Communications
in Computer and Information Science   1238

Commenced Publication in 2007
Founding and Former Series Editors:
Simone Diniz Junqueira Barbosa, Phoebe Chen, Alfredo Cuzzocrea,
Xiaoyong Du, Orhun Kara, Ting Liu, Krishna M. Sivalingam,
Dominik Ślęzak, Takashi Washio, Xiaokang Yang, and Junsong Yuan

More information about this series at http://www.springer.com/series/7899

Marie-Jeanne Lesot · Susana Vieira ·
Marek Z. Reformat · João Paulo Carvalho ·
Anna Wilbik · Bernadette Bouchon-Meunier ·
Ronald R. Yager (Eds.)

Information Processing and Management of Uncertainty in Knowledge-Based Systems

18th International Conference, IPMU 2020
Lisbon, Portugal, June 15–19, 2020
Proceedings, Part II

 Springer

Editors
Marie-Jeanne Lesot
LIP6-Sorbonne University
Paris, France

Susana Vieira
IDMEC, IST, Universidade de Lisboa
Lisbon, Portugal

Marek Z. Reformat
University of Alberta
Edmonton, AB, Canada

João Paulo Carvalho
INESC, IST, Universidade de Lisboa
Lisbon, Portugal

Anna Wilbik
Eindhoven University of Technology
Eindhoven, The Netherlands

Bernadette Bouchon-Meunier
CNRS-Sorbonne University
Paris, France

Ronald R. Yager
Iona College
New Rochelle, NY, USA

ISSN 1865-0929   ISSN 1865-0937 (electronic)
Communications in Computer and Information Science
ISBN 978-3-030-50142-6   ISBN 978-3-030-50143-3 (eBook)
https://doi.org/10.1007/978-3-030-50143-3

This Springer imprint is published by the registered company Springer Nature Switzerland AG
The registered company address is: Gewerbestrasse 11, 6330 Cham, Switzerland

Preface

We are very pleased to present you with the proceedings of the 18th International Conference on Information Processing and Management of Uncertainty in Knowledge-Based Systems (IPMU 2020), held during June 15–19, 2020. The conference was scheduled to take place in Lisbon, Portugal, at the Instituto Superior Técnico, University of Lisbon, located in a vibrant renovated area 10 minutes from downtown. Unfortunately, due to the COVID-19 pandemic and international travel restrictions around the globe, the Organizing Committee made the decision to make IPMU 2020 a virtual conference taking place as scheduled.

The IPMU conference is organized every two years. Its aim is to bring together scientists working on methods for the management of uncertainty and aggregation of information in intelligent systems. Since 1986, the IPMU conference has been providing a forum for the exchange of ideas between theoreticians and practitioners working in these areas and related fields. In addition to many contributed scientific papers, the conference has attracted prominent plenary speakers, including the Nobel Prize winners Kenneth Arrow, Daniel Kahneman, and Ilya Prigogine.

A very important feature of the conference is the presentation of the *Kampé de Fériet Award* for outstanding contributions to the field of uncertainty and management of uncertainty. Past winners of this prestigious award are Lotfi A. Zadeh (1992), Ilya Prigogine (1994), Toshiro Terano (1996), Kenneth Arrow (1998), Richard Jeffrey (2000), Arthur Dempster (2002), Janos Aczel (2004), Daniel Kahneman (2006), Enric Trillas (2008), James Bezdek (2010), Michio Sugeno (2012), Vladimir N. Vapnik (2014), Joseph Y. Halpern (2016), and Glenn Shafer (2018). This year, the recipient of the *Kampé de Fériet Award* is Barbara Tversky. Congratulations!

The IPMU 2020 conference offers a versatile and comprehensive scientific program. There were four invited talks given by distinguished researchers: Barbara Tversky (Stanford University and Columbia University, USA), Luísa Coheur (Universidade de Lisboa, Instituto Superior Técnico, Portugal), Jim Keller (University of Missouri, USA), and Björn Schuller (Imperial College London, UK). A special tribute was organized to celebrate the life and achievements of Enrique Ruspini who passed away last year. He was one of the fuzzy-logic pioneers and researchers who contributed enormously to the fuzzy sets and systems body of knowledge. Two invited papers are dedicated to his memory. We would like to thank Rudolf Seising, Francesc Esteva, Lluís Godo, Ricardo Oscar Rodriguez, and Thomas Vetterlein for their involvement and contributions.

The IPMU 2020 program consisted of 22 special sessions and 173 papers authored by researchers from 34 different countries. All 213 submitted papers underwent the thorough review process and were judged by at least three reviewers. Many of them were reviewed by more – even up to five – referees. Furthermore, all papers were examined by the program chairs. The review process respected the usual conflict-of-interest standards, so that all papers received multiple independent evaluations.

Organizing a conference is not possible without the assistance, dedication, and support of many people and institutions.

We are particularly thankful to the organizers of special sessions. Such sessions, dedicated to variety of topics and organized by experts, have always been a characteristic feature of IPMU conferences. We would like to pass our special thanks to Uzay Kaymak, who helped evaluate many special session proposals.

We would like to acknowledge all members of the IPMU 2020 Program Committee, as well as multiple reviewers who played an essential role in the reviewing process, ensuring a high-quality conference. Thank you very much for all your work and efforts.

We gratefully acknowledge the technical co-sponsorship of the IEEE Computational Intelligence Society and the European Society for Fuzzy Logic and Technology (EUSFLAT).

A huge thanks and appreciation to the personnel of Lisbon's Tourism Office 'Turismo de Lisboa' (www.visitlisboa.com) for their eagerness to help, as well as their enthusiastic support.

Our very special and greatest gratitude goes to the authors who have submitted results of their work and presented them at the conference. Without you this conference would not take place. Thank you!

We miss in-person meetings and discussions, yet we are privileged that despite these difficult and unusual times all of us had a chance to be involved in organizing the virtual IPMU conference. We hope that these proceedings provide the readers with multiple ideas leading to numerous research activities, significant publications, and intriguing presentations at future IPMU conferences.

April 2020

Marie-Jeanne Lesot
Marek Z. Reformat
Susana Vieira
Bernadette Bouchon-Meunier
João Paulo Carvalho
Anna Wilbik
Ronald R. Yager

Organization

General Chair

João Paulo Carvalho   INESC-ID, Instituto Superior Técnico,
                                  Universidade de Lisboa, Portugal

Program Chairs

Marie-Jeanne Lesot   LIP6, Sorbonne Université, France
Marek Z. Reformat   University of Alberta, Canada
Susana Vieira   IDMEC, Instituto Superior Técnico,
                                  Universidade de Lisboa, Portugal

Executive Directors

Bernadette   LIP6, CNRS, France
  Bouchon-Meunier
Ronald R. Yager   Iona College, USA

Special Session Chair

Uzay Kaymak   Technische Universiteit Eindhoven, The Netherlands

Publication Chair

Anna Wilbik   Technische Universiteit Eindhoven, The Netherlands

Sponsor and Publicity Chair

João M. C. Sousa   IDMEC, Instituto Superior Técnico,
                                  Universidade de Lisboa, Portugal

Web Chair

Fernando Batista   INESC-ID, Instituto Superior Técnico,
                                  Universidade de Lisboa, Portugal

Keeley Crockett	Manchester Metropolitan University, UK
Giuseppe D'Aniello	University of Salerno, Italy
Bernard De Baets	Ghent University, Belgium
Martine De Cock	University of Washington, USA
Guy De Tré	Ghent University, Belgium
Sébastien Destercke	CNRS, UMR Heudiasyc, France
Antonio Di Nola	University of Salerno, Italy
Scott Dick	University of Alberta, Canada
Didier Dubois	IRIT, RPDMP, France
Fabrizio Durante	Free University of Bozen-Bolzano, Italy
Krzysztof Dyczkowski	Adam Mickiewicz University, Poland
Zied Elouedi	Institut Supérieur de Gestion de Tunis, Tunisia
Francesc Esteva	IIIA-CSIC, Spain
Dimitar Filev	Ford Motor Company, USA
Matteo Gaeta	University of Salerno, Italy
Sylvie Galichet	LISTIC, Université de Savoie, France
Jonathan M. Garibaldi	University of Nottingham, UK
Lluis Godo	IIIA-CSIC, Spain
Fernando Gomide	University of Campinas, Brazil
Gil González-Rodríguez	University of Oviedo, Spain
Przemysław Grzegorzewski	Systems Research Institute, Polish Academy of Sciences, Poland
Lawrence Hall	University of South Florida, USA
Istvan Harmati	Széchenyi István Egyetem, Hungary
Timothy Havens	Michigan Technological University, USA
Francisco Herrera	University of Granada, Spain
Enrique Herrera-Viedma	University of Granada, Spain
Ludmila Himmelspach	Heirich Heine Universität Düsseldorf, Germany
Eyke Hüllemeier	Paderborn University, Germany
Michal Holčapek	University of Ostrava, Czech Republic
Janusz Kacprzyk	Systems Research Institute, Polish Academy of Sciences, Poland
Uzay Kaymak	Eindhoven University of Technology, The Netherlands
Jim Keller	University of Missouri, USA
Frank Klawonn	Ostfalia University of Applied Sciences, Germany
László T. Kóczy	Budapest University of Technology and Economics, Hungary
John Kornak	University of California, San Francisco, USA
Vladik Kreinovich	University of Texas at El Paso, USA
Ondrej Krídlo	University of P. J. Safarik in Kosice, Slovakia
Rudolf Kruse	University of Magdeburg, Germany
Christophe Labreuche	Thales R&T, France
Jérôme Lang	CNRS, LAMSADE, Université Paris-Dauphine, France
Anne Laurent	LIRMM, UM, France
Chang-Shing Lee	National University of Tainan, Taiwan

Martin Štěpnička	IRAFM, University of Ostrava, Czech Republic
Umberto Straccia	ISTI-CNR, Italy
Olivier Strauss	LIRMM, France
Michio Sugeno	Tokyo Institute of Technology, Japan
Eulalia Szmidt	Systems Research Institute, Polish Academy of Sciences, Poland
Marco Tabacchi	Università degli Studi di Palermo, Italy
Vicenc Torra	Maynooth University, Ireland
Linda C. van der Gaag	Utrecht University, The Netherlands
Barbara Vantaggi	Sapienza University of Rome, Italy
José Luis Verdegay	University of Granada, Spain
Thomas Vetterlein	Johannes Kepler University Linz, Austria
Susana Vieira	Universidade de Lisboa, Portugal
Christian Wagner	University of Nottingham, UK
Anna Wilbik	Eindhoven University of Technology, The Netherlands
Sławomir Zadrożny	Systems Research Institute, Polish Academy of Sciences, Poland

Additional Members of the Reviewing Committee

Raoua Abdelkhalek
Julien Alexandre Dit Sandretto
Zahra Alijani
Alessandro Antonucci
Jean Baratgin
Laécio C. Barros
Leliane N. Barros
Libor Behounek
María José Benítez Caballero
Kyle Bittner
Jan Boronski
Reda Boukezzoula
Ross Boylan
Andrey Bronevich
Petr Bujok
Michal Burda
Rafael Cabañas de Paz
Inma P. Cabrera
Tomasa Calvo
José Renato Campos
Andrea Capotorti
Diego Castaño
Anna Cena
Mihir Chakraborty

Yurilev Chalco-Cano
Manuel Chica
Panagiotis Chountas
Davide Ciucci
Frank Coolen
Maria Eugenia Cornejo Piñero
Cassio P. de Campos
Gert De Cooman
Laura De Miguel
Jean Dezert
J. Angel Diaz-Garcia
Graçaliz Dimuro
Paweł Drygaś
Hassane Essafi
Javier Fernandez
Carlos Fernandez-Basso
Juan Carlos Figueroa-García
Marcelo Finger
Tommaso Flaminio
Robert Fullér
Marek Gagolewski
Angel Garcia Contreras
Michel Grabisch
Karel Gutierrez

Allel Hadjali
Olgierd Hryniewicz
Miroslav Hudec
Ignacio Huitzil
Seong Jae Hwang
Atsushi Inoue
Vladimir Janis
Balasubramaniam Jayaram
Richard Jensen
Luis Jimenez Linares
Katarzyna Kaczmarek
Martin Kalina
Hiroharu Kawanaka
Alireza Khastan
Martins Kokainis
Ryszard Kowalczyk
Maciej Krawczak
Jiri Kupka
Serafina Lapenta
Ulcilea Leal
Antonio Ledda
Eric Lefevre
Nguyen Linh
Nicolas Madrid
Arnaud Martin
Denis Maua
Gilles Mauris
Belen Melian
María Paula Menchón
David Mercier
Arnau Mir
Soheyla Mirshahi
Marina Mizukoshi
Jiří Močkoř
Miguel Molina-Solana
Ignacio Montes
Serafin Moral
Tommaso Moraschini
Andreia Mordido
Juan Antonio Morente-Molinera
Fred Mubang
Vu-Linh Nguyen
Radoslaw Niewiadomski

Carles Noguera
Pavels Orlovs
Daniel Ortiz-Arroyo
Jan W. Owsinski
Antonio Palacio
Manuel J. Parra Royón
Jan Paseka
Viktor Pavliska
Renato Pelessoni
Barbara Pękala
Benjamin Quost
Emmanuel Ramasso
Eloisa Ramírez Poussa
Luca Reggio
Juan Vicente Riera
Maria Rifqi
Luis Rodriguez-Benitez
Guillaume Romain
Maciej Romaniuk
Francisco P. Romero
Clemente Rubio-Manzano
Aleksandra Rutkowska
Juan Jesus Salamanca Jurado
Teddy Seidenfeld
Mikel Sesma-Sara
Babak Shiri
Amit Shukla
Anand Pratap Singh
Damjan Skulj
Sotir Sotirov
Michal Stronkowski
Andrea Stupnánová
Matthias Troffaes
Dana Tudorascu
Leobardo Valera
Arthur Van Camp
Paolo Vicig
Amanda Vidal Wandelmer
Joaquim Viegas
Jin Hee Yoon
Karl Young
Hua-Peng Zhang

Special Session Organizers

Javier Andreu	University of Essex, UK
Michał Baczyński	University of Silesia in Katowice, Poland
Isabelle Bloch	Télécom ParisTech, France
Bernadette Bouchon-Meunier	LIP6, CNRS, France
Reda Boukezzoula	Université de Savoie Mont-Blanc, France
Humberto Bustince	Public University of Navarra, Spain
Tomasa Calvo	University of Alcalá, Spain
Martine Ceberio	University of Texas at El Paso, USA
Yurilev Chalco-Cano	University of Tarapacá at Arica, Chile
Giulianella Coletti	Università di Perugia, Italy
Didier Coquin	Université de Savoie Mont-Blanc, France
M. Eugenia Cornejo	University of Cádiz, Spain
Bernard De Baets	Ghent University, Belgium
Guy De Tré	Ghent University, Belgium
Graçaliz Dimuro	Universidade Federal do Rio Grande, Brazil
Didier Dubois	IRIT, Université Paul Sabatier, France
Hassane Essafi	CEA, France
Carlos J. Fernández-Basso	University of Granada, Spain
Javier Fernández	Public University of Navarra, Spain
Tommaso Flaminio	Spanish National Research Council, Spain
Lluis Godo	Spanish National Research Council, Spain
Przemyslaw Grzegorzewski	Warsaw University of Technology, Poland
Rajarshi Guhaniyogi	University of California, Santa Cruz, USA
Karel Gutiérrez Batista	University of Granada, Spain
István Á. Harmati	Széchenyi István University, Hungary
Michal Holčapek	University of Ostrava, Czech Republic
Atsushi Inoue	Eastern Washington University, USA
Balasubramaniam Jayaram	Indian Institute of Technology Hyderabad, India
Janusz Kacprzyk	Systems Research Institute, Polish Academy of Sciences, Poland
Hiroharu Kawanaka	Mie University, Japan
László T. Kóczy	Budapest University of Technology and Economics, Hungary
John Kornak	University of California, San Francisco, USA
Vladik Kreinovich	University of Texas at El Paso, USA
Henrik Legind Larsen	Legind Technologies, Denmark
Weldon Lodwick	Federal University of São Paulo, Brazil
Maria Jose Martín-Bautista	University of Granada, Spain
Sebastia Massanet	University of the Balearic Islands, Spain
Jesús Medina	University of Cádiz, Spain
Belén Melián-Batista	University of La Laguna, Spain
Radko Mesiar	Slovak University of Technology, Slovakia
Enrique Miranda	University of Oviedo, Spain

Ignacio Montes	University of Oviedo, Spain
Juan Moreno-Garcia	University of Castilla-La Mancha, Spain
Petra Murinová	University of Ostrava, Czech Republic
Vílem Novák	University of Ostrava, Czech Republic
David A. Pelta	University of Granada, Spain
Raúl Pérez-Fernández	University of Oviedo, Spain
Irina Perfilieva	University of Ostrava, Czech Republic
Henri Prade	IRIT, Université Paul Sabatier, France
Anca Ralescu	University of Cincinnati, USA
Eloísa Ramírez-Poussa	University of Cádiz, Spain
Luis Rodriguez-Benitez	University of Castilla-La Mancha, Spain
Antonio Rufian-Lizana	University of Sevilla, Spain
M. Dolores Ruiz	University of Granada, Spain
Andrea Stupnanova	Slovak University of Technology, Slovakia
Amanda Vidal	Czech Academy of Sciences, Czech Republic
Aaron Wolfe Scheffler	University of California, San Francisco, USA
Adnan Yazici	Nazarbayev University, Kazakhstan
Sławomir Zadrożny	Systems Research Institute Polish Academy of Sciences, Poland

List of Special Sessions

Fuzzy Interval Analysis

Antonio Rufian-Lizana	University of Sevilla, Spain
Weldon Lodwick	Federal University of São Paulo, Brazil
Yurilev Chalco-Cano	University of Tarapacá at Arica, Chile

Theoretical and Applied Aspects of Imprecise Probabilities

| Enrique Miranda | University of Oviedo, Spain |
| Ignacio Montes | University of Oviedo, Spain |

Similarities in Artificial Intelligence

| Bernadette Bouchon-Meunier | LIP6, CNRS, France |
| Giulianella Coletti | Università di Perugia, Italy |

Belief Function Theory and Its Applications

| Didier Coquin | Université de Savoie Mont-Blanc, France |
| Reda Boukezzoula | Université de Savoie Mont-Blanc, France |

Aggregation: Theory and Practice

Tomasa Calvo	University of Alcalá, Spain
Radko Mesiar	Slovak University of Technology, Slovakia
Andrea Stupnánová	Slovak University of Technology, Slovakia

Aggregation: Pre-aggregation Functions and Other Generalizations

Humberto Bustince   Public University of Navarra, Spain
Graçaliz Dimuro   Universidade Federal do Rio Grande, Brazil
Javier Fernández   Public University of Navarra, Spain

Aggregation: Aggregation of Different Data Structures

Bernard De Baets   Ghent University, Belgium
Raúl Pérez-Fernández   University of Oviedo, Spain

Fuzzy Methods in Data Mining and Knowledge Discovery

M. Dolores Ruiz   University of Granada, Spain
Karel Gutiérrez Batista University of Granada, Spain
Carlos J. Fernández-Basso University of Granada, Spain

Computational Intelligence for Logistics and Transportation Problems

David A. Pelta   University of Granada, Spain
Belén Melián-Batista   University of La Laguna, Spain

Fuzzy Implication Functions

Michał Baczyński   University of Silesia in Katowice, Poland
Balasubramaniam Jayaram Indian Institute of Technology Hyderabad, India
Sebastià Massanet   University of the Balearic Islands, Spain

Soft Methods in Statistics and Data Analysis

Przemysław Grzegorzewski Warsaw University of Technology, Poland

Image Understanding and Explainable AI

Isabelle Bloch   Télécom ParisTech, France
Atsushi Inoue   Eastern Washington University, USA
Hiroharu Kawanaka   Mie University, Japan
Anca Ralescu   University of Cincinnati, USA

Fuzzy and Generalized Quantifier Theory

Vilém Novák   University of Ostrava, Czech Republic
Petra Murinová   University of Ostrava, Czech Republic

Mathematical Methods Towards Dealing with Uncertainty in Applied Sciences

Irina Perfilieva   University of Ostrava, Czech Republic
Michal Holčapek   University of Ostrava, Czech Republic

Statistical Image Processing and Analysis, with Applications in Neuroimaging

John Kornak	University of California, San Francisco, USA
Rajarshi Guhaniyogi	University of California, Santa Cruz, USA
Aaron Wolfe Scheffler	University of California, San Francisco, USA

Interval Uncertainty

Martine Ceberio	University of Texas at El Paso, USA
Vladik Kreinovich	University of Texas at El Paso, USA

Discrete Models and Computational Intelligence

László T. Kóczy	Budapest University of Technology and Economics, Hungary
István Á. Harmati	Széchenyi István University, Hungary

Current Techniques to Model, Process and Describe Time Series

Juan Moreno-Garcia	University of Castilla-La Mancha, Spain
Luis Rodriguez-Benitez	University of Castilla-La Mancha, Spain

Mathematical Fuzzy Logic and Graded Reasoning Models

Tommaso Flaminio	Spanish National Research Council, Spain
Lluís Godo	Spanish National Research Council, Spain
Vílem Novák	University of Ostrava, Czech Republic
Amanda Vidal	Czech Academy of Sciences, Czech Republic

Formal Concept Analysis, Rough Sets, General Operators and Related Topics

M. Eugenia Cornejo	University of Cádiz, Spain
Didier Dubois	IRIT, Université Paul Sabatier, France
Jesús Medina	University of Cádiz, Spain
Henri Prade	IRIT, Université Paul Sabatier, France
Eloísa Ramírez-Poussa	University of Cádiz, Spain

Computational Intelligence Methods in Information Modelling, Representation and Processing

Guy De Tré	Ghent University, Belgium
Janusz Kacprzyk	Systems Research Institute, Polish Academy of Sciences, Poland
Adnan Yazici	Nazarbayev University, Kazakhstan
Sławomir Zadrożny	Systems Research Institute Polish Academy of Sciences, Poland

Contents - Part II

Aggregation: Pre-aggregation Functions and Other Generalizations of Monotonicity

Aggregation: Aggregation of Different Data Structures

Fuzzy Implication Functions

Fuzzy Interval Analysis

An Introduction to Differential Algebraic Equations Under Interval Uncertainty: A First Step Toward Generalized Uncertainty DAEs

Weldon Alexander Lodwick[1]([✉])[iD] and Marina Tuyako Mizukoshi[2]([✉])[iD]

[1] Department of Mathematical and Statistical Sciences, University of Colorado, 1201 Larimer Street, Denver, CO 80204, USA
Weldon.Lodwick@ucdenver.edu

[2] Instituto de Matemática e Estatística, Universidade Federal de Goiás, Campus II - Samambaia, Goiânia, GO 74690-900, Brazil
tuyako@ufg.br
https://clas.ucdenver.edu/mathematical-and-statistical-sciences,
http://www.ime.ufg.br

Abstract. This presentation introduces the theory leading to solution methods for differential algebraic equations (DAEs) under interval uncertainty in which the uncertainty is in the initial conditions of the differential equation and/or the entries of the coefficients of the differential equation and algebraic restrictions. While we restrict these uncertainties to be *intervals*, other types of uncertains like generalized uncertainties such as fuzzy intervals are done in a similar manner albeit leading to more complex analyses. Linear constant coefficient DAEs and then interval linear constant coefficient problems will illustrate both the theoretically challenges and solution approaches. The way the interval uncertainty is handled is novel and serves as a basis for more general uncertainty analysis.

Keywords: Interval analysis · Differential algebraic equations · Constraint interval

1 Introduction

This presentation introduces *interval* differential-algebraic equations. To our knowledge, the publication that is closest to our theoretical approach is [11], in which an interval arithmetic, they call ValEncIA, is used to analyzed interval DAEs. What is presented here and what is new is that we solve the interval DAE problem using the constraint interval representation (see [6,7]) to encode all interval initial condition and/or interval coefficients. It is shown that this representation has theoretical advantages not afforded to the usual interval representation. The coefficients and initial values are, for this presentation, constant

© Springer Nature Switzerland AG 2020
M.-J. Lesot et al. (Eds.): IPMU 2020, CCIS 1238, pp. 3–12, 2020.
https://doi.org/10.1007/978-3-030-50143-3_1

to maintain the article relatively short. However, the approach in which there are variable coefficients and/or initial values can easily be extend and this is pointed out. Since we transform (variable) interval coefficients to (variable) interval initial values, it is, in theory, a straight forward process, albeit more complex.

The constraint interval representation leads to useful numerical methods as will be demonstrated. The limitations of this publication prohibit the full development of the methods. The initial steps are indicated. What is presented here does not deal directly with what is called *validated methods* (see [9,10] for example). However, when the processes developed here are carried out in a system that accounts for numerical errors using outward directed rounding, for example, in INTLAB or CXSC, then the results will be validated. We restrict ourselves to what is called (see below) semi-explicit DAEs. That is, all problems are assumed to have been transformed to the semi-explicit form. However, our approach is much wider.

The general form of the *semi-explicit* DAE is

$$y' = F(y(t), t), \ y(t_0) = y_0 \tag{1}$$

$$G(y(t), t) = 0. \tag{2}$$

While the focus is incorporating (constant) interval uncertainties in (1), and (2), generalized uncertainties as developed in [8], can be analyzed in similar fashion. Note that the variable (and constant) interval coefficient/initial value is a type of generalized uncertainty and fits perfectly in the theory that is presented.

2   Definition and Properties

One can also think of the implicit ODE with an algebraic constraint,

$$F(y(t), y'(t), t) = 0, \ y(t_0) = y_0 \tag{3}$$
$$G(y(t), t) = 0, \tag{4}$$

as a semi-explicit DAE as follows Let

$$y'(t) = z(t)$$
$$F(y(t), z(t), t) = 0,$$
$$G(y(t), t) = 0,$$

This will increase the number of variables. However, this will not be the approach for this presentation. Our general form will assume that the DAE is in the semi-explicit form (1), (2).

Given an implicit differential equation (3), when $\frac{\partial F}{\partial y}$ is not invertible, that is, we do not have an explicit differential equation, at least theoretically, in the form (1), (2)), we can differentiate (4) to obtain

$$\frac{\partial G}{\partial y}(y(t), t)y'(t) + \frac{\partial G}{\partial t}(y(t), t) = 0. \tag{5}$$

If $\frac{\partial G}{\partial y}(y,t)$ is non-singular, then (5) can be solved explicitly for y' as follows.

$$y' = \left[\frac{\partial G}{\partial y}(y(t),t)\right]^{-1}\frac{\partial G}{\partial t}(y,t) \tag{6}$$

$$F\left(y(t),\left[\frac{\partial G}{\partial y}(y(t),t)\right]^{-1}\frac{\partial G}{\partial t}(y(t),t),t\right) = 0 \tag{7}$$

$$G(y(t),t) = 0 \tag{8}$$

and (6), (7), and (8) is in the form (1), (2) and the DAE is called an **index 1 DAE**.

This not being the case, that is, $\frac{\partial G}{\partial y}(y(t),t)$ is singular, then we have the form

$$F(y(t),y'(t),t) = 0,$$
$$\frac{\partial G}{\partial y}(y(t),t)y' + \frac{\partial G}{\partial t}(y(t),t) = 0,$$

which can be written as

$$H(y(t),y'(t),t) = 0$$

and we again differentiate with respect to y' and test for singularity. If the partial of H can be solved for y', then we have an **index-2 DAE**. This process can be continued, in principle, until (hopefully) y' as an explicit function of y and t is found.

DAEs arise in various contexts, in applications. We present two types of problems where DAEs arise - the simple pendulum and unconstrained optimal control, that illustrate the main issues associated with DAEs. Then we will show interval uncertainty in the DAEs using some of these examples. Our solution methods for the interval DAEs are based on what is developed for the examples.

3 Linear Constant Coefficient DAEs

Linear constant coefficient DAEs arise naturally in electrical engineering circuit problems as well as in some control theory problems. A portion of this theory, sufficient to understand our solution methods, is presented next. The linear constant coefficient DAE is defined as

$$Ax'(t) + Bx(t) = f, \quad y(t_0) = y_0, \tag{9}$$

where A and B are $m \times m$ matrices, $x(t)$ is a $m \times 1$ vector "state function" and f is a $m \times 1$ function vector, and

$$A = \begin{bmatrix} A_1 & A_2 \\ 0 & 0 \end{bmatrix}, B = \begin{bmatrix} B_1 & B_2 \\ B_3 & B_4 \end{bmatrix}, f = \begin{bmatrix} f_1 \\ f_2 \end{bmatrix},$$

so that

$$\begin{bmatrix} A_1 & A_2 \\ 0 & 0 \end{bmatrix} \begin{bmatrix} x_1'(t) \\ x_2'(t) \end{bmatrix} + \begin{bmatrix} B_1 & B_2 \\ B_3 & B_4 \end{bmatrix} \begin{bmatrix} x_1(t) \\ x_2(t) \end{bmatrix} = \begin{bmatrix} f_1 \\ f_2 \end{bmatrix}.$$

Note that this is indeed a semi-explicit DAE, where differential part is

$$x'(t) = F(x(t), t) = -A^{-1}(B)x(t) + f(t)$$

$$A = \begin{bmatrix} A_1 & A_2 \end{bmatrix}, B = \begin{bmatrix} B_1 & B_2 \end{bmatrix}, f = \begin{bmatrix} f_1 \\ f_2 \end{bmatrix},$$

where the matrix A is assumed to be invertible, and the algebraic part is

$$G(x(t), t) = B_3 x_1(t) + B_4 x_2(t) = f_2.$$

Example 1. Consider the linear constant coefficient DAE $Ax'(t) + Bx(t) = f$ where

$$A = \begin{bmatrix} 1 & 1 \\ 0 & 0 \end{bmatrix}, B = \begin{bmatrix} 0 & 0 \\ 2 & 1 \end{bmatrix}, f = \begin{bmatrix} t \\ e^t \end{bmatrix}.$$

The ODE part is

$$x_1'(t) + x_2'(t) = t \tag{10}$$

and the algebraic part is

$$G(x(t), t) = 2x_1(t) + x_2(t) = e^t. \tag{11}$$

Integrating the ODE part (10), we get

$$\int dx_1 + \int dx_2 = \int t \, dt,$$

$$x_1(t) + x_2(t) = \frac{1}{2}t^2 + c_1. \tag{12}$$

Solving (11) and (12)

$$2x_1(t) + x_2(t) = e^t$$

$$x_1(t) + x_2(t) = \frac{1}{2}t^2 + c_1$$

simultaneously, we get

$$x_1(t) = e^t - \frac{1}{2}t^2 - c_1, \tag{13}$$

$$x_2(t) = -e^t + t^2 + 2c_1. \tag{14}$$

Example 2. Consider the linear constant coefficient DAE

$$\begin{bmatrix} 1 & 0 \\ 0 & 0 \end{bmatrix} x'(t) + \begin{bmatrix} 1 & 1 \\ 1 & 1 \end{bmatrix} x(t) = \begin{bmatrix} t \\ e^t \end{bmatrix}.$$

The ODE part is

$$x_1'(t) + x_1(t) + x_2(t) = t \qquad (15)$$

and the algebraic constraint is

$$G(x(t), t) = x_1(t) + x_2(t) = e^t. \qquad (16)$$

Putting (16) into (15) and solving for x_1', we have

$$x_1'(t) = t - e^t$$

$$x_1(t) = \frac{1}{2}t^2 - e^t + c_1$$

$$x_2(t) = 2e^t - \frac{1}{2}t^2 - c_1.$$

Remark 1. 1) When, for the $m-$variable problem, the algebraic constraint can be substituted into the differential equation, and the differential equation, which is linear, is integrated, there are m equations in m unknowns. Interval uncertainty enters when the matrices A and/or B and/or the initial condition y_0 are intervals $A \in [A], B \in [B], y_0 \in [y_0]$. This is illustrated.

4 Illustrative Examples

Two DAE examples, beginning with the simple pendulum, are presented next.

4.1 The Simple Pendulum

Consider the following problem (see [4]) arising from a simple pendulum,

$$x''(t) = -\gamma x(t), \ x(t_0) = x_0, x'(t_0) = x_0'$$
$$y''(t) = -\gamma y(t) - g, \ y(t_0) = y_0, y'(t_0) = y_0' \qquad (17)$$
$$0 = x^2(t) + y^2(t) - L^2 \text{ (mechanical constraint) or}$$
$$0 = (x')^2 (t) + (y')^2 (t) - y(t)g \text{ (energy constraint)}$$

where g is the acceleration due to gravity, γ is the unknown tension in the string, and $L = 1$ is the length of the pendulum string. In this example, we will consider the unknown tension to be an interval $[\gamma] = [\underline{\gamma}, \overline{\gamma}]$ to model the uncertainty of the value of the tension. Moreover, we will also assume that the initial values are also intervals. We can restate (17) as a first order system where we will focus on the mechanical constraint omitting the energy constraint, as follows:

$$u_1(t) = x(t) \Rightarrow u_1'(t) = u_3(t), \ u_1(t_0) \in [(u_1)_0]$$
$$u_2(t) = y(t) \Rightarrow u_2'(t) = u_4(t), \ u_2(t_0) \in [(u_2)_0]$$
$$u_3(t) = x'(t) \Rightarrow u_3'(t) = -u_5(t) \cdot u_1, \ u_3(t_0) \in [(u_3)_0] \qquad (18)$$
$$u_4(t) = y'(t) \Rightarrow u_4'(t) = -u_5(t) \cdot u_2(t) - g, \ u_4(t_0) \in [(u_4)_0]$$
$$u_5(t) = \gamma \Rightarrow u_5'(t) = 0, \ u_5(t_0) \in [\gamma] = [\underline{\gamma}, \overline{\gamma}]$$
$$G(u_1(t), u_2(t), t) = u_1^2(t) + u_2^2(t) - 1 = 0.$$

Note that the uncertainty parameter γ is considered as a part of the differential equation and is constant (u_5) whose initial condition is an interval. This is in general the way constant interval (generalized) uncertainty in the parameters is handled. If the coefficient were in fact variable, $u_5'(t) \neq 0$ but a differential equation itself, for example,

$$u_5'(t) = h(t)$$

where $h(t)$ would define how the rate of change of the coefficient is varying with respect to time. Equation (18) is a standard semi-explicit real DAE and in this case, with interval initial conditions. How to solve such a system is presented in Sect. 5.

4.2   Unconstrained Optimal Control

We next present the transformation of unconstrained optimal control problems to a DAE. The form of the general unconstrained optimal control is the following.

$$\max_{u \in \Omega} J[u] = \int_0^1 L(x(u,t), u, t)dt$$

$$\text{subject to:} x'(u,t) = f(x, u, t) \tag{19}$$

$$x(u,0) = x_0.$$

$$x : \mathbb{R}^m \times \mathbb{R} \to \mathbb{R}^n, u : \mathbb{R} \to \mathbb{R}^m \tag{20}$$

When Ω is the set of all square integrable functions, then the problem becomes unconstrained. In this case we denote the constraint set Ω_0. The Pontryagin Maximization Principle utilizes the Hamiltonian function, which is defined for (19) as

$$H(x(t), \lambda, u, t) = \lambda(t)f(x(t), u, t) + L(x(t), u, t). \tag{21}$$

The function $\lambda(t)$ is called the co-state function and is a row vector ($1 \times n$). The co-state function can be thought of as the dynamic optimization equivalent to the Lagrange multiplier and is defined by the following differential equation:

$$\lambda'(t) = -\frac{\partial H(x(t), \lambda, u, t)}{\partial x} \tag{22}$$

$$\lambda(1) = 0.$$

Under suitable conditions (see [5]), the Pontryagin Maximization Principle (PMP) states that if there exists an (optimal) function $v(t)$ such that $J[v] \geq J[u], \forall u \in \Omega$, (the optimal control), then it maximizes the Hamiltonian with respect to the control, which in the case of unconstrained optimal control means that

$$\frac{\partial H(x(t), \lambda, v, t)}{\partial u} = 0, \ v \in \Omega_0, \tag{23}$$

where (23) is the algebraic constraint. Thus, the unconstrained optimal control problem, (19) together with the differential equation of the co-state, (22), and the PMP (23) results in a boundary valued DAE as follows:

$$x'(u,t) = f(x(t), u, t)$$
$$\lambda'(x(t), u, t) = -\frac{\partial H(x, \lambda, u, t)}{\partial x}$$
$$x(u, 0) = x_0$$
$$\lambda(x, u, 1) = 0$$
$$G(x(t), \lambda(t), t) = \frac{\partial H(x, \lambda, u, t)}{\partial u} = 0.$$

One example that is well studied is the linear quadratic optimal control problem (LQP), which is defined for the unconstrained optimal control problem as

$$\max_{u \in \Omega_0} J[u] = \frac{1}{2} \int_0^1 \left[x^T(t) Q x(t) + u^T(t) R u(t) \right] dt$$
$$\text{subject to:} x'(t) = A x(t) + B u(t) \tag{24}$$
$$x(u, 0) = x_0,$$

where $A_{n \times n}$ is an $n \times n$ real matrix, $B_{n \times m}$ is an $m \times n$ real matrix, $Q_{n \times n}$ is a real symmetric positive definite matrix, and $R_{m \times m}$ is a real invertible positive semi-definite matrix. For the LQP

$$H(x, u, \lambda, t) = \lambda(t) \left(A x(t) + B u(t) \right) + \frac{1}{2} \left(x^T(t) Q x(t) + u^T(t) R u(t) \right)$$
$$\lambda'(t) = -\frac{\partial H(x, \lambda, u, t)}{\partial x} = -\lambda(t) A + Q x(t), \quad \lambda(1) = 0,$$

remembering that λ is a row vector. The optimal control is obtained by solving

$$G(x(t), \lambda(t), t) = \frac{\partial H(x, \lambda, u, t)}{\partial u} = \lambda(t) B + R u(t) = 0$$
$$u(t) = -R^{-1} \lambda(t) B,$$

which, when put into the differential equations, yields

$$x'(t) = A x(t) - B R^{-1} \lambda(t) B, \quad x(0) = x_0$$
$$\lambda'(t) = Q x(t) - \lambda(t) A, \quad \lambda(1) = 0.$$

This results in the system

$$y'(t) = \begin{bmatrix} x'(t) \\ \lambda'(t) \end{bmatrix} = \begin{bmatrix} A & -B R^{-1} B \\ Q & A \end{bmatrix} \begin{bmatrix} x(t) \\ \lambda(t) \end{bmatrix}$$
$$x(0) = x_0, \lambda(1) = 0$$

The next section will consider DAEs with interval uncertainty in the coefficients and in the initial conditions.

5   Interval Uncertainty in DAEs

This section develops an interval solution method. The interval problem for non-linear equation considers interval coefficients as variables whose differential is zero and initial condition is the respective interval. For linear problems, it is sometimes advantageous to deal with the interval coefficients directly as illustrated next.

5.1   An Interval Linear Constant Coefficient DAE

Given the linear constant coefficient DAE $Ax'(t) + Bx(t) = f$, suppose that the coefficient matrices are interval matrices $[A]$ and $[B]$. That is,

$$[A] \, x'(t) + [B] \, x(t) = f.$$

Example 3. Consider Example 1, except we have interval entries

$$[A] = \begin{bmatrix} [A_1] & [A_2] \\ [A_3] & [A_4] \end{bmatrix}, B = \begin{bmatrix} [B_1] & [B_2] \\ [B_3] & [B_4] \end{bmatrix}, f = \begin{bmatrix} t \\ e^t \end{bmatrix},$$

where $[A_1] = [A_2] = [B_4] = [0.9, 1.1]$, $[A_3] = [A_4] = [B_1] = [B_2] = [0,0]$, and $[B_3] = [1.9, 2.1]$. The ODE part is

$$[0.9, 1.1] \, x_1'(t) + [0.9, 1.1] \, x_2'(t) = t \tag{25}$$

and the algebraic part is

$$G(x(t), t) = [1.9, 2.1] \, x_1(t) + [0.9, 1.1] \, x_{2(t)} = e^t. \tag{26}$$

Integrating (25) we have

$$[0.9, 1.1] \, x_1(t) + [0.9, 1.1] \, x_{2(t)} = \frac{1}{2} t^2 + c_1$$

which together with (26) forms the interval linear system

$$[0.9, 1.1] \, x_1(t) + [0.9, 1.1] \, x_2(t) = \frac{1}{2} t^2 + c_1 \tag{27}$$

$$[1.9, 2.1] \, x_1(t) + [0.9, 1.1] \, x_2(t) = e^t. \tag{28}$$

Using constraint interval (see [8] where any interval $[a, b]$ has the representation $[a, b] = a + \lambda(b - a)$), then

$$\begin{bmatrix} x_1(\vec{\lambda}) \\ x_2(\vec{\lambda}) \end{bmatrix} = \begin{bmatrix} 0.9 + 0.2\lambda_{11} & 0.9 + 0.2\lambda_{12} \\ 1.9 + 0.2\lambda_{21} & 0.9 + 0.2\lambda_{22} \end{bmatrix}^{-1} \begin{bmatrix} \frac{1}{2} t^2 + c_1 \\ e^t \end{bmatrix}$$

$$= \frac{1}{(0.9 + 0.2\lambda_{11})(0.9 + 0.2\lambda_{22}) - (0.9 + 0.2\lambda_{12})(1.9 + 0.2\lambda_{21})}$$

$$\times \begin{bmatrix} 0.9 + 0.2\lambda_{22} & -(0.9 + 0.2\lambda_{12}) \\ -(1.9 + 0.2\lambda_{21}) & 0.9 + 0.2\lambda_{11} \end{bmatrix} \begin{bmatrix} \frac{1}{2} t^2 + c_1 \\ e^t \end{bmatrix}$$

$$= \begin{bmatrix} -\dfrac{90.0c_1 - 90.0e^t - 20.0\lambda_{12}e^t + 20.0\lambda_{22}c_1 + 10.0t^2\lambda_{22} + 45.0t^2}{38.0\lambda_{12} - 18.0\lambda_{11} + 18.0\lambda_{21} - 18.0\lambda_{22} - 4.0\lambda_{11}\lambda_{22} + 4.0\lambda_{12}\lambda_{21} + 90.0} \\ \dfrac{190.0c_1 - 90.0e^t - 20.0\lambda_{11}e^t + 20.0\lambda_{21}c_1 + 10.0t^2\lambda_{21} + 95.0t^2}{38.0\lambda_{12} - 18.0\lambda_{11} + 18.0\lambda_{21} - 18.0\lambda_{22} - 4.0\lambda_{11}\lambda_{22} + 4.0\lambda_{12}\lambda_{21} + 90.0} \end{bmatrix},$$

where $\overrightarrow{\lambda} = (\lambda_{11}, \lambda_{12}, \lambda_{21}, \lambda_{22})$. Any instantiation of $\lambda_{ij} \in [0, 1]$ will yield a valid solution given the associated uncertainty. However, if one wishes to extract the interval containing $\begin{bmatrix} x_1(\overrightarrow{\lambda}) \\ x_2(\overrightarrow{\lambda}) \end{bmatrix}$, a global min/max over $0 \leq \lambda_{ij} \leq 1$ would need to be implemented.

Example 4. Consider the linear quadratic problem with interval initial condition for x,

$$\max J[u] = \frac{1}{2} \int_0^1 \left[x^2(t) + u^2(t) \right] dt$$

$$\text{subject to:} x'(t) = u(t), \ x(u, 0) = \left[\frac{1}{2}, \frac{3}{2} \right] = \frac{1}{2} + \gamma, \tag{29}$$

$$0 \leq \gamma \leq 1. \tag{30}$$

For this problem,

$$H(x(t), \lambda(t), u(t), t) = \lambda(t)u(t) - \frac{1}{2}x^2(t) - \frac{1}{2}u^2(t),$$

$$\lambda'(t) = -\frac{\partial H(x, \lambda, u, t)}{\partial x} = x(t), \ \lambda(1) = 0$$

$$G(x(t), \lambda(t), t) = \frac{\partial H(x(t), \lambda(t), v(t), t)}{\partial u} = \lambda(t) - v(t) = 0 \text{ or } v(t) = \lambda(t).$$

Thus

$$x'(t) = \lambda(t), \ x(0) = \frac{1}{2} + \gamma$$
$$\lambda'(t) = x(t), \ \lambda(1) = 0.$$

This implies that

$$x''(t) - x(t) = 0$$
$$x(t) = c_1 e^t + c_2 e^{-t}$$
$$\lambda(t) = c_1 e^t - c_2 e^{-t}$$

and with the initial conditions

$$x(t) = \frac{\frac{1}{2} + \gamma}{1 + e^2} e^t + \frac{e^2(\frac{1}{2} + \gamma)}{1 + e^2} e^{-t},$$

$$\lambda(t) = \frac{\frac{1}{2} + \gamma}{1 + e^2} e^t - \frac{e^2(\frac{1}{2} + \gamma)}{1 + e^2} e^{-t},$$

$$u_{opt}(t) = v(t) = \frac{\frac{1}{2} + \gamma}{1 + e^2} e^t - \frac{e^2(\frac{1}{2} + \gamma)}{1 + e^2} e^{-t}.$$

$$0 \leq \gamma \leq 1.$$

6   Conclusion

This study introduced the method to incorporate interval uncertainty in differential algebraic problems. Two examples of where DAEs under uncertainty arise were presented. Two solution methods with interval uncertainty for the linear problem and for the linear quadratic unconstrained optimal control problem were shown. Unconstrained optimal control problems lead to interval boundary-valued problems, which subsequent research will address. Moreover, more general uncertainties such as generalized uncertainties (see [8]), probability distributions, fuzzy intervals are the next steps in the development of a theory of DAEs under generalized uncertainties.

References

1. Ascher, U.M., Petzhold, L.R.: Computer Methods of Ordinary Differential Equations and Differential-Algebraic Equations. SIAM, Philadelphia (1998)
2. Brenan, K.E., Campbell, S.L., Petzhold, L.R.: Numerical Solution of Initial-Valued Problems in Differential-Algebraic Equations. SIAM, Philadelphia (1996)
3. Corliss, G.F., Lodwick, W.A.: Correct computation of solutions of differential algebraic control equations. Zeitschrift für Angewandte Mathematik nd Mechanik (ZAMM) 37–40 (1996). Special Issue "Numerical Analysis, Scientific Computing, Computer Science"
4. Corliss, G.F., Lodwick, W.A.: Role of constraints in validated solution of DAEs. Marquette University Technical Report No. 430, March 1996
5. Lodwick, W.A.: Two numerical methods for the solution of optimal control problems with computed error bounds using the maximum principle of pontryagin. Ph.D. thesis, Oregon State University (1980)
6. Lodwick, W.A.: Constrained interval arithmetic. CCM Report 138, February 1999
7. Lodwick, W.A.: Interval and fuzzy analysis: an unified approach. In: Hawkes, P.W. (ed.) Advances in Imagining and Electronic Physics, vol. 148, pp. 75–192. Academic Press, Cambridge (2007)
8. Lodwick, W.A., Thipwiwatpotjana, P.: Flexible and Generalized Uncertainty Optimization. SCI, vol. 696. Springer, Cham (2017). https://doi.org/10.1007/978-3-319-51107-8. ISSN 978-3-319-51105-4
9. Nedialkov, N.: Interval tools for ODEs and DAEs. In: 12th GAMM-IMACS International Symposium on Scientific Computing, Computer Arithmetic, and Validated Numerics (SCAN 2006) (2006)
10. Nedialkov, N., Pryce, J.: Solving DAEs by Taylor series (1), computing Taylor coefficients. BIT Numer. Math. **45**(3), 561–591 (2005)
11. Andreas, A., Brill, M., Günther, C.: A novel interval arithmetic approach for solving differential-algebraic equations with ValEncIA-IVP. Int. J. Appl. Math. Comput. Sci. **19**(3), 381–397 (2009)

Classification of Hyperbolic Singularities in Interval 3-Dimensional Linear Differential Systems

Marina Tuyako Mizukoshi[1]([✉])[iD], Alain Jacquemard[2]([✉]),
and Weldon Alexander Lodwick[3]([✉])[iD]

[1] Instituto de Matemática e Estatística, Universidade Federal de Goiás,
Campus II - Samambaia, Goiânia, GO 74690-900, Brazil
tuyako@ufg.br
[2] Université Bourgogne Franche-Comté, Institut de Mathématiques de Bourgogne,
UMR 5584, 21000 Dijon, France
Alain.Jacquemard@u-bourgogne.fr
[3] Department of Mathematical and Statistical Sciences, University of Colorado,
1201 Larimer Street, Denver, CO 80204, USA
Weldon.Lodwick@ucdenver.edu
http://www.ime.ufg.br, https://math.u-bourgogne.fr/,
https://clas.ucdenver.edu/mathematical-and-statistical-sciences

Abstract. We study the classification of the hyperbolic singularities to 3-dimensional interval linear differential equations as an application of interval eigenvalues using the Constraint Interval Arithmetic (CIA). We also present the ideas to calculate the interval eigenvalues using the standard interval arithmetic.

Keywords: Interval eigenvalues · Dynamical systems · Classification of singularities

1 Introduction

Many applied problems have uncertainties or inaccuracies due to data measurement errors, lack of complete information, simplification assumption of physical models, variations of the system, and computational errors. An encoding of uncertainty as intervals instead of numbers when applicable is an efficient way to address the aforementioned challenges.

When studying an interval problem we need to first understand what is the context. In this presentation, we are concerned if there exists dependence, independence or both in the parameters involved. In accordance to this context we need to choose the appropriated arithmetic. Where we have the total independence or dependence, we can use interval arithmetic or single level arithmetic

© Springer Nature Switzerland AG 2020
M.-J. Lesot et al. (Eds.): IPMU 2020, CCIS 1238, pp. 13–27, 2020.
https://doi.org/10.1007/978-3-030-50143-3_2

(SLA). If we are studying a problem where there is the independence as well as the dependence in parameters then constraint interval arithmetic (CIA) is a good choice.

We are interested in studying the interval eigenvalue problem associated with differential equations. This problem has many applications in the fields of mechanics and engineering. The first interval eigenvalues results were obtained by Deif [5], Deif and Rohn [15], Rohn [16]. Subsequently, approximation methods results were obtained by Qiu et al. [14], Leng et al. [10], Hladik [9] and Hladik et al. [6–8].

This presentation establishes conditions on the parameters of interval linear autonomous differential systems to classify the hyperbolic equilibrium point in 3-dimensions. Moreover a detailed study is given for an example using CIA along with a computational method for complex conjugate eigenvalues where we obtain the lower and upper bounds of the real eigenvalue.

2 Preliminaries

The following outlines the standard interval arithmetic WSMA (Warmus, Sunaga and Moore Arithmetic) and CIA (constraint interval arithmetic).

Let $\mathbf{x} = [\underline{x}\ \overline{x}]\ \mathbf{y} = [\underline{y}\ \overline{y}]$, be such that $\underline{x} \leq \overline{x}$ and $\underline{y} \leq \overline{y}$, then for WSMA arithmetic we have the following operations:

1. $\mathbf{x} + \mathbf{y} = [\underline{x} + \underline{y}\ \overline{x} + \overline{y}];$
2. $\mathbf{x} - \mathbf{y} = [\underline{x} - \overline{y}\ \overline{x} - \underline{y}];$
3. $\mathbf{x} \times \mathbf{y} = [min\{\underline{x}\underline{y}, \underline{x}\overline{y}, \overline{x}\underline{y}, \underline{x}\overline{y}\}\ max\{\underline{x}\underline{y}, \underline{x}\overline{y}, \overline{x}\underline{y}, \underline{x}\overline{y}\}];$
4. $\mathbf{x} \div \mathbf{y} = [min\{\underline{x} \div \underline{y}, \overline{x} \div \overline{y}, \underline{x} \div \overline{y}, \overline{x} \div \underline{y}\}\ max\{\underline{x} \div \underline{y}, \overline{x} \div \overline{y}, \underline{x} \div \overline{y}, \overline{x} \div \underline{y}\}], 0 \notin [\underline{y}\ \overline{y}].$

Remark: Note that in WSMA arithmetic $\mathbf{x} - \mathbf{x}$ is never 0 unless $\mathbf{x}$ is a real number (width zero) nor is $\mathbf{x} \div \mathbf{x} = 1$.

Definition 1. *[11] An interval $[\underline{x}\ \overline{x}]$ CI (constraint interval) representation is the real single-valued function $\boldsymbol{x}(\gamma) = \gamma\overline{x} + (1 - \gamma)\underline{x}, 0 \leq \gamma \leq 1$. Constraint interval arithmetic (CIA) is $\boldsymbol{z} = \boldsymbol{x} \circ \boldsymbol{y}$, where $\boldsymbol{z} = [\underline{z}\ \overline{z}] = \{z(\gamma_1, \gamma_2); z(\gamma_1, \gamma_2) = (\gamma_1\overline{x} + (1 - \gamma_1)\underline{x}) \circ (\gamma_2\overline{y} + (1 - \gamma_2)\underline{y}), 0 \leq \gamma_1 \leq 1, 0 \leq \gamma_2 \leq 1\}$, and $\underline{z} = min\{z(\gamma_1, \gamma_2)\}, \overline{z} = max\{z(\gamma_1, \gamma_2)\}, \circ \in \{+, -, \times, \div\}$.*

The set of $m \times n$ interval matrices will be denoted by $\mathbb{IR}^{m \times n}$. An interval matrix $\mathbf{A} = (\mathbf{A}_{ik})$ is interpreted as a set of real $m \times n$ matrices

$$\mathbf{A} = \{A \in \mathbb{R}^{m \times n}; A_{ik} \in (\mathbf{A}_{ik})\ \text{for}\ i = 1, \dots, m, k = 1, \dots, n\}.$$

Denote by $\mathbf{A} = [\underline{A}\ \overline{A}]$, where $\underline{A}$ and $\overline{A}$ are matrix whose entries are given by right and left sides of all intervals numbers $(\mathbf{a}_{ik}) \in \mathbf{A}$, respectively. In the CI context each element in $\mathbf{A}$ is given by $\mathbf{a}_{ij}(\gamma_{ij}) = \underline{a}_{ij} + \gamma_{ij}w_{\mathbf{a}_{ij}}, 1 \leq i \leq n, 1 \leq j \leq m, w_{\mathbf{a}_{ij}} = \overline{a}_{ij} - \underline{a}_{ij}, \gamma_i \in [0, 1]$.

Definition 2. *[5] Given* $[A] = [\underline{A} \ \overline{A}]$*, an interval matrix in* $\mathbb{IR}^{n \times n}$*, the set of eigenvalues is given by:*

$$\Lambda([A]) = \{\lambda; Ax = \lambda x, x \neq 0, A \in [A]\}.$$

In addition, we denote by $A_c = \frac{1}{2}(\underline{A} + \overline{A}), \Delta_A = \frac{1}{2}(\underline{A} - \overline{A})$*, the midpoint and the radius of* $[A]$*, respectively.*

In what follows we use the notation $\overrightarrow{\gamma}$ to mean the dependence of the choice of the values for $\gamma_{ij}, i, j = 1, \ldots, n$ in the interval $[0 \ 1]$.

Definition 3. *[12] Let be an interval matrix* $\boldsymbol{A} = [\underline{A} \ \overline{A}]$*, then the CI matrix is defined by*

$$A(\overrightarrow{\gamma}) = \begin{pmatrix} \underline{a}_{11} + \gamma_{11} w_{a_{11}} & \cdots & \underline{a}_{1n} + \gamma_{1n} w_{a_{1n}} \\ \cdots & \cdots & \cdots \\ \underline{a}_{n1} + \gamma_{n1} w_{a_{n1}} & \cdots & \underline{a}_{nn} + \gamma_{nn} w_{a_{nn}} \end{pmatrix} = \underline{A} + \Gamma \odot W,$$

where $\underline{A} = (\underline{a}_{ij}), W = (w_{a_{ij}}) = (\overline{a}_{ij} - \underline{a}_{ij}), \Gamma = (\gamma_{ij}), 0 \leq \gamma_{ij} \leq 1, for \ i = 1, \ldots, n$ *and* $j = 1, \ldots, n$ *and the symbol* $\odot$ *denotes componentwise multiplication. Then, we say that* $\lambda(\Gamma)$ *is an eigenvalue of* $A(\overrightarrow{\gamma})$ *if* $\exists x \neq 0 \mid A(\overrightarrow{\gamma})x = \lambda(\overrightarrow{\gamma})$ *i.e.,* $det(A(\overrightarrow{\gamma})) - \lambda(\overrightarrow{\gamma})I_n) = 0$*, where* I_n *is the identity matrix of order* n*.*

Remark: Here for each choice of matrix Γ we have a deterministic problem to calculate eigenvalues. We can get the interval eigenvalues by minimizing and maximizing $\lambda(\overrightarrow{\gamma})$ by varying all $\gamma_{ij}, i, j = 1, \ldots, n$ between 0 and 1.

To classify the equilibrium point in a 3-dimensional linear differential system, firstly we need to know how we can classify it in according the eigenvalues obtained of the matrix of the coefficients from linear differential system.

Consider a linear three-dimensional autonomous systems $X'(t) = AX(t), X(t), X'(t) \in M_{3 \times 1}(\mathbb{R}), A \in M_{3 \times 3}(\mathbb{R})$ of the form

$$\begin{cases} x'(t) = a_{11}x + a_{12}y + a_{13}z \\ y'(t) = a_{21}x + a_{22}y + a_{23}z \\ z'(t) = a_{31}x + a_{32}y + a_{33}z \end{cases} \tag{1}$$

where the a_{ij} are constants. Suppose that (1) satisfies the existence and uniqueness theorem. Given a matrix of order 3×3, we have the following possibilities for the real canonical forms:

$$\begin{pmatrix} \lambda_1 & 0 & 0 \\ 0 & \lambda_2 & 0 \\ 0 & 0 & \lambda_3 \end{pmatrix}, \begin{pmatrix} \lambda_1 & 1 & 0 \\ 0 & \lambda_1 & 0 \\ 0 & 0 & \lambda_3 \end{pmatrix}, \begin{pmatrix} \lambda_1 & 1 & 0 \\ 0 & \lambda_1 & 1 \\ 0 & 0 & \lambda_1 \end{pmatrix}, \begin{pmatrix} \alpha & -\beta & 0 \\ \beta & \alpha & 0 \\ 0 & 0 & \lambda_3 \end{pmatrix},$$

where the eigenvalue $\lambda = \alpha \pm i\beta$ with $\alpha = 0$ for pure imaginary, $\beta = 0$ for real case and both are different of zero for complex λ. If the singularities in matrix A are hyperbolic $\alpha \neq 0$ then we have the following possibilities:

Table 1. Classification of the hyperbolic flows in dimension 3.

Classification	Eigenvalues $\lambda_1, \lambda_2, \lambda_3$
Attractors (stable)	$\lambda = \alpha \pm i\beta(\alpha < 0)$ and $\lambda_3 < 0$ or $\lambda_1, \lambda_2, \lambda_3 < 0$
Saddle point	$\lambda = \alpha \pm i\beta, \alpha < 0(>0)$ and $\lambda_3 > 0(<0)$ or $\lambda_1, \lambda_2 < 0(>0)$ and $\lambda_3 > 0(<0)$
Repellors (unstable)	$\lambda = \alpha \pm i\beta(\alpha > 0)$ and $\lambda_3 > 0$ or $\lambda_1, \lambda_2, \lambda_3 > 0$

Remark: We are not interested in the cases $\alpha = 0$ and/or the real eigenvalue equal to zero since we cannot classify the equilibrium point.

3   Interval 3-Dimensional Linear Differential System

Given system (1) with the initial conditions, we can to consider the Initial Value Problem with uncertainty, where the initial conditions and/or coefficients are uncertainty. The behavior of the solution trajectories are not changed if only its initial condition has a small perturbation. For example, consider the system

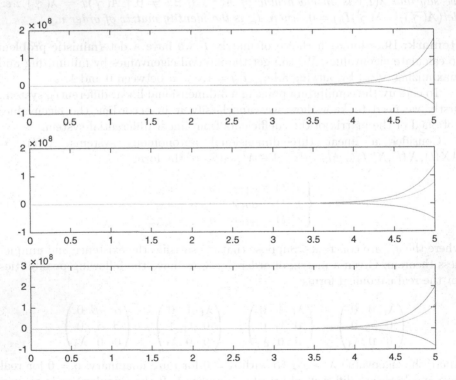

Fig. 1. The graph for $x(t), y(t), z(t)$ com initial conditions [0.8 0.8 0.8], [1 1 1] and [1.2 1.2 1.2], respectively.

$X'(t) = AX(t)$, where $A = \begin{pmatrix} 2 & 0 & -1 \\ 3 & 4 & 1 \\ 2 & 1 & 3 \end{pmatrix}$ with the initial conditions $\mathbf{x}_0 = \mathbf{y}_0 = \mathbf{z}_0 = [0.8 \; 1.2]$ then we have the following unstable trajectories (see Fig. 1).

Remark: When only the initial conditions are intervals then the eigenvalues do not change and the stability or instability is kept. But, if in (1) the entries in the matrix of coefficient vary, then we need to evaluate what will happen with the equilibrium point in (1).

Then, we consider in (1), A as an interval matrix. According to Definition 3, we have the following problem:

$$\begin{cases} x'(t) = (\underline{a}_{11} + \gamma_{11}w_{11})x(t) + (\underline{a}_{12} + \gamma_{12}w_{12})y(t) + (\underline{a}_{13} + \gamma_{13}w_{13})z(t) \\ y'(t) = (\underline{a}_{21} + \gamma_{21}w_{21})x(t) + (\underline{a}_{22} + \gamma_{22}w_{22})y(t) + (\underline{a}_{23} + \gamma_{23}w_{23})z(t) \\ z'(t) = (\underline{a}_{31} + \gamma_{31}w_{31})x(t) + (\underline{a}_{32} + \gamma_{32}w_{32})y(t) + (\underline{a}_{33} + \gamma_{33}w_{33})z(t), \end{cases} \quad (2)$$

where $\gamma_{ij} \in [0,1]$, $w_{ij} = \overline{a}_{ij} - \underline{a}_{ij}$, for $i,j = 1,2,3$.
Observe that system (2):

1. Has an unique equilibrium point at the origin $(0,0,0)$ if given the matrix

$$A(\overrightarrow{\gamma}) = \begin{pmatrix} \underline{a}_{11} + \gamma_{11}w_{11} & \underline{a}_{12} + \gamma_{12}w_{12} & \underline{a}_{13} + \gamma_{13}w_{13} \\ \underline{a}_{21} + \gamma_{21}w_{21} & \underline{a}_{22} + \gamma_{22}w_{22} & \underline{a}_{22} + \gamma_{23}w_{23} \\ \underline{a}_{31} + \gamma_{31}w_{31} & \underline{a}_{32} + \gamma_{32}w_{32} & \underline{a}_{33} + \gamma_{33}w_{33} \end{pmatrix},$$

   the $det(A(\overrightarrow{\gamma})) \neq 0, \forall \gamma_{ij} \in [0,1]$ for $i,j = 1,2,3$.

2. For $\overrightarrow{\gamma} = (\gamma_{11}, \gamma_{12}, \gamma_{13}, \gamma_{21}, \gamma_{22}, \gamma_{23}, \gamma_{31}, \gamma_{32}, \gamma_{33})$, the eigenvalues are obtained from the equation:

$$p(\lambda(\overrightarrow{\gamma})) = -\lambda^3(\overrightarrow{\gamma}) + a_2(\overrightarrow{\gamma})\lambda^2(\overrightarrow{\gamma}) - a_1(\overrightarrow{\gamma})\lambda(\overrightarrow{\gamma}) + a_0(\overrightarrow{\gamma}), \quad (3)$$

   where

   $a_2(\overrightarrow{\gamma}) = tr(A(\Gamma)) = \underline{a}_{11} + \gamma_{11}w_{11} + \underline{a}_{22} + \gamma_{22}w_{22} + \underline{a}_{33} + \gamma_{33}w_{33}$;

   $a_1(\overrightarrow{\gamma}) = [-(\underline{a}_{11} + \gamma_{11}w_{11}) - (\underline{a}_{22} + \gamma_{22}w_{22}) -](\underline{a}_{33} + \gamma_{33}w_{33}) + [\underline{a}_{31} + \gamma_{31}w_{31} + \underline{a}_{13} + \gamma_{13}w_{13}](\underline{a}_{32} + \gamma_{32}w_{32}) + (\underline{a}_{12} + \gamma_{12}w_{12})(\underline{a}_{13} + \gamma_{13}w_{13}) - (\underline{a}_{11} + \gamma_{11}w_{11})(\underline{a}_{22} + \gamma_{22}w_{22})$;

   $a_0(\overrightarrow{\gamma}) = [(\underline{a}_{11} + \gamma_{11}w_{11})(\underline{a}_{22} + \gamma_{22}w_{22}) - (\underline{a}_{12} + \gamma_{12}w_{12})(\underline{a}_{13} + \gamma_{13}w_{13})(\underline{a}_{21} + \gamma_{21}w_{21})](\underline{a}_{33} + \gamma_{33}w_{33}) + [[\underline{a}_{12} + \gamma_{12}w_{12} - \underline{a}_{11} - \gamma_{11}w_{11}](\underline{a}_{31} + \gamma_{31}w_{31}) + [\underline{a}_{21} + \gamma_{21}w_{21} - \underline{a}_{22} - \gamma_{22}w_{22}](\underline{a}_{13} + \gamma_{13}w_{13})](\underline{a}_{32} + \gamma_{32}w_{32})$.

Theorem 1. *If system (2) has a unique equilibrium point at $(0,0,0)$, then there is a matrix $\Gamma = (\gamma_{ij}), i,j = 1,2,3$ such that the equilibrium point is classified according to Table 1.*

Proof. Given the matrix $A(\overrightarrow{\gamma})$ in system (2), the nature of the equilibrium is defined according to the zeroes of the characteristic polynomial of $A(\overrightarrow{\gamma})$

$$-\lambda^3(\overrightarrow{\gamma}) + a_2(\overrightarrow{\gamma})\lambda^2(\overrightarrow{\gamma}) + a_1(\overrightarrow{\gamma})\lambda(\overrightarrow{\gamma}) + a_3(\overrightarrow{\gamma}) = 0, \quad (4)$$

The roots of the polynomial of order 3 in (4) are defined according to the following discriminant [4]:

$$\Delta(\overrightarrow{\gamma}) = 4p^3(\overrightarrow{\gamma}) + q^2(\overrightarrow{\gamma}), \tag{5}$$

where $q(\overrightarrow{\gamma}) = a_0(\overrightarrow{\gamma}) + \dfrac{a_2(\overrightarrow{\gamma})a_1(\overrightarrow{\gamma})}{3} + \dfrac{2}{27}a_2^3(\overrightarrow{\gamma})$ and $p(\overrightarrow{\gamma}) = -\dfrac{1}{3}a_1(\overrightarrow{\gamma}) - \dfrac{1}{9}a_2^2(\overrightarrow{\gamma})$. Then, we have:

1. If $\Delta(\overrightarrow{\gamma}) < 0$, (4) has 3 different real roots;
2. If $\Delta(\overrightarrow{\gamma}) > 0$, (4) has 1 real root and 2 complex (conjugate) roots;
3. If $\Delta(\overrightarrow{\gamma}) = 0$ and $p(\overrightarrow{\gamma}) < 0$ then (4) has 3 real roots, where two of them are equal;
4. If $p(\overrightarrow{\gamma}) = q(\overrightarrow{\gamma}) = 0$, then (4) has 3 equal real roots.

The analysis depends of the entries in matrix $A(\overrightarrow{\gamma})$ for $\gamma_{ij}, i, j = 1, 2, 3$.

1. First case: if all entries of matrix $A(\overrightarrow{\gamma})$ in system (2) are dependent, then $\gamma_{ij} = \gamma, \forall i, j = 1, 2, 3$ and the eigenvalues are obtained from the equation:
   $p(\lambda(\gamma)) = -\lambda^3(\gamma) + a_2(\gamma)\lambda^2(\gamma) - a_1(\gamma)\lambda(\gamma) + a_0(\gamma)$,
   where
   $a_2(\gamma) = tr(A(\gamma)) = \underline{a}_{11} + \underline{a}_{22} + \underline{a}_{33} + \gamma(w_{11} + w_{22} + w_{33})$;
   $a_1(\gamma) = \begin{vmatrix} \underline{a}_{11} + \gamma w_{11} & \underline{a}_{12} + \gamma w_{12} \\ \underline{a}_{21} + \gamma w_{21} & \underline{a}_{22} + \gamma w_{22} \end{vmatrix} + \begin{vmatrix} \underline{a}_{11} + \gamma_1 w_{11} & \underline{a}_{13} + \gamma_3 w_{13} \\ \underline{a}_{31} + \gamma_7 w_{31} & \underline{a}_{33} + \gamma_9 w_{33} \end{vmatrix} + \begin{vmatrix} \underline{a}_{22} + \gamma_5 w_{22} & \underline{a}_{22} + \gamma_6 w_{23} \\ \underline{a}_{32} + \gamma_8 w_{32} & \underline{a}_{33} + \gamma_9 w_{33} \end{vmatrix}$;
   $a_0(\gamma) = det(A(\gamma))$.

   Then, the classification can be obtained using the particular expression (5).
2. Second case: if the matrix $A(\overrightarrow{\gamma})$ is symmetric, then $\underline{a}_{ij} + \gamma_{ij}w_{ij} = \underline{a}_{ji} + \gamma_{ji}w_{ji}, \forall i \neq j, i, j = 1, 2, 3$ and $\overrightarrow{\gamma} = (\gamma_{11}, \gamma_{12}, \gamma_{13}, \gamma_{22}, \gamma_{23}, \gamma_{33})$. In this case, the eigenvalues are obtained from the equation:
   $p(\lambda(\overrightarrow{\gamma})) = -\lambda^3(\overrightarrow{\gamma}) + \lambda^2(\overrightarrow{\gamma})a_2(\overrightarrow{\gamma}) - \lambda(\overrightarrow{\gamma})a_1(\overrightarrow{\gamma}) + a_0(\overrightarrow{\gamma})$,
   where
   $a_2(\overrightarrow{\gamma}) = tr(A(\overrightarrow{\gamma})) = \underline{a}_{11} + \underline{a}_{22} + \underline{a}_{33} + \gamma_{11}w_{11} + \gamma_{22}w_{22} + \gamma_{33}w_{33}$;
   $a_1(\overrightarrow{\gamma}) = \begin{vmatrix} \underline{a}_{11} + \gamma_{11}w_{11} & \underline{a}_{12} + \gamma_{12}w_{12} \\ \underline{a}_{12} + \gamma_{12}w_{12} & \underline{a}_{22} + \gamma_{22}w_{22} \end{vmatrix} + \begin{vmatrix} \underline{a}_{11} + \gamma_{11}w_{11} & \underline{a}_{13} + \gamma_{13}w_{13} \\ \underline{a}_{13} + \gamma_{13}w_{13} & \underline{a}_{33} + \gamma_{33}w_{33} \end{vmatrix}$
   $+ \begin{vmatrix} \underline{a}_{22} + \gamma_{22}w_{22} & \underline{a}_{23} + \gamma_{23}w_{23} \\ \underline{a}_{23} + \gamma_{23}w_{23} & \underline{a}_{33} + \gamma_{33}w_{33} \end{vmatrix}$;
   $a_0(\overrightarrow{\gamma}) = det(A(\overrightarrow{\gamma}))$.
3. Third case: if all elements of the matrix are independent and the eigenvalues are obtained from Eq. (3). For each $\gamma_{ij} \in [0 \; 1], w_{ij} = \overline{a}_{ij} - \underline{a}_{ij}, i, j = 1, 2, 3$, we have the characteristic polynomial of degree 3.

Note that if in all cases we have $\gamma_{ij} = 0, i, j = 1, 2, 3$ we have the deterministic case and for each $\gamma, \gamma_{ij} \in [0 \; 1], w_{ij} = \overline{a}_{ij} - \underline{a}_{ij}, i, j = 1, 2, 3$, we need to get the sign of the roots for the characteristic polynomial of degree 3 to study the stability in (2). Here we want to choose $\gamma_{ij} \in [0 \; 1]$ so that there is an unique equilibrium [12].

Remark 1. For the square matrix of order 3, it is difficult to find, explicitly, the regions of the hypercube of dimension 9 that give a complete classification of the singularities of the characteristic matrix equation of system (2), even in the case of symmetry or the case of total dependence. Considering these, we will first describe the method called Cylindrical Algebraic Decomposition (CAD) used to find the lower and upper bound for real interval eigenvalue (see [1–3]).

To this end, we deal with a semi-algebraic set $S \subset \mathbb{R}^n$ which a finite union of sets defined by polynomial equations and inequalities with real coefficients. The CAD provides a partition of S into semi-algebraic pieces which are homeomorphic to $]0\ 1[^i$ for $i = 1, \ldots, n$. Moreover, classification problems involving such a set S, reduces to the computing of a finite number of sample points in each connected component, and then facing a polynomial optimization question. The algorithm is implemented in RAGlib (Real Algebraic Geometry library) of the software Maple. For example, if $S = \{P_1 = \ldots = P_n = 0, \lambda_1 > 0, \ldots, \lambda_m > 0\}$, the first step is a reduction to compute sample points in each component of S defined with non-strict inequalities. There is a connected component C_e of $S_e = \{P_1 = \ldots = P_n = 0, \lambda_1 \geq e, \ldots, \lambda_m \geq 0, 0 < e < e_0\}$ and a suitable e_0 that can be found using notions of critical values and asymptotic critical values. The next step of the algorithm addresses an algebraic problem.

In the next example, we analyze a particular 3-dimensional interval differential system when all entries in the matrix are dependent and independent via CI. For the independent case, firstly we find conditions to get 3, 2 and 1 real eigenvalues and one real and a pair of complex as was described in the proof of the Theorem 1. Besides, in Proposition 1 and 2 we find the real interval eigenvalue by using techniques from real algebraic geometry. The same method cannot be used to find the complex interval eigenvalue, since the $\leq$ real ordering is not longer available. Finally, we compare the values obtained with the Deif's method [5] and Rohn's method [16].

Example 1. Consider the system of the differential equations $X'(t) = A(\overrightarrow{\gamma})X(t)$, where $A(\overrightarrow{\gamma})$ is an interval matrix written as

$$A(\overrightarrow{\gamma}) = \begin{pmatrix} 2\gamma_{11} & 0 & -3 + 2\gamma_{12} \\ -1 + 4\gamma_{21} & -2 + 6\gamma_{22} & -1 + 2\gamma_{23} \\ -2 + 4\gamma_{31} & 1 & 1 + 2\gamma_{33} \end{pmatrix} \tag{6}$$

Here to simplify the notation, $\gamma_{11} = \gamma_1, \gamma_{12} = 0, \gamma_{13} = \gamma_2, \gamma_{21} = \gamma_3, \gamma_{22} = \gamma_4, \gamma_{23} = \gamma_5, \gamma_{31} = \gamma_6, \gamma_{32} = 0, \gamma_{33} = \gamma_7$ in such way $\overrightarrow{\gamma} = (\gamma_1, \gamma_2, \gamma_3, = \gamma_4, \gamma_5, \gamma_6, \gamma_7)$. $det(A(\overrightarrow{\gamma}) - \lambda I_3) = 0$ implies that the characteristic polynomial

$$P(\overrightarrow{\gamma}) = a_3(\overrightarrow{\gamma})\lambda^3\overrightarrow{\gamma}(\overrightarrow{\gamma}) + a_2(\overrightarrow{\gamma})\lambda^2(\overrightarrow{\gamma}) + a_1(\overrightarrow{\gamma})\lambda(\overrightarrow{\gamma}) + a_0(\overrightarrow{\gamma}) = 0, \tag{7}$$

where $a_3(\overrightarrow{\gamma}) = -1$;
$a_2(\overrightarrow{\gamma}) = -1 + 2\gamma_1 + 6\gamma_4 + 2\gamma_7$;
$a_1(\overrightarrow{\gamma}) = 7 + 2\gamma_1 - 4\gamma_2 - 6\gamma_4 - 12\gamma_1\gamma_4 + 2\gamma_5 - 12\gamma_6 + 8\gamma_2\gamma_6 + 4\gamma_7 - 4\gamma_1\gamma_7 - 12\gamma_4\gamma_7$;
$a_0(\overrightarrow{\gamma}) = 15 - 2\gamma_1 - 10\gamma_2 - 12\gamma_3 + 8\gamma_2\gamma_3 - 36\gamma_4 + 12\gamma_1\gamma_4 + 24\gamma_2\gamma_4 - 4\gamma_1\gamma_5 - 24\gamma_6 + 16\gamma_2\gamma_6 + 72\gamma_4\gamma_6 - 48\gamma_2\gamma_4\gamma_6 - 8\gamma_1\gamma_7 + 24\gamma_1\gamma_4\gamma_7$.

The solution of (7), $P(\overrightarrow{\gamma}) = 0$ for λ subject to $\overrightarrow{\gamma} = (\gamma_1, \ldots, \gamma_7), 0 \leq \gamma_i \leq 1, i = 1, \ldots, 7$ is obtained using the cube root formula

$$\lambda(\overrightarrow{\gamma}) = (P(\overrightarrow{\gamma}))^{\frac{1}{3}}, \overrightarrow{\gamma} = (\gamma_1, \ldots, \gamma_7). \tag{8}$$

Thus, the real interval eigenvalue is

$$[\lambda] = \left[\min_{0 \leq \gamma_i \leq 1} (P(\overrightarrow{\gamma}))^{\frac{1}{3}}, \min_{0 \leq \gamma_i \leq 1} (P(\overrightarrow{\gamma}))^{\frac{1}{3}} \right]. \tag{9}$$

Firstly, consider in (7) that all interval entries in matrix are dependent, then $\gamma_i = \gamma, \forall i = 1, \ldots, 7$ and, we have the following equation for the eigenvalues:
$$-\lambda^3(\gamma) + (1 - 10\gamma)\lambda^2(\gamma) + (-20\gamma^2 - 14\gamma + 7)\lambda(\gamma) + (-24\gamma^3 + 120\gamma^2 - 84\gamma + 15) = 0.$$
Thus, the eigenvalues are:

$$\lambda_{1,2}(\gamma) = \left(\frac{1}{2} \pm \frac{\sqrt{3}}{2} i \right) \left\{ \frac{2(\gamma - 1)\sqrt{-128\gamma^4 - 736\gamma^3 + 2052\gamma^2 - 1114\gamma + 69}}{\sqrt{27}} - \frac{224\gamma^3 - 780\gamma^2 + 726\gamma - 170}{27} \right\}^{1/3}$$

$$+ \left(\frac{\mp\sqrt{3}}{18}i - \frac{1}{18} \right) \sqrt[3]{\dfrac{(40\gamma^2 - 62\gamma + 22)^3}{\dfrac{2(\gamma - 1)\sqrt{-128\gamma^4 - 736\gamma^3 + 2052\gamma^2 - 1114\gamma + 69}}{\sqrt{27}} - \dfrac{224\gamma^3 - 780\gamma^2 + 726\gamma - 170}{27}}} - \frac{1 - 10\gamma}{3}$$

and $\lambda_3(\gamma)$ is

$$\left\{ \frac{2(\gamma - 1)\sqrt{-128\gamma^4 - 736\gamma^3 + 2052\gamma^2 - 1114\gamma + 69}}{\sqrt{27}} - \frac{224\gamma^3 - 780\gamma^2 + 726\gamma - 170}{27} \right\}^{1/3}$$

$$+ \sqrt[3]{\dfrac{40\gamma^2 - 62\gamma + 22}{\dfrac{2(\gamma - 1)\sqrt{-128\gamma^4 - 736\gamma^3 + 2052\gamma^2 - 1114\gamma + 69}}{\sqrt{27}} - \dfrac{224\gamma^3 - 780\gamma^2 + 726\gamma - 170}{27}}}$$

For $\gamma = 0, \lambda_3(0) = 3, \lambda_{1,2}(0) = -2 \pm i; \gamma = 1, \lambda_{1,2,3} = 3, \gamma = .07098016326$ 152795, $\lambda_3.07098016326152795)(= 2.89667, \lambda_{1,2} = -1.59343 \pm 0.886988i; \gamma = .6768941902714837$, $\lambda_{1,2}(.07098016326152795) = 2.53294 \pm 1.28951i, \lambda_3 = 0.703058$, and so on.

Analysing the graph of

$$\left\{ \frac{2(\gamma - 1)\sqrt{-128\gamma^4 - 736\gamma^3 + 2052\gamma^2 - 1114\gamma + 69}}{\sqrt{27}} - \frac{224\gamma^3 - 780\gamma^2 + 726\gamma - 170}{27} \right\}^{1/3},$$

we can conclude that for γ between 0.29 and 0.39 we have three real eigenvalues, for $\gamma = 1$ there is a unique triple real root and in other cases we have one real and two complex eigenvalues.

Secondly, consider $0 \leq \gamma_i \leq 1, i = 1, \ldots, 7$ are independent, then can be proved that for the real case, the min/max of the eigenvalues are obtained at a corner point on the boundary of the space of the parameters in a hypercube of the 7-dimension. Note that to obtain conditions for the complex eigenvalues is not easy, because the complex set is not an ordered set and we have an optimization problem to find the conditions for the parameters $\gamma_i, i = 1, \ldots, 7$. For all $\gamma_i \in \mathbb{R}^7$, the 3×3 matrix $A(\overrightarrow{\gamma})$ has at least one real eigenvalue, therefore for all $\overrightarrow{\gamma} \in \mathbb{R}^7$, one can define $\lambda_{min}(A(\overrightarrow{\gamma}))$

(*resp.* $\lambda_{max}(A(\overrightarrow{\gamma}))$) as the minimal (*resp.* maximal) real eigenvalue of $A(\overrightarrow{\gamma}) = A(\gamma_1, \ldots, \gamma_7)$. Let us now consider the compact set $\mathcal{B}_7 = [0,1]^7 \subset \mathbb{R}^7$.

We denote by $P = P(\lambda, \overrightarrow{\gamma})$ the characteristic polynomial of $A(\overrightarrow{\gamma})$. Then by considering the discriminant (5), where $q(\overrightarrow{\gamma}) = a_0(\overrightarrow{\gamma}) + \dfrac{a_2(\overrightarrow{\gamma})a_1(\overrightarrow{\gamma})}{3} + \dfrac{2}{27}a_2^3(\overrightarrow{\gamma})$ and $p(\overrightarrow{\gamma}) = -\dfrac{1}{3}a_1(\overrightarrow{\gamma}) - \dfrac{1}{3}a_2^2(\overrightarrow{\gamma})$ we have the following analysis.

1. There are three different real eigenvalues for $\triangle = [-162.963, 0]$. The left side is defined by parameters $\gamma_4 = \gamma_5 = \gamma_7 = 1, \gamma_1 = \gamma_2 = \gamma_3 = \gamma_6 = 0$ that define the characteristic equation $-\lambda^3 + 7\lambda^2 - 5\lambda - 21 = -(\lambda-3)(\lambda-5.31)(\lambda-1.32) = 0$ and the right side by $\gamma_1 = 0.64, \gamma_2 = 0.38, \gamma_3 = 0, \gamma_4 = .25, \gamma_5 = .62; \gamma_6 = 0.64, \gamma_7 = 0.33$ such that $-\lambda^3 + 2.44\lambda^2 - 1.6992\lambda - 0.2432 = -(\lambda - .198725)(\lambda - 0.941681)(\lambda - 1.29959) = 0$. We have that the equilibrium point in this case are saddle and repulsor, respectively.

2. There is one real and a complex conjugate for $\triangle = [0 \ 452]$ The left side is defined by parameters $\gamma_4 = 0.75, \gamma_5 = 0.12, \gamma_7 = .52, \gamma_1 = 0.89, \gamma_2 = 0.92, \gamma_3 = 0, \gamma_6 = .18$ that define the characteristic equation $-\lambda^3 + 6.32\lambda^2 - 12.4564\lambda + 7.8788 = -(\lambda - 3.18777)(\lambda - 1.56612 + 0.137312i)(\lambda - 1.56612 - 0.137312i) = 0$ and the right side by $\gamma_1 = \gamma_2 = \gamma_3 = \gamma_7 = 0, \gamma_4 = \gamma_5\gamma_6 = 1$, such that $-\lambda^3 + 5\lambda^2 - 9\lambda + 27 = -(\lambda - 0.321648 + 2.46858i)(\lambda - 0.321648 - 2.46858i)(\lambda - 4.3567) = 0$. Note that the equilibrium point in both cases are repulsing.

3. If $\triangle(\overrightarrow{\gamma}) = 0$ and $p(\overrightarrow{\gamma}) < 0$, then (7) has 3 real roots, where two them are equal. Then,

$$
\begin{aligned}
\triangle(\overrightarrow{\gamma}) = \Big(&(15 - 2\gamma_1 - 10\gamma_2 - 12\gamma_3 + 8\gamma_2\gamma_3 - 36\gamma_4 + 12\gamma_1\gamma_4 + 24\gamma_2\gamma_4 - 4\gamma_1\gamma_5 - 24\gamma_6 \\
& + 16\gamma_2\gamma_6 + 72\gamma_4\gamma_6 - 48\gamma_2\gamma_4\gamma_6 - 8\gamma_1\gamma_7 + 24\gamma_1\gamma_4\gamma_7 + \frac{1}{3}\Big(7 + 2\gamma_1 - 4\gamma_2 - 6\gamma_4 \\
& - 12\gamma_1\gamma_4 + 2\gamma_5 - 12\gamma_6 + 8\gamma_2\gamma_6 + 4\gamma_7 - 4\gamma_1\gamma_7 - 12\gamma_4\gamma_7\Big)\Big(-1 + 2\gamma_1 + 6\gamma_4 + 2\gamma_7\Big) \\
& + \frac{2}{27}\Big(-1 + 2\gamma_1 + 6\gamma_4 + 2\gamma_7\Big)^3\Big)^2 + 4\Big(-\frac{1}{3}\Big(7 + 2\gamma_1 - 4\gamma_2 - 6\gamma_4 - 12\gamma_1\gamma_4 + 2\gamma_5 \\
& - 12\gamma_6 + 8\gamma_2\gamma_6 + 4\gamma_7 - 4\gamma_1\gamma_7 - 12\gamma_4\gamma_7 + \frac{1}{3}\Big(-1 + 2\gamma_1 + 6\gamma_4 + 2\gamma_7\Big)^2\Big)^3\Big)
\end{aligned}
$$

and $p(\overrightarrow{\gamma}) < 0$. (7) has two equal roots if, and only if, (7) and its first derivative have the same roots. Then we have the condition $\gamma_i = 0, i = 1, 2, 7$. Moreover, if $\gamma_i = 0, i = 2, 5$, we have the condition $\gamma_6 < \dfrac{11}{16}$. Then for example $\gamma_6 = \dfrac{1}{2}$, then $\gamma_3 = \dfrac{43}{16}$ or $\dfrac{1}{6}$, with characteristic equation given by $-\dfrac{5}{27} + \lambda - \lambda^2 - \lambda^3 = -\Big(\lambda - \dfrac{1}{3}\Big)^2\Big(\lambda + \dfrac{5}{3}\Big) = 0$ and $1 + \lambda - \lambda^2 - \lambda^3 = -(\lambda+1)^2(\lambda-1) = 0$, respectively.

If $p(\overrightarrow{\gamma}) = q(\overrightarrow{\gamma}) = 0$, that is, $a_0(\overrightarrow{\gamma}) = \dfrac{25}{27}a_2^3(\overrightarrow{\gamma})$, (7) has one triple real root. The expression is

$$
\begin{aligned}
\gamma_2 = \Big(&- 215 + 102\gamma_1 - 150\gamma_1^2 + 100\gamma_1^3 + 162\gamma_3 + 711\gamma_4 - 1062\gamma_1\gamma_4 + 900\gamma_1^2\gamma_4 \\
& - 1350\gamma_4^2 + 2700\gamma_1\gamma_4^2 + 2700\gamma_4^3 + 54\gamma_1\gamma_5 + 324\gamma_6 - 972\gamma_4\gamma_6 + 75\gamma_7 - 192\gamma_1\gamma_7 \\
& + 300\gamma_1^2\gamma_7 - 900\gamma_4\gamma_7 + 1476\gamma_1\gamma_4\gamma_7 + 2700\gamma_4^2\gamma_7 - 150\gamma_7^2 + 300\gamma_1\gamma_7^2 + 900\gamma_4\gamma_7^2 \\
& + 100\gamma_7^3\Big)/27(-5 + 4\gamma_3 + 12\gamma_4 + 8\gamma_6 - 24\gamma_4\gamma_6),
\end{aligned}
$$

where $0 \leq \gamma_i \leq 1, i = 1, \ldots, 7$. For example, if $\gamma_i = 1, \forall i = 1, \ldots, 7, \lambda = 3$ is the unique eigenvalue.

Equation (5) is equivalent to $4q^3(\overrightarrow{\gamma}) + 27p^2(\overrightarrow{\gamma})$ and to $\dfrac{q^2}{4}(\overrightarrow{\gamma}) + \dfrac{p^3(\overrightarrow{\gamma})}{27}$, but to find the roots of (7) we need to use the last one [4], so that

$$\left[\frac{1}{4} \left(\frac{2}{27} a_2(\overrightarrow{\gamma})^2 + \frac{a_2(\overrightarrow{\gamma}) a_1(\overrightarrow{\gamma})}{3} + a_0(\overrightarrow{\gamma}) \right)^2 + \frac{1}{27} \left(\frac{-a_2(\overrightarrow{\gamma})^2}{3} - a_1 \right)^3 \right] \quad (10)$$

Our first goal is to estimate the extreme values of the multiple roots in $\mathcal{B}_7$. We begin by dealing with the multiple roots of the derivative $\frac{\partial P}{\partial \lambda}$ of the characteristic polynomial. We observe that the parameter γ_3 is not present in this derivative. Let us denote $\overrightarrow{\gamma} = (\gamma_1, \gamma_2, \gamma_4, \gamma_5, \gamma_6, \gamma_7)$. Let

$$P_1(\overrightarrow{\gamma}) = \frac{\partial P}{\partial \lambda}(\lambda, \overrightarrow{\gamma}) = 3\lambda^2 + 2a_2(\overrightarrow{\gamma})\lambda + a_1(\overrightarrow{\gamma}). \quad (11)$$

In the Eq. (11,) let us now consider the compact set $\overrightarrow{\gamma} \in \mathcal{B}_6 = [0, 1]^6 \subset \mathbb{R}^6$. Then, $P(\lambda, \overrightarrow{\gamma})$ has a double root in an interior point of $\mathcal{B}_6$ if for $\lambda = \dfrac{-a_2(\overrightarrow{\gamma}) \pm \sqrt{a_2^2(\overrightarrow{\gamma}) - 3a_1(\overrightarrow{\gamma})}}{3}$, we have $a_2^2(\overrightarrow{\gamma}) - 3a_1(\overrightarrow{\gamma}) = 0$ and in this case $\lambda = \dfrac{-a_2(\overrightarrow{\gamma})}{3}$.

The real eigenvalues can be shown to be in $[-3.39 \ 5.69]$ in accordance to the Propositions 1 and 2, which follow.

Third, Deif [5] considers the interval matrix $\begin{pmatrix} [0 \ 2] & [0 \ 0] & [-3 \ -1] \\ [-1 \ 3] & [-2 \ 4] & [-1 \ 1] \\ [-2 \ 2] & [1 \ 1] & [1 \ 3] \end{pmatrix}$ and found that $Re(\lambda) \in [0.2873 \ 4.7346]$ and $Im(\lambda) \in [0 \ 2.1754]$.

Fourth, by using the Rohn's Method outlined in [12], we found $Re(\lambda) \in [-2.70 \ 5.27]$ and $Im(\lambda) \in [-5.34544 \ 5.34544]$.

However, in the method using CI, the matrix $\begin{pmatrix} 0 & 0 & 0 \\ 3 & -2 & 1 \\ -2 & 1 & 1 \end{pmatrix}$ and $\begin{pmatrix} 2 & 0 & -3 \\ 3 & -2 & 1 \\ -2 & 1 & 1 \end{pmatrix}$ give us real eigenvalues in the interval $[-3.39 \ 5.69]$.

In the complex case, we find numerically, $Re(\lambda) \in [-2.11 \ 3.36429]$ for $\gamma_i = 0, \forall i \neq 5$ and $\gamma_1 = 0.94, \gamma_i = 1, i \neq 1, 5, \gamma_5 = 0$, respectively (Fig. 2).

The Propositions 1 and 2 proof that the real interval eigenvalue is $[-3.39 \ 5.69]$. Firstly, it is necessary to find a value Λ such that for all $\lambda > \Lambda, F_\lambda = \{\gamma \in int(\mathcal{B}_7) | \lambda_M(\gamma) = \lambda\}$ does not intersects the discriminant of $P(\lambda, \overrightarrow{\gamma})$.

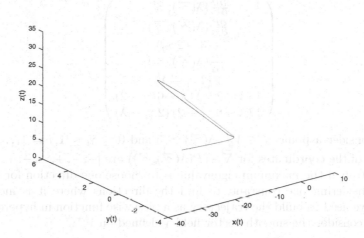

Fig. 2. Trajectories solutions for $\gamma_i = 3/5, 1/5, i = 1, \ldots, 7$ and $\gamma_i = 1/5, i \neq 6, \gamma_6 = 3/5$, respectively.

Proposition 1. λ_{max} *is the upper bound for the real interval eigenvalue in (7) and is the greatest root of* $\lambda^3 - 9\lambda^2 + 19\lambda - 1$, *obtained from the Eq. (7) for* $\overrightarrow{\gamma} = (1\ 0\ 0\ 1\ 1\ 0\ 1)$.

Proof. Let $V_5^+ = \{\overrightarrow{\gamma} \in \mathcal{B}_7 \,|\lambda(\overrightarrow{\gamma}) > 5\}$. First, we show that V_5^+ is non-empty. Note that for $\overrightarrow{\gamma}^0 = (1, 1, 0, 1, 1, 0, 1)$ we have $P(\lambda, \overrightarrow{\gamma}^0) = (\lambda - 5)(\lambda - 1)(\lambda - 3)$ such that $\lambda(\overrightarrow{\gamma}) = 5$. By taking $\overrightarrow{\gamma}^\varepsilon = (1 - c, 1\ \ 8\varepsilon, \varepsilon, 1 - \varepsilon, 1 - \varepsilon, 1 - \varepsilon), \varepsilon > 0$ sufficiently small, one gets $\lambda(\overrightarrow{\gamma}^\varepsilon) = 5 + \dfrac{\varepsilon}{4} + o(\varepsilon) > 5$ with $\overrightarrow{\gamma}^\varepsilon \in int(\mathcal{B}_7)$. Then, the maximum is not attained at an interior set.

We consider on V_5^+ a function $\lambda : \overrightarrow{\gamma} \mapsto \lambda(\overrightarrow{\gamma})$. Since $\lambda(\overrightarrow{\gamma})$ is smooth on V_5^+, $\frac{\partial P}{\partial \lambda(\overrightarrow{\gamma})}(\lambda(\overrightarrow{\gamma}), \overrightarrow{\gamma}) \neq 0$.

The gradient $\overrightarrow{\nabla}_{\overrightarrow{\gamma}}(P)$ of P with respect to $\overrightarrow{\gamma}$ is such that:

$$\overrightarrow{\nabla}_{\overrightarrow{\gamma}}(P(\lambda(\overrightarrow{\gamma}), \overrightarrow{\gamma})) = -\frac{\partial P}{\partial \lambda(\overrightarrow{\gamma})}(\lambda(\overrightarrow{\gamma}), \overrightarrow{\gamma}) \cdot \overrightarrow{\nabla}_{\overrightarrow{\gamma}}\lambda(\overrightarrow{\gamma}) \tag{12}$$

Moreover, $\lambda = \lambda(\overrightarrow{\gamma})$ is the greatest root of $P(\lambda, \overrightarrow{\gamma})$, which is a degree 3 polynomial with positive leading coefficient. Hence $\frac{\partial P}{\partial \lambda}(\lambda(\overrightarrow{\gamma}), \overrightarrow{\gamma}) > 0$. It follows that the respective coordinates of $\overrightarrow{\nabla}_{\overrightarrow{\gamma}}\lambda(\overrightarrow{\gamma})$ and $\overrightarrow{\nabla}_{\overrightarrow{\gamma}}P(\lambda(\overrightarrow{\gamma}), \overrightarrow{\gamma})$ have opposite signs.

The transposed gradient ${}^t\vec{\nabla}_{\vec{\gamma}}P(x,\vec{\gamma})$ is

$$
\begin{pmatrix}
\frac{\partial P}{\partial \gamma_1}(\lambda(\vec{\gamma}),\vec{\gamma}) \\
\frac{\partial P}{\partial \gamma_2}(\lambda(\vec{\gamma}),\vec{\gamma}) \\
4\,(3-2\,\gamma_2) \\
\frac{\partial P}{\partial \gamma_4}(\lambda(\vec{\gamma}),\vec{\gamma}) \\
2\,(2\,\gamma_1-\lambda) \\
4\,(3-2\,\gamma_2)\,(\lambda-6\,\gamma_4+2) \\
2\,(\lambda-6\,\gamma_4+2)\,(2\,\gamma_1-\lambda)
\end{pmatrix}
\tag{13}
$$

Let us consider a point $\vec{\gamma} \in V_5^+, \lambda(\vec{\gamma}) > 5$ and $0 \le \gamma_i \le 1, i = 1,\ldots,7$,then the signs of the coordinates for $\vec{\nabla}_{\vec{\gamma}}(P(\lambda(\vec{\gamma}),\vec{\gamma}))$ are $[-,-,+,-,-,+,-]$. The strategy to find the maximum eigenvalue is to choose one direction for the gradient, considering one constant, to find the directions where it is increasing, because we need to build the trajectory as a piecewise function in hypercube $\mathcal{B}_7$. Then we consider the smooth vector field χ defined on V_5^+,

$$
\chi(\vec{\gamma}) =
\begin{pmatrix}
0 \\
0 \\
4\,(3-2\gamma_2) \\
0 \\
2\,(2\gamma_1-\lambda(\vec{\gamma}) \\
4\,(3-2\gamma_2)\,(\xi(\Gamma)-6\gamma_4+2) \\
2\,(\xi(\Gamma)-6\gamma_4+2)\,(2\gamma_1-\lambda(\vec{\gamma})).
\end{pmatrix}
\tag{14}
$$

Let $\lambda(\vec{\gamma})^0 \in V_5^+ \cap int(\mathcal{B}_7)$ and $\varphi : t \mapsto \varphi(t)$ be the trajectory of χ such that $\varphi(0) = \lambda(\vec{\gamma})^0$. Along this trajectory $\xi(\varphi(t))$ strictly increases, hence $\varphi(t)$ remains in V_5^+. Let $\psi(\lambda(\vec{\gamma})) = \gamma_3, \gamma_6\,(1-\gamma_5)\,(1-\gamma_7)$. For all $\lambda(\vec{\gamma} \in int(\mathcal{B}_7)$, we have $sign(\frac{\partial \psi}{\partial \gamma_3}) = sign(\frac{\partial \psi}{\partial \gamma_6}) > 0$ and $sign(\frac{\partial \psi}{\partial \gamma_5}) = sign(\frac{\partial \psi}{\partial \gamma_7}) < 0$. In addition V_5^+, $\vec{\nabla}_{\lambda(\vec{\gamma})}(P(\lambda(\vec{\gamma}),\lambda(\vec{\gamma})))$ and $\vec{\nabla}_\Gamma\psi(\Gamma)$ have exactly the same signs, excepting where the sign does not change, as well as $\lambda(\lambda(\vec{\gamma}))$ and $\vec{\nabla}_{\lambda(\vec{\gamma})}\psi(\lambda(\vec{\gamma}))$.

It follows that $t \mapsto \psi(\varphi(t))$ decreases, so that it cannot be the minimum on $int(\mathcal{B}_7)$ and there exists a minimal $t^1 > 0$ such that $\psi(\varphi(t^1)) = 0$. Furthermore, if one puts $Z(\psi) = \{\lambda(\vec{\gamma}) \in \mathcal{B}_7 \mid \psi(\lambda(\vec{\gamma})) = 0\}$, this proves that $\lambda_{max} = \max\limits_{\lambda(\vec{\gamma}) \in (Z(\psi) \cap \mathcal{B}_7)} \{\lambda(\vec{\gamma})\}$.

We decrease/increase iteratively each coordinate $\{\gamma_3, \gamma_6, \gamma_5, \gamma_7\}$ to 0 or 1 accordingly to the (constant, non-zero) corresponding gradient coordinate sign. We finally get a point $\gamma^k \in \partial\mathcal{B}_7$ ($k \ge 1$) such that $\{\gamma_5^k = \gamma_7^k = 1, \gamma_3^k = \gamma_6^k = 0\}$ and $\lambda(\vec{\gamma}^k) > \lambda(\vec{\gamma}^0)$. Observe that $\gamma_1^k = \gamma_1^0, \gamma_2^k = \gamma_2^0, \gamma_4^k = \gamma_4^0$.

We now have to deal with the remaining free coordinates $(\gamma_1, \gamma_2, \gamma_4)$. So let us consider the polynomial

$$
\tilde{P}(\lambda, (\gamma_1, \gamma_2, \gamma_4)) = P(\lambda, (\gamma_1, \gamma_2, 0, \gamma_4, 1, 0, 1))
$$

with $(\gamma_1, \gamma_2, \gamma_4) \in \mathcal{B}_3 = [0,1]^3$. The problem we are facing is exactly the same as the original one for P. For all $h = (\gamma_1, \gamma_2, \gamma_4)$, denote by $\tilde{\lambda}(h) =$

$\lambda(\gamma_1, \gamma_2, 0, \gamma_4, 1, 0, 1))$ the greatest root of $\tilde{P}(\lambda, h)$. We introduce the non-empty semi-algebraic subset:

$$\tilde{V}_5^+ = \{(\gamma_1, \gamma_2, \gamma_4) \in \mathcal{B}_3 \mid (\gamma_1, \gamma_2, 0, \gamma_4, 1, 0, 1) \in V_5^+\}.$$

For all $h = (\gamma_1, \gamma_2, \gamma_4) \in \mathcal{B}_3$, and all $\lambda_1 \in \mathbb{R}$, the gradient $\overrightarrow{\nabla}_h \tilde{P}(\lambda_1, h)$ equals $\left(\frac{\partial P}{\partial \gamma_1}(\lambda_1, \lambda(\overrightarrow{\gamma})), \frac{\partial P}{\partial \gamma_2}(\lambda_1, \Gamma), \frac{\partial P}{\partial \gamma_4}(\lambda_1, \lambda(\overrightarrow{\gamma})) \right)$ where $\lambda(\overrightarrow{\gamma}) = (\gamma_1, \gamma_2, 0, \gamma_4, 1, 0, 1)$ $\in V_5^+$, then $\tilde{\lambda}((\overrightarrow{\gamma}) > 5$, such that

- $\frac{\partial \tilde{P}}{\partial \gamma_2}(\tilde{\xi}(h), h) > 0.$
- $\frac{\partial \tilde{P}}{\partial \gamma_1}(\tilde{\xi}(h), h) < 0.$
- $\frac{\partial \tilde{P}}{\partial \gamma_4}(\tilde{\xi}(h), h) < 0.$

We integrate $\overrightarrow{\nabla}_h \tilde{P}$ from a point $h^0 \in \mathcal{B}_3$ and get after at most two other iterations, a point h^* such that $\{\gamma_1 = \gamma_4 = 1, \gamma_2 = 0\}$, with $\tilde{\lambda}(h^*) > \tilde{\lambda}(h^0)$. To $h^* \in \mathcal{B}_3$ corresponds the unique point $\overrightarrow{\gamma}^* \in \mathcal{B}_7$ such that $\{\gamma_2 = \gamma_3 = \gamma_6 = 0, \gamma_1 = \gamma_4 = \gamma_5 = \gamma_7 = 1\}$. Moreover $\lambda(\overrightarrow{\gamma}^*) > \lambda((\overrightarrow{\gamma}^k) > \lambda(\overrightarrow{\gamma}^0)$.

This results in $P(\lambda, \overrightarrow{\gamma}^*) = \lambda^3 - 9\lambda^2 + 19\lambda - 1$ and one can compute explicitly the value $\lambda(\overrightarrow{\gamma}^*)$ which is realized at $\lambda_{max} = 5.69$.

Proposition 2. $\lambda_{min} = -3.39$ is the lower bound for the real interval eigenvalue in (7) and is the smallest root of $X^3 + X^2 - 9X - 3$, obtained from the Eq. (7) for $\overrightarrow{\gamma} = (0\,0\,1\,0\,1\,0\,0)$.

Proof. The proof is similar to the proof of Proposition 1.

Remark 2. Therefore, in this example we showed:

1. Given an equation

$$P_n(x) = x^a + a_{n-1}x^{n-1} + \ldots + a_1x + a_0 = 0, \tag{15}$$

   The discriminant $\Delta(P_n)$ of P_n is a polynomial in the indeterminates $(a_{n-1}, \ldots, a_1, a_0)$ with integer coefficients (explicitly computed as the determinant of a Sylvester matrix, see [13]). The set $\{(a_{n-1}, \ldots, a_1, a_0)|P_n(x) = 0 \text{ has root with multiplicity}\}$ is exactly the set $\{(a_{n-1}, \ldots, a_1, a_0)\,\Delta(P_n) = 0\}$. Then, varying the coefficients in (15) we can get what type of roots it has. In particular the condition on the discriminant for $n = 3$ defines the type of roots for the polynomial equation when $0 \leq \gamma_i \leq 1, i = 1, \ldots, 7$ in (7) are varying. In the real case, we have methods to find them, but in the complex case it is not easy to characterize them completely;

2. There are other methods to find the bounds for interval eigenvalue for (7), but it is not easy(in general not possible) to define the matrix by choosing the entries in interval matrix to get the corresponding eigenvalues. That is, once has the max/min eigenvalues the matrix that generated these eigenvalues is impossible to find. This is not true with the CI approach. We always know the matrix that generated the eigenvalues;

3. CIA gives us the option to choose the parameters for each eigenvalue in the way we can find the matrix explicitly, but may be a NP-hard procedure;
4. Many authors consider the interval matrix $[A] = [A_c - \triangle A \; A_c + \triangle A] = [\underline{A} \; \overline{A}]$. By CIA, we get $A_c - \triangle A$ for $\gamma_i = 0$ and $A_c + \triangle A$ for $\gamma_i = 1$.
5. A_c is obtained from CIA taking $\gamma_i = \dfrac{1}{2}$, but for the element $[-3 \; -1]$, $\triangle a_{13} = \dfrac{-1 - (-3)}{2} = 1$ and by CIA $a_{13}(\gamma_2) = -3 + 2\gamma_2$, then $-3 + 2\gamma_2 = 1$ if $\gamma_2 = 2$. This means that methods used by Deif [5], Rohn [16] and [9] are not equivalent. Note the elements $\triangle a_{13} = 1 \notin [-3 \; -1]$ neither $\triangle a_{32} = 0 \notin [1 \; 1]$.
6. Mathematica and Maple were used as tools to analyze and get some results.

4   Conclusion

This research outlined a method, involving semi algebraic sets theory, for which the stability of interval linear differential equations can be analyzed via constraint intervals. As a by product, a method for obtaining conditions about parameters to get real or complex eigenvalues of interval matrices of order 3×3 were developed.

References

1. Benedetti, R., Risler, J.J.: Real Algebraic and Semi-algebraic sets, Actualiteés mathematique. Herman, Berlin (1990)
2. Bonnard, B., Cotes, O., Faugère, J.C., Jacquemard, A., Rouot, M.S., Verron, T.: Algebraic geometric classification of the singular flow in the contrast imaging problem in nuclear magnetic resonance. https://hal.inria.fr/hal-01556806. Accessed 4 Feb 2020
3. Bonnard, B., Faugère, J.C., Jacquemard A., Safey El Din, M., Verron, T.: Determinantal sets, singularities and application to optimal control in medical imagery. In: Proceedings of the ISSAC, pp. 103–110 (2016)
4. Burnside, W.S., Panton, W.A.: The Theory of Equations: With an Introduction to the Theory of Binary Algebraic Forms. Dublin University Press Series. Longmans Green, London (1802)
5. Deif, A.S.: The interval eigenvalue problem. Z. Angew. Math. Mech. **71**, 61–64 (1991)
6. Hladík, M., Daney, D., Tsigaridas, E.: An algorithm for addressing the real interval eigenvalue problem. J. Comput. Appl. Math. **235**, 2715–2730 (2010)
7. Hladík, M., Daney, D., Tsigaridas, E.: A filtering method for the interval eigenvalue problem. Appl. Math. Comput. **217**, 5236–5242 (2011)
8. Hladík, M., Daney, D., Tsigaridas, E.: Bounds on real eigenvalues and singular values of interval matrices. SIAM J. Matrix Anal. Appl. **31**(4), 2116–2129 (2010)
9. Hladík, M.: Bounds on eigenvalues of real and complex interval matrices. Appl. Math. Comput. **219**(10), 5584–5591 (2013)
10. Leng, H., He, Z., Yuan, Q.: Computing bounds to real eigenvalues of real-interval matrices. Int. J. Numer. Methods Eng. **74**, 523–530 (2008)
11. Lodwick, W.A.: Constrained interval arithmetic. CCM Report 138, University of Colorado (USA), February 1999

12. Mizukoshi, M.T., Lodwick, W.A.: The interval eigenvalue problem using constraint interval analysis with an application to linear differential equations: a first step toward the fuzzy eigenvalue problem. Fuzzy Sets and Systems (2019, submitted)
13. Mignotte, M.: Mathematics for Computer Algebra. Springer, New York (1992). https://doi.org/10.1007/978-1-4613-9171-5
14. Qiu, Z., Müller, P.C., Frommer, A.: An approximate method for the standard interval eigenvalue problem of real non-symmetric interval matrices. Commun. Numer. Methods Eng. **17**, 239–251 (2001)
15. Rohn, J., Deif, A.: On the range of an interval matrix. Computing **47**, 373–377 (1992)
16. Rohn, J.: Interval matrices: singularity and real eigenvalues. SIAM J. Matrix Anal. Appl. **14**, 82–91 (1993)

New Results in the Calculus of Fuzzy-Valued Functions Using Mid-Point Representations

Luciano Stefanini[1]([✉]), Laerte Sorini[1], and Mina Shahidi[1,2]

[1] DESP - University of Urbino Carlo Bo, Urbino, Italy
{luciano.stefanini,laerte.sorini}@uniurb.it
[2] Department of Mathematics - IASBS Institute, Zanjan, Iran
minashahidi@iasbs.ac.ir

Abstract. We present new results in the calculus for fuzzy-valued functions of a single real variable. We adopt extensively the midpoint-radius representation of intervals in the real half-plane and show its usefulness in fuzzy calculus. Concepts related to convergence and limits, continuity, level-wise gH-differentiability of first and second orders have nice and useful midpoint expressions. Using mid-point representation of fuzzy-valued functions, partial orders and properties of monotonicity and convexity are discussed and analysed in detail. Periodicity is easy to represent and identify. Graphical examples and pictures accompany the presentation.

Keywords: Fuzzy-valued function · Midpoint representation ·
Monotonic fuzzy function · Convexity of fuzzy function · Periodic fuzzy function

1   Introduction to Intervals and Fuzzy Numbers

We denote by $\mathcal{K}_C$ the family of all bounded closed intervals in $\mathbb{R}$, i.e.,

$$\mathcal{K}_C = \left\{ [a^-, a^+] \mid a^-, a^+ \in \mathbb{R} \text{ and } a^- \leq a^+ \right\}.$$

To describe and represent basic concepts and operations for real intervals, the well-known *midpoint-radius* representation is very useful (see e.g. [2] and the references therein): for a given interval $A = [a^-, a^+]$, define the midpoint $\widehat{a}$ and radius $\widetilde{a}$, respectively, by

$$\widehat{a} = \frac{a^+ + a^-}{2} \text{ and } \widetilde{a} = \frac{a^+ - a^-}{2},$$

so that $a^- = \widehat{a} - \widetilde{a}$ and $a^+ = \widehat{a} + \widetilde{a}$. We denote an interval by $A = [a^-, a^+]$ or, in midpoint notation, by $A = (\widehat{a}; \widetilde{a})$; so

$$\mathcal{K}_C = \{(\widehat{a}; \widetilde{a}) \mid \widehat{a}, \widetilde{a} \in \mathbb{R} \text{ and } \widetilde{a} \geq 0\}.$$

© Springer Nature Switzerland AG 2020
M.-J. Lesot et al. (Eds.): IPMU 2020, CCIS 1238, pp. 28–42, 2020.
https://doi.org/10.1007/978-3-030-50143-3_3

When we refer to an interval $C \in \mathcal{K}_C$, its elements are denoted as c^-, c^+, $\widehat{c}$, $\widetilde{c}$, with $\widetilde{c} \geq 0$, $c^- \leq c^+$ and the interval by $C = [c^-, c^+]$ in extreme-point representation and by $C = (\widehat{c}; \widetilde{c})$, in midpoint notation.

Given $A = [a^-, a^+]$, $B = [b^-, b^+] \in \mathcal{K}_C$ and $\tau \in \mathbb{R}$, we have the following classical (Minkowski-type) addition, scalar multiplication and difference:

- $A + B = [a^- + b^-, a^+ + b^+]$,
- $\tau A = \{\tau a : a \in A\} = \begin{cases} [\tau a^-, \tau a^+], \text{ if } \tau \geq 0, \\ [\tau a^+, \tau a^-], \text{ if } \tau \leq 0 \end{cases}$,
- $-A = (-1)A = [-a^+, -a^-]$,
- $A - B = A + (-1)B = [a^- - b^+, a^+ - b^-]$.

Using midpoint notation, the previous operations, for $A = (\widehat{a}; \widetilde{a})$, $B = (\widehat{b}; \widetilde{b})$ and $\tau \in \mathbb{R}$ are:

- $A + B = (\widehat{a} + \widehat{b}; \widetilde{a} + \widetilde{b})$,
- $\tau A = (\tau \widehat{a}; |\tau| \widetilde{a})$,
- $-A = (-\widehat{a}; \widetilde{a})$,
- $A - B = (\widehat{a} - \widehat{b}; \widetilde{a} + \widetilde{b})$.

We denote the generalized Hukuhara difference (gH-difference in short) of two intervals A and B as $A \ominus_{gH} B = C \iff (A = B + C \text{ or } B = A - C)$; the gH-difference of two intervals always exists and, in midpoint notation, is equal to

$$A \ominus_{gH} B = (\widehat{a} - \widehat{b}; |\widetilde{a} - \widetilde{b}|) \subseteq A - B.$$

The gH-addition for intervals is defined by

$$A \oplus_{gH} B = A \ominus_{gH} (-B) = (\widehat{a} + \widehat{b}; |\widetilde{a} - \widetilde{b}|) \subseteq A + B.$$

If $A \in \mathcal{K}_C$, we will denote by $len(A) = a^+ - a^- = 2\widehat{a}$ the *length* of interval A. Remark that $\alpha A - \beta A = (\alpha + \beta)A$ only if $\alpha\beta \geq 0$ (except for trivial cases) and that $A \ominus_{gH} B = A - B$ or $A \oplus_{gH} B = A + B$ only if A or B are singletons.

For two intervals $A, B \in \mathcal{K}_C$ the Pompeiu–Hausdorff distance $d_H : \mathcal{K}_C \times \mathcal{K}_C \to \mathbb{R}_+ \cup \{0\}$ is defined by

$$d_H(A, B) = \max \left\{ \max_{a \in A} d(a, B), \max_{b \in B} d(b, A) \right\}$$

with $d(a, B) = \min_{b \in B} |a - b|$. The following properties are well known:

$$d_H(\tau A, \tau B) = |\tau| d_H(A, B), \forall \tau \in \mathbb{R},$$
$$d_H(A + C, B + C) = d_H(A, B), \text{ for all } C \in \mathcal{K}_C,$$
$$d_H(A + B, C + D) \leq d_H(A, C) + d_H(B, D).$$

It is known (see [4, 8]) that $d_H(A, B) = \|A \ominus_{gH} B\|$ where for $C \in \mathcal{K}_C$, the quantity $\|C\| = \max\{|c| ; c \in C\} = d_H(C, \{0\})$ is called the *magnitude* of C; an immediate property of the gH-difference for $A, B \in \mathcal{K}_C$ is

$$d_H(A, B) = 0 \iff A \ominus_{gH} B = 0 \iff A = B. \tag{1}$$

It is also well known that $(\mathcal{K}_C, d_H)$ is a complete metric space.

A fuzzy set on $\mathbb{R}$ is a mapping $u : \mathbb{R} \to [0,1]$. we denote its α-level set as $[u]_\alpha = \{x \in \mathbb{R} : \ u(x) \geq \alpha\}$ for any $\alpha \in [0,1]$. The $supp(u) = \{x \in \mathbb{R} | u(x) > 0\}$; 0-level set of u is defined by $[u]_0 = cl(supp(u))$ where $cl(M)$ means the closure of the subset $M \subset \mathbb{R}$.

A fuzzy set u on $\mathbb{R}$ is said to be a fuzzy number if:

(i) u is normal, i.e., there exists $x_0 \in \mathbb{R}$ such that $u(x_0) = 1$,
(ii) u is a convex fuzzy set (i.e. $u(tx + (1-t)y) \geq \min\{u(x), u(y)\}$, $\forall t \in [0,1]$, $x, y \in \mathbb{R}$),
(iii) u is upper semi-continuous on $\mathbb{R}$,
(iv) $cl\{x \in \mathbb{R} | u(x) > 0\}$ is compact.

Let $\mathbb{R}_F$ denote the family of fuzzy numbers. So, for any $u \in \mathbb{R}_F$ we have $[u]_\alpha \in \mathbb{K}_C$ for all $\alpha \in [0,1]$ and thus the α-levels of a fuzzy number are given by $[u]_\alpha = [u_\alpha^-, u_\alpha^+]$, $u_\alpha^-, u_\alpha^+ \in \mathbb{R}$ for all $\alpha \in [0,1]$. In midpoint notation, we will write $[u]_\alpha = (\widehat{u}_\alpha, \widetilde{u}_\alpha)$ where $\widehat{u}_\alpha = \frac{u_\alpha^+ + u_\alpha^-}{2}$ and $\widetilde{u}_\alpha = \frac{u_\alpha^+ - u_\alpha^-}{2}$ so that $u_\alpha^- = \widehat{u}_\alpha - \widetilde{u}_\alpha$ and $u_\alpha^+ = \widetilde{u}_\alpha + \widehat{u}_\alpha$. If $[u]_1$ is a singleton then we say that u is a fuzzy number. Triangular fuzzy numbers are a special type of fuzzy numbers which are well determined by three real numbers $a \leq b \leq c$, denoted by $u = < a, b, c >$, with α-levels $[u]_\alpha = [a + (b-a)\alpha, c - (c-b)\alpha]$ for all $\alpha \in [0,1]$.

It is well known that in terms of α-levels and taking into account the midpoint notation, for every $\alpha \in [0,1]$ $[u+v]_\alpha = [u]_\alpha + [v]_\alpha = [u_\alpha^- + v_\alpha^-, u_\alpha^+ + v_\alpha^+] = (\widehat{u}_\alpha + \widehat{v}_\alpha; \widetilde{u}_\alpha + \widetilde{v}_\alpha)$ and $[\lambda u]_\alpha = [\min\{\lambda u_\alpha^-, \lambda u_\alpha^+\}, \max\{\lambda u_\alpha^-, \lambda u_\alpha^+\}] = (\lambda \widehat{u}_\alpha; |\lambda| \widetilde{u}_\alpha)$.

The following LgH-difference is somewhat more general than the gH-difference:

Definition 1. *For given two fuzzy numbers u, ν, the level-wise generalized Hukuhara difference (LgH-difference, for short) of u, ν is defined as the set of interval-valued gH-differences*

$$u \ominus_{LgH} \nu = \{w_\alpha \in \mathcal{K}_C | w_\alpha = [u]_\alpha \ominus_{gH} [\nu]_\alpha \text{ for } \alpha \in [0,1]\},$$

that is, for each $\alpha \in [0,1]$, either $[u]_\alpha = [\nu]_\alpha + w_\alpha$ or $[\nu]_\alpha = [u]_\alpha - w_\alpha$.

1.1   Orders for Fuzzy Numbers

The LU-fuzzy partial order is well known in the literature. Let us recall that given $u, \nu \in \mathbb{R}_F$ and given $\alpha \in [0,1]$, heir α-levels are $u_\alpha = [u_\alpha^-, u_\alpha^+] \in \mathbb{K}_C$ and $\nu = [\nu_\alpha^-, \nu_\alpha^+] \in \mathbb{K}_C$, respectively.

Definition 2. *[7] Given $u, \nu \in \mathbb{R}_F$ and given $\alpha \in [0,1]$, we say that*

(i) *$u \precsim_{\alpha-LU} \nu$ if and only if $u_\alpha \precsim_{LU} \nu_\alpha$, that is, $u_\alpha^- \leq \nu_\alpha^-$ and $u_\alpha^+ \leq \nu_\alpha^+$,*
(ii) *$u \preceq_{\alpha-LU} \nu$ if and only if $u_\alpha \preceq_{LU} \nu_\alpha$,*
(iii) *$u \prec_{\alpha-LU} \nu$ if and only if $u_\alpha \prec_{LU} \nu_\alpha$.*

Correspondingly, the analogous LU-fuzzy orders can be obtained by

(a) $u \precsim_{LU} \nu$ *if and only if* $u \precsim_{\alpha-LU} \nu$ *for all* $\alpha \in [0,1]$,
(b) $u \preceq_{LU} \nu$ *if and only if* $u \preceq_{\alpha-LU} \nu$ *for all* $\alpha \in [0,1]$,
(c) $u \prec_{LU} \nu$ *if and only if* $u \prec_{\alpha-LU} \nu$ *for all* $\alpha \in [0,1]$.

The corresponding reverse orders are, respectively, $u \succsim_{LU} \nu \Longleftrightarrow \nu \precsim_{LU} u$, $u \succeq_{LU} \nu \Longleftrightarrow \nu \preceq_{LU} u$ and $u \succ_{LU} \nu \Longleftrightarrow \nu \prec_{LU} u$.

Using α-levels midpoint notation $u_\alpha = (\widehat{u}_\alpha; \widetilde{u}_\alpha)$, $\nu_\alpha = (\widehat{\nu}_\alpha; \widetilde{\nu}_\alpha)$ for all $\alpha \in [0,1]$, the partial orders (a) and (c) above can be expressed for all $\alpha \in [0,1]$ as

$$(LU_a) \begin{cases} \widehat{u}_\alpha \leq \widehat{\nu}_\alpha \\ \widetilde{\nu}_\alpha \leq \widetilde{u}_\alpha + (\widehat{\nu}_\alpha - \widehat{u}_\alpha) \quad \text{and} \\ \widetilde{\nu}_\alpha \geq \widetilde{u}_\alpha - (\widehat{\nu}_\alpha - \widehat{u}_\alpha) \end{cases}$$

$$(LU_c) \begin{cases} \widehat{u}_\alpha < \widehat{\nu}_\alpha \\ \widetilde{\nu}_\alpha < \widetilde{u}_\alpha + (\widehat{\nu}_\alpha - \widehat{u}_\alpha) \; ; \\ \widetilde{\nu}_\alpha > \widetilde{u}_\alpha - (\widehat{\nu}_\alpha - \widehat{u}_\alpha) \end{cases}$$

the partial order (b) can be expressed in terms of (LU_a) with the additional requirement that at least one of the inequalities is strict.

In the sequel, the results are expressed without proof because they are similar to the ones in [12] and [13]. For the family of intervals $u \ominus_{LgH} \nu$ we write $u \ominus_{LgH} \nu \precsim_{LU} 0$ (and similarly with other orders) to mean that $w_\alpha \precsim_{LU} 0$ for all $w_\alpha \in u \ominus_{LgH} \nu$.

Proposition 1. *Let* $u, \nu \in \mathbb{R}_\mathcal{F}$ *with* $u_\alpha = (\widehat{u}_\alpha; \widetilde{u}_\alpha)$, $\nu_\alpha = (\widehat{\nu}_\alpha; \widetilde{\nu}_\alpha)$ *for all* $\alpha \in [0,1]$. *We have*

(i.a) $u \precsim_{LU} \nu \Leftrightarrow u \ominus_{LgH} \nu \precsim_{LU} 0$,
(ii.a) $u \preceq_{LU} \nu \Leftrightarrow u \ominus_{LgH} \nu \preceq_{LU} 0$,
(iii.a) $u \prec_{LU} \nu \Leftrightarrow u \ominus_{LgH} \nu \prec_{LU} 0$,
(iv.a) $u \prec_{LU} \nu \Longrightarrow u \preceq_{LU} \nu \Longrightarrow u \precsim_{LU} \nu$,
(i.b) $u \succsim_{LU} \nu \Leftrightarrow u \ominus_{LgH} \nu \succsim_{LU} 0$,
(ii.b) $u \succeq_{LU} \nu \Leftrightarrow u \ominus_{LgH} \nu \succeq_{LU} 0$,
(iii.b) $u \succ_{LU} \nu \Leftrightarrow u \ominus_{LgH} \nu \succ_{LU} 0$,
(iv.b) $u \succ_{LU} \nu \Longrightarrow u \succeq_{LU} \nu \Longrightarrow u \succsim_{LU} \nu$.

We say that u and ν are LU-incomparable if neither $u \precsim_{LU} \nu$ nor $u \succsim_{LU} \nu$ and u and ν are α-LU-incomparable if neither $u \precsim_{\alpha-LU} \nu$ nor $u \succsim_{\alpha-LU} \nu$.

Proposition 2. *Let* $u, \nu \in \mathbb{R}_\mathcal{F}$ *with* $u_\alpha = (\widehat{u}_\alpha; \widetilde{u}_\alpha)$, $\nu_\alpha = (\widehat{\nu}_\alpha; \widetilde{\nu}_\alpha)$ *for all* $\alpha \in [0,1]$. *The following are equivalent:*

(i) u *and* ν *are* α-LU-incomparable;
(ii) $u_\alpha \ominus_{gH} \nu_\alpha$ *is not a singleton and* $0 \in int(u_\alpha \ominus_{gH} \nu_\alpha)$;
(iii) $|\widehat{u}_\alpha - \widehat{\nu}_\alpha| < |\widetilde{\nu}_\alpha - \widetilde{u}_\alpha|$ *for* $\alpha \in [0,1]$;
(iv) $u_\alpha \subset int(\nu_\alpha)$ *or* $\nu_\alpha \subset int(u_\alpha)$.

Proposition 3. *If u, ν, $w \in \mathbb{R}_{\mathcal{F}}$, then*

(i) $u \gtrsim_{LU} \nu$ *if and only if* $(u + w) \gtrsim_{LU} (v + w)$;

(ii-a) *If* $u + \nu \gtrsim_{LU} w$ *then* $u \gtrsim_{LU} (w \ominus_{LgH} \nu)$;

(ii-b) *If* $u + \nu \gtrsim_{LU} w$ *then* $u \gtrsim_{LU} (w \ominus_{LgH} \nu)$;

(iii) $u \gtrsim_{LU} \nu$ *if and only if* $(-\nu) \gtrsim_{LU} (-u)$.

2   Fuzzy-Valued Functions

A function $F : [a, b] \longrightarrow \mathbb{R}_F$ is said to be a fuzzy-valued function. For any $\alpha \in [0, 1]$, associated to F, we define the family of interval-valued functions $F_\alpha : [a, b] \longrightarrow \mathbb{K}_C$ given by $[F(x)]_\alpha = [f_\alpha^-(x), f_\alpha^+(x)]$ for all $\alpha \in [0, 1]$. In midpoint representation, we write $[F(x)]_\alpha = \left(\widehat{f}_\alpha(x); \widetilde{f}_\alpha(x) \right)$ where $\widehat{f}_\alpha(x) \in \mathbb{R}$ is the midpoint value of interval $[F(x)]_\alpha$ and $\widetilde{f}_\alpha(x) \in \mathbb{R}^+ \cup \{0\}$ is the nonnegative half-length of $F_\alpha(x)$:

$$\widehat{f}_\alpha(x) = \frac{f_\alpha^+(x) + f_\alpha^-(x)}{2} \text{ and}$$

$$\widetilde{f}_\alpha(x) = \frac{f_\alpha^+(x) - f_\alpha^-(x)}{2} \geq 0$$

so that

$$f_\alpha^-(x) = \widehat{f}_\alpha(x) - \widetilde{f}_\alpha(x) \text{ and } f_\alpha^+(x) = \widehat{f}_\alpha(x) + \widetilde{f}_\alpha(x).$$

Proposition 4. *Let $F : T \longrightarrow \mathbb{R}_F$ be a fuzzy-valued function and $x_0 \in T \subseteq \mathbb{R}$ be an accumulation point of T. If $\lim_{x \to x_0} F(x) = L$ with $L_\alpha = [l_\alpha^-, l_\alpha^+]$. Then $\lim_{x \to x_0} [F(x)]_\alpha = [l_\alpha^-, l_\alpha^+]$ for all α (uniformly in $\alpha \in [0, 1]$).*

In midpoint notation, let $[F(x)]_\alpha = (\widehat{f}_\alpha(x); \widetilde{f}_\alpha(x))$ and $L_\alpha = (\widehat{l}_\alpha; \widetilde{l}_\alpha)$ for all $\alpha \in [0, 1]$; then the limits and continuity can be expressed, respectively, as

$$\lim_{x \to x_0} [F(x)]_\alpha = L_\alpha \iff \begin{cases} \lim_{x \to x_0} \widehat{f}_\alpha(x) = \widehat{l}_\alpha \\ \lim_{x \to x_0} \widetilde{f}_\alpha(x) = \widetilde{l}_\alpha \end{cases} \tag{2}$$

and

$$\lim_{x \to x_0} [F(x)]_\alpha = [F(x_0)]_\alpha \iff \begin{cases} \lim_{x \to x_0} \widehat{f}_\alpha(x) = \widehat{f}_\alpha(x_0) \\ \lim_{x \to x_0} \widetilde{f}_\alpha(x) = \widetilde{f}_\alpha(x_0). \end{cases}$$

The following proposition connects limits to the order of fuzzy numbers. Analogous results can be obtained for the reverse partial order $\gtrsim_{LU}$.

Proposition 5. *Let $F, G, H : T \longrightarrow \mathbb{R}_F$ be fuzzy-valued functions and x_0 an accumulation point for T.*

(i) If $F(x) \lesssim_{LU} G(x)$ for all $x \in T$ in a neighborhood of x_0 and $\lim_{x \to x_0} F(x) = L \in \mathbb{R}_F$, $\lim_{x \to x_0} G(x) = M \in \mathbb{R}_F$, then $L \lesssim_{LU} M$;

(ii) If $F(x) \lesssim_{LU} G(x) \lesssim_{LU} H(x)$ for all $x \in T$ in a neighborhood of x_0 and $\lim_{x \to x_0} F(x) = \lim_{x \to x_0} H(x) = L \in \mathbb{R}_F$, then $\lim_{x \to x_0} G(x) = L$.

Similar results as in Propositions 4 and 5 are valid for the left limit with $x \longrightarrow x_0$, $x < x_0$ ($x \nearrow x_0$ for short) and for the right limit $x \longrightarrow x_0$, $x > x_0$ ($x \searrow x_0$ for short); the condition that $\lim_{x \to x_0} F(x) = L$ if and only if $\lim_{x \nearrow x_0} F(x) = L = \lim_{x \searrow x_0} F(x)$ is obvious.

The graphical representation of a fuzzy-valued function is then possible in terms either of the standard way, by picturing the level curves $y = f_\alpha^-(x)$ and $y = f_\alpha^+(x)$ in the plane (x, y), or, in the half-plane $(\widehat{z}; \widetilde{z})$, by plotting the parametric curves $\widehat{z} = \widehat{f}_\alpha(x)$ and $\widetilde{z} = \widetilde{f}_\alpha(x)$; Figs. 1 and 2 give an illustration of the two graphical alternatives for the (periodic, with period 2π) fuzzy function $F(x)$ having α-cuts defined by functions $\widehat{f}_\alpha(x) = 5\cos(x) - (\sqrt{2} - 1)\cos(5x)$ and $\widetilde{f}_\alpha(x) = (1 - 0.5\alpha)^2(1.5 + \sin(4x))$ for $x \in [0, 2\pi]$; only $n = 11$ α-cuts are pictured for uniform $\alpha \in \left\{ \dfrac{i-1}{10} | i = 1, 2, ..., n \right\}$ (see also Fig. 3).

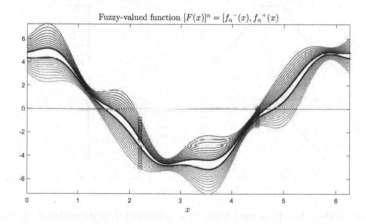

Fuzzy-valued function $[F(x)]^\alpha = [f_\alpha^-(x), f_\alpha^+(x)]$

Fig. 1. Level-wise endpoint graphical representation of the fuzzy-valued function with α-cuts $[F(x)]_\alpha = \left[\widehat{f}_\alpha(x) - \widetilde{f}_\alpha(x), \widehat{f}_\alpha(x) + \widetilde{f}_\alpha(x)\right]$ where $\widehat{f}_\alpha(x) = 5\cos(x) - (\sqrt{2} - 1)\cos(5x)$ and $\widetilde{f}_\alpha(x) = (1 - 0.6\alpha)^2(1.5 + \sin(4x))$ for $x \in [0, 2\pi]$. The core, intercepted by the black-colored curves, is the interval-valued function $x \longrightarrow [F(x)]_1 = [f_1^-(x), f_1^+(x)]$. The other α-cuts are represented by red-colored curves for the left extreme functions $f_\alpha^-(x)$ and blue-colored curves for the right extreme functions $f_\alpha^+(x)$. The marked points correspond to $x = 2.2$ and $x = 4.5$. (Color figure online)

In relation with the LgH-difference, we consider the concept of LgH-differentiability.

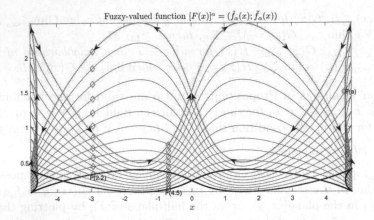

Fig. 2. Level-wise midpoint graphical representation in the half plane $(\widehat{z}; \widecheck{z})$ with $\widecheck{z} \geq 0$ as vertical axis, of the fuzzy-valued function $F(x)$, $x \in [0, 2\pi]$, described in Fig. 1. In this representation, each curve corresponds to a single α-cut (only $n = 11$ curves are pictured with uniform $\alpha = 0, 0.1, ..., 1$); the core corresponds to the black-colored curve, the support to the red-colored one. The arrows give the *direction* of x from initial 0 to final 2π. The marked points correspond to $x = 2.2$ and $x = 4.5$. (Color figure online)

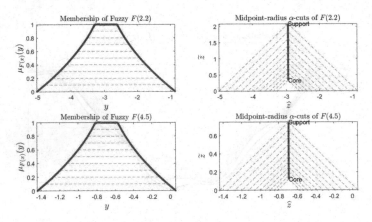

Fig. 3. Membership function and level-wise midpoint representations of two values $F(2.2)$ and $F(4.5)$ of the fuzzy-valued function $F(x)$ described in Fig. 1. In the midpoint representation, a vertical curve corresponds to the displacement of the $n = 11$ computed α-cuts; the red lines on the right pictures reconstruct the α-cuts. Remark that y and $\widehat{z}$ represent the same domain and that a linear vertical segment in the midpoint representation corresponds to a symmetric membership function having the same value of $\widehat{f}_\alpha(x)$ for all α. (Color figure online)

Definition 3. *[6] Let $x_0 \in]a, b[$ and h be such that $x_0 + h \in]a, b[$, then the level-wise gH-derivative (LgH-differentiable for short) of a function $F :]a, b[\rightarrow \mathbb{R}_F$ at*

x_0 is defined as the set of interval-valued gH-derivatives, if they exist,

$$F'_{LgH}(x_0)_\alpha = \lim_{h \to 0} \frac{1}{h} \left([F(x_0 + h)]_\alpha \ominus_{gH} [F(x_0)]_\alpha \right) \qquad (3)$$

if $F'_{LgH}(x_0)_\alpha$ is a compact interval for all $\alpha \in [0, 1]$, we say that F is level-wise generalized Hukuhara differentiable (LgH-differentiable for short) at x_0 and the family of intervals $\{F'_{LgH}(x_0)_\alpha | \alpha \in [0, 1]\}$ is the LgH-derivative of F at x_0, denoted by $F'_{LgH}(x_0)$.

Also, one-side derivatives can be considered. The right LgH-derivative of F at x_0 is $F'_{(r)gH}(x_0)_\alpha = \lim_{h \searrow 0} \frac{1}{h} \left([F(x_0 + h)]_\alpha \ominus_{gH} [F(x_0)]_\alpha \right)$ while to the left it is defined as $F'_{(l)LgH}(x_0)_\alpha = \lim_{h \nearrow 0} \frac{1}{h} \left([F(x_0 + h)]_\alpha \ominus_{gH} [F(x_0)]_\alpha \right)$. The LgH-derivative exists at x_0 if and only if the left and right derivatives at x_0 exist and are the same interval.

In terms of midpoint representation $[F(x)]_\alpha = \left(\widehat{f_\alpha}(x); \widetilde{f_\alpha}(x) \right)$, for all $\alpha \in [0, 1]$, we can write

$$\frac{[F(x + h)]_\alpha \ominus_{gH} [F(x)]_\alpha}{h} = \left(\widehat{\Delta}_{gH} F_\alpha(x, h); \widetilde{\Delta}_{gH} F_\alpha(x, h) \right), \text{ where}$$

$$\widehat{\Delta}_{gH} F_\alpha(x, h) = \frac{\widehat{f_\alpha}(x + h) - \widehat{f_\alpha}(x)}{h},$$

$$\widetilde{\Delta}_{gH} F_\alpha(x, h) = \left| \frac{\widetilde{f_\alpha}(x + h) - \widetilde{f_\alpha}(x)}{h} \right|$$

and taking the limit for $h \longrightarrow 0$, we obtain the LgH-derivative of F_α, if and only if the two limits $\lim_{h \longrightarrow 0} \frac{\widehat{f_\alpha}(x+h) - \widehat{f_\alpha}(x)}{h}$ and $\lim_{h \longrightarrow 0} \left| \frac{\widetilde{f_\alpha}(x+h) - \widetilde{f_\alpha}(x)}{h} \right|$ exist in $\mathbb{R}$; remark that the midpoint function $\widehat{f_\alpha}$ is required to admit the ordinary derivative at x. With respect to the existence of the second limit, the existence of the left and right derivatives $\widetilde{f}'_{(l)\alpha}(x)$ and $\widetilde{f}'_{(r)\alpha}(x)$ is required with $\left| \widetilde{f}'_{(l)\alpha}(x) \right| = \left| \widetilde{f}'_{(r)\alpha}(x) \right| = \widetilde{w}_\alpha(x) \geq 0$ (in particular $\widetilde{w}_\alpha(x) = \left| \widetilde{f_\alpha}'(x) \right|$ if $\widetilde{f_\alpha}'(x)$ exists) so that we have

$$F'_{LgH}(x)_\alpha = \left(\widehat{f_\alpha}'(x); \widetilde{w}_\alpha(x) \right) \qquad (4)$$

or, in the standard interval notation,

$$F'_{LgH}(x)_\alpha = \left[\widehat{f_\alpha}'(x) - \widetilde{w}_\alpha(x), \widehat{f_\alpha}'(x) + \widetilde{w}_\alpha(x) \right]. \qquad (5)$$

3 Monotonicity of Functions with Values in $(\mathbb{R}_F, LU)$

Monotonicity of fuzzy-valued functions has not been much investigated and this is partially due to the lack of unique meaningful definition of an order for fuzzy-valued functions. We can analyze monotonicity and, using the gH-difference, related characteristics of inequalities for fuzzy-valued functions.

Definition 4. *Let* $F \; : \; [a,b] \; \to \; \mathbb{R}_F$ *be fuzzy function, where* $[F(x)]_\alpha \; = \; \left(\widehat{f_\alpha}(x); \widetilde{f_\alpha}(x)\right)$ *for all* $\alpha \in [0,1]$. *We say that* F *is*

(a-i) $(\gtrsim_{LU})$-*nondecreasing on* $[a,b]$ *if* $x_1 < x_2$ *implies* $F(x_1) \gtrsim_{LU} F(x_2)$ *for all* $x_1, x_2 \in [a,b]$;

(a-ii) $(\gtrsim_{LU})$-*nonincreasing on* $[a,b]$ *if* $x_1 < x_2$ *implies* $F(x_2) \gtrsim_{LU} F(x_1)$ *for all* $x_1, x_2 \in [a,b]$;

(b-i) *(strictly)* $(\preceq_{LU})$-*increasing on* $[a,b]$ *if* $x_1 < x_2$ *implies* $F(x_1) \preceq_{LU} F(x_2)$ *for all* $x_1, x_2 \in [a,b]$;

(b-ii) *(strictly)* $(\preceq_{LU})$-*decreasing on* $[a,b]$ *if* $x_1 < x_2$ *implies* $F(x_2) \preceq_{LU} F(x_1)$ *for all* $x_1, x_2 \in [a,b]$;

(c-i) *(strongly)* $(\prec_{LU})$-*increasing on* $[a,b]$ *if* $x_1 < x_2$ *implies* $F(x_1) \prec_{LU} F(x_2)$ *for all* $x_1, x_2 \in [a,b]$;

(c-ii) *(strongly)* $(\prec_{LU})$-*decreasing on* $[a,b]$ *if* $x_1 < x_2$ *implies* $F(x_2) \prec_{LU} F(x_1)$ *for all* $x_1, x_2 \in [a,b]$.

If one of the six conditions is satisfied, we say that F *is monotonic on* $[a,b]$; *the monotonicity is strict if (b-i,b-ii) or strong if (c-i,c-ii) are satisfied.*

The monotonicity of $F : [a,b] \to \mathbb{R}_F$ can be analyzed also locally, in a neighborhood of an internal point $x_0 \in]a,b[$, by considering condition $F(x) \gtrsim_{LU} F(x_0)$ (or condition $F(x) \gtrsim_{LU} F(x_0)$) for $x \in]a,b[$ and $|x - x_0| < \delta$ with a positive small δ. We omit the corresponding definitions as they are analogous to the previous ones.

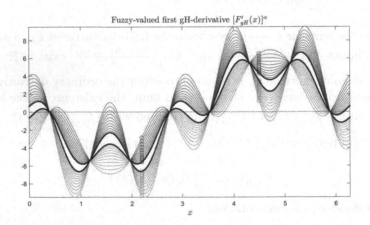

Fuzzy-valued first gH-derivative $[F'_{gH}(x)]^\alpha$

Fig. 4. Level-wise endpoint graphical representation of the fuzzy-valued LgH-derivative $[F'_{LgH}](x)]_\alpha = \left[\widehat{f}'_\alpha(x) - |\widetilde{f}'_\alpha(x)|, \widehat{f}'_\alpha(x) + |\widetilde{f}'_\alpha(x)|\right]$ where $\widehat{f}_\alpha(x) = 5cos(x) - (\sqrt{2} - 1)cos(5x)$ and $\widetilde{f}_\alpha(x) = (1 - 0.6\alpha)^2(1.5 + sin(4x))$ for $x \in [0, 2\pi]$. The core is intercepted by the black-colored curves. The other α-cuts are represented by red-colored curves for the left extreme functions and blue-colored curves for the right extreme functions. The marked points correspond to $x = 2.2$ and $x = 4.5$. (Color figure online)

Proposition 6. *Let* $F : [a, b] \to \mathbb{R}_F$ *be fuzzy function, where* $F_\alpha(x) = \Big(\widehat{f_\alpha}(x);$ *$\widetilde{f_\alpha}(x) \Big)$ *for* $\alpha \in [0, 1]$ *and* $x_0 \in]a, b[$. *Then*

(i) $F(x)$ is $(\precsim_{LU})$-nondecreasing at x_0 if and only if $\widehat{f_\alpha}(x)$ is nondecreasing, $\widetilde{f_\alpha}(x) - \widehat{f_\alpha}(x)$ is nonincreasing and $\widetilde{f_\alpha}(x) + \widehat{f_\alpha}(x)$ is nondecreasing at x_0 for all $\alpha \in [0, 1]$;

(ii) $F(x)$ is $(\precsim_{LU})$-nonincreasing at x_0 if and only if $\widehat{f_\alpha}(x)$ is nonincreasing, $\widetilde{f_\alpha}(x) - \widehat{f_\alpha}(x)$ is nondecreasing and $\widetilde{f_\alpha}(x) + \widehat{f_\alpha}(x)$ is nonincreasing at x_0 for all $\alpha \in [0, 1]$;

(iii) Analogous conditions are valid for strict and strong monotonicity.

Proposition 7. *Let* $F :]a, b[\to \mathbb{R}_F$ *be fuzzy function, where* $[F(x)]_\alpha = \Big(\widehat{f_\alpha}(x); \widetilde{f_\alpha}(x) \Big)$ *for all* $\alpha \in [0, 1]$ *and let F be LgH-differentiable at the internal points $x \in]a, b[$. Then*

(1) If F is $(\precsim_{LU})$-nondecreasing on $]a, b[$, then for all x, $F'_{LgH}(x) \succsim_{LU} 0$;
(2) If F is $(\precsim_{LU})$-nonincreasing on $]a, b[$, then for all x, $F'_{LgH}(x) \precsim_{LU} 0$.

In Fig. 4 we picture the fuzzy-valued first order LgH-derivative of a function $F(x)$ and in Fig. 5 we show graphically the membership functions of $F'_{LgH}(x)$ at two points $x = 2.2$ and $x = 4.5$. Observe from Fig. 5 that $F'_{LgH}(4.5)$ is positive in the $(\precsim_{LU})$ order and that $F'_{LgH}(2.2)$ is negative in the same order relation, denoting that $F(x)$ is locally strictly increasing around $x = 4.5$ and locally strictly decreasing around $x = 2.2$ (see also Fig. 4).

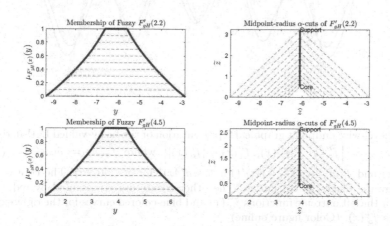

Fig. 5. Membership function and level-wise midpoint representations of two values $F'_{LgH}(2.2)$ and $F'_{LgH}(4.5)$ of the fuzzy-valued LgH-derivative of function $F(x)$ described in Fig. 1. In the midpoint representation, a vertical curve corresponds to the displacement of the $n = 11$ computed α-cuts; the red lines on the right pictures reconstruct the α-cuts. Remark that y and $\widehat{z}$ represent the same domain and that a linear vertical segment in the midpoint representation corresponds to a symmetric membership function. (Color figure online)

Analogous results are also immediate, relating strong (local) monotonicity of F to the "sign" of its left and right derivatives $F'_{(l)LgH}(x)$ and $F'_{(r)LgH}(x)$; at the extreme points of $[a, b]$, we consider only right (at a) or left (at b) monotonicity and right or left derivatives.

Proposition 8. *Let $F : [a, b] \rightarrow \mathbb{R}_F$ be fuzzy function, where $[F(x)]_\alpha = \left(\widehat{f_\alpha}(x); \widetilde{f_\alpha}(x)\right)$ for all $\alpha \in [0, 1]$ with left and/or right gH-derivatives at a point $x_0 \in [a, b]$. Then*

(i.a) if $0 \prec_{LU} F'_{(l)LgH}(x_0)$, then F is strongly ($\prec_{LU}$)-increasing on $[x_0 - \delta, x_0]$ for some $\delta > 0$ (here $x_0 > a$);

(i.b) if $0 \prec_{LU} F'_{(r)LgH}(x_0)$, then F is strongly ($\prec_{LU}$)-increasing on $[x_0, x_0 + \delta]$ for some $\delta > 0$ (here $x_0 < b$);

(ii.a) if $0 \succ_{LU} F'_{(l)LgH}(x_0)$, then F is strongly ($\prec_{LU}$)-decreasing on $[x_0 - \delta, x_0]$ for some $\delta > 0$ (here $x_0 > a$);

(ii.b) if $0 \succ_{LU} F'_{(r)LgH}(x_0)$, then F is strongly ($\prec_{LU}$)-decreasing on $[x_0, x_0 + \delta]$ for some $\delta > 0$ (here $x_0 < b$).

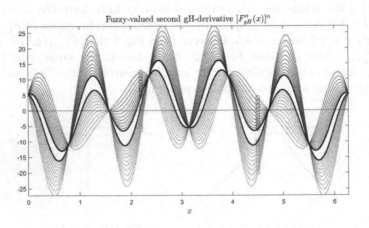

Fig. 6. Level-wise endpoint graphical representation of the fuzzy-valued LgH-derivative $[F''_{LgH}](x)]_\alpha = \left[\widehat{f}''_\alpha(x) - |\widetilde{f}''_\alpha(x)|, \widehat{f}''_\alpha(x) + |\widetilde{f}''_\alpha(x)|\right]$ where $\widehat{f}_\alpha(x) = 5cos(x) - (\sqrt{2} - 1)cos(5x)$ and $\widetilde{f}_\alpha(x) = (1 - 0.6\alpha)^2(1.5 + sin(4x))$ for $x \in [0, 2\pi]$. The core is intercepted by the black-colored curves. The other α-cuts are represented by red-colored curves for the left extreme functions $f^-_\alpha(x)$ and blue-colored curves for the right extreme functions $f^+_\alpha(x)$. (Color figure online)

4   Concavity and Convexity of Fuzzy-Valued Functions

We have three types of convexity, similar to the monotonicity and local extremum concepts.

Definition 5. *Let $F : [a,b] \to \mathbb{R}_F$ be a function and let $\precsim_{LU}$ be a partial order on $\mathbb{R}_F$. We say that*

(a-i)   *F is $(\precsim_{LU})$-convex on $[a,b]$ if and only if $F((1-\lambda)x_1 + \lambda x_2) \precsim_{LU}$ $(1-\lambda)F(x_1) + \lambda F(x_2)$, $\forall x_1, x_2 \in [a,b]$ and all $\lambda \in [0,1]$;*

(a-ii)   *F is $(\precsim_{\gamma^-,\gamma^+})$-concave on $[a,b]$ if and only if $F((1-\lambda)x_1 + \lambda x_2) \succsim_{LU}$ $(1-\lambda)F(x_1) + \lambda F(x_2)$, $\forall x_1, x_2 \in [a,b]$ and all $\lambda \in [0,1]$.*

(b-i)   *F is strictly $(\preceq_{LU})$-convex on $[a,b]$ if and only if $F((1-\lambda)x_1 + \lambda x_2) \preceq_{LU} (1-\lambda)F(x_1) + \lambda F(x_2)$, $\forall x_1, x_2 \in [a,b]$ and all $\lambda \in]0,1[$;*

(b-ii)   *F is strictly $(\preceq_{LU})$-concave on $[a,b]$ if and only if $F((1-\lambda)x_1 + \lambda x_2) \succeq_{LU} (1-\lambda)F(x_1) + \lambda F(x_2)$, $\forall x_1, x_2 \in [a,b]$ and all $\lambda \in]0,1[$.*

(c-i)   *F is strongly $(\prec_{LU})$-convex on $[a,b]$ if and only if $F((1-\lambda)x_1 + \lambda x_2) \prec_{LU} (1-\lambda)F(x_1) + \lambda F(x_2)$, $\forall x_1, x_2 \in [a,b]$ and all $\lambda \in]0,1[$;*

(c-ii)   *F is strongly $(\prec_{LU})$-concave on $[a,b]$ if and only if $F((1-\lambda)x_1 + \lambda x_2) \succ_{LU} (1-\lambda)F(x_1) + \lambda F(x_2)$, $\forall x_1, x_2 \in [a,b]$ and all $\lambda \in]0,1[$.*

The convexity of a function $[F]_\alpha = \left(\widehat{f_\alpha}; \widetilde{f_\alpha}\right)$ is related to the concavity of function $[-F]_\alpha = \left(-\widehat{f_\alpha}; \widetilde{f_\alpha}\right)$ for all $\alpha \in [0,1]$. It is easy to see that F is $(\precsim_{LU})$-convex if and only if $-F$ is $(\precsim_{LU})$-concave.

Proposition 9. *Let $F : [a,b] \to \mathbb{R}_F$ with $[F]_\alpha = \left(\widehat{f_\alpha}; \widetilde{f_\alpha}\right)$ for all $\alpha \in [0,1]$ and $\precsim_{LU}$ be a given partial order; then*

1. F is $(\precsim_{LU})$-convex if and only if $\widehat{f_\alpha}$ is convex, $\widetilde{f_\alpha} - \widehat{f_\alpha}$ is concave and $\widehat{f_\alpha} + \widetilde{f_\alpha}$ is convex;

2. F is $(\precsim_{LU})$-concave if and only if $\widehat{f_\alpha}$ is concave, $\widetilde{f_\alpha} - \widehat{f_\alpha}$ is convex and $\widehat{f_\alpha} + \widetilde{f_\alpha}$ is concave.

From Proposition 9, it is easy to see that F is $(\precsim_{LU})$-convex (or concave) if and only if f_α^- and f_α^+ are convex (or concave) for all $\alpha \in [0,1]$. Also, several ways to analyze $(\precsim_{LU})$-convexity (or concavity) in terms of the first or second derivatives of functions $\widehat{f_\alpha}$, $\widetilde{f_\alpha} - \widehat{f_\alpha}$ and $\widehat{f_\alpha} + \widetilde{f_\alpha}$ can be easily deduced.

Proposition 10. *Let $F :]a,b[\to \mathbb{R}_F$ with $[F]_\alpha = \left(\widehat{f_\alpha}; \widetilde{f_\alpha}\right)$ for all $\alpha \in [0,1]$. Real-valued functions $\widehat{f_\alpha}$ and $\widetilde{f_\alpha}$ are differentiable, then;*

1. If the first order derivatives $\widehat{f_\alpha'}$ and $\widetilde{f_\alpha'}$ exist, then:

(1-a) F is $(\precsim_{LU})$-convex on $]a,b[$ if and only if $\widehat{f_\alpha'}$, $\widetilde{f_\alpha'} - \widehat{f_\alpha'}$ and $\widetilde{f_\alpha'} + \widehat{f_\alpha'}$ are increasing (nondecreasing) for all $\alpha \in [0,1]$ on $]a,b[$;

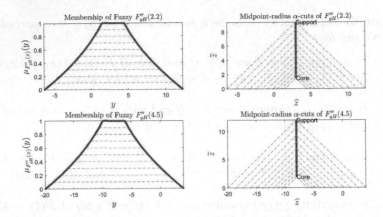

Fig. 7. Membership function and level-wise midpoint representations of two values $F''_{LgH}(2.1)$ and $F''_{LgH}(4.6)$ of the fuzzy-valued function $F(x)$ described in Fig. 1. In this midpoint representation, a vertical curve corresponds to the displacement of the $n = 11$ computed α-cuts; the red lines on the right pictures reconstruct the α-cuts. Remark that y and $\widehat{z}$ represent the same domain and that a linear vertical segment in the midpoint representation corresponds to a symmetric membership function, having the same value for all $\alpha \in [0, 1]$. (Color figure online)

(1-b)   F is $(\precsim_{LU})$-concave on $]a, b[$ if and only if $\widehat{f}'_\alpha$, $\widehat{f}'_\alpha - \widetilde{f}'_\alpha$ and $\widehat{f}'_\alpha + \widetilde{f}'_\alpha$ are decreasing (nonincreasing) for all $\alpha \in [0, 1]$ on $]a, b[$;

2. *If the second order derivatives $\widehat{f}''_\alpha$ and $\widetilde{f}''_\alpha$ exist and are continuous, then:*

(2-a)   F is $(\precsim_{LU})$-convex on $]a, b[$ if and only if $\widehat{f}''_\alpha \geq 0$, $\widehat{f}''_\alpha - \widetilde{f}''_\alpha \geq 0$ and $\widehat{f}''_\alpha + \widetilde{f}''_\alpha \geq 0$ for all $\alpha \in [0, 1]$ on $]a, b[$;

(2-b)   F is $(\precsim_{LU})$-concave on $]a, b[$ if and only if $\widehat{f}''_\alpha \leq 0$, $\widehat{f}''_\alpha - \widetilde{f}''_\alpha \leq 0$ and $\widehat{f}''_\alpha + \widetilde{f}''_\alpha \leq 0$ for all $\alpha \in [0, 1]$ on $]a, b[$.

In Fig. 6 we picture the fuzzy-valued second order LgH-derivative of a function $F(x)$ and in Fig. 7 we show graphically the membership functions of $F''_{LgH}(x)$ at two points $x = 2.2$ and $x = 4.5$. Observe from Fig. 7 that $F'_{LgH}(2.2)$ and $F'_{LgH}(4.5)$ are not positive nor negative in the $(\precsim_{LU})$ order, denoting that $F(x)$ is (locally) not convex nor concave around $x = 2.2$ or $x = 4.5$. See also Fig. 6 where regions of positive and negative second-order derivative can be identified.

Remark 1. Analogously to the relationship between the sign of second derivative and convexity for ordinary functions, we can establish conditions for convexity of fuzzy functions and the sign of the second order LgH-derivative $F''_{LgH}(x)$; for example, a sufficient condition for strong $\prec_{LU}$-convexity is the following (compare with Proposition 10):

1. If $F''_{LgH}(x_0) \prec_{LU} 0$ then $F(x)$ is strongly concave at x_0.
2. If $F''_{LgH}(x_0) \succ_{LU} 0$ then $F(x)$ is strongly convex at x_0.

5   An Outline of Periodic Fuzzy-Valued Functions

Definition 6. *A function $F : [a, b] \to \mathbb{R}_F$ is said to be periodic if, for some nonzero constant $T \in\ ]0, b - a[$, it occurs that $F(x + T) = F(x)$ for all $x \in [a, b]$ with $x + T \in [a, b]$ (i.e., for all $x \in [a, b - T]$). A nonzero constant T for which this is verified, is called a period of the function and if there exists a least positive constant T with this property, it is called the fundamental period.*

Obviously, if F has a period T, then this also implies that F_α for all $\alpha \in [0, 1]$ has a period T i.e., for all $\alpha \in [0, 1]$, $\widehat{f_\alpha}$ and $\widetilde{f_\alpha}$ are periodic with period T. On the other hand, the periodicity of functions F_α for all $\alpha \in [0, 1]$ does not necessarily imply the periodicity of F.

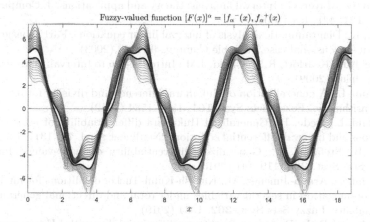

Fuzzy-valued function $[F(x)]^\alpha = [f_\alpha{}^-(x), f_\alpha{}^+(x)]$

Fig. 8. Level-wise endpoint graphical representation of the periodic fuzzy-valued function $[F(x)]_\alpha = \left[\widehat{f_\alpha}(x) - \widetilde{f_\alpha}(x), \widehat{f_\alpha}(x) + \widetilde{f_\alpha}(x) \right]$ where $\widehat{f_\alpha}(x) = 5cos(x) - (\sqrt{2} - 1)cos(5x)$ and $\widetilde{f_\alpha}(x) = (1 - 0.6\alpha)^2(1.5 + sin(4x))$ for $x \in [0, 6\pi]$. The period is $p = 2\pi$.

Proposition 11. *Let $F : [a, b] \to \mathbb{R}_F$ be a continuous function such that $[F(x)]_\alpha = (\widehat{f_\alpha}(x); \widetilde{f_\alpha}(x))$ for all $\alpha \in [0, 1]$ with $\widehat{f_\alpha}$ periodic of period $\widehat{T}$ and $\widetilde{f_\alpha}$ of period $\widetilde{T}$. Then it holds that:*

(1) if the periods $\widehat{T}$ and $\widetilde{T}$ are commensurable, i.e., $\frac{\widehat{T}}{\widetilde{T}} \in \mathbb{Q}$ ($\frac{\widehat{T}}{\widetilde{T}} = \frac{p}{q}$, such that p and q are coprime) then the function F is periodic of period $T = lcm(\widehat{T}, \widetilde{T})$, i.e,. T is the least common multiple between $\widehat{T}$ and $\widetilde{T}$ (i.e., $T = p\widetilde{T} = q\widehat{T}$);

(2) if the periods $\widehat{T}$ and $\widetilde{T}$ are not commensurable, i.e., $\frac{\widehat{T}}{\widetilde{T}} \notin \mathbb{Q}$, then function F is not periodic.

A periodic function of period $p = 2\pi$ is given in Fig. 8. It is the function reported in all figures above, extended to the domain $[0, 6\pi]$.

6   Conclusions and Further Work

We have developed new results to define monotonicity ans convexity or concavity for fuzzy-valued functions, in terms of the LU-partial order; similar results can be obtained for other types of partial orders. It appears that midpoint representation of the α-cuts is a useful tool to analyse and visualize properties of fuzzy-valued functions. In further work, we will analyse important concepts of (local) minimal and maximal points for fuzzy valued functions, by the use of first-order and second-order LgH-derivatives.

References

1. Alefeld, G., Mayer, G.: Interval analysis: theory and applications. J. Comput. Appl. Math. **121**, 421–464 (2000)
2. Kulpa, K.: Diagrammatic analysis of interval linear equations, Part I: basic notions and one-dimensional case. Reliable Comput. **9**, 1–20 (2003)
3. Moore, R.E., Kearfott, R.B., Cloud, J.M.: Introduction to Interval Analysis. SIAM, Philadelphia (2009)
4. Stefanini, L.: A generalization of Hukuhara difference and division for interval and fuzzy arithmetic. Fuzzy Sets Syst. **161**, 1564–1584 (2010)
5. Stefanini, L., Bede, B.: Generalized Hukuhara differentiability of interval-valued functions and interval differential equations. Nonlinear Anal. **71**, 1311–1328 (2009)
6. Bede, B., Stefanini, L.: Generalized differentiability of fuzzy-valued functions. Fuzzy Sets Syst. **230**, 119–141 (2013)
7. Stefanini, L., Arana-Jimenez, M.: Karush-Kuhn-Tucker conditions for interval and fuzzy optimization in several variables under total and directional generalized differentiability. Fuzzy Sets Syst. **362**, 1–34 (2019)
8. Stefanini, L., Bede, B.: Generalized fuzzy differentiability with LU-parametric representations. Fuzzy Sets Syst. **257**, 184–203 (2014)
9. Bede, B.: Mathematics of Fuzzy Sets and Fuzzy Logic. STUDFUZZ, vol. 295. Springer, Heidelberg (2013). https://doi.org/10.1007/978-3-642-35221-8
10. Stefanini, L., Bede, B.: A new gH-difference for multi-dimensional convex sets and convex fuzzy sets. Axioms **8**, 48 (2019). https://doi.org/10.3390/axioms8020048
11. Guerra, M.L., Stefanini, L.: A comparison index for interval ordering based on generalized Hukuhara difference. Soft Comput. **16**, 1931–1943 (2012)
12. Stefanini, L., Guerra, M.L., Amicizia, B.: Interval analysis and calculus for interval-valued functions of a single variable. Part I: partial orders, gH-derivative, monotonicity. Axioms **8**, 113 (2019). https://doi.org/10.3390/axioms8040113
13. Stefanini, L., Sorini, L., Amicizia, B.: Interval analysis and calculus for interval-valued functions of a single variable. Part II: external points, convexity, periodicity. Axioms **8**, 114 (2019). https://doi.org/10.3390/axioms8040114

On the Sum of Generalized Hukuhara Differentiable Fuzzy Functions

Yurilev Chalco-Cano[1]($\boxtimes$) (iD), A. Khastan[2](iD), and Antonio Rufián-Lizana[3]

[1] Departamento de Matemática, Universidad de Tarapacá, Casilla 7D, Arica, Chile
ychalco@uta.cl
[2] Department of Mathematics, Institute for Advanced Studies in Basic Sciences,
Zanjan, Iran
khastan@gmail.com
[3] Departamento de Estadística e Investigación Operativa, Universidad de Sevilla,
Sevilla, Spain

Abstract. In this article we present new results on the sum of gH-differentiable fuzzy functions. We give conditions so that the sum of two gH-differentiable fuzzy functions become gH-differentiable. We present also practical rules for obtaining the gH-derivative of the sum of fuzzy functions.

Keywords: Fuzzy functions · gH-differentiable fuzzy functions · Algebra of gH-differentiable fuzzy functions

1 Introduction

The generalized Hukuhara differentiability (gH-differentiability) for fuzzy functions is a very useful concept in the area of fuzzy mathematical analysis. This concept has been very important in the development of various topics into fuzzy theory, for instance, fuzzy differential equations and fuzzy optimization problems.

Obtaining the gH-derivative of an interval-valued function directly from the definition is a rather complex task. In contrast, the use of the (lateral) differentiability of its endpoint functions considerably simplifies the problem. In this direction, a characterization of the gH-differentiable fuzzy functions through of the (lateral) differentiability of its endpoint functions was obtained in [4].

Calculus for fuzzy functions is an important topic. Several properties of the gH-differentiable fuzzy functions have been obtained in [2–5]. In particular, some results on the algebra of gH-differentiable fuzzy functions have been presented in [1,3,5].

This article was partially supported by the project UTA-Mayor Number 4745-19.

M.-J. Lesot et al. (Eds.): IPMU 2020, CCIS 1238, pp. 43–50, 2020.
https://doi.org/10.1007/978-3-030-50143-3_4

In this paper we present some results on the sum of gH-differentiable fuzzy functions. We prove that the $(k)gH$-derivative is a linear operator. We show also that the sum of two $(k)gH$-differentiable fuzzy functions with different $k \in \{i, ii, iii, iv\}$ is not always gH-differentiable. We give conditions so that sum of fuzzy functions become gH-differentiable.

2 Preliminaries

We denote by $\mathbb{I}$ the family of all bounded closed intervals in $\mathbb{R}$, i.e.,

$$\mathbb{I} = \{A = [\underline{a}, \overline{a}] \; / \; \underline{a}, \overline{a} \in \mathbb{R} \text{ and } \underline{a} \leq \overline{a}\}.$$

On $\mathbb{I}$ we consider the algebraic operations "+" and "." by (see, e.g., [6])

$$[\underline{a}, \overline{a}] + [\underline{b}, \overline{b}] = [\underline{a} + \underline{b}, \overline{a} + \overline{b}] \quad \lambda \cdot [\underline{a}, \overline{a}] = [\min\{\lambda \underline{a}, \lambda \overline{a}\}, \max\{\lambda \underline{a}, \lambda \overline{a}\}].$$

Note that $(\mathbb{I}, +, \cdot)$ is a quasilinear space [7]. An interval $A = [\underline{a}, \overline{a}]$ with $\underline{a} < \overline{a}$ does not have an additive inverse. So, in the article [8] was introduced the generalized Hukuhara difference "$\ominus_{gH}$" by

$$[\underline{a}, \overline{a}] \ominus_{gH} [\underline{b}, \overline{b}] = [\min\{\underline{a} - \underline{b}, \overline{a} - \overline{b}\}, \max\{\underline{a} - \underline{b}, \overline{a} - \overline{b}\}].$$

The gH-difference is a complementary operation in the quasilinear space $(\mathbb{I}, +, \cdot)$. We note that the gH-difference $A \ominus_{gH} B$, of two intervals A and B, there is always.

Let $\mathbb{R}_{\mathcal{F}}$ denote the family of all fuzzy numbers. Let us denote by $[u]^\alpha = [\underline{u}_\alpha, \overline{u}_\alpha] \in \mathbb{R}$ the α-level set of u, for all $\alpha \in [0, 1]$.

Let $u, v \in \mathbb{R}_{\mathcal{F}}$, with α-levels represented by $[\underline{u}_\alpha, \overline{u}_\alpha]$ and $[\underline{v}_\alpha, \overline{v}_\alpha]$, respectively, and let $\lambda \in \mathbb{R}$. The addition $u + v$ and the scalar multiplication λu are defined via the α-levels by

$$[u + v]^\alpha = [(\underline{u + v})_\alpha, (\overline{u + v})_\alpha] = [\underline{u}_\alpha + \underline{v}_\alpha, \overline{u}_\alpha + \overline{v}_\alpha],$$

$$[\lambda u]^\alpha = [(\underline{\lambda u})_\alpha, (\overline{\lambda u})_\alpha] = [\min\{\lambda \underline{u}_\alpha, \lambda \overline{u}_\alpha\}, \max\{\lambda \underline{u}_\alpha, \lambda \overline{u}_\alpha\}], \quad (1)$$

for every $\alpha \in [0, 1]$. In this case $(\mathbb{R}_{\mathcal{F}}, +, \cdot)$ is a quasilinear space [7].

Definition 1. *([9]) Given two fuzzy numbers u, v, the generalized Hukuhara difference (gH-difference for short) is the fuzzy interval w, if it exists, such that*

$$u \ominus_{gH} v = w \Leftrightarrow \begin{cases} (i) \, u = v + w, \\ or \; (ii) \, v = u + (-1)w. \end{cases}$$

It may happen that there is not gH-difference of two fuzzy numbers [9].

If $u \ominus_{gH} v$ exists then, in terms of α-levels, we have

$$[u \ominus_{gH} v]^\alpha = [u]^\alpha \ominus_{gH} [u]^\alpha$$
$$= [\min\{\underline{u}_\alpha - \underline{v}_\alpha, \overline{u}_\alpha - \overline{v}_\alpha\}, \max\{\underline{u}_\alpha - \underline{v}_\alpha, \overline{u}_\alpha - \overline{v}_\alpha\}],$$

for all $\alpha \in [0, 1]$, where $[u]^\alpha \ominus_{gH} [u]^\alpha$ denotes the gH-difference between two intervals.

Given $u, v \in \mathbb{R}_{\mathcal{F}}$ we define the distance between u and v by

$$D(u, v) = \sup_{\alpha \in [0,1]} \max\{|\underline{u}_\alpha - \underline{v}_\alpha|, |\overline{u}_\alpha - \overline{v}_\alpha|\}.$$

3 Generalized Hukuhara Differentiable Fuzzy Functions

Henceforth, $T =]a, b[$ denotes an open interval in $\mathbb{R}$. A function $F : T \to \mathbb{R}_{\mathcal{F}}$ is said to be a fuzzy function. For each $\alpha \in [0, 1]$, associated with F, we define the family of interval-valued functions $F_\alpha : T \to \mathbb{I}$ given by $F_\alpha(x) = [f(x)]^\alpha$. For any $\alpha \in [0, 1]$, we denote

$$F_\alpha(x) = \left[\underline{f}_\alpha(x), \overline{f}_\alpha(x) \right].$$

Here, for each $\alpha \in [0, 1]$, the endpoint functions $\underline{f}_\alpha, \overline{f}_\alpha : T \to \mathbb{R}$ are called upper and lower functions of F, respectively.

The following definition is the well-known concept of generalized Hukuhara differentiable fuzzy functions (gH-differentiable fuzzy functions, for short) based on the gH-difference of fuzzy intervals.

Definition 2. *([3]) The gH-derivative of a fuzzy function $F : T \to \mathbb{R}_{\mathcal{F}}$ at $x_0 \in T$ is defined as*

$$F'(x_0) = \lim_{h \to 0} \frac{1}{h} \left[F(x_0 + h) \ominus_{gH} F(x_0) \right]. \tag{2}$$

If $F'(x_0) \in \mathbb{R}_{\mathcal{F}}$ satisfying (2) exists, we say that F is gH-differentiable at x_0.

Obtaining the gH-derivative of a fuzzy function via (2) is a rather complex problem. The following result characterizes the gH-differentiability of F in terms of the differentiability of its endpoint functions $\underline{f}_\alpha$ and $\overline{f}_\alpha$ which makes it more practice to calculate the gH-derivative.

Theorem 1. *([4]) Let $F : T \to \mathbb{R}_{\mathcal{F}}$ be a fuzzy function and $x \in T$. Then F is gH-differentiable at x if and only if one of the following four cases holds:*

(i) *$\underline{f}_\alpha$ and $\overline{f}_\alpha$ are differentiable at x, uniformly in $\alpha \in [0, 1]$, $(\underline{f}_\alpha)'(x)$ is monotonic increasing and $(\overline{f}_\alpha)'(x)$ is monotonic decreasing as functions of α and $(\underline{f}_1)'(x) \leq (\overline{f}_1)'(x)$. In this case,*

$$F'_\alpha(x) = \left[(\underline{f}_\alpha)'(x), (\overline{f}_\alpha)'(x) \right],$$

for all $\alpha \in [0, 1]$.

(ii) *$\underline{f}_\alpha$ and $\overline{f}_\alpha$ are differentiable at x, uniformly in $\alpha \in [0, 1]$, $(\underline{f}_\alpha)'(x)$ is monotonic decreasing and $(\overline{f}_\alpha)'(x)$ is monotonic increasing as functions of α and $(\overline{f}_1)'(x) \leq (\underline{f}_1)'(x)$. In this case,*

$$F'_\alpha(x) = \left[(\overline{f}_\alpha)'(x), (\underline{f}_\alpha)'(x) \right],$$

for all $\alpha \in [0, 1]$.

(iii) $(\underline{f}_\alpha)'_{+/-}(x)$ and $(\overline{f}_\alpha)'_{+/-}(x)$ exist uniformly in $\alpha \in [0,1]$, $(\underline{f}_\alpha)'_+(x) = (\overline{f}_\alpha)'_-(x)$ is monotonic increasing and $(\overline{f}_\alpha)'_+(x) = (\underline{f}_\alpha)'_-(x)$ is monotonic decreasing as functions of α and $(\underline{f}_1)'_+(x) \le (\overline{f}_1)'_+(x)$. In this case,

$$F'_\alpha(x) = \left[(\underline{f}_\alpha)'_+(x), (\overline{f}_\alpha)'_+(x)\right] = \left[(\overline{f}_\alpha)'_-(x), (\underline{f}_\alpha)'_-(x)\right],$$

for all $\alpha \in [0,1]$.

(iv) $(\underline{f}_\alpha)'_{+/-}(x)$ and $(\overline{f}_\alpha)'_{+/-}(x)$ exist uniformly in $\alpha \in [0,1]$, $(\underline{f}_\alpha)'_+(x) = (\overline{f}_\alpha)'_-(x)$ is monotonic decreasing and $(\overline{f}_\alpha)'_+(x) = (\underline{f}_\alpha)'_-(x)$ is monotonic increasing as functions of α and $(\overline{f}_1)'_+(x) \le (\underline{f}_1)'_+(x)$. In this case,

$$F'_\alpha(x) = \left[(\overline{f}_\alpha)'_+(x), (\underline{f}_\alpha)'_+(x)\right] = \left[(\underline{f}_\alpha)'_-(x), (\overline{f}_\alpha)'_-(x)\right],$$

for all $\alpha \in [0,1]$.

From Theorem 1 we can distinguish four cases. We say that an interval-valued function $F : T \to \mathbb{R}_\mathcal{F}$ is $(k)gH$-differentiable if case k in Theorem 1 holds, for $k \in \{i, ii, iii, iv\}$. Note that if $F : T \to \mathbb{R}_\mathcal{F}$ is gH-differentiable at x_0 in more than one case then F is gH-differentiable at x_0 in all four cases and $F'(x_0)$ is a trivial fuzzy number or singleton, i.e. $F'(x_0) = \chi_{\{a\}}$ for some $a \in \mathbb{R}$.

4   The Sum of gH-differentiable Fuzzy Functions

Given two fuzzy functions $F, G : T \to \mathbb{R}_\mathcal{F}$ the sum operation is defined by

$$(F + G)(x) = F(x) + G(x),$$

for all $x \in T$. Via the α-levels, with $F_\alpha(x) = \left[\underline{f}_\alpha(x), \overline{f}_\alpha(x)\right]$ and $G_\alpha(x) = \left[\underline{g}_\alpha(x), \overline{g}_\alpha(x)\right]$, for all $\alpha \in [0,1]$, we have

$$(F_\alpha + G_\alpha)(x) = F_\alpha(x) + G_\alpha(x) = \left[\underline{f}_\alpha(x) + \underline{g}_\alpha(x), \overline{f}_\alpha(x) + \overline{g}_\alpha(x)\right],$$

for all $x \in T$.

In this Section we present results on the gH-differentiability of $F + G$. We give rules for calculating it as well.

We start by showing that if F and G are gH-differentiable at x_0 with same type of gH-differentiability, then $F + G$ is also gH-differentiable at x_0 with the same type of gH-differentiability as F and G.

Theorem 2. *Let $F, G : T \to \mathbb{R}_\mathcal{F}$ be two fuzzy functions. If F and G are $(k)gH$-differentiable at x_0 for the same $k \in \{i, ii, iii, iv\}$ then $F + G$ is $(k)gH$-differentiable at x_0. Moreover,*

$$(F + G)'(x_0) = F'(x_0) + G'(x_0).$$

Proof. To prove the result we will consider separately the four possible cases for k. For $k \in \{1,2\}$, the results were proven in [3]. For case $k = iii$, we consider $F_\alpha(x) = \left[\underline{f}_\alpha(x), \overline{f}_\alpha(x)\right]$ and $G_\alpha(x) = \left[\underline{g}_\alpha(x), \overline{g}_\alpha(x)\right]$, for all $x \in T$, and assume that F and G are $(iii)gH$-differentiable fuzzy functions at x_0 then, from Theorem 1 and properties of lateral derivatives, the lateral derivatives $(\underline{f}_\alpha + \underline{g}_\alpha)'_{+/-}(x_0)$ and $(\overline{f}_\alpha + \overline{g}_\alpha)'_{+/-}(x_0)$ exist uniformly in $\alpha \in [0,1]$. In addition, $(\underline{f}_\alpha + \underline{g}_\alpha)'_+(x_0) = (\overline{f}_\alpha + \overline{g}_\alpha)'_-(x_0)$ is monotonic increasing and $(\overline{f}_\alpha + \overline{g}_\alpha)'_+(x_0) = (\underline{f}_\alpha + \underline{g}_\alpha)'_-(x_0)$ is monotonic decreasing as functions of α and $(\underline{f}_1 + \underline{g}_1)'_+(x_0) \leq (\overline{f}_1 + \overline{g}_1)'_+(x_0)$. Thus, from Theorem 1, part (iii), $F + G$ is a $(iii)gH$-differentiable fuzzy function at x. Moreover

$$
\begin{aligned}
\left[(F+G)'(x_0)\right]^\alpha &= \left[\left(\underline{f}_\alpha + \underline{g}_\alpha\right)'_+ (x_0), \left(\overline{f}_\alpha + \overline{g}_\alpha\right)'_+ (x_0)\right] \\
&= \left[\left(\underline{f}_\alpha\right)'_+ (x_0) + \left(\underline{g}_\alpha\right)'_+ (x_0), \left(\overline{f}_\alpha\right)'_+ (x_0) + \left(\overline{g}_\alpha\right)'_+ (x_0)\right] \\
&= \left[\left(\underline{f}_\alpha\right)'_+ (x_0), \left(\overline{f}_\alpha\right)'_+ (x_0)\right] + \left[\left(\underline{g}_\alpha\right)'_+ (x_0), \left(\overline{g}_\alpha\right)'_+ (x_0)\right] \\
&= [F'(x_0)]^\alpha + [G'(x_0)]^\alpha,
\end{aligned}
$$

for all $\alpha \in [0,1]$, and so

$$
(F+G)'(x_0) = F'(x_0) + G'(x_0).
$$

The proof for $k = iv$ is similar, so we omit it. □

From Theorem 2 it follows that the $(k)gH$-derivative is an additive operator, for $k \in \{i, ii, iii, iv\}$. Proving that the gH-derivative is a linear operator is our main objective. For this, given a fuzzy function $F : T \to \mathbb{R}_\mathcal{F}$ and a $\lambda \in \mathbb{R}$, we define the product $\lambda \cdot F$ by

$$
(\lambda \cdot F)(x) = \lambda \cdot F(x), \tag{3}
$$

for all $x \in T$.

Theorem 3. *Let $F : T \to \mathbb{R}_\mathcal{F}$ be a fuzzy function and let $\lambda \in \mathbb{R}$. If F is $(k)gH$-differentiable at x_0, for some $k \in \{i, ii, iii, iv\}$, then $\lambda \cdot F$ is also $(k)gH$-differentiable at x_0 and*

$$
(\lambda \cdot F)'(x_0) = \lambda \cdot F'(x_0).
$$

Proof. We consider $\lambda \in \mathbb{R}$ and $x \in T$, then from (1) and (3) we have

$$
\begin{aligned}
\left[(\lambda \cdot F)(x)\right]^\alpha &= [\lambda \cdot F(x)]^\alpha \\
&= \lambda \cdot [F(x)]^\alpha \\
&= \lambda \cdot \left[\underline{f}_\alpha(x), \overline{f}_\alpha(x)\right] \\
&= \begin{cases} \left[\lambda \underline{f}_\alpha(x), \lambda \overline{f}_\alpha(x)\right] & \text{if } \lambda \geq 0 \\[2mm] \left[\lambda \overline{f}_\alpha(x), \lambda \underline{f}_\alpha(x)\right] & \text{if } \lambda < 0. \end{cases}
\end{aligned}
$$

To prove the result we consider separately the four possible cases for k. We start considering $k = i$ and so, from Theorem 1, we have that the endpoint functions $\underline{f}_\alpha$ and $\overline{f}_\alpha$ are differentiable at x_0, uniformly in $\alpha \in [0,1]$, and satisfies the monotonicity conditions in relation to α as in item (i) of the Theorem 1. Now we consider two cases for λ: $\lambda \geq 0$ and $\lambda < 0$.

If $\lambda \geq 0$, taking into account item (i) of Theorem 1 and properties of monotonous functions, we have that $\lambda \underline{f}'_\alpha(x)$ is monotonic increasing and $\lambda \overline{f}'_\alpha(x)$ is monotonic decreasing as functions of α and $\lambda \underline{f}'_1(x) \leq \lambda \overline{f}'_1(x)$. Then, from Theorem 1, λF is an $(i)gH$-differentiable fuzzy function.

On other hand, if $\lambda < 0$, taking into account item (i) of Theorem 1 and properties of monotonous functions, we have that $\lambda \overline{f}'_\alpha(x)$ is monotonic increasing and $\lambda \underline{f}'_\alpha(x)$ is monotonic decreasing as functions of α and $\lambda \overline{f}'_1(x) \leq \lambda \underline{f}'_1(x)$. Then, from Theorem 1, λF is an $(i)gH$-differentiable fuzzy function.

In the same way we proof the cases for $k = ii, iii, iv$. $\qquad \square$

Summarizing, from Theorem 2 and Theorem 3 we have our main result.

Corollary 1. *The $(k)gH$-derivative is a linear operator, for each $k \in \{i, ii, iii, iv\}$.*

Remark 1. Theorem 3 and Corollary 1 correct Remark 31 in [3], where the authors assert that the $(i)gH$-derivative and $(ii)gH$-derivative are not linear in general.

Following with our study, what happen with the gH-differentiability of the sum of two $(k)gH$-differentiable fuzzy functions for different k?. In general, $F + G$ is not necessarily a gH-differentiable fuzzy function being F and G $(k)gH$-differentiable fuzzy functions with different k. In fact, Example 1 in [5], $F + G$ is not gH-differentiable being that F is $(ii)gH$-differentiable and G is $(iv)gH$-differentiable.

In the same way, if F is $(i)gH$-differentiable and G is $(ii)gH$-differentiable then $F + G$ is gH-differentiable, when F and G are interval-valued functions which are particular cases of fuzzy functions, (see [5]). More precisely.

Theorem 4. *([5]) Let $F, G : T \to \mathbb{I}$ be two interval-valued functions. If F is $(i)gH$-differentiable at x_0 and G is $(ii)gH$-differentiable at x_0 then $F + G$ is either $(i)gH$-differentiable or $(ii)gH$-differentiable at x_0. Moreover, we have*

$$(F + G)'(x_0) = F'(x_0) \ominus_{gH} (-1)G'(x_0). \tag{4}$$

However, in general Theorem 4 is not valid for fuzzy functions as we will show in the following example.

Example 1. We consider the fuzzy functions $F, G : (0,1) \to \mathbb{R}_\mathcal{F}$ defined via α-levels by

$$F_\alpha(x) = [\alpha, 2 - \alpha] \cdot x^2 \quad and \quad G_\alpha(x) = [0, 1 - \alpha^2] \cdot (1 - x).$$

The endpoint functions are

$$\underline{f}_\alpha(x) = \alpha x^2, \quad \overline{f}_\alpha(x) = (2-\alpha)x^2,$$

$$\underline{g}_\alpha(x) = 0, \quad \overline{g}_\alpha(x) = (1-\alpha^2)(1-x);$$

which are differentiable. Also, we have that F is $(i)gH$-differentiable and G is $(ii)gH$-differentiable. Now, the endpoint functions of the fuzzy function $F+G$ are

$$\underline{(F+G)}_\alpha(x) = \underline{f}_\alpha(x) + \underline{g}_\alpha(x) = \alpha x^2$$

and

$$\overline{(F+G)}_\alpha(x) = \overline{f}_\alpha(x) + \overline{g}_\alpha(x) = (2-\alpha)x^2 + (1-\alpha^2)(1-x).$$

Clearly, the endpoint functions are differentiable and

$$\underline{(F+G)}'_\alpha(x) = 2\alpha x$$

and

$$\overline{(F+G)}'_\alpha(x) = 2(2-\alpha)x - (1-\alpha^2) = 4x - 2\alpha x - 1 + \alpha^2.$$

Thus, the endpoint functions $\underline{(F+G)}_\alpha$ and $\overline{(F+G)}_\alpha$ do not satisfy any of the conditions of the Theorem 1. Therefore $F+G$ is not gH-differentiable.

From (4) we can see that for obtaining $(F+G)'(x)$ depends on the gH-difference. In fact, in Example 1 there is not the gH-difference $F'(x_0) \ominus_{gH} (-1)G'(x_0)$. The gH-difference of two fuzzy numbers does not always exists while the gH-difference between interval always exists [9]. So we have to include conditions on the existence of the gH-difference such as in the following result.

Theorem 5. *Let* $F, G : T \to \mathbb{R}_\mathcal{F}$ *be fuzzy functions. If* F *is* $(i)gH$*-differentiable at* x_0 *and* G *is* $(ii)gH$*-differentiable at* x_0 *and the* gH*-difference* $F'(x_0) \ominus_{gH} (-1)G'(x_0)$ *exists, then* $F+G$ *is* gH*-differentiable and*

$$(F+G)'(x_0) = F'(x_0) \ominus_{gH} (-1)G'(x_0). \tag{5}$$

Proof. Since the gH-difference $F'(x_0) \ominus_{gH} (-1)G'(x_0)$ exists, then one of the following cases holds:

$$Case(1) \begin{cases} len(F'(x_0)) \geq len(G'(x_0)), \ \forall \alpha \in [0,1], \\ \underline{f}'_\alpha(x_0) + \underline{g}'_\alpha(x_0) \text{ is monotonic increasing respect to } \alpha \text{ and} \\ \overline{f}'_\alpha(x_0) + \overline{g}'_\alpha(x_0) \text{ is monotonic decreasing respect to } \alpha. \end{cases}$$

$$Case(2) \begin{cases} len(F'(x_0)) \leq len(G'(x_0)), \ \forall \alpha \in [0,1], \\ \underline{f}'_\alpha(x_0) + \underline{g}'_\alpha(x_0) \text{ is monotonic decreasing respect to } \alpha \text{ and} \\ \overline{f}'_\alpha(x_0) + \overline{g}'_\alpha(x_0) \text{ is monotonic increasing respect to } \alpha. \end{cases}$$

If Case (1) holds then, from Theorem 1 item (i), $F+G$ is $(i)gH$-differentiable at x_0 and (5) holds. If Case (2) holds then, from Theorem 1 item (ii), $F+G$ is $(ii)gH$-differentiable at x_0 and (5) holds. Therefore, $F+G$ is $(i)gH$ or $(ii)gH$-differentiable at x_0 and (5) holds.   □

We note that a similar result to Theorem 5 was obtained in [1] to case of strongly generalized differentiable fuzzy functions.

5   Conclusions

In this article we have made a study on the sum of generalized Hukuhara differentiable fuzzy functions. The results obtained in this study should be very useful in fuzzy and interval optimization, fuzzy and interval differential equations and other topics in fuzzy and interval mathematical analysis.

References

1. Alikhani, R., Bahrami, F., Parvizi, S.: Differential calculus of fuzzy multi-variable functions and its applications to fuzzy partial differential equations. Fuzzy Sets Syst. **375**, 100–120 (2019)
2. Armand, A., Allahviranloo, T., Gouyandeh, Z.: Some fundamental results on fuzzy calculus. Iran. J. Fuzzy Syst. **15**, 27–46 (2018)
3. Bede, B., Stefanini, L.: Generalized differentiability of fuzzy-valued functions. Fuzzy Sets Syst. **230**, 119–141 (2013)
4. Chalco-Cano, Y., Rodríguez-López, R., Jiménez-Gamero, M.D.: Characterizations of generalized differentiable fuzzy functions. Fuzzy Sets Syst. **295**, 37–56 (2016)
5. Chalco-Cano, Y., Maqui-Huamán, G.G., Silva, G.N., Jiménez-Gamero, M.D.: Algebra of generalized Hukuhara differentiable interval-valued functions: review and new properties. Fuzzy Sets Syst. **375**, 53–69 (2019)
6. Moore, R.E.: Methods and Applications of Interval Analysis. Studies in Applied and Numerical Mathematics. Society for Industrial and Applied Mathematics, Philadelphia (1979)
7. Rojas-Medar, M.A., Jiménez-Gamero, M.D., Chalco-Cano, Y., Viera-Brandão, A.J.: Fuzzy quasilinear spaces and applications. Fuzzy Sets Syst. **152**, 173–190 (2005)
8. Stefanini, L., Bede, B.: Generalized Hukuhara differentiability of interval-valued functions and interval differential equations. Nonlinear Anal. **71**(34), 1311–1328 (2009)
9. Stefanini, L.: A generalization of Hukuhara difference and division for interval and fuzzy arithmetic. Fuzzy Sets Syst. **161**, 1564–1584 (2010)

Theoretical and Applied Aspects of Imprecise Probabilities

Imprecise Classification
with Non-parametric Predictive Inference

Serafín Moral[✉], Carlos J. Mantas, Javier G. Castellano,
and Joaquín Abellán

Department of Computer Science and Artifical Intelligence, University of Granada,
Granada, Spain
{seramoral,cmantas,fjgc,jabellan}@decsai.ugr.es

Abstract. In many situations, classifiers predict a set of states of a class
variable because there is no information enough to point only one state.
In the data mining area, this task is known as Imprecise Classification.
Decision Trees that use imprecise probabilities, also known as Credal
Decision Trees (CDTs), have been adapted to this field. The adaptation
proposed so far uses the Imprecise Dirichlet Model (IDM), a mathematical model of imprecise probabilities that assumes prior knowledge about
the data, depending strongly on a hyperparameter. This strong dependence is solved with the Non-Parametric Predictive Inference Model
(NPI-M), also based on imprecise probabilities. This model does not
make any prior assumption of the data and does not have parameters.
In this work, we propose a new adaptation of CDTs to Imprecise Classification based on the NPI-M. An experimental study carried out in this
research shows that the adaptation with NPI-M has an equivalent performance than the one obtained with the adaptation based on the IDM
with the best choice of the hyperparameter. Consequently, since the NPI-
M is a non-parametric approach, it is concluded that the NPI-M is more
appropriated than the IDM to be applied to the adaptation of CDTs to
Imprecise Classification.

Keywords: Imprecise classification · Credal decision trees · IDM ·
NPI-M · Imprece probabilities

1 Introduction

Supervised classification [15] aims to predict the value of a *class variable* associated with an instance, described by a set of *features* or *attributes*. This prediction
usually consists of a single value.

However, in many cases, there is no information available enough to point
only one state of the class variable. In these cases, it is more informative that
the classifier predicts a set of values of the class variable, which is known as an

Supported by the Spanish "Ministerio de Economía y Competitividad" and by "Fondo
Europeo de Desarrollo Regional" (FEDER) under Project TEC2015-69496-R.

© Springer Nature Switzerland AG 2020
M.-J. Lesot et al. (Eds.): IPMU 2020, CCIS 1238, pp. 53–66, 2020.
https://doi.org/10.1007/978-3-030-50143-3_5

imprecise prediction. Classifiers that make this type of predictions are known as *imprecise classifiers.*

When it is used an imprecise classifier, a set of class values might be obtained. It is composed of those states for which there is no another "better" one according to a criterion, which is called *dominance criterion.* The set of predicted states of the class variable is known as the set of *non-dominated states.*

In order to build an imprecise classifier, it is more suitable to apply models based on imprecise probabilities, instead of the ones that use the classical probability theory. In the literature, there are many mathematical theories associated with imprecise probabilities, such as belief functions, closed and convex sets of probability distributions (also called credal sets), probability intervals, etc [16].

In the literature, few methods for imprecise classification have been developed. The first one of them was the Naive Credal Classifier (NCC) [10, 24]. It uses the Imprecise Dirichlet Model (IDM) [22], a mathematical model of imprecise probabilities that makes statistical inferences from multinomial data, and the Naive Bayes assumption (all the attributes are independent given the class variable) to produce an imprecise classification.

In [4], it is proposed a new adaptation of the Credal Decision Trees (CDTs) [5], very simple and interpretable models, to Imprecise Classification. It is called Imprecise Credal Decision Tree (ICDT). In that work, it is shown, via an experimental analysis, that ICDT is a more informative method than NCC since it is more precise. In this work, we focus on the ICDT algorithm.

The ICDT proposed so far is based on the IDM. This model satisfies several principles which have been claimed to be desirable for inference, such as the *representation invariance principle* [22]. According to it, inferences on future events should be independent of the arrangement and labeling of the sample space. Nevertheless, IDM assumes previous knowledge about the data through a single hyperparameter s [22]. It is not a very desirable property because these assumptions are not always realistic.

For the previous reason, a Non Parametric model for Predictive Inference (NPI-M) was proposed in [8]. This model does not make any prior assumptions about the data. In addition, NPI-M is a nonparametric approach.

Both IDM and NPI-M have been applied to Decision Trees (DT) for precise classification in the literature [6, 18, 19]. When the IDM is applied to DTs, it has been shown that the performance has a strong dependence on the s parameter [19]. In [6], the NPI-M is shown to have always an equivalent performance to IDM with the standard s value when both models are applied to DTs.

For the previous reasons, in this work, we propose a new adaptation of CDTs to Imprecise Classification based on the NPI-M. It is called Imprecise Credal Decision Tree NPI (ICDT-NPI). It is similar to the already existing adaptation, but our proposed one is based on the NPI-M, instead of the IDM.

An extensive experimental research is carried out in this work. In it, we use the ICDT-NPI algorithm and the ICDT with different values of the s parameter for the IDM. This experimentation shows that, as in precise classification, the NPI-M provides equivalent results to the IDM with the best choice of the s

hyperparameter when both models are applied to the adaptation of CDTs to Imprecise Classification.

This paper is arranged as follows: The Imprecise Dirichlet Model and the Non-Parametric Predictive Inference Model are explained in Sects. 2 and 3, respectively. Section 4 describes the dominance criteria for Imprecise Classification used in this research. The adaptation of the Credal Decision Trees to Imprecise Classification is exposed in Sect. 5. Section 6 describes the main evaluation metrics that are used in imprecise classification. In Sect. 7, the experimental analysis is detailed. Conclusions are given in Sect. 8.

2   The Imprecise Dirichlet Model

Let us suppose that we have a dataset $\mathcal{D}$ with N instances. Let X be an attribute that takes values in $\{x_1, \cdots, x_t\}$.

The Imprecise Dirichlet Model (IDM) [22] is subsumed into the probability intervals theory [11]. According to this model, the variable X takes each one of its possible values x_i, $1 \leq i \leq t$ with a probability that belongs to the following interval:

$$I_i = \left\{ \left[\frac{n_i}{N+s}, \frac{n_i+s}{N+s} \right] \right\}, \forall i = 1, 2, \ldots, t, \tag{1}$$

being n_i the number of instances in $\mathcal{D}$ for which $X = x_i$, $\forall i = 1, 2, \ldots, t$ and $s > 0$ a given parameter of the model.

As it is shown in [1], this set of probability intervals is reachable and gives rise to the following closed and convex set of probability distributions, also called credal set:

$$\mathcal{P}^{\mathcal{D}}(X) = \left\{ p \mid \sum_{i=1}^{t} p(x_i) = 1, \ p(x_i) \in I_i, \quad \forall i = 1, 2, \ldots, t \right\}. \tag{2}$$

A crucial issue is the selection of the s hyperparameter. It is easy to observe that, if the s value is higher, then the intervals are wider. This parameter determines the speed of convergence of lower and upper probabilities as the size of the training set is larger. In [22], the values $s = 1$ and $s = 2$ are proposed.

3   Non-parametric Predictive Inference Model

Let X be a discrete variable whose set of possible values is $\{x_1, \cdots, x_T\}$. Let us suppose that there is a sample of N independent and identically distributed outcomes of X. Let n_i be the number of observations for which $X = x_i$, $\forall i = 1, 2, \ldots, T$. Let us assume that the first t observations have been observed, where $1 \leq t \leq T$, which implies that $n_i > 0$, $\forall i = 1, 2, \cdots, t$ and $n_i = 0$, $\forall i = t+1, \cdots, T$. Clearly, $\sum_{i=1}^{t} n_i = N$.

The Non-Parametric Predictive Inference Model (NPI-M) [8,9] utilizes a *probability wheel* representation of the data. On it, it is used a line from the center of the wheel to its boundary to represent each one of the observations. The wheel is partitioned into N slices with the same size. Each possible value can be represented only by a single sector of the wheel. This implies that two or more lines representing the same category must always be positioned next to each other on the wheel. The NPI-M is based on the circular $A_{(n)}$ assumption [9]. According to it, the probability that the next observation falls into any given slice is $\frac{1}{N}$. Thus, it must be decided which value of the X variable represents. If two lines that represent the same category border to a slice, that slice must be assigned to this value. Nevertheless, when a slice is bordered by two lines that represent different values, it can be assigned to one of the two categories associated with the slice's bordering lines, or to any value that has not been observed yet.

Let $A \subseteq \{x_1, x_2, \ldots, x_T\}$ be a subset of the set of possible values of the X variable. Let us denote $n_A = \sum_{x_i \in A} n_i$ the number of outcomes of X for which its value belongs to A and $r_A = |\{x_i \in A \mid n_i > 0, 1 \le i \le t\}|$ the number of possible values in A that have been already observed.

In order to determine the lower and upper probabilities of A, NPI-M considers all the possible configurations of the wheel. The difference between both probabilities is due to the non-observed categories. In [3], it is shown that the lower and upper probabilities of A are obtained as follows:

$$P_*(A) = \frac{n_A - \min(r_A, |\overline{A}|)}{N}, \quad P^*(A) = \frac{n_A + \min(|A|, t - r_A)}{N}. \tag{3}$$

As it can be seen, for singletons, $\{x_i\}$, $1 \le i \le T$, the lower and upper probabilities are given by:

$$P_*(\{x_i\}) = \max\left(\frac{n_i - 1}{N}, 0\right), \quad P^*(\{x_i\}) = \min\left(\frac{n_i + 1}{N}, 1\right).$$

Hence, it is disposed of the following set of probability intervals for singletons:

$$\mathcal{I} = \left\{ [l_i, u_i], \, l_i = \max\left(\frac{n_i - 1}{N}, 0\right), \, u_i = \min\left(\frac{n_i + 1}{N}, 1\right), \quad \forall i = 1, \ldots, T \right\}.$$

According to [11], this set of probability intervals corresponds to the following credal set:

$$\mathcal{P}(\mathcal{I}) = \{p \in \mathcal{P}(X) \mid p(x_i) \in [l_i, u_i], \quad \forall i = 1, 2 \ldots, T\}. \tag{4}$$

being $\mathcal{P}(X)$ the set of all probability distributions on the X variable, $l_i = \max\left(\frac{n_i - 1}{N}, 0\right)$, and $u_i = \min\left(\frac{n_i + 1}{N}, 1\right)$, $\forall i = 1, 2, \ldots, T$.

In [3], it was proved that the lower and upper probabilities associated with $\mathcal{P}(\mathcal{I})$ coincide with the lower and upper probabilities given by (3). Therefore, the lower and upper probabilities corresponding to the NPI-M can be extracted via the lower and upper probabilities for singletons, which produce a set of probability intervals, and, consequently, a credal set. Nevertheless, in [3], it is shown that, in this set, there are probability distributions that are not compatible with the NPI-M.

If we consider all the probability distributions belonging to $\mathcal{P}(\mathcal{I})$, it is obtained an approximated model, called Approximate Non-Parametric Predictive Inference Model (A-NPI-M) [3]. It utilizes the convex hull of the set of probability distributions compatible with the NPI-M. In this way, when the A-NPI-M is used, a set of difficult constraints is avoided and the exact model is simplified. In [3], it is shown that NPI-M and A-NPI-M have a similar behavior when both models are applied to CDTs. For these reasons, in this work, we consider the A-NPI-M.

4   Dominance Criteria in Imprecise Classification

In Imprecise Classification, it is used a *dominance criterion* to select the states of the class variable that are not "defeated" under that criterion by another. In order to do it, if we have a set of probability intervals, as in this research, we can use the bounds of the intervals.

Let c_i and c_j be two possible values of the class variable C. Two *dominance criteria* very used are the following:

1. Let $[l_i, u_i]$ and $[l_j, u_j]$ be the probability intervals on how c_i and c_j happen, respectively. It is said that there is *stochastic dominance* or *strong dominance* of c_i on c_j if, and only if, $l_i \geq u_j$.
2. Let suppose now that the probability of the class variable C is expressed by a non-empty credal set $\mathcal{P}$. It is said that there is credal dominance of c_i on c_j iff $p(C = c_i) \geq p(C = c_j)$, for all probability distribution $p \in \mathcal{P}$.

Credal dominance is a more significant criterion than stochastic dominance [24]. However, it is usually more difficult to verify. Under the IDM and the A-NPI-M, both dominance criteria are equivalent [2]. Hence, with both IDM and A-NPI-M, if we check that one state dominates stochastically to another, then we know that there is credal dominance of the first state on the second one. Therefore, with IDM, as well as with A-NPI-M, it is just necessary to consider the extreme values of the intervals to know the cases of credal dominance among the possible states of C.

5   Credal Decision Trees for Imprecise Classification

The adaptation of Credal Decision Tree algorithm (CDT) [5] to Imprecise Classification (ICDT) was proposed in [4]. It is called Imprecise Credal Decision Tree algorithm (ICDT).

As in CDTs, in ICDT, each node corresponds to an attribute or feature variable and there is a branch for each possible value of that attribute. When entering a feature in a node does not provide more information about the class variable according to a criterion, a terminal or leaf node is reached. Unlike in precise classification, where this leaf node is labeled with the most probable class value, in ICDTs, the leafs nodes are empty. When a new example is required to be classified, it is made a path from the root to a terminal node using the values of its attributes. Whereas in precise classification the most probable class value in that leaf node is assigned, in ICDTs, for each value of the class variable, it is obtained a probability interval.

The most important issue of the building process of the ICDT is the split criterion, i.e, the criterion utilized to select the attribute to split in each node. In ICDT, the split criterion is the same as the one used in CDT.

Let us suppose that $\mathcal{D}$ is a partition of the training set in a certain node. Let C be the class variable and let us suppose that $\{c_1, \ldots, c_k\}$ are its possible values. Let X be an attribute variable and let $\{x_1, x_2, \ldots, x_t\}$ be its possible values. Let us assume that $\mathcal{P}^{\mathcal{D}}(C)$ is the credal set on $\mathcal{D}$ associated with C corresponding to a model based on probability intervals[1] .

The split criterion utilized in the ICDT algorithm utilizes the maximum of the Shannon entropy [21] on $\mathcal{P}^{\mathcal{D}}(C)$:

$$H^*(\mathcal{P}^{\mathcal{D}}(C)) = \max \left\{ H(p) \mid p \in \mathcal{P}^{\mathcal{D}}(C) \right\} \qquad (5)$$

being H the Shannon entropy.

The maximum of entropy is a well-established measure on credal sets that satisfies good properties and behavior [16].

Hence, the split criterion used in ICDT is the Imprecise Information Gain (IIG) [5]. It is defined as follows:

$$IIG(C, X) = H^*(\mathcal{P}^{\mathcal{D}}(C)) - \sum_{i=1}^{t} P^{\mathcal{D}}(X = x_i) H^*(\mathcal{P}^{\mathcal{D}}(C \mid X = x_i)), \qquad (6)$$

where $P^{\mathcal{D}}(X)$ is the maximum of entropy on the credal set corresponding to the X attribute and $H^*(\mathcal{P}^{\mathcal{D}}(C \mid X = x_i))$ is the maximum of the entropy on the credal set associated with the C variable and with the partition of $\mathcal{D}$ composed by the instances of $\mathcal{D}$ that verify that $X = x_i$.

The main difference among the CDT and ICDT algorithms resides in the criterion utilized to classify an instance once a terminal node is reached. The CDT algorithm assigns the most frequent class value in that leaf. Nevertheless, the ICDT algorithm assigns a probability interval to each one of the possible values of the class variable using the relative frequencies in the leaf node and a model based on probability intervals. Then, a dominance criterion is used to obtain the set of non-dominated states. In this research, the models considered are the IDM and the A-NPI-M. Thus, since with these models stochastic and credal dominance are equivalent and the first one is much easier to verify, we use the stochastic dominance in this work.

The procedure to classify a new instance in the ICDT algorithm can be summarized in Fig. 1.

---

[1] In this work we will consider the A-NPI-M, unlike in [4], where the IDM is employed.

> Procedure **Classify_Instance_ICDT**(Built Tree $\mathcal{T}$, new instance **x**)
>
> 1. Apply **x** in $\mathcal{T}$ to reach a leaf node.
> 2. Obtain the probability intervals in this terminal node for **x**, based on relative frequencies using a model based on imprecise probabilities:
>
> $$\{[l_i, u_i], i = 1, \cdots, k\}.$$
>
> 3. Apply a dominance criterion to the above intervals to get a set of non-dominated states for **x**: $\{c_{i_1}, c_{i_2}, \cdots, c_{i_r}\}$, with $r \leq k$.

Fig. 1. Classification of a new instance in ICDT algorithm.

In this research, we compare the use of the A-NPI-M for the credal sets associated with the class variable, in the building process of the ICDT algorithm and to obtain the probability intervals for the class values in the terminal nodes, with the use of the IDM with different values of the s parameter.

6 Evaluation Metrics in Imprecise Classification

An evaluation measure for Imprecise Classification should take into consideration two points. The first one of them is if the prediction is right, i.e if the real class value is among the predicted ones. The second point is how informative is the predicted set of states, which is measured by its cardinality.

Several metrics only focus on one of the issues commented above, such as:

- **Determinacy:** It is the proportion of instances for which the classifier returns a single class value.
- **Single Accuracy:** It consists of the accuracy between the instances for which there is just one predicted state.
- **Set Accuracy:** It measures, on the instances for which there is more than one predicted state, the proportion of them for which its real state is among the predicted ones.
- **Indeterminacy Size:** It is the average size of the predicted states set.

As it can be observed, none of these metrics is suitable to measure the whole performance of an imprecise classifier.

In [10], it was proposed a measure to provide a global evaluation of an imprecise classifier, called *Discounted Accuracy measure* (DACC), defined as:

$$DACC = \frac{1}{N_{Test}} \sum_{i=1}^{N_{Test}} \frac{(correct)_i}{|U_i|}, \tag{7}$$

where N_{Test} is the number of instances of the test set; U_i is the predicted states set for the i-th instance; $|U_i|$ its cardinality; $(correct)_i$ is equal to 1 if U_i contains the real class value and 0 otherwise, $\forall i = 1, 2, \ldots, n$.

We shall denote k the number of class values.

It can be observed that DACC is an accuracy measure: it does not add any value for the erroneous predictions and, for the correct ones, the added value is "penalized" by the number of predicted states. The optimal value of $DACC$ is 1. It is achieved when there is always a single predicted state and all the predictions are right. If the classifier always predicts all the possible values of the class variable, the value of DACC is $\frac{1}{k}$. This value should be lower because in this case, the classifier is not informative.

In [4], a new metric for imprecise classification, MIC, was proposed. It penalizes the errors in a strict sense. When there is a correct prediction for an instance, MIC adds a value that depends on $\frac{|U_i|}{k}$. If the prediction for an instance is incorrect, MIC adds a constant value, which depends on k. More specifically, MIC is defined as follows:

$$MIC = \frac{1}{N_{Test}} \left(\sum_{i:Success} \log \frac{|U_i|}{k} + \frac{1}{k-1} \sum_{i:Error} \log k \right) \qquad (8)$$

As can be seen, the optimal value of MIC is $\log k$. It is reached when, for all the instances, only the real class value is predicted. Besides, when a classifier always returns as predicted states all the possible ones, i.e, when $|U_i| = k$, $\forall i = 1, \ldots, N_{Test}$, the value of MIC is equal to 0. It makes sense because, in this case, the classifier does not give any information.

7   Experimentation

7.1   Experimental Settings

Remark that, in this experimentation, the aim is to compare the use of the A-NPI-M versus the IDM in the ICDT algorithm, in the building process and for the selection of non-dominated states. For the reasons explained in Sect. 5, the stochastic dominance criterion is used. For evaluation, we use the DACC and MIC measures, as in [4].

Within this Section, we call ICDT-IDM to the Imprecise Decision Tree with the IDM and ICDT-NPI to the Imprecise Decision Tree with the A-NPI-M.

To compare the performance between both algorithms, as in the experimentation carried out in [4], 34 known datasets have been used. They can be downloaded from the *UCI Machine Learning repository* [17]. These datasets are diverse concerning the size of the set, the number of continuous and discrete attributes, the number of values per variable, the number of class values, etc. The most relevant characteristics of each dataset can be seen in Table 1.

As it was done in [4], in each dataset, missing values have been replaced with the mean value for continuous variables and with modal values for discrete attributes. After that, in each database, continuous attributes have been discretized using the Fayyad and Irani's discretization method [13].

Table 1. Data set description. Column "N" is the number of instances in the data sets, column "Feat" is the number of features or attribute variables, column "Num" is the number of numerical variables, column "Nom" is the number of nominal variables, column "k" is the number of cases or states of the class variable (always a nominal variable) and column "Range" is the range of states of the nominal variables of each data set.

Data set	N	Feat	Num	Nom	k	Range
anneal	898	38	6	32	6	2–10
arrhythmia	452	279	206	73	16	2
audiology	226	69	0	69	24	2–6
autos	205	25	15	10	7	2–22
balance-scale	625	4	4	0	3	–
car	1728	6	0	6	4	3–4
cmc	1473	9	2	7	3	2–4
dermatology	366	34	1	33	6	2–4
ecoli	366	7	7	0	7	–
flags	194	30	2	28	8	2–13
hypothyroid	3772	30	7	23	4	2–4
iris	150	4	4	0	3	–
letter	20000	16	16	0	26	–
lymphography	146	18	3	15	4	2–8
mfeat-pixel	2000	240	0	240	10	4–6
nursery	12960	8	0	8	4	2–4
optdigits	5620	64	64	0	10	–
page-blocks	5473	10	10	0	5	–
pendigits	10992	16	16	0	10	–
postop-patient-data	90	9	0	9	3	2–4
primary-tumor	339	17	0	17	21	2–3
segment	2310	19	16	0	7	–
soybean	683	35	0	35	19	2–7
spectrometer	531	101	100	1	48	4
splice	3190	60	0	60	3	4–6
sponge	76	44	0	44	3	2–9
tae	151	5	3	2	3	2
vehicle	946	18	18	0	4	–
vowel	990	11	10	1	11	2
waveform	5000	40	40	0	3	–
wine	178	13	13	0	3	–
zoo	101	16	1	16	7	2

We have used the Weka software [23] for this experimentation. We have started from the implementation of the ICDT-IDM algorithm given in this software and we have added the necessary methods to use the ICDT-NPI. For ICDT-IDM, three values of the s parameter have been used: $s = 1$, $s = 2$ and $s = 3$. The rest of the parameters used in both algorithms have been the ones given by default in Weka. This software has been also used for the preprocessing steps described above. We denote ICDT-IDMi to ICDT-IDM with $s = i$, for $i = 1, 2, 3$.

For each dataset, a 10-fold cross-validation procedure has been repeated 10 times.

For statistical comparisons, consistently with [12], we have used the following tests to compare more than two classifiers on a large number of datasets with a level of significance of $\alpha = 0.05$:

- **Friedman test** [14]: A non-parametric test that ranks the algorithms separately for each dataset (the best performing algorithm is assigned to the rank 1, the second-best, rank 2, and so on). The null hypothesis of the Friedman test is that all the algorithms have equivalent performance.
- When the null hypothesis of the Friedman test is rejected, all the algorithms are compared to each other by using the **Nemenyi test** [20].

For the statistical tests, the Keel software [7] has been used.

7.2   Results and Discussion

Tables 2 and 3 show, respectively, the main results corresponding to DACC and MIC measures. Specifically, these tables allow us to see the average values, the Friedman ranks, and the pairs of algorithms for which there are significant differences according to Nemenyi pos-hoc. We do not show the complete results here due to the limitations of space, they can be found in http://flanagan.ugr.es/IPMU2020.html.

Table 2. Summary of the results for the DACC measure. Column "Nemenyi" shows the algorithms in which the algorithm in the row performs significantly better according to the Nemenyi test.

Algorithm	Average	Friedman rank	Nemenyi
ICDT-NPI	0.7675	1.9118	ICDT-IDM3
ICDT-IDM1	**0.7763**	2.3382	ICDT-IDM3
ICDT-IDM2	0.7606	2.4853	–
ICDT-IDM3	0.7482	3.2647	–

As it can be observed, for both DACC and MIC metrics, the best average value is obtained by ICDT-IDM1, followed by ICDT-NPI, ICDT-IDM2, and

Table 3. Summary of the results for the MIC measure. Column "Nemenyi" shows the algorithms in which the algorithm in the row performs significantly better according to the Nemenyi test.

Algorithm	Average	Friedman rank	Nemenyi
ICDT-NPI	1.3414	1.9706	ICDT-IDM3
ICDT-IDM1	**1.3652**	2.4412	–
ICDT-IDM2	1.3334	2.5	–
ICDT-IDM3	1.3065	3.0882	–

ICDT-IDM3. In addition, for both evaluation measures, the ICDT-NPI algorithm obtains the best rank according to the Friedman test. Regarding ICDT-IDM, the higher is the value of the s parameter, the higher is the rank. The results of the Nemenyi post-hoc allow us to observe that the results obtained by ICDT-NPI are significantly better than the ones obtained by ICDT-IDM with the worst s value ($s = 3$) for both MIC and DACC metrics. Also, for DACC, ICDT-IDM with $s = 1$ performs significantly better than ICDT-IDM with $s = 3$. For both evaluation metrics, ICDT-NPI, ICDT-IDM1, and ICDT-IDM2 have an equivalent performance.

Hence, the performance of the ICDT-IDM algorithm depends on the choice of the s hyperparameter. Regarding ICDT-NPI, the results obtained for this algorithm are statistically equivalent to the ones obtained by ICDT-IDM with the best s parameter. Furthermore, ICDT-NPI performs significantly better than ICDT-IDM with the worst value of the s hyperparameter.

For a deeper analysis, Table 4 shows the average values of Determinacy, Single Accuracy, Set Accuracy and Indeterminacy size for each algorithm.

Table 4. Average results obtained for basic metrics by each algorithm. Best scores are marked in bold.

Algorithm	Determinacy	Single accuracy	Set accuracy	Indeterminacy size
ICDT-NPI	0.9002	**0.8237**	**0.9561**	7.9381
ICDT-IDM1	**0.9477**	0.8023	0.8844	**5.2955**
ICDT-IDM2	0.8985	0.8119	0.9168	5.9313
ICDT-IDM3	0.8666	0.8151	0.9218	6.1346

Firstly, ICDT-IDM1 achieves the highest average Determinacy. It means that the highest number of instances for which only one state is predicted is obtained with ICDT-IDM1. NPI-M obtains the second-highest value in Determinacy, followed by ICDT-IDM2 and ICDT-IDM3 for all the noise levels.

However, for the accuracy among the instances for which it is predicted just a state of the class variable (Single Accuracy), ICDT-IDM obtains the worst performance with $s = 1$. In the ICDT-IDM algorithm, the higher is the value of the s parameter, the better is the performance in Single Accuracy. The best Single Accuracy is obtained with ICDT-NPI.

Regarding Set Accuracy, which measures the average number of instances for which the real class value is between the predicted ones, the results are similar to the ones obtained in Single Accuracy: ICDT-IDM performs better as the value of the s parameter is higher and ICDT-NPI outperforms ICDT-IDM with the three s values considered.

The lowest value of the Indeterminacy size, which measures the average size of the non-dominated states, is achieved with ICDT-IDM1. Moreover, the lower is the s value for ICDT-IDM, the lower is the indeterminacy size. The highest average number of non-dominated states is obtained with ICDT-NPI.

Therefore, with ICDT-NPI, it is attained the best trade-off between predicting only one state and making correct predictions. This algorithm obtains the second-highest score in Determinacy and the best one in Single Accuracy, whereas ICDT-IDM1, which achieves the highest Determinacy, obtains the worst results in Single Accuracy. Besides, when there is more than one predicted state, in the ICDT algorithm, the size of the predicted states sets is larger as the value if the s value is higher and the largest set is obtained with the ICDT-NPI algorithm. Nevertheless, ICDT-NPI obtains the highest percentage of instances for which the real class value is predicted and, in the ICDT-IDM algorithm, this percentage is lower as the value of the s parameter is higher.

Summary of the Results: The ICDT-IDM algorithm predicts the real class value more frequently as long as the value of the s parameter is higher. However, if the s value is higher, then the predictions made by ICDT-IDM are less informative in the sense that the size of the predicted class values set is larger. With ICDT-NPI, although the size of the predicted states set is, on average, larger than with ICDT-IDM, it is achieved the best trade-off between predicting fewer states of the class variable and making correct predictions.

The results obtained with DACC and MIC measures allow us to deduce that ICDT-NPI performs equivalently to ICDT-IDM with the best choice of the s parameter. Moreover, the results obtained by ICDT-NPI are significantly better than the ones obtained by ICDT-IDM with the worst s value. Consequently, the NPI-M is more suitable to be applied to the adaptation of CDTs to Imprecise Classification, since the NPI-M is free of parameters.

8   Conclusions

In this work, we have dealt with the problem of Imprecise Classification. Specifically, we have considered the adaptation of the Credal Decision Trees, Decision Trees that use imprecise probabilities, to this field.

The adaptation of Credal Decision Trees to Imprecise Classification proposed so far was based on the Imprecise Dirichlet Model, a model based on imprecise probabilities that assumes prior knowledge about the data through a hyperparameter s. For this reason, in this research, a new adaptation of Credal Decision Trees to Imprecise Classification based on the Non-Parametric Predictive Inference Model has been presented. This model, which is also based on imprecise probabilities, solves the main drawback of the Imprecise Dirichlet Model: it does not make any prior assumption about the data; it is a non-parametric approach.

An experimental research carried out in this work has shown that the new adaptation of Credal Decision Trees to Imprecise Classification based on the Non-Parametric Predictive Inference Model has equivalent performance to the Imprecise Credal Decision Tree based on the Imprecise Dirichlet Model with the best s value. The results obtained by Imprecise Credal Decision Tree with Non-Parametric Predictive Inference Model are also significantly better than the ones obtained by Imprecise Credal Decision Tree with the Imprecise Dirichlet Model with the worst s value. Although with the Non-Parametric Predictive Inference Model the set of predicted class values is larger than with the Imprecise Dirichlet Model, with the first model it is achieved a better trade-off between making correct predictions and predicting fewer states of the class variable.

Therefore, taking into account that the Non-Parametric Predictive Inference Model is free of parameters, it can be concluded that this model is more suitable to be applied to the adaptations of Credal Decision Trees to Imprecise Classification than the Imprecise Dirichlet Model.

References

1. Abellán, J.: Uncertainty measures on probability intervals from the imprecise dirichlet model. Int. J. Gen. Syst. **35**(5), 509–528 (2006). https://doi.org/10.1080/03081070600687643
2. Abellán, J.: Equivalence relations among dominance concepts on probability intervals and general credal sets. Int. J. Gen. Syst. **41**(2), 109–122 (2012). https://doi.org/10.1080/03081079.2011.607449
3. Abellán, J., Baker, R.M., Coolen, F.P.: Maximising entropy on the nonparametric predictive inference model for multinomial data. Eur. J. Oper. Res. **212**(1), 112–122 (2011). https://doi.org/10.1016/j.ejor.2011.01.020
4. Abellán, J., Masegosa, A.R.: Imprecise classification with credal decision trees. Int. J. Uncertainty Fuzziness Knowl. Based Syst. **20**(05), 763–787 (2012). https://doi.org/10.1142/S0218488512500353
5. Abellán, J., Moral, S.: Building classification trees using the total uncertainty criterion. Int. J. Intell. Syst. **18**(12), 1215–1225 (2003). https://doi.org/10.1002/int.10143
6. Abellán, J., Baker, R.M., Coolen, F.P., Crossman, R.J., Masegosa, A.R.: Classification with decision trees from a nonparametric predictive inference perspective. Comput. Stat. Data Anal. **71**, 789–802 (2014). https://doi.org/10.1016/j.csda.2013.02.009
7. Alcalá-Fdez, J., et al.: KEEL: a software tool to assess evolutionary algorithms for data mining problems. Soft Comput. **13**(3), 307–318 (2009). https://doi.org/10.1007/s00500-008-0323-y

8. Coolen, F.P.A.: Learning from multinomial data: a nonparametric predictive alternative to the imprecise dirichlet model. In: Cozman, F.G., Nau, R., Seidenfeld, T. (eds.) ISIPTA 2005: Proceedings of the Fourth International Symposium on Imprecise Probabilities and their Applications, pp. 125–134 (2005). (Published by SIPTA)

9. Coolen, F., Augustin, T.: A nonparametric predictive alternative to the imprecise Dirichlet model: The case of a known number of categories. Int. J. Approximate Reasoning **50**(2), 217–230 (2009). https://doi.org/10.1016/j.ijar.2008.03.011. Special Section on The Imprecise Dirichlet Model and Special Section on Bayesian Robustness (Issues in Imprecise Probability)

10. Corani, G., Zaffalon, M.: Learning reliable classifiers from small or incomplete data sets: the naive credal classifier 2. J. Mach. Learn. Res. **9**, 581–621 (2008)

11. De Campos, L.M., Huete, J.F., Moral, S.: Probability intervals: a tool for uncertain reasoning. Int. J. Uncertainty Fuzziness Knowl. Based Syst. **02**(02), 167–196 (1994). https://doi.org/10.1142/S0218488594000146

12. Demšar, J.: Statistical comparisons of classifiers over multiple data sets. J. Mach. Learn. Res. **7**, 1–30 (2006)

13. Fayyad, U., Irani, K.: Multi-valued interval discretization of continuous-valued attributes for classification learning. In: Proceeding of the 13th International Joint Conference on Artificial Inteligence, pp. 1022–1027. Morgan Kaufmann (1993)

14. Friedman, M.: A comparison of alternative tests of significance for the problem of m rankings. Ann. Math. Stat. **11**(1), 86–92 (1940). https://doi.org/10.1214/aoms/1177731944

15. Hand, D.J.: Construction and Assessment of Classification Rules. Wiley, New York (1997)

16. Klir, G.J.: Uncertainty and Information: Foundations of Generalized Information Theory. Wiley (2006). https://doi.org/10.1002/0471755575

17. Lichman, M.: UCI machine learning repository (2013). http://archive.ics.uci.edu/ml

18. Mantas, C.J., Abellán, J.: Credal-C4.5: Decision tree based on imprecise probabilities to classify noisy data. Expert Syst. Appl. **41**(10), 4625–4637 (2014). https://doi.org/10.1016/j.eswa.2014.01.017

19. Mantas, C.J., Abellán, J., Castellano, J.G.: Analysis of Credal-C4.5 for classification in noisy domains. Expert Syst. Appl. **61**, 314–326 (2016). https://doi.org/10.1016/j.eswa.2016.05.035

20. Nemenyi, P.: Distribution-free multiple comparisons. Doctoral dissertation, Princeton University, New Jersey, USA (1963)

21. Shannon, C.E.: A mathematical theory of communication. Bell Syst. Tech. J. **27**(3), 379–423 (1948). https://doi.org/10.1002/j.1538-7305.1948.tb01338.x

22. Walley, P.: Inferences from multinomial data; learning about a bag of marbles (with discussion). J. Roy. Stat. Soc.: Ser. B (Methodol.) **58**(1), 3–57 (1996). https://doi.org/10.2307/2346164

23. Witten, I.H., Frank, E.: Data Mining: Practical Machine Learning Tools and Techniques. Morgan Kaufmann Series in Data Management Systems, 2nd edn. Morgan Kaufmann Publishers Inc., San Francisco (2005)

24. Zaffalon, M.: The naive credal classifier. J. Stat. Plann. Infer. **105**(1), 5–21 (2002). https://doi.org/10.1016/S0378-3758(01)00201-4. Imprecise Probability Models and their Applications

On the Elicitation of an Optimal Outer Approximation of a Coherent Lower Probability

Enrique Miranda[1]($\boxtimes$)(ID), Ignacio Montes[1](ID), and Paolo Vicig[2](ID)

[1] University of Oviedo, C-Federico García Lorca, 18, 33007 Oviedo, Spain
{mirandaenrique,imontes}@uniovi.es
[2] University of Trieste (DEAMS), 34123 Trieste, Italy
paolo.vicig@deams.units.it

Abstract. The process of outer approximating a coherent lower probability by a more tractable model with additional properties, such as 2- or completely monotone capacities, may not have a unique solution. In this paper, we investigate whether a number of approaches may help in eliciting a unique outer approximation: minimising a number of distances with respect to the initial model, or maximising the specificity of the outer approximation. We apply these to 2- and completely monotone approximating lower probabilities, and also to possibility measures.

Keywords: Coherent lower probabilities · 2-monotonicity · Belief functions · Possibility measures · Specificity

1 Introduction

The theory of imprecise probabilities [13] encompasses the different models that may be used as an alternative to probability theory in situations of imprecise or ambiguous information. Among them, we can find credal sets [7], coherent lower probabilities [13], belief functions [11] or possibility measures [15].

Within imprecise probabilities, one of the most general models are coherent lower and upper probabilities. However, this generality is at times harmed by the difficulties that arise when using them in practice. For example, there is no simple procedure for computing the extreme points of its associated credal set, and there is no unique coherent extension to gambles. These problems are solved when the coherent lower probability satisfies the additional property of 2-monotonicity [4,12], or that of complete monotonicity.

For this reason, in previous papers [9,10] we investigated the problem of transforming a coherent lower probability into a 2-monotone one that does not add information to the model while being as close as possible to it. This led us to the notion of undominated outer approximations, formerly introduced in [2]. In [9] we analysed the properties of the 2-monotone outer approximations, while in [10] we studied the completely monotone ones, considering in particular the

© Springer Nature Switzerland AG 2020
M.-J. Lesot et al. (Eds.): IPMU 2020, CCIS 1238, pp. 67–81, 2020.
https://doi.org/10.1007/978-3-030-50143-3_6

outer approximations in terms of necessity measures. In both cases, we found out that there may be an infinity of undominated outer approximations, and that their computation may be quite involved. Nevertheless, in the case of necessity measures, we proved that there are a finite number of undominated ones and we introduced a procedure for determining them.

Since in any case there is not a unique undominated outer approximation in terms of 2- or completely monotone lower probabilities or even in terms of necessity measures, in this paper we explore a number of possibilities that may help single out a unique undominated outer approximation. After introducing some preliminary notions in Sect. 2, formalising the idea of outer approximation and summarising the main properties from [9,10] in Sect. 3, in Sects. 4 and 5 we introduce and compare a number of different procedures to elicit an undominated outer approximation. We conclude the paper in Sect. 6 summarising the main contributions of the paper. Due to space limitations, proofs have been omitted.

2 Imprecise Probability Models

Consider an experiment taking values in a finite possibility space $\mathcal{X} = \{x_1, \ldots, x_n\}$. A *lower probability* on $\mathcal{P}(\mathcal{X})$ is a monotone function $\underline{P} : \mathcal{P}(\mathcal{X}) \to [0,1]$ satisfying $\underline{P}(\emptyset) = 0, \underline{P}(\mathcal{X}) = 1$. For every $A \subseteq \mathcal{X}$, $\underline{P}(A)$ is interpreted as a lower bound for the true (but unknown) probability of A. Any lower probability determines the *credal set* of probability measures that are compatible with it, given by $\mathcal{M}(\underline{P}) = \{P \mid P(A) \geq \underline{P}(A) \; \forall A \subseteq \mathcal{X}\}$. We say that $\underline{P}$ *avoids sure loss* when $\mathcal{M}(\underline{P})$ is non-empty, and that it is *coherent* when $\underline{P}(A) = \min\{P(A) \mid P \in \mathcal{M}(\underline{P})\}$ for every $A \subseteq \mathcal{X}$.

Associated with $\underline{P}$, we can consider its conjugate *upper probability*, given by $\overline{P}(A) = 1 - \underline{P}(A^c)$ for every $A \subseteq \mathcal{X}$. The value $\overline{P}(A)$ may be interpreted as an upper bound for the unknown probability of A, and it follows that $P \geq \underline{P}$ if and only if $P \leq \overline{P}$. This means that the probabilistic information given by a lower probability and its conjugate upper probability are equivalent, and so it suffices to work with one of them.

A coherent lower probability $\underline{P}$ is *k-monotone* if for every $1 \leq p \leq k$ and $A_1, \ldots, A_p \subseteq \mathcal{X}$ it satisfies $\underline{P}(\cup_{i=1}^p A_i) \geq \sum_{\emptyset \neq I \subseteq \{1,\ldots,p\}} (-1)^{|I|+1} \underline{P}(\cap_{i \in I} A_i)$. Of particular interest are the cases of *2-monotonicity* and *complete monotonicity*; the latter refers to those lower probabilities that are k-monotone for every k.

Any lower probability $\underline{P}$ can be represented in terms of a function called *Möbius inverse*, denoted by $m_{\underline{P}} : \mathcal{P}(\mathcal{X}) \to \mathbb{R}$, and defined by:

$$m_{\underline{P}}(A) = \sum_{B \subseteq A} (-1)^{|A \setminus B|} \underline{P}(B), \quad \forall A \subseteq \mathcal{X}.$$

Conversely, $m_{\underline{P}}$ allows to retrieve $\underline{P}$ using the expression $\underline{P}(A) = \sum_{B \subseteq A} m_{\underline{P}}(B)$. Moreover, $m_{\underline{P}}$ is the Möbius inverse associated with a 2-monotone lower probability $\underline{P}$ if and only if [3] $m_{\underline{P}}$ satisfies:

$$\sum_{A \subseteq \mathcal{X}} m_{\underline{P}}(A) = 1, \quad m_{\underline{P}}(\emptyset) = 0, \quad m_{\underline{P}}(\{x_i\}) \geq 0 \; \forall x_i \in \mathcal{X}, \qquad \text{(2monot.1)}$$

$$\sum_{\{x_i,x_j\}\subseteq B\subseteq A} m_{\underline{P}}(B) \geq 0, \quad \forall A \subseteq \mathcal{X},\ \forall x_i, x_j \in A,\ x_i \neq x_j, \qquad (2monot.2)$$

while it is associated with a completely monotone lower probability if and only if [11] $m_{\underline{P}}$ satisfies:

$$\sum_{A\subseteq\mathcal{X}} m_{\underline{P}}(A) = 1, \quad m_{\underline{P}}(\emptyset) = 0, \quad m_{\underline{P}}(A) \geq 0 \ \ \forall A \subseteq \mathcal{X}. \qquad (Cmonot.1)$$

Completely monotone lower probabilities are also connected to Dempster-Shafer Theory of Evidence [11], where they are called *belief functions*. In that case, the events with strictly positive mass are called *focal events*.

Another usual imprecise model is that of necessity and possibility measures. A possibility measure [6,15], denoted by Π, is a supremum-preserving function:

$$\Pi(\cup_{i\in I} A_i) = \sup_{i\in I} \Pi(A_i), \quad \forall A_i \subseteq \mathcal{X},\ i \in I.$$

In our finite framework, the above condition is equivalent to $\Pi(A \cup B) = \max\{\Pi(A), \Pi(B)\}$ for every $A, B \subseteq \mathcal{X}$. Every possibility measure is a coherent upper probability. Its conjugate lower probability, denoted by N and called *necessity measure*, is a completely monotone lower probability and its focal events are nested.

3   Outer Approximations of Coherent Lower Probabilities

Even if coherent lower probabilities are more general than 2-monotone ones, the latter have some practical advantages. For example, they can be easily extended to gambles [4] and the structure of their credal set can be easily determined [12]. Motivated by this, in [9] we proposed to replace a given coherent lower probability by a 2-monotone one that does not add information to the model while being as close as possible to the initial model. The first condition gives rise to the notion of *outer approximation*, and the second leads to the notion of *undominated* approximations. These concepts were formalised by Bronevich and Augustin [2]:

Definition 1. *Given a coherent lower probability $\underline{P}$ and a family $\mathcal{C}$ of coherent lower probabilities, $\underline{Q}$ is an* outer approximation *(OA, for short) of $\underline{P}$ if $\underline{Q} \leq \underline{P}$. Moreover, $\underline{Q}$ is* undominated *in $\mathcal{C}$ if there is no $\underline{Q}' \in \mathcal{C}$ such that $\underline{Q} \lneq \underline{Q}' \leq \underline{P}$.*

Similarly, given a coherent upper probability $\overline{P}$ and a family $\mathcal{C}$ of coherent upper probabilities, $\overline{Q} \in \mathcal{C}$ is an *outer approximation* of $\overline{P}$ if $\overline{Q}(A) \geq \overline{P}(A)$ for every $A \subseteq \mathcal{X}$, and it is called *non-dominating* in $\mathcal{C}$ if there is no $\overline{Q}' \in \mathcal{C}$ such that $\overline{Q} \gneq \overline{Q}' \geq \overline{P}$. It follows that $\overline{Q}$ is an outer approximation of $\overline{P}$ if and only if its conjugate $\underline{Q}$ is an outer approximation of the lower probability $\underline{P}$ conjugate of $\overline{P}$.

Let us consider now the families $\mathcal{C}_2$, $\mathcal{C}_\infty$ and $\mathcal{C}_\Pi$ of 2- and completely monotone lower probabilities and possibility measures. In [9,10] we investigated several properties of the undominated (non-dominating for $\mathcal{C}_\Pi$) outer approximations in these families. We showed that determining the set of undominated OA in $\mathcal{C}_2$ and $\mathcal{C}_\infty$ is not immediate, and that these sets are infinite in general. The problem is somewhat simpler for the outer approximations in $\mathcal{C}_\Pi$, even if in this case there is not a unique non-dominating OA either. In this paper, we discuss different procedures to elicit a unique OA.

Before we proceed, let us remark that we may assume without loss of generality that all singletons have strictly positive upper probability.

Proposition 1. *Let $\overline{P}, \overline{Q} : \mathcal{P}(\mathcal{X}) \to [0,1]$ be two coherent upper probabilities such that $\overline{P} \leq \overline{Q}$. Assume that $\overline{P}(\{x\}) = 0 < \overline{Q}(\{x\})$ for a given $x \in \mathcal{X}$, and let us define $\overline{Q}' : \mathcal{P}(\mathcal{X}) \to [0,1]$ by $\overline{Q}'(A) = \overline{Q}(A \setminus \{x\})$ for every $A \subseteq \mathcal{X}$. Then:*

1. *$\overline{P} \leq \overline{Q}' \leq \overline{Q}$.*
2. *If $\overline{Q}$ is k-alternating, so is $\overline{Q}'$.*
3. *If $\overline{Q}$ is a possibility measure, so is $\overline{Q}'$.*

The proposition above allows us to deduce the following:

Corollary 1. *Let $\overline{P} : \mathcal{P}(\mathcal{X}) \to [0,1]$ be a coherent upper probability and let $\overline{Q}$ be a non-dominating outer approximation of $\overline{P}$ in $\mathcal{C}_2$, $\mathcal{C}_\infty$ or $\mathcal{C}_\Pi$. If $\overline{P}(\{x\}) = 0$, then also $\overline{Q}(\{x\}) = 0$.*

As a consequence, we may assume without loss of generality that $\overline{P}(\{x\}) > 0$ for every $x \in \mathcal{X}$. This is relevant for the proofs of the results in Sect. 5.

4   Elicitation of an Outer Approximation in $\mathcal{C}_2$ and $\mathcal{C}_\infty$

From [9,10], the number of undominated OAs in $\mathcal{C}_2$ and $\mathcal{C}_\infty$ is not finite in general. In [9,10] we focused on those undominated OAs in $\mathcal{C}_2$ and $\mathcal{C}_\infty$ that minimise the BV-distance proposed in [1] with respect to the original coherent lower probability $\underline{P}$, given by $d_{BV}(\underline{P}, \underline{Q}) = \sum_{A \subseteq \mathcal{X}} |\underline{P}(A) - \underline{Q}(A)|$. It measures the amount of imprecision added to the model when replacing $\underline{P}$ by its OA $\underline{Q}$. Hence, its seems reasonable to minimise the imprecision added to the model.

Let $\mathcal{C}_2^{BV}(\underline{P})$ and $\mathcal{C}_\infty^{BV}(\underline{P})$ denote the set of undominated OAs in $\mathcal{C}_2$ and $\mathcal{C}_\infty$, respectively, that minimise the BV-distance with respect to $\underline{P}$. One advantage of focusing our elicitation to $\mathcal{C}_2^{BV}(\underline{P})$ and $\mathcal{C}_\infty^{BV}(\underline{P})$ is that these sets can be easily determined. Indeed, both $\mathcal{C}_2^{BV}(\underline{P})$ and $\mathcal{C}_\infty^{BV}(\underline{P})$ can be computed as the set of optimal solutions of a linear programming problem ([9, Prop. 1], [10, Prop. 3]). Hence, both sets are convex, and have an infinite number of elements in general. In the rest of the section we discuss different approaches to elicit an undominated OA within $\mathcal{C}_2^{BV}(\underline{P})$ and $\mathcal{C}_\infty^{BV}(\underline{P})$.

4.1   Approach Based on a Quadratic Distance

One possibility for obtaining a unique solution to our problem could be to use the quadratic distance, i.e., to consider the OA in $\mathcal{C}_2^{BV}$ or $\mathcal{C}_\infty^{BV}$ minimising

$$d_p(\underline{P}, \underline{Q}) = \sum_{A \subseteq \mathcal{X}} (\underline{P}(A) - \underline{Q}(A))^2. \tag{1}$$

Given $\delta_2^{BV} = \min_{Q \in \mathcal{C}_2} d_{BV}(\underline{P}, Q)$ and $\delta_\infty^{BV} = \min_{Q \in \mathcal{C}_\infty} d_{BV}(\underline{P}, Q)$, in $\mathcal{C}_2^{BV}$ we may set up the quadratic problem of minimising Eq. (1) subject to the constraints (2monot.1)–(2monot.2), and also to

$$\sum_{B \subseteq A} m_{\underline{Q}}(B) \leq \underline{P}(A) \quad \forall A \neq \emptyset, \mathcal{X}. \tag{OA}$$

$$\sum_{A \subseteq \mathcal{X}} \left(\underline{P}(A) - \sum_{B \subseteq A} m_{\underline{Q}}(B) \right) = \delta_2^{BV}. \tag{2monot-δ}$$

Analogously, in $\mathcal{C}_\infty^{BV}$ we can minimise Eq. (1) subject to (Cmonot.1), (OA) and:

$$\sum_{A \subseteq \mathcal{X}} \left(\underline{P}(A) - \sum_{B \subseteq A} m_{\underline{Q}}(B) \right) = \delta_\infty^{BV}. \tag{Cmonot-δ}$$

Proposition 2. *Let $\underline{P}$ be a coherent lower probability. Then:*

1. *The problem of minimising Eq. (1) subject to (2monot.1) $\div$ (2monot.2), (OA) and (2monot-δ) has a unique solution, which is an undominated OA of $\underline{P}$ in $\mathcal{C}_2^{BV}(\underline{P})$.*
2. *Similarly, the problem of minimising Eq. (1) subject to (Cmonot.1), (OA) and (Cmonot-δ) has a unique solution, which is an undominated OA of $\underline{P}$ in $\mathcal{C}_\infty^{BV}(\underline{P})$.*

The following example illustrates this result.

Example 1. Consider the coherent $\underline{P}$ given on $\mathcal{X} = \{x_1, x_2, x_3, x_4\}$ by [9, Ex.1]:

$$\underline{P}(A) = \begin{cases} 0 & \text{if } |A| = 1 \text{ or } A = \{x_1, x_2\}, \{x_3, x_4\}. \\ 1 & \text{if } A = \mathcal{X}. \\ 0.5 & \text{otherwise.} \end{cases}$$

For this coherent lower probability, $\delta_2^{BV} = \delta_\infty^{BV} = 1$, the sets $\mathcal{C}_2^{BV}(\underline{P})$ and $\mathcal{C}_\infty^{BV}(\underline{P})$ coincide and they are given by $\{\underline{Q}_\alpha \mid \alpha \in [0, 0.5]\}$, where:

$$\underline{Q}_\alpha(A) = \begin{cases} 0 & \text{if } |A| = 1 \text{ or } A = \{x_1, x_2\}, \{x_3, x_4\}. \\ \alpha & \text{if } A = \{x_1, x_4\}, \{x_2, x_3\}. \\ 0.5 - \alpha & \text{if } A = \{x_1, x_3\}, \{x_2, x_4\}. \\ 0.5 & \text{if } |A| = 3. \\ 1 & \text{if } A = \mathcal{X}. \end{cases}$$

Therefore, if among these $\underline{Q}_\alpha$ we minimise the quadratic distance with respect to $\underline{P}$, the optimal solution is $\underline{Q}_{0.25}$, in both $\mathcal{C}_2$ and $\mathcal{C}_\infty$. ♦

Note that the solution we obtain in Proposition 2 is *not* the OA that minimises the quadratic distance in $\mathcal{C}_2$ (or $\mathcal{C}_\infty$), but the one that minimises it in $\mathcal{C}_2^{BV}$ (or $\mathcal{C}_\infty^{BV}$).

While the quadratic distance is in our view the most promising approach in order to elicit a unique undominated OA of $\underline{P}$ in $\mathcal{C}_2$ and $\mathcal{C}_\infty$, it is not the only possibility. In the rest of the section we explore other approaches.

4.2   Approach Based on the Total Variation Distance

Instead of considering the quadratic distance, we may consider some extensions of the total variation distance [8, Ch.4.1] defined between lower probabilities:

$$d_1(\underline{P}_1, \underline{P}_2) = \max_{A \subseteq \mathcal{X}} |\underline{P}_1(A) - \underline{P}_2(A)|, \tag{2}$$

$$d_2(\underline{P}_1, \underline{P}_2) = \frac{1}{2} \sum_{x \in \mathcal{X}} |\underline{P}_1(\{x\}) - \underline{P}_2(\{x\})|, \tag{3}$$

$$d_3(\underline{P}_1, \underline{P}_2) = \sup_{P_1 \geq \underline{P}_1, P_2 \geq \underline{P}_2} \left(\max_{A \subseteq \mathcal{X}} |P_1(A) - P_2(A)| \right). \tag{4}$$

Thus, instead of minimising Eq. (1) we may consider the OA in $\mathcal{C}_2^{BV}(\underline{P})$ or $\mathcal{C}_\infty^{BV}(\underline{P})$ that minimises one of $d_i(\underline{P}, \underline{Q})$. However, none of d_1, d_2, d_3 determines a unique OA in $\mathcal{C}_2^{BV}(\underline{P})$ or $\mathcal{C}_\infty^{BV}(\underline{P})$, as we next show.

Example 2. Consider the coherent $\underline{P}$ in a four-element space given by:

A	$\underline{P}(A)$	$\underline{Q}_0(A)$	$\underline{Q}_1(A)$	A	$\underline{P}(A)$	$\underline{Q}_0(A)$	$\underline{Q}_1(A)$
$\{x_1\}$	0.1	0.1	0.1	$\{x_2, x_3\}$	0.3	0.3	0.2
$\{x_2\}$	0	0	0	$\{x_2, x_4\}$	0.4	0.3	0.4
$\{x_3\}$	0	0	0	$\{x_3, x_4\}$	0.4	0.3	0.3
$\{x_4\}$	0.3	0.3	0.3	$\{x_1, x_2, x_3\}$	0.5	0.5	0.5
$\{x_1, x_2\}$	0.1	0.1	0.1	$\{x_1, x_2, x_4\}$	0.6	0.6	0.6
$\{x_1, x_3\}$	0.3	0.2	0.3	$\{x_1, x_3, x_4\}$	0.7	0.7	0.7
$\{x_1, x_4\}$	0.6	0.6	0.5	$\{x_2, x_3, x_4\}$	0.6	0.6	0.6
				$\mathcal{X}$	1	1	1

In [10, Ex.1] we showed that $\mathcal{C}_\infty^{BV}(\underline{P})$ is given by $\left\{ \underline{Q}_0, \underline{Q}_1, \underline{Q}_\alpha \mid \alpha \in (0, 1) \right\}$, where $\underline{Q}_\alpha = \alpha \underline{Q}_0 + (1 - \alpha) \underline{Q}_1$. In all the cases it holds that:

$d_1(\underline{P}, \underline{Q}_0) = d_1(\underline{P}, \underline{Q}_1) = d_1(\underline{P}, \underline{Q}_\alpha) = 0.1 \quad \forall \alpha \in (0, 1)$.

$d_2(\underline{P}, \underline{Q}_0) = d_2(\underline{P}, \underline{Q}_1) = d_2(\underline{P}, \underline{Q}_\alpha) = 0 \quad \forall \alpha \in (0, 1)$.

$d_3(\underline{P}, \underline{Q}_0) = d_3(\underline{P}, \underline{Q}_1) = d_3(\underline{P}, \underline{Q}_\alpha) = 0.6 = \max_{A \subseteq \mathcal{X}} |\overline{P}(A) - \underline{Q}_\alpha(A)| \quad \forall \alpha \in (0, 1)$.

This means that none of d_1, d_2 and d_3 allows to elicit a unique OA from $\mathcal{C}_\infty^{BV}$. ◆

Consider now the undominated OAs in $\mathcal{C}_2$. First of all, note we may disregard d_2, because from [9, Prop.2] every undominated OA $\underline{Q}$ in $\mathcal{C}_2$ satisfies $\underline{Q}(\{x\}) =$

$\underline{P}(\{x\})$ for every $x \in \mathcal{X}$, and therefore $d_2(\underline{P}, Q) = 0$. The following example shows that d_1 and d_3 do not allow to elicit a unique undominated OA from $\mathcal{C}_2^{BV}(\underline{P})$, either.

Example 3. Consider now the coherent lower probability $\underline{P}$ given by:

A	$\underline{P}(A)$	$\underline{Q}_0(A)$	$\underline{Q}_1(A)$	A	$\underline{P}(A)$	$\underline{Q}_0(A)$	$\underline{Q}_1(A)$
$\{x_1\}$	0	0	0	$\{x_2, x_3\}$	0.3	0.3	0.3
$\{x_2\}$	0	0	0	$\{x_2, x_4\}$	0.3	0.2	0.3
$\{x_3\}$	0.1	0.1	0.1	$\{x_3, x_4\}$	0.4	0.4	0.3
$\{x_4\}$	0	0	0	$\{x_1, x_2, x_3\}$	0.5	0.5	0.5
$\{x_1, x_2\}$	0.3	0.2	0.2	$\{x_1, x_2, x_4\}$	0.6	0.6	0.6
$\{x_1, x_3\}$	0.3	0.3	0.3	$\{x_1, x_3, x_4\}$	0.6	0.6	0.6
$\{x_1, x_4\}$	0.2	0.2	0.2	$\{x_2, x_3, x_4\}$	0.6	0.6	0.6
				$\mathcal{X}$	1	1	1

It holds that $\mathcal{C}_2^{BV}(\underline{P}) = \mathcal{C}_\infty^{BV}(\underline{P}) = \left\{\underline{Q}_0, \underline{Q}_1, \underline{Q}_\alpha \mid \alpha \in (0,1)\right\}$, where $\underline{Q}_\alpha = \alpha \underline{Q}_0 + (1 - \alpha)\underline{Q}_1$. However:

$$d_1(\underline{P}, \underline{Q}_0) = d_1(\underline{P}, \underline{Q}_1) = d_1(\underline{P}, \underline{Q}_\alpha) = 0.1 \quad \forall \alpha \in (0,1), \quad \text{and}$$
$$d_3(\underline{P}, \underline{Q}_0) = d_3(\underline{P}, \underline{Q}_1) = d_3(\underline{P}, \underline{Q}_\alpha) = 0.5 = \max_{A \subseteq \mathcal{X}} |\overline{P}(A) - \underline{Q}_\alpha(A)|, \, \forall \alpha \in (0,1).$$

Thus, neither d_1 nor d_3 determines a unique undominated OA in $\mathcal{C}_2^{BV}(\underline{P})$. ◆

4.3   Approach Based on Measuring Specificity

When we consider the OAs of $\underline{P}$ in $\mathcal{C}_\infty$, we may compare them by measuring their *specificity*. We consider here the specificity measure defined by Yager [14], that splits the mass of the focal events among its elements.

Definition 2. *Let Q be a completely monotone lower probability on $\mathcal{P}(\mathcal{X})$ with Möbius inverse m_Q. Its* specificity *is given by*

$$S(\underline{Q}) = \sum_{\emptyset \neq A \subseteq \mathcal{X}} \frac{m_{\underline{Q}}(A)}{|A|} = \sum_{i=1}^{n} \frac{1}{i} \sum_{A : |A| = i} m_{\underline{Q}}(A).$$

Hence, we can choose an undominated OA in $\mathcal{C}_\infty^{BV}(\underline{P})$ with the greatest specificity. The next example shows that this criterion does not give rise to a unique undominated OA.

Example 4. Consider again Example 1, where $\mathcal{C}_\infty^{BV} = \left\{\underline{Q}_\alpha \mid \alpha \in [0, 0.5]\right\}$. The Möbius inverse of $\underline{Q}_\alpha$ is given by

$$m_{\underline{Q}_\alpha}(\{x_1, x_4\}) = m_{\underline{Q}_\alpha}(\{x_2, x_3\}) = \alpha, \; m_{\underline{Q}_\alpha}(\{x_1, x_3\}) = m_{\underline{Q}_\alpha}(\{x_2, x_4\}) = 0.5 - \alpha$$

and zero elsewhere. Hence, the specificity of $\underline{Q}_\alpha$ is

$$S(\underline{Q}_\alpha) = \frac{1}{2}(\alpha + \alpha + 0.5 - \alpha + 0.5 - \alpha) = 0.5,$$

regardless of the value of $\alpha \in [0, 0.5]$. We conclude that all the undominated OAs in $\mathcal{C}_\infty^{BV}(\underline{P})$ have the same specificity. ◆

5 Elicitation of an Outer Approximation in $\mathcal{C}_\Pi$

In [10, Sec.6] we showed that the set of non-dominating OAs in $\mathcal{C}_\Pi$ is finite and that we have a simple procedure for determining them. Given the conjugate upper probability $\overline{P}$ of $\underline{P}$, each permutation σ in the set S_n of all permutations of $\{1, 2, \ldots, n\}$ defines a possibility measure by:

$$\Pi_\sigma(\{x_{\sigma(1)}\}) = \overline{P}(\{x_{\sigma(1)}\}), \text{ and} \tag{5}$$

$$\Pi_\sigma(\{x_{\sigma(i)}\}) = \max_{A \in \mathcal{A}_{\sigma(i)}} \overline{P}(A \cup \{x_{\sigma(i)}\}), \text{ where for every } i > 1: \tag{6}$$

$$\mathcal{A}_{\sigma(i)} = \left\{ A \subseteq \{x_{\sigma(1)}, \ldots, x_{\sigma(i-1)}\} \mid \overline{P}(A \cup \{x_{\sigma(i)}\}) > \max_{x \in A} \Pi_\sigma(\{x\}) \right\}, \tag{7}$$

and $\Pi_\sigma(A) = \max_{x \in A} \Pi_\sigma(\{x\})$ for every $A \subseteq \mathcal{X}$. Then, the set of non-dominating OAs of $\overline{P}$ is $\{\Pi_\sigma \mid \sigma \in S_n\}$ (see [10, Prop.11, Cor.13]).

Next we propose a number of approaches to elicit a unique OA of $\overline{P}$ among the Π_σ determined by Eqs. (5) $\div$ (7). Note that the procedure above may determine the same possibility measure using different permutations. The next result is concerned with such cases, and will be helpful for reducing the candidate possibility measures.

Proposition 3. *Let $\{\Pi_\sigma \mid \sigma \in S_n\}$ be the set of non-dominating OAs of $\overline{P}$ in $\mathcal{C}_\Pi$. Consider $\sigma \in S_n$ and its associated Π_σ. Assume that $\exists i \in \{2, \ldots, n\}$ such that $\Pi_\sigma(\{x_{\sigma(i)}\}) \neq \overline{P}(\{x_{\sigma(1)}, \ldots, x_{\sigma(i)}\})$. Then, there exists $\sigma' \in S_n$ such that*

$$\Pi_\sigma(A) = \Pi_{\sigma'}(A) \quad \forall A \subseteq \mathcal{X} \text{ and}$$

$$\Pi_{\sigma'}(\{x_{\sigma'(j)}\}) = \overline{P}(\{x_{\sigma'(1)}, \ldots, x_{\sigma'(j)}\}) \quad \forall j = 1, \ldots, n. \tag{8}$$

5.1 Approach Based on the BV-Distance

Our first approach consists in looking for a possibility measure, among $\{\Pi_\sigma \mid \sigma \in S_n\}$, that minimises the BV-distance with respect to the original model. If we denote by N_σ the conjugate necessity measure of Π_σ, the BV-distance can be expressed by:

$$d_{BV}(\underline{P}, N_\sigma) = \sum_{A \subseteq \mathcal{X}} (\Pi_\sigma(A) - \overline{P}(A)) = \sum_{A \subseteq \mathcal{X}} \Pi_\sigma(A) - \sum_{A \subseteq \mathcal{X}} \overline{P}(A).$$

To ease the notation, for each $\sigma \in S_n$ we denote by $\vec{\beta}_\sigma$ the ordered vector determined by the values $\Pi_\sigma(\{x_{\sigma(i)}\})$, $i = 1, \ldots, n$, so that $\beta_{\sigma,1} \leq \ldots \leq \beta_{\sigma,n}$. Using this notation:

$$\sum_{A \subseteq \mathcal{X}} \Pi_\sigma(A) = \beta_{\sigma,1} + 2\beta_{\sigma,2} + \ldots + 2^{n-1}\beta_{\sigma,n} = \sum_{i=1}^{n} 2^{i-1}\beta_{\sigma,i}. \tag{9}$$

This means that, in order to minimise $d_{BV}(\underline{P}, N_\sigma)$, we must minimise Eq. (9). Our next result shows that if a dominance relation exists between $\vec{\beta}_\sigma$ and $\vec{\beta}_{\sigma'}$, this induces an order between the values in Eq. (9).

Lemma 1. *Let $\vec{\beta}_\sigma$ and $\vec{\beta}_{\sigma'}$ be two vectors associated with two possibility measures Π_σ and $\Pi_{\sigma'}$. Then $\vec{\beta}_\sigma \leq \vec{\beta}_{\sigma'}$ implies that $d_{BV}(\underline{P}, N_\sigma) \leq d_{BV}(\underline{P}, N_{\sigma'})$, and $\vec{\beta}_\sigma \lneq \vec{\beta}_{\sigma'}$ implies that $d_{BV}(\underline{P}, N_\sigma) < d_{BV}(\underline{P}, N_{\sigma'})$.*

This result may contribute to rule out some possibilities in S_n, as illustrated in the next example.

Example 5. Consider the following coherent conjugate lower and upper probabilities, as well as their associated possibility measures Π_σ and vectors $\vec{\beta}_\sigma$:

A	$\underline{P}(A)$	$\overline{P}(A)$	σ	$\Pi_\sigma(\{x_1\})$	$\Pi_\sigma(\{x_2\})$	$\Pi_\sigma(\{x_3\})$	$\vec{\beta}_\sigma$
$\{x_1\}$	0.25	0.4	$\sigma_1 = (1,2,3)$	0.4	0.8	1	$(0.4, 0.8, 1)$
$\{x_2\}$	0.2	0.5	$\sigma_2 = (1,3,2)$	0.4	1	0.8	$(0.4, 0.8, 1)$
$\{x_3\}$	0.2	0.5	$\sigma_3 = (2,1,3)$	0.8	0.5	1	$(0.5, 0.8, 1)$
$\{x_1, x_2\}$	0.5	0.8	$\sigma_4 = (2,3,1)$	1	0.5	0.75	$(0.5, 0.75, 1)$
$\{x_1, x_3\}$	0.5	0.8	$\sigma_5 = (3,1,2)$	0.8	1	0.5	$(0.5, 0.8, 1)$
$\{x_2, x_3\}$	0.6	0.75	$\sigma_6 = (3,2,1)$	1	0.75	0.5	$(0.5, 0.75, 1)$
$\mathcal{X}$	1	1					

Taking σ_1 and σ_3, it holds that $\vec{\beta}_{\sigma_1} \lneq \vec{\beta}_{\sigma_3}$, so from Lemma 1 $d_{BV}(\underline{P}, N_{\sigma_1}) < d_{BV}(\underline{P}, N_{\sigma_3})$. Hence, we can discard Π_{σ_3}. The same applies to $\vec{\beta}_{\sigma_1}$ and $\vec{\beta}_{\sigma_5}$, whence $d_{BV}(\underline{P}, N_{\sigma_1}) < d_{BV}(\underline{P}, N_{\sigma_5})$. ◆

In the general case, the set of all vectors $\vec{\beta}_\sigma$ is not totally ordered. Then, the problem of minimising the BV-distance is solved by casting it into a shortest path problem, as we shall now illustrate.

As we said before, the possibility measure(s) in $\{\Pi_\sigma : \sigma \in S_n\}$ that minimise the BV-distance to $\underline{P}$ are the ones minimising $\sum_{A \subseteq \mathcal{X}} \Pi_\sigma(A)$. In turn, this sum can be computed by means of Eq. (9), once we order the values $\Pi_\sigma(\{x_{\sigma(i)}\})$, for $i = 1, \ldots, n$. Our next result will be useful for this aim:

Proposition 4. *Let $\{\Pi_\sigma \mid \sigma \in S_n\}$ be the set of non-dominating OAs of $\overline{P}$ in $\mathcal{C}_\Pi$. Then $\sum_{A \subseteq \mathcal{X}} \Pi_\sigma(A) \leq \sum_{i=1}^{n} 2^{i-1} \overline{P}(\{x_{\sigma(1)}, \ldots, x_{\sigma(i)}\})$, and the equality holds if and only if Π_σ satisfies Eq. (8).*

From Proposition 3, if Π_σ does not satisfy Eq. (8) then there exists another permutation σ' that does so and such that $\Pi_\sigma = \Pi_{\sigma'}$. This means that we can find Π_σ minimising the BV-distance by solving a shortest path problem. For this aim, we consider the Hasse diagram of $\mathcal{P}(\mathcal{X})$, and if $x_i \notin A$, we assign the weight $2^{|A|}\overline{P}(A \cup \{x_i\})$ to the edge $A \to A \cup \{x_i\}$. Since these weights are positive, we can find the optimal solution using Dijkstra's algorithm [5]. In this diagram, there are two types of paths:

(a) Paths whose associated Π_σ satisfies Eq. (8); then $\sum_{A \subseteq \mathcal{X}} \Pi_\sigma(A)$ coincides with the value of the path.
(b) Paths whose associated Π_σ does not satisfy Eq. (8); then $\sum_{A \subseteq \mathcal{X}} \Pi_\sigma(A)$ shall be strictly smaller than the value of the path, and shall moreover coincide with the value of the path determined by another permutation σ', as established in Proposition 3. Then the shortest path can never be found among these ones.

As a consequence, the shortest path determines a permutation σ whose associated Π_σ satisfies Eq. (8). Moreover, this Π_σ minimises the BV-distance with respect to $\overline{P}$ among all the non-dominating OAs in C_∞. And in this manner we shall obtain all such possibility measures.

Example 6. Consider the coherent conjugate lower and upper probability $\underline{P}$ and $\overline{P}$ from Example 5. The following figure pictures the Hasse diagram with weights of the edges we discussed before:

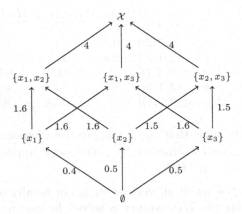

Solving the shortest path problem from $\emptyset$ to $\mathcal{X}$ using Dijkstra's algorithm, we obtain an optimal value of 6 that is attained with the following paths:

$$\emptyset \to \{x_1\} \to \{x_1, x_2\} \to \mathcal{X}, \qquad \emptyset \to \{x_1\} \to \{x_1, x_3\} \to \mathcal{X}.$$
$$\emptyset \to \{x_2\} \to \{x_2, x_3\} \to \mathcal{X}, \qquad \emptyset \to \{x_3\} \to \{x_2, x_3\} \to \mathcal{X}.$$

These four paths correspond to the permutations $\sigma_1 = (1, 2, 3)$, $\sigma_2 = (1, 3, 2)$, $\sigma_4 = (2, 3, 1)$ and $\sigma_6 = (3, 2, 1)$. Even if they induce four different possibility measures, all of them are at the same BV-distance to $\overline{P}$. Note also that the other two possibility measures are those that were discarded in Example 5. ◆

This example shows that with this approach we obtain the Π_σ at minimum BV-distance. It also shows that the solution is not unique, and that the vectors $\vec{\beta}_\sigma$ and $\vec{\beta}_{\sigma'}$ that are not pointwisely ordered may be associated with two different possibility measures Π_σ and $\Pi_{\sigma'}$ minimising the BV-distance (such as σ_1 and σ_6 in the example). Nevertheless, we can determine situations in which the BV-distance elicits one single Π_σ, using the following result:

Proposition 5. *Let $\underline{P}$ and $\overline{P}$ be coherent conjugate lower and upper probabilities. If there is a permutation $\sigma \in S_n$ satisfying*

$$\overline{P}\left(\{x_{\sigma(1)}, \ldots, x_{\sigma(j)}\}\right) = \min_{|A|=j} \overline{P}(A) \quad \forall j = 1, \ldots, n, \tag{10}$$

then Π_σ minimises the BV-distance.

As a consequence of this result, if there is only one permutation satisfying Eq. (10), this approach allows to elicit a unique undominated OA.

Example 7. Consider again $\underline{P}$ and $\overline{P}$ from Example 2. We can see that:

$$\overline{P}(\{x_2\}) = 0.3 = \min_{|A|=1} \overline{P}(A), \quad \overline{P}(\{x_2, x_3\}) = 0.4 = \min_{|A|=2} \overline{P}(A),$$

$$\overline{P}(\{x_1, x_2, x_3\}) = 0.7 = \min_{|A|=3} \overline{P}(A).$$

There is a (unique) chain of events satisfying Eq. (10), namely $\{x_2\} \subseteq \{x_2, x_3\} \subseteq \{x_1, x_2, x_3\} \subseteq \mathcal{X}$, that is associated with the permutation $\sigma = (2, 3, 1, 4)$. From Proposition 5, Π_σ is the unique undominated OA in $\mathcal{C}_\Pi$ minimising the BV-distance.   ◆

5.2 Approach Based on Measuring Specificity

Since any possibility measure is in particular the conjugate of a belief function, it is possible to compare them by means of specificity measures. In this section, we investigate which possibility measure(s) among $\{\Pi_\sigma \mid \sigma \in S_n\}$ are the most specific.

With each Π_σ in $\{\Pi_\sigma \mid \sigma \in S_n\}$, we consider its associated vector $\vec{\beta}_\sigma$. For possibility measures, the focal events A_i are nested: $A_i := \{x_{\sigma(n-i+1)}, \ldots, x_{\sigma(n)}\}$, with $m(A_i) = \beta_{\sigma,n-i+1} - \beta_{\sigma,n-i}$. Hence specificity simplifies to:

$$S(\Pi_\sigma) = 1 - \frac{\beta_{\sigma,n-1}}{2} - \frac{\beta_{\sigma,n-2}}{2 \cdot 3} - \cdots - \frac{\beta_{\sigma,1}}{n(n-1)}.$$

Thus, a most specific possibility measure will minimise

$$\frac{\beta_{\sigma,1}}{n(n-1)} + \frac{\beta_{\sigma,2}}{(n-1)(n-2)} + \ldots + \frac{\beta_{\sigma,n-1}}{2}. \tag{11}$$

Our first result is similar to Lemma 1, and allows to discard some of the possibility measures Π_σ.

Lemma 2. *Let $\vec{\beta}_\sigma$ and $\vec{\beta}_{\sigma'}$ be the vectors associated with the possibility measures Π_σ and $\Pi_{\sigma'}$. Then $\vec{\beta}_\sigma \leq \vec{\beta}_{\sigma'}$ implies that $S(\Pi_\sigma) \geq S(\Pi_{\sigma'})$ and $\vec{\beta}_\sigma \lneq \vec{\beta}_{\sigma'}$ implies that $S(\Pi_\sigma) > S(\Pi_{\sigma'})$.*

Example 8. Let us continue with Examples 5 and 6. In Example 5 we showed the possibility measures $\{\Pi_\sigma \mid \sigma \in S_n\}$ and their associated vectors $\vec{\beta}_\sigma$. As we argued in Example 5, $\vec{\beta}_{\sigma_1} \lneq \vec{\beta}_{\sigma_3}$, where $\sigma_1 = (1, 2, 3)$ and $\sigma_3 = (2, 1, 3)$. Hence from Lemma 2, $S(\Pi_{\sigma_1}) > S(\Pi_{\sigma_3})$, meaning that we can discard Π_{σ_3}. A similar reasoning allows us to discard Π_{σ_5}.   ◆

In order to find those possibility measures maximising the specificity, we have to minimise Eq. (11). Here we can make the same considerations as in Sect. 5.1.

Proposition 6. *Let $\{\Pi_\sigma \mid \sigma \in S_n\}$ be the set of non-dominating OAs of $\overline{P}$ in*

C_Π. *Then* $S(\Pi_\sigma) \geq 1 - \sum_{i=1}^{n-1} \dfrac{\overline{P}(\{x_{\sigma(1)}, \ldots, x_{\sigma(i)}\})}{(n-i)(n-i+1)}$, *and the equality holds if and*

only if Π_σ satisfies Eq. (8).

Moreover, from Proposition 3 we know that if Π_σ does not satisfy Eq. (8) then it is possible to find another permutation σ' that does so and such that $\Pi_\sigma = \Pi_{\sigma'}$.

This means that we can find the Π_σ maximising the specificity by solving a shortest path problem, similarly to what we did for the BV-distance. For this aim, we consider the Hasse diagram of $\mathcal{P}(\mathcal{X})$; if $x_i \notin A$, we assign the weight

$$\frac{\overline{P}(\{A \cup \{x_i\}\})}{(n - |A|)(n - |A| - 1)} \tag{12}$$

to the edge $A \to A \cup \{x_i\}$, and we give the fictitious weight 0 to $\mathcal{X} \setminus \{x_i\} \to \mathcal{X}$. In this diagram, there are two types of paths:

(a) Paths whose associated possibility measure Π_σ satisfies Eq. (8); then the value of Eq. (11) for Π_σ coincides with the value of the path.
(b) Paths whose associated possibility measure Π_σ does not satisfy Eq. (8); then the value of Eq. (11) for Π_σ is strictly smaller than the value of the path, and shall moreover coincide with the value of the path determined by another permutation σ', as established in Proposition 3. Then the shortest path can never be found among these ones.

As a consequence, if we find the shortest path we shall determine a permutation σ whose associated Π_σ satisfies Eq. (8), and therefore that maximises the specificity; and in this manner we shall obtain all such possibility measures.

Example 9. Consider again the running Examples 5, 6 and 8. In the next figure we can see the Hasse diagram of $\mathcal{P}(\mathcal{X})$ with the weights from Eq. (12).

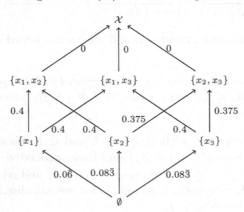

The optimal solutions of the shortest path problem are $\emptyset \to \{x_2\} \to \{x_2, x_3\} \to \mathcal{X}$ and $\emptyset \to \{x_3\} \to \{x_2, x_3\} \to \mathcal{X}$, which correspond to the permutations $\sigma_4 = (2, 3, 1)$ and $\sigma_6 = (3, 2, 1)$. ◆

These examples also show that the approach based on minimising the BV-distance and that based on maximising the specificity are not equivalent: in Example 5 we have seen that the possibility measures minimising the BV-distance are the ones associated with the permutations $(3, 1, 2)$ and $(3, 2, 1)$, while those maximising the specificity are the ones associated with $(2, 3, 1)$ and $(3, 2, 1)$.

To conclude this subsection, we establish a result analogous to Proposition 5.

Proposition 7. *Let $\underline{P}$ and $\overline{P}$ be coherent conjugate lower and upper probabilities. If there is a permutation σ satisfying Eq. (10), then Π_σ maximises the specificity.*

We arrive at the same conclusion of Proposition 5: if there is a unique permutation satisfying Eq. (10), then there is a unique possibility measure maximising the specificity; and in that case the chosen possibility measure maximises the specificity and at the same time minimises the BV-distance.

5.3   Approach Based on the Total Variation Distance

As we did in Sect. 4.2, we could elicit a possibility measure among $\{\Pi_\sigma \mid \sigma \in S_n\}$ by minimising one of the extensions of the TV-distance. When we focus on upper probabilities, the distances given in Eqs. (2) ÷ (4) can be expressed by:

$$d_1(\overline{P}_1, \overline{P}_2) = \max_{A \subseteq \mathcal{X}} |\overline{P}_1(A) - \overline{P}_2(A)|, \quad d_2(\overline{P}_1, \overline{P}_2) = \frac{1}{2} \sum_{x \in \mathcal{X}} |\overline{P}_1(\{x\}) - \overline{P}_2(\{x\})|,$$

$$d_3(\overline{P}_1, \overline{P}_2) = \sup_{P_1 \leq \overline{P}_1, P_2 \leq \overline{P}_2} \left(\max_{A \subseteq \mathcal{X}} |P_1(A) - P_2(A)| \right).$$

As in the case of $\mathcal{C}_2$ and $\mathcal{C}_\infty$, this approach is not fruitful:

Example 10. Consider our running Example 5. The values $d_i(\overline{P}, \Pi_j)$ are given by:

	Π_{σ_1}	Π_{σ_2}	Π_{σ_3}	Π_{σ_4}	Π_{σ_5}	Π_{σ_6}
$d_1(\overline{P}, \Pi_{\sigma_i})$	0.5	0.5	0.5	0.6	0.5	0.6
$d_2(\overline{P}, \Pi_{\sigma_i})$	0.4	0.4	0.45	0.425	0.45	0.425
$d_3(\overline{P}, \Pi_{\sigma_i})$	0.8	0.8	0.8	0.75	0.8	0.75

Thus, none of d_1, d_2 or d_3 allow to elicit a single possibility measure.   ◆

6   Conclusions

In this paper, we have explored a number of approaches to elicit a unique undominated OA of a given coherent lower probability. When the OA belongs to the families $\mathcal{C}_2$ and $\mathcal{C}_\infty$, we first focus on the ones minimising the BV-distance.

Among the approaches we have then considered, it seems that the best one is to consider the OA in $\mathcal{C}_2$ and $\mathcal{C}_\infty$ that minimises the quadratic distance: it singles out a unique undominated OA, while this is not the case when we minimise the TV-distance or maximise the specificity.

In the case of $\mathcal{C}_\Pi$, we know from [10] that there are at most $n!$ non-dominating OA of a coherent upper probability, and these are determined by Eqs. (5) ÷ (7). In order to elicit a unique possibility measure we have considered the approaches based on minimising the BV-distance, maximising the specificity and minimising the TV-distance. While none of them elicits a unique OA in general, we have given a sufficient condition for the uniqueness in Propositions 5 and 7. Moreover, we have seen that we can find the optimal OA according to the BV-distance and the specificity approaches by solving a shortest path problem.

In a future work, we intend to make a thorough comparison between the main approaches and to report on additional results that we have not included in this paper due to space limitations, such as the comparison between the OA in terms of the preservation of the preferences encompassed by the initial model, and the analysis of other particular cases of imprecise probability models, such as probability boxes.

Acknowledgements. The research reported in this paper has been supported by project PGC2018-098623-B-I00 from the Ministry of Science, Innovation and Universities of Spain and project IDI/2018/000176 from the Department of Employment, Industry and Tourism of Asturias.

References

1. Baroni, P., Vicig, P.: An uncertainty interchange format with imprecise probabilities. Int. J. Approx. Reason. **40**, 147–180 (2005). https://doi.org/10.1016/j.ijar.2005.03.001
2. Bronevich, A., Augustin, T.: Approximation of coherent lower probabilities by 2-monotone measures. In: Augustin, T., Coolen, F.P.A., Moral, S., Troffaes, M.C.M. (eds.) Proceedings of the Sixth ISIPTA, pp. 61–70 (2009)
3. Chateauneuf, A., Jaffray, J.Y.: Some characterizations of lower probabilities and other monotone capacities through the use of Möbius inversion. Math. Soci. Sci. **17**(3), 263–283 (1989). https://doi.org/10.1016/0165-4896(89)90056-5
4. Choquet, G.: Theory of capacities. Ann. l'Institut Fourier **5**, 131–295 (1954). https://doi.org/10.5802/aif.53
5. Dijkstra, E.: A note on two problems in connection with graphs. Numer. Math., 269–271 (1959) https://doi.org/10.1007/BF01386390
6. Dubois, D., Prade, H.: Possibility Theory. Plenum Press, New York (1988)
7. Levi, I.: The Enterprise of Knowledge. MIT Press, Cambridge (1980)
8. Levin, D.A., Peres, Y., Wilmer, E.: Markov Chains and Mixing Times. American Mathematical Society, Providence (2009)
9. Montes, I., Miranda, E., Vicig, P.: 2-monotone outer approximations of coherent lower probabilities. Int. J. Approx. Reason. **101**, 181–205 (2018). https://doi.org/10.1016/j.ijar.2018.07.004

10. Montes, I., Miranda, E., Vicig, P.: Outer approximating coherent lower probabilities with belief functions. Int. J. Approx. Reason. **110**, 1–30 (2019). https://doi.org/10.1016/j.ijar.2019.03.008
11. Shafer, G.: A Mathematical Theory of Evidence. Princeton University Press, Princeton (1976)
12. Shapley, L.S.: Cores of convex games. Int. J. Game Theory **1**, 11–26 (1971). https://doi.org/10.1007/BF01753431
13. Walley, P.: Statistical Reasoning with Imprecise Probabilities. Chapman and Hall, London (1991)
14. Yager, R.: Entropy and specificity in a mathematical theory of evidence. Int. J. Gen. Syst. **9**, 249–260 (1983). https://doi.org/10.1080/03081078308960825
15. Zadeh, L.A.: Fuzzy sets as a basis for a theory of possibility. Fuzzy Sets Syst. **1**, 3–28 (1978). https://doi.org/10.1016/S0165-0114(99)80004-9

Binary Credal Classification Under Sparsity Constraints

Tathagata Basu$^{(\boxtimes)}$ ⓘ, Matthias C. M. Troffaes ⓘ, and Jochen Einbeck ⓘ

Department of Mathematical Sciences, Durham University, Durham, UK
tathagata.basu@durham.ac.uk

Abstract. Binary classification is a well known problem in statistics. Besides classical methods, several techniques such as the naive credal classifier (for categorical data) and imprecise logistic regression (for continuous data) have been proposed to handle sparse data. However, a convincing approach to the classification problem in high dimensional problems (i.e., when the number of attributes is larger than the number of observations) is yet to be explored in the context of imprecise probability. In this article, we propose a sensitivity analysis based on penalised logistic regression scheme that works as binary classifier for high dimensional cases. We use an approach based on a set of likelihood functions (i.e. an imprecise likelihood, if you like), that assigns a set of weights to the attributes, to ensure a robust selection of the important attributes, whilst training the model at the same time, all in one fell swoop. We do a sensitivity analysis on the weights of the penalty term resulting in a set of sparse constraints which helps to identify imprecision in the dataset.

Keywords: Classification · High dimensional data · Imprecise probability

1 Introduction

Classification is a method for assigning a new object to a class or a group based on the observed features or attributes of the object. Classification is used in many applications such as pattern recognition for hand writing, disease treatment, facial recognition, chemical analysis, and so on. In general, a classifier can be seen as a function that maps a set of continuous or discrete variables into a categorical class variable. Constructing a classifier from random samples is an important problem in statistical inference. In our work, we will restrict ourselves to the case where there are only two classes to choose from, i.e. 'binary classification'.

Let C be a random variable that takes values in $\{0, 1\}$. Let a be a p-dimensional vector that denotes the attributes of an object and let $b =$

This work is funded by the European Commission's H2020 programme, through the UTOPIAE Marie Curie Innovative Training Network, H2020-MSCA-ITN-2016, Grant Agreement number 722734.

$(b_1, b_2, \ldots, b_p)^T$ denote the vector of regression coefficients. In a regression setting, we construct a classifier through a generalised linear model (GLM) as follows:

$$E(C \mid a) = h\left(a^T b\right) \tag{1}$$

where h acts as a 'link' function and E stands for expectation. We define

$$\pi(a) := E(C \mid a) = P(C = 1 \mid a). \tag{2}$$

Logistic regression is a well-used special case of the GLM, which is suitable
for classification with continuous attributes. Note that, for logistic regression, C
follows a Bernoulli distribution. However, in the high dimensional case i.e. when
the number of attributes is more than the number of observations $(p > n)$,
the performance of logistic regression is often not satisfactory. Apart from overfitting, numerical optimisation methods often converge to local solutions because
of multi-collinearity. Several techniques have been proposed to deal with this.
Generally, a penalty term is introduced in the negative log-likelihood, leading to
penalised logistic regression. A lasso-type penalty [16] is very popular because
of its variable selection property [15,21]. However, the lasso-type penalty can be
inconsistent. To tackle this, Zou [22] introduced an adaptive version of the lasso
for penalised logistic regression, which satisfies suitable asymptotic properties
[8] for variable selection, and leads to consistent estimates.

Several works related to classification can be found in the imprecise probability literature. Zaffalon [19] introduced the idea of the naive credal classifier
related to the imprecise Dirichlet model [18]. Bickis [3] introduced an imprecise
logit normal model for logistic regression. Corani and de Campos [5] proposed
the tree augmented naive classifier based on imprecise Dirichlet model. Paton
et al. [13,14] used a near vacuous set of priors for multinomial logistic regression.
Coz et al. [7] and Corani and Antonucci [4] investigated rejection based classifiers for attribute selection. However, high dimensional problems with automatic
attribute selection are yet to be tackled in the context of imprecise probability.

In this study, we propose a novel imprecise likelihood based approach for high
dimensional logistic regression problems. We use a set of sparsity constraints
through weights in the penalty term. Working with a *set* of weights relaxes the
assumption of preassigned weights and also helps to identify the behaviour of
the attributes, whereas sparsity constraints help in variable selection which is
essential for working with high dimensional problems. We use cross-validation
for model validation using different performance measures [6].

The paper is organised as follows. We first discuss some properties of
penalised logistic regression in Sect. 2. We discuss our sensitivity based classifier in Sect. 3. We discuss the model validation in Sect. 4, and we illustrate our
results using two datasets in Sect. 5. We conclude in Sect. 6.

Throughout the paper, capital letters denote random variables or estimators
that are dependent on any random quantity, and bold letters denote matrices.

2   Logistic Regressions for Sparse Problems

High dimensional regression is considered as sparse problem because of the small number of non-zero regression parameters. We often look for regularisation methods to achieve this sparsity/attribute selection. In this section, we discuss different penalised logistic regression schemes which are useful to attain a sparse model.

2.1   Penalised Logistic Regression (PLR)

Consider the generalised model in Eq. (1). For logistic regression, we use the following link function:

$$h(x) := \frac{\exp(x)}{1 + \exp(x)}. \tag{3}$$

We define a vector $C := (C_1, C_2, \ldots, C_n)^T$ denoting n observed classes such that, $C_i \in \{0, 1\}$. The C_i are thus Bernoulli random variables. Let $\boldsymbol{a} := [a_1, a_2, \ldots, a_n]$, with $a_i \in \mathbb{R}^p$, denote the observed attributes for n objects, so that $\boldsymbol{a}^T$ corresponds to the design matrix in the terminology of classical statistical modelling. It is easy to see that the negative log likelihood of the data is:

$$- \log(L(C, \boldsymbol{a}; b)) = \sum_{i=1}^{n} \left(- C_i \left(a_i^T b \right) + \log \left(1 + \exp(a_i^T b) \right) \right). \tag{4}$$

Therefore, the maximum likelihood estimate of the unknown parameter b is:

$$\hat{B}_{\mathrm{lr}} := \arg \min_b \{ - \log(L(C, \boldsymbol{a}; b)) \}. \tag{5}$$

Here, we denote estimates, such as $\hat{B}_{\mathrm{lr}}$, with capital letters because they are random variables (as they depend on C, which is random). The matrix of observed attributes $\boldsymbol{a}$ is denoted with a lower case letter, as it is customary to consider it as fixed and thereby non-random.

In high dimensional problems, we often seek for regularisation methods to avoid overfitting. We use penalised logistic regression (PLR) [15, 21] as a regularisation method which is defined by:

$$\hat{B}_{\mathrm{plr}}(\lambda) := \arg \min_b \{ - \log(L(C, \boldsymbol{a}; b)) + \lambda P(b) \}, \tag{6}$$

where $P(b)$ is a penalty function. We get sparse estimate for b when:

$$P(b) := \sum_{j=1}^{p} |b_j|^q \tag{7}$$

with $0 \leq q \leq 1$. However, for $q < 1$, the problem is non-convex (see Fig. 1) and the optimisation is computationally expensive. In contrast, for $q = 1$, the penalty is a lasso-type penalty [16], which is convex and easy to solve numerically. The value of λ is chosen through cross-validation, where λ acts as a tuning parameter. In Fig. 1, we show contours of different ℓ_q penalties for two variables.

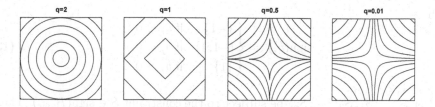

Fig. 1. Contour plots of (7) for different ℓ_q penalties.

2.2 Adaptive Penalised Logistic Regression (APLR)

The lasso type penalty in PLR can be inconsistent in variable selection and it is also not asymptotically unbiased. This issue can be resolved by assigning carefully chosen weights in the penalty term. This approach is known to be adaptive penalised logistic regression (APLR) [2,22].

Let $\hat{B} := (\hat{B}_1, \hat{B}_2, \cdots, \hat{B}_p)$ be any root-n consistent estimate for our logistic regression problem. Then, for any fixed $\gamma > 0$, the APLR [22] estimates are given by:

$$\hat{B}_{\text{aplr}}(\lambda, \gamma) := \arg \min_b \left(-\log(L(C, \boldsymbol{a}; b)) + \lambda \sum_{j=1}^{p} W_j(\gamma)|b_j| \right) \tag{8}$$

where

$$W_j(\gamma) := \frac{1}{|\hat{B}_j|^\gamma}. \tag{9}$$

Note that, for $\gamma = 0$, Eq. (8) becomes the regular penalised logistic regression with lasso penalty. Zou [22] showed that with these weights along with some mild regularity conditions, APLR follows desirable asymptotic properties for high dimensional problems [8].

Computation. For $\gamma > 0$, the objective function of APLR is given by:

$$J(b) := \left(\sum_{i=1}^{m} \left[-C_i \left(a_i^T b \right) + \log \left(1 + \exp(a_i^T b) \right) \right] + \lambda \sum_{j=1}^{p} W_j(\gamma)|b_j| \right), \tag{10}$$

where $W_j(\gamma)$ is given by Eq. (9). Now, for optimality Eq. (10) must satisfy Karush-Kuhn-Tucker condition. Therefore, we have,

$$0 \in \sum_{i=1}^{m} \left[-a_{ji}C_i + a_{ji}\frac{\exp(a_i^T b)}{1 + \exp(a_i^T b)} \right] + \lambda W_j(\gamma)\partial(|b_j|), \tag{11}$$

where, $\partial|b_j|$ is defined [12] as

$$\partial(|b_j|) = \text{sign}(b_j), \tag{12}$$

with

$$\text{sign}(b_j) := \begin{cases} \{-1\} & \text{if } b_j < 0 \\ [-1,1] & \text{if } b_j = 0 \\ \{1\} & \text{if } b_j > 0, \end{cases} \tag{13}$$

for $j = 1, 2, \cdots, p$.

Let $S := (S_1, S_2, \cdots, S_p)$ be subject to the constraint $S \in \text{sign}(\hat{B}_{\text{aplr}})$. Then, $\hat{B}_{\text{aplr}}$ satisfies the following:

$$\sum_{i=1}^{m} \left[-a_{ji} C_i + a_{ji} \frac{\exp\left(a_i^T \hat{B}_{\text{aplr}}(\lambda, \gamma)\right)}{1 + \exp\left(a_i^T \hat{B}_{\text{aplr}}(\lambda, \gamma)\right)} \right] = -\lambda W_j(\gamma) S_j \tag{14}$$

$$\sum_{i=1}^{m} a_{ji} \left[C_i - \frac{\exp\left(a_i^T \hat{B}_{\text{aplr}}(\lambda, \gamma)\right)}{1 + \exp\left(a_i^T \hat{B}_{\text{aplr}}(\lambda, \gamma)\right)} \right] = \lambda W_j(\gamma) S_j. \tag{15}$$

Now, let $h(\boldsymbol{a}^T \hat{B}) := \left(h\left(a_1^T \hat{B}\right), h\left(a_2^T \hat{B}\right), \cdots, h\left(a_n^T \hat{B}\right) \right)^T$, where h is the link function defined in Eq. (3). Then, we can write Eq. (15) as,

$$\boldsymbol{a} \left[C - h\left(\boldsymbol{a}^T \hat{B}_{\text{aplr}}(\lambda, \gamma)\right) \right] = \lambda W(\gamma) \cdot S \tag{16}$$

where '$\cdot$' denotes component wise multiplication. Note that Eq. (16) is not analytically solvable for $\hat{B}_{\text{aplr}}$. However, any sub-gradient based numerical optimisation method can be applied to solve it. Once we have the estimate, we can then define, for any new object with known attributes $a_* \in \mathbb{R}^p$ and unknown class C_*,

$$\hat{\Pi}(a_*, \lambda, \gamma) := P\left(C_* = 1 \mid a_*; \hat{B}_{\text{aplr}}(\lambda, \gamma)\right) = h\left(a_*^T \hat{B}_{\text{aplr}}(\lambda, \gamma)\right). \tag{17}$$

We can then for instance classify the object as 0 if $\hat{\Pi}(a_*, \lambda, \gamma) < 1/2$, as 1 if $\hat{\Pi}(a_*, \lambda, \gamma) > 1/2$, and as either if $\hat{\Pi}(a_*, \lambda, \gamma) = 1/2$. The parameter γ is often simply fixed (usually taken to be equal to 1), and λ is chosen through cross-validation, as with PLR.

Properties. For a sequence of n observations, where a_i is the attribute vector for the i-th observation, we now denote:

$$a_n := \boldsymbol{a} = [a_1, \cdots, a_n] \tag{18}$$

in order to make the dependence of this $p \times n$ matrix on n explicit. Define by $b^* := (b_1^*, \cdots, b_p^*)$ the vector of true regression coefficients. Assume the true model to be sparse, then without loss of generality $\mathcal{S} := \{j : b_j^* \neq 0\} = \{1, 2, \cdots, p_0\}$, where $p_0 < p$. Let $\phi(x) := \log(1 + \exp(x))$, then for any observation $a_i \in \mathbb{R}^p$ $(1 \leq i \leq n)$, we define the Fisher information matrix by:

$$I(b) := \phi''(a_i^T b) a_i a_i^T = \begin{bmatrix} I_{11} & I_{12} \\ I_{21} & I_{22} \end{bmatrix} \tag{19}$$

where, I_{11} is a $p_0 \times p_0$ matrix.

Regularity Conditions: We define the following regularity conditions for asymptotic properties of APLR.

C.1 Let $\lambda_n(\gamma)$ be a sequence such that, for $\gamma > 0$

$$\lim_{n\to\infty} \frac{\lambda_n(\gamma)}{\sqrt{n}} = 0 \quad \text{and} \quad \lim_{n\to\infty} \lambda_n(\gamma) \cdot n^{(\gamma-1)/2} = \infty. \tag{20}$$

For example, the above holds for $\lambda_n(\gamma) = n^{1/2-\gamma/4}$.

C.2 The Fisher information matrix is finite and positive definite.

C.3 Let there exist an open set $\mathcal{B} \subseteq \mathbb{R}^p$, such that $b^* \in \mathcal{B}$. Then for every $b \in \mathcal{B}$ and observation $a_i \in \mathbb{R}^p$ $(1 \leq i \leq n)$, there exists a function M so that

$$\left|\phi'''(a_i^T b)\right| \leq M(a_i) < \infty. \tag{21}$$

Let $\mathcal{S}_n = \{j : \hat{B}_{\text{aplr}, j} \neq 0\}$.

Theorem 1. *Under C.1–C.3, APLR estimates satisfy the following properties:*

P.1 Consistency in variable selection, i.e.

$$\lim_{n\to\infty} P(\mathcal{S}_n = \mathcal{S}) = 1 \tag{22}$$

P.2 Asymptotic normality, i.e.

$$\sqrt{n}\left(\hat{B}_{\text{aplr}, \mathcal{S}} - b_{\mathcal{S}}^*\right) \xrightarrow{d} N(0, I_{11}^{-1}) \tag{23}$$

Note, that here $\hat{B}_{\text{aplr}, \mathcal{S}}$ is dependent on both $\lambda_n(\gamma)$ and γ but we skip writing it for the sake of notation.

P.1 and P.2 are well known results for high dimensional problems and the proofs can be found in [22].

3   Imprecise Adaptive Penalised Logistic Regression

The use of data-driven weights in APLR makes APLR consistent in attribute selection, where the parameter γ is pre-assigned (usually equal to 1) or is estimated through cross-validation. However, high dimensional problems are sparse in nature, i.e. we have to deal with very limited information and therefore a single vector of weights is often proved to be sensitive and leads to misclassification, especially when the variability of the attributes is negligible with respect to each other. Sometimes, APLR may also perform poorly during model validation as, a single value of γ can provide two very different vectors of weights for two different parts of a single dataset. For instance, fixing $\gamma = 1$, essentially gives us the inverse of the absolute values of our estimates, which are generally sensitive to the data in sparse regime. So, we propose a sensitivity analysis of APLR over an interval of γ and obtain a non-determinate classifier. We call this method as imprecise adaptive penalised logistic regression or simply IAPLR. This allows the weights to vary in the order of γ providing us a set of sparse constraints of the form $\sum_{j=1}^{p} |b_j|/|\hat{B}_j|^{\gamma}$. This set of weight vectors allows the model to be flexible but consistent as we only rely on the data-driven weights.

The sensitivity analysis gives us a set of APLR estimates as a function of γ. We use this set of APLR estimates to obtain a set of estimated probabilities which are used for the decision making.

3.1 Decision Rule

Consider the APLR estimates defined by Eq. (8) and Eq. (9). As we described earlier, we perform a sensitivity analysis on the parameter γ. This gives us a set of estimated probabilities dependent on γ, such that $\gamma \in [\underline{\gamma}, \overline{\gamma}]$. We use the notion of credal dominance [19] for the decision criteria.

We can then for instance classify a new object with attributes $a_* \in \mathbb{R}^p$ as $\{0\}$ if $\hat{\Pi}(a_*, \lambda, \gamma) < 1/2$ for all $\gamma \in [\underline{\gamma}, \overline{\gamma}]$, as $\{1\}$ if $\hat{\Pi}(a_*, \lambda, \gamma) \geq 1/2$ for all $\gamma \in [\underline{\gamma}, \overline{\gamma}]$, and as $\{0, 1\}$ (i.e. indeterminate) otherwise. Note that our classifier now returns non-empty subsets of $\{0, 1\}$ rather than elements of $\{0, 1\}$, to allow indeterminate classifications to be expressed.

3.2 Prediction Consistency

We define the following:

$$a_{*,\mathcal{S}} := [a_{*,j}]_{j \in \mathcal{S}}, \tag{24}$$

i.e., $a_{*,\mathcal{S}}$ is a p_0-dimensional vector.

Theorem 2. Let $a_* \in \mathbb{R}^p$ such that $a_{*,\mathcal{S}}^T a_{*,\mathcal{S}} > 0$. Then for $\gamma > 0$ and under C.1–C.3, we have the following:

$$\sqrt{n}\left(\hat{\Pi}(a_*, \lambda_n(\gamma), \gamma) - \pi(a_*)\right) \xrightarrow{d} N\left(0, [\pi(a_*)(1 - \pi(a_*))]^2 a_{*,\mathcal{S}}^T I_{11}^{-1} a_{*,\mathcal{S}}\right) \tag{25}$$

where, I_{11} is the leading block matrix of the Fisher information matrix defined in Eq. (19).

Proof. We know that, under C.1–C.3 APLR estimates satisfies P.1. Therefore, as $n \to \infty$,

$$a_*^T \hat{B}_{\text{aplr}} = a_{*,\mathcal{S}_n}^T \hat{B}_{\text{aplr}, \mathcal{S}_n} = a_{*,\mathcal{S}}^T \hat{B}_{\text{aplr}, \mathcal{S}}. \tag{26}$$

Then from Eq. (26), we have,

$$\hat{\Pi}(a_*, \lambda_n(\gamma), \gamma) = h\left(a_{*,\mathcal{S}}^T \hat{B}_{\text{aplr}, \mathcal{S}}\right). \tag{27}$$

Now, by P.2, we know that $\hat{B}_{\text{aplr}, \mathcal{S}}$ is root-n consistent. Therefore,

$$\left(\hat{B}_{\text{aplr}, \mathcal{S}} - b_{\mathcal{S}}^*\right) = O_p(n^{-1/2}). \tag{28}$$

Following the approach of [1] for logistic regression problems, we apply Taylor's series expansion in Eq. (27) with respect to the true parameter $b_{\mathcal{S}}^*$. Then we have,

$$\hat{\Pi}(a_*, \lambda_n(\gamma), \gamma) = h\left(a_{*,\mathcal{S}}^T b_{\mathcal{S}}^*\right) + \left(\hat{B}_{\text{aplr},\,\mathcal{S}} - b_{\mathcal{S}}^*\right)^T \frac{\partial h\left(a_{*,\mathcal{S}}^T b_{\mathcal{S}}^*\right)}{\partial b_{\mathcal{S}}^*} + o_p(n^{-1/2}) \quad (29)$$

$$= \pi(a_*) + \left(\hat{B}_{\text{aplr},\,\mathcal{S}} - b_{\mathcal{S}}^*\right)^T \frac{\partial h\left(a_{*,\mathcal{S}}^T b_{\mathcal{S}}^*\right)}{\partial b_{\mathcal{S}}^*} + o_p(n^{-1/2}). \quad (30)$$

Here, $o_p(n^{-1/2})$ comes from the condition mentioned in Eq. (28). Now, re-arranging the terms we get,

$$\hat{\Pi}(a_*, \lambda_n(\gamma), \gamma) - \pi(a_*) = \left(\hat{B}_{\text{aplr},\,\mathcal{S}} - b_{\mathcal{S}}^*\right)^T \frac{\partial h\left(a_{*,\mathcal{S}}^T b_{\mathcal{S}}^*\right)}{\partial b_{\mathcal{S}}^*} + o_p(n^{-1/2}). \quad (31)$$

Now, from P.2 we have,

$$\sqrt{n}\left(\hat{B}_{\text{aplr},\,\mathcal{S}} - b_{\mathcal{S}}^*\right) \xrightarrow{d} N\left(0, I_{11}^{-1}\right). \quad (32)$$

Then, applying Eq. (32) in Eq. (30), we get

$$\sqrt{n}\left(\hat{\Pi}(a_*, \lambda_n(\gamma), \gamma) - \pi(a_*)\right) \xrightarrow{d} N\left(0, \left[\frac{\partial h\left(a_{*,\mathcal{S}}^T b_{\mathcal{S}}^*\right)}{\partial b_{\mathcal{S}}^*}\right]^T I_{11}^{-1} \frac{\partial h\left(a_{*,\mathcal{S}}^T b_{\mathcal{S}}^*\right)}{\partial b_{\mathcal{S}}^*}\right). \quad (33)$$

Now,

$$\frac{\partial h\left(a_{*,\mathcal{S}}^T b_{\mathcal{S}}^*\right)}{\partial b_{\mathcal{S}}^*} = \left[\frac{\exp\left(a_{*,\mathcal{S}}^T b_{\mathcal{S}}^*\right)\left(1 + \exp\left(a_{*,\mathcal{S}}^T b_{\mathcal{S}}^*\right)\right) - \exp\left(a_{*,\mathcal{S}}^T b_{\mathcal{S}}^*\right)^2}{\left(1 + \exp\left(a_{*,\mathcal{S}}^T b_{\mathcal{S}}^*\right)\right)^2}\right] a_{*,\mathcal{S}} \quad (34)$$

$$= \left[\frac{\exp\left(a_{*,\mathcal{S}}^T b_{\mathcal{S}}^*\right)}{\left(1 + \exp\left(a_{*,\mathcal{S}}^T b_{\mathcal{S}}^*\right)\right)^2}\right] a_{*,\mathcal{S}} \quad (35)$$

$$= h\left(a_{*,\mathcal{S}}^T b_{\mathcal{S}}^*\right)\left[1 - \frac{\exp\left(a_{*,\mathcal{S}}^T b_{\mathcal{S}}^*\right)}{1 + \exp\left(a_{*,\mathcal{S}}^T b_{\mathcal{S}}^*\right)}\right] a_{*,\mathcal{S}} \quad (36)$$

$$= h\left(a_{*,\mathcal{S}}^T b_{\mathcal{S}}^*\right)\left(1 - h\left(a_{*,\mathcal{S}}^T b_{\mathcal{S}}^*\right)\right) a_{*,\mathcal{S}} \quad (37)$$

$$= h\left(a_*^T b^*\right)\left(1 - h\left(a_*^T b^*\right)\right) a_{*,\mathcal{S}} \quad (38)$$

$$= \pi(a_*)\left(1 - \pi(a_*)\right) a_{*,\mathcal{S}}. \quad (39)$$

Therefore, using Eq. (39) in Eq. (33), we have

$$\sqrt{n}\left(\hat{\Pi}(a_*, \lambda_n(\gamma), \gamma) - \pi(a_*)\right) \xrightarrow{d} N\left(0, \left[\pi(a_*)\left(1 - \pi(a_*)\right)\right]^2 a_{*,\mathcal{S}}^T I_{11}^{-1} a_{*,\mathcal{S}}\right) \quad (40)$$

The result shows that for infinite data stream, the estimated probabilities will be equal to the true probability $\pi(a_*)$.

4 Model Validation

In our method, we perform a sensitivity analysis over γ. This gives us a set of estimated probabilities for each fixed value of λ. Depending on these values in this set, the predicted class will be either unique or both '0' and '1'. Therefore, the classical measures of accuracy will not be applicable in this context. So we use the following performance measures, proposed by Corani and Zaffalon [6] for Naive Credal Classifier (NCC).

4.1 Measures of Accuracy

We use cross-validation for model validation where λ is used as a tuning parameter. We consider the following performance measures [6,14] for credal classification.

Definition 1 (Determinacy). *Determincay is the performance measure that counts the percentage of classifications with unique output.*

Definition 2 (Single accuracy). *Single accuracy is accuracy of the classifications when the output is determinate.*

There are two other performance measures called *indeterminate output size* and *set accuracy*. However, in the context of binary credal classification, indeterminate output size is always equal to 2 and set accuracy is always equal to 1.

The above mentioned performance measures will be used for model validation but for the model selection, we first need to choose an optimal λ, i.e. a value of λ that maximises the performance of our model. For this purpose, we need to use a trade-off between determinacy and single accuracy. We use u_{65} utility on the discounted accuracy, as proposed by Zaffalon *et al.* [20]. We display u_{65} on the discounted accuracy measure in Table 1, where each row stands for predicted class and each column stands for the actual class.

Table 1. Discounted utility (u_{65}) table for binary credal classification

	{0}	{1}
{0}	1	0
{1}	0	1
{0, 1}	0.65	0.65

Note that, for binary credal classification, we can formulate this unified u_{65} accuracy measure in the following way:

$$\text{Accuracy} = \text{Determinacy} \times \text{Single accuracy} + 0.65 \times (1 - \text{Determinacy}) \quad (41)$$

4.2   Model Selection and Validation

We use nested loop cross-validation for model selection and validation. We first split the dataset $\mathcal{D}$ in 2 equal parts $\mathcal{D}_1$ and $\mathcal{D}_2$. We take $\mathcal{D}_1$ and split it in 5 equal parts. We use 4 of them to train our IAPLR model and use the remaining part for the selection of λ. We do this for each of the 5 parts to get an optimal λ based on the averaged performance measure. After obtaining the optimal λ though cross-validation, we validate our model with $\mathcal{D}_2$ for model validation.

We repeat the same for $\mathcal{D}_2$, we use $\mathcal{D}_2$ to obtain an optimal λ for model selection and then validate it using $\mathcal{D}_1$. By this way, we use each observation exactly once for testing. This also gives a comparison between these two models and gives us an idea of interactions between the observations.

5   Illustration

We use two different datasets for illustration. The Sonar dataset is a regular logistic regressional data while the LSVT dataset is high dimensional. In both cases, we normalise the attributes to avoid scaling issues and split the datasets in two equal parts $\mathcal{D}_{S,1}$, $\mathcal{D}_{S,2}$ (Sonar) and $\mathcal{D}_{L,1}$, $\mathcal{D}_{L,2}$ (LSVT). We first select our model using $\mathcal{D}_{S,1}$ ($\mathcal{D}_{L,1}$). We vary our set of weights through 20 different γ's ranging from 0.01 to 1. We take a grid of 50 λ values. We find optimal λ by 5-fold cross validation. We use this optimal λ for model selection.

We compare our results with the naive credal classifier (NCC) [19]. For this, we first categorise the attributes in 5 factors. We train our model in a grid of the concentration parameter s with 50 entries ranging from 0.04 to 2. We run a 5-fold cross-validation the choice of optimal s and this value of s for model selection. We also compare our result with naive Bayes classifier (NBC) [11] and APLR [2,22]. For APLR select the value of optimal λ through a 5-fold cross-validation. We use glmnet [9] for training APLR and IAPLR model. We validate our model using $\mathcal{D}_{S,2}$ ($\mathcal{D}_{L,2}$). We then select our model using $\mathcal{D}_{S,2}$ ($\mathcal{D}_{L,2}$) and validate using $\mathcal{D}_{S,1}$ ($\mathcal{D}_{L,1}$) to capture interaction between the observations.

We show a summary of our results in Table 2. The left most column denotes the training set. We show determinacy in the second column. In third and fourth column, we display the single accuracy and utility based (u_{65}) accuracy, respectively and in the right most column we display range of active attributes.

5.1   Sonar Dataset

We use the Sonar dataset [10] for the illustration of our method. The dataset consists of 208 observations on 60 attributes in the range of 0 to 1. Sonar signals are reflected by either a metallic cylinder or a roughly cylindrical rock, and the attributes represent the energy of the reflected signal within a particular frequency band integrated over time. We use these attributes to classify the types of the reflectors.

In the top row of Fig. 2, we show the cross validation plots with respect to λ. For $\mathcal{D}_{S,1}$, the optimal λ is found to be 0.039 and for $\mathcal{D}_{S,2}$ the value is equal to 0.087. We observe from Table 2, that IAPLR outperforms the rests in terms of derermincay and u_{65} utility measure. It also has a good agreement in model validation with respect to the datasets unlike NCC or NBC which are sensitive with respect to the training dataset. It performs an automatic variable selection like APLR. We show the selected variables in the left most column. For IAPLR, we have a range of active attributes unlike APLR, which is computed using $\gamma = 1$. We observe that for $\mathcal{D}_{S,1}$, the sparsity of the model is more sensitive than the sparsity of the model trained by $\mathcal{D}_{S,2}$. In the top row of Fig. 3, show the sensitivity of sparsity with respect to γ for the optimal value of λ obtained through cross-validation. We observe that for both partitions the method selects more attributes as the value of γ increases.

5.2   LSVT Dataset

We use the LSVT dataset [17] for the illustration with high dimensional data. The dataset consists of 126 observations on 310 attributes. The attributes are 310 different biomedical signal processing algorithms which are obtained through 126 voice recording signals of 14 different persons diagnosed with Parkinson's disease. The responses denote acceptable (1) vs unacceptable (2) phonation during LSVT rehabilitation.

In the bottom row of Fig. 2, we show the cross validation plots with respect to λ. For $\mathcal{D}_{L,1}$, the optimal λ is found to be 0.018 and for $\mathcal{D}_{L,2}$ the value is equal to 0.014. We observe from Table 2, that IAPLR outperforms the rests. It also has a good agreement in model validation with respect to the datasets unlike NCC,

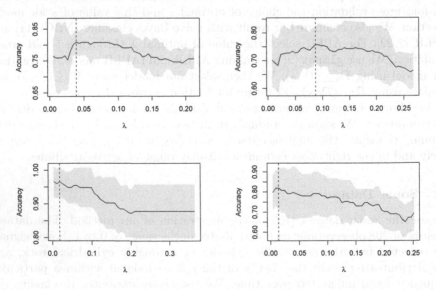

Fig. 2. Cross-validation curve with respect to the tuning parameter λ. The top row represents the results obtained for $\mathcal{D}_{S,1}$ (left), $\mathcal{D}_{S,2}$ (right) and the bottom row represents that of $\mathcal{D}_{L,1}$ (left), $\mathcal{D}_{L,2}$ (right).

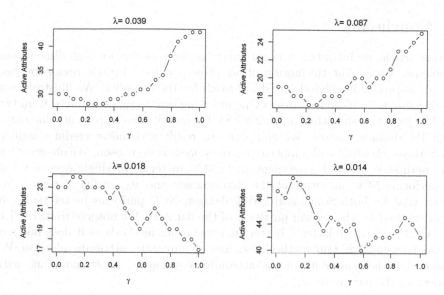

Fig. 3. Sensitivity of sparsity with respect to γ. The top row represents the results obtained for $\mathcal{D}_{S,1}$ (left), $\mathcal{D}_{S,2}$ (right) and the bottom row represents that of $\mathcal{D}_{L,1}$ (left), $\mathcal{D}_{L,2}$ (right).

Table 2. Summary of model selection and validation

Method	Training	Deter.	Single Acc.	u_{65}	Active
IAPLR ($\lambda = 0.039$)	$\mathcal{D}_{S,1}$	0.87	0.73	0.72	28–43
IAPLR ($\lambda = 0.087$)	$\mathcal{D}_{S,2}$	0.87	0.77	0.75	17–25
NCC ($s = 0.02$)	$\mathcal{D}_{S,1}$	0.77	0.68	0.67	–
NCC ($s = 0.56$)	$\mathcal{D}_{S,2}$	0.49	0.78	0.72	–
NBC	$\mathcal{D}_{S,1}$	–	–	0.59	–
NBC	$\mathcal{D}_{S,2}$	–	–	0.74	–
APLR ($\lambda = 0.104$)	$\mathcal{D}_{S,1}$	–	–	0.71	12
APLR ($\lambda = 0.189$)	$\mathcal{D}_{S,2}$	–	–	0.72	9
IAPLR ($\lambda = 0.018$)	$\mathcal{D}_{L,1}$	0.98	0.82	0.82	17–24
IAPLR ($\lambda = 0.014$)	$\mathcal{D}_{L,2}$	0.83	0.85	0.81	40–51
NCC ($s = 0.08$)	$\mathcal{D}_{L,1}$	0.14	0.78	0.67	–
NCC ($s = 0.04$)	$\mathcal{D}_{L,2}$	0.25	0.88	0.71	–
NBC	$\mathcal{D}_{L,1}$	–	–	0.51	–
NBC	$\mathcal{D}_{L,2}$	–	–	0.40	–
APLR ($\lambda = 0.052$)	$\mathcal{D}_{L,1}$	–	–	0.81	11
APLR ($\lambda = 0.285$)	$\mathcal{D}_{L,2}$	–	–	0.76	11

NBC and APLR. We notice that the sparsity levels are significantly different for different partitions of the dataset. We show the sparsity level in the bottom row of Fig. 3. We observe that for both partitions the method rejects more attributes as the value of γ increases.

6   Conclusion

In this article, we introduce a novel binary credal classifier for high dimensional problems. We exploit the notion of adaptive penalised logistic regression and use an imprecise likelihood based approach for the classifier. We illustrate our result using two different datasets One involving sonar signals bounced from two hard objects and the other involving LSVT rehabilitation of patients diagnosed with Parkinson's disease. We compare our result with naive credal classifier, naive Bayes classifier and adaptive penalised logistic regression. We observe that our method is in good agreement with NCC in terms of single accuracy but outperforms NCC in terms of the determinacy and u_{65} utility measure. We notice that for both Sonar and LSVT dataset, NCC performs better than any other method for the second partition of the datasets. We observe that, IAPLR or APLR performs relatively better than the other methods as it does not rely on the factorisation. Our method also does an automatic attribute selection. We notice that the sensitivity of the attribute selection is almost monotone with respect to the parameter γ.

References

1. Agresti, A.: Categorical Data Analysis. Wiley Series in Probability and Statistics. Wiley, Hoboken (2013). https://books.google.co.uk/books?id=UOrr47-2oisC
2. Algamal, Z.Y., Lee, M.H.: Penalized logistic regression with the adaptive lasso for gene selection in high-dimensional cancer classification. Expert Syst. Appl. **42**(23), 9326–9332 (2015). https://doi.org/10.1016/j.eswa.2015.08.016
3. Bickis, M.: The imprecise logit-normal model and its application to estimating hazard functions. J. Stat. Theory Pract. **3**(1), 183–195 (2009). https://doi.org/10.1080/15598608.2009.10411919
4. Corani, G., Antonucci, A.: Credal ensembles of classifiers. Comput. Stat. Data Anal. **71**, 818–831 (2014). https://doi.org/10.1016/j.csda.2012.11.010
5. Corani, G., de Campos, C.P.: A tree augmented classifier based on extreme imprecise Dirichlet model. Int. J. Approx. Reason. **51**(9), 1053–1068 (2010). https://doi.org/10.1016/j.ijar.2010.08.007
6. Corani, G., Zaffalon, M.: Learning reliable classifiers from small or incomplete data sets: the naive credal classifier 2. J. Mach. Learn. Res. **9**, 581–621 (2008)
7. José del Coz, J., Díez, J., Bahamonde, A.: Learning nondeterministic classifiers. J. Mach. Learn. Res. **10**, 2273–2293 (2009)
8. Fan, J., Li, R.: Variable selection via nonconcave penalized likelihood and its oracle properties. J. Am. Stat. Assoc. **96**(456), 1348–1360 (2001). http://www.jstor.org/stable/3085904
9. Friedman, J., Hastie, T., Tibshirani, R.: Regularization paths for generalized linear models via coordinate descent. J. Stat. Softw. **33**(1), 1–22 (2010). http://www.jstatsoft.org/v33/i01/
10. Gorman, R.P., Sejnowski, T.J.: Analysis of hidden units in a layered network trained to classify sonar targets. Neural Netw. **1**(1), 75–89 (1988). https://doi.org/10.1016/0893-6080(88)90023-8
11. Maron, M.E.: Automatic indexing: an experimental inquiry. J. ACM **8**(3), 404–417 (1961). https://doi.org/10.1145/321075.321084

12. Nesterov, Y.: Introductory Lectures on Convex Optimization: A Basic Course, 1st edn. Springer, Heidelberg (2014). https://doi.org/10.1007/978-1-4419-8853-9

13. Paton, L., Troffaes, M.C.M., Boatman, N., Hussein, M., Hart, A.: Multinomial logistic regression on Markov chains for crop rotation modelling. In: Laurent, A., Strauss, O., Bouchon-Meunier, B., Yager, R.R. (eds.) IPMU 2014. CCIS, vol. 444, pp. 476–485. Springer, Cham (2014). https://doi.org/10.1007/978-3-319-08852-5_49

14. Paton, L., Troffaes, M.C.M., Boatman, N., Hussein, M., Hart, A.: A robust Bayesian analysis of the impact of policy decisions on crop rotations. In: Augustin, T., Doria, S., Miranda, E., Quaeghebeur, E. (eds.) ISIPTA 2015: Proceedings of the 9th International Symposium on Imprecise Probability: Theories and Applications, Pescara, Italy, 20–24 July 2015, pp. 217–226. SIPTA, July 2015. http://dro.dur.ac.uk/15736/

15. Shevade, S., Keerthi, S.: A simple and efficient algorithm for gene selection using sparse logistic regression. Bioinformatics 19(17), 2246–2253 (2003). https://doi.org/10.1093/bioinformatics/btg308

16. Tibshirani, R.: Regression shrinkage and selection via the lasso. J. R. Stat. Soc.: Ser. B (Stat. Methodol.) 58(1), 267–288 (1996). http://www.jstor.org/stable/2346178

17. Tsanas, A., Little, M.A., Fox, C., Ramig, L.O.: Objective automatic assessment of rehabilitative speech treatment in Parkinson's disease. IEEE Trans. Neural Syst. Rehabil. Eng. 22(1), 181–190 (2014)

18. Walley, P.: Statistical Reasoning with Imprecise Probabilities. Monographs on Statistics & Applied Probability. Chapman & Hall/CRC , Taylor & Francis, New York (1991). https://books.google.co.uk/books?id=-hbvAAAAMAAJ

19. Zaffalon, M.: The Naive credal classifier. J. Stat. Plann. Infer. 105(1), 5–21 (2002). https://doi.org/10.1016/S0378-3758(01)00201 4. Imprecise Probability Models and their Applications

20. Zaffalon, M., Corani, G., Mauá, D.: Evaluating credal classifiers by utility-discounted predictive accuracy. Int. J. Approx. Reason. 53(8), 1282–1301 (2012). https://doi.org/10.1016/j.ijar.2012.06.022. Imprecise Probability: Theories and Applications (ISIPTA 2011)

21. Zhu, J., Hastie, T.: Classification of gene microarrays by penalized logistic regression. Biostatistics 5(3), 427–443 (2004). https://doi.org/10.1093/biostatistics/kxg046

22. Zou, H.: The adaptive lasso and its oracle properties. J. Am. Stat. Assoc. 101(476), 1418–1429 (2006). https://doi.org/10.1198/016214506000000735

Cautious Label-Wise Ranking
with Constraint Satisfaction

Yonatan-Carlos Carranza-Alarcon[✉], Soundouss Messoudi[✉],
and Sébastien Destercke[✉] [iD]

HEUDIASYC - UMR CNRS 7253, Université de Technologie de Compiègne,
57 avenue de Landshut, 60203 Compiegne Cedex, France
{yonatan-carlos.carranza-alarcon,soundouss.messoudi,
sebastien.destercke}@hds.utc.fr
https://www.hds.utc.fr/

Abstract. Ranking problems are difficult to solve due to their combinatorial nature. One way to solve this issue is to adopt a decomposition scheme, splitting the initial difficult problem in many simpler problems. The predictions obtained from these simplified settings must then be combined into one single output, possibly resolving inconsistencies between the outputs. In this paper, we consider such an approach for the label ranking problem, where in addition we allow the predictive model to produce cautious inferences in the form of sets of rankings when it lacks information to produce reliable, precise predictions. More specifically, we propose to combine a rank-wise decomposition, in which every sub-problem becomes an ordinal classification one, with a constraint satisfaction problem (CSP) approach to verify the consistency of the predictions. Our experimental results indicate that our approach produces predictions with appropriately balanced reliability and precision, while remaining competitive with classical, precise approaches.

Keywords: Label ranking problem · Constraint satisfaction · Imprecise probabilities

1 Introduction

In recent years, machine learning problems with structured outputs received an increasing interest. These problems appear in a variety of fields, including biology [33], image analysis [23], natural language treatment [5], and so on.

In this paper, we look at *label ranking (LR)*, where one has to learn a mapping from instances to rankings (strict total order) defined over a finite, usually limited number of labels. Most solutions to this problem reduce its initial complexity, either by fitting a probabilistic model (Mallows, Plackett-Luce [7]) with few parameters, or through a decomposition scheme. For example, ranking by pairwise comparison (RPC) [24] transforms the initial problem into binary problems. Constraint classification and log-linear models [13], as well as SVM-based

© Springer Nature Switzerland AG 2020
M.-J. Lesot et al. (Eds.): IPMU 2020, CCIS 1238, pp. 96–111, 2020.
https://doi.org/10.1007/978-3-030-50143-3_8

methods [30] learn, for each label, a (linear) utility function from which the ranking is deduced. Those latter approaches are close to other proposals [18] that
perform a label-wise decomposition.

In ranking problems, it may also be interesting [9,18] to predict partial rather
than complete rankings, abstaining to make a precise prediction in presence of
too little information. Such predictions can be seen as extensions of the reject
option [4] or of partial predictions [11]. They can prevent harmful decisions
based on incorrect predictions, and have been applied for different decomposition schemes, be it pairwise [10] or label-wise [18], always producing cautious
predictions in the form of partial order relations.

In this paper, we propose a new label ranking method, called LR-CSP, based
on a label-wise decomposition where each sub-problem intends to predict a set
of ranks. More precisely, we propose to learn for each label an imprecise ordinal
regression model of its rank [19], and use these models to infer a set of possible
ranks. To do this, we use imprecise probabilistic (IP) approaches are well tailored to make partial predictions [11] and represent potential lack of knowledge,
by describing our uncertainty by means of a convex set of probability distribu-
tions $\mathscr{P}$ [31] rather than by a classical single precise probability distribution $\mathbb{P}$.
An interesting point of our method, whose principle can be used with any set
of probabilities, is that it does not require any modification of the underlying
learning *imprecise* classifier, as long as the classifier can produce lower and upper
bounds $[\underline{P}, \overline{P}]$ over binary classification problems.

We then use CSP techniques on the set of resulting predictions to check
whether the prediction outputs are consistent with a global ranking (i.e. that
each label can be assigned a different rank).

Section 2 introduces the problem and our notations. Section 3 shows how
ranks can be predicted from imprecise probabilistic models and presents the
proposed inference method based on robust optimization techniques. Section 4
discusses related work. Finally, Sect. 5 is devoted to experimental evaluation
showing that our approach does reach a higher accuracy by allowing for partial
outputs, and remains quite competitive with alternative approaches to the same
learning problem.

2   Problem Setting

Multi-class problems consist in associating an instance $\mathbf{x}$ coming from an input
space $\mathcal{X}$ to a single label of the output space $\Lambda = \{\lambda_1, \ldots, \lambda_k\}$ representing
the possible classes. In label ranking, an instance $\mathbf{x}$ is no longer associated to a
unique label of Λ but to an order relation[1] $\succ_{\mathbf{x}}$ over $\Lambda \times \Lambda$, or equivalently to a
complete ranking over the labels in Λ. Hence, the output space is the set $\mathcal{L}(\Lambda)$
of complete rankings of Λ that contains $|\mathcal{L}(\Lambda)| = k!$ elements (i.e., the set of all
permutations). Table 1 illustrates a label ranking data set example with $k = 3$.

---

[1] A complete, transitive, and asymmetric relation.

Table 1. An example of label ranking data set $\mathbb{D}$

X_1	X_2	X_3	X_4	Y
107.1	25	Blue	60	$\lambda_1 \succ \lambda_3 \succ \lambda_2$
−50	10	Red	40	$\lambda_2 \succ \lambda_3 \succ \lambda_1$
200.6	30	Blue	58	$\lambda_2 \succ \lambda_1 \succ \lambda_3$
107.1	5	Green	33	$\lambda_1 \succ \lambda_2 \succ \lambda_3$
...	...	...	...	...

We can identify a ranking $\succ_{\mathbf{x}}$ with a permutation $\sigma_{\mathbf{x}}$ on $\{1, \ldots, k\}$ such that $\sigma_{\mathbf{x}}(i) < \sigma_{\mathbf{x}}(j)$ iff $\lambda_i \succ_{\mathbf{x}} \lambda_j$, as they are in one-to-one correspondence. $\sigma_{\mathbf{x}}(i)$ is the *rank* of label i in the order relation $\succ_{\mathbf{x}}$. As there is a one-to-one correspondence between permutations and complete rankings, we use the terms interchangeably.

Example 1. Consider the set $\Lambda = \{\lambda_1, \lambda_2, \lambda_3\}$ and the observation $\lambda_3 \succ \lambda_1 \succ \lambda_2$, then we have $\sigma_{\mathbf{x}}(1) = 2$, $\sigma_{\mathbf{x}}(2) = 3$, $\sigma_{\mathbf{x}}(3) = 1$.

The usual objective in label ranking is to use the training instances $\mathbb{D} = \{(\mathbf{x}_i, y_i) \mid i = 1, \ldots, n\}$ with $x_i \in \mathcal{X}$, $y_i \in \mathcal{L}(\Lambda)$ to learn a predictor, or a *ranker* $h : \mathcal{X} \to \mathcal{L}(\Lambda)$. While in theory this problem can be transformed into a multiclass problem where each ranking is a separate class, this is in practice undoable, as the number of classes would increase factorially with k. The most usual means to solve this issue is either to decompose the problem into many simpler ones, or to fit a parametric probability distribution over the ranks [7]. In this paper, we shall focus on a label-wise decomposition of the problem.

This rapid increase of $|\mathcal{L}(\Lambda)|$ also means that getting reliable, precise predictions of ranks is in practice very difficult as k increases. Hence it may be useful to allow the ranker to return partial but reliable predictions.

3   Label-Wise Decomposition: Learning and Predicting

This section details how we propose to reduce the initial ranking problem in a set of k label-wise problems, that we can then solve separately. The idea is the following: since a complete observation corresponds to each label being associated to a unique rank, we can learn a probabilistic model $p_i : K \to [0, 1]$ with $K = \{1, 2, \ldots, k\}$ and where $p_{ij} := p_i(j)$ is interpreted as the probability $P(\sigma(i) = j)$ that label λ_i has rank j. Note that $\sum_j p_{ij} = 1$.

A first step is to decompose the original data set $\mathbb{D}$ into k data sets $\mathbb{D}_j = \{(\mathbf{x}_i, \sigma_{\mathbf{x}_i}(j)) \mid i = 1, \ldots, n\}$, $j = 1, \ldots, k$. The decomposition is illustrated by Fig. 1. Estimating the probabilities p_{ij} for a label λ_i then comes down to solve an ordinal regression problem [27]. In such problems, the rank associated to a label is the one minimizing the expected cost $\mathbb{E}_{ij}$ of assigning label λ_i to rank j, that depends on p_{ij} and a distance $D : K \times K \to \mathbb{R}$ between ranks as follows:

$$\mathbb{E}_{ij} = \sum_{\ell=1}^{k} D(j, k) p_{ik}. \tag{1}$$

Common choices for the distances are the L_1 and L_2 norms, corresponding to

$$D_1(j,k) = |j - k| \quad \text{and} \quad D_2(j,k) = (j - k)^2. \tag{2}$$

Other choices include for instance the pinball loss [29], that penalizes asymmetrically giving a higher or a lower rank than the actual one. An interest of those in the imprecise setting we will adopt next is that it produces predictions in the form of intervals, i.e., in the sense that $\{1,3\}$ cannot be a prediction but $\{1,2,3\}$ can. In this paper, we will focus on the L_1 loss, as it is the most commonly considered in ordinal classification problems[2].

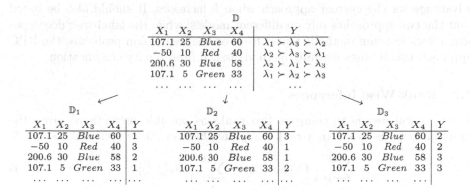

Fig. 1. Label-wise decomposition of rankings

3.1 Probability Set Model

Precise estimates for p_i issued from the finite data set $\mathbb{D}_k$ may be unreliable, especially if these estimates rely on little, noisy or incomplete data. Rather than relying on precise estimates in all cases, we propose to consider an imprecise probabilistic model, that is, to consider for each label λ_i a polytope (a convex set) $\mathscr{P}_i$ of possible probabilities. In our setting, a particularly interesting model are imprecise cumulative distributions [15], as they naturally encode the ordinal nature of rankings, and are a common choice in the precise setting [22]. They consist in providing bounds $[\underline{P}(A_\ell), \overline{P}(A_\ell)]$ on events $A_\ell = \{1, \dots, \ell\}$ and to consider the resulting set

$$\mathscr{P}_i = \left\{ p_i : \underline{P}_i(A_\ell) \le \sum_{j=1}^{\ell} p_{ij} \le \overline{P}_i(A_\ell), \sum_{j \in K} p_{ij} = 1 \right\}. \tag{3}$$

We will denote by $\underline{F}_{ij} = \underline{P}_i(A_j)$ and $\overline{F}_{ij} = \overline{P}_i(A_j)$ the given bounds. Table 2 provides an example of a cumulative distribution that could be obtained in a ranking problem where $k = 5$ and for a label λ_i. For other kinds of sets $\mathscr{P}_i$ we could consider, see [17].

---

[2] The approach easily adapts to the other losses.

Table 2. Imprecise cumulative distribution for λ_i

Rank j	1	2	3	4	5
$\overline{F}_{ij}$	0.15	**0.55**	*0.70*	***0.95***	1
$\underline{F}_{ij}$	**0.10**	*0.30*	***0.45***	0.80	1

This approach requires to learn k different models, one for each label. This is to be compared with the RPC [24] approach, in which $k(k-1)/2$ models (one for each pair of labels) have to be learned. There is therefore a clear computational advantage for the current approach when k increases. It should also be noted that the two approaches rely on different models: while the label-wise decomposition uses learning methods issued from ordinal regression problems, the RPC approach usually uses learning methods issued from binary classification.

3.2   Rank-Wise Inferences

The classical means to compare two ranks as possible predictions, given the probability p_i, is to say that rank ℓ is preferable to rank m (denoted $\ell \succ m$) iff

$$\sum_{j=1}^{k} D_1(j,m)p_{ij} \geq \sum_{j=1}^{k} D_1(j,\ell)p_{ij} \tag{4}$$

That is if the expected cost (loss) of predicting m is higher than the expected cost of predicting ℓ. The final prediction is then the rank that is not dominated or preferred to any other (with typically a random choice when there is some indifference between the top ranks).

When precise probabilities p_i are replaced by probability sets $\mathscr{P}_i$, a classical extension[3] of this rule is to consider that rank ℓ is preferable to rank m iff it is so for every probability in $\mathscr{P}_i$, that is if

$$\inf_{p_i \in \mathscr{P}_i} \sum_{j=1}^{k} (D_1(j,m) - D_1(j,\ell))p_{ij} \tag{5}$$

is positive. Note that under this definition we may have simultaneously $m \not\succ \ell$ and $\ell \not\succ m$, therefore there may be multiple undominated, incomparable ranks, in which case the final prediction is a set-valued one.

In general, obtaining the set of predicted values requires to solve Eq. (5) at most a quadratic number of times (corresponding to each pairwise comparison). However, it has been shown [16, Prop. 1] that when considering D_1 as a cost function, the set of predicted values corresponds to the set of possible medians within $\mathscr{P}_i$, which is straightforward to compute if one uses the generalized p-box [15] as an uncertainty model. Namely, if $\underline{F}_i, \overline{F}_i$ are the cumulative distributions for label λ_i, then the predicted ranks under D_1 cost are

$$\hat{R}_i = \left\{ j \in K : \underline{F}_{i(j-1)} \leq 0.5 \leq \overline{F}_{ij}, \quad \underline{F}_{i(0)} = 0 \right\}, \tag{6}$$

---

[3] Also, known as maximality criterion [31].

a set that is always non-empty and straightforward to obtain. Looking back at Table 2, our prediction would have been $\hat{R} = \{2, 3, 4\}$, as these are the three possible median values.

As for the RPC approach (and its cautious versions [9]), the label-wise decomposition requires to aggregate all decomposed models into a single (partial) prediction. Indeed, focusing only on decomposed models $\mathscr{P}_i$, nothing forbids to predict the same rank for multiple labels. In the next section, we discuss cautious predictions in the form of sets of ranks, as well as how to resolve inconsistencies.

3.3   Global Inferences

Once we have retrieved the different set-valued predictions of ranks for each label, two important questions remain:

1. Are those predictions consistent with the constraint that each label should receive a distinct rank?
2. If so, can we reduce the obtained predictions by integrating the aforementioned constraint?

Example 2. To illustrate the issue, let us consider the case where we have four labels $\lambda_1, \lambda_2, \lambda_3, \lambda_4$. Then the following predictions

$$\hat{R}_1 = \{1, 2\}, \ \hat{R}_2 = \{1, 2\}, \ \hat{R}_3 = \{1, 2\}, \ \hat{R}_4 = \{3, 4\}$$

are inconsistent, simply because labels $\lambda_1, \lambda_2, \lambda_3$ cannot be given simultaneously a different rank (note that pair-wisely, they are not conflicting). On the contrary, the following predictions

$$\hat{R}_1 = \{1, 2\}, \ \hat{R}_2 = \{1, 2, 3\}, \ \hat{R}_3 = \{2\}, \ \hat{R}_4 = \{1, 2, 3, 4\}$$

are consistent, and could also be reduced to the unique ranking

$$\hat{R}'_1 = \{1\}, \ \hat{R}'_2 = \{3\}, \ \hat{R}'_3 = \{2\}, \ \hat{R}'_4 = \{4\},$$

as the strong constraint $\hat{R}_3 = \{2\}$ propagates to all other predictions by removing λ_2 from them, which results in a new strong constraint $\hat{R}^*_1 = \{1\}$ that also propagates to all other predictions. This redundancy elimination is repeated as new strong constraints emerge until we get the unique ranking above.

Such a problem is well known in Constraint Programming [12], where it corresponds to the *alldifferent* constraint. In the case where all rank predictions are intervals, that is a prediction $\hat{R}_i$ contains all values between $\min \hat{R}_i$ and $\max \hat{R}_i$, efficient algorithms using the fact that one can concentrate on bounds alone exist, that we can use to speed up computations [28].

4   Discussion of Related Approaches

As said in the introduction, one of our main goals in this paper is to introduce a label ranking method that allows the ranker to partially abstain when it has insufficient information, therefore producing a corresponding set of possible rankings. We discuss here the usefulness of such rank-wise partial prediction (mainly w.r.t. approaches producing partial orders), as well as some related works.

4.1   Partial Orders vs Imprecise Ranks

Most existing methods [9,10] that propose to make set-valued or cautious predictions in ranking problems consider partial orders as their final predictions, that is pairwise relations $\succ_x$ that are transitive and asymmetric, but no longer necessarily complete. To do so, they often rely on decomposition approaches estimating preferences between each pairs of labels [24].

However, while a complete order can be equivalently described by the relation $\succ_x$ or by the rank associated to each label, this is no longer true when one considers partial predictions. Indeed, consider for instance the case where the set of rankings over three labels $\{\lambda_1, \lambda_2, \lambda_3\}$ we would like to predict is $S = \{\lambda_1 \succ \lambda_2 \succ \lambda_3, \lambda_1 \prec \lambda_2 \prec \lambda_3\}$, which could correspond to an instance where λ_2 is a good compromise, and where the population is quite divided about λ_1 and λ_3 that represent more extreme options.

While the set S can be efficiently and exactly represented by providing sets of ranks for each item, none of the information it contains can be retained in a partial order. Indeed, the prediction $\hat{R}_1 = \{1,3\}, \hat{R}_2 = \{2\}, \hat{R}_3 = \{1,3\}$ perfectly represents S, while representing it by a partial order would result in the empty relation (since for all pairs i, j, we have $\lambda_i \succ \lambda_j$ and $\lambda_j \succ \lambda_i$ in the set S).

We could find an example that would disadvantage a rank-wise cautious prediction over one using partial orders, as one representation is not more general than the other[4]. Yet, our small example shows that considering both approaches makes sense, as one cannot encapsulate the other, and vice-versa.

4.2   Score-Based Approaches

In a recent literature survey [30], we can see that there are many score-based approaches, already been studied and compared in [24], such as constraint classification, log-linear models, etc. Such approaches learn, from the samples, a function h_j for each label λ_j that will predict a strength $h_j(x^*)$ for a new instance. Labels are then ranked accordingly to their predicted strengths.

We will consider a typical example of such approaches, based on SVM, that we will call SVM label ranking (SVM-LR). Vembu and Gärtner [30] show that the SVM method [20] solving multi-label problems can be straightforwardly generalized to a label ranking problem. In contrast to our approach where each model is learned separately, SVM-LR fits all the functions at once, even if at prediction time they are evaluated independently. While this may account for label dependencies, this comes at a computational cost since we have to solve a quadratic optimization problem (i.e. the dual problem introduced in [20]) whose scale increases rapidly as the number of training samples and labels grows.

More precisely, the score functions $h_j(x^*) = \langle w_j \mid x^* \rangle$ are scalar products between a weight vector w_j and x^*. If α_{ijq} are coefficients that represent the

---

[4] In the sense that the family of subsets of ranking representable by one is not included in the other.

existence of either the preference $\lambda_q \succ_{\boldsymbol{x}_i} \lambda_j$ or $\lambda_j \succ_{\boldsymbol{x}_i} \lambda_q$ of the instance $\boldsymbol{x}_i$, $\boldsymbol{w}_j$ can be obtained from the dual problem in [20, Sect. 5] as follows:

$$
w_j = \frac{1}{2} \sum_{i=1}^{n} \left[\sum_{(j,q)\in E_i} \alpha_{ijq} - \sum_{(p,j)\in E_i} \alpha_{ipj} \right] \boldsymbol{x}_i \tag{7}
$$

where α_{ipq} are the weighted target values to optimize into the dual problem. E_i contains all preferences, i.e. $\{(p,q) \in E_i \iff \lambda_p \succ \lambda_q\}$, of the training instance $\boldsymbol{x}_i$.

It may seem at first that such approaches, once made imprecise, could be closer to ours. Indeed, the obtained models h_i after training also provide label-wise information. However, if we were to turn these method imprecise and obtain imprecise scores $[\underline{h}_i, \overline{h}_i]$, the most natural way to build a partial prediction would be to consider that $\lambda_i \succ \lambda_j$ when $\underline{h}_i > \overline{h}_j$, that is when the score of λ_i would certainly be higher than the one of λ_j. Such a partial prediction would be an interval order and would again not encompass the same family of subsets of rankings, as it would constitute a restricted setting compared to the one allowing for prediction any partial order.

5 Experiments

This section describes our experiments made to test if our approach is (1) competitive with existing ones and if (2) the partial predictions indeed provide more reliable inferences by abstaining on badly predicted ranks.

5.1 Data Sets

The data sets used in the experiments come from the UCI machine learning repository [21] and the Statlog collection [25]. They are synthetic label ranking data sets built either from classification or regression problems. From each original data set, a transformed data set $(\mathbf{x}_i, y_i)$ with complete rankings was obtained by following the procedure described in [8]. A summary of the data sets used in the experiments is given in Table 3. We perform 10×10-fold cross-validation procedure on all the data sets (c.f. Table 3).

5.2 Completeness/Correctness Trade-Off

To answer the question whether our method correctly identifies on which label it is desirable to abstain or to deliver a set of possible rankings, it is necessary to measure two aspects: how accurate and how precise the predictions are. Indeed, a good balance should be sought between informativeness and reliability of the predictions. For this reason, and similarly to what was proposed in the pairwise setting [9], we use a completeness and a correctness measure to assess the quality of the predictions. Given the prediction $\hat{R} = \{\hat{R}_i, i = 1, \ldots, k\}$, we propose as the completeness (CP) and correctness (CR) measure

$$
CP(\hat{R}) = \frac{k^2 - \sum_{i=1}^{k} |\hat{R}_i|}{k^2 - k} \quad \text{and} \quad CR(\hat{R}) = 1 - \frac{\sum_{i=1}^{k} \min_{\hat{r}_i \in \hat{R}_i} |\hat{r}_i - r_i|}{0.5k^2} \tag{8}
$$

Table 3. Experimental data sets

#	Data set	Type	#Inst	#Attributes	#Labels
a	authorship	classification	841	70	4
b	bodyfat	regression	252	7	7
c	calhousing	regression	20640	4	4
d	cpu-small	regression	8192	6	5
e	fried	regression	40768	9	5
f	glass	classification	214	9	6
g	housing	regression	506	6	6
h	iris	classification	150	4	3
i	pendigits	classification	10992	16	10
j	segment	classification	2310	18	7
k	stock	regression	950	5	5
l	vehicle	classification	846	18	4
m	vowel	classification	528	10	11
n	wine	classification	178	13	3

where CP is null if all $\hat{R}_i$ contains the k possible ranks and has value one if all $\hat{R}_i$ are reduced to singletons, whilst CR is equivalent to the Spearman Footrule when having a precise observation. Note that classical evaluation measures [36] used in an IP setting cannot be straightforwardly applied here, as they only extend the 0/1 loss and are not consistent with Spearman Footrule, and adapting cost-sensitive extensions [34] to the ranking setting would require some development.

5.3 Our Approach

As mentioned in Sect. 3, our proposal is to fit an *imprecise* ordinal regression model for every label-wise decomposition $\mathbb{D}_i$, in which the lower and upper bounds of the cumulative distribution $[\underline{F}_i, \overline{F}_i]$ must be estimated in order to predict the set of rankings (Eq. 6) of an unlabeled instance $\boldsymbol{x}^*$. In that regard, we propose to use an extension of Frank and Hall [22] method to imprecise probabilities, already studied in detail in [19].

Frank and Hall's method takes advantage of k ordered label values by transforming the original k-label ordinal problem to $k - 1$ binary classification sub-problems. Each estimates of the probability[5] $P_i(A_\ell) := F_i(\ell)$ where $A_\ell = \{1, \ldots, \ell\} \subseteq K$ and the mapping $F_i : K \to [0, 1]$ can be seen as a discrete cumulative distribution. We simply propose to make these estimates imprecise and to use bounds

---

[5] For readability, we here drop the condition of a new instance in all probabilities, i.e. $P_i(A_\ell) := P_i(A_\ell|\boldsymbol{x}^*)$.

$$\underline{P}_i(A_j) := \underline{F}_i(j) \quad \text{and} \quad \overline{P}_i(A_j) := \overline{F}_i(j)$$

which is indeed a generalized p-box model [15], as defined in Eq. (3).

To estimate these bounds, we use the naive credal classifier (NCC)[6] [35], which extends the classical naive Bayes classifier (NBC), as a base classifier. This classifier imprecision level is controlled through a hyper-parameter $s \in \mathbb{R}$. Indeed, the higher s, the wider the intervals $[\underline{P}_i(A_j), \overline{P}_i(A_j)]$. For $s = 0$, we retrieve the classical NBC with precise predictions, and for $s >>> 0$, the NCC model will make vacuous predictions (i.e. all rankings for every label).

However, the imprecision induced by a peculiar value of s differs from a data set to another (as show the values in Fig. 2), and it is essential to have an adaptive way to quickly obtain two values:

- the value s_{min} corresponding to the value with an average completeness close to 1, making the corresponding classifier close to a precise one. This value is the one we will use to compare our approach to standard, precise ones;
- the value s_{max} corresponding to the value with an average correctness close to 1, and for which the made predictions are almost always right. The corresponding completeness gives an idea of how much we should abstain to get strong guarantees on the prediction, hence of how "hard" is a given data set.

To find those values, we proceed with the following idea: we start from an initial interval of values $[\underline{s}, \overline{s}]$, and from target intervals $[\underline{CP}, \overline{CP}]$ and $[\underline{CR}, \overline{CR}]$, typically $[0.95, 1]$ of average completeness and correctness. Note that in case of inconsistent predictions, $\hat{R}_i = \emptyset$ and the completeness is higher than 1 (in such case, we consider $CR = 0$). For s_{min}, we will typically start from $\underline{s} = 0$ (for which $CP > 1$) and will consider a value $\overline{s}$ large enough for which $CP < 0.95$ (e.g., starting from $s = 2$ as advised in [32] and doubling s iteratively until $CP < 0.95$, as when s increases completeness decreases and correctness increases in average). We then proceed by dichotomy to find a value s_{min} for which average predictions are within interval $[\underline{CP}, \overline{CP}]$. We proceed similarly for s_{max}.

With s_{min} and s_{max} found, a last issue to solve is how to get intermediate values of $s \in [s_{min}, s_{max}]$ in order to get an adaptive evolution of completeness/correctness, as in Fig. 2. This is done through a simple procedure: first, we start by calculating the completeness/correctness for the middle value between s_{min} and s_{max}, that is for $(s_{min} + s_{max})/2$. We then compute the distance between all the pairs of completeness/correctness values obtained for consecutive s values, and add a new s point in the middle between the two points with the biggest Euclidean distance. We repeat the process until we get the number of s values requested, for which we provide completeness/correctness values.

---

[6] Bearing in mind that they can be replaced by any other imprecise classifiers, see [2,6].

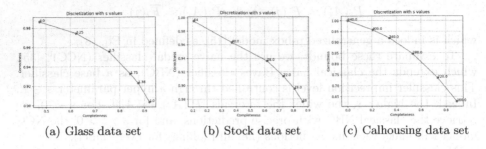

(a) Glass data set   (b) Stock data set   (c) Calhousing data set

Fig. 2. Evolution of the hyper-parameter s on glass, stock and calhousing data sets.

The Fig. 2 shows that the boundary values of the hyper-parameter of imprecision s actually significantly depend on the data set. Our approach enables us to find the proper "optimal" value s_{min} for each data set, which can be small (as in glass where $s_{min} = 1$) or big (as in calhousing where $s_{min} = 160$).

Figure 2 is already sufficient to show that our abstention method is working as expected, as indeed correctness increases quickly when we allow abstention, that is when completeness decreases. Figure 2(a) shows that for some data sets, one can have an almost perfect correctness while not being totally vacuous (as correctness of almost 1 is reached for a completeness slightly below 0.5, for a value $s = 4$), while this may not be the case for other more difficult data sets such as calhousing, for which one has to choose a trade-off between completeness and correctness to avoid fully vacuous predictions. Yet, for all data sets (only three being shown for lack of space), we witness a regular increase of correctness.

5.4   Comparison with Other Methods

A remaining question is to know whether our approach is competitive with other state-of-art approaches. To do this, we compare the results obtained on test data sets (in a 10×10 fold cross validation) between the results we obtain for $s = s_{min}$ and several methods. Those results are indeed the closest we can get to precise predictions in our setting. The methods to which we compare ourselves are the following:

- The ranking by pairwise comparisons (RPC), as implemented in [3];
- The Label ranking tree (LRT [8]), that adopt a local non-decomposed scheme;
- The SVM-LR approach that we already described in Sect. 4.2.

As the NCC deals with discrete attributes, we need to discretize continuous attributes in z intervals before training[7]. While z could be optimized, we use in this paper only two arbitrarily chosen levels of discretization $z = 5$ and $z = 6$ (i.e. LR-CSP-5 and LR-CSP-6 models) to compare our method against the others, for simplicity and because our goal is only to show competitiveness of our approach.

---

[7] Available in https://github.com/sdestercke/classifip.

As mentioned, we consider the comparison by picking the value s_{min}. By fixing this hyper-parameter regulating the imprecision level of our approach, we then compare the correctness measure (8) with the Spearman Footrule loss obtained for RCP and LRT methods, and implemented into existing software [3]. For the SVM-LR, of which we did not find an online implementation, we used a Python package[8], which solves a quadratic problem with known solvers [1] for little data sets, or a Frank-Wolfe algorithm for bigger data sets. In fact, Frank-Wolfe's algorithm almost certainly guarantees the convergence to the global minimum for convex surfaces and to a local minimum for non-convex surfaces [26].

A last issue to solve is how to handle inconsistency predictions, ones in which the *alldifferent* constraint would not find a precise or partial solution but an empty one. Here, such predictions are ignored, and our results consider correctness and Spearman footrule on consistent solutions only, as dealing with inconsistent predictions will be the object of future works.

5.5   Results

The average performances and their ranks in parentheses obtained in terms of the correctness (CR) measure are shown in Table 4(a) and 4(b), with discretization 5 and 6 respectively applied to our proposal method LR-CSP.

A Friedman test [14] on the ranks yields p-values of 0.00006176 and 0.0001097 for LR-CSP-5 and LR-CSP-6, respectively, thus strongly suggesting performance differences between the algorithms. The Nemenyi post-hoc test (see Table 5) further indicates that LR-CSP-5 (and LR-CSP-6) is significantly better than SVM-LR. Our approach also remains competitive with LRT and RPC.

Finally, recall that our method is also quite fast to compute, thanks to the simultaneous use of decomposition (requiring to build k classifiers), and of probability sets and loss functions offering computational advantages that make the prediction step very efficient. Also, thanks to the fact that our predictions are intervals, i.e. sets of ranks without holes in them, we can use very efficient algorithms to treat the *alldifferent* constraints [28].

Note also that our proposal discretized at $z = 6$ intervals gets more accurate predictions (and also indicate a little drop in the *p-value* of all comparisons of Table 5) what can suggest us that an optimal value of $\hat{z}$ may improve the prediction performance (all that remains, of course, hypothetical).

---

[8] Available in https://pypi.org/project/svm-label-ranking/.

Table 4. Average correctness accuracies (%) compared to LR-CSP-5 (left) and LR-CSP-6 (right)

	LR-CSP-5	LRT	RPC	SVM-LR		LR-CSP-6	LRT	RPC	SVM-LR
a	94.19 ± 1.31 (1)	91.49 ± 0.31 (3)	93.25 ± 0.25 (2)	64.75 ± 0.41 (4)	a	93.90 ± 0.69 (1)	91.53 ± 0.31 (3)	93.21 ± 0.23 (2)	64.42 ± 0.36 (4)
b	53.30 ± 3.81 (1)	41.56 ± 1.17 (4)	52.05 ± 0.42 (2)	52.56 ± 0.39 (3)	b	54.12 ± 3.73 (1)	41.70 ± 1.48 (4)	50.43 ± 0.39 (3)	51.10 ± 0.49 (2)
c	61.46 ± 0.92 (1)	58.33 ± 0.28 (2)	51.76 ± 0.01 (3)	38.48 ± 0.02 (4)	c	61.05 ± 0.80 (1)	58.37 ± 0.28 (2)	51.85 ± 0.02 (3)	38.45 ± 0.02 (4)
d	68.66 ± 0.63 (1)	60.96 ± 0.24 (3)	62.10 ± 0.04 (2)	47.08 ± 0.85 (4)	d	68.72 ± 1.42 (1)	60.76 ± 0.30 (3)	61.93 ± 0.04 (2)	46.71 ± 0.87 (4)
e	99.34 ± 0.07 (1)	91.29 ± 0.08 (3)	99.92 ± 0.01 (2)	84.19 ± 2.63 (4)	e	99.20 ± 0.07 (2)	91.26 ± 0.06 (3)	99.92 ± 0.01 (1)	84.18 ± 2.67 (4)
f	90.92 ± 3.48 (3)	91.75 ± 0.50 (1)	91.05 ± 0.18 (2)	87.16 ± 0.34 (4)	f	91.95 ± 2.90 (1)	91.59 ± 0.47 (2)	90.83 ± 0.24 (3)	85.68 ± 0.33 (4)
g	79.18 ± 1.98 (2)	84.55 ± 0.51 (1)	74.54 ± 0.15 (3)	69.54 ± 0.35 (4)	g	79.21 ± 3.37 (2)	85.09 ± 0.46 (1)	74.86 ± 0.16 (3)	70.16 ± 0.46 (4)
h	96.86 ± 3.72 (1)	96.77 ± 0.60 (2)	93.16 ± 0.56 (3)	88.39 ± 0.41 (4)	h	99.36 ± 1.28 (1)	97.16 ± 0.55 (2)	92.75 ± 0.58 (3)	87.39 ± 0.37 (4)
i	91.55 ± 0.24 (3)	95.15 ± 0.05 (1)	94.12 ± 0.01 (2)	58.66 ± 2.71 (4)	i	91.31 ± 0.14 (3)	95.14 ± 0.05 (1)	94.12 ± 0.01 (2)	58.75 ± 2.71 (4)
j	90.37 ± 0.46 (3)	96.19 ± 0.09 (1)	94.56 ± 0.02 (2)	66.39 ± 3.00 (4)	j	91.20 ± 0.85 (3)	96.11 ± 0.10 (1)	94.52 ± 0.03 (2)	66.25 ± 3.05 (4)
k	86.75 ± 1.59 (2)	91.46 ± 0.34 (1)	82.59 ± 0.06 (3)	74.49 ± 0.20 (4)	k	88.63 ± 1.53 (2)	91.64 ± 0.27 (1)	82.23 ± 0.08 (3)	75.20 ± 0.17 (4)
l	84.81 ± 2.13 (3)	88.07 ± 0.40 (2)	89.28 ± 0.17 (1)	82.30 ± 0.96 (4)	l	85.29 ± 1.91 (3)	88.03 ± 0.44 (2)	89.24 ± 0.14 (1)	81.93 ± 1.00 (4)
m	86.32 ± 2.34 (1)	85.36 ± 0.97 (2)	74.32 ± 0.06 (3)	66.60 ± 1.23 (4)	m	88.23 ± 1.00 (1)	84.40 ± 0.62 (2)	72.88 ± 0.06 (3)	65.41 ± 1.21 (4)
n	97.98 ± 2.89 (1)	91.75 ± 0.88 (4)	94.55 ± 0.62 (2)	94.53 ± 0.50 (3)	n	98.20 ± 1.19 (1)	91.80 ± 0.87 (4)	94.58 ± 0.61 (2)	94.56 ± 0.50 (3)
avg.	84.41 ± 1.83(1.72)	83.19 ± 0.46(2.14)	81.95 ± 0.18(2.28)	69.65 ± 1.00(3.86)	avg.	85.03 ± 1.49(1.64)	83.18 ± 0.45(2.21)	81.67 ± 0.19(2.36)	69.30 ± 1.02(3.79)

Table 5. Nemenyi post-hoc test: null hypothesis H_0 and p-value

#	H_0		LRT	RPC	SVM-LR
1	LR-CSP-5 =	0.8161	0.6452	**0.000066**
2	LR-CSP-6 =	0.6450	0.4600	**0.000066**

6 Conclusion and Perspectives

In this paper, we have proposed a method to make partial predictions in label ranking, using a label-wise decomposition as well as a new kind of partial predictions in terms of possible ranks. The experiments on synthetic data sets show that our proposed model (LR-CSP) produces reliable and cautious predictions and performs close to or even outperforms the existing alternative models.

This is quite encouraging, as we left a lot of room for optimization, e.g., in the base classifiers or in the discretization. However, while our method extends straightforwardly to partially observed rankings in training data when those are top-k rankings (considering for instance the rank of all remaining labels as $k+1$), it may be trickier to apply it to pairwise rankings, another popular way to get such data. Some of our future works will focus on that.

Acknowledgments. This work was carried out in the framework of the Labex MS2T and PreServe projects, funded by the French Government, through the National Agency for Research (Reference ANR-11-IDEX-0004-02 and ANR-18-CE23-0008).

References

1. Andersen, M., Dahl, J., Vandenberghe, L.: CVXOPT: a python package for convex optimization (2013). http://abel.ee.ucla.edu/cvxopt
2. Augustin, T., Coolen, F., de Cooman, G., Troffaes, M.: Introduction to Imprecise Probabilities. Wiley, Chichester (2014)
3. Balz, A., Senge, R.: WEKA-LR: a label ranking extension for weka (2011). https://cs.uni-paderborn.de/de/is/research/research-projects/software/weka-lr-a-label-ranking-extension-for-weka/
4. Bartlett, P., Wegkamp, M.: Classification with a reject option using a hinge loss. J. Mach. Learn. Res. **9**, 1823–1840 (2008)
5. Bordes, A., Glorot, X., Weston, J., Bengio, Y.: Joint learning of words and meaning representations for open-text semantic parsing. J. Mach. Learn. Res. Proc. Track **22**, 127–135 (2012)
6. Carranza-Alarcon, Y.C., Destercke, S.: Imprecise gaussian discriminant classification. In: International Symposium on Imprecise Probabilities: Theories and Applications, pp. 59–67 (2019)
7. Cheng, W., Dembczynski, K., Hüllermeier, E.: Label ranking methods based on the Plackett-Luce model. In: Proceedings of the 27th Annual International Conference on Machine Learning - ICML, pp. 215–222 (2010)
8. Cheng, W., Hühn, J., Hüllermeier, E.: Decision tree and instance-based learning for label ranking. In: Proceedings of the 26th Annual International Conference on Machine Learning - ICML 2009 (2009)

9. Cheng, W., Rademaker, M., De Baets, B., Hüllermeier, E.: Predicting partial orders: ranking with abstention. In: Balcázar, J.L., Bonchi, F., Gionis, A., Sebag, M. (eds.) ECML PKDD 2010. LNCS (LNAI), vol. 6321, pp. 215–230. Springer, Heidelberg (2010). https://doi.org/10.1007/978-3-642-15880-3_20
10. Cheng, W., Hüllermeier, E., Waegeman, W., Welker, V.: Label ranking with partial abstention based on thresholded probabilistic models. In: Advances in Neural Information Processing Systems, pp. 2501–2509 (2012)
11. Corani, G., Antonucci, A., Zaffalon, M.: Bayesian networks with imprecise probabilities: theory and application to classification. In: Holmes, D.E., Jain, L.C. (eds.) Data Mining: Foundations and Intelligent Paradigms, pp. 49–93. Springer, Heidelberg (2012). https://doi.org/10.1007/978-3-642-23166-7_4
12. Dechter, R.: Constraint Processing. Morgan Kaufmann, San Mateo (2003)
13. Dekel, O., Manning, C.D., Singer, Y.: Log-linear models for label ranking. In: Advances in Neural Information Processing Systems (2003)
14. Demšar, J.: Statistical comparisons of classifiers over multiple data sets. J. Mach. Learn. Res. 7, 1–30 (2006)
15. Destercke, S., Dubois, D., Chojnacki, E.: Unifying practical uncertainty representations: I. Generalized p-boxes. Int. J. Approximate Reasoning 49, 649–663 (2008)
16. Destercke, S.: On the median in imprecise ordinal problems. Ann. Oper. Res. 256(2), 375–392 (2016). https://doi.org/10.1007/s10479-016-2253-x
17. Destercke, S., Dubois, D.: Special cases. In: Introduction to Imprecise Probabilities, pp. 79–92 (2014)
18. Destercke, S., Masson, M.H., Poss, M.: Cautious label ranking with label-wise decomposition. Eur. J. Oper. Res. 246(3), 927–935 (2015)
19. Destercke, S., Yang, G.: Cautious ordinal classification by binary decomposition. In: Calders, T., Esposito, F., Hüllermeier, E., Meo, R. (eds.) ECML PKDD 2014. LNCS (LNAI), vol. 8724, pp. 323–337. Springer, Heidelberg (2014). https://doi.org/10.1007/978-3-662-44848-9_21
20. Elisseeff, A., Weston, J.: Kernel methods for multi-labelled classification and categorical regression problems. In: Advances in Neural Information Processing Systems, pp. 681–687. MIT Press, Cambridge (2002)
21. Frank, A., Asuncion, A.: UCI machine learning repository (2010). http://archive.ics.uci.edu/ml
22. Frank, E., Hall, M.: A simple approach to ordinal classification. In: De Raedt, L., Flach, P. (eds.) ECML 2001. LNCS (LNAI), vol. 2167, pp. 145–156. Springer, Heidelberg (2001). https://doi.org/10.1007/3-540-44795-4_13
23. Geng, X.: Multilabel ranking with inconsistent rankers. In: Proceedings of CVPR 2014 (2014)
24. Hüllermeier, E., Furnkranz, J., Cheng, W., Brinker, K.: Label ranking by learning pairwise preferences. Artif. Intell. 172, 1897–1916 (2008)
25. King, R., Feng, C., Sutherland, A.: StatLog: comparison of classification algorithms on large real-world problems. App. Artif. Intell. 9(3), 289–333 (1995)
26. Lacoste-Julien, S.: Convergence rate of Frank-Wolfe for non-convex objectives. arXiv preprint arXiv:1607.00345 (2016)
27. Li, L., Lin, H.T.: Ordinal regression by extended binary classification. In: Advances in Neural Information Processing Systems, pp. 865–872 (2007)
28. López-Ortiz, A., Quimper, C.G., Tromp, J., Van Beek, P.: A fast and simple algorithm for bounds consistency of the all different constraint. In: IJCAI, vol. 3, pp. 245–250 (2003)
29. Steinwart, I., Christmann, A., et al.: Estimating conditional quantiles with the help of the pinball loss. Bernoulli 17(1), 211–225 (2011)

30. Vembu, S., Gärtner, T.: Label ranking algorithms: a survey. In: Fürnkranz, J., Hüllermeier, E. (eds.) Preference Learning, pp. 45–64. Springer, Heidelberg (2010). https://doi.org/10.1007/978-3-642-14125-6_3
31. Walley, P.: Statistical Reasoning with Imprecise Probabilities. Chapman and Hall, New York (1991)
32. Walley, P.: Inferences from multinomial data: learning about a bag of marbles. J. Roy. Stat. Soc.: Ser. B (Methodol.) **58**(1), 3–34 (1996)
33. Weskamp, N., Hullermeier, E., Kuhn, D., Klebe, G.: Multiple graph alignment for the structural analysis of protein active sites. IEEE/ACM Trans. Comput. Biol. Bioinf. **4**(2), 310–320 (2007)
34. Yang, G., Destercke, S., Masson, M.H.: The costs of indeterminacy: how to determine them? IEEE Trans. Cybern. **47**(12), 4316–4327 (2016)
35. Zaffalon, M.: The naive credal classifier. J. Stat. Plann. Infer. **105**(1), 5–21 (2002)
36. Zaffalon, M., Corani, G., Mauá, D.: Evaluating credal classifiers by utility-discounted predictive accuracy. Int. J. Approximate Reasoning **53**(8), 1282–1301 (2012)

Approximating General Kernels by Extended Fuzzy Measures: Application to Filtering

Sébastien Destercke[1]([✉]), Agnès Rico[2], and Olivier Strauss[3]

[1] Sorbonne Universités, UTC, CNRS, Heudiasyc, 57 Avenue de Landshut, Compiègne, France
`sebastien.destercke@hds.utc.fr`
[2] ERIC, Université Claude Bernard Lyon 1, Lyon, France
`agnes.rico@univ-lyon1.fr`
[3] LIRMM, Université de Montpellier, Montpellier, France
`olivier.strauss@lirmm.fr`

Abstract. Convolution kernels are essential tools in signal processing: they are used to filter noisy signal, interpolate discrete signals, …. However, in a given application, it is often hard to select an optimal shape of the kernel. This is why, in practice, it may be useful to possess efficient tools to perform a robustness analysis, talking the form in our case of an imprecise convolution. When convolution kernels are positive, their formal equivalence with probability distributions allows one to use imprecise probability theory to achieve such an imprecise convolution. However, many kernels can have negative values, in which case the previous equivalence does not hold anymore. Yet, we show mathematically in this paper that, while the formal equivalence is lost, the computational tools used to describe sets of probabilities by intervals on the singletons still retain their key properties when used to approximate sets of (possibly) non-positive kernels. We then illustrate their use on a single application that consists of filtering a human electrocardiogram signal by using a low-pass filter whose order is imprecisely known. We show, in this experiment, that the proposed approach leads to tighter bounds than previously proposed approaches.

Keywords: Signal filtering · Probability intervals · Signed fuzzy measures · Interval-valued filtering

1 Introduction

Filtering a signal aims at removing some unwanted components or features, i.e. other signals or measurement noise. It is a common problem in both digital analysis and signal processing [5]. In this context, kernels are used for impulse response modelling, interpolation, linear and non-linear transformations, stochastic or band-pass filtering, etc. However, how to choose a particular kernel and its parameters to filter a given signal is often a tricky question.

© Springer Nature Switzerland AG 2020
M.-J. Lesot et al. (Eds.): IPMU 2020, CCIS 1238, pp. 112–123, 2020.
https://doi.org/10.1007/978-3-030-50143-3_9

A way to circumvent this difficulty is to filter the signal with a convex set of kernels, thus ending up with a set-valued signal, often summarized by lower/upper bounds. The set of kernels has to be convex due to the fact that, if two kernels are suitable to achieve the filtering, a combination of those two kernel should be suitable too. A key problem is then to propose a model to approximate this convex set of kernels in a reasonable way (i.e., without losing too much information) that will perform the set-valued filtering in an algorithmic efficient way guaranteeing the provided bounds (in the sense that the sets of signals obtained by filtering with each kernel of the set is contained within the bounds).

In the case of positive summative kernels, i.e., positive functions summing up to one, previous works used the formal equivalence between such kernels and probabilities at their advantage, and have proposed to use well-known probability set models as approximation tools. For example, maxitive [6] and cloudy [4] kernels respectively use possibility distributions and generalized p-boxes to model sets of kernels, and have used the properties of the induced lower measure on events to propose efficient filtering solutions from a computational standpoint.

However, when the kernel set to approximate contains functions that are not positive everywhere (but still sum up to one), this formal analogy is lost, and imprecise probabilistic tools cannot be used straightforwardly to model the set of kernels. Yet, recent works [9] have shown that in some cases such imprecise models can be meaningfully extended to accommodate negative values, while preserving the properties that makes them interesting for signal filtering (i.e., the guarantees of obtained bounds and the algorithmic efficiency). More formally, this means that we have to study the extension of Choquet-integral based digital filtering to the situation where kernels $\kappa : \mathscr{X} \to [-A, B] \subseteq \mathbb{R}$ can be any (bounded) function.

In this work, we show that this is also true for another popular model, namely probability intervals [1], that consists in providing lower/upper bound on singleton probabilities. In particular, while principle applied to sets of additive but possibly negative measures lead to a model inducing a non-monotone set function, called *signed measure*, we show that the Choquet integral of such a measure still leads to interesting bounds for the filtered signal, in the sense that these bounds are guaranteed and are obtained for specific additive measure dominated by the signed measure. Let us call this new kind of interval-valued kernels *imprecise kernels*.

The paper is structured as follows. Section 2 recalls the setting we consider, as well as a few preliminaries. We demonstrate in Sect. 3 that probability intervals can be meaningfully extended to accommodate sets of additive but non-positive measures. Section 4 shows how these results can be applied to numerical signal filtering.

2 Preliminaries: Filtering, Signed Kernels and Fuzzy Measures

We assume a finite space $\mathscr{X} = \{x_1, \ldots, x_n\}$ of n points, that is a subset of an infinite discrete space Ω that may be a discretization of a continuous space (e.g., the real line), and an observed signal whose values at these points are $f(x_1), \ldots, f(x_n)$, that can represent a time- or space-dependent record (EEG signal, sound, etc.). A kernel is here a bounded discrete function $\eta : \Omega \to [-A, B]$, often computed from a continuous kernel (corresponding, for example, to assumed filter impulse response). This kernel is such that $\sum_{x \in \Omega} \eta(x) = 1$, and for a given kernel we will denote b_η and a_η the sum of the positive and negative parts of the kernel, respectively. That is:

$$b_\eta = \sum_{x \in \Omega} \max(0, \eta(x)) \quad a_\eta = \sum_{x \in \Omega} \max(0, -\eta(x))$$

Filtering the signal f by using the kernel η consists of estimating the filtered signal $\hat{f}$ at each point x of $\mathcal{X}$ by:

$$\hat{f}(x) = \sum_{i=1}^{n} f(x_i) \eta(x_i - x) = \sum_{i=1}^{n} f(x_i) \eta^x(x_i), \tag{1}$$

where $\forall y \in \mathcal{X}, \eta^x(y) = \eta(y - x)$.

Since filtering f amounts to compute the value of $\hat{f}$ at each point of $\mathcal{X}$, let us simplify the previous formal statements by assuming that, at each point x of $\mathcal{X}$ exists a kernel $\kappa = \eta^x$. Note that the domain of κ can be restricted to $\mathcal{X}$ without any loss of generality. Moreover, $b_\kappa = \sum_{x \in \Omega} \max(0, \kappa(x)) = b_\eta$ and $a_\kappa = \sum_{x \in \Omega} \max(0, -\kappa(x)) = a_\eta$.

From any kernel κ, we can build a set function $\mu_\kappa : 2^{\mathcal{X}} \rightarrow [-a_\kappa, b_\kappa]$ such that $\mu_\kappa(A) = \sum_{x \in A} \kappa(x)$ for any $A \subseteq \mathcal{X}$. It is clear that we have $\mu_\kappa(\mathcal{X}) = 1$ and $\mu_\kappa(\emptyset) = 0$, but that we can have $\mu_\kappa(A) > \mu_\kappa(B)$ with $A \subset B$ (simply consider $A = \emptyset$ and $B = x$ where $\kappa(x) < 0$). Note also that if $A \cap B = \emptyset$, we keep $\mu_\kappa(A \cup B) = \mu_\kappa(A) + \mu_\kappa(B)$, hence the additivity of the measure.

Estimating the value of the signal $\hat{f}$ in a given point $x \in \mathcal{X}$ requires to compute a value $\mathscr{C}_\kappa$ that can be written as a weighted sum $\mathscr{C}_\kappa(f) = \sum_{i=1}^{n} f(x_i) \kappa(x_i)$. If we order and rank the values of f such that $f(x_{(1)}) \leq f(x_{(2)}) \leq \ldots \leq f(x_{(n)})$, this weighted sum can be rewritten

$$\hat{f}(x) = \mathscr{C}_\kappa(f) = \sum_{i=1}^{n} \left(f(x_{(i)}) - f(x_{(i-1)}) \right) \mu_\kappa(A_{(i)}) \tag{2}$$

where $A_{(i)} = \{x_{(i)}, \ldots, x_{(n)}\}$, $f(x_{(0)}) = 0$ and $A_{(1)} = \mathcal{X}$. One can already notice the similarity with the usual Choquet integral.

Example 1. Consider a kernel η that is a Hermite polynomial of degree 3, and its sampled version pictured in Fig. 1, with $a = -2$. From the picture, it is obvious that some of the values are negative.

$$\eta(x) = \begin{cases} (a+2)|x|^3 - (a+3)|x|^2 + 1 & \text{if } |x| \leq 1 \\ a|x|^3 - 5a|x|^2 + 8a|x| - 4a & \text{if } 1 \leq |x| \leq 2 \\ 0 & \text{else} \end{cases}$$

In this paper, we consider the case where the ideal η is ill-known, that is we only know that, for each $x \in \mathcal{X}$, η belongs to a convex set $\mathcal{N}$, which entails that κ belongs to a convex set $\mathcal{K}$. This set $\mathcal{N}$ (and $\mathcal{K}$) can reflect, for example, our uncertainty about which kernel should be ideally used (they can vary in shape, bandwidth, etc.). In

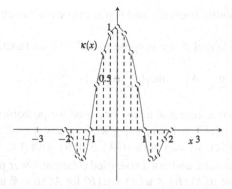

Fig. 1. Sampled Hermite polynomial of degree 3

the following section, we propose to approximate this set $\mathscr{K}$ of kernels that we deem suitable to filter f at point x by a measure $\underline{\mu}_{\mathscr{K}}$ that is close to a fuzzy measure: it shares with such measures the fact that $\underline{\mu}_{\mathscr{K}}(\emptyset) = 0$ and $\underline{\mu}_{\mathscr{K}}(\mathscr{X}) = 1$, and will be a standard fuzzy measure in the case where all $\kappa \in \mathscr{K}$ are positive. However, in the case of negative $\kappa \in \mathscr{K}$, we may have $\underline{\mu}_{\mathscr{K}}(A) < 0$ for some A, as well as have for some couple $A \subset B$ non-monotonicity in the form $\underline{\mu}_{\mathscr{K}}(A) > \underline{\mu}_{\mathscr{K}}(B)$.

3 Approximating Set of Kernels with Extended Probability Intervals

We now consider a set $\mathscr{K}$ of kernels κ defined on $\mathscr{X}$, that can be discretised versions of a set of continuous kernels. We make no assumptions about this set, except for the fact that each kernel $\kappa \in \mathscr{K}$ is bounded and such that $\sum_{x \in \mathscr{X}} \kappa(x) = 1$. If $\mathscr{K}$ contains a large amount of kernels, filtering the signal with each of them and getting the set answer $\{\mathscr{C}_\kappa | \kappa \in \mathscr{K}\}$ for each possible point of the signal can be untractable. Rather than doing so, we may search some efficient way to find some lower and upper bounds of $\{\mathscr{C}_\kappa | \kappa \in \mathscr{K}\}$ that are not too wide.

To achieve such a task, we propose here to use tools inspired by the imprecise probabilistic literature, namely non-additive set functions and the Choquet integral. To use such tools, we must first build a set-function approximating the set $\mathscr{K}$. To do so, we will extend probabilistic intervals [1] that consist in associating lower and upper bounds $[l(x), u(x)]$ to each atom, given a set of probabilities. In our case, for each elements $x \in \mathscr{X}$, we consider the interval-valued kernel

$$\overline{\underline{p}}(x) = [\underline{p}(x), \overline{p}(x)] \tag{3}$$

such that the bounds are given, for each $x \in \mathscr{X}$, as

$$\underline{p}(x) = \inf_{\kappa \in \mathscr{K}} \kappa(x) \text{ and } \overline{p}(x) = \sup_{\kappa \in \mathscr{K}} \kappa(x). \tag{4}$$

Clearly, both can be negative and are not classical probability intervals. Nevertheless, we will show that set-functions induced by these bounds enjoy properties similar to those of standard probability intervals, and hence can be used to efficiently approximate $\{\mathscr{C}_\kappa | \kappa \in \mathscr{K}\}$.

From the imprecise kernel $\overline{\rho}$, we propose to build the set function $\underline{\mu} : \mathscr{X} \to \mathbb{R}$ such that, for $A \subseteq \mathscr{X}$

$$\underline{\mu}_{\mathscr{X}}(A) = \max\{\sum_{x \in A} \underline{\rho}(x), 1 - \sum_{x \notin A} \overline{\rho}(x)\}, \tag{5}$$

which is formally the same equation as the one used for probability intervals. We still have that $\underline{\mu}(\emptyset) = 0$ and $\underline{\mu}(\mathscr{X}) = 1$ (as $\underline{\rho}(x) \leq \kappa(x)$ for any $\mu \in \mathscr{K}$, we necessarily have $\sum_{x \in \mathscr{X}} \underline{\rho}(x) \leq 1$). However, we can have $\underline{\mu}(A) > \underline{\mu}(B)$ with $A \subset B$, meaning that $\underline{\mu}$ is not a classical fuzzy measure, and not a so-called coherent lower probability[1]. It is also non-additive, as we have $\underline{\mu}(A \cup B) \neq \underline{\mu}(A) + \underline{\mu}(B)$ for $A \cap B = \emptyset$ in general.

Simply replacing μ_κ by $\underline{\mu}_{\mathscr{X}}$ in Eq. 2 gives us

$$\underline{\mathscr{C}}_{\mathscr{X}}(f) = \sum_{i=1}^{n} \left(f(x_{(i)}) - f(x_{(i-1)}) \right) \underline{\mu}_{\mathscr{X}}(A_{(i)}) \tag{6}$$

$$= \sum_{i=1}^{n} f(x_{(i)}) \left(\underline{\mu}(A_{(i)}) - \underline{\mu}(A_{(i+1)}) \right), \tag{7}$$

where $A_{(n+1)} = \emptyset$ and $f(x_0) = 0$.

In the positive case, $\underline{\mathscr{C}}_{\mathscr{X}}(f)$ is well-known to outer-approximate $\{\mathscr{C}_\kappa | \kappa \in \mathscr{K}\}$, as $\underline{\mathscr{C}}_{\mathscr{X}}(f) = \inf_{\mu_\kappa \geq \underline{\mu}_{\mathscr{X}}} \mathscr{C}_\kappa(f)$. In the rest of this section, we show that this is also true when the set $\mathscr{K}$ contains non-positive functions. In particular, we will show:

1. that Eq. (6) still provides a lower bound of $\inf_{\mu_\kappa \geq \underline{\mu}_{\mathscr{X}}} \mathscr{C}_\kappa(f)$ and,
2. that this lower bound corresponds to an infimum, meaning that it is obtained for a peculiar additive measure of $\mathscr{K}$.

To show the first point, we will first prove the following proposition concerning $\underline{\mu}$:

Proposition 1. *Given a set $\mathscr{K}$, we have for any $A \subseteq \mathscr{X}$*

$$\underline{\mu}_{\mathscr{X}}(A) \leq \inf_{\kappa \in \mathscr{K}} \mu_\kappa(A)$$

Proof. To prove it, consider a given $\kappa \in \mathscr{K}$, we have $\mu_\kappa(A) = \sum_{x \in A} \kappa(x) = 1 - \sum_{x \notin A} \kappa(x)$ since $\sum_{x \in \mathscr{X}} \kappa(x) = 1$. Now, from Eq. (5), we have either

- $\underline{\mu}_{\mathscr{X}}(A) = \sum_{x \in A} \underline{\rho}(x) \leq \sum_{x \in A} \kappa(x)$
- $\underline{\mu}_{\mathscr{X}}(A) = 1 - \sum_{x \notin A} \overline{\rho}(x) \leq 1 - \sum_{x \notin A} \kappa(x)$

and since this is true for any $\kappa \in \mathscr{K}$, we have the inequality.

Note that $\underline{\mu}$ is a tight measure on singletons, since $\underline{\mu}(x) = \underline{\rho}(x) = \inf_{\kappa \in \mathscr{K}} \kappa(x)$, hence any set-function higher than $\underline{\mu}$ on singletons would not be a lower envelope of $\mathscr{K}$. The fact that $\underline{\mathscr{C}}_{\mathscr{X}}(f) \leq \mathscr{C}_\kappa(f)$ for any $\kappa \in \mathscr{K}$ then simply follows from Proposition 1. If $\mathscr{K}$ reduces to a single kernel κ, then we find back the classical filtering result. Note

---

[1] The lower envelope of a set of probability measures.

that $\mathscr{C}_{\mathscr{H}}(f)$ is equivalent to filter f, that is to compute $\mathscr{C}_{\kappa}$, with the specific kernel $\kappa(x_{(i)}) = \underline{\mu}(A_{(i)}) - \underline{\mu}(A_{(i+1)})$. To prove that $\mathscr{C}_{\mathscr{H}}$ is a tight lower bound, it remains to show that such a kernel is within $\mathscr{H}$.

To show that the bound obtained by Eq. (6) is actually obtained by an additive measure dominated by $\underline{\mu}_{\mathscr{H}}$, we will first show that it still satisfies a convexity property.

Proposition 2. *Given a set $\mathscr{H}$ of kernels, the measure $\underline{\mu}_{\mathscr{H}}$ is 2-monotone and convex, as for every pair $A, B \subseteq \mathscr{X}$ we have*

$$\underline{\mu}_{\mathscr{H}}(A \cup B) + \underline{\mu}_{\mathscr{H}}(A \cap B) \geq \underline{\mu}_{\mathscr{H}}(A) + \underline{\mu}_{\mathscr{H}}(B)$$

Proof. We will mainly adapt the proof from [1, Proposition 5] to the case of non-positive kernels and signed measures, as its mechanism still works in this case.

A key element will be to show that for any two subsets C, D with $C \cap D = \emptyset$, there exists a **single** additive measure μ_{κ} with $\kappa \in \mathscr{H}$ such that

$$\underline{\mu}_{\mathscr{H}}(C) = \mu_{\kappa}(C) \text{ and } \underline{\mu}_{\mathscr{H}}(C \cup D) = \mu_{\kappa}(C \cup D), \tag{8}$$

as if we then take $C = A \cap B$ and $D = (A \cup B) \setminus (A \cap B)$ and choose κ so that it coincides on $\underline{\mu}_{\mathscr{H}}$ for events C, D, we do have

$$\underline{\mu}_{\mathscr{H}}(A \cap B) + \underline{\mu}_{\mathscr{H}}(A \cup B) = \mu_{\kappa}(A \cap B) + \mu_{\kappa}(A \cup B)$$
$$= \mu_{\kappa}(A) + \mu_{\kappa}(B)$$
$$\geq \underline{\mu}_{\mathscr{H}}(A) + \underline{\mu}_{\mathscr{H}}(B).$$

By Eq. (5), we know that

$$\underline{\mu}_{\mathscr{H}}(A) = \max\{\sum_{x \in A} \underline{\rho}(x), 1 - \sum_{x \notin A} \overline{\rho}(x)\}$$

which means that for any event A, we have two possibilities (the two terms of the max). This means four possibilities when considering C and $C \cup D$ together. Here, we will only show that Eq. (8) is true for one of those case, as the proofs for the other cases follow similar reasoning.

So let us consider the case where

$$\underline{\mu}_{\mathscr{H}}(C) = \sum_{x \in C} \underline{\rho}(x) \geq 1 - \sum_{x \notin C} \overline{\rho}(x),$$
$$\underline{\mu}_{\mathscr{H}}(C \cup D) = 1 - \sum_{x \notin C \cup D} \overline{\rho}(x) \geq \sum_{x \in C \cup D} \underline{\rho}(x).$$

Let us now consider the κ distribution such that $\kappa(x) = \underline{\rho}(x)$ if $x \in C$, $\kappa(x) = \overline{\rho}(x)$ if $x \in (C \cup D)^c$, that fits the requirements of Eq. (8) and so far satisfy the constraints on $\mathscr{H}$. To get an additive kernel whose weights sum up to one, we must still assign $\lambda = 1 - \sum_{x \in C} \underline{\rho}(x) - \sum_{x \in (C \cup D)^c} \overline{\rho}(x)$ mass over the singletons composing D. One can see that $\sum_{x \in D} \underline{\rho}(x) \leq \lambda \leq \sum_{x \in D} \overline{\rho}(x)$: for instance, that $\sum_{x \in D} \underline{\rho}(x) \leq \lambda$ immediately follows from the fact that in this sub-case $1 - \sum_{x \notin C \cup D} \overline{\rho}(x) \geq \sum_{x \in C \cup D} \underline{\rho}(x)$. This means that one can choose values $\kappa(x) \in [\underline{\rho}(x), \overline{\rho}(x)]$ for each $x \in D$ such that $\sum_{x \in D} \kappa(x) = \lambda$. So in this case we can build an additive $\kappa \in \mathcal{K}$ with $\sum \kappa(x) = 1$.

That a single additive $\kappa \in \mathcal{K}$ reaching the bounds $\underline{\mu}_{\mathcal{K}}(C)$ and $\underline{\mu}_{\mathcal{K}}(C \cup D)$ can be built in other sub-cases can be done similarly (we refer to [1, Proposition 5], as the proofs are analogous).

Hence $\mathscr{C}_{\mathcal{K}}$ is a signed Choquet integral with respect to the convex capacity $\underline{\mu}_{\mathcal{K}}$. We have $\underline{\mu}_{\mathcal{K}}(\emptyset) = 0$, so according to [10, Theorem 3], $\mathscr{C}_{\mathcal{K}}$ is the minimum of the integrals or expectations taken with respect to the additive measures dominating $\underline{\mu}_{\mathcal{K}}$, i.e., $\mathscr{C}_{\mathcal{K}}(f) = \min\{\mathscr{C}_\mu | \mu \in core(\underline{\mu}_{\mathcal{K}}\}$, where the core of a capacity is the set of additive set function that lie above the capacity everywhere. This allows us to state the following property.

Proposition 3. $\mathscr{C}_{\underline{\mu}_{\mathcal{K}}}(f) = \min\{\mathscr{C}_\mu | \mu \in core(\underline{\mu}_{\mathcal{K}})\}$.

We have therefore shown that, to approximate the result of filtering with any set of kernels (bounded and with no gain), it is still possible to use tools issued from imprecise probabilistic literature. However, it is even clearer in this case that such tools should not be interpreted straightforwardly as uncertainty models (as set-functions are not monotone, a property satisfied by standard fuzzy measures and coherent lower probabilities), but as convenient and efficient tools to perform robust filtering.

Remark 1. An upper bound $\overline{\mathscr{C}}_{\mathcal{K}}(f)$ can be obtained by using the conjugate capacity $\overline{\mu}_{\mathcal{K}}(A) = 1 - \underline{\mu}_{\mathcal{K}}(A^c)$ in Eq. (6). As $\overline{\mu}_{\mathcal{K}}$ is a concave capacity, we also have $\overline{\mathscr{C}}_{\mathcal{K}}(f) = \max\{\mathscr{C}_\mu | \mu \in anticore(\overline{\mu}_{\mathcal{K}})\}$, where the anticore of a capacity is the set of additive set function that lie below the capacity everywhere. Note that $\mu \in core(\underline{\mu}_{\mathcal{K}})$ is equivalent to $\mu \in anticore(\overline{\mu}_{\mathcal{K}})$.

Table 1. Imprecise kernel example

	x_1	x_2	x_3	x_4	x_5
$\overline{\rho}$	−0.1	0.3	0.9	0.3	−0.1
$\underline{\rho}$	−0.2	0.2	0.8	0.2	−0.2

Example 2. Consider the imprecise kernel $\overline{\rho}$ whose values on $\mathcal{X} = \{x_1, \ldots, x_5\}$ are given in Table 1. For such an imprecise kernel, we have for instance $\underline{\mu}_{\mathcal{K}}(\{x_1\}) = \max(-0.2, 1 - (1.4)) = -0.2$ and $\underline{\mu}_{\mathcal{K}}(\{x_1, x_5\}) = \max(-0.4, 1 - (1.5)) = -0.4$, showing that the defined measure is not monotonic with inclusion.

Finally, it should be noted that applying Eq. (6) does not require to evaluate our lower measure on every possible events, but only in a linear number of them (once function values have been ordered). Moreover, evaluating the value of this lower measure on any interval is quite straightforward given Eq. (5). So, even though the measure is non-additive (and not necessarily monotonic), evaluating the filtered values can be done quite efficiently.

4   Illustration of Signed Filtering on a Real Case

We now illustrate the use of our method on a real case scenario involving the filtering of human electrocardiogram (ECG) signals, using data initially collected to detect heart conditions under different settings [7]. ECG signals contain many types of noises - e.g. baseline wander, power-line interference, electromyographic (EMG) noise, electrode motion artifact noise, etc. Baseline wander is a low-frequency noise of around 0.5 to 0.6 Hz that is usually removed during the recording by a high-pass filtering of cut-off frequency 0.5 to 0.6 Hz. EMG noise, which is a high frequency noise of above 100 Hz, may be removed by a digital low-pass filter with an appropriate cut-off frequency. In [7] they propose to use a cut-off frequency of 45 Hz to preprocess the ECG signals. The noisy ECG signal to be filtered is presented in Fig. 2.

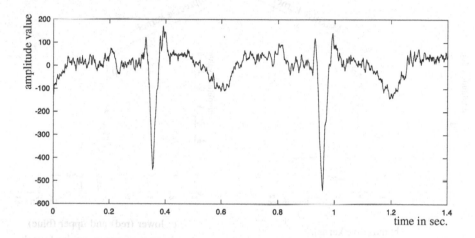

Fig. 2. The ECG signal to be low-pass filtered.

To prevent phase distortion in the bandpass, a Butterworth kernel should preferably be chosen. Moreover, since the signal has not to be processed on line, a symmetric Butterworth filter can be used, that is the combination of a causal and an anti-causal Butterworth kernel. Using such an even kernel prevents from phase delay.

In this experiment, we suppose that the 45 Hz cutoff frequency proposed in [7] is appropriate while the suitable order of the kernel is imprecisely known. Figure 3 presents the superposition of 13 kernels that are the impulse responses of the 13 symmetric lowpass Butterworth kernels of orders 1 to 13. Figure 3.a shows the superimposed kernels, that constitute the set $\mathcal{N}$ of kernels we have to approximate.

Applying our approximation to $\mathcal{N}$ provides the upper $(\overline{\rho})$ and lower $(\underline{\rho})$ bounds of the imprecise kernel that are pictured in Fig. 3.c, with the lower in red, the upper in blue. These bounds are simply obtained by computing $\underline{\rho} = \min_{n=1...13} \eta_n$ and $\overline{\rho} = \max_{n=1...13} \eta_n$ where η_n is the impulse response of the lowpass symmetric Butterworth kernel of order n with cutoff frequency equal to 45 Hz.

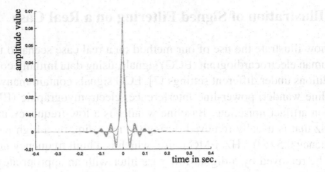

a: 13 superimposed symmetric Butterworth kernel impulse responses.

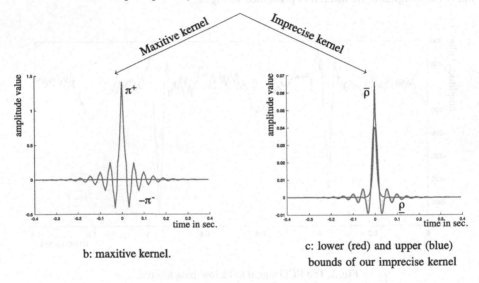

b: maxitive kernel.

c: lower (red) and upper (blue)
bounds of our imprecise kernel

Fig. 3. Original kernel family and corresponding imprecise models. (Color figure online)

To have a comparison point, we will also apply to the same signal the maxitive app-roach proposed in [9], where a signed kernel is approximated by a couple of extended possibility distributions (π^-, π^+). This couple of functions is computed in this way: $\pi^+ = \max_{n=1...13} \pi_n^+$, where π_n^+ is the most specific maxitive kernel that dominates $\eta_n^+ = \max(0, \eta_n)$ and $\pi^- = \max_{n=1...13} \pi_n^-$, where π_n^- is the most specific maxitive ker-nel that dominates $\eta_n^- = \max(0, -\eta_n)$ (see [9] Equation (4)). Figure 3.b plots π^+ (in blue) and $-\pi^-$ (in red). One can readily notice that, if their shape are similar, their boundary values are quite different (the imprecise maxitive kernel varying between -0.5 and 1.5, and our imprecise kernel between -0.01 and 0.07).

In Figs. 5 and 4, we have plotted the ECG signal of Fig. 2 filtered by the 13 kernels of Fig. 3.a, as well as the imprecise signal obtained by using the most specific signed maxitive kernel defined by the couple of functions (π^-, π^+), and the imprecise signal obtained by using the imprecise kernel plotted in Fig. 3.c, respectively. The upper bounds of the imprecise filtered signals are plotted in blue while their lower bounds is plotted in red.

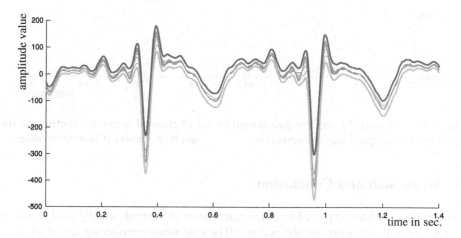

Fig. 4. The ECG signal being low-pass filtered by the 13 classical symmetric Butterworth (in cyan) and by the imprecise kernel (in red - lower and blue - upper). (Color figure online)

It seems obvious, by looking at Fig. 4, that the imprecise signal obtained by this new approach reaches our pursued goal, i.e. the obtained imprecise signal contains all the signals that would have been obtained by using the conventional approach. Moreover, the bounds are reasonably tight, which means that the core of the imprecise kernel is specific enough as an approximation. Indeed, non-parametric imprecise representation of kernels always leads to include unwanted kernels, and may lead to over-conservative bounds.

This is even more patent if we compare it to the signal bounds obtained with the maxitive approach, as this latter one leads to a less specific interval-valued signal. For instance, the values spanned by the interval-valued signal in our approach span from -500 to 200, and -800 to 500 for the maxitive approach. Another possible advantage of our approach is that the Krœnecker impulse is not necessarily included in the described set of kernels, while it is systematically included in a maxitive kernel, meaning that in this latter case the interval-valued signal always include the noisy original signal itself.

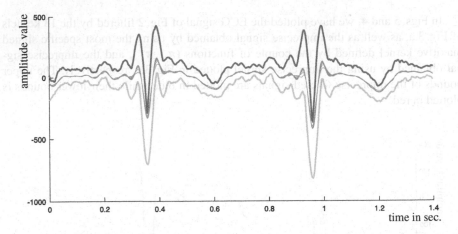

Fig. 5. The ECG signal being low-pass filtered by the 13 classical symmetric Butterworth (in cyan) and by the signed maxitive kernel (in red - lower and blue - upper). (Color figure online)

5   Discussion and Conclusion

In this paper, we have explored to which extent some of the tools usually used to model and reason with sets of probabilities can still be used when considering sets of additive measures that can be negative and fail the monotonicity condition. Such a situation happens, for instance, when filtering a signal.

We have proved that approximating such sets with interval-valued bounds on singletons by extending probability intervals still provides tools that allow on the one hand to use efficient algorithms, and on the other hand to get tight bounds (in the sense that obtained bounds are reached by specific additive measures). We have provided some preliminary experiments showing how our results could be used in filtering problems.

Future works could include the investigation of other imprecise probabilistic models that also offer computational advantages in the case of positive measures, such as using lower and upper bounds over sequences of nested events [3,8]. Complementarily, we could investigate whether computations with some parametric sets of signed kernels can be achieved exactly and efficiently without resorting to an approximation, as can be sometimes done for positive kernels [2].

References

1. De Campos, L.M., Huete, J.F., Moral, S.: Probability intervals: a tool for uncertain reasoning. Int. J. Uncertain. Fuzziness Knowl.-Based Syst. **2**(02), 167–196 (1994)
2. Dendievel, G., Destercke, S., Wachalski, P.: Density estimation with imprecise kernels: application to classification. In: Destercke, S., Denoeux, T., Gil, M.Á., Grzegorzewski, P., Hryniewicz, O. (eds.) SMPS 2018. AISC, vol. 832, pp. 59–67. Springer, Cham (2019). https://doi.org/10.1007/978-3-319-97547-4_9
3. Destercke, S., Dubois, D., Chojnacki, E.: Unifying practical uncertainty representations: I. Generalized p-boxes. Int. J. Approx. Reason. **49**(3), 649–663 (2008)

4. Destercke, S., Strauss, O.: Filtering with clouds. Soft. Comput. **16**(5), 821–831 (2012)
5. Jan, J.: Digital Signal Filtering, Analyses and Restoration. IET (2000)
6. Loquin, K., Strauss, O.: On the granularity of summative kernels. Fuzzy Sets Syst. **159**, 1952–1972 (2008)
7. Poree, F., Bansard, J.-Y., Kervio, G., Carrault, G.: Stability analysis of the 12-lead ECG morphology in different physiological conditions of interest for biometric applications. In: Proceedings of the 36th Annual Computers in Cardiology Conference (CinC), pp. 285–288 (2009)
8. Strauss, O., Destercke, S.: F-boxes for filtering. In: Proceedings of the 7th Conference of the European Society for Fuzzy Logic and Technology, EUSFLAT 2011, Aix-Les-Bains, France, 18–22 July 2011, pp. 935–942 (2011)
9. Strauss, O., Rico, A.: Where the domination of maxitive kernels is extended to signed values. In: 2019 IEEE International Conference on Fuzzy Systems (FUZZ-IEEE), pp. 1–6. IEEE (2019)
10. Waegenaere, A.D., Wakker, P.: Nonmonotonic Choquet integrals. J. Math. Econ. **36**, 45–60 (2001)

Metrical Approach to Measuring Uncertainty

Andrey G. Bronevich[1]([✉])[iD] and Igor N. Rozenberg[2][iD]

[1] National Research University Higher School of Economics,
Myasnitskaya 20, Moscow 101000, Russia
`brone@mail.ru`
[2] JSC "Research and Design Institute for Information Technology,
Signalling and Telecommunications on Railway Transport",
Orlikov per. 5, building 1, Moscow 107996, Russia
`I.Rozenberg@gismps.ru`

Abstract. Many uncertainty measures can be generated by the corresponding divergences, like the Kullback-Leibler divergence generates the Shannon entropy. Divergences can evaluate the information gain obtained by knowing a posterior probability distribution w.r.t. a prior one, or the contradiction between them. Divergences can be also viewed as distances between probability distributions. In this paper, we consider divergences that satisfy a weak system of axioms. This system of axioms does not guaranty additivity of divergences and allows us to consider, for example, the L_α-metric on probability measures as a divergence. We show what kind of uncertainty measures can be generated by such divergences, and how these uncertainty measures can be extended to credal sets.

Keywords: Uncertainty measures · Divergences · Credal sets

1 Introduction

In our experience, we deal with various types of uncertainty. Probability theory allows us to describe conflict in information, other uncertainty theories can generalize it admitting imprecision or non-specificity into models like the theory of imprecise probabilities [1,20] or the theory of belief functions [12,17]. We also need to merge information from different sources, and during this process, it is important to analyze how these sources are contradictory or conflicting. Therefore, we need to measure conflict and non-specificity within each source of information and to measure contradiction among information sources. In probability theory, there are many functionals for evaluating conflict called entropies [9,15,16] and there are many statistical distances called also divergences that can be used for measuring contradiction between probability models [9,16,18,19]. One can notice the visible interactions between various types of uncertainty like contradiction, conflict and non-specificity.

© Springer Nature Switzerland AG 2020
M.-J. Lesot et al. (Eds.): IPMU 2020, CCIS 1238, pp. 124–136, 2020.
https://doi.org/10.1007/978-3-030-50143-3_10

In this paper, we argue that a measure of contradiction (or divergence) is the basic one, and show how other measures of uncertainty can be expressed through it. The paper has the following structure. In Sect. 2, we give some notations and definitions related to probability measures and credal sets. In Sect. 3, we formulate a system of axioms for conflict measures without the requirement of their additivity. In Sect. 4, we introduce a system of axioms for divergences and illustrate them by examples like the Kullback-Leibler divergence, the Rényi divergence, and the g-divergence introduced in the paper. Section 5 is devoted to the question: how such uncertainty measures can be extended on credal sets. The paper finishes with the discussion of obtained results and conclusions.

2   Some Notations and Definitions

Let X be a finite non-empty set and let 2^X be the powerset of X. A set function $P : 2^X \to [0,1]$ is called a *probability measure* on 2^X if 1) $P(\emptyset) = 0$ and $P(X) = 1$; 2) $P(A \cup B) = P(A) + P(B)$ for all disjoint sets $A, B \in 2^X$. A function $p(x) = P(\{x\})$, $x \in X$, is called the *probability distribution*. We see that $P(A) = \sum_{x \in A} p(x)$ for every non-empty set $A \in 2^X$. We say that probabilities are *uniformly distributed* on X if $p(x) = 1/|X|$. The probability measure that corresponds to the uniform probability distribution is denoted by P_u. The set of all probability measures on 2^X is denoted by $M_{pr}(X)$, and we use the notation M_{pr} if if the basic set X can be chosen arbitrary.

We use the following operations on M_{pr}:

a) $P = aP_1 + (1 - a)P_2$ is the convex sum of $P_1, P_2 \in M_{pr}(X)$ with $a \in [0,1]$ if $P(A) = aP_1(A) + (1 - a)P_2(A)$ for all $A \in 2^X$;
b) let X and Y be non-empty finite sets, $\varphi : X \to Y$ be a mapping, and $P \in M_{pr}(X)$, then $P^\varphi \in M_{pr}(Y)$ is defined by $P^\varphi(B) = P(\varphi^{-1}(B))$, where $B \in 2^Y$ and $\varphi^{-1}(B) = \{x \in X | \varphi(x) \in B\}$.

We see that $M_{pr}(X)$ is a convex set and its extreme points are *Dirac measures*, such measures are defined by a probability distribution p_x, $x \in X$, for which $p_x(y) = 1$ if $x = y$, and $p_x(y) = 0$ otherwise. Let $P_x \in M_{pr}$ be a Dirac measure with the probability distribution p_x, then every $P \in M_{pr}(X)$ with a probability distribution p is represented as the convex sum of Dirac measures: $P = \sum_{x \in X} p(x)P_x$.

Clearly, we can identify every $P \in M_{pr}(X)$, where $X = \{x_1, ..., x_n\}$, with the point $(p(x_1), ..., p(x_n))$ in $\mathbb{R}^n$. A non-empty subset $\mathbf{P} \subseteq M_{pr}(X)$ is called a *credal set* [1] if this subset is closed and convex. Further $Cr(X)$ denotes the family of all possible credal sets in $M_{pr}(X)$. We reserve the notation Cr if the basic set X can be chosen arbitrary. In the theory of imprecise probabilities, the model uncertainty based on credal sets is the general one. Some credal sets can be defined by monotone measures. A set function $\mu : 2^X \to [0,1]$ is called a *monotone measure* [21] if 1) $\mu(\emptyset) = 0$ and $\mu(X) = 1$; 2) $\mu(A) \leqslant \mu(B)$ for every $A, B \in 2^X$ with $A \subseteq B$. A monotone measure μ on 2^X is called a *coherent upper probability* [1, 20] if there is a credal set $\mathbf{P} \in Cr(X)$ such that $\mu(A) = \sup\{P(A)|P \in \mathbf{P}\}$ for all $A \in 2^X$.

In the next, for the sake of notation simplicity, we will write $P(x)$ instead of $P(\{x\})$ omitting the figure brackets.

3   Axioms for a Conflict Measure on M_{pr}

In this paper, we will use axioms for a measure of conflict presented in [6]. Let us observe that in [6] such axioms have been formulated for belief functions and the authors show what it happens if the additivity axiom is dropped. We will formulate these axioms for probability measures.

A *measure of conflict* U_C is the functional $U_C : M_{pr} \to [0, +\infty)$ that satisfies the following axioms.

Axiom 1. $U_C(P) = 0$ for $P \in M_{pr}(X)$ iff $P = P_x$ for some $x \in X$.

Axiom 2. Let $\varphi : X \to Y$ be an injection, then $U_C(P^\varphi) = U_C(P)$ for every $P \in M_{pr}(X)$.

Axiom 3. Let $\varphi : X \to Y$ be an arbitrary mapping, then $U_C(P^\varphi) \leqslant U_C(P)$ for every $P \in M_{pr}(X)$.

Axiom 4. Let $P_1, P_2 \in M_{pr}(X)$ and $a \in [0, 1]$, then $U_C(aP_1 + (1 - a)P_2) \geqslant aU_C(P_1) + (1 - a)U_C(P_2)$.

Let us discuss the above axioms. Axiom 1 postulates that the conflict is equal to zero only for the information without uncertainty. Axiom 2 accumulates two known axioms for the Shannon entropy [16]. If $\varphi : X \to Y$ is the bijection, then Axiom 2 says that $U_C(P)$ does not depend on how elements of X are labeled. Let $\varphi : X \to Y$ be an injection such that $X \subseteq Y$ and $\varphi(x) = x$ for all $x \in X$, then P^φ has the following probability distribution: $P^\varphi(x) = P(x)$ if $x \in X$ and $P^\varphi(x) = 0$ otherwise. Thus, in this case Axiom 2 postulates that adding dummy elements to the set X has no influence on U_C values. Axiom 3 says that the conflict is not increases after a mapping. Notice that such a mapping can produce a loss of information, when two separate elements can map to the same element in Y. Axiom 4 shows the behavior of U_C after merging sources of information using the mixture rule of aggregation.

In [6], a reader can find the theorem that fully characterizes a system of functions that defines U_C. In this paper, we only give some examples of U_C discussed in [6].

Example 1. Let $f : [0, 1] \to [0, +\infty)$ be a concave function with the following properties:

1) $f(0) = f(1) = 0$;
2) $f(t)$ is strictly decreasing at $t = 1$.

Then $U_C(P) = \sum_{x \in X} f(P(x))$, where $P \in M_{pr}(X)$, is a measure of conflict on M_{pr}.

Some notable examples of this class of conflict measures are the Shannon entropy, when $f(t) = -t\log_2 t$, and the Gini index, when $f(t) = t(1-t)$.

Example 2. The functional U_C, defined by $U_C(P) = 1 - \max_{x \in X} P(x)$, where $P \in M_{pr}(X)$, is a conflict measure on M_{pr}.

We will establish later the connections between the conflict measure from Example 2 and the Rényi entropy of order infinity and other functionals for measuring conflict within belief functions.

4 Distances and Entropies in Probability Theory

Although, the Shannon entropy and that the Kullback-Leibler divergence (also called Kullback-Leibler distance or relative entropy) are very popular in probability theory, one can find many other functionals [15,16,18,19] that can be used to measure conflict within and between probability distributions. It is important to say that distances (or more exactly statistical distances) in probability theory are not distances as a reader would expect. They do not always obey the triangular inequality and can be non-symmetric. Such statistical distances measure the conflict (or contradiction) between a prior probability distribution of a random variable and its posterior distribution. We will also consider another possible interpretation of contradiction in Sect. 4. The aim of this section is to illustrate of how such distances or divergences can generate entropies or conflict measures on M_{pr}.

We postulate that a *statistical distance* or *divergence* is the functional $D \colon M_{pr} \times M_{pr} \to [0, +\infty]$ that satisfies the following axioms.

Axiom 5. $D(P_1, P_2) = 0$ for $P_1, P_2 \in M_{pr}(X)$ iff $P_1 = P_2$.

Axiom 6. $D(P_1, P_2) \in [0, +\infty)$ for $P_1, P_2 \in M_{pr}(X)$ if P_1 is absolutely continuous w.r.t. P_2.

Axiom 7. Let $\varphi : X \to Y$ be an injection, then $D(P_1^\varphi, P_2^\varphi) = D(P_1, P_2)$ for every $P_1, P_2 \in M_{pr}(X)$.

Axiom 8. Let P_u define the uniform probability distribution on 2^X, then

$$\sup_{P \in M_{pr}(X)} D(P, P_u) = \max_{x \in X} D(P_x, P_u).$$

Axiom 9. Let $\varphi_n : X \to X_n$ be an injection, in which $X = \{x_1, ..., x_m\}$, $X_n = \{x_1, ..., x_n\}$, $n > m$, $\varphi_n(x) = x$ for all $x \in X$, and let $P_u^{(n)} \in M_{pr}(X_n)$ define the uniform probability distribution on X_n. Then the functional

$$U_C(P) = \lim_{n \to \infty} \left(\sup_{P' \in M_{pr}(X_n)} D(P', P_u^{(n)}) - D(P^{\varphi_n}, P_u^{(n)}) \right), P \in M_{pr}(X),$$

is the conflict measure on M_{pr}.

Let us discuss the introduced axioms. Axiom 5 reflects the behavior of D like a distance or divergence. Axiom 6 allows us to cover the case of the Shannon entropy and the mentioned above interpretation of divergence. Axiom 7 is the similar to Axiom 2 for a measure of conflict. Axiom 8 postulates that the greatest distance between P_u (that symbolizes the highest uncertainty in $M_{pr}(X)$) and a $P \in M_{pr}(X)$ is achieved on P_x (the case, when uncertainty is absent). Axiom 9 establishes the main definition of U_C through D.

We will show several choices of divergences satisfying Axioms 5–9 described in the next subsections.

4.1   Kullback-Leibler Divergence

We remind that the *Kullback-Leibler divergence* (distance) [16,19] is defined for probability measures $P_1, P_2 \in M_{pr}(X)$ by

$$D_1(P_1, P_2) = \sum_{x \in X} P_1(x) \log_2 (P_1(x)/P_2(x)).$$

Computing U_C with $D = D_1$ and $|X| = n$, we get

$$\sup_{P \in M_{pr}(X)} D_1(P, P_u) = \log_2 n,$$

$$H_1(P) = \lim_{n \to \infty} \left[\log_2 n - \sum_{x \in X} P_1(x) \log_2 (n P_1(x)) \right] = - \sum_{x \in X} P_1(x) \log_2 P_1(x).$$

We see that H_1 is the *Shannon entropy*. We don't check Axioms 5–9 for D_1, because Axioms 5–7 are well-known properties of the Kullback-Leibler divergence [16,19], Axiom 8 follows from the fact that $D(P, P_u)$ is a convex function of P.

4.2   Rényi Divergence

The *Rényi divergence* [16,19] is the parametrical generalization of the Kullback-Leibler divergence with the parameter $\alpha \in [0, +\infty]$ defined as

$$D_\alpha(P_1, P_2) = \frac{1}{\alpha - 1} \log_2 \left(\sum_{x \in X} \frac{P_1^\alpha(x)}{P_2^{\alpha-1}(x)} \right) \quad \text{for } \alpha \neq 0, 1, +\infty.$$

For special cases, when $\alpha = 0, 1, +\infty$, D_α is defined by taking the limit on α. If $\alpha \to 1$, then we get the Kullback-Leibler divergence. Analogously,

$$D_0(P_1, P_2) = -\log_2 P_2(A), \text{ where } A = \{x \in X | P_1(x) > 0\},$$

$$D_\infty(P_1, P_2) = \max_{x \in X} \log_2 \frac{P_1(x)}{P_2(x)}.$$

The computation of U_C for $D = D_\alpha$ with $\alpha \neq 0, 1, +\infty$ gives us the result

$$H_\alpha(P) = -\frac{1}{\alpha - 1} \log_2 \left(\sum_{x \in X} P^\alpha(x) \right).$$

H_α is called the *Rényi entropy of order* α. Let us consider also the special cases $\alpha = 0$ and $\alpha = \infty$. Substituting D by D_0 in the expression of U_C, we get the result

$$H_0(P) = \log_2 |A|, \text{ where } A = \{x \in X | P(x) > 0\}.$$

In this case, H_0 is called the *Hartley entropy*. Computing U_C with $D = D_\infty$, we get result

$$H_\infty(P) = -\log_2 \left(\max_{x \in X} P(x) \right),$$

and H_∞ is called the *min-entropy*. Again, we do not check axioms, because they follow from the known properties of the Rényi divergence and the Rényi entropy.

4.3 g-divergence

Let $g : [0,1] \to [0,1]$ be a convex, strictly increasing and continuous function on $[0,1]$ such that $g(0) = 0$ and $g(1) = 1$. Then the *g-divergence* is the functional

$$D_g(P_1, P_2) = \sum_{x \in X} g(|P_1(x) - P_2(x)|),$$

where $P_1, P_2 \in M_{pr}(X)$. Let us compute the functional U_C generated by D_g. Assume that $X_n = \{x_1, ..., x_n\}$ and $P_u^{(n)}$ defines the uniform probability distribution on 2^X. Then $\sup\limits_{P' \in M_{pr}(X_n)} D(P', P_u^{(n)}) = D(P_x, P_u^{(n)})$ for every $x \in X_n$ and, for our case,

$$D_g(P_x, P_u) = g(1 - 1/n) + (n - 1)g(1/n).$$

Without decreasing generality, consider $P \in M_{pr}(X)$, where $X = \{x_1, ..., x_m\}$, such that $P(x_i) > 0$, $i = 1, ..., m$. Let us choose n such that $P(x_i) > 1/n$, $i = 1, ..., m$. Then

$$D_g(P^{\varphi_n}, P_u^{(n)}) = \sum_{i=1}^{m} g(P(x_i)) + (n - m)g(1/n),$$

and

$$U_C(P) = \lim_{n \to \infty} \left(g(1 - 1/n) + (n - 1)g(1/n) - \sum_{i=1}^{m} g(P(x_i)) - (n - m)g(1/n) \right)$$

$$= \lim_{n \to \infty} \left(g(1 - 1/n) + (m - 1)g(1/n) - \sum_{i=1}^{m} g(P(x_i)) \right) = 1 - \sum_{i=1}^{m} g(P(x_i)).$$

Let us denote

$$H_g(P) = 1 - \sum_{x \in X} g(P(x)),$$

where $P \in M_{pr}(X)$, and call it the *g-entropy*. Next result directly follows from Example 1.

Proposition 1. H_g *is a conflict measure on* M_{pr} *if the function* $f(t) = t - g(t)$, $t \in [0, 1]$, *is strictly decreasing at* $t = 1$.

Proof. Obviously, the expression of H_g can be rewritten using the function f as $H_g(P) = \sum_{x \in X} f(P(x))$. We see that $f(0) = 0$, $f(1) = 1$, f is a concave function on $[0, 1]$, i.e. the all necessary conditions are fulfilled for H_g to be a conflict measure.

Theorem 1. D_g *satisfies Axioms 5–9 on* $M_{pr} \times M_{pr}$ *if the conditions of Proposition 1 are fulfilled.*

Proof. Let us check axioms. We see that the truth of Axioms 5–7 follows from the definition of D_g. Let us check Axiom 8. It is easy to see that $D_g(P, P_u)$ is a convex function of P, i.e. $D_g(aP_1 + (1-a)P_2, P_u) \leqslant aD_g(P_1) + (1-a)D_g(P_2, P_u)$, for every $a \in [0, 1]$, and $P_1, P_2 \in M_{pr}(X)$. Thus, $\sup\limits_{P \in M_{pr}(X)} D_g(P, P_u)$ is achieved on extreme points of $M_{pr}(X)$, i.e.

$$\sup_{P \in M_{pr}(X)} D_g(P, P_u) = \max_{x \in X} D_g(P_x, P_u).$$

Axiom 9 follows from Proposition 1. The theorem is proved.

Let us analyze the range of D_g. We see that $D_g(P_1, P_2)$, where $P_1, P_2 \in M_{pr}(X)$, is convex on both arguments P_1 and P_2, therefore, the maximum is achieved on extreme points of $M_{pr}(X)$ and

$$\sup_{P_1, P_2 \in M_{pr}(X)} D_g(P_1, P_2) = \max_{x, y \in X} D_g(P_x, P_y) = 2.$$

In some cases, it is convenient that D_g should be normalized. Then we use the normalized $\underline{D}_g$ defined by

$$\underline{D}_g(P_1, P_2) = 0.5 \sum_{x \in X} g(|P_1(x) - P_2(x)|).$$

The value $\underline{D}_g(P_1, P_2) = 1$ manifests the highest contradiction between P_1 and P_2. As a rule, in information theory, entropies are normalized. This means that $H(P_u) = 1$ for an entropy $H : M_{pr} \to [0, +\infty)$, where P_u defines the uniform probability distribution on X and $|X| = 2$. Thus, we can introduce the normalized entropy $\underline{H}_g(P)$ as

$$\underline{H}_g(P) = \frac{1 - \sum\limits_{x \in X} g(P(x))}{1 - 2g(0.5)}.$$

4.4 Divergence and Entropy Based on L_α-metric

Let us consider the case, when $g(x) = x^\alpha$, where $\alpha > 1$, then we denote D_g by D_{L_α} and

$$\underline{D}_{L_\alpha}(P_1, P_2) = 0.5 \sum_{x \in X} |P_1(x) - P_2(x)|^\alpha, P_1, P_2 \in M_{pr}(X).$$

We use this notation, because $d_{L_\alpha}(P_1, P_2) = \left(2\underline{D}_{L_\alpha}(P_1, P_2)\right)^{1/\alpha}$ is the L_α-metric on M_{pr}. In this case, the corresponding $\underline{H}_g$ for $g(x) = x^\alpha$ is

$$\underline{H}_{L_\alpha}(P) = \frac{1 - \sum\limits_{x \in X} |P(x)|^\alpha}{1 - 2^{1-\alpha}}.$$

Consider the case $\alpha = 1$. We see that

$$\underline{D}_{L_1}(P_1, P_2) = 0.5 \sum_{x \in X} |P_1(x) - P_2(x)|,$$

and we can find the expression for $\underline{H}_{L_1}$ taking the limit for $\alpha \to 1$. Using l'Hôpital's rule, we get

$$\underline{H}_{L_1}(P) = \lim_{\alpha \to 1} \frac{-\sum\limits_{x \in X} \ln\left(P(x)\right)\left(P(x)\right)^\alpha}{2^{1-\alpha}\ln 2} = -\sum_{x \in X} P(x)\log_2 P(x).$$

Thus, $\underline{H}_{L_1}(P)$ is the Shannon entropy.

4.5   Concluding Remarks

We see that any g-divergence has the properties that any metric has, and the entropies generated by D_{L_α}, $\alpha \geqslant 1$, looks identical to the Rényi entropy H_α of order α:

$$H_\alpha(P) = -\frac{1}{\alpha - 1}\log_2\left(\sum_{x \in X} P^\alpha(x)\right), \underline{H}_{L_\alpha}(P) = \frac{1}{1 - 2^{1-\alpha}}\left(1 - \sum_{x \in X} P^\alpha(x)\right), \alpha > 0;$$

$$\underline{H}_{L_1}(P) = H_1(P) = -\sum_{x \in X} P(x)\log_2 P(x), \alpha = 0.$$

We see that $\underline{H}_{L_1} = H_1$ is the Shannon entropy. If we denote $t = \sum\limits_{x \in X} P^2(x)$, then $H_\alpha(P) = -\frac{1}{1-\alpha}\log_2 t$ and $\underline{H}_{L_\alpha}(P) = \frac{1}{1-2^{1-\alpha}}(1 - t)$. We see that both functions $\varphi_1(t) = -\frac{1}{1-\alpha}\log_2 t$ and $\varphi_2(t) = \frac{1}{1-2^{1-\alpha}}(1 - t)$ are decreasing on $[0, 1]$, therefore, H_α and $\underline{H}_{L_\alpha}$ similarly discriminate uncertainty. Formally, we can take the L_α-metric

$$d_{L_\alpha}(P_1, P_2) = \left(\sum_{x \in X} |P_1(x) - P_1(x)|^\alpha\right)^{1/\alpha}, \alpha \geqslant 1,$$

as a divergence. The divergence d_{L_α} generates the entropy

$$\underline{h}_{L_\alpha}(P) = \frac{1}{1 - 2^{\frac{1-\alpha}{\alpha}}}\left(1 - \left(\sum_{x \in X} P^\alpha(x)\right)^{1/\alpha}\right), \alpha > 1,$$

that is identical to the Shannon entropy if $\alpha \to 1$. We can also define $d_{L_\alpha}(P_1, P_2)$ and $\underline{h}_{L_\alpha}(P)$ if $\alpha \to +\infty$. Then

$$d_{L_\infty}(P_1, P_2) = \max_{x \in X} |P_1(x) - P_2(x)|, \underline{h}_{L_\infty}(P) = 2\left(1 - \max_{x \in X} P(x)\right).$$

5   Uncertainty Measures on Credal Sets

The main idea is to use divergences or distances on probability measures and express through them other measures of uncertainty. As it was established by many papers (see [2] and the references there), imprecise probabilities describe two types of uncertainty: non-specificity and conflict. A pure conflict is described by the classical probability theory, non-specificity corresponds to the choice of a probability model among possible ones. It is also important to evaluate the degree of total uncertainty that aggregates conflict and non-specificity. In previous sections, we have described the choice of two functionals $U_C : M_{pr} \to [0, +\infty)$ and $D : M_{pr} \times M_{pr} \to [0, +\infty]$. The functional U_C evaluates the amount conflict if uncertainty in information described by a probability measure, and D describes the contradiction between two sources of information in a probabilistic setting. It is important to distinguish two possible interpretations of divergence. The first one related to the Rényi and Kullback-Leibler divergences that evaluate the contradiction between a prior probability distribution and a posterior one. Thus, in this case D is not symmetrical, $D(P_1, P_2) = +\infty$ iff P_1 is not absolutely continuous w.r.t. P_2. If D is a g-divergence or the L_α-metric, then D can be used for evaluating difference between probability models, or for finding the degree of how information obtained from two sources is the same. Thus, values of U_C can be understood differently. In the case of the Rényi and Kullback-Leibler divergences, the part $D(P^{\varphi_n}, P_u^{(n)})$ of $U_C^{(1)}(P)$ gives us the amount of information gain after obtaining P^{φ_n} w.r.t. the most uncertain information described by $P_u^{(n)}$ and the part $\sup_{P' \in M_{pr}(X_n)} D(P', P_u^{(n)})$ gives us the maximal information gain, therefore, the difference $\sup_{P' \in M_{pr}(X_n)} D(P', P_u^{(n)}) - D(P^{\varphi_n}, P_u^{(n)})$ evaluates the deficiency of information in P^{φ_n}. If D is understood as a distance, then $\sup_{P' \in M_{pr}(X_n)} D(P', P_u^{(n)})$ is the distance between exact and the most uncertain information, and $D(P^{\varphi_n}, P_u^{(n)})$ characterizes how far P^{φ_n} is located from $P_u^{(n)}$.

Assume that $\mathbf{P} \in Cr(X)$ and $P \in M_{pr}(X)$. Let us introduce the following functionals:

$$\Phi_1(\mathbf{P}, P) = \inf\{D(P', P)|P' \in \mathbf{P}\}, \Phi_2(\mathbf{P}, P) = \sup\{D(P', P)|P' \in \mathbf{P}\}.$$

We see that $\Phi_1(\mathbf{P}, P)$ and $\Phi_2(\mathbf{P}, P)$ give the smallest and highest information gains if the posterior probability distribution is described by the credal set $\mathbf{P}$. According to the Laplace principle the prior information, described by P, has the highest uncertainty if $P = P_u$, where P_u is the uniform probability distribution on 2^X. Thus, $\Phi_1(\mathbf{P}, P_u)$ gives us the amount of information in $\mathbf{P}$. Using the same logic as before, we define the measure U_T of total uncertainty on Cr as

$$U_T(\mathbf{P}) = \lim_{n \to \infty} \left(\sup_{P' \in M_{pr}(X_n)} D(P', P_u^{(n)}) - \inf\{D(P^{\varphi_n}, P_u^{(n)}) | P \in \mathbf{P}\} \right),$$

where $\mathbf{P} \in Cr(X)$ and we use notations and definitions from Axiom 9. If $\mathbf{P} = \{P\}$, then the expression for $U_T(\mathbf{P})$ is simplified to

$$U_T(P) = \lim_{n \to \infty} \left(\sup_{P' \in M_{pr}(X_n)} D(P', P_u^{(n)}) - D(P^{\varphi_n}, P_u^{(n)}) \right),$$

i.e. U_T is the conflict measure U_C defined in Axiom 9. We see that U_T on Cr can be seen as an extension of U_T on M_{pr} by the following formula:

$$U_T(\mathbf{P}) = \sup\{U_T(P) | P \in \mathbf{P}\}.$$

Observe that the functional U_T is called the *upper or maximal entropy* [2, 14] if U_T is the Shannon entropy on M_{pr}. Let us analyze the functional

$$\Phi(\mathbf{P}) = \sup\{D(P, P_u) | P \in \mathbf{P}\} - \inf\{D(P, P_u) | P \in \mathbf{P}\},$$

where $\mathbf{P} \in Cr(X)$ and P_u is the uniform probability distribution on 2^X. We see that it characterizes the amount of uncertainty in choosing a true probability measure in $\mathbf{P}$. Thus, we can choose as a measure of non-specificity the functional

$$U_N(\mathbf{P}) = \lim_{n \to \infty} \left(\sup\{D(P^{\varphi_n}, P_u^{(n)}) | P \in \mathbf{P}\} - \inf\{D(P^{\varphi_n}, P_u^{(n)}) | P \in \mathbf{P}\} \right),$$

where $\mathbf{P} \in Cr(X)$ and we use notations and definitions from Axiom 9. Assume that the measure of total uncertainty aggregates non-specificity and conflict additively, i.e. for every $\mathbf{P} \in Cr$

$$U_C(\mathbf{P}) + U_N(\mathbf{P}) = U_T(\mathbf{P}),$$

where U_C is a measure of conflict on Cr. This assumption implies that

$$U_C(\mathbf{P}) = \lim_{n \to \infty} \left(\sup_{P' \in M_{pr}(X_n)} D(P', P_u^{(n)}) - \sup\{D(P^{\varphi_n}, P_u^{(n)}) | P \in \mathbf{P}\} \right),$$

and we can express $U_C(\mathbf{P})$ through its values on M_{pr} as

$$U_C(\mathbf{P}) = \inf\{U_C(P) | P \in \mathbf{P}\}.$$

Note that U_C is called the *minimal or lower entropy* [2, 14] on Cr if U_C is the Shannon entropy on M_{pr}.

6   Discussion and Conclusion

Let us notice that the axioms for the Rényi divergence can be found in [16]. Although, in this paper, the author consider these axioms as evident ones, the interpretation of some of them seems to be problematic, because they are based on so called generalized probability distributions that are not necessarily normalized. In our approach, divergences and entropies are not necessarily additive, that allows, for example, to use L_α-metrics as such divergences.

The results of the paper allow us to resolve some problems in the theory of belief functions. For example, in this theory [17] the conflict between two sources of information is evaluated using Dempster's rule of aggregation. If sources of information are described by probability measures $P_1, P_2 \in M_{pr}(X)$, then this evaluation is produced by the formula:

$$k(P_1, P_2) = 1 - \sum_{x \in X} P_1(x) P_2(x).$$

We see that $k(P, P) = 1 - \sum_{x \in X} P(x) P(x) = H_{L_2}(P)$ for $P \in M_{pr}(X)$, i.e. $k(P, P)$ is the Gini index that can be interpreted as entropy or conflict measure. Another representation

$$D_{L_2}(P_1, P_2) = \sum_{x \in X} (P_1(x) - P_2(x))^2 = -H_{L_2}(P_1) - H_{L_2}(P_2) + 2k(P_1, P_2)$$

implies that $k(P_1, P_2) = 0.5(H_{L_2}(P_1) + H_{L_2}(P_2) + D_{L_2}(P_1, P_2))$ consists of two parts: the part $0.5(H_{L_2}(P_1) + H_{L_2}(P_2))$ measures conflict within information sources and the part $0.5 D_{L_2}(P_1, P_2)$ measures conflict (contradiction) between information sources. Let us notice that this problem of disaggregating of k on two parts for belief functions is investigated in [11].

We also pay attention on using $\underline{h}_{L_\infty}$ for defining U_C on credal sets. In this case,

$$U_C(\mathbf{P}) = \inf\{\underline{h}_{L_\infty}(P) | P \in \mathbf{P}\} = 2\left(1 - \max_{x \in X} \sup\{P(x) | P \in \mathbf{P}\}\right).$$

If a credal set $\mathbf{P}$ is described by a coherent upper probability μ on 2^X, then the expression for U_C can be rewritten as $U_C(\mu) = 2\left(1 - \max_{x \in X} \mu(x)\right)$. Such a functional (without the factor 2) is proposed for measuring inner conflict in the theory of belief functions [10].

Formally, in this paper we propose to analyze uncertainty by choosing a divergence D, and then to take the compatible with D measures of uncertainty. Let us discuss our approach in detail if we choose d_{L_1} as divergence. We see that we can use d_{L_1} with the Shannon entropy. Let us also look at the following functional:

$$U'_C(P) = \min_{x \in X} D(P_x, P), \quad \text{where } P \in M_{pr}(X).$$

We see that $U'_C(P)$ evaluates the distance between P and the exact information without uncertainty, thus, it can be considered as a candidate for the measure of conflict. If we take $D = d_{L_1}$, then it is possible to show that

$$d_{L_1}(P_1, P_2) = 2\left(1 - \sum_{x \in X} \min\{P_1(x), P_2(x)\}\right).$$

This implies that

$$U'_C(P) = 2\min_{x \in X}(1 - P(x)) = 2\left(1 - \max_{x \in X} P(x)\right) = \underline{h}_{L_\infty}(P).$$

Doing in the same way for $D = D_\alpha$, we get that U'_C coincides in this case with the min-entropy.

Let us consider the functional

$$Con(\mathbf{P}_1, \mathbf{P}_2) = \inf\{D(P_1, P_2) | P_1 \in \mathbf{P}_1, P_2 \in \mathbf{P}_2\}, \mathbf{P}_1, \mathbf{P}_2 \in Cr(X).$$

We see that $Con(\mathbf{P}_1, \mathbf{P}_2) = 0$ iff $\mathbf{P}_1 \cap \mathbf{P}_2 = \emptyset$, and Con can be used for measuring contradiction between information sources described by credal sets. This measure of contradiction for $D = d_{L_1}$ is well-known in the theory of belief functions [3,8,13], its axiomatic can be found in [4,7] and its extension to imprecise probabilities based on generalized credal sets is given in [5,7].

Finalizing our paper, we can conclude that there is the variety of divergences and the corresponding uncertainty measures. The choice one of them can depend on the problem statement or on the complexity of realization, or on their additional desirable properties. We aware that in this paper we do not investigate in detail ways of evaluating conflict, non-specificity and contradiction in information presented by credal sets with the help of divergences. This can be the topic of our next research.

References

1. Augustin, T., Coolen, F.P.A., de Cooman, G., Troffaes, F.C.M. (eds.): Introduction to Imprecise Probabilities. Wiley Series in Probability and Statistics. Wiley, New York (2014)
2. Bronevich, A.G., Klir, G.J.: Measures of uncertainty for imprecise probabilities: an axiomatic approach. Int. J. Approx. Reason. **51**, 365–390 (2010)
3. Bronevich, A.G., Rozenberg, I.N.: The choice of generalized Dempster-Shafer rules for aggregating belief functions. Int. J. Approx. Reason. **56**(Part A), 122–136 (2015)
4. Bronevich, A.G., Rozenberg, I.N.: Conjunctive rules in the theory of belief functions and their justification through decisions models. In: Vejnarová, J., Kratochvíl, V. (eds.) BELIEF 2016. LNCS (LNAI), vol. 9861, pp. 137–145. Springer, Cham (2016). https://doi.org/10.1007/978-3-319-45559-4_14
5. Bronevich, A.G., Rozenberg, I.N.: Modelling uncertainty with generalized credal sets: application to conjunction and decision. Int. J. Gen. Syst. **27**(1), 67–96 (2018)

6. Bronevich, A.G., Lepskiy, A.E.: Clustering a body of evidence based on conflict measures. In: Štěpnička, M. (ed.) Proceedings of the 11th Conference of the European Society for Fuzzy Logic and Technology, vol. 1, pp. 328–333, Atlantis Press, Paris (2019). https://doi.org/10.2991/eusflat-19.2019.47

7. Bronevich, A.G., Rozenberg, I.N.: The contradiction between belief functions: its description, measurement, and correction based on generalized credal sets. Int. J. Approx. Reason. **112**, 119–139 (2019)

8. Cattaneo, M.E.G.V.: Combining belief functions issued from dependent sources. In: Bernard, J.-M., Seidenfeld, T., Zaffalon, M. (eds.), ISIPTA 2003: Proceedings in Informatics, vol. 18, pp. 133–147, Carleton Scientific, Waterloo (2003). https://doi.org/10.3929/ethz-a-004531249

9. Csiszár, I.: Axiomatic characterizations of information measures. Entropy **10**, 261–273 (2008)

10. Daniel, Milan: Conflicts within and between belief functions. In: Hüllermeier, Eyke, Kruse, Rudolf, Hoffmann, Frank (eds.) IPMU 2010. LNCS (LNAI), vol. 6178, pp. 696–705. Springer, Heidelberg (2010). https://doi.org/10.1007/978-3-642-14049-5_71

11. Daniel, Milan: Conflict between belief functions: a new measure based on their nonconflicting parts. In: Cuzzolin, Fabio (ed.) BELIEF 2014. LNCS (LNAI), vol. 8764, pp. 321–330. Springer, Cham (2014). https://doi.org/10.1007/978-3-319-11191-9_35

12. Dempster, A.P.: Upper and lower probabilities induced by multivalued mapping. Ann. Math. Stat. **38**, 325–339 (1967)

13. Destercke, S., Burger, T.: Toward an axiomatic definition of conflict between belief functions. IEEE Trans. Cybern. **43**(2), 585–596 (2013)

14. Klir, G.J.: Uncertainty and Information: Foundations of Generalized Information Theory. Wiley, Hoboken (2006)

15. Morales, D., Pardo, L., Vajda, I.: Uncertainty of discrete stochastic systems: general theory and statistical theory. IEEE Trans. Syst. Man Cybern. **26**, 1–17 (1996)

16. Rényi, A.: On measures of entropy and information. In: Proceedings of the Fourth Berkeley Symposium on Mathematical Statistics and Probability. Contributions to the Theory of Statistics, vol. 1, pp. 547–561, University of California Press, Berkeley (1961)

17. Shafer, G.: A Mathematical Theory of Evidence. Princeton University Press, Princeton (1976)

18. Taneja, I.J.: Generalized symmetric divergence measures and the probability of error. Commun. Stat. - Theory Methods **42**(9), 1654–1672 (2013)

19. Van Erven, T., Harremos, P.: Rényi divergence and Kullback-Leibler divergence. IEEE Trans. Inf. Theory **60**(7), 3797–3820 (2014)

20. Walley, P.: Statistical Reasoning with Imprecise Probabilities. Chapman and Hall, London (1991)

21. Wang, Z., Klir, G.J.: Generalized Measure Theory. Springer, Heidelberg (2009). https://doi.org/10.1007/978-0-387-76852-6

Conditioning and Dilation with Coherent Nearly-Linear Models

Renato Pelessoni[✉][iD] and Paolo Vicig[iD]

University of Trieste (DEAMS), 34123 Trieste, Italy
{renato.pelessoni,paolo.vicig}@deams.units.it

Abstract. In previous work [1] we introduced Nearly-Linear (NL) models, a class of neighbourhood models obtaining upper/lower probabilities by means of a linear affine transformation (with barriers) of a given probability. NL models are partitioned into more subfamilies, some of which are coherent. One, that of the Vertical Barrier Models (VBM), includes known models, such as the Pari-Mutuel, the ε-contamination or the Total Variation model as special instances. In this paper we study conditioning of coherent NL models, obtaining formulae for their natural extension. We show that VBMs are stable after conditioning, i.e. return a conditional model that is still a VBM, and that this is true also for the special instances mentioned above but not in general for NL models. We then analyse dilation for coherent NL models, a phenomenon that makes our *ex-post* opinion on an event A, after conditioning it on any event in a partition of hypotheses, vaguer than our *ex-ante* opinion on A.

Keywords: Conditioning · Coherent imprecise probabilities · Nearly-Linear models · Dilation

1 Introduction

Among special imprecise probability models, *neighbourhood models* [10, Sec. 4.6.5] obtain an upper/lower probability from a given (precise) probability P_0. One reason for doing this may be that P_0 is not considered fully reliable. Even when it is, $P_0(A)$ represents a *fair* price for selling event A, meaning that the buyer is entitled to receive 1 if A is true, 0 otherwise. A seller typically requires a higher price than $P_0(A)$, $\overline{P}(A) \geq P_0(A)$, for selling A. The upper probability $\overline{P}$, to be interpreted as an infimum selling price in the behavioural approach to imprecise probabilities [10], is often obtained as a function of P_0.

Recently, we investigated Nearly-Linear (NL) models [1], a relatively simple class of neighbourhood models. In fact, they derive upper and lower probabilities, $\overline{P}$ and $\underline{P}$ respectively, as linear affine transformations of P_0 with barriers, to prevent reaching values outside the $[0, 1]$ interval. As proven in [1], some NL models are coherent, while other ones ensure weaker properties. The most important coherent NL models are Vertical Barrier Models (VBM), including

© Springer Nature Switzerland AG 2020
M.-J. Lesot et al. (Eds.): IPMU 2020, CCIS 1238, pp. 137–150, 2020.
https://doi.org/10.1007/978-3-030-50143-3_11

several well-known models as special cases, such as the Pari-Mutuel, the Total Variation, the ε-contamination model, and others.

In this paper we explore the behaviour of coherent NL models when conditioning. Precisely, after recalling essential preliminary notions in Sect. 2, in Sect. 3 the events in the (unconditional) domain of $\underline{P}$, $\overline{P}$ are conditioned on an event B, and the lower/upper natural extensions $\underline{E}$, $\overline{E}$ of $\underline{P}$, $\overline{P}$ on the new (conditional) environment are computed. The natural extension is a standard inferential procedure in the theory of imprecise probabilities [10], which gives the least-committal coherent extension of a lower (upper) probability. Since $\underline{P}$ ($\overline{P}$) is 2-monotone (2-alternating) in a coherent NL model, $\underline{E}$ ($\overline{E}$) is given by easy-to-apply formulae and is 2-monotone (2-alternating) too. An interesting result is that VMBs are *stable* after conditioning: the conditional model after applying the natural extension is still a VBM. We show that this property extends also to the mentioned special VBMs: conditioning each special VBM model returns a special VBM model of the same kind. By contrast, the property does not hold for other NL models. In Sect. 4 we explore the phenomenon of dilation, where, given a partition of events $\mathcal{B}$, it happens for some event A that $\underline{E}(A|B) \leq \underline{P}(A) \leq \overline{P}(A) \leq \overline{E}(A|B)$, for all $B \in \mathcal{B} \setminus \{\emptyset\}$. This means that our *a posteriori* evaluations are vaguer than the *a priori* ones. We derive a number of conditions for dilation to occur or not to occur. Section 5 concludes the paper.

2   Preliminaries

In this paper we shall be concerned with coherent lower and upper probabilities, both conditional and unconditional. Coherent means in both cases *Williams-coherent* [11], in the structure-free version studied in [6]:

Definition 1. *Let $\mathcal{D} \neq \emptyset$ be an arbitrary set of conditional events. A conditional lower probability $\underline{P} : \mathcal{D} \to \mathbb{R}$ is coherent on $\mathcal{D}$ iff $\forall n \in \mathbf{N}, \forall s_0, s_1, \ldots, s_n \geq 0, \forall A_0|B_0, A_1|B_1, \ldots, A_n|B_n \in \mathcal{D}$, defining $\underline{G} = \sum_{i=1}^n s_i I_{B_i}(I_{A_i} - \underline{P}(A_i|B_i)) - s_0 I_{B_0}(I_{A_0} - \underline{P}(A_0|B_0))$, $B = \bigvee_{i=0}^n B_i$ (I_A: indicator of event A), it holds that $\max\{\underline{G}|B\} \geq 0$.*

A similar definition applies to upper probabilities. However, when considering simultaneously lower and upper probabilities, they will be *conjugate*, i.e.

$$\underline{P}(A|B) = 1 - \overline{P}(A^c|B). \tag{1}$$

Equation (1) lets us refer to lower (alternatively upper) probabilities only.

When $\mathcal{D}$ is made of unconditional events only, Definition 1 coincides with Walley's coherence [10]. In general, a (Williams-)coherent $\underline{P}$ on $\mathcal{D}$ has a coherent extension, not necessarily unique, on any set of conditional events $\mathcal{D}' \supset \mathcal{D}$.

The *natural extension* $\underline{E}$ of $\underline{P}$ on $\mathcal{D}'$ is the *least-committal* coherent extension of $\underline{P}$ to $\mathcal{D}'$, meaning that if Q is a coherent extension of $\underline{P}$, then $\underline{E} \leq Q$ on $\mathcal{D}'$. Further, $\underline{E} = \underline{P}$ on $\mathcal{D}$ iff $\underline{P}$ is coherent [6,10].

In this paper, we shall initially be concerned with *unconditional* lower probabilities ($\underline{P}(\cdot)$) and their conjugates ($\overline{P}(\cdot)$).

Coherence implies that [10, Sec. 2.7.4]

$$\text{if } A \Rightarrow B, \underline{P}(A) \leq \underline{P}(B), \overline{P}(A) \leq \overline{P}(B) \text{ (monotonicity)}$$
$$\underline{P}(A) + \overline{P}(B) \geq \underline{P}(A \vee B). \tag{2}$$

The domain $\mathcal{D}$ of $\underline{P}(\cdot)$, $\overline{P}(\cdot)$ will often be $\mathcal{A}(\mathbb{P})$, the set of events logically dependent on a given partition $\mathbb{P}$ (the powerset of $\mathbb{P}$, in set theoretic language).

A lower probability $\underline{P}$, coherent on $\mathcal{A}(\mathbb{P})$, is 2-*monotone* if $\underline{P}(A \vee B) \geq \underline{P}(A) + \underline{P}(B) - \underline{P}(A \wedge B)$, $\forall A, B \in \mathcal{A}(\mathbb{P})$. Its conjugate $\overline{P}$ is 2-*alternating*, meaning that $\overline{P}(A \vee B) \leq \overline{P}(A) + \overline{P}(B) - \overline{P}(A \wedge B)$, $\forall A, B \in \mathcal{A}(\mathbb{P})$.

2-monotone and 2-alternating coherent imprecise probabilities have some special properties [8–10]. In particular,

Proposition 1. ([9, Thm. 7.2],[10, Sec. 6.4.6]). *If $\underline{P}$ is a coherent 2-monotone lower probability on $\mathcal{A}(\mathbb{P})$ and $\overline{P}$ is its conjugate, given $B \in \mathcal{A}(\mathbb{P})$ such that $\underline{P}(B) > 0$, then, $\forall A \in \mathcal{A}(\mathbb{P})$,*

$$\overline{E}(A|B) = \frac{\overline{P}(A \wedge B)}{\overline{P}(A \wedge B) + \underline{P}(A^c \wedge B)} \tag{3}$$

$$\underline{E}(A|B) = \frac{\underline{P}(A \wedge B)}{\underline{P}(A \wedge B) + \overline{P}(A^c \wedge B)} \tag{4}$$

$\underline{E}$ *is 2-monotone ($\overline{E}$ is 2-alternating) on $\mathcal{A}(\mathbb{P})|B = \{A|B : A \in \mathcal{A}(\mathbb{P})\}$, where B is fixed. $\underline{E}$, $\overline{E}$ are conjugate.*

2.1 Nearly-Linear Models

Nearly-Linear models have been defined in [1], where their basic properties have been investigated.

Definition 2. *A Nearly-Linear Model is a couple $(\underline{P}, \overline{P})$ of conjugate lower and upper probabilities on $\mathcal{A}(\mathbb{P})$, where $\forall A \in \mathcal{A}(\mathbb{P}) \setminus \{\emptyset, \Omega\}$*

$$\underline{P}(A) = \min\{\max\{bP_0(A) + a, 0\}, 1\}, \tag{5}$$
$$\overline{P}(A) = \max\{\min\{bP_0(A) + c, 1\}, 0\} \tag{6}$$

and $\underline{P}(\emptyset) = \overline{P}(\emptyset) = 0$, $\underline{P}(\Omega) = \overline{P}(\Omega) = 1$.

In Eqs. (5), (6), P_0 is an assigned (precise) probability on $\mathcal{A}(\mathbb{P})$, while

$$b > 0, \ a \in \mathbb{R}, \ c = 1 - (a + b). \tag{7}$$

It has been shown in [1, Sec. 3.1] that NL models are partitioned into three subfamilies, with varying consistency properties. Here we focus on the coherent NL models, which are all the models in the VBM subfamily and some of the HBM (to be recalled next), while, within the third subfamily, $\underline{P}$ and $\overline{P}$ are coherent iff the cardinality of $\mathbb{P}$ is 2 (therefore we neglect these latter models).

Definition 3. *A* Vertical Barrier Model *(VBM) is a NL model where* (5), (6) *specialise into*

$$\underline{P}(A) = \max\{bP_0(A) + a, 0\}, \ \forall A \in \mathcal{A}(I\!P) \setminus \{\Omega\}, \ \underline{P}(\Omega) = 1 \qquad (8)$$

$$\overline{P}(A) = \min\{bP_0(A) + c, 1\}, \ \forall A \in \mathcal{A}(I\!P) \setminus \{\emptyset\}, \ \overline{P}(\emptyset) = 0 \qquad (9)$$

$$0 \leq a + b \leq 1, a \leq 0 \qquad (10)$$

and c is given by (7) *(hence c $\geq$ 0).*

In a Horizontal Barrier Model *(HBM)* $\underline{P}$, $\overline{P}$ *are given by Definition 2, hence by* (5), (6), (7) $\forall A \in \mathcal{A}(I\!P) \setminus \{\emptyset, \Omega\}$, *where a, b satisfy the constraints*

$$a + b > 1, \ b + 2a \leq 1 \qquad (11)$$

(implying a < 0, b > 1, c < 0).

Proposition 2. *([1])* $\underline{P}$, $\overline{P}$ *are coherent and 2-monotone, respectively 2-alternating in any VBM; in a HBM they are so iff* $\overline{P}$ *is subadditive (i.e.* $\overline{P}(A) + \overline{P}(B) \geq \overline{P}(A \vee B), \forall A, B \in \mathcal{A}(I\!P)$).

Thus, VBMs and (partly) HBMs ensure very good consistency properties, while being relatively simple transformations of an assigned probability P_0. Further, a VBM generalises a number of well-known models. Among them we find:

- if $a + b = 0$, the *vacuous lower/upper probability model* [10, Sec. 2.9.1]:

$$\underline{P}_V(A) = 0, \forall A \neq \Omega, \ \underline{P}_V(\Omega) = 1,$$
$$\overline{P}_V(A) = 1, \forall A \neq \emptyset, \ \overline{P}_V(\emptyset) = 0;$$

- if $a = 0$, $0 < b < 1$ (hence $c = 1 - b > 0$), the *ε-contamination model* or linear-vacuous mixture model [10, Sec. 2.9.2], here $b = 1 - \varepsilon$:

$$\underline{P}_\varepsilon(A) = (1 - \varepsilon)P_0(A), \ \forall A \neq \Omega, \ \underline{P}_\varepsilon(\Omega) = 1,$$
$$\overline{P}_\varepsilon(A) = (1 - \varepsilon)P_0(A) + \varepsilon, \ \forall A \neq \emptyset, \ \underline{P}_\varepsilon(\emptyset) = 0;$$

- if $b = 1 + \delta > 1$, $a = -\delta < 0$ (hence $c = 0$), the *Pari-Mutuel Model* [4,7], [10, Sec. 2.9.3]:

$$\underline{P}_{PMM}(A) = \max\{(1 + \delta)P_0(A) - \delta, 0\}, \qquad (12)$$
$$\overline{P}_{PMM}(A) = \min\{(1 + \delta)P_0(A), 1\};$$

- if $b = 1$, $-1 < a < 0$ (hence $c = -a$), the *Total Variation Model* [2, Sec. 3], [7, Sec. 3.2]:[1]

$$\underline{P}_{TVM}(A) = \max\{P_0(A) + a, 0\} \ \forall A \neq \Omega, \underline{P}_{TVM}(\Omega) = 1, \qquad (13)$$
$$\overline{P}_{TVM}(A) = \min\{P_0(A) - a, 1\} \ \forall A \neq \emptyset, \ \overline{P}_{TVM}(\emptyset) = 0.$$

---

[1] Note that $\underline{P}_{TVM}(A) \leq \overline{P}_{TVM}(A), \forall A$, since $a < 0$.

3   Conditioning Coherent Nearly-Linear Models

Given a coherent NL model $(\underline{P}, \overline{P})$ on $\mathcal{A}(I\!\!P)$ and an event $B \in \mathcal{A}(I\!\!P) \setminus \{\emptyset\}$, assumed or known to be true, we look for the natural extensions $\underline{E}(A|B), \overline{E}(A|B)$ of $\underline{P}, \overline{P}$ respectively, for any $A \in \mathcal{A}(I\!\!P)$. In other words, $\underline{P}, \overline{P}$ are extended on $\mathcal{A}(I\!\!P)|B$.

When $\underline{P}(B) = 0$, we determine $\underline{E}, \overline{E}$ quickly thanks to Proposition 3, which follows after a preliminary Lemma, stated without proof in a finite setting in [7].

Lemma 1. *Let* $\underline{P} : \mathcal{D} \to \mathbb{R}$ *be a coherent lower probability on* $\mathcal{D}$, *non-empty set of unconditional events, and* $B \in \mathcal{D}$, $B \neq \emptyset$ *such that* $\underline{P}(B) = 0$. *Let also* $\mathcal{S} = \{A_i|B\}_{i \in I}$ *be a set of events such that* $A_i \in \mathcal{D}$, $B \not\Rightarrow A_i$, $\forall i \in I$. *Then, the lower probability* $\underline{P}'$ *defined by*

$$\underline{P}'(E) = \underline{P}(E), \ \forall E \in \mathcal{D}, \ \underline{P}'(A_i|B) = 0, \ \forall i \in I,$$

is a coherent extension of $\underline{P}$ *on* $\mathcal{D} \cup \{A_i|B\}_{i \in I}$.

Proof. Firstly, since $A \wedge B \Rightarrow B$ and $\underline{P}(B) = 0$, we can extend in a unique way $\underline{P}$ on $\mathcal{D} \cup \{A_i \wedge B\}_{i \in I}$ preserving coherence, letting $\underline{P}(A_i \wedge B) = 0, \forall i \in I$.

To prove coherence of $\underline{P}'$, take $E_j \in \mathcal{D}$ $(j = 1, \ldots, n)$, $A_k|B \in \mathcal{S}$ $(k = 1, \ldots, m)$ and $n + m$ real coefficients s_j $(j = 1, \ldots, n)$, t_k $(k = 1, \ldots, m)$, such that at most one of them is negative, and define

$$\underline{G} = \sum_{j=1}^{n} s_j(I_{E_j} - \underline{P}'(E_j)) + \sum_{k=1}^{m} t_k I_B(I_{A_k} - \underline{P}'(A_k|B)).$$

According to Definition 1, we have to prove that $\max\{\underline{G}|H\} \geq 0$, where $H = \Omega$ if $n > 0$, $H = B$ otherwise. We distinguish two cases.

(a) Let $n = 0$, hence $\underline{G} = \sum_{k=1}^{m} t_k I_B I_{A_k}$. If $t_k \geq 0$, $\forall k = 1, \ldots, m$, $\underline{G} \geq 0$ and $\max\{\underline{G}|H\} = \max\{\underline{G}|B\} \geq 0$. Otherwise, if $t_{\overline{k}} < 0$ and $t_k \geq 0$, $\forall k = 1, \ldots, m$, $k \neq \overline{k}$, then $\max\{\underline{G}|H\} \geq \max\{t_{\overline{k}} I_B I_{A_{\overline{k}}}|B\} = \max\{t_{\overline{k}} I_{A_{\overline{k}}}|B\} = 0$, where the last equality holds because $A_{\overline{k}}^c \wedge B \neq \emptyset$ (since $B \not\Rightarrow A_{\overline{k}}$).

(b) If $n > 0$, $\underline{G} = \sum_{j=1}^{n} s_j(I_{E_j} - \underline{P}'(E_j)) + \sum_{k=1}^{m} t_k I_B I_{A_k} = \sum_{j=1}^{n} s_j(I_{E_j} - \underline{P}(E_j)) + \sum_{k=1}^{m} t_k(I_{A_k \wedge B} - \underline{P}(A_k \wedge B))$. Then, $\max\{\underline{G}|H\} = \max\{\underline{G}\} \geq 0$, applying Definition 1 to the coherent extension of $\underline{P}$ on $\mathcal{D} \cup \{A_i \wedge B\}_{i \in I}$. $\square$

Proposition 3. *Let* $\underline{P} : \mathcal{D} \to \mathbb{R}$ *be a coherent lower probability on* $\mathcal{D}$, *non-empty set of unconditional events, and* $B \in \mathcal{D}$, $B \neq \emptyset$ *such that* $\underline{P}(B) = 0$. *Then the natural extension* $\underline{E}$ *of* $\underline{P}$ *on* $\mathcal{D} \cup \{A_i|B\}_{i \in I}$, *where* $A_i \in \mathcal{D}$, $\forall i \in I$, *is given by* $\underline{E}(F) = \underline{P}(F), \forall F \in \mathcal{D}$ *and,* $\forall i \in I$, *by*

$$\underline{E}(A_i|B) = 1 \ if \ B \Rightarrow A_i, \ \underline{E}(A_i|B) = 0 \ otherwise. \tag{14}$$

Proof. Since $\underline{P}$ is coherent on $\mathcal{D}$, we have $\underline{E}(F) = \underline{P}(F), \forall F \in \mathcal{D}$.

Let $j \in I$. If $B \Rightarrow A_j$, $A_j|B = B|B$, hence coherence of $\underline{E}$ implies $\underline{E}(A_j|B) = 1$. If $B \not\Rightarrow A_j$, by Lemma 1, $\underline{P}$ can be extended on $A_j|B$, letting $\underline{P}(A_j|B) = 0$. Since $\underline{E}$ is the least-committal coherent extension of $\underline{P}$, we get $0 \leq \underline{E}(A_j|B) \leq \underline{P}(A_j|B) = 0$, hence $\underline{E}(A_j|B) = 0$. $\square$

Proposition 3 ensures that:

- $\underline{E}(A|B) = 0$ if $B \not\Rightarrow A$, $\underline{E}(A|B) = 1$ if $B \Rightarrow A$ (just take $A_i = A$ in (14));
- $\overline{E}(A|B) = 1$ if $B \not\Rightarrow A^c$, $\overline{E}(A|B) = 0$ if $B \Rightarrow A^c$ (just take $A_i = A^c$ in (14) and apply conjugacy).

Let us now *assume* $\underline{P}(B) > 0$.

Then, $\underline{E}$, $\overline{E}$ are given by the next

Proposition 4. *Let $(\underline{P}, \overline{P})$ be a coherent NL model on $\mathcal{A}(I\!P)$. For a given $B \in \mathcal{A}(I\!P)$ such that $\underline{P}(B) > 0$, we have that*

$$\overline{E}(A|B) = \begin{cases} 0 \text{ iff } \overline{P}(A \wedge B) = 0 \\ \dfrac{bP_0(A \wedge B) + c}{bP_0(B) + 1 - b} \ (\in]0, 1[) \text{ iff } \underline{P}(A^c \wedge B), \overline{P}(A \wedge B) \in]0, 1[\\ 1 \text{ iff } \underline{P}(A^c \wedge B) = 0 \end{cases} \quad (15)$$

$$\underline{E}(A|B) = \begin{cases} 1 \text{ iff } \overline{P}(A^c \wedge B) = 0 \\ \dfrac{bP_0(A \wedge B) + a}{bP_0(B) + 1 - b} \ (\in]0, 1[) \text{ iff } \underline{P}(A \wedge B), \overline{P}(A^c \wedge B) \in]0, 1[\\ 0 \text{ iff } \underline{P}(A \wedge B) = 0 \end{cases} \quad (16)$$

Proof. We derive first the expression (15) for $\overline{E}(A|B)$.

Since $\overline{P}$ is 2-alternating, we may apply Eq. (3). There, $\overline{E}$ depends on $\overline{P}(A \wedge B)$ and $\underline{P}(A^c \wedge B)$, which *cannot be simultaneously* 0: by (2), this would imply $0 = \underline{P}(A^c \wedge B) + \overline{P}(A \wedge B) \geq \underline{P}(B)$, hence $\underline{P}(B) = 0$. Further,

$$\begin{aligned} \underline{P}(A^c \wedge B) = 1 \to \overline{P}(A \vee B^c) = 0 \to \overline{P}(A \wedge B) = 0, \\ \overline{P}(A \wedge B) = 1 \to \underline{P}(A^c \vee B^c) = 0 \to \underline{P}(A^c \wedge B) = 0, \end{aligned} \quad (17)$$

using in both derivations conjugacy first, monotonicity then.

Thus, only the following exhaustive alternatives may occur:

(a) $\overline{P}(A \wedge B) = 0$
(b) $\underline{P}(A^c \wedge B) = 0$
(c) $\underline{P}(A^c \wedge B), \overline{P}(A \wedge B) \in]0, 1[$.

It is immediate from (3) that $\overline{E}(A|B) = 0$ iff $\overline{P}(A \wedge B) = 0$ and that $\overline{E}(A|B) = 1$ iff $\underline{P}(A^c \wedge B) = 0$. Otherwise, $\overline{E}(A|B) \in]0, 1[$. Precisely, from (3), (5), (6), (7)

$$\overline{E}(A|B) = \frac{bP_0(A \wedge B) + c}{bP_0(A \wedge B) + c + bP_0(A^c \wedge B) + a} = \frac{bP_0(A \wedge B) + c}{bP_0(B) + 1 - b}.$$

Turning now to $\underline{E}(A|B)$, its value in Eq. (16) may be obtained simply by conjugacy, using $\underline{E}(A|B) = 1 - \overline{E}(A^c|B)$ and (15). $\quad\square$

For the VBM, it is productive to write $\underline{E}(A|B)$, $\overline{E}(A|B)$ as follows:

Proposition 5. *Let* $(\underline{P}, \overline{P})$ *be a VBM on* $\mathcal{A}(I\!P)$. *For a given* $B \in \mathcal{A}(I\!P)$, *with* $\underline{P}(B) > 0$, *we have that*

$$\underline{E}(A|B) = \max\{b_B P_0(A|B) + a_B, 0\}, \forall A \in \mathcal{A}(I\!P) \setminus \{\Omega\}, \ \underline{E}(\Omega|B) = 1 \quad (18)$$

$$\overline{E}(A|B) = \min\{b_B P_0(A|B) + c_B, 1\}, \ \forall A \in \mathcal{A}(I\!P) \setminus \{\emptyset\}, \ \overline{E}(\emptyset|B) = 0 \quad (19)$$

$$a_B = \frac{a}{bP_0(B) + 1 - b}, \ b_B = \frac{bP_0(B)}{bP_0(B) + 1 - b}, \ c_B = 1 - (a_B + b_B). \quad (20)$$

Moreover, it holds that $b_B > 0$, $a_B \leq 0$, $0 < a_B + b_B \leq 1$.

Proof. Preliminarily, note that the denominator in (20) is positive. In fact, by assumption $\underline{P}(B) > 0$, meaning by (8) that $bP_0(B) + a > 0$. Using (7) and recalling that $c \geq 0$ in a VBM, it holds also that

$$bP_0(B) + 1 - b = bP_0(B) + a + c > 0.$$

Given this, let us prove (18) (the argument for (19) is analogous or could be also derived using conjugacy and will be omitted). For this, recalling (16), it is sufficient to establish that

(i) $b_B P_0(A|B) + a_B \leq 1$, with equality holding iff $\overline{P}(A^c \wedge B) = 0$;
(ii) $b_B P_0(A|B) + a_B \leq 0$ iff $\underline{P}(A \wedge B) = 0$;
(iii) $b_B P_0(A|B) + a_B = \dfrac{bP_0(A \wedge B) + a}{bP_0(B) + 1 - b} \in]0, 1[$ iff $\underline{P}(A \wedge B), \overline{P}(A^c \wedge B) \in]0, 1[$.

(i) Using (20) and the product rule at the first equality, non-negativity of b, c in the VBM at the inequality, (7) at the second equality, we obtain:

$$b_B P_0(A|B) + a_B = \frac{bP_0(A \wedge B) + a}{bP_0(B) + 1 - b}$$
$$\leq \frac{bP_0(A \wedge B) + bP_0(A^c \wedge B) + a + c}{bP_0(B) + 1 - b} = 1.$$

Moreover, $b_B P_0(A|B) + a_B = 1$ iff $bP_0(A^c \wedge B) + c = 0$, which is equivalent to $\overline{P}(A^c \wedge B) = 0$ by (9) (and since $bP_0(A^c \wedge B) + c \geq 0$).

(ii) Taking account of the first equality in the proof of (i) above and since $bP_0(B) + 1 - b > 0$, it ensues that

$$b_B P_0(A|B) + a_B \leq 0 \text{ iff } bP_0(A \wedge B) + a \leq 0 \text{ iff } \underline{P}(A \wedge B) = 0.$$

(iii) Immediate from (i), (ii) and recalling (17) (exchanging there A with A^c).

Elementary computations ensure that $b_B > 0$, $a_B \leq 0$, $0 < a_B + b_B \leq 1$. □

Proposition 5 proves an interesting feature of a VBM: when all events in $\mathcal{A}(I\!P)$ are conditioned on the same B, the resulting model is still a VBM. (Note the this holds also when $\underline{P}(B) = 0$: here Proposition 5 does not apply, but from Proposition 3 we obtain the vacuous lower/upper probabilities, a special VBM.)

We synthesise this property saying that a VBM is *stable under conditioning*.

A natural question now arises: does a HBM ensure an analogous property? Specifically, condition on B, with $\underline{P}(B) > 0$, the events of $\mathcal{A}(I\!P)$, which are initially given a HBM lower probability $\underline{P}$. Is it the case that the resulting $\underline{E}(\cdot|B)$ is determined by the equation

$$\underline{E}(A|B) = \min\{\max\{b_B P_0(A|B) + a_B, 0\}, 1\} \tag{21}$$

with a_B, b_B given by (20) and obeying the HBM constraints (11) (and similarly with $\overline{P}$, $\overline{E}(\cdot|B)$)?

The answer is negative: although $\underline{E}(A|B)$ may occasionally be obtained from (21), for instance—as is easy to check—when $\underline{P}(A \wedge B)$, $\overline{P}(A^c \wedge B) \in]0,1[$, $\forall A \in \mathcal{A}(I\!P) \setminus \{\emptyset, \Omega\}$, this is not true in general, as shown in the next example.

Example 1. Given $I\!P = \{\omega_1, \omega_2, \omega_3\}$, Table 1 describes the values of an assigned P_0, and of $\underline{P}$, $\overline{P}$ obtained by (5), (6) with $a = -10$, $b = 12$. Since a, b satisfy (11) and, as can be easily checked, $\overline{P}$ is subadditive, $(\underline{P}, \overline{P})$ is a coherent HBM by Definition 3 and Proposition 2. Now, take $A = \omega_2$, $B = \omega_1 \vee \omega_2$. From

Table 1. Data for Example 1

	ω_1	ω_2	ω_3	$\omega_1 \vee \omega_2$	$\omega_1 \vee \omega_3$	$\omega_2 \vee \omega_3$	$\emptyset$	Ω
P_0	$\frac{3}{10}$	$\frac{3}{5}$	$\frac{1}{10}$	$\frac{9}{10}$	$\frac{4}{10}$	$\frac{7}{10}$	0	1
$\underline{P}$	0	0	0	0.8	0	0	0	1
$\overline{P}$	1	1	0.2	1	1	1	0	1

(16), $\underline{E}(A|B) = 0$, because $\underline{P}(A \wedge B) = \underline{P}(\omega_2) = 0$. Yet, since $b_B P_0(A|B) + a_B = \frac{b P_0(\omega_2) + a}{b P_0(\omega_1 \vee \omega_2) + 1 - b} = 14 > 1$, Eq. (21) would let us mistakenly conclude that $\underline{E}(A|B) = 1$.

Thus, a coherent HBM differs from a VBM with respect to conditioning on some event B, being not stable.

It is interesting to note that not only the VBM, but also its special submodels listed in Sect. 2.1 are stable: conditioning one of them on B returns a model of the same kind. Let us illustrate this in some detail.

For the *linear-vacuous model* $(\underline{P}_V, \overline{P}_V)$, it is well-known [10] that $\underline{E}_V(A|B) = 0$ if $B \not\Rightarrow A$, while $\underline{E}_V(A|B) = 1$ if $B \Rightarrow A$. Note that this follows also from Proposition 3, since $\underline{P}_V(B) = 0$. By conjugacy, $\overline{E}_V(\cdot|B)$ is vacuous too.

With the *ε-contamination model*, its conditional model is again of the same type: from (18), (20), we get $a_B = 0$, $b_B = 1 - \varepsilon_B \in]0,1[$.

Turning to the *Total Variation Model* (TVM) and applying again (18), (20)

$$b_B = 1, a_B = \frac{a}{P_0(B)} < 0$$

$$\underline{E}_{TVM}(A|B) = \max\left\{P_0(A|B) + \frac{a}{P_0(B)}, 0\right\}$$

$$= \frac{1}{P_0(B)} \max\{P_0(A \wedge B) + a, 0\}. \tag{22}$$

At first sight, the conditional model differs from a TVM. However, we may easily write (13) in the form (22):

$$\underline{P}_{TVM}(A) = \frac{1}{P_0(\Omega)} \cdot \max\{P_0(A \wedge \Omega) + a, 0\}. \tag{23}$$

Comparing (22) and (23) we see that the TVM is stable too under conditioning: $\underline{P}_{TVM}, \underline{E}_{TVM}$ may be thought of as normalised on the P_0-probability of what is currently assumed to be true (Ω first, B then).

The conjugate of $\underline{E}_{TVM}(A|B)$ is $\overline{E}_{TVM}(A|B) = \frac{1}{P_0(B)} \min\{P_0(A \wedge B) - a, 1\}$. $\underline{E}_{TVM}(A|B), \overline{E}_{TVM}(A|B)$ determine a *credal set* (i.e. the set of probabilities P such that $\underline{E}_{TVM}(A|B) \leq P(A|B) \leq \overline{E}_{TVM}(A|B), \forall A \in \mathcal{A}(\mathbb{P})$) still made of all probabilities at a total variation distance[2] from P_0 not larger than $-a$ (> 0). This is like the unconditional TVM, the difference being that any A is replaced by $A \wedge B$, i.e. by what remains possible of A after assuming B true.

Conditioning the *Pari-Mutuel Model* (PMM) on B leads to similar conclusions: the conditional model is again of the PMM type.

Take for instance $\underline{P}_{PMM}(A)$, given by (12), where $b = 1 + \delta > 1, a = -\delta < 0$. From (18), (20) and recalling that $\underline{P}_{PMM}(B) = (1 + \delta)P_0(B) - \delta > 0$, we obtain

$$a_B = \frac{-\delta}{\underline{P}_{PMM}(B)} < 0, \; b_B = 1 + \frac{\delta}{\underline{P}_{PMM}(B)} > 1,$$

$$\underline{E}_{PMM}(A|B) = \max\{(1 + \delta_B)P_0(A|B) - \delta_B, 0\}, \; \delta_B = \frac{\delta}{\underline{P}_{PMM}(B)}.$$

Clearly, $a_B + b_B = 1$, and we may conclude that $\underline{E}_{PMM}(\cdot|B)$ is again a PMM, with δ replaced by δ_B. Note that the starting $\underline{P}_{PMM}(\cdot)$ may be written as

$$\underline{P}_{PMM}(A) = \max\left\{(1 + \delta)P_0(A|\Omega) - \frac{\delta}{\underline{P}_{PMM}(\Omega)}, 0\right\}.$$

Similarly, we obtain

$$\overline{E}_{PMM}(A|B) = \min\{(1 + \delta_B)P_0(A|B), 1\}.$$

Note that $\delta_B \geq \delta$, with equality holding iff $P_0(B) = 1$. As well known [7, 10], δ (hence δ_B) has the interpretation of a loading factor, which makes a subject

---

[2] On the total variation distance see e.g. [3, Sec. 4.1].

'sell' A $(A|B)$ for a selling price $\overline{P}_{PMM}(A)$ $(\overline{E}_{PMM}(A|B))$ higher than the 'fair price' $P_0(A)$ $(P_0(A|B))$. With respect to this kind of considerations, conditioning increases the loading factor, and the smaller $\underline{P}_{PMM}(B)$, the higher the increase. Next to this, conditioning makes both the seller and the buyer more cautious, in the sense that they restrict the events they would sell or buy.

From the seller's perspective, we can see this noting that $\overline{P}_{PMM}(A) < 1$ iff $P_0(A) < \frac{1}{1+\delta} = \overline{t}_\Omega$. Here $\overline{t}_\Omega$ is the threshold to ensure that selling A may be considered: when $\overline{P}(A) = 1$, the seller is practically sure that A will occur. On the other hand, s/he will find no rational buyer for such a price: in fact, the buyer should pay 1 to receive at most 1 if A occurs, 0 otherwise. After conditioning, $\overline{E}_{PMM}(A|B) < 1$ iff $P_0(A|B) < \frac{1}{1+\delta_B} = \overline{t}_B \leq \overline{t}_\Omega$. When $\overline{t}_B < \overline{t}_\Omega$, the seller may have the chance to negotiate A, but not $A|B$, for some events A.

Analogously, a subject 'buying' A $(A|B)$ will be unwilling to do so when $\underline{P}_{PMM}(A) = 0$ (when $\underline{E}_{PMM}(A|B) = 0$), which happens iff $P_0(A) \leq \frac{\delta}{1+\delta} = \underline{t}_\Omega$ (iff $P_0(A|B) \leq \frac{\delta_B}{1+\delta_B} = \frac{\delta}{\underline{P}_{PMM}(B)+\delta} = \underline{t}_B$). Here $\underline{t}_B \geq \underline{t}_\Omega$, and again conditioning makes the buyer more cautious, see also Fig. 1.

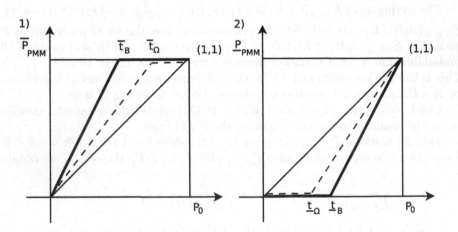

Fig. 1. 1) Plot of $\overline{P}_{PMM}$ against P_0 before (dashed bold line) and after (continuous bold line) conditioning. 2) The same for $\underline{P}_{PMM}$ against P_0.

4 Dilation with Coherent Nearly-Linear Models

Given a coherent NL model on $\mathcal{A}(I\!P)$, $B \in \mathcal{A}(I\!P) \setminus \{\emptyset\}$ and an event $A \in \mathcal{A}(I\!P)$, if we compute $\underline{E}(A|B)$, $\overline{E}(A|B)$ it may happen that

$$\underline{E}(A|B) \leq \underline{P}(A) \leq \overline{P}(A) \leq \overline{E}(A|B). \tag{24}$$

When Eq. (24) holds, the imprecision of our evaluation on A increases, or at least does not decrease, after assuming B true. Equation (24) is a condition preliminary to dilation, which we shall discuss later on. Next, we investigate when (24) holds.

Preliminarily, say that (24) *occurs trivially* when it holds and its three inequalities are equalities, and that A is an *extreme event* if either $\underline{P}(A) = \overline{P}(A) = 0$ or $\underline{P}(A) = \overline{P}(A) = 1$. When referring to extreme events in the sequel, we shall rule out $\emptyset$, Ω (for which no inference is needed).

Next, we investigate if Eq. (24) obtains when: $\underline{P}(B) = 0$ (Lemma 2); $\underline{P}(B) > 0$ and A is either an extreme event (Lemma 3) or non-extreme (Proposition 7).

Lemma 2. *Let $\underline{P}(B) = 0$. Then,*

(a_0) *If $B \not\Rightarrow A$ and $B \not\Rightarrow A^c$, (24) applies.*
(b_0) *If $B \Rightarrow A$, (24) occurs, trivially, iff $\underline{P}(A) = \overline{P}(A) = 1$.*
(c_0) *If $B \Rightarrow A^c$, (24) occurs, trivially, iff $\underline{P}(A) = \overline{P}(A) = 0$.*

Proof. As for (a_0), (24) holds because $\underline{E}(A|B) = 0$, $\overline{E}(A|B) = 1$ by Proposition 3. By the same Proposition 3, (b_0) and (c_0) follow easily.   □

Let us suppose now that $\underline{P}(B) > 0$. From (15), (16), and since $\overline{E}(\cdot|B) \geq \underline{E}(\cdot|B)$, the following alternatives may arise:

(a) $\overline{P}(A \wedge B) = 0$.
   Correspondingly, $\overline{E}(A|B) = \underline{E}(A|B) = 0$.
(b) $\underline{P}(A^c \wedge B), \overline{P}(A \wedge B) \in]0, 1[$; $\underline{P}(A \wedge B) = 0$.
   Here $\overline{E}(A|B) \in]0, 1[$, $\underline{E}(A|B) = 0$.
(c) $\underline{P}(A^c \wedge B), \overline{P}(A \wedge B) \in]0, 1[$; $\underline{P}(A \wedge B), \overline{P}(A^c \wedge B) \in]0, 1[$.
   Then, $\overline{E}(A|B), \underline{E}(A|B) \subset]0, 1[$.
(d) $\underline{P}(A^c \wedge B) = 0$; $\underline{P}(A \wedge B), \overline{P}(A^c \wedge B) \in]0, 1[$.
   Then, $\underline{E}(A|B) \in]0, 1[$, $\overline{E}(A|B) = 1$.
(e) $\overline{P}(A^c \wedge B) = 0$.
   Correspondingly, $\overline{E}(A|B) = \underline{E}(A|B) = 1$.
(f) $\underline{P}(A^c \wedge B) = \underline{P}(A \wedge B) = 0$.
   Here $\underline{E}(A|B) = 0$, $\overline{E}(A|B) = 1$.

Lemma 3. *If $\underline{P}(B) > 0$ and A is an extreme event, then (24) occurs trivially.*

Proof. If $\overline{P}(A) = \underline{P}(A) = 0$, then $\overline{P}(A \wedge B) = 0$ by monotonicity, and from (a) above $\overline{E}(A|B) = \underline{E}(A|B) = 0$ too. If $\overline{P}(A) = \underline{P}(A) = 1$, then $\overline{P}(A^c) = \underline{P}(A^c) = 0$, hence $\overline{P}(A^c \wedge B) = 0$ and $\overline{E}(A|B) = \underline{E}(A|B) = 1$ from (e) above.   □

We still have to establish whether (24) holds assuming that $\underline{P}(B) > 0$ and A is a non-extreme event. Trivially, (24) holds in case (f), while it does not in cases (a), (e). To see what happens in the remaining instances, let us prove that

Proposition 6. *Given a coherent NL model $(\underline{P}, \overline{P})$, let $B \in \mathcal{A}(I\!\!P)$, $\underline{P}(B) > 0$, and $A \in \mathcal{A}(I\!\!P)$, A non-extreme.*

(i) If $\underline{P}(A \wedge B), \overline{P}(A^c \wedge B) \in]0, 1[$, then

$$\underline{E}(A|B) \leq \underline{P}(A) \text{ iff either } P_0(B^c) = 0 \text{ or } \underline{P}(A) \leq P_0(A|B^c).$$

(ii) If $\underline{P}(A^c \wedge B), \overline{P}(A \wedge B) \in]0,1[$, *then*

$$\overline{E}(A|B) \geq \overline{P}(A) \text{ iff either } P_0(B^c) = 0 \text{ or } \overline{P}(A) \geq P_0(A|B^c).$$

Proof. (i) By monotonicity, $\underline{P}(A) \geq \underline{P}(A \wedge B) > 0$, and $\underline{P}(A) < 1$ because $\underline{P}(A) \leq \underline{P}(A \vee B^c) < 1$, the latter inequality holding since (by conjugacy) $\underline{P}(A \vee B^c) \in]0,1[$ iff $\overline{P}(A^c \wedge B) \in]0,1[$. Which is assumed to be true. Therefore, by (5), $\underline{P}(A) = bP_0(A) + a$, while $\underline{E}(A|B)$ is given by the second line in (16). Hence,

$$\underline{E}(A|B) \leq \underline{P}(A) \text{ iff } \frac{bP_0(A \wedge B) + a}{bP_0(B) + 1 - b} \leq bP_0(A) + a. \tag{25}$$

With some algebraic manipulations on the right-hand side of (25), noting that by assumption and (2), (5), (6), (7) we have that $0 < \underline{P}(B) \leq \overline{P}(A^c \wedge B) + \underline{P}(A \wedge B) = bP_0(B) + 1 - b$, we obtain

$$\frac{bP_0(A \wedge B) + a}{bP_0(B) + 1 - b} \leq bP_0(A) + a \text{ iff}$$

$$bP_0(A \wedge B) + a \leq b^2 P_0(A)(-P_0(B^c)) + ab(-P_0(B^c)) + bP_0(A) + a \text{ iff}$$

$$P_0(B^c)(bP_0(A)+a) \leq P_0(A) - P_0(A \wedge B) = P_0(A \wedge B^c) \text{ iff}$$

$$P_0(B^c) = 0 \text{ or } (P_0(B^c) > 0 \text{ and}) bP_0(A) + a \leq \frac{P_0(A \wedge B^c)}{P_0(B^c)} = P_0(A|B^c),$$

which proves (i).

(ii) It can be obtained in a very similar way or directly by conjugacy.   □

From Proposition 6, and recalling (15), (16), it follows straightforwardly that

Proposition 7. *Let* $(\underline{P}, \overline{P})$ *be a coherent NL model on* $\mathcal{A}(\mathbb{P})$, $B \in \mathcal{A}(\mathbb{P})$, $\underline{P}(B) > 0$. *For* $A \in \mathcal{A}(\mathbb{P})$, A *non-extreme, Eq. (24) holds iff*

$$\underline{P}(A^c \wedge B) = \underline{P}(A \wedge B) = 0 \text{ or } P_0(B^c) = 0 \text{ or } \underline{P}(A) \leq P_0(A|B^c) \leq \overline{P}(A). \tag{26}$$

The left (right) inequality does not apply if $\underline{E}(A|B) = 0$ *(if* $\overline{E}(A|B) = 1$*).*

The previous results pave the way to considerations on *dilation* with NL models. When discussing dilation [2,7], we consider a partition $\mathcal{B}$ of non-impossible events and say that (weak) *dilation* occurs, with respect to A and $\mathcal{B}$, when

$$\underline{E}(A|B) \leq \underline{P}(A) \leq \overline{P}(A) \leq \overline{E}(A|B), \ \forall B \in \mathcal{B}. \tag{27}$$

Recall that an event A is *logically non-independent* of a partition $\mathcal{B}$ iff $\exists B \in \mathcal{B} \setminus \{\emptyset\}$ such that either $B \Rightarrow A$ or $B \Rightarrow A^c$, *logically independent* of $\mathcal{B}$ otherwise.

We can now introduce several results concerning dilation.

Proposition 8. *Let* $(\underline{P}, \overline{P})$ *be a coherent NL model on* $\mathcal{A}(\mathbb{P})$, $A \in \mathcal{A}(\mathbb{P})$, $\mathcal{B} \subset \mathcal{A}(\mathbb{P}) \setminus \{\emptyset\}$ *a partition.*

(j) If A *is a non-extreme event logically non-independent of* $\mathcal{B}$, *dilation does not occur.*

(jj) If A is an extreme event logically independent of B, dilation occurs.

Proof. (j) Take $B \in \mathcal{B}$ such that either $B \Rightarrow A$ or $B \Rightarrow A^c$. If $\underline{P}(B) = 0$, (24) does not occur by Lemma 2, either (b_0) or (c_0). If $\underline{P}(B) > 0$, because of the logical non-independence hypothesis, either $A \wedge B = \emptyset$ or $A^c \wedge B = \emptyset$, hence either $B = A^c \wedge B$ or $B = A \wedge B$, respectively. This rules out case (f), as assuming it would imply here $\underline{P}(B) = 0$. Actually, recalling that A is non-extreme, case (a) or (e) applies, and (24) does not occur. Thus, the dilation condition (27) is not satisfied.

(jj) Follows from Lemma 2, (a_0) when $\underline{P}(B) = 0$, from Lemma 3 otherwise. □

Dilation is characterised for a non-extreme event A, logically independent of $\mathcal{B}$, as follows:

Proposition 9. *Given a coherent NL model $(\underline{P}, \overline{P})$ on $\mathcal{A}(I\!\!P)$, a partition $\mathcal{B} \subset \mathcal{A}(I\!\!P) \setminus \{\emptyset\}$ and $A \in \mathcal{A}(I\!\!P)$ non-extreme and logically independent of $\mathcal{B}$, dilation occurs (w.r.t. A, B) iff, $\forall B \in \mathcal{B}$ such that $\underline{P}(B) > 0$, (26) holds, where the left (right) inequality does not apply if $\underline{E}(A|B) = 0$ (if $\overline{E}(A|B) = 1$).*

Proof. We need not consider those $B \in \mathcal{B}$ such that $\underline{P}(B) = 0$, if any: by Lemma 2, (a_0) they ensure (24). For the others apply Proposition 7. □

We derive now an interesting sufficient condition for dilation with a VBM, extending an analogous property of a PMM [7, Corollary 2].

Proposition 10. *Dilation for a non-extreme event A, logically independent of partition $\mathcal{B}$, occurs in a VBM if*

$$P_0(A \wedge B) = P_0(A) \cdot P_0(B), \quad \forall B \in \mathcal{B}. \tag{28}$$

Proof. If $\underline{P}(B) = 0$, apply Lemma 2, (a_0). Otherwise, by Proposition 7, we have to check that, when $P_0(B^c) > 0$, $\underline{P}(A) \leq P_0(A|B^c) \leq \overline{P}(A)$ holds. Now, if $\underline{P}(B) > 0$, then necessarily by (5) $bP_0(B) + a > 0$, hence $P_0(B) > 0$, because $a \leq 0$ in a VBM. Further, if (28) holds, then $P_0(A \wedge B^c) = P_0(A) \cdot P_0(B^c)$, hence for those $B \in \mathcal{B}$ such that $P_0(B^c) > 0$, also $P_0(A|B^c) = P_0(A)$. Thus, the condition to check boils down to $\underline{P}(A) \leq P_0(A) \leq \overline{P}(A)$, which always applies for a VBM. □

Note that dilation occurs if event A in Proposition 10 is P_0-non-correlated with any $B \in \mathcal{B}$.

5 Conclusions

Among coherent NL models, VBMs ensure the property of being stable with conditioning, as also do several known submodels of theirs, such as the PMM. This implies also that results found in [5] on natural extensions of (unconditional) VBMs to gambles still apply here to conditional gambles $X|B$ defined on the conditional partition $I\!\!P|B = \{\omega|B : \omega \in I\!\!P\}$. By contrast, those HBMs that are coherent are generally not stable, thus these models confirm their weaker properties, in comparison with VBMs, already pointed out, from other perspectives, in [1]. Concerning dilation of A w.r.t. partition $\mathcal{B}$, we have seen that it may depend on more conditions, such as whether A is extreme or not, or it is logically independent of $\mathcal{B}$ or not, and we supplied several results. In future work, we plan to study the regular extension [10, Appendix J] of coherent NL models, determining how its being less conservative than the natural extension may limit extreme evaluations, as well as dilation. Concepts related to dilation and not presented here are also discussed in [2] for the ε-contamination and the Total Variation models, among others. The assumptions in [2] are usually less general than the present framework. The role of these notions within NL models is still to be investigated.

Acknowledgments. We are grateful to the referees for their stimulating comments and suggestions. We acknowledge partial support by the FRA2018 grant 'Uncertainty Modelling: Mathematical Methods and Applications'.

References

1. Corsato, C., Pelessoni, R., Vicig, P.: Nearly-Linear uncertainty measures. International Journal of Approximate Reasoning **114**, 1–28 (2019). https://doi.org/10.1016/j.ijar.2019.08.001
2. Herron, T., Seidenfeld, T., Wasserman, L.: Divisive Conditioning: Further Results on Dilation. Philosophy of Science **64**(3), 411–444 (1997). https://doi.org/10.1086/392559
3. Levin, D.A., Peres, Y.: Markov chains and mixing times. American Mathematical Society, Providence, R.I (2017). https://doi.org/10.1090/mbk/107
4. Montes, I., Miranda, E., Destercke, S.: Pari-mutuel probabilities as an uncertainty model. Information Sciences **481**, 550–573 (2019). https://doi.org/10.1016/j.ins.2019.01.005
5. Pelessoni, R., Vicig, P., Corsato, C.: Extending Nearly-Linear Models. In: De Bock, J., de Cooman, G., de Campos, C.P., Quaeghebeur, E., Wheeler, G. (eds.) PMLR, vol. 103, pp. 82–90 (2019)
6. Pelessoni, R., Vicig, P.: Williams coherence and beyond. International Journal of Approximate Reasoning **50**(4), 612–626 (2009). https://doi.org/10.1016/j.ijar.2008.10.002
7. Pelessoni, R., Vicig, P., Zaffalon, M.: Inference and risk measurement with the pari-mutuel model. International Journal of Approximate Reasoning **51**(9), 1145–1158 (2010). https://doi.org/10.1016/j.ijar.2010.08.005
8. Troffaes, M.C.M., de Cooman, G.: Lower previsions. Wiley series in probability and statistics, Wiley, Chichester, West Sussex, United Kingdom (2014)
9. Walley, P.: Coherent Lower (and upper) Probabilities. Technical Report 22, Department of Statistics, University of Warwick, Coventry (1981)
10. Walley, P.: Statistical reasoning with imprecise probabilities. No. 42 in Monographs on statistics and applied probability, Chapman and Hall, London; New York (1991)
11. Williams, P.: Notes on conditional previsions. International Journal of Approximate Reasoning **44**(3), 366–383 (2007). https://doi.org/10.1016/j.ijar.2006.07.019

Learning Sets of Bayesian Networks

Andrés Cano⑩, Manuel Gómez-Olmedo⑩, and Serafín Moral^(✉)⑩

Departamento de Ciencias de la Computación e Inteligencia Artificial,
18071 Granada, Spain
{acu,mgomez,smc}@decsai.ugr.es

Abstract. This paper considers the problem of learning a generalized credal network (a set of Bayesian networks) from a dataset. It is based on using the BDEu score and computes all the networks with score above a predetermined factor of the optimal one. To avoid the problem of determining the equivalent sample size (ESS), the approach also considers the possibility of an undetermined ESS. Even if the final result is a set of Bayesian networks, the paper also studies the problem of selecting a single network with some alternative procedures. Finally, some preliminary experiments are carried out with three small networks.

Keywords: Generalized credal networks · Learning · Likelihood regions · Probabilistic graphical models

1 Introduction

Probabilistic graphical models [17] and in particular Bayesian networks have been very successful for representing and reasoning in problems with several uncertain variables. The development of procedures to learn a Bayesian network from a dataset of observations [16] is one the most important reasons of this success. Usually, learning is carried out by selecting a score measuring the adequacy of a model given the data and optimizing it in the space of models. However, in most of the situations the selection of a single Bayesian network is not justified as there are many models that explain the data with a similar degree, being the selection of an optimal network a somewhat arbitrary choice [7]. For this reason, recently, there has been some effort in computing a set of possible models instead of selecting a single one [12]. The idea is to compute all the models with a score within a given factor of the optimal one. In this paper we will follow this line, but interpreting the result as a generalized credal network: a set of Bayesian networks which do not necessarily share the same graph [13]. The term credal network was introduced [6] for a set of Bayesian networks over a single graph (there is imprecision only in the parameters). The overall procedure is based on the general framework introduced in [15], where it is proposed a justification

This research was supported by the Spanish Ministry of Education and Science under project TIN2016-77902-C3-2-P, and the European Regional Development Fund (FEDER).

M.-J. Lesot et al. (Eds.): IPMU 2020, CCIS 1238, pp. 151–164, 2020.
https://doi.org/10.1007/978-3-030-50143-3_12

based on sets of desirable gambles [5,18,22] for the selection of a set of models instead of a single one, following the lines of Gärdenfors and Shalin [8] and α-cut conditioning by Cattaneo [2].

The basic criterion used for learning is the so called BDEu score [9]. This score needs a parameter, the equivalent sample size (ESS), which is usually arbitrarily selected in practice with a value between 1 and 10. However, there are results showing that the final network can have a dependence on the ESS, producing more dense networks when it is higher [4,14,21]. For this reason, our approach will also consider the possibility of imprecision due to an undetermined ESS.

The paper is organized as follows. Section 2 provides the basic theoretical framework for our problem. Section 3 describes the algorithms used in the computation. Section 4 is devoted to the experiments. Finally, the conclusions and future work are in Sect. 5.

2   Learning Imprecise Models

Given a set of variables, $\mathbf{X} = (X_1, \ldots, X_m)$, a Bayesian network [17] is a pair (G, β), where G is a directed acyclic graph such that each node represents a variable, and β is the set of parameters: a conditional probability distribution for each variable X_i given its parents in the graph, Pa_i, denoted as $P(X_i|Pa_i)$ or as $P_{(G,\beta)}(X_i|Pa_i)$ when we want to make reference to the associated model. It will be assumed that each variable X_i takes values on a finite set with K_i possible values. A generic value for variable X_i is denoted by x_i and a generic value for all the variables $\mathbf{X}$ is denoted as $\mathbf{x}$. An assignation of a concrete value to each variable in Pa_i is called a configuration and denoted as pa_i. The number of possible configurations of Pa_i is denoted by R_i. There is a joint probability distribution for variables $\mathbf{X}$ associated with a Bayesian network (G, β) that will be denoted as $P_{(G,\beta)}$ and that is equal to the product $\prod_{i=1}^{m} P_{(G,\beta)}(X_i|Pa_i)$.

We will consider that we have a set of full observations $\mathcal{D}$ for all the variables in $\mathbf{X}$. Given a graph G, n_{ijk} will denote the number of observations in $\mathcal{D}$ where $X_i = x_k$ and its parents Pa_i take the jth configuration, $n_{ij} = \sum_{k=1}^{K_i} n_{ijk}$, whereas n will be the total sample size. In the framework for learning proposed in [15], it is assumed that we have the following elements:

- A set of parameters Θ that corresponds to the space of possible decisions. In our case, Θ is the set of pairs (G, s), where G is a direct acyclic graph, and s is a possible ESS belonging to a finite set of values, S. For example, in our experiments we have considered $S = \{0.1, 0.5, 1.0, 2.0, 4.0, 6.0, 8.0, 10.0\}$. We assume a finite set instead of a continuous interval for computational reasons.
- A set of parameters B, and a conditional probability distribution $P(\beta|\theta)$ specifying the probability on B for each value of the parameter $\theta \in \Theta$. In our case the set B is the list of conditional probability distributions $P(X_i|Pa_i)$, where the probability values of the conditional distribution of X_i given the jth configuration of the parents are denoted by $\beta_{ij} = (\beta_{ij1}, \ldots, \beta_{ijK_i})$ (i.e. $\beta_{ijk} = P(X_i = x_k|Pa_i = pa_i^j)$). It is assumed that each β_{ij} follows an independent

Dirichlet distribution $D(s/(R_iK_i), \ldots, s/(R_iK_i))$. The set of all parameters β_{ij} will be denoted by β.

- A conditional distribution for an observation of the variables $(X_1, \ldots, X_m)$ given a pair $(\theta, \beta) \in \Theta \times B$ (in our case, given G, s and β). The probability of observing $X_1 = x_{k_1}, \ldots, X_m = x_{k_m}$ is the product $\prod_{i=1}^{m} \beta_{ijk_i} = \prod_{i=1}^{M} P(x_{k_i}|pa_i^j)$, where pa_i^j is the configuration of the parents compatible with the observation, and k_i represents the subscript of the observed value for $X_i = x_{k_i}$.

In this setting, a set of observations $\mathcal{O}$, defines a likelihood function L in Θ, which is given at the general case by,

$$L(\theta) = \int_B P(\beta|\theta)P(\mathcal{D}|\beta, \theta)d\beta. \tag{1}$$

In the particular case of learning generalized credal networks, we have that this likelihood is identical to the well known BDEu score [9] for learning Bayesian networks:

$$L(G, s) = BDEu(G, s) = P(\mathcal{D}|G) = \prod_{i=1}^{m} \prod_{j=1}^{R_i} \frac{\Gamma(s/R_i)}{\Gamma(n_{ij} + s/R_i)} \prod_{k=1}^{K_i} \frac{\Gamma(n_{ijk} + s/(R_iK_i))}{\Gamma(s/(R_iK_i))}. \tag{2}$$

The score for Pa_i as set of parents of X_i is the value:

$$\log(BDEu(Pa_i, s, X_i)) = \log\left(\prod_{j=1}^{R_i} \frac{\Gamma(s/R_i)}{\Gamma(n_{ij} + s/R_i)} \prod_{k=1}^{K_i} \frac{\Gamma(n_{ijk} + s/(R_iK_i))}{\Gamma(s/(R_iK_i))}\right). \tag{3}$$

It is immediate that $\log(BDEu(G, s)) = \sum_{i=1}^{m} \log(BDEu(Pa_i, s, X_i))$. Finally, a generalized uniform distribution on Θ is considered given in terms of a coherent set of desirable gambles [15]. When Θ is finite as in this case, associated credal set only contains the uniform probability, but when Θ is a continuous interval is quite different from the usual uniform density. Then a discounting is considered of this prior information on Θ given by a value $\epsilon \in [0, 1]$. This discounting is a generalization of the ϵ discounting of a belief function [20]. After the observations are obtained, the model is conditioned to them, obtaining a posterior information on Θ. It is assumed that the set of decisions is equivalent to the set of parameters Θ and the problem is solved by computing all un-dominated decisions under a 0–1 loss (details in [15]). Finally, in our case the set of un-dominated decisions is the set of parameters (G, s) such that:

$$L(G, s) = BDEu(G, s) \geq \alpha BDEu(\hat{G}, \hat{s}),$$

where $\alpha = \frac{1-\epsilon}{1+\epsilon} \in [0, 1]$ and $(\hat{G}, \hat{s})$ is the pair maximizing the likelihood $L(G, s)$ for $(G, s) \in \Theta$. The set of parameters satisfying the above inequality is denoted by H_L^α and defines what we shall call the set of possible models.

In the following, we will use the value $\alpha = \frac{1-\epsilon}{1+\epsilon} \in [0,1]$ which is computed as a continuous decreasing function from $[0,1]$ into $[0,1]$ and that determines the factor of the maximum entropy model which makes (G,S) non-dominated.

Given a parameter (G,s), the model for $\mathbf{X}$ is given by the Bayesian network $(G, \hat{\beta})$, where $\hat{\beta}$ is the Bayesian estimation of β (expected value of β given (G,s) and $\mathcal{D}$), and which can be computed in closed form by the well known expression:

$$\hat{\beta}_{ijk} = \frac{n_{ijk} + s/(R_i K_i)}{n_{ij} + s/(R_i)}. \tag{4}$$

The probability distribution associated with $(G, \hat{\beta})$ will be also denoted as $P_{(G,s)}(\mathbf{x})$[1]. Finally the set of possible models (a generalized credal network) is the set of Bayesian networks:

$$M_{\mathcal{D}}^{\alpha} = \{(G, \hat{\beta}) \mid (G,s) \in H_L^{\alpha}\}, \tag{5}$$

and where $\hat{\beta}$ is the set of parameters given by Eq. (4).

Though, in our opinion, the result of learning should be the set $M_{\mathcal{D}}^{\alpha}$, in some cases, it is interesting to select a single model. For example, we have carried out experiments in which we want to compare this approach to learning with a Bayesian procedure that always selects a single network. For this aim we have considered two approaches:

– *Maximum Entropy:* We select the pair $(G, \hat{\beta}) \in M_{\mathcal{D}}^{\alpha}$ maximizing the entropy, where the entropy of a model (G, β) is given by:

$$E(G, \beta) = -\sum_{\mathbf{x}} P_{(G,\beta)}(\mathbf{x}) \log P_{(G,\beta)}(\mathbf{x}). \tag{6}$$

– *Minimum of Maximum Kullback-Leibler Divergence:* If (G, β) and (G', β') are two models, then the Kullback-Leibler divergence of (G, β) to (G', β') is given by the expression:

$$KL((G', \beta'), (G, \beta)) = \sum_{\mathbf{x}} P_{(G',\beta')}(\mathbf{x}) \log \left(\frac{P_{(G',\beta')}(\mathbf{x})}{P_{(G,\beta)}(\mathbf{x})} \right). \tag{7}$$

Then, for each model $(G, \hat{\beta}) \in M_{\mathcal{D}}^{\alpha}$, the following value is computed:

$$MKL(G, \hat{\beta}) = \max\{KL((G', \hat{\beta}'), (G, \hat{\beta})) \mid (G', \hat{\beta}') \in M_{\mathcal{D}}^{\alpha}\}.$$

Finally, the model $(G, \hat{\beta}) \in M_{\mathcal{D}}^{\alpha}$ minimizing $MKL(G, \hat{\beta})$ is selected.

---

[1] In fact, this probability also depends on $\mathcal{D}$, but we do not include it to simplify the notation.

3   Algorithms

Given a set of observations $\mathcal{D}$ and a value of ϵ, our aim is to compute the set of Bayesian networks given by Eq. (5), where $\alpha = \frac{1-\epsilon}{1+\epsilon}$. For this we have taken as basis the pgmpy package in which basic procedures for inference and learning with Bayesian networks are implemented [1].

Our first algorithm AllScores(ESS, α) computes the set of possible parents as well as the logarithm of their scores for each variable X_i, for each sample size $s \in S$ and for a given value of α, being denoted this set as $Par(X_i, s, \alpha)$.

To do it, we compute the value of $\log(BDEu(Pa_i, s, X_i))$ following Eq. (3) for each set $Pa_i \subseteq \mathbf{X} \setminus \{X_i\}$, storing the pair $(Pa_i, \log(BDEu(Pa_i, s, X_i)))$, but taking into account the following pruning rules as in [12]:

- If $Pa_i \subset Pa'_i$ and $\log(BDEu(Pa_i, s, X_i)) > \log(BDEu(Pa'_i, s, X_i)) - log(\alpha)$, then Pa'_i is not added to $Par(X_i, s, \alpha)$ as there can not be a model in $M_{\mathcal{D}}^{\alpha}$ with this set of parents.
- If $Pa_i \subset Pa'_i$ and $\log(BDEu(Pa_i, s, X_i)) > \log(K_i) + R_i^+(Pa'_i) - \log(\alpha)$, where $R_i^+(Pa'_i)$ is the number of configurations of Pa'_i with $n_{ij} > 0$, then Pa'_i is not added to $Par(X_i, s, \alpha)$ and none of the supersets of Pa'_i is considered as possible set of parents for X_i.

Once AllScores(S, α) computes $Par(X_i, s, \alpha)$ for any $s \in S$ and any variable X_i, then an A* algorithm is applied to compute all the order relationships σ in $\{1, \ldots, m\}$ and values $s \in S$ such that there is a pair (G, s) in H_s^{α} with $Pa_{\sigma(i)} \subseteq \{X_{\sigma(1)}, \ldots, X_{\sigma(i-1)}\}$. For this we introduce two modifications of the algorithm proposed in [10]: a value α has been considered and we compute not only the order with the optimal value but also all the orders within a factor α of the optimal one.

A partial order σ_k is given by the values $(\sigma_k(1), \ldots, \sigma_k(k))$ for $k \leq m$ (only the first k values are specified). A partial order can be defined for values $k = 0, \ldots, m$. For $k = m$ we have a full order and for $k = 0$ the empty order. A graph G is compatible with a partial order σ_k when $Pa_{\sigma_k(i)} \subseteq \{X_{\sigma_k(1)}, \ldots, X_{\sigma_k(i-1)}\}$ for $i = 1, \ldots, k$. Given an ESS s and σ_k, it is possible to give an upper bound for the logarithm of the score of all the orders σ that are extensions of σ_k and which is given by,

$$
Bound(\sigma_k, s) = \sum_{i=1}^{k} Best(X_{\sigma(i)}, \{X_{\sigma(1)}, \ldots, X_{\sigma(i-1)}\}, s) +
$$
$$
\sum_{i=k+1}^{m} Best(X_{\sigma(i)}, \mathbf{X} \setminus \{X_{\sigma(i)}\}, s), \tag{8}
$$

where $Best(X_i, A, s)$ is the best score stored in $Par(X_i, s, \alpha)$, between those set of parents $Pa_i \subseteq A$, i.e. we select the parents compatible with partial order σ_k for variables $X_{\sigma_k(1)}, \ldots, X_{\sigma_k(k)}$ and the rest of the set of parents are chosen in an arbitrary way.

Our algorithm is applied to nodes $N(A_k, s, score, up)$, where A_k is a set $\{\sigma_k(1), \ldots, \sigma_k(k)\}$ for a partial order σ_k, s is a value in S, $score$ is the value of $Bound(\sigma_k, s)$, and up is a reference to the node $N(A_{k-1}, s, score', up')$ such that σ_k is obtained by extending partial order σ_{k-1} with the value $\sigma_k(k)$.

The A* algorithm is initiated with a priority queue with a node for each possible value of $s \in S$, $N(\emptyset, s, score, NULL)$, where $score$ is obtained by applying Eq. (8) to partial order σ_0 (empty partial order). The algorithm stores a value B which is the best score obtained so far for a complete order (which is equal to the score of the first complete order selected from the priority queue) and $H(A, s)$ which is the best score obtained so far for a node $N(A_k, s, score, up)$ where $A_k = A$. Let us note that for a node $N(A_k, s, score, up)$ it is always possible to recover its corresponding partial order σ_k as we have that $\sigma_k(k) = A_k \setminus A_{k-1}$ where A_{k-1} is the set appearing in node $N(A_{k-1}, s, score', up')$ referenced by up, and the rest of values can be recursively found by applying the same operation of the node $N(A_{k-1}, s, score', up')$.

The algorithm proceeds selecting the node with highest $score$ from the priority queue while the priority queue is not empty and $score \geq B + \log(\alpha)$. If this node is $N(A_k, s, score, up)$, then if it is complete ($A_k = \mathbf{X}$), the node is added to the set of solution nodes. In the case it is not complete then all the nodes $N(A_{k+1}, s, score', up')$ obtained by adding one variable X_l, in $\mathbf{X} \setminus A_k$ to A_k are computed, where up' points to former node $N(A_k, s, score, up)$ and the value of $score'$ is calculated taking into account that

$$score' = score + Best(X_l, A_k, s) - Best(X_l, \mathbf{X} \setminus \{X_{\sigma(k+1)}\}, s). \qquad (9)$$

The new node is added to the priority queue if and only if $score' \geq H(A_{k+1}, s) + \log(\alpha)$. In any case the value of $H(A_{k+1}, s)$ is updated if $score' > H(A_{k+1}, s)$.

Once A* is finished, we have a set of solution nodes $N(\mathbf{X}, score, s, up)$. For each one of these nodes we compute their associated order σ and then the order is expanded in a set of networks. Details are given in Algorithm 1. In that algorithm, $ParOrder(X_{\sigma(k)}, s, \alpha, \sigma)$ is the set of pairs $(Pa_{\sigma(k)}, t) \in Par(X_{\sigma(k)}, s, \alpha)$ such that $Pa_{\sigma(k)} \subseteq \{X_{\sigma(1)}, \ldots, X_{\sigma(k-1)}\}$, and t is $\log(BDEu(Pa_{\sigma(k)}, s, X_{\sigma(k)}))$.

The algorithm is initially called with a list L with a pair (G, u) where G is the empty graph, u is the value of $score$ in the solution node $N(\mathbf{X}, score, s, up)$ and with $k = 1$. It works in the following way: it considers pairs (G, u), where G is a partial graph (parents for variables $X_{\sigma(1)}, \ldots, X_{\sigma(k-1)}$ have been selected, but not for the rest of variables) and u is the best score that could be achieved if the optimal selection of parents is done for the variables $X_{\sigma(k)}, \ldots, X_{\sigma(m)}$. Then, the possible candidates for parents of variable $X_{\sigma(k)}$ are considered. If $Pa_{\sigma(k)}$ is a possible candidate set with a score of t and the optimal set of parents for this variable is T, then if this parent set is chosen, then $T - t$ is lost with respect to the optimal selection. If u was the previous optimal value, now it is $u' = u - T + t$. This set of parents can be selected only if $u' \geq B + \log(\alpha)$; in that case, the new graph G' obtained from G by adding links from $Pa_{\sigma(k)}$ to $X_{\sigma(k)}$ is considered

Algorithm 1. Computing the networks associated to an order and ESS s

Require: σ, an order of the variables
Require: α, the factor of the optimal score
Require: B, the best score of any network
Require: s, the ESS
Require: L a list of pairs (G, u) where G is a partial graph and u is the best score of a completion of G
Require: k the node to expand
Ensure: $LR = \{(G_1, \ldots, G_k\}$, a list of graphs with an admissible score compatible with σ

```
 1: procedure EXPAND(σ,α,B,s,L,k)
 2:     if k > m then
 3:         Let LR the list of graphs G such that (G, u) ∈ L
 4:         return LR
 5:     end if
 6:     Let L' equal to ∅
 7:     Let T = max{t : (Pa_σ(k), t) ∈ ParOrder(X_σ(k), s, α, σ)}
 8:     for (G, u) ∈ L do
 9:         Let Q be the set of pairs (Pa_σ(k), t) ∈ ParOrder(X_σ(k), s, α, σ) with
10:                 u − T + t ≥ B + log(α)
11:         for (Pa_σ(k), t) ∈ Q do
12:             Let G' the graph G expanded with links from Pa_σ(k) to X_σ(k)
13:             Let u' = u − T + t
14:             Add (G', u') to L'
15:         end for
16:     end for
17:     return EXPAND(σ,score,B,s,L',k + 1)
18: end procedure
```

with optimal value u'. The algorithm proceeds by expanding all the new partial graphs obtained this way, by assigning parents to the next variable, $X_{\sigma(k+1)}$.

Finally we compute the list of all the graphs associated to the result of the algorithm for any solution node $N(\mathbf{X}, score, s, up)$ with the corresponding value s. In this list, it is possible that the same graph is repeated with identical value of s (the same graph can be obtained with two different order of variables). To avoid repetitions a cleaning step is carried out in order to remove the repetitions of identical pairs (G, s). This is the final set of non-dominated set of parameters H_L^α. Finally the set of possible models M_D^α is the set of Bayesian networks $(G, \hat{\beta})$ that are computed for each pair $(G, s) \in H_L^\alpha$ where $\hat{\beta}$ has been obtained by applying Eq. (4).

The number of graphs compatible with an order computed by this algorithm can be very large. The size of L is initially equal to 1, and for each variable $X_{\sigma(k)}$ in $1, \ldots, m$, this number is increased in the different calls to EXPAND(σ,α,B,s,L,k). The increasing depends of the number of set of parents in Q (computed in lines 9–10 of the algorithm). If for each, (G, u), we denote by $NU(G, u)$ the cardinality of Q, then the new cardinality of L' is given by:

$$\sum_{(G,u)\in L} NU(G,u)$$

Observe that if $NU(G,u)$ is always equal to k, then the final number of networks is $O(k^m)$, and the complexity is exponential. However, in the experiments we have observed that this number is not very large (in the low size networks we have considered) as the cardinality of sets Q is decreasing for most of the pairs (G,u) when k increases, as the values u' associated with the new pairs (G',u') in L' are always lower than the value u in the pair (G,u) giving rise to them (see line 13 in the algorithm).

Above this, we have implemented some basic methods for computing the entropy of the probability distribution associated with a Bayesian network $E(G,\beta)$ and the Kullback-Leibler divergence from a model (G,β) to another one (G',β') given by $KL((G',\beta'),(G,\beta))$. For that, following [11, Theorem 8.5], we have implemented a function computing $ELL((G',\beta'),(G,\beta))$ given by:

$$ELL((G',\beta'),(G,\beta)) = \sum_{\mathbf{x}} P_{(G',\beta')}(\mathbf{x}) \log(P_{(G,\beta)}(\mathbf{x})).$$

For this computation, we take into account that $P_{(G,\beta)}(\mathbf{x}) = \prod_{i=1}^{m} P_{(G,\beta)}(x_i|pa_i)$, where pa_i is a generic configuration of the parents Pa_i of X_i in G, obtaining the following expression:

$$ELL((G',\beta'),(G,\beta)) = \sum_{i=1}^{m}\sum_{pa_i} P_{(G',\beta')}(x_i,pa_i) \log(P_{(G,\beta)}(x_i|pa_i)).$$

In this expression, $P_{(G,\beta)}(x_i|pa_i)$ is directly available in Bayesian network (G,β), but $P_{(G',\beta')}(x_i,pa_i)$ is not and have to be computed by means of propagation algorithms in Bayesian network (G',β'). This is done with a variable elimination algorithm for each configuration of the parents $Pa_i = pa_i$, entering it as evidence and computing the result for variable X_i without normalization. This provides the desired value $P_{(G',\beta')}(x_i,pa_i)$.

Finally, the values of entropy and Kullback-Leibler divergence are computed as follows:

$$E(G,\beta) = -ELL((G',\beta'),(G,\beta))$$
$$KL((G',\beta'),(G,\beta)) = ELL((G',\beta'),(G',\beta)) - ELL((G',\beta'),(G,\beta))$$

4   Experiments

To test the methods proposed in this paper we have carried out a series of experiments with 3 small networks obtained from the Bayesian networks repository in **bnlearn** web page [19]. The networks are: Cancer (5 nodes, 10 parameters), Earthquake (5 nodes, 10 parameters), and Survey (6 nodes, 21 parameters). The main reason for not using larger networks was the complexity associated to compute the Kullback-Leibler divergence for all the pairs of possible models. This is a really challenging problem, as if the number of networks is T, then $T(T-1)$ divergences must be computed, and each one of them, involves a significant number of propagation algorithms computing joint probability distributions. So, at this stage the use of large networks is not feasible to select the network with minimum maximum KL divergence to the rest of possible networks.

4.1 Experiment 1

In this case, we have considered a set of possible values for ESS, $S = \{0.1, 0.5, 1.0, 2.0, 4.0, 6.0, 8.0, 10.0\}$, and we have repeated 200 times the following sequence:

- A dataset of size 500 is simulated from the original network.
- The set of possible networks is computed with a value of $\alpha = e^{-0.6}$.
- The maximum entropy network (MEntropy), the minimum of maximum Kullback-Leibler divergence (MinMaxKL), and the maximum score network for all the sample sizes (Bayesian) are computed. For all of them the Kullback-Leibler divergence with the original one are also computed, as well as the maximum (MaxKL) and minimum divergence (MinKL) of all the possible models with the original one.

The means of the divergences of the estimated models can be seen in Table 1. We can observe as the usual method for learning Bayesian networks (considering the graph with highest score) gives rise to a network with a divergence between the maximum and minimum of the divergences of all the possible networks, and that the average is higher than the middle of the interval determined by the averages of the minimum and the maximum. This supports the idea that the Bayesian procedure somewhat makes an arbitrary selection among a set of networks that are all plausible given the data. This idea is also supported by Fig. 1 in which the density of the Bayesian, MinKL, and MaxKL divergences are depicted for each one of the networks[2]. On it we can see the similarities between the densities of these three values: of course the MinKL density is a bit biased to the left and MaxKL density to the right, being the Bayesian density in the middle, but with very small differences. This again supports the idea that all the computed models should be considered as result of the learning process.

When selecting a single model, we also show that our alternative methods based on considering a family of possible models and then selecting the one with maximum entropy or minimum of maximum of Kullback-Leibler divergence produce networks with a lower divergence on average to the original one than the usual Bayesian procedure. We have carried out a Friedman non-parametric test and in the three networks the differences are significant (p-values: 0.000006, 0.0159, 0.0023, for Cancer, Earthquake, and Survey networks, respectively). In a posthoc Friedman-Nemenyi test, the differences between MinMaxKL and

---

[2] Plotted with Python **seaborn** package.

Table 1. Means of divergences of estimated models and the original one

Network	Bayesian	MinKL	MaxKL	MEntroKL	MinMaxKL
Cancer	0.013026	0.011225	0.014126	0.012712	0.012270
Earthquake	0.017203	0.013132	0.019769	0.016784	0.016072
Survey	0.031459	0.028498	0.033899	0.031257	0.030932

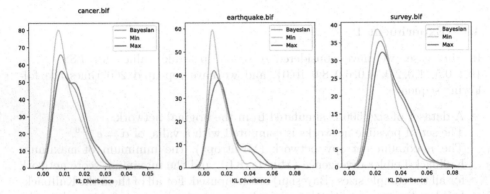

Fig. 1. Density for the Bayesian, minimum, and maximum Kullback-Leibler divergences.

Bayesian are significant in Cancer and Survey networks (p-values: 0.027, 0.017) but not in Cancer (p-value: 0.1187). The differences of MaxEntropy and the Bayesian procedure are not significant.

4.2   Experiment 2

In this case, we have a similar setting than in Experiment 1, but what we have measured is the number of networks that are obtained by our procedure (number of elements in $M_{\mathcal{D}}^{\alpha}$) and the distribution of the number of networks by each ESS $s \in S$. In Fig. 2 we can see the densities of the number of networks (left) and the figure with the averages of the networks by each $s \in S$. First, we can observe that the number of possible networks is low in average (below 5) for our selection of networks, α, and sample size, but that the right queue of the densities is somewhat large, existing cases in which the number of possible networks is 20 or more. With respect to the number of networks by value of ESS s, the most important fact is that the distribution of networks by ESS is highly dependent of the network, being the networks for Survey obtained with much higher values of s than in the case of Cancer or Earthquake. This result puts in doubt the usual practice of selecting a value of s when learning a Bayesian network without thinking that this does not have an effect in the final result.

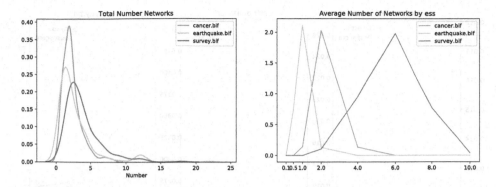

Fig. 2. Densities of the number of networks (left) and average number of networks by ESS (right).

4.3 Experiment 3

In this case, we compare the results of selecting a unique network by fixing a value of s (the optimal one for this value) with the result of selecting the parameter s and the network G optimizing the score. We again repeat a similar experiment to the other two cases, but we compute the networks: the Bayesian network, given by the pair (G, s) with highest score (the Bayesian approach in Experiment 1), and the best graph for each one of the values $s \in S$. For each one of the networks we compute the Kullback-Leibler divergence with the original one. The results of the averages of these divergences are depicted In Fig. 3 for each one of the networks. The dashed line represents the average of the divergence pair (G, s) with best score. On it, we can see that selecting the pair with best score is a good idea in Cancer network, as it produces an average divergence approximately equal to the best selection of value of s, but that is not the case of Earthquake and Survey networks, as there are many selections of s producing networks with lowest divergences than the divergence of the pair with best score. For this reason, is not always a good idea to select the equivalent sample size by using an empirical likelihood approach (the sample size giving rise to greatest likelihood). Other observation is that the shape of the densities of the divergences is quite different by network. For example, in Survey the lowest divergences are obtained with the highest values of s, while in Cancer a minimum is obtained for a low sample size of 2.

4.4 Experiment 4

In this case we have tested the evolution of the number of possible networks (elements in M_D^α) as a function of the sample size. For this aim, instead of fixing a sample size of 500, we have repeated the generation of a sample and the estimation of the models M_D^α for different samples sizes ($n = 400, 500, 1000, 2000, 5000, 10000, 20000, 40000$) and for each value of n we have compute the number of models in M_D^α (repeating it 200 times). Finally Fig. 4 shows the average number

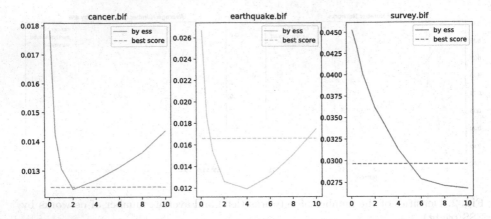

Fig. 3. Kulback Leibler of the best network and the best network by ESS s.

of models for each sample size. As it can be expected the number of models decreases when the sample size increases, very fast at the beginning and more slowly afterwards. In some cases, there are minor increasings in the average number of models when the sample size increases. We think that this is due to the fact that the density of the number of models has a long queue to the right, existing the possibility of obtaining some few cases with a high number of models. This fact can produce this small local irregularities.

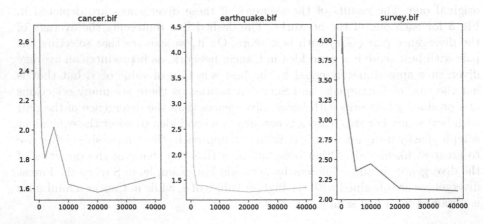

Fig. 4. Evolution of the number of possible networks as a function of the sample size.

5   Conclusions and Future Work

In this paper we have applied the general procedure proposed in Moral [15] to learn a generalized credal network (a set of possible Bayesian networks) given a dataset of observations. We have implemented algorithms for its computation and we have shown that the results applied to learning from samples simulated from small networks are promising. In particular, our main conclusion is that the usual procedure of selecting a network with the highest score does not make too much sense, when there is a set of networks that are equally plausible and that represents probability distributions with a similar divergence to the one associated with the true network. Even in this family, we can find networks using other alternative procedures with smaller divergences to the original one, as the case of considering the minimum of the maximum of Kullback-Leibler divergences in the family of possible models.

Our plans for future work are mainly related to making scalable the proposed procedures and algorithms. When the number of variables increases a direct application of the methods in this paper can be unfeasible. We could try to use more accurate bounds to prune A* search [3], but even so, the number of networks for a threshold could be too large to be computed. Experiments in this line are necessary. Then it would be convenient to develop approximations that could learn a set of significant networks from the full family of possible ones. Other line of research is to integrate several networks into a more compact representation: for example if a group of networks share the same structure with different probabilities try to represent it as a credal network with imprecision in the probabilities.

Other important task is to try to use the set of possible models to answer structural questions, as: is there a link from X_i to X_j? An obvious way to answer it is to see whether this link is present in all the networks of set of learned models, in none of the networks, or in some of them but not in all. In that case, the answer could be yes, no, or possibly. But a theoretical study justifying this or alternative decision rules would be necessary, as well as algorithms designed to answer these questions without an explicit construction of the full set of models.

References

1. Ankan, A., Panda, A.: pgmpy: Probabilistic graphical models using Python. In: Proceedings of the 14th Python in Science Conference (SCIPY 2015). Citeseer (2015). https://doi.org/10.25080/Majora-7b98e3ed-001
2. Cattaneo, M.E.G.V.: A continuous updating rule for imprecise probabilities. In: Laurent, A., Strauss, O., Bouchon-Meunier, B., Yager, R.R. (eds.) IPMU 2014. CCIS, vol. 444, pp. 426–435. Springer, Cham (2014). https://doi.org/10.1007/978-3-319-08852-5_44
3. Correia, A.H., Cussens, J., de Campos, C.P.: On pruning for score-based Bayesian network structure learning. arXiv preprint arXiv:1905.09943 (2019)
4. Correia, A.H.C., de Campos, C.P., van der Gaag, L.C.: An experimental study of prior dependence in Bayesian network structure learning. In: International Symposium on Imprecise Probabilities: Theories and Applications, pp. 78–81 (2019)
5. Couso, I., Moral, S.: Sets of desirable gambles: conditioning, representation, and precise probabilities. Int. J. Approximate Reasoning **52**(7), 1034–1055 (2011). https://doi.org/10.1016/j.ijar.2011.04.004

6. Cozman, F.: Credal networks. Artif. Intell. **120**, 199–233 (2000). https://doi.org/ 10.1016/S0004-3702(00)00029-1
7. Friedman, N., Koller, D.: Being Bayesian about network structure. A Bayesian approach to structure discovery in Bayesian networks. Mach. Learn. **50**, 95–125 (2003). https://doi.org/10.1023/A:1020249912095
8. Gärdenfors, P., Sahlin, N.E.: Unreliable probabilities, risk taking, and decision making. Synthese **53**(3), 361–386 (1982). https://doi.org/10.1007/BF00486156
9. Heckerman, D., Geiger, D., Chickering, D.M.: Learning Bayesian networks: The combination of knowledge and statistical data. Mach. Learn. **20**(3), 197–243 (1995). https://doi.org/10.1023/A:1022623210503
10. Karan, S., Zola, J.: Exact structure learning of Bayesian networks by optimal path extension. In: 2016 IEEE International Conference on Big Data (Big Data), pp. 48–55. IEEE (2016). https://doi.org/10.1109/BigData.2016.7840588
11. Koller, D., Friedman, N.: Probabilistic Graphical Models: Principles and Techniques. MIT press, Cambridge (2009)
12. Liao, Z.A., Sharma, C., Cussens, J., van Beek, P.: Finding all Bayesian network structures within a factor of optimal. In: Proceedings of the AAAI Conference on Artificial Intelligence, vol. 33, pp. 7892–7899 (2019). https://doi.org/10.1609/aaai. v33i01.33017892
13. Masegosa, A.R., Moral, S.: Imprecise probability models for learning multinomial distributions from data. Applications to learning credal networks. Int. J. Approximate Reasoning **55**(7), 1548–1569 (2014). https://doi.org/10.1016/j.ijar.2013.09. 019
14. Moral, S.: An empirical comparison of score measures for independence. In: Proceedings of the 10th IPMU International Conference, pp. 1307–1314 (2004)
15. Moral, S.: Learning with imprecise probabilities as model selection and averaging. Int. J. Approximate Reasoning **109**, 111–124 (2019). https://doi.org/10.1016/j. ijar.2019.04.001
16. Neapolitan, R.: Learning Bayesian Networks. Prentice Hall, Upper Saddle River (2004)
17. Pearl, J.: Probabilistic Reasoning with Intelligent Systems. Morgan & Kaufman, San Mateo (1988). https://doi.org/10.1016/C2009-0-27609-4
18. Quaeghebeur, E.: Desirability. In: Introduction to Imprecise Probabilities, chap. 1, pp. 1–27. Wiley (2014). https://doi.org/10.1002/9781118763117.ch1
19. Scutari, M.: Bayesian network repository of bnlearn (2007). https://www.bnlearn. com/bnrepository/
20. Shafer, G.: A Mathematical Theory of Evidence. Princeton University Press, Princeton (1976)
21. Silander, T., Kontkanen, P., Myllymäki, P.: On sensitivity of the MAP Bayesian network structure to the equivalent sample size parameter. In: Proceedings of the Twenty-Third Conference on Uncertainty in Artificial Intelligence, pp. 360–367. AUAI Press (2007)
22. Walley, P.: Towards a unified theory of imprecise probability. Int. J. Approximate Reasoning **24**, 125–148 (2000). https://doi.org/10.1016/S0888-613X(00)00031-1

A Study of the Set of Probability Measures Compatible with Comparative Judgements

Alexander Erreygers[1]([envelope])[iD] and Enrique Miranda[2][iD]

[1] Ghent University, Ghent, Belgium
alexander.erreygers@ugent.be
[2] University of Oviedo, Oviedo, Spain
mirandaenrique@uniovi.es

Abstract. We consider a set of comparative probability judgements over a finite possibility space and study the structure of the set of probability measures that are compatible with them. We relate the existence of some compatible probability measure to Walley's behavioural theory of imprecise probabilities, and introduce a graphical representation that allows us to bound, and in some cases determine, the extreme points of the set of compatible measures. In doing this, we generalise some earlier work by Miranda and Desterche on elementary comparisons.

Keywords: Comparative probabilities · Credal sets · Lower previsions · Sets of desirable gambles · Extreme points

1 Introduction

The elicitation of probability measures can be cumbersome in situations of imprecise or ambiguous information. To deal with this problem, a number of approaches have been put forward in the literature: we may work for instance with sets of probability measures, or *credal sets*; consider lower and/or upper bounds of the 'true' probability measure, representing the information in terms of *non-additive measures*; or model the information in terms of its behavioural implications. These different models are often referred to with the common term *imprecise probabilities* [1].

On the other hand, when the available information comes from expert judgements it may be easier to model it in terms of comparative assessments of the form 'event A is at least as probable as event B.' This leads to *comparative probabilities*, that were studied first by de Finetti [5] and later by other authors such as Koopman [9], Good [7] or Savage [15]. For a recent thorough overview, as well as an extensive philosophical justification and a summary of the most important results, we refer to [8].

© Springer Nature Switzerland AG 2020
M.-J. Lesot et al. (Eds.): IPMU 2020, CCIS 1238, pp. 165–179, 2020.
https://doi.org/10.1007/978-3-030-50143-3_13

In this paper, we consider a collection of comparative probability judgements over a finite possibility space and study the structure of the set of compatible probability measures. Specifically, we shall investigate in which cases this set is non-empty, the number of its extreme points and their features, and the properties of its associated lower probability. While most earlier work on comparative probabilities has mainly focused on the complete case—that is, the case where any two events are compared—ours is not the first study of the incomplete one; in this respect, the most influential works for this paper are those of Walley [17, Section 4.5] and Miranda and Destercke [11]. Our present contribution has the same goals as that of Miranda and Destercke, but our setting is more general: where they exclusively focused on the specific case of comparisons between elementary events, we generalise some of their results to the case of comparisons between arbitrary events.

The paper is organised as follows. We start with a formal introduction of comparative assessments in Sect. 2, and subsequently discuss the compatibility problem and show that it can be easily tackled using Walley's theory of lower previsions. From Sect. 3 on, we study the set of extreme points of the associated credal set. To that end, we introduce a graphical representation in Sect. 4; this representation allows us to determine the number of extreme points in a number of special cases in Sect. 5, where we also argue that this approach cannot be extended to the general case. We conclude in Sect. 6 with some additional comments and remarks. Due to space constraints, proofs have been omitted.

2   Comparative Assessments and Compatibility

Consider a finite possibility space $\mathscr{X}$ with cardinality n, and a (finite) number m of comparative judgements of the form 'event A is at least as likely as event B.' For ease of notation, we will represent the i-th judgement as a pair (A_i, B_i) of events—that is, subsets of the possibility space $\mathscr{X}$. Finally, we collect all m judgements in the comparative assessment

$$\mathscr{C} := \{(A_i, B_i) \colon i \in \{1, \ldots, m\}, A_i, B_i \subseteq \mathscr{X}\}.$$

Equivalently, the comparative judgements can be represented in terms of a (possibly partial) binary relation $\succeq$ on $2^{\mathscr{X}}$, the power set of the possibility space $\mathscr{X}$, with $A \succeq B$ being equivalent to $(A, B) \in \mathscr{C}$. Miranda and Destercke [11] exclusively dealt with comparative assessments that consist of comparative judgements that concern singletons, or equivalently, are a subset of $\{(\{x\}, \{y\}) \colon x, y \in \mathscr{X}\}$. We follow them in calling such comparative assessments *elementary*.

Throughout this contribution we will use a running example to illustrate much of the introduced concepts.

Running Example. Let $\mathscr{X} := \{1, 2, 3, 4\}$ and

$$\mathscr{C} := \{(\{1\}, \{2\}), (\{1, 2\}, \{3\}), (\{1, 3\}, \{4\}), (\{1, 2\}, \{4\})\}.$$

Clearly, the corresponding partial binary relation $\succeq$ is given by $\{1\} \succeq \{2\}$, $\{1,2\} \succeq \{3\}$, $\{1,3\} \succeq \{4\}$ and $\{1,2\} \succeq \{4\}$. ◆

Let $\Sigma_{\mathscr{X}}$ denote the set of all probability mass functions on $\mathscr{X}$. We follow the authors of [8,11,13,17] in considering the set

$$\mathscr{M}_{\mathscr{C}} := \left\{ p \in \Sigma_{\mathscr{X}} : (\forall (A, B) \in \mathscr{C}) \sum_{x \in A} p(x) \geq \sum_{x \in B} p(x) \right\} \tag{1}$$

of all probability mass functions that are compatible with the comparative judgements. Following Levi [10], we call $\mathscr{M}_{\mathscr{C}}$ the *comparative credal set*.

Given a set $\mathscr{C}$ of comparative judgements, we should first of all determine whether or not there is at least one compatible probability measure—that is, if the comparative credal set $\mathscr{M}_{\mathscr{C}}$ is non-empty. In the case of elementary judgements [11], this is trivial because the uniform probability distribution is compatible with any elementary comparative assessment. Unfortunately, when more elaborate judgements are allowed this is no longer the case, as is demonstrated by the following example.

Example 1. Consider the possibility space $\mathscr{X} := \{1, 2, 3\}$ and the comparative assessment $\mathscr{C} := \{(\{1\}, \{2, 3\}), (\{2\}, \{1, 3\}), (\{3\}, \{1, 2\})\}$. It follows immediately from these judgements that any compatible probability mass function p should satisfy $p(1) \geq 1/2$, $p(2) \geq 1/2$ and $p(3) \geq 1/2$. However, this is clearly impossible, whence $\mathscr{M}_{\mathscr{C}} = \emptyset$. ◆

2.1 Connection with Sets of Desirable Gambles

The existence of a compatible probability measure was characterized in [16, Theorem 4.1] in the case of complete comparative assessments and in [14, Proposition 4] and [13, Section 2] in the case of partial comparative assessments; see also [17, Section 4.5.2]. In this section, we use Walley's result to establish a connection with the theory of *sets of desirable gambles*, from which we shall derive a number of additional results. We refer to [17] for a detailed account of the theory.

A *gamble* f is a real-valued map on our finite possibility space $\mathscr{X}$. The set of all gambles on $\mathscr{X}$ is denoted by $\mathcal{L}$, and dominance between gambles is understood pointwise. Within $\mathcal{L}$ we may consider the subset $\mathcal{L}^+ := \{f \in \mathcal{L} : f \geq 0, f \neq 0\}$ of non-negative gambles, that in particular includes the *indicator* $\mathbb{I}_A$ of some event $A \subseteq \mathscr{X}$, taking value 1 on A and 0 elsewhere.

It is often convenient to think of a gamble f as an uncertain reward expressed in units of some linear utility scale: in case the outcome of our experiment is x, our subject receives the—possibly negative—pay-off $f(x)$. With this interpretation, our subject can specify a set of *almost desirable gambles* $\mathcal{K}$, being some set of gambles—or uncertain rewards—that she considers acceptable. Such a set $\mathcal{K}$ of almost desirable gambles can be extended to include gambles that are implied by rational behaviour; the least-committal of these extensions is the *natural extension* of $\mathcal{K}$, which is defined as $\mathcal{D}_{\mathcal{K}} := \overline{\mathrm{posi}(\mathcal{K} \cup \mathcal{L}^+)}$, where we consider the topological closure under the supremum norm and the posi operator is defined for any set of gambles $\mathcal{K}' \subseteq \mathcal{L}$ as

$$\mathrm{posi}(\mathcal{K}') := \left\{ \sum_{i=1}^{k} \lambda_i f_i : k \in \mathbb{N}, \lambda_i > 0, f_i \in \mathcal{K} \right\},$$

with $\mathbb{N}$ the set of natural numbers—that is, not including zero. We say that a set of almost desirable gambles $\mathcal{K}$ *avoids sure loss* if and only if $\max f \geq 0$ for all $f \in \mathcal{D}_{\mathcal{K}}$, and that it is *coherent* when $\mathcal{K} = \mathcal{D}_{\mathcal{K}}$. It turns out that $\mathcal{D}_{\mathcal{K}}$ is coherent if and only if $\mathcal{K}$ avoids sure loss, and that $\mathcal{K}$ avoids sure loss if and only if there exists a probability mass function p such that $\sum_{x \in \mathscr{X}} f(x)p(x) \geq 0$ for every $f \in \mathcal{K}$. As a consequence, the compatibility of $\mathscr{C}$ is equivalent to verifying that

$$\mathcal{K}_{\mathscr{C}} := \{\mathbb{I}_A - \mathbb{I}_B : (A, B) \in \mathscr{C}\} \tag{2}$$

avoids sure loss, which immediately leads to the following proposition—see also [14, Proposition 4], [13, Section 2] or [17, Lemma 3.3.2].

Proposition 1. *The comparative credal set $\mathscr{M}_{\mathscr{C}}$ is non-empty if and only if for every $\lambda : \{1, \ldots, m\} \to \mathbb{N} \cup \{0\}$, $\max \sum_{i=1}^{m} \lambda(i)(\mathbb{I}_{A_i} - \mathbb{I}_{B_i}) \geq 0$.*

Any set of gambles $\mathcal{K} \subseteq \mathcal{L}$ determines a *lower prevision* $\underline{P}_{\mathcal{K}}$ on $\mathcal{L}$ defined as

$$\underline{P}_{\mathcal{K}}(f) := \sup\{\mu \in \mathbb{R} : f - \mu \in \mathcal{K}\} \quad \text{for all } f \in \mathcal{L}$$

and a conjugate *upper prevision* $\overline{P}_{\mathcal{K}}$ defined as $\overline{P}_{\mathcal{K}}(f) := -\underline{P}_{\mathcal{K}}(-f)$ for all $f \in \mathcal{L}$. The lower prevision $\underline{P}_{\mathcal{K}}$ and its conjugate upper prevision $\overline{P}_{\mathcal{K}}$ are coherent if and only if $\mathcal{K}$ is coherent. Throughout this contribution, we let $\underline{P}_{\mathscr{C}}$ and $\overline{P}_{\mathscr{C}}$ denote the lower and upper previsions determined by $\mathcal{D}_{\mathcal{K}_{\mathscr{C}}}$. We can use $\underline{P}_{\mathscr{C}}$ to verify whether or not a comparative judgement is saturated and/or redundant:

Proposition 2. *Consider an assessment $\mathscr{C}$ such that $\mathscr{M}_{\mathscr{C}}$ is non-empty. If there is a comparative judgement $(A, B) \in \mathscr{C}$ such that $\mathbb{I}_A - \mathbb{I}_B \in \mathrm{posi}(\mathcal{K}_{\mathscr{C}} \setminus \{\mathbb{I}_A - \mathbb{I}_B\})$, then $\mathscr{M}_{\mathscr{C}} = \mathscr{M}_{\mathscr{C} \setminus \{(A,B)\}}$. If no $(A, B) \in \mathscr{C}$ satisfies this condition, then $\underline{P}_{\mathscr{C}}(\mathbb{I}_A - \mathbb{I}_B) = 0$ for every $(A, B) \in \mathscr{C}$ if and only if for every gamble $f \in \mathcal{K}_{\mathscr{C}}$, $f \notin \mathcal{D}_{\mathcal{K}_f}$ with $\mathcal{K}_f := \mathcal{K}_{\mathscr{C}} \setminus \{f\}$.*

This means that we should first analyse if each constraint (A_i, B_i) can be expressed as a positive linear combination of the other constraints in $\mathscr{C}$; if this is the case, we can remove (A_i, B_i) from our set of assessments. Once we have removed all these redundancies, any constraint that cannot be expressed as a linear combination of the other constraints together with trivial assessments of the type $(A, \emptyset)$ with $\emptyset \neq A \subseteq \mathscr{X}$ will be saturated by some $p \in \mathscr{M}_{\mathscr{C}}$ when the latter set is non-empty.

3   Bounding the Number of Extreme Points

It follows immediately from the properties of probability mass functions that the comparative credal set $\mathcal{M}_{\mathscr{C}}$ defined in Eq. (1) is a convex polytope as it is the intersection of $n + m + 1$ half spaces. It is well-known that if a convex polytope is non-empty, it is completely defined by its extreme points. A bound on the number of extreme points follows from McMullen's theorem [4]:

$$|\text{ext}(\mathcal{M}_{\mathscr{C}})| \leq \binom{m + 1 + \lfloor \frac{n}{2} \rfloor}{m + 1} + \binom{m + \lceil \frac{n}{2} \rceil}{m + 1}. \tag{3}$$

It is also possible to establish an upper bound on the number of extreme points that is independent on the number of comparative judgements; its proof is a relatively straightforward modification of the proofs of [3, Theorem 4.4] or [18, Theorem 5.13].

Proposition 3. *For any assessment $\mathscr{C}$, $|\text{ext}(\mathcal{M}_{\mathscr{C}})| \leq n! \, 2^n$.*

To give a sense of the absolute and relative performance of these bounds, we reconsider our running example.

Running Example. One can easily verify that the extreme points of the credal set $\mathcal{M}_{\mathscr{C}}$ are

$$p_1 := (1, 0, 0, 0), \qquad p_2 := (1/2, 1/2, 0, 0), \qquad p_3 := (1/2, 0, 1/2, 0),$$
$$p_4 := (1/2, 0, 0, 1/2), \qquad p_5 := (1/3, 1/3, 0, 1/3), \qquad p_6 := (1/3, 0, 1/3, 1/3),$$
$$p_7 := (1/4, 1/4, 1/2, 0), \qquad p_8 := (1/5, 1/5, 1/5, 2/5), \qquad p_9 := (1/6, 1/6, 1/3, 1/3).$$

Hence, $|\text{ext}(\mathcal{M}_{\mathscr{C}})| = 9$; the upper bounds on the number of extreme points of Eq. (3) and Proposition 3 are 27 and 384, respectively. ◆

On the other hand, the minimum number of extreme points of a non-empty comparative credal set $\mathcal{M}_{\mathscr{C}}$, regardless of the cardinality of the possibility space, is 1: if $\mathscr{X} := \{1, \ldots, n\}$ and $\mathscr{C} := \{(\{i\}, \{i+1\}) : i = 1, \ldots, n-1\} \cup \{(\{n\}, \{1\})\}$, then $\mathcal{M}_{\mathscr{C}}$ only includes the uniform distribution on $\mathscr{X}$, and as a consequence there is only one extreme point.

Our upper bound on the number of extreme points depends on the cardinality of the space n and the number m of comparative assessments; thus, the bound can be made tighter if we remove constraints that are redundant because they are implied by other constraints and the monotonicity and additivity properties of probability measures. For instance, we may assume without loss of generality that

$$(\forall (A, B) \in \mathscr{C}) \; A \neq \mathscr{X}, B \neq \emptyset, A \cap B = \emptyset. \tag{C0}$$

This allows us to bound the cardinality of $\mathscr{C}$:

Proposition 4. *If $\mathscr{C}$ satisfies (C0), then $m \leq 3^n - 2^{n+1} + 1$.*

Similarly, we may assume without loss of generality that any $(A, B) \in \mathscr{C}$ cannot be made redundant in the following senses:

$$(\nexists(A', B') \in \mathscr{C}, (A', B') \neq (A, B))\ A' \subseteq A, B' \supseteq B. \qquad \text{(monotonicity)}$$

$$(\nexists(A_1, B_1), (A_2, B_2) \in \mathscr{C}), A = A_1 \cup A_2, B = B_1 \cup B_2, A_1 \cap A_2 = \emptyset. \quad \text{(additiv.)}$$

$$(\nexists B_1, B_2 \in 2^{\mathscr{X}}, B_1 \supseteq B_2)\ (A, B_1) \in \mathscr{C}, (B_2, B) \in \mathscr{C}. \qquad \text{(transitivity)}$$

Nevertheless, it is more fruitful to detect redundant constraints using the theory of coherent lower previsions, as we did in Proposition 2. In this manner, given an initial (finite) set $\mathscr{C}$ of comparative assessments, we may proceed iteratively and remove all the redundant constraints, and then use Eq. (3) to bound the number of extreme points of the comparative credal set $\mathscr{M}_{\mathscr{C}}$.

4   A Graphical Approach

Essential for the results established in [11] is the representation of the elementary comparative assessments as a digraph. In the non-elementary case, such a graphical representation will also be helpful. Throughout this contribution we use the graph theoretic terminology as defined in [6]; we do allow ourselves one difference, however: we prefer to use *nodes* instead of vertices.

4.1   Representing the Comparative Assessment as a Graph

Miranda and Destercke [11] proposed a straightforward but powerful representation of the elementary comparative assessment $\mathscr{C}$ as a digraph: the atoms of the possibility space correspond to the nodes, and a directed edge is added from x to y for every $(\{x\}, \{y\}) \in \mathscr{C}$. The extreme points of the credal set are then obtained through the top subnetworks generated by certain sets of nodes [11, Theorem 1].

Because we do not limit ourselves to elementary comparative judgements, we cannot simply take over their construction. One straightforward generalisation of the aforementioned construction is to add a directed edge from x to y if there is a comparative judgement $(A, B) \in \mathscr{C}$ with $x \in A$ and $y \in B$. However, this approach is not terribly useful because there is loss of information: clearly, the digraph alone does not contain sufficient information to reconstruct the comparative judgements it represents. To overcome this loss of information and to end up with one-to-one correspondence, we borrow a trick from Miranda and Zaffalon [12] and add dummy nodes to our graph.

We represent the assessment $\mathscr{C}$ as a digraph $\mathcal{G}$ as follows. First, we add one node for every atom x in the possibility space $\mathscr{X}$. Next, for every comparison (A_i, B_i) in the assessment $\mathscr{C}$, we add an auxiliary node ξ_i, and we add a directed edge from every atom x in A_i to this auxiliary node ξ_i and a directed edge from the auxiliary node ξ_i to every atom y in B_i. Formally, the set of nodes is $\mathcal{N} := \mathscr{X} \cup \{\xi_1, \ldots, \xi_m\}$ and the set of directed edges is

$$\mathcal{E} := \bigcup_{i=1}^{m} \{(x, \xi_i) \colon x \in A_i\} \cup \{(\xi_i, y) \colon y \in B_i\}.$$

Running Example. The corresponding digraph $\mathcal{G}$ is depicted in Fig. 1.   ◆

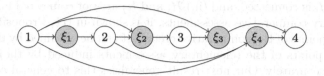

Fig. 1. The digraph $\mathcal{G}$ of the running example

Fix some node ν in the digraph $\mathcal{G}$. Following [11], we use $H(\nu)$ to denote the set that consists of the node ν itself and all of its predecessors, being those nodes ν' such that there is a directed path from ν' to ν. Following [2,11], for any subset N of the set of nodes $\mathcal{N}$, we let $H(N) := \cup_{\nu \in N} H(\nu)$ be the so-called *top subnetwork* generated by N. We will exclusively be concerned with the restriction of these top subnetworks to non-auxiliary nodes; therefore, we define $H'(x) := H(x) \cap \mathcal{X}$ for any x in $\mathcal{X}$ and $H'(A) := H(A) \cap \mathcal{X} = \cup_{x \in A} H'(x)$ for all $A \subseteq \mathcal{X}$.

Running Example. The top subnetwork of the node 1 is $H(1) = \{1\}$ and that of node 3 is $H(3) = \{1, \xi_1, 2, \xi_2, 3\}$. Hence, $H'(\{1,3\}) = \{1,2,3\}$.   ◆

4.2   Some Basic Observations

The following results are straightforward observations that follow almost immediately from our graphical representation $\mathcal{G}$ of the comparative assessment $\mathscr{C}$. The first lemma gives a useful sufficient condition for the existence of a compatible probability measure.

Lemma 1. *If the digraph $\mathcal{G}$ has a node with zero indegree , then $\mathscr{M}_{\mathscr{C}} \neq \emptyset$.*

To facilitate the statement of the following and future results, we introduce some additional notation. For any non-empty event $A \subseteq \mathcal{X}$, we denote the *uniform distribution over A* as p_A. In the particular case that the event A is the singleton $\{x\}$, we also speak of the *degenerate distribution on x*. The second lemma links atoms without predecessors with extreme points that are degenerate distributions.

Lemma 2. *If $x \in \mathcal{X}$ is a node with zero indegree, then the degenerate distribution $p_{\{x\}}$ on x is an extreme point of the comparative credal set $\mathscr{M}_{\mathscr{C}}$.*

Running Example. Observe that the node 1 is the only node with zero indegree. Then, Lemmas 1 and 2 imply that (i) the comparative credal set $\mathscr{M}_{\mathscr{C}}$ is non-empty; and (ii) the degenerate distribution on 1 is an extreme point because $p_1 = p_{\{1\}}$.   ◆

Our next result uses the well-known fact—see for instance [6, Sections 1.6 and 1.4.1]—that any digraph $\mathcal{H}$ can be uniquely decomposed into its *connected components*: the subdigraphs $\mathcal{H}_1, \ldots, \mathcal{H}_k$ such that (i) $\mathcal{H} = \cup_{i=1}^{k} \mathcal{H}_i$, (ii) each subdigraph $\mathcal{H}_i$ is connected, and (iii) $\mathcal{H}_i$ and $\mathcal{H}_j$ are not connected for any $i \neq j$. For elementary comparative assessments, it is shown in [11, Proposition 2] that the extreme points of the comparative credal set can be obtained by determining the extreme points of the (elementary assessments induced by the) connected components separately. Our next result generalises this to general comparative assessments.

Proposition 5. *Denote the connected components of the digraph $\mathcal{G}$ by $\mathcal{G}_1, \ldots, \mathcal{G}_k$. For every connected component $\mathcal{G}_i$, we denote its set of non-auxiliary nodes by $\mathscr{X}_i$ and we let $\mathscr{C}_i$ be the comparative assessment with possibility space $\mathscr{X}_i$ that is in one-to-one correspondence with $\mathcal{G}_i$. Then*

$$\mathrm{ext}(\mathscr{M}_{\mathscr{C}}) = \bigcup_{i=1}^{k} \{\mathrm{extend}(p_i) \colon p_i \in \mathrm{ext}(\mathscr{M}_{\mathscr{C}_i})\},$$

where $\mathrm{extend}(p_i)$ is the cylindrical extension of p_i to $\mathscr{X}$ that is obtained by assigning zero mass to $\mathscr{X} \setminus \mathscr{X}_i$.

Because of this result, without loss of generality we can restrict our attention in the remainder to digraphs $\mathcal{G}$ that are connected. It is also immediate to establish the following result.

Proposition 6. *Consider a set of comparative assessments $\mathscr{C}$, and let $\mathscr{C}' := \mathscr{C} \cup \{(A, B) \colon A \supset B\}$. If there is a cycle $A_1 \succeq A_2 \succeq A_3 \succeq \cdots \succeq A_k \succeq A_{k+1} = A_1$ in $\mathscr{C}'$, then for any $p \in \mathscr{M}_{\mathscr{C}}$, any $i, j \in \{1, \ldots, k\}$ such that $A_j \subset A_i$ and any $x \in A_i \setminus A_j$, $p(x) = 0$.*

In the language of Sect. 2, this means that $\overline{P}_{\mathscr{C}}(\mathbb{I}_{A_i \setminus A_{i+1}}) = 0$ if $A_{i+1} \subset A_i$, so any atom in $A_i \setminus A_{i+1}$ will always have zero mass. Hence, we can simplify the digraph $\mathcal{G}$ by removing nodes that are sure to have zero mass: (i) any atom in $A_i \setminus A_{i+1}$ with $A_{i+1} \subset A_i$; and (ii) if these removals result in the formation of one or more extra disconnected components, the entirety of those disconnected components that used to be connected exclusively by incoming directed edges from (the direct successors of) the previously removed atoms.

Remark 1. This graphical representation also allows us to simplify somewhat the study of the compatibility problem and the extreme points in the following manner. We define a relationship R between the elements of $\mathscr{X}$ as xRy if and only if there is a directed cycle going through x and y. It is easy to see that R is an equivalence relationship. Hence, we may consider the different equivalence classes and the directed edges between them that can be derived from $\mathcal{G}$, leading to a new acyclic digraph $\mathcal{G}'$ on the equivalence classes. Let $\mathcal{G}'_i$ denote the subdigraph associated with the i-th equivalence class and $\mathscr{C}_i$ the corresponding subset of comparative judgements. Observe that

- the set $\mathscr{M}_\mathscr{C}$ is non-empty if and only if at least one of the sets $\mathscr{M}_{\mathscr{C}_i}$ is non-empty, where the associated graph $\mathcal{G}'_i$ has no predecessors in $\mathcal{G}'$;
- if a subgraph $\mathcal{G}'_i$ is such that $\mathscr{M}_{\mathscr{C}_i}$ is empty, then for each of its successors $\mathcal{G}'_j$ any element of $\mathscr{M}_\mathscr{C}$ gives zero probability to the nodes in $\mathcal{G}'_j$.

This also allows us to remove redundant parts of the graph.   ♦

4.3 Acyclic Digraphs

If a digraph is free of directed cycles, then we call it *acyclic* [6, Section 4.2]. Any acyclic digraph has at least one node with zero indegree [6, Lemma 4.1]. Therefore, the following result is an immediate corollary of Lemma 1; alternatively, it is also a corollary of Propositions 8 and 10.

Corollary 1. *If the digraph $\mathcal{G}$ associated with the comparative assessment $\mathscr{C}$ is acyclic, then the associated comparative credal set $\mathscr{M}_\mathscr{C}$ is non-empty.*

On the other hand, a digraph is acyclic if and only if it has a topological ordering, sometimes also called an acyclic numbering [6, Proposition 4.1]. This necessary and sufficient condition allows us to establish the following result.

Proposition 7. *The digraph $\mathcal{G}$ associated with $\mathscr{C}$ is acyclic if and only if there is an ordering $x_1, \ldots, x_n$ of the atoms of the possibility space $\mathscr{X}$ such that*

$$(\forall (A, B) \in \mathscr{C})(\exists i \in \{1, \ldots, n-1\})\ A \subseteq \{x_1, \ldots, x_i\}\ and\ B \subseteq \{x_{i+1}, \ldots, x_n\}.$$

Running Example. It is easy to verify using Fig. 1 that the graph $\mathcal{G}$ is acyclic, and we have seen that the comparative credal set is non-empty. Furthermore, the ordering of Proposition 7 is clearly $1, 2, 3, 4$.   ♦

4.4 Strict Comparative Assessments

Our graphical representation also has implications when we consider a *strict* preference relation, where $A \succ B$ is to be interpreted as 'the event A is more likely than event B.' For a given set $\mathscr{C}$ of comparative judgements, we now consider the set

$$\mathscr{M}_\mathscr{C}^> := \left\{ p \in \Sigma_\mathscr{X} : (\forall (A, B) \in \mathscr{C}) \sum_{x \in A} p(x) > \sum_{y \in B} p(y) \right\}$$

of probability mass functions that are compatible with the strict comparative judgements. Since the set $\mathscr{M}_\mathscr{C}$ is a polytope, it follows that it is the closure of $\mathscr{M}_\mathscr{C}^>$, provided that this latter set is non-empty. In our case, we can prove something stronger: that $\mathscr{M}_\mathscr{C}^>$ is the topological interior of $\mathscr{M}_\mathscr{C}$.

Proposition 8. *For any comparative assessment $\mathscr{C}$, $\mathscr{M}_\mathscr{C}^> = \mathrm{int}(\mathscr{M}_\mathscr{C})$.*

In our next result, we establish a necessary and sufficient condition for $\mathscr{M}_\mathscr{C}^>$ to be non-empty.

Proposition 9. *Let $\mathscr{C}$ be a finite set of strict comparative assessments. Then the following are equivalent:*

(a) *$\mathscr{M}_{\mathscr{C}}^{\geq} \neq \emptyset$.*
(b) *Given the set $\mathcal{K}_{\mathscr{C}}$ defined by Eq. (2), $0 \notin \mathrm{posi}(\mathcal{K}_{\mathscr{C}} \cup \mathcal{L}^{+})$.*
(c) *For every $(A, B) \in \mathscr{C}$, $\overline{P}_{\mathscr{C}}(\mathbb{I}_A - \mathbb{I}_B) > 0$.*

In the case of elementary comparisons, it was established in [11, Lemma 1] that $\mathscr{M}_{\mathscr{C}}^{\geq}$ is non-empty if and only if the digraph $\mathscr{C}$ is acyclic. In the general case, the lack of directed cycles turns out to be sufficient as well.

Proposition 10. *Let $\mathscr{C}$ be a set of strict comparative assessments. If the associated digraph $\mathcal{G}$ is acyclic, then $\mathscr{M}_{\mathscr{C}}^{\geq} \neq \emptyset$.*

Quite remarkably and in contrast with the case of elementary probability comparisons, $\mathscr{M}_{\mathscr{C}}^{\geq}$ can be non-empty even though the digraph $\mathcal{G}$ has directed cycles. For example, if $\mathscr{X} = \{1, 2, 3\}$ and we make the assessments $(\{1, 2\}, \{3\})$ and $(\{3\}, \{1\})$, then the graph has a cycle involving 1 and 3; however, the probability mass function $(0.25, 0.45, 0.3)$ is compatible with the strict assessments.

On the other hand, a necessary condition for $\mathscr{M}_{\mathscr{C}}^{\geq}$ to be non-empty is that we cannot derive from $\mathscr{C}$ a cycle of the type $A_1 \succ A_2 \succ \cdots \succ A_k \succ A_1$. This is equivalent to the graph being acyclic in the case of elementary probability comparisons, and this is what leads to [11, Lemma 1]; however, the two conditions are not equivalent in the general case.

5   Extreme Points of the Comparative Credal Set

As we have often mentioned before, Miranda and Destercke [11] show that in the case of elementary comparative assessments, the extreme points of the comparative credal set can be determined using the graphical representation. More specifically, they show that:

E1. all the extreme points of $\mathscr{M}_{\mathscr{C}}$ correspond to uniform probability distributions [11, Lemma 2];
E2. if $C \subseteq \mathscr{X}$ is the support of an extreme point, then $C = H'(C)$ [11, Lemma 3];
E3. there are at most 2^{n-1} extreme points, and this bound is tight [11, Theorem 4].

Unfortunately, these observations do not hold in the case of non-elementary comparative assessments, as is illustrated by the following example.

Example 2. Let $\mathscr{X} := \{1, \ldots, 5\}$, and let $\mathscr{C}$ be given by

$$\mathscr{C} := \{(\{1, 4\}, \{5\}), (\{2, 4\}, \{1\}), (\{2, 5\}, \{1\}), (\{2, 3, 5\}, \{4\}), (\{2, 3\}, \{1\}),$$
$$(\{2, 4, 5\}, \{3\}), (\{1, 2, 3\}, \{4, 5\}), (\{3, 4\}, \{5\}), (\{1, 5\}, \{3\}),$$
$$(\{1, 3, 4, 5\}, \{2\}), (\{1, 3, 5\}, \{4\}), (\{3, 4, 5\}, \{1\})\}.$$

The 34 extreme points of $\mathscr{M}_{\mathscr{C}}$ are reported in Tab. 1. Note that $34 > 2^5 = 32$. ♦

Table 1. The extreme points of the comparative assessment in Example 2.

i	1	2	3	4	5	6	7	8	9	10	11	12	13	14	15	16	17
$p_i(1)$	$\frac{1}{3}$	$\frac{1}{3}$	$\frac{1}{4}$	$\frac{3}{8}$	$\frac{1}{7}$	$\frac{1}{5}$	$\frac{1}{5}$	0	$\frac{1}{6}$	$\frac{1}{4}$	$\frac{1}{4}$	$\frac{1}{3}$	$\frac{1}{4}$	$\frac{1}{4}$	$\frac{1}{3}$	$\frac{3}{14}$	$\frac{4}{11}$
$p_i(2)$	$\frac{1}{3}$	$\frac{1}{3}$	0	$\frac{2}{8}$	0	0	$\frac{1}{5}$	$\frac{1}{2}$	0	$\frac{1}{8}$	$\frac{1}{3}$	$\frac{1}{4}$	$\frac{1}{4}$	$\frac{1}{3}$	$\frac{1}{4}$	$\frac{1}{14}$	$\frac{3}{11}$
$p_i(3)$	0	$\frac{1}{3}$	$\frac{1}{4}$	$\frac{3}{8}$	$\frac{2}{7}$	$\frac{2}{5}$	$\frac{2}{5}$	0	$\frac{1}{3}$	0	$\frac{1}{8}$	0	0	$\frac{1}{4}$	$\frac{1}{6}$	$\frac{3}{14}$	$\frac{1}{11}$
$p_i(4)$	$\frac{1}{3}$	0	$\frac{1}{4}$	$\frac{1}{8}$	$\frac{1}{7}$	$\frac{1}{5}$	0	$\frac{1}{4}$	$\frac{1}{6}$	$\frac{3}{8}$	$\frac{3}{8}$	$\frac{1}{6}$	$\frac{1}{4}$	0	0	$\frac{2}{14}$	$\frac{1}{11}$
$p_i(5)$	0	0	$\frac{1}{4}$	$\frac{1}{8}$	$\frac{2}{7}$	$\frac{1}{5}$	$\frac{1}{5}$	$\frac{1}{4}$	$\frac{1}{3}$	$\frac{1}{8}$	$\frac{1}{8}$	$\frac{1}{6}$	$\frac{1}{4}$	$\frac{1}{4}$	$\frac{1}{6}$	$\frac{5}{14}$	$\frac{2}{11}$

i	18	19	20	21	22	23	24	25	26	27	28	29	30	31	32	33	34
$p_i(1)$	$\frac{3}{12}$	0	$\frac{1}{10}$	$\frac{1}{6}$	$\frac{1}{5}$	$\frac{2}{12}$	0	$\frac{1}{4}$	$\frac{1}{4}$	$\frac{1}{4}$	$\frac{1}{6}$	$\frac{1}{8}$	$\frac{1}{4}$	0	0	$\frac{1}{4}$	$\frac{1}{7}$
$p_i(2)$	$\frac{1}{12}$	$\frac{1}{4}$	0	0	$\frac{1}{5}$	$\frac{1}{12}$	$\frac{1}{2}$	$\frac{1}{2}$	$\frac{1}{2}$	$\frac{1}{2}$	$\frac{1}{2}$	$\frac{1}{2}$	$\frac{1}{2}$	$\frac{1}{3}$	$\frac{1}{2}$	$\frac{1}{2}$	$\frac{1}{7}$
$p_i(3)$	$\frac{2}{12}$	$\frac{1}{4}$	$\frac{4}{10}$	$\frac{1}{3}$	$\frac{1}{5}$	$\frac{3}{12}$	$\frac{1}{6}$	0	0	$\frac{1}{4}$	$\frac{1}{6}$	$\frac{1}{4}$	$\frac{1}{8}$	$\frac{1}{6}$	$\frac{1}{8}$	$\frac{1}{8}$	$\frac{3}{14}$
$p_i(4)$	$\frac{3}{12}$	$\frac{1}{4}$	$\frac{2}{10}$	$\frac{1}{3}$	$\frac{2}{5}$	$\frac{5}{12}$	$\frac{1}{6}$	$\frac{1}{4}$	$\frac{1}{8}$	0	0	0	0	$\frac{1}{3}$	$\frac{1}{4}$	$\frac{1}{4}$	$\frac{3}{7}$
$p_i(5)$	$\frac{4}{12}$	$\frac{1}{4}$	$\frac{3}{10}$	$\frac{1}{6}$	0	$\frac{1}{12}$	$\frac{1}{6}$	0	$\frac{1}{8}$	0	$\frac{1}{6}$	$\frac{1}{8}$	$\frac{1}{8}$	$\frac{1}{6}$	$\frac{1}{8}$	0	$\frac{1}{14}$

We learn from Example 2 that (E1) does not hold because p_4 is not a uniform distribution; (E2) does not hold because the support of p_1 is $\{1,2,4\}$ but $H'(\{1,2,4\}) = \{1,2,3,4,5\}$; and (E3) does not hold because there are more than $2^{5-1} = 16$ extreme points. In fact, we see that a comparative credal set can have more than 2^n extreme points. Consequently, we cannot use the strategy of [11, Algorithm 1]—that is, construct the possible supports and use the uniform distribution over them—to immediately determine the extreme points of the comparative credal set for some general comparative assessment. That being said, we have nevertheless identified some special cases other than the elementary one in which we *can* generate the extreme points procedurally.

5.1   Multi-level Partitions of Comparative Assessments

As a first special case, we consider a straightforward extension of [11] using a multi-level approach. At the core of this special case are some nested partitions of the possibility space and the restriction that the comparative judgements can only concern events that are on the same level of the nested partitions and belong to the same part of the partition in the previous level. We will here only explain the two-level case in detail; extending the approach to multiple levels is straightforward.

Let $C_1,\dots,C_k$ be a partition of the possibility space $\mathscr{X}$. A comparative assessment $\mathscr{C}$ is *two-level* over this partition if it can be partitioned as

$$\mathscr{C} = \mathscr{C}' \cup \bigcup_{i=1}^{k} \mathscr{C}_i,$$

with $\mathscr{C}' \subseteq \{(A,B) : A, B \in \{C_1,\dots,C_k\}\}$ and $\mathscr{C}_i \subseteq \{(x,y) : x,y \in C_i\}$ for all $i \in \{1,\dots,k\}$. Observe that if such a decomposition exists, then we can interpret $\mathscr{C}'$ as an elementary comparative assessment with possibility space $\mathscr{X}' := \{C_1, \ldots, C_k\}$ and, for all $i \in \{1, \ldots, k\}$, we can interpret $\mathscr{C}_i$ as an elementary comparative assessment with possibility space C_i. Hence, we may use the algorithm described in [11] to determine the extreme points of the comparative credal sets corresponding to these elementary comparative assessments, which we shall denote by $\mathscr{M}_{\mathrm{el}}, \mathscr{M}_{\mathrm{el},1}, \ldots, \mathscr{M}_{\mathrm{el},k}$, respectively. The following result establishes that we can combine these extreme points to obtain the extreme points of the original comparative credal set $\mathscr{M}_{\mathscr{C}}$.

Proposition 11. *Consider a comparative assessment $\mathscr{C}$ that is two-level over the partition $C_1, \ldots, C_k$ of the possibility space $\mathscr{X}$. Then $\mathrm{ext}(\mathscr{M}_{\mathscr{C}})$ is given by*

$$\{\mathrm{comb}(p, p_1, \ldots, p_k) \colon p \in \mathrm{ext}(\mathscr{M}_{\mathrm{el}}), (\forall i \in \{1, \ldots, k\}) \, p_i \in \mathrm{ext}(\mathscr{M}_{\mathrm{el},i})\},$$

where $\mathrm{comb}(p, p_1, \ldots, p_k)$ is the probability mass function defined for all $i \in \{1, \ldots, k\}$ and $x \in C_i$ as $\mathrm{comb}(p, p_1, \ldots, p_k)(x) := p(C_i)p_i(x)$.

Furthermore, as corollary of Proposition 11 and [11, Theorem 4] we obtain the following bound on the number of extreme points.

Corollary 2. *Consider a comparative assessment $\mathscr{C}$ that is two-level over some partition. Then $|\mathrm{ext}(\mathscr{M}_{\mathscr{C}})| \leq 2^{n-1}$.*

5.2 Acyclic Digraphs

Recall from Sect. 4.4 that the absence of cycles simplifies things if we are interested in the compatibility with *strict* comparative judgements. Hence, it does not seem all too far-fetched that determining the (number of) extreme points of the comparative credal set induced by a (non-strict) comparative assessment also simplifies under the absence of cycles. As will become clear in the remainder, this is only certainly so in some special cases.

First, we revisit the three main points of [11] that we recalled at the beginning of this section in the case of acyclic graphs. Our running example shows that also in the acyclic case (E1) does not hold because p_7, p_8 and p_9 are not uniform; (E2) does not hold because p_3 has support $C_3 := \{1, 3\}$ but $H'(C_3) = \{1, 2, 3\} \neq C_3$; and (E3) does not hold because there can be more than $2^{4-1} = 8$ extreme points. Furthermore, since different extreme points can have the same support—in our running example, this is the case for p_7, p_8 and p_9—there is no reason why the number of extreme points should be bounded above by 2^n. Nevertheless, and despite our rather extensive search, we have not succeeded in finding an example of a comparative assessment $\mathscr{C}$ with an acyclic digraph $\mathcal{G}$ that has a comparative credal set with more than 2^n extreme points. This is in contrast with the cyclic case, as we have shown in Example 2.

While the absence of cycles alone does not seem to allow us to efficiently determine the extreme points, there are two interesting special cases that permit us to do so. Essential to both these special cases is a specific class of subdigraphs of the digraph $\mathcal{G}$. To define this class, we first need to introduce two concepts from graph theory. The first concept is that of the *root* of a digraph $\mathcal{H}$: a node ν such that for any other node ν', there is a directed path from ν to ν'. The second concept is that of an *arborescence*: a digraph that has a root and whose underlying graph is a tree. We now call a subdigraph $\mathcal{G}'$ of the digraph $\mathcal{G}$ an *extreme arborescence* if (i) it is an arborescence whose root x^* has no predecessors in the digraph $\mathcal{G}$; and (ii) each of its auxiliary nodes has one direct predecessor and one direct successor.

Important to note here is that all extreme arborescences can be easily procedurally generated. In essence, one needs to (i) select a node x without predecessors in the original digraph $\mathcal{G}$; (ii) either stop or, if possible, (a) add one of the outgoing edges of x and the auxiliary node ξ in which it ends, (b) add one of the outgoing edges of ξ and the atom y such that y is not already in the arborescence; (iii) repeat step (ii) but with x being any of the atoms already in the arborescence.

Singular Assessments. The first special case of acyclic digraphs concerns representing digraphs where every atom has at most one direct predecessor. We call a comparative assessment $\mathcal{C}$ *singular* if $|\{(A, B) \in \mathcal{C}: x \in B\}| \leq 1$ for all $x \in \mathcal{X}$.

We see for instance that the comparative assessment in our running example is not singular, since 4 appears in both the assessments $(\{1,3\},\{4\})$ and $(\{1,2\},\{4\})$, while the comparative assessment $\mathcal{C} := \{(\{1\},\{2\}),(\{1,2\},\{3\}),(\{2,3\},\{4\})\}$, represented in Fig. 2, is.

Fig. 2. A singular digraph $\mathcal{G}$

The graph associated with a singular assessment need not be acyclic—for example, let $\mathcal{X} = \{1,2,3\}$ and consider the comparative judgements $(\{1\},\{2\})$, $(\{2\},\{3\})$ and $(\{3\},\{1\})$. In case it is, we can establish the following:

Theorem 1. *Consider a singular assessment $\mathcal{C}$ such that the associated digraph $\mathcal{G}$ is acyclic. Then every extreme point p of $\mathcal{M}_\mathcal{C}$ corresponds to a unique extreme arborescence $\mathcal{G}' \subseteq \mathcal{G}$ and vice versa, in the sense that p is the unique probability mass function that saturates the comparative constraints associated with the auxiliary nodes in $\mathcal{G}'$ and the non-negativity constraints associated with the atoms that are not in $\mathcal{G}'$.*

Because we can procedurally generate all extreme arborescences, it follows that we can use Theorem 1 to generate all extreme points of the comparative credal set. Another consequence of Theorem 1 is that we can establish a lower and upper bound on the number of extreme points in the singular case.

Theorem 2. *Consider a singular assessment $\mathcal{C}$ such that the associated digraph $\mathcal{G}$ is acyclic. Then $n \leq |\mathrm{ext}(\mathcal{M}_\mathcal{C})| \leq 2^{n-1}$.*

These lower and upper bounds are reached, as we can see from [11, Section 4.1].

Arborescences. Finally, we consider the case that the digraph $\mathcal{G}$ is an arborescence. Clearly, for this it is necessary that $\mathscr{C}$ is singular and that $|A| = 1$ for every $(A, B) \in \mathscr{C}$. As arborescences are special types of acyclic digraphs, we can strengthen Theorem 1 to be—in some sense—similar to [11, Theorem 1].

Theorem 3. *Consider and assessment $\mathscr{C}$ such that the associate graph is an arborescence. Then the set of extreme points of $\mathcal{M}_{\mathscr{C}}$ consists of the uniform distributions on $H'(C)$, where C is any set of atoms such that, for all $x, y \in C$, the closest common predecessor of x and y is a non-auxiliary node.*

We also observe that the bound on the number of extreme points established in Theorem 2 is still valid. To see that this result does not extend to all singular assessments, it suffices to take the extreme points of the assessment depicted in Fig. 2.

6 Conclusions

Although we find the results in this paper promising, there are some open problems that call for additional research, which should help towards making this model more operative for practical purposes. First and foremost, we would like to deepen the study of the acyclic case, and in particular to determine the number and the shape of the extreme points in other particular cases. In addition, a bound on the number of linearly independent constraints, in the manner hinted at in Sect. 3, should let us get a better bound on the number of extreme points. Finally, we should also look for graph decompositions that allow to work more efficiently with comparative judgements.

Acknowledgement. Enrique Miranda's research was supported by projects PGC2018-098623-B-I00 and IDI/2018/000176.

References

1. Augustin, T., Coolen, F., de Cooman, G., Troffaes, M. (eds.): Introduction to Imprecise Probabilities. Wiley Series in Probability and Statistics, Wiley (2014)
2. Cozman, F.G.: Credal networks. Artif. Intell. **120**(2), 199–233 (2000). https://doi.org/10.1016/S0004-3702(00)00029-1
3. Derks, J., Kuipers, J.: On the number of extreme points of the core of a transferable utility game. In: Borm, P., Peters, H. (eds.) Chapters in Game Theory. Theory and Decision Library C: (Series C: Game Theory, Mathematical Programming and Operations Research), pp. 83–97. Springer, Boston (2002). https://doi.org/10.1007/0-306-47526-X_4
4. Dyer, M.E.: The complexity of vertex enumeration methods. Math. Oper. Res. **8**(3), 381–402 (1983). https://doi.org/10.1287/moor.8.3.381
5. de Finetti, B.: Sul significato soggettivo della probabilità. Fundamenta Mathematicae **17**, 298–329 (1931). https://doi.org/10.4064/fm-17-1-298-329

6. Fournier, J.C.: Graphs Theory and Applications: With Exercises and Problems. ISTE & Wiley (2009)
7. Good, I.J.: Probability and the Weighting of Evidence. Griffin (1950)
8. Konek, J.: Comparative probabilities. In: The Open Handbook of Formal Epistemology, pp. 267–348. PhilPapers (2019)
9. Koopman, B.O.: The axioms and algebra of intuitive probability. Ann. Math. **41**(2), 269–292 (1940). https://doi.org/10.2307/1969003
10. Levi, I.: The Enterprise of Knowledge. MIT Press, Cambridge (1980)
11. Miranda, E., Destercke, S.: Extreme points of the credal sets generated by comparative probabilities. J. Math. Psychol. **64–65**, 44–57 (2015). https://doi.org/10.1016/j.jmp.2014.11.004
12. Miranda, E., Zaffalon, M.: Coherence graphs. Artif. Intell. **173**(1), 104–144 (2009). https://doi.org/10.1016/j.artint.2008.09.001
13. Regoli, G.: Comparative probability and robustness. Lect. Notes Monogr. Ser. **29**, 343–352 (1996)
14. Ríos Insua, D.: On the foundations of decision making under partial information. Theor. Decis. **33**, 83–100 (1992). https://doi.org/10.1007/BF00133984
15. Savage, L.J.: The Foundations of Statistics. Wiley, New York (1954)
16. Scott, D.: Measurement structures and linear inequalities. J. Math. Psychol. **1**, 233–247 (1964). https://doi.org/10.1016/0022-2496(64)90002-1
17. Walley, P.: Statistical Reasoning with Imprecise Probabilities. Chapman and Hall (1991)
18. Wallner, A.: Extreme points of coherent probabilities in finite spaces. Int. J. Approximate Reasoning **44**(3), 339–357 (2007). https://doi.org/10.1016/j.ijar.2006.07.017

Coherent and Archimedean Choice
in General Banach Spaces

Gert de Cooman$^{(\boxtimes)}$ (iD)

Foundations Lab for Imprecise Probabilities, Ghent University,
Technologiepark - Zwijnaarde 125, 9052 Zwijnaarde, Belgium
gert.decooman@ugent.be
https://users.ugent.be/~gdcooma

Abstract. I introduce and study a new notion of Archimedeanity for
binary and non-binary choice between options that live in an abstract
Banach space, through a very general class of choice models, called sets
of desirable option sets. In order to be able to bring horse lottery options
into the fold, I pay special attention to the case where these linear spaces
do not include all 'constant' options. I consider the frameworks of conservative inference associated with Archimedean choice models, and also
pay quite a lot of attention to representation of general (non-binary)
choice models in terms of the simpler, binary ones. The representation
theorems proved here provide an axiomatic characterisation of, amongst
other choice methods, Levi's E-admissibility and Walley–Sen maximality.

Keywords: Choice function · Coherence · Archimedeanity · Set of
desirable option sets.

1 Introduction

This paper is about rational decision making under uncertainty using choice
functions, along the lines established by Teddy Seidenfeld and colleagues [12].
What are the underlying ideas? A subject is to choose between options u, which
are typically uncertain rewards, and which live in a so-called option space $\mathcal{V}$.
Her choices are typically represented using a *rejection function* R or a *choice
function* C. For any finite set A of options, $R(A) \subseteq A$ contains those options
that our subject rejects from the option set A, and the remaining options in
$C(A) = A \setminus R(A)$ are the ones that are then still being considered. It is important
to note that $C(A)$ is not necessarily a singleton, so this approach allows for
indecision. Also, the *binary* choices are the ones where A has only two elements,
and I will not be assuming that these binary choices completely determine the
behaviour of R or C on option sets with more than two elements: I will be
considering choice behaviour that is not necessarily binary in nature.

My aim here is to present a theory of coherent and Archimedean choice
(functions), complete with conservative inference and representation results, for
very general option spaces: general Banach spaces that need not have constants.

© Springer Nature Switzerland AG 2020
M.-J. Lesot et al. (Eds.): IPMU 2020, CCIS 1238, pp. 180–194, 2020.
https://doi.org/10.1007/978-3-030-50143-3_14

For the basic theory of coherent choice (functions) on general linear option spaces but without Archimedeanity, I will rely fairly heavily on earlier work by Jasper De Bock and myself [6,7]. The present paper expands that work to include a discussion of a novel notion of Archimedeanity. Since this approach needs a notion of closeness, I will need to focus on option spaces that are Banach, but I still want to keep the treatment general enough so as to avoid the need for including constant options.

The reasons for working with option spaces that are general linear spaces are manifold, and were discussed at length in [7]. In summary, doing so allows us to deal with options that are gambles [14,19], i.e. bounded real-valued maps on some set of possible states $\mathcal{X}$, that are considered as uncertain rewards. These maps constitute a linear space $\mathcal{G}(\mathcal{X})$, closed under point-wise addition and point-wise multiplication with real numbers. But it also brings in, in one fell swoop, vector-valued gambles [15,23], polynomial gambles to deal with exchangeability [2,16], equivalence classes of gambles to deal with indifference [18], and abstract gambles defined without an underlying possibility space [21,22]. In all these cases, the space of options essentially includes all real constants—or constant gambles. But when we want our approach to also be able to deal generically with options that are horse lotteries, it is obvious that we need to consider option spaces that do not include all real constants.

In order to keep the length of this paper manageable, I have decided to focus on the mathematical developments only, and to keep the discussion fairly abstract. For a detailed exposition of the motivation for and the interpretation of the choice models discussed below, I refer to earlier joint papers by Jasper De Bock and myself [6,7]. I also recommend Jasper De Bock's most recent paper on Archimedeanity [5], as it contains a persuasive motivation for the new Archimedeanity condition, in the more restrictive and concrete context where options are gambles. I introduce binary choice models on abstract option spaces in Sect. 2, and extend the discussion to general—not necessarily binary—choice in Sect. 3. After these introductory sections, I focus on adding Archimedeanity to the picture. The basic representation tools that will turn out to be useful in this more restricted context, namely linear and superlinear bounded real functionals, are discussed in Sect. 4. The classical approach to Archimedeanity [10,11,19] for binary choice—which I will call essential Archimedeanity—is given an abstract treatment in Sect. 5. Sections 6 and 7 then deal with the new notion of Archimedeanity in the binary and general case, and discuss conservative inference and the representation of general Archimedean choice models in terms of binary (essentially) Archimedean ones. I conclude in Sect. 8 by stressing the relevance of my findings. Proofs for the results given below can be found in preprint version on arXiv [1].

2   Coherent Sets of Desirable Options

We begin by considering a linear space $\mathcal{V}$, whose elements u are called *options*, and which represent the objects that a subject can choose between. This *option space* $\mathcal{V}$ has some so-called *background ordering* $\succ$, which is 'natural' in that we assume that our subject's choices will always at least respect this ordering, even before she has started reflecting on her preferences. *This background ordering* $\succ$ *is taken to be a strict vector ordering on* u, so an irreflexive and transitive binary relation that is compatible with the addition and scalar multiplication of options.

We will assume that our subject's binary choices between options can be modelled by a so-called *set of desirable options* $D \subseteq \mathcal{V}$, where an option is called *desirable* when the subject strictly prefers it to the zero option. We will denote the set of all possible set of desirable option sets—all subsets of $\mathcal{V}$— by **D**. Of course, a set of desirable options D strictly speaking only covers the strict preferences $\triangleright$ between options u and 0: $u \triangleright 0 \Leftrightarrow u \in D$. For other strict preferences, it is assumed that they are compatible with the vector addition of options: $u \triangleright v \Leftrightarrow u - v \triangleright 0 \Leftrightarrow u - v \in D$.

We impose the following rationality requirements on a subject's strict preferences. A set of desirable options $D \in \mathbf{D}$ is called *coherent* [2,3,20] if it satisfies the following axioms:

D_1. $0 \notin D$;
D_2. if $u, v \in D$ and $(\lambda, \mu) > 0$, then $\lambda u + \mu v \in D$;
D_3. $\mathcal{V}_{\succ 0} \subseteq D$.

We will use the notation $(\lambda, \mu) > 0$ to mean that λ, μ are non-negative real numbers such that $\lambda + \mu > 0$. We denote the set of all coherent sets of desirable options by $\overline{\mathbf{D}}$.

$\overline{\mathbf{D}}$ is an intersection structure: for any non-empty family of sets of desirable options $D_i \in \overline{\mathbf{D}}$, $i \in I$, its intersection $\bigcap_{i \in I} D_i$ also belongs to $\overline{\mathbf{D}}$. This also implies that we can introduce a *coherent closure* operator $\mathrm{cl}_{\overline{\mathbf{D}}} \colon \mathbf{D} \to \overline{\mathbf{D}} \cup \{\mathcal{V}\}$ by letting $\mathrm{cl}_{\overline{\mathbf{D}}}(\mathcal{A}) := \bigcap \{D \in \overline{\mathbf{D}} \colon \mathcal{A} \subseteq D\}$ be the smallest—if any—coherent set of desirable options that includes $\mathcal{A}$. We call an assessment $\mathcal{A} \subseteq \mathcal{V}$ *consistent* if $\mathrm{cl}_{\overline{\mathbf{D}}}(\mathcal{A}) \neq \mathcal{V}$, or equivalently, if $\mathcal{A}$ is included in some coherent set of desirable options. The closure operator $\mathrm{cl}_{\overline{\mathbf{D}}}$ implements *conservative inference* with respect to the coherence axioms, in that it extends a consistent assessment $\mathcal{A}$ to the most conservative—smallest possible—coherent set of desirable options $\mathrm{cl}_{\overline{\mathbf{D}}}(\mathcal{A})$.

A coherent set of desirable options $\hat{D}$ is called *maximal* if none of its supersets is coherent: $(\forall D \in \overline{\mathbf{D}})(\hat{D} \subseteq D \Rightarrow \hat{D} = D)$. This turns out to be equivalent to the following so-called totality condition on $\hat{D}$ [2,3]:

D_T. for all $u \in \mathcal{V} \setminus \{0\}$, either $u \in \hat{D}$ or $-u \in \hat{D}$.

The set of all maximal sets of desirable options is denoted by $\overline{\mathbf{D}}_T$. These maximal elements can be used to represent all coherent sets of desirable options.

Theorem 1 ([2]). *For any* $D \in \mathbf{D}$, $\mathrm{cl}_{\overline{\mathbf{D}}}(D) = \bigcap \{\hat{D} \in \overline{\mathbf{D}}_T \colon D \subseteq \hat{D}\}$. *Hence, a consistent* D *is coherent if and only if* $D = \bigcap \{\hat{D} \in \overline{\mathbf{D}}_T \colon D \subseteq \hat{D}\}$.

Corollary 1. *A set of desirable options* $D \in \mathbf{D}$ *is coherent if and only if there is some non-empty* $\mathcal{D} \subseteq \overline{\mathbf{D}}_T$ *such that* $D = \bigcap \{\hat{D} \colon \hat{D} \in \mathcal{D}\}$. *The largest such set* $\mathcal{D}$ *is* $\{\hat{D} \in \overline{\mathbf{D}}_T \colon D \subseteq \hat{D}\}$.

For more details, and more 'constructive' expressions for $\mathrm{cl}_{\overline{\mathbf{D}}}$, see [2,7].

I also want to mention another, additional, rationality property, central in Teddy Seidenfeld's work [10–12], but introduced there in a form more appropriate for strict preferences between horse lotteries. We can get to the appropriate counterpart here when we introduce the $\mathrm{posi}(\cdot)$ operator, which, for any subset V of $\mathcal{V}$, returns the set of all positive linear combinations of its elements: $\mathrm{posi}(V) := \{\sum_{k=1}^{n} \lambda_k u_k \colon n \in \mathbb{N}, \lambda_k \in \mathbb{R}_{>0}, u_k \in V\}$. We call a set of desirable options $D \in \mathbf{D}$ *mixing* if it is coherent and satisfies the following mixingness axiom:

$\mathrm{D_M}$. for all finite subsets A of $\mathcal{V}$, if $\mathrm{posi}(A) \cap D \neq \emptyset$, then also $A \cap D \neq \emptyset$.

We denote the set of all mixing sets of desirable options by $\overline{\mathbf{D}}_{\mathrm{M}}$. They can be characterised as follows.

Proposition 1 ([15,17]). *Consider any set of desirable options $D \in \overline{\mathbf{D}}$ and let $D^c := \mathcal{V} \setminus D$. Then D is mixing if and only if $\mathrm{posi}(D^c) = D^c$, or equivalently, $D \cap \mathrm{posi}(D^c) = \emptyset$.*

They are therefore identical to the so-called *lexicographic* sets of desirable option sets introduced by Van Camp et al. [15,17]. For more details, see also [7,15,17].

3 Coherent Sets of Desirable Option Sets

We now turn from strict binary preferences—of one option u over another option v—to more general ones. The simplest way to introduce those more general choice models in the present context goes as follows. We call any finite subset A of $\mathcal{V}$ an *option set*, and we collect all such option sets into the set $\mathcal{Q}$. We call an option set A *desirable* to a subject if she assesses that at least one option in A is desirable, meaning that it is strictly preferred to 0. We collect a subject's desirable option sets into her *set of desirable option sets* K. We denote the set of all such possible sets of desirable option sets—all subsets of $\mathcal{Q}$—by $\mathbf{K}$.

The rationality requirements we will impose on such sets of desirable option sets turn out to be fairly natural generalisations of those for sets of desirable options. A set of desirable option sets $K \subseteq \mathcal{Q}$ is called *coherent* [7] if it satisfies the following axioms:

$\mathrm{K_0}$. if $A \in K$ then also $A \setminus \{0\} \in K$, for all $A \in \mathcal{Q}$;
$\mathrm{K_1}$. $\{0\} \notin K$;
$\mathrm{K_2}$. if $A_1, A_2 \in K$ and if, for all $u \in A_1$ and $v \in A_2$, $(\lambda_{u,v}, \mu_{u,v}) > 0$, then also $\{\lambda_{u,v} u + \mu_{u,v} v \colon u \in A_1, v \in A_2\} \in K$;
$\mathrm{K_3}$. if $A_1 \in K$ and $A_1 \subseteq A_2$, then also $A_2 \in K$, for all $A_1, A_2 \in \mathcal{Q}$;
$\mathrm{K_4}$. $\{u\} \in K$, for all $u \in \mathcal{V}_{\succ 0}$.

We denote the set of all coherent sets of desirable option sets by $\overline{\mathbf{K}}$.

A coherent set of desirable option sets K contains singletons, doubletons, and so on. Moreover, it also contains all supersets of its elements, by Axiom $\mathrm{K_3}$.

The singletons in K represent the binary choices, or in other words, the pure desirability aspects. We let $D_K := \{u \in \mathcal{V} : \{u\} \in K\}$ be the set of desirable options that represents the binary choices present in the model K. Its elements are the options that—according to K—are definitely desirable. But there may be elements A of K of higher cardinality that are minimal in the sense that K has none of its strict subsets. This means that our subject holds that at least one option in A is desirable, but her model holds no more specific information about which of these options actually are desirable. This indicates that the choice model K has non-binary aspects. If such is not the case, or in other words, if every element of K goes back to some singleton in K, meaning that $(\forall A \in K)(\exists u \in A)\{u\} \in K$, then we call the choice model K *binary*. With any $D \in \mathbf{D}$, our interpretation inspires us to associate a set of desirable option sets K_D, defined by $K_D := \{A \in \mathcal{Q} : A \cap D \neq \emptyset\}$. It turns out that a set of desirable option sets K is binary if and only if it has the form K_D, and the unique *representing* D is then given by D_K.

Proposition 2 ([7]). *A set of desirable option sets $K \in \mathbf{K}$ is binary if and only if there is some $D \in \mathbf{D}$ such that $K = K_D$. This D is then necessarily unique, and equal to D_K.*

The coherence of a binary set of desirable option sets is completely determined by the coherence of its corresponding set of desirable options.

Proposition 3 ([7]). *Consider any binary set of desirable option sets $K \in \mathbf{K}$ and let $D_K \in \mathbf{D}$ be its corresponding set of desirable options. Then K is coherent if and only if D_K is. Conversely, consider any set of desirable options $D \in \mathbf{D}$ and let K_D be its corresponding binary set of desirable option sets, then K_D is coherent if and only if D is.*

So the binary coherent sets of desirable option sets are given by $\{K_D : D \in \overline{\mathbf{D}}\}$, allowing us to call any coherent set of desirable option sets in $\overline{\mathbf{K}} \setminus \{K_D : D \in \overline{\mathbf{D}}\}$ *non-binary*. If we replace such a non-binary coherent set of desirable option sets K by its corresponding set of desirable options D_K, we lose information, because then necessarily $K_{D_K} \subset K$. Sets of desirable option sets are therefore more expressive than sets of desirable options. But our coherence axioms lead to a representation result that allows us to still use sets of desirable options, or rather, sets of them, to completely characterise *any* coherent choice model.

Theorem 2 ([7]). *A set of desirable option sets $K \in \mathbf{K}$ is coherent if and only if there is some non-empty set $\mathcal{D} \subseteq \overline{\mathbf{D}}$ of coherent sets of desirable options such that $K = \bigcap\{K_D : D \in \mathcal{D}\}$. The largest such set $\mathcal{D}$ is then $\overline{\mathbf{D}}(K) := \{D \in \overline{\mathbf{D}} : K \subseteq K_D\}$.*

It is also easy to see that $\overline{\mathbf{K}}$ is an intersection structure: if we consider any non-empty family of coherent sets of desirable options K_i, $i \in I$, then their intersection $\bigcap_{i \in I} K_i$ is still coherent. This implies that we can introduce a *coherent closure* operator $\mathrm{cl}_{\overline{\mathbf{K}}} : \mathbf{K} \to \overline{\mathbf{K}} \cup \{\mathcal{Q}\}$ by letting $\mathrm{cl}_{\overline{\mathbf{K}}}(\mathcal{A}) := \bigcap\{K \in \overline{\mathbf{K}} : \mathcal{A} \subseteq K\}$

be the smallest—if any—coherent set of desirable option sets that includes $\mathcal{A}$. We call an assessment $\mathcal{A} \subseteq \mathcal{Q}$ *consistent* if $\mathrm{cl}_{\overline{\mathbf{K}}}(\mathcal{A}) \neq \mathcal{Q}$, or equivalently, if $\mathcal{A}$ is included in some coherent set of desirable option sets. The closure operator $\mathrm{cl}_{\overline{\mathbf{K}}}$ implements *conservative inference* with respect to the coherence axioms, in that it extends a consistent assessment $\mathcal{A}$ to the most conservative—smallest possible—coherent set of desirable option sets $\mathrm{cl}_{\overline{\mathbf{K}}}(\mathcal{A})$. In combination with Theorem 2, this leads to the following important result.

Theorem 3 ([6,7]). *For any $K \in \mathbf{K}$, $\mathrm{cl}_{\overline{\mathbf{K}}}(K) = \bigcap \{K_D : D \in \overline{\mathbf{D}}(K)\}$. Hence, a consistent K is coherent if and only if $K = \bigcap \{K_D : D \in \overline{\mathbf{D}}(K)\}$.*

We can also lift the mixingness property from binary to general choice models, as Seidenfeld et al. have done [12].

K_M. if $B \in K$ and $A \subseteq B \subseteq \mathrm{posi}(A)$, then also $A \in K$, for all $A, B \in \mathcal{Q}$.

We call a set of desirable option sets $K \in \mathbf{K}$ *mixing* if it is coherent and satisfies K_M. The set of all mixing sets of desirable option sets is denoted by $\overline{\mathbf{K}}_M$, and it also constitutes an intersection structure. It therefore comes with its own mixing closure operator and associated conservative inference system.

The *binary* elements of $\overline{\mathbf{K}}_M$ are precisely the ones based on a mixing set of desirable options.

Proposition 4 ([7]). *For any set of desirable options $D \in \mathbf{D}$, K_D is mixing if and only if D is, so $K_D \in \overline{\mathbf{K}}_M \Leftrightarrow D \in \overline{\mathbf{D}}_M$.*

For general mixing sets of desirable option sets that are not necessarily binary, we still have a representation theorem analogous to Theorem 2.

Theorem 4 ([7]). *A set of desirable option sets $K \in \mathbf{K}$ is mixing if and only if there is some non-empty set $\mathcal{D} \subseteq \overline{\mathbf{D}}_M$ of mixing sets of desirable options such that $K = \bigcap \{K_D : D \in \mathcal{D}\}$. The largest such set $\mathcal{D}$ is then $\overline{\mathbf{D}}_M(K) := \{D \in \overline{\mathbf{D}}_M : K \subseteq K_D\}$.*

How can we connect this choice of model, sets of desirable option sets, to the rejection and choice functions that I mentioned in the Introduction, and which are much more prevalent in the literature? Their interpretation provides the clue. Consider any option set A, and any option $u \in A$. Then, with $A \ominus u := (A \setminus \{u\}) - u = \{v - u : v \in A, v \neq u\}$,

$$u \in R(A) \Leftrightarrow 0 \in R(A - u)$$
$$\Leftrightarrow \text{ there is some } v \in A \ominus u \text{ that is strictly preferred to } 0$$
$$\Leftrightarrow A \ominus u \in K.$$

4   Linear and Superlinear Functionals

Because the notions of essential Archimedeanity and Archimedeanity that I intend to introduce further on rely on an idea of openness—and therefore closeness—I will assume from now on that the option space $\mathcal{V}$ constitutes a *Banach space* with a norm $\|\cdot\|_{\mathcal{V}}$ and a corresponding topological closure operator Cl and interior operator Int. In this section, I have gathered a few useful definitions and basic results for linear and superlinear bounded real functionals on the space $\mathcal{V}$. These functionals generalise to our more general context the linear and coherent lower previsions defined by Peter Walley [19] on spaces of gambles.

A real functional $\Gamma\colon \mathcal{V} \to \mathbb{R}$ on $\mathcal{V}$ is called *bounded* if its *operator norm* $\|\Gamma\|_{\mathcal{V}^\circ} < +\infty$, where we let $\|\Gamma\|_{\mathcal{V}^\circ} := \sup_{u \in \mathcal{V} \setminus \{0\}} \frac{|\Gamma(u)|}{\|u\|_{\mathcal{V}}}$. We will denote by $\mathcal{V}^\circ$ the linear space of such bounded real functionals on $\mathcal{V}$.

This space can be topologised by the operator norm $\|\cdot\|_{\mathcal{V}^\circ}$, which leads to the so-called *initial topology* on $\mathcal{V}^\circ$. If we associate with any $u \in \mathcal{V}$ the so-called *evaluation functional* $u^\circ\colon \mathcal{V}^\circ \to \mathbb{R}$, defined by $u^\circ(\Gamma) := \Gamma(u)$ for all $\Gamma \in \mathcal{V}^\circ$, then u° is clearly a real linear functional on the normed linear space $\mathcal{V}^\circ$, whose operator norm $\sup_{\Gamma \in \mathcal{V}^\circ \setminus \{0\}} \frac{|u^\circ(\Gamma)|}{\|\Gamma\|_{\mathcal{V}^\circ}} \leq \sup_{\Gamma \in \mathcal{V}^\circ \setminus \{0\}} |u^\circ(\Gamma)| \frac{\|u\|_{\mathcal{V}}}{|\Gamma(u)|} = \|u\|_{\mathcal{V}} < +\infty$ is finite, which implies that u° is a continuous real linear functional on $\mathcal{V}^\circ$ [9, Section 23.1]. We will also retopologise $\mathcal{V}^\circ$ with the topology of pointwise convergence on $\mathcal{V}^\circ$, which is the weakest topology that makes all evaluation functionals u°, $u \in \mathcal{V}$ continuous. It is therefore weaker than the (so-called) *initial topology* induced by the norm $\|\cdot\|_{\mathcal{V}^\circ}$.

An interesting subspace of $\mathcal{V}^\circ$ is the linear space $\mathcal{V}^*$ of all *linear* bounded— and therefore continuous [9, Section 23.1]—real functionals on $\mathcal{V}$. We will also consider the set $\underline{\mathcal{V}}^*$ of all *superlinear* bounded real functionals $\underline{\Lambda}$ on $\mathcal{V}$, meaning that they are elements of $\mathcal{V}^\circ$ that are furthermore superadditive and non-negatively homogeneous:

$\underline{\Lambda}_1.$ $\underline{\Lambda}(u + v) \geq \underline{\Lambda}(u) + \underline{\Lambda}(v)$ for all $u, v \in \mathcal{V}$;   [superadditivity]
$\underline{\Lambda}_2.$ $\underline{\Lambda}(\lambda u) = \lambda \underline{\Lambda}(u)$ for all $u \in \mathcal{V}$ and all real $\lambda \geq 0$. [non-negative homogeneity]

Obviously, $\underline{\mathcal{V}}^*$ is a convex cone, and $\mathcal{V}^* \subseteq \underline{\mathcal{V}}^* \subseteq \mathcal{V}^\circ$.

With any $\underline{\Lambda} \in \mathcal{V}^\circ$ we can associate its *conjugate* (functional) $\overline{\Lambda}\colon \mathcal{V} \to \mathbb{R}$ defined by $\overline{\Lambda}(u) := -\underline{\Lambda}(-u)$ for all $u \in \mathcal{V}$. It is obviously also bounded. Clearly, a bounded real functional is linear if and only if it is superlinear and *self-conjugate*, i.e. equal to its conjugate.

Proposition 5. *Any* $\underline{\Lambda} \in \underline{\mathcal{V}}^*$ *is (uniformly) continuous.*

If we consider, for any $\underline{\Lambda} \in \underline{\mathcal{V}}^*$ its *set of dominating continuous linear functionals* $\mathcal{V}^*(\underline{\Lambda}) := \{\Lambda \in \mathcal{V}^*\colon (\forall u \in \mathcal{V})\underline{\Lambda}(u) \leq \Lambda(u)\}$, then a well-known version of the Hahn–Banach Theorem [9, Section 28.4, HB17] leads to the following representation result. An important condition for using this version is that $\underline{\Lambda}$ should be both superlinear and continuous.

Theorem 5 (Lower envelope theorem). *For all* $\underline{\Lambda} \in \underline{\mathcal{V}}^*$ *and* $u \in \mathcal{V}$ *there is some* $\Lambda \in \mathcal{V}^*(\underline{\Lambda})$ *such that* $\underline{\Lambda}(u) = \Lambda(u)$.

5   Essential Archimedeanity for Sets of Desirable Options

The background ordering $\succ$ on $\mathcal{V}$ introduced in Sect. 2 allows us to define convex cones of *positive* real functionals:

$$\mathcal{V}^*_{\succ 0} := \{\Lambda \in \mathcal{V}^* : (\forall u \in \mathcal{V}_{\succ 0})\Lambda(u) > 0\} \tag{1}$$

$$\underline{\mathcal{V}}^*_{\succ 0} := \{\underline{\Lambda} \in \underline{\mathcal{V}}^* : (\forall u \in \mathcal{V}_{\succ 0})\underline{\Lambda}(u) > 0\}. \tag{2}$$

Observe that $\mathcal{V}^*_{\succ 0} \subseteq \underline{\mathcal{V}}^*_{\succ 0}$. *We will implicitly assume from now on that the strict vector ordering $\succ$ is such that $\mathcal{V}^*_{\succ 0} \neq \emptyset$, which then of course also implies that $\underline{\mathcal{V}}^*_{\succ 0} \neq \emptyset$. We will also require for the remainder of this paper that $Int(\mathcal{V}_{\succ 0}) \neq \emptyset$: the background cone of positive options has a non-empty interior.*
    With any $\underline{\Lambda} \in \underline{\mathcal{V}}^*$, we can associate a set of desirable options as follows:

$$D_\Lambda := \underline{\Lambda}_{>0} = \{u \in \mathcal{V} : \underline{\Lambda}(u) > 0\}. \tag{3}$$

Also, given a set of desirable options $D \in \mathbf{D}$, we let

$$\mathcal{V}^*(D) := \{\Lambda \in \mathcal{V}^* : (\forall u \in D)\Lambda(u) > 0\} = \{\Lambda \in \mathcal{V}^* : D \subseteq D_\Lambda\} \tag{4}$$

$$\underline{\mathcal{V}}^*(D) := \{\underline{\Lambda} \in \underline{\mathcal{V}}^* : (\forall u \in D)\underline{\Lambda}(u) > 0\} = \{\underline{\Lambda} \in \underline{\mathcal{V}}^* : D \subseteq D_{\underline{\Lambda}}\} \tag{5}$$

where we used Eq. (3) for the second equalities. Clearly, $\mathcal{V}^*(D) \subseteq \underline{\mathcal{V}}^*(D)$. These sets are convex subcones of the convex cone $\underline{\mathcal{V}}^*$, and $\mathcal{V}^*(D)$ is also a convex cone in the dual linear space $\mathcal{V}^*$ of continuous real linear functionals on $\mathcal{V}$.
    Inspired by Walley's [19] discussion of 'strict desirability', we will call a set of desirable options $D \in \mathbf{D}$ *essentially Archimedean* if it is coherent and open.
    It turns out that there is a close connection between essentially Archimedean sets of desirable options and superlinear bounded real functionals. Before we can lay it bare in Propositions 6–7, we need to find a way to associate a superlinear bounded real functional with a set of desirable options $D \in \mathbf{D}$. There are a number of different ways to achieve this, but the following approach is especially productive. Since we assumed from the outset that $Int(\mathcal{V}_{\succ 0}) \neq \emptyset$, we can fix any $u_o \in Int(\mathcal{V}_{\succ 0})$. We use this special option u_o to associate with the set of desirable options D a specific (possibly extended) real functional $\underline{\Lambda}_{D,u_o}$ by letting

$$\underline{\Lambda}_{D,u_o}(u) := \sup\{\alpha \in \mathbb{R} : u - \alpha u_o \in D\} \text{ for all } u \in \mathcal{V}. \tag{6}$$

Proposition 6. *Assume that the set of desirable options D is coherent, then $\underline{\Lambda}_{D,u_o} \in \underline{\mathcal{V}}^*$. Moreover, $\underline{\Lambda}_{D,u_o}(u) \geq 0$ for all $u \in D$ and $\underline{\Lambda}_{D,u_o}(v) \leq 0$ for all $v \in D^c$. Finally, $D_{\underline{\Lambda}_{D,u_o}} = Int(D)$.*

Proposition 7. *A set of desirable options $D \in \mathbf{D}$ is essentially Archimedean if and only if there is some $\underline{\Lambda} \in \underline{\mathcal{V}}^*_{\succ 0}$ such that $D = D_{\underline{\Lambda}}$. In that case, we always have that $D = D_{\underline{\Lambda}_{D,u_o}}$, and therefore $\underline{\Lambda}_{D,u_o} \in \underline{\mathcal{V}}^*(D)$.*

For sets of desirable options that are essentially Archimedean and mixing, we have similar results in terms of *linear* rather than superlinear bounded real functionals.

Proposition 8. *If the set of desirable options D is mixing, then $\Lambda_{D,u_o} \in \mathcal{V}^*$, and we will then denote this linear bounded real functional by Λ_{D,u_o}. Moreover, $\Lambda_{D,u_o}(u) \geq 0$ for all $u \in D$ and $\Lambda_{D,u_o}(v) \leq 0$ for all $v \in D^c$.*

Proposition 9. *A set of desirable options $D \in \mathbf{D}$ is essentially Archimedean and mixing if and only if there is some $\Lambda \in \mathcal{V}_{\succ 0}^*$ such that $D = D_\Lambda$. In that case, we always have that $D = D_{\Lambda_{D,u_o}}$, and therefore $\Lambda_{D,u_o} \in \mathcal{V}^*(D)$.*

Proposition 7 has interesting implications, and it will be helpful to pay more attention to them, for a better understanding of what we are actually doing in Propositions 6–9. The essentially Archimedean sets of desirable options D are all those and only those for which there is some superlinear bounded real functional $\Lambda \in \underline{\mathcal{V}}^*$ such that $D = D_{\underline{\Lambda}}$. But Proposition 7 also guarantees that in this representation $D_{\underline{\Lambda}}$ for D, the superlinear bounded real functional $\underline{\Lambda}$ can always be replaced by the superlinear bounded real functional $\underline{\Lambda}_{D,u_o}$, as we know that $D = D_{\underline{\Lambda}} = D_{\underline{\Lambda}_{D,u_o}}$. The import of all this is that we can associate, with any $u_o \in Int(\mathcal{V}_{\succ 0})$, the following so-called *normalisation map*:

$$N_{u_o} \colon \underline{\mathcal{V}}^* \to \underline{\mathcal{V}}^* \colon \underline{\Lambda} \mapsto N_{u_o}\underline{\Lambda} := \underline{\Lambda}_{D_{\underline{\Lambda}},u_o}$$

where, after a few manipulations, we get

$$N_{u_o}\underline{\Lambda}(u) = \sup\{\alpha \in \mathbb{R} \colon \underline{\Lambda}(u - \alpha u_o) > 0\} \text{ for all } u \in \mathcal{V}. \tag{7}$$

It is the purport of Proposition 7 that if $\underline{\Lambda}$ represents an essentially Archimedean D in the sense that $D = D_{\underline{\Lambda}}$, then so does the version $N_{u_o}\underline{\Lambda}$, in the sense that also $D = D_{N_{u_o}\underline{\Lambda}}$.

6 Archimedeanity for Sets of Desirable Options

One of the drawbacks of working with essentially Archimedean sets of desirable options in Sect. 5, is that they do not constitute an intersection structure—and therefore do not come with a conservative inference method: an arbitrary intersection of essentially Archimedean sets of desirable options is no longer necessarily essentially Archimedean, simply because openness is not necessarily preserved under arbitrary intersections. To remedy this, we now turn to arbitrary intersections of essentially Archimedean models, which of course do constitute an intersection structure. We will see that these types of models also allow for a very elegant and general representation.

We will call a set of desirable options $D \in \overline{\mathbf{D}}$ *Archimedean* if it is coherent and if the following separation property is satisfied:

D_A. $(\forall u \notin D)(\exists \underline{\Lambda} \in \underline{\mathcal{V}}^*(D))\underline{\Lambda}(u) \leq 0,$

and we denote by $\overline{\mathbf{D}}_A$ the set of all coherent and Archimedean sets of desirable options. It is an immediate consequence of the Lower Envelope Theorem [Theorem 5] for superlinear bounded real functionals that this separation property is equivalent to

$\mathrm{D}_A^{\mathrm{p}}$. $(\forall u \notin D)(\exists \Lambda \in \mathcal{V}^*(D))\Lambda(u) \leq 0$,

which shows that the Archimedean sets of desirable option sets are in particular also evenly convex; see [4, Definition 1].

Since, by Proposition 7, all essentially Archimedean sets of desirable options have the form $D_{\underline{\Lambda}}$, Eq. (3) points to the fact that all essentially Archimedean models are also Archimedean:

$$(\forall \underline{\Lambda} \in \underline{\mathcal{V}}^*)D_{\underline{\Lambda}} \in \overline{\mathbf{D}}_A \text{ and in particular also } (\forall \Lambda \in \mathcal{V}^*)D_\Lambda \in \overline{\mathbf{D}}_A. \qquad (8)$$

$\overline{\mathbf{D}}_A$ is an intersection structure. Indeed, consider any non-empty family of Archimedean sets of desirable options D_i, $i \in I$ and let $D := \bigcap_{i \in I} D_i$, then we already know that D is coherent, so we only need to check that the separation condition D_A is satisfied. So consider any $u \notin D$, meaning that there is some $i \in I$ such that $u \notin D_i$. Hence there is some $\underline{\Lambda} \in \underline{\mathcal{V}}^*(D_i)$ such that $\underline{\Lambda}(u) \leq 0$. Since it follows from Eq. (5) that also $\underline{\Lambda} \in \underline{\mathcal{V}}^*(D)$, D is Archimedean.

That $\overline{\mathbf{D}}_A$ is an intersection structure also implies that we can introduce an *Archimedean closure* operator $\mathrm{cl}_{\overline{\mathbf{D}}_A}: \mathbf{D} \to \overline{\mathbf{D}}_A \cup \{\mathcal{V}\}$ by letting $\mathrm{cl}_{\overline{\mathbf{D}}_A}(\mathcal{A}) := \bigcap\{D \in \overline{\mathbf{D}}_A : \mathcal{A} \subseteq D\}$ be the smallest—if any—Archimedean set of desirable options that includes $\mathcal{A}$. We call an assessment $\mathcal{A} \subseteq \mathcal{V}$ *Archimedean consistent* if $\mathrm{cl}_{\overline{\mathbf{D}}_A}(\mathcal{A}) \neq \mathcal{V}$, or equivalently, if $\mathcal{A}$ is included in some Archimedean set of desirable options. The closure operator $\mathrm{cl}_{\overline{\mathbf{D}}_A}$ implements *conservative inference* with respect to the Archimedeanity axioms, in that it extends an Archimedean consistent assessment $\mathcal{A}$ to the most conservative—smallest possible—Archimedean set of desirable options $\mathrm{cl}_{\overline{\mathbf{D}}_A}(\mathcal{A})$.

Theorem 6. *For any set of desirable options $D \in \mathbf{D}$, $\mathrm{cl}_{\overline{\mathbf{D}}_A}(D) = \bigcap\{D_\Lambda : \Lambda \in \mathcal{V}^*(D)\} = \bigcap\{D_{\underline{\Lambda}} : \underline{\Lambda} \in \underline{\mathcal{V}}^*(D)\}$. Hence, an Archimedean consistent D is Archimedean if and only if $D = \bigcap\{D_\Lambda : \Lambda \in \mathcal{V}^*(D)\} = \bigcap\{D_{\underline{\Lambda}} : \underline{\Lambda} \in \underline{\mathcal{V}}^*(D)\}$.*

The following important representation theorem confirms that the essentially Archimedean sets of desirable options can be used to represent all Archimedean sets of desirable options via intersection.

Corollary 2. *For any set of desirable options $D \in \mathbf{D}$, the following statements are equivalent:*

(i) D is Archimedean;

(ii) there is some non-empty set $\underline{\mathcal{L}} \subseteq \underline{\mathcal{V}}_{\succ 0}^*$ of positive superlinear bounded real functionals such that $D = \bigcap\{D_{\underline{\Lambda}} : \underline{\Lambda} \in \underline{\mathcal{L}}\}$;

(iii) there is some non-empty set $\mathcal{L} \subseteq V_{\succ 0}^*$ of positive linear bounded real functionals such that $D = \bigcap \{D_\Lambda : \Lambda \in \mathcal{L}\}$.

The largest such set $\underline{\mathcal{L}}$ is $\underline{V}^*(D)$, and the largest such set $\mathcal{L}$ is $V^*(D)$.

The sets of functionals in these results theorem can of course also be replaced by $N_{u_o}(\underline{V}^*(D))$, $N_{u_o}(V^*(D))$, $N_{u_o}(\underline{\mathcal{L}})$ and $N_{u_o}(\mathcal{L})$ respectively, where u_o is any option in $Int(V_{\succ 0})$.

7   Archimedeanity for Sets of Desirable Option Sets

We now lift the discussion of Archimedeanity from binary to general choice models, that is, sets of desirable option sets. Given a set of desirable option sets $K \in \mathbf{K}$, we let

$$\underline{V}^*(K) := \{\underline{\Lambda} \in \underline{V}^* : (\forall A \in K)(\exists u \in A)\underline{\Lambda}(u) > 0\} = \{\underline{\Lambda} \in \underline{V}^* : K \subseteq K_{\underline{\Lambda}}\}, \quad (9)$$

where we used Eq. (3) and let

$$K_{\underline{\Lambda}} := K_{D_{\underline{\Lambda}}} = \{A \in Q : A \cap D_{\underline{\Lambda}} \neq \emptyset\} = \{A \in Q : (\exists u \in A)\underline{\Lambda}(u) > 0\}. \quad (10)$$

Similarly,

$$V^*(K) := \{\Lambda \in V^* : K \subseteq K_\Lambda\} \subseteq \underline{V}^*(K). \quad (11)$$

If we pick any $u_o \in Int(V_{\succ 0})$ and associate with it the normalisation map N_{u_o}, then since we know that $D_{\underline{\Lambda}} = D_{N_{u_o}\underline{\Lambda}}$ and $D_\Lambda = D_{N_{u_o}\Lambda}$, we also have that

$$K_{\underline{\Lambda}} = K_{N_{u_o}\underline{\Lambda}} \text{ and } K_\Lambda = K_{N_{u_o}\Lambda} \text{ for all } \underline{\Lambda} \in \underline{V}^* \text{ and } \Lambda \in V^*. \quad (12)$$

We will call a set of desirable option sets $K \in \mathbf{K}$ *Archimedean* if it is coherent and if the following separation property is satisfied:

K_A. $(\forall A \notin K)(\exists \underline{\Lambda} \in \underline{V}^*(K))(\forall u \in A)\underline{\Lambda}(u) \leq 0$,

and we denote by $\overline{\mathbf{K}}_A$ the set of all Archimedean sets of desirable option sets.

If we look at Proposition 7, we see that the essentially Archimedean sets of desirable options are all the $D_{\underline{\Lambda}}$, and Eq. (10) then tells us that the corresponding binary sets of desirable option sets $K_{\underline{\Lambda}}$ are all Archimedean:

$$(\forall \underline{\Lambda} \in \underline{V}^*)K_{\underline{\Lambda}} \in \overline{\mathbf{K}}_A. \quad (13)$$

But we can go further than this, and establish a strong connection between Archimedean sets of desirable option sets on the one hand, and Archimedean binary sets of desirable option sets on the other.

Proposition 10. *For any $D \in \mathbf{D}$, K_D is Archimedean if and only if D is.*

$\overline{\mathbf{K}}_A$ is an intersection structure. Indeed, consider any non-empty family of Archimedean sets of desirable option sets K_i, $i \in I$ and let $K := \bigcap_{i \in I} K_i$, then we already know that K is coherent, so we only need to show that the separation condition K_A is satisfied. So consider any $A \notin K$, meaning that there is some $i \in I$ such that $A \notin K_i$. Hence there is some $\underline{\Lambda} \in \underline{\mathcal{V}}^*(K_i)$ such that $\underline{\Lambda}(u) \le 0$ for all $u \in A$. Since it follows from Eq. (9) that also $\underline{\Lambda} \in \underline{\mathcal{V}}^*(K)$, we see that, indeed, K is Archimedean.

That $\overline{\mathbf{K}}_A$ is an intersection structure also implies that we can introduce an *Archimedean closure* operator $\mathrm{cl}_{\overline{\mathbf{K}}_A} : \mathbf{K} \to \overline{\mathbf{K}}_A \cup \{\mathcal{Q}\}$ by letting $\mathrm{cl}_{\overline{\mathbf{K}}_A}(A) := \bigcap \{K \in \overline{\mathbf{K}}_A : A \subseteq K\}$ be the smallest—if any—Archimedean set of desirable option sets that includes A. We call an assessment $K \subseteq \mathcal{Q}$ *Archimedean consistent* if $\mathrm{cl}_{\overline{\mathbf{K}}_A}(K) \ne \mathcal{Q}$, or equivalently, if K is included in some Archimedean set of desirable option sets.

Theorem 7. *For any set of desirable option sets* $K \in \mathbf{K}$, $\mathrm{cl}_{\overline{\mathbf{K}}_A}(K) = \bigcap \{K_{\underline{\Lambda}} : \underline{\Lambda} \in \underline{\mathcal{V}}^*(K)\}$. *Hence, an Archimedean consistent set of desirable option sets* K *is Archimedean if and only if* $K = \bigcap \{K_{\underline{\Lambda}} : \underline{\Lambda} \in \underline{\mathcal{V}}^*(K)\}$.

And here too, the following important representation theorem confirms that the positive superlinear bounded real functionals can be used to represent all Archimedean sets of desirable option sets.

Corollary 3. *A set of desirable option sets* $K \in K$ *is Archimedean if and only if there is some non-empty set* $\underline{\mathcal{L}} \subseteq \underline{\mathcal{V}}^*_{\succ 0}$ *of positive superlinear bounded real functionals such that* $K = \bigcap \{K_{\underline{\Lambda}} : \underline{\Lambda} \in \underline{\mathcal{L}}\}$. *The largest such set* $\underline{\mathcal{L}}$ *is* $\underline{\mathcal{V}}^*(K)$.

The sets of functionals in these results can also be replaced by $N_{u_o}(\underline{\mathcal{V}}^*(D))$ and $N_{u_o}(\underline{\mathcal{L}})$ respectively, where u_o is any option in $Int(\mathcal{V}_{\succ 0})$.

To conclude, let us see what happens if we also impose mixingness for sets of desirable option sets: what can we say about Archimedean and mixing sets of desirable option sets?

Proposition 11. *Consider any set of desirable option sets* $K \in \mathbf{K}$, *any* $\underline{\Lambda} \in \underline{\mathcal{V}}^*$ *and any* $u_o \in Int(\mathcal{V}_{\succ 0})$. *If* K *is mixing, then* $K \subseteq K_{\underline{\Lambda}}$ *implies that* $N_{u_o}\underline{\Lambda}$ *is linear. Hence,* $N_{u_o}(\underline{\mathcal{V}}^*(K)) = N_{u_o}(\mathcal{V}^*(K))$.

And as a sort of converse, the following result identifies the mixing and Archimedean binary sets of desirable option sets. It extends Proposition 9 from essential Archimedeanity to Archimedeanity.

Proposition 12. *Consider any* $u_o \in Int(\mathcal{V}_{\succ 0})$ *and any set of desirable options* $D \in \mathbf{D}$, *then* K_D *is mixing and Archimedean if and only if* $D = D_\Lambda$ *for some* $\Lambda \in \mathcal{V}^*_{\succ 0}$, *and we can always make sure that* $\Lambda(u_o) = 1$.

Corollary 4. *Consider any* $u_o \in Int(\mathcal{V}_{\succ 0})$, *then a set of desirable option sets* $K \in \mathbf{K}$ *is mixing and Archimedean if and only if there is some non-empty set* $\mathcal{L} \subseteq \mathcal{V}^*_{\succ 0}$ *of positive linear bounded real functionals* Λ, *with moreover* $\Lambda(u_o) = 1$, *such that* $K = \bigcap \{K_\Lambda : \Lambda \in \mathcal{L}\}$. *The largest such set* $\mathcal{L}$ *is* $N_{u_o}(\underline{\mathcal{V}}^*(K)) = N_{u_o}(\mathcal{V}^*(K))$.

192   G. de Cooman

8   Conclusions

The results presented here constitute the basis for a very general theory of binary
and non-binary choice. Its foundations are laid by the coherence axioms, which
can be made stronger by adding mixingness and Archimedeanity, separately or
jointly. For each of the sets of axioms thus formed, we get a conservative inference framework with corresponding closure operators, as well as representation
theorems that allow us to construe all coherent, Archimedean or mixing models—
as well as combinations of them—as intersections (infima) of specific types of
binary ones. These representations are especially interesting because they lead
to a complete axiomatic characterisation of various well-known decision making schemes. To give one example, the (coherent and) Archimedean and mixing
models are exactly the ones that correspond to decision making using Levi's
E-admissibility scheme [8,13] associated with general—not necessarily closed or
convex—sets of linear previsions. I believe such a characterisation—jointly with
the one in Jasper De Bock's paper [5]—is achieved here for the first time in its
full generality. And the theory is also flexible enough to allow for characterisations for a plethora of other schemes, amongst which Walley–Sen maximality
[13,19]. Indeed, for the binary choice models we get the decision making schemes
based on maximality for sets of desirable gambles (coherent binary models),
lexicographic probability orderings (mixing binary models), evenly convex sets
of positive superlinear bounded real functionals—lower previsions essentially—
(Archimedean binary models), and evenly convex sets of positive linear bounded
real functionals—linear previsions essentially—(Archimedean and mixing binary
models). And for their more general, non-binary counterparts we get, through
our representation theorems, schemes that are based on arbitrary intersections—
infima—of a whole variety of such binary cases.

What I haven't talked about here are the more constructive aspects of the
various conservative inference frameworks. The representation results in this
paper essentially allow us to express the closure operator that effects the conservative inference as an intersection of dominating special binary models, which
are not always easy (and in some cases even impossible) to identify construc-
tively. We therefore also need to look for other and more constructive ways of
tackling the conservative inference problem; early work on this topic seems to
suggest that this is not an entirely hopeless endeavour [7]. On a related note, the
Archimedeanity axioms D_A, D_A^p and K_A are similarly 'nonconstructive', as they
are based on the existence of (super)linear functionals that 'do certain things'.
For an equivalent approach to these axioms with a more constructive flavour,
and with gambles as options, I refer to Jasper De Bock's paper on this topic [5].

Finally, in a future paper I intend to use the results presented here to derive
similar axiomatic characterisations, conservative inference frameworks and representation theorems when the option space is a set of horse lotteries.

Acknowledgements. This work owes an intellectual debt to Teddy Seidenfeld, who
introduced me to the topic of choice functions. I have been working closely with Jasper
De Bock on diverse aspects of coherent choice, and my continuous discussions with him

have, as ever, been very helpful in finding the right questions to ask here. The basis for much of the work presented above was laid during a research stay as visiting professor at Durham University in the late spring of 2019. I am grateful to Durham University's Department of Mathematical Sciences for making that stay pleasant, fruitful, and possible.

References

1. De Cooman, G.: Coherent and Archimedean choice in general Banach spaces (2020). arXiv:2002.05461
2. De Cooman, G., Quaeghebeur, E.: Exchangeability and sets of desirable gambles. Int. J. Approximate Reasoning **53**(3), 363–395 (2012). Special issue in honour of Henry E. Kyburg, Jr
3. Couso, I., Moral, S.: Sets of desirable gambles: conditioning, representation, and precise probabilities. Int. J. Approximate Reasoning **52**(7), 1034–1055 (2011)
4. Daniilidis, A., Martinez-Legaz, J.E.: Characterizations of evenly convex sets and evenly quasiconvex functions. J. Math. Anal. Appl. **273**(1), 58–66 (2002)
5. De Bock, J.: Archimedean choice functions: an axiomatic foundation for imprecise decision making (2020). arXiv:2002.05196
6. De Bock, J., de Cooman, G.: A desirability-based axiomatisation for coherent choice functions. In: Destercke, S., Denoeux, T., Gil, M.Á., Grzegorzewski, P., Hryniewicz, O. (eds.) SMPS 2018. AISC, vol. 832, pp. 46–53. Springer, Cham (2019). https://doi.org/10.1007/978-3-319-97547-4_7
7. De Bock, J., De Cooman, G.: Interpreting, axiomatising and representing coherent choice functions in terms of desirability. In: International Symposium on Imprecise Probabilities: Theories and Applications, ISIPTA 2019, Proceedings, vol. 103, pp. 125–134 (2019)
8. Levi, I.: The Enterprise of Knowledge. MIT Press, London (1980)
9. Schechter, E.: Handbook of Analysis and Its Foundations. Academic Press, San Diego (1997)
10. Seidenfeld, T., Schervish, M.J., Kadane, J.B.: A representation of partially ordered preferences. Ann. Stat. **23**, 2168–2217 (1995). Reprinted in [11], pp. 69–129
11. Seidenfeld, T., Schervish, M.J., Kadane, J.B.: Rethinking the Foundations of Statistics. Cambridge University Press, Cambridge (1999)
12. Seidenfeld, T., Schervish, M.J., Kadane, J.B.: Coherent choice functions under uncertainty. Synthese **172**(1), 157–176 (2010). https://doi.org/10.1007/s11229-009-9470-7
13. Troffaes, M.C.M.: Decision making under uncertainty using imprecise probabilities. Int. J. Approximate Reasoning **45**(1), 17–29 (2007). https://doi.org/10.1016/j.ijar.2006.06.001
14. Troffaes, M.C.M., De Cooman, G.: Lower Previsions. Wiley, Hoboken (2014)
15. Van Camp, A.: Choice Functions as a Tool to Model Uncertainty. Ph.D. thesis, Ghent University, Faculty of Engineering and Architecture (2018)
16. Van Camp, A., De Cooman, G.: Exchangeable choice functions. Int. J. Approximate Reasoning **100**, 85–104 (2018)
17. Van Camp, A., De Cooman, G., Miranda, E.: Lexicographic choice functions. Int. J. Approximate Reasoning **92**, 97–119 (2018)
18. Van Camp, A., De Cooman, G., Miranda, E., Quaeghebeur, E.: Coherent choice functions, desirability and indifference. Fuzzy Sets Syst. **341**, 1–36 (2018)

19. Walley, P.: Statistical Reasoning with Imprecise Probabilities. Chapman and Hall, London (1991)
20. Walley, P.: Towards a unified theory of imprecise probability. Int. J. Approximate Reasoning **24**, 125–148 (2000)
21. Williams, P.M.: Notes on conditional previsions. Technical report, School of Mathematical and Physical Science, University of Sussex, UK (1975). Revised journal version: [21]
22. Williams, P.M.: Notes on conditional previsions. Int. J. Approximate Reasoning **44**, 366–383 (2007). Revised journal version of [21]
23. Zaffalon, M., Miranda, E.: Axiomatising incomplete preferences through sets of desirable gambles. J. Artif. Intell. Res. **60**, 1057–1126 (2017)

Archimedean Choice Functions
An Axiomatic Foundation for Imprecise Decision Making

Jasper De Bock[✉]

FLip, ELIS, Ghent University, Ghent, Belgium
jasper.debock@ugent.be

Abstract. If uncertainty is modelled by a probability measure, decisions are typically made by choosing the option with the highest expected utility. If an imprecise probability model is used instead, this decision rule can be generalised in several ways. We here focus on two such generalisations that apply to sets of probability measures: E-admissibility and maximality. Both of them can be regarded as special instances of so-called choice functions, a very general mathematical framework for decision making. For each of these two decision rules, we provide a set of necessary and sufficient conditions on choice functions that uniquely characterises this rule, thereby providing an axiomatic foundation for imprecise decision making with sets of probabilities. A representation theorem for Archimedean choice functions in terms of coherent lower previsions lies at the basis of both results.

Keywords: E-admissibility · Maximality · Archimedean choice functions · Decision making · Imprecise probabilities

1 Introduction

Decision making under uncertainty is typically carried out by combining an uncertainty model with a decision rule. If uncertainty is modelled by a probability measure, the by far most popular such decision rule is maximising expected utility, where one chooses the option—or makes the decision—whose expected utility with respect to this probability measure is the highest.

Uncertainty can also be modelled in various other ways though. The theory of imprecise probabilities, for example, offers a wide range of extensions of probability theory that provide more flexible modelling possibilities, such as differentiating between stochastic uncertainty and model uncertainty. The most straightforward such extension is to consider a set of probability measures instead of a single one, but one can also use interval probabilities, coherent lower previsions, sets of desirable gambles, belief functions, to name only a few.

For all these different types of uncertainty models, various decision rules have been developed, making the total number of possible combinations rather daunting. Choosing which combination of uncertainty model and decision rule to use is therefore difficult and often dealt with in a pragmatic fashion, by using a combination that one is familiar with, that is convenient or that is computationaly advantageous.

© Springer Nature Switzerland AG 2020
M.-J. Lesot et al. (Eds.): IPMU 2020, CCIS 1238, pp. 195–209, 2020.
https://doi.org/10.1007/978-3-030-50143-3_15

To motivate the use of a specific combination in a more principled way, one can also consider its properties. That is, one can aim to select an uncertainty model and decision rule whose resulting decisions satisfy the properties that one finds desirable for the decision problem at hand. In some cases, imposing a given set of properties as axioms can even rule out all combinations but one, thereby providing an axiomatic foundation for the use of a specific type of uncertainty model and decision rule. The famous work of Savage [9], for example, provides an axiomatic foundation for maximising expected utility with respect to a probability measure.

The main contributions of this paper are axiomatic foundations for three specific decision rules that make use of imprecise probability models [11]. The first two decision rules, called E-admissibility [7] and maximality [13], apply to a set of probability measures; they both reduce to maximising expected utility when this set contains only a single probability measure, but are otherwise fundamentally different. The third decision rule applies to sets of coherent lower previsions; it is more abstract then the other two, but includes both of them as special cases. This allows us to use our axiomatic foundation for the third rule as an intermediate step towards axiomatising E-admissibility and maximality.

To obtain our results, we make extensive use of choice functions [4, 10, 12]: a unifying framework for studying decision making. These choice functions require no reference to an uncertainty model or a decision rule, but are simply concerned with the decisions themselves, making them an excellent tool for comparing different methods. We will be especially interested in Archimedean choice functions, because all of the decision schemes that we consider are of this particular type.

In order to adhere to the page limit constraint, all proofs are omitted; they are available in the appendix of an extended online version [2].

2   Choice Functions and Uncertainty Models

A choice function C, quite simply, is a function that chooses. Specifically, for every finite set A of options, it returns a subset $C(A)$ of A. We here consider the special case where options are gambles: bounded real functions on some fixed state space $\mathcal{X}$. We let $\mathcal{L}$ be the set of all gambles on $\mathcal{X}$ and we use $\mathcal{Q}$ to denote the set of all finite subsets of $\mathcal{L}$, including the empty set. A choice function C is then a map from $\mathcal{Q}$ to $\mathcal{Q}$ such that, for all $A \in \mathcal{Q}$, $C(A) \subseteq A$.

If $C(A)$ contains only a single option u, this means that u is chosen from A. If $C(A)$ consists of multiple options, several interpretations can be adopted. On the one hand, this can be taken to mean that each of the options in $C(A)$ is chosen. On the other hand, $C(A)$ can also be regarded as a set of options among which it is not feasible to choose, in the sense that they are incomparable based on the available information; in other words: the elements of $A \setminus C(A)$ are rejected, but those in $C(A)$ are not necessarily 'chosen'. While our mathematical results further on do not require a philosophical stance in this matter, it will become apparent from our examples and interpretation that we have the later approach in mind.

A very popular class of choice functions—while not necessarily always called as such—are those that correspond to maximising expected utility. The idea there is to consider a probability measure P and to let $C(A)$ be the element(s) of A whose expected value—or utility—is highest. The probability measure in question is often taken to be countably additive, but we will not impose this restriction here, and impose only finite additivity. These finitely additive probability measures are uniquely characterised by their corresponding expectation operators, which are linear real functionals on $\mathscr{L}$ that dominate the infinum operator. We follow de Finetti in denoting these expectation operators by P as well, and in calling them linear previsions [6, 13].

Definition 1. *A linear prevision P on $\mathscr{L}$ is a map from $\mathscr{L}$ to $\mathbb{R}$ that satisfies*

$P_1.$ $P(u) \geq \inf u$ *for all* $u \in \mathscr{L}$;   boundedness
$P_2.$ $P(\lambda u) = \lambda P(u)$ *for all real* λ *and* $u \in \mathscr{L}$;   homogeneity
$P_3.$ $P(u+v) = P(u) + P(v)$ *for all* $u, v \in \mathscr{L}$.   additivity

We denote the set of all linear previsions on $\mathscr{L}$ by **P**.

For any such linear prevision—or equivalently, any finitely additive probability measure—the choice function obtained by maximising expected utility is defined by

$$C_P(A) := \big\{ u \in A : (\forall v \in A \setminus \{u\})\, P(u) \geq P(v) \big\} \text{ for all } A \in \mathscr{Q}. \tag{1}$$

It returns the options u in A that have the highest prevision—or expectation—$P(u)$.

However, there are also many situations in which it is not feasible to represent uncertainty by a single prevision or probability measure [13, Section 1.4.4]. In those cases, imprecise probability models can be used instead. The most straightforward such imprecise approach is to consider a non-empty set $\mathscr{P} \subseteq \mathbf{P}$ of linear previsions—or probability measures—as uncertainty model, the elements of which can be regarded as candidates for some 'true' but unknown precise model.

In that context, maximising expected utility can be generalised in several ways [11], of which we here consider two. The first is called E-admissibility; it chooses those options that maximise expected utility with respect to at least one precise model in $\mathscr{P}$:

$$C_{\mathscr{P}}^E(A) := \big\{ u \in A : (\exists P \in \mathscr{P})(\forall v \in A \setminus \{u\})\, P(u) \geq P(v) \big\} \text{ for all } A \in \mathscr{Q}. \tag{2}$$

The second generalisation is called maximality and starts from a partial order on the elements of A. In particular, for any two options $u, v \in A$, the option u is deemed better than v if its expectation is higher for every $P \in \mathscr{P}$. Decision making with maximality then consists in choosing the options u in A that are undominated in this order, in the sense that no other option $v \in A$ is better than u:

$$C_{\mathscr{P}}^M(A) := \big\{ u \in A : (\forall v \in A \setminus \{u\})(\exists P \in \mathscr{P})\, P(u) \geq P(v) \big\} \text{ for all } A \in \mathscr{Q}. \tag{3}$$

One can easily verify that $C_{\mathscr{P}}^E(A) \subseteq C_{\mathscr{P}}^M(A)$, making maximality the most conservative decisions rule of the two. Furthermore, in the particular case where $\mathscr{P}$ contains only a single linear prevision, they clearly coincide and both reduce to maximising expected utility. In all other cases, however, maximality and E-admissibility are different; see for example Proposition 3 in Sect. 9 for a formal statement.

One of the main aims of this paper is to characterise each of these two types of choice functions in terms of their properties. That is, we are looking for necessary and sufficient conditions under which a general choice function C is of the form $C_{\mathscr{P}}^{\mathrm{E}}$ or $C_{\mathscr{P}}^{\mathrm{M}}$, without assuming a priori the existence of a set of linear previsions $\mathscr{P}$. Such conditions will be presented in Sects. 7 and 8, respectively.

A crucial intermediate step in obtaining these two results will consist in finding a similar characterisation for choice functions that correspond to (sets of) coherent lower previsions [13], a generalisation of linear previsions that replaces additivity by the weaker property of superadditivity.

Definition 2. *A coherent lower prevision $\underline{P}$ on $\mathscr{L}$ is a map from $\mathscr{L}$ to $\mathbb{R}$ that satisfies*

$LP_1.$ $\underline{P}(u) \geq \inf u$ *for all* $u \in \mathscr{L}$;   boundedness
$LP_2.$ $\underline{P}(\lambda u) = \lambda \underline{P}(u)$ *for all real* $\lambda > 0$ *and* $u \in \mathscr{L}$;   positive homogeneity
$LP_3.$ $\underline{P}(u+v) \geq \underline{P}(u) + \underline{P}(v)$ *for all* $u, v \in \mathscr{L}$.   superadditivity

We denote the set of all coherent lower previsions on $\mathscr{L}$ by $\mathbf{P}$.

That linear previsions are a special case of coherent lower previsions follows trivially from their definitions. There is however also a more profound connection between both concepts: coherent lower previsions are minima of linear ones.

Theorem 1 [13, Section 3.3.3.]. *A real-valued map $\underline{P}$ on $\mathscr{L}$ is a coherent lower prevision if and only if there is a non-empty set $\mathscr{P} \subseteq \mathbf{P}$ of linear previsions such that*

$$\underline{P}(u) = \min\{P(u) : P \in \mathscr{P}\} \text{ for all } u \in \mathscr{L}.$$

Alternatively, coherent lower previsions can also be given a direct behavioural interpretation in terms of gambling, without any reference to probability measures or linear previsions [13,14].

Regardless of their interpretation, with any given non-empty set $\mathscr{P} \subseteq \underline{\mathbf{P}}$ of these coherent lower previsions, we can associate a choice function $C_{\mathscr{P}}$ in the following way:

$$C_{\mathscr{P}}(A) := \left\{u \in A : (\exists \underline{P} \in \mathscr{P})(\forall v \in A \setminus \{u\}) \, \underline{P}(v-u) \leq 0\right\} \text{ for all } A \in \mathscr{Q}. \qquad (4)$$

If the lower previsions in $\mathscr{P}$ are all linear, this definition reduces to E-admissibility, as can be seen by comparing Eqs. (2) and (4). What is far less obvious though, is that maximality is also special case of Eq. (4); see Theorem 7 in Sect. 8 for a formal statement. In that case, the sets $\mathscr{P}$ in Eqs. (4) and (3) may of course—and typically will—be different.

Because the choice functions that correspond to E-admissibility and maximality are both of the form $C_{\mathscr{P}}$, with $\mathscr{P}$ a set of coherent lower previsions, any attempt at characterising the former will of course benefit from characterising the latter. A large part of this paper will therefore be devoted to the development of necessary and sufficient conditions for a general choice function C to be of the form $C_{\mathscr{P}}$. In order to obtain such conditions, we will interpret choice functions in terms of (strict) desirability and establish a connection with so-called sets of desirable option sets. This interpretation will lead to a natural set of conditions that, as we will eventually see in Sect. 6, uniquely characterises choice functions of the form $C_{\mathscr{P}}$. We start with a brief introduction to desirability and sets of desirable option sets.

3   Coherent Sets of Desirable Option Sets

The basic notion on which our interpretation for choice functions will be based, and from which our axiomatisation will eventually be derived, is that of a desirable option: an option that is strictly preferred over the status quo [1, 8, 13]. In our case, where options are gambles $u \in \mathscr{L}$ on $\mathscr{X}$ and the status quo is the zero gamble, this means that the uncertain—and possibly negative—reward $u(x)$, whose actual value depends on the uncertain state $x \in \mathscr{X}$, is strictly preferred over the zero reward. In other words: gambling according to u is strictly preferred over not gambling at all.

We will impose the following three principles on desirable options, where we use '$(\lambda, \mu) > 0$' as a shorthand notation for '$\lambda \geq 0$, $\mu \geq 0$ and $\lambda + \mu > 0$'. The first two principles follow readily from the meaning of desirability. The third one follows from an assumption that rewards are expressed in a linear utility scale.

  d_1. 0 is not desirable;
  d_2. if $\inf u > 0$, then u is desirable;[1]
  d_3. if u, v are desirable and $(\lambda, \mu) > 0$, then $\lambda u + \mu v$ is desirable.

The notion of a desirable option gives rise to two different frameworks for modelling a subject's uncertainty about the value $x \in \mathscr{X}$. The first, which is well established, is that of sets of desirable options—or sets of desirable gambles. The idea there is to consider a set D that consists of options that are deemed desirable by a subject. If such a set is compatible with the principles d_1–d_3, it is called coherent.

Definition 3. *A set of desirable options $D \in \mathbf{D}$ is coherent if it satisfies:*

D_1. $0 \notin D$;
D_2. *if* $\inf u > 0$, *then* $u \in D$;
D_3. *if* $u, v \in D$ *and* $(\lambda, \mu) > 0$, *then* $\lambda u + \mu v \in D$.

We denote the set of all coherent sets of desirable options by $\overline{\mathbf{D}}$.

A more general framework, which will serve as our main workhorse in this paper, is that of sets of desirable option sets [3]. The idea here is to consider a set K of so-called desirable option sets A, which are finite sets of options that, according to our subject, are deemed to contain at least one desirable option. To say that $A = \{u, v\}$ is a desirable option set, for example, means that u or v is desirable. Crucially, the framework of sets of desirable option sets allows a subject to make this statement without having to

---

[1] There is no consensus on which properties to impose on desirability; the main ideas and results are always very similar though [1, 8, 13, 14]. In particular, d_2 is often strengthened by requiring that u is desirable as soon as $\inf u \geq 0$ and $u \neq 0$; we here prefer d_2 because it is less stringent and because it combines more easily with the notion of strict desirability that we will consider in Sect. 5. Annalogous comments apply to D_2 and K_2 further on.

specify—or know—which of the two options u or v is desirable. As explained in earlier work [4, Section 3],[2] it follows from d_1–d_3 that any set of desirable option sets K should satisfy the following axioms. If it does, we call K coherent.

Definition 4. *A set of desirable option sets $K \subseteq \mathcal{Q}$ is coherent if it satisfies:*

K_0. *if $A \in K$ then also $A \setminus \{0\} \in K$, for all $A \in \mathcal{Q}$;*
K_1. *$\{0\} \notin K$;*
K_2. *$\{u\} \in K$ for all $u \in \mathcal{L}$ with $\inf u > 0$;*
K_3. *if $A_1, A_2 \in K$ and if, for all $u \in A_1$ and $v \in A_2$, $(\lambda_{u,v}, \mu_{u,v}) > 0$, then also*

$$\{\lambda_{u,v} u + \mu_{u,v} v : u \in A_1, v \in A_2\} \in K;$$

K_4. *if $A_1 \in K$ and $A_1 \subseteq A_2$, then also $A_2 \in K$, for all $A_1, A_2 \in \mathcal{Q}$.*

We denote the set of all coherent sets of desirable option sets by $\overline{\mathbf{K}}$.

One particular way of obtaining a set of desirable option sets, is to derive it from a set of desirable options D, as follows:

$$K_D := \{A \in \mathcal{Q} : A \cap D \neq \emptyset\}. \tag{5}$$

One can easily verify that if D is coherent, then K_D will be as well [4, Proposition 8]. In general, however, sets of desirable option sets are more expressive than sets of desirable options. The link between both is provided by Theorem 2, which shows that a set of desirable option sets can be equivalently represented by a set of sets of desirable options.

Theorem 2 [4, Theorem 9]. *A set of desirable option sets K is coherent if and only if there is some non-empty set $\mathcal{D} \subseteq \overline{\mathbf{D}}$ such that $K = \bigcap \{K_D : D \in \mathcal{D}\}$.*

In practice, modelling a subject's uncertainty does not require her to specify a full coherent set of desirable option sets though. Instead, it suffices for her to provide an assessment $\mathscr{A} \subseteq \mathcal{Q}$, consisting of option sets A that she considers desirable. If such an assessment is consistent with coherence, meaning that there is at least one coherent set of desirable option sets K that includes $\mathscr{A}$, then this assessment can always be extended to a unique smallest—most conservative—coherent set of desirable option sets, called the natural extension of $\mathscr{A}$. This natural extension is given by

$$\text{Ex}(\mathscr{A}) := \bigcap \{K \in \overline{\mathbf{K}} : \mathscr{A} \subseteq K\} = \bigcap \{K_D : D \in \overline{\mathbf{D}}, \mathscr{A} \subseteq K_D\}, \tag{6}$$

as follows readily from Theorem 2. If $\mathscr{A}$ is not consistent with coherence, $\text{Ex}(\mathscr{A})$ is an empty intersection, which, by convention, we set equal to $\mathcal{Q}$.

---

[2] Reference [4] deals with the more general case where options take values in an abstract vector space $\mathscr{V}$, and where d_2 imposes that u should be desirable if $u \succ 0$, with $\succ$ an arbitrary but fixed strict vector ordering. Whenever we invoke results from [4], we are applying them for the special case where $\mathscr{V} = \mathscr{L}$ and $u \succ v \Leftrightarrow \inf(u - v) > 0$.

4 Strongly Archimedean Sets of Desirable Option Sets

That sets of desirable option sets can be used to axiomatise choice functions of the type $C_{\mathscr{P}}$, with $\mathscr{P}$ a set of coherent lower previsions, was already demonstrated in earlier work [4] for the specific case where $\mathscr{P}$ is closed with respect to pointwise convergence. A key step in that result consisted in strengthening the interpretation of K, replacing desirability with the stronger notion of strict desirability [13, Section 3.7.7]. We here repeat the reasoning that led to this result, before adapting it in Sect. 5 to get rid of the closure condition.

We call a desirable option u strictly desirable if there is some real $\varepsilon > 0$ such that $u - \varepsilon$ is desirable.[3] As a simple consequence of this definition and d_1–d_3, we find that

sd$_1$. 0 is not strictly desirable;
sd$_2$. if $\inf u > 0$, then u is strictly desirable;
sd$_3$. if u, v are strictly desirable and $(\lambda, \mu) > 0$, then $\lambda u + \mu v$ is strictly desirable;
sd$_4$. if u is strictly desirable, then $u - \varepsilon$ is strictly desirable for some real $\varepsilon > 0$.

By applying these principles to sets of desirable options, we arrive at the concept of a coherent set of strictly desirable options: a coherent set of desirable options D that is compatible with sd$_4$. What is particularly interesting about such sets is that they are in one-to-one correspondence with coherent lower previsions [4, 13], thereby allowing us to move from desirability to lower previsions as a first step towards choice functions of the form $C_{\mathscr{P}}$. The problem with coherent sets of strictly desirable options, however, is that they correspond to a single lower prevision $\underline{P}$, while we which to consider a set $\mathscr{P}$ of them. To achieve this, we again consider sets of desirable option sets, but now suitably adapted to strict desirability.

So consider any set of desirable option sets K and let us interpret it in terms of strict desirability. That A belongs to K then means that A contains at least one strictly desirable option. Given this interpretation, what properties should K satisfy? Since the principles sd$_1$–sd$_3$ are identical to d_1–d_3, K should clearly be coherent, meaning that it should satisfy K_0–K_4. Formalising the implications of sd$_4$ is more tricky though, as it can be done in several ways.

The first and most straighforward approach is to impose the following immediate translation of sd$_4$ to desirable option sets, where for all $A \in \mathscr{Q}$ and real ε:

$$A - \varepsilon := \{u - \varepsilon : u \in A\}$$

Definition 5. *A set of desirable option sets K is strongly Archimedean[4] if it is coherent and satisfies the following property:*

K_{SA}. *if $A \in K$, then also $A - \varepsilon \in K$ for some real $\varepsilon > 0$.*

---

[3] Walley's original notion of strict desirability [13, Section 3.7.7] is slightly different. In his version, if $\inf u = 0$ (but $u \neq 0$) then u should also be strictly desirable (but need not satisfy sd$_4$).
[4] In earlier work [4], we have referred to this property as Archimedeanity. With hindsight, however, we now prefer to reserve this terminology for the property in Definition 6.

The reasoning behind this axiom goes as follows. Since $A \in K$ is taken to mean that there is at least one $u \in A$ that is strictly desirable, it follows from sd$_4$ that there is some real $\varepsilon > 0$ such that $u - \varepsilon$ is strictly desirable. This implies that $A - \varepsilon$ contains at least one strictly desirable option. It therefore seems sensible to impose that $A - \varepsilon \in K$.

To explain the implications of this axiom, and how it is related to lower previsions, we need a way to link the latter to sets of desirable option sets. The first step is to associate, with any coherent lower prevision $\underline{P} \in \mathbf{P}$, a set of desirable option sets

$$K_{\underline{P}} := \{A \in \mathcal{Q} : (\exists u \in A)\underline{P}(u) > 0\}. \tag{7}$$

The coherence of this set can be easily verified [4, Propositions 8 and 24]. More generally, with any non-empty set $\mathcal{P}$ of coherent lower previsions, we associate a set of desirable option sets

$$K_{\mathcal{P}} := \bigcap\{K_{\underline{P}} : \underline{P} \in \mathcal{P}\}. \tag{8}$$

Coherence is again easily verified; it follows directly from the coherence of $K_{\underline{P}}$ and the fact that coherence is preserved under taking intersections. The final tool that we need to explain the implications of strong Archimedeanity, does the opposite; it starts with a coherent set of desirable option sets K, and associates a set of coherent lower previsions with it, defined by

$$\mathbf{P}(K) := \{\underline{P} \in \mathbf{P} : K \subseteq K_{\underline{P}}\}.$$

If K is strongly Archimedean, then as the following result shows, $\mathbf{P}(K)$ serves as a mathematically equivalent representation for K, from wich K can be recovered through Eq. (8). The representing set $\mathbf{P}(K)$ will then furthermore be closed with respect to the topology induced by pointwise convergence.

Theorem 3 [4, Theorem 28 and Proposition 24]. A set of desirable option sets K is strongly Archimedean if and only if there is some non-empty closed set $\mathcal{P} \subseteq \mathbf{P}$ of coherent lower previsions such that $K = K_{\mathcal{P}}$. Closure is with respect to pointwise convergence, and the largest such set $\mathcal{P}$ is then $\mathbf{P}(K)$.

If the representing coherent lower previsions in $\mathcal{P}$ or $\mathbf{P}(K)$ were linear, this result would already brings us very close to decision rules based on sets of linear previsions— or sets of probability measures. As we will see further on in Sect. 7, this can be achieved by imposing an additional axiom called mixingness. Before we do so, however, we will do away with the closure condition in Theorem 3, as it is overly restrictive. Imagine for example that we are modelling a subject's uncertainty about the outcome of a coin toss, and that she beliefs the coin to be unfair. In terms of probabilities, this would mean that her probability for heads is different from one half. Strong Archimeanity is not compatible with such an assessment, as the set of probability measures that satisfy this (strict) probability constraint is not closed. Our first main contribution will consist in resolving this issue, by suitably modifying the notion of strong Archimedeanity.

5 Archimedean Sets of Desirable Option Sets

At first sight, it may seem as if K_{SA} is the only way in which sd$_4$ can be translated to option sets. There is however also a second, far more subtle approach.

The crucial insight on which this second approach is based is that our interpretation in terms of strict desirability does not require ε to be known; it only imposes the existence of such an ε. Consider a subject whose uncertainty is represented by a set of desirable option sets K and let us adopt an interpretation in terms of strict desirability. This implies that the option sets $A \in K$ are option sets that, according to her beliefs, contain at least one strictly desirable option $u \in A$. As a consequence of sd$_4$, this implies that she beliefs that there is some real $\varepsilon > 0$ such that $u - \varepsilon$ is strictly desirable. Hence, she believes that there is some real $\varepsilon > 0$ such that $A - \varepsilon$ contains at least one strictly desirable option. Strong Archimedeanity, at that point, concludes that $A - \varepsilon \in K$. However, this is only justified if our subject knows ε. If she doesn't know ε, but only believes that there is such an ε, then there is no single $\varepsilon > 0$ for which she believes that $A - \varepsilon$ contains at least one strictly desirable option. Since the option sets in K are options sets for which our subject believes that they contain at least one strictly desirable option, it follows that $A \in K$ need not necessarily imply that $A - \varepsilon \in K$ for some $\varepsilon > 0$. Strong Archimedeanity is therefore indeed, as its name suggests, a bit too strong for our purposes.

So if we can't infer that $A - \varepsilon \in K$, what is it then that we can infer from $A \in K$ and sd$_4$? As explained above, the only thing that can be inferred is that for any $A \in K$, there is some $\varepsilon > 0$ such that $A - \varepsilon$ contains at least one strictly desirable option. Let us denote this specific epsilon by $\varepsilon(A)$. Crucially, we may not know—or rather, our subject may not know—the specific value of $\varepsilon(A)$. Nevertheless, any inferences we can make without knowing the specific value of $\varepsilon(A)$, can and should still be made. Our approach will therefore consist in finding out what inferences can be made for a specific choice of the $\varepsilon(A)$, to do this for every such choice, and to then only consider those inferences that can be made regardless of the specific choice of $\varepsilon(A)$.

To formalize this, we consider the set $\mathbb{R}_{>0}^K$ of all functions ε that associate a strictly positive real $\varepsilon(A) > 0$ with every option set A in K. As a consequence of our interpretation, we know that there is at least one $\varepsilon \in \mathbb{R}_{>0}^K$ such that, for every $A \in K$, $A - \varepsilon(A)$ contains a strictly desirable option.

Let us now assume for a minute that our subject does know for which specific ε in $\mathbb{R}_{>0}^K$ this is the case. In order to be compatible with sd$_1$–sd$_3$, the resulting assessment

$$K_\varepsilon := \{A - \varepsilon(A) : A \in K\} \tag{9}$$

should then be consistent with coherence, meaning that there is at least one coherent set of desirable option sets that includes K_ε. Whenever this is the case, then as explained in Sect. 3, we can use coherence to extend the assessment K_ε to the unique smallest coherent set of desirable option sets that incudes it: the natural extension $\mathrm{Ex}(K_\varepsilon)$ of K_ε. Based on the assessment K_ε and coherence, each of the option sets in $\mathrm{Ex}(K_\varepsilon)$ must necessarily contain a strictly desirable option. Hence, still assuming for the moment that our subject knows ε, it follows that every option set in $\mathrm{Ex}(K_\varepsilon)$ should belong to K.

Our subject may not know ε though; all we can infer from sd$_4$ is that there must be at least one ε for which the above is true. Let us denote this specific—but possibly unkown—ε by ε^*. Then as argued above, for every option set A in $\mathrm{Ex}(K_{\varepsilon^*})$, it follows from our interpretation that A should also belong to K. Since we don't know ε^*, however, we don't know for which option sets this is the case. What we can do though, is to consider those option sets $A \in \mathscr{Q}$ that belong to $\mathrm{Ex}(K_\varepsilon)$ for *every* possible $\varepsilon \in \mathbb{R}^K_{>0}$. For those option sets, regardless of whether we know ε^* or not, it trivially follows that $A \in \mathrm{Ex}(K_{\varepsilon^*})$, and therefore, that A should belong to K. Any coherent set of desirable option sets K that satisfies this property, we will call Archimedean.

Definition 6. *A set of desirable option sets K is* Archimedean *if it is coherent and satisfies the following property:*

K_A. *for any $A \in \mathscr{Q}$, if $A \in \mathrm{Ex}(K_\varepsilon)$ for all $\varepsilon \in \mathbb{R}^K_{>0}$, then also $A \in K$.*

Note that Archimedeanity also rules out the possibility that K_ε is inconsistent for all $\varepsilon \in \mathbb{R}^K_{>0}$, for this would imply that $K = \mathscr{Q}$, hence contradicting K_1.

By replacing strong Archimedeanity with Archimedeanity, the condition that the representing set $\mathscr{P}$ must be closed can be removed from Theorem 3, and we obtain a representation in terms of general sets of lower previsions.

Theorem 4. *A set of desirable option sets K is Archimedean if and only if there is some non-empty set $\mathscr{P} \subseteq \underline{\mathbf{P}}$ of coherent lower previsions such that $K = K_{\mathscr{P}}$. The largest such set $\mathscr{P}$ is then $\underline{\mathbf{P}}(K)$.*

The significance of this result is that it relates two very different things: sets of desirable option sets and sets of coherent lower previsions. While this may not yet be obvious, this is a major step in characterising choice functions of the form $C_{\mathscr{P}}$. In fact, we are nearly there. The only thing left to do is to connect choice functions with sets of desirable option sets. As we will explain in the next section, this connection comes quite naturally once we interpret choice functions in terms of (strict) desirability.

6 Archimedean Choice Functions

In order to provide choice functions with an interpretation, we need to explain what it means for an option u to be chosen from A, or alternatively, what it means for u to be rejected from A, in the sense that $u \notin C(A)$. We here adopt the latter approach. In particular, if our subject states that $u \notin C(A)$, we take this to mean that she is convinced that there is at least one other option v in $A \setminus \{u\}$ that is better that u, where 'v is better than u' is taken to mean that $v - u$ is strictly desirable, or equivalently, that there is a positive price $\varepsilon > 0$ for which paying ε to exchange the uncertain reward u for v is preferrable to the status quo. Note however that this interpretation does not not assume that our subject knows the specific ε and v for which this is the case.

Our interpretation has two implications for C. First, since $v - u = (v - u) - 0$, it immediately implies that C should be *translation invariant*, in the sense that

$$u \in C(A) \Leftrightarrow 0 \in C(A - u) \text{ for all } A \in \mathscr{Q} \text{ and } u \in A, \tag{10}$$

with $A - u := \{v - u : v \in A\}$. Second, for all $A \in \mathscr{Q}$ such that $0 \notin C(A \cup \{0\})$, it implies that A should contain at least one strictly desirable gamble. Indeed, if $0 \notin C(A \cup \{0\})$, then according to our interpretation, there is some $v \in (A \cup \{0\}) \setminus \{0\} \subseteq A$ such that

$v - 0 = v$ is strictly desirable. Hence, A indeed contains a strictly desirable option. For any choice function C, this leads us to consider the set of desirable option sets

$$K_C := \{ A \in \mathscr{Q} : 0 \notin C(A \cup \{0\}) \}. \tag{11}$$

According to our interpretation, each of the option sets in K_C contains at least one strictly desirable option. Following the discussion in Sect. 6, we will therefore require K_C to be Archimedean. When a choice function C satisfies both of the conditions that are implied by our interpretation, we call it Archimedean.

Definition 7. *A choice function C is Archimedean if K_C is Archimedean and C is translation invariant.*

Instead of deriving a set of desirable options sets K_C from a choice function C, we can also do the converse. That is, with any set of desirable option sets K, we can associate a choice function C_K, defined by

$$C_K(A) := \{ u \in A : A \ominus u \notin K \} \text{ for all } A \in \mathscr{Q}, \tag{12}$$

where $A \ominus u := \{ v - u : v \in A \setminus \{u\} \}$. Similarly to K_C, the expression for C_K is motivated by our interpretation. Indeed, for any option $u \in A$, the statement that $A \ominus u \in K$ means that $A \ominus u$ contains a strictly desirable option, so there is some $v \in A \setminus \{u\}$ such that $v - u$ is strictly desirable. This is exactly our interpretation for $u \notin C(A)$.

If a set of desirable option sets K is Archimedean, then C_K will be Archimedean as well. In fact, as our next result shows, every Archimedean choice function is of the form C_K, with K an Archimedean set of desirable option sets.

Proposition 1. *Let C be a choice function. Then C is Archimedean if and only if there is an Archimedean set of desirable option sets K such that $C = C_K$. This set K is then necessarily unique and furthermore equal to K_C.*

At this point, the hard work in characterising choice functions of the form $C_{\mathscr{P}}$ is done. Proposition 1 relates Archimedean choice functions to Archimedean sets of desirable option sets, while Theorem 4 relates Archimedean sets of desirable option sets to sets of coherent lower previsions $\mathscr{P}$. Combining both results, we find that a choice function is Archimedean if and only if it is of the form $C_{\mathscr{P}}$.

Theorem 5. *A choice function C is Archimedean if and only if there is a non-empty set $\mathscr{P} \subseteq \underline{\mathbf{P}}$ of coherent lower previsions such that $C = C_{\mathscr{P}}$. Whenever this is the case, the largest such set $\mathscr{P}$ is $\underline{\mathbf{P}}(K_C)$.*

Starting from this result, we will now proceed to axiomatise maximality and E-admissibility, by combining Archimedeanity with additional axioms.

7 Axiomatising E-Admissibility

Archimedeanity implies that a choice function is representable by a set of coherent lower previsions $\mathscr{P}$, in the sense that is of the form $C_{\mathscr{P}}$. As can be seen by comparing

Eqs. (4) and (2), this already brings us very close E-admissibility. Indeed, all that we need in order to obtain E-admissibility is for the lower previsions in $\mathscr{P}$ to be linear. That is, we would like the role of $\underline{P}(K)$ to be taken up by

$$\mathbf{P}(K) := \{P \in \mathbf{P} \colon K \subseteq K_P\}$$

instead. To achieve this, we impose a property called mixingness [4] on the Archimedean set of desirable option sets K_C that corresponds to C. For any option set A, this property considers the set

$$\mathrm{posi}(A) := \left\{ \sum_{i=1}^{n} \lambda_i u_i \colon n \in \mathbb{N}, \lambda_i > 0, u_i \in A \right\},$$

of all positive linear combinations of the elements in A, and requires that if any of these positive linear combinations—any mixture—is strictly desirable, then A itself should contain a strictly desirable option as well.

Definition 8. *A set of desirable option sets K is* mixing *if it satisfies*

K_M. *if $B \in K$ and $A \subseteq B \subseteq \mathrm{posi}(A)$, then also $A \in K$, for all $A, B \in \mathcal{Q}$;*

A choice function C is called mixing *if K_C is.*

As the following result shows, mixingness achieves exactly what we need: for any coherent set of desirable option sets K, it guarantees that the coherent lower previsions in $\underline{P}(K)$ are in fact linear.

Proposition 2. *Let K be a coherent set of desirable option sets that is mixing. Then for any $\underline{P} \in \underline{P}(K)$, we have that $\underline{P} \in \mathbf{P}$. Hence, $\underline{P}(K) = \mathbf{P}(K)$.*

By combining this result with Theorem 5, it follows that Archimedean choice functions that are mixing correspond to E-admissibility. The next result formalizes this and furthermore shows that the converse is true as well.

Theorem 6. *A choice function C is Archimedean and mixing if and only if there is a non-empty set $\mathscr{P} \subseteq \mathbf{P}$ of linear previsions such that $C = C_{\mathscr{P}}^{\mathrm{E}}$. The largest such set $\mathscr{P}$ is then $\mathbf{P}(K_C)$.*

8 Axiomatising Maximality

Having axiomatised E-admissibility, we now proceed to do the same for maximality. The link with Archimedeanity is not that obvious here though, because there is no immediate connection between Eqs. (4) and (3). Rather than focus on how to relate these two equations, we therefore zoom in on the properties of maximality itself. One such property, which is often used to illustrate the difference with E-admissibility, is that a choice function that corresponds to maximality is completely determined by its restriction to so-called *binary* choices—that is, choices between two options.

Definition 9. *A choice function C is binary if for all $A \in \mathscr{Q}$ and $u \in A$:*

$$u \in C(A) \Leftrightarrow (\forall v \in A \setminus \{u\})\, u \in C(\{u,v\})$$

Inspired by this observation, we impose binarity as an additional axiom, alongside Archimedeanity. As the following result shows, these two conditions are necessary and sufficient for a choice function C to be of the form $C_{\mathscr{P}}^{M}$, hence providing an axiomatisation for decision making with maximality.

Theorem 7. *A choice function C is Archimedean and binary if and only if there is a non-empty set $\mathscr{P} \subseteq \mathbf{P}$ of linear previsions such that $C = C_{\mathscr{P}}^{M}$. The largest such set $\mathscr{P}$ is then $\mathbf{P}(K_C)$.*

The formal proof of this result is rather technical, but the basic idea behind the sufficiency proof is nevertheless quite intuitive. First, for every $u \in C(A)$, the binarity of C implies that $u \in C(\{u,v\})$ for every $v \in A \setminus \{u\}$. For every such $v \in A \setminus \{u\}$, since C is Archimedean, Theorem 5 furthermore implies that there is a coherent lower prevision $\underline{P}$ such that $\underline{P}(v - u) \leq 0$. Because of Theorem 1, this in turn implies that there is a linear prevision such that $P(v - u) \leq 0$ and therefore also $P(u) \geq P(v)$. The challenging part consists in showing that $P \in \mathbf{P}(K_C)$ and establishing necessity.

9 An Axiomatisation for Maximising Expected Utility

As we have seen in Sects. 7 and 8, mixingness and binarity have quite a different effect on Archimedean choice functions. The former implies that they correspond to E-admissibility, while the latter leads to maximality. What is intriguing though is that the set of linear previsions $\mathscr{P}$ is twice the same. Indeed, as can be seen from Theorem 6 and 7, we may assume without loss of generality that this set is equal to $\mathbf{P}(K_C)$. For a choice function C that is mixing *and* binary, we therefore find that $C = C_{\mathscr{P}}^{E} = C_{\mathscr{P}}^{M}$, with $\mathscr{P} = \mathbf{P}(K_C)$. As the following result shows, this is only possible if $\mathscr{P}$ is a singleton.

Proposition 3. *Let $\mathscr{P} \subseteq \mathbf{P}$ be a non-empty set of linear previsions. Then $C_{\mathscr{P}}^{E} = C_{\mathscr{P}}^{M}$ if and only if $\mathscr{P} = \{P\}$ consists of a single linear prevision $P \in \mathbf{P}$.*

As a fairly immediate consequence, we obtain the following axiomatic characterisation of choice functions that correspond to maximising expected utility.

Theorem 8. *A choice function C is Archimedean, binary and mixing if and only if there is a linear prevision $P \in \mathbf{P}$ such that $C = C_P$.*

10 Conclusion and Future Work

The main conclusion of this work is that choice functions, when interpreted in terms of (strict) desirability, can provide an axiomatic basis for decision making with sets of probability models. In particular, we were able to derive necessary and sufficient conditions for a choice function to correspond to either E-admissibility or maximality.

As a byproduct, we also obtained a characterisation for choice functions that correspond to maximising expected utility.

The key concept on which these results were based is that of an Archimedean choice function, where Archimedeanity is itself a combination of several conditions. The first of these conditions is translation invariance; this condition is fairly simple and allows for a reduction from choice functions to sets of desirable option sets. The resulting set of desirable option sets should then satisfy two more conditions: coherence and K_A. Coherence is also fairly simple, because it follows directly from the principles of desirability. The condition K_A, however, is more involved, making it perhaps the least intuitive component of Archimedeanity.

The abstract character of K_A is not intrinsic to the property itself though, but rather to the framework on which it is imposed. In fact, the basic principle sd_4 on which K_A is based is very simple: if u is strictly desirable, then there must be some positive real ε such that $u - \varepsilon$ is strictly desirable as well. The reason why this simplicity does not translate to K_A is because we restrict attention to option sets that are finite. Consider for example an assessment of the form $\{u\} \in K$. This means that u is strictly desirable and therefore implies, due to sd_4, that the option set $\{u - \varepsilon \colon \varepsilon \in \mathbb{R}_{>0}\}$ contains at least one strictly desirable option. Hence, we should simply impose that this set belongs to K. This is not possible though because $\{u - \varepsilon \colon \varepsilon \in \mathbb{R}_{>0}\}$ is infinite, while our framework of sets of desirable option sets only considers finite option sets.

This situation can be remedied, and the axiom of Archimedeanity can be simplified, by developing and adopting a framework of sets of desirable options that allows for infinite option sets, and connecting it to a theory of choice funtions that chooses from possibly infinite option sets. Explaining how this works is beyond the scope and size of the present contribution though; I intend to report on those results elsewhere.

Acknowlegements. This work was funded by the BOF starting grant 01N04819 and is part of a larger research line on choice functions of Gert de Cooman and I [3,4]. Within this line of research, this contribution is one of two parallel papers on Archimedean choice functions, one by each of us. My paper—this one—deals with the case where options are bounded real-valued functions. It axiomatises Archimedean choice functions from the ground up by starting from (strict) desirability principles and proves that the resulting axioms guarantee a representation in terms of coherent lower prevision. The paper of Gert [5] defines Archimedeanity directly in terms of coherent lower previsions—superlinear functionals, actually, in his case—but considers the more general case where options live in an abstract Banach space; he also extends the concept of a coherent lower prevision to this more general context and discusses the connection with horse lotteries. We would like to combine our respective results in future work.

References

1. Couso, I., Moral, S.: Sets of desirable gambles: conditioning, representation, and precise probabilities. Int. J. Approximate Reasoning 52(7), 1034–1055 (2011)
2. De Bock, J.: Archimedean choice functions: an axiomatic foundation for imprecise decision making (2020). http://arxiv.org/abs/2002.05196

3. De Bock, J., de Cooman, G.: A desirability-based axiomatisation for coherent choice functions. In: Destercke, S., Denoeux, T., Gil, M.Á., Grzegorzewski, P., Hryniewicz, O. (eds.) SMPS 2018. AISC, vol. 832, pp. 46–53. Springer, Cham (2019). https://doi.org/10.1007/978-3-319-97547-4_7

4. De Bock, J., de Cooman, G.: Interpreting, axiomatising and representing coherent choice functions in terms of desirability. In: Proceedings of ISIPTA 2019. PMLR, vol. 103, pp. 125–134 (2019)

5. De Cooman, G.: Coherent and archimedean choice in general banach spaces (2020). http://arxiv.org/abs/2002.05461

6. de Finetti, B.: Teoria delle Probabilità. Einaudi, Turin (1970)

7. Levi, I.: On indeterminate probabilities. Jo. Philos. 71(13), 391–418 (1974)

8. Quaeghebeur, E., de Cooman, G., Hermans, F.: Accept & reject statement-based uncertainty models. Int. J. Approximate Reasoning 57, 69–102 (2015)

9. Savage, L.J.: The Foundations of Statistics, 2nd edn. Dover, New York (1972)

10. Seidenfeld, T., Schervish, M.J., Kadane, J.B.: Coherent choice functions under uncertainty. Synthese 172(1), 157–176 (2010)

11. Troffaes, M.C.M.: Decision making under uncertainty using imprecise probabilities. Int. J. Approximate Reasoning 45(1), 17–29 (2007)

12. Van Camp, A.: Choice Functions as a Tool to Model Uncertainty. Ph.D. thesis, Ghent University, Faculty of Engineering and Architecture (2018)

13. Walley, P.: Statistical Reasoning with Imprecise Probabilities. Chapman and Hall, London (1991)

14. Williams, P.M.: Notes on conditional previsions. Int. J. Approximate Reasoning 44, 366–383 (2007)

Dynamic Portfolio Selection Under Ambiguity in the ϵ-Contaminated Binomial Model

Paride Antonini[1] , Davide Petturiti[1](✉) , and Barbara Vantaggi[2]

[1] Dip. Economia, University of Perugia, Perugia, Italy
paride.antonini@studenti.unipg.it, davide.petturiti@unipg.it
[2] Dip. MEMOTEF, "La Sapienza" University of Rome, Rome, Italy
barbara.vantaggi@uniroma1.it

Abstract. Investors often need to look for an optimal portfolio acting under ambiguity, as they may not be able to single out a unique real-world probability measure. In this paper a discrete-time dynamic portfolio selection problem is studied, referring to an ϵ-contaminated binomial market model and assuming investors' preferences are consistent with the Choquet expected utility theory. We formulate the portfolio selection problem for a CRRA utility function in terms of the terminal wealth, and provide a characterization of the optimal solution in the case stock price returns are uniformly distributed. In this case, we further investigate the effect of the contamination parameter ϵ on the optimal portfolio.

Keywords: Ambiguity · Optimal portfolio · ϵ-Contamination model · Choquet integral

1 Introduction

The standard approach to uncertainty in portfolio selection models is to refer to a unique real-world probability measure **P**, usually estimated from data. Nevertheless, sometimes, due to the presence of unobserved variables or since the information is partial, misspecified or "imprecise", it is no longer possible to handle uncertainty through a single probability measure. This has led to the development of models taking care of ambiguity, i.e., coping with a class of probability measures.

In recent years a growing interest has been addressed towards ambiguity in decision and economic literature (see, e.g., the survey papers [5] and [6]). Mainly, the aim is to propose models generalizing the classical subjective expected utility model in a way to avoid paradoxical situations like that discussed in [4], showing preference patterns under ambiguity that are not consistent with the expected utility paradigm. In particular, referring to the Anscombe-Aumann framework, the seminal papers [17] and [7] incorporate ambiguity in decisions, respectively, either through a non-additive uncertainty measure or through a class of probability measures. The two approaches are generally not equivalent but in case

© Springer Nature Switzerland AG 2020
M.-J. Lesot et al. (Eds.): IPMU 2020, CCIS 1238, pp. 210–223, 2020.
https://doi.org/10.1007/978-3-030-50143-3_16

of a 2-monotone capacity [19] they reduce to Choquet expected utility. Choquet expected utility has been investigated also in a von Neumann-Morgenstern-like framework, by referring to "objective" completely monotone capacities modeling generalized lotteries [2,10,14].

In this paper, we formalize a dynamic portfolio selection problem under ambiguity, referring to an ϵ-contamination [9] of a binomial market model [3] and to a CRRA utility function [18]. Thanks to the completeness of the market, we formulate the dynamic portfolio selection in terms of the final wealth. Moreover, since the lower envelope of the ϵ-contamination class, obtained from the real-world probability $\mathbf{P}$, is a completely monotone capacity, we assume that the investors' preferences are consistent with the Choquet expected utility theory. The problem amounts to finding a final wealth maximizing the corresponding Choquet expected utility functional. We stress that, due to the properties of the Choquet integral [8], an agent which is a Choquet expected utility maximizer is actually an expected utility maximinimizer, with respect to the ϵ-contamination class.

Notice that, referring to a one-period model, a portfolio selection problem under ambiguity via maximinimization has been faced in [15,16]. Moreover, our problem is analogous to that formulated in Appendix C of [11], the latter differing from ours for working in a continuous setting and considering a distortion of the probability $\mathbf{P}$.

We show that our problem can be reduced to a family of linearly constrained concave problems, indexed by the set of all permutations of the sample space. Moreover, the initial problem is proved to have a unique optimal solution. Focusing on the special case of uniformly distributed stock returns (determining a uniform probability distribution on the sample space) we provide a characterization of the optimal solution relying only on a number of permutations equal to the cardinality of the sample space. Yet in the uniform case, we study the effect of the contamination parameter ϵ on the optimal portfolio, showing the presence of a threshold above which the optimal self-financing strategy reduces to 0 for all times: this highlights that with such values of ϵ the ambiguity in the model is so high to make the risk-free portfolio the most suitable choice.

This paper differs from the approach used in [12], where among their results, the authors characterize a generalization of the random walk by replacing a probability with a capacity on the branches of a tree, for each time and by imposing dynamic consistency (or rectangularity). We stress that dynamic consistency is a mathematically helpful property in order to capture a form of martingale property, but it is a strong limitation for a non-additive measure.

2 Dynamic Portfolio Selection in the Binomial Model

We refer to the multi-period binomial model introduced by Cox, Ross and Rubinstein [3]. Such model considers a perfect (competitive and frictionless) market under no-arbitrage, where two basic securities are traded: a non-dividend-paying stock and a risk-free bond.

The evolution of the prices of such securities is expressed by the stochastic processes $\{S_0, \ldots, S_T\}$ and $\{B_0, \ldots, B_T\}$, with $T \in \mathbb{N}$, $S_0 = s > 0$, $B_0 = 1$, and for $t = 1, \ldots, T$,

$$\frac{S_t}{S_{t-1}} = \begin{cases} u, \text{ with probability } p \\ d, \text{ with probability } 1 - p \end{cases} \quad \text{and} \quad \frac{B_t}{B_{t-1}} = (1+r),$$

where $u > d > 0$ are the "up" and "down" stock price coefficients, r is the risk-free interest rate over each period, satisfying $u > (1+r) > d$, and $p \in (0,1)$ is the probability of an "up" movement for the stock price.

All the processes we consider are defined on a filtered probability space $(\Omega, \mathcal{F}, \{\mathcal{F}_t\}_{t=0,\ldots,T}, \mathbf{P})$, where $\Omega = \{1, \ldots, 2^T\}$ and $\mathcal{F}_t$ is the algebra generated by random variables $\{S_0, \ldots, S_t\}$, for $t = 0, \ldots, T$, with $\mathcal{F}_0 = \{\emptyset, \Omega\}$ and $\mathcal{F}_T = \mathcal{F} = \wp(\Omega)$, where the latter denotes the power set of Ω. As usual $\mathbf{E}^{\mathbf{P}}$ denotes the expected value with respect to $\mathbf{P}$.

The process $\{S_0, \ldots, S_T\}$ is a multiplicative binomial process where the returns $\frac{S_1}{S_0}, \ldots, \frac{S_T}{S_{T-1}}$ are i.i.d. random variables, and the probability $\mathbf{P}$, usually said *real-world probability*, is completely singled out by the parameter p, indeed, for $t = 0, \ldots, T$ and $k = 0, \ldots, t$, we have that

$$\mathbf{P}(S_t = u^k d^{t-k} s) = \frac{t!}{k!(t-k)!} p^k (1-p)^{t-k}.$$

We consider a *self-financing strategy* $\{\theta_0, \ldots, \theta_{T-1}\}$, that is an adapted process where each θ_t is the (random) number of shares of stock to buy (if positive) or short-sell (if negative) at time t up to time $t + 1$. Such strategy determines an adapted wealth process $\{V_0, \ldots, V_T\}$, where, for $t = 0, \ldots, T - 1$,

$$V_{t+1} = (1+r)V_t + \theta_t S_t \left(\frac{S_{t+1}}{S_t} - (1+r) \right). \tag{1}$$

This market model is said to be *complete*, i.e., there is a unique probability measure $\mathbf{Q}$ on $\mathcal{F}$ equivalent to $\mathbf{P}$, usually said *risk-neutral probability*, such that the discounted wealth process of any self-financing strategy is a martingale under $\mathbf{Q}$:

$$\frac{V_t}{(1+r)^t} = \mathbf{E}_t^{\mathbf{Q}} \left[\frac{V_T}{(1+r)^T} \right],$$

for $t = 0, \ldots, T$, where $\mathbf{E}_t^{\mathbf{Q}}[\cdot] = \mathbf{E}^{\mathbf{Q}}[\cdot | \mathcal{F}_t]$ and $\mathbf{E}^{\mathbf{Q}}$ is the expectation with respect to $\mathbf{Q}$. Notice that $\mathbf{E}_0^{\mathbf{Q}}$ can be identified with $\mathbf{E}^{\mathbf{Q}}$. In particular, completeness implies that every payoff $V_T \in \mathbb{R}^{\Omega}$ depending only on the stock price history can be replicated by a dynamic self-financing strategy $\{\theta_0, \ldots, \theta_{T-1}\}$ and its unique no-arbitrage price at time $t = 0$ is

$$V_0 = \frac{\mathbf{E}^{\mathbf{Q}}[V_T]}{(1+r)^T}.$$

Notice that the process $\{S_0, \ldots, S_T\}$ continues to be a multiplicative binomial process also under $\mathbf{Q}$, the latter being completely characterized by the parameter

$$q = \frac{(1+r) - d}{u - d} \in (0, 1),$$

which replaces p. In particular, every state $i \in \Omega$ is identified with the path of the stock price evolution corresponding to the T-digit binary expansion of number $i - 1$, in which zeroes are interpreted as "up" movements and ones as "down" movements. If in such representation there are k "up" movements and $T - k$ "down" movements, then (we avoid braces to simplify writing)

$$\mathbf{P}(i) = p^k (1-p)^{T-k} \quad \text{and} \quad \mathbf{Q}(i) = q^k (1-q)^{T-k}.$$

In the market model we have just introduced, the real-world probability measure $\mathbf{P}$ is assumed to encode the beliefs of an investor. Assuming that investor's preferences are consistent with the expected utility theory, given a utility function u with $\operatorname{dom} u = \mathbb{R}_{++} = (0, +\infty)$, usually assumed to be sufficiently regular, strictly increasing and strictly concave, and an initial wealth $V_0 > 0$, the dynamic portfolio selection consists in determining a self-financing strategy $\{\theta_0, \ldots, \theta_{T-1}\}$ resulting in a final wealth $V_T \in \mathbb{R}_{++}^\Omega$ solving

$$\max_{\theta_0, \ldots, \theta_{T-1}} \mathbf{E}^{\mathbf{P}}[u(V_T)].$$

Due to the dynamic consistency property of the conditional expected value operator, this problem can be efficiently solved through dynamic programming [18].

In this paper we suppose that our investor has ambiguous beliefs, meaning that, rather than having a single probability $\mathbf{P}$, he/she actually considers a class of probabilities $\mathcal{P}$. In this case, the choice of $\{\theta_0, \ldots, \theta_{T-1}\}$ should take into account all the probabilities in $\mathcal{P}$, by adopting a suitable criterion of choice, justified by a normative criterion of rationality.

3   Modeling Ambiguity Through the ϵ-Contamination Model

Here we consider the binomial market model (see Sect. 2), introducing ambiguity to get an ϵ-*contaminated binomial model*. Given the real-world probability $\mathbf{P}$ defined on $\mathcal{F}$ (which is completely singled out by p) and $\epsilon \in (0, 1)$, the corresponding ϵ-*contamination model* (see, e.g., [9, 19]) is the class of probability measures on $\mathcal{F}$ defined as

$$\mathcal{P}_{p,\epsilon} = \{\mathbf{P}' = (1 - \epsilon)\mathbf{P} + \epsilon\mathbf{P}'' : \mathbf{P}'' \text{ is a probability measure on } \mathcal{F}\},$$

whose lower envelope $\nu_{p,\epsilon} = \min \mathcal{P}_{p,\epsilon}$ is a completely monotone normalized capacity defined on $\mathcal{F}$ such that

$$\nu_{p,\epsilon}(A) = \begin{cases} (1 - \epsilon)\mathbf{P}(A), & \text{if } A \neq \Omega, \\ 1, & \text{if } A = \Omega. \end{cases}$$

For every permutation σ of Ω, we can define a probability measure $\mathbf{P}^\sigma$ on $\mathcal{F}$, whose value on the singletons (we avoid braces to simplify writing) is

$$\mathbf{P}^\sigma(\sigma(i)) = \nu_{p,\epsilon}(E_i^\sigma) - \nu_{p,\epsilon}(E_{i+1}^\sigma),$$

where $E_i^\sigma = \{\sigma(i), \ldots, \sigma(2^T)\}$, for all $i \in \Omega$, and $E_{2^T+1}^\sigma = \emptyset$. It turns out that the class $\mathcal{P}_{p,\epsilon}$ is the convex hull of probabilities $\mathbf{P}^\sigma$'s.

We consider a CRRA utility function defined, for $\gamma > 0$, $\gamma \neq 1$, as

$$u_\gamma(x) = \frac{x^{1-\gamma}}{1-\gamma}, \quad \text{for } x > 0.$$

For every random variable $V_T \in \mathbb{R}_{++}^\Omega$, we can define the functional

$$\mathbf{CEU}_{\gamma,p,\epsilon}[V_T] = \oint u_\gamma(V_T) \, d\nu_{p,\epsilon},$$

where the integral on the right side is of Choquet type. In particular, since u_γ is strictly increasing, for all $V_T \in \mathbb{R}_{++}^\Omega$, if σ is a permutation of Ω such that $V_T(\sigma(1)) \leq \ldots \leq V_T(\sigma(2^T))$ then the functional $\mathbf{CEU}_{\gamma,p,\epsilon}$ can be expressed (see [8]) as follows

$$\mathbf{CEU}_{\gamma,p,\epsilon}[V_T] = \sum_{k=1}^{2^T} \mathbf{P}^\sigma(k) \frac{(V_T(k))^{1-\gamma}}{1-\gamma}. \tag{2}$$

Proposition 1. *The functional* $\mathbf{CEU}_{\gamma,p,\epsilon}$ *is concave, that is, for all* $V_T, V_T' \in \mathbb{R}_{++}^\Omega$ *and all* $\alpha \in [0,1]$, *it holds*

$$\mathbf{CEU}_{\gamma,p,\epsilon}[\alpha V_T + (1-\alpha)V_T'] \geq \alpha \mathbf{CEU}_{\gamma,p,\epsilon}[V_T] + (1-\alpha)\mathbf{CEU}_{\gamma,p,\epsilon}[V_T'].$$

Proof. Since u_γ is (strictly) concave, for all $i \in \Omega$ and $\alpha \in [0,1]$, we have

$$u_\gamma(\alpha V_T(i) + (1-\alpha)V_T'(i)) \geq \alpha u_\gamma(V_T(i)) + (1-\alpha)u_\gamma(V_T'(i)).$$

Hence, since $\nu_{p,\epsilon}$ is completely monotone and so 2-monotone, Theorems 4.24 and 4.35 in [8] imply (by monotonicity and concavity of the Choquet integral)

$$\mathbf{CEU}_{\gamma,p,\epsilon}[\alpha V_T + (1-\alpha)V_T'] = \oint u_\gamma(\alpha V_T + (1-\alpha)V_T') \, d\nu_{p,\epsilon}$$

$$\geq \oint [\alpha u_\gamma(V_T) + (1-\alpha)u_\gamma(V_T')] \, d\nu_{p,\epsilon}$$

$$\geq \alpha \oint u_\gamma(V_T) \, d\nu_{p,\epsilon} + (1-\alpha)\oint u_\gamma(V_T') \, d\nu_{p,\epsilon}$$

$$= \alpha \mathbf{CEU}_{\gamma,p,\epsilon}[V_T] + (1-\alpha)\mathbf{CEU}_{\gamma,p,\epsilon}[V_T'].$$

By Theorem 4.39 in [8], we also have that

$$\mathbf{CEU}_{\gamma,p,\epsilon}[V_T] = \min_{\mathbf{P}' \in \mathcal{P}_{p,\epsilon}} \mathbf{E}^{\mathbf{P}'}[u_\gamma(V_T)],$$

therefore, $\mathbf{CEU}_{\gamma,p,\epsilon}$ is a *lower expected utility* and maximizing it we are actually applying a *maximin criterion* of choice.

Given an initial wealth $V_0 > 0$, our aim is to select a self-financing strategy $\{\theta_0, \ldots, \theta_{T-1}\}$ resulting in a final wealth $V_T \in \mathbb{R}_{++}^{\Omega}$, solving

$$\max_{\theta_0, \ldots, \theta_{T-1}} \mathbf{CEU}_{\gamma, p, \epsilon}[V_T].$$

Taking into account the completeness of the market, the above problem can be rewritten maximizing over the final wealth V_T

$$\text{maximize } \mathbf{CEU}_{\gamma, p, \epsilon}[V_T]$$

subject to:

$$\begin{cases} \mathbf{E}^{\mathbf{Q}}[V_T] - (1+r)^T V_0 = 0, \\ V_T \in \mathbb{R}_{++}^{\Omega}. \end{cases} \tag{3}$$

By Proposition 1, the objective function in (3) is a concave function on $\mathbb{R}_{++}^{\Omega}$, subject to a linear constraint, therefore every local maximum is a global maximum and, further, the set of global maxima is convex [1].

Since by (2) the computation of $\mathbf{CEU}_{\gamma, p, \epsilon}[V_T]$ depends on a permutation σ such that the values of V_T are increasingly ordered, the above problem (3) can be decomposed in a family of optimization problems, each indexed by a permutation σ of $\Omega = \{1, \ldots, 2^T\}$, corresponding to a possible ordering of V_T:

$$\text{maximize } \left[\sum_{k=1}^{2^T} \mathbf{P}^{\sigma}(k) \frac{(V_T(k))^{1-\gamma}}{1-\gamma} \right]$$

subject to:

$$\begin{cases} \sum_{k=1}^{2^T} \mathbf{Q}(k) V_T(k) - (1+r)^T V_0 = 0, \\ V_T(\sigma(i-1)) - V_T(\sigma(i)) \leq 0, \quad \text{for all } i \in \Omega \setminus \{1\}, \\ V_T \in \mathbb{R}_{++}^{\Omega}. \end{cases} \tag{4}$$

Denoting by V_T^{σ} an optimal solution related to permutation σ, an optimal solution V_T^* for (3) can be found by selecting V_T^{σ} for a permutation σ where the objective function is maximum. Notice that there can be more permutations where the maximum is attained.

Proposition 2. *The following statements hold:*

(i) for every permutation σ, problem (4) has an optimal solution and such optimal solution is unique;

(ii) problem (3) has an optimal solution and such optimal solution is unique.

Proof. (i). The subset $\mathcal{V}^{\sigma}$ of $\mathbb{R}^{\Omega}$ satisfying the equality and inequality constraints in (4) is closed, and $\mathcal{V}^{\sigma} \cap \mathbb{R}_{++}^{\Omega} \neq \emptyset$, since $V_T = (1+r)^T V_0 \in \mathcal{V}^{\sigma} \cap \mathbb{R}_{++}^{\Omega}$. Proceeding as in the proof of Theorem 2.12 in [13], we have that problem (4) has an optimal solution V_T^{σ}, and such optimal solution is unique since the objective function in (4) is strictly concave [1], as u_{γ} is strictly concave.

(ii). Statement *(i)* implies that problem (3) has an optimal solution V_T^* obtained by selecting V_T^σ for a permutation σ where the objective function is maximum. Suppose there are two permutations σ, σ' attaining the maximum and such that $V_T^\sigma \neq V_T^{\sigma'}$. Since the set of optimal solutions of (3) is convex, this implies that, for all $\alpha \in [0,1]$, $V_T^\alpha = \alpha V_T^\sigma + (1-\alpha) V_T^{\sigma'}$ is an optimal solution of (3). So, since the number of permutations of Ω is finite, we can find an α^* such that $V_T^{\alpha^*}$ solves (4) for a permutation σ'' but $V_T^{\alpha^*} \neq V_T^{\sigma''}$, reaching a contradiction. Finally, this implies that problem (3) has a unique optimal solution.

3.1   Characterization of the Case $p = \frac{1}{2}$

Here we focus on the special case $p = \frac{1}{2}$, for which the probability distribution $\mathbf{P}^\sigma$ on Ω is such that

$$\mathbf{P}^\sigma(\sigma(1)) = \frac{1 + (2^T - 1)\epsilon}{2^T} \quad \text{and} \quad \mathbf{P}^\sigma(\sigma(i)) = \frac{1 - \epsilon}{2^T}, \quad \text{for all } i \in \Omega \setminus \{1\}.$$

The definition of $\mathbf{P}^\sigma$ shows that only the value $\sigma(1)$ of every permutation σ deserves attention: all the permutations σ, σ' such that $\sigma(1) = \sigma'(1)$ can be considered to be equivalent. Therefore, we can restrict to 2^T arbitrary permutations $\sigma_1, \ldots, \sigma_{2^T}$ satisfying $\sigma_h(1) = h$, for all $h \in \Omega$.

In this case, the family of optimization problems (4) can be reduced to a family of optimization problems indexed by the permutations $\sigma_1, \ldots, \sigma_{2^T}$:

$$\text{maximize} \left[\sum_{k=1}^{2^T} \mathbf{P}^{\sigma_h}(k) \frac{(V_T(k))^{1-\gamma}}{1-\gamma} \right]$$

subject to:

$$\begin{cases} \sum_{k=1}^{2^T} \mathbf{Q}(k) V_T(k) - (1+r)^T V_0 = 0, \\ V_T(\sigma_h(1)) - V_T(\sigma_h(i)) \leq 0, \quad \text{for all } i \in \Omega \setminus \{1\}, \\ V_T \in \mathbb{R}_{++}^\Omega. \end{cases} \tag{5}$$

Analogously to problem (4), for every permutation σ_h, problem (5) has a unique optimal solution $V_T^{\sigma_h}$. Then, the optimal solution V_T^* for (3) can be found by selecting a $V_T^{\sigma_h}$ for a permutation σ_h (possibly non-unique) where the objective function is maximum. This shows that, when $p = \frac{1}{2}$, problem (3) can be solved by solving 2^T problems (5).

Below we provide a characterization of the optimal solution of problem (3). In what follows, let σ be a permutation of Ω and $I \subseteq \Omega \setminus \{1\}$. If $I \neq \emptyset$, for all $i \in I$, consider the constants

$$A_i^{\sigma, I} = \frac{\mathbf{P}^\sigma(\sigma(i))}{\sum\limits_{k \in I \cup \{1\}} \mathbf{P}^\sigma(\sigma(k))} \left[\left(\frac{\mathbf{Q}(\sigma(i))}{\mathbf{P}^\sigma(\sigma(i))} - \frac{\mathbf{Q}(\sigma(1))}{\mathbf{P}^\sigma(\sigma(1))} \right) \left(\sum_{k \in (I \cup \{1\}) \setminus \{i\}} \mathbf{P}^\sigma(\sigma(k)) \right) \right.$$

$$\left. + \sum_{k \in I \setminus \{i\}} \left(\mathbf{P}^\sigma(\sigma(k)) \left(\frac{\mathbf{Q}(\sigma(1))}{\mathbf{P}^\sigma(\sigma(1))} - \frac{\mathbf{Q}(\sigma(k))}{\mathbf{P}^\sigma(\sigma(k))} \right) \right) \right],$$

where summations over an empty set are intended to be 0.

In turn, the above constants are used to define the following weights:

$$\lambda_1^{\sigma,I} = \left(\frac{(1+r)^T V_0}{\sum\limits_{k=1}^{2^T} \mathbf{Q}(\sigma(k)) \left(\frac{1}{\mathbf{P}^{\sigma}(\sigma(k))} \left(\mathbf{Q}(\sigma(k)) + \mathbf{1}_{\{1\}}(k) \left(\sum\limits_{i \in I} A_i^{\sigma,I} \right) - \sum\limits_{i \in I} \mathbf{1}_{\{i\}}(k) A_i^{\sigma,I} \right) \right)^{-\frac{1}{\gamma}}} \right)^{-\gamma},$$

where $\mathbf{1}_{\{i\}}$ is the indicator function of the singleton $\{i\}$, while, for all $i \in I$, set

$$\lambda_i^{\sigma,I} = A_i^{\sigma,I} \lambda_1^{\sigma,I},$$

and $\lambda_i^{\sigma,I} = 0$, for all $i \in \Omega \setminus (I \cup \{1\})$. Notice that, if $I = \emptyset$, then the two inner summations involving I in the definition of $\lambda_1^{\sigma,I}$ are set to 0: in this case, only $\lambda_1^{\sigma,I}$ will be non-null.

Theorem 1. *For $p = \frac{1}{2}$ and $T \geq 1$, a random variable $V_T \in \mathbb{R}_{++}^{\Omega}$ is the optimal solution of problem (3) if and only if there is a permutation σ of Ω and a subset $I \subseteq \Omega \setminus \{1\}$ inducing the weights $\lambda_i^{\sigma,I}$'s such that the following conditions hold:*

(i) $\lambda_i^{\sigma,I} \geq 0$, for all $i \in I$;

(ii) $V_T(\sigma(1)) = \left(\frac{1}{\mathbf{P}^{\sigma}(\sigma(1))} \left(\mathbf{Q}(\sigma(1)) \lambda_1^{\sigma,I} + \sum\limits_{k=2}^{2^T} \lambda_k^{\sigma,I} \right) \right)^{-\frac{1}{\gamma}}$, for all $i \in \Omega \setminus \{1\}$,

$$V_T(\sigma(i)) = \left(\frac{1}{\mathbf{P}^{\sigma}(\sigma(i))} \left(\mathbf{Q}(\sigma(i)) \lambda_1^{\sigma,I} - \lambda_i^{\sigma,I} \right) \right)^{-\frac{1}{\gamma}};$$

(iii) $V_T(\sigma(1)) \leq V_T(\sigma(i))$, for all $i \in \Omega \setminus (I \cup \{1\})$;

and there is no other permutation σ' of Ω and no other subset $I' \subseteq \Omega \setminus \{1\}$ determining the weights $\lambda^{\sigma',I'}$'s and the random variable $V_T' \in \mathbb{R}_{++}^{\Omega}$ satisfying (i)–(iii) such that $\mathbf{CEU}_{\gamma,p,\epsilon}[V_T'] > \mathbf{CEU}_{\gamma,p,\epsilon}[V_T]$.

Proof. Let σ be a permutation of Ω. We first show that $V_T \in \mathbb{R}_{++}^{\Omega}$ is the optimal solution of the corresponding problem (5) if and only if there is a subset $I \subseteq \Omega \setminus \{1\}$ inducing weights $\lambda_i^{\sigma,I}$'s satisfying (i)–(iii). Denote by $f(V_T) = f(V_T(1), \ldots, V_T(2^T))$ and $g(V_T) = g(V_T(1), \ldots, V_T(2^T)) = 0$ the objective function and the equality constraint in (5), which are, respectively, (strictly) concave and linear. We also have that all inequality constraints in (5) are linear, therefore, the Karush-Kuhn-Tucker (KKT) conditions are necessary and sufficient in this case [1]. Define the Lagrangian function

$$L(V_T, \lambda_1, \ldots, \lambda_{2^T}) = f(V_T) - \lambda_1 g(V_T) - \sum_{k=2}^{2^T} \lambda_k (V_T(\sigma(1)) - V_T(\sigma(k))),$$

for which it holds

$$\frac{\partial L}{\partial V_T(\sigma(1))} = \mathbf{P}^\sigma(\sigma(1))(V_T(\sigma(1)))^{-\gamma} - \lambda_1 \mathbf{Q}(\sigma(1)) - \sum_{k=2}^{2^T} \lambda_k,$$

$$\frac{\partial L}{\partial V_T(\sigma(i))} = \mathbf{P}^\sigma(\sigma(i))(V_T(\sigma(i)))^{-\gamma} - \lambda_1 \mathbf{Q}(\sigma(i)) + \lambda_i, \quad \text{for all } i \in \Omega \setminus \{1\}.$$

Imposing the KKT conditions, we look for $V_T \in \mathbb{R}_{++}^\Omega, \lambda_1 \in \mathbb{R}, \lambda_2, \ldots, \lambda_{2^T} \geq 0$ such that $\frac{\partial L}{\partial V_T(\sigma(k))} = 0$, for all $k \in \Omega$, $g(V_T) = 0$, $V_T(\sigma(1)) \leq V_T(\sigma(i))$ and $\lambda_i(V_T(\sigma(1)) - V_T(\sigma(i))) = 0$, for all $i \in \Omega \setminus \{1\}$.

By $\frac{\partial L}{\partial V_T(\sigma(k))} = 0$, for all $k \in \Omega$, we derive

$$V_T(\sigma(1)) = \left(\frac{1}{\mathbf{P}^\sigma(\sigma(1))} \left(\lambda_1 \mathbf{Q}(\sigma(1)) + \sum_{k=2}^{2^T} \lambda_k \right) \right)^{-\frac{1}{\gamma}},$$

$$V_T(\sigma(i)) = \left(\frac{1}{\mathbf{P}^\sigma(\sigma(i))} \left(\lambda_1 \mathbf{Q}(\sigma(i)) - \lambda_i \right) \right)^{-\frac{1}{\gamma}}, \quad \text{for all } i \in \Omega \setminus \{1\}.$$

Moreover, by the complementary slackness conditions $\lambda_i(V_T(\sigma(1)) - V_T(\sigma(i))) = 0$, for all $i \in \Omega \setminus \{1\}$, there must exist $I \subseteq \Omega \setminus \{1\}$ such that $\lambda_i = 0$, for all $i \in \Omega \setminus (I \cup \{1\})$, while, for all $i \in I$, $V_T(\sigma(1)) - V_T(\sigma(i)) = 0$.

The case $I = \emptyset$ is trivial, thus suppose $I \neq \emptyset$. For every $i \in I$, equation $V_T(\sigma(1)) - V_T(\sigma(i)) = 0$ holds if and only if

$$\left(\frac{\mathbf{Q}(\sigma(1))}{\mathbf{P}^\sigma(\sigma(1))} - \frac{\mathbf{Q}(\sigma(i))}{\mathbf{P}^\sigma(\sigma(i))} \right) \lambda_1 + \left(\frac{1}{\mathbf{P}^\sigma(\sigma(1))} + \frac{1}{\mathbf{P}^\sigma(\sigma(i))} \right) \lambda_i + \frac{1}{\mathbf{P}^\sigma(\sigma(1))} \sum_{k \in I \setminus \{i\}} \lambda_k = 0.$$

Choose an enumeration of $I \cup \{1\} = \{i_1, i_2, \ldots, i_n\}$ with $i_1 = 1$. Then the above equations give rise to the homogeneous linear system $A\mathbf{x} = \mathbf{0}$, whose unknown is the column vector $\mathbf{x} = [\lambda_1 \ \lambda_{i_2} \ \cdots \ \lambda_{i_n}]^T \in \mathbb{R}^{n \times 1}$ and whose coefficient matrix is $A = [\mathbf{q}|B] \in \mathbb{R}^{(n-1) \times n}$ where

$$\mathbf{q} = \left[\left(\frac{\mathbf{Q}(\sigma(1))}{\mathbf{P}^\sigma(\sigma(1))} - \frac{\mathbf{Q}(\sigma(i_2))}{\mathbf{P}^\sigma(\sigma(i_2))} \right) \left(\frac{\mathbf{Q}(\sigma(1))}{\mathbf{P}^\sigma(\sigma(1))} - \frac{\mathbf{Q}(\sigma(i_3))}{\mathbf{P}^\sigma(\sigma(i_3))} \right) \cdots \left(\frac{\mathbf{Q}(\sigma(1))}{\mathbf{P}^\sigma(\sigma(1))} - \frac{\mathbf{Q}(\sigma(i_n))}{\mathbf{P}^\sigma(\sigma(i_n))} \right) \right]^T,$$

$$B = \begin{bmatrix} \left(\frac{1}{\mathbf{P}^\sigma(\sigma(1))} + \frac{1}{\mathbf{P}^\sigma(\sigma(i_2))} \right) & \frac{1}{\mathbf{P}^\sigma(\sigma(1))} & \cdots & \frac{1}{\mathbf{P}^\sigma(\sigma(1))} \\ \frac{1}{\mathbf{P}^\sigma(\sigma(1))} & \left(\frac{1}{\mathbf{P}^\sigma(\sigma(1))} + \frac{1}{\mathbf{P}^\sigma(\sigma(i_3))} \right) & \cdots & \frac{1}{\mathbf{P}^\sigma(\sigma(1))} \\ \vdots & \vdots & \cdots & \vdots \\ \frac{1}{\mathbf{P}^\sigma(\sigma(1))} & \frac{1}{\mathbf{P}^\sigma(\sigma(1))} & \cdots & \left(\frac{1}{\mathbf{P}^\sigma(\sigma(1))} + \frac{1}{\mathbf{P}^\sigma(\sigma(i_n))} \right) \end{bmatrix}.$$

Hence, $B \in \mathbb{R}^{(n-1) \times (n-1)}$ can be decomposed in the sum of two matrices $C, D \in \mathbb{R}^{(n-1) \times (n-1)}$, where C is a constant matrix with all entries equal to $\frac{1}{\mathbf{P}^\sigma(\sigma(1))}$ and D is the diagonal matrix whose diagonal contains the elements

$\frac{1}{\mathbf{P}^\sigma(\sigma(i_2))}, \ldots, \frac{1}{\mathbf{P}^\sigma(\sigma(i_n))}$. Since the determinant does not change subtracting the first row from all other rows, we can consider the matrix

$$
\begin{bmatrix}
\left(\frac{1}{\mathbf{P}^\sigma(\sigma(1))} + \frac{1}{\mathbf{P}^\sigma(\sigma(i_2))}\right) & \frac{1}{\mathbf{P}^\sigma(\sigma(1))} & \cdots & \frac{1}{\mathbf{P}^\sigma(\sigma(1))} \\
-\frac{1}{\mathbf{P}^\sigma(\sigma(i_2))} & \frac{1}{\mathbf{P}^\sigma(\sigma(i_3))} & \cdots & 0 \\
\vdots & \vdots & \cdots & \vdots \\
-\frac{1}{\mathbf{P}^\sigma(\sigma(i_2))} & 0 & \cdots & \frac{1}{\mathbf{P}^\sigma(\sigma(i_n))}
\end{bmatrix}.
\tag{6}
$$

Applying the Laplace expansion along the first row we get that

$$
\det B = \det(C + D) = \frac{\sum_{j=1}^{n} \mathbf{P}^\sigma(\sigma(i_j))}{\prod_{j=1}^{n} \mathbf{P}^\sigma(\sigma(i_j))},
$$

and since $\det B \neq 0$, we have that $\operatorname{rank} B = n - 1$ and the system admits non-trivial solutions, depending on one real parameter that we identify with λ_1. Now, apply Cramer's rule to the reduced system $B\mathbf{y} = -\lambda_1 \mathbf{q}$ with unknown the column vector $\mathbf{y} = [\lambda_{i_2} \cdots \lambda_{i_n}]^T \in \mathbb{R}^{(n-1)\times 1}$.

For $j = 2, \ldots, n$, denote by B_{j-1} the matrix obtained by substituting the $(j-1)$-th column of B with the vector $-\lambda_1 \mathbf{q}$. Applying the Laplace expansion along the $(j-1)$-th column of B_{j-1} and noticing that all minors can be transformed (by swapping rows and keeping track of sign changes) in the sum of a constant matrix and a diagonal matrix (possibly with a zero on the diagonal), we have

$$
\det B_{j-1} = \left[\left(\frac{\mathbf{Q}(\sigma(i_j))}{\mathbf{P}^\sigma(\sigma(i_j))} - \frac{\mathbf{Q}(\sigma(1))}{\mathbf{P}^\sigma(\sigma(1))} \right) \frac{\sum_{k\in(I\cup\{1\})\setminus\{i_j\}} \mathbf{P}^\sigma(\sigma(k))}{\prod_{k\in(I\cup\{1\})\setminus\{i_j\}} \mathbf{P}^\sigma(\sigma(k))} \right.
$$
$$
\left. + \sum_{k\in I\setminus\{i_j\}} \left(\left(\frac{\mathbf{Q}(\sigma(1))}{\mathbf{P}^\sigma(\sigma(1))} - \frac{\mathbf{Q}(\sigma(k))}{\mathbf{P}^\sigma(\sigma(k))} \right) \frac{1}{\prod_{s\in(I\cup\{1\})\setminus\{i_j,k\}} \mathbf{P}^\sigma(\sigma(s))} \right) \right] \lambda_1
$$

therefore, $\lambda_{i_j} = \frac{\det B_{j-1}}{\det B} = A_{i_j}^{\sigma,I} \lambda_1$.

Substituting in $g(V_T) = 0$ the expressions of $V_T(\sigma(k))$, for all $k \in \Omega$, and λ_i, for all $i \in \Omega \setminus \{1\}$, we get for λ_1 the expression of $\lambda_1^{\sigma,I}$, thus λ_i coincides with $\lambda_i^{\sigma,I}$. Hence, $V_T \in \mathbb{R}_{++}^\Omega$ is the optimal solution for the problem (5) if and only if there exists $I \subseteq \Omega \setminus \{1\}$ inducing weights $\lambda_i^{\sigma,I}$'s satisfying (i)–(iii).

For every $h \in \Omega$, let σ_h be a permutation of Ω such that $\sigma_h(1) = h$ and denote by $V_T^{\sigma_h}$ the optimal solution of problem (5) for σ_h. By the definition of the Choquet integral [8], problem (3) is equivalent to maximizing $\mathbf{CEU}_{\gamma,p,\epsilon}$ over optimal solutions of the family of problems (4), indexed by all permutations of Ω. In turn, since, for every permutations σ, σ' of Ω such that $\sigma(1) = \sigma'(1)$, it holds $\mathbf{P}^\sigma = \mathbf{P}^{\sigma'}$, such maximization can be reduced to maximizing $\mathbf{CEU}_{\gamma,p,\epsilon}$ over optimal solutions of the family of problems (5), indexed by permutations $\sigma_1, \ldots, \sigma_{2^T}$ of Ω. This finally proves the theorem.

Theorem 1 allows to find the analytic expression of the optimal solution of problem (3) in the case $p = \frac{1}{2}$, reducing it to 2^T combinatorial optimization problems, each corresponding to a permutation σ_h of Ω such that $\sigma_h(1) = h$, for all $h \in \Omega$.

Example 1. Consider a two-period model with $T = 2$, $\gamma = 2$, $\epsilon = \frac{1}{50}$, $p = \frac{1}{2}$, $u = 2$, and $d = \frac{1}{2}$. Suppose $S_0 = €100$, $V_0 = €10$, and the risk-free interest rate is $r = 5\%$ over every period. This implies that $q = \frac{11}{30}$ while, for every permutation σ of $\Omega = \{1, 2, 3, 4\}$, it holds $\mathbf{P}^\sigma(\sigma(1)) = \frac{53}{200}$ and $\mathbf{P}^\sigma(\sigma(i)) = \frac{49}{200}$.

For $\sigma_1 = \langle 1, 2, 3, 4 \rangle$ the only subset of $\Omega \setminus \{1\}$ satisfying *(i)–(iii)* is $I = \{4\}$ with:

- $\lambda_1^{\sigma_1, I} = 0.00822164$, $\lambda_2^{\sigma_1, I} = 0$, $\lambda_3^{\sigma_1, I} = 0$, $\lambda_4^{\sigma_1, I} = 0.00118255$,
- $V_2^{\sigma_1}(1) = 10.7623$, $V_2^{\sigma_1}(2) = 11.328$, $V_2^{\sigma_1}(3) = 11.328$, $V_2^{\sigma_1}(4) = 10.7623$,
- $\mathbf{CEU}_{\gamma, p, \epsilon}[V_2^{\sigma_1}] = -0.0906436$.

For $\sigma_2 = \langle 2, 1, 3, 4 \rangle$ the only subset of $\Omega \setminus \{1\}$ satisfying *(i)–(iii)* is $I = \{4\}$ with:

- $\lambda_1^{\sigma_2, I} = 0.00803641$, $\lambda_2^{\sigma_2, I} = 0$, $\lambda_3^{\sigma_2, I} = 0$, $\lambda_4^{\sigma_2, I} = 0.000778428$,
- $V_2^{\sigma_2}(1) = 15.0585$, $V_2^{\sigma_2}(2) = 10.0101$, $V_2^{\sigma_2}(3) = 11.4578$, $V_2^{\sigma_2}(4) = 10.0101$,
- $\mathbf{CEU}_{\gamma, p, \epsilon}[V_2^{\sigma_2}] = -0.0886014$.

For $\sigma_3 = \langle 3, 1, 2, 4 \rangle$ the only subset of $\Omega \setminus \{1\}$ satisfying *(i)–(iii)* is $I = \{4\}$ with:

- $\lambda_1^{\sigma_3, I} = 0.00803641$, $\lambda_2^{\sigma_3, I} = 0$, $\lambda_3^{\sigma_3, I} = 0$, $\lambda_4^{\sigma_3, I} = 0.000778428$,
- $V_2^{\sigma_3}(1) = 15.0585$, $V_2^{\sigma_3}(2) = 11.4578$, $V_2^{\sigma_3}(3) = 10.0101$, $V_2^{\sigma_3}(4) = 10.0101$,
- $\mathbf{CEU}_{\gamma, p, \epsilon}[V_2^{\sigma_3}] = -0.0886014$.

For $\sigma_4 = \langle 4, 1, 2, 3 \rangle$ the only subset of $\Omega \setminus \{1\}$ satisfying *(i)–(iii)* is $I = \emptyset$ with:

- $\lambda_1^{\sigma_4, I} = 0.0079751$, $\lambda_2^{\sigma_4, I} = 0$, $\lambda_3^{\sigma_4, I} = 0$, $\lambda_4^{\sigma_4, I} = 0$,
- $V_2^{\sigma_4}(1) = 15.1162$, $V_2^{\sigma_4}(2) = 11.5017$, $V_2^{\sigma_4}(3) = 11.5017$, $V_2^{\sigma_4}(4) = 9.1017$,
- $\mathbf{CEU}_{\gamma, p, \epsilon}[V_2^{\sigma_4}] = -0.0879255$.

A simple inspection shows that the maximum value of $\mathbf{CEU}_{\gamma, p, \epsilon}$ is obtained for σ_4, therefore we take $V_2^* = V_2^{\sigma_4}$.

Using the martingale property with respect to $\mathbf{Q}$ of the wealth process $\{V_0^*, V_1^*, V_2^*\}$, that is $V_1^* = \frac{\mathbf{E}_1^\mathbf{Q}[V_2^*]}{1+r}$ and $V_0^* = V_0 = \frac{\mathbf{E}_0^\mathbf{Q}[V_2^*]}{(1+r)^2}$, we can recover the optimal self-financing strategy $\{\theta_0, \theta_1\}$ through (1):

Ω	V_0^*	V_1^*	V_2^*	θ_0	θ_1
1	€10	€12.2162	€15.1162	0.018	0.012
2	€10	€12.2162	€11.5017	0.018	0.012
3	€10	€9.5064	€11.5017	0.018	0.032
4	€10	€9.5064	€9.1017	0.018	0.032

4   The Effect of ϵ on the Optimal Portfolio

Now we investigate the effect of ϵ on the optimal portfolio. Figures 1 and 2 show the optimal value of $\mathbf{CEU}_{\gamma,p,\epsilon}$ for $V_0 = €10$, $u \in \{1.2, 1.4, 1.6, 1.8, 2\}$, $d = \frac{1}{u}$, $r = 5\%$, $\gamma = 2$, $p = \frac{1}{2}$ and ϵ ranging in $[0, 1)$ with step 0.01, for $T = 2$ and $T = 3$, respectively. In particular, $\epsilon = 0$ stands for absence of ambiguity.

In both figures we can see that, for increasing ϵ, the optimal value of $\mathbf{CEU}_{\gamma,p,\epsilon}$ decreases until reaching a constant value that corresponds to $u_\gamma((1 + r)^T V_0)$. It actually holds that, once the optimal value of $\mathbf{CEU}_{\gamma,p,\epsilon}$ reaches $u_\gamma((1 + r)^T V_0)$, the optimal portfolio results in the final risk-free wealth $V_T^* = (1 + r)^T V_0$. In such cases, the corresponding self-financing strategy is such that $\theta_t = 0$, for $t = 0, \ldots, T - 1$, i.e., the optimal portfolio consists of only a risk-free bond investment for all the periods.

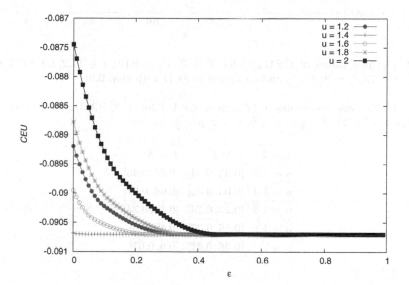

Fig. 1. Optimal values of $\mathbf{CEU}_{\gamma,p,\epsilon}$ for $T = 2$, $V_0 = €10$, $u \in \{1.2, 1.4, 1.6, 1.8, 2\}$, $d = \frac{1}{u}$, $r = 5\%$, $\gamma = 2$, $p = \frac{1}{2}$ and ϵ ranging in $[0, 1)$ with step 0.01.

Figures 1 and 2 highlight the existence of a value $\epsilon^*(T, \gamma, p, u, d, r, V_0)$ above which the optimal self-financing strategy reduces to 0 for all times. In a sense, we may think at $\epsilon^*(T, \gamma, p, u, d, r, V_0)$ as a threshold above which the ambiguity incorporated in the real-world probability $\mathbf{P}$ through the ϵ-contamination is so high to make the risk-free portfolio the most suitable choice. The following Table 1 reports the intervals containing $\epsilon^*(T, \gamma, p, u, d, r, V_0)$ for the parameter setting of Figs. 1 and 2. Let us stress that a numerical approximation of $\epsilon^*(T, \gamma, p, u, d, r, V_0)$ can be achieved by applying a suitable bisection algorithm, while its analytic expression will be the aim of future research.

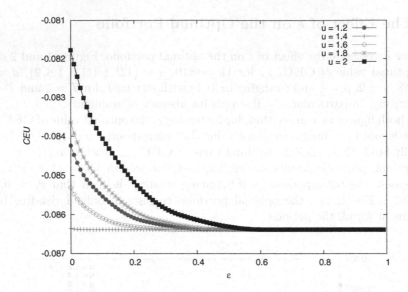

Fig. 2. Optimal values of $\mathbf{CEU}_{\gamma,p,\epsilon}$ for $T = 3$, $V_0 = €10$, $u \in \{1.2, 1.4, 1.6, 1.8, 2\}$, $d = \frac{1}{u}$, $r = 5\%$, $\gamma = 2$, $p = \frac{1}{2}$ and ϵ ranging in $[0, 1)$ with step 0.01.

Table 1. Intervals containing $\epsilon^*(T, \gamma, p, u, d, r, V_0)$ for $T \in \{2, 3\}$, $V_0 = €10$, $u \in \{1.2, 1.4, 1.6, 1.8, 2\}$, $d = \frac{1}{u}$, $r = 5\%$, $\gamma = 2$, $p = \frac{1}{2}$.

$\gamma = 2$	$T = 2$	$T = 3$
$u = 1.2$	$[0.33, 0.34]$	$[0.45, 0.46]$
$u = 1.4$	$[0.04, 0.05]$	$[0.06, 0.07]$
$u = 1.6$	$[0.23, 0.24]$	$[0.33, 0.34]$
$u = 1.8$	$[0.36, 0.37]$	$[0.49, 0.5]$
$u = 2$	$[0.46, 0.47]$	$[0.6, 0.61]$

5 Conclusions and Future Works

In this paper we formulate a dynamic portfolio selection problem under ambiguity by referring to an ϵ-contaminated binomial market model and a CRRA utility function. We provide a characterization of the optimal solution in the case the stock price returns are uniformly distributed and investigate the effect of the contamination parameter ϵ on the optimal portfolio. It turns out that, in the uniform case, one can find a threshold for ϵ above which the optimal portfolio reduces to a risk-free investment on every period. An analytic characterization of such threshold is reserved for future research. Further, a characterization of the optimal solution in the general case remains open, as well as the generalization to other classes of utility functions. Another line of investigation is the design of efficient algorithms to solve the family of concave optimization problems to which the initial problem can be reduced, since, being indexed by the permutations of the sample space, they constitute a computational challenge.

References

1. Boyd, S., Vandenberghe, L.: Convex Optimization. Cambridge University Press, Cambridge (2004)
2. Coletti, G., Petturiti, D., Vantaggi, B.: Rationality principles for preferences on belief functions. Kybernetika 51(3), 486–507 (2015)
3. Cox, J., Ross, S., Rubinstein, M.: Option pricing: a simplified approach. J. Finan. Econ. 7(3), 229–263 (1979)
4. Ellsberg, D.: Risk, ambiguity, and the savage axioms. Q. J. Econ. 75(4), 643–669 (1961)
5. Etner, J., Jeleva, M., Tallon, J.M.: Decision theory under ambiguity. J. Econ. Surv. 26(2), 234–270 (2012)
6. Gilboa, I., Marinacci, M.: Ambiguity and the Bayesian paradigm. In: Arló-Costa, H., Hendricks, V.F., van Benthem, J. (eds.) Readings in Formal Epistemology. SGTP, vol. 1, pp. 385–439. Springer, Cham (2016). https://doi.org/10.1007/978-3-319-20451-2_21
7. Gilboa, I., Schmeidler, D.: Maxmin expected utility with non-unique prior. J. Math. Econ. 18(2), 141–153 (1989)
8. Grabisch, M.: Set Functions, Games and Capacities in Decision Making. Theory and Decision Library C. Springer, Heidelberg (2016). https://doi.org/10.1007/978-3-319-30690-2
9. Huber, P.J.: Robust Statistics. Wiley, New York (1981)
10. Jaffray, J.Y.: Linear utility theory for belief functions. Oper. Res. Lett. 8(2), 107–112 (1989)
11. Jin, H., Yu Zhou, X.: Behavioral portfolio selection in continuous time. Math. Finan. 18(3), 385–426 (2008)
12. Kast, R., Lapied, A., Roubaud, D.: Modelling under ambiguity with dynamically consistent Choquet random walks and Choquet-Brownian motions. Econ. Model. 38, 495–503 (2014)
13. Pascucci, A., Runggaldier, W.: Financial Mathematics: Theory and Problems for Multi-period Models. Springer, Heidelberg (2009). https://doi.org/10.1007/978-88-470-2538-7
14. Petturiti, D., Vantaggi, B.: Modeling agent's conditional preferences under objective ambiguity in Dempster-Shafer theory. Int. J. Approximate Reasoning 119, 151–176 (2020)
15. Pflug, G., Wozabal, D.: Ambiguity in portfolio selection. Quant. Finan. 7(4), 435–442 (2007)
16. Pflug, G.C., Pohl, M.: A review on ambiguity in stochastic portfolio optimization. Set-Valued Variational Anal. 26(4), 733–757 (2018)
17. Schmeidler, D.: Subjective probability and expected utility without additivity. Econometrica 57(3), 571–587 (1989)
18. Černý, A.: Mathematical Techniques in Finance: Tools for Incomplete Markets, 2nd edn. Princeton University Press, Princeton (2009)
19. Walley, P.: Statistical Reasoning with Imprecise Probabilities. Chapman and Hall, London (1991)

Limit Behaviour of Upper and Lower Expected Time Averages in Discrete-Time Imprecise Markov Chains

Natan T'Joens[✉] and Jasper De Bock

FLip, Ghent University, Ghent, Belgium
{natan.tjoens,jasper.debock}@ugent.be

Abstract. We study the limit behaviour of upper and lower bounds on expected time averages in imprecise Markov chains; a generalised type of Markov chain where the local dynamics, traditionally characterised by transition probabilities, are now represented by sets of 'plausible' transition probabilities. Our main result is a necessary and sufficient condition under which these upper and lower bounds, called upper and lower expected time averages, will converge as time progresses towards infinity to limit values that do not depend on the process' initial state. Remarkably, our condition is considerably weaker than those needed to establish similar results for so-called limit—or steady state—upper and lower expectations, which are often used to provide approximate information about the limit behaviour of time averages as well. We show that such an approximation is sub-optimal and that it can be significantly improved by directly using upper and lower expected time averages.

Keywords: Imprecise Markov chain · Upper expectation · Upper transition operator · Expected time average · Weak ergodicity

1 Introduction

Markov chains are probabilistic models that can be used to describe the uncertain dynamics of a large variety of stochastic processes. One of the key results within the field is the point-wise ergodic theorem. It establishes a relation between the long-term time average of a real-valued function and its limit expectation, which is guaranteed to exist if the Markov chain is ergodic.[1] For this reason, limit expectations and limit distributions have become central objects of interest. Of course, if one is interested in the long-term behaviour of time averages, one could also study the expected values of these averages directly. This is not often done though, because the limit of these expected time averages coincides with the

---

[1] The term ergodicity has various meanings; sometimes it refers to properties of an invariant measure, sometimes it refers to properties such as irreducibility (with or without aperiodicity), regularity, ... Our usage of the term follows conventions introduced in earlier work [2,8] on imprecise Markov chains; see Sects. 2 and 4.

© Springer Nature Switzerland AG 2020
M.-J. Lesot et al. (Eds.): IPMU 2020, CCIS 1238, pp. 224–238, 2020.
https://doi.org/10.1007/978-3-030-50143-3_17

aforementioned limit expectations, which can straightforwardly be obtained by solving a linear eigenproblem [10].

We here consider a generalisation of Markov chains, called imprecise Markov chains [2,4,9], for which the considerations above are not necessarily true. Imprecise Markov chains are sets of traditional ("precise") probabilistic models, where the Markov property (history independence) and time-homogeneity apply to the collection of precise models as a whole, but not necessarily to the individual models themselves. Imprecise Markov chains therefore allow one to incorporate model uncertainty about the numerical values of the transition probabilities that make up a Markov chain, but also, and more importantly, about structural assumptions such as time-homogeneity and the Markov property. For such an imprecise Markov chain, one is then typically interested in obtaining tight upper and lower bounds on inferences for the individual constituting models. The operators that represent these upper and lower bounds are respectively called upper and lower expectations.

Just like traditional Markov chains can have a limit expectation, an imprecise Markov chain can have limit upper and lower expectations. There are necessary and sufficient conditions for their existence [8] as well as an imprecise variant of the point-wise ergodic theorem [2]. An important difference with traditional Markov chains however, is that upper and lower bounds on expectations of time averages—we will call these upper and lower expected time averages—may not converge to limit upper and lower expectations. Nevertheless, because they give conservative bounds [11, Lemma 57], and because they are fairly easy to compute, limit upper and lower expectations are often used as descriptors of the long-term behaviour of imprecise Markov chains, even if one is actually interested in time averages. This comes at a cost though: as we illustrate in Sect. 4, both inferences can differ greatly, with limit expectations providing far too conservative bounds.

Unfortunately, apart from some experiments in [11], little is known about the long-term behaviour of upper and lower expected time averages in imprecise Markov chains. The aim of this paper is to remedy this situation. Our main result is an accessibility condition that is necessary and sufficient for upper and lower expected time averages to converge to a limit value that does not depend on the process' initial state; see Sect. 7. Remarkably, this condition is considerably weaker than the ones required for limit lower and upper expectations to exist.

Technical proofs are relegated to the appendix of an extended online version [12]. This is particularly true for the results in Sect. 7, where the main text provides an informal argument that aims to provide intuition.

2   Markov Chains

We consider an infinite sequence $X_0 X_1 X_2 \cdots$ of uncertain states, where each state X_k at time $k \in \mathbb{N}_0 := \mathbb{N} \cup \{0\}$ takes values in some finite set $\mathcal{X}$, called the *state space*. Such a sequence $X_0 X_1 X_2 \cdots$ will be called a *(discrete-time) stochastic process*. For any $k, \ell \in \mathbb{N}_0$ such that $k \leq \ell$, we use $X_{k:\ell}$ to denote the finite subsequence $X_k \cdots X_\ell$ of states that takes values in $\mathscr{X}^{\ell-k+1}$. Moreover, for any $k, \ell \in \mathbb{N}_0$ such that $k \leq \ell$ and any $x_{k:\ell} \in \mathscr{X}^{\ell-k+1}$, we use $X_{k:\ell} = x_{k:\ell}$ to denote the event that $X_k = x_k \cdots X_\ell = x_\ell$. The uncertain dynamics of a stochastic process are then typically described by probabilities of the form $\mathrm{P}(X_{k+1} = x_{k+1} | X_{0:k} = x_{0:k})$, for any $k \in \mathbb{N}_0$ and any $x_{0:k+1} \in \mathscr{X}^{k+2}$. They represent beliefs about which state the process will be in at time $k+1$ given that we know that it was in the states $x_0 \cdots x_k$ at time instances 0 through k. Additionally, our beliefs about the value of the initial state X_0 can be represented by probabilities $\mathrm{P}(X_0 = x_0)$ for all $x_0 \in \mathscr{X}$. The local probability assessments $\mathrm{P}(X_{k+1} = x_{k+1} | X_{0:k} = x_{0:k})$ and $\mathrm{P}(X_0 = x_0)$ can now be combined to construct a global probability model P that describes the dynamics of the process on a more general level. This can be done in various ways; one of the most common ones being a measure-theoretic approach where countable additivity plays a central role. For our purposes however, we will only require finite additivity. Regardless, once you have such a global probability model P, it can then be used to define expectations and make inferences about the uncertain behaviour of the process.

For any set A, let us write $\mathscr{L}(A)$ to denote the set of all real-valued functions on A. Throughout, for any $a \in A$, we use $\mathbb{I}_a$ to denote the *indicator* of a: the function in $\mathscr{L}(A)$ that takes the value 1 in a and 0 otherwise. We will only be concerned with (upper and lower) expectations of *finitary functions*: functions that depend on the state of the process at a finite number of time instances. So if f is finitary, we can write $f = g(X_{0:k})$ for some $k \in \mathbb{N}_0$ and some $g \in \mathscr{L}(\mathscr{X}^{k+1})$. Note that finitary functions are bounded; this follows from their real-valuedness and the fact that $\mathscr{X}$ is finite. The expectation of a finitary function $f(X_{0:k})$ conditional on some event $X_{0:\ell} = x_{0:\ell}$ simply reduces to a finite weighted sum:

$$\mathrm{E}_{\mathrm{P}}(f(X_{0:k}) | X_{0:\ell} = x_{0:\ell}) = \sum_{x_{\ell+1:k} \in \mathscr{X}^{k-\ell}} f(x_{0:k}) \prod_{i=\ell}^{k-1} \mathrm{P}(X_{i+1} = x_{i+1} | X_{0:i} = x_{0:i}).$$

A particularly interesting case arises when studying stochastic processes that are described by a probability model P that satisfies

$$\mathrm{P}(X_{k+1} = y \,|\, X_{0:k} = x_{0:k}) = \mathrm{P}(X_{k+1} = y \,|\, X_k = x_k),$$

for all $k \in \mathbb{N}_0$, all $y \in \mathscr{X}$ and all $x_{0:k} \in \mathscr{X}^{k+1}$. This property, known as the *Markov property*, states that given the present state of the process the future behaviour of the process does not depend on its history. A process of this type is called a *Markov chain*. We moreover call it *(time) homogeneous* if additionally $\mathrm{P}(X_{k+1} = y \,|\, X_k = x) = \mathrm{P}(X_1 = y \,|\, X_0 = x)$, for all $k \in \mathbb{N}_0$ and all $x, y \in \mathscr{X}$. Hence, together with the assessments $\mathrm{P}(X_0 = x_0)$, the dynamics of a homogeneous Markov chain are fully characterised by the probabilities $\mathrm{P}(X_1 = y \,|\, X_0 = x)$. These probabilities are typically gathered in a *transition matrix* T; a row-stochastic $|\mathscr{X}| \times |\mathscr{X}|$ matrix T that is defined by $T(x, y) := \mathrm{P}(X_1 = y \,|\, X_0 = x)$ for all $x, y \in \mathscr{X}$. This matrix representation T is particularly convenient because it can be regarded as a linear operator from

$\mathscr{L}(\mathscr{X})$ to $\mathscr{L}(\mathscr{X})$, defined for any $k \in \mathbb{N}_0$, any $f \in \mathscr{L}(\mathscr{X})$ and any $x \in \mathscr{X}$ by

$$Tf(x) := \sum_{y \in \mathscr{X}} f(y)\mathrm{P}(X_{k+1} = y \mid X_k = x) = \mathrm{E}_{\mathrm{P}}(f(X_{k+1}) \mid X_k = x).$$

More generally, we have that $\mathrm{E}_{\mathrm{P}}(f(X_{k+\ell}) \mid X_k = x) = T^\ell f(x)$ for all $k \in \mathbb{N}_0$, all $\ell \in \mathbb{N}_0$ and all $x \in \mathscr{X}$. Then, under some well-known accessibility conditions [8, Proposition 3], the expectation $T^\ell f(x)$ converges for increasing ℓ towards a constant $\mathrm{E}_\infty(f)$ independently of the initial state x. If this is the case for all $f \in \mathscr{L}(\mathscr{X})$, the homogeneous Markov chain will have a steady-state distribution, represented by the limit expectation E_∞, and we call the Markov chain *ergodic*. The expectation E_∞ is in particular also useful if we are interested in the limit behaviour of expected time averages. Indeed, let $\overline{f}_k(X_{\ell:\ell+k}) := {}^1\!/(k+1) \sum_{i=\ell}^{\ell+k} f(X_i)$ be the time average of some function $f \in \mathscr{L}(\mathscr{X})$ evaluated at the time instances ℓ through $k + \ell$. Then, according to [11, Theorem 38], the limit of the expected average $\lim_{k \to +\infty} \mathrm{E}_{\mathrm{P}}(\overline{f}_k(X_{0:k}))$ coincides with the limit expectation $\mathrm{E}_\infty(f)$. One of the aims of this paper is to explore to which extent this remains true for imprecise Markov chains.

3   Imprecise Markov Chains

If the basic probabilities $\mathrm{P}(X_{k+1} | X_{0:k} = x_{0:k})$ that describe a stochastic process are imprecise, in the sense that we only have partial information about them, then we can still model the process' dynamics by considering a set $\mathscr{T}_{x_{0:k}}$ of such probabilities, for all $k \in \mathbb{N}_0$ and all $x_{0:k} \in \mathscr{X}^{k+1}$. This set $\mathscr{T}_{x_{0:k}}$ is then interpreted as the set of all probability mass functions $\mathrm{P}(X_{k+1} | X_{0:k} = x_{0:k})$ that we deem "plausible". We here consider the special case where the sets $\mathscr{T}_{x_{0:k}}$ satisfy a Markov property, meaning that $\mathscr{T}_{x_{0:k}} = \mathscr{T}_{x_k}$ for all $k \in \mathbb{N}_0$ and all $x_{0:k} \in \mathscr{X}^{k+1}$. Similar to the precise case, the sets $\mathscr{T}_x$, for all $x \in \mathscr{X}$, can be gathered into a single object: the set $\mathscr{T}$ of all row stochastic $|\mathscr{X}| \times |\mathscr{X}|$ matrices T such that, for all $x \in \mathscr{X}$, the probability mass function $T(x, \cdot)$ is an element of $\mathscr{T}_x$. A set $\mathscr{T}$ of transition matrices defined in this way is called *separately specified* [9]. For any such set $\mathscr{T}$, the corresponding *imprecise Markov chain under epistemic irrelevance* $\mathscr{P}_{\mathscr{T}}^{\mathrm{ei}}$ [3] is the set of all (precise) probability models P such that $\mathrm{P}(X_{k+1} | X_{0:k} = x_{0:k}) \in \mathscr{T}_{x_k}$ for all $k \in \mathbb{N}_0$ and all $x_{0:k} \in \mathscr{X}^{k+1}$. The values of the probabilities $\mathrm{P}(X_0 = x_0)$ will be of no importance to us, because we will focus solely on (upper and lower) expectations conditional on the value of the initial state X_0.

Clearly, an imprecise Markov chain $\mathscr{P}_{\mathscr{T}}^{\mathrm{ei}}$ also contains non-homogeneous, and even non-Markovian processes. So the Markov property does in this case not apply to the individual probability assessments, but rather to the sets $\mathscr{T}_{x_{0:k}}$. The model $\mathscr{P}_{\mathscr{T}}^{\mathrm{ei}}$ is therefore a generalisation of a traditional Markov chain where we allow for model uncertainty about, on the one hand, the mass functions $\mathrm{P}(X_{k+1} | X_{0:k} = x_{0:k})$ and, on the other hand, about structural assumptions such as the Markov and time-homogeneity property. However, there are also types of imprecise Markov chains that do impose some of these properties. For a given set $\mathscr{T}$, the *imprecise Markov chain under complete independence* $\mathscr{P}_{\mathscr{T}}^{\mathrm{ci}}$ is the subset of $\mathscr{P}_{\mathscr{T}}^{\mathrm{ei}}$ that contains all, possibly non-homogeneous, Markov chains in $\mathscr{P}_{\mathscr{T}}^{\mathrm{ei}}$ [11]. The *imprecise Markov chain under repetition independence* $\mathscr{P}_{\mathscr{T}}^{\mathrm{ri}}$ is the subset of $\mathscr{P}_{\mathscr{T}}^{\mathrm{ei}}$ containing all homogeneous Markov chains [11]. Henceforth, we let $\mathscr{T}$ be some fixed, arbitrary set of transition matrices that is separately specified.

Now, for any probability model P in the imprecise Markov chain $\mathscr{P}_{\mathscr{T}}^{\mathrm{ei}}$, we can again consider the corresponding expectation operator E_{P}. The *upper* and *lower expectation* are then respectively defined as the tightest upper and lower bound on this expectation:

$$\overline{\mathrm{E}}_{\mathscr{T}}^{\mathrm{ei}}(f|A) := \sup_{\mathrm{P} \in \mathscr{P}_{\mathscr{T}}^{\mathrm{ei}}} \mathrm{E}_{\mathrm{P}}(f|A) \quad \text{and} \quad \underline{\mathrm{E}}_{\mathscr{T}}^{\mathrm{ei}}(f|A) := \inf_{\mathrm{P} \in \mathscr{P}_{\mathscr{T}}^{\mathrm{ei}}} \mathrm{E}_{\mathrm{P}}(f|A),$$

for any finitary function f and any event A of the form $X_{0:k} = x_{0:k}$. The operators $\overline{\mathrm{E}}_{\mathscr{T}}^{\mathrm{ei}}$ and $\underline{\mathrm{E}}_{\mathscr{T}}^{\mathrm{ei}}$ are related by conjugacy, meaning that $\underline{\mathrm{E}}_{\mathscr{T}}^{\mathrm{ei}}(\cdot|\cdot) = -\overline{\mathrm{E}}_{\mathscr{T}}^{\mathrm{ei}}(-\cdot|\cdot)$, which allows us to focus on only one of them; upper expectations in our case. The lower expectation $\underline{\mathrm{E}}_{\mathscr{T}}^{\mathrm{ei}}(f|A)$ of a finitary function f can then simply be obtained by considering the upper expectation $-\overline{\mathrm{E}}_{\mathscr{T}}^{\mathrm{ei}}(-f|A)$.

In a similar way, we can define the upper expectations $\overline{\mathrm{E}}_{\mathscr{T}}^{\mathrm{ci}}$ and $\overline{\mathrm{E}}_{\mathscr{T}}^{\mathrm{ri}}$ and the lower expectations $\underline{\mathrm{E}}_{\mathscr{T}}^{\mathrm{ci}}$ and $\underline{\mathrm{E}}_{\mathscr{T}}^{\mathrm{ri}}$ as the tightest upper and lower bounds on the expectations corresponding to the models in $\mathscr{P}_{\mathscr{T}}^{\mathrm{ci}}$ and $\mathscr{P}_{\mathscr{T}}^{\mathrm{ri}}$, respectively. Since $\mathscr{P}_{\mathscr{T}}^{\mathrm{ri}} \subseteq \mathscr{P}_{\mathscr{T}}^{\mathrm{ci}} \subseteq \mathscr{P}_{\mathscr{T}}^{\mathrm{ei}}$, we have that $\overline{\mathrm{E}}_{\mathscr{T}}^{\mathrm{ri}}(f|A) \leq \overline{\mathrm{E}}_{\mathscr{T}}^{\mathrm{ci}}(f|A) \leq \overline{\mathrm{E}}_{\mathscr{T}}^{\mathrm{ei}}(f|A)$ for any finitary function f and any event A of the form $X_{0:k} = x_{0:k}$.

As we have mentioned before, imprecise Markov chains generalise traditional Markov chains by incorporating different types of model uncertainty. The corresponding upper (and lower) expectations then allow us to make inferences that are robust with respect to this model uncertainty. For a more detailed discussion on the motivation for and interpretation behind these and other types of so-called imprecise probability models, we refer to [1,5,14].

Within the context of imprecise Markov chains, we will be specifically concerned with two types of inferences: the upper and lower expectation of a function at a single time instant, and the upper and lower expectation of the time average of a function. For imprecise Markov chains under epistemic irrelevance and under complete independence, both of these inferences coincide [11, Theorem 51 & Theorem 52]. For any $f \in \mathscr{L}(\mathscr{X})$ and any $x \in \mathscr{X}$, we will denote them by

$$\overline{\mathrm{E}}_k(f|x) = \overline{\mathrm{E}}_{\mathscr{T}}^{\mathrm{ei}}(f(X_k)|X_0 = x) = \overline{\mathrm{E}}_{\mathscr{T}}^{\mathrm{ci}}(f(X_k)|X_0 = x)$$

$$\text{and} \quad \overline{\mathrm{E}}_{\mathrm{av},k}(f|x) = \overline{\mathrm{E}}_{\mathscr{T}}^{\mathrm{ei}}(\overline{f}_k(X_{0:k})|X_0 = x) = \overline{\mathrm{E}}_{\mathscr{T}}^{\mathrm{ci}}(\overline{f}_k(X_{0:k})|X_0 = x),$$

respectively, where the dependency on $\mathscr{T}$ is implicit. The corresponding lower expectations can be obtained through conjugacy: $\underline{\mathrm{E}}_k(f|x) = -\overline{\mathrm{E}}_k(-f|x)$ and $\underline{\mathrm{E}}_{\mathrm{av},k}(f|x) = -\overline{\mathrm{E}}_{\mathrm{av},k}(-f|x)$ for all $f \in \mathscr{L}(\mathscr{X})$ and all $x \in \mathscr{X}$. In the remainder, we will omit imprecise Markov chains under repetition independence from the discussion. Generally speaking, this type of imprecise Markov chain is less studied within the field of imprecise probability because of its limited capacity to incorporate model uncertainty. Indeed, it is simply a set of time-homogeneous precise Markov chains and therefore only allows for model uncertainty about the numerical values of the transition probabilities. Moreover, as far as we know, a characterisation for the ergodicity of such Markov chains—a central topic in this paper—is currently lacking. We therefore believe that this subject demands a separate discussion, which we defer to future work.

4 Transition Operators, Ergodicity and Weak Ergodicity

Inferences of the form $\overline{\mathbb{E}}_k(f|x)$ were among the first ones to be thoroughly studied in imprecise Markov chains. Their study was fundamentally based on the observation that $\overline{\mathbb{E}}_k(f|x)$ can be elegantly rewritten as the k-th iteration of the map $\overline{T}\colon \mathscr{L}(\mathscr{X}) \to \mathscr{L}(\mathscr{X})$ defined by

$$\overline{T}h(x) := \sup_{T \in \mathscr{T}} Th(x) = \sup_{T(x,\cdot) \in \mathscr{T}_x} \sum_{y \in \mathscr{X}} T(x,y)h(y),$$

for all $x \in \mathscr{X}$ and all $h \in \mathscr{L}(\mathscr{X})$. Concretely, $\overline{\mathbb{E}}_k(f|x) = [\overline{T}^k f](x)$ for all $x \in \mathscr{X}$ and all $k \in \mathbb{N}_0$ [4, Theorem 3.1]. The map $\overline{T}$ therefore plays a similar role as the transition matrix T in traditional Markov chains, which is why it is called the *upper transition operator* corresponding to the set $\mathscr{T}$.

In an analogous way, inferences of the form $\overline{\mathbb{E}}_{\mathrm{av},k}(f|x)$ can be obtained as the k-th iteration of the map $\overline{T}_f\colon \mathscr{L}(\mathscr{X}) \to \mathscr{L}(\mathscr{X})$ defined by $\overline{T}_f h := f + \overline{T}h$ for all $h \in \mathscr{L}(\mathscr{X})$. In particular, if we let $\tilde{m}_{f,0} := f = \overline{T}_f(0)$ and

$$\tilde{m}_{f,k} := f + \overline{T}\tilde{m}_{f,k-1} = \overline{T}_f \tilde{m}_{f,k-1} \text{ for all } k \in \mathbb{N}, \tag{1}$$

then it follows from [11, Lemma 41] that $\overline{\mathbb{E}}_{\mathrm{av},k}(f|x) = \frac{1}{k+1}\tilde{m}_{f,k}(x)$ for all $x \in \mathscr{X}$ and all $k \in \mathbb{N}_0$. Applying Eq. (1) repeatedly, we find that for all $x \in \mathscr{X}$:

$$\overline{\mathbb{E}}_{\mathrm{av},k}(f|x) = \frac{1}{k+1}\tilde{m}_{f,k}(x) = \frac{1}{k+1}[\overline{T}_f^k \tilde{m}_{f,0}](x) = \frac{1}{k+1}[\overline{T}_f^{k+1}(0)](x). \tag{2}$$

The same formula can also be obtained as a special case of the results in [13].

These expressions for $\overline{\mathbb{E}}_k(f|x)$ and $\overline{\mathbb{E}}_{\mathrm{av},k}(f|x)$ in terms of the respective operators $\overline{T}$ and $\overline{T}_f$ are particularly useful when we aim to characterise the limit behaviour of these inferences. As will be elaborated on in the next section, there are conditions on $\overline{T}$ that are necessary and sufficient for $\overline{\mathbb{E}}_k(f|x)$ to converge to a limit value that does not depend on the process' initial state $x \in \mathscr{X}$. If this is the case for all $f \in \mathscr{L}(\mathscr{X})$, the imprecise Markov chain is called *ergodic* and we then denote the constant limit value by $\overline{\mathbb{E}}_\infty(f) := \lim_{k \to +\infty} \overline{\mathbb{E}}_k(f|x)$. Similarly, we call an imprecise Markov chain *weakly ergodic* if, for all $f \in \mathscr{L}(\mathscr{X})$, $\lim_{k \to +\infty} \overline{\mathbb{E}}_{\mathrm{av},k}(f|x)$ exists and does not depend on the initial state x. For a weakly ergodic imprecise Markov chain, we denote the common limit value by $\overline{\mathbb{E}}_{\mathrm{av},\infty}(f) := \lim_{k \to +\infty} \overline{\mathbb{E}}_{\mathrm{av},k}(f|x)$. In contrast with standard ergodicity, weak ergodicity and, more generally, the limit behaviour of $\overline{\mathbb{E}}_{\mathrm{av},k}(f|x)$, is almost entirely unexplored. The aim of this paper is to remedy this situation. The main contribution will be a necessary and sufficient condition for an imprecise Markov chain to be weakly ergodic. As we will see, this condition is weaker than those needed for standard ergodicity, hence our choice of terminology. The following example shows that this difference already becomes apparent in the precise case.

Example 1. Let $\mathscr{X} = \{a, b\}$, consider any function $f = \begin{bmatrix} f_a \\ f_b \end{bmatrix} \in \mathscr{L}(\mathscr{X})$ and assume that $\mathscr{T}$ consists of a single matrix $T = \begin{bmatrix} 0 & 1 \\ 1 & 0 \end{bmatrix}$. Clearly, $\overline{T}$ is not ergodic because $\overline{T}^{(2\ell+1)} f = T^{(2\ell+1)} f = \begin{bmatrix} 0 & 1 \\ 1 & 0 \end{bmatrix} f = \begin{bmatrix} f_b \\ f_a \end{bmatrix}$ and $\overline{T}^{(2\ell)} f = \begin{bmatrix} 1 & 0 \\ 0 & 1 \end{bmatrix} f = \begin{bmatrix} f_a \\ f_b \end{bmatrix}$ for all $\ell \in \mathbb{N}_0$. $\overline{T}$ is weakly ergodic though, because

$$\overline{T}_f^{(2\ell)}(0) = \ell \begin{bmatrix} f_a + f_b \\ f_a + f_b \end{bmatrix} \text{ and } \overline{T}_f^{(2\ell+1)}(0) = f + \overline{T}\,\overline{T}_f^{(2\ell)}(0) = f + \ell \begin{bmatrix} f_a + f_b \\ f_a + f_b \end{bmatrix},$$

for all $\ell \in \mathbb{N}_0$, which implies that $\overline{\mathrm{E}}_{\mathrm{av},\infty}(f) := \lim_{k \to +\infty} \overline{T}_f^k(0)/k = (f_a + f_b)/2$ exists. ◊

Notably, even if an imprecise Markov chain is ergodic (and hence also weakly ergodic) and therefore both $\overline{\mathrm{E}}_\infty(f)$ and $\overline{\mathrm{E}}_{\mathrm{av},\infty}(f)$ exist, these inferences will not necessarily coincide. This was first observed in an experimental setting [11, Section 7.6], but the differences that were observed there were marginal. The following example shows that these differences can in fact be very substantial.

Example 2. Let $\mathscr{X} = \{a, b\}$, let $\mathscr{T}_a$ be the set of all probability mass functions on $\mathscr{X}$ and let $\mathscr{T}_b := \{p\}$ for the probability mass function $p = (p_a, p_b) = (1, 0)$ that puts all mass in a. Then, for any $f = \begin{bmatrix} f_a \\ f_b \end{bmatrix} \in \mathscr{L}(\mathscr{X})$, we have that

$$\overline{T}f(x) = \begin{cases} \max f & \text{if } x = a; \\ f_a & \text{if } x = b, \end{cases} \quad \text{and} \quad \overline{T}^2 f(x) = \begin{cases} \max \overline{T} f = \max f & \text{if } x = a; \\ \overline{T}f(a) = \max f & \text{if } x = b. \end{cases}$$

It follows that $\overline{T}^k f = \max f$ for all $k \geq 2$, so the limit upper expectation $\overline{\mathrm{E}}_\infty(f)$ exists and is equal to $\max f$ for all $f \in \mathscr{L}(\mathscr{X})$. In particular, we have that $\overline{\mathrm{E}}_\infty(\mathbb{I}_b) = 1$. On the other hand, we find that $\overline{T}_{\mathbb{I}_b}^{(2\ell)}(0) = \ell$ and $\overline{T}_{\mathbb{I}_b}^{(2\ell+1)}(0) = \mathbb{I}_b + \overline{T}\,\overline{T}_{\mathbb{I}_b}^{(2\ell)}(0) = \begin{bmatrix} \ell \\ \ell+1 \end{bmatrix}$ for all $\ell \in \mathbb{N}_0$. This implies that the upper expectation $\overline{\mathrm{E}}_{\mathrm{av},\infty}(\mathbb{I}_b) := \lim_{k \to +\infty} \overline{T}_{\mathbb{I}_b}^k(0)/k$ exists and is equal to $1/2$. This value differs significantly from the limit upper expectation $\overline{\mathrm{E}}_\infty(\mathbb{I}_b) = 1$.

In fact, this result could have been expected simply by taking a closer look at the dynamics that correspond to $\mathscr{T}$. Indeed, it follows directly from $\mathscr{T}$ that, if the system is in state b at some instant, then it will surely be in a at the next time instant. Hence, the system can only reside in state b for maximally half of the time, resulting in an upper expected average that converges to $1/2$. These underlying dynamics have little effect on the limit upper expectation $\overline{\mathrm{E}}_\infty(\mathbb{I}_b)$ though, because it is only concerned with the upper expectation of $\mathbb{I}_b$ evaluated at a single time instant. ◊

Although we have used sets $\mathscr{T}$ of transition matrices to define imprecise Markov chains, it should at this point be clear that, if we are interested in the inferences $\overline{\mathrm{E}}_k(f|x)$ and $\overline{\mathrm{E}}_{\mathrm{av},k}(f|x)$ and their limit values, then it suffices to specify $\overline{\mathscr{T}}$. In fact, we will henceforth forget about $\mathscr{T}$ and will assume that $\overline{T}$ is a *coherent* upper transition operator on $\mathscr{L}(\mathscr{X})$, meaning that it is an operator from $\mathscr{L}(\mathscr{X})$ to $\mathscr{L}(\mathscr{X})$ that satisfies

C1. $\min h \le \overline{T}h \le \max h$ [boundedness];
C2. $\overline{T}(h+g) \le \overline{T}h + \overline{T}g$ [sub-additivity];
C3. $\overline{T}(\lambda h) = \lambda \overline{T}h$ [non-negative homogeneity],

for all $h, g \in \mathscr{L}(\mathscr{X})$ and all real $\lambda \ge 0$ [5,14,15], and we will regard $\overline{\mathrm{E}}_k(f|x)$ and $\overline{\mathrm{E}}_{\mathrm{av},k}(f|x)$ as objects that correspond to $\overline{T}$. Our results and proofs will never rely on the fact that $\overline{T}$ is derived from a set $\mathscr{T}$ of transition matrices, but will only make use of C1–C3 and the following two properties that are implied by them [14, Section 2.6.1]:

C4. $\overline{T}(\mu + h) = \mu + \overline{T}h$ [constant additivity];
C5. if $h \le g$ then $\overline{T}h \le \overline{T}g$ [monotonicity],

for all $h, g \in \mathscr{L}(\mathscr{X})$ and all real μ. This can be done without loss of generality because an upper transition operator $\overline{T}$ that is defined as an upper envelope of a set $\mathscr{T}$ of transition matrices—as we did in Sect. 4—is always coherent [14, Theorem 2.6.3]. Since properties such as ergodicity and weak ergodicity can be completely characterised in terms of $\overline{T}$, we will henceforth simply say that $\overline{T}$ itself is (weakly) ergodic, instead of saying that the corresponding imprecise Markov chain is.

5   Accessibility Relations and Topical Maps

To characterise ergodicity and weak ergodicity, we will make use of some well-known graph-theoretic concepts, suitably adapted to the imprecise Markov chain setting; we recall the following from [4] and [8]. The *upper accessibility graph* $\mathscr{G}(\overline{T})$ corresponding to $\overline{T}$ is defined as the graph with vertices $x_1 \cdots x_n \in \mathscr{X}$, where $n := |\mathscr{X}|$, with an edge from x_i to x_j if $\overline{T}\mathbb{I}_{x_j}(x_i) > 0$. For any two vertices x_i and x_j, we say that x_j is *accessible* from x_i, denoted by $x_i \to x_j$, if $x_i = x_j$ or if there is a directed path from x_i to x_j, which means that there is a sequence $x_i = x'_0, x'_1, \cdots, x'_m = x_j$ of vertices, with $m \in \mathbb{N}$, such that there is an edge from $x'_{\ell-1}$ to x'_ℓ for all $\ell \in \{1, \cdots, m\}$. We say that two vertices x_i and x_j *communicate* and write $x_i \leftrightarrow x_j$ if both $x_i \to x_j$ and $x_j \to x_i$. The relation $\leftrightarrow$ is an equivalence relation (reflexive, symmetric and transitive) and the equivalence classes are called *communication classes*. We call the graph $\mathscr{G}(\overline{T})$ *strongly connected* if any two vertices x_i and x_j in $\mathscr{G}(\overline{T})$ communicate, or equivalently, if $\mathscr{X}$ itself is a communication class. Furthermore, we say that $\overline{T}$ (or $\mathscr{G}(\overline{T})$) has a *top class* $\mathcal{R}$ if

$$\mathcal{R} := \{x \in \mathscr{X} : y \to x \text{ for all } y \in \mathscr{X}\} \ne \emptyset.$$

So, if $\overline{T}$ has a top class $\mathcal{R}$, then $\mathcal{R}$ is accessible from any vertex in the graph $\mathscr{G}(\overline{T})$. As a fairly immediate consequence, it follows that $\mathcal{R}$ is a communication class that is *maximal* or *undominated*, meaning that $x \not\to y$ for all $x \in \mathcal{R}$ and all $y \in \mathcal{R}^c$. In fact, it is the only such maximal communication class.

Having a top class is necessary for $\overline{T}$ to be ergodic, but it is not sufficient. Sufficiency additionally requires that the top class $\mathcal{R}$ satisfies [8, Proposition 3]:

E1. $(\forall x \in \mathcal{R})(\exists k^* \in \mathbb{N})(\forall k \geq k^*) \ \min \overline{T}^k \mathbb{I}_x > 0$ [Regularity];
E2. $(\forall x \in \mathcal{R}^c)(\exists k \in \mathbb{N}) \ \overline{T}^k \mathbb{I}_{\mathcal{R}^c}(x) < 1$ [Absorbing].

We will say that $\overline{T}$ is *top class regular* (TCR) if it has a top class that is regular, and analogously for *top class absorbing* (TCA). Top class regularity represents aperiodic behaviour: it demands that there is some time instant $k^* \in \mathbb{N}$ such that all of the elements in the top class $\mathcal{R}$ are accessible from each other in k steps, for any $k \geq k^*$. In the case of traditional Markov chains, top class regularity suffices as a necessary and sufficient condition for ergodicity [4,10]. However, in the imprecise case, we need the additional condition of being top class absorbing, which ensures that the top class will eventually be reached. It requires that, if the process starts from any state $x \in \mathcal{R}^c$, the lower probability that it will ever transition to $\mathcal{R}$ is strictly positive. We refer to [4] for more details. From a practical point of view, an important feature of both of these accessibility conditions is that they can be easily checked in practice [8].

The characterisation of ergodicity using (TCR) and (TCA) was strongly inspired by the observation that upper transition operators are part of a specific collection of order-preserving maps, called *topical maps*. These are maps $F \colon \mathbb{R}^n \to \mathbb{R}^n$ that satisfy

T1. $F(\mu + h) = \mu + Fh$ [constant additivity];
T2. if $h \leq g$ then $F(h) \leq F(g)$ [monotonicity],

for all $h, g \in \mathbb{R}^n$ and all $\mu \in \mathbb{R}$. To show this, we identify $\mathscr{L}(\mathscr{X})$ with the finite-dimensional linear space $\mathbb{R}^n$, with $n = |\mathscr{X}|$; this is clearly possible because both are isomorph. That every coherent upper transition operator is topical now follows trivially from C4 and C5. What is perhaps less obvious, but can be derived in an equally trivial way, is that the operator $\overline{T}_f$ is also topical. This allows us to apply results for topical maps to $\overline{T}_f$ in order to find necessary and sufficient conditions for weak ergodicity.

6   A Sufficient Condition for Weak Ergodicity

As a first step, we aim to find sufficient conditions for the existence of $\overline{\mathbb{E}}_{\mathrm{av},\infty}(f)$. To that end, recall from Sect. 4 that if $\overline{\mathbb{E}}_{\mathrm{av},\infty}(f)$ exists, it is equal to the limit $\lim_{k \to +\infty} \overline{T}_f^k(0)/k$. Then, since $\overline{T}_f$ is topical, the following lemma implies that it is also equal to $\lim_{k \to +\infty} \overline{T}_f^k h/k$ for any $h \in \mathscr{L}(\mathscr{X})$.

Lemma 1 [7, Lemma 3.1]. *Consider any topical map $F \colon \mathbb{R}^n \to \mathbb{R}^n$. If the limit $\lim_{k \to +\infty} F^k h/k$ exists for some $h \in \mathbb{R}^n$, then the limit exists for all $h \in \mathbb{R}^n$ and they are all equal.*

Hence, if $\lim_{k \to +\infty} \overline{T}_f^k h/k$ converges to a constant vector μ for some $h \in \mathscr{L}(\mathscr{X})$, then $\overline{\mathrm{E}}_{\mathrm{av},\infty}(f)$ exists and is equal to μ. This condition is clearly satisfied if the map $\overline{T}_f$ has an (additive) eigenvector $h \in \mathscr{L}(\mathscr{X})$, meaning that $\overline{T}_f^k h = h + k\mu$ for some $\mu \in \mathbb{R}$ and all $k \in \mathbb{N}_0$. In that case, we have that $\overline{\mathrm{E}}_{\mathrm{av},\infty}(f) = \mu$, where μ is called the eigenvalue corresponding to h.

To find conditions that guarantee the existence of an eigenvector of $\overline{T}_f$, we will make use of results from [6] and [7]. There, accessibility graphs are defined in a slightly different way: for any topical map $F \colon \mathbb{R}^n \to \mathbb{R}^n$, they let $\mathscr{G}'(F)$ be the graph with vertices $v_1, \cdots, v_n$ and an edge from v_i to v_j if $\lim_{\alpha \to +\infty}[F(\alpha \mathbb{I}_{v_j})](v_i) = +\infty$. Subsequently, for such a graph $\mathscr{G}'(F)$, the accessibility relation $\cdot \to \cdot$ and corresponding notions (e.g. 'strongly connected', 'top class', ...) are defined as in Sect. 5. If we identify the vertices $v_1, \cdots, v_n$ in $\mathscr{G}'(\overline{T})$ and $\mathscr{G}'(\overline{T}_f)$ with the different states $x_1, \cdots, x_n$ in $\mathscr{X}$, this can in particular be done for the topical maps $\overline{T}$ and $\overline{T}_f$. The following results show that the resulting graphs coincide with the one defined in Sect. 5.

Lemma 2. *For any two vertices x and y in $\mathscr{G}'(\overline{T})$, there is an edge from x to y in $\mathscr{G}'(\overline{T})$ if and only if there is an edge from x to y in $\mathscr{G}(\overline{T})$.*

Proof. Consider any two vertices x and y in the graph $\mathscr{G}'(\overline{T})$. Then there is an edge from x to y if $\lim_{\alpha \to +\infty}[\overline{T}(\alpha \mathbb{I}_y)](x) = +\infty$. By non-negative homogeneity [C3], this is equivalent to the condition that $\lim_{\alpha \to +\infty} \alpha[\overline{T}\mathbb{I}_y](x) = +\infty$. Since moreover $0 \le \overline{T}\mathbb{I}_y \le 1$ by C1, this condition reduces to $\overline{T}\mathbb{I}_y(x) > 0$. $\qquad\square$

Corollary 1. *The graphs $\mathscr{G}'(\overline{T}_f)$, $\mathscr{G}'(\overline{T})$ and $\mathscr{G}(\overline{T})$ are identical.*

Proof. Lemma 2 implies that $\mathscr{G}'(\overline{T})$ and $\mathscr{G}(\overline{T})$ are identical. Moreover, that $\mathscr{G}'(\overline{T}_f)$ is equal to $\mathscr{G}'(\overline{T})$, follows straightforwardly from the definition of $\overline{T}_f$. $\qquad\square$

In principle, we could use this result to directly obtain the desired condition for the existence of an eigenvector from [6, Theorem 2]. However, [6, Theorem 2] is given in a multiplicative framework and would need to be reformulated in an additive framework in order to be applicable to the map $\overline{T}_f$; see [6, Section 2.1]. This can be achieved with a bijective transformation, but we prefer to not do so because it would require too much extra terminology and notation. Instead, we will derive an additive variant of [6, Theorem 2] directly from [6, Theorem 9] and [6, Theorem 10].

The first result establishes that the existence of an eigenvector is equivalent to the fact that trajectories are bounded with respect to the Hilbert semi-norm $\|\cdot\|_{\mathrm{H}}$, defined by $\|h\|_{\mathrm{H}} := \max h - \min h$ for all $h \in \mathbb{R}^n$.

Theorem 1 [6, Theorem 9]. *Let $F \colon \mathbb{R}^n \to \mathbb{R}^n$ be a topical map. Then F has an eigenvector in $\mathbb{R}^n$ if and only if $\{\|F^k h\|_{\mathrm{H}} : k \in \mathbb{N}\}$ is bounded for some (and hence all) $h \in \mathbb{R}^n$.*

That the boundedness of a single trajectory indeed implies the boundedness of all trajectories follows from the non-expansiveness of a topical map with respect to the Hilbert semi-norm [6]. The second result that we need uses the notion of a *super-eigenspace*, defined for any topical map F and any $\mu \in \mathbb{R}$ as the set $S^\mu(F) := \{h \in \mathbb{R}^n : Fh \leq h + \mu\}$.

Theorem 2 [6, Theorem 10]. *Let $F \colon \mathbb{R}^n \to \mathbb{R}^n$ be a topical map such that the associated graph $\mathscr{G}'(F)$ is strongly connected. Then all of the super-eigenspaces are bounded in the Hilbert semi-norm.*

Together, these theorems imply that any topical map $F \colon \mathbb{R}^n \to \mathbb{R}^n$ for which the graph $\mathscr{G}'(F)$ is strongly connected, has an eigenvector. The connection between both is provided by the fact that trajectories cannot leave an eigenspace. The following result formalises this.

Theorem 3. *Let $F \colon \mathbb{R}^n \to \mathbb{R}^n$ be a topical map such that the associated graph $\mathscr{G}'(F)$ is strongly connected. Then F has an eigenvector in $\mathbb{R}^n$.*

Proof. Consider any $h \in \mathbb{R}^n$ and any $\mu \in \mathbb{R}$ such that $\max(Fh - h) \leq \mu$. Then $Fh \leq h + \mu$, so $h \in S^\mu(F)$. Now notice that $F(Fh) \leq F(h + \mu) = Fh + \mu$ because of T1 and T2, which implies that also $Fh \in S^\mu(F)$. In the same way, we can also deduce that $F^2 h \in S^\mu(F)$ and, by repeating this argument, that the whole trajectory corresponding to h remains in $S^\mu(F)$. This trajectory is bounded because of Theorem 2, which by Theorem 1 guarantees the existence of an eigenvector. □

In particular, if $\mathscr{G}'(\overline{T}_f)$ is strongly connected then $\overline{T}_f$ has an eigenvector, which on its turn implies the existence of $\overline{\mathrm{E}}_{\mathrm{av},\infty}(f)$ as explained earlier. If we combine this observation with Corollary 1, we obtain the following result.

Proposition 1. *An upper transition operator $\overline{T}$ is weakly ergodic if the associated graph $\mathscr{G}(\overline{T})$ is strongly connected.*

Proof. Suppose that $\mathscr{G}(\overline{T})$ is strongly connected. Then, by Corollary 1, $\mathscr{G}'(\overline{T}_f)$ is also strongly connected. Hence, since $\overline{T}_f$ is a topical map, Theorem 3 guarantees the existence of an eigenvector of $\overline{T}_f$. As explained in the beginning of this section, this implies by Lemma 1 that $\overline{\mathrm{E}}_{\mathrm{av},\infty}(f)$ exists, so we indeed find that $\overline{T}$ is weakly ergodic. □

In the remainder of this paper, we will use the fact that $\overline{T}$ is coherent—so not just topical—to strengthen this result. In particular, we will show that the condition of being strongly connected can be replaced by a weaker one: being top class absorbing. It will moreover turn out that this property is not only sufficient, but also necessary for weak ergodicity.

7   Necessary and Sufficient Condition for Weak Ergodicity

In order to gain some intuition about how to obtain a more general sufficient condition for weak ergodicity, consider the case where $\overline{T}$ has a top class $\mathcal{R}$ and the process' initial state x is in $\mathcal{R}$. Since $\mathcal{R}$ is a maximal communication class, the process surely remains in $\mathcal{R}$ and hence, it is to be expected that the time average of f will not be affected by the dynamics of the process outside $\mathcal{R}$. Moreover, the communication class $\mathcal{R}$ is a strongly connected component, so one would expect that, due to Proposition 1, the upper expected time average $\overline{E}_{av,k}(f|x)$ converges to a constant that does not depend on the state $x \in \mathcal{R}$. Our intuition is formalised by the following proposition. Its proof, as well as those of the other statements in this section, are available in the appendix of [12].

Proposition 2. *For any maximal communication class S and any $x \in S$, the upper expectation $\overline{E}_{av,k}(f|x)$ is equal to $\overline{E}_{av,k}(f\mathbb{I}_S|x)$ and converges to a limit value. This limit value is furthermore the same for all $x \in S$.*

As a next step, we want to extend the domain of convergence of $\overline{E}_{av,k}(f|x)$ to all states $x \in \mathcal{X}$. To do so, we will impose the additional property of being top class absorbing (TCA), which, as explained in Sect. 5, demands that there is a strictly positive (lower) probability to reach the top class $\mathcal{R}$ in a finite time period. Once in $\mathcal{R}$, the process can never escape $\mathcal{R}$ though. One would therefore expect that as time progresses—as more of these finite time periods go by—this lower probability increases, implying that the process will eventually be in $\mathcal{R}$ with practical certainty. Furthermore, if the process transitions from $x \in \mathcal{R}^c$ to a state $y \in \mathcal{R}$, then Proposition 2 guarantees that $\overline{E}_{av,k}(f|y)$ converges to a limit and that this limit value does not depend on the state y. Finally, since the average is taken over a growing time interval, the initial finite number of time steps that it took for the process to transition from x to y will not influence the time average of f in the limit. This leads us to suspect that $\overline{E}_{av,k}(f|x)$ converges to the same limit as $\overline{E}_{av,k}(f|y)$. Since this argument applies to any $x \in \mathcal{R}^c$, we are led to believe that $\overline{T}$ is weakly ergodic. The following result confirms this.

Proposition 3. *Any $\overline{T}$ that satisfies (TCA) is weakly ergodic.*

Conversely, suppose that $\overline{T}$ does not satisfy (TCA). Then there are two possibilities: either there is no top class or there is a top class but it is not absorbing. If there is no top class, then it can be easily deduced that there are at least two maximal communication classes S_1 and S_2. As discusssed earlier, the process cannot escape the classes S_1 and S_2 once it has reached them. So if it starts in one of these communication classes, the process' dynamics outside this class are irrelevant for the behaviour of the resulting time average. In particular, if we let f be the function that takes the constant value c_1 in S_1 and c_2 in S_2, with $c_1 \neq c_2$, then we would expect that $\overline{E}_{av,k}(f|x) = c_1$ and $\overline{E}_{av,k}(f|y) = c_2$ for all $k \in \mathbb{N}_0$, any $x \in S_1$ and any $y \in S_2$. In fact, this can easily be formalised by means of Proposition 2. Hence, $\overline{E}_{av,\infty}(f|x) = c_1 \neq c_2 = \overline{E}_{av,\infty}(f|y)$, so the upper transition operator $\overline{T}$ cannot be weakly ergodic.

Proposition 4. *Any weakly ergodic $\overline{T}$ has a top class.*

Finally, suppose that there is a top class $\mathcal{R}$, but that it is not absorbing. This implies that there is an $x \in \mathcal{R}^c$ and a compatible precise model such that the process is guaranteed to remain in $\mathcal{R}^c$ given that it started in x. If we now let

$f = \mathbb{I}_{\mathcal{R}^c}$, then conditional on the fact that $X_0 = x$, the expected time average of f corresponding to this precise model is equal to 1. Furthermore, since $f \leq 1$, no other process can yield a higher expected time average. The upper expected time average $\overline{\mathrm{E}}_{av,k}(f|x)$ is therefore equal to 1 for all $k \in \mathbb{N}_0$. However, using Proposition 2, we can also show that $\overline{\mathrm{E}}_{av,k}(f|y) = 0$ for any $y \in \mathcal{R}$ and all $k \in \mathbb{N}_0$. Hence, $\overline{\mathrm{E}}_{av,\infty}(f|x) = 1 \neq 0 = \overline{\mathrm{E}}_{av,\infty}(f|y)$, which precludes $\overline{T}$ from being weakly ergodic.

Proposition 5. *Any weakly ergodic $\overline{T}$ that has a top class satisfies (TCA).*

Together with Propositions 3 and 4, this allows us to conclude that (TCA) is a necessary and sufficient condition for weak ergodicity.

Theorem 4. *$\overline{T}$ is weakly ergodic if and only if it is top class absorbing.*

8 Conclusion

The most important conclusion of our study of upper and lower expected time averages is its final result: that being top class absorbing is necessary and sufficient for weak ergodicity; a property that guarantees upper and lower expected time averages to converge to a limit value that does not depend on the process' initial state. In comparison with standard ergodicity, which guarantees the existence of a limit upper and lower expectation, weak ergodicity thus requires less stringent conditions to be satisfied. We illustrated this difference in Example 1, where we considered a(n imprecise) Markov chain that satisfies (TCA) but not (TCR).

Apart from the fact that their existence is guaranteed under weaker conditions, the inferences $\overline{\mathrm{E}}_{av,\infty}(f)$ are also able to provide us with more information about how time averages might behave, compared to limit expectations. To see why, recall Example 2, where the inference $\overline{\mathrm{E}}_{av,\infty}(\mathbb{I}_b) = 1/2$ significantly differed from $\overline{\mathrm{E}}_\infty(\mathbb{I}_b) = 1$. Clearly, the former was more representative for the limit behaviour of the time average of $\mathbb{I}_b$. As a consequence of [11, Lemma 57], a similar statement holds for general functions. In particular, it implies that $\overline{\mathrm{E}}_{av,\infty}(f) \leq \overline{\mathrm{E}}_\infty(f)$ for any function $f \in \mathscr{L}(\mathscr{X})$. Since both inferences are upper bounds, $\overline{\mathrm{E}}_{av,\infty}(f)$ is therefore at least as informative as $\overline{\mathrm{E}}_\infty(f)$.

In summary then, when it comes to characterising long-term time averages, there are two advantages that (limits of) upper and lower expected time averages have over conventional limit upper and lower expectations: they exist under weaker conditions and they are at least as (and sometimes much more) informative.

That said, there is also one important feature that limit upper and lower expectations have, but that is currently still lacking for upper and lower expected time averages: an (imprecise) point-wise ergodic theorem [2, Theorem 32]. For the limit upper and lower expectations of an ergodic imprecise Markov chain, this result states that

$$\underline{\mathrm{E}}_\infty(f) \leq \liminf_{k \to +\infty} \overline{f}_k(X_{0:k}) \leq \limsup_{k \to +\infty} \overline{f}_k(X_{0:k}) \leq \overline{\mathrm{E}}_\infty(f),$$

with lower probability one. In order for limit upper and lower expected time averages to be the undisputed quantities of interest when studying long-term time averages, a similar result would need to be obtained for weak ergodicity, where the role of $\overline{E}_\infty(f)$ and $\underline{E}_\infty(f) := -\overline{E}_\infty(-f)$ is taken over by $\overline{E}_{av,\infty}(f)$ and $\underline{E}_{av,\infty}(f) := -\overline{E}_{av,\infty}(-f)$, respectively. If such a result would hold, it would provide us with (strictly almost sure) bounds on the limit values attained by time averages that are not only more informative as the current ones, but also guaranteed to exist under weaker conditions. Whether such a result indeed holds is an open problem that we would like to address in our future work.

A second line of future research that we would like to pursue consists in studying the convergence of $\overline{E}_{av,k}(f|x)$ in general, without imposing that the limit value should not depend on x. We suspect that this kind of convergence will require no conditions at all.

References

1. Augustin, T., Coolen, F.P., de Cooman, G., Troffaes, M.C.: Introduction to Imprecise Probabilities. Wiley, Chichester (2014)
2. de Cooman, G., De Bock, J., Lopatatzidis, S.: Imprecise stochastic processes in discrete time: global models, imprecise Markov chains, and ergodic theorems. Int. J. Approximate Reasoning **76**, 18–46 (2016)
3. de Cooman, G., Hermans, F., Antonucci, A., Zaffalon, M.: Epistemic irrelevance in credal nets: the case of imprecise Markov trees. Int. J. Approximate Reasoning **51**(9), 1029–1052 (2010)
4. de Cooman, G., Hermans, F., Quaeghebeur, E.: Imprecise Markov chains and their limit behaviour. Prob. Eng. Inf. Sci. **23**(4), 597–635 (2009)
5. de Cooman, G., Troffaes, M.C.: Lower Previsions. Wiley, Chichester (2014)
6. Gaubert, S., Gunawardena, J.: The Perron-Frobenius theorem for homogeneous, monotone functions. Trans. Am. Math. Soc. **356**(12), 4931–4950 (2004)
7. Gunawardena, J.: From max-plus algebra to nonexpansive mappings: a nonlinear theory for discrete event systems. Theor. Comput. Sci. **293**(1), 141–167 (2003)
8. Hermans, F., de Cooman, G.: Characterisation of ergodic upper transition operators. Int. J. Approximate Reasoning **53**(4), 573–583 (2012)
9. Hermans, F., Škulj, D.: Stochastic processes. In: Augustin, T., Coolen, F.P., de Cooman, G., Troffaes, M.C. (eds.) Introduction to Imprecise Probabilities, pp. 258–278. Wiley, Chichester (2014)
10. Kemeny, J.G., Snell, J.L.: Finite Markov chains. Undergraduate Text in Mathematics. Springer, New York (1976)
11. Lopatatzidis, S.: Robust Modelling and Optimisation in Stochastic Processes using Imprecise Probabilities, with an Application to Queueing Theory. Ph.D. thesis, Ghent University (2017)
12. T'Joens, N., De Bock, J.: Limit behaviour of upper and lower expected time averages in discrete-time imprecise markov chains (2020). http://arxiv.org/abs/2002. 05661
13. T'Joens, N., Krak, T., Bock, J.D., Cooman, G.: A Recursive Algorithm for Computing Inferences in Imprecise Markov Chains. In: Kern-Isberner, G., Ognjanović, Z. (eds.) ECSQARU 2019. LNCS (LNAI), vol. 11726, pp. 455–465. Springer, Cham (2019). https://doi.org/10.1007/978-3-030-29765-7_38

14. Walley, P.: Statistical Reasoning with Imprecise Probabilities. Chapman and Hall, London (1991)
15. Williams, P.M.: Notes on conditional previsions. Int. J. Approximate Reasoning **44**(3), 366–383 (2007)

Similarities in Artificial Intelligence

An Interval-Valued Divergence for Interval-Valued Fuzzy Sets

Susana Díaz[1], Irene Díaz[2], and Susana Montes[1]

[1] Department of Statistics and O.R., University of Oviedo, Oviedo, Spain
{diazsusana,montes}@uniovi.es
[2] Department of Computer Science, University of Oviedo, Oviedo, Spain
sirene@uniovi.es

Abstract. Characterizing the degree of similarity or difference between two sets is a very important topic, since it has many applications in different areas, including image processing or decision making. Several studies have been done about the comparison of fuzzy sets and its extensions, in particular for interval-valued fuzzy sets. However, in most of the cases, the results of the comparison is just a number. In order to avoid this reduction of the information, we have introduced a measure for comparing two interval-valued fuzzy sets such that it is an interval itself, which can be reduced to a number if it is necessary. Thus, a richer class of measures is now considered.

Keywords: Interval-valued fuzzy set · Order between intervals · Dissimilarity · Divergence measure

1 Introduction

In many real-life situations, we have to compare two objects, opinions, etc having an incomplete or not precise information about them. One of the most extended approach to model these situations is to consider fuzzy sets and compare them when necessary. Comparison represents an important topic in fuzzy sets such that several measures for comparing fuzzy sets have been proposed in literature. A general study was done by Bouchon-Meunier et al. in 1996 [5]. Since then, a lot of measures have been introduced, some of them as constructive definitions, that is, specific formulas (see, among many others, [1,30,31,34]) and some of them by means of axiomatic definitions (see [12,20,23] for instance).

The necessity of dealing with imprecision in real world problems has been a long-term research challenge that has originated different extensions of the fuzzy sets. Interval–valued fuzzy sets are one of the most challenging extensions.

Authors would like to thank for the support of Spanish Ministry of Science and Technology project TIN-2017-87600-P (I. Diaz), Spanish Ministry of Science and Technology project PGC2018-098623-B-I00 (S. Diaz and S. Montes) and FICYT Project IDI/2018/000176 (S. Diaz, I. Diaz and S. Montes).

© Springer Nature Switzerland AG 2020
M.-J. Lesot et al. (Eds.): IPMU 2020, CCIS 1238, pp. 241–249, 2020.
https://doi.org/10.1007/978-3-030-50143-3_18

They were introduced independently by Zadeh [33], Grattan-Guiness [17], Jahn [19] and Sambuc [24] in the seventies. Interval-valued fuzzy sets can be useful to deal with situations where the classical fuzzy tools are not so efficient as, for instance, when there is not an objective procedure to select the crisp membership degrees. This extension has attracted very quickly the attention of many researchers, since they could see the high potential of them for different applications. Thus, for instance, Sambuc [24] used them in medical diagnosis in thyrodian pathology, Bustince [6] and Gozalczany [16] in approximate reasoning and Cornelis et al. [11] and Turksen [29] in logic.

Based on their utility, several concepts, tools and trends related to this extension have to be studied. In particular, we are specially interested on the measures of similarity, or their dual measures of the difference, between interval-valued fuzzy sets. There are a lot of papers related to this topic in the literature (see e.g. [2,13,32,35]).

Another very related concept is the notion of intuitionistic fuzzy set, introduced by Atanassov [3] about ten years later than interval-valued fuzzy sets. Despite the semantic differences, it was proven by many authors that intuitionistic fuzzy sets and interval-valued fuzzy sets are equipollent ([4,14]), that is, there is a bijection function mapping one onto the other. Thus, the measures of comparison between intuitionistic fuzzy sets (see, for instance, [21,22]) could propose us a first idea about the way to compare two interval-valued fuzzy sets, although they cannot be directly used, as it was shown in [10,25,26].

Many of the previously introduced measures represent the result of the comparison as a value in the real line. However, if we are dealing with interval-valued fuzzy sets, even the total similarity of incomplete descriptions does not guarantee the total similarity of the described objects. In order to solve this problem, a similarity described by means of a range of values could be more appropriate. However, this is not the usual case and, as far as we know, it has been only considerer in a few number of works (see [28,36]).

Although interval-valued fuzzy sets could be compared by means of distances, the most usual case in the literature is the use of dissimilarity measures. However, it was noticed in [22] that these measures could not be appropriate in some cases, for some counterintuitive examples, when we are comparing Atanassov intuitionistic fuzzy sets. To avoid this problem, it is necessary to introduce a measure of comparison with stronger properties than dissimilarities. Thus, the main aim of this work is to obtain a first idea on dissimilarity and divergence measures which are able to compare two interval-valued fuzzy sets assuming values in the set of closed real intervals and to relate both concepts.

This paper is organized as follows. In Sect. 2, some basic concepts are introduced and the notation is fixed for the remaining parts of the paper. Section 3 is devoted to the new definition of divergence measures between interval-valued fuzzy sets and dissimilarities. Finally, some conclusions and open problems are drawn in Sect. 4.

2   Basic Concepts

Let X denote the universe of discourse. An interval-valued fuzzy subset of X is a mapping $A : X \to L([0,1])$ such that $A(x) = [\underline{A}(x), \overline{A}(x)]$, where $L([0,1])$ denotes the family of closed intervals included in the unit interval $[0,1]$. Thus, an interval-valued fuzzy set A is totally characterized by two mapping, $\underline{A}$ and $\overline{A}$, from X into $[0,1]$ such that $\underline{A}(x) \leq \overline{A}(x), \forall x \in X$. These maps represent the lower and upper bound of the corresponding intervals. Let us notice that if $\underline{A}(x) = \overline{A}(x), \forall x \in X$, then A is a classical fuzzy sets. The collection of all the interval-valued fuzzy sets in X is denoted by $IVFS(X)$ and the subset formed by all the fuzzy sets in X is denoted by $FS(X)$.

Several operations have been considered to this concept in the literature. We will consider now the most usual ones, since they are the most usual particular case of the general operations defined by means of t-norms and t-conorms defined on $L([0,1])$ (see e.g. [15]). Thus, for any $A, B \in IVFS(X)$, we have that:

- The intersection of A and B is the interval-valued fuzzy set defined by $\underline{A \cap B}(x) = \min\{\underline{A}(x), \underline{B}(x)\}$ and $\overline{A \cap B}(x) = \min\{\overline{A}(x), \overline{B}(x)\}$ for any $x \in X$.
- The union of A and B is the interval-valued fuzzy set defined by $\underline{A \cup B}(x) = \max\{\underline{A}(x), \underline{B}(x)\}$ and $\overline{A \cup B}(x) = \max\{\overline{A}(x), \overline{B}(x)\}$ for any $x \in X$.
- The complement of A is the interval-valued fuzzy set defined by $\underline{A^c}(x) = 1 - \overline{A}(x)$ and $\overline{A^c}(x) = 1 - \underline{A}(x)$ for any $x \in X$.
- A is a subset of B if, and only if, $\underline{A}(x) \leq \underline{B}(x)$ and $\overline{A}(x) \leq \overline{B}(x)$ for any $x \in X$. Thus, in fact, we are saying that A is included in B if the membership degree of any element in X is an interval lower than or equal to the interval representing the membership degree of B, when we use the usual lattice-ordering between closed intervals of the real line. We are going to consider here this order but, of course, any other order between intervals (see, for instance, [7]) could be considered to define the inclusion.

Another important partial order on $IVFS(X)$ could be considered:

$$A \sqsubseteq B \text{ iff } A(x) \subseteq B(x), \forall x \in X$$

for the usual inclusion between intervals.

A first intuitive ways to compare two interval-valued fuzzy sets, based on the Hausdorff distance (see [18]), could be:

$$d_H(A, B) = \sum_{x \in X} \max\{|\underline{A}(x) - \underline{B}(x)|, |\overline{A}(x) - \overline{B}(x)|\}$$

where X is a finite universe.

Note that in all the above mentioned families of measures, the degree of difference between two sets is just a number. In the following some axiomatic definitions about how to compare two sets while keeping the original information as much as possible are introduced.

3   Interval-Valued Measures for the Difference for Interval-Valued Fuzzy Sets

The measure of the difference between two interval-valued fuzzy sets is defined axiomatically on the basis of some natural properties:

- It is non negative and symmetric.
- It becomes zero when the two sets are fuzzy and equal, since the equality of two interval-valued fuzzy sets does not imply they are equal.

Every definition of divergence or dissimilarity satisfies the two previous properties. However, divergences and dissimilarities differ in a third axiom, the one representing the idea that the closer sets, the less differences.

On the other hand, when intervals are considered their width should be also considered as an evidence of uncertainty. Thus,

- For a wider interval, the uncertainty is greater and therefore the width of the difference with another interval should be bigger.

The usual way to compare two sets is by means of dissimilarities. If we adapt the previous ideas to our purposes, we could define a dissimilarity between interval-valued fuzzy sets as follows.

Definition 1. *[28] A map $D : IVFS(X) \times IVFS(X) \to L([0,1])$ is a dissimilarity on $IVFS(X)$ if for any $A, B, C \in IVFS(X)$ the following conditions are fulfilled:*

(Dis1) $D(A,B) = [0,0]$ iff $A, B \in FS(X)$ and $A = B$.
(Dis2) $D(A,B) = D(B,A)$.
(Dis3) if $A \subseteq B \subseteq C$, then $D(A,B) \leq D(A,C)$ and $D(B,C) \leq D(A,C)$.
(Dis4) if $B \sqsubseteq C$, then $D(A,B) \subseteq D(A,C)$.

Example 1. From [28] we can obtain several examples of dissimilarity measures:

- The trivial dissimilarity:

$$D_0(A,B) = \begin{cases} [0,0] & \text{if } A, B \in FS(X), A = B, \\ [0,1] & \text{otherwise.} \end{cases}$$

- The dissimilarity induced by a numerical distance:

$$D_1(A,B) = \frac{1}{|X|} \sum_{x \in X} \left[\inf_{a \in A(x), b \in B(x)} |a - b|, \sup_{a \in A(x), b \in B(x)} |a - b| \right]$$

- The dissimilarity induced by the numerical trivial dissimilarity:

$$D_2(A,B) = \begin{cases} [0,0] & \text{if } A, B \in FS(X), A = B, \\ [0,1] & \text{if } A \neq B \text{ and } A(x) \cap B(x) \neq \emptyset, \forall x \in X, \\ [1,1] & \text{if } \exists x \in X \text{ such that } A(x) \cap B(x) = \emptyset. \end{cases}$$

Although dissimilarities are the usual way to compare two sets, they are based on a partial order on $IVFS(X)$, so one of the main properties is only required for a few number of elements in $IVFS(X)$, what represents an important drawback. In fact, we could find some counterintuitive dissimilarities based on the numerical measure given by Chen [9]. Thus, to avoid this circumstance, the third axiom could be replaced as follows.

Definition 2. *A map* $D : IVFS(X) \times IVFS(X) \to L([0,1])$ *is a divergence on* $IVFS(X)$ *if for any* $A, B, C \in IVFS(X)$ *the following conditions are fulfilled:*

(Dis1) $D(A, B) = [0, 0]$ *iff* $A, B \in FS(X)$ *and* $A = B$.
(Dis2) $D(A, B) = D(B, A)$
(Div3) $D(A \cap C, B \cap C) \le D(A, B)$ *and* $D(A \cup C, B \cup C) \le D(A, B)$
(Dis4) *if* $B \sqsubseteq C$, *then* $D(A, B) \subseteq D(A, C)$.

A fuzzy set can be seen as a particular case of interval-valued fuzzy set such that the width of the membership interval is always zero. Thus, another logical requirement would be that when we restrict a divergence measure to the set formed by all the fuzzy sets in X, we obtain a divergence measure in the sense introduced in [23], although now the value could be an interval instead of just a number. This requirement was not added, since it is a consequence of the other four axioms, as we can see at the following proposition.

Proposition 1. *Let* D *be a divergence measure in* $IVFS(X)$. *Then the map* $D|_{FS(X)}$ *defined by*

$$D|_{FS(X)}(A, B) = D(A', B')$$

with $A'(x) = [A(x), A(x)]$ *and* $B'(x) = [B(x), B(x)]$ *for any* $x \in X$ *and for any* $A, B \in FS(X)$, *is a divergence measure in* $FS(X)$.

The same proposition could be proven for dissimilarities. It is clear that both concepts are very related. In fact, the family of divergence measures is really a subfamily of the dissimilarities, as we can see from the following proposition.

Proposition 2. *If a map* $D : IVFS(X) \times IVFS(X) \to L([0,1])$ *is a divergence on* $IVFS(X)$, *then it is a dissimilarity on* $IVFS(X)$.

Thus, any example of divergence is an example of dissimilarity. However, the converse is not true, as we can see at the following example.

Example 2. Let us consider the map $D_3 : IVFS(X) \times IVFS(X) \to L([0,1])$ defined by

$$D_3(A, B) = \begin{cases} [0, 0] & \text{if } A, B \in FS(X), A = B, \\ [0, 0.5] & \text{if } A \ne B \text{ and } A \ne X, B \ne X, \\ [0, 1] & \text{otherwise.} \end{cases}$$

First we check that D_3 is a dissimilarity. It is trivial that Axioms (Dis1) and (Dis2) are fulfilled. For Axiom (Dis3) we are going to consider three cases:

- If $D(A, C) = [0, 1]$, then it is trivial.
- If $D(A, C) = [0, 0.5]$, then $A \neq C$, $A \neq X$ and $C \neq X$. Since $B \subseteq C$, then $B \neq X$. Thus, $D(A, B) = [0, 0]$ or $D(A, B) = [0, 0.5]$ and the same happens for $D(B, C)$.
- If $D(A, C) = [0, 0]$, then $A = C$ and $A, C \in FS(X)$. Then, $B = A$ and therefore $D(A, B) = D(B, C) = [0, 0]$.

Finally, for Axiom (Dis4), we have again three cases:

- If $D(A, C) = [0, 1]$, then it is trivial.
- If $D(A, C) = [0, 0.5]$, then $A \neq C$, $A \neq X$ and $C \neq X$. Since $B \sqsubseteq C$, then $B \neq X$. Thus, $D(A, B) = [0, 0]$ or $D(A, B) = [0, 0.5]$.
- If $D(A, C) = [0, 0]$, then $A = C$ and $A, C \in FS(X)$. Since $B \sqsubseteq C$, then $B = C$ and therefore $D(A, B) = D(A, C)$.

Thus, D_3 is a dissimilarity, but it is not a divergence since for $X = \{x, y\}$, if we consider the sets A, B defined by $A(x) = [1, 1]$, $A(y) = [0, 1]$, $B(x) = [0, 1]$ and $B(y) = [1, 1]$, we have that $A \cup B = X$. Then, $D_3(A, B) = [0, 0.5]$ but $D_3(A \cup B, B \cup B) = D_3(X, B) = [0, 1]$ which is not lower than or equal to $D_3(A, B)$. Then Axiom (Div3) is not fulfilled by D_3.

Example 3. Trivial dissimilarity D_0 in Example 1 is neither a divergence as $D_0(A, B) = [0, 0]$ when $A, B \in FS(X), A = B$, but $D_0(A \cup C, B \cup C)$ is not necessary $[0, 0]$. For example, if $X = \{x\}$, $A(x) = B(x) = \{0.5\}$ and $C(x) = [0.2, 0.6]\}$, then $A \cup C(x) = [0.2, 0.5]$ and $B \cup C(x) = [0.2, 0.5]$. $A \cup C(x), B \cup C(x) \notin FS(X)$, then $D_0(A \cup C, B \cup C) = [0, 1]$ and thus axiom (Div3) is not satisfied.

As the previous result shows, divergences are particular cases of dissimilarities, with specific properties. Next proposition show several interesting properties satisfied by divergences defined by means of a range of values.

Proposition 3. *Let* $D : IVFS(X) \times IVFS(X) \to L([0, 1])$ *be a divergence on* $IVFS(X)$. *Then,* $\forall A, B \in IVFS(X)$

1. $[0, 0] \leq D(A, B)$.
2. $D(A \cap B, B) \leq D(A, A \cup B)$.
3. $D(A \cap B, B) \leq D(A, B)$.
4. $D(A \cap B, B) \leq D(A \cap B, A \cup B)$.
5. $D(B, A \cup B) \leq D(A \cap B, A \cup B)$.

From these measures we could obtain the ones defined just by a number if, for instance, we aggregate the lower and upper bound of the divergence measure in order to obtain just a number. This could be done by any average function as the arithmetic mean or the median.

4   Concluding Remarks

In this work we have introduce a way to compare two interval-valued fuzzy sets such that the value is itself an interval instead of the usual numbers. Dissimilarities and the particular case of divergences are defined and some properties are studied. Of course, this is just a first approach to this topic and a lot of work is still pending. In particular, we would like to study in detail the importance of the width of the interval, in a similar way as it was done in [8,27], since distances, dissimilarities and divergences are concepts clearly related. Apart from that, we would like to consider a more general definition of the concepts of intersection, union and complement based on general t-norms and study the behaviour of the proposed measures in that general case. We would also like to develop methods for building divergence measures.

References

1. Anthony, M., Hammer, P.L.: A boolean measure of similarity. Discrete Appl. Math. **154**(16), 2242–2246 (2006)
2. Arefi, M., Taheri, S.M.: Weighted similarity measure on interval-valued fuzzy sets and its application to pattern recognition. Iranian J. Fuzzy Syst. **11**(5), 67–79 (2014)
3. Atanassov, K.: Intuitionistic fuzzy sets. Fuzzy Sets Syst. **20**, 87–96 (1986)
4. Atanassov, K., Gargov, G.: Interval valued intuitionistic fuzzy sets. Fuzzy Sets Syst. **31**, 343–349 (1989)
5. Bouchon-Meunier, B., Rifqi, M., Bothorel, S.: Towards general measures of comparison of objects. Fuzzy Sets Syst. **84**, 143–153 (1996)
6. Bustince, H.: Indicator of inclusion grade for interval-valued fuzzy sets. Application to approximate reasoning based on interval-valued fuzzy sets. Int. J. Approximate Reasoning **23**, 137–209 (2000)
7. Bustince, H., Fernandez, J., Kolesarova, A., Mesiar, R.: Generation of linear orders for intervals by means of aggregation functions. Fuzzy Sets Syst. **220**(1), 69–77 (2013)
8. Bustince, H., Marco-Detcharta, C., Fernández, J., Wagner, C., Garibaldi, J.M., Takáč, Z.: Similarity between interval-valued fuzzy sets taking into account the width of the intervals and admissible orders. Fuzzy Sets Syst. (in press). https://doi.org/10.1016/j.fss.2019.04.002
9. Chen, S.M.: Measures of similarity between vague sets. Fuzzy Sets Syst. **74**(2), 217–223 (1995)
10. Chen, T.-Y.: A note on distances between intuitionistic fuzzy sets and/or interval-valued fuzzy sets based on the Hausdorff metric. Fuzzy Sets Syst. **158**(22), 2523–2525 (2007)
11. Cornelis, C., Deschrijver, G., Kerre, E.E.: Implication in intuitionistic fuzzy and interval-valued fuzzy set theory: construction, classification, application. Int. J. Approximate Reasoning **35**, 55–95 (2004)
12. Couso, I., Montes, S.: An axiomatic definition of fuzzy divergence measures. Int. J. Uncertainty Fuzziness Knowl. Based Syst. **16**(1), 1–17 (2008)
13. Deng, G., Song, L., Jiang, Y., Fu, J.: Monotonic similarity measures of interval-valued fuzzy sets and their applications. Int. J. Uncertainty Fuzziness Knowl. Based Syst. **25**(4), 515–544 (2017)

14. Deschrijver, G., Kerre, E.E.: On the relationship between some extensions of fuzzy set theory. Fuzzy Sets Syst. **133**, 227–235 (2013)
15. Deschrijver, G., Král, P.: On the cardinalities of interval-valued fuzzy sets. Fuzzy Sets Syst. **158**, 1728–1750 (2007)
16. Gorzalczany, M.B.: A method of inference in approximate reasoning based on interval-valued fuzzy sets. Fuzzy Sets Syst. **21**, 1–17 (1987)
17. Grattan-Guinness, I.: Fuzzy membership mapped onto intervals and many-valued quantities. Math. Logic Q. **22**(1), 149–160 (1976)
18. Grzegorzewski, P.: Distances between intuitionistic fuzzy sets and/or interval-valued fuzzy sets based on the Hausdorff metric. Fuzzy Sets Syst. **148**, 319–328 (2004)
19. Jahn, K.U.: Intervall-wertige Mengen. Math. Nach. **68**, 115–132 (1975)
20. Kobza, V., Janis, V., Montes, S.: Generalizated local divergence measures. J. Intell. Fuzzy Syst. **33**, 337–350 (2017)
21. Montes, I., Pal, N.R., Janis, V., Montes, S.: Divergence measures for intuitionistic fuzzy sets. IEEE Trans. Fuzzy Syst. **23**(2), 444–456 (2015)
22. Montes, I., Janis, V., Pal, N.R., Montes, S.: Local divergences for Atanassov intuitionistic fuzzy sets. IEEE Trans. Fuzzy Syst. **24**(2), 360–373 (2016)
23. Montes, S., Couso, I., Gil, P., Bertoluzza, C.: Divergence measure between fuzzy sets. Int. J. Approximate Reasoning **30**(2), 91–105 (2002)
24. Sambuc, R.: Fonctions ϕ-floues. Application l'aide au diagnostic en pathologie thyroidienne, These de Doctorar en Merseille (1975)
25. Szmidt, E., Kacprzyk, J.: Intuitionistic fuzzy sets - two and three term representations in the context of a Hausdorff distance. Acta Universitatis Matthiae Belii. Series Mathematics. **19**, 53–62 (2011)
26. Szmidt, E., Kacprzyk, J.: A perspective on differences between Atanassov's intuitionistic fuzzy sets and interval-valued fuzzy sets. In: Torra, V., Dahlbom, A., Narukawa, Y. (eds.) Fuzzy Sets, Rough Sets, Multisets and Clustering. SCI, vol. 671, pp. 221–237. Springer, Cham (2017). https://doi.org/10.1007/978-3-319-47557-8_13
27. Takáč, Z., Bustince, H., Pintor, J.M., Marco-Detchart, C., Couso, I.: Width-based interval-valued distances and fuzzy entropies. IEEE Access **7**, 14044–14057 (2019)
28. Torres-Manzanera, E., Kral, P., Janis, V., Montes, S.: Uncertainty-aware dissimilarity measures for interval-valued fuzzy sets. Int. J. Uncertainty Fuzziness Knowl. Based Syst. (in press)
29. Turksen, I.B., Zhong, Z.: An approximate analogical reasoning schema based on similarity measures and interval-valued fuzzy sets. Fuzzy Sets Syst. **34**, 323–346 (1990)
30. Valverde, L., Ovchinnikov, S.: Representations of T-similarity relations. Fuzzy Sets Syst. **159**(17), 211–220 (2008)
31. Wilbik, A., Keller, J.M.: A fuzzy measure similarity between sets of linguistic summaries. IEEE Trans. Fuzzy Syst. **21**(1), 183–189 (2013)
32. Wu, C., Luo, P., Li, Y., Ren, X.: A new similarity measure of interval-valued intuitionistic fuzzy sets considering its hesitancy degree and applications in expert systems. Math. Prob. Eng. **2014**, 1–16 (2014)
33. Zadeh, L.: The concept of a linguistic variable and its application to approximate reasoning-I. Inf. Sci. **8**, 199–249 (1975)
34. Zhang, C., Fu, H.: Similarity measures on three kinds of fuzzy sets. Pattern Recogn. Lett. **27**(12), 1307–1317 (2006)

35. Zhang, H., Zhang, W., Mei, C.: Entropy of interval-valued fuzzy sets based on distance and its relationship with similarity measure. Knowl. Based Syst. **22**, 449–454 (2009)
36. Zywica, P., Stachowiak, A.: Uncertainty-aware similarity measures - properties and construction method. In: Proceedings of 11th Conference of the European Society for Fuzzy Logic and Technology (EUSFLAT2019), Atlantis Studies in Uncertainty Modelling, vol. 1, pp. 512–519. Atlantis Press (2019). https://doi.org/10.2991/eusflat-19.2019.71

The Fuzzy Processing of Metaphors

Charles Tijus[✉] ⓘ

Cognitions Humaine et Artificielle, Université Paris 8, Saint Denis,
93526 Saint Denis, France
tijus@lutin-userlab.fr

Abstract. The question under investigation about verbal metaphor is what kind of thinking and reasoning can help catching the metaphorical target category (e.g., octopus meaning citrus press, child meaning son or daughter, pizzas meaning people) from a source category (children, pizzas or octopuses); knowing that this kind of "logically false" way of talking/understanding and of reasoning appears to be the most prominent kind of human of thinking. We are reviewing some of the prominent work of Bernadette Bouchon-Meunier with her team to evaluate how much the fuzzy logic computation of metaphorical reasoning and schemes can be used to model the human computation of metaphors.

Keywords: Thinking · Reasoning · Similarity · Analogy · Metaphor

1 Introducing Metaphor as One of Main Ways of Thinking and Reasoning

Let's imagine three situations

1) A grandmother and her grandson

   - The grandson: *Granny, is Dad your child?*
   - Grandmother: *Yes, he is my child*
   - The grandson: *Yes but yet, dad is no longer a child!*

2) A waiter of a restaurant and a group of 3 friends who have lunch together in this restaurant

   - The first client says to the waiter: *I'm going to have a pizza!*
   - The second client: *me too!*
   - The third client: *I too will have a pizza.*
   - (after the meal, the customers pay and leave)
   - The waiter shouting: *hey you! The 3 pizzas, you're forgetting your change!*

3) Philippe Starck, who is a designer, invented his famous citrus press that has an octopus shape, named *Juicy Salif*, by drawing on the paper tablecloth of an Italian restaurant, while tasting octopus seasoned with lemon. People that see *Juicy Salif* are usually saying *"oh this is an octopus!"*.

© Springer Nature Switzerland AG 2020
M.-J. Lesot et al. (Eds.): IPMU 2020, CCIS 1238, pp. 250–257, 2020.
https://doi.org/10.1007/978-3-030-50143-3_19

Homonyms are words that are spelled the same and sound the same way but have different meanings. In these three situations, either child, pizza or octopus has two different meanings although they are not homonyms:

- one meaning relates to a known category of things (*children, pizzas or octopuses*) and is properly used (*child used by the grandson as being a boy or girl from the time of birth until he or she is an adult; pizza as a large circle of flat bread baked with cheese, tomatoes, and so on or octopus as a sea creature with a soft, oval body and tentacles*), while
- the other meaning relates to a different category (*child used by the grandmother as being a son or daughter of any age, pizzas to denote customers, octopus to represent a citrus press*).

Thus, the question under investigation is what kind of thinking and reasoning can help catching the metaphorical (*e.g., octopus meaning citrus press*), pragmatic (*e.g., child meaning son or daughter*) or metaphorical target category (*e.g., pizzas meaning people*) from a source category (*children, pizzas or octopuses*); knowing that this kind of "*logically false*" way of talking/understanding and of reasoning appears to be the most prominent kind of human thinking [1]. From now, we label "metaphor", from the Latin "*metaphora*", itself from the Greek "μεταφορά", a figure of style based on attributive categorization [2]. It designates a target thing (*e.g., some specific people being called as "pizzas"*) with the name of another category (*pizza*) as a source for attributing to the target thing some of the features of the source category (*have a pizza for lunch*). Thus, instead of saying "*Hey you people that were eating pizzas*", they can be named and recognized as the meanings of "*pizzas*" in a sentence such as "*hey you, the pizzas, you have forgotten your money*".

2   The Logic Behind Metaphor Thinking

Associations, - such as Juliet is my sun (she brings me joy and shine) in the Romeo's diary of Shakespeare's Romeo and Juliet-, are in fact based on some fuzzy resemblance, on analogy or metaphor (upon, according to/so to speak) [1]. Analogy is usually based on similarity of things and on a computation based on "partial agreement, likeness or proportion between things", i.e. for transfer of learning. Although such metaphorically-based reasoning is providing sentences that do not have an evidential support and are logically "false", this kind of reasoning appears to be the most prominent kind of human thinking [1].

Although Metaphors seem to be produced through metaphorical reasoning based on comparison (e.g., "Peter is a dog"): Peter, who is a troubadour, sings to beg for a little love. Droopy, who is a dog, barks to beg for food. Peter's voice sounds like a barking: a proposition such as "Jean is barking" is an analogy, while "Peter is a dog" is a metaphorical proposition [2] based on attributive categorization.

Unlike deduction which provides "true" conclusions, the human inferences which produce a metaphor (major premise: a person had a meal, minor premise: the meal was pizza, conclusion: this person is a pizza) concludes with a false expression (not literal; not legally comparable). In such way of thinking, the things that are under cognitive investigation (e.g. the people forgetting their change in a restaurant) that are used to produce metaphors are of inter-domain constraints (people are not food). Although they are of different categories, generally of exclusive domains, inference is made by starting from an example of a given category and ends with another different category.

First, note the process is done from target-to-source to produce a metaphor (to produce Juliette is my sun: from Juliette, Shakespeare had to find a source of resemblance) whereas to understand the resemblance of the source, the cognitive inquiry of the listener is a process from source-to-target (to understand that Juliette is my sun: from the sun, the listener must find the resemblance that is to be to attributed to Juliette) [3].

Note also that if the metaphor is based on implication (x is a kind of y), the source (that serve as vehicle to transmit meanings) and the target (that is the topic) cannot swap: "Peter is a dog", the reverse is not true: dogs are not like John [2]. This is somewhat an asymmetric similarity measure on sets, not to compare a variant to a prototype [4], but a prototype to many variants.

In fact, there is a scientific and technological problem that is very difficult to solve: the what and the how of thought processes that are capable of producing and understanding metaphors. What type of calculation linking a target category to a source category belonging to another domain could be able to support reasonings based on imperfection, imprecision and approximation, graduality, imprecision, blurring, uncertainty and plausibility.

It seems to us that significant advances in the metaphor resolution stem from the work of Bernadette Bouchon-Meunier with her team: the calculation in fuzzy logic of reasoning and metaphorical schemes [5, 6].

3   The Cognitive Psychology of Metaphorical Thinking

There are psychological models of producing and of understanding metaphors. The model of Ortony [7] tries to apprehend the statements and the metaphors and proposes that the comparative statements are not necessarily true: their value of truth depends on the fact that the terms which they compare can be effectively assigned common predicates. Thus, the difference between literal comparisons and non-literal comparisons would arise from the fact that the terms linked in the literal comparisons have one or more significant predicates in common ((*e.g., John is an Irish man, comparing Jogn to Irish people*)). In non-literal comparisons, the common predicate, if a salient one, is only marginal for the topic (*e.g., John is a fox, comparing John to foxes*). Thus, non-literal statements induce graduality. Ortony is also following Tversky by taking into account the features of both topic and vehicle and explains that a "feature is as an attribute or a predicate in a rather general sense: feature X is something that we know about X"» and that according to Tversky the degree of similarity between two objects is a weighting function of the common features according to the features which distinguish them. Ortony argued that the essence of metaphoricity is salience imbalance; an imbalance that can be enhanced by attribute inequality. His proposals predict "*that nonliteral similarity statements will tend to be much less reversible than literal similarity statements and that in cases in which reversals still result in meaningful comparisons, the meaning change will be greater for similes than for literals*".

Thus, Ortony attempts to explain the way in which people understand metaphors by proposing *"the model of the salient imbalance"*. This model focuses on the fact that we must assign properties to the terms constituting the statements in order to be able to understand and interpret the metaphors. According to Ortony, there are a large number of properties that fight against each other (*there is a confrontation of properties*) and it is the most salient property that will allow to have a representation of a given object. Glucksberg and Keysar [8] offer their concept of understanding metaphor, which is based on a process of categorization. The authors believe that when an individual reads a metaphor, he activates the vehicle category (*i.e. the term that gives metaphorical meaning*) which allows him to have a representation of the situation. For example, the statement *"Sophie is an ice cube"* is a metaphor made up of a vehicle (*i.e. ice cube*) which qualifies a topic (*i.e. Sophie*). To understand the latter, the individual activates the category to which the ice cube vehicle belongs (*i.e. cold things*) before assigning the properties of the ice cube to the Sophie topical. Through their studies on the treatment of metaphorical statements, [7, 8] highlight the attribution of properties to objects via the transfer of knowledge. These studies show, moreover, that regardless of the differences in strategies implemented to process the statements, it is the transfer of knowledge which is at the origin of the recovery of information initiating the processing strategy.

Counterfactual reasoning, which is conventionally used in psychology as a diagnostic tool for evaluating creative potential, can help stem the phenomenon of metaphorical thought. According to [9–11] counterfactual reasoning makes it possible to transfer properties from one object to another object by an alternative attribution of categorized properties. This type of reasoning based on an alternative reality [12] implemented in our mind [1] allows a specific transfer of property between objects allowing not to refer to the prototype of the object to be considered but to what the object is not and could have been. For example, someone saying on Monday a counterfactual such as "If Paris was London" might imagine Parisian Museums being open on Mondays.

4 The Fuzzy Logic Computation of Metaphors

Since Zadeh's pioneering work on modifiers and on the one of Bouchon-Meunier and Colleagues [13, 14] which exploit the distribution of defined categories in a single universe, the mathematical transformation can be applied to different kinds of features, the concept of fuzzy relations being a generalization of the concept of relation, corresponding again to a fuzzy subset. In contrast to the preceding case concerning modifiers, this fuzzy set will not be constructed from a fuzzy set but from a value for two parts: the relation part of a specific feature and a category part. It is then possible to define mechanisms such that from a given feature and category, the fuzzy set representing the utterance can be constructed.

A major advance made by Bernadette Bouchon-Meunier that can be used for processing metaphor is the use of the main principle of fuzzy logic: that is to say the graduality of the membership function. Thus, the analogy as a metaphor (Juliet is the sun of Romeo) are fuzzy sets of particular types: analogy being calculated as a particular case of metaphor, with relations of transitivity, asymmetry and irreflexivity. This is due to her insight that the main principle of fuzzy logic, i.e. the graduality of membership function, might be a central core of metaphorical thinking and reasoning. There is a graduality of membership function that can be used both for "John is a man" and "John is a cat"; certainty (John is a man) being a special case of uncertainty (John is a cat). So, as crisp sets, that are representing precise and certain descriptions of objects, are to be regarded as particular cases of fuzzy sets [10], tautology as identity (*the sun is the sun*), analogy as comparison (*the nuclear of the atom is as the sun of the solar system*) and antilogy as metaphorical contradiction (*Juliet is the sun although Juliet is not the sun*) are all fuzzy sets of special kinds: identity being computed as a particular case of analogy, analogy being computed as a particular case of metaphor, with transitivity, asymmetry and irreflexivity relations [2].

As for analogy, concerning metaphorical reasoning, there are inter-domain relationships: a target object [T], that is to say the subject/tenor, is studied from the point of view of a source which is used as a vehicle (V) for the transmission of meanings [2]. Having comparisons between the same category (*tautology*), between different categories in the same domain (*analogy*), between different categories between domains or between inter-domain categories (*metaphor*), T and V can be calculated according to the evaluation of their "*Proximity*" through fuzzy modifiers to measure their similarity [6]. Thus, the proximity of two objects (*i.e. the moon and the sun*) as semantic distance (*very close, close, far, very far*) can be described through fuzzy sets which can manage the approximation. Indeed, the approximation of the proximity depends on the contextual role of T and V (*the moon-sun proximity in "the moon is a sun" having a different value in "the sun is a moon"*), according to the context (*the moon proximity: sun in "tonight the moon is a sun" (is very bright in the night) being of a different value in "today at noon the moon is a sun"*) being very large; depending on the motif (*the moon-sun proximity in "in your drawing; the moon is a sun "being of another value" in the sky, the moon is a sun"*).

In the past, most of the cognitive models of metaphor understanding have adopted the approach according to which metaphor is an implicit comparison: understanding a metaphor "X (*topic*) is Y (*vehicle*)" consists in converting it into a simile "*X (the topic) is like Y (the vehicle)*". This comparison-based model of metaphor understanding is a mechanism of property matching. This is why these models are confronted with the problem of measuring the similarity of properties as well as with the problem of calculating the distance between properties, which makes a simile literal or unliteral - metaphoric.

As a difference with analogies, metaphors appear to be assertions that have a not reversible structure: *if New York is a big apple, a big apple is not New York*. Metaphor is inter-domains and the source is not a particular thing, but a category of things that serve to make the Target inherit the properties of the category. If the understanding of metaphor is based on the similarity computation of T and V; the comparison would be symmetrical and reversible: T can be compared to V and V can be compared to T. When two objects, situations or domains are comparable, either one or the other can serve as a source or as target. However, Metaphor is a particular case of comparison. Most of source-target comparison are oriented: a target can imply a source, the reverse being not true.

There is a clear distinction between the objects of the physical world and the categories that humans use to represent them, in order to think, speak and communicate about them. Categories as sources of understanding can be used literally as in deduction and induction, but also in vague terms in metaphors. Like, we recommend that in both cases, the categorization of a target as being of a source type provides invisible features. As semantic relationships, metaphors are based on categorization. They activate a category and the functions attached to it. They also activate the superordinate categories and the functionalities attached to them.

For metaphors understanding [5] in which a source (*for example, the sun*) is a vehicle for transferring meanings to the target subject (*for example, Juliet*), the literature is about the notion that vehicles in metaphors are attributive categories. This categorization-based approach is mainly the Glucksberg's class inclusion model: a metaphorical statement of the type "*X is Y*" is solved by searching for the category, (represented by the term *Y*), that provides source properties potentially relevant to the topic target X. The general assumption underlying Glucksberg's approach is that understanding the metaphor involves including the topic in the source vehicle category and assigning it properties in that category that are consistent with what is known about the topic.

Following Glucksberg, we assume that the interpretation is constructed online and that knowledge of the subject comes into play at an early stage of processing by constraining the selection of functionalities. Bouchon-Meunier [15] noted that crisp sets representing precise descriptions and certain objects could be considered as special cases of fuzzy sets. Likewise, we argue that deduction, induction, abduction, induction, analogy and verbal metaphor are a special case of metaphorical thinking, from the least fuzzy-certain to the most fuzzy-uncertain; analogy being a special case of metaphor.

Bouchon-Meunier [11] proposed fuzzy reasoning based on prototypes for the fabrication and resolution of analogies. A fuzzy prototype of a category makes it possible to generate the typicity and the set of relevant objects and can therefore be used to match the source to the target; as Tversky suggests, the degree of typicity depends both on the resemblance to other objects of the same category and on the dissimilarity with objects of other categories. Thus, the metaphorical question that arises is "does the target gradually satisfy the prototype of the source category?" These are solutions for different modes of reasoning, including analogy and metaphor. Bouchon-Meunier and Valverde [12] were also tackling the problem of representing the similarities involved in metaphorical reasoning and used fuzzy relations to compare situations. Because vagueness involves various forms of reasoning, from real literal sentences to false non-literal sentences, it is a powerful method of calculation for human thinking and reasoning which is primarily reasoning based on metaphor and analogy. In addition, analogical and metaphorical sentences often include modifiers that Bouchon-Meunier and Marsala put at the core of fuzzy systems that can be interpreted.

5   The Fuzzy Logic Computation of Metaphors

In short, according to Bouchon-Meunier [12], we use fuzzy relations to compare situations that can be used to model natural analogy: resemblance relations can be used to define a kind of metaphorical scheme compatible with approximate reasoning in fuzzy logic, with measures of satisfiability, likeness and inclusion. These fuzzy relationships can be seen as measures of a categorization process devoted to analogy and metaphor in order to convey knowledge from source to target.

Analogies and metaphors are common modes of thinking and reasoning, although based on a false categorization: "this lawyer is really a shark". So there is a powerful use of approximation and imprecision by the brain using analogies and metaphors through fuzzy inference. According to Bouchon-Meunier [11], progressive reasoning can be obtained using linguistic modifiers, the link between gradual reasoning and metaphorical reasoning corresponding to the use of a relation between the variations of X and the variations of Y expressed in progressive knowledge to deduce a value of Y from a given value of X. Thus Bouchon-Meunier and her collaborators introduced a general framework representing the analogy, on the basis of a link between variables and measures of comparison between values of variables. This metaphorical scheme is a common description of several forms of reasoning used in fuzzy control or in the management of knowledge-based systems, such as deductive reasoning, inductive reasoning or prototypical reasoning, progressive reasoning.

References

1. Lakoff, G., Johnson, M.: Metaphors We Live By. University of Chicago Press, Chicago (2008)
2. Glucksberg, S.: The psycholinguistics of metaphor. Trends Cogn. Sci. **7**(2), 92–96 (2003)
3. Tijus, C.: Analogy. In: A Fuzzy Dictionary of Fuzzy Modelling. Common Concepts and Perspectives. Tribute to Bernadette Bouchon-Meunier (in press)
4. Tversky, A.: Features of similarity. Psychol. Rev. **84**, 327–352 (1977)
5. Bouchon-Meunier, B., Ramdani, M., Valverde, L.: Fuzzy logic, inductive and analogical reasoning. In: Ralescu, A.L. (ed.) FLAI 1993. LNCS, vol. 847, pp. 38–50. Springer, Heidelberg (1994). https://doi.org/10.1007/3-540-58409-9_4
6. Bouchon-Meunier, B., Marsala, C.: Fuzzy modifiers at the core of interpretable fuzzy systems. In: Tamir, D.E., Rishe, N.D., Kandel, A. (eds.) Fifty Years of Fuzzy Logic and its Applications. SFSC, vol. 326, pp. 51–63. Springer, Cham (2015). https://doi.org/10.1007/978-3-319-19683-1_4
7. Ortony, A.: Beyond literal similarity. Psychol. Rev. **86**(3), 161–180 (1979)
8. Glucksberg, S.: Keysar, B: Understanding metaphorical comparisons: beyond similarity. Psychol. Rev. **97**(1), 3–18 (1990)
9. Tijus, C., Barcenilla, J., Rougeaux, M., Jouen, F.: Open innovation and prospective ergonomics for smart clothes. In: 2nd International Conference on Ergonomics in Design (2014)
10. Candau, C., Ventalon, G., Barcenilla, J., Tijus, C.: Analogy and metaphors in images. In: Laurent, A., Strauss, O., Bouchon-Meunier, B., Yager, R.R. (eds.) IPMU 2014. CCIS, vol. 443, pp. 135–142. Springer, Cham (2014). https://doi.org/10.1007/978-3-319-08855-6_14

11. Vivian, R., Brangier, E., Bornet, C.: How design the future hydrogen users' needs? A contribution of prospective ergonomics. In: Rebelo, F., Soares, M. (eds.) AHFE 2017. AISC, vol. 588, pp. 400–410. Springer, Cham (2018). https://doi.org/10.1007/978-3-319-60582-1_40

12. Tijus, C.: Contextual categorization and cognitive phenomena. In: Akman, V., Bouquet, P., Thomason, R., Young, R. (eds.) CONTEXT 2001. LNCS (LNAI), vol. 2116, pp. 316–329. Springer, Heidelberg (2001). https://doi.org/10.1007/3-540-44607-9_24

13. Bouchon-Meunier, B., Yao, J.: Linguistic modifiers and gradual membership to a category. Int. J. Intell. Syst. 7, 25–36 (1992)

14. Brouard, C., Bouchon-Meunier, B., Tijus, C.: Modelling action in verbal command context with fuzzy subsets and semantic networks. In: Twentieth Annual Meeting of the Cognitive Science Society, pp. 179–184 (1998)

15. Bouchon-Meunier, B., Rifqi, M., Bothorel, S.: Towards general measures of comparison of objects. Fuzzy Set. Syst. 84(2), 143–153 (1996). Author, F.: Article title. Journal 2(5), 99–110 (2016)

A Measurement Theory Characterization of a Class of Dissimilarity Measures for Fuzzy Description Profiles

Giulianella Coletti[1] , Davide Petturiti[2(✉)] ,
and Bernadette Bouchon-Meunier[3]

[1] Dip. Matematica e Informatica, University of Perugia, Perugia, Italy
giulianella.coletti@unipg.it
[2] Dip. Economia, University of Perugia, Perugia, Italy
davide.petturiti@unipg.it
[3] LIP6, Sorbonne Université, Paris, France
bernadette.bouchon-meunier@lip6.fr

Abstract. We consider objects associated with a fuzzy set-based representation. By using a classic method of measurement theory, we characterize dissimilarity relations agreeing with a particular class of fuzzy dissimilarity measures. Dissimilarity measures in the considered class are those only depending on the attribute-wise distance of fuzzy description profiles. In particular, we analyze the subclass of weighted Manhattan dissimilarity measures.

Keywords: Dissimilarity relation · Fuzzy description profiles · Axioms · Weighted Manhattan dissimilarity measure

1 Introduction

Many situations in daily life or in science need to distinguish between several objects. Therefore, similarity and dissimilarity measures are important to evaluate a degree of resemblance between two or more objects. Many similarity/dissimilarity measures are available in the literature and the choice of one of them is done each time two images, cases, objects, situations, texts or data must be compared.

We are interested in measuring the similarity/dissimilarity of general objects characterized by a profile formed by a finite number of attributes or features. Then any object is identified by a vector which is binary (if the features can only belong or not to the object) or, as it has been more recently preferred, with elements in $[0,1]$ (if a partial degree of membership is accepted). In this setting, one can simply compare the vectors, rather than the objects themselves. For that,

The first and second authors are members of GNAMPA-INdAM. The second author was supported by University of Perugia, FRB 2019, project "Modelli per le decisioni economiche e finanziarie in condizioni di ambiguità ed imprecisione".

© Springer Nature Switzerland AG 2020
M.-J. Lesot et al. (Eds.): IPMU 2020, CCIS 1238, pp. 258–268, 2020.
https://doi.org/10.1007/978-3-030-50143-3_20

many papers on this subject present in the literature (see for instance [5,7,10]) discuss about the opportunity of considering as dissimilarity measure a (pseudo) distance function or a measure of comparison, as studied in [1] generalizing Tversky's contrast model [11] (for a general parametrised form, see [5]). Usually the comparison is made for a particular environment and "a posteriori" (on the basis of the obtained results), focusing on one or more specific properties.

With the purpose to provide conscious reasons to use a particular similarity measure on the basis of the semantics behind this choice, in [2-4] we characterized two classes of similarity measures using the paradigm of measurement theory. These classes are very large and contain as particular cases almost all the known measures in the sense of Tversky's contrast model and its generalizations. The study starts from the concept of comparative similarity and provides the conditions related to the primitive idea of similarity which are accepted when choosing a measure of this class.

The aim of this paper is to make an equivalent study for dissimilarity measures which only depend on the distances between the degrees of membership of two objects, related to any feature. This class contains many distances (for instance, the Manhattan distance, the Minkowski distance, and the weighted Euclidean distance). The particular subclass of the weighted Manhattan distance (depending on a vector of parameters) is then considered and is completely characterized by the comparative dissimilarity agreeing with one element of such class.

2   Preliminaries

Let $\mathcal{H}$ be a set of p attributes h_k, ($k \in I = \{1, \ldots, p\}$), each of which being present in an object with a degree of membership $\mu_k(\cdot) \in [0, 1]$.

Assume that the attributes in $\mathcal{H}$ are ordered increasingly according to indices in I. Let $\mathcal{Y} = [0, 1]^p$ be the set of all *fuzzy description profiles*: objects are identified through vectors $X = (x_1, \ldots, x_p) \in \mathcal{Y}$, where $x_i \in [0, 1]$ expresses the degree of membership of attribute i in the considered object. In other words, fuzzy description profiles in $\mathcal{Y}$ can be regarded as membership functions of fuzzy subsets of the ordered set $\mathcal{H}$ of p attributes. Let us stress that $\mathcal{Y}$ is endowed with the partial order relation $\leq$ defined component-wise.

Since the attributes can be expressed by a vague characterization, we can regard each of them as a fuzzy subset of a corresponding hidden variable. So each $X \in \mathcal{Y}$ is a projection of the Cartesian product of p possibly fuzzy subsets of p variables. For instance if the attributes h_1 and h_2 represent a person as "old" and "fat", every $X = (x_1, x_2)$ is a projection of the Cartesian product of the fuzzy sets "old" and "fat" of variables "age" and "weight", both taking values in $\mathbb{R}$.

We denote by $\mathcal{X} \subset \mathcal{Y}$ the set of *crisp description profiles*, i.e., $\mathcal{X} = \{0, 1\}^p$, and for any $X \in \mathcal{Y}$, we consider the *support* $s_X = \{i : x_i > 0\}$, so, in particular, $\underline{0}$ is the fuzzy description profile with $s_X = \emptyset$. More generally, if $\varepsilon \in [0, 1]$, then $\underline{\varepsilon}$ denotes the element of $\mathcal{Y}$ whose components are all equal to ε.

For every $0 \leq \delta \leq x_i$ and $0 \leq \eta \leq 1 - x_i$ we denote by $x_i^{-\delta}$ the value $x_i - \delta$, and by x_i^{η} the value $x_i + \eta$, and consider the elements of $\mathcal{Y}$: $X_k^{-\delta} = (x_1, \ldots, x_k^{-\delta}, \ldots, x_p)$, and $X_k^{\eta} = (x_1, \ldots, x_k^{\eta}, \ldots, x_p)$.

Given $X, Y \in \mathcal{Y}$, we denote by $U = |X - Y|$ the element of $\mathcal{Y}$ whose k-th component is $u_k = |x_k - y_k|$, while $W = X^c$ is the element of $\mathcal{Y}$, whose k-th component is $w_k = 1 - x_k$, which is referred to as the *complement* of X.

Let us now consider a *comparative dissimilarity* that is a binary relation $\precsim$ on $\mathcal{Y}^2$, with the following meaning: for all $X, Y, X', Y' \in \mathcal{Y}$, $(X, Y) \precsim (X', Y')$ means that X is no more dissimilar to Y than X' is dissimilar to Y'.

The relations $\sim$ and $\prec$ are then induced by $\precsim$ in the usual way: $(X, Y) \sim (X', Y')$ stands for $(X, Y) \precsim (X', Y')$ and $(X', Y') \precsim (X, Y)$, while $(X, Y) \prec (X', Y')$ stands for $(X, Y) \precsim (X', Y')$ and not $(X', Y') \precsim (X, Y)$.

If $\precsim$ is assumed to be complete, then $\sim$ and $\prec$ are the symmetric and the asymmetric parts of $\precsim$, respectively.

Definition 1. *Let $\precsim$ be a comparative dissimilarity and $D : \mathcal{Y}^2 \to \mathbb{R}$ a dissimilarity measure. We say that D **represents** $\precsim$ if and only if, for all $(X, Y), (X', Y') \in \mathcal{Y}^2$, it holds that*

$$\begin{cases} (X, Y) \precsim (X', Y') \Longrightarrow D(X, Y) \leq D(X', Y'), \\ (X, Y) \prec (X', Y') \Longrightarrow D(X, Y) < D(X', Y'). \end{cases}$$

As is well-known, if $\precsim$ is complete, the above conditions can be summarized as follows:

$$(X, Y) \precsim (X', Y') \Longleftrightarrow D(X, Y) \leq D(X', Y').$$

3   Basic Axioms

Given a comparative dissimilarity relation $\precsim$ on $\mathcal{Y}^2$, in the following we propose a set of axioms that reveal to be necessary and sufficient for $\precsim$ to have a dissimilarity measure representation inside a suitable class of dissimilarity measures.

(FD0) $\precsim$ is a weak order on $\mathcal{Y}^2$ (i.e., it is a complete and transitive binary relation on $\mathcal{Y}^2$).

We note that the completeness of relation $\precsim$ can be removed and required only in some specific cases: we assume it for simplicity.

The next axiom requires the comparative degree of dissimilarity to be independent of the common increment of presence/absence of the features in the objects of a pair. In fact, what is discriminant is the difference between the membership degrees of each feature.

(FD1) For all $X, Y \in \mathcal{Y}$, for all $k \in I$, for all $\varepsilon \leq \min(x_k, y_k)$, it holds:

$$(X, Y) \sim (X_k^{-\varepsilon}, Y_k^{-\varepsilon}).$$

The next example shows a situation of three pairs that the axioms **(FD0)** and **(FD1)** require to be equivalent.

Example 1. Let us consider a comparative dissimilarity among apartments in New York, described by the following attributes:

- a = it is small;
- b = it is centrally located;
- c = it has a modern kitchen;
- d = it has a panoramic view;
- e = it is near a metro station.

Given the fuzzy description profiles below, axioms **(FD0)** and **(FD1)** require that one must retain $(X, Y) \sim (X', Y') \sim (X'', Y'')$.

$\mathcal{H}$	a	b	c	d	e
X	0.5	0.4	0.9	0.6	0.1
Y	0.4	0.8	0.3	0.8	0.2
X'	0.3	0.2	0.6	0.2	0.05
Y'	0.2	0.6	0.2	0.4	0.15
X''	0.1	0	0.6	0	0
Y''	0	0.4	0	0.2	0.1

The next axiom is a local strong form of symmetry.

(FD2) For all $X, Y \in \mathcal{Y}$, for all $k \in I$, denoting $X'_k = (x_1, \ldots, y_k, \ldots, x_p)$ and $Y'_k = (y_1, \ldots, x_k, \ldots, y_p)$, it holds:

$$(X, Y) \sim (X'_k, Y'_k).$$

Example 2. Refer to the features in Example 1 and consider the fuzzy description profiles below.

$\mathcal{H}$	a	b	c	d	e
X''	0.1	0	0.6	0	0
Y''	0	0.4	0	0.2	0.1
X'''	0.1	0.4	0.6	0.2	0.1
Y'''	0	0	0	0	0

Accepting axioms **(FD0)** and **(FD2)** implies to set $(X'', Y'') \sim (X''', Y''')$.

As the following proposition shows, for a transitive complete relation, local symmetry implies symmetry. We note that the transitivity is necessary and that the converse does not hold.

Proposition 1. *Let $\precsim$ a comparative dissimilarity on $\mathcal{Y}^2$. If $\precsim$ satisfies axioms* **(FD0)** *and* **(FD2)***, then, for every $X, Y \in \mathcal{Y}$ one has: $(X, Y) \sim (Y, X)$.*

Proof. The proof trivially follows by applying at most p times **(FD2)** and **(FD0)**. $\square$

The next proposition shows that under axioms **(FD0)** and **(FD1)**, all pairs of identical fuzzy description profiles are equivalent.

Proposition 2. *Let $\precsim$ be a comparative dissimilarity on $\mathcal{Y}^2$. If $\precsim$ satisfies axioms* **(FD0)** *and* **(FD1)**, *then for every $X \in \mathcal{Y}$ one has:* $(\underline{1},\underline{1}) \sim (X,X) \sim (\underline{0},\underline{0})$.

Proof. For every $X \in \mathcal{Y}$, in particular $X = \underline{1}$, apply p times axiom **(FD1)** taking $\varepsilon = x_k$ and then use **(FD0)**. $\square$

The following axiom is a boundary condition. It provides a natural left limitation: "the elements of each pair (X,Y) are at least dissimilar to each other as an element of the pair is from itself". On the other hand, for the right limitation it is not enough to refer to any pair (X, X^c) formed by a profile and its complement, but it is required that the profile X is crisp, or equivalently that the supports of X and X^c are disjoint.

(FD3)
> a) for every $X, Y \in \mathcal{Y}$,
> > $(X,X) \precsim (X,Y)$ and $(Y,Y) \precsim (X,Y)$,
> b) for every $X \in \mathcal{X}$ and $Y \in \mathcal{Y}$,
> > 1) $(X,Y) \precsim (X,X^c)$,
> > 2) if $(Y,Z) \precsim (Y,Y^c)$ for any $Z \in \mathcal{Y}$, then $Y \in \mathcal{X}$.

The following is a monotonicity axiom.

(FD4) For all $X, Y \in \mathcal{Y}$, for all $k \in I$, such that $x_k \leq y_k$, for all $0 < \varepsilon \leq x_k$ and $0 < \eta \leq 1 - y_k$, it holds:

$$(X,Y) \precsim (X_k^{-\varepsilon}, Y) \quad \text{and} \quad (X,Y) \precsim (X, Y_k^{\eta}).$$

The following Theorem 1 shows that the introduced axioms are necessarily satisfied by any comparative dissimilarity agreeing with a dissimilarity measure, taking into account the distances of the degree of membership of each feature in the compared fuzzy description profiles. The same axioms become necessary and sufficient together with the following structural axiom **(Q)**, known as Debreu's condition [6], which assures the representability of a weak order $\precsim$ by a real function.

(Q) There is a countable $\prec$-dense set $\mathcal{Z} \subseteq \mathcal{Y}^2$ (i.e., for all $(X,Y), (X',Y') \in \mathcal{Y}^2$, with $(X,Y) \prec (X',Y')$, there exists $(X'',Y'') \in \mathcal{Z}$, such that $(X,Y) \prec (X'',Y'') \prec (X',Y')$).

We need to precise that we adopt as fuzzy inclusion the classic concept introduced by Zadeh [12], as follows: if X, Y are fuzzy sets of a set $\mathcal{C} = \{z_1, \ldots, z_n\}$, then

$$X \subseteq Y \text{ if and only if } \mu_X(z_k) \leq \mu_Y(z_k) \text{ for every } z_k \in \mathcal{C}. \qquad (1)$$

In particular, in our case, we have that $X \subseteq Y$ if and only if every component of X is smaller than or equal to the same component of Y, i.e. every attribute is no more present in X than it is in Y. So, $X \subseteq Y$ can be written as $X \leq Y$, where the inequality is component-wise.

Theorem 1. *Let $\precsim$ be a comparative dissimilarity relation on $\mathcal{Y}^2$. Then, the following statements are equivalent:*

(i) $\precsim$ satisfies **(FD0)**–**(FD4)** *and* **(Q)***;*

(ii) there exists a function (unique under increasing transformations) $\Phi : \mathcal{Y}^2 \rightarrow [0,1]$ representing $\precsim$ in the sense of Definition 1 and a function $\varphi : \mathcal{Y} \rightarrow \mathbb{R}$ such that:

    a) $Z \subseteq Z' \Longrightarrow \varphi(Z) \leq \varphi(Z')$, for every $Z, Z' \in \mathcal{Y}$;

    b) $\varphi(\underline{0}) = 0$, and $\varphi(\underline{1}) = 1$;

    c) for all $X, Y \in \mathcal{Y}$:

$$\Phi(X,Y) = \Phi(|X - Y|, \underline{0}) = \varphi(|X - Y|).$$

Proof. We first prove that *(i)* $\Longrightarrow$ *(ii)*. Axioms **(FD0)** and **(Q)** are sufficient conditions for the existence of a function $\Phi : \mathcal{Y}^2 \rightarrow [0,1]$ representing $\precsim$ [8]. Now, applying at most p times **(FD1)** with $\varepsilon = \min(x_i, y_i)$ and at most $\frac{p}{2}$ times **(FD2)** we get, by **(FD0)**, that $(X,Y) \sim (|X - Y|, \underline{0})$. Then, since Φ represents $\precsim$ we have $\Phi(X,Y) = \Phi(|X - Y|, \underline{0})$. Thus it is sufficient to define, for all $Z \in \mathcal{Y}$, $\varphi(Z) = \Phi(Z, \underline{0})$.

We now prove the validity of statement *a)* of condition *(ii)*. Let $Z = |X - Y| \subseteq Z' = |X' - Y'|$. Among the pairs respectively equivalent to $(Z, \underline{0})$ and $(Z', \underline{0})$ there are two pairs in the hypotheses of axiom **(FD4)** for at least one index. So, from **(FD4)** condition *a)* follows. Condition *b)* follows by axiom **(FD3)**, by considering that $|X - Y| = \underline{0}$ is obtained if and only if $X = Y$ and $|X - Y| = \underline{1}$ is obtained if and only if $X = Y$ and $Y = X^c$ and $X \in \mathcal{X}$. Then it is sufficient to recall that Φ is unique under increasing transformations.

Let us consider now the implication *(ii)* $\Longrightarrow$ *(i)*. Every binary relation $\precsim$ representable by a real function satisfies axiom **(FD0)** and **(Q)** [8]. We must prove that $\precsim$ satisfies axioms **(FD1)**–**(FD4)**: taking into account representability of $\precsim$ by Φ we deduce that condition *c)* in *(ii)* implies **(FD1)** and **(FD2)** whereas condition *a)* implies **(FD3)**. To prove axiom **(FD4)** it is sufficient to consider that Φ assigns 1 to all and only the elements of the equivalence class of $(\underline{1}, \underline{0})$, i.e. only to the pairs (X, X^c) with $X \in \mathcal{X}$. Similarly, Φ assigns 0 to all and only the elements of the equivalence class of $(\underline{0}, \underline{0})$, i.e. to the pairs (X, X) for every $X \in \mathcal{Y}$. $\qquad\square$

4 Representation by a Weighted Manhattan Distance

Condition *(ii)* of Theorem 1 identifies a too wide and therefore too general class of functions. In the following we will study the relations representable by the elements of a particular subclass of functions Φ that is the class of the weighted Manhattan distances.

Definition 2. *A* **weighted Manhattan distance** *is a function $D_{\alpha} : \mathcal{Y}^2 \rightarrow \mathbb{R}$ parameterized by $\boldsymbol{\alpha} = (\alpha_1, \dots, \alpha_p)$ with $\alpha_k \geq 0$ and $\sum_{k=1}^{p} \alpha_k = 1$, defined, for every $X, Y \in \mathcal{Y}$, as*

$$D_{\alpha}(X,Y) = \sum_{k=1}^{p} \alpha_k |x_k - y_k|.$$

The next axiom will represent "the constraint accepted" to obtain that the function representing our comparative dissimilarity $\precsim$ belongs to this particular subclass.

4.1 Rationality Principle

(R) For all $n \in \mathbb{N}$, for all $(X_1, Y_1), \ldots, (X_n, Y_n), (X_1', Y_1'), \ldots, (X_n', Y_n') \in \mathcal{Y}^2$ with $(X_i, Y_i) \precsim (X_i', Y_i')$, $i = 1, \ldots, n$, for all $\lambda_1, \ldots, \lambda_n > 0$ it holds:

$$\sum_{i=1}^{n} \lambda_i |X_i' - Y_i'| \leq \sum_{i=1}^{n} \lambda_i |X_i - Y_i| \implies (X_i, Y_i) \sim (X_i', Y_i'), \ i = 1, \ldots, n.$$

The above axiom has an easy interpretation. First of all it requires to evaluate as equally dissimilar every pairs (X, Y) and $(|X - Y|, \underline{0})$. Moreover, it asserts that if you have n pairs (X_i, Y_i) of fuzzy profiles and you judge the elements of each of them no more dissimilar than those of other n pairs (X_i', Y_i'), with at least a strict comparison, combining in a positive linear combination the first and the second you cannot obtain two vectors Z and Z' such that the fuzzy description profiles $W = \frac{Z}{\sum_{i=1}^{n} \lambda_i}$ and $W' = \frac{Z'}{\sum_{i=1}^{n} \lambda_i}$ satisfy $W' \leq W$.

In the next example we provide a comparative dissimilarity assessment which violates the above rationality principle.

Example 3. Referring to the features in Example 1, let us consider the following profiles:

$\mathcal{H}$	a	b	c	d	e
X_1	1/2	1	1/4	3/4	1/10
Y_1	1/2	2/3	1/4	1/2	1/10
X_2	1	1	1/6	1/2	1/2
Y_2	1/2	1	1/6	1/2	1/2
X_3	2/3	2/3	1/3	0	1/4
Y_3	1/6	1	1/3	0	1/4
X_4	1/8	1	1/2	1/2	0
Y_4	1/8	1	1/3	1/4	0
X_5	1	1/8	0	0	1
Y_5	1/2	1/8	0	1/4	1
X_6	0	1/2	1/3	0	0
Y_6	0	1/6	1/3	0	2/3
X_7	1/4	1/6	1/6	1	1/3
Y_7	1/4	1/6	0	1	1
X_8	0	1/3	1/4	3/4	0
Y_8	1/2	0	1/4	1/2	0

Suppose now to assign the following reasonable relation: $(X_1, Y_1) \prec (X_2, Y_2)$, $(X_3, Y_3) \prec (X_4, Y_4)$, $(X_5, Y_5) \prec (X_6, Y_6)$, $(X_7, Y_7) \prec (X_8, Y_8)$.

It is easy to prove that the relation violates axiom **(R)**. By trivial computations, taking all λ_i's equal to 1, one obtains

$$|X_1 - Y_1| + |X_3 - Y_3| + |X_5 - Y_5| + |X_7 - Y_7| = (1, 2/3, 1/6, 1/2, 2/3) =$$
$$|X_2 - Y_2| + |X_4 - Y_4| + |X_6 - Y_6| + |X_8 - Y_8|.$$

The next theorem shows that, for a complete $\precsim$, condition **(R)** implies all the axioms from **(FD0)** to **(FD4)**. Nevertheless, since condition **(R)** deals with finite sets of pairs, condition **(Q)** is not guaranteed to hold, so **(R)** does not assure representability of $\precsim$ on the whole $\mathcal{Y}^2$.

Theorem 2. *Let $\precsim$ be a complete comparative dissimilarity relation on $\mathcal{Y}^2$ satisfying **(R)**. Then $\precsim$ is transitive and axioms **(FD1)**–**(FD4)** hold.*

Proof. To prove transitivity suppose that we have $(X, Y) \precsim (X', Y')$, $(X', Y') \precsim (X'', Y'')$ and $(X'', Y'') \prec (X, Y)$. Then we have

$$|X - Y| + |X' - Y'| + |X'' - Y''| = |X' - Y'| + |X'' - Y''| + |X - Y|,$$

contradicting **(R)**. The proof of the other cases is similar.

To prove **(FD1)** and **(FD2)** it is sufficient to note that $|X - Y| = |X_k^{-\varepsilon} - Y_k^{-\varepsilon}| = |X_k' - Y_k'|$. Condition *a)* of **(FD3)** follows immediately from the inequality $|\underline{0}| = |X - X| = |Y - Y| \leq |X - Y|$. To prove condition *b)* let us consider that for all and only $X \in \mathcal{X}$ one has $|X - X^c| = \underline{1} \geq |X - Y|$. Similar considerations prove **(FD4)**. $\qquad\square$

4.2 Representability Theorems

In the following we consider nontrivial relations $\precsim$, i.e., we assume that there exist $(X, Y), (X', Y')$ with $(X, Y) \prec (X', Y')$.

Theorem 3. *Let $\precsim$ be a nontrivial complete comparative dissimilarity relation on a finite $\mathcal{F} \subset \mathcal{Y}^2$. Then, the following statements are equivalent:*

*(i) $\precsim$ satisfies **(R)**;*
(ii) there exists a weight vector $\boldsymbol{\alpha} = (\alpha_1, \ldots, \alpha_p)$ with $\alpha_k \geq 0$ and $\sum_{k=1}^{p} \alpha_k = 1$ such that, for all $(X, Y), (X'Y') \in \mathcal{F}$, it holds that

$$(X, Y) \precsim (X', Y') \iff D_\alpha(X, Y) \leq D_\alpha(X', Y').$$

Proof. Since $\mathcal{F}$ is finite, the binary relation $\precsim$ amounts to a finite number of comparisons. Consider the sets

$$\mathcal{S} = \{((X, Y), (X', Y')) \in \mathcal{Y}^2 : (X, Y) \prec (X', Y')\},$$
$$\mathcal{W} = \{((X, Y), (X', Y')) \in \mathcal{Y}^2 \setminus \mathcal{S} : (X, Y) \precsim (X', Y')\},$$

with $s = \operatorname{card} \mathcal{S}$ and $w = \operatorname{card} \mathcal{W}$, and fix two enumerations $\mathcal{S} = \{((X_j, Y_j), (X_j', Y_j'))\}_{j \in J}$ and $\mathcal{W} = \{((X_h, Y_h), (X_h', Y_h'))\}_{h \in H}$ with $J = \{1, \ldots, s\}$ and $H = \{1, \ldots, w\}$.

Condition *(ii)* is equivalent to the solvability of the following linear system

$$\begin{cases} A\beta > 0, \\ B\beta \geq 0, \\ \beta \geq 0, \end{cases}$$

with unknown $\beta \in \mathbb{R}^{p \times 1}$, and $A \in \mathbb{R}^{s \times p}$ and $B \in \mathbb{R}^{w \times p}$, where the s rows of A are the vectors $|X'_j - Y'_j| - |X_j - Y_j|$, for all $j \in J$, while the w rows of B are the vectors $|X'_h - Y'_h| - |X_h - Y_h|$, for all $h \in H$. Indeed, if we have a weight vector α satisfying *(ii)*, then setting $\beta = \alpha^T$ we get a solution of the above system. On the converse, if β is a solution of the above system, then defining $\alpha_k = \frac{\beta_k}{\sum_{i=1}^{p} \beta_i}$, we get a weight vector α satisfying *(ii)*.

By the Motzkin theorem of the alternative [9], the solvability of the above system is equivalent to the non-solvability of the following system

$$\begin{cases} \mu A + \nu B \leq 0, \\ \mu, \nu \geq 0, \\ \mu \neq 0, \end{cases}$$

with unknowns $\mu \in \mathbb{R}^{1 \times s}$ and $\nu \in \mathbb{R}^{1 \times w}$. In particular, the first inequality reduces to

$$\sum_{j \in J} \mu_j (|X'_j - Y'_j| - |X_j - Y_j|) + \sum_{h \in H} \nu_h (|X'_h - Y'_h| - |X_h - Y_h|) \leq 0,$$

thus the non-solvability of the above system is equivalent to condition **(R)**. □

Remark 1. We note that, in the hypotheses of previous theorem, if $(X, Y), (X', Y') \in \mathcal{F}$ and it holds $|X - Y| < |X' - Y'|$ then it must be $(X, Y) \prec (X', Y')$. In particular, if $(\underline{0}, \underline{0}), (\underline{1}, \underline{0}) \in \mathcal{F}$, then it must be $(\underline{0}, \underline{0}) \prec (\underline{1}, \underline{0})$.

Consider now the case where $\precsim$ is a nontrivial complete relation on $\mathcal{Y}^2$. In this case, axiom **(R)** is not sufficient to assure representability of $\precsim$ by a dissimilarity measure D_α on the whole $\mathcal{Y}^2$. Indeed, axiom **(R)** guarantees representability only on every finite subset $\mathcal{Y}^2$, by virtue of Theorem 3. This implies that the parameter α characterizing D_α depends on the particular finite subset $\mathcal{F}$. To remedy this problem it is necessary to introduce a further axiom which requires that in each equivalence class $(|X - Y|, \underline{0})$ there must be one pair $(\underline{\varepsilon}, \underline{0})$.

(FD5) For all $(X, Y) \in \mathcal{Y}^2$ there exists $\varepsilon \in [0, 1]$, such that $(X, Y) \sim (\underline{\varepsilon}, \underline{0})$.

Theorem 4. *Let $\precsim$ be a nontrivial complete comparative dissimilarity relation on $\mathcal{Y}^2$. Then, the following statements are equivalent:*

*(i) $\precsim$ satisfies **(R)** and **(FD5)**;*
(ii) there exists a weight vector $\alpha = (\alpha_1, \ldots, \alpha_p)$ with $\alpha_k \geq 0$ and $\sum_{k=1}^{p} \alpha_k = 1$ such that, for all $(X, Y), (X'Y') \in \mathcal{Y}^2$, it holds

$$(X, Y) \precsim (X', Y') \iff D_\alpha(X, Y) \leq D_\alpha(X', Y').$$

Moreover, the weight vector $\boldsymbol{\alpha}$ is unique.

Proof. The implication *(ii)* $\Longrightarrow$ *(i)* is easily proven, therefore we only prove *(i)* $\Longrightarrow$ *(ii)*.

For every finite $\mathcal{F} \subset \mathcal{Y}^2$ such that the restriction of $\precsim$ to $\mathcal{F}$ is nontrivial, Theorem 3 implies the existence of a weight vector $\boldsymbol{\alpha}^{\mathcal{F}} = (\alpha_1^{\mathcal{F}}, \ldots, \alpha_p^{\mathcal{F}})$ with $\alpha_k^{\mathcal{F}} \geq 0$ and $\sum_{k=1}^p \alpha_k^{\mathcal{F}} = 1$, such that the corresponding $D_{\boldsymbol{\alpha}^{\mathcal{F}}}$ represents the restriction of $\precsim$ to $\mathcal{F}$. Notice that, by Remark 1 every finite subset of $\mathcal{Y}^2$ containing $(\underline{0}, \underline{0})$ and $(\underline{1}, \underline{0})$ meets nontriviality, as it must be $(\underline{0}, \underline{0}) \prec (\underline{1}, \underline{0})$.

Next, axiom **(FD5)** implies that, for all $(X, Y) \in \mathcal{Y}^2$, there exists $\varepsilon_{(X,Y)} \in [0, 1]$, such that $(X, Y) \sim (\varepsilon_{(X,Y)}, \underline{0})$. In particular, denoting by E_k the element of $\mathcal{Y}$ whose k-th component is 1 and the others are 0, we have that there exists $\alpha_k \in [0, 1]$ such that $(E_k, \underline{0}) \sim (\underline{\alpha_k}, \underline{0})$. Now, for every $(X, Y), (X', Y') \in \mathcal{Y}^2$ we consider the finite subset of $\mathcal{Y}^2$

$$\mathcal{F} = \{(X, Y), (X', Y'), (\varepsilon_{(X,Y)}, \underline{0}), (\varepsilon_{(X',Y')}, \underline{0}),$$
$$(E_1, \underline{0}), \ldots, (E_p, \underline{0}), (\underline{\alpha_1}, \underline{0}), \ldots, (\underline{\alpha_p}, \underline{0}), (\underline{0}, \underline{0}), (\underline{1}, \underline{0})\}.$$

By the previous point we have that there is a weight vector $\boldsymbol{\alpha}^{\mathcal{F}} = (\alpha_1^{\mathcal{F}}, \ldots, \alpha_p^{\mathcal{F}})$ with $\alpha_k^{\mathcal{F}} \geq 0$ and $\sum_{k=1}^p \alpha_k^{\mathcal{F}} = 1$ such that

$$(X, Y) \precsim (X', Y') \Longleftrightarrow D_{\boldsymbol{\alpha}^{\mathcal{F}}}(X, Y) \leq D_{\boldsymbol{\alpha}^{\mathcal{F}}}(X', Y'),$$
$$(X, Y) \sim (\varepsilon_{(X,Y)}, \underline{0}) \Longleftrightarrow D_{\boldsymbol{\alpha}^{\mathcal{F}}}(X, Y) = D_{\boldsymbol{\alpha}^{\mathcal{F}}}(\varepsilon_{(X,Y)}, \underline{0}) = \varepsilon_{(X,Y)},$$
$$(X', Y') \sim (\varepsilon_{(X',Y')}, \underline{0}) \Longleftrightarrow D_{\boldsymbol{\alpha}^{\mathcal{F}}}(X', Y') = D_{\boldsymbol{\alpha}^{\mathcal{F}}}(\varepsilon_{(X',Y')}, \underline{0}) = \varepsilon_{(X',Y')}.$$

Moreover, for all $k = 1, \ldots, p$, we have

$$(E_k, \underline{0}) \sim (\underline{\alpha_k}, \underline{0}) \Longleftrightarrow D_{\boldsymbol{\alpha}^{\mathcal{F}}}(E_k, \underline{0}) = \alpha_k^{\mathcal{F}} = \alpha_k = D_{\boldsymbol{\alpha}^{\mathcal{F}}}(\underline{\alpha_k}, \underline{0}),$$

thus we get

$$\varepsilon_{(X,Y)} = \sum_{k=1}^p \alpha_k |x_k - y_k| \quad \text{and} \quad \varepsilon_{(X',Y')} = \sum_{k=1}^p \alpha_k |x'_k - y'_k|.$$

Hence, there exists a weight vector $\boldsymbol{\alpha} = (\alpha_1, \ldots, \alpha_p)$ with $\alpha_k \geq 0$ and $\sum_{k=1}^p \alpha_k = 1$ such that, for all $(X, Y), (X'Y') \in \mathcal{Y}^2$, it holds that

$$(X, Y) \precsim (X', Y') \Longleftrightarrow D_{\boldsymbol{\alpha}}(X, Y) \leq D_{\boldsymbol{\alpha}}(X', Y'),$$

and such weight vector is unique. Indeed, suppose there exists $\boldsymbol{\alpha}' = (\alpha'_1, \ldots, \alpha'_p)$ with $\alpha'_k \geq 0$ and $\sum_{k=1}^p \alpha'_k = 1$, such that $D_{\boldsymbol{\alpha}'}$ represents $\precsim$ on the whole $\mathcal{Y}^2$ and $\boldsymbol{\alpha}' \neq \boldsymbol{\alpha}$. For $k = 1, \ldots, p$, it holds that

$$(E_k, \underline{0}) \sim (\underline{\alpha_k}, \underline{0}) \Longleftrightarrow D_{\boldsymbol{\alpha}'}(E_k, \underline{0}) = \alpha'_k = \alpha_k = D_{\boldsymbol{\alpha}'}(\underline{\alpha_k}, \underline{0}),$$

reaching in this way a contradiction. $\qquad\square$

5   Conclusions

In this paper we characterize comparative dissimilarities on fuzzy description profiles, representable by elements of a class of dissimilarity measures only depending on the attribute-wise distance. This very large class contains all L^p distances and, in particular, the weighted Manhattan distances. Then, we characterize comparative dissimilarities representable by the latter subclass. Our aim for future research is to provide a characterization of comparative dissimilarities representable by other distinguished subclasses.

References

1. Bouchon-Meunier, B., Rifqi, M., Bothorel, S.: Towards general measures of comparison of objects. Fuzzy Sets Syst. **84**, 143–153 (1996)
2. Coletti, G., Bouchon-Meunier, B.: Fuzzy similarity measures and measurement theory. In: IEEE International Conference on Fuzzy Systems (FUZZ-IEEE), New Orleans, LA, USA, pp. 1–7 (2019)
3. Coletti, G., Bouchon-Meunier, B.: A study of similarity measures through the paradigm of measurement theory: the classic case. Soft. Comput. **23**(16), 6827–6845 (2019). https://doi.org/10.1007/s00500-018-03724-3
4. Coletti, G., Bouchon-Meunier, B.: A study of similarity measures through the paradigm of measurement theory: the fuzzy case. Soft Computing (under revision), in press. https://doi.org/10.1007/s00500-020-05054-9
5. De Baets, B., Janssens, S., Meyer, H.D.: On the transitivity of a parametric family of cardinality-based similarity measures. Int. J. Approximate Reasoning **50**(1), 104–116 (2009)
6. Debreu, P.: Representation of Preference Ordering by a Numerical Function. Wiley, New York (1954)
7. Goshtasby, A.: Similarity and dissimilarity measures. In: Image Registration. Advances in Computer Vision and Pattern Recognition, pp. 7–66. Springer, London (2012). https://doi.org/10.1007/978-1-4471-2458-0_2
8. Krantz, D., Luce, R., Suppes, P., Tversky, A.: Foundations of Measurement. Academic Press, New York (1971)
9. Mangasarian, O.: Nonlinear Programming, Classics in Applied Mathematics, vol. 10. SIAM (1994)
10. Prasetyo, H., Purwarianti, A.: Comparison of distance and dissimilarity measures for clustering data with mix attribute types. In: 2014 The 1st International Conference on Information Technology, Computer, and Electrical Engineering, pp. 276–280 (2014)
11. Tversky, A.: Features of similarity. Psychol. Rev. **84**, 327–352 (1977)
12. Zadeh, L.: Fuzzy sets. Inf. Control **8**, 338–353 (1965)

Learning Tversky Similarity

Javad Rahnama and Eyke Hüllermeier[✉]

Heinz Nixdorf Institute and Institute of Computer Science, Paderborn University,
Paderborn, Germany
javad.rahnama@uni-paderborn.de, eyke@upb.de

Abstract. In this paper, we advocate Tversky's ratio model as an appropriate basis for computational approaches to semantic similarity, that is, the comparison of objects such as images in a semantically meaningful way. We consider the problem of learning Tversky similarity measures from suitable training data indicating whether two objects tend to be similar or dissimilar. Experimentally, we evaluate our approach to similarity learning on two image datasets, showing that is performs very well compared to existing methods.

Keywords: Similarity · Machine learning · Semantic features · Image data

1 Introduction

Similarity is an important cognitive concept and a key notion in various branches of artificial intelligence, including case-based reasoning [26], information retrieval [5], machine learning [6], data analysis [30] and data mining [4], among others. Specified as a numerical (real-valued) function on pairs of objects, it can be applied in a rather generic way for various problems and purposes [21]. In particular, similarity is of great interest for structured objects such as text and images. In such domains, similarity measures are normally not defined by hand but learned from data, i.e., they are automatically extracted from sample data in the form of objects along with information about their similarity.

In the image domain, numerous methods of that kind have been proposed, based on different types of feature information, including visual [12,37] and semantic features [24], and exploiting different types of measurements, such as non-metric [5,15] and metric ones [19]. Most common is a geometrical model of similarity that relies on an embedding of objects as points in a suitable (often high-dimensional) vector space, in which similarity between objects is inversely related to their distance. As popular examples of (dis-)similarity measures in this field, let us mention the Euclidean and the Cosine distance [25].

Somewhat surprisingly, another concept of similarity, the one put forward by Tversky [35], appears to be less recognized in this field, in spite of its popularity in psychology and cognitive science. Tversky argued that the geometrical model of similarity is not fully compatible with human perception, and empirically

© Springer Nature Switzerland AG 2020
M.-J. Lesot et al. (Eds.): IPMU 2020, CCIS 1238, pp. 269–280, 2020.
https://doi.org/10.1007/978-3-030-50143-3_21

demonstrated that human similarity assessment does not obey all properties implied by this model, such as minimality, symmetry, and triangle inequality [36]. Due to its cognitive plausibility, we hypothesize that Tversky similarity could provide a suitable basis for mimicking human similarity assessment in domains such as *art images* [20]. For example, an art historian will probably find the copy of a painting more similar to the original than the original to the copy, thereby violating symmetry. Not less importantly, Tversky similarity is a potentially interpretable measure, which does not merely produce a number, but is also able to explain why two objects are deemed more or less similar. In other words, it is able to fill the *semantic gap* between the extracted visual information and a human's interpretation of an image. This feature of *explainability* is of critical importance in many applications [29]. Last but not least, the similarity also exhibits interesting theoretical properties; for example, see [9] for a recent analysis from the perspective of measurement theory.

In this paper, we elaborate on the potential of Tversky similarity as an alternative to existing measures, such as Euclidean and Cosine distance, with a specific focus on the image domain. In particular, we consider the problem of learning Tversky similarity from data, i.e., tuning its parameters on the basis of suitable training data, a problem that has not received much attention so far. As a notable exception we mention [2], where the problem of learning the importance of features (and feature combinations) for a generalized Jaccard measure (a special case of the Tversky measures) on the basis of training data is addressed, so as to achieve optimal performance in similarity-based (nearest neighbor) classification; to this end, the authors make use of stochastic optimization techniques (particle swarm optimization and differential evolution). On the other side, it is also worth noticing that the problem of learning Tversky similarity is in general different from the use of the Tversky measure as a loss function in machine learning [1,27]. Here, the measure serves as a means to accomplish a certain goal, namely to learn a classification model (e.g., a neural network) that achieves a good compromise between precision and recall, whereas is our case, the measure corresponds to the sought model itself.

The paper is organized as follows. After a brief review of related work in the next section, we recall Tversky's notion of similarity based on the ratio model in Sect. 3. In Sect. 4, we then propose a method for learning this similarity from suitable training data. Finally, we present experimental results in Sect. 5, prior to concluding the paper in Sect. 6.

2   Related Work

2.1   Image Similarity

Image similarity and its quantitative assessment in terms of similarity measures strongly depends on the image representation. Numerous approaches have been presented to extract different types of representations based on visual and semantic features of images [12,22]. Most of the state-of-the-art methods for extracting visual features are based on deep neural networks, which produce a variety of features, ranging from low level features in the early layers to more abstract features in the last layers [18,39]. In general, however, similarity is a complex concept that can not be derived from visual features alone. Therefore, some studies have exploited the use of prior knowledge as well as intermediate or high-level representation of features to capture similarity in specific applications [3,11,12].

The similarity or distance on individual features is commonly combined into an overall assessment using measures such as weighted Euclidean or weighted Cosine distance, whereas Tversky similarity is much less used in the image domain in this regard. There are, however, a few notable exceptions. For example, Tversky similarity is used in the context of image retrieval in [28] and for measuring the similarity between satellite images in [33].

2.2 Metric Learning

As already said, the notion of similarity is closely connected to the notion of distance. Whether relationships between objects are expressed in terms of similarity or distance is often a matter of taste, although small differences in the respective mathematical formalizations also exist. In the literature, distance seems to be even a bit more common than similarity. In particular, there is large body of literature on distance (metric) learning [19]. Most distance learning methods focus on tuning the weights of attributes in the Mahalanobis distance or weighted Euclidean distance. As training information, these methods typically use constraints on pairwise or relative relations among objects [7,16].

3   Tversky Similarity

Tversky suggested that humans perceive the similarity between objects based on contrasting (the measure of) those features they have in common and those on which they differ [35]. Moreover, he suggested that more attention is paid to the shared than to the distinctive features. Thus, in his (feature-matching) model, an object is represented as a set of meaningful features, and similarity is defined based on suitable set operations.

3.1 Formal Definition

More formally, consider a set of objects $\mathcal{X}$ and a finite set of features $\mathcal{F} = \{f_1, \ldots, f_m\}$. Each feature is considered as a binary predicate, i.e., a mapping $f_i : \mathcal{X} \longrightarrow \{0, 1\}$, where $f_i(x) = 1$ is interpreted as the presence of the ith feature for the object x, and $f_i(x) = 0$ as the absence. Thus, each object $x \in \mathcal{X}$ can be represented by the subset of features it exhibits:

$$F(x) = \{f_i \mid f_i(x) = 1, \, i \in \{1, \ldots, m\}\} \subseteq \mathcal{F}$$

Tversky similarity, in the form of the so-called ratio model, is then defined as a function $S_{\alpha,\beta} : \mathcal{X}^2 \longrightarrow [0,1]$ as follows:

$$S_{\alpha,\beta}(x,y) = \frac{g(|A \cap B|)}{g(|A \cap B|) + \alpha\, g(|A \setminus B|) + \beta\, g(|B \setminus A|)}, \tag{1}$$

where $A = F(x)$, $B = F(y)$, and g is a non-negative, increasing function $\mathbb{N}_0 \longrightarrow \mathbb{N}_0$; in the simplest case, g is the identity measuring set cardinality. According to (1), Tversky similarity puts the number of features that are shared by two objects in relation to the number of relevant features, where relevance means that a feature is present in at least one of the two objects—features that are absent from both objects are completely ignored, which is an important property of Tversky similarity, and distinguishes it from most other similarity measures.

The coefficients $0 \leq \alpha, \beta \leq 1$ in (1) control the impact of the distinctive features as well as the asymmetry of the measure. The larger these coefficients, the more important the distinctive feature are. For $\alpha = \beta$, the similarity measure is symmetric, i.e., $S_{\alpha,\beta}(x,y) = S_{\alpha,\beta}(y,x)$ for all $x,y \in \mathcal{X}$, for $\alpha \neq \beta$ it is asymmetric. Important special cases include the Jaccard coefficient ($\alpha = \beta = 1$) and the Dice similarity ($\alpha = \beta = 1/2$).

3.2   Feature Weighting

The Tversky similarity (TS) measure (1) implicitly assumes that all features have the same importance, which might not always be true. In fact, g is only a function of the cardinality of feature subsets, but ignores the concrete elements of these subsets. In other words, only the number of shared and distinctive features is important, no matter what these features are.

A natural generalization of (1) is a weighted variant of Tversky similarity (WTS), in which each feature f_i is weighted by some $w_i \in [0,1]$:

$$S_{\alpha,\beta,w}(x,y) = \frac{\sum_{i=1}^m w_i f_i(x) f_i(y)}{\sum_{i=1}^m w_i \Big(f_i(x) f_i(y) + \alpha f_i(x) \bar{f}_i(y) + \beta \bar{f}_i(x) f_i(y) \Big)}, \tag{2}$$

with $\bar{f}_i(\cdot) = 1 - f_i(\cdot)$. Thus, in addition to α and β, this version of Tversky similarity is now parametrized by a weight vector $w = (w_1, \ldots, w_m)$.

3.3   Semantic Similarity

As already said, we believe that Tversky similarity may offer a semantically meaningful and cognitively plausible tool for object comparison, especially in the image domain—provided the f_i are "semantic features", that is, features with a meaningful semantic interpretation. In the image domain, such features are normally not given right away. Instead, meaningful (high-level) features have to be extracted from "low-level" feature information on the pixel level.

Thus, what we envision is an approach to image similarity based on a two-level architecture, in which semantic features are extracted from images in the

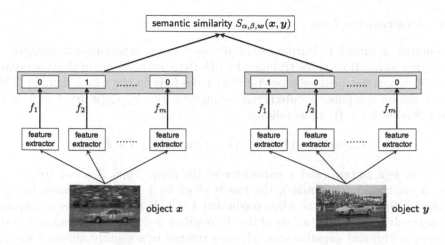

Fig. 1. Two-level architecture for semantic similarity.

first step, and these features are then used to specify similarity in the second step (cf. Fig. 1). While the features could be predefined by a human domain expert, the feature extraction itself should be accomplished automatically based on suitable image processing techniques.

For example, a human expert may know that, in a certain context, the presence of water on an image is a relevant feature. Water might then be included as a feature f_i in $\mathcal{F}$. The function f_i itself is a mapping that assumes an image (represented as a set of pixels) as an input and returns 0 or 1 as an output, depending on whether water is visible on the image or not. It could be realized, for example, by a neural network that has been trained to recognize water on images. Obviously, feature descriptors obtained in this way are never perfect, so that feature descriptions $F(x)$ will be "noisy" in practice.

4   Learning Tversky Similarity

In this section, we address the problem of *learning* the Tversky similarity, that is, tuning the parameters of the measure so as the optimally adapt it to a concrete application at hand. To this end, we assume suitable training data to be given, which informs about the similarity or dissimilarity between objects. More specifically, we assume training data of the following form:

$$\mathcal{D} = \left\{ (F(\boldsymbol{x}_n), F(\boldsymbol{y}_n), s_n) \right\}_{n=1}^{N} \subset \mathcal{X} \times \mathcal{X} \times \{0, 1\} \tag{3}$$

Each training example is a triplet $(F(\boldsymbol{x}_n), F(\boldsymbol{y}_n), s_n)$, where $s_n \in \{0, 1\}$ indicates whether the two objects $\boldsymbol{x}_n$ and $\boldsymbol{y}_n$ are considered similar or not. Thus, instead of precise numerical similarity degrees, the feedback is binary and only provides a rough indication. Note that the objects in the training data are already represented in terms of a feature description, i.e., we disregard the step of feature extraction here and completely focus on specifying the Tversky similarity.

4.1 Contrastive Loss

A common approach to learning similarity or distance functions is to minimize a suitable loss function on the training data $\mathcal{D}$. Here, we make use of the contrastive loss [7], which has been employed quite successfully in the image domain. This loss function compares a "predicted" similarity $\hat{s} = S_{\alpha,\beta,w}(x,y) \in [0,1]$ with given feedback $s \in \{0,1\}$ as follows:

$$L_m(s,\hat{s}) = s(1 - \hat{s}) + (1 - s) \max\left(m - 1 + \hat{s}, 0\right), \qquad (4)$$

where m is a margin (and a parameter of the loss). Thus, if $s = 1$ (objects x and y are considered similar), the loss is given by $1 - \hat{s}$ and increases linearly with the distance from the ideal prediction $\hat{s} = 1$. If $s = 0$, the loss is specified analogously, except that a loss of 0 only requires $\hat{s} \leq 1 - m$ (instead of $\hat{s} = 0$). Thus, positive and negative examples are treated in a slightly different way: A high (predicted) similarity on negative pairs is penalized less strongly than a low similarity on positive pairs. This could be meaningful, for example, when the feedback s is obtained from class membership in a classification or clustering context (as will be the case in our experiments) with many classes, some of which may (necessarily) overlap to a certain extent.

4.2 Loss Minimization

Training the weighted Tversky similarity (WTS) essentially consists of minimizing the total loss on the training data, i.e., finding

$$(\alpha^*, \beta^*, w^*) = \underset{\alpha,\beta,w}{\operatorname{argmin}} \sum_{n=1}^{N} L_m(s_n, S_{\alpha,\beta,w}(x_n, y_n)) \qquad (5)$$

Besides, we consider two restricted variants of this problem: Learning the (unweighted) TS measure comes down to omitting the weight vector w from (5), and enforcing symmetry to optimizing a single parameter α instead of two parameters α and β.

Our learning algorithm (see Algorithm 1) is quite simple and essentially based on gradient descent optimization. Gradients for the parameter updates are computed on mini-batches. To avoid overfitting of the training data, we take a part of the data for validation and apply an early stopping technique. More specifically, we stop the training process as soon as the accuracy on the validation data decreases by more than 0.01 in 20 consecutive iterations.

5 Experiments

In this section, we present the results of our experimental study. Before doing so, we first describe the data we used for the experiments, the experimental setting, and the baseline methods we compared with.

Algorithm 1. Learning Algorithm

Input: maximum iteration MT, batch size B, margin m, training data $\mathcal{D}$
Output: α, β, w

1: split $\mathcal{D}$ into training data $\mathcal{D}_{train}$ and validation data $\mathcal{D}_{val}$
2: randomly initialize parameters α, β, w
3: **while** ($t < MT$) and (stopping criterion = false)
4:     sample mini-batch uniformly at random from similar and dissimilar pairs
5:     compute $S_{\alpha, \beta, w}$ according to (2)
6:     update parameters by minimizing (5)
7:     test stopping condition on $\mathcal{D}_{val}$
8: **end while**

5.1   Data

Since the extraction of semantic features (cf. Sect. 3.3) is beyond the scope of this paper, we collected two image datasets for which this information, or at least information that can be interpreted as such, is already given.

The **a-Pascal-VOC2008** data [14] consists a total of 4340 images, which are split into 2113 training and 2227 test images. Each image is labeled with one of 32 class categories, such as "horse". Moreover, each image is described by 64 additional binary attributes, which we interpret as semantic features. Mostly, these features characterize the presence of certain objects on the image, such as saddle, tail, or snout.

The **Sun attributes** dataset [23] includes 14340 images equally distributed over 717 categories, such as "forest road" or "mountain path". Each image is also described by 102 attributes, such as "snow", "clouds", "flowers", etc., which describe images in a semantically meaningful way. These features are attached by 3 human annotators, and we assume the presence of each attribute in an image if it is attached by at least one of them.

5.2   Experimental Setting

We split the Sun attribute dataset into training data, validation data, and test data with a ratio of 70/10/20. The validation data is used to fine-tune parameters like the margin m of the contrastive loss. After each iteration of the training procedure (cf. Algorithm 1), the model is evaluated on validation data, and the model with highest performance is stored. We consider the same setting for the a-Pascal-VOC dataset, except for adopting the predefined train/test split. For this data, we extract 10% of the training data for validation.

Training examples for similarity learning (i.e., triplets of the form (x, y, s)) are extracted from the set of images belonging to the training data as follows: A pair of images x, y is considered as similar ($s = 1$) if both images have the same class label, and otherwise as dissimilar ($s = 0$). To facilitate learning and have balanced batches, similar and dissimilar pairs of images are sampled with the same probability of 0.5 (and uniformly within the set of all similar pairs and all dissimilar pairs, respectively).

Once a similarity measure S (or distance measure) has been learned, its performance is evaluated on the test data. To this end, we generate data in the form of triplets (x, y, s) in the same way as we did for training. The similarity measure is then used as a threshold classifier $(x, y) \mapsto [\![S(x, y) > t]\!]$ with a threshold t tuned on the validation data, and evaluated in terms of its classification rate (percentage of correct predictions) and F1 measure. Eventually, the average performance is determined over a very large set of randomly selected triplets (4M in the case of the a-Pascal-VOC2008 data and 6M in the case of Sun attributes), so that the estimation error is essentially 0.

5.3   Methods

We train both variants of the Tversky similarity, the unweighted (TS) and the weighted one (WTS). Since the "ground-truth" similarity in our data is symmetric by construction, we restrict to the respective symmetric versions ($\alpha = \beta$). We train unweighted Tversky similarity and weighted Tversky similarity using stochastic gradient descent with Nestrov's momentum [31] (learnin rate 0.01) and Adagrad [13] (learning rate 0.01), respectively.

As baselines for comparison, we learn two common similarity/distance metrics, namely the weighted Euclidean distance and weighted Cosine distance. Both distances are trained using Adam optimization [17] (with learning rate 0.001), in very much the same ways as the Tversky similarity. To prevent overfitting, the contrastive loss is combined with L1 and L2 regularization terms.

Moreover, we include LMNN [38] and a modified Siamese network (Siamese Net Semantic) based on [15]. The modified Siamese network consists of two blocks: two feature extraction blocks with shared weights for a pair of objects and a non-metric similarity block. A feature extraction block maps the input semantic feature vector into an embedding space, in which similar objects have small and dissimilar objects a larger distance. The non-metric similarity block predicts the similarity score that indicates the degree of similarity or dissimilarity of the input pair.

Since the number of input (semantic) features for the feature extraction block is relatively low in our experiments with the Sun attribute dataset, we only use three fully connected layers of size [64, 32, 16] and activation functions [Relu, Relu, Sigmoid], respectively. Correspondingly, the dimensions of two fully connected layers are [32, 16] with [Relu, Sigmoid] as activation functions for the a-Pascal-VOC2008 dataset. The non-metric network consists of an L1 distance part and two fully connected layers. In the L1 distance part, we calculate the L1 distance between features of the object pair, produced by the feature extraction blocks. The dimensions of two fully connected layers are [16, 1] with [Relu, Sigmoid] as activation functions, respectively. The parameters of the modified Siamese network are learned by minimizing a combination of two common loss functions using back propagation and the Adam optimization method [17] with learning rate 0.01. The contrastive loss is exploited to pull the representation of similar images closer and push the dissimilar ones apart. Moreover, since the evaluation is eventually done on a binary classification task, we combine the contrastive loss with the cross-entropy loss to improve the classification accuracy.

To show the effectiveness of our method in obtaining semantic similarity among images, we also train the modified Siamese network based on only visual features (Siamese Net Visual). The inputs of this network are the original images, and the output is a similarity prediction that indicates whether the two input images are similar or not. In the feature extraction block, we extract high-level features from the pre-trained Inception-V3 [32] followed by a flatten layer, batch normalization, and two fully connected layers with dimensions [256, 128] and activation functions [Relu, Sigmoid]. We also use the same non-metric similarity block and optimization method as explained above with a learning rate 0.01.

5.4 Results

Table 1. Performance on the a-Pascal-VOC2008 dataset.

Method	Classification rate	F1 measure
Weighted Euclidean	0.65	0.74
Weighted Cosine	0.81	0.78
LMNN [38]	0.78	0.81
Siamese Net Semantic [15]	0.82	**0.84**
Tversky (TS)	0.82	0.82
Weighted Tversky (WTS)	**0.83**	**0.84**

Table 2. Performance on the Sun attribute dataset.

Method	Classification rate	F1 measure
Weighted Euclidean	0.76	0.77
Weighted Cosine	0.78	0.80
LMNN [38]	0.79	0.81
Siamese Net Semantic [15]	0.79	0.81
Siamese Net Visual [15]	0.74	0.78
Tversky (TS)	0.80	0.81
Weighted Tversky (WTS)	**0.81**	**0.83**

The results are summarized in Tables 1 and 2. As can be seen, Tversky similarity performs very well and is highly competitive. Somewhat surprisingly, the performance of the simple unweighted version of Tversky similarity is already extremely strong. It can still be improved a bit by the weighted version, but not very much.

The Euclidean (and likewise the Cosine) distance performs quite poorly. As a possible explanation, note that the Euclidean distance is not able to ignore presumably irrelevant features that occur in none of the two objects (images) at hand. Since the distance between the corresponding feature values is 0, such features contribute to the similarity of the objects (e.g., the simultaneous absence of, say, trees on two images contributes to their similarity). This is certainly a strength of the Tversky similarity (and part of its motivation).

6   Conclusion

This paper presents first steps toward learning Tversky similarity, i.e., machine learning methods for tuning the parameters of Tversky's similarity model to training data collected in a concrete application context. To the best of our knowledge, such methods do not exist so far, in spite of the popularity of Tversky's model. The experimental results we obtained for image data so far, even if preliminary, are quite promising.

There are various directions for extending the approach presented in this paper, which ought to be addressed in future work:

- Our learning algorithm implements a rather plain solution and essentially applies a general purpose learning technique (a gradient-based optimization) to a specific loss function. More sophisticated methods, specifically tailored to the Tversky similarity and exploiting properties thereof, promise to improve efficiency and perhaps even boost performance. In this regard, we plan to elaborate on different ideas, such as alternating optimization and correlation analysis to judge feature importance.
- In practical applications, other types of training data may become relevant. An interesting example is relative similarity information of the form "object x is more similar to y than to z". Devising methods for learning from data of that kind is another important topic of future work.
- Likewise, other loss functions and performance metrics should be considered, both for training and evaluation. Specifically relevant are ranking measures from information retrieval, because similarity is often used for the purpose of object retrieval (e.g., ranking images stored in a database in decreasing order of their similarity to a query image).
- Further generalizations of the Tversky similarity itself could be considered as well, for example using fuzzy instead of binary features [8,10,34].

Acknowledgements. This work was supported by the German Research Foundation (DFG) under grant HU 1284/22-1. The authors strongly profited from discussions with the collaborators Ralph Ewerth, Hubertus Kohle, Stefanie Schneider, and Matthias Springstein.

References

1. Abraham, N., Khan, N.: A novel focal Tversky loss function with improved attention U-net for lesion segmentation. In: Proceedings ISBI, IEEE International Symposium on Biomedical Imaging (2019)

2. Baioletti, M., Coletti, G., Petturiti, D.: Weighted attribute combinations based similarity measures. In: Greco, S., Bouchon-Meunier, B., Coletti, G., Fedrizzi, M., Matarazzo, B., Yager, R.R. (eds.) IPMU 2012. CCIS, vol. 299, pp. 211–220. Springer, Heidelberg (2012). https://doi.org/10.1007/978-3-642-31718-7_22
3. Baltrušaitis, T., Ahuja, C., Morency, L.P.: Multimodal machine learning: a survey and taxonomy. IEEE Trans. Pattern Anal. Mach. Intell. **41**(2), 423–443 (2018)
4. Bouchon-Meunier, B., Rifqi, M., Lesot, M.-J.: Similarities in fuzzy data mining: from a cognitive view to real-world applications. In: Zurada, J.M., Yen, G.G., Wang, J. (eds.) WCCI 2008. LNCS, vol. 5050, pp. 349–367. Springer, Heidelberg (2008). https://doi.org/10.1007/978-3-540-68860-0_17
5. Chechik, G., Sharma, V., Shalit, U., Bengio, S.: Large scale online learning of image similarity through ranking. J. Mach. Learn. Res. **11**, 1109–1135 (2010)
6. Chen, Y., Garcia, E.K., Gupta, M.R., Rahimi, A., Cazzanti, L.: Similarity-based classification: concepts and algorithms. J. Mach. Learn. Res. **10**, 747–776 (2009)
7. Chopra, S., Hadsell, R., Le Cun, Y.: Learning a similarity metric discriminatively, with application to face verification. In: IEEE Computer Society Conference on Computer Vision and Pattern Recognition (CVPR), vol. 1, pp. 539–546 (2005)
8. Coletti, G., Bouchon-Meunier, B.: Fuzzy similarity measures and measurement theory. In: Proceedings FUZZ-IEEE, International Conference on Fuzzy Systems, New Orleans, LA, USA, pp. 1–7 (2019)
9. Coletti, G., Bouchon-Meunier, B.: A study of similarity measures through the paradigm of measurement theory: the classic case. Soft. Comput. **23**(16), 6827–6845 (2019)
10. Coletti, G., Petturiti, D., Vantaggi, B.: Fuzzy weighted attribute combinations based similarity measures. In: Antonucci, A., Cholvy, L., Papini, O. (eds.) ECSQARU 2017. LNCS (LNAI), vol. 10369, pp. 364–374. Springer, Cham (2017). https://doi.org/10.1007/978-3-319-61581-3_33
11. Datta, R., Joshi, D., Li, J., Wang, J.Z.: Image retrieval: ideas, influences, and trends of the new age. ACM Comput. Surv. **40**(2), 1–60 (2008)
12. Deselaers, T., Ferrari, V.: Visual and semantic similarity in ImageNet. In: IEEE Conference on Computer Vision and Pattern Recognition (CVPR), pp. 1777–1784. IEEE (2011)
13. Duchi, J., Hazan, E., Singer, Y.: Adaptive subgradient methods for online learning and stochastic optimization. J. Mach. Learn. Res. **12**, 2121–2159 (2011)
14. Farhadi, A., Endres, I., Hoiem, D., Forsyth, D.: Describing objects by their attributes. In: IEEE Conference on Computer Vision and Pattern Recognition (CVPR), pp. 1778–1785. IEEE (2009)
15. Garcia, N., Vogiatzis, G.: Learning non-metric visual similarity for image retrieval. Image Vis. Comput. **82**, 18–25 (2019)
16. Hermans, A., Beyer, L., Leibe, B.: In defense of the triplet loss for person re-identification. arXiv preprint arXiv:1703.07737 (2017)
17. Kingma, D.P., Ba, J.: Adam: a method for stochastic optimization. arXiv preprint arXiv:1412.6980 (2014)
18. Krizhevsky, A., Sutskever, I., Hinton, G.E.: ImageNet classification with deep convolutional neural networks. In: Advances in Neural Information Processing Systems (NIPS), pp. 1097–1105 (2012)
19. Kulis, B., et al.: Metric learning: a survey. Found. Trends® Mach. Learn. **5**(4), 287–364 (2013)
20. Lang, S., Ommer, B.: Attesting similarity: supporting the organization and study of art image collections with computer vision. Digit. Scholarsh. Humanit. **33**(4), 845–856 (2018)

21. Lesot, M., Rifqi, M., Benhadda, H.: Similarity measures for binary and numerical data: a survey. Int. J. Knowl. Eng. Soft Data Paradigms **1**(1), 63–84 (2009)
22. Liu, Y., Zhang, D., Lu, G., Ma, W.Y.: A survey of content-based image retrieval with high-level semantics. Pattern Recogn. **40**(1), 262–282 (2007)
23. Patterson, G., Xu, C., Su, H., Hays, J.: The sun attribute database: beyond categories for deeper scene understanding. Int. J. Comput. Vision **108**(1–2), 59–81 (2014)
24. Pedersen, T., Patwardhan, S., Michelizzi, J.: Wordnet::Similarity - measuring the relatedness of concepts. In: Proceedings AAAI, vol. 4, pp. 25–29 (2004)
25. Qian, G., Sural, S., Gu, Y., Pramanik, S.: Similarity between Euclidean and cosine angle distance for nearest neighbor queries. In: ACM Symposium on Applied Computing, pp. 1232–1237 (2004)
26. Richter, M.M.: On the notion of similarity in case-based reasoning. In: Della Riccia, G., Kruse, R., Viertl, R. (eds.) Proceedings of the ISSEK94 Workshop on Mathematical and Statistical Methods in Artificial Intelligence. ICMS, vol. 363, pp. 171–183. Springer, Vienna (1995). https://doi.org/10.1007/978-3-7091-2690-5_12
27. Salehi, S.S.M., Erdogmus, D., Gholipour, A.: Tversky loss function for image segmentation using 3D fully convolutional deep networks. In: Wang, Q., Shi, Y., Suk, H.-I., Suzuki, K. (eds.) MLMI 2017. LNCS, vol. 10541, pp. 379–387. Springer, Cham (2017). https://doi.org/10.1007/978-3-319-67389-9_44
28. Santini, S., Jain, R.: Similarity measures. IEEE Trans. Pattern Anal. Mach. Intell. **21**(9), 871–883 (1999)
29. Smeulders, A.W., Worring, M., Santini, S., Gupta, A., Jain, R.: Content-based image retrieval at the end of the early years. IEEE Trans. Pattern Anal. Mach. Intell. **22**(12), 1349–1380 (2000)
30. Strehl, A., Ghosh, J., Mooney, R.: Impact of similarity measures on web-page clustering. In: Workshop on Artificial Intelligence for Web Search (AAAI), vol. 58, p. 64 (2000)
31. Sutskever, I., Martens, J., Dahl, G., Hinton, G.: On the importance of initialization and momentum in deep learning. In: International Conference on Machine Learning, pp. 1139–1147 (2013)
32. Szegedy, C., Vanhoucke, V., Ioffe, S., Shlens, J., Wojna, Z.: Rethinking the inception architecture for computer vision. In: IEEE Conference on Computer Vision and Pattern Recognition (CVPR), pp. 2818–2826 (2016)
33. Tang, H., Maitre, H., Boujemaa, N.: Similarity measures for satellite images with heterogeneous contents. In: Proceedings Urban Remote Sensing, pp. 1–9. IEEE (2007)
34. Tolias, Y.A., Panas, S.M., Tsoukalas, L.H.: Generalized fuzzy indices for similarity matching. Fuzzy Sets Syst. **120**(2), 255–270 (2001)
35. Tversky, A.: Features of similarity. Psychol. Rev. **84**(4), 327–352 (1977)
36. Tversky, A., Gati, I.: Similarity, separability, and the triangle inequality. Psychol. Rev. **89**(2), (1982)
37. Wang, J., et al.: Learning fine-grained image similarity with deep ranking. In: IEEE Conference on Computer Vision and Pattern Recognition (CVPR), pp. 1386–1393 (2014)
38. Weinberger, K.Q., Saul, L.K.: Distance metric learning for large margin nearest neighbor classification. J. Mach. Learn. Res. **10**, 207–244 (2009)
39. Yue-Hei Ng, J., Yang, F., Davis, L.S.: Exploiting local features from deep networks for image retrieval. In: IEEE Conference on Computer Vision and Pattern Recognition Workshops (CVPR), pp. 53–61 (2015)

Belief Function Theory and Its Applications

Belief Functions for the Importance Assessment in Multiplex Networks

Alexander Lepskiy[1] and Natalia Meshcheryakova[1,2]

[1] National Research University Higher School of Economics,
20 Myasnitskaya Ulitsa, 101000 Moscow, Russia
alex.lepskiy@gmail.com
[2] V.A. Trapeznikov Institute of Control Sciences of Russian Academy of Science,
65 Profsoyuznaya Ulitsa, 117342 Moscow, Russia
natamesc@gmail.com
https://www.hse.ru/en/staff/natamesc

Abstract. We apply Dempster-Shafer theory in order to reveal important elements in undirected weighted networks. We estimate cooperation of each node with different groups of vertices that surround it via construction of belief functions. The obtained intensities of cooperation are further redistributed over all elements of a particular group of nodes that results in pignistic probabilities of node-to-node interactions. Finally, pairwise interactions can be aggregated into the centrality vector that ranks nodes with respect to derived values. We also adapt the proposed model to multiplex networks. In this type of networks nodes can be differently connected with each other on several levels of interaction. Various combination rules help to analyze such systems as a single entity, that has many advantages in the study of complex systems. In particular, Dempster rule takes into account the inconsistency in initial data that has an impact on the final centrality ranking. We also provide a numerical example that illustrates the distinctive features of the proposed model. Additionally, we establish analytical relations between a proposed measure and classical centrality measures for particular graph configurations.

Keywords: Belief functions · Network analysis · Centrality measures

1 Introduction

Dempster-Shafer theory of belief functions [1,2] is a widely used tool to measure belief or conflict between elements in a considered system [1,2]. Recently it has also found use in the field of social network analysis [3]. Social networks represent interactions that are met between people, countries, in transportation systems, etc.

One of the core problems in network science is the detection of central elements. In [4] a modified evidential centrality and evidential semi-local centrality in weighted network are proposed. The measures use the combination of "high",

© Springer Nature Switzerland AG 2020
M.-J. Lesot et al. (Eds.): IPMU 2020, CCIS 1238, pp. 283–296, 2020.
https://doi.org/10.1007/978-3-030-50143-3_22

"low" and "(high, low)" probabilities of the influence based on weighted and unweighted degrees of nodes via Dempster's rule. In [5] the same rule is applied in order to combine different node-to-node interactions in a network. The proposed measures that are able to detect social influencers were applied to Twitter data.

The theory of belief functions can be also adapted to the problem of community detection, i.e. the partition of nodes into tightly connected groups. For instance, in [6] the author proposed a novel method based on local density measures assigned to each node that are further used for the detection density peaks in a graph.

In the frame of the recent work we mostly focus on the problem of the detection of the most influential as well as the most affected elements in networks. The knowledge about the position of nodes plays a significant role in understanding of structural properties of complex systems.

There exist several networking approaches that aim to assess the importance of nodes in graphs. The first class of the methods refers to classical centrality measures [7]. It includes degree centrality measure that prioritizes over nodes with the largest number of neighbors or with the largest sum of incoming/outcoming weights. The eigenvector group of centralities, that includes eigenvector centrality itself, Bonacich, PageRank, Katz, Hubs and Authorities, Alpha centrality, etc., takes into account the importance of neighbors of a node, i.e. the centrality of a vertex depends on centralities of the adjacent nodes [8–12]. Closeness and betweenness centralities consider the distance between nodes and the number of the shortest paths that go through nodes in a network [13,14].

Another class of measures, that detect the most important elements, employs cooperative game theoretical approach. It includes the estimation of Myerson values, that is similar to Shapley-Shubik index calculation [15]. It also requires the introduction of nodes set functions, that can vary depending on the problem statement. In [16] the Hoede–Bakker index is adjusted to the estimation of the influence elements in social networks. In [17] Long-Range Interaction Centrality (LRIC) is proposed, that estimates node-to-node influence with respect to individual attributes of nodes, the possibility of the group influence and indirect interactions through intermediate nodes.

However, all the approaches described above are designed for so-called monoplex networks and require adaptation to complex structures with many types of interactions between adjacent nodes (so-called multilayer networks [18]). In recent years multilayer networks became one of the central topics in the field of network science. A multilayer network where the set of nodes (or a part of nodes) remains the same through all layers is called multiplex network, which is the object of the research in this work.

There exist several ways for the assessment of central elements in multiplex networks. Firstly, one can calculate centralities for each layer separately and further aggregate the obtained values through all considered networks. Secondly, one can aggregate connections between pairs of nodes to obtain monoplex network and then apply centrality measures to a new weighted graph. The modification of classical centrality measures to interconnected multilayer networks is described in [18,19]. In [20] social choice theory rules are applied to multiplex networks in order to detect key elements.

However, the final results for these approaches are calculated from the secondary data. In this work we propose a novel technique of the key elements assessment. We construct a mapping between each node and sets of other nodes, which is a mass function. In case of several layers we combine mass functions on each layer to a unique function that can be used for the centrality estimation in the whole system. The key advantages of our approach are that we take into account interactions with different groups of nodes and we are able to estimate node-to-node influence within the whole network structure. We also take into account the consistency on connections on different network layers.

This paper is organized as follows: in Sect. 2 we describe some basic concepts from belief functions theory. In Sect. 3 we propose a centrality measure for one-layer network and apply it to a toy example. In Sect. 4 we develop an approach to elucidate important elements in networks with several layers. In the same Section we apply the proposed method to two-layers network. Section 5 contains a discussion of our approach as well as conclusion to the work.

2   Background to the Theory of Belief Functions

In this Section we will remind some basic definitions and notions from Dempster-Shafer theory of belief functions [1,2] that are further employed in this work.

Let X be a finite set that is called *frame of discernment* and 2^X is a set of all subsets of X. Function $m : 2^X \rightarrow [0;1]$ that meets the requirements of normalization condition, i.e. $m(\emptyset) = 0$ and $\sum_{A \in 2^X} m(A) = 1$, is called *basic probability assignment* or a *mass function*. All $A \in 2^X$ such that $m(A) > 0$ are called *focal elements* and the family of all focal elements is called the *body of evidence*.

Mass function m can be associated with two set functions namely a belief function denoted by $g(A) = \sum_{B \subset A} m(B)$ and a plausibility function denoted $\bar{g}(A) = \sum_{B:A \cap B \neq \emptyset} m(B)$, that is dual to belief function $g(A)$. These two functions can be considered as lower and upper bounds for the probability estimation of event $A : g(A) \leq P(A) \leq \bar{g}(A), A \in 2^X$. The value of function $g(A)$ reflects the belief level to the fact that $x \in A \subseteq X$, where x from X. We denote by $Bel(X)$ a set of all belief functions g on set X.

Belief function g can be also represented as a convex combination of categorical belief functions $\eta_B(A) = \begin{cases} 1, B \subseteq A \\ 0, B \not\subseteq A \end{cases}, B \in 2^X \setminus \{\emptyset\}$ with $\{m(B)\}$ multipliers: $g(A) = \sum_B m(B)\eta_B(A)$. Note that η_X describes vacuous evidence that $x \in X$. Thus, we call this function as vacuous belief function. Additionally, mass function $m(A)$ can be also expressed from belief function g with Möbius transformation as $m(A) = \sum_{B \subset A} (-1)^{|A \setminus B|} g(B)$.

In this work we mainly focus on combination techniques adopted from Dempster-Shafer theory. By combination we mean some operator $R : Bel(X) \times$

$Bel(x) \to Bel(X)$ that transforms two belief functions into one belief function. We denote by $m = m_1 \otimes_R m_2$ the combinations of two mass functions m_1 and m_2 under rule R.

There exist various combination rules that are widely used in the theory and applications of belief functions. For instance, Dempster rule [1], that is regarded as the pioneered and the most popular combination technique in Dempster-Shafer theory, is calculated as follows:

$$m(A) = (m_1 \otimes_D m_2)(A) = \frac{1}{1 - K} \sum_{B \cap C = A} m_1(B) \cdot m_2(C) \qquad (1)$$

for all $A \neq \emptyset$ and $m(\emptyset) = 0$, where $K = \sum_{B \cap C = \emptyset} m_1(B) \cdot m_2(C)$. Parameter $K = K(m_1, m_2) \in [0; 1]$ indicates the level of conflict between two evidences. If $K = 1$ then the level of conflict is the highest and rule (1) is not applicable in this case.

Another combination technique that is similar to Demster rule is Yager combination rule [21] that is defined as

$$m(A) = (m_1 \otimes_Y m_2)(A) = \sum_{B \cap C = A} m_1(B) \cdot m_2(C) \qquad (2)$$

for all $A \neq \emptyset$, $m(\emptyset) = 0$ and $m(X) = K + m_1(X) \cdot m_2(X)$. According to this rule, the value of conflict K is reallocated among the mass of ignorance $m(X)$.

Other combination rules are also described in [22], some generalizations can be found in [23,24], axiomatics and the description of conflict rules are reviewed in [25–28].

Additionally, discounted technique proposed in [1] can be applied to mass functions in case when various sources of information that are determined by their belief functions have different levels of reliability or different priority. Discounting of mass functions can be performed with the help of parameter $\alpha \in [0; 1]$ as follows:

$$m^\alpha(A) = (1 - \alpha)m(A) \text{ for } A \neq X \text{ and } m^\alpha(X) = (1 - \alpha)m(X) + \alpha.$$

If $\alpha = 0$ then the source of information is regarded as thoroughly reliable and $m^\alpha(A) = m(A) \ \forall A \in 2^X$. Conversely, if $\alpha = 1$ then $m^\alpha(X) = 1$ and the related belief function is vacuous.

3   Centrality Assessment with Belief Functions

In this Section we describe a graph model with one layer of interaction as well as the construction of centrality measure based on a mass function for a network.

We consider connected graph as tuple $G = (V, E, W)$, where $V = \{v_1, ..., v_n\}$ is a set of nodes, $|V| = n$, and $E = \{e(v_i, v_j)\}$ as a set of edges. For the simplicity, we associate v_k with number k, $k = 1, ..., n$ and denote $e(v_i, v_j)$ as e_{ij}. In this work we consider undirected network, i.e. $e_{ij} \in E$ implies that $e_{ji} \in E$. We also analyze weighted networks, i.e. each edge e_{ij} in network G associates with weight $w_{ij} \in W$. Without loss of generality, we assume that all weights $w_{ij} \in [0; 1]$ and $w_{ij} = 0$ implies that $e_{ij} \notin E$. Weight w_{ij} between nodes v_i and v_j indicates the degree of interaction between corresponding nodes.

Our main focus is to range nodes with respect their importance in a network. We assume that a node is considered to be pivotal if it actively interacts with other nodes in a graph. In our analysis we take into account the connections with distant nodes as well as the cooperation with group of other nodes. More precisely, we suppose that centrality of a node depends on relative aggregated weight of adjacent subgraphs to the considered node. At the same time, the aggregated weight of a subgraph can be estimated with the help of monotonic measures including such measures as belief functions.

We consider a family of belief functions $g_k = \sum_B m_k(B)\eta_B$ for all vertices $v_k \in V$ in network G. Let $N_k^{(p)}$ be a p-neighborhood of node v_k, i.e. a set of nodes of graph G whose distance from node v_k is at most p edges and $v_k \notin N_k^{(p)}$. We denote by $|W| = \sum_{i<j} w_{ij}$ the total sum of all weights in a graph. Next, we define mass function m_k of node v_k in connected graph G as follows:

$$m_k(A) = \frac{1}{|W|} \begin{cases} w_{ik}, & \text{if } A = \{v_i\} \subseteq N_k^{(1)}, \\ \gamma_{ij}^{(k)} \cdot w_{ij}, & \text{if } A = \{v_i, v_j\} \subseteq N_k^{(p)} \wedge e_{ij} \in E, \\ |W| - \sum_{v_i \in N_k^{(1)}} w_{ik} - \sum_{\substack{v_i, v_j \in N_k^{(p)} \\ e_{ij} \in E}} \gamma_{ij}^{(k)} \cdot w_{ij}, & \text{if } A = V, \\ 0, & \text{otherwise,} \end{cases} \quad (3)$$

where $N_k^{(p)} \neq \emptyset$ and $\gamma_{ij}^{(k)} \in [0; 1]$ is a discount factor that decreases the importance of the connection of node v_k with distant nodes. This coefficient can be determined in the following way:

$$\gamma_{ij}^{(k)} = \frac{1}{1 + \min\{d_{ik}, d_{jk}\}}, \quad (4)$$

where d is a distance between corresponding nodes. A mass function of the k-th node reaches the higher values on single nodes that are adjacent to the k-th node and the lower values on the pairs of connected nodes that both belongs to the p-neighborhood of node k. Thus, the value of mass function (3) on one- and two-element sets is proportional to the weights on corresponding edges and inversely proportional to the distance to a considered node. Belief function g_k aggregates the obtained mass functions and corresponding weights over all nodes and edges that are contained in p-neighborhood of the k-th node. Other characteristics can be also taken into account as weighted path between nodes, the joint intensity of the connections along the considered path, etc.

It can be seen that the proposed mass function $m_k : 2^V \to [0; 1]$ satisfies the normalization condition: $m_k(\emptyset) = 0$, $\sum_{A \in 2^V} m_k(A) = 1$. Thus, we can regard m_k as basic probability assignments and $g_k = \sum_B m_k(B)\eta_B$ are belief functions of V.

Similar measures for the nodes influence assessment in networks are proposed in [29].

The proposed mass function $m_k(A)$ characterizes the distribution of pure interaction of node v_k with a set of closely located nodes from A, i.e. we exclude the interactions with other subsets of A in $m_k(A)$. The value $m_k(V)$ indicates the level of ignorance of the interactions between node v_k and distant nodes in the graph outside $N_k^{(p)}$.

The pairwise interaction between nodes can be estimated by the reallocation of $m_k(A)$ among all nodes in A. This can be done with the help of so-called pignistic probabilities proposed in [30]:

$$Bet_{v_k}(\{u\}) = \sum_{A: u \in A} \frac{m_k(A)}{|A|}. \tag{5}$$

It has been known that pignistic probabilities for belief function $g_k = \sum_B m_k(B)\eta_B$ defined on V coincide with Shapley values [31] that are widely used in cooperative game theory and are calculated as follows:

$$Bet_{v_k}(\{u\}) = \sum_{\substack{A \subseteq V \\ u \in A}} \frac{(n-|A|)!(|A|-1)!}{n!} (g_k(A) - g_k(A \setminus \{u\})).$$

Value $Bet_{v_k}(\{u\})$ indicates the fraction of interaction of node v_k with node u. Hence, the value

$$q_v = \sum_{u \in V} Bet_v(\{u\}), \ \forall v \in V \tag{6}$$

shows the total cooperation with node v in graph G. Consequently, the ranking of q values demonstrates the importance on nodes with respect to their activity in the considered graph.

We also note that if graph G has several connected components then the proposed analysis is provided for each component separately. The size of each component can be taken into account in order not to overestimate the interactions in small groups.

We illustrate the proposed model on a toy example represented on Fig. 1.

For the graph on Fig. 1 $|W| = 3.2$. Let us estimate the belief functions for each node according to formulas (3) and (4) taking into account 2-neighborhood of each node:

$$g_1 = \frac{9}{32}\eta_{\{v_2\}} + \frac{3}{32}\eta_{\{v_2,v_4\}} + \frac{3}{32}\eta_{\{v_2,v_5\}} + \frac{5}{96}\eta_{\{v_4,v_5\}} + \frac{23}{48}\eta_V,$$

$$g_2 = \frac{9}{32}\eta_{\{v_1\}} + \frac{3}{16}\eta_{\{v_4\}} + \frac{3}{16}\eta_{\{v_5\}} + \frac{3}{64}\eta_{\{v_3,v_4\}} + \frac{3}{64}\eta_{\{v_3,v_5\}} + \frac{5}{64}\eta_{\{v_4,v_5\}} + \frac{11}{64}\eta_V,$$

$$g_3 = \frac{3}{32}\eta_{\{v_4\}} + \frac{3}{32}\eta_{\{v_5\}} + \frac{3}{32}\eta_{\{v_2,v_4\}} + \frac{3}{32}\eta_{\{v_2,v_5\}} + \frac{5}{64}\eta_{\{v_4,v_5\}} + \frac{35}{64}\eta_V,$$

$$g_4 = \frac{3}{16}\eta_{\{v_2\}} + \frac{3}{32}\eta_{\{v_3\}} + \frac{5}{32}\eta_{\{v_5\}} + \frac{9}{64}\eta_{\{v_1,v_2\}} + \frac{3}{32}\eta_{\{v_2,v_5\}} + \frac{3}{64}\eta_{\{v_3,v_5\}} + \frac{9}{32}\eta_V,$$

$$g_5 = \frac{3}{16}\eta_{\{v_2\}} + \frac{3}{32}\eta_{\{v_3\}} + \frac{5}{32}\eta_{\{v_4\}} + \frac{9}{64}\eta_{\{v_1,v_2\}} + \frac{3}{32}\eta_{\{v_2,v_4\}} + \frac{3}{64}\eta_{\{v_3,v_4\}} + \frac{9}{32}\eta_V.$$

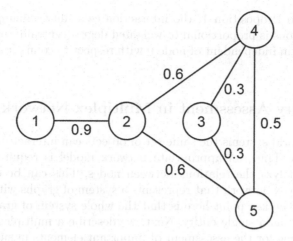

Fig. 1. Graph for the numerical example.

Hence, the matrix of pignistic probabilities with $Bet_{v_k}(\{v_i\})$ values in the k-th row and the i-th column is the following:

$$
Bet = \begin{pmatrix}
0.096 & 0.471 & 0.096 & 0.169 & 0.169 \\
0.316 & 0.034 & 0.081 & 0.284 & 0.284 \\
0.109 & 0.203 & 0.109 & 0.289 & 0.289 \\
0.127 & 0.361 & 0.173 & 0.056 & 0.283 \\
0.127 & 0.361 & 0.173 & 0.283 & 0.056
\end{pmatrix}
$$

Finally, the vector of interactions $q = (q_1, ..., q_5)$ is equal to (0.774, 1.43, 0.633, 1.081, 1.081). As the result, the final ranking of centrality vector arranges nodes in the following order: $v_2 \succ v_4 = v_5 \succ v_1 \succ v_3$. The same ranking can be obtained with eigenvector centrality measure, which confirms the consistency of the proposed approach.

It can be also proved that if we consider a 1-neighborhood of a node that, in its turn, induces an acyclic subgraph (i.e. a star subgraph) then the centrality value of a node according to formulas (3)–(6) is associated with a degree centrality measure. More precisely, it can be formulated as follows.

Proposition 1. *If centrality value q_v for node $v \in V$ in graph $G = (V, E, W)$, $|V| = n$, $n \geq 2$ is calculated with regard to 1-neighborhood $N_v^{(1)}$ of node v and a subgraph constructed on nodes $N_v^{(1)} \cup \{v\}$ is a star graph with center v then the interaction value of node v is equal to*

$$
q_v = \frac{n(v)}{|W|} + 1 - \frac{2}{n},
$$

where $n(v) = \sum_{u:e(v,u)\in E} w(v,u)$ is a weighted degree of node v, $n = |V|$.

According to Proposition 1, the interaction centrality value q_v constructed via belief functions is proportional to weighted degree centrality of node v up to the constant term independent of node v with respect to configuration described above.

4   Centrality Assessment in Multiplex Network

In many real-world systems the same set of objects can interact with each other in several ways. Thus, an appropriate network model is required in order to qualitatively analyze the relations between nodes. This can be done with the help of a multiplex network that represents a system of graphs with many layers of interaction. The key point here is that the whole system of graphs should be analyzed jointly as a single entity. Next, we describe a multiplex graph model and the technique for the assessment of important elements in such systems.

We consider a multiplex graph as the family of networks $G = (V, E^{(s)}, W^{(s)})$ with the common set of nodes $V = \{v_1, ..., v_n\}$ but distinct set of edges $E^{(s)} = \{e^{(s)}(v_i, v_j)\}$, where $s = \{1, ..., l\}$ indicates a particular network-layer in a multiplex graph with l levels of interaction. As it is stated above, we associate node v_k with number k, $k = 1, ..., n$ and denote $e^{(s)}(v_i, v_j)$ as $e_{ij}^{(s)}$. The other notations remain the same adjusted to layer s.

In order to obtain central elements in the whole system we can estimate the important elements at each layer separately, for example, with the help of the approach described above. As the result, we derive a family of vectors $q^{(s)}$ that can be aggregated into a single ranking. However, we consider all layers independently of each other, which means that multiplex system loses its significance. Instead, we consider another approach.

Firstly, we calculate the set of mass functions $\left\{m_k^{(s)}\right\}_{k=1}^{n}$ for each layer s, $s = 1, ..., l$. Further, this set can be put together into an aggregated mass function with the help of combination rule R as $m_k = m_k^{(1)} \otimes_R ... \otimes_R m_k^{(l)}$. Note that discounted coefficients α can be applied at this stage as well. Finally, we can derive centralities for all elements in the considered system as it is described in formulas (5) and (6).

In order to demonstrate the whole idea let us investigate the following example. Assume that the system of nodes from the graph on Fig. 1 also interacts on another layer as it is shown on Fig. 2.

The sum of weights for the second graphs is $|W| = 4.2$. Hence, the belief functions for each node of the graph on Fig. 2 with respect to 2-neighborhood of nodes are following:

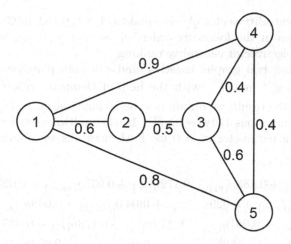

Fig. 2. The second graph for the numerical example.

$$g_1^{(2)} = \frac{1}{7}\eta_{\{v_2\}} + \frac{3}{14}\eta_{\{v_4\}} + \frac{4}{21}\eta_{\{v_5\}} + \frac{5}{84}\eta_{\{v_2,v_3\}} + \frac{1}{21}\eta_{\{v_3,v_4\}}$$
$$+ \frac{1}{14}\eta_{\{v_3,v_5\}} + \frac{1}{21}\eta_{\{v_4,v_5\}} + \frac{19}{84}\eta_V,$$

$$g_2^{(2)} = \frac{1}{7}\eta_{\{v_1\}} + \frac{5}{42}\eta_{\{v_3\}} + \frac{3}{28}\eta_{\{v_1,v_4\}} + \frac{2}{21}\eta_{\{v_1,v_5\}} + \frac{1}{21}\eta_{\{v_3,v_4\}}$$
$$+ \frac{1}{14}\eta_{\{v_3,v_5\}} + \frac{2}{63}\eta_{\{v_4,v_5\}} + \frac{97}{252}\eta_V,$$

$$g_3^{(2)} = \frac{5}{42}\eta_{\{v_2\}} + \frac{2}{21}\eta_{\{v_4\}} + \frac{1}{7}\eta_{\{v_5\}} + \frac{1}{14}\eta_{\{v_1,v_2\}} + \frac{3}{28}\eta_{\{v_1,v_4\}}$$
$$+ \frac{2}{21}\eta_{\{v_1,v_5\}} + \frac{1}{21}\eta_{\{v_4,v_5\}} + \frac{9}{28}\eta_V,$$

$$g_4^{(2)} = \frac{3}{14}\eta_{\{v_1\}} + \frac{2}{21}\eta_{\{v_3\}} + \frac{2}{21}\eta_{\{v_5\}} + \frac{1}{14}\eta_{\{v_1,v_2\}} + \frac{2}{21}\eta_{\{v_1,v_5\}}$$
$$+ \frac{5}{84}\eta_{\{v_2,v_3\}} + \frac{1}{14}\eta_{\{v_3,v_5\}} + \frac{25}{84}\eta_V,$$

$$g_5^{(2)} = \frac{4}{21}\eta_{\{v_1\}} + \frac{1}{7}\eta_{\{v_3\}} + \frac{2}{21}\eta_{\{v_4\}} + \frac{1}{14}\eta_{\{v_1,v_2\}} + \frac{3}{28}\eta_{\{v_1,v_4\}}$$
$$+ \frac{5}{84}\eta_{\{v_2,v_3\}} + \frac{1}{21}\eta_{\{v_3,v_4\}} + \frac{2}{7}\eta_V.$$

The matrix of pignistic probabilities $Bet_k^{(2)}(\{v_i\})$ for Numerical example 2 is equal to

$$Bet^{(2)} = \begin{pmatrix} 0.045 & 0.218 & 0.135 & 0.307 & 0.295 \\ 0.321 & 0.077 & 0.256 & 0.170 & 0.176 \\ 0.201 & 0.219 & 0.064 & 0.237 & 0.279 \\ 0.357 & 0.125 & 0.220 & 0.060 & 0.238 \\ 0.337 & 0.123 & 0.253 & 0.230 & 0.057 \end{pmatrix}$$

As the result, centrality vector $q^{(2)}$ is equal to $(1.261, 0.762, 0.928, 1.006, 1.045)$, that ranks nodes in the following order: $v_1 \succ v_5 \succ v_4 \succ v_3 \succ v_2$, that also coincides with eigenvector centrality ranking.

We now bring two graphs together and calculate pairwise aggregation of belief functions $g_k^{(1)}$ and $g_k^{(2)}$ with the help of Dempster rule D according to formula (1). As the result, we obtain new belief functions $g_k = D(g_k^{(1)}, g_k^{(2)})$ that indicate the interactions between nodes in the multiplex structure. Assuming that discounting parameter $\alpha = 0$ for both networks we obtain the following belief functions:

$$g_1 = 0.291\eta_{\{v_2\}} + 0.186\eta_{\{v_4\}} + 0.172\eta_{\{v_5\}} + 0.037\eta_{\{v_2,v_3\}} + 0.027\eta_{\{v_2,v_4\}}$$
$$+ 0.027\eta_{\{v_2,v_5\}} + 0.029\eta_{\{v_3,v_4\}} + 0.044\eta_{\{v_3,v_5\}} + 0.048\eta_{\{v_4,v_5\}} + 0.139\eta_V,$$
$$g_2 = 0.0318\eta_{\{v_1\}} + 0.051\eta_{\{v_3\}} + 0.174\eta_{\{v_4\}} + 0.178\eta_{\{v_5\}} + 0.025\eta_{\{v_1,v_4\}}$$
$$+ 0.023\eta_{\{v_1,v_5\}} + 0.039\eta_{\{v_3,v_4\}} + 0.047\eta_{\{v_3,v_5\}} + 0.053\eta_{\{v_4,v_5\}} + 0.092\eta_V,$$
$$g_3 = 0.116\eta_{\{v_2\}} + 0.167\eta_{\{v_4\}} + 0.208\eta_{\{v_5\}} + 0.045\eta_{\{v_1,v_2\}} + 0.068\eta_{\{v_1,v_4\}}$$
$$+ 0.060\eta_{\{v_1,v_5\}} + 0.035\eta_{\{v_2,v_4\}} + 0.035\eta_{\{v_2,v_5\}} + 0.063\eta_{\{v_4,v_5\}} + 0.203\eta_V,$$
$$g_4 = 0.148\eta_{\{v_1\}} + 0.144\eta_{\{v_2\}} + 0.119\eta_{\{v_3\}} + 0.211\eta_{\{v_5\}} + 0.103\eta_{\{v_1,v_2\}}$$
$$+ 0.038\eta_{\{v_1,v_5\}} + 0.024\eta_{\{v_2,v_3\}} + 0.040\eta_{\{v_2,v_5\}} + 0.053\eta_{\{v_3,v_5\}} + 0.120\eta_V,$$
$$g_5 = 0.138\eta_{\{v_1\}} + 0.143\eta_{\{v_2\}} + 0.145\eta_{\{v_3\}} + 0.208\eta_{\{v_4\}} + 0.102\eta_{\{v_1,v_2\}}$$
$$+ 0.044\eta_{\{v_1,v_4\}} + 0.024\eta_{\{v_2,v_3\}} + 0.039\eta_{\{v_2,v_4\}} + 0.042\eta_{\{v_3,v_4\}} + 0.116\eta_V.$$

Hence, the matrix of pignistic probabilities of aggregated belief functions is

$$Bet = \begin{pmatrix} 0.028 & 0.364 & 0.083 & 0.266 & 0.259 \\ 0.361 & 0.018 & 0.113 & 0.251 & 0.257 \\ 0.127 & 0.214 & 0.041 & 0.291 & 0.328 \\ 0.243 & 0.251 & 0.181 & 0.024 & 0.301 \\ 0.234 & 0.249 & 0.201 & 0.293 & 0.023 \end{pmatrix}$$

Finally, the vector of centrality values q for the multiplex network is equal to $(0.992, 1.097, 0.618, 1.125, 1.168)$, that ranks nodes as $v_5 \succ v_4 \succ v_2 \succ v_1 \succ v_3$. If we average over eigenvector centrality values then we obtain the following ordering: $v_1 \succ v_5 \succ v_4 \succ v_2 \succ v_3$, that almost agrees with the ordering of the second graph.

As we can see, the results obtained by two approaches differ significantly. This can be explained by the choice of aggregation rule (1) that takes into account the possible disagreement in initial data. It means that if the connections diverse for some node on different levels of interaction it leads to the decrease of the importance of this node in aggregated ranking. This fact can be seen by the example of node 1 that has only one strong connection in the first graph and three connections in the second graph. Despite the fact that this node is ranked as the first for the second network it has low position in aggregated ranking.

Another important point is that stability through all layers is encouraged by higher ranks. For instance, node 5 is ranked as the first one for both layers separately. However, in aggregated ranking it takes the first place as this node demonstrates more stable connections through both layers as well as node 4.

Additionally, if we consider a 1-neighborhood of nodes in multiplex acyclic graphs with two layers then the following propositions concerning the aggregated interaction centrality value can be proved.

Proposition 2. *If centrality value q_v for node $v \in V$ in acyclic graph $G = (V, E^{(s)}, W^{(s)})$, $s = 1, 2$, $|V| = n$, $n \geq 2$ is calculated with regard to 1-neighborhood $N_v^{(1)}$ of node v and the Dempster combination rule (1) is used then the aggregated interaction value of node v is equal to*

$$q_v = \frac{1}{|W^{(1)}||W^{(2)}|} \sum_{u \in V} \frac{w^{(1)}(u, v)w^{(2)}(u, v)}{1 - K_u}$$

$$+ \frac{1}{|W^{(1)}|} \sum_{u \in V} \frac{w^{(1)}(u, v)}{1 - K_u} \left(1 - \frac{n^{(2)}(u)}{|W^{(2)}|}\right)$$

$$+ \frac{1}{|W^{(2)}|} \sum_{u \in V} \frac{w^{(2)}(u, v)}{1 - K_u} \left(1 - \frac{n^{(1)}(u)}{|W^{(1)}|}\right) + B(G),$$

where $n^{(s)}(v) = \sum_{u:e(v,u) \in E^{(s)}} w^{(s)}(v, u)$ is a weighted degree of node v in graph G on layer s, $s = 1, 2$; $K_u = \frac{1}{|W^{(1)}||W^{(2)}|} \sum_{\substack{x,y \in V: x \neq y, \\ e(u,x) \in E^{(1)}, \\ e(u,y) \in E^{(2)}}} w^{(1)}(u, x)w^{(2)}(u, y)$

is the level of conflict of node $u \in V$; $B(G) = \frac{1}{n} \sum_{u \in V} \frac{1}{1-K_u} \left(1 - \frac{n^{(1)}(u)}{|W^{(1)}|}\right)$ $\left(1 - \frac{n^{(2)}(u)}{|W^{(2)}|}\right)$ is a constant that is independent of node v.

If we apply Yager combination rule (2) then this expression is simplified.

Proposition 3. *If centrality value q_v for node $v \in V$ in acyclic graph $G = (V, E^{(s)}, W^{(s)})$, $s = 1, 2$, $|V| = n$, $n \geq 2$ is calculated with regard to 1-neighborhood $N_v^{(1)}$ of node v and the Yager combination rule (2) is used then the aggregated interaction value of node v is equal to*

$$q_v = \frac{n^{(1)}}{|W^{(1)}|} + \frac{n^{(2)}}{|W^{(2)}|}$$

$$+ \frac{1}{|W^{(1)}||W^{(2)}|} \sum_{u \in V} \left(w^{(1)}(u, v)w^{(2)}(u, v) - w^{(1)}(u, v)n^{(2)}(u) - w^{(2)}(u, v)n^{(1)}(u)\right)$$

$$+ C(G),$$

where $C(G) = \frac{1}{n} \sum_{u \in V} \left(K_u + \left(1 - \frac{n^{(1)}(u)}{|W^{(1)}|}\right) \left(1 - \frac{n^{(2)}(u)}{|W^{(2)}|}\right)\right)$ is a constant that is independent of node v.

It can be seen that in the last case the aggregated interaction centrality value of node v is represented as a sum of normalized weighted degree centralities on each layer and the value that represents the interactions of v's neighbors through different layers.

5 Conclusion

In this work we propose a new approach for the nodes importance assessment in multiplex networks. We apply Dempster-Shafer theory in order to reveal key elements in undirected weighted graphs as well as to aggregate interactions between nodes into the total ranking.

We assess the cooperation of nodes with different subgroups of vertices by evaluating the corresponding belief functions. The proposed model of belief function takes into account the cooperation with nodes and the groups of nodes that are located in a given neighborhood of a considered vertex. Further, the obtained intensities of cooperation with different groups of nodes are redistributed among the participants of these groups that results in node-to-node intensity of cooperation. The obtained values show real interactions between nodes with respect to distant connections, that is already informative and cannot be derived with most of the classical centrality measures. Finally, we aggregate the pairwise interaction into the final centrality vector that gives the ranking of nodes and helps to reveal key elements in a network.

If nodes cooperate with each other on different levels of interactions then we apply a combination rule to mass functions obtained for different layers of a multiplex structure. In particular, Dempster rule takes into account the disagreement in data on different levels of interaction, i.e. it rewards nodes that have consistent connections through all layers and reduces the importance of nodes with unstable links. Additionally, other combination rules can be used as well in order to estimate the aggregated centralities in multilayer networks. These rules may consider the reliability, the inconsistency, the uncertainty, etc. of a network structure in various ways.

It is important to note that the proposed approach can be easily adapted to directed networks. We also emphasize that parameters of the introduced functions such as discounting coefficients, the radius of nodes neighborhood, etc. can be tuned in accordance with the problem statement.

It is also shown that the proposed methods give comparable results for one-layer networks and take into account specifications of a multilayer structure. In further research we aim to improve the proposed approach and apply the developed methods to real networks.

Acknowledgements. The article was prepared within the framework of the Basic Research Program at the National Research University Higher School of Economics (HSE University) and supported within the framework of a subsidy by the Russian Academic Excellence Project '5–100'. This work is also supported by the Russian Foundation for Basic Research under grant No 18-01-00877a.

References

1. Dempster, A.P.: Upper and lower probabilities induced by a multivalued mapping. Ann. Math. Stat. **38**(2), 325–339 (1967)
2. Shafer, G.A.: Mathematical Theory of Evidence. Princeton University Press, Princeton (1976)
3. Ben Dhaou, S., Kharoune, M., Martin, A., Ben Yaghlane, B.: Belief approach for social networks. In: Cuzzolin, F. (ed.) BELIEF 2014. LNCS (LNAI), vol. 8764, pp. 115–123. Springer, Cham (2014). https://doi.org/10.1007/978-3-319-11191-9_13
4. Gao, C., Wei, D., Hu, Y., Mahadevan, S., Deng, Y.: A modified evidential methodology of identifying influential nodes in weighted networks. Phys. A **392**(21), 5490–5500 (2013)
5. Jendoubi, S., Martin, A.: A reliability-based approach for influence maximization using the evidence theory. In: Bellatreche, L., Chakravarthy, S. (eds.) DaWaK 2017. LNCS, vol. 10440, pp. 313–326. Springer, Cham (2017). https://doi.org/10.1007/978-3-319-64283-3_23
6. Zhou, K., Pan, Q., Martin, A.: Evidential community detection based on density peaks. In: Destercke, S., Denoeux, T., Cuzzolin, F., Martin, A. (eds.) BELIEF 2018. LNCS (LNAI), vol. 11069, pp. 269–277. Springer, Cham (2018). https://doi.org/10.1007/978-3-319-99383-6_33
7. Newman, M.E.J.: Networks: An Introduction. Oxford University Press, Oxford (2010)
8. Bonacich, P.: Power and centrality: a family of measures. Am. J. Sociol. **92**(5), 1170–1182 (1987)
9. Brin, S., Page, L.: The anatomy of a large-scale hypertextual Web search engine. Comput. Netw. **30**, 107–117 (1998)
10. Katz, L.: A new status index derived from sociometric index. Psychometrika **18**, 39–43 (1953)
11. Kleinberg, J.M.: Authoritative sources in a hyperlinked environment. J. ACM **46**(5), 604–632 (1999)
12. Bonacich, P., Lloyd, P.: Eigenvector-like measures of centrality for asymmetric relations. Soc. Netw. **23**, 191–201 (2001)
13. Bavelas, A.: Communication patterns in task-oriented groups. J. Acoust. Soc. Am. **22**(6), 725–730 (1950)
14. Freeman, L.C.: A set of measures of centrality based upon betweenness. Sociometry **40**, 35–41 (1977)
15. Myerson, R.B.: Graphs and cooperation in games. Math. Oper. Res. **2**(3), 225–229 (1977)
16. Grabisch, M., Rusinowska, A.: A model of influence in a social network. Theor. Decis. **69**(1), 69–96 (2008)
17. Aleskerov, F., Meshcheryakova, N., Shvydun, S.: Power in network structures. In: Kalyagin, V.A., Nikolaev, A.I., Pardalos, P.M., Prokopyev, O.A. (eds.) NET 2016. SPMS, vol. 197, pp. 79–85. Springer, Cham (2017). https://doi.org/10.1007/978-3-319-56829-4_7
18. Kivela, M., Arenas, A., Barthelemy, M., Gleeson, J.P., Moreno, Y., Porter, M.A.: Multilayer networks. J. Complex Netw. **2**(3), 203–271 (2014)
19. De Domenico, M., Solé-Ribalta, A., Omodei, E., Gómez, S., Arenas, A.: Centrality in interconnected multilayer networks. arXiv:1311.2906 (2013)
20. Shvydun, S.: Influence assessment in multiplex networks using social choice rules. Procedia Comput. Sci. **139**, 182–189 (2018)

21. Yager, R.: On the Dempster-Shafer Framework and New Combination Rules. Inf. Sci. **41**, 93–137 (1987)
22. Sentz, K., Ferson, S.: Combination of evidence in Dempster-Shafer theory. Report SAND 2002–0835, Sandia National Laboratories (2002)
23. Bronevich, A., Rozenberg, I.: The choice of generalized Dempster-Shafer rules for aggregating belief functions. Int. J. Approximate Reasoning **56**, 122–136 (2015)
24. Lepskiy, A.: General schemes of combining rules and the quality characteristics of combining. In: Cuzzolin, F. (ed.) BELIEF 2014. LNCS (LNAI), vol. 8764, pp. 29–38. Springer, Cham (2014). https://doi.org/10.1007/978-3-319-11191-9_4
25. Destercke, S., Burger, T.: Toward an axiomatic definition of conflict between belief functions. IEEE Trans. Syst. Man Cybern. **43**(2), 585–596 (2013)
26. Bronevich, A., Lepskiy, A., Penikas, H.: The application of conflict measure to estimating incoherence of analyst's forecasts about the cost of shares of russian companies. Procedia Comput. Sci. **55**, 1113–1122 (2015)
27. Bronevich, A., Rozenberg, I.: The contradiction between belief functions: its description, measurement, and correction based on generalized credal sets. Int. J. Approximate Reasoning **112**, 119–139 (2019)
28. Lepskiy, A.: On the conflict measures agreed with the combining rules. In: Destercke, S., Denoeux, T., Cuzzolin, F., Martin, A. (eds.) BELIEF 2018. LNCS (LNAI), vol. 11069, pp. 172–180. Springer, Cham (2018). https://doi.org/10.1007/978-3-319-99383-6_22
29. Ren, J., Wang, C., He, H., Dong, J.: Identifying influential nodes in weighted network based on evidence theory and local structure. Int. J. Innovative Comput. Inf. Control **11**(5), 1765–1777 (2015)
30. Smets, P., Kennes, R.: The transferable belief model. Artif. Intell. **66**, 191–243 (1994)
31. Shapley, L.: A value for n-person games. In: Contributions to the Theory of Games. II (28) in Annals of Mathematics Studies, pp. 307–317. Princeton University Press (1953)

Correction of Belief Function to Improve the Performances of a Fusion System

Didier Coquin[✉] ⓘ, Reda Boukezzoula ⓘ, and Rihab Ben Ameur

LISTIC, University of Savoie Mont Blanc, Annecy le vieux, 74944 Annecy, France
didier.coquin@univ-smb.fr

Abstract. Our application concerns the fusion of classifiers for the recognition of trees from their leaves, in the framework of belief functions theory. In order to improve the rate of good classification it is necessary to correct Bayesian mass functions. This correction will be done from the meta-knowledge which is estimated from the confusion matrix. The corrected mass functions considerably improve the recognition rate based on the decisions provided by the classifiers.

Keywords: Belief functions theory · Mass correction ·
Meta-knowledge · Fusion of classifiers

1 Introduction

Tree species recognition is the problem of identifying the species of a given tree. This task may be easy for a botanist who has strong knowledge about trees. However, a novice or a lover of the trees universe may have difficulties. This paper is part of the ReVeRiES project that seeks to develop a mobile application that recognizes tree species without needing an internet connection. Indeed, the application will be used in nature, forests, mountains where an internet connection is not available. In the context of tree species recognition, we treat a real problem since the photos of leaves are taken in the natural environment. The processing of this type of data is complicated because of their specificities due firstly to the nature of the objects to be recognized (inter-species similarity and intra-species variability) and secondly to the environment. Errors can be accumulated during the pre-fusion process. The merit of the fusion is to take into account all the imperfections that can taint the available data and try to model them well. The fusion is more effective if the data are well modeled. The theory of belief functions represents one of the best theoretical frameworks able to manage and represent uncertainty, inaccuracy, conflict, etc. This theory is important because of its wealth of tools to manage the various sources of imperfections as well as the specificities of the available data. In the framework of this theory, it is possible to model the data through the construction of mass functions. It is also possible to manage the computational complexity thanks to the approximations allowing to reduce the number of focal elements. Conflict

© Springer Nature Switzerland AG 2020
M.-J. Lesot et al. (Eds.): IPMU 2020, CCIS 1238, pp. 297–311, 2020.
https://doi.org/10.1007/978-3-030-50143-3_23

being one of the most present sources of imperfections, can be dealt through the selection of the best combination rule. By merging sources of information with different degrees of reliability, the least reliable source may affect the data issued from the most reliable one. One of the solutions for this problem is to try to improve the performances of the least reliable source. Thus, by merging with other sources, it will provide useful information and will in turn contribute in improving the performance of the fusion system. The performance improvement of an information source can be affected through the **correction of mass functions**. In this context, the correction can be made based on measures of the relevance or sincerity of the studied source [14]. The confusion matrices present a data source from which meta-knowledge characterizing the state of a source can be extracted [17]. In this paper, the proposed fusion system is a hierarchical fusion system set up within the framework of belief function theory. It allows to merge data from leaves and provides the user with a list of the most likely species while respecting the educational purpose of the application. The computational complexity of this fusion system is quite small allowing, in the long term, to implement the application on a Smartphone.

The paper is organized in the following way: Sect. 2 describes the fusion system. Section 3 presents the different mechanisms for correcting the mass functions. A correction method is proposed in Sect. 4 which we have applied to our fusion system. Section 5 shows the improvement brought by this correction on real data.

2   Fusion System

The sub-classification strategy consists on recognizing a leaf through its botanical properties. The application consists of taking a photo of a leaf with a smartphone and the system extracts morphological characteristics (the lobe shape $\overrightarrow{A_F}$ described by the Polygonal model, the Margin $\overrightarrow{A_C}$, the Base and the Apex $\overrightarrow{A_{BS}}$). The fusion system is composed of three classifiers based on random forests that have been previously trained on learning images. As outputs of each classifier S_i, we have a distribution of probabilities $P_j[e_i]$ in the species referential with $e_i \in \Omega = 72$ the number of species [1]. The fusion system, shown in Fig. 1, consists of 3 steps:

1. **Mass functions construction**: the passage towards the belief functions theory requires the construction of mass functions based on the distributions of probabilities.
2. **Mass functions approximation**: to reduce the computational complexity.
3. **Mass functions combination**: in the framework of the belief functions theory, several combination rules exist. Each one allows to manage one of more sources of imperfections.

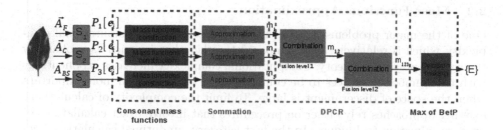

Fig. 1. Hierarchical fusion system for combining leaves data characteristics

2.1 Mass Functions Construction

Let $\Omega = \{w_1, w_2, ..., w_N\}$ be a frame of discernment composed of N exclusive and exhaustive hypothesis w_i. The set of all subsets A of Ω is called the power set of Ω and is denoted $2^\Omega = \{A | A \subseteq \Omega\} = \{\emptyset, w_1, w_2, ..., w_N, w_1 \cup w_2, ..., \Omega\}$. A mass function on Ω is a mapping m from the power set 2^Ω, to the interval $[0, 1]$ such that:

$$\sum_{A \subseteq \Omega} m(A) = 1 \tag{1}$$

Subsets A of Ω such that $m(A) > 0$ are called *focal sets* of m. A mass function is said to be: *vacuous* if Ω is its only focal set, in which case it is denoted by m_Ω; *inconsistent* if $\emptyset$ is its only focal set, in which case it is denoted by $m_\emptyset$; *dogmatic* if Ω is not a focal set; *normal* if $\emptyset$ is not a focal set.

Mass functions construction from probability distributions is an important step in the fusion process. Indeed, the mass functions must be sufficiently representative of all available information. In our case, the data obtained at the output of the classifiers are uncertain because the sources have different degrees of reliability and are sometimes conflicting. It is therefore important that the mass functions can represent all these imperfections and take into account as much information as possible [1]. In our case, we have a large discernment framework ($\Omega = 72$ assumptions). To construct 2^{72} focal elements and to give a mass to all possible combinations would lead to significant computational complexity. In the literature, there are several methods for constructing a mass function from a probability distribution, the best known of which are: bayesian mass functions and consonant mass functions [2]. We have studied and analyzed these two methods in depth. Following these studies, we found that by building bayesian mass functions, we obtain precise and certain focal elements. The obtained mass functions are representative enough of the available information and the focal elements are singletons which causes the elimination of certain species. Concerning consonant mass functions, we noticed that this method leads to the construction of large focal elements (made up of several species). It has the advantage of allowing a better presentation of all available information.

2.2   Mass Functions Approximation

One of the major problems of belief function theory is the computational complexity which is relatively high compared to other theories such as probability theory and possibility theory. Computational complexity increases exponentially with the number of sources to be combined, and becomes even more important when the space of discernment is large. To limit the complexity of calculations, existing approaches rely either on procedures that involve exact calculations or on approximation techniques. In the first category, an optimal calculation algorithm has been proposed by Kennes in [9]. This algorithm makes it possible to considerably optimize the number of addition and multiplication operations when calculating mass functions and to reduce the complexity of the calculations. The disadvantage of this method is that the number of focal elements is not reduced since it is a procedure that performs exact calculations. This can be problematic in our case because we have to manage a large area of discernment and perform the same calculation for several sources of information. The second category of approaches is composed of mass functions approximation procedures. The mass functions approximation procedures allow the reduction of the number of focal elements based on different selection criteria. In particular, we focused on the most commonly used procedures: the "Inner" and "Outer" approximations [4], the $k - l - x$ [21] and the "Summation method" [11]. The analysis, as well as the experiments carried out, have led to the choice of the summation approximation as the most suitable method for our case since it offers the best relationship between computational complexity and preservation of a maximum of useful information.

"Summation" approximation consists of keeping the $k - 1$ focal elements having the largest mass. The union of the remaining focal elements forms the kth focal element whose sum of their masses is transferred to the element k. The application of this approximation in our specific context is interesting. Unlike the $k - l - x$ approximation, even focal elements with low mass will be kept. Thus, a maximum of information will be transferred allowing more flexibility to the fusion process. In addition, this approximation is simple, and it does not require a lot of calculation operations which reduces the processing time.

2.3   Mass Functions Combination

To better manage the imperfections of the data, several combination rules have been proposed as part of the theory of belief functions. Each rule has been put in place to handle one or more aspects of imperfections. Given the limited and variable performance of the data sources to be combined, we consider that we have to deal with unreliable sources of data. According to Lefevre et al. [10] the combination rules proposed in the literature represent solutions to manage the different sources of conflict. They divide these combination rules into two categories:

1. the first category groups together the combination rules requiring that the sources of information to combine are reliable.

2. the second category groups combination rules where at least one of the sources of information is reliable without necessarily knowing which one.

In our case, the main problematic situation we face is the following:

- no intersection between 2 focal elements issued from two sources. In that case, our need is to transfer mass to the elements involved in the conflict.
- two or more sources propose the true species. In that case, our need is to transfer mass to the intersection.
- no intersection between the species proposed by the different sources (total conflict). In that case, our need is to transfer mass to the partial ignorance.

After the study and analysis of the different existing combination rules: Dempster combination rule [18], Dubois and Prade combination rule [6], Mixte combination rule, Florea and Jousselme combination rule [8], PCR6 [12] and DPCR combination rule [13], we found that the **DPCR combination rule** is the best compromise between a good recognition rate of the first species and the processing time per species (see Table 1). Indeed, it transfers the mass to the intersection, the elements involved in the conflict as well as to partial ignorance. It also offers the best compromise between classification ratio and computational complexity. DPCR is called the rule of proportional redistribution of the weakened conflict, this rule was proposed by Martin [13]. It is an extension of the $PCR5$ combination rule. It consists in applying a discounting procedure to transfer part of the partial conflict to partial ignorance (the union).

Table 1. Processing time per species and recognition rate according to the combination rule

Combination rule	Processing time per species	Recognition rate
Florea	522 ms	57.1%
PCR6	163 ms	51%
Dubois et Prade	199 ms	57.2%
Mixte	634 ms	56%
DPCR	305 ms	58.3%

2.4 Decision Making Criteria

Since the objective of the application is to present a list of the top 10 most probable species, we are interested in the decision criterion "the maximum of pignistic probability" or BetP which is proposed by Smets in [20]. These decision-making criteria makes it possible to transform the mass functions into measures of probabilities for all $w_k \in \Omega$, but in our case, we work with bayesian bba and in a such a case, we have the same results. Thus, we can to provide the user with an ordered list of the most likely species and not an ordered list of the most likely sets of species. The criterion makes it possible to lift the inaccuracy which can taint the data since we work with consonant mass functions. The dashed colored lines in the Fig. 1 represents the different choices made for the different blocks of the fusion system, which nave been previously explained. The three sources S_i are three feature vectors (the Polygonal model, the Margin, the Base and the Apex) that are used as input for the three classifiers (Random Forest).

3   Belief Function Corrections

The theory of belief functions provides a flexible framework for modeling and managing information uncertainty. Several tools are proposed in the literature, they allow to modify or correct some of the information based on additional information on the relevance or sincerity of a source.

3.1   Discounting Based on Degree of Relevance

Historically, discounting is the first correction tool proposed in the literature [19]. It is based on the degree of relevance of an information source S to weaken the mass of information it provides [16]. The discounting was developed as part of the Transferable Belief Model (TBM) [20]. Suppose that an information source S provides information represented by a mass function m_S and assume that β with β in [0, 1], is the degree of relevance relating to this source S. The discounting is done as follows:

$$m(A) = \beta m_S(A) + (1 - \beta)m_\Omega(A), \forall A \subseteq \Omega \qquad (2)$$

Discounting therefore consists of weakening the initial mass function according to its degree of relevance and giving more weight to ignorance. The reverse operation of the discounting, called, de-discounting was introduced in [5]. If after making the discounting it turns out that the information on the relevance of the source was false, the de-discounting allows to find $m_S(A)$ from $m(A)$. This correction method uses the overall degree of relevance to weaken the mass functions. However, this measure is insufficient since the degree of relevance may vary from one class to another. Given the quality of the data, it would be more useful to apply contextual correction methods where the weakening of a mass function takes into account each context.

3.2   Contextual Discounting Based on the Degree of Relevance

In [15], Mercier et al. studied contextual discounting based on the degree of contextual relevance of an information source. Its application requires the presence of additional information on the relevance of an information source conditionally in different contexts $\mathcal{A}$ of Ω such as the union of subsets $\mathcal{A}$ form a partition of Ω. Suppose that β_A, with β_A in [0, 1], is the degree of relevance of a source S in a context A and the union of context elements $\mathcal{A}$ form the $\mathcal{A}$ partition of Ω.

The belief m obtained by taking contextual elements $\mathcal{A}$ into account is defined as follows:

$$m = m_S \bigcup_{A \in \mathcal{A}} A_{\beta_A} \qquad (3)$$

where $\bigcup_{A \in \mathcal{A}}$ represents the canonical disjunctive decomposition, A_{β_A} represents a simple mass function having as focal elements $\emptyset$ and A whose corresponding masses are β_A and $1 - \beta_A$ respectively. A more general case of contextual discounting is discussed in [14]. Its application is simpler and more general since the formation of a partition by A context elements is no longer required. Thus, this correction tool can be applied even if the A context elements do not form a partition of Ω. Suppose β_A, with β_A the degree of relevance given to a source of information in a context A and that $\mathcal{A}$ is the set of contexts for which we know contextually the degree of relevance of the information source. The resulting mass function by taking into account these context elements is calculated as follows:

$$m = m_S \bigcup \{ \bigcap_{A \in \mathcal{A}} (\overline{A})^{1 - \beta_A} \} \qquad (4)$$

where $\bigcap_{A \in \mathcal{A}}$ represents the canonical conjunctive decomposition. The contextual discounting discussed in this section is based on contextual elements that provide information about the contextual relevance of the information source. This technique of mass function correction is interesting in our case since the relevance of our classifiers is variable from one species to another. With this technique, it would be possible to introduce additional information on the contextual relevance of each source of information to each species. Contextual discounting can also be applied to contextually modify a mass function based on another measure of reliability: sincerity.

3.3   Correction Techniques Based on the Degree of Sincerity

a) Discounting based on the degree of sincerity
It is possible to estimate the sincerity or not of a source of information by assuming that it is reliable. Thus, the correct decisions and incorrect decisions of the source respectively correspond to a sincere or insincere behavior of the source of information. Suppose a source of information is sincere with a β degree and not sincere with a $1 - \beta$ degree. According to [7], the correction of mass functions by taking into account meta-knowledge about the sincerity of a source can be done in 3 different ways:

- by weakening the initial mass while giving mass to ignorance as follows:

$$m = \beta m_S + (1 - \beta) m_\Omega \qquad (5)$$

- by weakening the initial mass given to A ($\forall A \subseteq \Omega$) while tuning more mass to its complement $\overline{A}$ ($\forall \overline{A} \subseteq \Omega$) as follows:

$$m = \beta m_S + (1 - \beta) \overline{m_S} \qquad (6)$$

- by weakening the initial mass by a factor of $\beta = 1 - 2\epsilon$, where ϵ is the classification error rate in the previous Equation, and guarantying mass to ignorance.

According to the degree of sincerity of the source of information, the initial mass is weakened while granting mass to ignorance as shown by the Eqs. 5. The Eq. 6 presents a slightly different way since the weakening of the initial mass is coupled to a mass transfer to $\overline{A}$ which is the complement set of the focal element A in question. This approach is very interesting because it allows a better preservation of the specificity of the information. In the next section, we are particularly interested in this approach and will propose a method to contextually define the complement set $\overline{A}$ of A.

b) The contextual discounting based on the degree of sincerity
The definition of different forms of non-veracity in [17] made it possible to define a new form of contextual discounting based on the degree of sincerity. Suppose the source S is negatively untrue in $\overline{A}$. This new form of contextual discounting is defined as follows:

$$m = m_S \bigcup_{A \in \mathcal{A}} A_{\beta_A} \tag{7}$$

c) Contextual reinforcement based on the degree of sincerity
In [14], Mercier and et al. define a new correction mechanism called contextual reinforcement. Suppose the source S is positively untrue in A. This new form of contextual reinforcement is defined as follows:

$$m = m_S \bigcap_{A \in \mathcal{A}} A_{\beta_A} \tag{8}$$

These two new forms of discounting represented in the Eqs. 7 and 8 are interesting because the discounting takes into account the contextual sincerity of the source of information. Its application requires the setting up of assumptions in relation to the sincerity of the source. So we need meta-knowledge to assert if the source is negatively untrue in A or if it is positively untrue in $\overline{A}$.

3.4 Estimation of Knowledge from a Confusion Matrix

The application of the correction mechanisms requires knowledge of the state of the information source to be processed. This knowledge, called "meta-knowledge", is used to characterize the state of a data source: relevant or not, sincere or not. This meta-knowledge is generally uncertain and often presented by mass functions [17]. In the context of classifiers fusion, we are interested in the estimation of meta-knowledge from the confusion matrices obtained for each classifier. A confusion matrix $M_i = (n_{kl})_{k \in \{1,...,K\}, l \in \{1,...,K\}}$ is a table characterizing the performance of a data source S_i on a test set of K classes (Table 2). Each line k in the confusion matrix represents the decision made in favor of w_{lk}. Each column l represents the truth w_l. Note that $n_{k.} = \sum_{l=1}^{K} n_{kl}$ represents the number of objects in w_k and $n_{.l} = \sum_{k=1}^{K} n_{kl}$ is the number of objects in w_l. The total number of objects is $n = \sum_{k=1}^{K} \sum_{l=1}^{K} n_{kl}$.

Table 2. Illustration of a confusion matrix

Verity decision	$w_1 \ldots w_l \ldots w_K$
w_1	$n_{11} \quad \cdots \quad n_{1K}$
$\vdots$	$\ddots$
w_k	$\vdots \quad n_{kl} \quad \vdots$
$\vdots$	$\ddots$
w_K	$n_{K1} \ldots n_{KK}$

3.5 Estimation of Contextual Relevance of a Source from the Confusion Matrix

In [7], Elouedi et al. define two ways to calculate the contextual relevance for each context: using the percentage of correct classifications for each context, and using a distance to determine the reliability rate of a source in the different contexts. The contextual weakening of a mass function based on its degree of contextual relevance is interesting. In our case, the application of this approach amounts to disjunctively combining the initial mass function with the 72 contextual elements that can be highlighted by the confusion matrix. This is very complicated from a computational point of view since it involves making 72 combinations for each mass function and for each source of information.

So, in the following we will use the confusion matrix to estimate the sincerity of the source.

3.6 Estimation of a Source Sincerity from the Confusion Matrix

It is possible to estimate the sincerity of a source always using a confusion matrix. In [17], Pichon et al. suggest the three approaches that can be used: generation of a Bayesian mass function characterizing the sincerity of a source, generation of a mass function from the pignistic probability, or generation of a mass function to characterize the sincerity of a source based on the work of Dempster [3].

As we mentioned, we are interested in the weakening of the mass function as proposed in Eq. 10. Indeed, in some applications, it is easy to define the set $\overline{A}$ complement of A. This is even simpler when it comes to a small discernment space or where relevant knowledge is available (such as expert opinion). In our case, we can assimilate the set $\overline{A}$ complement of A to the set of all the classes belonging to the discernment space Ω but not belonging to A. Assume that A is formed of 10 species, a mass will be transferred to the complement set $\overline{A}$ composed of 62 species. The specificity of the information can be lost. For all these reasons, and considering the quality of the processed data as well as the available information, we were interested in defining the $\overline{A}$ complement set for each context.

4 The Proposed Correction Method

The proposed correction method consists in reinforcing, contextually, the mass attributed to some classes. This correction requires retrieving information from the test database's confusion matrix to define the complement set $\overline{A}_i$ of each context c_i. The proposed approach consists of the following 3 steps:

- **construction of the Bayesian mass functions** from the probability distributions obtained at the output of a classifier
- **extraction of contextual meta-knowledges** from the confusion matrix. These meta-knowledges allow to define a complement set $\overline{A}_i$ for each context c_i
- **contextual use of meta-knowledge** to correct the Bayesian mass functions

4.1 Building Bayesian Mass Functions

At the output of a classifier, we obtain a probability distribution $P_i(e_k)$ between the species. To move from probability theory to belief function theory, we construct Bayesian mass functions and transfer all available information to it as follows:

$$m_i\{e_k\} = P_i(e_k), \qquad \qquad \forall e_k \in \Omega \qquad (9)$$
$$m_i\{X\} = 0, \qquad \qquad if \ |X| \neq 1 \qquad (10)$$

where X is any set containing more than one species. Mass m_i equal to the probability P_i is given to each element e_k of Ω having a non-zero probability. A zero mass is given to all subsets.

4.2 Meta-knowledge Extraction

The purpose of this step is to use the confusion matrix to define, for each context c_i, the complement set $\overline{A}_i$. In our case, there are 72 contexts corresponding to the 72 species forming the discernment space Ω. In a confusion matrix, each line k represents the decision made in favor of w_k. Each column l represents the truth w_l. By traversing the matrix of confusion line by line without considering the diagonal, it is possible to extract, for a decision w_k, all the species in which the truth is found. This set is the complement set $\overline{A}_k$ for a context c_k corresponding to w_k decision.

Example 2: Let's consider the confusion matrix shown in the Table 3.

Let's analyze the confusion matrix line by line without taking into account the diagonal. The first line corresponds to the decision e_1. We notice through the third column that the classifier decides e_1 whereas it is e_3. We thus consider, that for the context c_1 corresponding to the decision e_1, the complement set is $\overline{A}_1 = \{e_3\}$. So, the idea is to assign a mass to e_3 each time the classifier decides e_1.

Table 3. Example of confusion matrix

Truth decision	e_1	e_2	e_3
e_1	3	0	2
e_2	1	4	2
e_3	1	0	3

The second line corresponds to the decision e_2. We notice through the first column that once on 5 the classifier decides e_2 when it is e_1 and through the third column that the classifier decides e_2 when it is e_3. So a mass will be given to e_1 and e_3 each time the classifier decides e_2.

The third line corresponds to the decision e_3. According to the first column, once out of 5 the classifier decides e_3 whereas it is e_1.

By going through the confusion matrix line by line, and following the classifier's decision, the three complementary sets corresponding respectively to a decision in favor of e_1, e_2 and e_3 are obtained:

$$L(decision = e_1) = L_1 = \overline{A_1} = \{e_3\}$$
$$L(decision = e_2) = L_2 = \overline{A_2} = \{e_1, e_3\} \qquad (11)$$
$$L(decision = e_3) = L_3 = \overline{A_3} = \{e_1\}$$

4.3 Using Meta-knowledge for Bayesian Mass Function Correction

Following the extraction of the complementary sets of each context, the correction of the mass functions is carried out by applying the approach presented in the Eq. 10. Applying this equation, must not change the Bayesian form of the mass function so that we can then build consonant mass functions. The information available will therefore be represented in the best possible way. The initial mass function will be weakened by β and a mass equal to $(1 - \beta)/|\overline{A_i}|$ is given to each element of the complement set. Depending on how the degree of sincerity β is calculated, the correction can be done in two ways:

1. β = overall degree of sincerity of the information source.
2. β = degree of contextual sincerity. In this case, the value of β is extracted from the diagonal of the confusion matrix for the 72 assumptions.

To weaken a mass function, the idea is to contextually vary the value of β. Thus, if the classifier decides that it is the species e_2, the contextual value of β is that corresponding to the degree of sincerity of the classifier relative to the species e_2. It is important to note that in our case, we consider that the decision of a classifier corresponds to the species with the maximum mass. Note that for most species, the context value of β is zero. Thus, in these cases, the initial mass function is multiplied by zero and a null mass is given to the complement set. In other words, the mass function becomes zero.

5   Experiments and Results: Application of the Mass Function Correction Method

5.1   Dataset Description

In the data set, there is 5104 leaves photos unequally distributed on 72 species. In fact, we have some species that are well presented in the dataset. However, we have some other species that are less presented and some species that have only two exemplary. We used 2572 leaves photos for training and the rest for testing.

5.2   What About Degree of Sincerity: Global or Contextual?

When the degree of sincerity is global, the value of β is the same for all contexts, while when the degree of sincerity is contextual, the value of β will depend on the context. We notice that in many contexts the value of β is null: 21 contexts for the Polygonal model classifier S_1, 22 contexts for the Margin classifier S_2 and 34 contexts for the Base and Apex classifier S_3. Even though contextual values of β usually allow contextual truthfulness to be taken into account in each case, in our case this approach is useless since, in most contexts, β have a null value.

5.3   Impact of the Proposed Method on the Classifier Performances

In order to evaluate the performance of the classifiers before and after the correction of the mass functions, the idea is to use the 10-fold cross-validation procedure. Meta-knowledge is extracted from 9/10th samples, and the effectiveness of the correction method is tested on the remaining 10. This operation is repeated 10 times, and each time the configuration and test samples are randomly drawn. The end result is the average value of the 10 results corresponding to the 10 repetitions. Figure 2 shows the results obtained before and after the correction of the mass functions of the test database. Since botanists like to see similar species, we have presented the results of the good classification rate according to the number N of species selected. We note a significant improvement in classifier performance for both modalities. Nevertheless, through Table 4 representing the

Table 4. Standard deviation of the different classifiers without and with the correction of mass functions, N is the number of returned species

Modality	Characteristics	Standard deviation					
		Without correction			With correction		
		N = 1	N = 5	N = 10	N = 1	N = 5	N = 10
Leaf	S_1: Polygonal model	4.4604	4.7424	4.3574	4.4326	6.0894	2.9342
	S_2: Margin	3.7012	4.3951	5.0837	5.3637	6.5911	4.4604
	S_3: Base and Apex	5.4614	3.9912	3.2008	3.8133	3.8328	3.2924

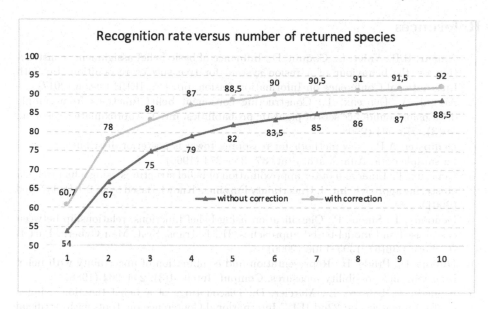

Fig. 2. Recognition rate without and with mass correction.

standard deviation for each classifier, we notice that the standard deviation is often important. Indeed, by randomly pulling the configuration data, the sets formed can be quite representative of the variations existing in the data as they can be weakly representative of these variations. The quality of the extracted meta-knowledge, therefore, varies according to the relevance of the configuration database, which subsequently influences the ability of the fusion system to generalize.

6 Conclusion

In this paper, we presented the different mechanisms for correcting mass functions and applied the most appropriate ones for our application. We have shown that the use of the confusion matrix allows us to define meta-knowledge useful for the correction of mass functions. We applied this technique to our merging system and showed that it significantly improved the rate of good classification. Work is in progress on the addition of the bark modality from which we can extract characteristics such as texture, color, ... which would allow us to differentiate between neighboring species. The bark-derived characteristics could serve as a meta-knowledge that we will use to resolve ambiguities when there are significant conflicts on species associated with leaf recognition. This would improve the performance of this system.

Acknowledgements. This work is part of ReVeRIES project supported by the French National Agency for Research with the reference ANR-15-CE38-0004.

References

1. Ameur, R.B., Valet, L., Coquin, D.: Influence of basic belief assignments construction on the behaviour of a fusion system for tree species recognition. In: 20th International Conference on Information Fusion, pp. 1–8. IEEE Fusion (2017)
2. Aregui, A., Denœux, T.: Constructing consonant belief functions from sample data using confidence sets of pignistic probabilities. Int. J. Approximate Reasoning 49(3), 575–594 (2008)
3. Dempster, A.P.: New methods for reasoning towards posterior distributions based on sample data. Ann. Math. Stat. 37, 355–374 (1966)
4. Denœux, T.: Inner and outer approximation of belief structures using a hierarchical clustering approach. Int. J. Uncertainty Fuzziness Knowl. Based Syst. 9(04), 437–460 (2001)
5. Denoeux, T., Smets, P.: Classification using belief functions: relationship between case-based and model-based approaches. IEEE Trans. Syst. Man Cybern. Part B (Cybern.) 36(6), 1395–1406 (2006)
6. Dubois, D., Prade, H.: Representation and combination of uncertainty with belief functions and possibility measures. Comput. Intell. 4(3), 244–264 (1988)
7. Elouedi, Z., Lefevre, E., Mercier, D.: Discountings of a belief function using a confusion matrix. In: 22nd IEEE International Conference on Tools with Artificial Intelligence (ICTAI), vol. 1, pp. 287–294. IEEE (2010)
8. Florea, M.C., Jousselme, A.L., Bossé, É., Grenier, D.: Robust combination rules for evidence theory. Inf. Fusion 10(2), 183–197 (2009)
9. Kennes, R.: Computational aspects of the mobius transformation of graphs. IEEE Trans. Syst. Man Cybern. 22(2), 201–223 (1992)
10. Lefevre, E., Colot, O., Vannoorenberghe, P.: Belief function combination and conflict management. Inf. fusion 3(2), 149–162 (2002)
11. Lowrance, J.D., Garvey, T.D., Strat, T.M.: A framework for evidential-reasoning systems. In: Yager, R.R., Liu, L. (eds.) Classic works of the Dempster-Shafer theory of belief functions. Studies in Fuzziness and Soft Computing, pp. 419–434. Springer, Heidelberg (2008). https://doi.org/10.1007/978-3-540-44792-4_16
12. Martin, A., Osswald, C.: A new generalization of the proportional conflict redistribution rule stable in terms of decision. In: Advances and Applications of DSmT for Information Fusion: Collected Works, vol. 2, pp. 69–88 (2006)
13. Martin, A., Osswald, C.: Toward a combination rule to deal with partial conflict and specificity in belief functions theory. In: 10th International Conference on Information Fusion, pp. 1–8. IEEE (2007)
14. Mercier, D., Lefèvre, É., Delmotte, F.: Belief functions contextual discounting and canonical decompositions. Int. J. Approximate Reasoning 53(2), 146–158 (2012)
15. Mercier, D., Quost, B., Denœux, T.: Refined modeling of sensor reliability in the belief function framework using contextual discounting. Inf. Fusion 9(2), 246–258 (2008)
16. Pichon, F., Dubois, D., Denoeux, T.: Relevance and truthfulness in information correction and fusion. Int. J. Approximate Reasoning 53(2), 159–175 (2012)
17. Pichon, F., Mercier, D., Lefevre, É., Delmotte, F.: Proposition and learning of some belief function contextual correction mechanisms. Int. J. Approximate Reasoning 72, 4–42 (2016)
18. Shafer, G.: A Mathematical Theory of Evidence, vol. 1. Princeton University Press, Princeton (1976)

19. Smets, P.: Belief functions: the disjunctive rule of combination and the generalized Bayesian theorem. Int. J. Approximate Reasoning **9**(1), 1–35 (1993)
20. Smets, P., Kennes, R.: The transferable belief model. Artif. Intell. **66**(2), 191–234 (1994)
21. Tessem, B.: Approximations for efficient computation in the theory of evidence. Artif. Intell. **61**(2), 315–329 (1993)

Evaluation of Probabilistic Transformations for Evidential Data Association

Mohammed Boumediene[1,2(⊠)] and Jean Dezert[3]

[1] Signals and Images Laboratory, USTO, Mohamed Boudiaf, Oran, Algeria
[2] National Institute of Telecommunications and ICT, Abdelhafid Boussouf, Oran, Algeria
mboumediene@inttic.dz
[3] The French Aerospace Lab, ONERA - DTIS, 92320 Palaiseau, France
jean.dezert@onera.fr

Abstract. Data association is one of the main tasks to achieve in perception applications. Its aim is to match the sensor detections to the known objects. To treat such issue, recent research focus on the evidential approach using belief functions, which are interpreted as an extension of the probabilistic model for reasoning about uncertainty. The data fusion process begins by quantifying sensor data by belief masses. Thereafter, these masses are combined in order to provide more accurate information. Finally, a probabilistic approximation of these combined masses is done to make-decision on associations. Several probabilistic transformations have been proposed in the literature. However, to the best of our knowledge, these transformations have been evaluated only on simulated examples. For this reason, the objective of this paper is to benchmark most of interesting probabilistic transformations on real-data in order to evaluate their performances for the autonomous vehicle perception problematic.

Keywords: Data association · Evidential theory · Belief functions · Probabilistic transformation

1 Introduction

Multiple Target Tracking (MTT) is important in perception applications (autonomous vehicle, surveillance, etc.). The MTT system is usually based on two main steps: data association and tracking. The first step associates detected objects in the perceived scene, called *targets*, to known objects characterized by their predicted *tracks*. The second step estimates the *track* states over time typically thanks to Kalman Filters [1], or improved state estimation techniques (like particle filters, etc). Nevertheless, bad associations provide wrong *track* estimation and then leads to false perception results.

© Springer Nature Switzerland AG 2020
M.-J. Lesot et al. (Eds.): IPMU 2020, CCIS 1238, pp. 312–326, 2020.
https://doi.org/10.1007/978-3-030-50143-3_24

The data association problem is usually resolved by Bayesian theory [1,2]. Several methods have been proposed as the Global Nearest Neighbor (GNN) method, the Probabilistic Data Association Filter (PDAF), and the Multiple Hypothesis Tracking (MHT) [3,12,22]. However, the Bayesian theory doesn't manage efficiently data imperfection due to the lack of knowledge we can have on sensor quality, reliability, etc. To circumvent this drawback, the Evidential theory [9,25] appears as an interesting approach because of its ability to model and deal with epistemic uncertainty. Its provides a theoretical framework to manage ignorance and data imperfection.

Several evidential data association approaches have been proposed [6,10,20, 23] in the framework of belief functions. Rombaut [23] uses the Evidential theory to measure the confidence of the association between perceived and known obstacles. To manage efficiently objects appearance and disappearance, Gruyer and Cherfaoui [15] propose the bi-directional data association. The first direction concerns the *target-to-track* pairings which provides a good way to manage the appearance of the new *tracks*. The second direction concerns the *track-to-target* pairings and then manage disappearance of *tracks*. This approach has been extended by Mercier *et al.* [20] to track vehicles by using a global optimization to make assignment decisions. To reduce the complexity for real-time applications, a local optimization has been used [5,6]. For all these methods, the data fusion process begins by defining belief masses from sensor information and prior knowledge. These masses represent the belief and ignorance on the assignment hypotheses. Thereafter, the masses are combined in order to provide a complete information of the considered problem. Finally, to make a decision, the belief masses are classically approximated by a probability measure thanks to a chosen probabilistic transformation.

For data association applications, the widely used probabilistic transformation (i.e. approximation) is the pignistic transformation [5,6,17,20]. This transformation is based on a simple mapping process from belief to probability domain. However, several published works criticize the pignistic transformation and propose generalized and/or alternative transformations [7,8,11,19,21,30]. To our knowledge, the proposed transformations have been evaluated by their authors only on simulated examples. The main objective of this paper is to compare these transformations on real-data in order to determine which one is well-suited for assignment problems.

The rest of the paper is structured as follows. Section 2 recalls the basics of belief functions and their uses in data association problems. In Sect. 3, the most appealing probabilistic transformations are presented and compared on the well-known KITTI public database in Sect. 4. Finally, Sect. 5 concludes the paper.

2   Belief Functions for Data Association

To select "best" associations, the data fusion process consists in four steps: modeling, estimation, combination and decision-making. This section presents their definitions and principles.

2.1   Basic Fundamentals

The Belief Functions (BF) have been introduced by Shafer [25] based on Dempster's researches [9]. They offer a theoretical framework for reasoning about uncertainty. Let's consider a problem where we have an exhaustive list of hypotheses (H_j) which are mutually exclusive. They define a so-called *frame of discernment* Θ:

$$\Theta = \bigcup_{j=1}^{k} \{H_j\} \text{ with } H_i \cap H_j = \emptyset \qquad (1)$$

The power set 2^{Θ} is the set of all subsets of Θ, that is:

$$2^{\Theta} = \{\emptyset, H_1, ..., H_k, ..., \{H_1, H_2, H_3\}, ..., \Theta\} \qquad (2)$$

The proposition $A = \{H_1, H_2, H_3\}$ represents the disjunction meaning that either H_1 or H_2 or H_3 can be the solution to the problem under concern. In other words, A represents a partial ignorance if A is the disjunction of several elements of Θ. The union of all hypotheses Θ represents the total ignorance and $\emptyset$ is the empty set that represents the impossible solution (interpreted usually as the conflicting information).

The truthfulness of each proposition $A \in 2^{\Theta}$ issued from source j is modeled by a *basic belief assignment* (bba) $m_j^{\Theta}(A)$:

$$m_j^{\Theta} : 2^{\Theta} \to [0, 1], \sum_{A \in 2^{\Theta}} m_j^{\Theta}(A) = 1 \qquad (3)$$

Thereafter, the different *bbas* (m_j^{Θ}) are combined which provides a global knowledge of the considered problem. Several rules of combination have been proposed [29], the conjunctive operator is widely used in many rules proposed in the literature for the combination of sources of evidence. For instance, Shafer [25] did propose Dempster's rule of combination below which is nothing but the normalized version of the conjunctive rule [26]:

$$\begin{cases} m_{DS}^{\Theta}(A) = \frac{1}{1-K} \sum_{A_1 \cap ... \cap A_p = A} \prod_{j=1}^{p} m_j^{\Theta}(A_j) \\ m_{DS}^{\Theta}(\emptyset) = 0, \end{cases} \qquad (4)$$

where K is a normalized coefficient:

$$K = \sum_{A_1 \cap ... \cap A_p = \emptyset} \prod_{j=1}^{p} m_j^{\Theta}(A_j). \qquad (5)$$

Finally, in order to make decisions in Θ, a probabilistic approximation of the combined *bbas* ($m_{DS}^{\Theta}(A)$) is usually done. The upper and the lower bounds of the unknown probability $P(A)$ are defined by the belief $Bel(A)$ and the plausibility $Pl(A)$ functions given respectively by:

$$\begin{cases} Bel(A) = \sum_{B \subseteq A} m_{DS}^{\Theta}(B) \\ Pl(A) \ = \sum_{B \cap A \neq \emptyset} m_{DS}^{\Theta}(B) \end{cases} \tag{6}$$

2.2   Belief Modeling

The data association problem can be analyzed from two points of view: *target-to-track* and *track-to-target* association. Consequently, two frames of discernment are defined: $\Theta_{i,\cdot}$ and $\Theta_{\cdot,j}$, $i = 1, ..., n$, with n the number of targets, and $j = 1, ..., m$, with m the number of tracks:

$$\begin{aligned} \Theta_{i,\cdot} &= \left\{ Y_{(i,1)}, Y_{(i,2)}, ..., Y_{(i,m)}, Y_{(i,*)} \right\} \\ \Theta_{\cdot,j} &= \left\{ X_{(1,j)}, X_{(2,j)}, ..., X_{(n,j)}, X_{(*,j)} \right\} \end{aligned} \tag{7}$$

where $\Theta_{i,\cdot}$ is composed of the m possible target(i)-to-track(j) associations denoted $Y_{(i,j)}$. The hypothesis of appearance is represented by $Y_{(i,*)}$[1]. $\Theta_{\cdot,j}$ contains the n possible track(j)-to-target(i) associations denoted $X_{(i,j)}$, and $X_{(*,j)}$ is the track disappearance.

2.3   Basic Belief Assignment

For *target-to-track* assignment, three *bba's* are used to answer the question "Is target X_i associated with track Y_j?":

- $m_j^{\Theta_{i,\cdot}}(Y_{(i,j)})$: belief in "$X_i$ is associated with Y_j",
- $m_j^{\Theta_{i,\cdot}}(\overline{Y_{(i,j)}})$: belief in "$X_i$ is not associated with Y_j"[2],
- $m_j^{\Theta_{i,\cdot}}(\Theta_{i,\cdot})$: the degree of ignorance.

The recent benchmark [4] on huge real data shows that the most suited model is the non-antagonist model [14, 23] which is defined as follows:

$$m_j^{\Theta_{i,\cdot}}(Y_{(i,j)}) = \begin{cases} 0 & , I_{i,j} \in [0, \tau] \\ \Phi_1(I_{i,j}) & , I_{i,j} \in [\tau, 1] \end{cases} \tag{8}$$

$$m_j^{\Theta_{i,\cdot}}(\overline{Y_{(i,j)}}) = \begin{cases} \Phi_2(I_{i,j}) & , I_{i,j} \in [0, \tau] \\ 0 & , I_{i,j} \in [\tau, 1] \end{cases} \tag{9}$$

---

[1] $Y_{(i,*)}$ refers to the fact that no track is assigned to the target(i).
[2] $\overline{Y_{(i,j)}}$ defines the complementary hypothesis of $Y_{(i,j)}$,
$\overline{Y_{(i,j)}} = \{Y_{(i,1)}, ..., Y_{(i,j-1)}, Y_{(i,j+1)}, ..., Y_{(i,m)}, Y_{(i,*)}\}$.

$$m_j^{\Theta_{i,\cdot}}(\Theta_{i,\cdot}) = 1 - m_j^{\Theta_{i,\cdot}}(Y_{(i,j)}) - m_j^{\Theta_{i,\cdot}}(\overline{Y_{(i,j)}}), \tag{10}$$

where $0 < \tau < 1$ represents the impartiality of the association process and $I_{i,j} \in [0,1]$ is an index of similarity between X_i and Y_j. $\Phi_1(.)$ and $\Phi_2(.)$ are two cosine functions defined by:

$$\begin{cases} \Phi_1(I_{i,j}) = \frac{\alpha}{2}\left[1 - \cos(\pi\frac{I_{i,j}-\tau}{\tau})\right] \\ \Phi_2(I_{i,j}) = \frac{\alpha}{2}\left[1 + \cos(\pi\frac{I_{i,j}}{\tau})\right], \end{cases} \tag{11}$$

where $0 < \alpha < 1$ is the reliability factor of the data source. In the same manner, belief masses are generated for the *track-to-target* assignment.

2.4 Belief Combination

Based on Dempster's rule (4), the combined masses $m^{\Theta_{i,\cdot}}$ (and $m^{\Theta_{\cdot,j}}$) over $2^{\Theta_{i,\cdot}}$ (and $2^{\Theta_{\cdot,j}}$) can be computed as follows [24]:

$$m^{\Theta_{i,\cdot}}\left(Y_{(i,j)}\right) = K \cdot m_j^{\Theta_{i,\cdot}}\left(Y_{(i,j)}\right) \prod_{\substack{a=1 \\ a \neq j}}^{m} \alpha_{(i,a)}$$

$$m^{\Theta_{i,\cdot}}(\{Y_{(i,j)}, \ldots, Y_{(i,l)}, Y_{(i,*)}\}) = K \cdot \gamma_{(i,(j,\ldots,l))} \prod_{\substack{a=1 \\ a \neq j \\ \cdots\cdots \\ a \neq l}}^{m} \beta_{(i,a)} \tag{12}$$

$$m^{\Theta_{i,\cdot}}\left(Y_{(i,*)}\right) = K \cdot \prod_{a=1}^{m} \beta_{(i,a)}$$

$$m^{\Theta_{i,\cdot}}(\Theta_{i,\cdot}) = K \cdot \prod_{a=1}^{m} m_a^{\Theta_{i,\cdot}}(\Theta_{i,\cdot})$$

with:

$$\begin{cases} \alpha_{(i,a)} = 1 - m_a^{\Theta_{i,\cdot}}\left(Y_{(i,a)}\right) \\ \beta_{(i,a)} = m_a^{\Theta_{i,\cdot}}\left(\overline{Y_{(i,a)}}\right) \\ \gamma_{(i,(j,\ldots,l))} = m_j^{\Theta_{i,\cdot}}(\Theta_{i,\cdot}) \ldots m_l^{\Theta_{i,\cdot}}(\Theta_{i,\cdot}) \\ K = \left[\prod_{a=1}^{m}\alpha_{(i,a)} + \sum_{a=1}^{m} m_a^{\Theta_{i,\cdot}}\left(Y_{(i,a)}\right)\prod_{\substack{b=1 \\ b \neq a}}^{m}\alpha_{(i,b)}\right]^{-1} \end{cases}$$

2.5 Decision-Making

Finally, the probabilities matrix $P_{i,\cdot}$ ($P_{\cdot,j}$) is obtained by using a probabilistic transformation. Table 1 presents the $P_{i,\cdot}$ matrix where each line defines the association probabilities of the target X_i with all tracks Y_j. $P_{i,\cdot}(Y_{(i,*)})$ represents the appearance probability of X_i.

Table 1. Probabilities of target-to-track associations

$P_{i,.}(.)$	Y_1	$\dots$	Y_m	Y_*
X_1	$P_{1,.}(Y_{(1,1)})$	$\dots$	$P_{1,.}(Y_{(1,m)})$	$P_{1,.}(Y_{(1,*)})$
X_2	$P_{2,.}(Y_{(2,1)})$	$\dots$	$P_{2,.}(Y_{(2,m)})$	$P_{2,.}(Y_{(2,*)})$
$\vdots$	$\vdots$	$\vdots$	$\vdots$	$\vdots$
X_n	$P_{n,.}(Y_{(n,1)})$	$\dots$	$P_{n,.}(Y_{(n,m)})$	$P_{n,.}(Y_{(n,*)})$

The association decisions are made by using a global or a local optimization strategy. The Joint Pignistic Probability (JPP) [20] selects associations that maximize the probability product. However, this global optimization is time-consuming and can select doubtful local associations. To cope these drawbacks, local optimizations have been proposed as the Local Pignistic Probability (LPP). Interested readers in the benchmark of these algorithms can refer to [17,18].

3   Probabilistic Transformations

The generalized formula of the probabilistic transformation can be defined as follows:

$$P_{i,.}\left(Y_{(i,j)}\right) = m^{\Theta_{i,.}}\left(Y_{(i,j)}\right) + \sum_{\substack{A \in 2^{\Theta_{i,.}} \\ Y_{(i,j)} \subset A}} T(Y_{(i,j)}, A) \cdot m^{\Theta_{i,.}}(A), \tag{13}$$

where A represents the partial/global ignorance about the association of target X_i and $T(Y_{(i,j)}, A)$ represents the rate of the ignorance mass $m^{\Theta_{i,.}}(A)$ which is transferred to singleton $Y_{(i,j)}$.

Several probabilistic transformations have been proposed in the literature. In this section, only the most interesting ones are presented.

3.1   Pignistic Probability

The pignistic transformation denoted by $BetP$ and proposed by Smets [27,28] is still widely used for evidential data association applications [6,14,16,20]. This transformation redistributes equitably the mass of ignorance on singletons as follows:

$$T_{BetP_{i,.}}(Y_{(i,j)}, A) = \frac{1}{|A|}, \tag{14}$$

where $|A|$ represents the cardinality of the subset A. However, the pignistic transformation (14) ignores the *bbas* of singletons which can be considered as a crude commitment. $BetP$ is easy to implement because it has a low complexity due to its simple redistribution process.

3.2   Dezert-Smarandache Probability

Besides of the cardinality, Dezert-Smarandache Probability $(DSmP)$ transformation [11] considers the values of masses when transferring ignorance on singletons:

$$T_{DSmP_{i,.}}(Y_{(i,j)}, A) = \frac{m^{\Theta_{i,.}}(Y_{(i,j)}) + \epsilon}{\sum\limits_{Y_{(i,k)} \subset A} m^{\Theta_{i,.}}(Y_{(i,k)}) + \epsilon \cdot |A|} \tag{15}$$

The value of the tuning parameter $\epsilon \geq 0$ is used to adjust the effect of focal element's cardinality in the proportional redistribution, and to make $DSmP$ defined and computable when encountering zero masses. Typically, one takes $\epsilon = 0.001$. The smaller ϵ, the better approximation of probability measure we get [11]. $DSmP$ allows to obtain in general a higher Probabilistic Information Content (PIC) [31] than $BetP$ because it uses more information than $BetP$ for its establishment. The PIC indicates the level of the available knowledge to make a correct decision. $PIC = 0$ indicates that no knowledge exists to take a correct decision.

3.3   MultiScale Probability

The Multiscale Probability $(MulP)$ transformation [19] highlights the proportion of each hypothesis in the frame of discernment by using a difference function between belief and plausibility:

$$T_{MulP_{i,.}}(Y_{(i,j)}, A) = \frac{(Pl^{\Theta_{i,.}}(Y_{(i,j)}) - Bel^{\Theta_{i,.}}(Y_{(i,j)}))^q}{\sum\limits_{Y_{(i,k)} \subset A} (Pl^{\Theta_{i,.}}(Y_{(i,k)}) - Bel^{\Theta_{i,.}}(Y_{(i,k)}))^q}, \tag{16}$$

where $q \geq 0$ is a factor used to amend the proportion of the difference $(Pl(\cdot) - Bel(\cdot))$. However, the $T_{MulP_{i,.}}$ is not defined $(\frac{0}{0})$ when $m(\cdot)$ is a Bayesian mass $(Pl(\cdot) = Bel(\cdot))$.

3.4 Sudano's Probabilities

Sudano proposes several alternatives to $BetP$ as the Proportional Plausibility ($PrPl$) and the Proportional Belief ($PrBel$) transformations [11,30]. Those latter redistribute respectively the ignorance mass according to the normalized plausibility and belief functions:

$$T_{PrPl_{i,.}}(Y_{(i,j)}, A) = \frac{Pl^{\Theta_{i,.}}(Y_{(i,j)})}{\sum_{Y_{(i,k)} \subseteq A} Pl^{\Theta_{i,.}}(Y_{(i,k)})} \tag{17}$$

$$T_{PrBel_{i,.}}(Y_{(i,j)}, A) = \frac{Bel^{\Theta_{i,.}}(Y_{(i,j)})}{\sum_{Y_{(i,k)} \subseteq A} Bel^{\Theta_{i,.}}(Y_{(i,k)})} \tag{18}$$

3.5 Pan's Probabilities

Other proportional transformations have been proposed in [21]. Those transformations assume that the *bba* are proportional to a function $S(\cdot)$ which is based on the belief and the plausibility:

$$T_{PrBP_{i,.}}(Y_{(i,j)}, A) = \frac{S(i,j)}{\sum_{Y_{(i,k)} \subseteq A} S(i,k)}, \tag{19}$$

where different definitions of S have been proposed:

$$\begin{cases} PrBP1_{i,.} : S(i,j) = Pl^{\Theta_{i,.}}(Y_{(i,j)}) \cdot Bel^{\Theta_{i,.}}(Y_{(i,j)}) \\ PrBP2_{i,.} : S(i,j) = Bel^{\Theta_{i,.}}(Y_{(i,j)}) \cdot (1 - Pl^{\Theta_{i,.}}(Y_{(i,j)}))^{-1} \\ PrBP3_{i,.} : S(i,j) = Pl^{\Theta_{i,.}}(Y_{(i,j)}) \cdot (1 - Bel^{\Theta_{i,.}}(Y_{(i,j)}))^{-1} \end{cases} \tag{20}$$

4 Results

This section presents a benchmark of the probabilistic transformations in the framework of the object association system for autonomous vehicles. The aim is to assign detected objects in the scene (*targets*) to known ones (*tracks*). The transformations have been evaluated on real data.

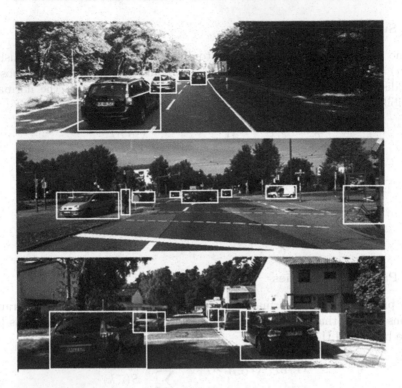

Fig. 1. Examples of images provided by KITTI [4].

The KITTI dataset[3] provides 21 sequences recorded from cameras mounted on a moving vehicle on urban roads [13]. To our knowledge, no comparison of probabilistic transformations has been done on real data where more than 30000 associations have been observed. These latter cover different road scenario as shown in Fig. 1. For this work, detections are defined only by 2D bounding box in the image plane as presented in Fig. 1.

4.1 Experimental Setting

The assignment information are based on the distance between objects in the image plane. For that, the distance $d_{i,j}$ is defined as follows:

$$d_{i,j} = \frac{1}{2}(d_{i,j}^{right} + d_{i,j}^{left}), \tag{21}$$

where $d_{i,j}^{right}$ (resp. $d_{i,j}^{left}$) is the Euclidean distance between bottom-right (resp. top-left) corners of the bounding boxes of target X_i (detected object) and track Y_j (known object) as presented in Fig. 2.

---

[3] http://www.cvlibs.net/datasets/kitti/eval_tracking.php.

Fig. 2. The illustration of the distances $d_{i,j}^{right}$ and $d_{i,j}^{left}$ [4].

The parameters of the *bba* model (11) are: $\alpha = 0.9$ and $\tau = 0.5$. The index of similarity is defined as follows:

$$I_{i,j} = \begin{cases} 1 - \frac{d_{i,j}}{D} & , if \ d_{i,j} < D \\ 0 & , otherwise, \end{cases} \tag{22}$$

where D is the limit distance for association which is determined heuristically, e.g. $D = 210$ in this work.

The tuning parameters $\epsilon = 0.001$ and $q = 5$ for $DSmP$ and $MulP$ transformations respectively. The LPP algorithm has been used as optimization strategy in the decision-making step.

4.2   Comparison of Probabilistic Transformations

All discussed transformations are characterized by an equivalent complexity except the pignistic transformation. $BetP$ is computed directly from combined masses which leads to a lower computational time.

To compare the performance of the probabilistic transformations presented previously, the object association system is evaluated by the True Associations Rate (TAR):

$$TAR = \frac{\sum_t True \ Association_t}{\sum_t Ground \ Truth_t}, \tag{23}$$

where t is the frame index.

Table 2 compares association outcomes of the system based on different probabilistic transformations. Only *target-to-track* association results have been presented in Table 2 due to the lack of space. However, from *track-to-target* association results, similar comments/conclusions hold. The penultimate row of Table 2 shows the weighted average of TAR value based on all sequences which is given by:

$$TAR_{avg} = \sum_{i=0}^{20} w_i TAR_i \qquad (24)$$

where TAR_i is the TAR value of the i-th sequence, and where the weight w_i is $w_i = n_i / \sum_{i=0}^{20} n_i$ and n_i being the number of associations of the i-th sequence. For instance, $TAR_{avg} = 0.9852$ (or 98.52%) for the *BetP* transformation, etc. The last row of Table 2 represents the weighted standard deviation (σ_w) of association scores defined as follows:

$$\sigma_w = \sqrt{\sum_{n=0}^{20} w_i (TAR_i - TAR_{avg})^2} \qquad (25)$$

Table 2. *target-to-track* associations score (in %) obtained by different probabilistic transformations.

Seq. $n°$	nb of Frame	nb of Ass.	BetP	DSmP	PrPl	PrBel	MulP	PrBP1	PrBP2	PrBP3
Seq. 0	154	675	**99.41**	99.11	**99.41**	99.26	**99.41**	**99.41**	99.26	99.11
Seq. 1	447	2643	**97.50**	96.71	97.43	96.03	97.47	95.88	95.50	97.39
Seq. 2	233	668	**99.70**	**99.70**	**99.70**	97.75	**99.70**	98.65	97.46	**99.70**
Seq. 3	144	337	**99.41**	**99.41**	**99.41**	98.81	**99.41**	99.11	98.81	**99.41**
Seq. 4	314	545	89.72	93.39	93.21	92.29	90.09	92.29	91.56	**93.76**
Seq. 5	297	925	98.59	97.51	98.16	96.00	**99.46**	96.32	95.24	97.95
Seq. 6	270	474	**100**	**100**	**100**	98.95	**100**	98.95	98.73	**100**
Seq. 7	800	2084	**97.60**	96.55	97.17	95.11	97.55	95.25	94.63	97.02
Seq. 8	390	492	99.19	98.78	99.19	97.76	**99.39**	97.76	97.56	99.19
Seq. 9	802	2888	**98.44**	97.33	98.10	97.09	98.37	97.13	96.92	97.82
Seq. 10	294	541	98.71	98.34	98.71	97.78	**99.26**	98.15	98.89	98.34
Seq. 11	373	3001	**99.37**	98.77	99.30	99.30	99.33	99.27	99.27	99.23
Seq. 12	78	67	**100**	**100**	**100**	**100**	**100**	**100**	**100**	**100**
Seq. 13	340	617	93.35	**95.62**	94.00	93.35	93.19	93.35	92.06	94.00
Seq. 14	106	374	89.04	89.57	88.50	88.77	88.50	**89.84**	89.04	88.77
Seq. 15	376	1249	**99.28**	**99.28**	**99.28**	99.04	**99.28**	98.80	98.80	**99.28**
Seq. 16	209	1872	**97.54**	96.63	97.44	96.69	**97.54**	96.85	96.90	97.38
Seq. 17	145	486	**99.18**	98.35	**99.18**	96.71	**99.18**	97.33	96.91	**99.18**
Seq. 18	339	1130	**99.82**	98.41	99.65	98.94	**99.82**	99.03	98.94	99.29
Seq. 19	1059	4968	**99.42**	98.73	**99.42**	97.83	99.36	97.89	97.36	99.34
Seq. 20	837	4673	**99.68**	98.35	99.59	98.35	99.64	98.20	98.10	99.42
All Seq.	8007	30709	**98.52**	97.85	**98.47**	97.35	**98.52**	97.40	97.10	98.37
	std. dev.	σ_w	1.38	**1.05**	1.22	1.26	1.39	1.21	1.36	1.18

The obtained results show that $PrBel$, $PrBP1$, and $PrBP2$ provide the worst mean associations scores ($\leq 97.40\%$) with the largest standard deviation (1.36%) for $PrBP2$. It can be explained by the fact that these transformations are based on the Bel function which is a pessimistic measurement. The rest of the transformations provide rates of correct association (i.e. scores) >98.40% which represents a gain of +1%. The best mean score $\approx 98.50\%$ is given by $BetP$, $PrPl$, and $MultP$ transformations. Based only on the mean score criterion, $BetP$ seems more interesting because it provides better scores on 15 sequences from 21 as illustrated in Fig. 3. In addition, $BetP$ is based on a very simple transferring process of uncertainty which makes $BetP$ a good choice for real-time applications. However, this apparent advantage of $BetP$ needs to be seen in relative terms because $BetP$ also generates a quite large standard deviation of 1.38%, which clearly indicate that $BetP$ is not very precise. $PrPl$ and $MultP$ are also characterized by a relatively high standard deviation (1.22% and 1.39%). On the other hand, the lower standard deviation 1.05% is given by $DSmP$ transformation with a good association score = 97.85%. This transformation performs well in term of PCI criteria which leads to make correct decisions [11]. Consequently, $DSmP$ is an interesting alternative to $BetP$ for the data association process in autonomous vehicle perception system.

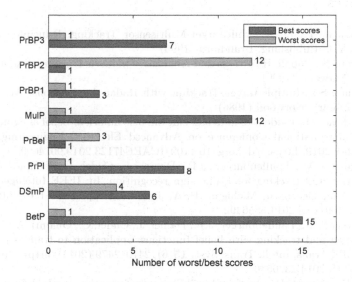

Fig. 3. The number of worst/best scores obtained by each probabilistic transformation on 21 sequences; e.g. $PrBel$ provides three worst scores (sequences 3, 10, and 17) and only one best score on sequence 12.

5   Conclusion

An evaluation of several probabilistic transformations for evidential data association has been presented in this paper. These transformations approximate the belief masses by a probability measure in order to make association decisions. The widely used probabilistic approximation is the pignistic transformation. However, several published studies criticize the choice of this method of approximation and propose generalized transformations.

We did compare the performances of these probabilistic transformations on real-data in order to determine which one is more suited for assignment problems in the context of autonomous vehicle navigation based on real datasets. The obtained results based on the well-known KITTI dataset show that the pignistic transformation provides one of the better scores. However, it provides a quite large standard deviation contrary to $DSmP$ transformation which provides the lowest standard deviation. In addition, $DSmP$ procures a nearly similar association score to that given by $BetP$. Consequently, $DSmP$ can be a good alternative to $BetP$ for the autonomous vehicle perception problematic requiring a bit more computational power with respect to $BetP$.

References

1. Bar-Shalom, Y., Li, X.: Multitarget-Multisensor Tracking: Principles and Techniques. YBS Publishing, Brandford (1995)
2. Blackman, S., Popoli, R.: Design and Analysis of modern Tracking Systems. Artech House, Norwood (1999)
3. Blackman, S.: Multiple-target Tracking with Radar Applications. Artech House Radar Library, Norwood (1986)
4. Boumediene, M.: Evidential data association: benchmark of belief assignment models. In: International Conference on Advanced Electrical Engineering, Algeria, November 2019. https://doi.org/10.1109/ICAEE47123.2019.9014699
5. Boumediene, M., Lauffenburger, J.P., Daniel, J., Cudel, C.: Coupled detection, association and tracking for traffic sign recognition. In: IEEE Intelligent Vehicle Symposium, Dearborn, Michigan, USA, June 2014, pp. 1402–1407. https://doi.org/10.1109/IVS.2014.6856492
6. Boumediene, M., Lauffenburger, J.P., Daniel, J., Cudel, C., Ouamri, A.: Multi-ROI association and tracking with belief functions: application to traffic sign recognition. IEEE Trans. Intell. Trans. Syst. **15**(6), 2470–2479 (2014). https://doi.org/10.1109/TITS.2014.2320536
7. Cobb, B., Shenoy, P.: On the plausibility transformation method for translating belief function models to probability models. IJAR **41**(3), 314–330 (2006)
8. Cuzzolin, F.: On the properties of the intersection probability, submitted to the Annals of Mathematics and AI (2007)
9. Dempster, A.P.: Upper and lower probabilities induced by a multiple valued mapping. Ann. Math. Statistics **38**, 325–339 (1967)
10. Denœux, T., El Zoghby, N., Cherfaoui, V., Jouglet, A.: Optimal object association in the dempster-shafer framework. IEEE Trans. Cybern. **44**(22), 2521–2531 (2014). https://doi.org/10.1109/TCYB.2014.2309632

11. Dezert, J., Smarandache, F.: A new probabilistic transformation of belief mass assignment. In: International Conference on Information Fusion (FUSION), Cologne, Germany , pp. 1410–1417. (2008)
12. Fortmann, T.E., Bar-Shalom, Y., Scheffe, M.: Sonar tracking of multiple targets using joint probabilistic data association. IEEE J. Oceanic Eng. 8(3), 173–184 (1983). https://doi.org/10.1109/JOE.1983.1145560
13. Geiger, A., Lenz, P., Urtasun, R.: Are we ready for autonomous driving? The KITTI vision benchmark suite. In: Conference on Computer Vision and Pattern Recognition (CVPR), Rhode Island, USA (2012). https://doi.org/10.1109/CVPR. 2012.6248074
14. Gruyer, D., Demmel, S., Magnier, V., Belaroussi, R.: Multi-hypotheses tracking using the Dempster-Shafer theory, application to ambiguous road context. Inf. Fusion 29, 40–56 (2016)
15. Gruyer, D., Berge-Cherfaoui, V.: Multi-objects association in perception of dynamical situation. In: Proceedings of the Fifteenth Conference on Uncertainty in Artificial Intelligence, UAI 1999, pp. 255–262. Morgan Kaufmann Publishers Inc., San Francisco (1999)
16. Hachour, S., Delmotte, F., Mercier, D.: Comparison of credal assignment algorithms in kinematic data tracking context. In: Laurent, A., Strauss, O., Bouchon-Meunier, B., Yager, R.R. (eds.) IPMU 2014. CCIS, vol. 444, pp. 200–211. Springer, Cham (2014). https://doi.org/10.1007/978-3-319-08852-5_21
17. Lauffenburger, J.P., Boumediene, M.: Adaptive credal multi-target assignment for conflict resolving. In: International Conference on Information Fusion (FUSION), Heidelberg, Germany, July 2016, pp. 1578–1584
18. Lauffenburger, J.P., Daniel, J., Boumediene, M.: Traffic sign recognition: benchmark of credal object association algorithms. In: International Conference on Information Fusion (FUSION), Salamanca, Spain, July 2014, pp. 1–7
19. Li, M., Lu, X., Zhang, Q., Deng, Y.: Multiscale probability transformation of basic probability assignment. Math. Prob. Eng. 2014, 6 (2014)
20. Mercier, D., Lefèvre, E., Jolly, D.: Object association with belief functions, an application with vehicles. Inf. Sci. 181(24), 5485–5500 (2011). https://doi.org/10. 1016/j.ins.2011.07.045
21. Pan, W., Yang, H.: New methods of transforming belief functions to pignistic probability functions in evidence theory. In: International Workshop on Intelligent Systems and Applications, China, May 2009
22. Reid, D.: An algorithm for tracking multiple targets. IEEE Trans. Autom. Control 24(6), 843–854 (1979). https://doi.org/10.1109/TAC.1979.1102177
23. Rombaut, M.: Decision in multi-obstacle matching process using Dempster-Shafer's theory. In: International Conference on Advances in Vehicle Control and Safety, Amiens, France, pp. 63–68 (1998)
24. Royère, C., Gruyer, D., Cherfaoui, V.: Data association with believe theory. In: International Conference on Information Fusion (FUSION), Paris, France, July 2000. https://doi.org/10.1109/IFIC.2000.862698
25. Shafer, G.: A Mathematical Theory of Evidence. Princeton University Press, Princeton (1976)
26. Smets, P.: The combination of evidence in the transferable belief model. IEEE Trans. Pattern Anal. Mach. Intell. 12, 447–458 (1990)
27. Smets, P., Kennes, R.: The transferable belief model. Artif. Intell. 66(2), 191–234 (1994). https://doi.org/10.1016/0004-3702(94)90026-4
28. Smets, P.: Decision making in the tbm: the necessity of the pignistic transformation. Int. J. Approximate Reasoning 38, 133–147 (2005)

29. Smets, P.: Analyzing the combination of conflicting belief functions. Inf. Fusion **8**(4), 387–412 (2007). https://doi.org/10.1016/j.inffus.2006.04.003
30. Sudano, J.: Pignistic probability transforms for mixes of low- and high-probability events. In: International Conference on Information Fusion, Montreal, August 2001
31. Sudano, J.: The system probability information content (pic) relationship to contributing components, combining independent multi-source beliefs, hybrid and pedigree pignistic probabilities. In: International Conference on Information Fusion, Annapolis, July 2002

A Belief Classification Approach Based on Artificial Immune Recognition System

Rihab Abdelkhalek[✉] and Zied Elouedi[✉]

LARODEC, Institut Supérieur de Gestion de Tunis, Université de Tunis,
Tunis, Tunisia
rihab.abdelkhalek@gmail.com, zied.elouedi@gmx.com

Abstract. Artificial Immune Recognition Systems (AIRS) are supervised classification methods inspired by the immune system metaphors.
They enjoy a great popularity in the filed of machine learning by achieving good and competitive classification results. Nonetheless, while these
approaches work properly under a certain framework, they present some
weaknesses basically related to their inability to deal with uncertainty.
This is considered as an important challenge in real-world classification
problems. Furthermore, using traditional AIRS approaches, all memory
cells are considered with the same importance during the classification
process which may affect the final generated results. To tackle these
issues, we propose in this paper a new AIRS approach under the belief
function framework. Our approach tends to handle the uncertainty pervaded in the classification stage while taking into account the number of
training antigens represented by each memory cell. The performance of
the proposed evidential AIRS approach is validated on real-world data
sets and compared to state of the art AIRS under certain and uncertain
frameworks.

Keywords: Machine learning · Classification · Artificial Immune
Recognition Systems · Uncertainty · Belief function theory

1 Introduction

Artificial Immune Recognition System (AIRS) [1] is considered as a popular
supervised classification method fully inspired by the biological immune system
metaphors. It allows us to make suitable decisions related to various types of
problems. Yet, the AIRS technique has achieved better and competitive classification results when compared to other well-established classification techniques
like Naive Bayes, decision trees and artificial neural networks classifiers [2]. That
is why, it has attracted a great deal of attention in different areas such as pattern recognition, computer virus detection, anomaly detection, optimization and

© Springer Nature Switzerland AG 2020
M.-J. Lesot et al. (Eds.): IPMU 2020, CCIS 1238, pp. 327–340, 2020.
https://doi.org/10.1007/978-3-030-50143-3_25

robotics [3]. Different AIRS approaches have been proposed in the literature aiming to improve the classification accuracy [4]. However, while these approaches work properly under a certain framework, they present some weaknesses basically related to their inability to deal with uncertainty. In fact, uncertainty may undoubtedly spread throughout the classification process, which may deeply affect the classification results. Thus, managing uncertainty in AIRS systems becomes a widespread interest.

In this context, various theories like the fuzzy set and the possibility theories [5,6] can be adopted to deal with imperfection within AIRS. The belief function theory [7,8] is considered among the most appropriate and powerful theories for representing and reasoning under uncertainty. It presents a flexible and rich framework for handling imperfection in different levels ranging from the complete ignorance to the total certainty. Indeed, some works have focused on the problem of managing uncertainty in the field of AIRS, either by means of fuzzy set theory [10–13] or possibility theory [14]. For instance, a new medical decision making system conceived by fuzzy-AIRS was proposed in [10] for the task of lymph diseases diagnosis. In such work, the resource allocation mechanism of AIRS was replaced by fuzzy resource allocation in order to improve its classification performance. However, Golzari et al. [11] performed a statistical study that proved that the integration of fuzzy resource allocation causes significant improvement in a minority of data sets. On this basis, they assumed that investigating the more accurate fuzzy memberships, fuzzy rules and fuzzy values may increase the accuracy in more data sets. Later, authors in [12] proposed a fuzzy weighted pre-processing strategy before the application of AIRS which is based on sample means of each feature values. More specifically, they defined two membership functions known as input and output membership functions which are allowed to allocate a new feature value for each feature in accordance with its old value. An extended version of AIRS with Fuzzy-KNN has been proposed in [13] where classes memberships have been assigned as a function of the vectors distance from the K-nearest neighbors. On the other hand, authors in [14] have introduced a new classification technique combining the AIRS method within the possibility theory, where the training instances have been represented via possibility distributions. More recently, a very preliminary work has been proposed under the belief function theory [15] where the K-nearest antigens are considered as different pieces of evidence contributing to the final class assignment. Nonetheless, in such AIRS approaches, all memory cells are considered with the same importance during the classification process which may affect the generated results. Actually, an improved version of AIRS, called AIRS3, has been proposed in this context. Despite its good performance, such approach is not able to deal with the uncertainty pervaded in the final classification results. That is why, we propose in this paper a new AIRS3 approach under the belief function framework. Our aim in this work is not only to handle the uncertainty pervaded throughout the classification process, but also to take into account the number of training antigens represented by each memory cell.

The remainder of this paper is organized as follows: Sect. 2 recalls the belief function theory basic concepts. Section 3 introduces the Artificial Immune Recognition System. Our proposed Evidential AIRS approach is described in Sect. 4. Section 5 illustrates its experimental results conducted on real-world data sets. Finally, the paper is concluded in Sect. 6.

2   Belief Function Theory

The belief function theory, also called evidence theory, is a flexible and rich framework for dealing with uncertainty [7,8]. In this section, we recall its basic concepts and operations as interpreted in the Transferable Belief Model [9].

2.1   Frame of Discernment

In the belief function theory, a problem domain is represented by a finite set of elementary events called the frame of discernment, denoted by Θ, which contains hypotheses concerning the given problem [9] such that: $\Theta = \{\theta_1, \theta_2, \cdots, \theta_n\}$. In fact, all the possible values that each subset of Θ can take is called the power set of Θ and denoted by 2^Θ, where $2^\Theta = \{A : A \subseteq \Theta\}$.

2.2   Basic Belief Assignment

A basic belief assignment (*bba*) is an expression of the belief committed to the elements of the frame of discernment Θ [8]. It is a mapping function such that:

$$m : 2^\Theta \to [0,1] \quad and \quad \sum_{A \subseteq \Theta} m(A) = 1 \tag{1}$$

Each mass $m(A)$, called a basic belief mass (*bbm*), quantifies the degree of belief exactly assigned to the event A of Θ.

2.3   Simple Support Function

The *bba* which has at most one focal element aside from the frame of discernment Θ is called simple support function. It is defined as follows:

$$m(X) = \begin{cases} w & \text{if } X = \Theta \\ 1 - w & \text{if } X = A \text{ for some } A \subseteq \Theta \\ 0 & \text{otherwise} \end{cases} \tag{2}$$

where A is the focal element and $w \in [0,1]$.

2.4 Dempster's Rule of Combination

Considering two bba's m_1 and m_2 induced from two reliable and independent information sources, the evidence can be combined using Dempster's rule of combination defined as:

$$(m_1 \oplus m_2)(A) = k. \sum_{B,C \subseteq \Theta : B \cap C = A} m_1(B) \cdot m_2(C) \tag{3}$$

$$where \quad (m_1 \oplus m_2)(\varnothing) = 0 \quad and \quad k^{-1} = 1 - \sum_{B,C \subseteq \Theta : B \cap C = \varnothing} m_1(B) \cdot m_2(C)$$

2.5 Decision Making

To make decisions, beliefs can be represented by pignistic probabilities such as:

$$BetP(A) = \sum_{B \subseteq \Theta} \frac{|A \cap B|}{|B|} \frac{m(B)}{(1 - m(\varnothing))} \quad for \ all \ A \in \Theta \tag{4}$$

3 The Artificial Immune Recognition System

Artificial Immune Recognition System (AIRS) is a resource limited supervised learning algorithm inspired by immune metaphors [1]. In what follows, we recall the two improved versions of AIRS namely AIRS2 [19] and AIRS3 [20].

3.1 AIRS2 Method

Two major phases characterize the AIRS2 method namely, the learning-reduction procedure and the classification procedure.

The Learning-Reduction Procedure
The learning-reduction procedure represents the main phase in the AIRS algorithm. The input data present the training set T where each object is considered as an antigen following the same representation as an antibody [20]. Each antigen is represented by a set of attributes values and class values. The output of this procedure is a reduced data set called memory cell pool (MC) containing memory cells, which are later used in the classification process. This phase is divided into four stages: initialization step, memory cell identification and Artificial Recognition Balls (ARB) generation, competition for resources and development of a candidate memory cell, and memory cell introduction.

1. Initialization
The initialization step can be considered as a pre-processing stage and it is performed on the following three steps:

- **Data normalization**

  This step consists in normalizing all numeric attributes of the training data in order to make sure that the Euclidean distance between two training items is in the range of [0,1]. The normalized Euclidean distance is also used as an affinity measure between two cells. It is computed as follows:

$$affinity\ (ag_1, ag_2) = \sqrt{\sum_{i=1}^{m}(ag_{1i} - ag_{2i})^2} \qquad (5)$$

  Where ag_1 and ag_2 are the two attribute vectors representing two cells (data samples) and m is the number of attributes.

- **Computing the affinity threshold**

  After normalization has been performed, the $affinity\_threshold$ is then computed based on Eq. (6).

$$affinity\_threshold = \frac{\sum_{i=1}^{n}\sum_{j=i+1}^{n} affinity\ (ag_i, ag_j)}{\frac{n \cdot (n-1)}{2}} \qquad (6)$$

  Where n is the number of antigens in the training set, ag_i and ag_j are receptively the ith and j^{th} antigens and $affinity\ (ag_i, ag_j)$ represents the affinity measure between the two antigens ag_i and ag_j.

- **MC and AB initializations**

  The final stage is the initialization of the memory cell pool (MC) and the ARB pool. This is done by randomly selecting 0 or more antigens from the training set to be included in the MC and ARB sets.

2. **Memory cell identification and ARB generation**

   At this level, only one antigen from the training instances is required in the training process and the following steps outlined below are applied.

   - **mc_match identification**

     For each memory cell in the MC pool having the same class as the training antigen ag, we calculate the stimulation according to Eq. (7) below:

$$Stim(ag, mc) = 1 - affinity(ag, mc) \qquad (7)$$

     The cell having the greatest stimulation is selected as the best match memory cell denoted as mc_match. It will be employed in the affinity maturation process. If there is no mc in the MC pool having the same class as the training antigen, this antigen will be integrated directly to MC.

   - **New ARBs generation**

     Once the mc_match is selected, this memory cell is used to generate a number of mutated clones added to ARB pool. The number of clones is proportional to the affinity between mc_match and the presented antigen ag and it is computed as follows:

$$mc\_match \cdot Num_{clones} = ColanalRate \cdot HyperColanlRate \cdot Stim(ag, mc\_match) \qquad (8)$$

3. Competition for resources and development of a candidate memory cell

At this point, a set of $ARBs$ which includes mc_match and mutations of mc_match is considered. We aim in this phase to extract a candidate memory cell ($mc\_candidate$) which better recognizes the given antigen ag. To this end, a computation of the normalized stimulation between each ARB and the antigen is first performed and a finite number of resources is then allocated to each ARB such as:

$$ARB.resources = ClonalRate \cdot Stim\_normalized(ag, ARB) \qquad (9)$$

Each class has a restricted number of resources allowed to be allocated. If the total number of resources of all the $ARBs$ overcomes the allowed limit, the surplus resources are eliminated from the weakest ARB and then empty $ARBs$ will be deleted from the system. Therefore, only $ARBs$ with the highest stimulation level are able to compete for resources. After this competition, a testing process of the stopping criterion is performed. This latter is achieved if the mean affinity value for all the existing $ARBs$ with the antigen surpasses the $Stimulation\_threshold$ which is a pre-defined parameter. Otherwise, $ARBs$ will be mutated and the new mutated clones will be included into the ARB pool. Then, the competition for resources processes will be held and only survived $ARBs$ are going to move through the clonal expansion and the mutation procedures until the stopping criterion is reached. The number of clones corresponding to each ARB is measured using the following equation:

$$ARB.num\_Clones = Stim(ag, ARB) \cdot ClonalRate \qquad (10)$$

Finally, the ARB having the highest stimulation level will be picked up as the candidate memory cell ($mc\_candidate$).

4. Memory cell introduction

This step corresponds to the final stage of the training process where the memory cell pool is revised. If the affinity between the extracted $mc\_candidate$ and the antigen ag is higher than that of the mc_match, the $mc\_candidate$ will be integrated to MC, becoming a long-lived memory cell. Moreover, if the affinity measure between mc_match and $mc\_candidate$ is also lower than the product of the $affinity\_threshold$ and $affinity\_threshold\_scalar$, mc_match will be replaced by $mc\_candidate$ in the set of memory cells.

The Classification Step

Once the training phase is achieved, the resulting MC pool will be used for the classification process. In this context, the K-Nearest Neighbors (KNN) technique is adopted and the test antigen is classified by the majority vote of the K nearest memory cells.

3.2 AIRS3 Method

In order to improve the performance of AIRS2, a new version called AIRS3 has been proposed in [20]. The main idea of the AIRS3 is to add a new component allowing to keep the number of represented antigens (numRepAg) for each memory cell in the resulting memory cell pool. This number is preserved during the training phase and will be used in the classification stage. An extended version of the K-Nearest Neighbors is then applied in the classification process to take into account $numRepAg$. In this version, the K value becomes the number of training antigens represented by some memory cells in MC instead of the number of selected memory cells. The sum of $numRepAg$ of all chosen cells must be equal to K. Thereafter, for each class label, the sum of $numRepAg$ of all the selected cells having the same class label will be computed. Finally, the new unlabeled antigen will be assigned to the class with the highest sum of $numRepAg$.

The two standard versions of AIRS2 and AIRS3 show a good performance under a certain context. However, these two AIRS approaches are not able to deal with the uncertainty pervaded in the final classification results. Hence, we propose a new version of AIRS3, where we embrace the belief function theory for handling such uncertainty while taking into account the number of training antigens represented by each memory cell. The proposed approach is called the Evidential AIRS3, that we introduce in the next section.

4   Evidential AIRS3 Approach

Inspired by the standard AIRS3 method, we propose a new classification version of AIRS under the belief function framework. Our main goal is to improve the classification performance of the existing AIRS approaches under certain and uncertain frameworks. For this purpose, we opt for the Evidential K-Nearest Neighbors (EKNN) [16,17] in order to take into account the uncertain aspect of the final classes assignment. Instead of the standard K-Nearest Neighbors [18] commonly used in AIRS, the EKNN formalism is adapted within the AIRS3 where different belief function tools come into play. The proposed approach, that we denote by Evidential AIRS3 (EAIRS3), is based on two consecutive reduction phases. The first one is performed during the learning-reduction procedure where we obtain the first reduced memory cell pool (MC pool). Otherwise, the second reduction is performed during the selection of the best-stimulated samples, taking into account the number of represented antigens by each memory cell denoted by $numRepAg$. Hence, we get in the end a new reduced MC poll called R-MC pool. The obtained R-MC pool will be considered as the input for the EKNN and the classification task will be performed accordingly.

Figure 1 below illustrates the flowchart of the Evidential AIRS3 approach.

4.1   Notations

In order to better explain our contribution, let us first define and clarify the basic notations to be used in our approach.

- $\Theta = \{c_1, c_2, \cdots, c_M\}$: The frame of discernment including a finite set of classes where M refers to the number of classes.

- n: The number of the obtained memory cells in the R-MC pool.
- R-MC $= \{ (mc_{(1)}, c_1), \cdots , (mc_{(n)}, c_n) \}$: The reduced memory cell pool including the available information of n selected memory cells and their corresponding class labels c_i.
- $m(\{c_p\}|mc_{(i)})$: The basic belief mass assigned to the class c_p.
- $I_n = \{I_1, \cdots , I_n\}$: represents the indexes of the n memory cells in the R-MC pool.

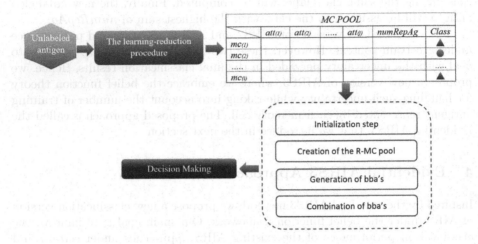

Fig. 1. Evidential AIRS3 process

4.2 Initialization Step

The classification process of the Evidential AIRS3 is based on two parameters, α and $\gamma = (\gamma_1, \cdots , \gamma_p)$ which will be used in the next stages. Therefore, the first step consists in initializing and computing these two parameters. We assign to the parameter α the value 0.95 as mentioned in the EKNN formalism [16]. In fact, this value is highly recommended in order to have good results and to obtain a best classification accuracy. Otherwise, in order to compute the γ_p parameter, we extract the set of the training samples having the same class C_p and we calculate the inverse of the mean distance between each pair of antigens in this set. This computation is executed based on the normalized Euclidean distance defined by:

$$Dist\,(ag_1, ag_2) = \sqrt{\sum_{i=1}^{dim}(ag_{1i} - ag_{2i})^2} \tag{11}$$

Where ag_1 and ag_2 are the two attribute vectors representing two data samples and dim is the number of attributes. In the end of this step, we store the values of γ_p computed for each class in a vector γ. This latter will be further employed in the creation of the basic belief assignment (bba).

4.3   Creation of the R-MC Pool

The second phase of the Evidential AIRS3 approach is the collection of the best-stimulated antigens while taking into account the number of represented antigens $(numRepAg)$ by each memory cell in the MC pool. That is why, we calculate first the similarity between the unlabeled antigen and the whole samples in the MC pool. According to the computed distances, we have to pick out the memory cells having the lowest distances. Unlike the evidential approach proposed in [15], the selection procedure is achieved when the sum of the $numRepAg$ corresponding to the selected cells is equal to K. Hence, the result of this step is a new reduced MC pool, named R-MC pool, containing the nearest neighbors to the test antigen having in total: $\sum numRepAg = K$.

4.4   Generation of bba's

Traditional methods in this step, generate the basic belief assignments $(bba$'s$)$ of all the K-nearest neighbors which could be too complicated for a large value of K. In our approach, we aim to alleviate this problem by generating only the bba's of the obtained cells in the R-MC pool. In other words, these extracted cells will be the pieces of evidence on which we will rely on the classification and the decision making processes. By exploring the R-MC pool obtained in the previous step, we have to generate the basic belief assignment (bba) for the n selected memory cells such that:

$$m(.|mc_{(i)}) = \begin{cases} m(\{c_p\}|mc_{(i)}) = \alpha\phi_p(d_{(i)}) \\ m(\Theta|mc_{(i)}) = 1 - \alpha\phi_p(d_{(i)}) \\ m(\{E\}|mc_i) = 0, \ \forall E \in 2^\Theta \setminus \{\Theta, \{c_p\}\} \end{cases} \quad (12)$$

Where $d_i = d(ag, mc_{(i)})$ is the euclidean distance between the antigen ag and each memory cell $mc_{(i)} \in I_n$, c_p is the class of $mc_{(i)}$ (i.e. $c_i = c_p$), ϕ_p is a decreasing function verifying:

$$\phi_p(0) = 1 \ and \ \lim_{d_{(i)} \to \infty} \phi_p(d_{(i)}) = 0. \quad (13)$$

A popular choice for the decreasing function $\phi_p(d_{(i)})$ is:

$$\phi_p(d_{(i)}) = exp^{-(\gamma_p \cdot (d_{(i)})^2)} \quad (14)$$

We recall here that γ_p and α are the two parameters already computed in the first step.

4.5   Combination of *bba*'s

Once the *bba*'s are generated for each memory cell in the R-MC pool, we combine these *bba*'s using the Dempster's rule of combination as illustrated in Fig. 2 such that:

$$m = m(.|mc_{(1)}) \oplus ... \oplus m(.|mc_{(n)}) \tag{15}$$

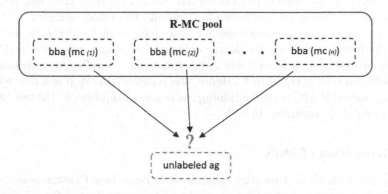

Fig. 2. Combination of the n *bba*'s in R-MC pool

In order to get the final *bba* corresponding to the unlabeled antigen, the evidence of the n memory cells are aggregated using the following expression:

$$m(\{c_p\}) = \frac{1}{N}(1 - \prod_{i \in I_{n,p}} (1 - \alpha\phi_p(d_{(i)})) \cdot \prod_{r \neq p} \prod_{I \in I_{n,r}} (1 - \alpha\phi_r(d_{(i)}))) \qquad \forall p \in \{1, \cdots, M\} \tag{16}$$

$$m(\Theta) = \frac{1}{N} \prod_{r=1}^{M} (1 - \prod_{I \in I_{n,r}} (1 - \alpha\phi_r(d_{(i)})))$$

Where N is a normalized factor [16].

4.6   Decision Making

The last step of our approach is the decision making process where we should provide an appropriate class to the test antigen. That is why, we opt for the pignistic transformation function, denoted by *BetP*, which allows us to create a probability function from the n generalized basic belief assignments. Once the pignistic probability distribution is derived, the test antigen is assigned to the class having the maximum degree of pignistic probability.

5   Experimental Study

In order to evaluate the performance of our Evidential AIRS3 approach, we performed various experiments on real world data sets selected from the U.C.I repository. In these tests, we compare the Evidential AIRS3 with the two standard AIRS methods under a certain framework namely, AIRS2 and AIRS3. Furthermore, two other AIRS approaches working under uncertainty have been compared namely the Fuzzy AIRS2 and the Evidential AIRS2.

5.1   The Framework

Five real data sets have been used in our experiments namely, Cryotherapy (C), Wine (W), Fertility (F), Somerville Happiness Survey (SHS) and Pima Indians Diabetes (PID). The specifications of each data set are shown in Table 1 below, where # nbInstances is the number of antigens, # nbAttributes represents the number of attributes and # nbclass is the number of class labels of a data set.

Table 1. The characteristics of used data sets

Data set	#nbInstances	#nbAttributes	#nbClass
C	90	6	2
W	178	13	3
F	100	9	2
SHS	143	6	2
PID	768	8	2

For all these data sets, we used the following parameter values for all the diverse versions of AIRS employed in our comparison: AIRS2, AIRS3, Fuzzy AIRS2, Evidential AIRS2 (EAIRS2), and Evidential AIRS3 (EAIRS3):

Clonal rate = 10, Mutation rate = 0.4, HyperClonal Rate = 2, Number of resources = 200, Stimulation threshold = 0.3, Affinity threshold scalar = 0.2.

In addition, we have tested with different values of K = [3, 5, 7, 8, 9, and 10].

5.2   Evaluation Criterion

In order to evaluate our approach, we rely on one of the most frequent metrics which is the Percent Correct Classification (PCC). This evaluation metric computes the classification accuracy using the following expression:

$$PCC = \frac{Number\ of\ correctly\ classified\ instances}{Total\ number\ of\ classified\ instances} \tag{17}$$

Therefore, during our experiments, we opt to the cross-validation technique in order to assess the performance of our method. More specifically, we choose to use the 10-fold cross-validation where we estimate the efficiently of our approach by averaging the accuracies derived in all the 10 cases of cross validation.

5.3   Experimental Results

Considering different versions of AIRS, we compare the accuracy of our method to the traditional AIRS2, AIRS3, Fuzzy AIRS2 and EAIRS2. Our comparison is based on the PCC of the several used K-values. The experimental results through the different K values are illustrated in Fig. 3.

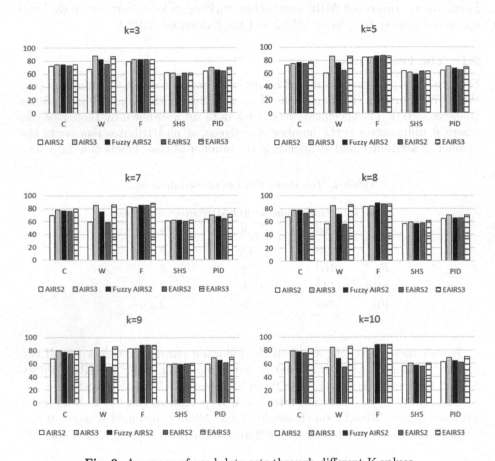

Fig. 3. Accuracy of used data sets through different K-values

Accordingly, we observe that our approach EAIRS3, outperforms the other versions of AIRS in most data sets through the various selected K-values. For example, for the data set W with K = 10, the PCC reaches 85.95% with EAIRS3, while it corresponds respectively to 54.35% with AIRS2, 84.18% with AIRS3, 68.01% with Fuzzy AIRS2 and 55.03% with EAIRS2.

In Table 2, we represent the comparison results of the average PCC obtained for each data set with the different employed K-values.

We can notice through the results shown in Table 2, that EAIRS3 surpasses all the other versions of AIRS in term of classification accuracy for all the used

Table 2. The average PCC (%) of evidential AIRS3 VS AIRS2, AIRS3, Fuzzy AIRS2 and EAIRS2.

Data sets	AIRS2	AIRS3	Fuzzy AIRS2	EAIRS2	EAIRS3
C	68.62	77.44	76.78	74.82	**78.51**
W	59.00	85.41	73.90	60.80	**86.13**
F	82.66	82.91	86.30	86.36	**86.63**
SHS	59.75	60.57	58.23	59.54	**61.40**
PID	62.99	69.67	65.88	63.69	**70.03**

data sets. If we take for example the data set F, our EAIRS3 achieves the best result of 86.63% compared to 82.66% for AIRS2, 82.91% for AIRS3, 86.3% for Fuzzy AIRS2 and 86.36% for EAIRS2. These results prove the greatest performance of our approach over all the other traditional AIRS in term of classification accuracy.

6 Conclusion

In this paper, we have proposed a new AIRS approach under the belief function theory. Our Evidential AIRS3 approach allows not only to deal with the uncertainty pervaded in the classification process, but also to take into account the number of training antigens represented by each memory cell. Experimental results have shown an improvement of the classification performance over state of the art AIRS methods under certain an uncertain frameworks.

References

1. Watkins, A.: A resource limited artificial immune classifier. In: The 2002 Congress on Evolutionary Computation, pp. 926–931 (2002)
2. Polat, K., Sahan, S., Kodaz, H., Günes, S.: A new classification method for breast cancer diagnosis: feature selection artificial immune recognition system (FS-AIRS). In: Wang, L., Chen, K., Ong, Y.S. (eds.) ICNC 2005. LNCS, vol. 3611, pp. 830–838. Springer, Heidelberg (2005). https://doi.org/10.1007/11539117_117
3. Castro, L.N., De Castro, L.N., Timmis, J.: Artificial Immune Systems: A New Computational Intelligence Approach. Springer, London (2002)
4. Brownlee, J.: Artificial immune recognition system (AIRS)-a review and analysis. Center for Intelligent Systems and Complex Processes (CISCP), Faculty of Information and Communication Technologies (ICT), Swinburne University of Technology, Victoria, Australia (2005)
5. Zadeh, L.A.: Fuzzy sets. Inf. Control 8(3), 338–353 (1965)
6. Tzvieli, A.: Possibility theory: an approach to computerized processing of uncertainty. J. Am. Soc. Inf. Sci. 41(2), 153–154 (1990)
7. Dempster, A.P.: A generalization of Bayesian inference. J. Roy. Stat. Soc. Ser. B (Methodol.) 30, 205–232 (1968)

8. Shafer, G.: A Mathematical Theory of Evidence. Princeton University Press, New Jersey (1976)
9. Smets, P.: The transferable belief model for quantified belief representation. In: Smets, P. (ed.) Quantified Representation of Uncertainty and Imprecision. HDRUMS, vol. 1, pp. 267–301. Springer, Dordrecht (1998). https://doi.org/10.1007/978-94-017-1735-9_9
10. Polat, K., Gunes, S.: Automated identification of diseases related to lymph system from lymphography data using artificial immune recognition system with fuzzy resource allocation mechanism (fuzzy-AIRS). Biomed. Signal Process. Control **1**(4), 253–260 (2006)
11. Golzari, S., Doraisamy, S., Sulaiman, M.N., Udzir, N.I.: Effect of fuzzy resource allocation method on AIRS classifier accuracy. J. Theor. Appl. Inf. Technol. **5**, 18–24 (2005)
12. Polat, K., Gunes, S.: Principles component analysis, fuzzy weighting pre-processing and artificial immune recognition system based diagnostic system for diagnosis of lung cancer. Expert Syst. Appl. **34**(1), 214–221 (2008)
13. Chikh, M.A., Saidi, M., Settouti, N.: Diagnosis of diabetes diseases using an artificial immune recognition system2 (AIRS2) with fuzzy k-nearest neighbor. J. Med. Syst. **36**(5), 2721–2729 (2012)
14. Hentech, R., Jenhani, I., Elouedi, Z.: Possibilistic AIRS induction from uncertain data. Soft Comput. **20**(1), 3–17 (2015). https://doi.org/10.1007/s00500-015-1627-3
15. Lahsoumi, A., Elouedi, Z.: Evidential artificial immune recognition system. In: Douligeris, C., Karagiannis, D., Apostolou, D. (eds.) KSEM 2019. LNCS (LNAI), vol. 11775, pp. 643–654. Springer, Cham (2019). https://doi.org/10.1007/978-3-030-29551-6_57
16. Denoeux, T.: A k-nearest neighbor classification rule based on Dempster-Shafer theory. IEEE Trans. Syst. Man Cybern. **25**, 804–813 (1995)
17. Zouhal, L.M., Denoeux, T.: An evidence theoretic kNN rule with parameter optimization. IEEE Trans. Syst. Man Cybern. **28**(2), 263–271 (1998)
18. Dasarathy, B.V.: Nearest Neighbour NN Norms: NN Pattern Classification Techniques. IEEE Computer Society Press, Los Alamitos (1991)
19. Watkins, A., Timmis, J.: Artificial immune recognition system (AIRS): revisions and refinements. In: AISB 2004 Convention, p. 18 (2004)
20. Jenhani, I., Elouedi, Z.: Re-visiting the artificial immune recognition system: a survey and an improved version. Artif. Intell. Rev. **42**(4), 821–833 (2012). https://doi.org/10.1007/s10462-012-9360-0

Evidential Group Spammers Detection

Malika Ben Khalifa[1,2]([✉]), Zied Elouedi[1], and Eric Lefèvre[2]

[1] Institut Supérieur de Gestion de Tunis, LARODEC,
Université de Tunis, Tunis, Tunisia
malikabenkhalifa2@gmail.com, zied.elouedi@gmx.fr
[2] Univ. Artois, EA 3926, Laboratoire de Génie Informatique et d'Automatique
de l'Artois (LGI2A), 62400 Béthune, France
eric.lefevre@univ-artois.fr

Abstract. Online reviews are considered as one of the most prevalent reference indicators for people to evaluate the quality of different products or services before purchasing. Since these reviews affect the buying decision of customers and control the success of the different e-commerce websites, the activity of fake reviews posting is more and more increasing. These fraudulent reviews are posted by a large number of spammers who try to promote or demote target products or companies. The reviewers spammers generally work collaboratively under group of spammers to take control of reviews given to some products, which seriously damage the review system. To deal with this issue, we propose a novel method aim to detect group spammers while relying on various group spamming behavioral indicators. Our approach is based on the K-nearest neighbors algorithm under the belief function theory to treat the uncertainty in the used behavioral indicators. Our method succeeds in distinguishing between genuine and fraudulent group of reviewers. It was tested on two large real datasets extracted from yelp.com.

Keywords: Fake reviews · Group spammers · Uncertainty · Belief function theory · Evidential KNN · E-commerce

1 Introduction

Products, brands, hotels, restaurant, cities, places to visit and all services are now identified through a rating score which is generally the average score of the different reviews given by customers. Such rating score or reviews become one of the most influenced source on consumer's purchase decisions. We can assume that, online reviews nowadays control e-commerce and even international commerce. To increase their market share and to stay ahead of their competitors, companies and business try to over qualify their products by posting fake positive reviews, and even by posting fake negative reviews to damage their competitors' e-reputation. Those who post these fake reviews are called fake reviewers or review spammers, and the products being spammed are called target products.

© Springer Nature Switzerland AG 2020
M.-J. Lesot et al. (Eds.): IPMU 2020, CCIS 1238, pp. 341–353, 2020.
https://doi.org/10.1007/978-3-030-50143-3_26

As the commercialization of these fraudulent activities, such spammers are organized to collaboratively write fake reviews in order magnify the effect of review manipulation. Such review group spammers are more frequently occurred to control the sentiment of the target products. They are even more harmful than individual review spammers' cause over their ability to deviate the overall rating in a short time interval and with different reviewers profiles to mislead spammer detection tools. The review spam detection issue attracts significant researchers during the last years. The main objective of these methods is to distinguish between fake and genuine reviews in order to protect ensure a safe environment and an equitable concurrence between companies. These research can be classified into three categories [11]; review spam detection, review spammer detection and group spammer detection. Several approaches are based on the review spam text information as the semantic and linguistic aspects [7,17]. Moreover, there are different methods which try to detect spammers through graph based aspects [1,8,23]. Others, detect spammers while relying on the spammers behavioral indicators [12,15,18], and on the brust patterns as new indicators [9]. These approaches give significant results in the spam reviews detection field. Recently, there were increasingly research interests in group spamming detection aspects cause of their powerful manipulation thanks to their huge reviewers' members. The first study was introduced by Mukherjee et al. [14], in which they rely on the Frequent Itemset Mining (FIM) technique to generate candidate review spammers groups. This technique considers reviewers as items and products as transactions. Through the FIM technique and by initializing the minimum support count to 3, they can spot at least 2 reviewers, while each reviewer review at least 3 common products. Many techniques rely on these candidate groups and propose different computing frameworks to evaluate the suspicion of each candidate spammer groups. Such that, in [14] authors proposed an iterative computing GSRank to rank candidate groups which spots the relationship among candidate groups, target products and individual reviewers. Xu et al. [27] introduce a statistical model based on the EM algorithm to calculate the collusiveness of each group member from one FIM candidate group at least. Another proposed method in [24] relies on FIM method to capture bicliques or sub-bicliques candidates them check them to detect real collusion groups through group spam indicators. Moreover, Xu et al. [28] use FIM to find groups of reviewers who have reviewed various common products. They introduce a KNN-based method and a graph based classification method to predict the fake or not fake labels for each reviewer belonging to at least one FIM candidate group. They evaluated the effectiveness of the used group spammer indicators on a large Chinese review websites. The KNN-method proves its performance in this study. Some other recent works in this aspect do not rely on the FIM techniques such in [25], where authors propose a top-down computing framework (GGSpam) to detect review spammer groups by exploiting the topological structure of the underlying reviewer graph. We can also cite [26] which propose an unsupervised approach named LDA-based group spamming detection in product reviews (GSLDA) which adapt Latent Dirichlet Allocation (LDA) in order to bound the closely related group spammers into small cluster of reviewers and extracts high suspicious reviewers groups from each LDA-clusters. These proposed methods achieve also significant results. The spam review detection issue can be considerate as one of the most uncertain challenging problem due to the ambiguity provided by the spammers and the group spammers to mislead the detection systems. Nevertheless, the previous proposed methods did not take into consideration the uncertain aspect while trying to detect group spammers. We think that ignoring such uncertainty may deeply affect the quality of detection. For these reasons, we propose a novel method aims to detect group spammers based on the FIM technique to generate candidate group and also on the different group spammer indicators, using the K-nearest neighbors' algorithm within the belief function theory. This theory has shown its robustness in this field through our previous methods which achieve significant results [2–4]. Furthermore, the use of the Evidential K-NN has been based on its robustness in the real world classification problems under uncertainty. We seek to involve imprecision in the Group spammers behaviors indicators which are considered as the fundamental interest in our approach since they are used as features for the Evidential K-NN. In such way, our method predicts the labels spammers or not spammers reviewers (belonging or not to the FIM candidate groups).

This paper is structured as follows: In the first section, we present the basic concepts of the belief function theory and the Evidential K-nearest neighbors, then we elucidate the proposed method in Sect. 2. Section 3 is consecrated for the experimental results and we finish with a conclusion and some future work.

2   Belief Function Theory

In section, we elucidate the fundamentals of the belief function theory as well as the Evidential K-nearest neighbors classifier.

2.1   Basic Concepts

The belief function theory, called also the Dempster Shafer theory, is one of the powerful theories that handles uncertainty in different tasks. It was introduced by Shafer [20] as a model to manage beliefs.

In this theory, a given problem is represented by a finite and exhaustive set of different events called the frame of discernment Ω. 2^{Ω} is the power set of Ω that includes all possible hypotheses and it is defined by: $2^{\Omega} = \{A : A \subseteq \Omega\}$.

A basic belief assignment (bba) named also a belief mass represents the degree of belief given to an element A. It is defined as a function m^{Ω} from 2^{Ω} to $[0, 1]$ such that:

$$\sum_{A \subseteq \Omega} m^{\Omega}(A) = 1. \tag{1}$$

A focal element A is a set of hypotheses with positive mass value $m^{\Omega}(A) > 0$.

Several types of bba's have been proposed [21] in order to model special situations of uncertainty. Here, we present some special cases of bba's:

- The certain *bba* represents the state of total certainty and it is defined as follows: $m^{\Omega}(\{\omega_i\}) = 1$ and $\omega_i \in \Omega$.
- The categorical *bba* has a unique focal element A different from the frame of discernment defined by: $m^{\Omega}(A) = 1, \forall A \subset \Omega$ and $m^{\Omega}(B) = 0, \forall B \subseteq \Omega$ $B \neq A$.
- Simple support function: In this case, the *bba* focal elements are $\{A, \Omega\}$. A simple support function is defined as the following equation:

$$m^{\Omega}(X) = \begin{cases} w & \text{if } X = \Omega \\ 1 - w & \text{if } X = A \text{ for some } A \subset \Omega \\ 0 & \text{otherwise} \end{cases} \qquad (2)$$

where A is the focus and $w \in [0,1]$.

Belief Function

The belief function, denoted *bel*, includes all the basic belief masses given to the subsets of A. It quantifies the total belief committed to an event A by assigning to every subset A of Ω the sum of belief masses committed to every subset of A. *bel* is represented as follows:

$$bel(A) = \sum_{\emptyset \neq B \subseteq \Omega} m^{\Omega}(B) \qquad (3)$$

$$bel(\emptyset) = 0 \qquad (4)$$

Combination Rules

Several combination rules have been proposed in the framework of belief functions to aggregate a set of *bba*'s provided by pieces for evidence from different experts. Let m_1^{Ω} and m_2^{Ω} two *bba*'s modeling two distinct sources of information defined on the same frame of discernment Ω. In what follows, we elucidate the combination rules related to our approach.

1. *Conjunctive rule*: It was introduces in [22], denoted by $\bigcirc$ and defined as:

$$m_1^{\Omega} \bigcirc m_2^{\Omega}(A) = \sum_{B \cap C = A} m_1^{\Omega}(B) m_2^{\Omega}(C) \qquad (5)$$

2. *Dempster's rule of combination*: This combination rule is a normalized version of the conjunctive rule [5]. It is denoted by $\oplus$ and defined as:

$$m_1^{\Omega} \oplus m_2^{\Omega}(A) = \begin{cases} \frac{m_1^{\Omega} \bigcirc m_2^{\Omega}(A)}{1 - m_1^{\Omega} \bigcirc m_2^{\Omega}(\emptyset)} & \text{if } A \neq \emptyset, \forall A \subseteq \Omega, \\ 0 & \textit{otherwise.} \end{cases} \qquad (6)$$

Decision Process

The belief function framework provides numerous solutions to make decision. Within the Transferable Belief Model (TBM) [22], the decision process is performed at the pignistic level where *bba's* are transformed into the pignistic probabilities denoted by $BetP$ and defined as:

$$BetP(B) = \sum_{A \subseteq \Omega} \frac{|A \cap B|}{|A|} \frac{m^{\Omega}(A)}{(1 - m^{\Omega}(\emptyset))} \quad \forall B \in \Omega \qquad (7)$$

2.2   Evidential K-Nearest Neighbors

The Evidential K-Nearest Neighbors (EKNN) [6] is one of the best known classification methods based in the belief function framework. It performs the classification over the basic crisp KNN method thanks to its ability to offer a credal classification of the different objects. This credal partition provides a richer information content of the classifier's output.

Notations

- $\Omega = \{C_1, C_2, ..., C_N\}$: The frame of discernment containing the N possible classes of the problem.
- $X_i = \{X_1, X_2, ..., X_m\}$: The object X_i belonging to the set of m distinct instances in the problem.
- A new instance X to be classified.
- $N_K(X)$: The set of the K-Nearest Neighbors of X.

EKNN Method

The main objective of the EKNN is to classify a new object X based on the information given by the training set. A new instance X to be classified must be allocated to one class of the $N_K(X)$ founded on the selected neighbors. Nevertheless, the knowledge that a neighbor X_i belongs to class C_q may be deemed $d(X, X_i)$ as a piece of evidence that raises the belief that the object X to be classified belongs to the class C_q. For this reason, the EKNN technique deals with this fact and treats each neighbor as a piece of evidence that support some hypotheses about the class of the pattern X to be classified. In fact, the more the distance between X and X_i is reduces, the more the evidence is strong. This evidence can be illustrated by a simple support function with a *bba* such that:

$$m_{X,X_i}(\{C_q\}) = \alpha_0 \exp^{-(\gamma_q^2 d(X,X_i)^2)} \tag{8}$$

$$m_{X,X_i}(\Omega) = 1 - \alpha_0 \exp^{-(\gamma_q^2 d(X,X_i)^2)} \tag{9}$$

Where:

- α_0 is a constant that has been fixed in 0.95.
- $d(X, X_i)$ represents the Euclidean distance between the instance to be classified and the other instances in the training set.
- γ_q assigned to each class C_q has been defined as a positive parameter. It represents the inverse of the mean distance between all the training instances belonging to the class C_q.

After the generation of the different *bba's* by the K-nearest neighbors, they can be combined through the Dempster combination rule as follows:

$$m_X = m_{X,X_1} \oplus ... \oplus m_{X,X_K} \tag{10}$$

where $\{1, ..., K\}$ is the set including the indexes of the K-Nearest Neighbors.

3   Proposed Method

The idea behind our method is to take into account the uncertain aspect in order to improve detecting the group spammer reviewers. For that, we propose a novel approach based on FIM techniques to generate candidate groups, different group spammers indicators and we rely on the Evidential K-nearest neighbors which is famous classifier under the belief function framework. In the remainder of this section, we will elucidate the different steps of our proposed approach; in the first step we will construct the different group spammers from data and we model the group spammers indicators which will be used as features in our method. In the second step, we detail the applying of the EKNN in which we present the initialization and learning phase. Finally, we distinguish between the group spammers and the innocent reviewers through the classification phase.

3.1   Step1: Pre-processing Phase

Spammers who get paid to post fake reviews can not just writing one review for a single product because they would not make enough money that way. Rather, they post various reviews for many products. That's why, we use Frequent pattern mining, that can find them working together on multiple products, to construct candidate spammer groups. Then, we elucidate the different group spammers indicators [14] which can control the candidate spammers behaviors and to find out whether these groups behave strangely.

1- Construction of Spammer Groups From Data
To create a dataset that holds sufficient colluders for evaluation, the first task is to search for the places where colluders would probably be found. A good way to achieve this is to use frequent itemset mining (FIM). In such context, reviewer IDs are regarded as items, each transaction is the set of reviewer IDs who have reviewed a particular product.

Through FIM, groups of reviewers who have reviewed multiple common products can be found. Here we use maximal frequent itemset mining (MFIM) to discover groups with maximal size since we focus on the worst spamming activities in our dataset.

2-Group Spammer Indicators
In our method, we rely on these different group spammers indicators:

- Time Window (TW): Reviewers in a spammer group usually work together in order to post fake reviews for a target product in a short time interval.
- Group Deviation (GD): Members of group spammers are generally give either very high (5*) or very low (1*) ratings to the products. The same products typically are also reviewed by other genuine reviewers. Group spammers generally deviate in their ratings by a significant amount from the mean review ratings score. Therefore, the bigger the deviation, the worse the group is.
- Group Content Similarity (GCS): Members of group spammers usually copy reviews among themselves. Therefore, the products or the services which are victims of such group spamming can have many reviews with similar content.

- Member Content Similarity (MCS): The members of a group may not know one another. Each of them just copy or modify his/her own previous reviews. If multiple members of the group do this, the group is more likely to be a spammer group.
- Early Time Frame (ETF): One damaging group spam activity is to strike right after a product is launched or is made available for reviewing. The purpose is to make a big impact and to take control of the sentiment on the product.
- Ratio of Group Size (RGS): The ratio of the group size and the total number of reviewers for the product is also a good indicator of spamming. In one extreme (the worst case), the group members are the only reviewers of the product, which is very damaging.
- Group Size (GS): The group size itself also tells something quite interesting. If a group is large, then the probability of members happening to be in the group by chance is small. Furthermore, the larger the group, the more devastating is its effect.
- Support Count (SC): Support count is the number of products for which the group has worked on together. If a group has a very high support count, it is clearly alarming.

3.2 Step2: Evidential KNN Application

After applying the FIM algorithm with fixed parameter settings, we note that the suspicious groups are found to be highly similar with each other in term of members, reviewed product also similar ratings. This is because of the dense reviewer product bipartite graph that can decomposed into many small pieces of fully connected sub-graphs (groups). This may have many overlapped nodes (members and products) and such small sub-graphs may dilute the effectiveness of some indicators. However, being used properly, these tiny groups may be favorable to detect colluders in a novel way.

That's why, we propose to rely on the Evidential KNN-based method to detect colluders by utilizing the similarities between such groups. Let $\{g_j\}$ $j = 1..m$ a set of groups and $\{R_i\}$ $i = 1..n$ be a set of reviewers with each associated with an vector a_i of attributes which are the different group spammer indicators mentioned above. Note that each reviewer may belong to multiple groups.

By modeling the colluder detection problem as a binary classification problem our goal is to assign each reviewer R_i with a class label $\Omega = \{S, \bar{S}\}$ where S represents the class of the spammers reviewers and $\bar{S}$ contains the class of the not spammers (innocent) reviewers.

The idea is that given a set of groups, the reviewers who belong to "similar" groups may be more likely to have the same class labels. Thus the class label of a reviewer R_i can be determined commonly by a set of k reviewers who belong to groups most "similar" to the groups R_i belongs to.

1-Initialization and Learning Phase

When applying the Evidential K-NN classifier, we start by initializing the parameters α_0 et γ_0 to be used in the learning phase. The α_0 is fixed to 0.95, as mentioned in the EKNN algorithm [6]. To ensure the γ_{Ii} computation performance, first of all we must find reviewers belonging to different groups are having separately exclusive group spammers indicators. We measure the pairwise similarity of two groups which consists of three measurements as follows:

Common Member Ratio
It measures the Jaccard similarity of the sets of members of two groups:

$$S_{cm} = \frac{|M_i \cap M_j|}{|M_i \cup M_j|} \tag{11}$$

Where M_i and M_j are the member sets of groups g_i and g_j.

Common Product Ratio
It is computed as the sum of the number of products (hotel/ restaurant) of the same brand reviewed by each group, divided by the sum of the number of products reviewed by each group:

$$S_{cp} = max_{b \in B} \frac{(P_{b,i}) + (P_{b,j})}{P_i + P_j} \tag{12}$$

where B is the set of common brands reviewed by both groups g_i and g_j. $P_{b,i}$ (respectively $P_{b,j}$) is the set of the products with brand b reviewed by group g_i (respectively g_j), and P_i (respectively P_j) is the set of the products reviewed by group g_i (respectively g_j).

Common Rating Deviation
It computes the deviation between the average ratings given to the products of common restaurant/hotel reviewed by two groups:

$$S_{crd} = \frac{1}{1 + \sqrt{\frac{1}{|B|} \sum_{b \in B} (\bar{r}_{b,i} - \bar{r}_{b,j})^2}} \tag{13}$$

where $\bar{r}_{b,i}$ (respectively $\bar{r}_{b,j}$) is the average rating given to the products with brand b by group g_i (respectively g_j). Accordingly, the pairwise similarity of two groups is defined as the weighted average of the above components:

$$Sg_{i,j} = \frac{w_k S_k}{\sum w_k S_k} \tag{14}$$

where $S_k \in \{S_{cp}, S_{crd}, S_{cm}\}$ and w_k is a non negative weight for S_k where $\sum_k w_k = 1$.
After defining the pairwise similarity of two groups, the pairwise similarity of two reviewers is computed by taking the average over the pairwise similarity of each pair of their respective groups:

$$d(R_i, R_j) = \frac{\sum_{k \in G_i} \sum_{l \in G_j} Sg_{k,l}}{|G_i||G_j|} \tag{15}$$

where G_i and G_j are the set of groups that have reviewer R_i and R_j respectively. Then, we must select a set of reviewers and for each reviewer R_j in the database, we measure its distance with the target reviewer R_i. Given a target reviewer, we have to select its K-most similar neighbors, by choosing only the K reviewers having the smallest distances values that is calculated through the pairwise similarity of two reviewers calculated above.

2-Classification Phase

In this part, we aim to classify the target reviewer R_i into spammer or not where our frame of discernment $\Omega = \{S, \bar{S}\}$.

The bba's Generation

Each reviewer R_I provides a piece of evidence that represents our belief about the class that he belongs. However, this information does not offer certain knowledge about the class. In the belief function framework, this case is represented by simple support functions, where only a part of belief is assigned to $\omega_i \in \Omega$ and the rest is committed to Ω. Consequently, we obtain bba as follows:

$$m_{R_i,R_j}(\{\omega_i\}) = \alpha_{R_i} \qquad (16)$$

$$m_{R_i,R_j}(\Omega) = 1 - \alpha_{R_i} \qquad (17)$$

Where R_i is the target reviewer and R_j is its similar reviewer that $j = \{1..K\}$, $\alpha_{R_i} = \alpha_0 \exp^{(-\gamma_{Ii} d(R_i, R_j))}$, α_0 and γ_{Ii} are two parameters and $d(R_i, R_j)$ is the distance between the two reviewers R_i and R_j measured above.

In our situation, each neighbor of the target reviewer has two possible hypotheses. It can be near to a spammer reviewer in which his the committed belief is assigned to the spammer class S and the rest is given Ω. On the contrary, it can be similar to an innocent reviewer where the committed belief is allocated to the not spammer class $\bar{S}$ and the rest to the whole frame Ω. We treat the K-most similar reviewers independently where each one is represented by a bba. Hence, K various bba's can be created for each reviewer.

The bba's Combination

After the bba's generation for each reviewer R_i, we detail how to aggregate these bba's in order to get the final belief concerning the reviewer classification. We combine these bba's through the Dempster combination rule to obtain the whole bba that represent the evidence of the K-nearest Neighbors regarding the class of the reviewer. Hence, this global mass function m is calculated as such:

$$m_{R_i} = m_{R_i,R_1} \oplus m_{R_i,R_2} \oplus .... \oplus m_{R_i,R_K} \qquad (18)$$

Decision Making

In order to determine the membership of the reviewer R_i to one of the classes of Ω, we apply the pignistic probability $BetP$. Therefore, the classification decision is made either the reviewer is a spammer or innocent. For this, we select the class that has the grater value of $BetP$ as the final classification.

4   Experimentation and Results

The evaluation in the fake reviews detection problem was always a challenging issue due to the unavailability of the true real world growth data and variability of the features also the classification methods used by the different related works which can lead to unsafe comparison in this field.

Data Description

In order to test our method performance, we use two datasets collected from yelp.com. These datasets represent the more complete, largest, the more diversified and general purpose labeled datasets that are available today for the spam review detection field. They are labeled through the classification based on the yelp filter which has been used in various previous works [2,10,16,19,26] as ground truth in favor of its efficient detection algorithm based on experts judgment and on various behavioral features. Table 1 introduces the datasets content where the percentages indicate the filtered fake reviews (not recommended) also the spammers reviewers.

The YelpNYC dataset contains reviews of restaurants located in New York City; the Zip dataset is bigger than the YelpNYC datasets, since it includes businesses in different regions of the U.S., such that New Jersey, New York, Vermont and Connecticut. The strong points of these datasets are:

- The high number of reviews per user, which facilities to modeling of the behavioral features of each reviewer.
- The divers kinds of entities reviewed, i.e., hotels and restaurants.
- Above all, the datasets hold just basic information, such as the content, label, rating, and date of each review, connected to the reviewer who generated them. Thanks to the over-specific information, we can generalize the proposed method to different review sites.

Table 1. Datasets description

Datasets	Reviews (filtered %)	Reviewers (Spammer %)	Services (Restaurant or hotel)
YelpZip	608,598 (13.22%)	260,277 (23.91%)	5,044
YelpNYC	359,052 (10.27%)	160,225 (17.79%)	923

Evaluation Criteria

We rely on the three following criteria to evaluate our method performance: Accuracy, precision and recall, they can be defined as Eqs. 19, 20, 21 respectively where TP, TN, FP, FN denote True Positive, True Negative, False Positive and False Negative respectively.

$$Accuracy = \frac{(TP + TN)}{(TP + TN + FP + FN)} \tag{19}$$

$$Precision = \frac{TP}{(TP + FP)} \qquad (20)$$

$$Recall = \frac{TP}{(TP + FN)} \qquad (21)$$

Experimental Results

First of all, we apply the frequent itemset mining FIM, where I is the set of all reviewer ids in our two datasets. Each transaction is the set of the reviewer ids who have reviewed a particular hotel or restaurant. Thus, each hotel or restaurant generates a transaction of reviewer ids. By mining frequent itemsets, we find groups of reviewers who have reviewed multiple restaurants or hotels together. Then, we rely on the Maximal Frequent Itemset Mining (MFIM) to spot groups with maximal size in order to focus on the worst spamming activities. In the YelpZip dataset we found 74,364 candidate groups and 50,050 candidate groups for the YelpNYC dataset. We use $k = 3$ for our proposed approach.

Trying to ensure a safe comparison, we compare our method named Evidential Group Spammers Detection (EGSD) with two previous works in which authors rely on the FIM technique to generate the candidate groups and almost the same features used in our work. The first method introduced in [13] Detecting Group Review Spam (DGRS) used the FIM to generate candidate groups then computed the different indicators value and use the SVM rank algorithm to rank them, the other method proposed in [14] we focus on the Ranking Group Spam algorithm (GSRank) which rely on an iterative algorithm to effectively rank the group spammers. The results are reported in the Table 2.

Table 2. Comparative results

Evaluation Criteria	Accuracy			Precision			Recall		
Methods	DGRS	GSRank	EGSD	DGRS	GSRank	EGSD	DGRS	GSRank	EGSD
YelpZip	65%	78%	**85%**	70%	76%	**83.5%**	71%	74%	**86%**
YelpNYC	60%	74%	**84.3%**	62%	76.5%	**83.55%**	61.3%	77.2%	**85%**

Our method achieves the best performance detection according to accuracy, precision and recall over-passing the compared methods. We record at best an accuracy improvement over 10% in both YelpZip and YelpNYC data-sets compared to DGRS and over 7% compared to GSRank.

5   Conclusion

In this work, we tackle the group spammer review detection problem which become a real issue to the online rating systems and we propose a novel approach that aims to distinguish between the spammer and the innocent reviewers while taking into account the uncertainty in the different suspicious behavioral group spammer indicators. Experimental study on a real-world datasets against several state-of-the-art approaches verifies the effectiveness and efficiency of our method. Our proposed approach can be useful for different reviews sites in various fields. As future work, we aim to introduce other features to further improve the detection.

References

1. Akoglu, L., Chandy, R., Faloutsos, C.: Opinion fraud detection in online reviews by network effects. In: Proceedings of the Seventh International Conference on Weblogs and Social Media, ICWSM, vol. 13, pp. 2–11 (2013)
2. Ben Khalifa, M., Elouedi, Z., Lefèvre, E.: Fake reviews detection based on both the review and the reviewer features under belief function theory. In: The 16th International Conference Applied Computing (AC 2019), pp. 123–130 (2019)
3. Ben Khalifa, M., Elouedi, Z., Lefèvre, E.: Spammers detection based on reviewers' behaviors under belief function theory. In: Wotawa, F., Friedrich, G., Pill, I., Koitz-Hristov, R., Ali, M. (eds.) IEA/AIE 2019. LNCS (LNAI), vol. 11606, pp. 642–653. Springer, Cham (2019). https://doi.org/10.1007/978-3-030-22999-3_55
4. Ben Khalifa, M., Elouedi, Z., Lefèvre, E.: Multiple criteria fake reviews detection using belief function theory. In: Abraham, A., Cherukuri, A.K., Melin, P., Gandhi, N. (eds.) ISDA 2018 2018. AISC, vol. 940, pp. 315–324. Springer, Cham (2020). https://doi.org/10.1007/978-3-030-16657-1_29
5. Dempster, A.P.: Upper and lower probabilities induced by a multivalued mapping. Ann. Math. Stat. **38**, 325–339 (1967)
6. Denoeux, T.: A K-nearest neighbor classification rule based on Dempster-Shafer theory. IEEE Trans. Syst. Man Cybern. **25**(5), 804–813 (1995)
7. Deng, X., Chen, R.: Sentiment analysis based online restaurants fake reviews hype detection. Web Technol. Appl., 1–10 (2014)
8. Fayazbakhsh, S., Sinha, J.: Review spam detection: a network-based approach. Final Project Report: CSE 590 (Data Mining and Networks) (2012)
9. Fei, G., Mukherjee, A., Liu, B., Hsu, M., Castellanos, M., Ghosh, R.: Exploiting burstiness in reviews for review spammer detection. In: Proceedings of the Seventh International Conference on Weblogs and Social Media, ICWSM, vol. 13, pp. 175–184 (2013)
10. Fontanarava, J., Pasi, G., Viviani, M.: Feature analysis for fake review detection through supervised classification. In: Proceedings of the International Conference on Data Science and Advanced Analytics, pp. 658–666 (2017)
11. Heydari, A., Tavakoli, M., Ismail, Z., Salim, N.: Leveraging quality metrics in voting model based thread retrieval. Int. J. Comput. Electr. Autom. Control Inf. Eng. **10**(1), 117–123 (2016). World Academy of Science, Engineering and Technology
12. Lim, P., Nguyen, V., Jindal, N., Liu, B., Lauw, H.: Detecting product review spammers using rating behaviors. In: Proceedings of the 19th ACM International Conference on Information and Knowledge Management, pp. 939–948 (2010)
13. Mukherjee, A., Liu, B., Wang, J., Glance, N., Jindal, N.: Detecting group review spam. In: Proceedings of the 20th International Conference on World Wide Web, WWW 2011, Hyderabad, India. ACM (2011). 978-1-4503-0637-9/11/03
14. Mukherjee, A., Liu, B., Glance, N.: Spotting fake reviewer groups in consumer reviews. In: Proceedings of the 21st International Conference on World Wide Web, pp. 191–200. ACM, New York (2012)

15. Mukherjee, A., Kumar, A., Liu, B., Wang, J., Hsu, M., Castellanos, M.: Spotting opinion spammers using behavioral footprints. In: Proceedings of the ACM International Conference on Knowledge Discovery and Data Mining, pp. 632–640 (2013)
16. Mukherjee, A., Venkataraman, V., Liu, B., Glance, N.: What Yelp fake review filter might be doing. In: Proceedings of the Seventh International Conference on Weblogs and Social Media, ICWSM, pp. 409–418 (2013)
17. Ong, T., Mannino, M., Gregg, D.: Linguistic characteristics of shill reviews. Electron. Commer. Res. Appl. **13**(2), 69–78 (2014)
18. Pan, L., Zhenning, X., Jun, A., Fei, W.: Identifying indicators of fake reviews based on spammer's behavior features. In: Proceedings of the IEEE International Conference on Software Quality, Reliability and Security Companion, QRS-C, pp. 396–403 (2017)
19. Rayana, S., Akoglu, L.: Collective opinion spam detection: bridging review networks and metadata. In: Proceedings of the 21th International Conference on Knowledge Discovery and Data Mining, ACM SIGKDD, pp. 985–994 (2015)
20. Shafer, G.: A Mathematical Theory of Evidence, vol. 1. Princeton University Press, Princeton (1976)
21. Smets, P.: The canonical decomposition of a weighted belief. In: Proceedings of the Fourteenth International Joint Conference on Artificial Intelligence, pp. 1896–1901 (1995)
22. Smets, P.: The transferable belief model for quantified belief representation. In: Smets, P. (ed.) Quantified Representation of Uncertainty and Imprecision. HDRUMS, vol. 1, pp. 267–301. Springer, Dordrecht (1998). https://doi.org/10.1007/978-94-017-1735-9_9
23. Wang, G., Xie, S., Liu, B., Yu, P.S.: Review graph based online store review spammer detection. In: Proceedings of 11th International Conference on Data Mining, ICDM, pp. 1242–1247 (2011)
24. Wang, Z., Hou, T., Song, D., Li, Z., Kong, T.: Detecting review spammer groups via bipartite graph projection. Comput. J. **59**(6), 861–874 (2016)
25. Wang, Z., Gu, S., Zhao, X., Xu, X.: Graph-based review spammer group detection. Knowl. Inf. Syst. **55**(3), 571–597 (2017). https://doi.org/10.1007/s10115-017-1068-7
26. Wang, Z., Gu, S., Xu, X.: GSLDA: LDA-based group spamming detection in product reviews. Appl. Intell. **48**(9), 3094–3107 (2018). https://doi.org/10.1007/s10489-018-1142-1
27. Xu, C., Zhang, J.: Towards collusive fraud detection in online reviews. In: 2015 IEEE International Conference on Data Mining, ICDM Atlantic City, pp. 1051–1056 (2015)
28. Xu, C., Zhang, J., Chang, K., Long. C.: Uncovering collusive spammers in Chinese review websites. In: Proceedings of the 22nd ACM International Conference on Conference on Information and Knowledge Management, pp 979–988. ACM, New York (2013)

Dempster-Shafer Theory: How Constraint Programming Can Help

Alexandros Kaltsounidis[(⊠)] and Isambo Karali[(⊠)]

Department of Informatics and Telecommunications,
National and Kapodistrian University of Athens, Athens, Greece
{A.Kaltsounidis,izambo}@di.uoa.gr

Abstract. Dealing with uncertainty has always been a challenging topic in the area of knowledge representation. Nowadays, as the internet provides a vast platform for knowledge exploitation, the need becomes even more imminent. The kind of uncertainty encountered in most of these cases as well as its distributed nature make Dempster-Shafer (D-S) Theory to be an appropriate framework for its representation. However, we have to face the drawback of the computation burden of Dempster's rule of combination due to its combinatorial behavior. Constraint Programming (CP) has proved to be an efficient tool in cases where results have to satisfy some specified properties and pruning of the computation space can be achieved. As D-S theory measures' computation fulfills this requirement, CP seems a promising framework to employ for this purpose. In this paper, we present our approach to use CP to compute the *belief* and *plausibility* measures of D-S Theory and Dempster's rule of combination as well as the results of the effort. As it was expected, the results are quite promising and in many cases impressive.

Keywords: Dempster-Shafer theory · Uncertainty · Constraint Programming · ECLiPSe prolog

1 Introduction

In the area of Knowledge Representation there are many frameworks whose purpose is to model the raw information available so as to create a meaningful and useful inference. Unfortunately, knowledge is not always expressed if terms of indisputable facts. It can be either uncertain or vague or both. Uncertainty may be the product of incomplete or unreliable knowledge and it raises a significant confusion as to how it should be treated. A lot of research has been done in the area and different approaches have been suggested. Efforts combine results from computer science, statistics, game theory and philosophy [1]. These include, but are not limited to, probability, possibility theory [2], probabilistic reasoning such as Bayesian Networks [3], Non-monotonic reasoning [4] and Dempster-Shafer (D-S) theory [1, 5–7]. In case of imprecise or vague information, its modeling has been somehow interrelated with Fuzzy Sets and Fuzzy Logic [8, 9].

The Internet, the World Wide Web (WWW) and the ongoing evolution of the Semantic Web (SW) [10], provide an enormous information store for knowledge extraction and exploitation. The need for an efficient framework able to reason under

© Springer Nature Switzerland AG 2020
M.-J. Lesot et al. (Eds.): IPMU 2020, CCIS 1238, pp. 354–367, 2020.
https://doi.org/10.1007/978-3-030-50143-3_27

uncertainty is more urgent than ever. Apart from the above mentioned problems, frequently missing information, the size of the available data and its distributed nature provide some special characteristics. As an example, consider the case of the online traveling sites which provide information about hotels and a user's need to be provided with some accommodation suggestion according to some personal criteria. Hotel information is distributed among the sites, information is stored heterogeneously and some of it may be missing from some sites. Regarding dealing with the uncertainty in the Web, a detailed discussion can be found in [11].

Dempster-Shafer (D-S) theory, also referred to as the *theory of belief functions* or *evidence theory* is able to deal with ignorance and missing information (i.e. epistemic uncertainty). It is a generalization of the Bayesian theory of subjective probabilities and as such it allows for a greater degree of flexibility. For compositionality purposes, it also offers a rule for combining evidence from different independent sources [12]. The Theory has already been adopted in WWW or SW environments even as a tool for inexact knowledge modelling [13–17]. Some criticism has been done regarding when to use Dempster's Rule of Combination, as it might produce counter intuitive results in some cases [18, 19]. However, the Theory is widely used, especially in practical situations relative to data fusion [20]. According to [21], whether Dempster's rule of combination is appropriate depends on the problem's characteristics. Although D-S theory is a valuable framework for handling uncertainty with the current traits of knowledge representation, the computation of Dempster's rule puts a significant barrier to the theory being used more in practice because of its high complexity. It has been proved that computing Dempster's rule of combination is a #P-complete problem [22] and so it cannot be performed in an acceptable time when the data grow significantly, although methods of computing Dempster's rule in a more efficient way have been proposed [23]. To overcome this obstacle of complexity, many researchers have resorted to approximation algorithms [23–27]. In general, their efforts fall into two categories. The first one contains algorithms that make use of Monte Carlo, or similar random methodologies, to compute a solution [23]. The second category [24–27] consists of algorithms which alter the input data, in order to create an easier to compute problem. This is carried out by disregarding facts that have little evidence.

In order to avoid losing accuracy and be able to use Dempster's Rule of Combination as is, our approach to the complexity of the rule is to use Constraint Programming for its computation. Constraint programming (CP) [28] is a programming paradigm where relationships between the variables of the problem's computational space are stated in the form of constraints. In the case of Dempster's Rule of Combination, we utilize constraints to compute only the combinations that have a nonempty intersection (we can extract the normalization factor out of these). CP is also used to avoid evaluating redundant combinations, when computing *belief* and *plausibility* of a specific set by setting constrains on subset and non-empty intersection relationships, respectively. In order to evaluate the method, we employed ECLiPSe Prolog [29] using its set constraints solver [30, 31]. We created a program that performs Dempster's rule of combination on any number of mass functions using constraints and we compared it with a *generate-and-test* implementation. The latter evaluates every possible focal point combination. For comparison reasons, both programs ran on a number of random test cases that we created by using a variable number of: i) mass functions, ii) focal points per mass function, and iii) elements of the Universe. The *constrain-and-generate* program outperformed the *generate-and-test* one in the tests. As expected, the time the constraint program needed was relative to the number of combinations which have a non-empty intersection, whereas the generate-and-test program's time was related to the total number of combinations. When computing the *belief* (or *plausibility*) for a set *A*, the time needed by the constraint program was relative to the number of combinations whose intersection is a subset of (or intersects with) *A*. Thus, for the *constraint-and-generate* method no redundant sets are generated.

It is worth mentioning that Constraint Systems (CS), the formalism that models constraint problems, and D-S theory have already been related since many years ago. CS have been employed to model the uncertainty expressed by belief functions in a variety of early works, for instance [32]. In addition, recently, Constraint Satisfaction Problems (CSPs) have been extended with uncertainty. In [33], a unifying CSP extension based on the D-S theory is presented that supports uncertainty, soft and prioritized constraints and preferences over the solutions.

In this paper, we present our approach in using CP to reduce the computation time for D-S theory measures, namely *belief* and *plausibility*, as well as Dempster's rule of combination and summarize our results. The paper is structured as follows. The necessary background on D-S theory will be recalled in Sect. 2. In Sect. 3 the complexity of Dempster's rule as well as related work will be discussed. A brief description of Constraint Programming and Logic Programming will be given in Sect. 4, as constraint programming within logic programming is used in our prototype implementation. Our approach, its implementation and test results will be presented in Sect. 5. Section 6 will then conclude this paper.

2 Dempster Shafer Theory

Dempster-Shafer (D-S) Theory is a framework designed for reasoning under uncertainty, which allows to combine evidence from different independent sources. The latter is achieved by Dempster's rule of combination. This rule produces common shared belief between multiple sources and distributes non-shared belief through a normalization factor. Dempster's rule of combination is a powerful and useful tool and one of the reasons why the D-S theory has been so widely spread in many areas in computer science, from artificial intelligence to databases as it allows the Theory to deal with distributed sources.

D-S theory considers a universe, *U*, of mutually exclusive events. A basic *mass function*, *m*, assigns subjective probabilities to subsets of *U*. Then, on top of *m*, *belief* and *plausibility* functions are defined for any subset of *U* assigning lower and upper likelihoods to each of them. Composition of different *mass functions* on the same universe is achieved using *Dempster's Rule of Combination* considering independent sources. More formally, these measures are defined as follows.

2.1 Formal Definition

Let U be the Universe, i.e. the set containing all possible states of a system under consideration. The power set 2^U of U, is the set of all subsets of U, including the empty set $\emptyset$.

Mass Function: A mass function (also called *basic probability assignment* or *basic belief assignment*) is the most fundamental function for assigning degrees of belief to a set of one or more hypotheses. Formally, a mass function m is a function from 2^U to $[0, 1]$: $m : 2^U \longrightarrow [0, 1]$ with the following properties:

1. The mass of the empty set is equal to zero $m(\emptyset) = 0$, and
2. The masses of the elements of the power set must sum to a total of one.
$$\sum_{A \in 2^U} m(A) = 1$$

Belief Function: The *belief* (also known as *support*) of a set $A \in 2^U$, denoted by $bel(A)$ expresses the total amount of belief committed to A and is defined as:

$$bel(A) = \sum_{B \subseteq A} m(B), \qquad \forall A \subseteq U.$$

Plausibility Function: The *plausibility* for a set $A \in 2^U$, denoted by $pl(A)$ expresses the amount of belief not committed to the complement of A and thus declares how plausible A is. It is computed as:

$$pl(A) = \sum_{B \cap A \neq \emptyset} m(B), \qquad \forall A \subseteq U.$$

Dempster's Rule of Combination: The theory of evidence also handles the problem of how to combine evidence from different independent sources.

Let $m_1, m_2, \ldots, m_n$ be mass functions defined over the same Universe U. Then using Dempster's rule of combination we can compute a new mass function that incorporates the evidence of $m_1, m_2, \ldots, m_n$. We use $\oplus$ to denote the operator of Dempster's rule of combination. Then the combination of $m_1, m_2, \ldots, m_n$ is called the *joint mass* $m_{1,2,\ldots,n} \equiv m_1 \oplus m_2 \oplus \ldots \oplus m_n$ and is defined as:

$$m_1 \oplus m_2 \oplus \ldots \oplus m_n = \overset{n}{\underset{i=1}{\oplus}} m_i(A) = \frac{1}{1-K} \sum_{B_1,\ldots,B_n \mid \cap_{i=1}^n B_i = A} m_1(B_1) \cdot \ldots \cdot m_n(B_n), \forall A \subseteq U, A \neq \emptyset$$

where,

$$K = \sum_{B_1,\ldots,B_n \mid \cap_{i=1}^n B_i = \emptyset} m_1(B_1) \cdot \ldots \cdot m_n(B_n).$$

Here, K is a normalization constant that accounts for the products of mass values corresponding to the empty intersections of focal points. It can be considered as a measure of conflict between the mass functions.

3   Complexity of Dempster's Rule of Combination – Approximation Algorithms

Dempster's Rule of Combination has exponential complexity. In [22], it was shown that evaluating the rule is a #P-complete problem.

To handle the complexity problem, many have resorted to approximation algorithms to compute the rule. An approximation algorithm is an efficient algorithm that finds approximate solutions for the desired problem with provable guarantees on the distance of the evaluated solution to the exact one. Towards this, a lot of work has been done into trying to reduce the size of input (i.e. the number of focal points of each mass function), with some of the most well-known methods being the Bayesian approximation [24], the k-l-x method [25], the summarization method [26], and the D1 Approximation [27]. There has also been effort into developing algorithms to directly compute an approximate value for Dempster's rule's result using Monte Carlo models and/or Markov Chain models [23].

When a lot of information is available, i.e. the size of the problem is big, we can resort to approximation algorithms, as a loss of accuracy can be tolerated, for the sake of a significant reduction in time needed for the computation. On the other hand, when we have only a few sources of evidence to combine, and/or a small number of focal points per mass function, an approximation can be deceptive, thus we have to compute the exact solution.

To give a better understanding of the problem, we can summarize it as follows. Given a frame of discernment U of size $|U|$, a mass function m can have up to $2^{|U|}$ ($2^{|U|} - 1$ to be precise, as the empty set cannot be a focal point) focal points. Given a mass function m, the computation of the *belief*, or the *plausibility* function for m is linear to the number of focal points of m (as computation of sums). Note, however, that the combination of two mass functions through Dempster's rule of combination requires the computation of up to $2^{2|U|}$ intersections. To generalize, let n be the number of mass functions $m_1, \ldots, m_n$, the computation of the *joint mass* $m_{1,\ldots,n} = \bigoplus_{i=1}^{n} m_i$ needs up to $2^{n|U|}$ intersections to be computed. Thus the worst case complexity of Dempster's rule is $O(2^{n|U|})$. To be precise, say we know the number of focal elements of each mass function, i.e. let $q_1, \ldots, q_{n-}$ be the number of focal points of each mass function $m_1, \ldots, m_n$ respectively, then the complexity of the rule of combination is $\Theta(Q)$, where Q is the product $Q = q_1 \cdot q_2 \cdot \ldots \cdot q_n$.

When Constraint Programming is involved while computing Dempster's rule of combination, we consider only the number of combinations that result in a desired non-empty set. Let A be a set and Q_A the number of combinations whose intersection is a subset of A, then a method utilizing Constraint Programming would need time proportional to Q_A, where obviously $Q_A < Q$, where Q as above.

In the following section we shall give a short description of constraint programming as well as logic programming to allow a better understanding of the approach that we follow.

4   Constraint Programming and Logic Programming

Constraint Programming (CP) [28] is a powerful paradigm that can be used to solve a variety of problems referred to as Constraint Satisfaction Problems (CSP). In CP a declarative description of the problem is given by the programmer and a constraint solver is used to find an acceptable solution to it. More precisely, a CSP can be described as a set of Variables, each associated with a domain of values, and a set of Constraints over subsets of these Variables. A solution to the problem is an assignment of values to the Variables so that all Constraints are satisfied. Constraint Programming uses constraint propagation to reduce the search space allowing for solving the problem faster. However, the time that the constraint propagation process needs must be taken into account. We are concerned with CSPs where Variables are sets of integers. It is shown in [34] that a CSP with Set Variables with at least one binary constraint (i.e. a constraint that involves two Variables) has exponential complexity.

Logic Programming [35], as the name suggests is based on the idea "that logic can be used as a programming language". A logic program (i.e. a program written in a logic programming language) is a sequence of sentences in logical form, each of which expresses a fact or a rule for a given domain. More precisely, a logic program consists of clauses named *Horn Clauses*. Horn clauses can have the form of a *fact*, e.g. *likes (mary,john)*, denoting that "*mary likes john*", or a *rule*, e.g. *parent(x,y)∧male(x)→ father(x,y)*, denoting that for any unknown individuals x and y, in case x is *parent* of y and x is *male* then x is *father* of y. Horn clauses can also have the form of a *goal*, e.g. *father(x,mary)→*. In this case, the goal is said to *be satisfied* if there is an individual x that is *father* of *mary*. In particular, Prolog [36] is a practical logic programming language based on Horn clauses.

Constraint Programming can be hosted by a Logic Programming language. Then it is referred to as Constraint Logic Programming [37, 38].

We will be working with ECLiPSe Prolog, a software system implementing Prolog that also offers libraries for Constraint Programming. We will be using ECLiPSe's *ic* library that supports finite domain constraints, as well as the *ic_sets* library which implements constraints over the domain of finite sets of integers and cooperates with ic.

The constraint propagation algorithm that ECLiPSe's solver uses for Set Variables has a complexity $O(ld + (e - l)dd')$, where l is the number of inclusion constraints, e the number of total constraints, d the sum of cardinalities of the largest domain bounds and d' their difference [30]. More about ECLiPSe Prolog can be found in [29].

5 Constraint Programming for Computing Dempster's Rule of Combination

To face the complexity problem that Dempster's Rule of Combination introduces, we use Constraint Programming to perform the computation. The idea behind using CP is to reduce the number of computations needed to evaluate the Rule to the ones that are absolutely necessary. Constraint Logic Programming (CLP) has been chosen for the sake of simplicity and prototype experimentation. The Complexity of the Rule is directly related to the number of focal points each mass function has, as we have to evaluate all combinations of focal points. Each combination of focal points might either intersect to a set or not. By using Constraint Programming, we enumerate only those combinations of focal points whose intersection is not empty. We know beforehand that this method does not improve the computational class of the Rule, as in worst case, all combinations intersect, but nevertheless it might reduce the number of computations needed, and thus the time needed, to compute Dempster's Rule of Combination.

In addition, we use Constraint Programming to compute *belief/plausibility* for a given set. Recall that *belief* of a set $A \subseteq U$, where U is the Universe, is the sum of the mass values of all subsets of A. When computing the *belief* of a set A, given mass functions $m_1, \ldots, m_n$, we use constraints so that only the desired combinations of focal points, i.e. those that intersect to a subset of A, are created. In the general case, we need also to compute the normalization constant, K, in order to normalize the value, which means that every intersecting combination (or every non-intersecting combination) will have to be evaluated anyway, and thus the number of combinations to be evaluated is not reduced. However, even in the special case where we know that $K = 0$, that is there is no conflict between the different sources of evidence, we trust that constraints will be proved useful for computing the *belief* of a set A.

5.1 Implementation

We implemented both algorithms, i.e. the *generate-and-test* and the *constrain-and-generate*, in *ECLiPSe Prolog*, so that we accomplish a more fair time comparison. The generate-and-test algorithm evaluates every possible combination of focal points and, then, the unnecessary results are discarded. The *constrain-and-generate* one exploits CP and constraints are set in order to generate only combinations with the desired properties. In the following figures, the fundamental predicates for both algorithms are presented (Figs. 1 and 2).

```
%compute(-A, - Val, + Hyper)
compute(A,Val, Hyper) : -
        pick(1, Sets),                    % generate a combination
        m_intersection(Sets, A),          % find intersection
        not_empty(A),                     % check if intersection is empty
        subset(A, Hyper),                 % check if intersection is subset of Hyper
        values(1, Sets, Vals),            % get mass values
        compute_val(Vals, Val).           % multiply them
```

Fig. 1. generate-and-test

```
%compute(-A, - Val, + Hyper)
compute(A, Val, Hyper) : -
        ic_sets : subset(A, Hyper),            % subset constraint
        not_empty(A),                          % non empty constraint
        new_focal_points(Sets),                % declaration of variables
        ic_sets : all_intersection(Sets, A),   % constrain A to be the intersection of variables
        constrain_membership(Sets),            % generate a combination
        values(1, Sets, Vals),                 % get mass values
        compute_val(Vals, Val).                % multiply them
```

Fig. 2. constrain-and-generate

The predicate *compute/3* is used to compute every combination of focal points whose intersection is a non-empty subset of *Hyper* for both the *generate-and-test* and the *constrain-and-generate* algorithms. This predicate is called while computing the *belief/plausibility* of a desired set, as well as the *joint mass*.

5.2   Test Results

In order to compare both methods and examine whether the use of Constraint Programming reduces the time needed for the computation we created a number of random test cases on which we ran both programs. When creating random sets, we experimented with different values of the following parameters: i) the cardinality of Universe, ii) the number of mass functions, and iii) the number of focal points per mass function (for simplicity this is the same for every mass function). Both methods were run on each test case and the time needed for the execution was recorded.

The values of many parameters were highly influenced from values used in [8]. A Universe of size $|U| = 20$ was assumed as a basis for most of the tests. As focal points are created at random, choosing a small Universe results in mass functions sharing the same focal points (high "density"), whereas picking a large Universe will result in mass functions with different focal points (low "density"). We found that the value $|U| = 20$ keeps a good balance between a "sparse" and a "dense" case.

In the following, the first part is concerned with comparing the two methods for evaluating Dempster's Rule of Combination for every possible set so as to compute the *joint mass*. Next, we focus on computing *belief* for a specific set.

Computing the Joint Mass

To demonstrate the time gain of the constrain-and-generate algorithm as discussed in Sect. 3, we created a number of random test cases with fixed parameters and recorded the number of combinations that intersect and time needed for the computation for both the *generate-and-test*, and the *constrain-and-generate* algorithms. Notice that the time needed for the *constrain-and-generate* algorithm is proportional to the number of intersecting combinations. On the other hand, the *generate-and-test* algorithm always evaluates all possible combinations and its run-time is unrelated to the number of intersecting combinations, but depends on the total number of possible combinations. Some sample test cases are presented below (Figs. 3, 4, 5 and Tables 1, 2).

Table 1. $|U| = 20$, number of focal points per mass function = 3

Number of mass functions	generate-and-test (s)	constrain-and-generate (s)
10	0.58	0.22
11	1.77	1.28
12	5.39	0.52
13	8.45	0.20
14	21.55	0.66
15	113.41	0.86

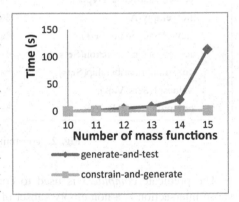

Fig. 3. $|U| = 20$, number of focal points per mass function = 3

Table 2. $|U| = 20$, number of mass functions = 10

Number of focal points per mass function	generate-and-test (s)	constrain-and-generate (s)
3	0.58	0.22
4	7.94	4.97
5	66.70	8.23
6	249.36	56.89

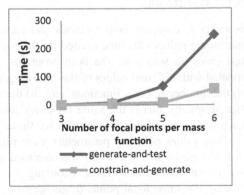

Fig. 4. $|U| = 20$, number of mass functions = 10

Table 3. $|U| = 20$, number of mass functions = 5, number of focal points per mass function = 10

Possible combinations	Intersecting combinations	constrain-and-generate (s)
100000	22240	2.17
	25204	2.36
	36753	3.05
	39085	3.16
	41821	3.84
	48379	3.98
	53184	4.68

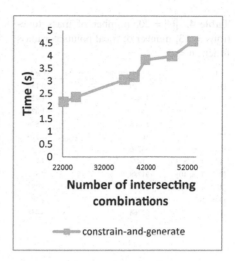

Fig. 5. $|U| = 20$, number of mass functions = 5, number of focal points per mass function = 10

Table 3 does not contain time results for the generate-and-test algorithm as it did not execute within an acceptable time.

Computing Belief

In this section, we discuss the results of the algorithms regarding computing *belief* of a specific set. Note that, similar results hold for computing *plausibility*, as we can compute *plausibility* (resp. *belief*) for a set by computing the *belief* (resp. *plausibility*) of its complement. Recall that when computing the *belief* of a set the normalization factor K must be evaluated in order to normalize the result. Note that, computing K is of the same complexity as computing the *joint mass* and the benefit of using CP in this case was discussed in the previous paragraph. So, we considered a case where $K = 0$ in order to examine the extra benefit by using CP in the case of belief evaluation.

We worked as before, creating random test cases so as to compare both methods. The number of all possible combinations is q^n, where n is the number of mass functions, and q is the number of focal points for each mass function. We have already mentioned that the *generate-and-test (g-t)* algorithm has to evaluate all q^n combinations and then discard those that do not intersect to the desired set, i.e. a subset of the set under consideration, whereas the *constrain-and-generate (c-g)* one sets constraints to be satisfied, so as to avoid evaluating every possible combination. We tested with different sizes of Universe, number of focal points per mass function, and number of mass functions. The run-time for both algorithms and the number of combinations that had to be evaluated for each set were recorded. The last set in each trial was the Universe itself. In all test cases for sets different from the Universe, improvement was noticed on run-time. A sample result is shown below.

Table 4. $|U| = 20$, number of mass functions = 13, number of focal points per mass function = 3

Set	Combinations	g-t (s)	c-g (s)
[1, 7, 9]	0		0.156
[4, 5, 9, 10]	14848		1.422
[2, 3, 6, 9, 10]	15360		1.625
[2, 6, 8]	15888		1.828
[2, 4–7, 10]	16128	4.438	1.719
[4, 8–10]	30736		3.313
[1, 2, 4, 5, 8–10]	31488		3
[2, 3, 7, 8, 10]	31788		3.5
[1, 3, 4, 6–8, 10]	31808		3.234
[1–8, 10]	32256		4.391
U	32768		4.438

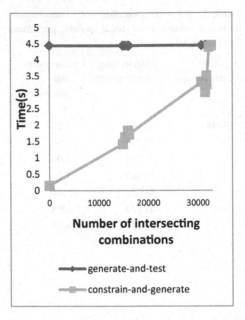

Fig. 6. $|U| = 20$, number of mass functions = 13, number of focal points per mass function = 3

Table 4 and Fig. 6 highlight the importance of constraints for computing *belief* for sets that are smaller than the Universe, or, otherwise stated, sets that are formed out by fewer combinations than the Universe.

To sum up, from the test results verified that Constraint Programming can help to reduce the time needed for computing Dempster's Rule of Combination whether we wish to compute the *joint mass* or *belief/plausibility* for a set. How much we can benefit from CP, depends deeply on the number of intersections that must be evaluated, which, unfortunately, is unknown prior to the computation. In general, the *constrain-and-generate's* performance is remarkable in cases where empty intersections exist, as it executes the calculation much faster than the *generate-and-test* method.

6   Conclusions and Future Work

Dempster-Shafer theory remains one of the most expressive frameworks for reasoning under uncertainty. However, the high complexity of Dempster's rule of combination imposes a significant restriction to its application. This becomes worse, considering the tremendous amount of data available. The method that we proposed to overcome this problem makes use of Constraint Programming to optimize the evaluation of Dempster's rule by computing only the appropriate combinations. The conclusion that can be drawn from the empirical testing that we performed is that the *constrain-and-generate* algorithm utilizing Constraint Programming needs considerable less time to compute the *joint mass*, or *belief/plausibility* of a specific set, than the *generate-and-test* algorithm. Therefore, such a method allows the computation of Dempster's rule of combination to be performed in an acceptable time even for more complex cases.

However, concerning the implementation platform, we have to make some remarks. *ECLiPSe Prolog* is one of the first systems to embed Constraint Programming libraries. While working with the *ic_sets* library, though, we encountered some anomalies. In cases where the combinations that have to be evaluated are significantly large, even the generation part of the *constrain-and-generate* method takes too much time, even more than the time needed by the simple *generate-and-test*. As far as we know, this could be an overhead of the library itself, or the way it handles some constraints.

As we have already pointed out, Constraint Programming allowed us to avoid generating non intersecting combinations. Notice that we can compute the normalization factor, *K*, by generating either all intersecting combinations or all non-intersecting ones. An idea that we would like to exploit is the concurrent computation of the intersecting and non-intersecting combinations in order to evaluate the normalization factor. This approach should ensure the fastest termination in all cases.

As mentioned, our implementation provides us with a prototype experimentation. It would be worthwhile to compare the use of constraint programming with approximation methods using data from a real-life application so as the comparison depicts the benefit we would gain in real world. Moreover, we believe it would be meaningful to compare our method with methods using the Fast Mobius Transformation [39–41]. In this case, another CP platform may be considered.

Acknowledgements. This research was partially funded by the National and Kapodistrian University of Athens Special Account of Research Grants no 13233.

References

1. Shafer, G.: A mathematical theory of evidence turns 40. Int. J. Approx. Reasoning **79**, 7–25 (2016)
2. Dubois, D., Prade, H.: Possibility theory, probability theory and multiple-valued logics: a clarification. Ann. Math. Artif. Intell. **32**(1), 35–66 (2001)
3. Pearl, J.: Fusion, propagation, and structuring in belief networks. Artif. Intell. **29**(3), 241–288 (1986)
4. McDermott, D., Doyle, J.: Non-monotonic logic I. Artif. Intell. **13**(1–2), 41–72 (1980)
5. Dempster, A.P.: Upper and lower probabilities induced by a multivalued mapping. Ann. Math. Stat. **38**(2), 325–339 (1967)
6. Shafer, G.: A Mathematical Theory of Evidence. Princeton University Press, Princeton (1976)
7. Shafer, G.: Perspectives on the theory and practice of belief functions. Int. J. Approximate Reasoning **4**(5–6), 323–362 (1990)
8. Zadeh, L.A.: Fuzzy Sets, Fuzzy Logic, and Fuzzy Systems. World Scientific Press, Singapore (1996)

9. Zadeh, L.A.: Fuzzy sets. Inf. Control **8**(3), 338–353 (1965)
10. World Wide Web Consortium. Semantic Web. https://www.w3.org/standards/semanticweb/. Accessed 13 Feb 2020
11. Laskey, K.J., Laskey, K.B., Costa, P.C.G., Kokar, M.M., Martin, T., Lukasiewicz, T.: Uncertainty reasoning for the world wide web. W3C Incubator Group Report, Technical report (2008)
12. Shafer, G.: Dempster's rule of combination. Int. J. Approx. Reasoning **79**, 26–40 (2016)
13. Lukasiewicz, T., Straccia, U.: Managing uncertainty and vagueness in description logics for the semantic web. Web Semant. **6**(4), 291–308 (2008)
14. Ortega, F.B.: Managing vagueness in ontologies. Ph.D. dissertation, Universidad de Granada (2008)
15. Stoilos, G., Stamou, G.B., Tzouvaras, V., Pan, J.Z., Horrocks, I.: Fuzzy OWL: uncertainty and the semantic web. In: Proceedings of the OWLED*05 Workshop on OWL: Experiences and Directions, Galway, Ireland (2005)
16. Karanikola, L., Karali, I.: Towards a Dempster-Shafer fuzzy description logic - handling imprecision in the semantic web. IEEE Trans. Fuzzy Syst. **26**(5), 3016–3026 (2018)
17. Karanikola, L., Karali, I.: Semantic web and ignorance: Dempster-Shafer description logics. In: FLAIRS Conference 2017, pp. 68–73 (2017)
18. Zadeh, L.A.: A simple view of the Dempster-Shafer theory of evidence and its implications for the rule of combination. AI Mag. **7**(2), 86–90 (1986)
19. Pei, W.: A defect in Dempster-Shafer theory. In: Proceedings of the 10th Conference on Uncertainty in Artificial Intelligence, Seattle, pp. 560–566 (1994)
20. Khan, N., Anwar, S.: Time-domain data fusion using weighted evidence and Dempster-Shafer combination rule: application in object classification. Sensors (Basel) **19**(23), 5187 (2019). https://doi.org/10.3390/s19235187
21. S. Mckeever, J. Ye, A Comparison of Evidence Fusion Rules for Situation Recognition in Sensor-Based Environments. In: Communications in Computer and Information Science. pp. 163–175 (2013). https://doi.org/10.1007/978-3-319-04406-4_16
22. Orponnen, P.: Dempster's rule of combination is #P-complete. Artif. Intell. **44**(1–2), 245–253 (1990)
23. Wilson, N.: Algorithms for Dempster-Shafer theory. In: Kohlas, J., Moral, S. (eds.) Handbook of Defeasible Reasoning and Uncertainty Management Systems, vol. 5, pp. 421–475. Springer, Dordrecht (2000). https://doi.org/10.1007/978-94-017-1737-3_10
24. Voorbraak, F.: A computationally efficient approximation of Dempster-Shafer theory. Int. J. Man Mach. Stud. **30**(5), 525–536 (1989)
25. Tessem, B.: Approximations for efficient computation in the theory of evidence. Artif. Intell. **61**(2), 315–329 (1993)
26. Lowrance, J.D., Garvey, T.D., Strat, T.M.: A framework for evidential-reasoning systems. In: 5th National Conference on Artificial Intelligence, Menlo Park, California, pp. 896–903 (1986)
27. Bauer, M.: Approximation algorithms and decision making in the Dempster-Shafer theory of evidence – an empirical study. Int. J. Approximate Reasoning **17**(2–3), 217–237 (1997)
28. Mayoh, B., Tyugu, E., Penjam, J.: Constraint Programming. Springer, Heidelberg (1993). https://doi.org/10.1007/978-3-642-85983-0
29. The ECLiPSe Constraint Programming System. https://eclipseclp.org/. Accessed 13 Feb 2020
30. Gervet, C: Conjunto: constraint logic programming with finite set domains. In: ILPS (1994)
31. Eclipse conjunto library. http://eclipseclp.org/doc/bips/lib/conjunto/index.html. Accessed 13 Feb 2020

32. Kohlas, J., Monney, P.-A.: Propagating belief functions through constraint systems. Int. J. Approximate Reasoning **5**(5), 433–461 (1991)
33. Rouahi, A., Ben Salah, K., Ghédira, K.: Belief constraint satisfaction problems. In: 12th International Conference of Computer Systems and Applications (AICCSA), Marrakech, pp. 1–4 (2015)
34. Aiken, A., Kozen, D., Vardi, M., Wimmers, E.: The complexity of set constraints. In: Börger, E., Gurevich, Y., Meinke, K. (eds.) CSL 1993. LNCS, vol. 832, pp. 1–17. Springer, Heidelberg (1994). https://doi.org/10.1007/BFb0049320
35. Lloyd, J.W.: Foundations of Logic Programming. Springer-Verlag, Heidelberg (1984). https://doi.org/10.1007/978-3-642-96826-6
36. Colmerauer, A., Kanoui, H.: Philippe Roussel et Robert Pasero, Un système de communication homme-machine en Français, rapport de recherche. Groupe de recherche en Intelligence Artificielle, Marseille (1973)
37. Jaffar, J., Lassez, J.L.: Constraint logic programming. In: Proceedings of the 14th ACM SIGACT-SIGPLAN Symposium on Principles of Programming Languages, Munch, West Germany, pp. 111–119 (1987)
38. Van Hentenryck, P.: Constraint Satisfaction in Logic Programming. MIT Press, Cambridge (1989)
39. Kennes, R., Smets, P.: Computational aspects of the Mobius transformation. In: Proceedings of the Sixth Annual Conference on Uncertainty in Artificial Intelligence, pp. 401–416. Elsevier Science Inc., USA (1990)
40. Kennes, R., Smets, P.: Fast algorithms for Dempster-Shafer theory. In: Uncertainty in Knowledge Bases. IPMU 1990, Paris, France, pp. 14-26 (1990)
41. Chaveroche, M., Davoine, F., Cherfaoui, V.: Efficient Möbius transformations and their applications to D-S theory. In: 13th International Conference on Scalable Uncertainty Management, Compiègne, France, pp. 390–403 (2019)

Bayesian Smoothing of Decision Tree Soft Predictions and Evidential Evaluation

Nicolas Sutton-Charani(✉) ⓘ

EuroMov Digital Health in Motion, Univ Montpellier, IMT Mines Alès, Alès, France
nicolas.sutton-charani@mines-ales.fr

Abstract. As for many classifiers, decision trees predictions are naturally probabilistic, with a frequentist probability distribution on labels associated to each leaf of the tree. Those probabilities have the major drawback of being potentially unreliable in the case where they have been estimated from a limited number of examples. Empirical Bayes methods enable the updating of observed probability distributions for which the parameters of the *prior* distribution are estimated from the data. This paper presents an approach of smoothing decision trees predictive binary probabilities with an empirical Bayes method. The update of probability distributions associated with tree leaves creates a correction concentrated on small-sized leaves, which improves the quality of probabilistic tree predictions. The amplitude of these corrections is used to generate predictive belief functions which are finally evaluated through the ensemblist extension of three evaluation indexes of predictive probabilities.

Keywords: Smoothing · Correction · Predictive probabilities · Decision trees · Bayesian empirical methods · Predictive belief functions · Uncertain evaluation

1 Introduction

Even if the predictions provided by classifiers are generally considered in a *precise* or *crisp* form, they are often initially computed as *soft* predictions through probability distributions, most probable labels being used as *hard* predictions at the final predictive or decision-making step. Decision trees are basic classifiers and regressors that are at the basis of many famous supervised learning algorithms (random forest, XGBoost, etc.). Once a tree built, the proportions of labels contained in each leaf are used to compute these predictive probabilities. Small leaves, *i.e.* containing only a small number of examples, therefore produce unreliable predictive probabilities since they are computed from a limited amount of data. Those unreliable leaves are usually cut in a post-pruning step in order to avoid overfitting, but due to the complexity pruning methodologies often involve, users tend to choose pre-pruning strategies, *i.e.* set more conservative stopping criteria, instead.

In the classic machine learning literature some work focus on classifiers predictive probabilities *calibration* in order to make them smoother or to correct

© Springer Nature Switzerland AG 2020
M.-J. Lesot et al. (Eds.): IPMU 2020, CCIS 1238, pp. 368–381, 2020.
https://doi.org/10.1007/978-3-030-50143-3_28

some intrinsic biases typical of different predictive models [16]. These approaches often involve the systematic application of mathematical functions requiring a set of dedicated data at the stage of calibration and sometimes requiring heavy computations in terms of complexity [20] or only considering a subset of the learning data [27]. Statistical approaches such as Laplacian or additive smoothing provide tools that are known to correct estimators in order to increase the impact of classes for which only few or even no data are available. Those techniques have been largely used in Natural Language Processing [12] and Machine learning [21,23]. From a Bayesian point of view, Laplacian smoothing equates to using a non-informative prior, such as the uniform distribution, for updating the expection of a Dirichlet distribution. Other works based on evidential models enable local adjustments during the learning phase of decision trees. In these methodologies class labels estimates are carried out on small subsets of data independently to the dataset global distribution [7].

This works aims at providing basic adjustments of decision trees outputs based on the tree structure and the global distribution of the learning data without involving any additional complexity. The approach presented in this article allows the correction of classification trees predictive probabilities in the case of binary classes. To achieve this, an empirical Bayesian method taking into account the whole learning sample as prior knowledge is applied and results in the adjustments of the probabilities associated with leaves relatively to their size. Unlike Bayesian smoothing which takes benefits only from the size of sub-samples corresponding to leaves, empirical Bayesian smoothing uses the whole distribution of labels in regards to leaves which can be viewed as a rich piece of information and is therefore legitimate to be incorporated in any predictive evidential modelling. To this extent, the ranges of the resulting estimates corrections are finally used to generate predictive belief functions by discounting the leaves predictive probabilities which can be finally evaluated following extensions of existing evaluation metrics. It should be noticed that this work is out of the scope of learning decision trees from uncertain data as in [6,25,26]. In this paper all the learning data are precise, it is at the prediction step that the uncertainty is modelled by frequentist probabilities wich are smoothed and transformed into belief functions from their correction ranges.

After recalling the necessary background in Sect. 2, the proposed approach is described in detail in Sect. 3. Taking into account the predictive probability adjustments amplitude enables, in Sect. 4, the formalisation of an evidential generative model and the extension of three evaluation metrics of predictive probabilities to the case of uncertain evidential predictions. In Sect. 6, a first set of experiments shows the contribution of the methodology on the one hand in a pragmatic point of view in terms of predictive performance and on the other hand by the flexibility of decision-making it offers.

2   Basics

The learning of a decision tree corresponds to a recursive partitioning of the attributes space aiming at separating labels as well as possible (classification) or to decrease their variance (regression) [1]. The learning data are thus distributed in the different tree leaves, which are then associated with probability distributions over class variables according to the proportions of labels in the examples they contain. In this article we restrict ourselves to the case of binary classes noted $\{1, 0\}$ or $\{+, -\}$.

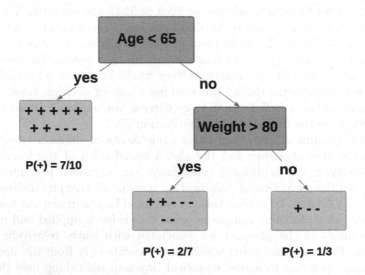

Fig. 1. Predictive probabilities

Fig. 1 is an example of a decision tree in which any example *younger* than 65 years old will have a positive class probability estimated by $P(+) = \frac{7}{10}$, for any *older* example (than 65 years old) and *heavier* than 80 kg we will have $P(+) = \frac{2}{7}$ and finally for examples older than 65 years old and weighing less than 80 kg we will predict $P(+) = \frac{1}{3}$. This last probability is here estimated from only 3 examples, it is therefore natural to consider it relatively unreliable (in comparison to the two others based on respectively 10 and 7 examples).

Various stopping criteria can be used during the learning of a decision tree depending on the structure of the tree (leaves number, depth, etc.) or in terms of information (impurity gain, variance). In order to avoid overfitting, pruning strategies are generally implemented to limit the number of leaves (which reduces variance and complexity).

2.1 Empirical Bayesian Methods

Bayesian inference is an important field of Statistics which consists in using some *prior* knowledge in order to update the estimations computed on data (according to Bayes theorem). This updates often result in predictive probabilities whose quality directly depends on the prior information. While Bayesian priors are generally constituted of probability distributions that the user subjectively express about the phenomenon of interest (expert opinions are often at the basis of the prior's modelling), for the Bayesian empirical methods [4, 22], the parameters of these prior distributions are estimated from the data. Some authors consider these methods as approximations of hierarchical Bayesian models [19].

Considering a sample of a binary variable $y = (y_1, ..., y_n) \in \{0, 1\}^n$, for all subset $y^* \subseteq y$, a natural (and frequentist) estimate of the probability of label 1, i.e. of $p^* = P(Y = 1 | Y \in y^*)$, is its observed frequency $\widetilde{p}^* = \frac{|\{y_i \in y^* : y_i = 1\}|}{|y^*|}$. This estimator makes implicitly the assumption that the sample y^* is large enough to estimate the probability of label 1 by a random draw in it.

In Bayesian statistics, for the binomial model corresponding to y^*'s generation, a *prior* knowledge about p^* is usually modelled in terms of probability distribution by $p^* \sim Beta(\alpha, \beta)$ (*flexible* prior) and the model update from the data y^* results in a *posterior* distribution $p_1^* y^* \sim Beta(\alpha + n_1^*, \beta + n^* - n_1^*)$ with $n_1^* = |\{y_i \in y^* : y_i = 1\}|$ and $n^* = |y^*|$.

The Bayesian (*posterior*) estimator of p^* is finally computed as its conditional expectation given y^*:

$$\widehat{p}^* = \mathbb{E}[p^* | y^*] = \frac{n_1^* + \alpha}{n^* + \alpha + \beta} \tag{1}$$

By doing so, $\widetilde{p}^*$ is shifted toward its global expectation $E[\widetilde{p}^*] = \frac{\alpha}{\alpha + \beta}$ and this shift's range relies on the value of n^*.

Whereas in standard Bayesian works the *prior*'s parameters are often set by an expert according to his knowledge, i.e. subjectively, or by default in a non-informative form as with Laplacian smoothing ($p^* \sim Beta(1, 1)$ is equivalent to $p^* \sim \mathcal{U}[0, 1]$), in empirical Bayesian approaches they are estimated from the whole sample y (by likelihood maximisation) with the hypothesis that $p \sim Beta(\alpha, \beta)$. One direct consequence is that the smaller the size of y^*, the greater the amplitude of the shift of p^* toward its global expectation on y. The Bayesian estimator of this approach, illustrated for the number of successful baseball shots per player in [2], is here applied to the probabilistic predictions attached with the leaves of binary classification trees.

2.2 Evaluation of Predictive Probabilities

Even if the evaluation of a classifier is often done from precise or *crisp* predictions by comparing them to the real class labels through different metrics (*e.g.* accuracy, precision, recall, etc.) it can however be done at the level of predictive probabilities, thus upstream. Three metrics for evaluating binary probabilistic predictions are presented hereinafter.

Let $y = (y_1, ..., y_n) \in \{1, 0\}^n$ be the *true* labels of a given sample of size n and $p = (p_1, ..., p_n) = [P(Y_1 = 1), ..., P(Y_n = 1)]$ the class predictive probabilities of label $\{1\}$ according to a given predictive model M, applied to the sample x. The Table 1 summarizes the definitions of the *log-loss*, the Brier score and the area under the *ROC* curve (*AUC*). The log-loss can be interpreted as a Kullback–Leibler divergence between p and y that takes into account p's entropy and is therefore called *cross-entropy* by some authors. The Brier score is defined as the mean squared difference between p and y. The log-loss and Brier score thus measure the difference between observations and predicted probabilities, penalizing the probabilities of the least probable labels. The ROC curve is a standard measure of a binary classifier's predictive power, it represents the rate of true positives or *sensibility* (*i.e.* the proportion of positive examples that are predicted as positive) as a function of the rate of false positives or $1 - specificity$ (*i.e.* the proportion of negative examples that are predicted as positive), we have $ROC : sensibility(1 - specificity)$. The AUC area under the ROC curve is a well known indicator of the quality of probabilistic predictions.

The three metrics thus defined lie in $[0, 1]$ and a good binary classifier will be characterized by log-loss and Brier score values close to 0 and an AUC value close to 1. It should be noted, however, that these three metrics are defined for standard uncertain predictions, *i.e.* probabilistic predictions.

Table 1. Evaluation metrics of binary probabilistic predictions $(p_1, ...p_n)$ with respect to true labels $(y_1, ...y_n)$

Name	Definition
Log-loss	$-\frac{1}{n} \sum_{i=1}^{n} \left[y_i \log(p_i) + (1 - y_i) \log(1 - p_i) \right]$
Brier score	$\frac{1}{n} \sum_{i=1}^{n} (y_i - p_i)^2$
AUC	$\int_0^1 ROC(p, y, \lambda) d\lambda$

3 Empirical Bayesian Correction of Decision Trees Predictive Probabilities

The approach presented in this paper consists in correcting the predictive probabilities of a binary classification tree containing H leaves denoted $l_1, ..., l_H$ with an empirical Bayesian method. It is assumed that the proportion of label $\{1\}$ within the leaves of the tree follows a $Beta(\alpha, \beta)$ distribution (without setting *a priori* the values of the parameters α and β). We can notice that the Beta distribution is both a special case of the Dirichlet distribution (widely used in Bayesian statistics) and a generalization of the uniform distribution $(\mathcal{U}_{[0,1]} = Beta(1, 1))$ which is supposed to model *non-informative* prior knowledge.

Once the hypothesis of the Beta law is formulated, its parameters α and β are estimated from the set $\{p_1^1, ..., p_1^H\}$ of label proportions $\{1\}$ within the H leaves of the considered tree. In order to penalize small leaves, we will consider an artificial sample denoted E containing each p_1^h proportion repeated a number of times equal to the size of the considered leaf, i.e. to the number of examples it contains.

We have $E = (\underbrace{p_1^1, ..., p_1^1}_{|l_1| \text{ times}}, ..., \underbrace{p_1^H, ..., p_1^H}_{|l_H| \text{ times}})$.

The estimation of the parameters α and β can then be performed on E by likelihood maximisation following different approaches (moments, least squares, etc.). This step makes α and β's values take into account the leaves sizes through an artificial weighting that is defined according to their sizes. By doing so, the information of the tree structure is used through the leaves partitioning and their sizes to define the prior or the empirical Bayes model. The probabilities $\tilde{p}_1^h$ of the leaves $l_1, ..., l_H$ are finally corrected according to the Eq. (1):

$$\tilde{p}_1^h = \frac{n_1^h}{n^h} \rightsquigarrow \tilde{p}_1^h = \frac{n_1^h + \alpha}{n^h + \alpha + \beta} \tag{2}$$

with n^h and n_1^h denoting respectively the size of the leaf l_h (i.e. the number of examples it contains) and the number of examples it contains that have the label $\{1\}$.

It should be recalled that other works [20,27] allow a *calibration* of the predictive probabilities by different approaches using either only the distributions of the examples contained in the leaves independently from one another, or the distribution of the whole learning sample but applying systematic transformations based on estimations requiring many computations (often obtained by cross-validation). The method proposed in this paper uses both the whole distribution of the training data (once distributed in the different leaves of a tree) and remains very simple in terms of complexity, the α and β parameters of Eq. (2) being estimated only once for the whole sample and then used locally on leaves according to their sizes at the prediction step. Nevertheless the range of these probabilistic correction represent a piece of information by itself that should be incorporated into the leaves in order to express a confidence level on themselves. In the next section an evidential generative model based on these correction ranges is presented.

4 Generation of Predictive Belief Functions

The uncertainty expressed in the predictive probabilities of a classifier is mainly *aleatory*. It is based on the mathematical model underlying the classifier and on frequentist estimates. The knowledge of the empirical Bayesian adjustment and its range is a piece of information in itself that can allow epistemic uncertainty to be incorporated into the leaves predictive probabilities. It is indeed natural to consider unreliable the predictive probabilities that are estimated on a small number of examples (and therefore highly adjusted).

Unlike many works on belief functions generation where uncertain data are used in evidential likelihoods applied to parametric models [9,14] or random generation is extended to mass function [3], the context of this article is restricted to the case of discounting leaves predictive probabilities according to the range of their empirical Bayesian adjustment. As a first approach, we propose to generate a belief function from a predictive probability using its empirical Bayesian correction range $|\tilde{p} - \hat{p}|$ as unreliability indicator ($\tilde{p}$ and $\hat{p}$ denoting respectively standard frequentist and empirical Bayesian adjusted estimates of p), assigning its weight to ignorance and substracting it uniformly to singletons probabilities. The resulting belief function can be viewed as a type of evidential discounted predictive probabilities. This modeling relies on the hypothesis that the more a predictive probability is corrected (*i.e.* the smaller the subsample considered), the less reliable it is.

$$\begin{cases} m(\{1,0\}) = |\tilde{p} - \hat{p}| \\ m(\{1\}) \ = \tilde{p} - \frac{|\tilde{p}-\hat{p}|}{2} \\ m(\{0\}) \ = 1 - \tilde{p} - \frac{|\tilde{p}-\hat{p}|}{2} \end{cases} \tag{3}$$

In order to make as few assumptions as possible, once the mass of ignorance is defined, the mass of the two classes are symmetrically discounted. We can notice that we have $Bel(\{0\}) = 1 - Pl((\{1\})$ and $Bel(\{1\}) = 1 - Pl((\{0\})$.

Remark: This belief function modelling can also be written in terms of imprecise probability: $p \in [p^-, p^+]$ with

$$\begin{cases} p^- = \tilde{p} - \frac{|\tilde{p}-\hat{p}|}{2} \\ p^+ = \tilde{p} + \frac{|\tilde{p}-\hat{p}|}{2} \end{cases} \tag{4}$$

The generative model (3–4) can be interpreted as a reliability modelling of the corrected leaves predictive probabilities. The output nature of a classification tree can thus be considered through imprecise probabilities. In the next section, an imprecise evaluation model is presented that keeps an evidential uncertainty level until final outputs of decision trees evaluation.

5   Imprecise Evaluation

Evidential predictions evaluation remains a challenging task, some works have been presented based on evidential likelihood maximisation of evaluation model parameters (error rate or accuracy) [25] but in case of a predictive belief function the simplest evaluation solution is to convert it into standard probability with the pignistic transform for instance. The main drawback of such practice is that all the information contained in the uncertain modelling of the predictions is lost but it has the pragmatic advantage of providing *crisp* evaluation metrics that can be easily interpreted.

In order to keep the predictive uncertainty or imprecision provided by the model (4) until the stage of the classifiers evaluation, it is possible to consider the metrics defined in Table 1 in an ensemblist or *intervalist* perspective. Indeed, to a set of predictive probabilities naturally corresponds a set of values taken by these evaluation metrics. An imprecise probability $[p^-, p^+]$ computed according to the model (4) will thus be evaluated imprecisely by an interval defined as follows:

$$eval([p^-, p^+], y) = \left[\min_{p \in [p^-, p^+]} eval(p, y), \max_{p \in [p^-, p^+]} eval(p, y) \right]$$

where *eval* is one of the metrics defined in Table 1.

This type of evaluation approach involves that, the smaller the leaves of the evaluated decision tree, the more the probabilities associated with these leaves will be adjusted and the more imprecise the evaluation of these trees will be (*i.e.* the wider the intervals obtained). This approach is therefore a means of propagating epistemic uncertainty about the structure of the tree into its evaluation metrics. It should be noted, however, that this solution requires an effective browse of the entire $[p^-, p^+]$ interval, which implies a significant computational cost.

6 Experiments

In this section, a set of experiments is implemented to illustrate the practical interest of the empirical Bayesian correction model presented in this paper. Using six benchmark datasets from the UCI[1] and Kaggle[2] sites, 10-fold cross-validations are carried out with, for each fold of each data set, the learning of decision trees corresponding to different complexities on the nine other folds and an evaluation on the fold in question using the three precise metrics defined in Table 1. The 'cp' parameter (of the 'rpart' function in R) allows to control the trees complexity, it represents the minimal relative information gain of each considered cut during the learning of the trees. Trees denoted 'pruned' are learned with a maximum complexity ($cp = 0$) and then pruned according to the classical *cost-complexity* criteria approach of the *CART* algorithm [1]. The evaluations consist of the precise metrics computations presented in Table 1 as well as their uncertain extensions defined in Sect. 5. These steps are repeated 150 times in order to make the results robust to fold random generation and only the mean evaluations of the trees predictive probabilities are represented here. The codes used for the implementation of all the experiments presented below are available at https://github.com/lgi2p/empiricalBayesDecisionTrees.

Table 2 represents the characteristics of the different datasets used in terms of number of examples (n), number of attributes or predictor variables (J) and number of class labels (K). The Tables 3, 4, 5, 6, 7 and 8 contain the mean evaluations computed for each dataset and for each tree type, over all 150 cross-validations performed, before and after empirical Bayesian smoothing. Figures 2 and 3 illustrate the distributions of these results with evaluation intervals for the log-loss on the 'banana' dataset and for the Brier score on the 'bankLoan' dataset with respect to the supports of the corresponding uncertain evaluations (only the extreme points of the uncertain metrics are represented).

The log-loss of corrected trees is almost always lower than that of original trees. This increase in performance is clear for large trees (learnt with a low

---

[1] https://archive.ics.uci.edu/ml/datasets.html.
[2] https://www.kaggle.com/datasets.

Table 2. Datasets dimensions

	n	J	K
banana	5300	2	2
bankLoan	5000	12	2
banknote	1372	4	2
mammo	830	5	2
pima	768	8	2
ticTacToe	958	9	2

Table 3. Log-loss before smoothing

dataset\cp	0	0.001	0.005	0.01	0.05	pruned
banana	0.410	0.351	0.355	0.419	0.425	0.359
bankLoan	1	0.665	0.568	0.571	0.571	0.566
banknote	0.330	0.329	0.322	0.331	0.347	0.329
mammo	0.444	0.437	0.414	0.409	0.450	0.416
pima	0.886	0.884	0.742	0.655	0.565	0.630
ticTacToe	0.221	0.221	0.213	0.210	0.549	0.214

Table 4. Log-loss after smoothing

dataset\cp	0	0.001	0.005	0.01	0.05	pruned
banana	0.287	0.300	0.334	0.418	0.424	0.358
bankLoan	0.640	0.582	0.568	0.571	0.571	0.566
banknote	0.173	0.174	0.180	0.225	0.346	0.328
mammo	0.419	0.419	0.410	0.406	0.450	0.416
pima	0.557	0.557	0.545	0.541	0.564	0.630
ticTacToe	0.188	0.188	0.188	0.191	0.550	0.214

Table 5. Brier score before smoothing

dataset\cp	0	0.001	0.005	0.01	0.05	pruned
banana	0.084	0.086	0.097	0.129	0.131	0.086
bankLoan	0.229	0.204	0.190	0.192	0.192	0.189
banknote	0.040	0.040	0.043	0.057	0.098	0.041
mammo	0.130	0.130	0.125	0.123	0.140	0.126
pima	0.188	0.188	0.184	0.183	0.189	0.186
ticTacToe	0.061	0.061	0.061	0.062	0.187	0.062

Table 6. Brier score after smoothing

dataset\cp	0	0.001	0.005	0.01	0.05	pruned
banana	0.083	0.086	0.097	0.129	0.131	0.086
bankLoan	0.219	0.198	0.190	0.192	0.192	0.189
banknote	0.040	0.040	0.042	0.057	0.098	0.041
mammo	0.129	0.129	0.125	0.123	0.140	0.126
pima	0.183	0.183	0.179	0.179	0.188	0.186
ticTacToe	0.060	0.060	0.060	0.062	0.187	0.061

Table 7. AUC before smoothing

dataset\cp	0	0.001	0.005	0.01	0.05	pruned
banana	0.946	0.936	0.920	0.872	0.867	0.937
bankLoan	0.778	0.772	0.747	0.722	0.720	0.755
banknote	0.843	0.838	0.821	0.799	0.777	0.827
mammo	0.852	0.848	0.835	0.819	0.789	0.839
pima	0.839	0.836	0.826	0.812	0.771	0.821
ticTacToe	0.861	0.858	0.850	0.838	0.758	0.846

Table 8. AUC after smoothing

dataset\cp	0	0.001	0.005	0.01	0.05	pruned
banana	0.948	0.937	0.920	0.872	0.867	0.937
bankLoan	0.781	0.773	0.747	0.722	0.720	0.755
banknote	0.845	0.840	0.822	0.799	0.777	0.827
mammo	0.854	0.850	0.836	0.819	0.789	0.838
pima	0.842	0.839	0.828	0.813	0.771	0.821
ticTacToe	0.864	0.860	0.851	0.839	0.758	0.845

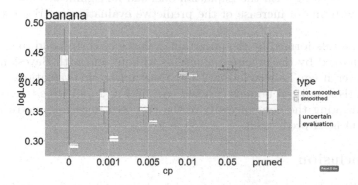

Fig. 2. Precise and uncertain log-loss as a function of the complexity

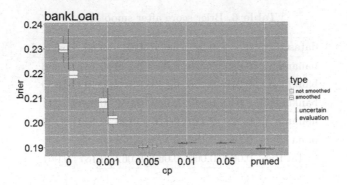

Fig. 3. Precise and uncertain Brier score as a function of the complexity

value of the hyper-parameter '*cp*' and thus containing many leaves), slightly less visible for small trees and very limited for pruned trees. This difference in terms of performance gain in regards to the tree size is probably due to the fact that the bigger the trees, the smaller are their leaves. Indeed, if the examples of the same learning data are spread into a higher number of leaves, their number inside the leaves have to be lower (in average). This conclusion shows that empirical Bayesian adjustment makes sense especially for complex models. The same phenomenon of gain in performance proportional to tree size is globally observable for the Brier score and the AUC index but in smaller ranges.

Other works have already illustrated the quality increase of decision trees predictions by smoothing methods [5,17,18] but none of these neither illustrated nor explained the link between pruning and those increase ranges based on overfitting intuition. Moreover, using the learning sample labels distribution as prior knowledge is a new proposal that has not yet been studied in the context of decision trees leaves, especially in order to generate predictive belief functions (and their evaluation counterpart). Even if it was not in the scope of this article, some experiments have been carried on in order to compare the empirical Bayesian smoothing with the Laplacian one, and no significant differences were observed in terms of increase of the predictive evaluation metrics used in this paper.

The intervals formed by the uncertain evaluations correspond roughly to the intervals formed by the precise evaluations without and with Bayesian smoothing. However, it can be seen in Fig. 2 and 3 that uncertain evaluations sometimes *exit* from these natural bounds (pruned trees in Fig. 2 and *large* trees in Fig. 3), thus highlighting the non-convexity of the proposed uncertain metrics for evaluating evidential predictions.

7   Conclusion

The empirical Bayesian correction model presented in this paper for decision trees predictive probabilities is of interest in terms of predictive performance, and this interest is particularly relevant for large trees. The fact that Bayesian corrections hardly improve the performance of *small* (i.e. with low complexity) or pruned trees suggests that the Bayesian correction represents an alternative to pruning. By shifting the predictive probabilities of small leaves to their global averages (i.e. calculated within the total learning sample) it reduces the phenomenon of overfitting.

In this paper Bayesian corrections are only performed at the predictive stage, i.e. at the leaf level. Adopting the same approach throughout the learning process is possible by proceeding in the same way at the purity gain computation level (i.e. for all the considered cuts). It would be interesting to compare the approach presented in this work with the one of [7] which pursues the same goal (penalizing small leaves) based on an evidential extension of purity gain computation where a mass of $\frac{1}{n+1}$ is assigned to *ignorance* in the impurity measure presented in [15] that combines *variability* and *non-specificity* computed on belief functions. It is important to note that the approach previously mentioned is based on the distribution of examples within the leaves individually, in case of unbalanced learning samples they do not allow correction in the direction of the general distribution as it is the case with the empirical Bayesian model.

The approaches of predictive belief functions generation based on the use of evidential likelihood [9,13] also represent an interesting alternative to which it will be important to compare oneself both in terms of predictive performances and with respect to the underlying semantics. In the same vein, the approach presented in this paper for the binary classification context could be extended to the multi-class case following the approaches used in [24]. More generally, all classifiers whose learning at the predictive stage involves frequentist probability computations could potentially benefit from this type of Bayesian correction. Ensemble learning approaches could be enhanced by empirical Bayesian corrections at different levels. When their single classifiers are decision trees, it could be straightforward to correct trees with the correction model presented in this article. A global adjustment could be also be achieved at the aggregation phase with the same type of artificial sample creation as for tree correction (leaves level could be extended to classifiers level).

The model for generating predictive belief functions and especially the extension of evaluation metrics to the credibility context proposed in this work could be greatly enriched by a more refined modeling of the uncertainty resulting from the initial predictive probabilities and their corrections. For example, it would be possible to use the distances proposed in [11] in order to make the representation of uncertain assessment metrics more complex beyond simple intervals. It would also be desirable to directly estimate the bounds of the uncertain evaluation intervals without having to browse them effectively from simulation-based approaches such as in [8] or by optimization results such as in [10].

References

1. Breiman, L., Friedman, J., Stone, C.J., Olshen, R.A.: Classification and Regression Trees. Chapman & Hall, Boca Raton (1984)

2. Brown, L.P.: Empirical Bayes in-season prediction of baseball batting averages. Ann. Appl. Stat. **2**(1), 113–152 (2008)
3. Burger, T., Destercke, S.: How to randomly generate mass functions. Int. J. Uncertainty Fuzziness Knowl. Based Syst. **21**, 645–673 (2013)
4. Casella, G.: An introduction to empirical Bayes data analysis. Am. Stat. **39**(5), 83–87 (1985)
5. Chawla, N.V.: Many are better than one: improving probabilistic estimates from decision trees. In: Quiñonero-Candela, J., Dagan, I., Magnini, B., d'Alché-Buc, F. (eds.) MLCW 2005. LNCS (LNAI), vol. 3944, pp. 41–55. Springer, Heidelberg (2006). https://doi.org/10.1007/11736790_4
6. Elouedi, Z., Mellouli, K., Smets, P.: Belief decision trees: theoretical foundations. Int. J. Approximate Reasoning **28**(2–3), 91–124 (2001)
7. Denœux, T., Bjanger, M.: Induction of decision trees from partially classified data using belief functions. In: International Conference on Systems, Man And Cybernetics (SMC 2000), vol. 4, pp. 2923–2928 (2000)
8. Denœux, T., Masson, M.-H., Hébert, P.-A.: Non-parametric rank-based statistics and significance tests for fuzzy data. Fuzzy Sets Syst. **153**(1), 1–28 (2005)
9. Denœux, T.: Likelihood-based belief function: justification and some extensions to low-quality data. Int. J. Approximate Reasoning **55**(7), 1535–1547 (2014)
10. Destercke, S., Strauss, O.: Kolmogorov-Smirnov test for interval data. In: Information Processing and Management of Uncertainty (IPMU), pp. 416–425 (2014)
11. Jousselme, A.-L., Maupin, P.: Distances in evidence theory: comprehensive survey and generalizations. Int. J. Approximate Reasoning **53**(2), 118–145 (2012)
12. Jurafsky, D., Martin, J.H.: Speech and Language Processing: An Introduction to Natural Language Processing, Computational Linguistics, and Speech Recognition. Prentice Hall PTR, Upper Saddle River (2000)
13. Kanjanatarakul, O., Sriboonchitta, S., Denœux, T.: Forecasting using belief functions: an application to marketing econometrics. Int. J. Approximate Reasoning **55**(5), 1113–1128 (2014)
14. Kanjanatarakul, O., Denœux, T., Sriboonchitta, S.: Prediction of future observations using belief functions: a likelihood-based approach. Int. J. Approximate Reasoning **72**, 71–94 (2016)
15. Klirr, G.J.: Uncertainty and Information: Foundations of gEneralized Information Theory. Wiley- IEEE Press, New York (2013)
16. Kuhn, Max, Johnson, Kjell: Applied Predictive Modeling. Springer, New York (2013). https://doi.org/10.1007/978-1-4614-6849-3
17. Ferri, C., Flach, P.A., Hernández-Orallo, J.: Decision trees for ranking: effect of new smoothing methods, new splitting criteria and simple pruning methods. Mathematics (2003)
18. Margineantu, D.D., Dietterich, T.G.: Improved class probability estimates from decision tree models. In: Denison, D.D., Hansen, M.H., Holmes, C.C., Mallick, B., Yu, B. (eds.) Nonlinear Estimation and Classification, pp. 173–188. Springer, New York (1989). https://doi.org/10.1007/978-0-387-21579-2_10
19. Maritz, J.S., Lwin, T.: Applied Predictive Modeling, 2nd edn. Chapman and Hall, London (1989)
20. Niculescu-Mizil, A., Caruana, R.: Predicting good probabilities with supervised learning. In: Proceedings of the 22nd International Conference on Machine Learning (ICML 2005), pp. 625–632 (2005)
21. Osher, S.J., Wang, B., Yin, P., Luo, X., Pham, M., Lin, A.T.: Laplacian smoothing gradient descent. In: International Conference on Learning Representations (ICLR) 2019 Conference - Blind Submission (2019)

22. Robbins, H.: An Empirical Bayes Approach to Statistics. University of California Press **1**, 157–163 (1956)
23. Sucar, L.E.: Probabilistic Graphical Models. ACVPR. Springer, London (2015). https://doi.org/10.1007/978-1-4471-6699-3
24. Sutton-Charani, N., Destercke, S., Denœux, T.: Classification trees based on belief functions. In: Proceedings of the 2nd International Conference on Belief Functions (BELIEF 2012) (2012)
25. Sutton-Charani, N., Destercke, S., Denœux, T.: Training and evaluating classifiers from evidential data: application to E2M tree pruning. In: Proceedings of the 3rd International Conference on Belief Functions (BELIEF 2014) (2014)
26. Trabelsi, A., Elouedi, Z., Lefevre, E.: Handling uncertain attribute values in decision tree classifier using the belief function theory. In: Dichev, C., Agre, G. (eds.) AIMSA 2016. LNCS (LNAI), vol. 9883, pp. 26–35. Springer, Cham (2016). https://doi.org/10.1007/978-3-319-44748-3_3
27. Zadrozny, B., Elkan, C.: Obtaining calibrated probability estimates from decision trees and naive Bayesian classifiers. In: Proceedings of the Eighteenth International Conference on Machine Learning (ICML 2001) (2001)

On Solutions of Marginal Problem in Evidence Theory

Jiřina Vejnarová(✉)

Czech Academy of Sciences, Institute of Information Theory and Automation,
Pod Vodárenskou Věží 4, Prague, Czech Republic
vejnar@utia.cas.cz

Abstract. Recently introduced marginal problem – which addresses the question of whether or not a common extension exists for a given set of marginal basic assignments – in the framework of evidence theory is recalled. Sets of solutions are studied in more detail and it is shown, by a simple example, that their structure is much more complicated (i.e. the number of extreme vertices of the convex set of solutions is substantially greater) than that in an analogous problem in probabilistic framework. The concept of product extension of two basic assignments is generalized (via operator of composition) to a finite sequence of basic assignments. This makes possible not only to express the extension, if it exists, in a closed form, but also enables us to find the sufficient condition for the existence of an extension of evidential marginal problem. Presented approach is illustrated by a simple example.

Keywords: Marginal problem · Extension · Product extension

1 Introduction

The marginal problem is one of the most challenging problem types in probability theory. It addresses the question of whether or not a common extension exists for a given set of marginal distributions. The challenges lie not only in a wide range of the relevant theoretical problems (probably the most important among them is to find conditions for the existence of a solution to this problem), but also in its applicability to various problems of artificial intelligence [11], statistics [4] and computer tomography [6]. The importance of this problem type is emphasized by the fact that from 1990 a series of conferences with this topic takes place every three years. During the last two decades it has also been studied in other frameworks, for example, in possibility theory [9] and quantum mathematics [7].

In [10] we introduced an evidential marginal problem analogous to that encountered in the probabilistic framework. We demonstrated the similarities between these frameworks concerning necessary conditions and convexity of sets of solutions. We also introduced the concept of product extension of two marginal basic assignments.

The support of Grant No. GAČR 19-04579S is gratefully acknowledged.

M.-J. Lesot et al. (Eds.): IPMU 2020, CCIS 1238, pp. 382–393, 2020.
https://doi.org/10.1007/978-3-030-50143-3_29

This paper is a natural continuation of the above-mentioned one. We study the convex sets of solutions in more details and reveal one substantial difference concerning the complexity of sets of all extensions between probabilistic and evidential frameworks. Nevertheless, the main attention is devoted to the generalization of product extension to finite sets of marginal basic assignments. This generalization is realized via operator of composition introduced in [5], which makes possible to express the product extension of a sequence of marginal basic assignments in an elegant way. This finally allows us to find sufficient condition for the existence of evidential marginal problem solution.

The paper is organised as follows: after a brief overview of necessary concepts and notation (Sect. 2) we start Sect. 3 by a motivation example, then we recall the evidential marginal problem (and what was done in [10]) and we also study convex sets of extensions. In Sect. 4 we deal with generalization of product extension to finite sets of basic assignments and find a sufficient condition for the existence of an extension. Finally, before the Conclusions we present a solution of the motivation example via perfect sequence model.

2 Basic Concepts and Notation

In this section we will, as briefly as possible, recall basic concepts from evidence theory [8] concerning sets and set functions.

2.1 Set Projections and Extension

For an index set $N = \{1, 2, \ldots, n\}$, let $\{X_i\}_{i \in N}$ be a system of variables, each X_i having its values in a finite set $\mathbf{X}_i$. In this paper we will deal with a *multidimensional frame of discernment*

$$\mathbf{X}_N = \mathbf{X}_1 \times \mathbf{X}_2 \times \ldots \times \mathbf{X}_n,$$

and its *subframes* (for $K \subseteq N$)

$$\mathbf{X}_K = \underset{i \in K}{\times} \mathbf{X}_i.$$

Throughout this paper, X_K will denote a group of variables $\{X_i\}_{i \in K}$ when dealing with groups of variables on these subframes.

For $M \subset K \subseteq N$ and $A \subset \mathbf{X}_K$, we denote by $A^{\downarrow M}$ a *projection* of A into $\mathbf{X}_M$:

$$A^{\downarrow M} = \{y \in \mathbf{X}_M \mid \exists x \in A : y = x^{\downarrow M}\},$$

where, for $M = \{i_1, i_2, \ldots, i_m\}$,

$$x^{\downarrow M} = (x_{i_1}, x_{i_2}, \ldots, x_{i_m}) \in \mathbf{X}_M.$$

In addition to the projection, in this text we will also need its inverse operation that is usually called a cylindrical extension. The *cylindrical extension* of $A \subset \mathbf{X}_K$ to $\mathbf{X}_L$ ($K \subset L$) is the set

$$A^{\uparrow L} = \{x \in \mathbf{X}_L : x^{\downarrow K} \in A\} = A \times \mathbf{X}_{L \setminus K}.$$

A more complex instance is to make a common extension of two sets, which will be called a *join* [1]. By a *join* of two sets $A \subseteq \mathbf{X}_K$ and $B \subseteq \mathbf{X}_L$ $(K, L \subseteq N)$, we will understand a set

$$A \bowtie B = \{x \in \mathbf{X}_{K \cup L} : x^{\downarrow K} \in A \ \& \ x^{\downarrow L} \in B\}.$$

Let us note that, for any $C \subseteq \mathbf{X}_{K \cup L}$, it naturally holds $C \subseteq C^{\downarrow K} \bowtie C^{\downarrow L}$, but generally $C \neq C^{\downarrow K} \bowtie C^{\downarrow L}$.

Let us also note that if K and L are disjoint, then the join of A and B is just their Cartesian product, $A \bowtie B = A \times B$, and if $K = L$ then $A \bowtie B = A \cap B$. If $K \cap L \neq \emptyset$ and $A^{\downarrow K \cap L} \cap B^{\downarrow K \cap L} = \emptyset$ then $A \bowtie B = \emptyset$ as well. Generally, $A \bowtie B = A^{\uparrow K \cup L} \cap B^{\uparrow K \cup L}$, i.e., a join of two sets is the intersection of their cylindrical extensions.

2.2 Set Functions

In evidence theory [8], two dual measures are used to model the uncertainty: belief and plausibility measures. Each of them can be defined with the help of another set function called a *basic (probability* or *belief) assignment m* on $\mathbf{X}_N$, i.e.,

$$m : \mathscr{P}(\mathbf{X}_N) \longrightarrow [0, 1],$$

where $\mathscr{P}(\mathbf{X}_N)$ is the power set of $\mathbf{X}_N$, and

$$\sum_{A \subseteq \mathbf{X}_N} m(A) = 1.$$

Furthermore, we assume that $m(\emptyset) = 0$.[1] A set $A \in \mathscr{P}(\mathbf{X}_N)$ is a *focal element* if $m(A) > 0$.

For a basic assignment m on $\mathbf{X}_K$ and $M \subset K$, a *marginal basic assignment* of m on $\mathbf{X}_M$ is defined (for each $A \subseteq \mathbf{X}_M$) by the equality

$$m^{\downarrow M}(A) = \sum_{\substack{B \subseteq \mathbf{X}_K \\ B^{\downarrow M} = A}} m(B). \tag{1}$$

In this paper we will confine ourselves to basic assignments.

3 Marginal Problem and Its Solutions

In this section we first recall what we understand under evidential marginal problem, then we briefly discuss the necessary condition for the existence of a solution and finally study in more detail convex sets of its solutions.

---

[1] This assumption is not generally accepted, e.g., in [2] it is omitted.

3.1 Motivation Example—Poll Opinion

Let us consider the following situation. We have obtained poll opinion contained in the left part of Table 1 expressing the dependence of preferences P with values A and B on sex S (M, F). Some voters are still undecided, we denote this fact by $\{A, B\}$.

From last census we can get data concerning education E (primary—P, high school—H and university—U) and place of residence R (town—T and countryside—C) both of them with respect to sex. They are contained in the middle and right part of of Table 1, respectively. In both cases some values, denoted by $\{P, H, U\}$ and $\{T, C\}$ are missing.

Throughout this example we will express values of basic assignments in the percentage form in order to avoid to small numbers in the joint model.

Table 1. Motivation example: poll opinion and census.

S & P	percentage	S & E	percentage	S & R	percentage
M, A	19	M, P	24	M, T	22
M, B	23	M, H	15	M, C	25
$M, \{A, B\}$	7	M, U	7	$M, \{T, C\}$	2
F, A	25	$M, \{P, H, U\}$	3	F, T	27
F, B	16	F, P	18	F, C	23
$F, \{A, B\}$	10	F, H	23	$F, \{T, C\}$	1
		F, U	8		
		$F, \{P, H, U\}$	2		

Our aim is to get more detailed information about the voters by integrating information from all these tables together. Undecided voters and missing data suggest that this problem has to be modeled by evidence theory. In Sect. 5 one can find solution via method presented in this paper, more precisely in Sect. 4.

3.2 Definition

The evidential marginal problem was, analogous to probability theory, introduced in [10] as follows: Let us assume that X_i, $i \in N$, $1 \leq |N| < \infty$ are finitely-valued variables, $\mathscr{K}$ is a system of nonempty subsets of N and

$$\mathscr{S} = \{m_K, K \in \mathscr{K}\} \tag{2}$$

is a family of basic assignments, where each m_K is a basic assignment on $\mathbf{X}_K$.

The problem we are interested in is the existence of an *extension*, i.e., a basic assignment m on $\mathbf{X}$ whose marginals are basic assignments from $\mathscr{S}$; or, more generally, the set

$$\mathscr{E} = \{m : m^{\downarrow K} = m_K, K \in \mathscr{K}\} \tag{3}$$

is of interest.

In the above-mentioned paper we proved that the necessary condition for the existence of an extension of basic assignments from $\mathscr{S}$ is their pairwise projectivity. Having two basic assignments m_1 and m_2 on $\mathbf{X}_K$ and $\mathbf{X}_L$, respectively ($K, L \subseteq N$), we say that these assignments are *projective* if

$$m_1^{\downarrow K \cap L} = m_2^{\downarrow K \cap L}.$$

If set $\mathscr{S}$ consists of just two basic assignments this condition is also sufficient, but in more general case it need not be, as we showed in [10].

3.3   Sets of Solutions

In [10] we started to study also sets of all solutions of a marginal problem. We proved that, analogous to probabilistic framework, the set is convex. Nevertheless, as suggests the following simple example, resulting convex set is much more complicated.

Example 1. Consider, for $i = 1, 2$, two basic assignments m_i on $\mathbf{X}_i = \{a_i, b_i\}$, specified in Table 2. Our task is to find basic assignments on $\mathbf{X}_1 \times \mathbf{X}_2$ which are extreme points of the set $\mathscr{E}(m_1, m_2)$. We found the 23 extreme points[2] summarized in Table 3. Let us note, that any of basic assignments m^{2-7}, m^{9-14} and m^{16-21} stands for six basic assignments, where three focal elements are fixed (as suggested in the table) and the fourth one is one set A from

$$\mathscr{A} = \{\mathbf{X}_1 \times \mathbf{X}_2 \setminus \{(a_1, a_2)\}, \mathbf{X}_1 \times \mathbf{X}_2 \setminus \{(a_1, b_2)\}, \mathbf{X}_1 \times \mathbf{X}_2 \setminus \{(b_1, a_2)\},$$
$$\mathbf{X}_1 \times \mathbf{X}_2 \setminus \{(b_1, b_2)\}, \{(a_1, a_2), (b_1, b_2)\}, \{(a_1, b_2), (b_1, a_2)\}\},$$

with the values contained in the last column.

Table 2. Example 1: basic assignments m_1 and m_2.

$A \subseteq \mathbf{X}_1$	$m_1(A)$	$A \subseteq \mathbf{X}_2$	$m_2(A)$
$\{a_1\}$	0.2	$\{a_2\}$	0.6
$\{b_1\}$	0.3	$\{b_2\}$	0
$\mathbf{X}_1$	0.5	$\mathbf{X}_2$	0.4

On the other hand, although they are numerous, they are quite "nice", as any of them has only four focal elements in contrary to the basic assignments from the interior of $\mathscr{E}$, any of which has twelve focal elements (from possible 15 in case of two binary variables). An example of these basic assignments is m^i, presented in last line of the table, a convex combination of $m^1, \ldots, m^{22}$, where, in contrary to $m^1, \ldots, m^{22}$ each set from $\mathscr{A}$ is a focal element with the same value (as indicated in the table).

From this point of view a deeper study of extreme points of solutions seems to be of importance. We intend to concentrate to this problem in the future research.

---

[2] Let us note that in analogous case of two binary variables in probabilistic framework the resulting convex set has two extreme points.

Table 3. Example 1: extreme points of $\mathscr{E}(m_1, m_2)$.

	$\{(a_1,a_2)\}$	$\{(b_1,a_2)\}$	$\{a_1\} \times \mathbf{X}_2$	$\{b_1\} \times \mathbf{X}_2$	$\mathbf{X}_1 \times \{a_2\}$	$\mathbf{X}_1 \times \mathbf{X}_2$	A
m^1	0.2	0.3	0	0	0.1	0.4	0
m^{2-7}	0.2	0.3	0	0	0.1	0	0.4
m^8	0.2	0	0	0.3	0.4	0.1	0
m^{9-14}	0.2	0	0	0.3	0.4	0	0.1
m^{15}	0	0.3	0.2	0	0.3	0.2	0
m^{16-21}	0	0.3	0.2	0	0.3	0	0.2
m^{22}	0	0.1	0.2	0.2	0.5	0	0
m^{23}	0.1	0	0.1	0.3	0.5	0	0
m^i	0.1	0.205	0.1	0.095	0.295	0.07	0.0225

4 Product Extension of Sets of Marginals

In this section we will first recall the concept of product extension [10] as well as the composition operator [5] of two basic assignments. We will show that the former is a special case of the latter. Then we use this relationship in order to generalize product extension to finite sets of basic assignments and finally to find a sufficient condition for the existence of an extension.

4.1 Product Extensions

Dempster's rule of combination [8] (and its various modifications as e.g. [3]) is the usual way to combine, in the framework of evidence theory, information from different sources. It is quite natural that several attempts were done to use it in order to merge information expressed by marginal basic assignments. Nevertheless none of them was able to keep both marginals even in case of projective basic assignments. We discussed this problem in more detail in [10].

Instead of the use of Dempster's rule we suggested to use product extension of marginal basic assignments defined as follows.

Definition 1. *Let m_1 and m_2 be projective basic assignments on $\mathbf{X}_K$ and $\mathbf{X}_L$ ($K, L \subseteq N$), respectively. We will call basic assignment m on $\mathbf{X}_{K \cup L}$ product extension of m_1 and m_2 if for any $A = A^{\downarrow K} \bowtie A^{\downarrow L}$*

$$m(A) = \frac{m_1^{\downarrow K}(A^{\downarrow K}) \cdot m_2^{\downarrow L}(A^{\downarrow L})}{m_1^{\downarrow K \cap L}(A^{\downarrow K \cap L})}, \tag{4}$$

whenever the right-hand side is defined, and $m(A) = 0$ otherwise.

Let us note that the expression on the right-hand side of (4) is only seemingly asymmetric, as m_1 and m_2 are supposed to be projective. Therefore, it is irrelevant which marginal is used in the denominator.

From the following theorem proven in [10] one can easily see that product extension is superior to Dempster's rule as concerns keeping marginals.

Theorem 1. *Let m_1 and m_2 be two projective basic assignments on $\mathbf{X}_K$ and $\mathbf{X}_L$ ($K, L \subseteq N$), respectively, and m be their product extension. Then*

$$m^{\downarrow K}(B) = m_1(B),$$
$$m^{\downarrow L}(C) = m_2(C)$$

for any $B \in \mathbf{X}_K$ and $C \in \mathbf{X}_L$, respectively.

4.2 Composition Operator

Now, let us recall the concept of composition operator, introduced in [5].

Definition 2. *For two arbitrary basic assignments m_1 on $\mathbf{X}_K$ and m_2 on $\mathbf{X}_L$ a composition $m_1 \triangleright m_2$ is defined for all $A \subseteq \mathbf{X}_{K \cup L}$ by one of the following expressions:*

[a] *if $m_2^{\downarrow K \cap L}(A^{\downarrow K \cap L}) > 0$ and $A = A^{\downarrow K} \bowtie A^{\downarrow L}$ then*

$$(m_1 \triangleright m_2)(A) = \frac{m_1(A^{\downarrow K}) \cdot m_2(A^{\downarrow L})}{m_2^{\downarrow K \cap L}(A^{\downarrow K \cap L})};$$

[b] *if $m_2^{\downarrow K \cap L}(A^{\downarrow K \cap L}) = 0$ and $A = A^{\downarrow K} \times \mathbf{X}_{L \setminus K}$ then*

$$(m_1 \triangleright m_2)(A) = m_1(A^{\downarrow K});$$

[c] *in all other cases*

$$(m_1 \triangleright m_2)(A) = 0.$$

The purpose of the composition operator is to integrate information from two sources expressed by basic assignments m_1 and m_2. The resulting basic assignment $m_1 \triangleright m_2$ has to keep all the information contained in the first basic assignment, and as much as possible from the second one.

The definition completely fulfills this requirement. The only focal elements are those obtained via [a] or [b]. Both of them keep the first marginal, i.e. all the information contained in m_1.

In case [a] "as much as possible from the second basic assignment" is obtained via multiplication by m_2 (divided by its marginal). If m_1 and m_2 are projective, then also all the information from m_2 is kept[3] (cf. Theorem 1 and Lemma 1).

In case [b] no information about $A^{\downarrow K \cap L}$ is available, so $m_1 \triangleright m_2$ is the least specific basic assignment with marginal m_1—its vacuous extension.

The following lemma reveals, that product extension of two basic assignments is a special case of composition of two basic assignments.

Lemma 1. *Let m_1 and m_2 be projective basic assignments. Then $(m_1 \triangleright m_2)(A)$ is equal to their product extension defined by (4).*

Proof. First, let us note that when m_1 and m_2 are projective, case [b] is not applied. Therefore, $(m_1 \triangleright m_2)(A)$ is computed via [a], which is equal to the right hand side of (4) (due to projectivity of m_1 and m_2), whenever it is defined. Otherwise, by [c] $(m_1 \triangleright m_2)(A) = 0$.

---

[3] Let us note that if K and L are disjoint, then m_1 and m_2 are trivially projective.

4.3 Iterative Application of Composition Operator

Composing of $m_1, m_2, \ldots, m_n$ defined on $\mathbf{X}_{K_1}, \ldots, \mathbf{X}_{K_n}$, respectively, together by multiple application of the operator of composition, one gets multidimensional basic assignments on $\mathbf{X}_{K_1 \cup K_2 \cup \ldots \cup K_n}$. However, since the operator of composition is neither commutative nor associative (cf. [5]), we have to specify what "composing them together" means.

To avoid using too many brackets let us make the following convention. Whenever we put down the expression $m_1 \triangleright m_2 \triangleright \ldots \triangleright m_n$ we will understand that the operator of composition is performed successively from left to right:[4]

$$m_1 \triangleright m_2 \triangleright \ldots \triangleright m_n = (\ldots ((m_1 \triangleright m_2) \triangleright m_3) \triangleright \ldots) \triangleright m_n.$$

Therefore, it is obvious that ordering of basic assignments in question is substantial for the properties of the resulting model. Let us demonstrate it by the following simple example.

Example 2. Consider three variables X_1, X_2 and X_3 with values in $\mathbf{X}_1, \mathbf{X}_2$ and $\mathbf{X}_3$, respectively, $\mathbf{X}_i = \{a_i, b_i\}, i = 1, 2, 3$. Let m_1, m_2 and m_3 be defined as shown in Table 4.

Table 4. Example 2: basic assignments m_1, m_2 and m_3.

$A \subseteq \mathbf{X}_1 \times \mathbf{X}_2$	$m_1(A)$	$A \subseteq \mathbf{X}_2 \times \mathbf{X}_3$	$m_2(A)$	$A \subseteq \mathbf{X}_1 \times \mathbf{X}_3$	$m_3(A)$
$\{(a_1, a_2)\}$	0.5	$\{(a_2, a_3)\}$	0.5	$\{a_1\} \times \mathbf{X}_3$	0.5
$\mathbf{X}_1 \times \mathbf{X}_2$	0.5	$\mathbf{X}_2 \times \mathbf{X}_3$	0.5	$\mathbf{X}_1 \times \{a_3\}$	0.5

These basic assignments are pairwise projective (any one-dimensional marginal has just two focal elements, namely $\{a_i\}$ and $\mathbf{X}_i$), but common extension does not exist (as we already showed in [10]). One can judge that application of the operator of composition to different orderings of these three basic assignments will lead to different joint basic assignments on $\mathbf{X}_1 \times \mathbf{X}_2 \times \mathbf{X}_3$. And it is the case. Each of these composed basic assignments has again only two focal elements, namely

$$\{(a_1, a_2, a_3)\}, \mathbf{X}_1 \times \mathbf{X}_2 \times \mathbf{X}_3$$

for m_1, m_2, m_3 and m_2, m_1, m_3,

$$\{(a_1, a_2)\} \times \mathbf{X}_3, \mathbf{X}_1 \times \mathbf{X}_2 \times \{a_3\}$$

for m_1, m_3, m_2 and m_3, m_1, m_2 and, finally,

$$\mathbf{X}_1 \times \{(a_2, a_3)\}, \{a_1\} \times \mathbf{X}_2 \times \mathbf{X}_3$$

for m_2, m_3, m_1 and m_3, m_2, m_1.

In the next section we will deal with special kind of sequences of basic assignments.

---

[4] Naturally, if we want to change the ordering in which the operators are to be performed we will do so with the help of brackets.

4.4 Perfect Sequences and Sufficient Condition

When representing knowledge in a specific area of interest, a special role is played by the so-called *perfect sequences*, i.e., sequences $m_1, m_2, \ldots, m_n$, for which

$$m_1 \triangleright m_2 = m_2 \triangleright m_1,$$

$$m_1 \triangleright m_2 \triangleright m_3 = m_3 \triangleright (m_1 \triangleright m_2),$$

$$\vdots$$

$$m_1 \triangleright m_2 \triangleright \ldots \triangleright m_n = m_n \triangleright (m_1 \triangleright \ldots \triangleright m_{n-1}).$$

The following theorem proven in [5] explains why perfect sequences are useful for marginal problem solution.

Theorem 2. *A sequence* $m_1, m_2, \ldots, m_n$ *is perfect if and only if all* $m_1, m_2, \ldots, m_n$ *are marginal basic assignments of the multidimensional basic assignment* $m_1 \triangleright m_2 \triangleright \ldots \triangleright m_n$*:*

$$(m_1 \triangleright m_2 \triangleright \ldots \triangleright m_n)^{\downarrow K_i} = m_i,$$

for all $i = 1, \ldots, n$.

Let us note that this theorem expresses necessary and sufficient condition for the existence of an extension of basic assignments from $\mathscr{S}$, however, this condition is hardly verifiable as it is obvious from the definition of the perfect sequence. Nevertheless, we can formulate a sufficient condition expressed by Theorem 3. Before doing it, let us recall the well-known running intersection property and a lemma (proven in [5]) necessary to prove Theorem 3.

We say that $K_1, K_2, \ldots, K_n$ meets the *running intersection property* iff

$$\forall i = 2, 3, \ldots, n \ \exists j (1 \le j < i) \qquad \text{such that} \quad K_i \cap (K_1 \cup \ldots \cup K_{i-1}) \subseteq K_j.$$

Lemma 2. *A sequence* $m_1, m_2, \ldots, m_n$ *is perfect iff the pairs of basic assignments* m_i *and* $(m_1 \triangleright \ldots \triangleright m_{i-1})$ *are projective, i.e. if*

$$m_i^{\downarrow K_i \cap (K_1 \cup \ldots \cup K_{i-1})} = (m_1 \triangleright \ldots \triangleright m_{i-1})^{\downarrow K_i \cap (K_1 \cup \ldots \cup K_{i-1})},$$

for all $i = 2, 3, \ldots, n$.

Theorem 3. *Let* $\mathscr{S}$ *be a system of pairwise projective basic assignments on* $\mathbf{X}_K$, $K \in \mathscr{K}$*, and let* $\mathscr{K}$ *be ordered in such a way that* $K_1, K_2, \ldots, K_n$ *meets running intersection property. Then* $\mathscr{E} \neq \emptyset$.

Proof. We have to prove that the sequence of basic assignments $m_1, m_2, \ldots, m_n$ on $\mathbf{X}_{K_1}, \mathbf{X}_{K_2}, \ldots, \mathbf{X}_{K_n}$, respectively, is perfect. Then, according to Theorem 2 ($m_1 \triangleright m_2 \triangleright \ldots \triangleright m_n) \in \mathscr{E}$.

Due to Lemma 2 it is enough to show that for each $i = 2, \ldots, n$ basic assignment m_i and the composed assignment $m_1 \triangleright \ldots \triangleright m_{i-1}$ are projective. Let us prove it by induction.

For $i = 2$ the required projectivity is guaranteed by the fact that we assume pairwise projectivity of all $m_1, \ldots, m_n$. So we have to prove it for general $i > 2$ under the assumption that the assertion holds for $i-1$, which means (due to Theorem 2) that all $m_1, m_2, \ldots, m_{i-1}$ are marginal to $m_1 \triangleright \ldots \triangleright m_{i-1}$. Since we assume that $K_1, \ldots, K_n$ meets the running intersection property, there exists $j \in \{1, 2, \ldots i-1\}$ such that $K_i \cap (K_1 \cup \ldots \cup K_{i-1}) \subseteq K_j$. Therefore $(m_1 \triangleright \ldots \triangleright m_{i-1})^{\downarrow K_i \cap (K_1 \cup \ldots \cup K_{i-1})}$ and $m_j^{\downarrow K_i \cap (K_1 \cup \ldots \cup K_{i-1})}$ are the same marginals of $m_1 \triangleright \ldots \triangleright m_{i-1}$ and therefore they have to equal to each other:

$$(m_1 \triangleright \ldots \triangleright m_{i-1})^{\downarrow K_i \cap (K_1 \cup \ldots \cup K_{i-1})} = m_j^{\downarrow K_i \cap (K_1 \cup \ldots \cup K_{i-1})}.$$

However we assume that m_i and m_j are projective and therefore also

$$(m_1 \triangleright \ldots \triangleright m_{i-1})^{\downarrow K_i \cap (K_1 \cup \ldots \cup K_{i-1})} = m_i^{\downarrow K_i \cap (K_1 \cup \ldots \cup K_{i-1})}.$$

Nevertheless, there exist a big "grey zone" of problems satisfying necessary condition but not the sufficient one. In this case the answer can be negative (as in Example 2) or positive (as in the following example).

Example 3. Consider again three variables X_1, X_2 and X_3 with values in X_1, X_2 and X_3, respectively, $X_i = \{a_i, b_i\}, i = 1, 2, 3$. Let m_1, m_2 and m_3' be defined as shown in Table 5. The only difference with Example 2 consists in using m_3' instead of m_3.

Table 5. Example 3: basic assignments m_1, m_2 and m_3'.

$A \subseteq X_1 \times X_2$	$m_1(A)$	$A \subseteq X_2 \times X_3$	$m_2(A)$	$A \subseteq X_1 \times X_3$	$m_3'(A)$
$\{(a_1, a_2)\}$	0.5	$\{(a_2, a_3)\}$	0.5	$\{(a_1, a_3)\}$	0.5
$X_1 \times X_2$	0.5	$X_2 \times X_2$	0.5	$X_1 \times X_3$	0.5

These basic assignments are again pairwise projective and the running intersection property for $\{1, 2\}, \{2, 3\}$ and $\{1, 3\}$ (for any ordering of these sets) does not again hold. However, in this case a common extensions exist; product extension can be found in Table 6.

Table 6. Example 3: product extension of m_1, m_2 and m_3'.

$A \subseteq X_1 \times X_2 \times X_3$	$m(A)$
$\{(a_1, a_2, a_3)\}$	0.5
$X_1 \times X_2 \times X_3$	0.5

The difference between Examples 2 and 3 consists in the fact that m_1, m_2 and m_3' is perfect (although running intersection property is no satisfied) in contrary to m_1, m_2 and m_3.

From this example one can see, that perfectness does not depend only on relations among different subframes (or their index sets), on which low-dimensional basic assignments are defined, but also of the individual structure of any of them.

5 Poll Opinion—Solution

It can be easily seen from Table 1 that m_1, m_2 and m_3 are pairwise projective, as

$$m_1(M) = m_2(M) = m_3(M) = 0.49$$

and

$$m_1(F) = m_2(F) = m_3(F) = 0.51.$$

Since the sequence $\{S, O\}, \{S, E\}, \{S, R\}$, satisfies RIP, m_1, m_2 and m_3 form a perfect sequence. Application of composition operator to m_1 and m_2 and then to $m_1 \triangleright m_2$ and m_3 gives rise to the joint basic assignment contained in Table 7.

Table 7. Poll opinion—solution

		Male			Female		
		T	C	{C,T}	T	C	{C,T}
$P = A$	$E = P$	4.19	4.75	0.38	4.67	3.98	0.17
	$E = H$	2.61	2.97	0.24	5.99	5.1	0.22
	$E = A$	1.21	1.38	0.11	2.07	1.76	0.08
	$E = \{P, H, U\}$	0.54	0.61	0.05	0.53	0.45	0.02
$P = B$	$E = P$	5.08	5.78	0.46	2.96	2.53	0.11
	$E = H$	3.15	3.58	0.29	3.82	3.26	0.14
	$E = A$	1.49	1.69	0.13	1.33	1.13	0.05
	$E = \{P, H, U\}$	0.63	0.71	0.06	0.32	0.27	0.01
$P = \{A, B\}$	$E = P$	1.54	1.74	0.14	1.86	1.57	0.07
	$E = H$	0.94	1.07	0.09	2.39	2.04	0.09
	$E = A$	0.45	0.51	0.04	0.85	0.72	0.03
	$E = \{P, H, U\}$	0.18	0.2	0.02	0.21	0.18	0.01

This table contains all focal elements of the basic assignment $m_1 \triangleright m_2 \triangleright m_3$, i.e only 72 from possible almost 17 million. Therefore it is evident, that despite the fact that evidential models are super-exponentially complex, compositional models belong (among them) to those with lower complexity. However, complexity of the resulting model strongly depends on the complexity of input basic assignments as well as on their number.

From this table one can obtain by simple marginalization marginal tables of the relationship between some variables of interest — e.g. preference and education.

6 Conclusions

We have recalled an evidential marginal problem introduced in [10] in a way analogous to a probability setting, where marginal probabilities are substituted by marginal basic assignments.

We studied the structure of the sets of extensions of the problem in more detail and realized that it is much more complicated than that in analogous probabilistic case. This will be topic of our future research.

We also generalized concept of product extension to a finite set of basic assignments using the operator of composition introduced in [5]. The result of this effort is not only the closed form of an extension of a finite set of basic assignments (if it exists), but also the sufficient condition for the existence of such an extension. The obtained results are illustrated by a simple practical example.

References

1. Beeri, C., Fagin, R., Maier, D., Yannakakis, M.: On the desirability of acyclic database schemes. J. Assoc. Comput. Mach. **30**(3), 479–513 (1983)
2. Yaghlane, B.B., Smets, P., Mcllouli, K.: Belief functions independence: I. The marginal case. Int. J. Approximate Reasoning **29**(1), 47–70 (2002)
3. Yaghlane, B., Smets, P., Mellouli, K.: Belief functions independence: II. The conditional case. Int. J. Approximate Reasoning **31**(1–2), 31–75 (2002)
4. Janžura, M.: Marginal problem, statistical estimation, and Möbius formula. Kybernetika **43**(5), 619–631 (2007)
5. Jiroušek, R., Vejnarová, J.: Compositional models and conditional independence in evidence theory. Int. J. Approximate Reasoning **52**, 316–334 (2011)
6. Pougaza, D.-B., Mohammad-Djafaria, A., Berchera, J.-F.: Link between copula and tomography. Pattern Recogn. Lett. **31**(14), 2258–2264 (2010)
7. Schilling, C.: The quantum marginal problem. In: Proceedings of the Conference QMath 12, Berlin (2013)
8. Shafer, G.: A Mathematical Theory of Evidence. Princeton University Press, Princeton (1976)
9. Vejnarová, J.: On possibilistic marginal problem. Kybernetika **43**(5), 657–674 (2007)
10. Vejnarová, J.: On marginal problem in evidence theory. In: Cuzzolin, F. (ed.) Belief Functions: Theory and Applications, pp. 331–338 (2014)
11. Vomlel, J.: Integrating inconsistent data in a probabilistic model. J. Appl. Non-Class. Logics **14**(3), 367–386 (2004)

Handling Mixture Optimisation Problem Using Cautious Predictions and Belief Functions

Lucie Jacquin[1]([✉]), Abdelhak Imoussaten[1], and Sébastien Destercke[2]

[1] EuroMov Digital Health in Motion, Univ Montpellier, IMT Mines Ales, Ales, France
{lucie.jacquin,Abdelhak.imoussaten}@mines-ales.fr
[2] Sorbonne universités, UTC, CNRS, Heudiasyc, 57 Avenue de Landshut, Compiègne, France
sebastien.destercke@hds.utc.fr

Abstract. Predictions from classification models are most often used as final decisions. Yet, there are situations where the prediction serves as an input for another constrained decision problem. In this paper, we consider such an issue where the classifier provides imprecise and/or uncertain predictions that need to be managed within the decision problem. More precisely, we consider the optimisation of a mix of material pieces of different types in different containers. Information about those pieces is modelled by a mass function provided by a cautious classifier. Our proposal concerns the statement of the optimisation problem within the framework of belief function. Finally, we give an illustration of this problem in the case of plastic sorting for recycling purposes.

Keywords: Belief functions · Sum rule of mass functions · Mixture optimisation · Plastic sorting

1 Introduction

Mixing materials in the right amount is a common problem in many industries. Depending on the desired properties, the mixture must meet certain constraints on the proportions of each material. In the case where the mixing is done progressively, one must know, at each step, the materials present in the piece to be added and the materials present in the existing mixture in order to check if the new mixture respects the proportion constraints. This problem can be encountered in several applications; when refining crude-oil into useful petroleum products, one has to manage the mixture of different hydrocarbon products; when recycling plastic, the portion of some material type should not exceed some thresholds; when producing different types of wood paneling, each type of paneling is made by gluing and pressing together a different mixture of pine and oak chips; etc. The work presented in this paper is motivated by the problem of plastic sorting for recycling purposes, that will serve as a running and illustrative example

© Springer Nature Switzerland AG 2020
M.-J. Lesot et al. (Eds.): IPMU 2020, CCIS 1238, pp. 394–407, 2020.
https://doi.org/10.1007/978-3-030-50143-3_30

of our proposal. More precisely, we have to assign plastic pieces issued from a deposit to various containers, knowing that pieces can be of different materials, and that each container should satisfy some constraints w.r.t. the proportion of materials it contains. Our goal is then to find the sorting optimizing the recycling process.

As sorting plastic manually is time and cost-consuming, automatic processing machines are now put in place, with several sensors (e.g., infra-red cameras) installed to recognize the material of a plastic piece. The obtained signal is then processed by automatic model learned from pieces labelled in favourable conditions (see [8] for more details). Of course, as real conditions are much less favourable, there may be a lot of uncertainties regarding the actual material of on-line processed pieces, which explains the need for reliable yet precise enough classifiers [2,8,11,15]. In our setting, we consider that such classifiers returns mass functions modelling our knowledge about the material type.

A classical tool to perform optimization under uncertainty is stochastic optimization. We will extend such a setting to belief functions, first by considering the Choquet integral instead of the classical expectation as an objective function, and second by replacing the probability measure by the pair belief/plausibility measures. As we add pieces to a given container, we will also have to compute the global uncertainty of a container by adding mass functions of different weights. To do so, we will adapt the technique proposed in [7] for general intervals to the case of discrete proportions.

The paper is organised as follows. The problem is formalized as a stochastic optimisation problem in Sect. 2. Section 3 gives some reminders about belief functions, summing operation of mass functions, cautious prediction, and Choquet integral. In Sect. 4, the optimisation problem of pieces sorting is formalized in the framework of belief functions. The illustration concerning plastic sorting is presented in Sect. 5.

2   Stochastic Optimisation Problem Formalisation

We consider a deposit of scrap plastic, crude-oil, wood, etc., with a total physical weight W. This weight represent a set of pieces that will be put in C containers depending on the composition of each piece. In the end, each container c will contain a weight of material w_c^{end}, with $\sum_{c=1}^{C} w_c^{end} = W$. The n types of materials are represented by the set $S = \{s_1, \ldots, s_n\}$, and we denote by $\theta_i^{c,end}$ the proportion of material s_i present in the container at the end of the sorting.

Since pieces are supposed to be on conveyor belts, the optimisation process will be performed step-wisely, deciding for each new piece in which container it should go. Doing so, the final step, i.e., *end*, gives the proportions $\theta_1^{c,end}, \theta_2^{c,end}, \ldots, \theta_n^{c,end}$ in each container and the weights $w_1^{end}, \ldots, w_n^{end}$ can be deduced by weighting each container. To avoid complicating notations, we omit the time or step reference in the optimisation problem. The optimisation problem can be set as follows:

$$\max_{c\in\{1,\dots,C\}} \quad g_c(s_f) \tag{1a}$$

$$\text{subject to} \quad h_c(\theta_1^c,\dots,\theta_n^c) \leq 0,\ c = 1,\dots,C, \tag{1b}$$

$$\sum_{i=1}^{n} \theta_i^c = 1,\ c = 1,\dots,C \tag{1c}$$

where:

- The objective function (1a) is such that $g : S \to \mathbb{R}^+$, with $g_c(s_i)$ the gain obtained if a material of type s_i is added to container c;
- θ_i^c is the proportion of material type s_i in the container c after adding the new piece to it,
- The constraints (1b) are expressed using function $h_c : [0,1]^n \to [-1,1]$. They are of the form $h_{c,A}(\theta_1,\dots,\theta_n) = \sum_{i\in A} \theta_i - \alpha_c \leq 0$ with $A \subseteq S$, meaning that the proportion of materials of types A should not exceed α_c in container c;
- The constraint (1c) means simply that proportions sum up to 1.

The deterministic version of this problem is easy to solve, but becomes more complicated if the piece f composition is uncertain, for instance given by a probability mass function (pmf) $p(.|f)$ over S. The optimisation becomes then stochastic, and (1a) is replaced by

$$\max_{c\in\{1,\dots,C\}} \quad \mathbb{E}_{p(.|f)}[g_c] \tag{2}$$

where $\mathbb{E}_{p(.|f)}$ is the expectation w.r.t. $p(.|f)$. Remark then that $p(.|f)$ can be converted to a pmf over the discrete subset of proportions $\{(1,0,\dots,0),\dots,(0,\dots,1)\}$ of $[0,1]^n$. Indeed, to check to which extent constraints are satisfied, we will need to compute probabilities over proportions. We denote by $p_c \oplus p(.|f)$ the result of adding the current probabilistic proportions p_c of the container with $p(.|f)$, accounting for the current weight of the container and the weight of f.

The constraints (1b) are then replaced by chance constraints

$$\mathbb{P}_{f,c}(h_{c,A}(\theta_1^c,\dots,\theta_n^c) \leq 0) \geq \eta,\ c = 1,\dots,C. \tag{3}$$

where $\mathbb{P}_{f,c}$ is the measure induced from $p_c \oplus p(.|f)$, and η is typically close to 1. Finally the stochastic optimisation problem is the following

$$\max_{c\in\{1,\dots,C\}} \quad \mathbb{E}_{p(.|f)}[g_c] \tag{4a}$$

$$\text{subject to} \quad \mathbb{P}_{f,c}(h_{c,A}(\theta_1^c,\dots,\theta_n^c) \leq 0) \geq \eta,\ c = 1,\dots,C, \tag{4b}$$

$$\sum_{i=1}^{n} \theta_i^c = 1,\ c = 1,\dots,C. \tag{4c}$$

However, it may be the case that pieces uncertainty is too severe to be modelled by probabilities, in which case more general models, such as belief functions, should be used. In the next sections, we discuss an extension of Eqs. (2)–(3) for such uncertainty models.

3   Reminders

3.1   Belief Functions

Belief functions [12,14] are uncertainty models that combine probabilistic and set-valued uncertainty representations, therefore providing an expressive and flexible framework to represent different kinds of uncertainty. Beyond probabilities and sets, they also extend possibility theory [5].

Given a space $\mathcal{X}$ with elements x, the basic tools used within belief function theory is the *mass function*, also called *basic belief assignment* (bba), is a set function $m : 2^{\mathcal{X}} \to [0,1]$ satisfying

$$m(\emptyset) = 0 \text{ and } \sum_{A \subseteq \mathcal{X}} m(A) = 1.$$

The elements $A \in 2^{\mathcal{X}}$ such that $m(A) > 0$ are called focal elements and they form a set denoted $\mathbb{F}$. $(m, \mathbb{F})$ is called body of evidence.

The *belief function* $Bel : 2^{\mathcal{X}} \to [0,1]$ is a set function that measures how much an event A is implied by our information such that

$$Bel(A) = \sum_{B \subseteq \mathcal{X}, B \subseteq A} m(B).$$

The *plausibility function* $Pl : 2^{\mathcal{X}} \to [0,1]$ is a set function that measures how much an event A is consistent with our information such that

$$Pl(A) = \sum_{B \subseteq \mathcal{X}, B \cap A \neq \emptyset} m(B).$$

Note that when focal elements are singletons x, we have $Bel = Pl$ and retrieve probabilities.

3.2   Sum Operation on Imprecise Proportion

Let us denote the unit simplex by $\mathbb{U} = \{(\theta_1, \ldots, \theta_n) \in [0,1]^n : \sum_{i=1}^{n} \theta_i = 1\}$.

Let us consider two sets of pieces sf^1 and sf^2 made of materials among $S = \{s_1, \ldots, s_n\}$ with physical masses w^1 and w^2. The information about the material type proportions in sf^1 and sf^2 are given respectively by the bodies of evidence $(m^1, \mathbb{F}^1)$ and $(m^2, \mathbb{F}^2)$ defined over $\mathbb{U}$, with discrete focal elements in a finite number. A focal element in $\mathbb{F}^1$ (resp. $\mathbb{F}^2$) is in the form $J = J_1 \times \ldots \times J_n$ (resp. $K = K_1 \times \ldots \times K_n$) where J_i (resp. K_i), $i \in \{1, \ldots, n\}$ is an imprecise information about the proportion of s_i in sf^1 (resp. sf^2).

The information resulting from adding sf^2 with sf^1 is a mass function denoted $m^{1 \oplus 2}$ and defined as follows for $I \subset \mathbb{U}$ [7]:

$$m^{1 \oplus 2}(I) = \sum_{\substack{J \in \mathbb{F}^1, K \in \mathbb{F}^2 \\ I = J \boxplus K}} m^1(J) \cdot m^2(K). \tag{5}$$

where $\mathbb{F}^{1\oplus 2}$ is a finite set made of discrete subsets of $\mathbb{U}$ resulting from summing proportion in $\mathbb{F}^1$ and $\mathbb{F}^2$; the total weight associated to the mixture is $w^1 + w^2$ and $\boxplus$ is defined for two focal elements $J \in \mathbb{F}^1$ and $K \in \mathbb{F}^2$ as follows:

$$J \boxplus K = I_1 \times \ldots \times I_n, \text{ with } I_i = \{\frac{w^1\, x + w^2\, y}{w^1 + w^2}, x \in J_i, y \in K_i\}.$$

Note that in case where imprecise information are convex sets, e.g. intervals, only the lower and the upper bounds of the intervals are involved in the determination of $J \boxplus K$ [7].

Example 1. Let us consider the case where $S = \{s_1, s_2, s_3, s_4\}$ and sf^1 and sf^2 are both composed of a single piece each with weight $1\,\text{kg}$. In Table 1, we give an example of two bodies of evidence for these two sets of pieces. The focal elements presented in Table 1 have the following meaning: J_1 means that sf^1 is a pure material of type s_1 or s_2, and J_2 means that sf^1 is a pure material of type s_2, and similarly for K_1, K_2.

Table 1. Bodies of evidence.

sf^1 ($w^1 = 1\,\text{kg}$)		sf^2 ($w^2 = 1\,\text{kg}$)	
$\mathbb{F}^1$	m^1	$\mathbb{F}^2$	m^2
$J_1 = \{0,1\} \times \{0,1\} \times \{0\} \times \{0\}$	0.5	$K_1 = \{0,1\} \times \{0,1\} \times \{0\} \times \{0\}$	0.6
$J_2 = \{0\} \times \{1\} \times \{0\} \times \{0\}$	0.5	$K_2 = \{1\} \times \{0\} \times \{0\} \times \{0\}$	0.4

The obtained mass function when mixing sf^1 and sf^2 is given by its body of evidence $(\{I_1, I_2, I_3, I_4\}, m^{1\oplus 2})$ as follows:

$$I_1 = J_1 \boxplus K_1 = \{0, \frac{1}{2}, 1\} \times \{0, \frac{1}{2}, 1\} \times \{0\} \times \{0\}, m^{1\oplus 2}(I_1) = 0.3,$$

$$I_2 = J_1 \boxplus K_2 = \{\frac{1}{2}, 1\} \times \{0, \frac{1}{2}\} \times \{0\} \times \{0\}, \qquad m^{1\oplus 2}(I_2) = 0.2,$$

$$I_3 = J_2 \boxplus K_1 = \{0, \frac{1}{2}\} \times \{\frac{1}{2}, 1\} \times \{0\} \times \{0\}, \qquad m^{1\oplus 2}(I_3) = 0.3,$$

$$I_4 = J_2 \boxplus K_2 = \{\frac{1}{2}\} \times \{\frac{1}{2}\} \times \{0\} \times \{0\}, \qquad m^{1\oplus 2}(I_4) = 0.2.$$

3.3   Inference from Imprecise Proportions

The set A_α of vector proportions that satisfy $\sum_{i \in A} \theta_i \leq \alpha$ is of interest in our problem because it allows expressing constraints containers must respect, as indicate Eq. (1b). Thus we need to make inferences over such event. Given focal elements $I = I_1 \times \ldots \times I_n$, in case where $I_i = [\ell_i, u_i]$ are intervals it was shown [7] that

$$I \subseteq A_\alpha \Leftrightarrow \min(\sum_{s_i \in A} u_i, 1 - \sum_{s_i \notin A} \ell_i,) \leq \alpha$$

$$\mathcal{I} \cap A_\alpha \neq \emptyset \Leftrightarrow \max(\sum_{s_i \in A} \ell_i, 1 - \sum_{s_i \notin A} u_i) \leq \alpha$$

In the discrete case where $I_i = \{\tau_1, \tau_2, \ldots, \tau_{|I_i|}\}$, $\tau_i \in [0,1]$, the two previous formulae remain valid when considering $u_i = \max\limits_{t=1,\ldots,|I_i|} \tau_t$ and $\ell_i = \min\limits_{t=1,\ldots,|I_i|} \tau_t$.

3.4 Cautious Predictions

In our case, belief functions will be produced by classifiers that will be learned from a set of examples/pieces $f_1, \ldots, f_l$ having m features $X_1, \ldots, X_m$ having received a label in S. Given a new object f, this classifier will output a mass $m(.|f)$ as a prediction.

Such classifiers are indeed useful in our application, as they provide more reliable information, and can account for many defects, such as the missingness of some feature X_j for f (due to a broken sensor), or the fact that measurements are done by industrial on line machine device instead of laboratory measurements, meaning that variability in measurement due to atmospheric disturbances, ageing of plastics, black or dark-coloured materials, etc. lead to reducing the quality of the spectrum obtained from plastic pieces. In this situation, classifier producing point prediction, i.e., single element from S as prediction, will make to many errors to provide a reliable sorting. Instead of point prediction classifiers, we will use classifiers providing cautious predictions in form of a posterior mass function over S [8], but the approach could apply to other such classifiers [3,4,11]. It should be stressed that in our case, one could prefer to put a good plastic in a low price container rather than ruining a high price container by violating constraints (1b), so being cautious by accounting for imperfectness of information is essential.

3.5 Choquet Integral

The Choquet integral [9] is an integral that applies to non-additive measures, often referred as fuzzy measures [10]. Since a Belief function defined over a space S is such a fuzzy measure[1], we can apply the Choquet integral to it in the following way: given a vector of real positive values $y = (y_1, \ldots, y_n) \in \mathbb{R}^{+n}$, its Choquet integral w.r.t. Bel is defined as

$$C_{Bel}(y) = \sum_{i=1}^{n} (y_{\sigma(i)} - y_{\sigma(i-1)}) Bel(\{s_{\sigma(i)}, s_{\sigma(i+1)}, \ldots, s_{\sigma(n)}\}) \tag{6}$$

where $0 = y_{\sigma(0)} \leq y_{\sigma(1)} \leq y_{\sigma(2)} \leq \cdots \leq y_{\sigma(n)}$ (σ is a permutation over $\{1, \ldots, n\}$).

If $Bel = Pl$, then Eq. (6) is simply the standard expectation operator. Otherwise, it can be interpreted as the lower expectation taken over all probabilities $Bel \leq P \leq Pl$, i.e., all probabilities bounded by our imprecise knowledge.

---

[1] It is such that $Bel(\emptyset) = 0$, $Bel(S) = 1$ and is montonic, i.e., $A \subseteq B \rightarrow Bel(A) \leq Bel(B)$.

4   Optimisation Problem Statement in the Framework of Belief Function

We now provide an equivalent of the optimisation problem ingredients (4a)–(4c) in the framework of belief function. We consider all the previous ingredients, except that now the information about a new piece to add to a container is given by a mass function $m(.|f)$ defined over $S = \{s_1, \ldots, s_n\}$, and our information about the proportions of materials in a given container c is also given by a mass function m^c bearing on $\mathbb{U}$. As before, one can easily go from a mass $m(.|f)$ on S to a mass on $\mathbb{U}$ (see Example 1 for an illustration).

4.1   The Objective Function

The expected value in the objective function (2) can be replaced by the Choquet integral based on the belief function $Bel(.|f)$. As in Sect. 2, we will only be interested to model in the objective function the potential gain of adding the new piece f to one of the container, without bothering about the container current proportions, as those will be treated in the constraints. If g is the overall gain of a container containing materials of a specified kind $\overline{A}$, where elements $A \subset S$ are considered as impurities whose percentage should not exceed α_c, we simply consider the function $g_c(s) = g(s)$ for $s \in \overline{A}$, and $g_c(s) = \alpha_c \cdot g(s)$.

Example 2. Consider four material types $S = \{s_1, \ldots, s_4\}$ and three containers. Table 2 presents an example of gains obtained when adding piece f to each container. We consider that container 1 is dedicated to s_1 and other type proportions should not exceed α_1; container 2 is dedicated to s_2 and s_3 (deemed compatible for recycling) and other type proportions should not exceed α_2; container 3 is the garbage bin, so $\alpha_3 = 1$.

Table 2. Container gains.

	s_1	s_2	s_3	s_4
Container 1	100\$	$\alpha_1 . 100\$$	$\alpha_1 . 100\$$	$\alpha_1 . 100\$$
Container 2	$\alpha_2 . 100\$$	100\$	100\$	$\alpha_2 . 100\$$
Container 3	1\$	1\$	1\$	1\$

The example of Table 2 shows that the larger the threshold, the higher the gain when adding impurities to a container.

Still denoting by $g_c(s_i)$ the gain obtained if the real type of the added piece to the container c is s_i, Eq. (2) becomes:

$$\max_{c \in \{1, \ldots, C\}} C_{Bel(.|f)}(g_c(s_1), \ldots, g_c(s_n)) \tag{7}$$

The objective function (7) is an expected value based on Choquet integral where gains are weighted related to our belief on the material type of the new piece including imprecise information. Let denotes $x_{(1)} = \min\limits_{i=1,n} g_c(s_i)$, ..., $x_{(n)} = \max\limits_{i=1,n} g_c(s_i)$ such as $x_{(1)} \leq x_{(2)} \leq \cdots \leq x_{(n)}$, then this expected value guarantee $x_{(1)}$ surely and adds to it the gaps $x_{(i)} - x_{(i-1)}$ weighted by $Bel(\{s_{(i)}, \ldots, s_{(n)}\}|f)$.

Example 3. Let us consider a mass function $m(.|f)$ with the following body of evidence $(\{\{s_1\}, \{s_1, s_2\}\}, (0.2, 0.8))$. The resulting $Bel(.|f)$ is given in Table 3.

Table 3. Belief function.

	$\emptyset$	$\{s_1\}$	$\{s_2\}$	$\{s_1, s_2\}$	$\{s_3\}$	$\{s_1, s_3\}$	$\{s_2, s_3\}$	$\{s_1, s_2, s_3\}$
$Bel(.|f)$	0	0.2	0	1	0	0.2	0	1

	$\{s_4\}$	$\{s_1, s_4\}$	$\{s_2, s_4\}$	$\{s_1, s_2, s_4\}$	$\{s_3, s_4\}$	$\{s_1, s_3, s_4\}$	$\{s_2, s_3, s_4\}$	$\{s_1, s_2, s_3, s_4\}$
$Bel(.|f)$	0	0.2	0	1	0	0.2	0	1

If we consider $\alpha_1 = 0.25$ and $\alpha_2 = 0.3$ in Table 2, we obtain the gains in Table 4. In this case, without considering constraints, f should go in container 1.

Table 4. Container gains.

	s_1	s_2	s_3	s_4	Expected gain
Container 1	100\$	25\$	25\$	25\$	$25 + 75\ Bel(\{s_1\}|f) = 40$
Container 2	30\$	100\$	100\$	30\$	$30 + 70\ Bel(\{s_2, s_3\}|f) = 30$
Container 3	1\$	1\$	1\$	1\$	1

4.2 The Constraints

Let us consider that the physical weight of f is w^f and the physical weight of the current pieces in the container c is w^c. The formula (5) gives us the new mass function $m^{f \oplus c}$ when adding the piece f to the container c. The constraints in (3) check whether impurities in containers are not too high. However, we must now replace he probability measure $\mathbb{P}^{f \oplus c}$ in this constraint is by the pair $(Bel^{f \oplus c}, Pl^{f \oplus c})$. One may reasonably requires the degree of certainty that a constraint is satisfied to be very high, and the degree of plausibility of this same constraint to be satisfied to be close to 1. Such a reasoning can be applied by replacing the constraint (3) by two constraints:

$$Bel^{f \oplus c}(h_c(\theta_1, \ldots, \theta_n) \leq 0) > \eta_c, \quad c = 1, \ldots, C, \tag{8a}$$

$$Pl^{f \oplus c}(h_c(\theta_1, \ldots, \theta_n) \leq 0) \sim 1, \quad c = 1, \ldots, C. \tag{8b}$$

where $\eta_c \in]0,1]$ are enough large. Note that such ideas are not new, and have been for instance recently applied to the travelling salesman problem [6].

Example 4. If we go back to the Example 2, the considered constraints for each container can be given as follows:
Container 1:

$$Bel^{f \oplus c}(\sum_{i \neq 1} \theta_i \leq \alpha_1) > \eta_1, \quad Pl^{f \oplus c}(\sum_{i \neq 1} \theta_i \leq \alpha_1) \sim 1,$$

Container 2:

$$Bel^{f \oplus c}(\sum_{i \neq 2,3} \theta_i \leq \alpha_2) > \eta_2, \quad Pl^{f \oplus c}(\sum_{i \neq 2,3} \theta_i \leq \alpha_2) \sim 1,$$

Container 3:

$$Bel^{f \oplus c}(\sum_{i \neq 4} \theta_i \leq \alpha_3) > \eta_3, \quad Pl^{f \oplus c}(\sum_{i \neq 4} \theta_i \leq \alpha_3) \sim 1.$$

Let us denote A_α the set of vector proportions that satisfy $\sum_{i \in A} \theta_i \leq \alpha$. In Sect. 3.3 we give the way to determine $Bel(A_\alpha)$ and $Pl(A_\alpha)$ that are required to check the constraints (8a) and (8b).

Finally, we have the following optimisation problem to decide in each container a piece f should be added:

$$\max_{c \in \{1,\dots,C\}} \quad C_{Bel(.|f)}(g_c(s_1), \dots, g_c(s_n)) \tag{9a}$$

$$\text{subject to} \quad Bel^{f \oplus c}(h_c(\theta_1^c, \dots, \theta_n^c) \leq 0) > \eta_c, \; c = 1, \dots, C, \tag{9b}$$

$$Pl^{f \oplus c}(h_c(\theta_1^c, \dots, \theta_n^c) \leq 0) \sim 1, \; c = 1, \dots, C, \tag{9c}$$

$$\sum_{i=1}^{n} \theta_i^c = 1, c = 1, \dots, C. \tag{9d}$$

To solve the optimisation problem (9a)–(9d) one needs to assess (9a) for each container for the finite number of pieces in the deposit. Complexity issues arise when the number of pieces is very large. Indeed, the number of focal elements involved when determining $Bel^{f \oplus c}$ (9b) and $Pl^{f \oplus c}$ (9c) become exponential, yet one can easily solve this issue by considering approximations (e.g., deleting focal elements of very small mass).

5   Illustration

In this section we present an application concerning plastic sorting where the pieces of a deposit should be separated by types of materials in different containers prior to recycling due to some physico-chemical reasons related to non-miscibility. Optical sorting devices are used to automatically sort the pieces. As

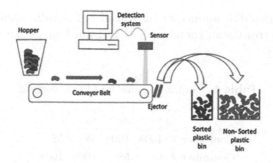

Fig. 1. Example of sorting device

it is shown in Fig. 1 borrowed from [1], pieces of plastics arrive continuously on a conveyor belt before being recorded by an infra-red camera. However, the on line acquired information is subject to several issues inducing the presence of imprecision on one hand, i.e. some features information are not precise enough to draw clear distinctions between the materials type, and uncertainty on the other hand, i.e. due to the reliability of information caused by atmospheric disturbance, etc (please refer to [8] for more details). Two sources of information are used to collect data. The first source of data is the Attenuated Total Reflection (ATR) which gives excellent quality of spectra that allows experts to label pieces easily. The second source is the optical device which provides spectra of lesser quality. Since small quantity of badly sorted plastics can lead to high decreases of impact resistance [13] and of monetary value, impurities should be limited. Thus, experts have defined tolerance threshold on the proportions of impurities.

In this illustration we propose a sorting procedure based on the optimisation problem in (9a)–(9d). The cautious classification is provided using the evidential classifier proposed in [8].

Let us recap the procedure performed to sort each fragment f:

- Estimate the resulting composition of each container c if we add f to it as a mass function $m^{f \oplus c}$ using the sum operation defined in Sect. 3.2.
- Select the containers verifying the constraints (9b) and (9c).
- Compare the objective function (9a) for the selected container.
- Update the evidence about the chosen container.

5.1 Data Presentation

Let us consider a plastic waste deposit composed of 25 pieces of four material types s_1, s_2, s_3, s_4. All the pieces have the weight $w = 1$. Each piece should be sent to one of the three containers dedicated for specific material types. The first container is dedicated to plastic types s_1, s_2 and the proportions of impurities, i.e s_3, s_4, should not exceed $\alpha_1 = 0.05$. The second container is dedicated to plastic types s_3, s_4, and the proportions of impurities, i.e s_1, s_2, should not exceed $\alpha_2 = 0.05$. The third container is actually the reject option, thus all types of plastics are considered as impurities (or considered as valid materials), but there is no need to control them, thereby $\alpha_3 = 1$. Table 5 gives the gains considered for the containers.

Table 5. Container gains for plastic sorting.

	s_1	s_2	s_3	s_4
Container 1	100\$	100\$	5\$	5\$
Container 2	5\$	5\$	100\$	100\$
Container 3	1\$	1\$	1\$	1\$

The database used for the experimentation are 23365 industrially acquired spectra. Each example of the database is composed of its 154-dimension features and its ATR label.

5.2 Simulations

The evidential classifier proposed in [8] has been trained on the 11747 examples and applied on the testing set, i.e., 11618 other examples. We obtained 11618 mass functions $m(.|f_1), \ldots, m(.|f_{11618})$. In order to evaluate the sorting procedure, we tested the performances on 40 simulations of fragment streams. The simulation of a stream was done by selecting randomly indexes orders of testing fragments $f_1, \ldots, f_{11618}$. For computational reasons, we stopped the sorting procedure at the 25th fragment for each simulation. Note that the complexity of the sorting procedure is exponential, i.e., $\mathcal{O}((2^{|S|})^{nb \ of \ pieces})$ [7]. Figure 2, 3 and 4 show respectively the evolution of the weight of materials in the two first containers, the belief that the constraints are respected and the real proportions of impurities. Each curves represents one simulation and we keep the same color in all the figures. The thresholds are set to $\eta_1 = \eta_2 = 0.6$.

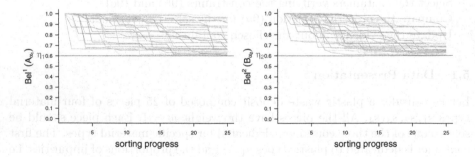

Fig. 2. Evolution of the weight of materials in container 1 and 2.

In Fig. 2 we observe that the choice between the two first containers is balanced. As we can see in Fig. 3, the constraints defined in (9b) are always respected. Using the testing labels we can evaluate the real proportions of impurities.

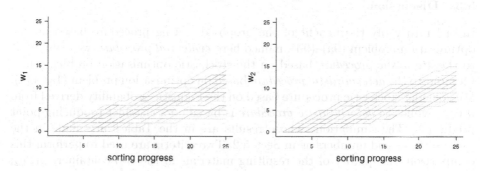

Fig. 3. Evolution of the belief that the constraints are respected in containers 1 and 2.

In Fig. 4, we observe that the proportion of impurities are most of the times below the required threshold except for a few simulations where mistakes are made for the first pieces added in container 1 and 2. Since at the beginning of the sorting, there are only few pieces, the mistakes have a high impact on the proportions. After checking, it turned out that the mass functions provided for these examples were not accurate. In order to evaluate the quality of the resulting sorted material, we introduce the score q_c as the percentage of simulations respecting impurities proportions constraints at the end of the sorting in the container c. With the proposed approach we obtain $q_1 = 77.5\%$ and $q_2 = 62.5\%$. This is significantly higher than the required levels 60%, which is in-line with the fact that we are acting cautiously. In terms of gains, the average gain obtained in the simulations is 1901.475$ while the optimal would have been 2500$, in

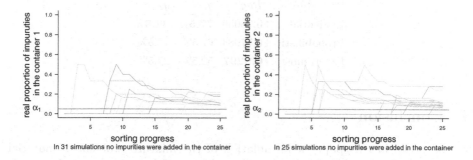

Fig. 4. Evolution of real proportions of impurities in containers 1 and 2.

the ideal case where all pieces are sorted in the correct container. However, this would only have been possible if we had perfect classification results, something that is unlikely.

5.3   Discussion

In order to verify the benefit of the proposed sorting procedure based on the optimisation problem (9a)–(9d), named here *evidential procedure*, we compare it to the *stochastic procedure* based on the stochastic optimisation problem (4a)–(4c) and to the *deterministic procedure* based on optimisation problem (1a)–(1c). We consider stochastic procedure based on the Pignistic probability derived from $m(.|f)$ while the *deterministic procedure* is based on a classifier producing point prediction. The simulations whose results are in the Table 6 are made in the same settings and numbers as in Sect. 5.2. Two criteria are used to perform this comparison: the quality of the resulting materials in the two containers q_1, q_2; the rate of average gain obtained on all simulations, denoted Rag.

What we see here is that not accounting for uncertainty, or considering a less expressive model (i.e., probabilities) do indeed bring a better average gain, but fails to meet the constraints imposed to the containers for them to be usable at all. Indeed, the *evidential procedure* achieves high quality of the sorting material while the two other procedures do not respect the required constraints on the containers composition. This could be solved by considering more penalizing gains in case of bad sorting for the *deterministic procedure* and *stochastic procedure*, yet this would complexify the procedure. Thus the *evidential procedure* seems preferable for applications where constraints on impurities are strong, i.e. very small α or when the confidence level required for the application is high, i.e., η closer to 1. When such requirements are not necessary, we would advice the use of an alternative procedure less computationally demanding.

Table 6. Comparison with alternative procedures

Procedures	Rag	q_1	q_2
Evidential	0.76059	**77.5%**	**62.5%**
Probabilistic	0.77984	67.5%	57.5%
Deterministic	**0.9297**	52.5%	27.5%

6   Conclusion

We proposed in this paper a formulation of the mixture problem of material types in the framework of belief functions. The usefulness of this work is illustrated using the sorting procedure of plastic material. A stepwise approach is proposed to avoid the complicated complete resolution. As perspectives for this work, one should optimise the stepwise summing of mass functions in on line sorting procedure by controlling the focal elements generated at each step in order to overcome the exponential complexity. Furthermore, one may relax the constraints on impurities at each step by requiring them only at the end of the sorting procedure.

References

1. Beigbeder, J., Perrin, D., Mascaro, J.F., Lopez-Cuesta, J.M.: Study of the physico-chemical properties of recycled polymers from waste electrical and electronic equipment (WEEE) sorted by high resolution near infrared devices. Resour. Conserv. Recycl. **78**, 105–114 (2013)
2. Côme, E., Oukhellou, L., Denoeux, T., Aknin, P.: Learning from partially supervised data using mixture models and belief functions. Pattern Recogn. **42**(3), 334–348 (2009)
3. Corani, G., Zaffalon, M.: Learning reliable classifiers from small or incomplete data sets: the naive credal classifier 2. J. Mach. Learn. Res. **9**, 581–621 (2008)
4. Del Coz, J.J., Díez, J., Bahamonde, A.: Learning nondeterministic classifiers. J. Mach. Learn. Res. **10**, 2273–2293 (2009)
5. Dubois, D., Prade, H.: Possibility theory. In: Meyers, R. (eds.) Computational Complexity. Springer, New York (2012). https://doi.org/10.1007/978-1-4614-1800-9
6. Helal, N., Pichon, F., Porumbel, D., Mercier, D., Lefèvre, É.: The capacitated vehicle routing problem with evidential demands. Int. J. Approximate Reasoning **95**, 124–151 (2018)
7. Jacquin, L., Imoussaten, A., Destercke, S., Trousset, F., Montmain, J., Perrin, D.: Manipulating focal sets on the unit simplex: application to plastic sorting. In: 2020 IEEE International Conference on Fuzzy Systems (FUZZ-IEEE). IEEE (2020)
8. Jacquin, L., Imoussaten, A., Trousset, F., Montmain, J., Perrin, D.: Evidential classification of incomplete data via imprecise relabelling: application to plastic sorting. In: Ben Amor, N., Quost, B., Theobald, M. (eds.) SUM 2019. LNCS (LNAI), vol. 11940, pp. 122–135. Springer, Cham (2019). https://doi.org/10.1007/978-3-030-35514-2_10
9. Labreuche, C., Grabisch, M.: The Choquet integral for the aggregation of interval scales in multicriteria decision making. Fuzzy Sets Syst. **137**(1), 11–26 (2003)
10. Murofushi, T., Sugeno, M.: An interpretation of fuzzy measures and the choquet integral as an integral with respect to a fuzzy measure. Fuzzy Sets Syst. **29**(2), 201–227 (1989)
11. Nguyen, V.L., Destercke, S., Masson, M.H., Hüllermeier, E.: Reliable multi-class classification based on pairwise epistemic and aleatoric uncertainty. In: International Joint Conference on Artificial Intelligence (2018)
12. Shafer, G.: A Mathematical Theory of Evidence, vol. 42. Princeton University Press, Princeton (1976)
13. Signoret, C., Caro-Bretelle, A.S., Lopez-Cuesta, J.M., Ienny, P., Perrin, D.: Mir spectral characterization of plastic to enable discrimination in an industrial recycling context: II. Specific case of polyolefins. Waste Manage. **98**, 160–172 (2019)
14. Smets, P., Kennes, R.: The transferable belief model. Artif. Intell. **66**(2), 191–234 (1994)
15. Zaffalon, M.: The naive credal classifier. J. Stat. Plann. Infer. **105**(1), 5–21 (2002)

Aggregation: Theory and Practice

A Note on Aggregation of Intuitionistic Values

Anna Kolesárová[1] and Radko Mesiar[2(✉)]

[1] Faculty of Chemical and Food Technology, Slovak University of Technology,
Radlinského 9, 812 37 Bratislava, Slovakia
anna.kolesarova@stuba.sk
[2] Faculty of Civil Engineering, Slovak University of Technology,
Radlinského 11, 810 05 Bratislava, Slovakia
radko.mesiar@stuba.sk

Abstract. Atanassov's intuitionistic fuzzy set theory is based on the lattice L^* of intuitionistic values and their aggregation. There are lots of works dealing with this topic, but mostly considering some particular cases. In this contribution, we offer a rather general view on aggregation of intuitionistic values with transparent proofs of several properties which significantly shorten the related proofs for particular cases known from the literature.

Keyword: Aggregation function, Intuitionistic values, Representable aggregation function, t-norm, t-conorm

1 Introduction

The first connectives/aggregation functions for intuitionistic fuzzy logic/theory were proposed by Atanassov [1,2]. They were related to the basic aggregation functions on $[0, 1]$, i.e., minimum, maximum, product and probabilistic sum of aggregation functions. Later, Deschrijver and Kerre [11] introduced a class of intuitionistic connectives based on t-norms on $[0, 1]$ and the corresponding dual t-conorms. Xu [19] introduced several functions dealing with intuitionistic values, including aggregation functions based on the standard product, such as intuitionistic fuzzy weighted averaging (IFWA) operator. Similar functions/aggregation functions dealing with intuitionistic values but based on the Einstein t-norm (a particular t-norm from the Hamacher family [16]) were proposed in [18] and consequently mentioned in several other papers. In papers [15,17], Hamacher's product or Dombi's t-norms were considered for deriving aggregation functions acting on intuitionistic values.

In all mentioned cases, one can find long and tedious proofs of some properties of the introduced functions, in particular, the proofs of the basic fact that (in some cases) they are aggregation functions on the lattice L^* of intuitionistic values.

© Springer Nature Switzerland AG 2020
M.-J. Lesot et al. (Eds.): IPMU 2020, CCIS 1238, pp. 411–418, 2020.
https://doi.org/10.1007/978-3-030-50143-3_31

The main aim of this contribution is to propose a unified approach to intuitionistic aggregation based on strict t-norms and provide short and transparent proofs of important properties of the introduced functions. As a byproduct, our results can help to the researchers who work in intuitionistic fuzzy set theory to avoid introducing "new" intuitionistic aggregation functions whose only novelty is that they are based on t-norms which have not been explicitly mentioned in the literature yet. Moreover, we show several techniques for making proofs of new results in aggregation domain shorter and more transparent.

2   Preliminaries

We expect that the readers are familiar with the basics of fuzzy set theory [20] and intuitionistic fuzzy set theory [1–3]. Similarly, the basic knowledge concerning triangular norms [16] and aggregation functions [4,6,13] is expected. Here we only recall some most relevant basic notions.

Note that the concept of Atanassov's intuitionistic fuzzy sets was introduced to generalize the concept of Zadeh's fuzzy sets [20], and the original definition [1] of an Atanassov intuitionistic fuzzy set (AIFS for short) in a universe X can be recalled as follows:

Definition 2.1. *Let $X \neq \emptyset$ be any set. An Atanassov intuitionistic fuzzy set A in X is defined as a set of triplets*

$$A = \{(x, \mu_A(x), \nu_A(x)) \mid x \in X\},$$

where $\mu_A, \nu_A \colon X \to [0,1]$, and for all $x \in X$, also $\mu_A(x) + \nu_A(x) \in [0,1]$.

Note that, for each element $x \in X$, $\mu_A(x)$ and $\nu_A(x)$ are called the membership and the non-membership degrees of x to A, respectively, and the pair $(\mu_A(x), \nu_A(x))$ is called an Atanassov intuitionistic fuzzy value [18].

Denote by L^* the set of all possible intuitionistic values, i.e.,

$$L^* = \{(a, b) \in [0,1]^2 \mid a + b \leq 1\}.$$

In particular, the set L^* with the partial order $\leq_{L^*}$, given by

$$(a, b) \leq_{L^*} (c, d) \text{ if and only if } a \leq c \text{ and } b \geq d,$$

form a bounded lattice with the top element $\mathbf{1}_{L^*} = \top = (1, 0)$ and bottom element $\mathbf{0}_{L^*} = \bot = (0, 1)$.

Then an Atanassov intuitionistic fuzzy set A in a universe X can be seen as a mapping $A \colon X \to L^*$. To define basic operations of union and intersection of AIFSs, Atanassov [1] has proposed to apply pointwisely some appropriate binary functions mapping $(L^*)^2$ onto L^*. Due to their associativity, the n-ary forms of these operations are uniquely defined.

The first intuitionistic connectives proposed in [1] are just the join $\vee_{L^*}$ and meet $\wedge_{L^*}$, inducing the union and intersection of AIFSs. Clearly,

$$(a, b) \vee_{L^*} (c, d) = (a \vee c, b \wedge d), \tag{1}$$

and
$$(a, b) \wedge_{L^*} (c, d) = (a \wedge c, b \vee d). \tag{2}$$

As another type of operations on L^*, Atanassov [1] proposed the operations given by
$$(a, b) \oplus (c, d) = (a + c - ac, bd), \tag{3}$$
and
$$(a, b) \otimes (c, d) = (ac, b + d - bd). \tag{4}$$

Obviously, all four operations $\vee_{L^*}, \wedge_{L^*}, \oplus$ and $\otimes$ are binary aggregation functions on L^* [5,18], i.e., they are non-decreasing in each variable with respect to the order $\leq_{L^*}$, and they aggregate $\top$ and $\top$ into $\top$ and also $\perp$ and $\perp$ into $\perp$. Moreover, they are commutative and associative, $\top$ is a neutral element of $\wedge_{L^*}$ and $\otimes$, whereas $\perp$ is a neutral element of $\vee_{L^*}$ and $\oplus$. Hence, $\wedge_{L^*}$ and $\otimes$ can be seen as t-norms, and $\vee_{L^*}$ and $\oplus$ as t-conorms on L^*. In all four cases we see the connection with the standard t-norms T_M and T_P and the standard t-conorms S_M and S_P, defined on the real unit interval $[0, 1]$ by

$$T_M(a, b) = \min\{a, b\}, \quad T_P(a, b) = ab,$$
$$S_M(a, b) = \max\{a, b\}, \quad S_P(a, b) = a + b - ab.$$

For more details see, e.g., [16]. Formally, $\vee_{L^*} \approx (S_M, T_M)$, i.e.,

$$(a, b) \vee_{L^*} (c, d) = (S_M(a, c), T_M(b, d)).$$

Further, $\wedge_{L^*} \approx (T_M, S_M)$, and thus it can be seen as a dual operation to $\vee_{L^*}$ on L^*. Similarly, $\oplus \approx (S_P, T_P)$ and $\otimes \approx (T_P, S_P)$ are dual operations.

This observation has been generalized by Deschrijver and Kerre [11]. For an arbitrary t-norm $T: [0, 1]^2 \to [0, 1]$ and the corresponding dual t-conorm $S: [0, 1]^2 \to [0, 1]$, satisfying, for each $(a, b) \in [0, 1]^2$, $S(a, b) = 1 - T(1 - a, 1 - b)$, the mappings $\mathbf{T}_T, \mathbf{S}_T : (L^*)^2 \to L^*$,

$$\mathbf{T}_T \approx (T, S) \quad \text{and} \quad \mathbf{S}_T \approx (S, T), \tag{5}$$

are called a representable t-norm and a t-conorm on L^*, respectively.

For possible applications of AIFS theory some other operations/aggregation functions have also been proposed. For example, Xu [19] has proposed to define the product ".", $\cdot :]0, \infty[\times L^* \to L^*$ by

$$\lambda \cdot (a, b) = (1 - (1 - a)^\lambda, b^\lambda) \tag{6}$$

and the power on the same domain by

$$(a, b)^\lambda = (a^\lambda, 1 - (1 - b)^\lambda). \tag{7}$$

Based on these functions, in [19] some new n-ary aggregation functions on L^* were proposed and studied, for example, $IFWA_{\mathbf{w}}$ (intuitionistic fuzzy weighted average) given by

$$IFWA_{\mathbf{w}}((a_1, b_1), \dots, (a_n, b_n)) = \left(1 - \prod_{i=1}^{n} (1 - a_i)^{w_i}, \prod_{i=1}^{n} b^{w_i} \right), \tag{8}$$

where $\mathbf{w} = (w_1, \ldots, w_n) \in [0,1]^n$, $\sum_{i=1}^{n} w_i = 1$, is a weighting vector. Evidently, formulas (6), (7) and (8) are based on the product t-norm. Similarly, modifying formulas (6), (7) and (8), Wang and Liu [18] have proposed functions dealing with values from L^* based on the Einstein t-norm $T_E \colon [0,1]^2 \to [0,1]$ given by $T_E(a,b) = \frac{ab}{2-a-b+ab}$ and its dual, Einstein t-conorm, $S_E \colon [0,1]^2 \to [0,1]$, $S_E(a,b) = \frac{a+b}{1+ab}$ (note that the formula defining S_E is just the Einstein formula for summing relative velocities). For example, the product $\cdot_E \colon \,]0, \infty[\times L^* \to L^*$ they have defined as follows:

$$\lambda \cdot_E (a,b) = \left(\frac{(1+a)^\lambda - (1-a)^\lambda}{(1+a)^\lambda + (1-a)^\lambda}, \frac{2b^\lambda}{(2-b)^\lambda + b^\lambda} \right). \tag{9}$$

3 Representable Aggregation Functions on L^*

All till now discussed aggregation functions on L^* have been based on some aggregation function $A \colon [0,1]^n \to [0,1]$ and its dual $A^d \colon [0,1]^n \to [0,1]$,

$$A^d(a_1, \ldots, a_n) = 1 - A(1 - a_1, \ldots, 1 - a_n)$$

and defined as follows: $\mathbf{A} \colon (L^*)^n \to L^*$, $\mathbf{A} \approx (A, A^d)$,

$$\mathbf{A}((a_1, b_1), \ldots, (a_n, b_n)) = (A(a_1, \ldots, a_n), A^d(b_1, \ldots, b_n)). \tag{10}$$

It is obvious that for any aggregation functions $A, B \colon [0,1]^n \to [0,1]$ such that $B \leq A^d$, the mapping $\mathbf{A} \colon (L^*)^n \to L^*$, $\mathbf{A} \approx (A, B)$,

$$\mathbf{A}((a_1, b_1), \ldots, (a_n, b_n)) = (A(a_1, \ldots, a_n), B(b_1, \ldots, b_n)) \tag{11}$$

is an aggregation function on L^*. Moreover, $\mathbf{A}$ inherits all common properties of A and B. In particular, if both A and B are associative (symmetric/idempotent or if they have a neutral element/an annihilator) then $\mathbf{A} \approx (A, B)$ also possesses this property. Such types of intuitionistic aggregation functions are called representable intuitionistic aggregation functions [10,12]. They also cover intuitionistic aggregation functions mentioned in (1)–(5) and (8).

As an example when $B \neq A^d$ consider $A = T_P$ and $B = S_M$. Clearly, $S_M \leq T_P^d = S_P$ and thus $\mathbf{T} \approx (T_P, S_M)$ given by

$$\mathbf{T}((a,b),(c,d)) = (ac, b \vee d)$$

is a representable t-norm on L^*.

Note that there are also non-representable aggregation functions on L^*, including t-norms. For example, consider the mapping $\mathbf{T} \colon (L^*)^2 \to L^*$ given by

$$\mathbf{T}((a,b),(c,d)) = (a \wedge c, (b \vee (1-c)) \wedge ((1-a) \vee d)).$$

Then $\mathbf{T}$ is a t-norm on L^* which is not representable. For more examples see [10].

4   Main Results

Formulas (3), (4), (6), (7) and (8) are related to the product t-norm. Similarly, (9) and several other formulas in [15] and [18] are related to the Hamacher and Einstein t-norms, respectively, and formulas based on Dombi's t-norms have been discussed in [17]. In all mentioned cases, a strict t-norm $T\colon [0,1]^2 \to [0,1]$ and its dual t-conorm $S\colon [0,1]^2 \to [0,1]$ have been considered. The main advantage of strict t-norms is the fact that they are generated by an additive generator $t\colon [0,1] \to [0,\infty]$ which is a decreasing bijection, i.e.,

$$T(a,b) = t^{-1}(t(a) + t(b)).$$

To have a unique correspondence between strict t-norms and their additive generators, one can always choose the generator satisfying $t(0.5) = 1$. Then the dual t-conorm S is generated by an additive generator $s\colon [0,1] \to [0,\infty]$, $s(x) = t(1-x)$,

$$S(a,b) = s^{-1}(s(a) + s(b)) = 1 - t^{-1}(t(1-a) + t(1-b)).$$

Clearly, a representable intuitionistic t-norm $\mathbf{T}_T \approx (T,S)$ and t-conorm $\mathbf{S}_T \approx (S,T)$ in their n-ary form are given by

$$\mathbf{T}_T((a_1,b_1),\dots,(a_n,b_n)) = \left(t^{-1}\left(\sum_{i=1}^{n} t(a_i)\right), 1 - t^{-1}\left(\sum_{i=1}^{n} t(1-b_i)\right)\right)$$

and

$$\mathbf{S}_T((a_1,b_1),\dots,(a_n,b_n)) = \left(1 - t^{-1}\left(\sum_{i=1}^{n} t(1-a_i)\right), t^{-1}\left(\sum_{i=1}^{n} t(b_i)\right)\right).$$

Now, fix a strict t-norm T having an additive generator t, and introduce the mappings:

$$\lambda \cdot_T (a,b) = \left(1 - t^{-1}\left(\lambda t(1-a)\right), t^{-1}\left(\lambda t(b)\right)\right), \ \lambda > 0, \ (a,b) \in L^*, \tag{12}$$

$$(a,b)^{\lambda_T} = \left(t^{-1}\left(\lambda t(a)\right), 1 - t^{-1}\left(\lambda t(1-b)\right)\right), \ \lambda > 0, \ (a,b) \in L^*, \tag{13}$$

and for all $(a_1,b_1),\dots,(a_n,b_n) \in L^*$,

$$IFWA_{\mathbf{w}}^T((a_1,b_1),\dots,(a_n,b_n)) = \mathbf{S}_T(w_1 \cdot_T (a_1,b_1),\dots, w_n \cdot_T (a_n,b_n)). \tag{14}$$

Then we have

$$IFWA_{\mathbf{w}}^T((a_1,b_1),\dots,(a_n,b_n)) = \left(1 - t^{-1}\left(\sum_{i=1}^{n} w_i t(1-a_i)\right), t^{-1}\left(\sum_{i=1}^{n} w_i t(b_i)\right)\right), \tag{15}$$

and the following results hold:

Theorem 4.1. *Let $T\colon [0,1]^2 \to [0,1]$ be a strict t-norm with an additive generator t. Then*

(i) $\mathbf{S}_T((a,b),(a,b)) = 2 \cdot_T (a,b)$;
(ii) $\mathbf{T}_T((a,b),(a,b)) = (a,b)^{2_T}$;
(iii) $IFWA_{\mathbf{w}}^T$ is an idempotent representable n-ary aggregation function on L^*, which is $\oplus_T$-additive where $\oplus_T = S_T$.

Proof: The proofs of (i) and (ii) are only a matter of simple calculations, e.g.,

$$\mathbf{S}_T((a,b),(a,b)) = \left(1 - t^{-1}\left(2t(1-a)\right), t^{-1}\left(2t(b)\right)\right) = 2 \cdot_T (a,b).$$

To prove (iii), first observe that for any $b_1,\ldots,b_n \in [0,1]$ and a weighting vector $\mathbf{w}$, the formula

$$t^{-1}\left(\sum_{i=1}^n w_i t(b_i)\right) = T_{\mathbf{w}}(b_1,\ldots,b_n)$$

defines a weighted t-norm $T_{\mathbf{w}}$, see [7]. Obviously, for the dual t-conorm S to T and $a_1,\ldots,a_n \in [0,1]$, the formula

$$1 - t^{-1}\left(\sum_{i=1}^n w_i t(1-a_i)\right) = S_{\mathbf{w}}(a_1,\ldots,a_n) = T_{\mathbf{w}}^d(a_1,\ldots,a_n)$$

gives the weighted t-conorm $S_{\mathbf{w}}$, and the weighted aggregation functions $T_{\mathbf{w}}$ and $S_{\mathbf{w}}$ based on the same weighting vector are dual. Then $IFWA_{\mathbf{w}}^T \approx (S_{\mathbf{w}}, T_{\mathbf{w}})$ is a representable aggregation function on L^*. As both $S_{\mathbf{w}}$ and $T_{\mathbf{w}}$ are idempotent, $IFWA_{\mathbf{w}}^T$ is also idempotent.

Finally, verification of the $\oplus_T$-additivity of $IFWA_{\mathbf{w}}^T$, i.e., the property

$$IFWA_{\mathbf{w}}^T\left((a_1,b_1)\oplus_T(c_1,d_1),\ldots,(a_n,b_n)\oplus_T(c_n,d_n)\right)$$
$$= IFWA_{\mathbf{w}}^T\left((a_1,b_1),\ldots,(a_n,b_n)\right)\oplus_T IFWA_{\mathbf{w}}^T\left((c_1,d_1),\ldots,(c_n,d_n)\right)$$

is a matter of an easy computation. $\square$

Note that the proofs of these results realized for particular t-norms known from the literature, see, e.g. [18,19], are rather long and non-transparent.

Remark 4.1. Nilpotent t-norms [16] are also generated by additive generators. In this case, an additive generator is a decreasing bijection $t\colon [0,1] \to [0,1]$, and then

$$T(a,b) = t^{-1}\left(\min\{1, t(a) + t(b)\}\right),$$

and for the corresponding dual t-conorm S we have

$$S(a,b) = t^{-1}\left(\min\{1, t(1-a) + t(1-b)\}\right).$$

Surprisingly, nilpotent t-norms are rarely applied in intuitionistic fuzzy set theory (IFST), possibly except the Łukasiewicz t-norm T_L, $T_L(a,b) = \max\{0, a + b - 1\}$ and its dual t-conorm S_L, $S_L(a,b) = \min\{1, a + b\}$. The corresponding intuitionistic t-norm $\mathbf{T}_L \approx (T_L, S_L)$ and t-conorm $\mathbf{S}_L \approx (S_L, T_L)$ form the basis of probability theory in IFST, see [14] and several Riečan's works [8,9]. Based on a nilpotent t-norm with additive generator t, one can introduce the related product, power and weighted mean, paraphrasing formulas (12), (13) and (15). Considering T_L and its additive generator $t_L \colon [0,1] \to [0,1]$, $t_L(x) = 1 - x$, we obtain:

$$\lambda \cdot_{T_L} (a,b) = \left(1 - t_L^{-1}\left(\min\{1, \lambda t_L(1-a)\}\right), t_L^{-1}\left(\min\{1, \lambda t_L(b)\}\right)\right)$$
$$= \left(\min\{1, \lambda a\}, \max\{0, 1 - \lambda(1-b)\}\right);$$
$$(a,b)^{\lambda_{T_L}} = \left(\max\{0, 1 - \lambda(1-a)\}, \min\{1, \lambda b\}\right),$$

and

$$IFWA_{\mathbf{w}}^{T_L}((a_1, b_1), \ldots, (a_n, b_n)) = \left(\sum_{i=1}^{n} w_i a_i, \sum_{i=1}^{n} w_i b_i\right),$$

i.e., $IFWA_{\mathbf{w}}^{T_L}$ is an intuitionistic fuzzy weighted arithmetic mean. Note that considering continuous t-norms and their weighted forms discussed in [7], one can introduce more general types of weighted aggregation functions also in the intuitionistic framework.

5 Concluding Remarks

We have introduced a rather general approach to functions dealing with intuitionistic values from L^*, in particular to functions based on continuous Archimedean t-norms and their additive generators. Our results significantly generalize the related results based on special strict t-norms (e.g., the standard product, Einstein's t-norm, Dombi's t-norms) known from the literature. The main advantage of our approach is its transparentness and obtaining short general proofs of the results in contrast to the long and non-transparent proofs of results in particular cases. We believe that our paper will contribute not only to the theoretical basis of IFST, but in particular, it can make intuitionistic fuzzy sets more efficient for applications. Moreover, it can also help in related domains, such as picture fuzzy sets, neutrosophic fuzzy sets, etc.

Acknowledgement. The authors kindly acknowledge the support of the project APVV-18-0052 and the grant VEGA 1/0614/18.

References

1. Atanassov, K.T.: Intuitionistic fuzzy sets. Fuzzy Sets Syst. **20**(1), 87–96 (1986). https://doi.org/10.1016/S0165-0114(86)80034-3
2. Atanassov, K.T.: New operations defined over the intuitionistic fuzzy sets. Fuzzy Sets Syst. **61**(2), 137–142 (1994). https://doi.org/10.1016/0165-0114(94)90229-1
3. Atanassov, K.T.: Intuitionistic Fuzzy Sets: Theory and Applications. Physica-Verlag, Heidelberg/New York (1999). https://doi.org/10.1007/978-3-7908-1870-3
4. Beliakov, G., Pradera, A., Calvo, T.: Aggregation Functions: A Guide for Practitioners. Springer, Heidelberg (2007). https://doi.org/10.1007/978-3-540-73721-6

5. Beliakov, G., Bustince, H., Goswami, D.P., Mukherjee, U.K., Pal, N.R.: On averaging operators for Atanassov's intuitionistic fuzzy sets. Inf. Sci. **181**(60), 1116–1124 (2011). https://doi.org/10.1016/j.ins.2010.11.024

6. Calvo, T., Kolesárová, A., Komorníková, M., Mesiar, R.: Aggregation operators: properties, classes and construction methods. In: Calvo, T., Mayor, G., Mesiar, R. (eds.) Aggregation Operators, pp. 3–107. Physica-Verlag, Heidelberg (2002). https://doi.org/10.1007/978-3-7908-1787-4

7. Calvo, T., Mesiar, R.: Weighted triangular norms-based aggregation operators. Fuzzy Sets Syst. **137**(1), 3–10 (2003). https://doi.org/10.1016/S0165-0114(02)00428-1

8. Ciungu, L., Riečan, B.: General form of probabilities on IF-sets. In: Di Gesù, V., Pal, S.K., Petrosino, A. (eds.) WILF 2009. LNCS (LNAI), vol. 5571, pp. 101–107. Springer, Heidelberg (2009). https://doi.org/10.1007/978-3-642-02282-1_13

9. Ciungu, L., Riečan, B.: Representation theorem for probabilities on IFS-events. Inf. Sci. **180**, 793–798 (2010). https://doi.org/10.1016/j.ins.2009.11.003

10. Deschrijver, G., Cornelis, C., Kerre, E.: On the representation of intuitionistic fuzzy t-norms and t-conorms. IEEE Trans. Fuzzy Syst. **12**(1), 45–61 (2004). https://doi.org/10.1109/TFUZZ.2003.822678

11. Deschrijver, G., Kerre, E.: A generalization of operators on intuitionistic fuzzy sets using triangular norms and conorms. Notes Intuitionistic Fuzzy Sets **8**(1), 19–27 (2002)

12. Deschrijver, G., Kerre, E.: Aggregation operators in interval-valued fuzzy and Atanassovs intuitionistic fuzzy set theory. In: Bustince, H., et al. (eds.) Fuzzy Sets and Their Extensions: Representation, Aggregation and Models. Studies in Fuzziness and Soft Computing, vol. 220, pp. 183–203. Springer, Heidelberg (2008). https://doi.org/10.1007/978-3-540-73723-0-10

13. Grabisch, M., Marichal, J.-L., Mesiar, R., Pap, E.: Aggregation Functions. Cambridge University Press, Cambridge (2009)

14. Grzegorzewski, P., Mrowka, E.: Probability of intuitionistic fuzzy events. In: Grzegorzewski, P., Hryniewicz, O., Gil, M.A. (eds.) Soft Methods in Probability, Statistics and Data Analysis. Advances in Intelligent and Soft Computing, vol. 16, pp. 105–115. Physica, Heidelberg (2002). https://doi.org/10.1007/978-3-7908-1773-7-8

15. Huang, J.-Y.: Intuitionistic fuzzy Hamacher aggregation operators and their application to multiple attribute decision making. J. Intell. Fuzzy Syst. **27**(1), 505–513 (2014). https://doi.org/10.3233/IFS-131019

16. Klement, E.P., Mesiar, R., Pap, E.: Triangular Norms. Kluwer, Dordrecht (2000). https://doi.org/10.1007/978-94-015-9540-7

17. Liu, P.D., Liu, J.L., Chen, S.M.: Some intuitionistic fuzzy Dombi Bonferroni mean operators and their application to multi-attribute group decision making. J. Oper. Res. Soc. **69**, 1–24 (2018). https://doi.org/10.1057/s41274-017-0190-y

18. Wang, W., Liu, X.: Intuitionistic fuzzy information aggregation using Einstein operations. IEEE Trans. Fuzzy Syst. **20**(5), 923–938 (2012). https://doi.org/10.1109/TFUZZ.2012.2189405

19. Xu, Z.S.: Intuitionistic fuzzy aggregation operators. IEEE Trans. Fuzzy Syst. **15**(6), 1179–1187 (2007). https://doi.org/10.1109/TFUZZ.2006.890678

20. Zadeh, L.A.: Fuzzy sets. Inf. Control **8**(3), 338–353 (1965). https://doi.org/10.1016/S0019-9958(65)90241-X

BIOWA Operators

Andrea Stupňanová[1]([⊠])[ID] and LeSheng Jin[2][ID]

[1] Faculty of Civil Engineering, Slovak University of Technology in Bratislava,
Radlinského 11, 810 05 Bratislava, Slovak Republic
andrea.stupnanova@stuba.sk
[2] Business School, Nanjing Normal University, Nanjing, China
jls1980@163.com

Abstract. Based on bi-capacities and related Choquet integral introduced by Grabisch and Labreuche, a new generalization of OWA operators, namely BIOWA operators are introduced. Our approach is exemplified by several examples. Bi-capacities leading to the standard Yager's OWA operators on real line are completely characterized.

Keywords: Aggregation function · Bi-capacity · Choquet integral · OWA operator

1 Introduction

Bi-capacities arise as a natural generalization of capacities (fuzzy measures, normed monotone measures) in the context of decision making where underlying scales are bipolar. They, together with the related Choquet integral, are able to capture a wide variety of decision behaviours, encompassing models such as Cumulative Prospect Theory [9]. The related Choquet integral generalizes the standard extensions of the original Choquet integral, such as the asymmetric Choquet integral [3] or the symmetric (Šipoš) integral [10]. Hence we expect possible applications in all areas dealing with the aggregation of data from bipolar scales, such as expert systems [8].

In what follows, for an arbitrary $n > 0$ denote $N = \{1, \dots, n\}$. Then a set function $\mu : 2^N \to [0, 1]$ is capacity if it is increasing and fulfils boundary conditions $\mu(\emptyset) = 0$ and $\mu(N) = 1$. The capacity μ is symmetric if

$$\mu(A) = \mu(\pi(A)) \text{ for any } A \in 2^N \text{ and permutation } \pi : N \to N,$$

$\pi(A) = \{\pi(i) | i \in A\}$. This symmetry holds if and only if $\mu(A) = g(\text{card}A)$, where $g : N \cup \{0\} \to [0, 1]$ is an increasing function fulfilling $g(0) = 0$ and $g(n) = 1$.

Let $f : N \to \mathbb{R}^+$ and μ be a capacity on N. The Choquet integral [2] of f with respect to μ is defined by

$$\mathbf{Ch}_\mu(f) := \int_0^\infty \mu(\{i | f(i) \geq t\}) \mathrm{d}t = \sum_{i=1}^n f(\tau(i)) \cdot [\mu(A_{\tau,i}) - \mu(A_{\tau,i+1})], \quad (1)$$

© Springer Nature Switzerland AG 2020
M.-J. Lesot et al. (Eds.): IPMU 2020, CCIS 1238, pp. 419–425, 2020.
https://doi.org/10.1007/978-3-030-50143-3_32

where $\tau : \{1, \ldots, n\} \to \{1, \ldots, n\}$ is any permutation such that $f(\tau(1)) \leq \cdots \leq f(\tau(n))$, $A_{\tau,i} = \{\tau(i), \ldots, \tau(n)\}$ and $A_{\tau,n+1} := \emptyset$. If μ is a symmetric capacity generated by g, then (1) can be rewritten into

$$\mathbf{Ch}_\mu(f) = \sum_{i=1}^{n} f(\tau(i)) \cdot [g(n - i + 1) - g(n - i)]. \tag{2}$$

The Choquet integral (2) is a symmetric n-ary aggregation function on $\mathbb{R}^+$, i.e., for any permutation π it hold $\mathbf{Ch}_\mu(f \circ \pi) = \mathbf{Ch}_\mu(f)$, where $(f \circ \pi)(i) = f(\pi(i))$. Formula (2) can be applied also for inputs from $\mathbb{R}$ (corresponding to the asymmetric Choquet integral [7]).

Moreover it is enough to put $w_i^g = g(n - i + 1) - g(n - i)$ to see that

$$\mathbf{Ch}_\mu(f) = \mathrm{OWA}_{\mathbf{w}^g}(f),$$

i.e., we get an OWA operator introduced by Yager [11]. Following [7], the Choquet integral $\mathbf{Ch}_\mu$ coincides with OWA operator if and only if $\mathbf{Ch}_\mu$ is a symmetric aggregation function, or equivalently, if and only if μ is a symmetric capacity, $\mu(A) = g(\mathrm{card}A)$. In our contribution we apply and study the same ideas on bi-capacity-based Choquet integral [5].

The structure of this contribution is as follows. In the next section, some necessary preliminaries are given. In Sect. 3 we define a new generalization of OWA operators, BIOWA operators, based on symmetric bi-capacities. Several examples are illustrating our new approach. Finally, some concluding remarks are given.

2 Preliminaries

Grabisch and Labreuche [5] have introduced bi-capacities $\nu : 3^N \to [-1, 1]$ and related Choquet integrals [6] $\mathcal{C}_\nu : \mathbb{R}^n \to \mathbb{R}$. Recall that $3^N = \{(A, B)|A, B \in 2^N, A \cap B = \emptyset\}$. Then bi-capacity is defined as follows.

Definition 1. *The function* $\nu : 3^N \to [-1, 1]$ *is a bi-capacity whenever*

- $\nu(N, \emptyset) = 1, \nu(\emptyset, \emptyset) = 0$ *and* $\nu(\emptyset, N) = -1$,
- ν *is increasing in the first coordinate and decreasing in the second coordinate.*

More, the Choquet integral $\mathcal{C}_\nu : \mathbb{R}^n \to \mathbb{R}$ is defined by

$$\mathcal{C}_\nu(f) = \sum_{i=1}^{n} |f(\sigma(i))| \cdot \Big[\nu\Big(A_{\sigma,i} \cap N_f^+, A_{\sigma,i} \cap N_f^-\Big) \tag{3}$$
$$- \nu\Big(A_{\sigma,i+1} \cap N_f^+, A_{\sigma,i+1} \cap N_f^-\Big)\Big],$$

where $\sigma : N \to N$ is an arbitrary permutation such that

$$|f(\sigma(1))| \leq \cdots \leq |f(\sigma(n))|,$$

$N_f^+ = \{i \in N | f(i) \geq 0\}$ and $N_f^- = N \setminus N_f^+$.

Note that $A_{\sigma,i} = \{\sigma(i), \ldots, \sigma(n)\}$ for $i = 1, \ldots, n$ and if $f \geq 0$ then $\mathcal{C}_\nu(f) = \mathbf{Ch}_\mu(f)$, where the right-hand side is the standard Choquet integral with respect to the capacity $\mu : 2^N \to [0,1], \mu(A) = \nu(A, \emptyset)$. Similarly, if $f \leq 0$ then $\mathcal{C}_\nu(f) = -\mathbf{Ch}_\gamma(-f)$, where $\gamma : 2^N \to [0,1]$ is the capacity given by $\gamma(B) = -\nu(\emptyset, B)$.

3   BIOWA Operator

In our contribution, we discuss symmetric Choquet integrals with respect to bi-capacities, which can be seen as an important generalization of standard OWA operators on $\mathbb{R}$ [11]. We will define a bi-capacity based OWA operator, BIOWA operator, as a symmetric Choquet integral $\mathcal{C}_\nu$, i.e. for any permutation π and a real function $f : N \to \mathbb{R}$ it holds $\mathcal{C}_\nu(f) = \mathcal{C}_\nu(f \circ \pi)$. Clearly

$$|f \circ \pi \circ \pi^{-1} \circ \sigma(1)| \leq \cdots \leq |f \circ \pi \circ \pi^{-1} \circ \sigma(n)|,$$

so, for $f \circ \pi$, the role of ordering permutation is done by $\pi^{-1} \circ \sigma$, similarly $N_{f \circ \pi^{-1}}^+ = \pi^{-1}(N_f^+), N_{f \circ \pi^{-1}}^- = \pi^{-1}(N_f^-)$. Based on the fact that $\mathcal{C}_\nu(1_A - 1_B) = \nu(A, B)$ for any disjoint $A, B \subseteq N$, the symmetry of the Choquet integral $\mathcal{C}_\nu$ is equivalent to the symmetry of the related bi-capacity ν, $\nu(A, B) = \nu(\pi^{-1}(A), \pi^{-1}(B))$ for any permutation π. We denote $\tilde{N} = \{(i, j) | i, j \in N \cup \{0\}, i + j \leq n\}$.

Lemma 1. *A bi-capacity* $\nu : 3^N \to [-1, 1]$ *is symmetric if and only if*

$$\nu(A, B) = h(\mathrm{card}A, \mathrm{card}B),$$

where $h : \tilde{N} \to [-1, 1]$

- *satisfies 3 boundary conditions* $h(n, 0) = 1, h(0, 0) = 0, h(0, n) = -1$,
- *is increasing in the first coordinate and*
- *is decreasing in the second coordinate.*

Example 1. Among several interesting examples of functions h, we mention here the next:

(i) $h_{\mathrm{ad}}(a, b) = \dfrac{a - b}{n}$ (the additive h)

(ii) $h_0(a, b) = \begin{cases} 1 & \text{if} \quad a = n, \\ -1 & \text{if} \quad b = n, \\ 0 & \text{otherwise}; \end{cases}$

(iii) $h^*(a, b) = \begin{cases} 1 & \text{if} \quad a > 0, \\ 0 & \text{if} \quad b < n, a = 0, \quad \text{(the greatest } h) \\ -1 & \text{if} \quad b = n; \end{cases}$

$$(\text{iv}) \ h_*(a,b) = \begin{cases} 1 & \text{if } a = n, \\ 0 & \text{if } a < n, b = 0, \\ -1 & \text{if } b < n; \end{cases} \qquad (\text{the smallest } h)$$

$$(\text{v}) \ h_1(a,b) = \begin{cases} 1 & \text{if } a > 0, b = 0, \\ -1 & \text{if } b > 0, a = 0, \\ 0 & \text{otherwise}; \end{cases}$$

(vi) $h_g(a,b) = g(a) - 1 + g(n - b)$, where $g : N \cup \{0\} \to [0,1]$ is an increasing function fulfilling boundary conditions $g(0) = 0, g(n) = 1$. Note that if $g(i) = \frac{i}{n}$ then $h_g = h_{\text{ad}}$.

In the formula for the Choquet integral $\mathcal{C}_\nu$ (3), the absolute values $|f(\sigma(i))|$ of aggregated inputs are multiplied by weights

$$w_i^f = \left[\nu\left(A_{\sigma,i} \cap N_f^+, A_{\sigma,i} \cap N_f^- \right) - \nu\left(A_{\sigma,i+1} \cap N_f^+, A_{\sigma,i+1} \cap N_f^- \right) \right].$$

For $f : N \to \mathbb{R}^n$, we define function $s_f : N \to \{(1,0),(0,1)\}$ by

$$s_f(i) := \begin{cases} (1,0) & \text{if } \sigma(i) \in N_f^+, \\ (0,1) & \text{if } \sigma(i) \in N_f^-, \end{cases} \qquad (4)$$

where $\sigma : n \to n$ is the unique permutation such that $|f(\sigma(1))| \leq \cdots \leq |f(\sigma(n))|$, and if for some $i < j$ it holds $|f(\sigma(i))| = |f(\sigma(j))|$ then either $\mathrm{sign} f(\sigma_i) = -1$ and $\mathrm{sign} f(\sigma_j) = 1$, or $\mathrm{sign} f(\sigma_i) = \mathrm{sign} f(\sigma_j)$ and $\sigma(i) < \sigma(j)$.

Consider a symmetric bi-capacity ν described by a function h. Then

$$w_i^f = h\left(\sum_{j=i}^{n} s_f(j) \right) - h\left(\sum_{j=i+1}^{n} s_f(j) \right), \qquad (5)$$

with convention that $\sum_{j=n+1}^{n} s_f(j) = (0,0)$. Then $\mathcal{C}_\nu(f) = \sum_{i=1}^{n} w_i^f |f(\sigma(i))|$. As we can see from (5), for a fixed h the weights w_i^f depend on the function s_f only, and thus we have, in general, 2^n possible weighting vectors $\mathbf{w}_s = (w_1^s, \ldots, w_n^s)$,

$$w_i^s = h\left(\sum_{j=i}^{n} s(j) \right) - h\left(\sum_{j=i+1}^{n} s(j) \right). \qquad (6)$$

Now, we can define the BIOWA operator.

Definition 2. *Consider a symmeric bi-capacity $\nu : 3^N \to [-1,1]$ described by a generating function h. Then BIOWA operator $\mathrm{BIOWA}_h : \mathbb{R}^n \to \mathbb{R}$ is defined as*

$$\mathrm{BIOWA}_h(f) = \mathcal{C}_\nu(f) = \sum_{i=1}^{n} w_i^s |f(\sigma(i))|, \qquad (7)$$

where the weighting vector $\mathbf{w}_s$ is defined by (6) and $s = s_f$ is given by (4).

Obviously, properties of the bi-polar capacities-based Choquet integrals are herited by the BIOWA operators. In particular, each BIOWA operator is an idempotent, continuous, positively homogenous and symmetric aggregation function on $\mathbb{R}$ [6].

Example 2. Let us consider $N = \{1, 2, 3, 4, 5\}$, and $f : (1, 2, 3, 4, 5) \rightarrow (-4, -1, 2, 3, 0)$. Then $\sigma = (5, 2, 3, 4, 1), N_f^+ = \{3, 4, 5\}, N_f^- = \{1, 2\}$ and $s_f : (1, 2, 3, 4, 5) \rightarrow ((1, 0), (0, 1), (1, 0), (1, 0), (0, 1))$. Then the weighting vector $\mathbf{w}_s$ is given by

$$w_1^s = h(3, 2) - h(2, 2)$$
$$w_2^s = h(2, 2) - h(2, 1)$$
$$w_3^s = h(2, 1) - h(1, 1)$$
$$w_4^s = h(1, 1) - h(0, 1)$$
$$w_5^s = h(0, 1)$$

and the corresponding BIOWA operator is

$$\begin{aligned}
\mathrm{BIOWA}_h(f) &= w_1^s \cdot 0 + w_2^s \cdot 1 + w_3^s \cdot 2 + w_4^s \cdot 3 + w_5^s \cdot 4 \\
&= h(2, 2) - h(2, 1) + 2(h(2, 1) - h(1, 1)) + 3(h(1, 1) - h(0, 1)) + 4h(0, 1) \\
&= h(2, 2) + h(2, 1) + h(1, 1) + h(0, 1).
\end{aligned}$$

For $h = h_{\mathrm{ad}}$ we get

$$\mathrm{BIOWA}_{h_{\mathrm{ad}}}(f) = \frac{1}{5} - \frac{1}{5} = 0 = \frac{1}{5} \sum_{i=1}^{5} f(i).$$

Note that:

- if $f \geq 0$ (i.e., $N_f^+ = N, N_f^- = \emptyset$), then the standard $\mathrm{OWA}_{\mathbf{w}^+}$ is recovered, where $\mathbf{w}^+ = (w_1^+, \ldots, w_n^+)$, $w_i^+ = h(n - i + 1, 0) - h(n - i, 0)$,

$$\mathrm{BIOWA}_h(f) = \mathrm{OWA}_{\mathbf{w}^+}(f) = \sum_{i=1}^{n} w_i^+ f(\tau(i)),$$

where $f(\tau(1)) \leq \cdots \leq f(\tau(n))$.
- if $f < 0$ (i.e., $N_f^+ = \emptyset, N_f^- = N$), again an OWA operator is recovered, but related to the weighting vector $\mathbf{w}^-$, $w_i^- = h(0, i - 1) - h(0, i)$, $\sum_{i=1}^{5} w_i = 1$, and then

$$\mathrm{BIOWA}_h(f) = \mathrm{OWA}_{\mathbf{w}^-}(f) = \sum_{i=1}^{n} w_i^- f(\tau(i)).$$

where $f(\tau(1)) \leq \cdots \leq f(\tau(n))$. Note that this formula holds also if $f \leq 0$.

These two facts justify the name BIOWA operator.

The BIOWA operator $\text{BIOWA}_h : \mathbb{R}^n \to \mathbb{R}$ is a standard OWA operator of Yager if and only if $h(a,b) = h_g(a,b)$ from the Example 1 (vi), (i.e., g generates weighting vector $\mathbf{w} = (w_1, \ldots, w_n)$, $w_i = g(n-i+1) - g(n-i)$). Then

$$\text{BIOWA}_{h_g}(f) = \sum_{i=1}^n f(\tau(i)) \cdot (g(n-i+1) - g(n-i)),$$

where $\tau : N \to N$ is a permutation such that $f(\tau(1)) \leq \cdots \leq f(\tau(n))$;

Example 3.

(i) Consider $h = h_{\text{ad}}$. Then

$$\text{BIOWA}_{h_{\text{ad}}}(f) = \frac{1}{n} \sum_{i=1}^n f(i) = \text{AM}(f) \quad \text{(arithmetic mean)}.$$

(ii) Consider $h = h_0$. Then

$$\text{BIOWA}_{h_0}(f) = \text{med}_0(f(1), \ldots, f(n)) = \text{med}\left(\min_{i \in N} f(i), 0, \max_{i \in N} f(i)\right) \quad \text{(0-median)},$$

for more details about 0-median see [1].

(iii) Consider $h = h^*$. Then

$$\text{BIOWA}_{h^*}(f) = \max\{f(i) | i \in N\}.$$

(iv) Consider $h = h_*$. Then

$$\text{BIOWA}_{h_*}(f) = \min\{f(i) | i \in N\}.$$

(v) Consider $h = h_1$. Then

$$\text{BIOWA}_{h_1}(f) = \max\{f(i) | f(i) \geq 0\} + \min\{f(i) | f(i) \leq 0\}.$$

If $n = 2$, $f(1) = x$, $f(2) = y$, then

$$\text{BIOWA}_{h_1}(f) = \begin{cases} \max\{x,y\} & \text{if} \quad x, y \geq 0, \\ \min\{x,y\} & \text{if} \quad x, y < 0, \\ x + y & \text{otherwise.} \end{cases}$$

Remark 1. If we define dual function to h by $h^d(a,b) = -h(b,a)$ then

$$\text{BIOWA}_{h^d}(f) = -\text{BIOWA}_h(-f);$$

Note that $h^d_{\text{ad}} = h_{\text{ad}}$ and $h^d_0 = h_0$, thus $\text{BIOWA}_{h_{\text{ad}}}$ and BIOWA_{h_0} are homogenous. This is not the case of h^* and h_*, as $h^d_* = h^* \neq h_*$.

4 Concluding Remarks

Based on symmetric bi-capacities and related Choquet integral introduced by Grabisch and Labreuche, we have introduced and discussed BIOWA operators. These operators act just as standard OWA operator of Yager once all inputs have the same sign (but such two OWA operators may differ in weights). Symmetric bi-capacities yielding standard OWA operators being in coincidence with the related BIOWA operator are completely characterized.

We believe that the introduced BIOWA operators will find numerous applications in all areas where real data are successfully processed by means of OWA operators [4], as well as in all areas dealing with the aggregation of data from bipolar scales, such as expert systems, see [8].

Acknowledgments. The support of the grants APVV-17-0066 and VEGA 1/0006/19 is kindly announced.

References

1. Calvo, T., Kolesárová, A., Komorníková M., Mesiar, R.: Aggregation operators: properties, classes and construction methods. In: Calvo, T., Mayor, G., Mesiar, R. (eds.) Aggregation Operators: New Trends and Applications. Studies in Fuzziness and Soft Computing, vol. 97, pp. 3–104. Physica, Heidelberg (2002). https://doi.org/10.1007/978-3-7908-1787-4_1
2. Choquet, G.: Theory of capacities. Ann. Inst. Fourier **5**, 131–295 (1953/1954)
3. Dennenberg, D.: Non-Additive Measure and Integral. Kluwer Academic Publishers, Dordrecht (1994)
4. Emrouznejad, A., Marra, M.: Ordered weighted averaging operators 1988–2014: a citation-based literature survey. Int. J. Intell. Syst. **29**, 994–1014 (2014). https://doi.org/10.1002/int.21673
5. Grabisch, M., Labreuche, C.: Bi-capacities-I: definition, Möbius transform and interaction. Fuzzy Sets Syst. **151**, 211–236 (2005). https://doi.org/10.1016/j.fss.2004.08.012
6. Grabisch, M., Labreuche, C.: Bi-capacities-II: the Choquet integral. Fuzzy Sets Syst. **151**, 237–256 (2005)
7. Grabisch, M., Marichal, J.-L., Mesiar, R., Pap, E.: Aggregation Functions. Cambridge University Press, Cambridge (2009)
8. Heckerman, D., Shortliffe, E.: From certainty factors to belief networks. Artif. Intell. Med. **4**(1), 35–52 (1992). https://doi.org/10.1016/0933-3657(92)90036-O
9. Tversky, A., Kahneman, D.: Advances in prospect theory: cumulative representation of uncertainty. J. Risk Uncertainty **5**(4), 297–323 (1992). https://doi.org/10.1007/BF00122574
10. Šipoš, J.: Integral with respect to a pre-measure. Math. Slovaca **29**, 141–155 (1979)
11. Yager, R.R.: On ordered weighted averaging aggregation operators in multicriteria decisionmaking. IEEE Trans. Syst. Man Cybern. **18**(1), 183–190 (1988). https://doi.org/10.1109/21.87068

On Compatibility of Two Approaches to Generalization of the Lovász Extension Formula

Ľubomíra Horanská[✉] [iD]

Institute of Information Engineering, Automation and Mathematics, Faculty of
Chemical and Food Technology, Slovak University of Technology in Bratislava,
Radlinského 9, Bratislava 1, 812 37 Bratislava, Slovakia
lubomira.horanska@stuba.sk

Abstract. We present a method of generalization of the Lovász extension formula combining two known approaches - the first of them based on
the replacement of the product operator by some suitable binary function
F and the second one based on the replacement of the minimum operator by a suitable aggregation function A. We propose generalization by
simultaneous replacement of both product and minimum operators and
investigate pairs (F, A) yielding an aggregation function for all capacities.

Keywords: Aggregation function · Choquet integral · Capacity ·
Möbius transform

1 Introduction

Aggregation of several values into a single value proves to be useful in many fields,
e.g., multicriteria decision making, image processing, deep learning, fuzzy systems etc. Using the Choquet integral [3] as a mean of aggregation process allows
to capture relations between aggregated data through so-called fuzzy measures
[9]. This is the reason of the nowadays interest in generalizations of the Choquet
integral, for a recent state-of-art see, e.g., [4].

In our paper we focus on generalizations of the Choquet integral expressed
by means of the so-called Möbius transform, which is also known as Lovász
extension formula, see (2) below. Recently, two different approaches occured - in
the first one the Lovász extension formula is modified by replacing of the product
operator by some suitable binary function F and the second one is based on the
replacement of the minimum operator by a suitable aggregation function A. We
study the question, when these two approaches can be used simultaneously and
we investigate the functional $I_{F,A}^m$ obtained in this way.

The paper is organized as follows. In the next section, some necessary preliminaries are given. In Sect. 3, we propose the new functional $I_{F,A}^m$ and exemplify
the instances, when the obtained functional is an aggregation function for all

Supported by the grants VEGA 1/0614/18 and VEGA 1/0545/20.

capacities and when it is not. Section 4 contains results concerning the binary case. Finally, some concluding remarks are given.

2   Preliminaries

In this section we recall some definitions and results which will be used in the sequel. We also fix the notation, mostly according to [5], wherein more information concerning the theory of aggregation functions can be found.

Let $n \in \mathbb{N}$ and $N = \{1, \cdots, n\}$.

Definition 1. *A function $A \colon [0,1]^n \to [0,1]$ is an (n-ary) aggregation function if A is monotone and satisfies the boundary conditions $A(0, \ldots, 0) = 0$ and $A(1, \ldots, 1) = 1$.*

We denote the class of all n-ary aggregations functions by $\mathcal{A}_{(n)}$.

Definition 2. *An aggregation function $A \in \mathcal{A}_{(n)}$ is*

- *conjunctive, if $A(\mathbf{x}) \leq \min_{i \in N} x_i$ for all $\mathbf{x} \in [0,1]^n$,*
- *disjunctive, if $A(\mathbf{x}) \geq \max_{i \in N} x_i$ for all $\mathbf{x} \in [0,1]^n$.*

Definition 3. *A set function $m \colon 2^N \to [0,1]$ is a capacity if $m(C) \leq m(D)$ whenever $C \subseteq D$ and m satisfies the boundary conditions $m(\emptyset) = 0$, $m(N) = 1$.*

We denote the class of all capacities on 2^N by $\mathcal{M}_{(n)}$.

Definition 4. *The set function $M_m \colon 2^N \to \mathbb{R}$, defined by*

$$M_m(I) = \sum_{K \subseteq I} (-1)^{|I \setminus K|} m(K)$$

for all $I \subseteq N$, is called Möbius transform corresponding to a capacity m.

Möbius transform is invertible by means of the so-called Zeta transform:

$$m(A) = \sum_{B \subseteq A} M_m(B), \tag{1}$$

for every $A \subseteq N$.

Denote $\mathcal{R}_n \subsetneq \mathbb{R}$ the range of the Möbius transform. The bounds of the Möbius transform have recently been studied by Grabisch et al. in [6].

Definition 5. *Let $m \colon 2^N \to [0,1]$ be a capacity and $\mathbf{x} = (x_1, \ldots, x_n) \in [0,1]^n$. Then the Choquet integral of $\mathbf{x}$ with respect to m is given by*

$$\mathbf{Ch}_m(\mathbf{x}) = \int_0^1 m(\{i \in N | x_i \geq t\}) \, dt,$$

where the integral on the right-hand side is the Riemann integral.

Proposition 1. Let $m : 2^N \to [0,1]$ and $\mathbf{x} \in [0,1]^n$. Then the discrete Choquet integral can be expressed as:

$$\mathbf{Ch}_m(\mathbf{x}) = \sum_{\emptyset \neq B \subseteq N} \left(M_m(B) \cdot \min_{i \in B} x_i \right). \tag{2}$$

Formula (2) is also known as the *Lovász extension* formula [8].

Now we recall two approaches to generalization of the formula (2). The first one is due to Kolesárová et al. [7] and is based on replacing the minimum operator in (2) by some other aggregation function in the following way:

Let $m \in \mathcal{M}_{(n)}$ be a capacity, $A \in \mathcal{A}_{(n)}$ be an aggregation function. Define $F_{m,A} \colon [0,1]^n \to \mathbb{R}$ by

$$F_{m,A}(x_1, \ldots, x_n) = \sum_{B \subseteq N} M_m(B) A(\mathbf{x}_B), \tag{3}$$

where $(\mathbf{x}_B)_i = x_i$ whenever $i \in B$ and $(\mathbf{x}_B)_i = 1$ otherwise. The authors focused on characterization of aggregation functions A yielding, for all capacities $m \in \mathcal{M}_{(n)}$, an aggregation function $F_{m,A}$ extending the capacity m, i.e., on such A that $F_{m,A} \in \mathcal{A}_{(n)}$ and $F_{m,A}(\mathbf{1}_B) = m(B)$ for all $B \subseteq N$ (here $\mathbf{1}_B$ stands for the indicator of the set B).

Remark 1. There was shown in [7] that (among others) all copulas are suitable to be taken in rôle of A in (3). For instance, taking $A = \Pi$, where $\Pi(\mathbf{x}) = \prod_{i=1}^n x_i$ is the product copula, we obtain the well-known Owen multilinear extension (see [10]).

The second approach occured recently in [2] and is based on replacing the product of $M_m(A)$ and minimum operator in the formula (2) by some function $F \colon \mathbb{R} \times [0,1] \to \mathbb{R}$ in the following way:

Let $m \in \mathcal{M}_{(n)}$, $F \colon \mathbb{R} \times [0,1] \to \mathbb{R}$ be a function bounded on $[0,1]^2$. Define the function $\mathfrak{I}_m^F \colon [0,1]^n \to \mathbb{R}$ by

$$\mathfrak{I}_m^F(\mathbf{x}) = \sum_{\emptyset \neq B \subseteq N} F(M_m(B), \min_{i \in B}\{x_i\}). \tag{4}$$

The authors focused on functions F yielding an aggregation function $\mathfrak{I}_m^F$ for all capacities $m \in \mathcal{M}_{(n)}$.

Remark 2. It was shown in [2] that all functions F yielding for all $m \in \mathcal{M}_{(n)}$ aggregation functions $\mathfrak{I}_m^F$ with a given diagonal section $\delta \in \mathcal{A}_{(1)}$ are exactly those of the form

$$F(u,v) = u\,h(v) + \frac{\delta(v) - h(v)}{2^n - 1}, \tag{5}$$

where $h \colon [0,1] \to \mathbb{R}$ is a function satisfying

$$-\frac{\delta(y) - \delta(x)}{2^n - 2} \leq h(y) - h(x) \leq \delta(y) - \delta(x),$$

for all $(x, y) \in [0, 1]^2$, such that $x < y$.

However, there is no full characterization of all functions F yielding an aggregation function $\mathcal{I}_m^F$ for every $m \in \mathcal{M}_{(n)}$ in [2].

3   Double Generalization of the Lovász Extension Formula

Let $F \colon \mathbb{R} \times [0, 1] \to [0, 1]$ be a function bounded on $[0, 1]^2$, A be an aggregation function $A \in \mathcal{A}_{(n)}$, m be a capacity $m \in \mathcal{M}_{(n)}$. We define the function $\mathcal{I}_{F,A}^m \colon [0, 1]^n \to \mathbb{R}$ as

$$\mathcal{I}_{F,A}^m(\mathbf{x}) = \sum_{\emptyset \neq B \subseteq N} F(M_m(B), A(\mathbf{x}_B)), \tag{6}$$

where $(\mathbf{x}_B)_i = x_i$ whenever $i \in B$ and $(\mathbf{x}_B)_i = 1$ otherwise.

Lemma 1. *Let $F \colon \mathbb{R} \times [0, 1] \to \mathbb{R}$ be a function bounded on $[0, 1]^2$ and $c \in \mathbb{R}$. Let $F_c \colon \mathbb{R} \times [0, 1] \to \mathbb{R}$ be a function defined by*

$$F_c(x, y) = F(x, y) + c\left(x - \frac{1}{2^n - 1}\right).$$

Then, that for any $m \in \mathcal{M}_{(n)}$, it holds $\mathcal{I}_{F,A}^m(\mathbf{x}) = \mathcal{I}_{F_c,A}^m(\mathbf{x})$ for all $\mathbf{x} \in [0, 1]^n$.

Proof. Since $\displaystyle\sum_{\emptyset \neq B \subseteq N} c\left(M_m(B) - \frac{1}{2^n - 1}\right) = 0$, the result follows.

Consequently, one can consider $F(0, 0) = 0$ with no loss of generality (compare with Proposition 3.1 in [2]).

Let us define

$$\mathcal{F}_0 = \{F \colon \mathbb{R} \times [0, 1] \to \mathbb{R} \mid F(0, 0) = 0 \text{ and } F \text{ is bounded on } [0, 1]^2\}$$

Definition 6. *A function $F \in \mathcal{F}_0$ is I-compatible with an aggregation function $A \in \mathcal{A}_{(n)}$ iff $\mathcal{I}_{F,A}^m \in \mathcal{A}_{(n)}$ for all $m \in \mathcal{M}_{(n)}$.*

Note that, according to Remark 1, the product operator $\Pi(u, v) = uv$ is I-compatible with every copula. Next, according to Remark 2, all binary functions of the form (5) are I-compatible with $A = \min$.

Example 1. Let $F(u, v) = \frac{v}{2^n - 1}$, $A \in \mathcal{A}_{(n)}$ be a conjunctive aggregation function. We have

$$\mathcal{I}_{F,A}^m(\mathbf{x}) = \frac{1}{2^n - 1} \sum_{\emptyset \neq B \subseteq N} A(\mathbf{x}_B).$$

Clearly, it is a monotone function and $\mathcal{I}_{F,A}^m(\mathbf{1}) = 1$. Moreover, conjunctivity of A gives $\mathcal{I}_{F,A}^m(\mathbf{0}) = 0$. Thus, $\mathcal{I}_{F,A}^m$ is an aggregation function for all capacities $m \in \mathcal{M}_{(n)}$ and therefore F is I-compatible with every conjunctive aggregation function $A \in \mathcal{A}_{(n)}$.

Example 2. Let $f: [0,1] \to [0,1]$ be a nondecreasing function such that $f(0) = 0$ and $f(1) = 1$, i.e., $f \in \mathcal{A}_{(1)}$. Let $F(u,v) = (2 - 2^n)u + f(v)$. Then F is I-compatible with every disjunctive aggregation function $A \in \mathcal{A}_{(n)}$. Indeed, disjunctivity of A implies $A(\mathbf{x}_B) = 1$ for all $x \in [0,1]^n$, $\emptyset \neq B \subsetneq N$. Then, using (1), we obtain

$$\mathcal{I}_{F,A}^m(\mathbf{x}) = (2 - 2^n) \sum_{\emptyset \neq B \subseteq N} M_m(B) + \sum_{\emptyset \neq B \subseteq N} f(A(\mathbf{x}_B))$$

$$= 2 - 2^n + f(A(\mathbf{x})) + \sum_{\emptyset \neq B \subsetneq N} f(A(\mathbf{x}_B))$$

$$= 2 - 2^n + f(A(\mathbf{x})) + 2^n - 2 = f(A(\mathbf{x})),$$

which is an aggregation function for all $m \in \mathcal{M}_{(n)}$.

On the other hand, for $n > 1$, F is not I-compatible with the minimal aggregation function A_* defined as $A_*(\mathbf{x}) = 1$ if $\mathbf{x} = \mathbf{1}$ and $A_*(\mathbf{x}) = 0$ otherwise, since in this case $\mathcal{I}_{F,A_*}^m(\mathbf{x}) = 2 - 2^n$ for all $\mathbf{x} \neq \mathbf{1}$. Note that for $n = 1$ we obtain $\mathcal{I}_{F,A_*}^m = A_*$.

For a measure $m \in \mathcal{M}_{(2)}$ let us denote $m(\{1\}) = a$ and $m(\{2\}) = a$.

Example 3. Let $n = 2$. Let $F(u,v) = u v^{u+1}$, $A(x,y) = \max\{x + y - 1, 0\}$. Then

$$\mathcal{I}_{F,A}^m(x,y) = \begin{cases} ax^{a+1} + by^{b+1} + (1 - a - b)(x + y - 1)^{2-a-b} & \text{if } x + y \geq 1 \\ ax^{a+1} + by^{b+1} & \text{otherwise} \end{cases},$$

which is an aggregation function for all $m \in \mathcal{M}_{(2)}$, thus F is I-compatible with A. However, taking a disjunctive aggregation function in rôle of A, we obtain

$$\mathcal{I}_{F,A}^m(x,y) = a + b + (1 - a - b)A(x,y)^{2-a-b},$$

which is not an aggregation function for all capacities up to the minimal one ($a = b = 0$). Hence, F is not I-compatible with any disjunctive aggregation function.

4   Binary Case

Let $n = 2$. Then the function $\mathcal{I}_{F,A}^m$ defined by (6) can be expressed as

$$\mathcal{I}_{F,A}^m(x,y) = F(a, A(x,1)) + F(b, A(1,y)) + F(1 - a - b, A(x,y)). \qquad (7)$$

Proposition 2. *Let $F \in \mathcal{F}_0$, $A \in \mathcal{A}_{(2)}$. Then F is I-compatible with A iff the following conditions are satisfied*

(i) *There exist constants $k, \kappa \in \mathbb{R}$ such that for any $u \in \mathcal{R}_2 = [-1,1]$ it holds*
$F(u, A(0,1)) = F(u, A(1,0)) = k(u - \frac{1}{2})$
$F(u, 0) = ku,$
$F(u, 1) = \kappa u + \frac{1-\kappa}{3}.$

(ii) For all $x, x', y, y' \in [0,1]$ such that $x \le x'$ and $y \le y'$ it holds

$$F(a, A(x',1)) - F(a, A(x,1)) + F(1-a-b, A(x',y)) - F(1-a-b, A(x,y)) \ge 0$$

and

$$F(b, A(1,y')) - F(b, A(1,y)) + F(1-a-b, A(x,y')) - F(1-a-b, A(x,y)) \ge 0,$$

for any $a, b \in [0,1]$.

Proof. It can easily be checked that conditions (i) ensure boundary conditions $\mathcal{I}_{F,A}^m(0,0) = 0$ and $\mathcal{I}_{F,A}^m(1,1) = 1$. To show necessity, let us consider the following equation:

$$\mathcal{I}_{F,A}^m(0,0) = F(a, A(0,1)) + F(b, A(1,0)) + F(1-a-b, A(0,0)) = 0,$$

for all $a, b \in [0,1]$.

Denoting $F(u, A(0,1)) = f(u)$, $F(u, A(1,0)) = h(u)$ and $F(u,0) = g(u)$, for $u \in [-1,1]$, the previous equation takes form

$$f(a) + h(b) + g(1 - a - b) = 0, \tag{8}$$

for all $a, b \in [0,1]$. Following techniques used for solving Pexider's equation (see [1]), we can put $a = 0$ and $b = 0$ respectively, obtaining

$$f(0) + h(b) + g(1 - b) = 0,$$
$$f(a) + h(0) + g(1 - a) = 0.$$

Thus, for any $t \in [0,1]$, we have

$$f(0) + h(t) + g(1 - t) = 0,$$
$$f(t) + h(0) + g(1 - t) = 0.$$

Consequently

$$h(t) = f(t) + f(0) - h(0), \tag{9}$$
$$g(t) = -f(1 - t) - h(0). \tag{10}$$

Therefore, formula (8) turns into

$$f(a + b) = f(a) + f(b) + f(0) - 2h(0),$$

for all $a, b \in [0,1]$. Now, denoting $\varphi(t) = f(t) + f(0) - 2h(0)$, we get

$$\varphi(a + b) = \varphi(a) + \varphi(b), \tag{11}$$

which is the Cauchy equation. Taking $a = b = 0$, we get $\varphi(0) = 0$. Therefore, putting $a = t, b = -t$, we get $\varphi(t) = -\varphi(-t)$, i.e., φ is an odd function. Since we suppose F to be bounded on $[0,1]^2$, according to Aczél [1], all solutions of the

Eq. (11) on the interval $[-1,1]$ can be expressed as $\varphi(t) = kt$, for some $k \in \mathbb{R}$. Therefore,
$$f(t) = kt - f(0) + 2h(0),$$

for all $t \in [-1,1]$, which for $t = 0$ gives $f(0) = h(0)$. Denoting $f(0) = c$, by (9) and (10) we obtain

$$f(t) = kt + c,$$
$$h(t) = kt + c,$$
$$g(t) = -f(1-t) - h(0) = kt - k - 2c.$$

Since by assumption $g(0) = F(0,0) = 0$, we have $c = -\frac{k}{2}$ and consequently,

$$f(t) = h(t) = k(t - \frac{1}{2}), \tag{12}$$

$$g(t) = kt, \tag{13}$$

for all $t \in [-1,1]$ as asserted.

The second boundary condition for $\mathcal{I}_{F,A}^m$ gives

$$\mathcal{I}_{F,A}^m(1,1) = F(a, A(1,1)) + F(b, A(1,1)) + F(1 - a - b, A(1,1)) = 1,$$

for all $a, b \in [0,1]$. As A is an aggregation function, it holds $A(1,1) = 1$. Denoting $F(u,1) = \psi(u)$ we obtain

$$\psi(a) + \psi(b) + \psi(1 - a - b) = 1,$$

for all $a, b \in [0,1]$. Similarly as above, this equation can be transformed into the Cauchy equation (see also [2]) having all solutions of the form $\psi(t) = \kappa t + \frac{1-\kappa}{3}$, for $\kappa \in \mathbb{R}$ and $t \in [-1,1]$.

The conditions (ii) are equivalent to monotonicity of $\mathcal{I}_{F,A}^m$, which completes the proof.

Considering aggregation functions satisfying $A(0,1) = A(1,0) = 0$ (e.g., all conjunctive aggregation functions are involved in this subclass), the conditions in Proposition 2 ensuring the boundary conditions of $\mathcal{I}_{F,A}^m$ can be simplified in the following way.

Corollary 1. *Let $F \in \mathcal{F}_0$, $A \in \mathcal{A}_{(2)}$ be an aggregation function with $A(0,1) = A(1,0) = 0$. Then the following holds:*

(i) $I_{F,A}^m(0,0) = 0$ iff $F(u,0) = 0$ for any $u \in \mathcal{R}_2$,
(ii) $I_{F,A}^m(1,1) = 1$ iff there exist a constant $\kappa \in \mathbb{R}$ such that $F(u,1) = \kappa u + \frac{1-\kappa}{3}$ for any $u \in \mathcal{R}_2$. Moreover, if F is I-compatible with A, then $\kappa \in [-\frac{1}{2}, 1]$.

Proof. We have $F(u, A(0,1)) = F(u, A(1,0)) = F(u,0)$ for all $u \in \mathcal{R}_n$. The conditions (i) in Proposition 2 yield $k(u - \frac{1}{2}) = ku$, and consequently $k = 0$, thus $F(u,0) = 0$ for all $u \in \mathcal{R}_n$ as asserted.

Supposing that F is I-compatible with A and considering nondecreasingness of $\mathcal{I}_{F,A}^m$ in the first variable, we obtain

$$0 \leq \mathcal{I}_{F,A}^m(0,0) - \mathcal{I}_{F,A}^m(1,0)$$
$$= F(a, A(0,1)) - F(a, A(1,1)) + F(1 - a - b, A(0,0)) - F(1 - a - b, A(1,0))$$
$$= F(a,0) - F(a,1) + F(1 - a - b, 0) - F(1 - a - b, 0)$$
$$= F(a,0) - F(a,1),$$

for all $a \in [0,1]$.

Hence,

$$0 = F(u,0) \leq F(u,1) = \kappa u + \frac{1 - \kappa}{3}$$

for all $u \in [0,1]$ and consequently $-\frac{1}{2} \leq \kappa \leq 1$, which completes the proof.

Considering aggregation functions satisfying $A(0,1) = A(1,0) = 1$ (e.g., all disjunctive aggregation functions are involved in this subclass), the conditions in Proposition 2 ensuring the boundary conditions of $I_{F,A}^m$ can be simplified in the following way.

Corollary 2. *Let $F \in \mathcal{F}_0$, $A \in \mathcal{A}_{(2)}$ be an aggregation function with $A(0,1) = A(1,0) = 1$. Then the following holds:*

(i) $I_{F,A}^m(0,0) = 0$ iff $F(u,0) = -2u$ for any $u \in \mathcal{R}_2$,
(ii) $I_{F,A}^m(1,1) = 1$ iff $F(u,1) = -2u + 1$ for any $u \in \mathcal{R}_2$,

Proof. Since $A(0,1) = A(1,0) = 1$, the Eq. (8) takes form

$$f(a) + f(b) + g(1 - a - b) = 0,$$

for all $a, b \in [0,1]$. Taking $b = 1 - a$ and considering $g(0) = 0$ we obtain

$$f(a) = -f(1 - a),$$

for all $a \in [0,1]$, and thus $f(\frac{1}{2}) = 0$. Proposition 2(i) yields

$$F\left(\frac{1}{2}, 1\right) = \frac{\kappa}{2} + \frac{1 - \kappa}{3} = 0,$$

thus $\kappa = -2$, and consequently formulae (12),(13) imply the assertion.

5   Conclusion

We have introduced a new functional $I_{F,A}^m$ generalizing the Lovász extension formula (or the Choquet integral expressed in terms of Möbius transform) using simultaneously two known approaches. We have investigated when the obtained functional is an aggregation function for all capacities and exemplified positive and negative instances. In case of the binary functional we have found a characterization of all pairs (F, A) which are I-compatible, i.e., yielding an aggregation function $I_{F,A}^m$ for all capacities m. In our future reasearch we will focus on the characterization of all I-compatible pairs (F, A) in general n-ary case. Another interesting unsolved problem is the problem of giving back capacity, i.e., characterization of pairs (F, A) satisfying $I_{F,A}^m(\mathbf{1}_E) = m(E)$ for all $E \subseteq N$.

Acknowledgments. The support of the grant VEGA 1/0614/18 and VEGA 1/0545/20 is kindly acknowledged.

References

1. Aczél, J.: Lectures on Functional Equations and Their Applications. Academic Press, Cambridge (1966)
2. Bustince, H., Fernandez, J., Horanská, Ľ., Mesiar, R., Stupňanová, A.: A generalization of the Choquet integral defined in terms of the Möbius transform. IEEE Trans. Fuzzy Syst. https://doi.org/10.1109/TFUZZ.2019.2933803
3. Choquet, G.: Theory of capacities. Ann. de l'Institut Fourier **5**, 131–295 (1953–54)
4. Dimuro, G.P., et al.: The state-of-art of the generalizations of the Choquet integral: from aggregation and pre-aggregation to ordered directionally monotone functions. Inf. Fusion **57**, 27–43 (2020)
5. Grabisch, M., Marichal, J.-L., Mesiar, R., Pap, E.: Aggregation Functions. Cambridge University Press, Cambridge (2009)
6. Grabisch, M., Miranda, P.: Exact bounds of the Möbius inverse of monotone set functions. Discrete Appl. Math. **186**, 7–12 (2015)
7. Kolesárová, A., Stupňanová, A., Beganová, J.: Aggregation-based extensions of fuzzy measures. Fuzzy Sets Syst. **194**, 1–14 (2012)
8. Lovász, L.: Submodular function and convexity. In: Bachem, A., Korte, B., Grötschel, M. (eds.) Mathematical Programming: The State of the Art, pp. 235–257. Springer, Heidelberg (1983). https://doi.org/10.1007/978-3-642-68874-4_10
9. Murofushi, T., Sugeno, M., Machida, M.: Non-monotonic fuzzy measures and the Choquet integral. Fuzzy Sets Syst. **64**(1), 73–86 (1994)
10. Owen, G.: Multilinear extensions of games. In: Roth, A.E. (ed.) The Shapley Value, pp. 139–151. Cambridge University Press, Cambridge (1988)

The Formalization of Asymmetry in Disjunctive Evaluation

Miroslav Hudec[1](✉) ⓘ and Radko Mesiar[2] ⓘ

[1] Faculty of Economics, VSB – Technical University of Ostrava,
Ostrava, Czech Republic
miroslav.hudec@vsb.cz
[2] Faculty of Civil Engineering, Slovak University of Technology, Bratislava, Slovakia
mesiar@math.sk

Abstract. The main property of disjunction is substitutability, i.e., the fully satisfied predicate substitutes the rejected one. But, in many real–world cases disjunction is expressed as the fusion of full and optional alternatives, which is expressed as *OR ELSE* connective. Generally, this logical connective provides a solution lower than or equal to the *MAX* operator, and higher than or equal to the projection of the full alternative, i.e., the solution does not go below any averaging function and above *MAX* function. Therefore, the optional alternative does not influence the solution when it is satisfied with a degree lower than the degree of full alternative. In this work, we propose further generalization by other disjunctive functions in order to allow upward reinforcement of asymmetric disjunction. Finally, the obtained results are illustrated and discussed.

Keywords: Asymmetric disjunction · Averaging functions · Probabilistic sum · Łukasiewicz t–conorm · Generalization · Upward reinforcement

1 Introduction

One of the key properties of disjunction is commutativity, i.e., the order of predicates is irrelevant. The full substitutability means that at least one predicate should be met [7]. In technical systems, it is a desirable property, because if one unit fails, the others substitute its functionality. But, in evaluating records from a large data set, a larger number of records might be selected and moreover a significant proportion of them might get the ideal evaluation score. An example is searching for a house which has *spacious basement* **or** *spacious attic* **or** *suitable tool–shed* for storing the less–frequently used items. Such query might lead to the over–abundant answer problem. This problem was discussed in [3], where several solutions have been proposed. People often consider disjunction as the left–right order of predicates (alternatives), that is, the first predicate is the full alternative, whereas the other ones are less relevant options [10,13], or formally P_1 *OR ELSE* P_2 *OR ELSE* P_3.

© Springer Nature Switzerland AG 2020
M.-J. Lesot et al. (Eds.): IPMU 2020, CCIS 1238, pp. 435–446, 2020.
https://doi.org/10.1007/978-3-030-50143-3_34

In the above example, predicates might not be the equal substitutes for a particular house buyer. In such cases, the substitutability should be more restrictive. If an entity fails to meet the first predicate, but meets the second one, it is still considered as a solution, but as a non–ideal one (the evaluation degree should be lower than the ideal value, usually denoted as 1). This observation leads to the intensified disjunction. The literature offers two approaches: bipolar and asymmetric. The bipolar form of *OR ELSE* connective consists of the positive pole, which expresses the perfect values (full alternative) and the negative pole expressing the acceptable values. This approach has been examined in, e.g., [4,12].

In this work, the focus is on the latter. Asymmetric disjunction has been proposed by Bosc and Pivert [2], where authors suggested weighted arithmetic mean for reducing the strength of substitutability of the optional alternative. The axiomatization of averaging functions for covering a larger scale of possibilities for disjunctive asymmetric behaviour is proposed by Hudec and Mesiar [9]. This work goes further by examining possibilities for replacing *MAX* function by any disjunctive function and therefore extending asymmetric disjunction to cover diverse managerial evaluation tasks based on the disjunctive principle of human reasoning.

The structure of paper is as follows: Sect. 2 provides the preliminaries of aggregation functions. Section 3 is dedicated to the formalization of averaging and disjunctive functions in *OR ELSE*, whereas Sect. 4 discusses the results, and emphasizes strengths and weak points. Finally, Sect. 5 concludes the paper.

2   Preliminaries of Aggregation Functions

Disjunction belongs to the large class of aggregation functions, i.e., functions $A : [0,1]^n \rightarrow [0,1]$ which are monotone and satisfy the boundary conditions $A(0,...,0) = 0$ and $A(1,...,1) = 1$, $n \in \mathbb{N}$. The standard classification of aggregation functions is due to Dubois and Prade [6]. Namely, conjunctive aggregation functions are characterized by $A(\mathbf{x}) \leq \min(\mathbf{x})$, disjunctive by $A(\mathbf{x}) \geq \max(\mathbf{x})$, averaging by $\min(\mathbf{x}) \leq A(\mathbf{x}) \leq \max(\mathbf{x})$, and remaining aggregation functions are called mixed, where $\mathbf{x}$ is a vector, $\mathbf{x} = (x_1,...,x_n)$.

In this work, we denote by $\mathcal{A}v$ averaging aggregation functions, and by $\mathcal{D}is$ disjunctive aggregation functions. More, $\mathcal{A}v_2$ is the set of bivariate averaging functions, whereas $\mathcal{D}is_2$ represents the set of bivariate disjunctive functions. Note that if the arity of a considered aggregation is clear (mostly $n = 2$), we will use notation $\mathcal{A}v$ instead of $\mathcal{A}v_2$, and $\mathcal{D}is$ instead of $\mathcal{D}is_2$.

The extremal elements of $\mathcal{A}v_2$ are *MAX* (which is also called Zadeh's disjunction, *OR* operator) and *MIN* (Zadeh's conjunction, *AND* operator). To characterize the disjunctive (conjunctive) attitude of members of $\mathcal{A}v_2$, one can consider the *ORNESS* measure $ORNESS : \mathcal{A}v_2 \rightarrow [0,1]$ given by

$$ORNESS(A) = 3 \cdot \int_0^1 \int_0^1 A(x,y)\, dy\, dx - 1 \tag{1}$$

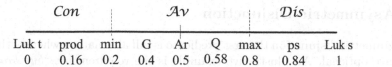

Con				Av		Dis		
Luk t	prod	min	G	Ar	Q	max	ps	Luk s
0	0.16	0.2	0.4	0.5	0.58	0.8	0.84	1

Legend:
Luk t – Łukasiewicz t-norm, prod – product t-norm, min – *MIN* function,
G – geometric mean, Ar – arithmetic mean, Q – quadratic mean, max – *MAX*
function, ps – probabilistic sum t-conorm, Luk s – Łukasiewicz t-conorm,
for $x = 0.8$ and $y = 0.2$.

Fig. 1. An illustrative example of aggregation functions.

Analogously, the *ANDNESS* measure $ANDNESS : Av_2 \to [0,1]$ characterizes the conjunctive attitude by

$$ANDNESS(A) = 2 - 3 \cdot \int_0^1 \int_0^1 A(x,y)\, dy\, dx \tag{2}$$

More details regarding these measures can be found in, e.g., [8,11]. Obviously,
the disjunctive attitude of *MAX* is equal to 1, whereas the conjunctive attitude
of *MAX* is 0. The opposite holds for *MIN*.

The arithmetic mean is an element of Av having the full compensation effect,
or neutrality [7] (*ORNESS* and *ANDNESS* measures are equal to 0.5). Therefore, the remaining elements of Av have either partial disjunctive or partial
conjunctive behaviour.

The extremal elements of Dis are *MAX* and drastic sum

$$S_{DS}(x,y) = \begin{cases} \max(x,y) & \text{for } \min(x,y) = 0 \\ 1 & \text{otherwise} \end{cases}$$

where *MAX* is the only idempotent element. The other elements have upward
reinforcement property [1], like probabilistic sum

$$S_P(x,y) = x + y - xy$$

and Łukasiewicz t–conorm

$$S_L(x,y) = \min 1, x + y$$

used in this work.

Analogously, the extremal elements of the set of conjunctive functions (Con)
are drastic product and *MIN*, where *MIN* is the only idempotent element. The
other elements have downward reinforcement property.

These observations for the symmetric case are illustrated in Fig. 1, where
predicate P_1 is satisfied with 0.8 ($x = 0.8$) and predicate P_2 with 0.2 ($y = 0.2$).
Generally, P_1 and P_2 can be any kind of predicates (elementary, compound,
quantified, etc.). Just as reminder, geometric mean is $G(x,y) = \sqrt{xy}$, arithmetic
mean is $W = \frac{1}{2}(x + y)$ and quadratic mean is $Q(x,y) = \sqrt{0.5x^2 + 0.5y^2}$.

3   Asymmetric Disjunction

In asymmetric disjunction the first predicate is full alternative, whereas the other ones are optional. An illustrative example is the requirement: *"buy broccoli or else cauliflower"* [9]. If we find neither broccoli nor cauliflower, the score is 0. If we find both, the score is 1. If we find only broccoli, the score is 1. Finally, if we find only cauliflower, the score should be less than 1, but better than 0. Thus, the last option should be managed by an element of $\mathcal{A}v_2$, whereas the other options by an element of $\mathcal{D}is_2$ and therefore we should aggregate both cases.

Contrary, in the frame of two–valued logic, the left–right order of predicates has been solved by the Qualitative Choice Logic [5]. In that approach, when first predicate is satisfied, the solution gets value 1; when second is satisfied, the solution gets value 2; etc. If not a single predicate is satisfied, the solution is 0. The problem arise when it is integrated into a complex predicate like: P_1 *AND* P_2 *AND* (P_3 *OR ELSE* P_4 *OR ELSE* P_5)., i.e., the overall solution is expected to be in the unit interval.

3.1   The Formalization of Asymmetric Disjunction

Bosc and Pivert [2] proposed the following six axioms in order to formalize *OR ELSE* operator D, where x and y are the values of predicates P_1 and P_2, respectively:

A1 D is more drastic than *OR* operator: $D(x, y) \leq \max(x, y)$, i.e. we are crossing the border between averaging and disjunctive functions.

A2 D is softer than when only P_1 appears, because P_2 opens new choices: $D(x, y) \geq x$.

A3 D is an increasing function in its first argument.

A4 D is an increasing function in its second argument.

A5 D has asymmetric behaviour, i.e. $D(x, y) \neq D(y, x)$ for some $(x, y) \in [0, 1]^2$.

A6 D is equivalent to x *OR ELSE* (x *OR* y): $D(x, y) = D(x, x \vee y)$.

Note that, for the simplicity, sometimes we use the lattice connectives notation $\vee = MAX$ and $\wedge = MIN$.

The operator which meets these axioms is expressed by the function

$$D_A(x, y) = \max(x, A(x, y)) \tag{3}$$

where $A \in \mathcal{A}v_2$.

As a typical example of *OR ELSE* operator (3), Bosc and Pivert [2] have proposed a parametrized class of functions

$$D_{BPk}(x, y) = \max(x, k \cdot x + (1 - k)y) \tag{4}$$

where $k \in {]}0, 1]$ (i.e., A is the weighted arithmetic mean W_k) and BP stands for the Bosc-Pivert operator. For the asymptotic extremal value $k = 0$, we get the disjunction expressed by the *MAX* function: $\max(x, y)$. For $k = 0.5$ and $y \geq x$ we get the non–weighted arithmetic mean W, i.e., $D_{BP0.5}^W(x, y) = \max(x, \frac{x+y}{2})$.

Another example is [9]

$$D_{BPk}^G(x,y) = \max(x, G_k(x,y)) \tag{5}$$

where G_k is the weighted geometric mean and $k \in]0,1]$. Analogously to (4), we write

$$D_{BPk}^G(x,y) = \max(x, x^k \cdot y^{(1-k)}) \tag{6}$$

where $k \in]0,1]$. For $k = 0.5$ we get $D_{BP0.5}^G(x,y) = \max(x, \sqrt{xy})$

Similarly, we can use the other elements of $\mathcal{A}v$, e.g., quadratic mean, where for $k = 0.5$ we get

$$D_{BP0.5}^Q(x,y) = \max(x, \sqrt{0.5x^2 + 0.5y^2}) \tag{7}$$

The asymmetric disjunction considers all $\mathcal{A}v$ elements including the extremal elements MIN and MAX. Obviously,

$$D_{BP0.5}^{MAX} = \max(x, \max(0.5x, 0.5y)) = \begin{cases} \max(x, 0.5x) = x & \text{for } x \geq y \\ \max(x, 0.5y) & \text{for } x < y \end{cases}$$

The analogous observation holds for the extremal element MIN.

Recently, Hudec and Mesiar [9] proposed the axiomatization of asymmetric conjunction and disjunction for continuous as well as non–continuous cases (P_1 OR ELSE P_2, but when P_2 has a high satisfaction degree it becomes the full alternative, i.e., *broccoli or else cauliflower, but if cauliflower is very ripe, then it becomes the full alternative*), and discussed requirements for associative behaviour. In all above cases $D = MAX$. The next example illustrates semantics of diverse elements of $\mathcal{A}v$.

Example. Let a house buyer has raised conditions regarding the storage space for the less–frequently or seasonally used items by the condition: *spacious basement or else spacious attic*. Here x (resp. y) stands for the intensity of spaciousness for basement (resp. attic).

The following observations illustrate the suitable elements of $\mathcal{A}v$ for several decision–making requirements:

- The *ORNESS* measure for MAX is 1, i.e., we have the full substitutability of alternatives, $A = MAX$, or disjunction $x \vee y$.
- The *ORNESS* measure for arithmetic mean W is 0.5 (regardless of the number of predicates), i.e., we model the basic case when attic is less suitable alternative to basement, $A = W$.
- The *ORNESS* measure for geometric mean G is 0.33 (for two predicates), so we are able to model the situation for an elderly buyer and a quite steep ladder, to decrease the relevance for the optional alternative, and even reject house having no basement (all heavy items must be stored in attic), $A = G$.
- The *ORNESS* measure for quadratic mean Q is 0.62 (for two predicates), thus we model the situation for a younger buyer and a less steep stairs, to increase the relevance for the optional alternative and to still keep attic as a less convenient than basement, $A = Q$.

The solution is shown in Table 1 for several hypothetical houses, where the solution is lower than or equal to the MAX operator, and higher than or equal to the projection of the full alternative. Observe that $H2$ and $H8$ have the same suitability degree, but $H8$ might be considered as a better option.

Table 1. *OR ELSE* connective expressed by the continuous Bosc–Pivert operators for arithmetic mean, geometric mean and quadratic mean, where $k = 0.5$.

House	x	y	$A = W$ (4)	$A = G$ (6)	$A = Q$ (7)
H1	0.80	0.80	0.80	0.80	0.80
H2	0.80	0.20	0.80	0.80	0.80
H3	0.20	0.80	0.50	0.40	0.583
H4	1	0.50	1	1	1
H5	0.50	1	0.75	0.707	0.791
H6	0	1	0.50	0	0.707
H7	0.10	0.90	0.50	0.30	0.64
H8	0.80	0.78	0.80	0.80	0.80
H9	0.90	1	0.95	0.949	0.951
H10	0.34	1	0.67	0.58	0.75
H11	0.33	1	0.65	0.548	0.74
H12	0.70	0.10	0.70	0.70	0.70

□

3.2   The Generalization of Asymmetric Disjunction

Axioms A1 and A2 ensure for any *OR ELSE* operator D its idempotency, i.e., $D(x, x) = x$ for all $x \in [0, 1]$. The question is, whether we can apply other disjunctive functions than MAX in (3), e.g., probabilistic sum or Łukasiewicz t–conorm, or whether the idempotency is mandatory.

Therefore, a general form of (3) is

$$D_{H,A}(x, y) = H(x, A(x, y)) \qquad (8)$$

where $A \in \mathcal{A}v$ and $H \in \mathcal{D}is$, i.e., $H : [0, 1]^2 \to [0, 1]$ is a disjunctive aggregation function.

Observe that the idempotency of D in (8) is equivalent to $H = MAX$, i.e., to the original approach proposed by Bosc and Pivert [2].

This structure keeps the asymmetry in the most cases, but not in general. So, e.g., if H is the second projection (i.e., one keeps the second argument) and A is symmetric, then also $D_{H,A}$ given by (8) is symmetric. The same claim is valid if H is symmetric and A is the second projection. The following examples support this claim:

$$D_{H,MAX}(x,y) = \begin{cases} H(x,x) & \text{if } x \geq y \\ H(x,y) & \text{for } x < y \end{cases}$$

$$D_{H,MIN}(x,y) = \begin{cases} H(x,y) & \text{if } x \geq y \\ H(x,x) & \text{for } x < y \end{cases}$$

$$D_{SP,G}(x,y) = x + \sqrt{xy} - x \cdot \sqrt{xy}$$

$$D_{SP,A}(x,y) = x + (1-x) \cdot A(x,y)$$

An interesting class of our operators introduced in (8) is generated by a generator $g : [0,1] \rightarrow [0,\infty]$, g being an increasing bijection. Then H given by $H(x,y) = g^{-1}(g(x) + g(y))$ is a strict t–conorm, and A_k, $k \in [0,1]$ given by $A_k(x,y) = g^{-1}(k \cdot g(x) + (1-k) \cdot g(y))$ is a weighted quasi–geometric mean. The related operator $D_{H,A_k} = D_{g,k}$ is then given by $D_{g,k}(x,y) = g^{-1}((1+k) \cdot g(x) + (1-k) \cdot g(y))$.

For $g(x) = -\log(1-x)$ the related strict t–conorm is the probabilistic sum S_P and then $D_{g,k}(x,y) = 1 - (1-x)^{1+k} \cdot (1-y)^{1-k}$.

For the extremal cases we obtain $D_{g,0} = S_P = H$ and $D_{g,1}(x,y) = 2x - x^2 = S_P(x,x)$.

As another example, consider $g(x) = x/(1-x)$. Then $H = S_H$ is the t–conorm dual to the Hamacher product, and then $D_{g,k}(x,y) = ((1+k)x + (1-k)y - 2xy)/(1+kx-ky-xy)$, and $D_{g,0} = S_H$, $D_{g,1}(x,y) = 2x/(1+x) = S_H(x,x)$.

In general, $D_{H,A}(x,y) \geq x$. Thus, the newly introduced operators $D_{H,A}$ allow to increase the value x (of the first argument), i.e., $x \leq D_{H,A}(x,y) \leq 1$. Consequently

$$0 \leq D_{H,A}(x,y) - x \leq 1 - x \tag{9}$$

The minimal compensation $0 = D_{H,A}(x,y) - x$ (for any $x,y \in [0,1]$) is obtained if and only if H is the first projection, $H(x,y) = x$, and $A \in \mathcal{A}v_2$ is arbitrary, or $H = MAX$ and A is the first projection.

On the other hand, the maximal compensation $1 - x = D_{H,A}(x,y)$ cannot be attained if $x = y = 0$, as then, for any H and A, $D(0,0) = H(0, A(0,0)) = H(0,0) = 0$. However, if we insist that for any $(x,y) \neq (0,0)$ the compensation $D_{H,A}(x,y) - x = 1 - x$ is maximal, then necessarily $D_{H,A} = A^*$ is the greatest aggregation function given by $A^*(x,y) = 1$ whenever $(x,y) \neq (0,0)$ and $A^*(0,0) = 0$.

Obviously, we obtain $D_{H,A} = A^*$ whenever $A = A^*$. A similar claim is valid if $H = A^*$ and $A(0,y) > 0$ for any $y > 0$. Complete proofs will be added into the full version of this contribution.

For the probabilistic sum S_p (an Archimedean t–conorm) we have

$$D_{S_p,A}(x,y) = x + A(x,y) - x \cdot A(x,y) \tag{10}$$

and then $D_{S_p,A}(x,x) = 2x - x^2$ for an arbitrary $A \in \mathcal{A}v_2$.

For the Łukasiewicz t–conorm S_L (a nilpotent t–conorm) we have

$$D_{S_L,A}(x,y) = \min(1, x + A(x,y)) \tag{11}$$

and then $D_{S_L,A}(x,x) = \min(1, 2x)$ for an arbitrary $A \in \mathcal{A}v_2$.

Table 2. *OR ELSE* connective expressed by (10) for arithmetic mean, geometric mean and quadratic mean, where $k = 0.5$.

House	x	y	$A = W$ (4)	$A = G$ (6)	$A = Q$ (7)	S_L*
H1	0.80	0.80	0.96	0.96	0.96	0.96
H2	0.80	0.20	0.90	0.88	0.9166	0.84
H3	0.20	0.80	0.60	0.52	0.6665	0.84
H4	1	0.50	1	1	1	1
H5	0.50	1	0.875	0.8535	0.8953	1
H6	0	1	0.50	0	0.7701	1
H7	0.10	0.90	0.55	0.37	0.6763	0.91
H8	0.80	0.78	0.958	0.9579	0.9581	0.956
H9	0.90	1	0.995	0.9949	0.9951	1
H10	0.34	1	0.782	0.725	0.833	1
H11	0.33	1	0.755	0.683	0.817	1
H12	0.70	0.10	0.82	0.779	0.85	0.73

*to compare with the symmetric case $S_p = x + y - x \cdot y$

For the same data as in Table 1, we have the solution for $H = S_p$ shown in Table 2, whereas for $H = S_L$ the solution is in Table 3.

Observe that houses *H2* and *H8* are now distinguishable (Table 2), that is, *H8* is preferred due to significantly higher value y. Further, the differences among averaging functions for *H8* are almost negligible (due to high values of x and y). For $H = MAX$ and $H = S_p$ the optional alternative influences solution also when $y > x$, but does not become the full alternative (see, *H9*, *H10*, *H11* in Tables 1 and 2).

The feature of Łukasiewicz t–conorm is reflected in the evaluation. Houses, which significantly meet full and optional alternatives get value 1 and become undistinguishable, see Table 3. For $H = S_L$ the optional alternative influences solution also when $y > x$ and moreover becomes the full alternative, especially for $A = Q$, compare *H10* and *H11*.

Theoretically, MAX in (3) can be replaced by any disjunctive function. In such cases, the asymmetric disjunction is more flexible allowing optional alternative to influence the solution in all cases, including when $y < x$. But, we should keep solution equal to 1 when both alternatives, or only full alternative assign value 1. When only optional alternative gets value 1, the solution should be less than the ideal satisfaction, thus nilpotent t–conorm functions are not suitable. Further, various averaging functions emphasize or reduce the relevance of optional alternatives as was illustrated in example in Sect. 3.1. Therefore, the proposed aggregation covers diverse managerial needs in evaluation tasks.

Table 3. *OR ELSE* connective expressed by (11) for arithmetic mean, geometric mean and quadratic mean, where $k = 0.5$.

House	x	y	$A = W$ (4)	$A = G$ (6)	$A = Q$ (7)	S_L*
H1	0.80	0.80	1	1	1	1
H2	0.80	0.20	1	1	1	1
H3	0.20	0.80	0.70	0.60	0.78	1
H4	1	0.50	1	1	1	1
H5	0.50	1	1	1	1	1
H6	0	1	0.50	0	0.71	1
H7	0.10	0.90	0.60	0.40	0.74	1
H8	0.80	0.78	1	1	1	1
H9	0.90	1	1	1	1	1
H10	0.34	1	1	0.923	1	1
H11	0.33	1	0.95	0.848	1	1
H12	0.70	0.10	1	0.96	1	0.80

*to compare with the symmetric case $S_L = \min(1, x + y)$

4 Discussion

In the literature, we find that the general models of substitutability should not go below the neutrality, or arithmetic mean W [7]. The asymmetric disjunction proposed by Bosc and Pivert [2] is inside this frame (*ORNESS* of W is 0.5 regardless of weights). It also does not go above *MAX*. The asymmetric disjunction proposed by Hudec and Mesiar [9] goes below W, i.e., $A \in Av\backslash\{MIN, MAX\}$, but not above *MAX* (3) to cover further users expectations, and meets axioms A1–A6. So, these approaches do not support upward reinforcement.

This model considers the whole classes of Av and Dis Fig. 1, i.e., above *MAX* and idempotency for upwardly reinforcing evaluated items as disjunction do (see, Table 2, e.g., *H1*, *H2* and *H8*). In the case of (10), i.e., $H = S_p$, we cannot reach solution 1 when $x < 1$ and $y = 1$. Thus, the optional alternative influences solution even when $y < x$, but does not become the full alternative.

On the other hand, by Eq. (11) the optional alternative in certain situations becomes a full one, namely when $x + A(x, y) \geq 1 \land y > x$. Observe that for $A = W_k$, we get $D(x, y) = \min(1, x + kx + (1 - k)y)$ and therefore for $x > 1/(1 + k)$, the solution is equal to 1, regardless of value y. This situation is plotted in Fig. 2. For instance, for $k = 0.5$, we have $D_{S_L, W}(0.7, 0.1) = 1$. But, by symmetric disjunction we have $S_L(0.7, 0.1) = 0.8$. For the completeness Table 3 shows the solution for disjunction S_L.

Theoretically, for $H = S_L$ we have asymmetric aggregation, but from the perspective of human logic evaluation of optional alternative it is questionable. The solution might be penalization when $y > x$ and the solution is equal to 1

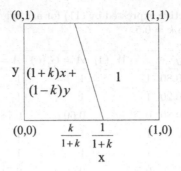

Fig. 2. Asymmetric disjunction by $H = S_L$ and $A = W_{0.5}$.

(when the optional alternative becomes the full one). In general, if $D(x, y) = S_L(x, A(x, y))$, then considering (9) we assign $D(x, y) - x = \min((1 - x, A(x, y))$.

The answer to the question: "is the idempotency mandatory?" is as follows: The idempotency property may be too restrictive for general purpose of asymmetric disjunction and therefore other functions than MAX in (3) may be appropriate to cover diverse requirements in evaluation tasks. We have proven that for $H = S_p$ and $A \in Av$ we have an upwardly reinforced asymmetric disjunction. Therefore, $H = S_p$ can be directly used. But, it does not hold for $H = S_L$ where we should adopt penalization. The drastic sum has the theoretical meaning of an upper bound of $\mathcal{D}is$ without significant applicability. Hence, the same observation holds for asymmetric disjunction when $H = S_{DS}$.

Axioms A2–A6 are still valid. Axiom A1 should be relaxed for the generalized asymmetric disjunction. Generally, we can apply Archimedean t–conorms for H, but for the nilpotent one, we should consider penalty. The work in this field should continue in generalization of the other connectives and in evaluation, which of them correlate with the human reasoning.

In idempotent disjunction, i.e., MAX function, lower values than the maximal one do not influence solution. In all other functions, lower values somehow influence the solution. This holds for the symmetric case. We offered this option for the asymmetric case. In cases when lower values of optional alternative should be considered, we need this approach. For instance, in aforementioned requirement: *spacious basement or else spacious attic*, clearly, a house of 0.7 spaciousness of basement and 0.4 of attic is better than a house of values (0.7 and 0.3), which is better than (0.4, 0.7).

Considering the afore results, we propose to entitle this operator as *Compensatory OR ELSE*, due to the integrated asymmetry and compensative effect.

It is worth noting that the asymmetric disjunction can be combined with the other logic aggregations. For instance, consider buying a house. A buyer might pose the following requirement: *size around 200 m^2 AND short distance to work AND (spacious basement OR ELSE spacious attic) AND (**most of the following requirements**: {short distance to theatre, short distance to train station, low population density, detached garage, etc.} satisfied)*. In the first step, we should calculate matching degrees of asymmetric disjunction and quantified aggregation, while in the second step we calculate the overall matching degree.

5 Conclusion

It is a challenging task to cover the diverse needs for disjunctively polarized decision making and evaluation tasks. Hence, practice searches for the robust mathematical solutions to cover the whole range of disjunctive aggregation, including the asymmetric case of full and optional predicates. In order to contribute, the theoretical part of this work has recognized and formalized requirements for asymmetric disjunction, which are illustrated on the illustrative examples. The answer to the question *whether the idempotency is mandatory* is the following: The idempotency property may be too restrictive for general purpose of asymmetric disjunction and therefore other functions than MAX may be appropriate. We have proven that for $H = S_p$ and $A \in \mathcal{A}v$ we get an upwardly reinforced asymmetric disjunction. Therefore, $H = S_p$ can be directly used. But, it does not hold for $H = S_L$, where we should adopt penalization to keep the solution equal to 1 when both alternatives, or only the full alternative is satisfied with value 1.

In the everyday decision making tasks and database queries, asymmetric disjunction could appear as the whole condition. On the other hand, in the complex managerial evaluation of entities asymmetric disjunction is just a part of the overall criterion as is illustrated in Sect. 4. The topics for the future work should include extending this study for the weighted asymmetric disjunction case, testing on real–world data sets, and examining the consistency with the disjunctive managerial decision making and evaluation.

Acknowledgments. This paper was supported by SGS project No. SP2020/125. Also the support of the project APVV 18-0052 is kindly announced.

References

1. Beliakov, G., Pradera, A., Calvo, T.: Aggregation Functions: A Guide for Practitioners. Springer, Heidelberg (2007). https://doi.org/10.1007/978-3-540-73721-6
2. Bosc, P., Pivert, O.: On four noncommutative fuzzy connectives and their axiomatization. Fuzzy Sets Syst. **202**, 42–60 (2012)
3. Bosc, P., Hadjali, A., Pivert, O.: Empty versus overabundant answers to exible relational queries. Fuzzy Sets Syst. **159**, 1450–1467 (2008)
4. Bosc, P., Pivert, O.: On a fuzzy bipolar relational algebra. Inf. Sci. **219**, 1–16 (2013)
5. Brewka, G., Benferhat, S., Berre, D.L.: Qualitative choice logic. Artif. Intell. **157**, 203–237 (2004)
6. Dubois, D., Prade, H.: On the use of aggregation operations in information fusion processes. Fuzzy Sets Syst. **142**, 143–161 (2004)

7. Dujmović, J.: Soft Computing Evaluation Logic: The LSP Decision Method and its Applications. Wiley-IEEE Computer Society, Hoboken (2018)
8. Dujmović, J.: Weighted conjunctive and disjunctive means and their application in system evaluation. Ser. Math. Phys. **461/497**, 147–158 (1974). Univ. Beograd. Publ. Elektrotechn. Fak, Belgrade
9. Hudec, M., Mesiar, R.: The axiomatization of asymmetric disjunction and conjunction. Inf. Fusion **53**, 165–173 (2020)
10. Karttunen, L.: Presupposition and linguistic context. Theor. Linguist. **1**, 181–193 (1974)
11. Kolesárová, A., Mesiar, R.: Parametric characterization of aggregation functions. Fuzzy Sets Syst. **160**, 816–831 (2009)
12. Lietard, L., Nouredine, T., Rocacher, D.: Fuzzy bipolar conditions of type "or else". In: 20th IEEE International Conference on Fuzzy Systems (FUZZ-IEEE 2011), pp. 2546–2551, Taiwan (2011)
13. Singh, R.: On the interpretation of disjunction: asymmetric, incremental, and eager for inconsistency. Linguist. Philos. **31**, 245–260 (2008). https://doi.org/10.1007/s10988-008-9038-x

Fuzzy Inference System as an Aggregation Operator - Application to the Design of a Soil Chemical Quality Index

Denys Yohana Mora-Herrera[1], Serge Guillaume[2(✉)] [ID], Didier Snoeck[3], and Orlando Zúñiga Escobar[1]

[1] Universidad del Valle, Cali, Colombia
{denys.mora,orlando.zuniga}@correounivalle.edu.co
[2] ITAP, Univ Montpellier, INRAE, Montpellier SupAgro, Montpellier, France
serge.guillaume@inrae.fr
[3] CIRAD, Montpellier, France
didier.snoeck@cirad.fr

Abstract. Fuzzy logic is widely used in linguistic modeling. In this work, fuzzy logic is used in a multicriteria decision making framework in two different ways. First, fuzzy sets are used to model an expert preference relation for each of the individual information sources to turn raw data into satisfaction degrees. Second, fuzzy rules are used to model the interaction between sources to aggregate the individual degrees into a global score. The whole framework is implemented as an open source software called *GeoFIS*. The potential of the method is illustrated using an agronomic case study to design a soil chemical quality index from expert knowledge for cacao production systems. The data come from three municipalities of Tolima department in Colombia. The output inferred by the fuzzy inference system was used as a target to learn the weights of classical numerical aggregation operators. Only the *Choquet Integral* proved to have a similar modeling ability, but the weights would have been difficult to set from expert knowledge without learning.

Keywords: Fusion · Multicriteria · Preference · Decision

1 Introduction

Fuzzy logic is widely used as an interface between symbolic and numerical spaces, allowing the implementation of human reasoning in computers. Fuzzy inference systems are well known for their ability for linguistic modeling and approximate reasoning. They are used in this work in the framework of multicriteria decision making to model expert knowledge.

Complex systems, such as agricultural production systems, are characterized by several interrelated dimensions, for instance agronomic, social and economic. The production process includes different steps performed systematically, from the plant selection to the commercialization. Decisions are made at each step of

© Springer Nature Switzerland AG 2020
M.-J. Lesot et al. (Eds.): IPMU 2020, CCIS 1238, pp. 447–456, 2020.
https://doi.org/10.1007/978-3-030-50143-3_35

this process that may degrade or support the sustainability of the production system.

Decision making usually involve several, may be conflicting, attributes. Multicriteria decision analysis (MCDA) has been indicated as the appropriate set of tools to perform assessments of sustainability, by considering different sustainability spheres, perspectives, stakeholders, values, uncertainties and intra and inter-generational considerations [2]. In this article five MCDA methods are reviewed on the basis of ten criteria that they should satisfy to properly handle problems concerning sustainability. The methods are from three main families: i) utility-based theory: multi attribute utility theory (MAUT) and analytical hierarchy process (AHP), ii) outranking relation theory: elimination and choice expressing the reality (ELECTRE) and preference ranking organization method for enrichment of evaluations (PROMETHEE), iii) the sets of decision rules theory: dominance based rough set approach (DRSA). The latter uses crisp *'if ... then'* rules where the premise compares for each criterion its satisfaction degree to a threshold and the conclusion is a category or a set of categories.

In a review about MCDA applied to forest and other natural resource management [7] a special attention is paid to methods that deal with uncertainty. The possible causes of uncertainty are analyzed and the study reports how the methods are adapted to manage some dimension of uncertainty. A fuzzy multiple objective linear programming method is mentioned [8].

Agricultural production systems involve agronomic, social, cultural, institutional, economic and other natural elements that are interrelated. Cacao production has been studied for many years. It is grown in climatic, economic and social uncertain contexts, then more efforts by farmers (time, money or land) do not always produce more quality, quantity, profitability nor a better life quality for farmers. So, it is a dynamic and complex system characterized by nonlinear relationships dependent of local contexts. Moreover data is not enough and knowledge is needed to turn data into valuable agronomic information, for instance to make a decision about fertilization from a soil content analysis.

The challenge is to design a tool that includes the available scientific knowledge to help farmers in decision making. This tool is designed as an indicator with three main components: agronomic, economic and social, and final product quality. For sustainable and competitive cacao production the fertility status of the soil is an important variable that is generally related to cacao agronomic yield. A subsystem of the agricultural quality part is analyzed in this work: the soil chemical quality index.

The implementation is achieved by the means of an open source software, called *GeoFIS*[1], a platform for spatial data processing with a decision support perspective. The fusion module uses fuzzy logic in two different contexts: first to turn raw data into satisfaction degrees and second to model variable interaction using linguistic rules.

---

[1] https://www.geofis.org.

Section 2 describes the data fusion framework, a subsystem of the soil chemical quality index is studied in Sect. 3. In Sect. 4 the FIS is compared with classical numerical operators. Finally the main conclusions are summarized in Sect. 5.

2   Data Fusion and Multicriteria Decision Making

The process of data fusion for decision making is driven by expert knowledge. Information fusion is done with a specific goal, for instance risk level evaluation or variable application rate in agriculture. The selection of the relevant and available information sources is done by the decision maker. Then the next step is to evaluate what could be the level of decision, e.g. risk or rate, according to each of the sources for a given entity defined by its spatial coordinates. The final step comes down to aggregate these partial levels, or degrees, to make the final decision. The aggregation *function* models the decision maker preferences: Are some attributes more important than others? How to combine conflicting information sources?

The whole framework can be illustrated as follows:

$$(a_1, \ldots, a_n) \ , \ (b_1, \ldots, b_n) \ \xrightarrow{\mathbf{A}} \ f(a) \ , \ f(b)$$

$$\uparrow \qquad\qquad\qquad c \downarrow$$

$$(x_1, \ldots, x_n) \ , \ (y_1, \ldots, y_n) \qquad \precsim (a, b)$$

There are two steps to formalize expert knowledge and preferences for the decision process. The first one deals with each individual variable, or information source. The second one addresses the interaction between sources.

The first step aims to turn, for each individual variable, x_i or y_j, raw data into **satisfaction degrees**, a_i or b_j. This is done by defining a preference relation through a fuzzy set. The scale is the unit interval, $[0, 1]$ with zero meaning the criterion is totally not satisfied and that it is fully satisfied with one. Then the degrees are aggregated using the f function to compute a global score, $f(a)$ or $f(b)$.

2.1   Numerical Operators

The most popular techniques to aggregate commensurable degrees are numerical operators. The main families of such operators, with suitable properties, are the Weighted Arithmetic Mean (*WAM*), the Ordered Weighted Average (*OWA*) and the Choquet Integral (*CI*).

Let X be the set of sources to aggregate: $X = \{a_1, \ldots, a_n\}$.

The *WAM* aggregation is recalled in Eq. 1.

$$WAM(a_1, \ldots, a_n) = \sum_{i=1}^{n} w_i a_i \tag{1}$$

with $w_i \in [0,1]$ and $\sum_{i=1}^{n} w_i = 1$. The weights are assigned to the sources of information.

The OWA is computed as shown in Eq. 2.

$$OWA(a_1, \ldots, a_n) = \sum_{i=1}^{n} w_i a_{(i)} \tag{2}$$

where (.) is a permutation such as $a_{(1)} \leq \cdots \leq a_{(n)}$.

In this case, the degrees are ordered and the weights are assigned to the locations in the distribution, from the minimum to the maximum, whatever the information sources.

These operators are easy to use, the number of parameters is the number of information sources to aggregate[2], but their modeling ability is limited. The *Choquet Integral* [1] proved to be useful in multicriteria decision making [3]. It is computed according to Eq. 3.

$$CI(a_1, \ldots, a_n) = \sum_{i=1}^{n} (a_{(i)} - a_{(i-1)}) w(A_{(i)}) \tag{3}$$

where (.) is a the permutation previously defined with $a_{(0)} = 0$ and $A_{(i)} = \{(i), \ldots, (n)\}$, meaning the set of sources with a degree $a \geq a_{(i)}$.

The weights must fulfill the two following conditions:

1. Normalization: $w(\emptyset) = 0$, $w(X) = 1$
2. Monotonicity condition: $\forall A \subset B \subset X$, $w(A) \leq w(B)$.

The weights are not only defined for each of the information sources but for all their possible combinations. Specific configurations include WAM and OWA modeling. The *Choquet Integral* is equivalent to a WAM when the sum of the weights assigned to the individual sources is one and when the weight of any coalition is the sum of the weights of its individual components. In this case the measure is additive. It is equivalent to an OWA when the weight of a coalition only depends on its size: for instance all the subsets with two elements have the same weight. This kind of measure is said to be symmetric. In the general case, the aggregation of n information sources requires $2^n - 2$ coefficients. These are usually set by learning algorithms [9].

2.2 Fuzzy Inference Systems

A fuzzy inference system usually requires more parameters than the former numerical aggregators but, in this particular case of data fusion, the design can be simplified as all the input variables are satisfaction degrees that share the same scale, the unit interval, and the same meaning. Strong fuzzy partitions

---

[2] Only $(n-1)$ parameters have to be defined for n sources as their sum is 1.

with regular grids are used to ensure semantic integrity. The inference system is also automatically defined: the membership degrees are aggregated using the minimum operator to compute the matching degree of each rule, the rule conclusions are aggregated using the Sugeno operator in the crisp case or using the centroid operator if fuzzy. More details are available in the *FisPro* documentation[3].

The unique parameter left to the user is the number of linguistic terms for each input variable. In this work it was set at 2, *Low* and *High*, for all of them. With two linguistic labels by variable, the number of rules is 2^n, i.e. the number of coefficients required by the *Choquet Integral*. A rule describes a local context the expert domain, the decision maker, is able to understand. In this way, the rule conclusions are easier to define than the *Choquet Integral* coefficients.

2.3 Implementation

The fusion module is implemented as an open source software in the *GeoFIS* program. The data must be co-located, i.e. a record includes the spatial coordinates of the cell, from a pixel to a zone, and the corresponding attributes.

The available functions to turn raw data into degrees are of the following shapes:

SemiTrapInf ⬛: low values are preferred
SemiTrapSup ⬛: high values are preferred
Trapezoidal ⬛: about an interval
Triangular ⬛: about a value

Three aggregation operators are currently available: *WAM*, *OWA* and a fuzzy inference system (FIS) including linguistic rules.

For *WAM* and *OWA* the weights can be learned provided a co-located target is available. Rule conclusions can also be learned using the *FisPro* software [4].

Rule conclusion can be either a linguistic term, fuzzy output, or a real value, crisp output. Using a fuzzy output, it would be necessary to define as many labels as different suitable rule conclusions. As a crisp conclusion may take any value in the output range, it allows for more versatility.

The output should also range in the unit interval. This constraint ensures the output can feed a further step of the process as shown in Fig. 1.

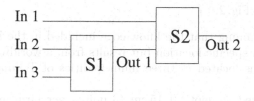

Fig. 1. A hierarchical structure.

---
[3] https://www.fispro.org.

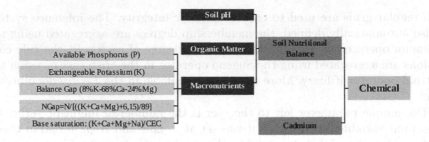

Fig. 2. The chemical subsystem.

In this way, the intermediate systems can be kept small, making their design and interpretation easier.

The *GeoFIS* program includes a distance function based on a fuzzy partition that allows for integrating expert knowledge into distance calculations [5] as well other functionalities, such as a zone delineation algorithm [10]. An illustration of its potential use in Precision Agriculture can be found in [6].

3   Case Study

Different indicators had been designed as guidance for characterizing and improving agricultural system sustainability, one of them is *Soil Fertility*. It is defined by the Soil Science Society of America as[4]: "The quality of a soil that enables it to provide nutrients in adequate amounts and in proper balance for the growth of specified plants or crops."

The chemical index structure is shown in Fig. 2. The *Soil Nutritional Balance* is analyzed in this work. This subsystem include three input variables:

– *Soil pH* depends of the nature of the soil and controls the chemical processes that take place in the soil; specifically, the availability of nutrients. Moreover, it is a parameter that is very easy to measure in the field. The preference relation is defined as: ⌁ 5, 5.5, 6.5, 7.5 [11].
– *Organic matter* brings Carbon that is an important macronutrient, its presence is very regular and usual in cacao plantations and, usually, it is not a limiting criterion. The preference relation is defined as: ⟋ 3, 5 [11].
– *Macronutrients*: this variable is calculated from a combination of soil chemical data as shown in Fig. 2.

It is worth mentioning that the knowledge included in the index is generic: it is not related to a specific location but results from a worldwide analysis.

The study area is located in three municipalities of Tolima department in Colombia.

Soils were sampled at depth 0–15 cm (3 points per farm and 3 repetitions in each one). An agronomist surveyed the crops and filled out field register

---

[4] https://www.soils.org/publications/soils-glossary.

about nutritional, productivity and plant health stage. An economist applied a structured interview to farmers to collect information about their agronomic and post-harvest practices and socioeconomic conditions. The fieldwork was achieved between December 2018 and January 2019.

Table 1. The *Soil Nutritional Balance* rule base.

	Organic matter	Soil pH	Macronutrients	Conclusion
1	Low	Low	Low	0.0
2	Low	Low	High	0.2
3	Low	High	Low	0.4
4	Low	High	High	0.6
5	High	Low	Low	0.3
6	High	Low	High	0.5
7	High	High	Low	0.7
8	High	High	High	1.0

Even if the main interactions are known, there are various ways of modeling them. Rule conclusions were valued taking account hierarchical importance of each input variable and their contribution to aggregated variable when it is satisfied individually and when it is satisfied in combination with others. Always considering the premise that all is more than the sum its parts. The rule base for the *Soil Nutritional Balance* is given in Table 1. The most important input variable is *Soil pH* because it controls the chemical processes that take place in the soil, so when only this variable is satisfied, *High*, then the conclusion is 0.4, rule number 3. If only *Organic matter* or *Macronutrients* is *High* then the respective rule conclusions are 0.3 and 0.2. For any combination of two variables with *High*, the rule conclusion is set at the sum of conclusions when only one of them is *High*. For instance, rule number 4 involves the *High* label for *Soil pH* and *Macronutrients* and the conclusion is $0.6 = 0.4 + 0.2$. However, when the three variables are satisfied to a *High* level the conclusion is higher than the sum of the individual contributions, 0.9, to highlight their positive interaction on the crop agricultural quality. The corresponding rule conclusion, number 8, is set at 1.

4   Comparison with Numerical Aggregation Operators

The output of the fuzzy system for the 10 farms from the Chaparral municipality is used to learn the weights of the *WAM, OWA* and *Choquet Integral*. The weights of the *WAM* and the *OWA* were learned using a least square minimization procedure under two constraints for the weights: they must be positive and their sum should be 1. The *pnnls* function from the *lsei R* package was used. In a preprocessing step, the degrees for each farm were sorted in an increasing order to learn the *OWA* weights.

The data are given in Table 2.

Table 2. Comparison of the aggregation operators using the 10 farms from Chaparral.

	OM	pH	MacN	FIS	WAM	OWA	CI
1	1	0.5	0.696	0.655	0.731	0.730	0.655
2	0.87	0.76	0.414	0.586	0.666	0.660	0.607
3	0	1	0.337	0.467	0.437	0.447	0.463
4	0	1	0.669	0.534	0.561	0.520	0.533
5	0.525	1	0.470	0.675	0.653	0.680	0.687
6	0.86	1	0.484	0.770	0.764	0.759	0.749
7	0.355	1	0.559	0.652	0.632	0.641	0.651
8	0	1	0.773	0.555	0.599	0.543	0.555
9	0	0.6	0.380	0.309	0.328	0.307	0.315
10	0	1	0.614	0.523	0.540	0.508	0.522
R^2				-	0.912	0.936	0.993

The degrees to aggregate are in the first three columns for *Organic matter* (*OM*), *Soil pH* (*pH*) and *Macronutrients* (*MacN*), followed by the output inferred by the fuzzy system (*FIS*). The remaining columns give the score for the *Soil Nutritional Balance* for the three aggregation operators tested. The determination coefficient between the operators and the *FIS* target are in the last row.

As expected, the *WAM* yields the poorest result. The weights for *OM*, *pH* and *MacN* are: 0.317, 0.312 and 0.371. The score for the first farm is similar to the one for farm #6 and higher than the one for farm #5. This result is not expected as soil of farm #1 has a less suitable *Soil pH* than farm #6. Farm #2 is also assigned a high score, higher than the ones of farms #5 and #7. This is not expected as the degrees of *Soil pH* and, to a lower extent, of *Macronutrients* are better for farms #5 and #7. The exception of *Organic matter* should not have been sufficient to get a higher score.

The weights for the *OWA*, from the minimum to the maximum, are: 0.407, 0.220 and 0.373. The results are slightly improved compared to the *WAM*, but the same comments about the score of farm #1 can be made. Farm #2 also gets an higher score than farm #7.

The *Choquet Integral* gives the best results and the farms are ranked in the same order. This is not surprising as the model requires more coefficients, $2^n - 2$ as the empty coalition is assigned a zero value and the whole set a one. The weights yielded by the HLMS algorithm [9] are shown in Table 3.

This table shows that the coefficients of the optimal *Choquet Integral* are really different from the fuzzy rule conclusions reported in Table 1. Two weights, *OM* and *MacN*, are set at zero. The first one has been optimized by the algorithm but this is not the case for the weight assigned to *MacN*. This is explained

Table 3. The weights of the *Choquet Integral* learned by the HLMS algorithm using the 10 farms from Chaparral.

OM	pH	MacN	OM-pH	OM-MacN	pH-MacN
0.000	0.392	0.000	0.558	0.794	0.603

by the fact that this degree is never the highest in the training set. This is an identified drawback of the algorithm, some values are not handled by the algorithm, depending on the data: they are called untouched coefficients in [9]. Any value lower or equal to 0.603, the minimum value for a coalition that includes $MacN$, would be acceptable.

A zero value for OM does not mean this variable is not used: the weight was put on coalitions that include OM. In the two cases, the weight of the set is higher that the sum of the weights of its elements. For instance, the OM-pH set is given a 0.558 weight, higher than the sum $0 + 0.392$.

Even if the *Choquet Integral* proved to have an important modeling ability, its tuning remains difficult without learning algorithms.

5   Conclusion

In multicriteria decision making various kinds of operators can be used. Some are easy to use but have a limited modeling ability, such as the Weighted Arithmetic Mean. Others are more efficient but require a more important number of parameters whose setting may be difficult. This is the case for the *Choquet Integral*. This work shows that fuzzy logic can be used in two key steps of the aggregation process. First, fuzzy membership functions are used to model individual preferences and to turn raw data into satisfaction degrees for each of the information sources. Second, fuzzy inference systems, that implement linguistic reasoning, are suitable to model variable interaction and collective behavior in local contexts. Linguistic rules are easy to design for domain experts as they naturally use linguistic reasoning.

In the general case fuzzy inference systems require a lot of parameters to define the input partitions and the inference operators. In the particular case of data aggregation all the input variable are satisfaction degrees with common scale and common meaning. This leads to a automatic setting of inputs using a strong fuzzy partition with two linguistic terms, *Low* and *High*. As a consequence, only the rule conclusions have to be specifically defined by the user. This is the way expert knowledge about variable interaction is modeled.

This framework is implemented as an open source software called *GeoFIS* (See footnote 1). This is a strong asset as software support availability is a key factor for a method to be adopted.

The proposal was used to design a soil chemical quality index for cacao crop. It has a hierarchical structure with intermediate outputs easy to analyze. The membership functions were defined according to the available scientific knowledge. Even if the main interactions are known, there are several ways of modeling them. The results were easy to analyze and consistent with the field observations.

The output inferred by the fuzzy system was used as a target to learn the weights of alternative numerical aggregation operators. The most simple ones, *WAM* and *OWA*, yielded poor results. Only the *Choquet Integral* proved able to fit the target. The weights defined by the learning algorithm proved that the expert tuning of the *Choquet Integral* would have been difficult.

Fuzzy inference systems thanks to their proximity with natural language and expert reasoning are a good alternative framework for modeling preferences and multicriteria decision making.

References

1. Choquet, G.: Theory of capacities. Ann. de l'Institut Fourier **5**, 131–295 (1954). https://doi.org/10.5802/aif.53
2. Cinelli, M., Coles, S.R., Kirwan, K.: Analysis of the potentials of multi criteria decision analysis methods to conduct sustainability assessment. Ecol. Ind. **46**, 138–148 (2014). https://doi.org/10.1016/j.ecolind.2014.06.011. http://www.sciencedirect.com/science/article/pii/S1470160X14002647
3. Grabisch, M., Labreuche, C.: A decade of application of the choquet and sugeno integrals in multi-criteria decision aid. 4OR **6**(1), 1–44 (2008). https://doi.org/10.1007/s10288-007-0064-2
4. Guillaume, S., Charnomordic, B.: Learning interpretable fuzzy inference systems with FisPro. Inf. Sci. **181**, 4409–4427 (2011). https://doi.org/10.1016/j.ins.2011.03.025
5. Guillaume, S., Charnomordic, B., Loisel, P.: Fuzzy partitions: a way to integrate expert knowledge into distance calculations. Inf. Sci. **245**, 76–95 (2013). https://doi.org/10.1016/j.ins.2012.07.045
6. Leroux, C., et al.: GeoFIS: an open source, decision-support tool for precision agriculture data. Agriculture **8**(6), 73 (2018). https://doi.org/10.3390/agriculture8060073
7. Mendoza, G., Martins, H.: Multi-criteria decision analysis in natural resource management: a critical review of methods and new modelling paradigms. For. Ecol. Manag. **230**(1), 1–22 (2006). https://doi.org/10.1016/j.foreco.2006.03.023. http://www.sciencedirect.com/science/article/pii/S0378112706002258
8. Mendoza, G.A., Bare, B.B., Zhou, Z.: A fuzzy multiple objective linear programming approach to forest planning under uncertainty. Agric. Syst. **41**(3), 257–274 (1993). https://doi.org/10.1016/0308-521X(93)90003-K. http://www.sciencedirect.com/science/article/pii/0308521X9390003K. Multiple Criteria Analysis in Agricultural Systems
9. Murillo, J., Guillaume, S., Tapia, E., Bulacio, P.: Revised HLMS: a useful algorithm for fuzzy measure identification. Inf. Fusion **14**(4), 532–540 (2013). https://doi.org/10.1016/j.inffus.2013.01.002
10. Pedroso, M., Taylor, J., Tisseyre, B., Charnomordic, B., Guillaume, S.: A segmentation algorithm for the delineation of management zones. Comput. Electron. Agric. **70**(1), 199–208 (2010). https://doi.org/10.1016/j.compag.2009.10.007
11. Rojas, F., Sacristán Sánchez, E.J.: Guía ambiantal para el cultivo del cacao. Ministerio de Agricultura y Desarrollo Rural y Federación Nacional de Cacaoteros, Colombia (2013)

Necessary and Possible Interaction Between Criteria in a General Choquet Integral Model

Paul Alain Kaldjob Kaldjob, Brice Mayag[✉], and Denis Bouyssou[✉]

University Paris-Dauphine, PSL Research University, LAMSADE, CNRS, UMR 7243, Place du Maréchal de Lattre de Tassigny, 75775 Paris, Cedex 16, France
{paul-alain.kaldjob-kaldjob,brice.mayag,denis.bouyssou}@dauphine.fr

Abstract. This paper deals with interaction between criteria in a general Choquet integral model. When the preference of the Decision Maker (DM) contains no indifference, we first give a necessary and sufficient condition for them to be representable by a Choquet integral model. Using this condition, we show that it is always possible to choose from the numerical representations, one relatively for which all the Shapley interaction indices are strictly positive. We illustrate our results with an example.

Keywords: Interaction index · General Choquet integral model · Shapley interaction indices

1 Introduction

In Multiple Criteria Decision Making (MCDM), the independence hypothesis of preferences is often considered to be restrictive. Thus, several other models that do not require the independence hypothesis have been developed, including the Choquet integral model. The Choquet integral model assumes that the criteria has been constructed so as to be commensurate. Here, we are not concerned with the commensurability hypothesis and therefore we assume that the criteria have been constructed to be commensurate.

In [6] we find two necessary and sufficient conditions for a preferential information on set of binary alternatives to be represented by a 2-additive Choquet integral model (i.e., a Choquet integral model using a 2-additive capacity). This result is extended by using a representation of the general model based on a set of generalized binary alternatives. We extend this result. Indeed, our first result gives a necessary and sufficient condition for preferential information on generalized binary alternatives to be representable by a general model of Choquet integral. In [5] it is proven that in the framework of binary alternatives, if the preferential information contains no indifference, and is representable by a 2-additive Choquet model, then we can choose among these representations one for which all Shapley interaction indices between two criteria are strictly

© Springer Nature Switzerland AG 2020
M.-J. Lesot et al. (Eds.): IPMU 2020, CCIS 1238, pp. 457–466, 2020.
https://doi.org/10.1007/978-3-030-50143-3_36

positive. We extend also this result. Indeed, under the conditions of our first result, we shows that in the framework of generalized binary alternatives, if the preference information contains no indifference, it is always possible to represent it by a general Choquet integral model which all interaction indices are strictly positive.

This paper is organized as follows. After having recalled in the second section some basic elements on the model of the Choquet integral in MCDM, in the third section, we talk about of concept of necessary and possible interaction introduced in [5]. Then, in the fourth section, we expose our two results. Indeed, we begin by giving a necessary and sufficient condition for preferential information on the set of generalized binary alternatives containing no indifference to be represented by Choquet integral model. Under this condition, we show that it is always possible to represent this preferential information by a Choquet integral model where all interaction indices are strictly positive. We illustrate our results with an example, and we end with a conclusion.

2   Notations and Definitions

2.1   The Framework

Let X be a set of alternatives evaluate on a set of n criteria $N = \{1, 2, ..., n\}$. The set of all alternatives X is assumed to be a subset of a Cartesian product $X_1 \times X_2 \times ... \times X_n$.

The criteria are recoded numerically using, for all $i \in N$, a function u_i from X_i into $\mathbb{R}$. Using these functions allows to assume that the various recoded criteria are "commensurate" and, hence, the application of the Choquet integral model is meaningful [4].

For all $x = (x_1, ..., x_n) \in X$, we will sometimes write $u(x)$ as a shorthand for $(u_1(x_1), ..., u_n(x_n))$.

We assume the DM is able to identify on each criterion $i \in N$ two reference levels 1_i and 0_i:

- the level 0_i in X is considered as a neutral level and we set $u_i(0_i) = 0$;
- the level 1_i in X is considered as a good level and we set $u_i(1_i) = 1$.

For a subset $S \subseteq N$ we define the element $a_S = (1_S; 0_{-S})$ of X such that $a_i = 1_i$ if $i \in S$ and $a_i = 0_i$ otherwise. We suppose that for all $S \subseteq N$, $a_S \in X$.

2.2   Choquet Integral

The Choquet integral [3] in an aggregation function known in MCDM as a tool generalizing the arithmetic mean. It is based on the notion of capacity μ defined as a function from the powerset 2^N into $[0, 1]$ such that:
$\mu(\emptyset) = 0$,
$\mu(N) = 1$,
$\forall S, T \in 2^N, [S \subseteq T \implies \mu(S) \leq \mu(T)]$ (monotonicity).

For an alternative $x = (x_1, ..., x_n) \in X$, the expression of the Choquet integral w.r.t. a capacity μ is given by:

$$C_\mu(u(x)) = C_\mu(u_1(x_1), ..., u_n(x_n))$$

$$= \sum_{i=1}^{n} \left[u_{\sigma(i)}(x_{\sigma(i)}) - u_{\sigma(i-1)}(x_{\sigma(i-1)}) \right] \mu(\{\sigma(i), ..., \sigma(n)\})$$

Where σ is a permutation on N such that $u_{\sigma(1)}(x_{\sigma(1)}) \leq u_{\sigma(2)}(x_{\sigma(2)}) \leq ... \leq u_{\sigma(n)}(x_{\sigma(n)})$ and $u_{\sigma(0)}(x_{\sigma(0)}) = 0$.

Our work is based on the set $\mathcal{B}$ defined as following.

Definition 1. *The set of **generalized binary alternatives** is defined by $\mathcal{B} = \{a_S = (1_S, 0_{-S}) : S \subseteq N\}$.*

Remark 1. For all $S \subseteq N$, we have $C_\mu(a_S) = \mu(S)$.

The idea is to ask to the DM its preferences by comparing some elements of $\mathcal{B}$. We obtain the binary relations P and I defined as follows.

Definition 2. *An **ordinal preference information** $\{P, I\}$ on $\mathcal{B}$ is given by:*

$P = \{(x, y) \in \mathcal{B} \times \mathcal{B}: DM \text{ strictly prefers } x \text{ to } y\}$,
$I = \{(x, y) \in \mathcal{B} \times \mathcal{B}: DM \text{ is indifferent between } x \text{ and } y\}$.

We add to this ordinal preference information a relation M modeling the monotonicity relations between binary alternatives, and allowing us to ensure the satisfaction of the monotonicity condition $[S \subseteq T \implies \mu(S) \leq \mu(T)]$.

Definition 3. *For all $a_S, a_T \in \mathcal{B}$, $a_S M a_T$ if [not($a_S(P \cup I)a_T$) and $S \supseteq T$].*

Remark 2. $a_S M a_T \implies C_\mu(a_S) \geq C_\mu(a_T)$.

Definition 4. *There exists a **strict cycle** of $(P \cup M)$ if there exists the elements $x_0, x_1 ..., x_r$ of $\mathcal{B}$ such that $x_0(P \cup M)x_1(P \cup M)...(P \cup M)x_r(P \cup M)x_0$ and for a least one $i \in \{0, ..., r-1\}$, $x_i P x_{i+1}$.*

2.3   General Interaction Index

Definition 5. *The general interaction [2] index w.r.t. a capacity μ is defined by:*
$\forall A \subseteq N$,

$$I_A^\mu = \sum_{K \subseteq N \setminus A} \frac{(n-k-a)! k!}{(n-a+1)!} \sum_{L \subseteq A} (-1)^{a-\ell} \mu(K \cup L)$$

where $\ell = |L|$, $k = |K|$ and $a = |A| \geq 2$.

Remark 3. Given a capacity μ and $A \subseteq N$, we can rewrite the general interaction index as follows

$$I_A^\mu = \sum_{K \subseteq N \setminus A} \frac{(n-k-a)!k!}{(n-a+1)!} \Delta_A^K$$

where $\ell = |L|$, $a = |A|$ and $k = |K|$ with $\Delta_A^K = \sum_{L \subseteq A} (-1)^{a-\ell} \mu(K \cup L)$.

The following lemma gives a decomposition of Δ_A^K (we assume that 0 is an even number).

Lemma 1. $\forall A \subseteq N, \forall K \subseteq N \setminus A$,

$$\Delta_A^K = \sum_{\substack{p=0, \\ p \text{ even}}}^{a} \left[\sum_{\substack{L \subseteq A, \\ \ell = a-p}} \mu(K \cup L) - \sum_{\substack{L \subseteq A, \\ \ell = a-p-1}} \mu(K \cup L) \right].$$

Proof. We will reason according to the parity of a.

– If a is even.

$$\Delta_A^K = \sum_{L \subseteq A} (-1)^{a-\ell} \mu(K \cup L)$$

$$\Delta_A^K = \left[\sum_{\substack{L \subseteq A, \\ \ell = a}} \mu(K \cup L) - \sum_{\substack{L \subseteq A, \\ \ell = a-1}} \mu(K \cup L) \right] + \left[\sum_{\substack{L \subseteq A, \\ \ell = a-2}} \mu(K \cup L) - \sum_{\substack{L \subseteq A, \\ \ell = a-3}} \mu(K \cup L) \right] + \dots$$

$$+ \left[\sum_{\substack{L \subseteq A, \\ \ell = 2}} \mu(K \cup L) - \sum_{\substack{L \subseteq A, \\ \ell = 1}} \mu(K \cup L) \right] + \left[\sum_{\substack{L \subseteq A, \\ \ell = 0}} \mu(K \cup L) - \sum_{\substack{L \subseteq A, \\ \ell = -1}} \mu(K \cup L) \right]$$

where $\sum_{\substack{L \subseteq A, \\ \ell = -1}} \mu(K \cup L) = 0$.

$$\Delta_A^K = \sum_{\substack{p=0, \\ p \text{ even}}}^{a} \left[\sum_{\substack{L \subseteq A, \\ \ell = a-p}} \mu(K \cup L) - \sum_{\substack{L \subseteq A, \\ \ell = a-p-1}} \mu(K \cup L). \right]$$

– If a is odd.

$$\Delta_A^K = \sum_{L \subseteq A} (-1)^{a-\ell} \mu(K \cup L)$$

$$\Delta_A^K = \left[\sum_{\substack{L \subseteq A, \\ \ell = a}} \mu(K \cup L) - \sum_{\substack{L \subseteq A, \\ \ell = a-1}} \mu(K \cup L) \right] + \left[\sum_{\substack{L \subseteq A, \\ \ell = a-2}} \mu(K \cup L) - \sum_{\substack{L \subseteq A, \\ \ell = a-3}} \mu(K \cup L) \right] + \dots$$

$$+\left[\sum_{\substack{L\subseteq A,\\ \ell=1}}\mu(K\cup L)-\sum_{\substack{L\subseteq A,\\ \ell=0}}\mu(K\cup L)\right]$$

$$\Delta_A^K=\sum_{\substack{p=0,\\ p\ even}}^{a-1}\left[\sum_{\substack{L\subseteq A,\\ \ell=a-p}}\mu(K\cup L)-\sum_{\substack{L\subseteq A,\\ \ell=a-p-1}}\mu(K\cup L)\right]$$

$$\Delta_A^K=\sum_{\substack{p=0,\\ p\ even}}^{a}\left[\sum_{\substack{L\subseteq A,\\ \ell=a-p}}\mu(K\cup L)-\sum_{\substack{L\subseteq A,\\ \ell=a-p-1}}\mu(K\cup L)\right]$$

since a is odd but p is even.   □

Remark 4. Let N be the set of criteria, $A\subseteq N$, $K\subseteq N\setminus A$ and $1\le t\le a$. We have:

$$\sum_{\substack{L\subseteq A,\\ \ell=t}}\sum_{i\in L}\mu(K\cup L\setminus\{i\})=C_{a-t+1}^1\sum_{\substack{L\subseteq A,\\ \ell=t-1}}\mu(K\cup L)\ \text{where}\ C_n^p=\frac{n!}{p!(n-p)!}.$$

Let us illustrate Remark 4 with this example.

Example 1. Let $N=\{1,2,3,4\}$, $A=\{1,2,3\}$ and $K\subseteq N\setminus A=\{4\}$. We have:

$$\sum_{\substack{L\subseteq A,\\ \ell=2}}\sum_{i\in L}\mu(K\cup L\setminus\{i\})=\sum_{i\in\{1,2\}}\mu(K\cup\{1,2\}\setminus\{i\})+\sum_{i\in\{1,3\}}\mu(K\cup\{1,3\}\setminus\{i\})$$

$$+\sum_{i\in\{2,3\}}\mu(K\cup\{2,3\}\setminus\{i\})$$

$$=2\mu(K\cup\{1\})+2\mu(K\cup\{2\})+2\mu(K\cup\{3\})$$

$$=2\sum_{\substack{L\subseteq A,\\ \ell=1}}\mu(K\cup L)$$

with $2=C_{3-2+1}^1$.

3   Necessary and Possible Interaction

Once the DM compares a number a alternatives in terms of strict preferences (P) or indifference (I), the following definition tells us when this ordinal preference information is representable by Choquet integral model.

Definition 6. *An ordinal preference information $\{P,I\}$ on $\mathcal{B}$ is representable by a Choquet integral model if we can find a capacity μ such that: For all $S,T\subseteq N$,*

$$a_S P a_T\implies C_\mu(u(a_S))>C_\mu(u(a_T));$$
$$a_S I a_T\implies C_\mu(u(a_S))=C_\mu(u(a_T)).$$

The set of all capacities that can be used to represent the ordinal preference information $\{P,I\}$ at hand will be denoted $C_{Pref}(P,I)$. When there is no ambiguity on the underlying ordinal preference information, we will simply write C_{Pref}.

The following definition of necessary and possible interactions will be central in the rest of this text. This definition is given in [5].

Definition 7. *Let A be a subset of N. We say that:*

1. *There exists a possible positive (resp. null, negative) interaction among the elements of A if there exists a capacity $\mu \in C_{Pref}$ such that $I_A^\mu > 0$ (resp. $I_A^\mu = 0$, $I_A^\mu < 0$);*
2. *There exists a necessary positive (resp. null, negative) interaction among the elements of A if $I_A^\mu > 0$ (resp. $I_A^\mu = 0$, $I_A^\mu < 0$) for all capacity $\mu \in C_{Pref}$.*

Remark 5. Let be A a subset of criteria.

- If there exists a necessary positive (resp. null, negative) interaction among the elements of A, then there exists a possible positive (resp. null, negative) interaction among the elements of A.
- If there is no necessary positive (resp. null, negative) interaction among the elements of A, then there exists a possible negative or null (resp. positive or negative, positive or null) interaction among the elements of A.

Let A be a subset of N and $\{P, I\}$, an ordinal preference information. We can have a possible but not necessary interaction, what makes the interpretation difficult because dependent on the capacity chosen into $C_{Pref}(P, I)$. Indeed, the interpretation of the interaction only makes sense in the case of the necessary interaction.

4   Results

The following proposition gives a necessary and sufficient condition for an ordinal preference information on $\mathcal{B}$ containing no indifference to be representable by a Choquet integral model.

Proposition 1. *Let $\{P, I\}$ be an ordinal preference information on $\mathcal{B}$ such that $I = \emptyset$.*
$\{P, I\}$ is representable by a Choquet integral if and only if the binary relation $(P \cup M)$ contains no strict cycle.

Proof. **Necessity.** Suppose that the ordinal preference information $\{P, I\}$ on $\mathcal{B}$ is representable by a Choquet integral. So there exists a capacity μ such that $\{P, I\}$ is representable by C_μ.

If $P \cup M$ contains a strict cycle, then there exists $x_0, x_1, ..., x_r$ on $\mathcal{B}$ such that $x_0(P \cup M)x_1(P \cup M)...(P \cup M)x_r(P \cup M)x_0$ and there exists $x_i, x_{i+1} \in \{x_0, x_1, ..., x_r\}$ such that $x_i P x_{i+1}$. Since $\{P, I\}$ is representable by C_μ, therefore $C_\mu(x_0) \geq ... \geq C_\mu(x_i) > C_\mu(x_{i+1}) \geq ... \geq C_\mu(x_0)$, then $C_\mu(x_0) > C_\mu(x_0)$. Contradiction.
So, $P \cup M$ contains no strict cycle.

Sufficiency. Assume that $(P \cup M)$ contains no strict cycle, then there exists $\{\mathcal{B}_0, \mathcal{B}_1, ..., \mathcal{B}_m\}$ a partition of $\mathcal{B}$, builds by using a suitable topological sorting on $(P \cup M)$ [1].

We construct a partition $\{\mathcal{B}_0, \mathcal{B}_1, ..., \mathcal{B}_m\}$ as follows:
$\mathcal{B}_0 = \{x \in \mathcal{B} : \forall y \in \mathcal{B}, not[x(P \cup M)y]\}$,
$\mathcal{B}_1 = \{x \in \mathcal{B} \setminus \mathcal{B}_0 : \forall y \in \mathcal{B} \setminus \mathcal{B}_0, not[x(P \cup M)y]\}$,
$\mathcal{B}_i = \{x \in \mathcal{B} \setminus (\mathcal{B}_0 \cup ... \cup \mathcal{B}_{i-1}) : \forall y \in \mathcal{B} \setminus (\mathcal{B}_0 \cup ... \cup \mathcal{B}_{i-1}), not[x(P \cup M)y]\}$, for
all $i = 1; 2; ...; m$.

Let us define the mapping $f : \mathcal{B} \longrightarrow \mathbb{R}$ and $\mu : 2^N \longrightarrow [0, 1]$ as follows: for
$\ell \in \{0, 1, ..., m\}$,

$$\forall x \in \mathcal{B}_\ell, \ f(\phi(x)) = \begin{cases} 0 & if \quad \ell = 0, \\ (2n)^\ell & if \quad \ell \in \{1, 2, ..., m\}. \end{cases}$$

$\mu(S) = \frac{f_S}{\alpha}$, where $f_S = f(\phi(a_S))$ and $\alpha = f_N = (2n)^m$.
Let $a_S, a_T \in \mathcal{B}$ such that $a_S P a_T$. Show that $C_\mu(a_S) > C_\mu(a_T)$.
As $I = \emptyset$, then $\mathcal{B}_0 = \{a_0\}$ and $\mathcal{B}_m = \{a_N\}$.

- If $T = \emptyset$, then $a_T \in \mathcal{B}_0$ and $a_S \in \mathcal{B}_r$ with $r \geq 1$.
  We have $C_\mu(a_S) = \dfrac{(2n)^r}{\alpha} > 0 = \mu(\emptyset) = C_\mu(a_T)$.
- If $\emptyset \subsetneq T$, since $a_S, a_T \in \mathcal{B}$, and $\{\mathcal{B}_0, \mathcal{B}_1, ..., \mathcal{B}_m\}$ is a partition of $\mathcal{B}$ (with
  $\mathcal{B}_0 = \{a_0\}$), then there exists $r, q \in \{1, ..., m\}$ such that $a_S \in \mathcal{B}_r$, $a_T \in \mathcal{B}_q$.
  Therefore $C_\mu(a_S) = \mu(S) = \dfrac{f_S}{\alpha} = \dfrac{(2n)^r}{\alpha}$, $C_\mu(a_T) = \mu(T) = \dfrac{f_T}{\alpha} = \dfrac{(2n)^q}{\alpha}$.
  Moreover $a_S P a_T$, then $r > q$, so $(2n)^r > (2n)^q$, therefore $\dfrac{(2n)^r}{\alpha} > \dfrac{(2n)^q}{\alpha}$, i.e.,
  $C_\mu(a_S) > C_\mu(a_T)$.

In both cases, $C_\mu(a_S) > C_\mu(a_T)$. Therefore $\{P, I\}$ is representable by C_μ.   □

Given the ordinal preference information $\{P, I\}$ on $\mathcal{B}$, under the previous conditions, the following proposition shows that: it is always possible to choose in $C_{Pref}(P, I)$, one capacity allowing all the interaction indices are strictly positive. This result shows that positive interaction is always possible into all subsets of criteria in general Choquet integral model.

Proposition 2. *Let $\{P, I\}$ be an ordinal preference information on $\mathcal{B}$ such that $I = \emptyset$, and $(P \cup M)$ containing no strict cycle.*
There exists a capacity μ such that C_μ represents $\{P, I\}$ and for all $A \subseteq N, I_A^\mu > 0$.

Proof. To show that $I_A^\mu > 0$, we will prove that for all $K \subseteq N \setminus A$, $\displaystyle\sum_{L \subseteq A} (-1)^{a-\ell} \mu(K \cup L) > 0$.

The partition $\{\mathcal{B}_0, ..., \mathcal{B}_m\}$ of $\mathcal{B}$ and the capacity μ are built as in proof of Proposition 1. Since $I = \emptyset$, then we have $\mathcal{B}_0 = \{a_0\}$ and $\mathcal{B}_m = \{a_N\}$.

Consider capacity μ define by: $\mu(S) = \frac{f_S}{\alpha}$, where $f_S = f(\phi(a_S))$ and $\alpha = f_N = (2n)^m$.

Let $K \subseteq N \setminus A$. According to the previous Lemma 1 we have

$$\sum_{L\subseteq A}(-1)^{a-\ell}\mu(K\cup L) = \sum_{\substack{p=0,\\p\ even}}^{a}\left[\sum_{\substack{L\subseteq A,\\\ell=a-p}}\mu(K\cup L) - \sum_{\substack{L\subseteq A,\\\ell=a-p-1}}\mu(K\cup L)\right]$$

Let $L\subseteq A$, $|L| = a-p$ with $p\in\{0,1,...,a\}$ and even number.

As $K\cup L \supsetneq K\cup L\setminus\{i\}$ for all $i\in L$, then $a_{K\cup L}(P\cup M)a_{K\cup L\setminus\{i\}}$, hence there exists $q\in\{1,2,...,m\}$ such that $a_{K\cup L}\in\mathcal{B}_q$ and $\forall i\in L$, there exists $r_i\in\{0,1,2,...,m\}$ such that $a_{K\cup L\setminus\{i\}}\in\mathcal{B}_{r_i}$ with $r_i\leq q-1$.

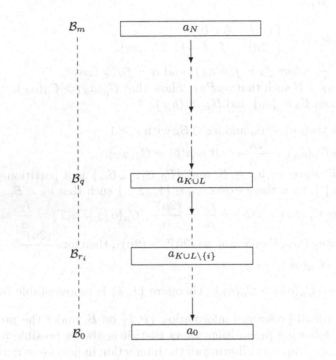

Fig. 1. An illustration of the elements $\mathcal{B}_m$, $\mathcal{B}_q$, $\mathcal{B}_{r_i}$, and $\mathcal{B}_0$ such that $m > q > r_i > 0$ with $i\in N$.

Then $\mu(K\cup L) = (2n)^q = (2n)(2n)^{q-1}$.

$$\sum_{i\in L}\mu(K\cup L\setminus\{i\}) = \sum_{i\in L}(2n)^{r_i} \leq \sum_{i\in L}(2n)^{q-1} = l(2n)^{q-1}$$

As $2n > l$, then $\mu(K\cup L) > \sum_{i\in L}\mu(K\cup L\setminus\{i\})$, hence $\sum_{\substack{L\subseteq A,\\l=a-p}}\mu(K\cup L) >$

$\sum_{\substack{L\subseteq A,\\l=a-p}}\sum_{i\in L}\mu(K\cup L\setminus\{i\})$ According to Remark 4 (with $t = a-p$), we have

$$\sum_{\substack{L\subseteq A,\\\ell=a-p}}\sum_{i\in L}\mu(K\cup L\setminus\{i\}) = C_{p+1}^1\sum_{\substack{L\subseteq A,\\\ell=a-p-1}}^{a}\mu(K\cup L) > \sum_{\substack{L\subseteq A,\\\ell=a-p-1}}^{a}\mu(K\cup L) \text{ since }$$

$C_{p+1}^1 = p+1 > 1$.

So $\displaystyle\sum_{\substack{L\subseteq A,\\ \ell=a-p}} \mu(K\cup L) > \sum_{\substack{L\subseteq A,\\ \ell=a-p-1}} \mu(K\cup L)$, i.e., $\displaystyle\sum_{\substack{L\subseteq A,\\ \ell=a-p}} \mu(K\cup L) - \sum_{\substack{L\subseteq A,\\ \ell=a-p-1}} \mu(K\cup L) > 0$.

We then have $\displaystyle\sum_{\substack{p=0,\\ p\ even}}^{a} \left[\sum_{\substack{L\subseteq A,\\ \ell=a-p}} \mu(K\cup L) - \sum_{\substack{L\subseteq A,\\ \ell=a-p-1}} \mu(K\cup L)\right] > 0$.

We have just proved that For all $K \subseteq N \setminus A$, $\displaystyle\sum_{L\subseteq A}(-1)^{a-\ell}\mu(K\cup L) > 0$.

We can therefore conclude that $I_A^\mu > 0$.   □

The following example illustrates our two results.

Example 2. $N = \{1,2,3,4\}$, $P = \{(a_{23},a_1),(a_{234},a_{123}),(a_2,a_{13})\}$.

The ordinal preference information $\{P,I\}$ contains no indifference and the binary relation $(P\cup M)$ contains no strict cycle, so $\{P,I\}$ is representable by a Choquet integral model.

A suitable topological sorting on $(P\cup M)$ is given by:
$\mathcal{B}_0 = \{a_0\}$; $\mathcal{B}_1 = \{a_1,a_3,a_4\}$; $\mathcal{B}_2 = \{a_{13},a_{14},a_{34}\}$; $\mathcal{B}_3 = \{a_2\}$; $\mathcal{B}_4 = \{a_{12},a_{23},a_{24}\}$;
$\mathcal{B}_5 = \{a_{123},a_{124},a_{134}\}$; $\mathcal{B}_6 = \{a_{234}\}$ and $\mathcal{B}_7 = \{a_N\}$.

The ordinal preference information $\{P,I\}$ is representable by a following capacity μ:

S	$\emptyset$	$\{1\}$	$\{2\}$	$\{3\}$	$\{4\}$	$\{1,2\}$	$\{1,3\}$	$\{1,4\}$	$\{2,3\}$	$\{2,4\}$	$\{3,4\}$	$\{1,2,3\}$	$\{1,2,4\}$	$\{1,3,4\}$	$\{2,3,4\}$	N
$8^7\times\mu(S)$	0	8	8^3	8	8	8^4	8^2	8^2	8^4	8^4	8^2	8^5	8^5	8^5	8^6	8^7

Considering the previous capacity, the corresponding interaction indices are given by the following table:

A	$\{1,2\}$	$\{1,3\}$	$\{1,4\}$	$\{2,3\}$	$\{2,4\}$	$\{3,4\}$	$\{1,2,3\}$	$\{1,2,4\}$	$\{1,3,4\}$	$\{2,3,4\}$	N
$8^7\times I_A^\mu$	611.33	611.66	688.16	726.16	726.16	726.33	899	899	612.83	612.83	612.83

We can see that $I_A^\mu > 0$, $\forall A \subseteq N$ such that $\mid A \mid \geq 2$.

5   Conclusion

This article deals with the notion of interaction within a subset of criteria of any size, in the Choquet integral model. Our first result gives a necessary and sufficient condition for ordinal preference information on generalized binary alternatives to be representable by a general model of Choquet integral. This extends theorem 1, Page 305 [6].

Under condition of our first result, our second result shows that in the framework of generalized binary alternatives, if the ordinal preference information contains no indifference, it is possible to represent it by a general Choquet integral model which all Shapley interaction indices between a subset of criteria are strictly positive. This extends theorem 2, Page 10 [5].

The subject of this paper offer several avenues for future research. In fact, It would be interesting as in [5] to provide a linear program to test the necessary interaction outside the framework of generalized binary alternatives. Otherwise we will examine the case where $I \neq \emptyset$. We are also interested in duality. Indeed, is it always possible to build a capacity relative to which all the interaction indices will be strictly negative? It would finally be interesting to study the case of bipolar scales. We are already investigating some of these research avenues.

References

1. Gondran, M., Minoux, M.: Graphes et algorithmes, 3e edn. Eyrolles, Paris (1995)
2. Grabisch, M.: k-order additive discrete fuzzy measures and their representation. Fuzzy Sets Syst. **92**(2), 167–189 (1997)
3. Grabisch, M.: Set Functions, Games and Capacities in Decision Making, vol. 46. Springer, Cham (2016). https://doi.org/10.1007/978-3-319-30690-2
4. Grabisch, M., Labreuche, C., Vansnick, J.-C.: On the extension of pseudo-Boolean functions for the aggregation of interacting criteria. Eur. J. Oper. Res. **148**, 28–47 (2003)
5. Mayag, B., Bouyssou, D.: Necessary and possible interaction between criteria in a 2-additive Choquet integral model. Eur. J. Oper. Res. **283**, 308–320 (2019)
6. Mayag, B., Grabisch, M., Labreuche, C.: A representation of preferences by the Choquet integral with respect to a 2-additive capacity. Theory Decis. **71**(3), 297–324 (2011). https://doi.org/10.1007/s11238-010-9198-3

Construction of Nullnorms Based on Closure and Interior Operators on Bounded Lattices

Gül Deniz Çaylı[✉]

Department of Mathematics, Faculty of Science, Karadeniz Technical University, 61080 Trabzon, Turkey
guldeniz.cayli@ktu.edu.tr

Abstract. In this paper, we introduce two rather effective methods for constructing new families of nullnorms with a zero element on the basis of the closure operators and interior operators on a bounded lattice under some additional conditions. Our constructions can be seen as a generalization of the ones in [28]. As a by-product, two types of idempotent nullnorms on bounded lattices are obtained. Several interesting examples are included to get a better understanding of the structure of new families of nullnorms.

Keywords: Bounded lattice · Construction method · Closure operator · Interior operator · Nullnorm

1 Introduction

The definitions of t-operators and nullnorms on the unit interval were introduced by Maş et al. [20] and Calvo et al. [4], respectively. In [21], it was shown that nullnorms coincide with t-operators on the unit interval $[0, 1]$ since both of them have identical block structures on $[0, 1]^2$. As a generalization of t-norms and t-conorms on the unit interval, nullnorms have a zero element s derived from the whole domain, regardless of whether t-norms and t-conorms have zero elements 0 and 1, respectively. In particular, a nullnorm is a t-norm if $s = 0$ while it is a t-conorm if $s = 1$. These operators are effective in various fields of applications, such as expert systems, fuzzy decision making and fuzzy system modeling. They are interesting also from theoretical point of view. For more studies about t-norms, t-conorms, nullnorms and related operators on the unit interval, it can be referred to [5, 10, 14, 17, 19, 22–24, 26, 27, 29].

In recent years, the study of nullnorms on bounded lattices was initiated by Karaçal et al. [18]. They demonstrated the presence of nullnorms with a zero element on the basis of a t-norm and a t-conorm on a bounded lattice. Notice that the families of nullnorms obtained in [18] are not idempotent, in general. For this reason, Çaylı and Karaçal [6] introduced a method showing the presence idempotent nullnorms on bounded lattices such that there is only

© Springer Nature Switzerland AG 2020
M.-J. Lesot et al. (Eds.): IPMU 2020, CCIS 1238, pp. 467–480, 2020.
https://doi.org/10.1007/978-3-030-50143-3_37

one element incomparable with the zero element. Moreover, they proposed that there does not need to exist an idempotent nullnorm on a bounded lattice. After then, Wang et al. [28] and Çaylı [7] presented some methods for constructing idempotent nullnorms on a bounded lattice with some additional conditions on theirs zero element. Their methods can be viewed as a generalization of the proposed method in [6]. On the contrary to the approaches in [6,7,28] based on the only infimum t-norm on $[s,1]^2$ and supremum t-conorm on $[0,s]^2$, in [8,9] by using an arbitrary t-norm on $[s,1]^2$ and an arbitrary t-conorm on $[0,s]^2$, it was described some construction methods for nullnorms on a bounded lattice L having a zero element s with some constraints.

In general topology, by considering a nonempty set A and the set $\wp(A)$ of all subsets of A, the closure operator (resp. interior operator) on $\wp(A)$ is defined as an expansive, isotone and idempotent map $cl : \wp(A) \to \wp(A)$ (resp. a contractive, isotone and idempotent map $int : \wp(A) \to \wp(A)$). Both of these operators can be used for constructing topologies on A in general topology [15]. More precisely, a one-to-one correspondence from the set of all topologies on A to the set of all closure (interior) operators on $\wp(A)$. That is, any topology on a nonempty set can induce the closure (interior) operator on its underlying powerset. It should be pointed out that closure and interior operators can be defined on a lattice $(\wp(A), \subseteq)$ of all subsets of a set A with set union as the join and set intersection as the meet. Hence, Everett [16] extended the closure operator (resp. interior operator) on $\wp(A)$ to a general lattice L where the condition $cl(\emptyset) = \emptyset$ (resp. $int(A) = A$) is omitted.

The main aim of this paper is to present some methods for yielding new families of nullnorms with a zero element by means of closure operators and interior operators on a bounded lattice. The remainder of this paper is organized as follows. In Sect. 2, we recall some preliminary details about bounded lattices and nullnorms, interior and closure operators on them. In Sect. 3, considering a bounded lattice L, we propose two new methods for generating nullnorms with a zero element based on the presence of closure operators $cl : L \to L$ and interior operators $int : L \to L$. We note that our constructions are a generalization of the constructions in [28]. We also provide some corresponding examples showing that our constructions actually create new types of nullnorms on bounded lattices different from those in [28]. It should be pointed out that our methods need some sufficient and necessary conditions to generate a nullnorm on a bounded lattice. As a by product, two classes of idempotent nullnorms on bounded lattices are obtained when taking the closure operator $cl : L \to L$ as $cl(x) = x$ for all $x \in L$ and the interior operator $int : L \to L$ as $int(x) = x$ for all $x \in L$. Furthermore, we exemplify that we cannot force new nullnorms to coincide with another predefined t-conorm $[0,s]^2$ and t-norm $[s,1]^2$ apart from the t-conorm $S_\vee : [0,s]^2 \to [0,s]$ defined by $S_\vee(x,y) = x \vee y$ for all $x,y \in [0,s]$ and the t-norm $T_\wedge : [s,1]^2 \to [s,1]$ defined by $T_\wedge(x,y) = x \wedge y$. Finally, some concluding remarks are added.

2 Preliminaries

In this part, some basic results about bounded lattices and nullnorms, closure and interior operators on them are recalled.

A lattice L is a nonempty set with the partial order $\leq$ where any two elements $x, y \in L$ have a smallest upper bound (called join or supremum), denoted by $x \vee y$ and a greatest lower bound (called meet or infimum), denoted by $x \wedge y$. For $a, b \in L$, we use the notation $a < b$ where $a \leq b$ and $a \neq b$. Moreover, we use the notation $a \parallel b$ to denote that a and b are incomparable. For $s \in L\backslash\{0,1\}$, we denote $D_s = [0,s] \times [s,1] \cup [s,1] \times [0,s]$ and $I_s = \{x \in L \mid x \parallel s\}$.

A bounded lattice $(L, \leq, \wedge, \vee)$ is a lattice having the top and bottom elements, which are written as 1 and 0, respectively, that is, there are two elements $1, 0 \in L$ such that $0 \leq x \leq 1$ for all $x \in L$.

Given $a, b \in L$ with $a \leq b$, the subinterval $[a, b]$ is a sublattice of L defined by $[a, b] = \{x \in L \mid a \leq x \leq b\}$.

The subintervals $]a, b]$, $[a, b[$ and $]a, b[$ are defined by $]a, b] = \{x \in L \mid a < x \leq b\}$, $[a, b[= \{x \in L \mid a \leq x < b\}$ and $]a, b[= \{x \in L \mid a < x < b\}$ (see [1]).

Definition 1 ([6–9,18]). *Let L be a bounded lattice. A binary operation $F : L \times L \to L$ is called a nullnorm on L if, for any $x, y, z \in L$, it satisfies the following properties:*

(1) $F(x, z) \leq F(y, z)$ for $x \leq y$.
(2) $F(x, F(y, z)) = F(F(x, y), z)$.
(3) $F(x, y) = F(y, x)$.
(4) There exists an element $s \in L$ called the zero element, such we obtain $F(x, 0) = x$ for all $x \leq s$ and $F(x, 1) = x$ for all $x \geq s$.

F is called an idempotent nullnorm on L whenever $F(x, x) = x$ for all $x \in L$. We note that a triangular norm T (t-norm for short) on L is a special case of nullnorm with $s = 0$ whereas a triangular conorm S (t-conorm for short) on L is a special case of nullnorm with $s = 1$ (see [2,3]).

Proposition 1 ([13,18]). *Let L be a bounded lattice and F be a nullnorm on L with the zero element $s \in L\backslash\{0,1\}$. Then the following statements hold:*

i) $F|[0,s]^2 : [0,s]^2 \to [0,s]$ is a t-conorm on $[0,s]^2$.
ii) $F|[s,1]^2 : [s,1]^2 \to [s,1]$ is a t-norm on $[s,1]^2$.

Proposition 2 ([9]). *Let L be a bounded lattice and F be an idempotent nullnorm on L with the zero element $s \in L\backslash\{0,1\}$. Then the following statements hold:*

i) $F(x, y) = x \vee y$ for all $(x, y) \in [0, s]^2$.
ii) $F(x, y) = x \wedge y$ for all $(x, y) \in [s, 1]^2$.
iii) $F(x, y) = x \vee (y \wedge s)$ for all $(x, y) \in [0, s] \times I_s$.
iv) $F(x, y) = y \vee (x \wedge s)$ for all $(x, y) \in I_s \times [0, s]$.
v) $F(x, y) = x \wedge (y \vee s)$ for all $(x, y) \in [s, 1] \times I_s$.
vi) $F(x, y) = y \wedge (x \vee s)$ for all $(x, y) \in I_s \times [s, 1]$.

Definition 2 ([11,12,16]). *Let L be a lattice. A mapping $cl : L \to L$ is called a closure operator if, for any $x, y \in L$, it satisfies the following properties:*

(1) Expansion: $x \leq cl(x)$.
(2) Preservation of join: $cl(x \vee y) = cl(x) \vee cl(y)$.
(3) Idempotence: $cl(cl(x)) = cl(x)$.

For a closure operator $cl : L \to L$ and for any $x, y \in L$, we obtain that $cl(cl(x)) \leq cl(x)$ and $cl(x) \leq cl(y)$ whenever $x \leq y$.

Definition 3 ([11,12,25]). *Let L be a lattice. A mapping $int : L \to L$ is called an interior operator if, for any $x, y \in L$, it satisfies the following properties:*

(1) Contraction: $int(x) \leq x$.
(2) Preservation of meet: $int(x \wedge y) = int(x) \wedge int(y)$.
(3) Idempotence: $int(int(x)) = int(x)$.

For an interior operator $int : L \to L$ and for any $x, y \in L$, we obtain that $int(x) \leq int(int(x))$ and $int(x) \leq int(y)$ whenever $x \leq y$.

3   Construction Methods for Nullnorms

In this section, considering a bounded lattice L, we introduce two methods to construct the classes of nullnorms $F_{cl} : L \times L \to L$ and $F_{int} : L \times L \to L$ with the zero element on the basis of the closure operator $cl : L \to L$ and interior operator $int : L \to L$, respectively. We note that our constructions require some sufficient and necessary conditions on the bounded lattice and the closure (interior) operator. These conditions play an effective role in our constructions, and they yield a nullnorm on a bounded lattice in only particular cases. We also present some illustrative examples to have a better understanding of the structures of new constructions.

Theorem 1. *Let L be a bounded lattice and $s \in L \backslash \{0, 1\}$ such that $a \wedge s = b \wedge s$ and $a \vee s = b \vee s$ for all $a, b \in I_s$. Given a closure operator $cl : L \to L$ such that $cl(p) \vee cl(q) \in I_s$ for all $p, q \in I_s$, the following function $F_{cl} : L \times L \to L$ defined by*

$$
F_{cl}(x, y) = \begin{cases}
x \vee y & if\ (x, y) \in [0, s]^2, \\
x \wedge y & if\ (x, y) \in [s, 1]^2, \\
s & if\ (x, y) \in D_s, \\
y \wedge (x \vee s) & if\ (x, y) \in I_s \times [s, 1], \\
x \wedge (y \vee s) & if\ (x, y) \in [s, 1] \times I_s, \\
y \vee (x \wedge s) & if\ (x, y) \in I_s \times [0, s], \\
x \vee (y \wedge s) & if\ (x, y) \in [0, s] \times I_s, \\
cl(x) \vee cl(y) & if\ (x, y) \in I_s^2
\end{cases}
\tag{1}
$$

is a nullnorm on L with the zero element s.

Theorem 2. *Let L be a bounded lattice, $s \in L \backslash \{0,1\}$ such that $a \wedge s = b \wedge s$ for all $a,b \in I_s$ and $cl : L \to L$ be a closure operator. If the function F_{cl} defined by the formula (1) is a nullnorm on L with the zero element s, then there holds $p \vee s = q \vee s$ and $cl(p) \vee cl(q) \in I_s$ for all $p,q \in I_s$.*

Proof. Let the function F_{cl} defined by the formula (1) be a nullnorm on L with the zero element s.

Given $p,q \in I_s$, from the monotonicity of F_{cl}, we have $q \leq cl(q) \leq cl(p) \vee cl(q) = F_{cl}(p,q) \leq F_{cl}(p,1) = p \vee s$ and $p \leq cl(p) \leq cl(p) \vee cl(q) = F_{cl}(p,q) \leq F_{cl}(1,q) = q \vee s$. Then $q \vee s \leq p \vee s$ and $p \vee s \leq q \vee s$. Hence, it holds $p \vee s = q \vee s$ for any $p,q \in I_s$.

Assume that $cl(p) \vee cl(q) \in [0,s]$. Then we have $p \vee q \leq cl(p) \vee cl(q) \leq s$, i.e, $p \leq s$. This is a contradiction. So, $cl(p) \vee cl(q) \in [0,s]$ does not hold. Supposing that $cl(p) \vee cl(q) \in]s,1]$, we have $F_{cl}(0, F_{cl}(p,q)) = s$ and $F_{cl}(F_{cl}(0,p),q) = F_{cl}(p \wedge s, q) = (p \wedge s) \vee (q \wedge s) = p \wedge s$. From the associativity of F_{cl}, it holds $s = p \wedge s$, i.e., $s \leq p$. This is a contradiction. So, $cl(p) \vee cl(q) \in]s,1]$ does not hold. Therefore, we have $cl(p) \vee cl(q) \in I_s$ for any $p,q \in I_s$.

Consider a bounded lattice L, $s \in L \backslash \{0,1\}$ such that $a \wedge s = b \wedge s$ for all $a,b \in I_s$ and a closure operator $cl : L \to L$. We observe that the conditions $p \vee s = q \vee s$ and $cl(p) \vee cl(q) \in I_s$ for all $p,q \in I_s$ are both sufficient and necessary to yield a nullnorm on L with the zero element s of the function F_{cl} defined by the formula (1). In this case, one can ask whether the condition $a \wedge s = b \wedge s$ for all $a,b \in I_s$ is necessary to be a nullnorm on L with the zero element s of F_{cl}. We firstly give an example to show that in Theorem 1, the condition $a \wedge s = b \wedge s$ for all $a,b \in I_s$ cannot be omitted, in general.

Example 1. Consider the bounded lattice $L_1 = \{0,v,s,r,u,t,1\}$ depicted by Hasse diagram in Fig. 1. Define the closure operator $cl : L_1 \to L_1$ by $cl(0) = cl(v) = v$, $cl(r) = cl(u) = cl(t) = t$ and $cl(s) = cl(1) - 1$. Notice that $p \vee s = q \vee s$ and $cl(p) \vee cl(q) \in I_s$ for all $p,q \in I_s$ but $t \wedge s = v \neq 0 = r \wedge s$ for $r,t \in I_s$. In this case, by using the approach in Theorem 1, we have $F_{cl}(0, F_{cl}(u,r)) = F_{cl}(0, cl(u) \vee cl(r)) = F_{cl}(0,t) = t \wedge s = v$ and $F_{cl}(F_{cl}(0,u),r) = F_{cl}(u \wedge s, r) = F_{cl}(0,r) - r \wedge s = 0$. Then, F_{cl} is not associative for the indicated closure operator on L_1. Therefore, F_{cl} is not a nullnorm on L_1 which does not satisfy the condition $a \wedge s = b \wedge s$ for all $a,b \in I_s$.

From Example 1, we observe that the condition $a \wedge s = b \wedge s$ for all $a,b \in I_s$ is sufficient in Theorem 1. Taking into account the above mentioned question, we state that this condition is not necessary in Theorem 1. In order to show this fact, we provide an example of a bounded lattice violating this condition on which the function F_{cl} defined by the formula (1) is a nullnorm with the zero element s.

Example 2. Consider the lattice L_1 depicted in Example 1 and the closure operator $cl : L_1 \to L_1$ defined by $cl(x) = x$ for all $x \in L_1$. If we apply the construction in Theorem 1, then we obtain the function $F_{cl} : L_1 \times L_1 \to L_1$ given as in Table 1.

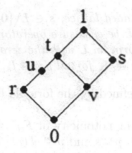

Fig. 1. Lattice L_1

It is easy to check that F_{cl} is a nullnorm with the zero element s for the chosen closure operator on L_1.

Table 1. Nullnorm F_{cl} on L_1

F_{cl}	0	v	s	r	u	t	1
0	0	v	s	0	0	v	s
v	v	v	s	v	v	v	s
s	s	s	s	s	s	s	s
r	0	v	s	r	u	t	1
u	0	v	s	u	u	t	1
t	v	v	s	t	t	t	1
1	s	s	s	1	1	1	1

In view of Theorems 1 and 2, when taking the closure operator $cl : L \to L$ as $cl(x) = x$ for all $x \in L$, we have the following Corollary 1 which shows the presence of idempotent nullnorms on L with the zero element $s \in L \backslash \{0,1\}$.

Corollary 1. *Let L be a bounded lattice and $s \in L \backslash \{0,1\}$ such that $a \wedge s = b \wedge s$ for all $a, b \in I_s$. Then, the following function $F_1 : L \times L \to L$ defined by*

$$
F_1(x,y) = \begin{cases}
x \vee y & if \ (x,y) \in [0,s]^2, \\
x \wedge y & if \ (x,y) \in [s,1]^2, \\
s & if \ (x,y) \in D_s, \\
y \wedge (x \vee s) & if \ (x,y) \in I_s \times [s,1], \\
x \wedge (y \vee s) & if \ (x,y) \in [s,1] \times I_s, \\
y \vee (x \wedge s) & if \ (x,y) \in I_s \times [0,s], \\
x \vee (y \wedge s) & if \ (x,y) \in [0,s] \times I_s, \\
x \vee y & if \ (x,y) \in I_s^2
\end{cases}
$$

is an idempotent nullnorm on L with the zero element s if and only if $p \vee s = q \vee s$ and $p \vee q \in I_s$ for all $p, q \in I_s$.

Remark 1. Let L be a bounded lattice, $s \in L\backslash\{0,1\}$, $a \wedge s = b \wedge s$ and $a \vee s = b \vee s$ for all $a, b \in I_s$. Consider a closure operator $cl : L \to L$ such that $cl(p) \vee cl(q) \in I_s$ for all $p, q \in I_s$. It should be pointed out that the classes of the nullnorms $V_\vee$ introduced in [28, Theorem 2] and F_{cl} defined by the formula (1) in Theorem 1 are different from each other. F_{cl} differs from $V_\vee$ on the domain $I_s \times I_s$. The value of F_{cl} is $cl(x) \vee cl(y)$ whereas $V_\vee$ has the value $x \vee y$ when $(x, y) \in I_s \times I_s$. Both of them have the same value on all remainder domains. From Corollary 1, we can easily observe that the nullnorm F_{cl} coincides with the nullnorm $V_\vee$ when defining the closure operator $cl : L \to L$ by $cl(x) = x$ for all $x \in L$. More precisely, our construction in Theorem 1 encompass as a special case the construction of $V_\vee$ in [28, Theorem 2]. Furthermore, the nullnorms F_{cl} and $V_\vee$ do not have to coincide with each other. In the following, we present an example to illustrative the correctness of this argument.

Example 3. Given the lattice $L_2 = \{0, u, m, n, s, v, p, q, t, r, 1\}$ characterized by Hasse diagram in Fig. 2, it is clear that $a \wedge s = b \wedge s$ and $a \vee s = b \vee s$ for all $a, b \in I_s$. Take the closure operator $cl : L_2 \to L_2$ as $cl(0) = cl(v) = v$, $cl(u) = cl(n) = cl(s) = cl(m) = u$, $cl(t) = cl(q) = q$, $cl(p) = cl(r) = r$ and $cl(1) = 1$. Then, by use of the approaches in Theorem 1 and [28, Theorem 2], respectively, the nullnorms F_{cl}, $V_\vee : L_2 \times L_2 \to L_2$ are defined in Tables 2 and 3, respectively. These nullnorms are different from each other since $F_{cl}(p, t) = r \neq p = V_\vee(p, t)$ for $p, t \in L_2$.

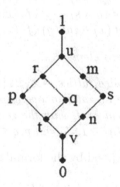

Fig. 2. Lattice L_2

Theorem 3. *Let L be a bounded lattice and $s \in L\backslash\{0,1\}$ such that $a \wedge s = b \wedge s$ and $a \vee s = b \vee s$ for all $a, b \in I_s$. Given an interior operator $int : L \to L$ such that $int(p) \wedge int(q) \in I_s$ for all $p, q \in I_s$, the following function $F_{int} : L \times L \to L$ defined by*

Table 2. Nullnorm F_{cl} on L_2

F_{cl}	0	v	n	s	t	p	q	r	m	u	1
0	0	v	n	s	v	v	v	v	s	s	s
v	v	v	n	s	v	v	v	v	s	s	s
n	n	n	n	s	n	n	n	n	s	s	s
s	s	s	s	s	s	s	s	s	s	s	s
t	v	v	n	s	q	r	q	r	m	u	u
p	v	v	n	s	r	r	r	r	m	u	u
q	v	v	n	s	q	r	q	r	m	u	u
r	v	v	n	s	r	r	r	r	m	u	u
m	s	s	s	s	m	m	m	m	m	m	m
u	s	s	s	s	u	u	u	u	m	u	u
1	s	s	s	s	u	u	u	u	m	u	1

Table 3. Nullnorm V_V on L_2

V_V	0	v	n	s	t	p	q	r	m	u	1
0	0	v	n	s	v	v	v	v	s	s	s
v	v	v	n	s	v	v	v	v	s	s	s
n	n	n	n	s	n	n	n	n	s	s	s
s	s	s	s	s	s	s	s	s	s	s	s
t	v	v	n	s	t	p	q	r	m	u	u
p	v	v	n	s	p	p	r	r	m	u	u
q	v	v	n	s	q	r	q	r	m	u	u
r	v	v	n	s	r	r	r	r	m	u	u
m	s	s	s	s	m	m	m	m	m	m	m
u	s	s	s	s	u	u	u	u	m	u	u
1	s	s	s	s	u	u	u	u	m	u	1

$$
F_{int}(x,y) = \begin{cases}
x \vee y & if\ (x,y) \in [0,s]^2, \\
x \wedge y & if\ (x,y) \in [s,1]^2, \\
s & if\ (x,y) \in D_s, \\
y \wedge (x \vee s) & if\ (x,y) \in I_s \times [s,1], \\
x \wedge (y \vee s) & if\ (x,y) \in [s,1] \times I_s, \\
y \vee (x \wedge s) & if\ (x,y) \in I_s \times [0,s], \\
x \vee (y \wedge s) & if\ (x,y) \in [0,s] \times I_s, \\
int(x) \wedge int(y) & if\ (x,y) \in I_s^2
\end{cases}
\tag{2}
$$

is a nullnorm on L with the zero element s.

Theorem 4. *Let L be a bounded lattice, $s \in L \setminus \{0,1\}$ such that $a \vee s = b \vee s$ for all $a,b \in I_s$, $int : L \to L$ be an interior operator. If the function F_{int} defined by the formula (2) is a nullnorm on L with the zero element s, then there holds $p \wedge s = q \wedge s$ and $int(p) \wedge int(q) \in I_s$ for all $p,q \in I_s$.*

Proof. Let the function F_{int} defined by the formula (2) be a nullnorm on L with the zero element s.

Given $p,q \in I_s$, from the monotonicity of F_{int}, we have $q \geq int(q) \geq int(p) \wedge int(q) = F_{int}(p,q) \geq F_{int}(p,0) = p \wedge s$ and $p \geq int(p) \geq int(p) \wedge int(q) = F_{int}(p,q) \geq F_{int}(0,q) = q \wedge s$. In this case, we obtain $q \wedge s \leq p \wedge s$ and $p \wedge s \leq q \wedge s$. So, it holds $p \wedge s = q \wedge s$ for any $p,q \in I_s$.

Assume that $int(p) \wedge int(q) \in [s,1]$. Then we have $s \leq int(p) \wedge int(q) \leq p \wedge q$. That is, $s \leq p$ which is a contradiction. Hence, $int(p) \wedge int(q) \in [s,1]$ cannot hold. Suppose that $int(p) \wedge int(q) \in [0,s[$. Then, it is obtained that $F_{int}(1, F_{int}(p,q)) = s$ and $F_{int}(F_{int}(1,p),q) = F_{int}(p \vee s, q) = (p \vee s) \wedge (q \vee s) = p \vee s$. Since F_{int} is associative, we get $s = p \vee s$, i.e., $p \leq s$ which is a contradiction. Hence, $int(p) \wedge int(q) \in [0,s[$ cannot hold. Therefore, it holds $int(p) \wedge int(q) \in I_s$ for any $p,q \in I_s$.

Consider a bounded lattice L, $s \in L \backslash \{0, 1\}$ such that $a \vee s = b \vee s$ for all $a, b \in I_s$ and an interior operator $int : L \to L$. It should be pointed out that the conditions $p \wedge s = q \wedge s$ and $int\,(p) \wedge int\,(q) \in I_s$ for all $p, q \in I_s$ are both sufficient and necessary to generate a nullnorm on L with the zero element s of the function F_{int} defined by the formula (2). Then a natural question arises: is it necessary the condition $a \vee s = b \vee s$ for all $a, b \in I_s$ to be a nullnorm on L with the zero element s of F_{cl}. At first, by the following example, we demonstrate that in Theorem 3, the condition $a \vee s = b \vee s$ for all $a, b \in I_s$ cannot be omitted, in general.

Example 4. Consider the bounded lattice $L_3 = \{0, s, k, n, t, m, 1\}$ with the lattice diagram shown in Fig. 3. Define the interior operator $int : L_3 \to L_3$ by $int\,(x) = x$ for all $x \in L_3$. It holds $p \wedge s = q \wedge s$ and $int\,(p) \wedge int\,(q) \in I_s$ for all $p, q \in I_s$, however, $t \vee s = m \neq 1 = n \vee s$ for $n, t \in I_s$. Then, by applying the method in Theorem 3, we obtain $F_{cl}\,(1, F_{cl}\,(k, n)) = F_{cl}\,(1, int\,(k) \wedge int\,(n)) = F_{cl}\,(1, t) = t \vee s = m$ and $F_{cl}\,(F_{cl}\,(1, k), n) = F_{cl}\,(k \vee s, n) = F_{cl}\,(1, n) = n \vee s = 1$. In that case, F_{int} is not associative for the indicated interior operator on L_3. Hence, F_{int} is not a nullnorm on L_3 violating the condition $a \vee s = b \vee s$ for all $a, b \in I_s$.

By Example 4, we observe that the condition $a \vee s = b \vee s$ for all $a, b \in I_s$ is sufficient in Theorem 3. Moreover, we answer the above question so that this is not a necessary condition in Theorem 3. In order to illustrative this observation, we provide an example of a bounded lattice violating this condition on which the function F_{int} defined by the formula (2) is a nullnorm on L with the zero element s.

Example 5. Consider the bounded lattice $L_4 = \{0, s, m, k, t, 1\}$ with the lattice diagram shown in Fig. 4. Notice that $t \vee s = m \neq 1 = k \vee s$ for $k, t \in I_s$. Define the interior operator $int : L_4 \to L_4$ by $int\,(x) = x$ for all $x \in L_4$. Then, by applying the construction in Theorem 3, we obtain the function $F_{int} : L_4 \times L_4 \to L_4$ given as in Table 4. It can be easily seen that F_{int} is a nullnorm on L_4 with the zero element s.

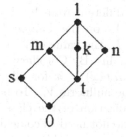

Fig. 3. Lattice L_3

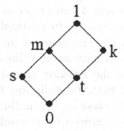

Fig. 4. Lattice L_4

Table 4. Nullnorm F_{int} on L_4

F_{int}	0	s	t	k	m	1
0	0	s	0	0	s	s
s	s	s	s	s	s	s
t	0	s	t	t	m	m
k	0	s	t	k	m	1
m	s	s	m	m	m	m
1	s	s	m	1	m	1

Taking into consideration Theorems 3 and 4, if we choose the interior operator $int : L \rightarrow L$ as $int\,(x) = x$ for all $x \in L$, then we get the following Corollary 2 which shows the presence of idempotent nullnorms on L with the zero element $s \in L \backslash \{0,1\}$.

Corollary 2. *Let L be a bounded lattice and $s \in L \backslash \{0,1\}$ such that $a \vee s = b \vee s$ for all $a, b \in I_s$. Then, the following function $F_2 : L \times L \rightarrow L$ defined by*

$$F_2\,(x,y) = \begin{cases} x \vee y & if \ (x,y) \in [0,s]^2, \\ x \wedge y & if \ (x,y) \in [s,1]^2, \\ s & if \ (x,y) \in D_s, \\ y \wedge (x \vee s) & if \ (x,y) \in I_s \times [s,1], \\ x \wedge (y \vee s) & if \ (x,y) \in [s,1] \times I_s, \\ y \vee (x \wedge s) & if \ (x,y) \in I_s \times [0,s], \\ x \vee (y \wedge s) & if \ (x,y) \in [0,s] \times I_s, \\ x \wedge y & if \ (x,y) \in I_s^2 \end{cases} \tag{3}$$

is an idempotent nullnorm on L with the zero element s if and only if $p \wedge s = q \wedge s$ and $p \wedge q \in I_s$ for all $p, q \in I_s$.

Remark 2. Let L be a bounded lattice, $s \in L \backslash \{0,1\}$, $a \wedge s = b \wedge s$ and $a \vee s = b \vee s$ for all $a, b \in I_s$. Consider an interior operator $int : L \rightarrow L$ such that $int\,(p) \wedge int\,(q) \in I_s$ for all $p, q \in I_s$. We note that F_{int} defined by the formula (2) in Theorem 3 can create new type of nullnorm different from $V_\wedge$ described in [28, Theorem 2]. In particular, F_{int} differs from $V_\wedge$ on the domain $I_s \times I_s$. While F_{int} has the value of $int\,(x) \wedge int\,(y)$ on $I_s \times I_s$, the value of $V_\wedge$ is $x \wedge y$. Both of them have same value on all remainder domains. By Corollary 2, when considering the interior operator $int : L \rightarrow L$ as $int\,(x) = x$ for all $x \in L$, we observe that the nullnorm F_{int} coincides with the nullnorm $V_\wedge$. To be more precise, the class of the nullnorm F_{int} is a generalization of the class of the nullnorm $V_\wedge$. Moreover, the nullnorms F_{int} and $V_\wedge$ do not need to coincide with each other. We provide the following example to demonstrate this observation.

Example 6. Consider the lattice L_2 with the given order in Fig. 2 and the interior operator $int : L_2 \rightarrow L_2$ defined by $int(0) = 0$, $int\,(v) = v$, $int\,(p) = int\,(q)$

Table 5. Nullnorm F_{int} on L_2

F_{int}	0	v	n	s	t	p	q	r	m	u	1
0	0	v	n	s	v	v	v	v	s	s	s
v	v	v	n	s	v	v	v	v	s	s	s
n	n	n	n	s	n	n	n	n	s	s	s
s	s	s	s	s	s	s	s	s	s	s	s
t	v	v	n	s	t	t	t	t	m	u	u
p	v	v	n	s	t	t	t	t	m	u	u
q	v	v	n	s	t	t	t	t	m	u	u
r	v	v	n	s	t	t	t	r	m	u	u
m	s	s	s	s	m	m	m	m	m	m	m
u	s	s	s	s	u	u	u	u	m	u	u
1	s	s	s	s	u	u	u	u	m	u	1

Table 6. Nullnorm $V_\wedge$ on L_2

$V_\wedge$	0	v	n	s	t	p	q	r	m	u	1
0	0	v	n	s	v	v	v	v	s	s	s
v	v	v	n	s	v	v	v	v	s	s	s
n	n	n	n	s	n	n	n	n	s	s	s
s	s	s	s	s	s	s	s	s	s	s	s
t	v	v	n	s	t	t	t	t	m	u	u
p	v	v	n	s	t	p	t	p	m	u	u
q	v	v	n	s	t	t	q	q	m	u	u
r	v	v	n	s	t	p	q	r	m	u	u
m	s	s	s	s	m	m	m	m	m	m	m
u	s	s	s	s	u	u	u	u	m	u	u
1	s	s	s	s	u	u	u	u	m	u	1

$= int(t) = t$, $int(m) = m$, $int(r) = r$, $int(n) = int(s) = n$ and $int(u) = int(1) = u$. Then, by use of the approaches in Theorem 3 and [28, Theorem 2], respectively, the nullnorms F_{int}, $V_\wedge : L_2 \times L_2 \to L_2$ are defined in Tables 5 and 6, respectively. These nullnorms are different from each other since $F_{int}(q,r) = t \neq q = V_\wedge(q,r)$ for $q, r \in L_2$.

Remark 3. It should be noted that the restriction of the nullnorms F_{cl} and F_{int} on $[0,s]^2$ is the t-conorm $S_\vee : [0,s]^2 \to [0,s]$ defined by $S_\vee(x,y) = x \vee y$ for all $x, y \in [0,s]$. However, F_{cl} and F_{int} do not need to coincide with another predefined t-conorm except for the t-conorm $S_\vee$ on $[0,s]^2$. To illustrate this argument, considering the lattice $L_5 - \{0,k,t,m,s,n,1\}$ according to the lattice diagram shown in Fig. 5, we assume that the restriction of the nullnorms F_{cl} and F_{int} on $[0,s]^2$ is the t-conorm $S : [0,s]^2 \to [0,s]$ given as in Table 7. Then, by applying the construction approaches in Theorems 1 and 3, we have $F_{cl}(F_{cl}(t,k),n) = m$ $(F_{int}(F_{int}(t,k),n) = m)$ and $F_{cl}(t,F_{cl}(k,n)) = s$ $(F_{int}(t,F_{int}(k,n)) = s)$. Since F_{cl} and F_{int} do not satisfy associativity property, we cannot force F_{cl} and F_{int} to coincide with another prescribed t-conorm except for the t-conorm $S_\vee$ on $[0,s]^2$.

Similarly, we note that the restriction of the nullnorms F_{cl} and F_{int} on $[s,1]^2$ is the t-norm $T_\wedge : [s,1]^2 \to [s,1]$ defined by $T_\wedge(x,y) = x \wedge y$ for all $x, y \in [s,1]$. However, F_{cl} and F_{int} do not need to coincide with another predefined t-norm except for the t-norm $T_\wedge$ on $[s,1]^2$. To illustrate this observation, we consider the lattice $L_6 = \{0,k,s,t,r,m,n,1\}$ according to the lattice diagram shown in Fig. 6 and assume that the restriction of the nullnorms F_{cl} and F_{int} on $[s,1]^2$ is the t-norm $T : [s,1]^2 \to [s,1]$ given as in Table 8. Then, by means of the construction approaches in Theorems 1 and 3, we have $F_{cl}(F_{cl}(m,n),k) = r$ $(F_{int}(F_{int}(m,n),k) = r)$ and $F_{cl}(m,F_{cl}(n,k)) = t$ $(F_{int}(m,F_{int}(n,k)) = t)$.

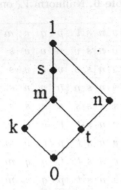

Fig. 5. Lattice L_5

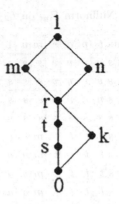

Fig. 6. Lattice L_6

Table 7. T-conorm S on $[0, s]^2$

S	0	k	t	m	s
0	0	k	t	m	s
k	k	k	m	m	s
t	t	m	s	s	s
m	m	m	s	s	s
s	s	s	s	s	s

Table 8. T-norm T on $[s, 1]^2$

T	s	t	r	m	n	1
s	s	s	s	s	s	s
t	s	t	t	t	t	t
r	s	t	t	t	t	r
m	s	t	t	r	r	m
n	s	t	t	r	r	n
1	s	t	r	m	n	1

Since F_{cl} and F_{int} do not satisfy associativity property, we cannot force F_{cl} and F_{int} to coincide with another prescribed t-norm except for the t-norm $T_\wedge$ on $[s, 1]^2$.

4   Concluding Remarks

Following the characterization of nullnorms on the real unit interval $[0, 1]$, the structure of nullnorms concerning algebraic structures on bounded lattices has attracted researchers' attention. The definition of nullnorms was extended to bounded lattices by Karaçal et al. [18]. They also demonstrated the presence of nullnorms based on a t-norm and a t-conorm on bounded lattices. Some further methods for constructing nullnorms (in particular, idempotent nullnorms) on a bounded lattice were introduced in the papers [2, 6–9, 28]. In this paper, we continued to investigate the methods for obtaining new classes of nullnorms on bounded lattices with the zero element different from the bottom and top elements. More particularly, by using the existence of closure operators and interior operators on a bounded lattice L, we proposed two different construction methods for nullnorms on L with the zero element $s \in L\backslash\{0,1\}$ under some additional conditions. We also pointed out that our constructions encompass as a special case the ones in [28]. The proposed constructions for nullnorms both in this paper and in [28] coincide with the supremum t-conorm $S_\vee$ on $[0,s]^2$ and the infimum t-norm $T_\wedge$ on $[s,1]^2$. We demonstrated that these classes of nullnorms do not need to coincide with another predefined t-conorm except for $S_\vee$ on $[0,s]^2$ and t-norm except for $T_\wedge$ on $[s,1]^2$. Moreover, some specific examples were presented to illustrate more clearly new methods of nullnorms on bounded lattices.

References

1. Birkhoff, G.: Lattice Theory. American Mathematical Society Colloquium Publishers, Providence (1967)
2. Bodjanova, S., Kalina, M.: Nullnorms and T-operators on bounded lattices: coincidence and differences. In: Medina, J., et al. (eds.) IPMU 2018. CCIS, vol. 853, pp. 160–170. Springer, Cham (2018). https://doi.org/10.1007/978-3-319-91473-2_14
3. Bodjanova, S., Kalina, M.: Uninorms on bounded lattices with given underlying operations. In: Halaš, R., Gagolewski, M., Mesiar, R. (eds.) AGOP 2019. AISC, vol. 981, pp. 183–194. Springer, Cham (2019). https://doi.org/10.1007/978-3-030-19494-9_17
4. Calvo, T., De Baets, B., Fodor, J.: The functional equations of Frank and Alsina for uninorms and nullnorms. Fuzzy Sets Syst. **120**, 385–394 (2001)
5. Calvo, T., Kolesárová, A., Komorníková, M., Mesiar, R.: Aggregation operators: properties, classes and construction methods. In: Calvo, T., et al. (eds.) Aggregation Operators, New Trends and Applications, pp. 3–104. Physica, Heidelberg (2002). https://doi.org/10.1007/978-3-7908-1787-4_1
6. Çaylı, G.D., Karaçal, F.: Idempotent nullnorms on bounded lattices. Inf. Sci. **425**, 153–164 (2018)
7. Çaylı, G.D.: Construction methods for idempotent nullnorms on bounded lattices. Appl. Math. Comput. **366**, 124746 (2020)
8. Çaylı, G.D.: Nullnorms on bounded lattices derived from t-norms and t-conorms. Inf. Sci. **512**, 1134–1154 (2020)
9. Çaylı, G.D.: Some results about nullnorms on bounded lattices. Fuzzy Sets Syst. **386**, 105–131 (2020)
10. Drewniak, J., Drygaś, P., Rak, E.: Distributivity between uninorms and nullnorms. Fuzzy Sets Syst. **159**, 1646–1657 (2008)
11. Drossos, C.A.: Generalized t-norm structures. Fuzzy Sets Syst. **104**, 53–59 (1999)
12. Drossos, C.A., Navara, M.: Generalized t-conorms and closure operators. In: Proceedings of EUFIT 1996, Aachen, pp. 22–26 (1996)
13. Drygaś, P.: Isotonic operations with zero element in bounded lattices. In: Atanassov, K., et al. (eds.) Soft Computing Foundations and Theoretical Aspect, EXIT Warszawa, pp. 181–190 (2004)
14. Drygaś, P.: A characterization of idempotent nullnorms. Fuzzy Sets Syst. **145**, 455–461 (2004)
15. Engelking, R.: General Topology. Heldermann Verlag, Berlin (1989)
16. Everett, C.J.: Closure operators and Galois theory in lattices. Trans. Am. Math. Soc. **55**, 514–525 (1944)

17. Grabisch, M., Marichal, J.L., Mesiar, R., Pap, E.: Aggregation Functions. Cambridge University Press, Cambridge (2009)
18. Karaçal, F., İnce, M.A., Mesiar, R.: Nullnorms on bounded lattices. Inf. Sci. **325**, 227–236 (2015)
19. Li, G., Liu, H.W., Su, Y.: On the conditional distributivity of nullnorms over uninorms. Inf. Sci. **317**, 157–169 (2015)
20. Mas, M., Mayor, G., Torrens, J.: T-operators. Int. J. Uncertain Fuzziness Knowl. Based Syst. **7**, 31–50 (1996)
21. Mas, M., Mayor, G., Torrens, J.: The distributivity condition for uninorms and t-operators. Fuzzy Sets Syst. **128**, 209–225 (2002)
22. Mas, M., Mayor, G., Torrens, J.: The modularity condition for uninorms and t-operators. Fuzzy Sets Syst. **126**, 207–218 (2002)
23. Mas, M., Mesiar, R., Monserrat, M., Torrens, J.: Aggregation operators with annihilator. Int. J. Gen. Syst. **34**, 1–22 (2005)
24. Mesiar, R., Kolesárová, A., Stupnanová, A.: Quo vadis aggregation. Int. J. Gen. Syst. **47**, 97–117 (2018)
25. Ouyang, Y., Zhang, H.P.: Constructing uninorms via closure operators on a bounded lattice. Fuzzy Sets Syst. https://doi.org/10.1016/j.fss.2019.05.006
26. Qin, F., Wang, Y.M.: Distributivity between semi-t-operators and Mayor's aggregation operators. Inf. Sci. **336–337**, 6–16 (2016)
27. Xie, A., Liu, H.: On the distributivity of uninorms over nullnorms. Fuzzy Sets Syst. **211**, 62–72 (2013)
28. Wang, Y.M., Zhan, H., Liu, H.W.: Uni-nullnorms on bounded lattices. Fuzzy Sets Syst. **386**, 132–144 (2020)
29. Zong, W., Su, Y., Liu, H.W., De Baets, B.: On the structure of 2-uninorms. Inf. Sci. **467**, 506–527 (2016)

General Grouping Functions

Helida Santos[1]([✉])[iD], Graçaliz P. Dimuro[1,5][iD], Tiago C. Asmus[2,5][iD],
Giancarlo Lucca[3][iD], Eduardo N. Borges[1][iD], Benjamin Bedregal[4][iD],
José A. Sanz[5,6][iD], Javier Fernández[5,6][iD], and Humberto Bustince[5,6,7][iD]

[1] Centro de Ciências Computacionais, Universidade Federal do Rio Grande,
Rio Grande, Brazil
{helida,gracalizdimuro,eduardoborges}@furg.br
[2] Instituto de Matemática, Estatística e Física, Universidade Federal do Rio Grande,
Rio Grande, Brazil
tiagoasmus@furg.br
[3] Programa de Pós-Graduação em Modelagem Computacional,
Universidade Federal do Rio Grande, Rio Grande, Brazil
giancarlo.lucca@furg.br
[4] Departamento de Informática e Matemática Aplicada,
Universidade Federal do Rio Grande do Norte, Natal, Brazil
bedregal@dimap.ufrn.br
[5] Departamento de Estadistica, Informatica y Matematicas,
Universidad Publica de Navarra, Pamplona, Spain
{joseantonio.sanz,fcojavier.fernandez,bustince}@unavarra.es
[6] Institute of Smart Cities, Universidad Publica de Navarra, Pamplona, Spain
[7] King Abdulaziz University, Jeddah, Saudi Arabia

Abstract. Some aggregation functions that are not necessarily associative, namely overlap and grouping functions, have called the attention of many researchers in the recent past. This is probably due to the fact that they are a richer class of operators whenever one compares with other classes of aggregation functions, such as t-norms and t-conorms, respectively. In the present work we introduce a more general proposal for disjunctive n-ary aggregation functions entitled general grouping functions, in order to be used in problems that admit n dimensional inputs in a more flexible manner, allowing their application in different contexts. We present some new interesting results, like the characterization of that operator and also provide different construction methods.

Keywords: Grouping functions · n-dimensional grouping functions ·
General grouping functions · General overlap functions

1 Introduction

Overlap functions are a kind of aggregation functions [3] that are not required to be associative, and they were introduced by Bustince et al. in [4] to measure the degree of overlapping between two classes or objects. Grouping functions

M.-J. Lesot et al. (Eds.): IPMU 2020, CCIS 1238, pp. 481–495, 2020.
https://doi.org/10.1007/978-3-030-50143-3_38

are the dual notion of overlap functions. They were introduced by Bustince et al. [5] in order to express the measure of the amount of evidence in favor of either of two alternatives when performing pairwise comparisons [1] in decision making based on fuzzy preference relations [6]. They have also been used as the disjunction operator in some important problems, such as image thresholding [17] and the construction of a class of implication functions for the generation of fuzzy subsethood and entropy measures [13].

Overlap and grouping functions have been largely studied since they are richer than t-norms and t-conorms [18], respectively. Regarding, for instance, some properties like idempotency, homogeneity, and, mainly, the self-closeness feature with respect to the convex sum and the aggregation by generalized composition of overlap/grouping functions [7,8,10,12]. For example, there is just one idempotent t-conorm (namely, the maximum t-conorm) and two homogeneous t-conorms (namely, the maximum and the probabilistic sum of t-conorms). On the contrary, there are uncountable numbers of idempotent, as well as homogenous, grouping functions [2,11]. For comparisons among properties of grouping functions and t-conorms, see [2,5,17]

However, grouping functions are bivariate functions. Since they may be non associative, they can only be applied in bi-dimensional problems (that is, when just two classes or objects are considered). In order to solve this drawback, Gómez et al. [16] introduced the concept of n-dimensional grouping functions, with an application to fuzzy community detection.

Recently, De Miguel et al. [20] introduced general overlap functions, by relaxing some boundary conditions, in order to apply to an n-ary problem, namely, fuzzy rule based classification systems, more specifically, in the determination of the matching degree in the fuzzy reasoning method. Then, inspired on the paper by De Miguel et al. [20], the objective of this present paper is to introduce the concept of general grouping functions, providing their characterization and different construction methods. The aim is to define the theoretical basis of a tool that can be used to express the measure of the amount of evidence in favor of one of multiple alternatives when performing n-ary comparisons in multi-criteria decision making based on n-ary fuzzy heterogeneous, incomplete preference relations [14,19,26], which we let for future work.

The paper is organized as follows. Section 2 presents some preliminary concepts. In Sect. 3, we define general grouping functions, studying some properties. In Sect. 4, we study the characterization of general grouping functions, providing some construction methods. Section 5 is the Conclusion.

2   Preliminary Concepts

In this section, we highlight some relevant concepts used in this work.

Definition 1. *A function* $N \colon [0,1] \to [0,1]$ *is a fuzzy negation if it holds:* (N1) *N is antitonic, i.e.* $N(x) \leq N(y)$ *whenever* $y \leq x$ *and* (N2) $N(0) = 1$ *and* $N(1) = 0.$

Definition 2. *[3] An n-ary aggregation function is a mapping $A\colon [0,1]^n \to [0,1]$ satisfying: (A1) $A(0,\ldots,0) = 0$ and $A(1,\ldots,1) = 1$; (A2) increasingness: for each $i \in \{1,\ldots,n\}$, if $x_i \leq y$ then $A(x_1,\ldots,x_n) \leq A(x_1,\ldots,x_{i-1},y,x_{i+1},\ldots,x_n)$.*

Definition 3. *An n-ary aggregation function $A\colon [0,1]^n \to [0,1]$ is called conjunctive if, for any $\vec{x} = (x_1,\ldots,x_n) \in [0,1]^n$, it holds that $A(\vec{x}) \leq \min(\vec{x}) = \min\{x_1,\ldots,x_n\}$. And A is called disjunctive if, for any $\vec{x} = (x_1,\ldots,x_n) \in [0,1]^n$, it holds that $A(\vec{x}) \geq \max(\vec{x}) = \max\{x_1,\ldots,x_n\}$.*

Definition 4. *[4] A binary function $O\colon [0,1]^2 \to [0,1]$ is said to be an overlap function if it satisfies the following conditions, for all $x,y,z \in [0,1]$:*

(O1) *$O(x,y) = O(y,x)$;*
(O2) *$O(x,y) = 0$ if and only if $x = 0$ or $y = 0$;*
(O3) *$O(x,y) = 1$ if and only if $x = y = 1$;*
(O4) *if $x \leq y$ then $O(x,z) \leq O(y,z)$;*
(O5) *O is continuous;*

Definition 5. *[5] A binary function $G\colon [0,1]^2 \to [0,1]$ is said to be a grouping function if it satisfies the following conditions, for all $x,y,z \in [0,1]$:*

(G1) *$G(x,y) = G(y,x)$;*
(G2) *$G(x,y) = 0$ if and only if $x = y = 0$;*
(G3) *$G(x,y) = 1$ if and only if $x = 1$ or $y = 1$;*
(G4) *If $x \leq y$ then $G(x,z) \leq G(y,z)$;*
(G5) *G is continuous;*

For all properties and related concepts on overlap functions and grouping functions, see [2,5,7,9,10,21,23–25].

Definition 6. *[22] A function $G\colon [0,1]^2 \to [0,1]$ is a 0-grouping function if the second condition in Definition 5 is replaced by: (G2') If $x = y = 0$ then $G(x,y) = 0$. Analogously, a function $G\colon [0,1]^2 \to [0,1]$ is a 1-grouping function if the third condition in Definition 5 is replaced by: (G3') If $x = 1$ or $y = 1$ then $G(x,y) = 1$.*

Both notions were extended in several ways and some of them are presented in the following definitions.

Definition 7. *[15] An n-ary function $\mathcal{G}\colon [0,1]^n \to [0,1]$ is called an n-dimensional grouping function if for all $\vec{x} = (x_1,\ldots,x_n) \in [0,1]^n$:*

1. *$\mathcal{G}$ is commutative;*
2. *$\mathcal{G}(\vec{x}) = 0$ if and only if $x_i = 0$, for all $i = 1,\ldots,n$;*
3. *$\mathcal{G}(\vec{x}) = 1$ if and only if there exists $i \in \{1,\ldots,n\}$ with $x_i = 1$;*
4. *$\mathcal{G}$ is increasing;*
5. *$\mathcal{G}$ is continuous.*

Definition 8. *[20]* *A function* $\mathcal{GO}\colon [0,1]^n \to [0,1]$ *is said to be a general over-lap function if it satisfies the following conditions, for all* $\vec{x} = (x_1,\ldots,x_n) \in [0,1]^n$:

$(\mathcal{GO}1)$ $\mathcal{GO}$ *is commutative;*

$(\mathcal{GO}2)$ *If* $\prod_{i=1}^n x_i = 0$ *then* $\mathcal{GO}(\vec{x}) = 0$;

$(\mathcal{GO}3)$ *If* $\prod_{i=1}^n x_i = 1$ *then* $\mathcal{GO}(\vec{x}) = 1$;

$(\mathcal{GO}4)$ $\mathcal{GO}$ *is increasing;*

$(\mathcal{GO}5)$ $\mathcal{GO}$ *is continuous.*

3   General Grouping Functions

Following the ideas given in [20], we can also generalize the idea of general grouping functions as follows.

Definition 9. *A function* $\mathcal{GG}\colon [0,1]^n \to [0,1]$ *is called a general grouping function if the following conditions hold, for all* $\vec{x} = (x_1,\ldots,x_n) \in [0,1]^n$:

$(\mathcal{GG}1)$ $\mathcal{GG}$ *is commutative;*

$(\mathcal{GG}2)$ *If* $\sum_{i=1}^n x_i = 0$ *then* $\mathcal{GG}(\vec{x}) = 0$;

$(\mathcal{GG}3)$ *If there exists* $i \in \{1,\ldots,n\}$ *such that* $x_i = 1$ *then* $\mathcal{GG}(\vec{x}) = 1$;

$(\mathcal{GG}4)$ $\mathcal{GG}$ *is increasing;*

$(\mathcal{GG}5)$ $\mathcal{GG}$ *is continuous.*

Note that $(\mathcal{GG}2)$ is the same of saying that 0 is an anhilator of the general grouping function $\mathcal{GG}$.

Proposition 1. *If* $\mathcal{G}\colon [0,1]^n \to [0,1]$ *is an* n-*dimensional grouping function, then* $\mathcal{G}$ *is also a general grouping function.*

Proof. Straighforward.   □

From this proposition, we can conclude that the concept of general grouping functions is a generalization of n-dimensional grouping functions, which on its turn is a generalization of the concepts of 0-grouping functions and 1-grouping functions.

Example 1. 1. Every grouping function $G\colon [0,1]^2 \to [0,1]$ is a general grouping function, but the converse does not hold.
2. The function $\mathcal{GG}(x,y) = \min\{1, 2 - (1-x)^2 - (1-y)^2\}$ is a general grouping function, but it is not a bidimensional grouping function, since $\mathcal{GG}(0.5, 0.5) = 1$.
3. Consider $G(x,y) = \max\{1 - (1-x)^p, 1 - (1-y)^p\}$, for $p > 0$ and $S_{\mathcal{L}}(x,y) = \min\{1, x+y\}$. Then, the function $\mathcal{GG}^{S_{\mathcal{L}}}(x,y) = G(x,y)S_{\mathcal{L}}(x,y)$ is a general grouping function.

4. Take any grouping function G, and a continuous t-conorm S. Then, the generalization of the previous item is the binary general grouping function given by: $GG(x, y) = G(x, y)S(x, y)$
5. Other examples are:

$$Prod\_S\_Luk(x_1, \ldots, x_n) = \left(1 - \prod_{i=1}^{n}(1 - x_i)\right) * \left(\min\left\{\sum_{i=1}^{n} x_i, 1\right\}\right)$$

$$GM\_S\_Luk(x_1, \ldots, x_n) = \left(1 - \sqrt[n]{\prod_{i=1}^{n}(1 - x_i)}\right) * \left(\min\left\{\sum_{i=1}^{n} x_i, 1\right\}\right).$$

The generalization of the third item of Example 1 can be seen as follows.

Proposition 2. *Take any grouping function G, and any t-conorm S. Then, the binary general grouping function given by: $GG(x, y) = G(x, y)S(x, y)$.*

Proposition 3. *Let $F: [0, 1]^n \to [0, 1]$ be a commutative and continuous aggregation function. Then the following statements hold:*

(i) *If F is disjunctive, then F is a general grouping function.*
(ii) *If F is conjunctive, then F is neither a general grouping function nor an n-dimensional grouping function.*

Proof. Consider a commutative and continuous aggregation function $F: [0, 1]^n \to [0, 1]$. It follows that:
(i) Since F is commutative ($GG1$), continuous ($GG5$) and clearly increasing ($GG4$), then it remains to prove the following:
($GG2$) Suppose that $\sum_{i=1}^{n} x_i = 0$. Then, since F is an aggregation function, it holds that $F(0, \ldots, 0) = 0$.
($GG3$) Suppose that, for some $\vec{x} - (x_1, \ldots, x_n) \subset [0, 1]^n$, there exists $i \in \{1, \ldots, n\}$ such that $x_i = 1$. Then, since F is disjunctive, then $F(\vec{x}) \geq \max\{x_1, \ldots, 1, \ldots, x_n\} = 1$, which means that $F(\vec{x}) = 1$.
(ii) Suppose that F is a conjunctive aggregation function and it is either a general grouping function or an n-dimensional grouping function. Then, by either ($GG3$) or ($G3$), if for some $\vec{x} = (x_1 \ldots, x_n) \in [0, 1]^n$, there exists $i \in \{1, \ldots, n\}$ such that $x_i = 1$, then $F(\vec{x}) = 1$. Take $\vec{x} = (1, 0 \ldots, 0)$, it follows that $F(1, 0 \ldots, 0) = 1 = \max\{1, 0 \ldots, 0\} \nleq 0 = \min\{1, 0 \ldots, 0\}$, which is a contradiction with the conjunctive property. Thus, one concludes that F is neither a general grouping function nor an n-dimensional grouping function. □

We say that an element $a \in [0, 1]$ is a neutral element of GG if for each $x \in [0, 1]$, $GG(x, \underbrace{a, \ldots, a}_{(n-1)}) = x$.

Proposition 4. *Let $GG: [0, 1]^n \to [0, 1]$ be a general grouping function with a neutral element $a \in [0, 1]$. Then, $a = 0$ if and only if GG satisfies, for all $\vec{x} = (x_1, \ldots, x_n) \in [0, 1]^n$, the following condition:*

$(\mathcal{GG}2')$ If $\mathcal{GG}(\vec{x}) = 0$, then $\sum_{i=1}^{n} x_i = 0$.

Proof. ($\Rightarrow$) Suppose that (i) the neutral element of $\mathcal{GG}$ is $a = 0$ and (ii) $\mathcal{GG}(x_1, \ldots, x_n) = 0$. Then, by (i), one has that, for each $x_1 \in [0,1]$, it holds that $x_1 = \mathcal{GG}(x_1, 0 \ldots, 0)$. By (ii) and since $\mathcal{GG}$ is increasing, it follows that

$$x_1 = \mathcal{GG}(x_1, 0 \ldots, 0) \leq \mathcal{GG}(x_1, \ldots, x_n) = 0.$$

Similarly, one shows that $x_2, \ldots, x_n = 0$, that is $\sum_{i=1}^{n} x_i = 0$.

($\Leftarrow$) Suppose that $\mathcal{GG}$ satisfies $(\mathcal{GG}2')$ and that $\mathcal{GG}(x_1, \ldots, x_n) = 0$, for $(x_1, \ldots, x_n) \in [0,1]^n$. Then, by $(\mathcal{GG}2')$, it holds that $\sum_{i=1}^{n} x_i = 0$. Since a is the neutral element of $\mathcal{GG}$, one has that $\mathcal{GG}(0, a, \ldots, a) = 0$, which means that $a = 0$, by $(\mathcal{GG}2')$.   □

Remark 1. Observe that the result stated by Proposition 4 does not mean that when a general grouping function has a neutral element, then it is necessarily equal to 0. In fact, for each $a \in (0,1)$, the function $\mathcal{GG}: [0,1]^n \to [0,1]$, for all $\vec{x} = (x_1 \ldots, x_n) \in [0,1]^n$, defined by:

$$\mathcal{GG}(\vec{x}) = \begin{cases} \min\{\vec{x}\}, & \text{if } \max\{\vec{x}\} \leq a \\ \max\{\vec{x}\}, & \text{if } \min\{\vec{x}\} \geq a \\ \frac{\min\{\vec{x}\} + \max\{\vec{x}\}(1 - \min\{\vec{x}\}) - a}{1 - a}, & \text{if } \min\{\vec{x}\} < a < \max\{\vec{x}\} \end{cases}$$

is a general grouping function with a as neutral element.

Proposition 5. *If 0 is the neutral element of a general grouping function* $\mathcal{GG}: [0,1]^n \to [0,1]$ *and $\mathcal{GG}$ is idempotent, then $\mathcal{GG}$ is the maximum.*

Proof. Since $\mathcal{GG}$ is idempotent and increasing in each argument, then one has that for all $\vec{x} = (x_1, \ldots, x_n) \in [0,1]^n$: **(1)** $\mathcal{GG}(x_1, \ldots, x_n) \leq \mathcal{GG}(\max(\vec{x}), \ldots, \max(\vec{x})) = \max\{\vec{x}\}$. Then we have that $x_k = \max\{\vec{x}\}$ for some $k = 1, \ldots, n$; so we have $x_k = \mathcal{GG}(0, \ldots, x_k, \ldots, 0) \leq \mathcal{GG}(x_1, \ldots, x_k, \ldots, x_n)$ and then **(2)** $\mathcal{GG}(x_1, \ldots, x_n) \geq x_k = \max\{\vec{x}\}$. So, from **(1)** and **(2)** one has that $\mathcal{GG}(x_1, \ldots, x_n) = \max\{\vec{x}\}$, for each $\vec{x} \in [0,1]^n$.   □

3.1 General Grouping Functions on Lattices

Following a similar procedure described in [20] for general overlap functions on lattices, it is possible to characterize general grouping functions. In order to do that, first we introduce some properties and notations.

Let us denote by $\mathfrak{G}^n$ the set of all general grouping functions. Define the ordering relation $\leq_{\mathfrak{G}^n} \in \mathfrak{G}^n \times \mathfrak{G}^n$, for all $\mathcal{GG}_1, \mathcal{GG}_2 \in \mathfrak{G}^n$ by:

$$\mathcal{GG}_1 \leq_{\mathfrak{G}^n} \mathcal{GG}_2 \Leftrightarrow \mathcal{GG}_1(\vec{x}) \leq \mathcal{GG}_2(\vec{x}), \text{ for all } \vec{x} = (x_1, \ldots, x_n) \in [0,1]^n.$$

The supremum and infimum of two arbitrary general grouping functions $\mathcal{GG}_1, \mathcal{GG}_2 \in \mathfrak{G}^n$ are, respectively, the general grouping functions $\mathcal{GG}_1 \vee \mathcal{GG}_2, \mathcal{GG}_1 \wedge \mathcal{GG}_2 \in \mathfrak{G}^n$, defined, for all $\vec{x} = (x_1, \ldots, x_n) \in [0,1]^n$ by: $\mathcal{GG}_1 \vee \mathcal{GG}_2(\vec{x}) = \max\{\mathcal{GG}_1(\vec{x}), \mathcal{GG}_2(\vec{x})\}$ and $\mathcal{GG}_1 \wedge \mathcal{GG}_2(\vec{x}) = \min\{\mathcal{GG}_1(\vec{x}), \mathcal{GG}_2(\vec{x})\}$.

The following result is immediate:

Theorem 1. *The ordered set $(\mathfrak{G}^n, \leq_{\mathfrak{G}^n})$ is a lattice.*

Remark 2. Note that the supremum of the lattice $(\mathfrak{G}^n, \leq_{\mathfrak{G}^n})$ is given, for all $\vec{x} = (x_1, \ldots, x_n) \in [0,1]^n$, by:

$$GG_{\text{sup}}(\vec{x}) = \begin{cases} 0, & \text{if } \sum_{i=1}^{n} x_i = 0 \\ 1, & \text{otherwise.} \end{cases}$$

On the other hand, the infimum of $(\mathfrak{G}^n, \leq_{\mathfrak{G}^n})$ is given, for all $\vec{x} = (x_1, \ldots, x_n) \in [0,1]^n$, by:

$$GG_{\text{inf}}(\vec{x}) = \begin{cases} 1, & \text{if } \exists i \in \{1, \ldots, n\} : x_i = 1 \\ 0, & \text{otherwise.} \end{cases}$$

However, neither GG_{sup} nor GG_{inf} are general grouping functions, since they are not continuous. Thus, in the lattice $(\mathfrak{G}^n, \leq_{\mathfrak{G}^n})$ there is no bottom neither top elements. Then, similarly to general overlap functions, the lattice $(\mathfrak{G}^n, \leq_{\mathfrak{G}^n})$ is not complete.

4   Characterization of General Grouping Functions and Construction Methods

In this section we provide a characterization and some constructions methods for general grouping functions.

Theorem 2. *The mapping $\mathcal{GG}: [0,1]^n \to [0,1]$ is a general grouping function if and only if*

$$\mathcal{GG}(\vec{x}) = \frac{f(\vec{x})}{f(\vec{x}) + h(\vec{x})} \tag{1}$$

for some $f, h: [0,1]^n \to [0,1]$ the following properties hold, for all $\vec{x} \in [0,1]^n$:

(i) *f and h are commutative;*
(ii) *f is increasing and h is decreasing.*
(iii) *If $\sum_{i=1}^{n} x_i = 0$, then $f(\vec{x}) = 0$.*
(iv) *If there exists $i \in \{1, \ldots, n\}$ such that $x_i = 1$, then $h(\vec{x}) = 0$.*
(v) *f and h are continuous.*
(vi) *$f(\vec{x}) + h(\vec{x}) \neq 0$ for any $\vec{x} \in [0,1]^n$.*

Proof. It follows that:

($\Rightarrow$) Suppose that $\mathcal{GG}$ is a general grouping function, and take $f(\vec{x}) = \mathcal{GG}(\vec{x})$ and $h(\vec{x}) = 1 - f(\vec{x})$. Then one always have $f(\vec{x}) + h(\vec{x}) \neq 0$, and so Equation (1) is well defined. Also, conditions (i)–(v) trivially hold.

($\Leftarrow$) Consider $f, h \colon [0,1]^n \to [0,1]$ satisfying conditions (i)–(v). We will show that $\mathcal{GG}$ defined according to Eq. (1) is a general grouping function. It is immediate that $\mathcal{GG}$ is commutative ($\mathcal{GG}1$) and continuous ($\mathcal{GG}5$). To prove ($\mathcal{GG}2$), notice that if $\sum_{i=1}^{n} x_i = 0$ then $f(\vec{x}) = 0$ and thus $\mathcal{GG}(\vec{x}) = 0$. Now, let us prove that ($\mathcal{GG}3$) holds. For that, observe that if there exists $i \in \{1, \ldots, n\}$ such that $x_i = 1$, then $h(\vec{x}) = 0$, and, thus, it is immediate that $\mathcal{GG}(\vec{x}) = 1$. The proof of ($\mathcal{GG}4$) is similar to [20, Theorem 3]. $\qquad\square$

Example 2. Observe that Theorem 2 provides a method for constructing general grouping functions. For example, take the maximum powered by p, defined by:

$$\max^p(\vec{x}) = \max_{1 \leq i \leq n}\{x_i^p\},$$

with $p > 0$. So, if we take the function $T\max_\alpha^p \colon [0,1]^n \to [0,1]$, called α-truncated maximum powered by p, given, for all $\vec{x} \in [0,1]^n$ and $\alpha \in (0,1)$, by:

$$T\max_\alpha^p(\vec{x}) = \begin{cases} 0, & \text{if } \max^p(\vec{x}) \leq \alpha \\ \max^p(\vec{x}), & \text{if } \max^p(\vec{x}) > \alpha \end{cases} \qquad (2)$$

then it is clear that $T\max_\alpha^p$ is not continuous. However, one can consider the function $CT\max_{\alpha,\epsilon}^p \colon [0,1]^n \to [0,1]$, called the continuous truncated maximum powered by p, for all $\vec{x} \in [0,1]^n$, $\alpha \in [0,1]$ and $\epsilon \in (0,\alpha]$, which is defined by:

$$CT\max_{\alpha,\epsilon}^p(\vec{x}) = \begin{cases} 0, & \text{if } \max^p(\vec{x}) \leq \alpha - \epsilon \\ \frac{\alpha}{\epsilon}(\max^p(\vec{x}) - (\alpha - \epsilon)), & \text{if } \alpha - \epsilon < \max^p(\vec{x}) < \alpha \\ \max^p(\vec{x}), & \text{if } \max^p(\vec{x}) \geq \alpha. \end{cases} \qquad (3)$$

Observe that taking $f = CT\max_{\alpha,\epsilon}^p$, then f satisfies conditions (i)–(iii) and (v) in Theorem 2. Now, take $h(\vec{x}) = \min_{1 \leq i \leq n}\{1 - x_i\}$, which satisfies conditions (i)–(ii) and (iv)–(v) required in Theorem 2. Thus, this assures that

$$\mathcal{GG}(\vec{x}) = \frac{CT\max_{\alpha,\epsilon}^p(\vec{x})}{CT\max_{\alpha,\epsilon}^p(\vec{x}) + \min_{1 \leq i \leq n}\{1 - x_i\}}$$

is a general grouping function.

Remark 3. Observe that the maximum powered by p is an n-dimensional grouping function [15] and that $CT\max_{\alpha,\epsilon}^p$ is a general grouping function. However, $CT\max_{\alpha,\epsilon}^p$ is not an n-dimensional grouping function, for $\alpha - \epsilon > 0$, since $CT\max_{\alpha,\epsilon}^p(\alpha - \epsilon, \ldots, \alpha - \epsilon) = 0$.

Corollary 1. *Consider the functions $f, h: [0,1]^n \to [0,1]$ and let $\mathcal{GG}: [0,1]^n \to [0,1]$ be a general grouping function constructed according to Theorem 2, and taking into account functions f and h. Then $\mathcal{GG}$ is idempotent if and only if, for all $x \in [0,1)$, it holds that:*

$$f(x, \ldots, x) = \frac{x}{1-x} h(x, \ldots, x).$$

Proof. ($\Rightarrow$) If $\mathcal{GG}$ is idempotent, then by Theorem 2 it holds that:

$$\mathcal{GG}(x, \ldots, x) = \frac{f(x, \ldots, x)}{f(x, \ldots, x) + h(x, \ldots, x)} = x.$$

It follows that: $f(x, \ldots, x) = x(f(x, \ldots, x) + h(x, \ldots, x))$

$$(1-x)f(x, \ldots, x) = x\, h(x, \ldots, x)$$
$$f(x, \ldots, x) = \frac{x}{1-x}\, h(x, \ldots, x).$$

($\Leftarrow$) It is immediate. $\square$

Example 3. Take the function $\alpha\beta$-truncated maximum powered by p, $T\max_{\alpha\beta}^p$: $[0,1]^n \to [0,1]$, for all $\vec{x} \in [0,1]^n$; $\alpha, \beta \in (0,1)$ and $\alpha < \beta$, defined by:

$$T\max_{\alpha\beta}^p(\vec{x}) = \begin{cases} 0, & \max^p(\vec{x}) \leq \alpha \\ \max^p(\vec{x}), & \alpha < \max^p(\vec{x}) < \beta \\ 1, & \max^p(\vec{x}) \geq \beta \end{cases}$$

It is clear that $T\max_{\alpha\beta}^p$ is not continuous. However, we can define its continuous version, $CT\max_{\alpha\beta,\epsilon\delta}^p: [0,1]^n \to [0,1]$, for all $\vec{x} \in [0,1]^n$; $\alpha \in [0,1)$; $\beta, \epsilon, \delta \in (0,1]$; $\alpha + \epsilon, \beta - \delta \in (0,1)$ and $\alpha + \epsilon < \beta - \delta$, as follows:

$$CT\max_{\alpha\beta,\epsilon\delta}^p(\vec{x}) = \begin{cases} 0, & \max^p(\vec{x}) \leq \alpha \\ \frac{1-(\alpha+\epsilon)}{\epsilon}(\alpha - \max^p(\vec{x})), & \alpha < \max^p(\vec{x}) < \alpha + \epsilon \\ 1 - \max^p(\vec{x}), & \alpha + \epsilon \leq \max^p(\vec{x}) \leq \beta - \delta \\ 1 - (\beta - \delta) - \frac{\beta-\delta}{\delta}(\beta - \delta - \max^p(\vec{x})), & \beta - \delta < \max^p(\vec{x}) < \beta \\ 1, & \max^p(\vec{x}) \geq \beta \end{cases}$$

Observe that $CT\max_{\alpha\beta,\epsilon\delta}^p$ satisfies conditions ($\mathcal{GG}1$)-($\mathcal{GG}5$) from Definition 9, and then it is a general grouping function. But, whenever $\alpha \neq 0$ or $\beta \neq 1$, then $CT\max_{\alpha\beta,\epsilon\delta}^p$ is not an n-dimensional grouping function, once $CT\max_{\alpha\beta,\epsilon\delta}^p(\alpha - \epsilon, \ldots, \alpha - \epsilon) = 0$, for $\alpha - \epsilon > 0$, because $\max^p(\alpha - \epsilon, \ldots, \alpha - \epsilon) = \alpha - \epsilon < \alpha$.

The following Theorem generalizes Example 3 providing a construction method for general grouping functions from truncated n-dimensional grouping functions.

Theorem 3. *Consider $\alpha \in [0,1)$; $\beta, \epsilon, \delta \in (0,1]$; $\alpha + \epsilon, \beta - \delta \in (0,1)$ and $\alpha < \beta$, $\alpha + \epsilon < \beta - \delta$. Let $\mathcal{G}$ be an n-dimensional grouping function, whose $\alpha\beta$-truncated version is defined, for all $\vec{x} = (x_1, \ldots, x_n) \in [0,1]^n$, by:*

$$
T\mathcal{G}_{\alpha\beta}(\vec{x}) = \begin{cases} 0, & \mathcal{G}(\vec{x}) \leq \alpha \\ \mathcal{G}(\vec{x}), & \alpha < \mathcal{G}(\vec{x}) < \beta \\ 1, & \mathcal{G}(\vec{x}) \geq \beta \end{cases}
$$

Then, the continuous version of $T\mathcal{G}_{\alpha\beta}$, for all $\vec{x} = (x_1, \ldots, x_n) \in [0,1]^n$, is given by:

$$
CT\mathcal{G}_{\alpha\beta,\epsilon\delta}(\vec{x}) = \begin{cases} 0, & \mathcal{G}(\vec{x}) \leq \alpha \\ \frac{1-(\alpha+\epsilon)}{\epsilon}(\alpha - \mathcal{G}(\vec{x})), & \alpha < \mathcal{G}(\vec{x}) < \alpha + \epsilon \\ 1 - \mathcal{G}(\vec{x}), & \alpha + \epsilon \leq \mathcal{G}(\vec{x}) \leq \beta - \delta \\ 1 - (\beta - \delta) - \frac{\beta-\delta}{\delta}(\beta - \delta - \mathcal{G}(\vec{x})), & \beta - \delta < \mathcal{G}(\vec{x}) < \beta \\ 1, & \mathcal{G}(\vec{x}) \geq \beta \end{cases}
$$

and it is a general grouping function. Besides that, whenever $\alpha = 0$ and $\beta = 1$, then $CT\mathcal{G}_{\alpha\beta,\epsilon\delta}$ is an n-dimensional grouping function.

The following proposition shows a construction method of general grouping functions that generalizes Example 1(4).

Proposition 6. *Let $\mathcal{G} \colon [0,1]^n \to [0,1]$ be an n-dimensional grouping function and let $F \colon [0,1]^n \to [0,1]$ be a commutative and continuous aggregation function such that, for all $\vec{x} = (x_1, \ldots, x_n) \in [0,1]^n$, if there exists $i \in \{1, \ldots, n\}$ such that $x_i = 1$, then $F(\vec{x}) = 1$. Then $\mathcal{GG}_{\mathcal{G}F}(\vec{x}) = \mathcal{G}(\vec{x})F(\vec{x})$ is a general grouping function.*

Proof. It is immediate that $\mathcal{GG}_{\mathcal{G}F}$ is well defined, $(\mathcal{GG}1)$ commutative, $(\mathcal{GG}4)$ increasing and $(\mathcal{GG}5)$ continuous, since $\mathcal{G}$, F and the product operation are commutative, increasing and continuous. To prove $(\mathcal{GG}2)$, whenever $\sum_{i=1}^{n} x_i = 0$, then by $(\mathcal{G}2)$, it holds that $\mathcal{G}(\vec{x}) = 0$, and, thus, $\mathcal{GG}_{\mathcal{G}F}(\vec{x}) = \mathcal{G}(\vec{x})F(\vec{x}) = 0$. For $(\mathcal{GG}3)$, whenever there exists $i \in \{1, \ldots, n\}$ such that $x_i = 1$, then, by $(\mathcal{G}3)$, one has that $\mathcal{G}(\vec{x}) = 1$, and, by the property of F, it holds that $F(\vec{x}) = 1$. It follows that: $\mathcal{GG}_{\mathcal{G}F}(\vec{x}) = \mathcal{G}(\vec{x})F(\vec{x}) = 1$. $\square$

The following result is immediate.

Corollary 2. *Let $\mathcal{GH} \colon [0,1]^n \to [0,1]$ be a general grouping function and let $F \colon [0,1] \to [0,1]$ be a commutative and continuous aggregation function such that, for all $\vec{x} = (x_1, \ldots, x_n) \in [0,1]^n$, if there exists $i \in \{1, \ldots, n\}$ such that $x_i = 1$, then $F(\vec{x}) = 1$. Then $\mathcal{GG}_{\mathcal{GH},F}(\vec{x}) = \mathcal{GH}(\vec{x})F(\vec{x})$ is a general grouping function.*

Note that $\mathfrak{G}^n$ is closed with respect to some aggregation functions, as stated by the following results, which provide construction methods of general grouping functions.

Theorem 4. *Consider* $M\colon [0,1]^2 \to [0,1]$. *For any* $\mathcal{GG}_1, \mathcal{GG}_2 \in \mathfrak{G}^n$, *define the mapping* $M_{\mathcal{GG}_1,\mathcal{GG}_2}\colon [0,1]^n \to [0,1]$, *for all* $\vec{x} \in [0,1]^n$, *by:*

$$M_{\mathcal{GG}_1,\mathcal{GG}_2}(\vec{x}) = M(\mathcal{GG}_1(\vec{x}), \mathcal{GG}_2(\vec{x})).$$

Then, $M_{\mathcal{GG}_1,\mathcal{GG}_2} \in \mathfrak{G}^n$ *if and only if* M *is a continuous aggregation function.*

Proof. It follows that:
($\Rightarrow$) Suppose that $M_{\mathcal{GG}_1,\mathcal{GG}_2} \in \mathfrak{G}^n$. Then it is immediate that M is continuous and increasing (A2). Now consider $\vec{x} = (x_1,\ldots,x_n) \in [0,1]^n$ and suppose that $\sum_{i=1}^{n} x_i = 0$. Then, by ($\mathcal{GG}2$), one has that: $M_{\mathcal{GG}_1,\mathcal{GG}_2}(\vec{x}) = M(\mathcal{GG}_1(\vec{x}), \mathcal{GG}_2(\vec{x})) = 0$ and $\mathcal{GG}_1(\vec{x}) = \mathcal{GG}_2(\vec{x}) = 0$. Thus, it holds that $M(0,0) = 0$. Now, consider $\vec{x} = (x_1,\ldots,x_n) \in [0,1]^n$, such that there exists $i \in \{1,\ldots,n\}$ such that $x_i = 1$. Then, by ($\mathcal{GG}3$), one has that: $M_{\mathcal{GG}_1,\mathcal{GG}_2}(\vec{x}) = M(\mathcal{GG}_1(\vec{x}), \mathcal{GG}_2(\vec{x})) = 1$ and $\mathcal{GG}_1(\vec{x}) = \mathcal{GG}_2(\vec{x}) = 1$. Therefore, it holds that $M(1,1) = 1$. This proves that M also satisfies (A1), and, thus, M is a continuous aggregation function.
($\Leftarrow$) Suppose that M is a continuous aggregation function. Then it is immediate that $M_{\mathcal{GG}_1,\mathcal{GG}_2}$ is ($\mathcal{GG}1$) commutative, ($\mathcal{GG}4$) increasing and ($\mathcal{GG}5$) continuous. For ($\mathcal{GG}2$), consider $\vec{x} = (x_1,\ldots,x_n) \in [0,1]^n$ such that $\sum_{i=1}^{n} x_i = 0$. Then, by ($\mathcal{GG}2$), one has that $\mathcal{GG}_1(\vec{x}) = \mathcal{GG}_2(\vec{x}) = 0$. It follows that: $M_{\mathcal{GG}_1,\mathcal{GG}_2}(\vec{x}) = M(\mathcal{GG}_1(\vec{x}), \mathcal{GG}_2(\vec{x})) = M(0,0) = 0$, by (A1), since M is an aggregation function. Finally, for ($\mathcal{GG}3$) consider that there exists $i \in \{1,\ldots,n\}$ such that $x_i = 1$ for some $\vec{x} = (x_1,\ldots,x_n) \in [0,1]^n$. Then, it holds that $\mathcal{GG}_1(\vec{x}) = \mathcal{GG}_2(\vec{x}) = 1$. It follows that: $M_{\mathcal{GG}_1,\mathcal{GG}_2}(\vec{x}) = M(\mathcal{GG}_1(\vec{x}), \mathcal{GG}_2(\vec{x})) = M(1,1) = 1$, by (A1), since M is an aggregation function. This proves that $M_{\mathcal{GG}_1,\mathcal{GG}_2} \in \mathfrak{G}^n$. $\qquad\square$

Example 4. In the sense of Theorem 4, $\mathfrak{G}^n$ is closed under any bidimensional overlap functions, grouping functions and continuous t-norms and t-conorms [18].

Corollary 3. *Consider* $M\colon [0,1]^2 \to [0,1]$. *For any* n-*dimensional grouping functions* $\mathcal{G}_1, \mathcal{G}_2\colon [0,1]^n \to [0,1]$, *define the mapping* $M_{\mathcal{G}_1,\mathcal{G}_2}\colon [0,1]^n \to [0,1]$, *for all* $\vec{x} = (x_1,\ldots,x_n) \in [0,1]^n$, *by:*

$$M_{\mathcal{G}_1,\mathcal{G}_2}(\vec{x}) = M(\mathcal{G}_1(\vec{x}), \mathcal{G}_2(\vec{x})).$$

Then, $M_{\mathcal{G}_1,\mathcal{G}_2} \in \mathfrak{G}^n$ *if and only if* M *is a continuous aggregation function.*

Proof. It follows from Theorem 4, since any n-dimensional grouping function is a general grouping function. $\qquad\square$

Theorem 4 can be easily extended for n-ary functions $M^n\colon [0,1]^n \to [0,1]$:

Theorem 5. *Consider* $M^n\colon [0,1]^n \to [0,1]$. *For any* $\mathcal{GG}_1,\ldots,\mathcal{GG}_n \in \mathfrak{G}^n$, *define the mapping* $M_{\mathcal{GG}_1,\ldots,\mathcal{GG}_n}\colon [0,1]^n \to [0,1]$, *for all* $\vec{x} \in [0,1]^n$, *by:*

$$M_{\mathcal{GG}_1,\ldots,\mathcal{GG}_n}(\vec{x}) = M^n(\mathcal{GG}_1(\vec{x}), \ldots, \mathcal{GG}_n(\vec{x})).$$

Then, $M_{\mathcal{GG}_1,\ldots,\mathcal{GG}_n} \in \mathfrak{G}^n$ *if and only if* $M^n : [0,1]^n \to [0,1]$ *is a continuous n-ary aggregation function.*

Proof. Analogous to the proof of Theorem 4.   □

This result can be extended for n-dimensional grouping functions.

Corollary 4. *Consider* $M^n : [0,1]^n \to [0,1]$ *and fr any n-dimensional grouping functions* $\mathcal{G}_1,\ldots,\mathcal{G}_n$ *define the mapping* $M_{\mathcal{G}_1,\ldots,\mathcal{G}_n} : [0,1]^n \to [0,1]$, *for all* $\vec{x} = (x_1,\ldots,x_n) \in [0,1]^n$, *by:*

$$M_{\mathcal{G}_1,\ldots,\mathcal{G}_n}(\vec{x}) = M^n(\mathcal{G}_1(\vec{x}),\ldots,\mathcal{G}_n(\vec{x})).$$

Then, $M_{\mathcal{G}_1,\ldots,\mathcal{G}_n}$ *is a general grouping function if and only if* $M^n : [0,1]^n \to [0,1]$ *is a continuous n-ary aggregation function.*

Corollary 5. *Let* $\mathcal{GG}_1,\ldots,\mathcal{GG}_m : [0,1]^n \to [0,1]$ *be general grouping functions and consider weights* $w_1,\ldots,w_m \in [0,1]$ *such that* $\sum_{i=1}^{m} w_i = 1$. *Then the convex sum* $\mathcal{GG} = \sum_{i=1}^{m} w_i \mathcal{GG}_i$ *is also a general grouping function.*

Proof. Since the weighted sum is a continuous commutative n-ary aggregation function, the result follows from Theorem 5.   □

It is possible to obtain general grouping functions from the generalized composition of general grouping functions and aggregation functions satisfying specific conditions:

Theorem 6. *Let* $\mathcal{GG}_2 : [0,1]^n \to [0,1]$ *be a general grouping function and let the n-ary aggregation functions* $A_1,\ldots,A_n : [0,1]^n \to [0,1]$ *be continuous, commutative and disjunctive. Then, the function* $\mathcal{GG}_1 : [0,1]^n \to [0,1]$ *defined, for all* $\vec{x} = (x_1,\ldots,x_n) \in [0,1]^n$, *by:* $\mathcal{GG}_1(\vec{x}) = \mathcal{GG}_2(A_1(\vec{x}),\ldots,A_n(\vec{x}))$ *is a general grouping function.*

Proof. Since $\mathcal{GG}_2, A_1,\ldots,A_n$ are commutative, increasing and continuous functions, then $\mathcal{GG}_1$ satisfies conditions $(\mathcal{GG}1)$, $(\mathcal{GG}4)$ and $(\mathcal{GG}5)$. So, it remains to prove:

$(\mathcal{GG}2)$ Let $\vec{x} = (x_1,\ldots,x_n) \in [0,1]^n$ be such that $\sum_{i=1}^{n} x_i = 0$. Then, since A_1 is disjunctive, we have that $A_1(\vec{x}) \geq \max(\vec{x}) = 0$, that is $A_1(\vec{x}) = 0$. Equivalently, one obtains $A_2(\vec{x}),\ldots,A_n(\vec{x}) = 0$. Thus, since $\mathcal{GG}_2$ is a general grouping function, one has that $\mathcal{GG}_1(\vec{x}) = \mathcal{GG}_2(A_1(\vec{x}),\ldots,A_n(\vec{x})) = \mathcal{GG}_2(0,\ldots,0) = 0$.

$(\mathcal{GG}3)$ Suppose that, for some $\vec{x} = (x_1,\ldots,x_n) \in [0,1]^n$, there exists $i \in \{1,\ldots,n\}$ such that $x_i = 1$. So, since A_1 is disjunctive then $A_1(\vec{x}) \geq \max(\vec{x}) = 1$, that is $A_1(\vec{x}) = 1$. Since $\mathcal{GG}_2$ is a general grouping function, it follows that $\mathcal{GG}_1(\vec{x}) = \mathcal{GG}_2(A_1(\vec{x}),\ldots,A_n(\vec{x})) = \mathcal{GG}_2(1, A_2(\vec{x}),\ldots,A_n(\vec{x})) = 1$.   □

Next proposition uses the n-duality property.

Proposition 7. *Consider a continuous fuzzy negation* $N \colon [0,1] \to [0,1]$ *and a general overlap function* $\mathcal{GO} \colon [0,1]^n \to [0,1]$, *then for all* $\vec{x} = (x_1, \ldots, x_n) \in [0,1]^n$:

$$\mathcal{GG}(\vec{x}) = N(\mathcal{GO}(N(x_1), \ldots, N(x_n))) \tag{4}$$

is a general grouping function. Reciprocally, if $\mathcal{GG} \colon [0,1]^n \to [0,1]$ *is a general grouping function, then for all* $\vec{x} = (x_1, \ldots, x_n) \in [0,1]^n$:

$$\mathcal{GO}(\vec{x}) = N(\mathcal{GG}(N(x_1), \ldots, N(x_n))) \tag{5}$$

is a general overlap function.

Proof. Since we have a continuous fuzzy negation and bearing in mind that general overlap functions and general grouping functions are commutative, increasing and continuous functions according to Definition 8 and Definition 9, respectively, then $\mathcal{GO}$ and $\mathcal{GG}$ satisfy conditions $(\mathcal{GO}1), (\mathcal{GG}1)$; $(\mathcal{GO}4), (\mathcal{GG}4)$ and $(\mathcal{GO}5), (\mathcal{GG}5)$. So, it remains to prove:

$(\mathcal{GG}2)$ For Eq. (4), take $x_i = 0$, for all $i \in \{1, \ldots, n\}$. Therefore,

$$\mathcal{GG}(\vec{x}) = N(\mathcal{GO}(N(0), \ldots, N(0))) \overset{N2}{=} N(\mathcal{GO}(1, \ldots, 1)) \overset{\mathcal{GO}3}{=} N(1) \overset{N2}{=} 0.$$

$(\mathcal{GG}3)$ If there exists a $x_i = 1$, for some $i \in \{1, \ldots, n\}$, then

$$
\begin{aligned}
\mathcal{GG}(\vec{x}) &= N(\mathcal{GO}(N(x_1), \ldots, N(1), \ldots, N(x_n))) \\
&\overset{N2}{=} N(\mathcal{GO}(N(x_1), \ldots, 0, \ldots, N(x_n))) \\
&\overset{\mathcal{GO}2}{=} N(0) \overset{N2}{=} 1.
\end{aligned}
$$

$(\mathcal{GO}2)$ Similarly, for Eq. (5), take a $x_i = 0$ for some $i \in \{1, \ldots, n\}$. So,

$$
\begin{aligned}
\mathcal{GO}(\vec{x}) &= N(\mathcal{GG}(N(x_1), \ldots, N(0), \ldots, N(x_n))) \\
&\overset{N2}{=} N(\mathcal{GG}(N(x_1), \ldots, 1, \ldots, N(x_n))) \\
&\overset{\mathcal{GG}3}{=} N(1) \overset{N2}{=} 0.
\end{aligned}
$$

$(\mathcal{GO}3)$ Now, consider that $x_i = 1$, for all $i \in \{1, \ldots, n\}$. Then,

$$\mathcal{GO}(\vec{x}) = N(\mathcal{GG}(N(1), \ldots, N(1))) \overset{N2}{=} N(\mathcal{GG}(0, \ldots, 0)) \overset{\mathcal{GG}2}{=} N(0) \overset{N2}{=} 1.$$

$\square$

5 Conclusions

In this paper, we first introduced the concept of general grouping functions and studied some of their properties. Then we provided a characterization of general grouping functions and some construction methods.

The theoretical developments presented here allow for a more flexible approach when dealing with decision making problems with multiple alternatives. Immediate future work is concerned with the development of an application in multi-criteria decision making based on n-ary fuzzy heterogeneous, incomplete preference relations.

Acknowledgments. Supported by CNPq (233950/2014-1, 307781/2016-0, 301618/2019-4), FAPERGS (17/2551 - 0000872-3, 19/2551 - 0001279-9, 19/ 2551 - 0001660-3), PNPD/CAPES (464880 /2019-00) and the Spanish Ministry of Science and Technology (PC093 - 094TFIPDL, TIN2016-81731-REDT, TIN2016-77356-P (AEI/FEDER, UE)).

References

1. Barzilai, J.: Consistency measures for pairwise comparison matrices. J. Multi-Criteria Decis. Anal. **7**(3), 123–132 (1998). https://doi.org/doi.org/10.1002/(SICI)1099-1360(199805)7:3%3C123::AID-MCDA181%3E3.0.CO;2-8
2. Bedregal, B.C., Dimuro, G.P., Bustince, H., Barrenechea, E.: New results on overlap and grouping functions. Inf. Sci. **249**, 148–170 (2013). https://doi.org/10.1016/j.ins.2013.05.004
3. Beliakov, G., Pradera, A., Calvo, T.: Aggregation Functions: A Guide for Practitioners. Springer, Berlin (2007). https://doi.org/10.1007/978-3-540-73721-6
4. Bustince, H., Fernandez, J., Mesiar, R., Montero, J., Orduna, R.: Nonlinear analysis: theory, overlap functions. Methods Appl. **72**(3–4), 1488–1499 (2010)
5. Bustince, H., Pagola, M., Mesiar, R., Hüllermeier, E., Herrera, F.: Grouping, overlaps, and generalized bientropic functions for fuzzy modeling of pairwise comparisons. IEEE Trans. Fuzzy Syst. **20**(3), 405–415 (2012). https://doi.org/10.1109/TFUZZ.2011.2173581
6. Chiclana, F., Herrera, F., Herrera-Viedma, E.: Integrating multiplicative preference relations in a multipurpose decision-making model based on fuzzy preference relations. Fuzzy Sets Syst. **122**(2), 277–291 (2001). https://doi.org/10.1016/S0165-0114(00)00004-X
7. Dimuro, G.P., Bedregal, B.: Archimedean overlap functions: the ordinal sum and the cancellation, idempotency and limiting properties. Fuzzy Sets Syst. **252**, 39–54 (2014). https://doi.org/10.1016/j.fss.2014.04.008
8. Dimuro, G.P., Bedregal, B.: On residual implications derived from overlap functions. Inf. Sci. **312**, 78–88 (2015). https://doi.org/10.1016/j.ins.2015.03.049
9. Dimuro, G.P., Bedregal, B., Bustince, H., Mesiar, R., Asiain, M.J.: On additive generators of grouping functions. In: Laurent, A., Strauss, O., Bouchon-Meunier, B., Yager, R.R. (eds.) IPMU 2014. CCIS, vol. 444, pp. 252–261. Springer, Cham (2014). https://doi.org/10.1007/978-3-319-08852-5_26
10. Dimuro, G.P., Bedregal, B., Fernandez, J., Sesma-Sara, M., Pintor, J.M., Bustince, H.: The law of O-conditionality for fuzzy implications constructed from overlap and grouping functions. Int. J. Approximate Reasoning **105**, 27–48 (2019). https://doi.org/10.1016/j.ijar.2018.11.006
11. Dimuro, G.P., Bedregal, B., Santiago, R.H.N.: On (G, N)-implications derived from grouping functions. Inf. Sci. **279**, 1–17 (2014). https://doi.org/10.1016/j.ins.2014.04.021
12. Dimuro, G.P., Bedregal, B., Bustince, H., Asiáin, M.J., Mesiar, R.: On additive generators of overlap functions. Fuzzy Sets Syst. **287**, 76–96 (2016). https://doi.org/10.1016/j.fss.2015.02.008. Theme: Aggregation Operations
13. Dimuro, G.P., Bedregal, B., Bustince, H., Jurio, A., Baczyński, M., Miś, K.: QL-operations and QL-implication functions constructed from tuples (O, G, N) and the generation of fuzzy subsethood and entropy measures. Int. J. Approximate Reasoning **82**, 170–192 (2017). https://doi.org/10.1016/j.ijar.2016.12.013
14. Fodor, J., Roubens, M.: Fuzzy Preference Modelling and Multicriteria Decision Support. Kluwer Academic Publisher, Dordrecht (1994)

15. Gómez, D., Rodríguez, J.T., Montero, J., Bustince, H., Barrenechea, E.: n-Dimensional overlap functions. Fuzzy Sets Syst. **287**, 57–75 (2016). https://doi.org/10.1016/j.fss.2014.11.023. Theme: Aggregation Operations
16. Gómez, D., Rodríguez, J.T., Montero, J., Yáñez, J.: Fuzzy community detection based on grouping and overlapping functions. In: 2015 Conference of the International Fuzzy Systems Association and the European Society for Fuzzy Logic and Technology (IFSA-EUSFLAT-15). Atlantis Press, Paris (2015). https://doi.org/10.2991/ifsa-eusflat-15.2015.215
17. Jurio, A., Bustince, H., Pagola, M., Pradera, A., Yager, R.: Some properties of overlap and grouping functions and their application to image thresholding. Fuzzy Sets Syst. **229**, 69–90 (2013). https://doi.org/10.1016/j.fss.2012.12.009
18. Klement, E., Mesiar, R., Pap, E.: Triangular Norms, Trends in Logic - Studia Logica Library, vol. 8. Kluwer Academic Publishers, Dordrecht (2000)
19. Lourenzutti, R., Krohling, R.A., Reformat, M.Z.: Choquet based topsis and todim for dynamic and heterogeneous decision making with criteria interaction. Inf. Sci. **408**, 41–69 (2017). https://doi.org/10.1016/j.ins.2017.04.037
20. Miguel, L.D., et al.: General overlap functions. Fuzzy Sets Syst. **372**, 81–96 (2019). https://doi.org/10.1016/j.fss.2018.08.003. Theme: Aggregation Operations
21. Qiao, J., Hu, B.Q.: On the distributive laws of fuzzy implication functions over additively generated overlap and grouping functions. IEEE Tran. Fuzzy Syst. **26**(4), 2421–2433 (2018). https://doi.org/10.1109/TFUZZ.2017.2776861
22. Qiao, J., Hu, B.Q.: On interval additive generators of interval overlap functions and interval grouping functions. Fuzzy Sets Syst. **323**, 19–55 (2017). https://doi.org/10.1016/j.fss.2017.03.007
23. Qiao, J., Hu, B.Q.: On the migrativity of uninorms and nullnorms over overlap and grouping functions. Fuzzy Sets Syst. **346**, 1–54 (2018). https://doi.org/10.1016/j.fss.2017.11.012
24. Qiao, J., Hu, B.Q.: On generalized migrativity property for overlap functions. Fuzzy Sets Syst. **357**, 91–116 (2019). https://doi.org/10.1016/j.fss.2018.01.007
25. Qiao, J., Hu, B.Q.: On homogeneous, quasi-homogeneous and pseudo-homogeneous overlap and grouping functions. Fuzzy Sets Syst. **357**, 58–90 (2019). https://doi.org/10.1016/j.fss.2018.06.001. Theme: Aggregation Functions
26. Ureña, R., Chiclana, F., Morente-Molinera, J., Herrera-Viedma, E.: Managing incomplete preference relations in decision making: a review and future trends. Inf. Sci. **302**, 14–32 (2015). https://doi.org/10.1016/j.ins.2014.12.061

The Necessary and Possible Importance Relation Among Criteria in a 2-Additive Choquet Integral Model

Brice Mayag[1]([✉]) and Bertrand Tchantcho[2]

[1] University Paris-Dauphine, PSL Research university, LAMSADE, CNRS, UMR 7243, Place du Maréchal de Lattre de Tassigny, 75775 Paris, cedex 16, France
brice.mayag@dauphine.fr
[2] MASS Laboratory, Ecole Normale Supérieure de Yaoundé, BP 47, Yaoundé, Cameroon
btchantcho@yahoo.fr

Abstract. In the context of the representation of a preference information by a 2-additive Choquet integral, we introduce the necessary and possible importance relations allowing to compare the Shapley values of two criteria. We present some sufficient conditions, using a set of binary alternatives, to get a necessary importance relation among two criteria.

Keywords: MCDA · Binary alternatives · Shapley value · Choquet integral · Necessary relations

1 Introduction

In Operational Research domain, Multiple Criteria Decision Making (MCDM) is a scientific field which tries to represent the preferences of a Decision Maker (DM) over a set of alternatives evaluated on a set of criteria often contradictory. To represent a preference information of a DM, allowing some dependencies or interactions among criteria, a 2-additive Choquet integral model, a generalization of the well-known arithmetic mean, is usually elaborated.

The 2-additive Choquet integral is a particular case of the Choquet integral [1–3,7], an aggregation function based on the notion of capacity or fuzzy measure. The identification of the capacity leads to the computation of two important parameters of the 2-additive integral model: the interaction index [10] related to only two criteria and the importance of each criterion (corresponding to the Shapley value [12]).

We assume that the DM can expresses his preferences through two binary relations on the set of alternatives: a strict and an indifference preference information. There exist some characterization about the representation of such information by a 2-additive Choquet integral, especially when the set of binary alternatives is considered [6,7,9]. A binary action is a fictitious alternative which

© Springer Nature Switzerland AG 2020
M.-J. Lesot et al. (Eds.): IPMU 2020, CCIS 1238, pp. 496–509, 2020.
https://doi.org/10.1007/978-3-030-50143-3_39

takes either the neutral value **0** for all criteria, or the neutral value **0** for all criteria except for one or two criteria for which it takes the satisfactory value **1**.

Under these hypotheses, we try to analyze, in this paper, the comparison of the importance index of two given criteria. To do so, we introduce the notions of necessary and possible importance relations among two criteria. A criterion i is possibly more important than criterion j, if there exists a compatible 2-additive capacity, representing the preference information given by the DM, such that the Shapley value associated to i is greater than the one associated to j. If this conclusion is made for all the compatible capacities, then i is judged necessarily more important than j. The concept of necessary and possible relations were introduced for the alternatives in a robust ordinal regression approach [5], and extended for the interactions indices in the 2-additive Choquet integral model [9]. We give also some sufficient conditions in order to obtain a necessary importance relation among criteria, in the framework of binary alternatives.

The paper is organized as follows. The next section presents the basic material we need on the 2-additive Choquet integral. The new notions of necessary and possible relations among criteria are introduced in Sect. 3, after a motivating example based on hospitals rankings, in a real-world situation. Our results are presented in Sect. 4 and we end the paper by giving some perspectives of this work.

2   A Choquet Integral w.r.t. a 2 Additive Capacity

Let X be a finite set of alternatives evaluated on a set of n criteria $N = \{1, \ldots, n\}$. The notation 2^N refers to the set of all subsets of N. The set of attributes is denoted by $X_1, \ldots, X_n$. An alternative x is presented by $x = (x_1, \ldots, x_n)$ where $x_i \in X_i$, $i = 1, \ldots, n$.

The notion of interaction among criteria is more simple and understandable, in MCDA, when it concerns only two criteria. That is why the Choquet integral w.r.t. a 2-additive capacity [6,7], also called 2-additive Choquet, was proposed in order to take into account the type of interaction between two criteria. This aggregation function, considered as a fuzzy integral, is based on the concept of *capacity or fuzzy measure* μ defined as a set function from the powerset of criteria 2^N to $[0, 1]$ such that:

1. $\mu(\emptyset) = 0$
2. $\mu(N) = 1$
3. $\forall A, B \in 2^N, \ [A \subseteq B \Rightarrow \mu(A) \le \mu(B)]$ (monotonicity).

We associate to each capacity another set function called the *Möbius transform* $m^\mu : 2^N \to \mathbb{R}$ defined by

$$m^\mu(T) := \sum_{K \subseteq T} (-1)^{|T \setminus K|} \mu(K), \forall T \in 2^N. \tag{1}$$

A capacity μ on N satisfying the following two conditions:

- For all subset T of N such that $|T| > 2$, $m^\mu(T) = 0$;
- There exists a subset B of N such that $|B| = 2$ and $m^\mu(B) \neq 0$.

is said to be *2-additive*.

In the sequel, we use, for a capacity μ and its Möbius transform m^μ, the following notations: $\mu_i := \mu(\{i\})$, $\mu_{ij} := \mu(\{i,j\})$, $m_i^\mu := m^\mu(\{i\})$, $m_{ij}^\mu := m^\mu(\{i,j\})$, for all $i,j \in N$, $i \neq j$. Whenever we use i and j together, it always means that they are different.

Given an alternative $x := (x_1, ..., x_n)$ of X, the 2-additive Choquet integral of x is expressed as follows [4]:

$$C_\mu(u(x)) = \sum_{i=1}^{n} \phi_i \, u_i(x_i) - \frac{1}{2} \sum_{\{i,j\} \subseteq N} I_{ij} \, |u_i(x_i) - u_j(x_j)| \tag{2}$$

where

- For all $i \in N$, $u_i : X_i \to \mathbb{R}_+$ is a marginal utility function associated to the attribute X_i;
- $u(x) = (u_1(x_1), \dots, u_n(x_n))$ for $x = (x_1, ..., x_n) \in X$;
- $I_{ij}^\mu = \mu_{ij} - \mu_i - \mu_j$ is the interaction index between the two criteria i and j [2,10];
- $\phi_i^\mu = \sum_{K \subseteq N \setminus i} \dfrac{(n - |K| - 1)!|K|!}{n!} (\mu(K \cup i) - \mu(K)) = \mu_i + \dfrac{1}{2} \sum_{j \in N, j \neq i} I_{ij}^\mu$ is

defined as the importance of criterion i and it corresponds to the Shapley value of i w.r.t. μ [11].

Equation (2) proves that the 2-additive Choquet integral is a generalization of the weighted sum. Indeed, when there is no interaction among criteria, the Shapley value ϕ_i^μ is the weight associated to the criterion i. There is another expression of $C_\mu(u(x))$, related to the coefficients of the Möbius transform of μ, given by:

$$C_\mu(u_1(x_1), \dots, u_n(x_n)) = \sum_{i \in N} m_i^\mu \, u_i(x_i) + \sum_{i,j \in N} m^\mu(\{i,j\}) \, \min(u_i(x_i), u_j(x_j)) \tag{3}$$

We assume that the DM expresses his preferences on X by giving a strict preference relation P and an indifference relation I on X. We say that the preference information $\{P, I\}$ on X is representable by a 2-additive Choquet integral if we have: for all $x, y \in X$,

$$\begin{cases} x \, P \, y \Longrightarrow C_\mu(u(x)) > C_\mu(u(y)) \\ x \, I \, y \Longrightarrow C_\mu(u(x)) = C_\mu(u(y)) \end{cases} \tag{4}$$

3   The Importance Relation Among Criteria Is Not Stable

3.1   A Motivating Example

Let us consider eight hospitals (see Table 1), specialized in weight loss surgery[1], and evaluated on four criteria given by the French magazine "Le Point[2]" [8] (see their evaluations in Table 1 below):

- *Criterion 1 - Activity*: number of procedures performed during one year. Since a hospital has a good score on activity then its teams are more trained and often have good results. Therefore this criterion has to be maximized.
- *Criterion 2 - Notoriety*: It corresponds to the reputation and attractiveness of the hospital. It is a percentage of patients treated in the hospital but living in another French administrative department. The more the percentage increases, more the hospital is attractive.
- *Criterion 3 - Average Length Of Stay (ALOS)*: a mean calculated by dividing the sum of inpatient days by the number of patients admissions with the same diagnosis-related group classification. If a hospital is more organized in terms of resources then its ALOS score should be low.
- *Criterion 4 - Technicality*: this particular indicator measures the ratio of procedures performed with an efficient technology compared to the same procedures performed with obsolete technology. The higher the percentage is, the more the team is trained in advanced technologies or complex surgeries.

Table 1. Evaluations of eight hospitals on activity, Notoriety, ALOS and Technicality.

	1- Activity	2- Notoriety	3- ALOS	4- Technicality
Hospital 1 ($H1$)	200	65	3.5	85
Hospital 2 ($H2$)	450	60	4	75
Hospital 3 ($H3$)	450	50	2.5	55
Hospital 4 ($H4$)	350	50	3.5	85
Hospital 5 ($H5$)	350	55	2	75
Hospital 6 ($H6$)	150	65	2.5	80
Hospital 7 ($H7$)	200	55	2	55
Hospital 8 ($H8$)	150	60	4	80

Based on its expertise, the DM (a team of some specialists on weight loss surgery) provides the following preferences where P refers to the strict preference relation:

$$H1 \ P \ H2; \quad H3 \ P \ H4; \quad H5 \ P \ H6; \quad H8 \ P \ H7. \tag{5}$$

---

[1] http://en.wikipedia.org/wiki/Bariatric_surgery.

[2] https://www.lepoint.fr/sante/le-palmares-des-hopitaux-et-cliniques-methodologie-21-08-2019-2330873_40.php.

Based on these preferences, he also asks himself the following questions which seem reasonable:

- Is the criterion Activity more important than the criterion Notoriety?
- Is the criterion Activity more important than the criterion ALOS?
- ...

First of all, let us try to model his preferences by an additive model. The four preferences could be representable by an arithmetic mean model, w.r.t. a vector of weights (w_1, w_2, w_3, w_4) associated to the four criteria, if the following system is feasible:

$$
\begin{cases}
H1 \ P \ H2 \Rightarrow u_1(200)w_1 + u_2(65)w_2 + u_3(3.5)w_3 + u_4(85)w_4 > \\
\qquad u_1(450)w_1 + u_2(60)w_2 + u_3(4)w_3 + u_4(75)w_4 \\
\\
H3 \ P \ H4 \Rightarrow u_1(450)w_1 + u_2(50)w_2 + u_3(2.5)w_3 + u_4(55)w_4 > \\
\qquad u_1(350)w_1 + u_2(50)w_2 + u_3(3.5)w_3 + u_4(85)w_4 \\
\\
H5 \ P \ H6 \Rightarrow u_1(350)w_1 + u_2(55)w_2 + u_3(2)w_3 + u_4(75)w_4 > \\
\qquad u_1(150)w_1 + u_2(65)w_2 + u_3(2.5)w_3 + u_4(80)w_4 \\
\\
H8 \ P \ H7 \Rightarrow u_1(200)w_1 + u_2(55)w_2 + u_3(2)w_3 + u_4(55)w_4 > \\
\qquad u_1(150)w_1 + u_2(60)w_2 + u_3(4)w_3 + u_4(80)w_4
\end{cases}
\tag{6}
$$

It is not difficult to see that the first three equations in this system lead to $[u_1(200) - u_1(150)]w_1 + [u_2(55) - u_2(60)]w_2 + [u_3(2) - u_3(4)]w_3 + [u_4(55) - u_4(80)]w_4 > 0$ contradicting the last equation. Therefore the arithmetic mean is not able to model the DM preferences (5).

To prove that these preferences are modeled by a 2-additive Choquet integral, we assume that the marginal utility functions are constructed by using the following monotone normalization formula of the criterion i, where U_i (respectively L_i) represents a upper bound (respectively a lower bound) associated to the values of X_i: Given a hospital $h = (h_1, h_2, h_3, h_4)$,

$$
\begin{cases}
u_i(h_i) = \dfrac{h_i}{U_i} & \text{if } i \text{ is to be maximized } (criteria\ 1,\ 2 \text{ and } 4) \\
\\
u_i(h_i) = 1 - \dfrac{h_i}{L_i} & \text{if } i \text{ is to be minimized } (criterion\ 3)
\end{cases}
\tag{7}
$$

By choosing $U_1 = 500, U_2 = U_4 = 100$ and $L_3 = 5$, the obtained utility functions, associated to each hospital, are given by the Table 2.

Table 3 below presents five 2-additive capacities allowing to represent the preferences (5) by 2-additive Choquet integral.

These results show that the importance index of Activity is more important than Notoriety when the capacity of the Parameters 1, 4 and 5 are chosen. Conversely the importance index of the criterion Notoriety is more important than Activity for the capacity of Parameters 2 and 3. Hence, based on the preference

Table 2. Utility functions of eight hospitals on Activity, Notoriety, ALOS and Technicality.

	1- Activity	2- Notoriety	3- ALOS	4- Technicality
Hospital 1 ($H1$)	0.4	0.65	0.3	0.85
Hospital 2 ($H2$)	0.9	0.60	0.2	0.75
Hospital 3 ($H3$)	0.9	0.50	0.5	0.55
Hospital 4 ($H4$)	0.7	0.50	0.3	0.85
Hospital 5 ($H5$)	0.7	0.55	0.6	0.75
Hospital 6 ($H6$)	0.3	0.65	0.5	0.80
Hospital 7 ($H7$)	0.4	0.55	0.6	0.55
Hospital 8 ($H8$)	0.3	0.60	0.2	0.80

giving by the Dean, it is not easy to conclude about the importance of the criterion Activity compared to the criterion Notoriety. We have similar conclusions with Activity and ALOS.

To overcome this limits we introduce in the next section the new notion of necessary and possible importance relation among two criteria.

3.2 Necessary and Possible Importance of Criteria

Let $\{P, I\}$ be a preference information on X representable by a 2-additive Choquet integral. We denote by $\mathcal{C}_2^{\{P,I\}}$ the set of all the 2-additive capacities compatible with $\{P, I\}$.

Definition 1. *Given two different criteria i and j. We say that:*

- *i is possibly more important than j, if there exists $\mu \in \mathcal{C}_2^{\{P,I\}}$ such that $\phi_i^\mu > \phi_j^\mu$.*
- *i is necessarily more important than j, if for all $\mu \in \mathcal{C}_2^{\{P,I\}}$, we have $\phi_i^\mu > \phi_j^\mu$.*
- *i and j are possibly equally important if there exists $\mu \in \mathcal{C}_2^{\{P,I\}}$ such that $\phi_i^\mu = \phi_j^\mu$.*
- *i and j are necessarily equally important if for all $\mu \in \mathcal{C}_2^{\{P,I\}}$, we have $\phi_i^\mu = \phi_j^\mu$.*

Using these definitions, we can conclude that, in our previous example, the criterion Activity is not necessary important than Notoriety. The converse is also true. There exists only a possible importance relation among these two criteria. Now, let us give some sufficient conditions allowing to get the necessary importance relation among two given criteria when the DM expresses his preferences on a set of binary alternatives.

Table 3. Five capacities compatible with the preferences (5)

	Par.1	Par.2	Par.3	Par.4	Par.5
$C_\mu(H1)$	0.5713	0.5723	0.5850	0.6413	0.5695
$C_\mu(H2)$	0.5613	0.5560	0.5750	0.5929	0.5415
$C_\mu(H3)$	0.5582	0.5531	0.5325	0.6513	0.6925
$C_\mu(H4)$	0.5482	0.5431	0.5134	0.6413	0.5695
$C_\mu(H5)$	0.6259	0.6228	0.5812	0.6931	0.6735
$C_\mu(H6)$	0.5483	0.5509	0.5712	0.6831	0.545
$C_\mu(H7)$	0.4920	0.4935	0.5117	0.5476	0.4735
$C_\mu(H8)$	0.5020	0.5035	0.5217	0.5724	0.494
μ_1	0.1175	0.1056	0.0712	0.3438	0.4799
μ_2	0.0948	0.0956	0.0612	0.01	0.01
μ_3	0	0	0	0	0
μ_4	0.2243	0.2175	0.0812	0.6206	0.4899
μ_{12}	0.2124	0.2012	0.1812	0.3538	0.4899
μ_{13}	0.3864	0.3762	0.2062	0.3593	0.99
μ_{14}	0.3418	0.3231	0.1525	0.6208	0.4899
μ_{23}	0.0948	0.0956	0.0724	0.0099	0.01
μ_{24}	0.6135	0.6237	0.7337	0.6206	0.4899
μ_{34}	0.2243	0.2175	0.0812	0.9844	0.4899
ϕ_1^μ	0.2520	0.2409	0.1631	0.1796	0.495
ϕ_2^μ	0.2420	0.2509	0.3868	0.005	0.0049
ϕ_3^μ	0.1344	0.1353	0.073	0.1896	0.255
ϕ_4^μ	0.3714	0.3728	0.3768	0.6256	0.245
I_{12}^μ	0	0	0.0487	0	0
I_{13}^μ	0.2689	0.2706	0.135	0.0155	0.51
I_{14}^μ	0	0	0	-0.3438	-0.4799
I_{23}^μ	0	0	0.0112	0	0
I_{24}^μ	0.2943	0.3106	0.5912	-0.0099	-0.0099
I_{34}^μ	0	0	0	0.3638	0

4   Sufficient Conditions Using the Set of Binary Alternatives

4.1   Preference Information on the Set of Binary Alternatives

In this section, we suppose that the DM is able to identify on each criterion i two reference levels: 1_i (the satisfactory or good level) and 0_i (the neutral level). These references are usually used in the elicitation of the parameters of the Choquet integral (see [3,4]). We set for convenience $u_i(1_i) = 1$ and $u_i(0_i) = 0$.

We ask to the DM, a preference information on a reference subset $\mathcal{B}$ of X, called the set of binary alternatives and defined by

$$\mathcal{B} = \{\mathbf{0}_N,\ (\mathbf{1}_i, \mathbf{0}_{N-i}),\ (\mathbf{1}_{ij}, \mathbf{0}_{N-ij}),\ i, j \in N, i \neq j\} \subseteq X$$

where

- $\mathbf{0}_N = (\mathbf{1}_\emptyset, \mathbf{0}_N) =: a_0$ is an action considered neutral on all criteria.
- $(\mathbf{1}_i, \mathbf{0}_{N-i}) =: a_i$ is an action considered satisfactory on criterion i and neutral on the other criteria.
- $(\mathbf{1}_{ij}, \mathbf{0}_{N-ij}) =: a_{ij}$ is an action considered satisfactory on criteria i and j and neutral on the other criteria.

The following remark shows that the use of the binary alternatives can help to the determination of the 2-additive capacity.

Remark 1. *Let μ be a 2-additive capacity. We have*

- $C_\mu(u(a_0)) = 0;$
- $C_\mu(u(a_i)) = \mu_i = \phi_i^\mu - \dfrac{1}{2} \displaystyle\sum_{l \in N \setminus \{i\}} I_{il}^\mu;$
- $C_\mu(u(a_{ij})) = \mu_{ij} = \phi_i^\mu + \phi_j^\mu - \dfrac{1}{2} \displaystyle\sum_{l \in N \setminus \{i,j\}} (I_{il} + I_{jl}).$

We introduce the relation M modeling the natural monotonicity constraints $\mu_{ij} \geq \mu_i \geq 0$, $i, j \in N$ for a 2-additive capacity μ. Let $x, y \in \mathcal{B}$, $x\, M\, y$ if one of the following two conditions is satisfied:

1. $y = a_0$ and $not(x\ (P \cup I)\ a_0)$,
2. $\exists i, j \in N$ such that $[x = a_{ij},\ y = a_i]$ and $not[x\ (P \cup I)\ y]$.

Definition 2. *Given two binary alternatives x and y,*

- *The notation $x\ TC\ y$ means that there is a path from x to y, i.e., there exists $x_1, x_2, \ldots, x_p \in \mathcal{B}$ such that $x = x_1\ (P \cup I \cup M)\ x_2\ (P \cup I \cup M) \cdots (P \cup I \cup M)\ x_{p-1}\ (P \cup I \cup M)\ x_p = y$.*
- *A path of $(P \cup I \cup M)$ from x to x is called a cycle of $(P \cup I \cup M)$.*
- *$x\ TC_P\ y$ denotes a path from x to y containing a strict preference P.*

It is proven in [7] that, when the indifference relation is empty, the relation P is representable by a 2-additive Choquet integral if and only if the relation $(P \cup M)$ contains no strict cycle, i.e., a cycle containing an element of P.

4.2 Our Results When $I = \emptyset$

Lemma 1. *Let μ be a 2-additive capacity on a finite set of n criteria N. Let i, j be two different criteria. We have:*

$$\phi_i^\mu - \phi_j^\mu = \frac{1}{2}\left[(4 - n)(\mu_i - \mu_j) + \sum_{l \in N \setminus \{i,j\}} (\mu_{il} - \mu_{jl})\right]. \tag{8}$$

Proof. Let μ be a 2-additive capacity on N and $i, j \in N$. The expression of the importance of these criteria w.r.t. μ are

$$\phi_i^\mu = \mu_i + \frac{1}{2} \sum_{l \in N \setminus \{i\}} I_{il}^\mu$$

$$\phi_j^\mu = \mu_j + \frac{1}{2} \sum_{k \in N \setminus \{j\}} I_{jk}^\mu.$$

Then we have:

$$\phi_i^\mu - \phi_j^\mu = \mu_i + \frac{1}{2} I_{ij}^\mu + \frac{1}{2} \sum_{l \in N \setminus \{i,j\}} I_{il}^\mu - \mu_j - \frac{1}{2} I_{ij}^\mu - \frac{1}{2} \sum_{k \in N \setminus \{i,j\}} I_{jk}^\mu$$

$$= \mu_i + \frac{1}{2} \sum_{l \in N \setminus \{i,j\}} I_{il}^\mu - \mu_j - \frac{1}{2} \sum_{l \in N \setminus \{i,j\}} I_{jl}^\mu$$

$$= \mu_i - \mu_j + \frac{1}{2} \sum_{l \in N \setminus \{i,j\}} (\mu_{il} - \mu_i - \mu_l) - \frac{1}{2} \sum_{l \in N \setminus \{i,j\}} (\mu_{jl} - \mu_j - \mu_l)$$

$$= \mu_i - \mu_j + \frac{1}{2} \sum_{l \in N \setminus \{i,j\}} (\mu_{il} - \mu_i - \mu_l - \mu_{jl} + \mu_j + \mu_l)$$

$$= \mu_i - \mu_j + \frac{1}{2} \sum_{l \in N \setminus \{i,j\}} \mu_{il} - \frac{1}{2} \sum_{l \in N \setminus \{i,j\}} \mu_i - \frac{1}{2} \sum_{l \in N \setminus \{i,j\}} \mu_{jl} + \frac{1}{2} \sum_{l \in N \setminus \{i,j\}} \mu_j$$

$$= \frac{1}{2} \left[(4 - n)(\mu_i - \mu_j) + \sum_{l \in N \setminus \{i,j\}} (\mu_{il} - \mu_{jl}) \right]$$

$\square$

This Lemma will help us to prove our proposition of sufficient conditions to obtain the necessary importance relation of two given criteria. As the number $4 - n$ appears in Eq. (8), we examine three cases of these sufficient conditions: $n = 3$, $n = 4$ and $n \geq 5$.

Proposition 1. *Let P be a strict preference relation on $\mathcal{B}$ representable by a 2-additive Choquet integral.*

1. *Case $N = \{i, j, k\}$.*

$$If \begin{cases} a_i \; TC_P \; a_j \\ and \\ a_{ik} \; TC_P \; a_{jk} \end{cases} or \begin{cases} a_i \; TC_P \; a_{jk} \\ and \\ a_{ik} \; TC_P \; a_j \end{cases}, \text{ then we have } \phi_i^\mu > \phi_j^\mu, \text{ for all } \mu \in \mathcal{C}_2^{\{P\}},$$

i.e., criterion i is necessary more important than criterion j.

2. *Case $N = \{i, j, k, l\}$.*

$$If \begin{cases} a_{ik} \; TC_P \; a_{jk} \\ and \\ a_{il} \; TC_P \; a_{jl} \end{cases} or \begin{cases} a_{ik} \; TC_P \; a_{jl} \\ and \\ a_{il} \; TC_P \; a_{jk} \end{cases}, \text{ then we have } \phi_i^\mu > \phi_j^\mu, \text{ for all } \mu \in \mathcal{C}_2^{\{P\}},$$

i.e., criterion i is necessary more important than criterion j.

3. *Case $|N| \geq 5$.*
   Let $i, j \in N$. We set $N \setminus \{i, j\} = \{l_1, l_2, \ldots, l_{n-2}\}$.

   If there exists a permutation σ on $N \setminus \{i, j\}$ such that
   $$\begin{cases} a_j \ TC_P \ a_i \\ \quad and \\ a_{il_t} \ TC_P \ a_{j\sigma(l_t)}, t = 1, \ldots, n-2 \end{cases}, \ then \ we \ have \ \phi_i^\mu > \phi_j^\mu, \ for \ all \ \mu \in \mathcal{C}_2^{\{P\}},$$
   i.e., criterion i is necessary more important than criterion j.

Proof. 1. Case $N = \{i, j, k\}$.

From Lemma 1 we have $\phi_i^\mu - \phi_j^\mu = \dfrac{1}{2}\left[(\mu_i - \mu_j) + (\mu_{ik} - \mu_{jk})\right] = \dfrac{1}{2}\left[(\mu_i - \mu_{jk}) + (\mu_{ik} - \mu_j)\right]$.

Let $\mu \in \mathcal{C}_2^{\{P\}}$.

If $\begin{cases} a_i \ TC_P \ a_j \\ \quad and \\ a_{ik} \ TC_P \ a_{jk} \end{cases}$ then we have $\begin{cases} \mu_i > \mu_j \\ \quad and \\ \mu_{ik} > \mu_{jk} \end{cases}$. Hence $\phi_i^\mu > \phi_j^\mu$.

The proof is similar if $\begin{cases} a_i \ TC_P \ a_{jk} \\ \quad and \\ a_{ik} \ TC_P \ a_j \end{cases}$

2. Case $N = \{i, j, k, l\}$.

From Lemma 1 we have $\phi_i^\mu - \phi_j^\mu = \dfrac{1}{2}\left[(\mu_{ik} - \mu_{jk}) + (\mu_{il} - \mu_{jl})\right] = \dfrac{1}{2}\left[(\mu_{ik} - \mu_{jl}) + (\mu_{il} - \mu_{jk})\right]$.

Let $\mu \in \mathcal{C}_2^{\{P\}}$.

If $\begin{cases} a_{ik} \ TC_P \ a_{jk} \\ \quad and \\ a_{il} \ TC_P \ a_{jl} \end{cases}$ then we have $\begin{cases} \mu_{ik} > \mu_{jk} \\ \quad and \\ \mu_{il} > \mu_{jl} \end{cases}$. Hence $\phi_i^\mu > \phi_j^\mu$.

The proof is similar if $\begin{cases} a_{ik} \ TC_P \ a_{jl} \\ \quad and \\ a_{il} \ TC_P \ a_{jk}. \end{cases}$

3. Case $|N| \geq 5$.
   Let $i, j \in N$. Let $\mu \in \mathcal{C}_2^{\{P\}}$.

   If there exists a permutation σ on $N \setminus \{i, j\} = \{l_1, l_2, \ldots, l_{n-2}\}$ such that
   $$\begin{cases} a_j \ TC_P \ a_i \\ \quad and \\ a_{il_t} \ TC_P \ a_{j\sigma(l_t)}, t = 1, \ldots, n-2 \end{cases}, \ then \ we \ have$$
   $$\begin{cases} \mu_j > \mu_i \\ \quad and \\ \mu_{il_t} > \mu_{j\sigma(l_t)}, t = 1, \ldots, n-2. \end{cases}$$

Since $\phi_i^\mu - \phi_j^\mu = \frac{1}{2}\left[(4-n)(\mu_i - \mu_j) + \sum_{l \in N \setminus \{i,j\}} (\mu_{il} - \mu_{jl})\right]$ can be rewritten

$$\phi_i^\mu - \phi_j^\mu = \frac{1}{2}\left[(4-n)(\mu_i - \mu_j) + \sum_{h=1}^{n-2}(\mu_{il_h} - \mu_{jl_{\sigma(h)}})\right]$$

then we have $\phi_i^\mu > \phi_j^\mu$.

$\square$

Example 1. *We suppose that the DM in our previous example on hospitals expresses his preferences on the following set of binary alternatives:*

$$\mathcal{B} = \{a_0, a_1, a_2, a_3, a_4, a_{12}, a_{13}, a_{14}, a_{23}, a_{24}, a_{34}\}$$

. he provides these two strict preferences:

- $a_{13}\, P\, a_{24}$: *a satisfactory hospital on Activity and ALOS is strictly preferred to a satisfactory hospital on Notoriety and Technicality.*
- $a_{14}\, P\, a_{23}$: *a satisfactory hospital on Activity and Technicality is strictly preferred to a satisfactory hospital on Notoriety and ALOS.*

Using Proposition 1 for $|N| = 4$, we can conclude that the criterion Activity is necessary important than Notoriety.

Example 2. *Let $N = \{1, 2, 3, 4, 5\}$ and $P = \{(a_2, a_1); (a_{13}, a_{25}), (a_{14}, a_{23}), (a_{15}, a_{24})\}$.*

It is not difficult to see that the conditions given in Proposition 1, for $|N| = 5$, are satisfied. The permutation σ used here is $\sigma(3) = 5$, $\sigma(4) = 3$ and $\sigma(5) = 4$.

Therefore criterion 1 is necessary more important than criterion 2, even if the DM prefers a_2 to a_1.

Definition 3. *Let $\{P, I\}$ be a preference information on $\mathcal{B}$ representable by a 2-additive Choquet integral. Let $i, j \in N$.*

*j is p-**dominated** (possibly dominated) by i if there exists $l_0 \in N \setminus \{i, j\}$ such that the following conditions are satisfied:*

1. *$a_{il_0}\, TC_P\, a_0$;*
2. *for all $k \in N \setminus \{i, j\}$, $not(a_{jk}\, TC\, a_{il_0})$.*

This property ensures to have an element a_{il_0} not dominated by any element related to the criterion j. The next proposition shows that, in this case, we always have i possibly important than j.

Proposition 2. *Let P be a strict preference relation on $\mathcal{B}$ representable by a 2-additive Choquet integral. Let $i, j \in N$.*

If j is p-dominated by i, then there exists $\mu \in \mathcal{C}_2^{\{P\}}$ such that $\phi_i^\mu > \phi_j^\mu$.

In other words, if j is p-dominated by i then the criterion j is not necessary more important than i.

Proof. Let $i, j \in N$. We suppose that j is p-dominated by i, i.e., there exists $l_0 \in N \setminus \{i, j\}$ such that for all $k \in N \setminus \{i, j\}$, $not(a_{jk} \, TC \, a_{il_0})$ and $a_{il_0} \, TC_P \, a_0$.

We add to the binary relation $(P \cup M)$, another binary relation T on $\mathcal{B}$ defined by: for all $x, y \in \mathcal{B}$,

$$x \, T \, y \Leftrightarrow \begin{cases} x = a_{il_0}, \; y = a_{jk}, \; k \in N \setminus \{i, j\} \\ \text{and} \\ not(x(P \cup M)y) \end{cases}$$

Since P is representable by a 2-additive Choquet integral, $(P \cup M \cup T)$ contains no strict cycle. Then there exists a partition $\{\mathcal{B}_0, \mathcal{B}_1, ..., \mathcal{B}_m\}$ of $\mathcal{B}$, build by using an appropriate topological sorting on $(P \cup M \cup T)$, as the one detailed in Sect. 5.2. of [7].

Therefore there exists $p \in \{1, ..., m\}$ (since $a_{il_0} \, TC_P \, a_0$) such that $a_{il_0} \in \mathcal{B}_p$ and each element a_{jk}, $k \in N \setminus \{i, j\}$ belongs to a set $\mathcal{B}_{q_k}$ with $q_k \in \{0, ..., m\}$, $q_k < p$. We have also $a_{il_0} \, (P \cup M) \, a_i$ and $a_{jk} \, (P \cup M) \, a_j$. Hence $a_i \in \mathcal{B}_r$, $r < p$ and $a_j \in \mathcal{B}_{r'}$, $r' < p$, $r, r' \in \{0, ..., m\}$.

Let us define the mapping $f : \mathcal{B} \to \mathbb{R}$ and $\mu : 2^N \to [0, 1]$ as follows: For $l \in \{0, ..., m\}$,

$$\forall x \in \mathcal{B}_l, \; f(x) = \begin{cases} 0 & \text{if } l = 0 \\ (2n)^l & \text{otherwise.} \end{cases} \tag{9}$$

$$\begin{cases} \mu_{\emptyset} = 0 \\ \mu_i = \frac{f_i}{\alpha}, \; \forall i \in N \\ \mu_{ij} = \frac{f_{ij}}{\alpha}, \; \forall i, j \in N \\ \mu(K) = \displaystyle\sum_{\{i,j\} \subseteq K} \mu_{ij} - (|K| - 2) \sum_{i \in K} \mu_i, \; \forall K \subseteq N, |K| > 2. \end{cases} \tag{10}$$

where $f_i := f(a_i)$, $f_{ij} := f(a_{ij})$ and $\alpha = \displaystyle\sum_{\{i,j\} \subseteq N} f_{ij} - (n - 2) \sum_{i \in N} f_i$.

The capacity μ, defined like this, is 2-additive (see Proposition 7 in Sect. 5.3. of [7]). Since $p > r$ and $p > p_k$, $k \in N \setminus \{i, j\}$, we have

$$(2n)^p \geq (2n)(2n)^{p-1} \geq n(2n)^r + n \sum_{k \in N \setminus \{i,j\}} (2n)^{p_k} \geq (n - 4)(2n)^r + \sum_{k \in N \setminus \{i,j\}} (2n)^{p_k}$$

i.e.,

$$\mu_{il_0} \geq (n - 4)\mu_i + \sum_{k \in N \setminus \{i,j\}} \mu_{jk}.$$

Hence $\phi_i^\mu \geq \phi_j^\mu$ since

$$\phi_i^\mu - \phi_j^\mu = \frac{1}{2}\left[(4-n)(\mu_i - \mu_j) + \sum_{l \in N \setminus \{i,j\}} (\mu_{il} - \mu_{jl})\right]$$

$$= \frac{1}{2}\left[(n-4)\mu_j - (n-4)\mu_i + \sum_{l \in N \setminus \{i,j\}} \mu_{il} - \sum_{l \in N \setminus \{i,j\}} \mu_{jl}\right]$$

$$= \frac{1}{2}\left[(n-4)\mu_j \sum_{l \in N \setminus \{i,j,l_0\}} \mu_{il} + \mu_{il_0} - (n-4)\mu_i - \sum_{l \in N \setminus \{i,j\}} \mu_{jl}\right]$$

□

Example 3 *In the previous Example 1 related to the hospitals, we had a_{13} P a_{24} and a_{14} P a_{23}.*

Using these preferences, it is not difficult to check that the criterion 4 is p-dominated by the criterion Notoriety, criterion 3.

Indeed, if we choose $l_0 = 1$, then we have $not(a_{14}\ TC\ a_{13})$, $not(a_{24}\ TC\ a_{13})$ and $not(a_{34}\ TC\ a_{13})$.

As indicated in the previous proof, the binary relation T is added to $(P \cup M)$ as follows: $a_{13}T a_{14}$ and $a_{13}T a_{34}$.

Hence the partition of $\mathcal{B}$, obtained after the topological sorting on $(P \cup I \cup T)$ given in the previous proof, leads to the following 2-additive capacity μ:

- $\mathcal{B}_0 = \{a_0\} \longrightarrow 0$
- $\mathcal{B}_1 = \{a_1, a_2, a_3, a_4\} \longrightarrow \dfrac{(2 \times 4)^1}{\alpha} = \dfrac{8}{4800}$
- $\mathcal{B}_2 = \{a_{12}, a_{23}, a_{24}, a_{34}\} \longrightarrow \dfrac{(2 \times 4)^2}{\alpha} = \dfrac{64}{4800}$
- $\mathcal{B}_3 = \{a_{14}\} \longrightarrow \dfrac{(2 \times 4)^3}{\alpha} = \dfrac{512}{4800}$
- $\mathcal{B}_4 = \{a_{13}\} \longrightarrow \dfrac{(2 \times 4)^4}{\alpha} = \dfrac{4096}{4800}$

where $\alpha = 8^4 + 8^3 + 4 \times 8^2 - 2 \times 8 = 4800$

Therefore we have
$$\phi_3^\mu - \phi_4^\mu = \frac{1}{2}\left[(\mu_{13} - \mu_{14}) + (\mu_{23} - \mu_{24})\right] = \frac{1}{2}\left[\left(\frac{4096}{4800} - \frac{512}{4800}\right) + \left(\frac{64}{4800} - \frac{64}{4800}\right)\right] > 0.$$

5 Conclusion

We introduced the concepts of necessary and possible importance relation among the criteria. These notions allow to have a robust interpretation when the parameters of the 2-additive Choquet integral are inferred from a preference information given by the DM. Of course, in the direct elicitation process, the DM may give directly a preference over the Shapley value of two criteria. In this case, the necessary relation is obviously obtained. The main perspective will be the complete characterization of the necessary and possible importance relations. We provided here some sufficient conditions as a first step of this future work.

References

1. Choquet, G.: Theory of capacities. In: Annales de l'Institut Fourier, vol. 5, pp. 131–295 (1953)
2. Grabisch, M.: k-order additive discrete fuzzy measures and their representation. Fuzzy Sets Syst. **92**, 167–189 (1997)
3. Grabisch, M., Labreuche, C.: Fuzzy measures and integrals in MCDA. Multiple Criteria Decision Analysis: State of the Art Surveys. ISORMS, vol. 78, pp. 563–604. Springer, New York (2005). https://doi.org/10.1007/0-387-23081-5_14
4. Grabisch, M., Labreuche, C.: A decade of application of the Choquet and Sugeno integrals in multi-criteria decision aid. 4OR **6**, 1–44 (2008). https://doi.org/10.1007/s10288-007-0064-2
5. Greco, S., Mousseau, V., Slowinski, R.: Ordinal regression revisited: multiple criteria ranking using a set of additive value functions. Eur. J. Oper. Res. **51**(2), 416–436 (2008)
6. Mayag, B., Grabisch, M., Labreuche, C.: A characterization of the 2-additive Choquet integral through cardinal information. Fuzzy Sets Syst. **184**(1), 84–105 (2011)
7. Mayag, B., Grabisch, M., Labreuche, C.: A representation of preferences by the Choquet integral with respect to a 2-additive capacity. Theory Decis. **71**(3), 297–324 (2011)
8. Mayag, B.: A 2-additive Choquet integral model for french hospitals rankings in weight loss surgery. In: Carvalho, J.P., Lesot, M.-J., Kaymak, U., Vieira, S., Bouchon-Meunier, B., Yager, R.R. (eds.) IPMU 2016. CCIS, vol. 610, pp. 101–116. Springer, Cham (2016). https://doi.org/10.1007/978-3-319-40596-4_10
9. Mayag, B., Bouyssou, D.: Necessary and possible interaction between criteria in a 2 additive Choquet integral model. Eur. J. Oper. Res. **283**(1), 308–320 (2020)
10. Murofushi, T., Soneda, S.: Techniques for reading fuzzy measures (III): interaction index. In: 9th Fuzzy System Symposium, Sapporo, Japan, pp. 693–696, May 1993. (In Japanese)
11. Shapley, L.S.: A value for n-person games. In: Kuhn, H.W., Tucker, A.W. (eds.) Contributions to the Theory of Games, Annals of Mathematics Studies, vol. II, no. 28, pp. 307–317. Princeton University Press (1953)
12. Shapley, L.S.: A value for n-person games. In: Contributions to the Theory of games, pp. 307–317 (1953)

Measuring Polarization: A Fuzzy Set Theoretical Approach

Juan Antonio Guevara[1]([⊠]) [iD], Daniel Gómez[1,4] [iD], José Manuel Robles[2] [iD], and Javier Montero[3] [iD]

[1] Facultad de Estudios Estadísticos, Universidad Complutense de Madrid,
Madrid, Spain
juanguev@ucm.es, dagomez@estad.ucm.es
[2] Facultad de Ciencias Económicas y Empresariales,
Universidad Complutense de Madrid, Madrid, Spain
jmrobles@ccee.ucm.es
[3] Facultad de Ciencias Matemáticas, Universidad Complutense de Madrid,
Madrid, Spain
monty@mat.ucm.es
[4] Instituto Universitario de Evaluación Sanitaria,
Universidad Complutense de Madrid, Madrid, Spain

Abstract. The measurement of polarization has been studied over the last thirty years. Despite the different applied approaches, since polarization concept is complex, we find a lack of consensus about how it should be measured. This paper proposes a new approach to the measurement of the polarization phenomenon based on fuzzy set. Fuzzy approach provides a new perspective whose elements admit degrees of membership. Since reality is not black and white, a polarization measure should include this key characteristic. For this purpose we analyze polarization metric properties and develop a new risk of polarization measure using aggregation operators and overlapping functions. We simulate a sample of $N = 391315$ cases across a 5-likert-scale with different distributions to test our measure. Other polarization measures were applied to compare situations where fuzzy set approach offers different results, where membership functions have proved to play an essential role in the measurement. Finally, we want to highlight the new and potential contribution of fuzzy set approach to the polarization measurement which opens a new field to research on.

Keywords: Polarization · Fuzzy set · Ordinal variation

1 Introduction

Polarization is one of the most studied concepts in social sciences, specially over the last few years due to the recent growth of polarization episodes in

Supported by PGC2018-096509-B-I00 national project.

different scenarios. The concept of polarization has been studied in social sciences from different perspectives [1,6,14,16–18]. As can be seen after a fast literature review there is not an universal and accepted measure. As a consequence, there is not a well-defined consensus in the literature about what is the true nature of polarization.

Nevertheless, one of the most cited and used polarization measures was defined in the economics framework. Wolfson (1992) and Esteban and Ray (1994) among others, were some of the first authors in measuring polarization [5–7,19]. These polarization measures are strongly linked to the concept of inequality. Since then, a growing number of diverse polarization measures has arisen, most of them based on the idea given in [6] and [7], although there are others that have also had a significant impact [1,4,16,17].

It is important to remark that in the classical definition of Esteban and Ray of 1994 (and any other polarization measure based on Esteban and Ray) concepts like identification, membership, alienation and at the end aggregation are included in respective formulas.

Some of the previous concepts allow graduation and are vague in nature. For example, the way in which an individual feels identified with a group can be modeled by a fuzzy membership function. These functions represents the individual's membership degree in a given group.

Focusing in the idea of Esteban and Ray in the bipolar case (i.e there are two extreme situations), in this paper we propose a new polarization measure expressed in terms of fuzzy membership functions. These functions are aggregated by adequate aggregation operators to obtain a final polarization score.

2 Preliminaries

2.1 Polarization Measures

Polarization literature is essentially divided in two main approaches. First, those measures which only admit the existence of two groups, where the maximum polarization values are found in those cases where the group size is equal. According to this point of view, polarization follows a bimodal distribution (e.g.: Reynal-Querol, 2001 [17]). Otherwise, there are approaches which accept the presence of multiple groups. So that, those measures which take into account such diversity, are closer to terms like dispersion and variation. Since the more different values the more polarized is a population, the measure is moderated by the existence of two main groups with significant size. In this section, we focus our attention in measures based on diversity. Furthermore, we use the IOV index [2] as a reference measure in following sections because of its closeness to the concept of polarization.

– Esteban and Ray (1994): Being one of the first polarization measures proposed, in [6] it is defined the ER polarization measure. This measure was proposed because of the need of measuring the polarization concept, where inequality indices do not fit to this task. Esteban and Ray aimed to establish a difference between polarization and inequality proposing three main basics of a polarization measure. So that, must be:

1. a) high degree of homogeneity within groups.
2. b) high degree of heterogeneity between groups.
3. c) few number of groups with significant size.

To assess this, given a population of N individuals that take values X along a given numeric variable, in [6] the measurement of polarization is based on the *effective antagonism approach*. This is also called the IA *approach* that contains two concepts: *identification* (I) and *alienation* (a). The first one, reflects the degree in which a given individual feels closeness with the group that he/she belongs to. Otherwise, a shows the absolute distance between two individuals in terms of income.

Finally, the authors proposed the next polarization measure:

$$P(\pi, \mathbf{y}) = \sum_{i=1}^{n} \sum_{j=1}^{n} \pi_i \pi_j \mathbf{T}(\mathbf{I}(\pi_i), a(\delta(y_i, y_j))) \tag{1}$$

Where *identification* (I) is a function which depends on π_i group's relative size, *alienation* (a) reflects the absolute distance between the groups y_i and y_j. Regarding these two key aspects, *effective antagonism* is the product between I and a. It is worth mentioning that in the following years other authors have adapted ER measure (i.e. [4,15]). This model assume a symmetric alienation between individuals. Authors also established an asymmetric model in [6].

According to Eq. (1), the most used and simplified version of this formula (that we have denoted as P_S here) are reformulated for those cases in which the only information available about the population N is the variable $X = \{Xu, \, u \in N\}$ with its relative distribution: $\{(x_1, \pi_1), \ldots, (x_n, \pi_n)\}$. So that, the following assumptions are made:

- The population N is partitioned into groups according to different values of X. An individual $u \in N$ that takes the value x_i belongs to the group $y_i = \{u \in N \, X_u = x_i\}$.
- The relative frequency of the group y_i is denoted by π_i.
- The identification felt by one individual u to his/her group y_i depend by the relative size of that group. In fact, we have that this value is $I(u) = \pi_i^{\alpha}$. The value of α should be greater than 1 (see [6]).
- The value of $\delta((y_i, y_j))$ that represents the discrepancy between these two groups (y_i and y_j) is the absolute difference between the values that they takes in the variable X, (i.e $\delta(y_i, y_j) = |x_i - x_j|$).
- The alienation function a is the identification function.
- Finally, the function T is the product operator between the I and a.

Taking this into account, previous expression is commonly used as follow:

$$P_S(X, \alpha) = \sum_{i=1}^{n} \sum_{j=1}^{n} \pi_i \pi_j \left(\pi_i^{\alpha} |x_i - x_j|\right). \tag{2}$$

Example 1. Let us have a given population N with $|N| = 1500$. Let the variable X be a variable that takes values $\{1, 2, 3\}$ with a relative frequency of 0.3, 0.5 and 0.2. Then:

$$P_S(X, \alpha = 2) = (0.3 * 0.5 * 0.3^2) + (0.3 * 0.2 * 0.3^2 * 2) + \dots (0.2 * 0.5 * 0.2^2).$$

Let us recall again that the definition of Esteban and Ray of polarization (and of course its simplified version P_S) assumes certain hypothesis that it is necessary to remark. From now on, we will denote by C_i, the group that is denoted as y_i by Esteban and Rey.

1. The different values of income present at X determine how many K groups are. So that, given a set of responses X with a finite domain $D_X = \{x_i, i = 1 \dots K\}$, the class of groups $\mathcal{C} = \{C_1, \dots, C_K\}$ are perfectly defined as $C_i = \{j \in N, x_j = x_i\}$.
2. Individuals can be only assigned to one group, since the $\mathcal{C}$ is a partition of N. Also let us observe that $|C_l| = N\pi_l$.
3. Given a group C_l, if we denote $\mu_{Cl}(i)$ as the degree to which the individual i feels identification to the group C_l, this value is assumed to be:

$$\mu_{Cl}(i) = \begin{cases} \pi_l^\alpha & if \ x_i = x_l, \\ 0 & otherwise. \end{cases}$$

Taking previous considerations into account, formula (2) can be viewed as

$$P_S(X, \alpha) - \frac{1}{|N|} \sum_{i,j \in N} I_{ij} \tag{3}$$

where $I_{i,j} = \mathbf{T}(\mu_{Ci}(i), a(\delta(C_i, C_j)))$ is the effective antagonism felt by person i to individual j that is not symmetric. The effective antagonism of two individuals is the aggregation T of two values: the identification of individual i with the group to which he/she belongs ($\mu_{Ci}(i)$) and the alienation a of the discrepancy of the groups to which individuals i and j belong $\delta(C_i, C_j)$.

Example 2. Let $X = \{1, 2, 1, 3, 3, 4, 5, 5\}$ be a population of $N = 8$ individuals. Following the assumptions of Esteban and Ray we have 5 groups that are perfectly identified with those individuals that takes values in $D_X = \{1, 2, 3, 4, 5\}$. So the relative frequency are $\pi = (\frac{2}{8}, \frac{1}{8}, \frac{2}{8}, \frac{1}{8}, \frac{2}{8})$.
In order to obtain the final polarization score, we have to sum for each pair of relative frequencies the value $\mathbf{T}(I(i, \in C_i), a(\delta(C_i, C_j)))$. Now we are going to see how this expression performs starting with two individuals i, j. Let us assume that $i = 1$ and $j = 7$. Then their values x_1 and x_7 are 1 and 5 respectively. Assuming that $I(i, C_i) = \pi_i^\alpha$, $a(u) = u$, $T(u, v) = uv$ we have that the expression corresponding to the individuals 1 and 7 is

$$I_{1,7} = \pi_1^\alpha |5 - 1|.$$

If we chose $\alpha = 2$ for a clearer example, we have that $I_{1,7} = \frac{16}{64} = \frac{1}{4}$.

- IOV Blair and Lacy. Ordinal index variation.

The concept of polarization have been confusing frequently with variation. Variability, dispersion and variance are key concepts in Statistics, and they are main argument to describe both the distribution of random variables and to describe the observed values of a statistical variable. In the last context, according to [12] the measurement of dispersion is usually associated to continuous statistical variables. When the dispersion has to be measured in ordinal variables (like for example a Likert-scale) the common approach is to convert the ordinal estimation into a numerical one by assigning numerical values to each ordinal variable category. Afterwards it is then possible to use a classical dispersion measure. But some authors [3,8,9] have pointed out that this procedure can lead to misunderstanding and misinterpretation of the measurement results.

Hence, some ordinal dispersion measures have been defined [3,8,9] to properly deal with ordinal statistical variables instead of forcing the use of classical measures (such as entropy, standard deviation, variance or quasivariance) that do take into account such ordinal characteristic.

Although other measures could be alternatively used within our ordinal framework, in this paper we will focus on the ordinal dispersion measure defined by Berry and Mielke [2], usually called as IOV.

Given an ordinal variable with values $X = \{L_1, \ldots, L_n\}$ and a relative frequency vector $f = (f_1, \ldots, f_n)$, the ordinal dispersion measure IOV is defined as:

$$IOV = \sum_{i=1}^{n-1} \sum_{j=i+1}^{n} f_i f_j (j - i). \tag{4}$$

2.2 Aggregation Operators: Overlapping and Grouping Functions

Fuzzy set were introduced by Zadeh in 1965 [20], with the idea of sets with a continuum grades of membership, instead of the classical dual (yes/no) membership. Thus, as [13] remark in their work:

Definition 1. *A fuzzy set $\widetilde{A}$ over the domain X is defined as*

$$\widetilde{A} = \{(x, \mu_A(x))\ \ x \in X\}$$

where μ_A represents a membership degree function, i.e. $\mu_A : X \longrightarrow [0,1]$.

Aggregation Operators (AO) are one of the hottest disciplines in information sciences. AO appears in a natural way when the soft information has to be aggregated. At the beginning, AO were defined to aggregate values from membership functions associated to fuzzy set (see [20]). A key concept for the development of this paper is that of aggregation function.

Definition 2. *A function $A : [0,1]^n \to [0,1]$ is said to be an n-ary aggregation function if the following conditions hold:*

(A1) *A is increasing in each argument: for each $i \in \{1, \ldots, n\}$, if $x_i \leq y$, then*
$$A(x_1, \ldots, x_n) \leq A(x_1, \ldots, x_{i-1}, y, x_{i+1}, \ldots, x_n);$$
(A2) *A satisfies the boundary conditions: $A(0, \ldots, 0) = 0$ and $A(1, \ldots, 1) = 1$.*

It is important to emphasize that previous definition can be extended into a more general framework allowing to deal with more general objects than values into $[0, 1]$.

Given two degrees of membership $x = \mu_A(c)$ and $y = \mu_B(c)$ of an object c into classes A and B, an overlap function is supposed to yield the degree z up to which the object c belongs to the intersection of both classes. Particularly, an overlap function was defined in [10] as a particular type of bivariate aggregation function characterized by a set of commutative, natural boundary and monotonicity properties. These authors extended the bivariate aggregation function to a n-dimensional case.

Definition 3. *A function $O : [0, 1]^2 \rightarrow [0, 1]$ is said to be an overlap function if the following conditions hold:*

(O1) *O is commutative;*
(O2) *$O(x, y) = 0$ if and only if $xy = 0$;*
(O3) *$O(x, y) = 1$ if and only if $xy = 1$;*
(O4) *O is increasing in each argument;*
(O5) *O is continuous.*

Grouping functions are supposed to yield the degree z up to which the combination (grouping) of the two classes A and B is supported, that is, the degree up to which either A or B (or both) hold.

Definition 4. *A function $G: [0, 1]^2 \rightarrow [0, 1]$ is said to be a grouping function if the following conditions hold:*

(G1) *G is commutative;*
(G2) *$G(x, y) = 0$ if and only if $x = y = 0$;*
(G3) *$G(x, y) = 1$ if and only if $x = 1$ or $y = 1$;*
(G4) *G is increasing in each argument;*
(G5) *G is continuous.*

Overlap functions are particularly useful. Furthermore, their applicability can be extended to community detection problems [11] or even to edge detection methods in the field of computer vision.

3 A New Polarization Measure from a Fuzzy Set Perspective: The One-Dimensional and Bipolar Case

In this section we are focused on the case in which the only available information of a given population is a one-dimensional variable X.

This variable X could be incomes (as it is assumed in ER approach) or even opinions. Now, let us assume that this variable X, presents two poles X_A

and X_B. In this situation we will say that the variable X is bipolar or present two extreme values. Furthermore, we assume that the communication between individual for those extreme poles is broken, and thus, polarization does exist.

The only information we need to assume for the measure here propose is that we are able to measure the identification (or membership or closeness) of each individual with both extreme values/poles. Let us denote by μ_{X_A}, μ_{X_B} the two membership degree functions that represent the membership degree. $\mu_{X_A}, \mu_{X_B} : N \longrightarrow [0,1]$ are functions and for each $i \in N$, $\mu_{X_A}(i)$ and $\mu_{X_B}(i)$ represent the membership degree of individual i to the classes of *extreme opinion* X_A and *extreme opinion* X_B respectively (Fig. 1).

For this bipolar case, in which we assume the existence of two radical/extreme or poles opinions and we don't have a-priori groups, we understand that polarization is associated when the following two situations appear:

1. a) A significant part of population is close to the pole X_A.
2. b) A significant part of population is close to the pole X_B.

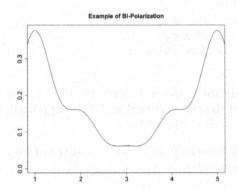

Fig. 1. A membership function of a bi-polarized population.

Also, as it happen with the ER case, we are going to assume that we are able to measure the discrepancy between these two poles or extreme situations by $\delta(X_A, X_B)$.

Finally, the polarization measure that we propose here can be expressed as the sum of the aggregation of three important concepts and could be understood as the risk of polarization. Let us remark that polarization appears when two groups break their relationships and also their communication.

We consider the risk of polarization between two individuals (e.g.: i, j) as the possibility of these two situations:

– How individual i is close to the extreme position X_A and j is close to the other pole X_B.
– How individual i is close to the extreme position X_B and j is close to the other pole X_A.

So that, if we assume that polarization appears in the last two situations we propose:

$$JDJ(X) = \sum_{i,j \in N, i \leq j} \varphi\left(\phi(\mu_{X_A}(i), \mu_{X_B}(j)), (\phi(\mu_{X_B}(i), \mu_{X_A}(j)))\right) \qquad (5)$$

where $\phi : [0,1]^2 \longrightarrow [0,1]$ is an overlapping aggregation operator and $\varphi : [0,1]^2 \longrightarrow [0,1]$ is a grouping function.

Example 3. To a better understand of the previous formula, let us analyze for a given pair of individuals $i, j \in N$ how the value is computed.

1. **Case 1. Individual i is close to the pole A and j is close to the pole B.** High polarization. In this case, we have that $\mu_{A,B}(i) = (\mu_A(i), \mu_B(i)) = (1, \beta)$ and $\mu_{A,B}(j) = (\mu_A(i), \mu_B(i)) = (\gamma, 1)$. Then we have to aggregate by a grouping function (φ) two values: the degree to which i belong to A and j to B $a = \phi(\mu_{X_A}(i), \mu_{X_B}(j)) = \phi(1,1) = 1$ and the degree to which i belong to B and j to A $b = \phi(\mu_{X_B}(i), \mu_{X_A}(j)) = \phi(0,0) = 0$.
Finally, we have to verify which one of these two facts are true by a grouping function $\varphi(1,0) = 1$, since this is a case with high polarization.

2. **Case 2. Individuals i and j are in the middle of poles A and B.** Here we propose the following case: $\mu_{A,B}(i) = (\mu_A(i), \mu_B(i)) = (0.5, 0.5)$ and $\mu_{A,B}(j) = (\mu_A(i), \mu_B(i)) = (0.5, 0.5)$. We have two individuals in the middle of the distribution. Then we have to aggregate by a grouping function the φ the results of two values: the degree to which i belong to A and j to B $a = \phi(\mu_{X_A}(i), \mu_{X_B}(j)) = \phi(0.5, 0.5) = 0.5$ and the degree to which i belong to B and j to A $b = \phi(\mu_{X_B}(i), \mu_{X_A}(j)) = \phi(0.5, 0.5) = 0.5$.
Finally, we have $\varphi(0.5, 0.5) = 0.5$. Since this is a case with medium polarization.

3. **Case 3. Individual i is close to pole A and j is also close to pole B.** $\mu_{A,B}(i) = (\mu_A(i), \mu_B(i)) = (1, 0)$ and $\mu_{A,B}(j) = (\mu_A(j), \mu_B(j)) = (1, 0.5)$. We have two individuals close to the pole A. No polarization case. Then we have to aggregate by a grouping function the φ the results of two value: the degree to which i belong to A and j to B $a = \phi(\mu_A(i), \mu_B(j)) = \phi(1, 0) = 0$ and the degree to which i belong to B and j to A $b = \phi(\mu_B(i), \mu_A(j)) = \phi(0, 1) = 0$.
Finally we have $\varphi(0, 0) = 0$. Since this is a case with low polarization.

Remark 1. Also let us note that since ϕ is an aggregation function, any increment of the three component will increase the polarization values for a fixed $i, j \in N$.

Remark 2. Those situations where a population X is partitioned into k groups, there will be as much groups as values are in the variable X (i.e. $D_X = \{X_1, \ldots, X_K\}$).
The previous bi-polarization index will be

$$JDJ(X) = \sum_{i=1,K} \sum_{j=1,K} \pi_i \pi_j \phi(\mu_{X_A}(i), \mu_{X_B}(j)) \qquad (6)$$

So that, the difference with the other polarization index is the way in which $I(i,j)$ are measured for each pair of the groups C_i, C_j.

4   A Comparison Between Polarization Measures in a 5-Liker Scale

Let us analyze the case in which we have a population with N individuals that takes values on a discrete/ordinal variable X with domain $D_x = \{1, 2, 3, 4, 5\}$. In order to build the JDJ polarization measure defined in the previous section, we need to chose the grouping, overlapping and membership functions that we are going to use. For simplicity, the grouping function that we have chosen is the Maximum aggregation operator. Furthermore, we are going to study two well-known overlapping functions: the minimum and the product. Finally we are going to analyze a triangular membership function μ^T (see Fig. 2):

$$(\mu_1^T(x), \mu_5^T(x)) = (1 - \frac{x-1}{4}, \frac{x-1}{4}) \tag{7}$$

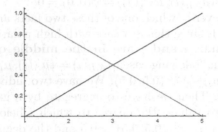

Fig. 2. Membership degree μ_1^T, μ_5^T for the two extreme poles $X_A = 1$, $X_B = 5$.

It is possible to consider different triangular membership functions too. For example, we can reduce the triangular membership function domain if we want to force the μ_1 function to zero for those values in x greater than a as well as to force the μ_5 function to be zero for those values in x lower than b.

For this experiment we have always considered $\varphi(u, v) = Max\{u, v\}$ and $\phi(x, y, z) = Min\{x, y, z\}$ or $\phi(x, y, z) = xyz$. We have also considered two options for JDJ polarization index denoted by $JDJ_{Max-Min-Tri(a=3,b=3)}$ and $JDJ_{Max-Pro-Tri}$.

In the following experiment, we reproduce a sample of $N = 391315$ cases and their different relative frequency distributions along a 5-Likert Scale with the same probability.

Therefore, three polarization measures were applied and compared for each case: Esteban and Ray measure (ER), and the last two index mentioned above $JDJ\_Max - Pro - Tri$ (JDJ_Pro) and $JDJ\_Max - Min - Tri(a = 3, b = 3)$ (JDJ_MIN). Furthermore, an index of ordinal variance (IOV) is applied. Thus, next table shows some descriptive statistics for the measures applied (Table 1).

As the following histograms suggest (see Fig. 3), it is worth mentioning the opposite skewed between ER and JDJ_Pro measures, finding lower ER values of polarization than JDJ_Pro. In fact, this is a key aspect between both measures

Table 1. Descriptive statistics for IOV, ER, JDJ_Pro_T and $JDJ\_MIN\_T\_a3b3$.

	Mean	SD	Median	Trimmed	Mad	Min	Max	Range	Skew	Kurtosis
IOV	0.60	0.17	0.60	0.60	0.19	0.00	1.00	1.00	−0.15	−0.49
ER	0.37	0.10	0.35	0.36	0.10	0.00	1.00	1.00	0.80	1.17
JDJ_Pro_T	0.75	0.13	0.76	0.76	0.12	0.00	1.00	1.00	−0.85	1.23
$JDJ\_MIN\_T\_a3b3$	0.28	0.18	0.27	0.27	0.20	0.00	1.00	1.00	0.41	−0.37

that underlies a difference between their conceptual and model properties that we explain below. Otherwise, low values of polarization are shown by JDJ_MIN. It is due to the membership function of JDJ_MIN.

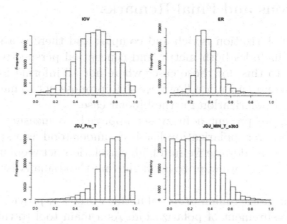

Fig. 3. Underlying frequency distribution for IOV, ER, JDJ_Pro_T and JDJ_MIN.

In Fig. 4 we show the relationship between each polarization measure and the IOV values grouped by deciles. We can see a natural tendency to find the higher polarization the higher IOV values. In fact, correlation between IOV and all polarization measures can be found in the figure below, where JDJ_Pro shows the highest correlation value (0.87), JDJ_MIN has a correlation value of 0.843 and ER shows the lower (0.78). Otherwise, we can see in ER measure a significant portion of medium values of polarization in the first decile of IOV, finding a lack of stability in this scenario.

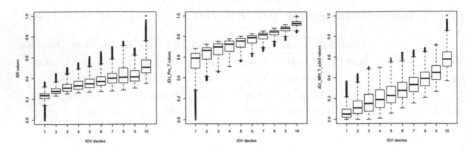

Fig. 4. Box plot displaying a distribution of polarization measures (ER, JDJ_Pro and JDJ_MIN_T) for each decile of IOV.

5   Conclusions and Final Remarks

The concept of polarization is rich and complex and there is a need to find an approach which includes both metric and conceptual perspectives at the same time. According to this, for those cases where not all information are available in data (such as communication flow), we shall propose not to mean polarization itself but a risk of polarization for the bipolar case.

In this work, we present a fuzzy set approach to measure the risk of bipolarization. Moreover, polarization has been understood as a synonym of variation. Regarding this, despite we find high correlation between ordinal variation and polarization values, we want to highlight that these small discrepancies make the difference.

Otherwise, as another main proposal in this work, is to provide a new methodology on the measurement of polarization. As a main tool to this new point of view, fuzzy set provides the appropriate resources. In one hand, in daily life people does not only feel identified with one single group but to some others too. Although, this duality is not a strict dichotomy but a long spectrum of nuances. Reality is fuzzy itself. As an example, an individual can be a strong supporter of a given political party but being identified with some contrary party proposals as well. In other hand, from a metric building perspective, using aggregation operators and membership functions, fuzzy set approach allows to pursue this philosophy. The membership functions used in this work are just a general example to apply this methodology. Along the different 391315 populations for a 5 likert scale, we have seen how the membership function determines the model behaviour. Specially for both measures proposed here, whose different membership functions reflect different results. Other membership functions more adequate are up to being develop for being applied.

Specifically, this key aspect has two main consequences: a) the frequency or bias to show high or low values of each polarization measure and b) those specific scenarios where high or low values should appear. It is important the equivalence between this membership functions and reality (e.g.: in those cases where individuals get clustered into two antagonistic groups, a given polarization measure should offer its highest values).

To conclude, we suggest for some directions for future research. Regarding membership functions, we consider as an important task to research about which membership function is more adequate for a given scenario. Furthermore, to develop new polarization measure incorporating a multi-dimensional case with multiples features. Moreover, including more theoretical polarization concepts like communication flow is needed to build an adequate polarization measure.

References

1. Apouey, B.: Measuring health polarization with self-assessed health data. Health Econ. **16**(9), 875–894 (2007)
2. Berry, K.J., Mielke Jr., P.W.: Indices of ordinal variation. Percept. Mot. Skills **74**(2), 576–578 (1992)
3. Blair, J., Lacy, M.: From the sage social science collections. Rights reserved. Sociol. Methods Res. **28**(3), 251–280 (2000)
4. Duclos, J.Y., Esteban, J., Ray, D.: Polarization: concepts, measurement, estimation. Econometrica **72**(6), 1737–1772 (2004). https://doi.org/10.1111/j.1468-0262.2004.00552.x
5. Esteban, J., Ray, D.: On the measurement of polarization, Boston university, institute for economic development, working paper 18 (1991)
6. Esteban, J.M., Ray, D.: On the measurement of polarization. Econ.: J. Econ. Soc. **62**(4), 819–851 (1994)
7. Foster, J.E., Wolfson, M.C.: Polarization and the decline of the middle class: Canada and the US mimeo. Vanderbilt University (1992)
8. Franceschini, F., Galetto, M., Varetto, M.: Qualitative ordinal scales: the concept of ordinal range. Qual. Eng. **16**(4), 515–524 (2004). https://doi.org/10.1081/QEN-120038013
9. Gadrich, T., Bashkansky, E.: Ordanova: Analysis of ordinal variation. J. Stat. Plan. Infer. **142**(12), 3174–3188 (2012). https://doi.org/10.1016/j.jspi.2012.06.004
10. Gómez, D., Rodriguez, J.T., Montero, J., Bustince, H., Barrenechea, E.: n-dimensional overlap functions. Fuzzy Sets Syst. **287**, 57–75 (2016)
11. Gomez, D., Rodríguez, J.T., Yanez, J., Montero, J.: A new modularity measure for fuzzy community detection problems based on overlap and grouping functions. Int. J. Approximate Reasoning **74**, 88–107 (2016)
12. Martínez, N., Gómez, D., Olaso, P., Rojas, K., Montero, J.: A novel ordered weighted averaging weight determination based on ordinal dispersion. Int. J. Intell. Syst. **34**(9), 2291–2315 (2019)
13. Montero, J., González-del Campo, R., Garmendia, L., Gómez, D., Rodríguez, J.T.: Computable aggregations. Inf. Sci. **460**, 439–449 (2018)
14. Morales, A.J., Borondo, J., Losada, J.C., Benito, R.M.: Measuring political polarization: Twitter shows the two sides of venezuela. Chaos: Interdiscip. J. Nonlinear Sci. **25**(3), 033114 (2015). https://doi.org/10.1063/1.4913758
15. Permanyer, I., D'Ambrosio, C.: Measuring social polarization with ordinal and categorical data. J. Publ. Econ. Theory **17**(3), 311–327 (2015)
16. Permanyer, I.: The conceptualization and measurement of social polarization. J. Econ. Inequality **10**(1), 45–74 (2012). https://doi.org/10.1007/s10888-010-9143-2
17. Reynal-Querol, M.: Ethnic and religious conflicts, political systems and growth. Ph.D. thesis, London School of Economics and Political Science (University of London) (2001)

18. Wang, Y.Q., Tsui, K.Y.: Polarization orderings and new classes of polarization indices. J. Publ. Econ. Theory **2**(3), 349–363 (2000). https://doi.org/10.1111/1097-3923.00042
19. Wolfson, M.C.: When inequalities diverge. Am. Econ. Rev. **84**(2), 353–358 (1994)
20. Zadeh, L.A.: Fuzzy sets. Inf. control **8**(3), 338–353 (1965)

New Methods for Comparing Interval-Valued Fuzzy Cardinal Numbers

Barbara Pękala[1][(✉)], Jarosław Szkoła[1], Krzysztof Dyczkowski[2], and Tomasz Piłka[2]

[1] Institute of Computer Sciences, University of Rzeszow, Rzeszów, Poland
{bpekala,jszkola}@ur.edu.pl
[2] Faculty of Mathematics and Computer Science, Adam Mickiewicz University,
Poznań, Poland
{chris,pilka}@amu.edu.pl

Abstract. In this contribution the concept how to solve the problem of comparability in the interval-valued fuzzy setting and its application in medical diagnosis is presented. Especially, we consider comparability of interval-valued fuzzy sets cardinality, where order of its elements is most important. We propose an algorithm for comparing interval-valued fuzzy cardinal numbers (IVFCNs) and we evaluate it in a medical diagnosis decision support system.

Keywords: Interval-valued Fuzzy Set (IVFS) · Interval-valued fuzzy cardinal number (IVFCN) · IV-order · Comparing IVFCN · Medical diagnosis

1 Introduction

Many new methods and theories behaving imprecision and uncertainty have been proposed since fuzzy sets were introduced by Zadeh [26]. Extensions of classical fuzzy set theory, intuitionistic fuzzy sets [3] and interval-valued fuzzy sets [22, 25] are very useful in dealing with imprecision and uncertainty (cf. [5] for more details). In this setting, different proposals for comparability relations between interval-valued fuzzy sets have been proposed (e.g. [23, 30]). However, the motivation of the present paper is to propose a new methods to comparability between interval-valued fuzzy sets (and their special type which are interval-valued fuzzy cardinal numbers) taking into account the widths of the intervals. We assume that the precise membership degree of an element in a given set is a number included in the membership interval. For such interpretation, the width of the membership interval of an element reflects the lack of precise membership degree of that element. Hence, the fact that two elements have the

This work was partially supported by the Centre for Innovation and Transfer of Natural Sciences and Engineering Knowledge of University of Rzeszów, Poland, the project RPPK.01.03.00-18-001/10.

same membership intervals does not necessarily mean that their corresponding membership values are the same. This is why we have taken into account the importance of the notion of width of intervals while proposing new algorithms.

Additionally, these developments are made according to the standard partial order between intervals, but also with respect to admissible orders [6], which are linear.

The paper is organized as follows. In Sect. 2 basic information on interval-valued setting are recalled. Especially, orders in interval setting and aggregation operators based on them are considered. Afterwards, in Sect. 3 we propose algorithm to compare interval values and finally in Sect. 4 we present mentioned methodology for comparing IVFCNs in the decision making used in the Ova-Expert system (intelligent decision support system for the diagnosis of ovarian tumors) (see [12, 13, 29]).

2   Preliminaries

Firstly, we recall some facts from interval-valued fuzzy set theory.

2.1   Orders in the Interval-Valued Fuzzy Settings

Definition 1 (cf. [22, 25]). *An interval-valued fuzzy set IVFS $\widetilde{A}$ in X is a mapping $\widetilde{A} : X \to L^I$ such that $\widetilde{A}(x) = [\underline{A}(x), \overline{A}(x)] \in L^I$ for $x \in X$, where*

$$\widetilde{A} \cap \widetilde{B} = \{\langle x, [\min\{\underline{A}(x), \underline{B}(x)\}, \min\{\overline{A}(x), \overline{B}(x)\}]\rangle : x \in X\},$$

$$\widetilde{A} \cup \widetilde{B} = \{\langle x, [\max\{\underline{A}(x), \underline{B}(x)\}, \max\{\overline{A}(x), \overline{B}(x)\}]\rangle : x \in X\}$$

and

$$L^I = \{[\underline{x}, \overline{x}] : \underline{x}, \overline{x} \in [0, 1], \underline{x} \le \overline{x}\}.$$

The well-known classical monotonicity (partial order) for intervals is of the form

$$[\underline{x}, \overline{x}] \le_{L^I} [\underline{y}, \overline{y}] \Leftrightarrow \underline{x} \le \underline{y}, \overline{x} \le \overline{y},$$

where $[\underline{x}, \overline{x}] <_{L^I} [\underline{y}, \overline{y}] \Leftrightarrow [\underline{x}, \overline{x}] \le_{L^I} [\underline{y}, \overline{y}]$ and $(\underline{x} < \underline{y}$ or $\overline{x} < \overline{y})$.

In L^I the operations joint and meet are defined respectively

$$[\underline{x}, \overline{x}] \vee [\underline{y}, \overline{y}] = [\max(\underline{x}, \underline{y}), \max(\overline{x}, \overline{y})],$$

$$[\underline{x}, \overline{x}] \wedge [\underline{y}, \overline{y}] = [\min(\underline{x}, \underline{y}), \min(\overline{x}, \overline{y})].$$

Note that the structure $(L^I, \vee, \wedge)$ is a complete lattice, with the partial order $\le_{L^I}$, where

$$\mathbf{1} = [1, 1] \quad and \quad \mathbf{0} = [0, 0]$$

are the greatest and the smallest element of $(L^I, \le_{L^I})$, respectively.

We are interested in extending the partial order $\le_{L^I}$ to a linear order, solving the problem of existence of incomparable elements. We recall the notion of an admissible order, which was introduced in [6] and studied, for example, in [2] and [27]. The linearity of the order is needed in many applications of real problems, in order to be able to compare any two interval data [7].

Definition 2 (cf. [6]). *An order $\leq_{Adm}$ in L^I is called admissible if it is linear and satisfies that for all $x, y \in L^I$, such that $x \leq_{L^I} y$, then $x \leq_{Adm} y$.*

Proposition 1 (cf. [6]). *Let $B_1, B_2 : [0,1]^2 \rightarrow [0,1]$ be two continuous aggregation functions, such that, for all $x = [\underline{x}, \overline{x}], y = [\underline{y}, \overline{y}] \in L^I$, the equalities $B_1(\underline{x}, \overline{x}) = B_1(\underline{y}, \overline{y})$ and $B_2(\underline{x}, \overline{x}) = B_2(\underline{y}, \overline{y})$ hold if and only if $x = y$. If the order $\leq_{B_{1,2}}$ on L^I is defined by $x \leq_{B_{1,2}} y$ if and only if*

$$B_1(\underline{x}, \overline{x}) < B_1(\underline{y}, \overline{y}) \quad or \quad (B_1(\underline{x}, \overline{x}) = B_1(\underline{y}, \overline{y}) \quad and \quad B_2(\underline{x}, \overline{x}) \leq B_2(\underline{y}, \overline{y})),$$

then $\leq_{B_{1,2}}$ is an admissible order on L^I.

Example 1 (cf. [6]). The following are special cases of admissible linear orders on L^I:

– The Xu and Yager order:

$$[\underline{x}, \overline{x}] \leq_{XY} [\underline{y}, \overline{y}] \Leftrightarrow \underline{x} + \overline{x} < \underline{y} + \overline{y} \text{ or } (\overline{x} + \underline{x} = \overline{y} + \underline{y} \text{ and } \overline{x} - \underline{x} \leq \overline{y} - \underline{y}).$$

– The first lexicographical order (with respect to the first variable), $\leq_{Lex1}$ defined as:

$$[\underline{x}, \overline{x}] \leq_{Lex1} [\underline{y}, \overline{y}] \Leftrightarrow \underline{x} < \underline{y} \text{ or } (\underline{x} = \underline{y} \text{ and } \overline{x} \leq \overline{y}).$$

– The second lexicographical order (with respect to the second variable), $\leq_{Lex2}$ defined as:

$$[\underline{x}, \overline{x}] \leq_{Lex2} [\underline{y}, \overline{y}] \Leftrightarrow \overline{x} < \overline{y} \text{ or } (\overline{x} = \overline{y} \text{ and } \underline{x} \leq \underline{y}).$$

– The $\alpha\beta$ order, $\leq_{\alpha\beta}$ defined as:

$$[\underline{x}, \overline{x}] \leq_{\alpha\beta} [\underline{y}, y] \Leftrightarrow K_\alpha(\underline{x}, \overline{x}) < K_\alpha(\underline{y}, \overline{y}) \text{ or}$$
$$(K_\alpha(\underline{x}, \overline{x}) = K_\alpha(\underline{y}, \overline{y}) \text{ and } K_\beta(\underline{x}, \overline{x}) \leq K_\beta(\underline{y}, \overline{y}))$$

for some $\alpha \neq \beta \in [0,1]$ and $x, y \in L^I$, where $K_\alpha : [0,1]^2 \rightarrow [0,1]$ is defined as $K_\alpha(x, y) = \alpha x + (1 - \alpha)y$.

The orders $\leq_{XY}$, $\leq_{Lex1}$ and $\leq_{Lex2}$ are special cases of the order $\leq_{\alpha\beta}$ with $\leq_{0.5\beta}$ (for $\beta > 0.5$), $\leq_{1,0}$, $\leq_{0,1}$, respectively. The orders $\leq_{XY}$, $\leq_{Lex1}$, $\leq_{Lex2}$, and $\leq_{\alpha\beta}$ are admissible linear orders $\leq_{B_{1,2}}$ defined by pairs of aggregation functions, namely weighted means. In the case of the orders $\leq_{Lex1}$ and $\leq_{Lex2}$, the aggregations that are used are the projections P_1, P_2 and P_2, P_1, respectively.

Remark 1. In the later part we will use the notation $\leq$ both for the partial or admissible linear order, with $\mathbf{0}$ and $\mathbf{1}$ as minimal and maximal element of L^I, respectively. Notation $\leq_{L^I}$ will be used while the results for the admissible linear orders will be used with the notation $\leq_{Adm}$.

2.2   Interval-Valued Aggregation Functions

Let us now recall the concept of an interval-valued aggregation function, or an aggregation function on L^I, which is an important notion for many applications. We consider interval-valued aggregation functions both with respect to $\leq_{L^I}$ and $\leq_{Adm}$. In many papers we may find the study of properties and possible applications of interval-valued operators/aggregation functions (e.g. [4,5,7,14,18,20]).

Definition 3 (cf. [16,27]). *An operation $\mathcal{A} : (L^I)^n \to L^I$ is called an interval-valued aggregation function if it is increasing with respect to the order $\leq$ (partial or total) and*

$$\mathcal{A}(\underbrace{0,...,0}_{n\times}) = 0, \quad \mathcal{A}(\underbrace{1,...,1}_{n\times}) = 1.$$

A special class of interval-valued aggregation functions is the one formed by the so-called representable interval-valued aggregation functions.

Definition 4 (cf. [9,11]). *An interval-valued aggregation function $\mathcal{A} : (L^I)^n \to L^I$ is said to be representable if there exist aggregation functions $A_1, A_2 : [0,1]^n \to [0,1]$ such that*

$$\mathcal{A}(x_1, \ldots, x_n) = [A_1(\underline{x}_1, \ldots \underline{x}_n), A_2(\overline{x}_1, \ldots, \overline{x}_n)]$$

for all $x_1, \ldots, x_n \in L^I$, provided that $A_1 \leq A_2$.

Example 2. Lattice operations $\wedge$ and $\vee$ on L^I are examples of representable aggregation functions on L^I with respect to the partial order $\leq_{L^I}$, with $A_1 = A_2 = \min$ in the first case and $A_1 = A_2 = \max$ in the second one. However, $\wedge$ and $\vee$ are not interval-valued aggregation functions with respect to $\leq_{Lex1}$, $\leq_{Lex2}$ or $\leq_{XY}$.

The following are other examples of representable interval-valued aggregation functions with respect to $\leq_{L^I}$.

– The representable arithmetic mean:

$$\mathcal{A}_{mean}([\underline{x}, \overline{x}], [\underline{y}, \overline{y}]) = [A_{mean}(\underline{x}, \underline{y}), A_{mean}(\overline{x}, \overline{y})] = [\frac{x+y}{2}, \frac{\overline{x}+\overline{y}}{2}].$$

– The representable geometric mean:

$$\mathcal{A}_{gmean}([\underline{x}, \overline{x}], [\underline{y}, \overline{y}]) = [A_{gmean}(\underline{x}, \underline{y}), A_{gmean}(\overline{x}, \overline{y})] = [\sqrt{\underline{x}\underline{y}}, \sqrt{\overline{x}\overline{y}}].$$

Representability is not the only possible way to build interval-valued aggregation functions with respect to $\leq_{L^I}$. Moreover, we may built interval-valued aggregation functions with respect to the other orders, i.e. $\leq_{Adm}$.

Let $A : [0,1]^2 \to [0,1]$ be an aggregation function.

- The function $\mathcal{A}_1 : (L^I)^2 \to L^I$, where

$$\mathcal{A}_1(x,y) = \begin{cases} [1,1], & if \ (x,y) = ([1,1],[1,1]), \\ [0,A(\underline{x},\overline{y})], & otherwise \end{cases}$$

is a non-representable interval-valued aggregation function with respect to $\leq_{L^I}$.

- The function $\mathcal{A}_2 : (L^I)^2 \to L^I$ ([19]), where

$$\mathcal{A}_2(x,y) = \begin{cases} [1,1], & if \ (x,y) = ([1,1],[1,1]) \\ [0,A(\underline{x},y)], & otherwise \end{cases}$$

is non-representable interval-valued aggregation functions with respect to $\leq_{Lex1}$.

- $\mathcal{A}_{mean}$ is an aggregation function with respect to $\leq_{\alpha\beta}$ (cf. [2]).
- The following function

$$\mathcal{A}_\alpha(x,y) = [\alpha\underline{x} + (1-\alpha)\underline{y}, \alpha\overline{x} + (1-\alpha)\overline{y}]$$

is an interval-valued aggregation function on L^I with respect to $\leq_{Lex1}, \leq_{Lex2}$ and $\leq_{XY}$ for $x,y \in L^I$ and $\alpha \in [0,1]$ (cf. [27]).

3   Subsethood Measure

Subsethood, or inclusion, measures have been studied mainly from constructive and axiomatic approaches and have been introduced successfully into the theory of fuzzy sets and their extensions. Many researchers have tried to relax the rigidity of Zadeh's definition of subsethood to get a soft approach which is more compatible with the spirit of fuzzy logic. For instance, Zhang and Leung (1996) defended that quantitative methods were the main approaches in uncertainty inference, a key problem in artificial intelligence, so they presented a generalized definition for subsethood measures, called *including degrees*.

3.1   Precedence Indicator

We use the notion of an interval subsethood measure for a pair of intervals with the partial and admissible orders and the width of intervals introduced and examined in [21].

Definition 5. *A function* $\mathrm{Prec} : (L^I)^2 \to L^I$ *is said to be a **precedence indicator** if it satisfies the following conditions for any* $a,b,c \in L^I$

P1 *if* $a = 1_{L^I}$ *and* $b = 0_{L^I}$, *then* $\mathrm{Prec}(a,b) = 0_{L^I}$,
P2 *if* $a < b$, *then* $\mathrm{Prec}(a,b) = 1_{L^I}$ *for any* $a,b \in L^I$,
P3 $\mathrm{Prec}(a,a) = [1 - w(a), 1]$ *for any* $a \in L^I$,
P4 *if* $a \leq b \leq c$ *and* $w(a) = w(b) = w(c)$, *then* $\mathrm{Prec}(c,a) \leq \mathrm{Prec}(b,a)$ *and* $\mathrm{Prec}(c,a) \leq \mathrm{Prec}(c,b)$, *for any* $a,b,c \in L^I$, *where*

$$w(a) = \overline{a} - \underline{a}. \tag{1}$$

The following construction method is based on the aggregation and negation functions which play important rule in many applications (e.g. [4,9,11,14]) and is presented in the next theorem. Recall that an interval-valued fuzzy negation N_{IV} is an antytonic operation that satisfies $N_{IV}(0_{L^I}) = 1_{L^I}$ and $N_{IV}(1_{L^I}) = 0_{L^I}$ ([1,10]).

Proposition 2 ([21]). *For $a, b \in L^I$ the operation $\mathrm{Prec}_{\mathcal{A}} : (L^I)^2 \to L^I$ is the precedence indicator*

$$\mathrm{Prec}_{\mathcal{A}}(a, b) = \begin{cases} [1 - w(a), 1], & a = b, \\ 1_{L^I}, & a < b, \\ \mathcal{A}(N_{IV}(a), b), & otherwise \end{cases}$$

for $a, b \in L^I$ and the interval-valued fuzzy negation N_{IV}, such that

$$N_{IV}(a) = [N(\overline{a}), N(\underline{a})] \le [1 - \overline{a}, 1 - \underline{a}],$$

where N is a fuzzy negation and $\mathcal{A}$ is a representable interval-valued aggregation such that $\mathcal{A} \le \vee$.

Using the construction methods from Proposition 2 we obtain the following examples.

Example 3. The following function is an interval subsethood measure with respect to $\le_{L^I}$:

$$\mathrm{Prec}_{\mathcal{A}_{mean L^I}}(x, y) = \begin{cases} [1 - w(x), 1], & x = y, \\ 1, & x <_{L^I} y, \\ [\frac{1 - \overline{x} + y}{2}, \frac{1 - \underline{x} + \overline{y}}{2}], & otherwise, \end{cases}$$

where $N_{IV}(x) = [1 - \overline{x}, 1 - \underline{x}]$.
Moreover, the following function is a subsethood measure with respect to $\le_{Lex2}$:

$$\mathrm{Prec}_{\mathcal{A}_{mean Lex2}}(x, y) = \begin{cases} [1 - w(x), 1], & x = y, \\ 1, & x <_{Lex2} y, \\ [\frac{y}{2}, \frac{1 - \overline{x} + \overline{y}}{2}], & otherwise. \end{cases}$$

Using the interval-valued aggregation function $\mathcal{A}_\alpha$ for $\alpha \in [0, 1]$, we get the subsethood measure

$$\mathrm{Prec}_{\mathcal{A}_\alpha Lex2}(x, y) =$$
$$\begin{cases} [1 - w(x), 1], & x = y, \\ 1, & x <_{Lex2} y, \\ [(1 - \alpha)\underline{y}, & \\ \quad \alpha(1 - \overline{x}) + (1 - \alpha)\overline{y}], & otherwise, \end{cases}$$

where

$$N_{IV}(x) = \begin{cases} 1, & x = 0, \\ [0, 1 - \overline{x}], & otherwise \end{cases}$$

is an interval-valued fuzzy negation with respect to $\le_{Lex2}$.

Remark 2 (cf. [27]). The aggregation $\mathcal{A}_\alpha$ preserves the width of the intervals of the same width.

Another construction method, which is inspired by the construction presented for generalization of the subsethood measure at paper [21], presents the following proposition.

Proposition 3. *The operation*

$$\text{Prec}_w(a,b) = \begin{cases} 1_{L^I}, & a < b, \\ [1 - \max(w(a), r(a,b)), 1 - r(a,b)], & else \end{cases}$$

is the precedence indicator with respect to $\leq$, where for $a, b \in L^I$
$r(a,b) = \max\{|\underline{a} - \underline{b}|, |\overline{a} - \overline{b}|\}$.

3.2 Interval-Valued Fuzzy Cardinal Numbers (IVFCNs)

In this section we briefly introduce main ideas about cardinalities of IVFSs. More details can be found in the monographs [12,24]. Such numbers are of great importance in solving decision problems in which uncertainty occurs (see [8,15,28]). In further part we will use the following notations:

- For given fuzzy set A a symbol $[A]_i$ is defined as:

$$[A]_i := \bigvee \{t \in (0,1]: |A_t| \geq i\} \text{ for } i \in \mathbb{N}.$$

- Function $f : [0,1] \rightarrow [0,1]$ is called cardinality pattern if it meets the following conditions:
  1. is nondecreasing i.e. $\forall_{a,b\in[0,1]} f(a) \leq f(b)$ if $a \leq b$,
  2. and meets limit conditions $f(0) = 0$ i $f(1) = 1$.
- Symbol $\cap_T$ means the triangular norm and N the fuzzy negation.

Generalized Fuzzy Cardinal Numbers

1. **Generalized cardinal number FGCount** is interpreted as a degree to which fuzzy set A has at least k elements

$$FG_f(k) := f([A]_1) \cap_T f([A]_2) \cap_T \ldots \cap_T f([A]_k) \text{ for } k \in \mathbb{N}.$$

2. **Generalized cardinal number FLCount** is interpreter as a degree to which A includes at most k elements

$$FL_f(k) := N(f([A]_{k+1})) \cap_T N(f([A]_{k+2})) \cap_T \cdot \cap_T N(f([A]_n)) \text{ for } k \in \mathbb{N}.$$

3. **Generalized cardinal number FECount** expresses the degree to which A has exactly k elements where

$$FE_f(k) := f([A]_1) \cap_T f([A]_2) \cap_T \ldots \cap_T f([A]_k) \cap_T$$

$$N(f([A]_{k+1})) \cap_T N(f([A]_{k+2})) \cap_T \ldots \cap_T N(f([A]_n)) \text{ for } k \in \mathbb{N}.$$

FE_f is the intersection of FG_f and FL_f. It may be perceived as the **'actual'** **generalized cardinal number** of a fuzzy set A.

Fuzzy Cardinality of IVFS. Cardinalities of interval-valued fuzzy sets are defined in a natural manner using cardinalities of fuzzy sets described in previous section.

For a finite interval-valued fuzzy set $\widetilde{A} = [\underline{A}, \overline{A}]$ fuzzy type cardinalities are defined as interval-valued fuzzy sets in $\mathbb{N}$ (see [24]).

Definition 6 (cf. [12,24]).

1. *f-FGCount of IVFS $\widetilde{A}$ for a given cardinality pattern f is defined as:*

$$\widetilde{FG}_f(\widetilde{A}) = [FG_f(\underline{A}), FG_f(\overline{A})], \tag{2}$$

i.e. for $k \in \mathbb{N}$:

$$\widetilde{FG}_f(\widetilde{A})(k) = [FG_f(\underline{A})(k), FG_f(\overline{A})(k)] = [f([\underline{A}]_k), f([\overline{A}]_k)], \tag{3}$$

where $FG_f(\underline{A})$ and $FG_f(\overline{A})$ are the fuzzy cardinalites defined in previous section.

2. *f-FLCount of IVFS $\widetilde{A}$ for a given cardinality pattern f is defined as:*

$$\widetilde{FL}_f(\widetilde{A}) = [FL_f(\overline{A}), FL_f(\underline{A})], \tag{4}$$

i.e. for $k \in \mathbb{N}$:

$$\widetilde{FL}_f(\widetilde{A})(k) = [FL_f(\underline{A})(k), FL_f(\overline{A}(k)] = [1 - f([\overline{A}]_{k+1}), 1 - f([\underline{A}]_k)], \tag{5}$$

where $FL_f(\underline{A})$ and $FL_f(\overline{A})$ are the fuzzy cardinalities defined in previous section.

3. *f-FECount of IVFS $\widetilde{A}$ for a given cardinality pattern f is defined as:*

$$\widetilde{FE}_f(\widetilde{A}) = \widetilde{FG}_f(\widetilde{A}) \cap \widetilde{FL}_f(\widetilde{A}), \tag{6}$$

i.e. for $k \in \mathbb{N}$:

$$\widetilde{FE}_f(\widetilde{A})(k) = [f([\underline{A}]_k) \wedge (1 - f([\overline{A}]_{k+1})), f([\overline{A}]_k) \wedge (1 - f([\underline{A}]_{k+1})]. \tag{7}$$

To simplify the notations, f-FECount of an IVFS will be denoted by $\widetilde{\sigma}$ and we will call it Interval-Valued Fuzzy Cardinal Number (in short IVFCN).

Comparability Algorithm of IVFCNs. In many decision-making applications, an important problem to solve is comparing the cardinalities of IVFSs.

To define algorithm more formally, we need to introduce some basics notations.

Definition 7. *Representative $Rep(x) \in \mathbb{R}$ of an interval $x \in L^I$ for $\alpha \in [0, 1]$ is defined as:*

$$Rep(x) = K_\alpha(x) = \underline{x} + \alpha * w(x), \tag{8}$$

where w fulfills (1).

In a special case value of representative of an interval can lead to middle or bounds of interval:

1. if $a = 0$ then $Rep(x) = \underline{x}$ - lower bound of the interval;
2. if $a = 1$ then $Rep(x) = \overline{x}$ - upper bound of the interval;
3. if $a = 0.5$ then $Rep(x) = (\underline{x} + \overline{x})/2$ - middle of the interval.

To order the IVFCNs we propose the following method inspired on [17] used in the next section for group decision making.

Algorithm for Ordering of IVFCNs

Step 1 (Inputs). Input data: interval-valued fuzzy cardinal numbers: $X_i \in IVFCNs$, $i = \{1, \ldots, n\}$, $n \in \mathbb{N}$

$$X_i = \{[\underline{x}_k, \overline{x}_k], \quad where \quad k = \{1, \ldots, m\} \; m \in \mathbb{N}\}.$$

Step 2 (Representatives). For input data with intervals data, we need the following process (representatives (see (8))):

$$Rep(x_k) = \underline{x}_k + \alpha * w(x_k), \quad where \quad \alpha \in [0, 1]$$

and we obtain $\mathbf{x}_i = \{Rep(x_k)\}$.

Step 3 (Aggregations). For m elements support we construct aggregations matrix with difference aggregation functions Let $\mathbf{F} = (F_1, \ldots F_m)$ be a sequence of m aggregation functions, $F_i : [0, 1]^m \rightarrow [0, 1]$, then we calculate

$$F_1(\mathbf{x}_1), \ldots, F_m(\mathbf{x}_1)$$
$$\ldots$$
$$F_1(\mathbf{x}_n), \ldots, F_m(\mathbf{x}_n)$$

for $\mathbf{x}_1, \ldots \mathbf{x}_n$ and we calculate for each $1 \leq i, j \leq k$ the measure of connectivity for pairs of values $\mathbf{x}_i$ and $\mathbf{x}_j$, $i \neq j$,

$$CON(\mathbf{x}_i, \mathbf{x}_j) = \sum_{1 \leq l \leq m} (F_l(\mathbf{x}_i(t)) - F_l(\mathbf{x}_j(t))), \quad 1 \leq t \leq n.$$

Step 4 (Selection). For each $1 \leq i, j \leq n$ we find

$$\max_{1 \leq i, j \leq n} CON(\mathbf{x}_i, \mathbf{x}_j) = CON(\mathbf{x}_z, \mathbf{x}_w), \quad 1 \leq z, w \leq n;$$

The element $\mathbf{x}_z$ is chosen as most appropriate.
We repeat Step Selection by omitting the wining values $\mathbf{x}_z$ in the next iteration. If $\max_{1 \leq i, j \leq n} CON(\mathbf{x}_i, \mathbf{x}_j) = CON(\mathbf{x}_z, \mathbf{x}_w) = CON(\mathbf{x}_{z'}, \mathbf{x}_{w'}), 1 \leq z, z', w, w' \leq n$, then we find

$$C_1 := \sum_{1 \leq w \leq n} (\max(0, CON(\mathbf{x}_z, \mathbf{x}_w)))$$

and
$$C_2 := \sum_{1 \leq w' \leq n} (\max((0, CON(\mathbf{x}_{z'}, \mathbf{x}_{w'}))).$$

If $C_1 > C_2$ then the element $\mathbf{x}_z$ is chosen as the most appropriate otherwise the element $\mathbf{x}_{z'}$ is chosen as most appropriate. Otherwise if both C_1 and C_2 are equal than $\mathbf{x}_z = \mathbf{x}_{z'}$.

As a consequence we obtain the sequence: $\mathbf{x}_z \succeq \mathbf{x}_{z'} \succeq \ldots$ and thus $\mathbf{X}_z \succeq \mathbf{X}_{z'} \succeq \ldots$. In particular, if we compare only two IVFCNs, i.e two sequences of representatives $\mathbf{x}, \mathbf{y}$, then $\mathbf{x} \succeq \mathbf{y}$ if $CON(\mathbf{x}, \mathbf{y}) > CON(\mathbf{y}, \mathbf{x})$ else if $CON(\mathbf{x}, \mathbf{y}) < CON(\mathbf{y}, \mathbf{x})$, then $\mathbf{x} \preceq \mathbf{y}$, otherwise they are equivalent.

4 Application in Decision Making

The presented methodology for comparing IVFCNs can be applied in the decision model used in the OvaExpert system. OvaExpert is an intelligent decision support system for the diagnosis of ovarian tumors. The system was developed as a result of joint research of two Polish research centers: the Division of Gynecologic Surgery of the Poznan University of Medical Sciences and the Department of Imprecise Information Processing Methods, Faculty of Mathematics. More detailed information about the system can be found in [12].

Figure 1 presents diagram showing OvaExpert counting approach for making decisions. This method of decision making utilised in the system is based on voting strategy with counting. On input system gets incomplete information about patient. In Step 1 many diagnostic models are computed and it results as IVFS of decisions. Then in Step 2 two IVFSNs are computed (which represents positive and negative diagnosis). And finally in Step 3 comparison of this two IVFSNs resulting decision. This step utilize methods presented earlier in this work.

4.1 Decision Making Algorithm Based on Bipolar Voting Strategy

The idea behind decision algorithm is to use bipolar perspective on IVFS. Because such an IVFS contains information on uncertainty level, it carries both information supporting and rejecting the decision. This property of IVFS is used in decision algorithm. The basic idea behind this algorithm consists of a couple of steps:

- As an input we have two IVFS's P and C (representing number of decision's D_{pro} and D_{contra} supporting given decision):
    $P = \sigma(D_{pro})$ - representing the number of decisions 'for';
    $C = \sigma(D_{contra})$ - representing the number of decisions 'against';
- To make decisions, we must choose a set that is more numerous e.g. decide if (or vice versa):
$$P < C.$$

For equivalency of P and C (see Algorithm. Ordering of IVFCNs) then we do not make decisions.

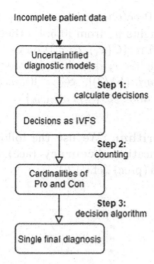

Fig. 1. OvaExpert counting approach for making medical decision

4.2 Results and Discussion

The presented algorithm have been tested on real medical data. These data described 388 cases of patients diagnosed and treated in the Division of Gynecological Surgery, Poznan University of Medical Sciences, between 2005 and 2015. Out of them 61% have been diagnosed as suffering from benign tumours and 39% as suffering from malign tumours. Moreover, 56% of patients had full diagnostic (no test required by diagnostic scales was missing), 40% had significant amounts of missing data varying from (0%, 50%], and for the remaining ones 50% of data was missing. Detailed description of data used for evaluation can be found in [12].

The goal of evaluation was to select a decision algorithm that would best classify malignity cases with the top possible decisiveness.

We tested the algorithm for the weighted average functions of F_i in $\mathbf{F}$ (see Step 3 of Algorithm. Ordering of IVFCNs).

(Weighted Average). We will use the following case of aggregation functions, i.e. arithmetic weight mean $AW_w : ([0, 1])^m \rightarrow [0, 1]$

$$AW_w(x_1, ..., x_m) = \frac{\sum_{i=1}^m w_i x_i}{\sum_{i=1}^m w_i}, \tag{9}$$

where the adequate vectors $\mathbf{w} = (w_1, ...w_m)$ we generate in the following way: In the first step divide a sum of supports of both IVFCNs $S = supp(P) \cup supp(C)$ into two equal parts S_1 and S_2. In the second step compare precedence indicators from both parts separately in the following way:

If $\mathrm{Prec}(P(x_i), C(x_i)) \leq \mathrm{Prec}(C(x_i), P(x_i))$ on S_1

then we generate weights w_i from $[0.5, 1]$ otherwise from $[0, 0.5]$;

If $\mathrm{Prec}(P(x_i), C(x_i)) > \mathrm{Prec}(C(x_i), P(x_i))$ on S_2

then we generate weights w_i from $[0.5, 1]$ otherwise from $[0, 0.5]$, where: $S_1 = [min(S), (max(S) - min(S)/2)]$, $S_2 = [(max(S) - min(S))/2, max(S)]$ and $\mathrm{Prec} \in \{\mathrm{Prec}_{\mathcal{A}}, \mathrm{Prec}_w\}$ and $\leq \in \{\leq_{LI}, \leq_{Adm}\}$.

Evaluation of the Algorithm. We use the following notations for evaluate the results of the classification: **Accuracy** (acc), **Sensitivity** (sen), **Specificity** (spec) and **Precision** (prec) (cf. [12]).

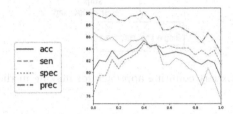

Fig. 2. Impact of α selection on decision quality

The Table 1 below presents analysis of different parameters of α (different methods to calculate representatives). We check values of α by different criteria, i.e., the best acc, sen, spec, prec, respectively. We present the best results obtained in series of tests for Prec_w and Xu and Yager linear order. We observe that we obtained comparable results to [12]. With comparison to OEA baseline model (which is current decision method in OvaExpert system) we see that new algorithm give the better results of specificity, but very similar values of other performance measures (like sensitivity and accurency). Figure 2 shows the impact of α selection on decision quality. So proposed algorithms are interesting from point of view possible applications and in other data also may be better methods to compare IV values.

Table 1. The best evaluation results for different α

F/	crit. of α	the best α	acc	sen	spec	prec
	acc/sen/prec	0.4	**85.31**	**84.88**	85.96	**90.12**
	spec	0	80.34	76.21	**86.79**	90.0

To obtain the best acc and sen values, the α values should be in the range of $[0.15, 0.5]$, which indicates a strong relationship between these values and the designated representatives.

For spec and prec values, the initial range of $[0, 0.45]$ is relatively neutral, only after exceeding the value of 0.45 for α, the above variables have a decreasing trend. To obtain the best results for all parameters, the recommended alpha values should be in the range $[0.15, 0.45]$.

5   Conclusions and Future Plans

In this presentation, we discuss possible algorithms for compare in interval-valued fuzzy setting, where these notions with widths of intervals involved. Moreover, new algorithms of comparing and ranking cardinalities of IVFS were applied in decision making algorithm. In future we will test presented algorithms for other types data. Moreover, we are currently working on several other types of ranking methods that uses aggregation functions.

References

1. Asiain, M.J., Bustince, H., et al.: Negations with respect to admissible orders in the interval-valued fuzzy set theory. IEEE Trans. Fuzzy Syst. **26**(2), 556–568 (2018)
2. Asiain, M.J., et al.: About the use of admissible order for defining implication operators. In: Carvalho, J.P., Lesot, M.-J., Kaymak, U., Vieira, S., Bouchon-Meunier, B., Yager, R.R. (eds.) IPMU 2016. CCIS, vol. 610, pp. 353–362. Springer, Cham (2016). https://doi.org/10.1007/978-3-319-40596-4_30
3. Atanassov, K.T.: Intuitionistic fuzzy sets. Fuzzy Sets. Syst. **20**(1), 87–96 (1986)
4. Bentkowska, U., Pękala, B., et al.: N-reciprocity property for interval-valued fuzzy relations with an application to group decision making problems in social networks. Int. J. Uncertain. Fuzziness Knowl. Based Syst. **25**(Suppl. 1), 43–72 (2017)
5. Bustince, H., Barrenechea, E., et al.: A historical account of types of fuzzy sets and their relationships. IEEE Trans. Fuzzy Syst. **24**(1), 179–194 (2016)
6. Bustince, H., Fernandez, J., et al.: Generation of linear orders for intervals by means of aggregation functions. Fuzzy Sets Syst. **220**, 69–77 (2013)
7. Bustince, H., Galar, M., et al.: A new approach to interval-valued choquet integrals and the problem of ordering in interval-valued fuzzy set applications. IEEE Trans. Fuzzy Syst. **21**(6), 1150–1162 (2013)
8. Dąbrowski, A., Matczak, P., et al.: Identification of experimental and control areas for CCTV effectiveness assessment-the issue of spatially aggregated data. ISPRS Int. J. Geo-Inf. **7**(12), 471 (2018)
9. Deschrijver, G.: Quasi-arithmetic means and OWA functions in interval-valued and Atanassov's intuitionistic fuzzy set theory. In: Proceedings of the 7th Conference of the European Society for Fuzzy Logic and Technology, pp. 506–513. Atlantis Press (2011)
10. Deschrijver, G., Cornelis, C.: Representability in interval-valued fuzzy set theory. Int. J. Uncertain. Fuzziness Knowl. Based Syst. **15**(03), 345–361 (2007)
11. Deschrijver, G., Kerre, E.E.: Implicators based on binary aggregation operators in interval-valued fuzzy set theory. Fuzzy Sets Syst. **153**(2), 229–248 (2005)
12. Dyczkowski, K.: Intelligent Medical Decision Support System Based on Imperfect Information. The Case of Ovarian Tumor Diagnosis. Studies in Computational Intelligence. Springer, Cham (2018). https://doi.org/10.1007/978-3-319-67005-8

13. Dyczkowski, K., Wójtowicz, A., Żywica, P., Stachowiak, A., Moszyński, R., Szubert, S.: An intelligent system for computer-aided ovarian tumor diagnosis. In: Filev, D., et al. (eds.) Intelligent Systems 2014. AISC, vol. 323, pp. 335–343. Springer, Cham (2015). https://doi.org/10.1007/978-3-319-11310-4_29

14. Elkano, M., Sanz, J.A., Galar, M., Pękala, B., Bentkowska, U., Bustince, H.: Composition of interval-valued fuzzy relations using aggregation functions. Inf. Sci. **369**, 690–703 (2016)

15. Jasiulewicz-Kaczmarek, M., Żywica, P.: The concept of maintenance sustainability performance assessment by integrating balanced scorecard with non-additive fuzzy integral. Eksploatacja i Niezawodność **20** (2018)

16. Komorníková, M., Mesiar, R.: Aggregation functions on bounded partially ordered sets and their classification. Fuzzy Sets Syst. **175**(1), 48–56 (2011)

17. Miguel, L.D., Sesma-Sara, M., et al.: An algorithm for group decision making using n-dimensional fuzzy sets, admissible orders and OWA operators. Inf. Fusion **37**, 126–131 (2017)

18. Pękala, B.: Operations on interval matrices. In: Kryszkiewicz, M., Peters, J.F., Rybinski, H., Skowron, A. (eds.) RSEISP 2007. LNCS (LNAI), vol. 4585, pp. 613–621. Springer, Heidelberg (2007). https://doi.org/10.1007/978-3-540-73451-2_64

19. Pękala, B.: Uncertainty Data in Interval-Valued Fuzzy Set Theory: Properties, Algorithms and Applications. Studies in Fuzziness and Soft Computing, vol. 367. Springer, Cham (2019). https://doi.org/10.1007/978-3-319-93910-0

20. Pękala, B., Bentkowska, U., et al.: Operators on intuitionistic fuzzy relations. In: 2015 IEEE International Conference on Fuzzy Systems (FUZZ-IEEE), pp. 1–8. IEEE (2015)

21. Pękala, B., Bentkowska, U., et al.: Interval subsethood measures with respect to uncertainty for the interval-valued fuzzy setting. Int. J. Comput. Intell. Syst. **13**, 167–177 (2020)

22. Sambuc, R.: Fonctions ϕ-floues: Application á l'aide au diagnostic en pathologie thyroidienne. Ph.D. thesis, Faculté de Médecine de Marseille (1975). (in French)

23. Sesma-Sara, M., Miguel, L.D., Pagola, M., Burusco, A., Mesiar, R., Bustince, H.: New measures for comparing matrices and their application to image processing. Appl. Math. Model. **61**, 498–520 (2018)

24. Wygralak, M.: Intelligent Counting under Information Imprecision. Studies in Fuzziness and Soft Computing, vol. 292. Springer, Heidelberg (2013). https://doi.org/10.1007/978-3-642-34685-9

25. Zadeh, L.: The concept of a linguistic variable and its application to approximate reasoning-I. Inf. Sci. **8**(3), 199–249 (1975)

26. Zadeh, L.A.: Fuzzy sets. Inf. control **8**(3), 338–353 (1965)

27. Zapata, H., Bustince, H., et al.: Interval-valued implications and interval-valued strong equality index with admissible orders. Int. J. Approx. Reason. **88**, 91–109 (2017)

28. Żywica, P.: Modelling medical uncertainties with use of fuzzy sets and their extensions. In: Medina, J., Ojeda-Aciego, M., Verdegay, J.L., Perfilieva, I., Bouchon-Meunier, B., Yager, R.R. (eds.) IPMU 2018. CCIS, vol. 855, pp. 369–380. Springer, Cham (2018). https://doi.org/10.1007/978-3-319-91479-4_31

29. Żywica, P., Dyczkowski, K., et al.: Development of a fuzzy-driven system for ovarian tumor diagnosis. Biocybern. Biomed. Eng. **36**(4), 632–643 (2016)

30. Żywica, P., Stachowiak, A., Wygralak, M.: An algorithmic study of relative cardinalities for interval-valued fuzzy sets. Fuzzy Sets Syst. **294**, 105–124 (2015)

Aggregation Functions Transformed by 0 - 1 Valued Monotone Systems of Functions

Martin Kalina$^{(\boxtimes)}$ (iD)

Faculty of Civil Engineering, Department of Mathematics,
Slovak University of Technology in Bratislava, Radlinského 11,
810 05 Bratislava, Slovakia
kalina@math.sk

Abstract. In the paper Jin et al. [8] the authors introduced a generalized phi-transformation of aggregation functions. This is a kind of two-step aggregation. This transformation was further developed in Jin et al. [9] into a Generalized-Convex-Sum-Transformation. A special case of the proposed Generalized-Convex-Sum-Transformation is the well-known *-product, also known as the Darsow product of copulas. This approach covers also the discrete Choquet integral. In this paper we study the monotone systems of functions, particularly the case when functions in these systems are just two-valued.

Keywords: Aggregation function · Copula · Generalized-convex-sum transformation · Monotone system of functions

1 Introduction

Jin et al. in [8] introduced a generalized φ-transformation of aggregation functions. This method is based on a so-called parametrized chain and an aggregation function F. The original aggregation function A is transformed into $A_{\langle F,\mathbf{c}\rangle}$, where $\mathbf{c}$ is a vector-function and $F(\mathbf{c}(t)) = t$. This method was modified by the same authors in [9] into a generalized-convex-sum-transformation. A special case of this generalized-convex-sum-transformation is the well-known *-product, also known as Darsow product, see [4]. As it is shown in [9], this method generalizes the discrete Choquet integral. The transformation is based on systems of monotone functions as follows (we illustrate here the transformation of binary aggregation functions)

$$A_{\mathbb{F}}(x,y) = \int\limits_0^1 A(f_x(t), g_y(t))\mathrm{d}t,$$

where $\mathbb{F}$ is the pair of monotone systems of functions $\{f_x\}_{x\in[0,1]}$, $\{g_y\}_{y\in[0,1]}$. This means that the particular choice of the pair $\mathbb{F}$ of monotone systems of functions

© Springer Nature Switzerland AG 2020
M.-J. Lesot et al. (Eds.): IPMU 2020, CCIS 1238, pp. 537–548, 2020.
https://doi.org/10.1007/978-3-030-50143-3_42

$\{f_x\}_{x\in[0,1]}$, $\{g_y\}_{y\in[0,1]}$ influences the resulting transform $A_{\mathbb{F}}$ of the (in this case) binary aggregation function A.

In this contribution we will study the case when the systems of functions $\{f_x\}_{x\in[0,1]}$, $\{g_y\}_{y\in[0,1]}$ are two-valued, i.e., $f_x(t) \in \{0,1\}$ and $g_y(t) \in \{0,1\}$ for all $x, y, t \in [0,1]$.

After recalling some preliminary notions and results in Sect. 2, in Sect. 3 we provide the results of our study. Finally, conclusions are given in Sect. 4.

2   Preliminaries

In this section we recall some basic definitions and known facts on aggregation functions. In the second part we provide basic idea of the generalized-convex-sum-transformation that was introduced in [9].

2.1   Basic Definitions and Known Facts

In this contribution we will deal with (n-ary) aggregation function on $[0,1]$. For more details including definitions and discussion concerning examples and properties of aggregation functions we recommend [1,2,7,10,12].

Some distinguished families of n-ary aggregation functions are given in the following definition.

Definition 1 ([7]). *An n-ary aggregation function A is said to be*

(1) an n-ary semi-copula if $e = 1$ is its neutral element,
(2) a t-norm if it is an associative and symmetric semi-copula,
(3) dual to a semi-copula if $e = 0$ is a neutral element,
(4) a t-conorm if it is associative and symmetric and dual to a semi-copula,
(5) an n-ary quasi-copula if it is a 1-Lipschitz semi-copula, i.e.,

$$|A(x_1,\dots,x_n) - A(y_1,\dots,y_n)| \le \sum_{i=1}^{n} |x_i - y_i|.$$

Definition 2 ([12]). *An n-ary aggregation function $C_n : [0,1]^n \to [0,1]$ is said to be an n-ary copula if it is an n-ary semi-copula which is n-increasing, i.e., if for all $\mathbf{x}^{(0)} \in [0,1]^n$ and $\mathbf{x}^{(1)} \in [0,1]^n$ such that $\mathbf{x}^{(0)} \ge \mathbf{x}^{(1)}$ the following holds*

$$\sum_{(i_1,\dots,i_n)\in\mathcal{I}} (-1)^{\sum_{k=1}^{n} i_k} C_n(x_1^{(i_1)},\dots,x_n^{(i_n)}) \ge 0, \tag{1}$$

where $\mathcal{I} = \{0,1\}^n$.

Lemma 1 ([12]). *Every copula C is 1-Lipschitz.*

Proposition 1 ([12]). *Let C be a binary copula (or a quasi-copula). Then for every $(x,y) \in [0,1]^2$*

$$\max(0, x + y - 1) \le C(x,y) \le \min(x,y). \tag{2}$$

Let us remark that the functions related to the lower and upper bound occurring in inequality (2) are denoted by

$$W(x,y) = \max(0, x + y - 1), \quad M(x,y) = \min(x,y) \tag{3}$$

and are called *the lower and upper Fréchet-Hoeffding bounds*, respectively. In the theory of t-norms the function W is usually denoted by T_L and M is denoted by T_M and are called the Łukasiewicz and minimum t-norm, respectively.

2.2 Generalized-Convex-Sum-Transformation

In this section we briefly recall some definitions and results from [9] explaining the construction method using transformation of a given aggregation function A by an n-tuple of monotone systems of functions $\mathbb{F}_n$. The notion of a monotone system of functions is crucial in the construction method in question.

Definition 3. *Let $\mathcal{F} = \{f_x\}_{x \in [0,1]}$ be a family of functions such that*

1. for every $x \in [0,1]$ $f_x : [0,1] \to [0,1]$ is a Lebesgue integrable function,
2. $f_{x_1} \le f_{x_2}$ for $x_1 \le x_2$,
3. for all $z \in [0,1]$ $f_0(z) = 0$ and $f_1(z) = 1$.

Then $\mathcal{F}$ is called a Monotone System of Functions, *MSF for brevity.*

Example 1. For every $x \in [0,1]$ let $f_x : [0,1] \to [0,1]$ and $g_x : [0,1] \to [0,1]$ be defined by

$$f_x(t) = \begin{cases} 0 & \text{for } x < 1, \\ 1 & \text{for } x = 1, \end{cases} \qquad g_x(t) = \begin{cases} 0 & \text{for } x = 0, \\ 1 & \text{for } x > 0. \end{cases}$$

Then $\mathcal{F} = \{f_x\}_{x \in [0,1]}$ and $\mathcal{G} = \{g_x\}_{x \in [0,1]}$ are the least and the greatest Monotone Systems of Functions, respectively.

Definition 4. *Let $\mathcal{F}^{(i)} = \left\{ f_x^{(i)} \right\}_{x \in [0,1]}$, $i = 1, 2, \ldots, n$, be MSF (Definition 3). Then $\mathbb{F}_n = (\mathcal{F}^{(1)}, \mathcal{F}^{(2)}, \ldots, \mathcal{F}^{(n)})$ is called an n-tuple of the Monotone Systems of Functions, n-MSF for brevity.*

Lemma 2 ([9]). *Let $A : [0,1]^n \to [0,1]$ be any Lebesgue integrable n-ary aggregation function and $\mathbb{F}_n = (\mathcal{F}^{(1)}, \mathcal{F}^{(2)}, \ldots, \mathcal{F}^{(n)})$ be an arbitrary n-MSF. Let a function $A_{\mathbb{F}_n} : [0,1]^n \to [0,1]$ be given by*

$$A_{\mathbb{F}_n}(x_1, \ldots, x_n) = \int_0^1 A\big(f_{x_1}^{(1)}(t), \ldots, f_{x_n}^{(n)}(t)\big) \, \mathrm{d}t, \tag{4}$$

where $f_{x_1}^{(1)} \in \mathcal{F}^{(1)}, \ldots, f_{x_n}^{(n)} \in \mathcal{F}^{(n)}$. Then $A_{\mathbb{F}_n}$ is an aggregation function.

Definition 5. *An n-ary aggregation function $A_{\mathbb{F}_n}$ defined by formula (4), where $\mathbb{F}_n = (\mathcal{F}^{(1)}, \ldots, \mathcal{F}^{(n)})$ is an n-tuple of monotone systems of functions, is said to be a* Generalized-Convex-Sum-Transform *of A by $\mathbb{F}_n$, or a* GCS-transform *in short.*

Definition 6. *(i) Let* $\mathcal{F} = \{f_x\}_{x \in [0,1]}$ *be an MSF. If* $\int\limits_0^1 f_x(t)\,\mathrm{d}t = x$ *is fulfilled for all* $x \in [0,1]$ *then* $\mathcal{F}$ *is called* a Standard Monotone System of Functions, *or SMSF for brevity.*

(ii) Let $\mathcal{F}^{(i)} = \left\{f_x^{(i)}\right\}_{x \in [0,1]}$, $i = 1, 2, \ldots, n$, *be an SMSF.*

Then $\mathbb{F}_n = (\mathcal{F}^{(1)}, \mathcal{F}^{(2)}, \ldots, \mathcal{F}^{(n)})$ *is called an* n*-tuple of the Standard Monotone Systems of Functions, or* n*-SMSF for brevity.*

Remark 1. It follows directly from Definition 3 and 6 that the set containing all the MSF and SMSF is convex.

Example 2 ([9]). Consider an SMSF $\mathcal{F} = \{f_x\}_{x \in [0,1]}$ such that

$$f_x(t) = \begin{cases} 1 & \text{for } x \in]0,1] \text{ and } t \in [0,x], \\ 0 & \text{otherwise.} \end{cases} \tag{5}$$

Denote $x_{(0)} = 0$ and $A(f_{x_1}(t), \ldots, f_{x_n}(t)) = A(\mathbf{1}_{\{(i),\ldots,(n)\}})$ for $t \in \left]x_{(i-1)}, x_{(i)}\right]$. Then for an arbitrary $n \geq 2$, an arbitrary n-ary aggregation function A and the n-tuple $\mathbb{F}_n = (\mathcal{F}, \ldots, \mathcal{F})$ the following holds

$$
\begin{aligned}
A_{\mathbb{F}_n}(\mathbf{x}) &= \int_0^1 A(f_{x_1}(t), \ldots, f_{x_n}(t))\mathrm{d}t \\
&= \int_0^{x_{(1)}} A(\mathbf{1}_{\{(1),\ldots,(n)\}})\mathrm{d}t + \int_{x_{(1)}}^{x_{(2)}} A(\mathbf{1}_{\{(2),\ldots,(n)\}})\mathrm{d}t + \cdots + \\
&\quad + \int_{x_{(n-1)}}^{x_{(n)}} A(\mathbf{1}_{\{(n)\}})\mathrm{d}t \\
&= x_{(1)}m_A(\{(1),\ldots,(n)\}) + (x_{(2)} - x_{(1)})m_A(\{(2),\ldots,(n)\}) + \cdots + \\
&\quad + (x_{(n)} - x_{(n-1)})m_A(\{(n)\}) = \mathrm{Ch}_{m_A}(\mathbf{x}),
\end{aligned}
$$

where $m_A : 2^{\{1,\ldots,n\}} \to [0,1]$ is a capacity given by $m_A(E) = A(\mathbf{1}_E)$, $(\cdot) : \{1,\ldots,n\} \to \{1,\ldots,n\}$ is any permutation satisfying $x_{(1)} \leq x_{(2)} \leq \cdots \leq x_{(n)}$, and Ch_{m_A} is the Choquet integral (see [3,6]) with respect to the capacity m_A. This fact shows that our construction method can be seen as a significant extension of the Choquet integrals.

3   0 - 1 Valued Standard Monotone Systems of Functions

In the rest of the paper we will consider only standard monotone systems of functions $\mathcal{F} = \{f_x\}_{x \in [0,1]}$ that fulfill the constraint $f_x(t) \in \{0,1\}$ for all $x \in [0,1]$ and $t \in [0,1]$. We will call them *0-1-valued monotone systems of functions*, abbreviation 0-1-SMSF.

Let $\mathcal{F} = \{f_x\}_{x \in [0,1]}$ and $\mathcal{G} = \{g_x\}_{x \in [0,1]}$ be 0-1-valued monotone systems of functions and $\mathbb{F} = (\mathcal{F}, \mathcal{G})$ be a pair of 0-1-SMSF. Define $C^{\mathbb{F}} : [0,1]^2 \to [0,1]$ by

$$C^{\mathbb{F}}(x,y) = \int f_x(t)g_y(t)\mathrm{d}t. \tag{6}$$

Let A be a binary aggregation function. Then, obviously

$$A_{\mathbb{F}}(x,y) = \int_0^1 A(f_x(t), g_y(t))dt$$
$$= \left(x - C^{\mathbb{F}}(x,y)\right) A(1,0) + \left(y - C^{\mathbb{F}}(x,y)\right) A(0,1) + C^{\mathbb{F}}(x,y).$$

In other words, knowing $C^{\mathbb{F}}$ we know the result of the GCS-transform $A_{\mathbb{F}}$ of A.

We will focus our attention only to binary aggregation functions and their GCS-transforms.

Proposition 2. *Let $\mathcal{F} = \{f_x\}_{x \in [0,1]}$ and $\mathcal{G} = \{g_x\}_{x \in [0,1]}$ be arbitrary 0-1-SMSF and $\mathbb{F} = (\mathcal{F}, \mathcal{G})$. Then, $C^{\mathbb{F}}$ defined by formula (6) is a copula.*

Proof. Directly by Definitions 3 and 6 we have that $C^{\mathbb{F}}$ is a semi-copula. Let us prove the two-increasingness. Assume $x_1 \geq x_2$ and $y_1 \geq y_2$ be arbitrary elements of $[0,1]$. Then for all $t \in [0,1]$ the following holds

$$f_{x_1}(t) \geq f_{x_2}(t), \quad g_{y_1}(t) \geq g_{y_2}(t). \tag{7}$$

These imply

$$C^{\mathbb{F}}(x_1, y_1) - C^{\mathbb{F}}(x_2, y_1) = \int_0^1 (f_{x_1}(t) - f_{x_2}(t))g_{y_1}(t)dt \tag{8}$$

$$C^{\mathbb{F}}(x_1, y_2) - C^{\mathbb{F}}(x_2, y_2) = \int_0^1 (f_{x_1}(t) - f_{x_2}(t))g_{y_2}(t)dt. \tag{9}$$

Formulae (8) and (9) imply

$$(C^{\mathbb{F}}(x_1, y_1) - C^{\mathbb{F}}(x_2, y_1)) - (C^{\mathbb{F}}(x_1, y_2) - C^{\mathbb{F}}(x_2, y_2))$$
$$= \int_0^1 (f_{x_1}(t) - f_{x_2}(t))(g_{y_1}(t) - g_{y_2}(t))dt,$$

and by inequalities (7),

$$\int_0^1 (f_{x_1}(t) - f_{x_2}(t))(g_{y_1}(t) - g_{y_2}(t))dt \geq 0.$$

$\square$

Definition 7. *Let $\mathcal{F} = \{f_x\}_{x \in [0,1]}$ and $\mathcal{G} = \{g_x\}_{x \in [0,1]}$ be arbitrary 0-1-SMSF and $\mathbb{F} = (\mathcal{F}, \mathcal{G})$. Then we denote $C^{\mathbb{F}}$ the copula given by Eq. (6) and we say that $C^{\mathbb{F}}$ is generated by $\mathbb{F}$.*

The following example illustrates some 0-1-SMSF and the copulas they generate.

Example 3. Denote $\tilde{\mathcal{F}} = \{f_x\}_{x \in [0,1]}$ where f_x are given by formula (5). Further, set $\mathcal{G}_1 = \{{}^1 g_x\}_{x \in [0,1]}$, $\mathcal{G}_2 = \{{}^2 g_x\}_{x \in [0,1]}$, $\mathcal{G}_3 = \{{}^3 g_x\}_{x \in [0,1]}$, $\mathcal{G}_4 = \{{}^4 g_x\}_{x \in [0,1]}$,

$\mathcal{G}_5 = \{{}^5 g_x\}_{x \in [0,1]}$, where ${}^i g_x$ are given by formulae (10), (11), (12), (13) and (14), respectively:

$$
{}^1 g_x(t) = \begin{cases} 1 & \text{if } t \in [0, \frac{x}{2}] \cup [1 - \frac{x}{2}, 1], \\ 0 & \text{otherwise,} \end{cases} \tag{10}
$$

$$
{}^2 g_x(t) = \begin{cases} 1 & \text{if } t \in [\frac{1-x}{2}, \frac{1+x}{2}], \\ 0 & \text{otherwise,} \end{cases} \tag{11}
$$

$$
{}^3 g_x(t) = \begin{cases} 1 \text{ for: } & x \leq \frac{1}{2} \text{ and } t \in [0, x], \\ & x \in \left] \frac{1}{2}, \frac{3}{4} \right[\text{ and } t \in \left[0, \frac{1}{2}\right] \cup \left[\frac{3}{2} - x, 1\right], \\ & t \geq \frac{3}{4} \text{ and } t \in \left[0, x - \frac{1}{4}\right] \cup \left[\frac{3}{4}, 1\right], \\ 0 & \text{otherwise,} \end{cases} \tag{12}
$$

$$
{}^4 g_x(t) = \begin{cases} 1 \text{ for: } & x \leq \frac{1}{2} \text{ and } t \in [0, x^2] \cup [1 - x + x^2, 1], \\ & x \in \left] \frac{1}{2}, \frac{3}{4} \right[\text{ and } t \in \left[0, \frac{1}{4}\right] \cup \left[\frac{3}{4}, 1\right] \cup \left[1 - x, \frac{1}{2}\right], \\ & t \geq \frac{3}{4} \text{ and } t \in \left[0, x - \frac{1}{4}\right] \cup \left[\frac{3}{4}, 1\right], \\ 0 & \text{otherwise,} \end{cases} \tag{13}
$$

$$
{}^5 g_x(t) = \begin{cases} 1 & \text{for } t \in \left[0, \frac{x}{2}\right] \cup \left[\frac{1}{2}, \frac{1}{2}(2 - x)\right], \\ 0 & \text{otherwise.} \end{cases} \tag{14}
$$

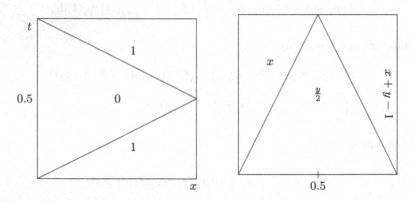

Fig. 1. Left the layout of $\mathcal{G}_1$, right the copula $C^{\mathbb{F}^{(1)}}$ where $\mathbb{F}^{(1)} = (\tilde{\mathcal{F}}, \mathcal{G}_1)$

The copula sketched in the right part of Fig. 1 is the so-called tent copula (see, e.g., [4]).

The following lemma is straightforward.

Lemma 3. *Let* $\mathcal{F} = \{f_x\}_{x \in [0,1]}$ *and* $\mathcal{G} = \{g_x\}_{x \in [0,1]}$ *be arbitrary 0-1-SMSF. Then*

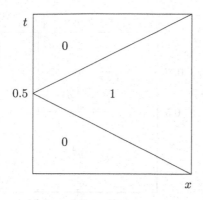 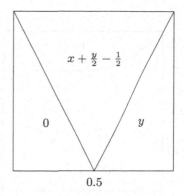

Fig. 2. Left the layout of $\mathcal{G}_2$, right the copula $C^{\mathbb{F}^{(2)}}$ where $\mathbb{F}^{(2)} = (\tilde{\mathcal{F}}, \mathcal{G}_2)$.

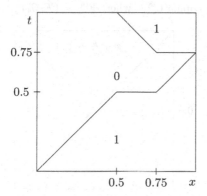 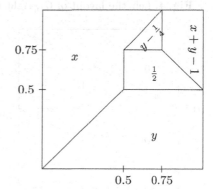

Fig. 3. Left the layout of $\mathcal{G}_3$, right the copula $C^{\mathbb{F}^{(3)}}$ where $\mathbb{F}^{(3)} = (\tilde{\mathcal{F}}, \mathcal{G}_3)$

(a) $C^{\mathbb{F}} = M$ for $\mathbb{F} = (\mathcal{F}, \mathcal{F})$,

(b) if $\mathbb{F} = (\mathcal{F}, \mathcal{G})$ generates a copula $C^{\mathbb{F}}$ then $\overline{\mathbb{F}} = (\mathcal{G}, \mathcal{F})$ generates the copula $C^{\overline{\mathbb{F}}}$ such that $C^{\overline{\mathbb{F}}}(x, y) = C^{\mathbb{F}}(y, x)$,

(c) $C^{\hat{\mathbb{F}}} = W$ for $\hat{\mathbb{F}} = (\mathcal{F}, \mathcal{F}^d)$, where $\mathcal{F}^d = \{f_x^d\}_{x \in [0,1]}$ and $f_x^d(t) = 1 - f_{1-x}(t)$.

Proposition 3. *Let $\tilde{\mathcal{F}}$ be the 0-1-SMSF defined by formula (5). There exists a pair of 0-1-SMSF $\mathbb{G} = (\mathcal{G}^{(1)}, \mathcal{G}^{(2)})$ and the copula $C^{\mathbb{G}}$ generated by $\mathbb{G}$ such that for arbitrary 0-1-SMSF $\mathcal{F} = \{f_x\}_{x \in [0,1]}$ and the pair of 0-1-SMSF $\mathbb{F} = (\tilde{\mathcal{F}}, \mathcal{F})$ we have*

$$C^{\mathbb{G}} \neq C^{\mathbb{F}}.$$

Proof. Choose $(x, y) \in [0, 1]^2$. Then

$$C^{\mathbb{F}}(x, y) = \int_0^x f_y(t) \mathrm{d}t.$$

There are three possibilities for the result of $C^{\mathbb{F}}(x, y)$.

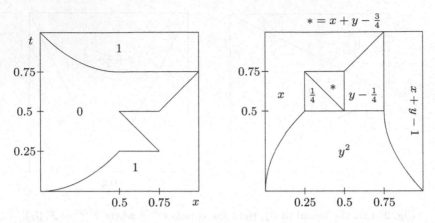

Fig. 4. Left the layout of $\mathcal{G}_4$, right the copula $C^{\mathbb{F}^{(4)}}$ where $\mathbb{F}^{(4)} = (\tilde{\mathcal{F}}, \mathcal{G}_4)$

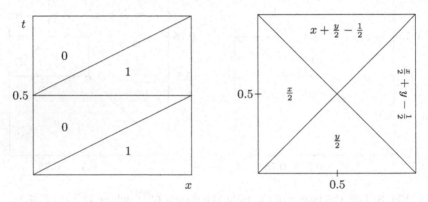

Fig. 5. Left the layout of $\mathcal{G}_5$, right the copula $C^{\mathbb{G}}$ where $\mathbb{G} = (\mathcal{G}_1, \mathcal{G}_5)$

1. $g_y(t) = 0$ almost everywhere on the interval $[0, x]$, i.e., $\lambda(\{t \in [0, x]; g_y(t) \neq 0\}) = 0$, where λ is the Lebesgue measure. In this case $C^{\mathbb{F}}(x, y) = 0$.
2. There exists $\varphi(y) \in [0, 1]$ such that $f_y(t) = 1$ almost everywhere on the interval $[0, \varphi(y)]$ and, if $x > \varphi(y)$ $f_y(t) = 0$ almost everywhere on $[\varphi(y), x]$. Then

$$C^{\mathbb{F}}(x, y) = \begin{cases} x & \text{if } x \leq \varphi(y), \\ \varphi(y) & \text{if } x > \varphi(y). \end{cases}$$

3. There exists a countable (finite or infinite) system of pairwise disjoint intervals $\{[a_i, b_i]\}_{i \in I}$ such that $[a_i, b_i] \cap [0, x] \neq \emptyset$, $f_y(t) = 1$ almost everywhere for $t \in \bigcup_{i \in I}$ and $f_y(t) = 0$ almost everywhere for $t \in [0, x] \setminus \bigcup_{i \in I}$. There are two possibilities:
   - $b_i \leq x$ for all $i \in I$, then

$$C^{\mathbb{F}}(x, y) = \sum_{i \in I} (b_i - a_i),$$

– there exists $j \in I$ such that $b_j > x$, then

$$C^{\mathbb{F}}(x,y) = \sum_{i \in I} (b_i - a_i) - (b_j - x).$$

This gives the following – setting $\mathcal{G}^{(1)} = \mathcal{G}_1$ and $\mathcal{G}^{(2)} = \mathcal{G}_5$, where $\mathcal{G}_1$ and $\mathcal{G}_5$ are defined by formulae (10) and (14), respectively, and $\mathbb{G} = (\mathcal{G}^{(1)}, \mathcal{G}^{(2)})$ (see also Fig. 5), then the copula $C^{\mathcal{G}}$ cannot be generated by any pair of 0-1-SMSF $\mathbb{F} = (\tilde{\mathcal{F}}, \mathcal{F})$. □

Remark 2. Analysing the proof of Proposition 3 we see that if a copula has a component that is non-linear in the first variable, then it cannot be generated by a pair of 0-1-SMSF $(\tilde{\mathcal{F}}, \mathcal{F})$ where $\tilde{\mathcal{F}}$ is the 0-1-SMSF defined by formula (5). Another consequence that can be derived is that the copula $C^{\mathbb{G}}$ sketched on Fig. 5 cannot be generated by a pair of 0-1-SMSF $(\tilde{\mathcal{F}}, \mathcal{F})$ nor by $(\mathcal{F}, \tilde{\mathcal{F}})$.

A characterization by the second mixed partial derivatives of copulas generated by pairs of 0-1-SMSF is contained in the following proposition.

Proposition 4. *Let $\mathbb{F}$ be a pair of 0-1-SMSF and $C^{\mathbb{F}}$ the copula generated by $\mathbb{F}$. Then*

$$\frac{\partial^2 C^{\mathbb{F}}}{\partial x \partial y}(x,y) = 0 \tag{15}$$

for all $(x,y) \in [0,1]^2$ where the second mixed partial derivative exists.

Durante et al. [5] have shown that if formula (15) holds for a copula $C^{\mathbb{F}}$ then $C^{\mathbb{F}}$ still may have a density.

Finally, we show how we can construct an arbitrary shuffle of M by a pair of 0-1-SMSF. The family of all shuffles of min is very important, since, as it is proven in [11], this family is dense in the system of all bivariate copulas.

A geometrical visualisation of a shuffle of min is quite straightforward. We choose a natural number $n > 1$, a system of nods $0 = a_0 < a_1 < \cdots < a_n = 1$ and cut the minimum copula parallel to the y-axis into n strips using those nods. Then we shuffle the strips (this means we choose a permutation Π : $\{1, 2, \ldots, n\} \to \{1, 2, \ldots, n\}$) and paste them together in the permuted order. If we denote $b_i = a_i - a_{i-1}$ for $i \in \{1, 2, \ldots, n\}$ then choosing a permutation $\Pi : \{1, 2, \ldots, n\} \to \{1, 2, \ldots, n\}$, a shuffle of min is given by $S = \langle n, (b_i)_{i=1}^n, \Pi \rangle$.

Example 4. Set $n = 5$, the permutation Π by $(2, 1, 4, 3, 5)$ and the nodes are given by $(0, 0.3, 0.4, 0.7, 0.9, 1)$. Then the corresponding 0-1-SMSF $\mathcal{G}_6 = \{^6 g_x\}_{x \in [0,1]}$ is given by formula (16) and the shuffle copula is then generated by the pair of 0-1-SMSF $\mathbb{F}_6 = (\tilde{\mathcal{F}}, \mathcal{G}_6)$ and displayed in Fig. 6.

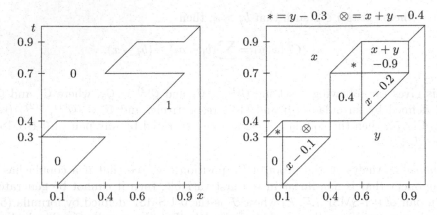

Fig. 6. Left the layout of $\mathcal{G}_6$, right the copula $C^{\mathbb{F}^{(6)}}$ where $\mathbb{F}^{(6)} = (\tilde{\mathcal{F}}, \mathcal{G}_6)$

The explicit formula for 0-1-SMSF $\mathcal{G}_6 = \{^6 g_x\}_{x \in [0,1]}$ is given by

$$
^6 g_x(t) = \begin{cases}
1 \text{ for:} & x \in [0, 0.1] \text{ and } t \in [0.3, x + 0.3], \\
& x \in [0.1, 0.4] \text{ and } t \in [0, x - 0.1] \cup [0.3, 0.4], \\
& x \in [0.4, 0.6] \text{ and } t \in [0, 0.4] \cup [0.7, x + 0.3], \\
& x \in [0.6, 0.9] \text{ and } t \in [0, x - 0.2] \cup [0.7, 0.9], \\
& x \in [0.9, 1] \text{ and } t \in [0, x], \\
0 & \text{otherwise.}
\end{cases}
\tag{16}
$$

Shuffles of min are in [11] described in a more general way than we have illustrated by Example 4. Namely, the shuffles can be combined with flips (a flip of the minimum copula M is W). This combination is sketched in Fig. 3. The following proposition gives a characterization of all shuffles (possibly combined with flips) of min as special cases of copulas generated by a pair of 0-1-SMSF. In this case a shuffle of min is given by $\langle n, (b_i)_{i=1}^n, \Pi, (m_i)_{i=1}^n \rangle$, where b_i denotes the width of the i-th strip, Π is a permutation of $\{1, 2, \ldots, n\}$ and $m_i = 1$ ($m = 0$) if we use the flip (if we do not use the flip) in the square $[a_{i-1}, a_i]^2$, where $a_i = \sum_{j=1}^i b_j$. We skip the proof of Proposition 5 since the construction there is just a generalization of formulae (12) and (16).

Proposition 5. *Set $n > 1$ a natural number, for $i \in \{1, \ldots, n\}$ let $b_i > 0$ be such that $\sum_{i=1}^n b_i = 1$, $m_i \in \{0, 1\}$ and $\Pi : \{1, \ldots, n\} \to \{1, \ldots, n\}$ be a permutation.*

Denote $a_0 = \tilde{a}_0 = 0$ and for $i \in \{1, \ldots, n\}$ $a_i = \sum_{j=1}^i b_j$ and $\tilde{a}_i = \sum_{j=1}^i b_{\Pi^{-1}(j)}$.

The shuffle of min given by $\langle n, (b_i)_{i=1}^n, \Pi, (m_i)_{i=1}^n \rangle$, is generated by the pair of 0-1-SMSF $\mathbb{F} = (\tilde{\mathcal{F}}, \mathcal{H})$, where $\mathcal{H} = \{h_x\}_{x \in [0,1]}$ is defined by

$$h_x(t) = \begin{cases} 1 & for\ x \in [0, \tilde{a}_1]\ and\ t \in A_1, \\ & x \in [\tilde{a}_{i-1}, \tilde{a}_i]\ and\ t \in A_i \cup \bigcup_{j=1}^{i-1} [\tilde{a}_{j-1}, \tilde{a}_j],\ for\ 2 \leq i \leq n, \\ 0 & otherwise, \end{cases} \quad (17)$$

where, for $i \in \{1, 2, \ldots, n\}$,

$$A_i = \begin{cases} [a_{\Pi}^{-1}(i) + \tilde{a}_{i-1} - x, a_{\Pi^{-1}(i)}] & if\ m = 1, \\ [a_{\Pi}^{-1}(i) - 1, x - (a_{\Pi^{-1}(i)} - \tilde{a}_i)] & if\ m = 0. \end{cases}$$

4   Conclusion

This paper contributes to a study of Generalized-Convex-Sum Transformation of (binary) aggregation functions. Particularly, we have studied copulas that can be generated by pairs of 0-1-SMSF (see formula (6)). Though, the 0-1-SMSF $\tilde{\mathcal{F}}$ given by formula (5) is, in a sense, a basic 0-1-SMSF, when we like to generate all possible copulas by pairs of 0-1-SMSF, it is not sufficient to consider only those pairs where one of the 0-1-SMSF is $\tilde{\mathcal{F}}$. We have also shown that every shuffle of the minimum copula (possibly combined with flips) can be generated by a pair of 0-1-SMSF and in Proposition 5 we have written down an explicit formula for such 0-1-SMSF.

Acknowledgements. The work on this paper has been supported from the Science and Technology Assistance Agency under the contract no. APVV-18-0052, and by the Slovak Scientific Grant Agency VEGA no. 1/0006/19 and 2/0142/20.

References

1. Beliakov, G., Bustince Sola, H., Calvo Sánchez, T.: A Practical Guide to Averaging Functions. SFSC, vol. 329. Springer, Cham (2016). https://doi.org/10.1007/978-3-319-24753-3
2. Beliakov, G., Pradera, A., Calvo, T.: Aggregation Functions: A Guide for Practitioners. Springer, Heidelberg (2007). https://doi.org/10.1007/978-3-540-73721-6
3. Choquet, G.: Theory of capacities. Ann. Inst. Fourier **5**, 131–295 (1953–1954)
4. Darsow, W.F., Nguyen, B., Olsen, T.E.: Copulas and Markov processes. Ill. J. Math. **36**(4), 600–642 (1992)
5. Durante, F., Sánchez, J.F., Sempi, C.: A note on the notion of singular copula. Fuzzy Sets Syst. **211**, 120–122 (2013). https://doi.org/10.1016/j.fss.2012.04.005
6. Grabisch, M.: Fuzzy integral in multicriteria decision making. Fuzzy Sets Syst. **69**(3), 279–298 (1995)
7. Grabisch, M., Marichal, J.L., Mesiar, R., Pap, E.: Aggregation Functions. Cambridge University Press, Cambridge (2009)
8. Jin, L., Mesiar, R., Kalina, M., Špirková, J., Borkotokey, S.: Generalized phi-transformations of aggregation functions. Fuzzy Sets Syst. **372**, 124–141 (2019). https://doi.org/10.1016/j.fss.2018.09.016

9. Jin, L., Mesiar, R., Kalina, M., Špirková, J., Borkotokey, S.: New transformations of aggregation functions based on monotone systems of functions. Int. J. Approximate Reasoning **118**, 79–95 (2020). https://doi.org/10.1016/j.ijar.2019.12.004
10. Klement, E.P., Mesiar, R., Pap, E.: Triangular Norms. Springer, Heidelberg (2000). https://doi.org/10.1007/978-94-015-9540-7
11. Mikusinski, P., Sherwood, H., Taylor, M.D.: Shuffles of min. Stochastica **XIII–1**, 61–74 (1992)
12. Nelsen, R.B.: An Introduction to Copulas. Springer Series in Statistics. Springer, New York (2006). https://doi.org/10.1007/0-387-28678-0

Aggregation: Pre-aggregation Functions and Other Generalizations of Monotonicity

Analyzing Non-deterministic Computable Aggregations

Luis Garmendia[1]([✉]) [iD], Daniel Gómez[2]([✉]) [iD], Luis Magdalena[3]([✉]) [iD],
and Javier Montero[4]([✉]) [iD]

[1] Facultad de Informática, Universidad Complutense de Madrid,
Madrid, Spain
lgarmend@fdi.ucm.es
[2] Facultad de Estudios Estadísticos, Universidad Complutense de Madrid,
Madrid, Spain
dagomez@estad.ucm.es
[3] ETSI Informáticos, Universidad Politécnica de Madrid, Madrid, Spain
luis.magdalena@upm.es
[4] Facultad de Ciencias Matemáticas, Universidad Complutense de Madrid,
Madrid, Spain
monty@mat.ucm.es

Abstract. Traditionally, the term aggregation is associated with an aggregation function, implicitly assuming that any aggregation process can be represented by a function. However, the concept of computable aggregation considers that the core of the aggregation processes is the program that enables it. This new concept of aggregation introduces the scenario where the aggregation can even be non-deterministic. In this work, this new class of aggregation is formally defined, and some desirable properties related with consistency, robustness and monotonicity are proposed.

Keywords: Aggregation · Computable aggregation · Nondeterministic aggregation

1 Introduction

Aggregation is a fundamental part of any decision, compression or summarization process of complex information [1–4]. For many years, aggregation has become one of the most relevant topics in soft computing, with multiple applications to decision making, artificial intelligence, data science, and image processing among many others. Aggregation processes have been associated in literature with aggregation functions. An aggregation process was usually represented by means of a function or a family of functions so that the aggregation result associated to a vector of elements was obtained through the image of the vector by the function.

Supported by PGC2018-096509-B-I00 research national project.

However, this association between aggregation processes and functions was broken in [5] with the definition of computable aggregations. In that paper, the authors put the emphasis in the programs enabling the aggregation processes, not necessarily being expressed in funct terms of functions. In that sense, the core of an aggregation process is the program allowing its computation, and not only the function that describes it. Given a function that describes an aggregation process, it can be implemented in many ways and the way in which this process is carried out is relevant. This new idea of aggregation, allow us to classify classical aggregation operators according to their algorithmic complexity [5], or to classify aggregation functions according to new ideas of recursivity [6,7] in terms of the programs instead of the functions.

The rupture between functions and aggregation operators opens the domain of aggregation processes to a field not yet analyzed in this discipline: non-deterministic computable aggregations. Aggregation processes where the same input can produce different outputs. This type of aggregation is very common in statistics, where, due to the volume of information to be processed, it is frequent to choose a representative sample on which the aggregation is operated. Obviously, replicating the process does not imply obtaining the same sample and consequently the result can change. Obviously, these types of aggregation processes can never be modeled by functions, as a result of the intrinsic definition of function. But leaving the well known arena of functions, studying the desirable properties for these new concept of aggregations represents a significant challenge.

The present paper focuses on the definition of properties related to robustness, stability, boundary conditions, and monotony, on nondeterministic computable aggregations.

2   Preliminaries

2.1   Aggregation Operators

Aggregation is a fundamental part of science. The process of aggregating the information is a key tool for most knowledge based systems. In general, we can say that aggregation has the aim of merging different pieces of information to come to a conclusion or a decision. Several research communities consider this kind of tools, such as multi-criteria community, decision-sensor fusion community, decision making community, data mining community, among many others.

Although this is not a necessary assumption, aggregation operators [4,8–11] are associated to the use of membership functions, and this is the reason why they are usually defined as follows:

Definition 1 [12]. *An aggregation operator is a mapping* $Ag : [0,1]^n \to [0,1]$ *that satisfies:*

1. $Ag(0,0...,0) = 0$ *and* $Ag(1,1,..,1) = 1$.
2. *Ag is monotonic.*

Let us observed that this definition presents an aggregation operator as a function that is linked to the value of n. There is a different function for each n. Other authors (see [12] for example) present an aggregation operator as a function that considers any cardinality for the set of items to be aggregated, defining Ag as a function $Ag : \cup_n [0,1]^n \to [0,1]]$. It is also possible (see for example [13–15]) to define the concept of family of aggregation functions as a set $\{Ag_n\}$ assuming some additional constraints for the relations between functions Ag_k, Ag_l of different cardinality.

The original definition of aggregation functions work on the unit interval. This classical definition has been extended to a more general class of situations replacing the lattice $[0,1]$ into a more general scenario T. Another extension of aggregation operators was done by relaxing the monotonicity or boundary conditions. One of these new classes of aggregation operators are the pre-aggregation functions in which the concept of directional monotonicity was introduced [16]. These pre-aggregations could be extended replacing the unit interval for a general T as it is done with the classical aggregation functions. We can find some interesting studies in this line in [17].

It is even possible in some cases to define aggregation processes going beyond functions by considering methods that do not match with the concept of function. To analyze this option let us first remind the concept of Computable aggregation, as well as that of function.

2.2 Computable Aggregation

In [5], it was introduced the concept of computable aggregation focusing on the idea that in aggregation processes we should pay our attention in the program that makes possible the aggregation instead of a generic function that has not yet been implemented.

Would it make sense to talk about an aggregation process where the considered function could not be implemented?. From the actual definition of aggregation operator the answer is yes of course. Or even more, given the same aggregation function, it is not equivalent at all to implement it in one way or another. And therefore, it would make sense to analyze the properties of the implementation (program), although from the functional point of view both implementations coincide. The main contribution in [5] was to separate the strong association that existed between "aggregation processes" and explicit functions.

In order to formally introduce computable aggregations it is necessary first to introduce what we understand by a program, a list and/or an algorithm.

Definition 2. *A list L is an abstract data type (ADT) that represents a sequence of values. A list can be defined by its behavior, and its implementation must provide at least the following operations: test whether a list is empty, add a value, remove a value, and compute the length of a list (number of elements).*

A list can be defined under a template data type. For example, a list $L<[0,1]>$ is a list of values in the space $X = [0,1]$.

Another definition that we need to give here is what we understand by algorithm and computer program. These concepts are necessary to formally define computable aggregations.

Definition 3 [18]. *In mathematics and computer science, an* **algorithm** *is a self-contained step-by-step set of operations to be performed.*

Algorithms can be expressed in many kinds of notation, including natural languages, pseudocode, flowcharts, drakon-charts, programming languages or control tables.

Definition 4 [19]. *A computer program, or simply a* **program**, *is a sequence of instructions written to perform a specified task on a computer.*

Now, we are able to define the concept of a computable aggregation as it was defined in [5].

Definition 5. *Let $L<T>$ be a finite and non-empty list of n elements with type T. A* **computable aggregation** *is a program P that transform the list $L<T>$ into an element of T.*

3   Non Deterministic Computable Aggregations

It is important to emphasize that the computable aggregation paradigm broadens the set of possible aggregation methods, being in this case not limited to those methods for which there is a function that explains the aggregation process.

In [7], it was demonstrated that any classical aggregation process (aggregation operators, pre-aggregations, fusion functions or extended fusion functions) can be implemented by an algorithm and by a program. However the opposite is not always true.

As was mentioned in [5], given an aggregation operator Ag in any of the previous settings, it is possible to build a program P that for any list L of T, $Ag_{|L|}(L) = P(L)$.

But let us note that the opposite is not always true. Given a program P, there exist some situations in which it is not possible to build a function associated to the program P.

Example 1. Let's consider a population X of size n ($X = \{x_1, \ldots, x_n\}$). Suppose that the objective of the aggregation process is to estimate the average value of X. If n is too large considering the available time for computing the aggregation, a reasonable solution would be to estimate the average value through statistical sampling. In such a situation, k elements of the population will be drawn at random to further calculate the average, i e., given the set $\{x_1, ..x_n\}$, we choose $\{x_{i1}, \ldots, x_{ik}\}$ at random, computing the arithmetic mean of these k elements. Obviously, this aggregation process can't be defined by means of an explicit function since the same input does not always produce the same output. This is a computable aggregation that can't be modeled as an aggregation function, and it is important to integrate such a situation in aggregation processes.

From previous example we can distinguish between those computable aggregations in which there exist an explicit function, from those in which this explicit function does not exist. Formally we can partition the set of computable aggregation programs into two groups. In order to establish this partition we will introduce first the concept of deterministic algorithm and program.

It is important to notice that in this framework, the concept of determinism is approached in two different ways:

- A determinism that only considers input/output relations.
- A determinism that considers the whole process, including the internal states.

With the first approach, a deterministic algorithm will be an algorithm which, given a particular input, will always produce the same output. On the other hand, the second approach assumes also that the underlying machine always passes through the same sequence of states.

From the point of view of computer programs as implementation of aggregation processes, our interest relies on input-output relations. Consequently we will focus on the first conception when defining a deterministic program.

Definition 6. *A program P is deterministic, or repeatable, if it produces the very same output when given the same input no matter how many times it is run.*

It would be possible adding to this definition of deterministic program, a concept similar to the *same sequence of states* considered for the deterministic algorithm. In fact, according to [20], in a deterministic program there is at most one instruction to be executed *next*, so that from a given initial state only one execution sequence is generated. This could be a good option in case we were interested in the verification of a program, being the execution process a key aspect. But it is obviously too restrictive for our purpose, since we are only interested in the result of the Program. Consequently, we will consider Definition 6 as our conception of a deterministic program.

No matter which of the two possible definitions we consider, when applying a deterministic program P we can refer to the output produced for a particular input (a list L), as $P(L)$. That is because the same input (L) will always produce the same output ($P(L)$), generating a mapping.

Obviously, a program or an algorithm are non-deterministic when they do not match with the previous definitions.

It is clear according to the definitions that (when having the same approach to determinism) a non-deterministic algorithm will always produce a non-deterministic program. Consequently, when analyzing if a computable aggregation is deterministic or non-deterministic we should simply consider the corresponding Program.

Definition 7. *A computable aggregation P over the set T is non deterministic if and only if the program P is non deterministic.*

From this definition we will say that a computable aggregation P is deterministic when the program P is deterministic, implying that the underlying algorithm is also deterministic.

Let us denote by $\mathcal{P}_\mathcal{D}$ the class of deterministic computable aggregations and by $\mathcal{P}_{\mathcal{ND}}$ the set of non deterministic computable aggregations.

From now on, we will analyze the case in $T = [0, 1]$.

3.1   Some Non Deterministic Computable Aggregations

It is obvious that in many cases the non-deterministic behavior of a program relates to an inappropriate coding that generates an unexpected problem. This is usually the case when we have a deterministic algorithm that being wrongly programed produces a non-deterministic output. But this is not the kind of non-determinism we are interested in. Our interest relies on programs describing aggregation processes that are intrinsically non-deterministic, processes where, as an example, random or probabilistic decisions are involved.

In this subsection, we will define some interesting cases of non deterministic computable aggregations (NDCAs).

Definition 8. *Given a value $p \in (0, 1]$, and given a family of aggregation operators $\{Ag_n : [0,1]^n \longrightarrow [0,1], \ n \geq 2\}$, let us define the computable aggregation $P_{Ag,p}$ as the two steps program that for a given list $l = (x_1, \ldots, x_n) \in L{<}[0,1]{>}$ performs the following actions:*

- **Step 1.** *To reduce the list l into another list l_p of lower (or equal) dimension by randomly erasing the elements of the list with probability $1 - p$.*
- **Step 2.** *To return the value $Ag_{|l_p|}(l_p)$ if $|l_p| \geq 2$ and 0 otherwise.*

Note that the computable aggregation $P_{Ag,p=1}$ coincides with the program associated with the family of aggregation operators $\{Ag_n, \ n \geq 2\}$ and is deterministic since the list $l_p = l$ when $p = 1$.

Obviously, $P_{Ag,p}$ as described in Definition 8 is not only a computational aggregation, it is in fact a generic approach that induces as many different computable aggregation as the possible families of aggregation operators $\{Ag_n\}$. In particular, we will analyze in this paper three cases: the arithmetic mean, maximum and minimum aggregation operators. From now on, we will denote the three of them as: $P_{M,p}$, $P_{Max,p}$ and $P_{Min,p}$ described as the sample mean or average, the sample maximum and the sample minimum respectively.

Another way to build a class of Non deterministic computable aggregations should be to fix a value k, and randomly select k elements from the list, as those to be aggregated. Given a list of n elements of $[0, 1]$, $l = \{x_1, \ldots, x_m\} \in {<}[0,1]{>}$, let us denote by Sel_k a program that randomly chooses a sample (without resampling) of k elements of the list if $m \geq k$, and maintains the same list in other case.

Definition 9. *Given a family of aggregation operators $\{Ag_n \ n \geq 2\}$, the computable aggregation $P_{Ag,k}$, with $k \leq n$, is defined as the program that for a given* *list l of m elements, first applies the procedure $l' = Sel_k(l)$, and then computes the value $Ag_{|l'|}(l')$.*

It is very easy to see that the computable aggregation $P_{Ag,k}$ is also a non-deterministic computable aggregation (being deterministic when $k \geq m$). The main difference with the previous computable aggregation is that here the dimension of the list to be aggregated is upper bounded by k, consequently the family of aggregation operators is also bounded in dimension, no matter the dimensions of the list to be aggregated.

Definition 10. *Given a normal distribution $N(\mu, \sigma)$ and a family of aggregation operators $\{Ag_n : [0,1]^n \longrightarrow [0,1], n > 2\}$, we define the **noisy computable aggregation** $P_{N(\mu,\sigma),Ag}$ as the program that for any list $l \in <[0,1]>$, with $|l| \leq n$, returns $P_{N(\mu,\sigma),Ag} = Ag(l_N)$, where l_N is the list generated by replacing each element in l with the same element after being modified by adding noise generated by the normal distribution (truncated to 0 or 1 in case that the resulting value was out of the $[0,1]$ interval).*

Definition 11. *The **pure random computable aggregation** P_{PR} is a program that for any list $l \in <[0,1]>$ returns a random value in $[0,1]$.*

Definition 12. *The **bounded random computable aggregation** P_{BR} is a program that for any list $l \in <[0,1]>$ returns a random value in the interval $[Min(l), Max(l)]$.*

Proposition 1. *The computable aggregation operators: $P_{M,p}$, $P_{Max,p}$, $P_{Min,p}$ $P_{Ag,k}$, $P_{N(\mu,\sigma),Ag}$, P_{PR}, P_{BR} are non deterministic computable aggregation operators if $\sigma > 0$ and $p < 1$.*

An example of C++ program P implementing $P_{M,p}$ and $P_{Max,p}$ is described in Fig. 1.

```
double aggSampleAVG(PopulationData d, double prob) {
    int n = 0;
    double sum = 0;
    for (int i = 0; i < PopulationSize; i++)
        if (generateUniform(0,1) < prob) {
            sum += d[i];
            n++;
        }
    return sum / n;
}
double aggSampleMax(PopulationData d, double prob) {
    double max = 0;
    for (int i = 0; i < PopulationSize; i++)
        if (generateUniform(0, 1) < prob)
            if(max< d[i]) max= d[i];
    return max;
}
```

Fig. 1. An implementation of $P_{M,p}$ and $P_{Max,p}$ in C++.

4   Exploring Non Deterministic Computable Aggregations

Given a computable aggregation P that aggregates a list $l \in L<T>$, let us denote by $\mathcal{P}(l)$ the theoretical distribution after all possible realizations of the program P over the fixed list l. Obviously, if $P \in \mathcal{P}_\mathcal{D}$, the associate $\mathcal{P}(l)$ for any list will be a single value. For non deterministic programs, we will have here a probability distribution $\mathcal{P}(l)$ for each fixed value of l.

In general, given a non deterministic computable aggregation P, it is not possible to know the theoretical distribution $\mathcal{P}(l)$. Nevertheless, we could try to approximate it by making many realizations of $P(l)$. In the following definition we distinguish between the empirical and theoretical distribution.

Definition 13. *Given a computable aggregation P and given a list $l \in L<T>$, the distribution of results obtained after n executions of the program P over the list l, will be referred us the empirical distribution with size n of the program P over the list l, represented by $DP_{n,l}$.*

Proposition 2. *Given a deterministic computable aggregation P, the following holds:*

$$DP_{n,l} = \mathcal{P}(l)$$

To analyze the aggregation process previously described and implemented by an NDCA we have several components to consider. There is a list $l \in L<T>$ of elements to be aggregated. There is also a family of aggregation operators ($\{Ag_n\}$) underlying in the non deterministic aggregation process. Finally, two distributions describe the aggregation process: the theoretical distribution $\mathcal{P}(l)$ and the empirical distribution ($DP_{n,l}$). It is obvious that the interactions and relations among these elements will describe and characterize an NDCA.

Some of the questions to be considered are obvious, as the relations between the properties of both distributions (the theoretical and the empirical). But it could also be interesting to consider potential relations between some properties of the list l (analyzed as a distribution) and the corresponding properties of the theoretical/empirical distribution. Another important question could be to compare the result produced by the deterministic underlying aggregation, that is $Ag_{|l|}(l)$, and some properties of the distributions (mean, median, etc) generated by the NDCA.

As said before, it is in general not possible to know the theoretical distribution generated by a non deterministic computable aggregation P when applied on a list l ($\mathcal{P}(l)$). We will replace it with an empirical distribution ($DP_{n,l}$), and we need both to have similar properties.

In that sense we can define the concept of robustness of an NDCA as a measure of the similarity of both distributions (theoretical and empirical).

Definition 14 Robust. *A non deterministic computable aggregation P is said to be Robust, if and only if, for any $l \in L<[0,1]>$, for any $t \in [0,1]$, and for* *any $\epsilon > 0$, there exists n_0 such that the absolute difference between the empirical distribution function*

$$\widehat{DP_{l,n}}(t) = \frac{number\ of\ elements\ in DP_{l,n} \leq t}{n}$$

and the real distribution function $F_{P(l)}(t)$ is lower than ϵ for $n \geq n_0$.

4.1 Empirically Exploring Non Deterministic Computable Aggregations

Given a non deterministic computable aggregation $P = agg$ and a list l, Fig. 2 presents a $C++$ program generating the empirical distribution $DP_{n,l}$ for a list l with length 1000.

```
//Population size
const int PopulationSize = 10000;
// Array (List) of population data
typedef double PopulationData[PopulationSize];
// Times to execute an aggregation to a Population Data
const int AggExecutionTimes = 1000;
// Array (List) to store the ExecutionTimes outputs
typedef double ExecutionsOutputData[AggExecutionTimes];
//Aggregation: List<[0, 1]> -> [0, 1]
double agg(PopulationData d);
// Input: Population Data  Output: ExecutionsOutputData
void generateAggExecutions(PopulationData l,ExecutionsOutputData o){
    for (int i = 0; i < AggExecutionTimes; i++) {
        o[i] = agg(l);
    }
}
```

Fig. 2. An implementation of $Dagg_{l,n}$ in C++.

Figure 3 presents the empirical distribution $(DP_{n,l})$ for some of the previously defined NDCAs[1], with $n = 1000$ and being l a list of 10000 random values[2] generated either using a uniform(0,1) distribution or a Normal(0.5,0.1) bounded in $[0, 1]$[3].

Note that the sample and the noisy mean NDCA have a normal distribution and the pure random and bounded pure random NDCA have uniform distribution. Note also that the sample min and the sample max NDCA have more dispersion in the $N(0.5, 0.1$ population generated list that with the $U(0, 1)$ population generated list.

---

[1] The C++ program could be downloaded from https://github.com/lgarmend/NonDeterministicComputableAggregations/.

[2] The generated population is available in file GeneratedPopulationData.txt, at https://github.com/lgarmend/NonDeterministicComputableAggregations/blob/master/GeneratedPopulationData.txt.

[3] The executions lists are saved in file GeneratedExecutionsList.txt for the cases of uniform and Normal population distribution.

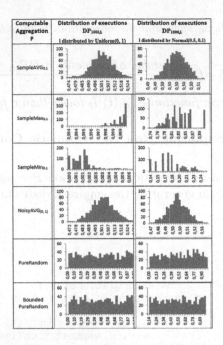

Fig. 3. Empirical distribution for some NDCAs

5   Characterizing Non Deterministic Computable Aggregations

Three properties were initially considered in aggregation operators theory: two boundary conditions plus monotonicity property. Taking into account now that for a non-deterministic computable aggregation the output for a fixed input does not necessarily have to be a single-point value, in this section we will try to extend (or at least give possible extensions) of how these concepts could be generalized in the context of non deterministic computable aggregations.

In the following definition, we present the pure case in which a computable aggregation satisfies the boundary conditions and also the idempotent property. Firs at all, let us denote by l_a a list with all elements equal to a, i.e $l_a = (a, \dots, a)$.

Definition 15 *A computable aggregation P over the set $T = [0, 1]$ is idempotent if and only if $P(l_a) = a$ for all $a \in [0, 1]$.*

Definition 16 *A computable aggregation P over the set $T = [0, 1]$ satisfies the two classical boundary conditions if and only the following holds:*

- $P(l_0) = 0$.
- $P(l_1) = 1$.

Let us note that if a computable aggregation is idempotent then it satisfies the two boundary condition.

Proposition 3 *If the family $\{Ag_n\}$ of aggregation operators is idempotent, then the following holds:*

- *The computable aggregation $P_{Ag,p}$ is idempotent.*
- *The computable aggregation $P_{Ag,k}$ is idempotent.*

As a consequence of the previous proposition the following corollary holds.

Corollary 1. *The computable aggregations $P_{M,p}$, $P_{Max,p}$, $P_{Min,p}$, $P_{M,k}$, $P_{Max,k}$, $P_{Mix,k}$ are idempotent.*

Proposition 4. *If the family $\{Ag_n\}$ of aggregation operators satisfies the boundary conditions then the following holds:*

- *The computable aggregation $P_{Ag,p}$ satisfies the boundary conditions.*
- *The computable aggregation $P_{Ag,k}$ satisfies the boundary conditions.*

As a consequence of the previous proposition the following corollary holds.

Corollary 2. *The computable aggregations $P_{M,p}$, $P_{Max,p}$, $P_{Min,p}$, $P_{M,k}$, $P_{Max,k}$, $P_{Mix,k}$ satisfy the boundary conditions.*

Proposition 5. *The computable aggregations P_{BR}, P_{PR}, $P_{N(\mu,\sigma)}$ are non idempotent and do not satisfy the boundary conditions.*

In previous section we introduced robustness as the way to establish that both the theoretical and the empirical distributions are similar. This idea will allow us to work on the basis of $DP_{l,n}$, considering that in most cases the theoretical distribution is unknown.

We will introduce now the idea of consistency of an NDCA, as the property that considers how close is the behavior of P, with that of Ag, the underlying aggregation process considered by the NDCA.

Definition 17 Consistent in ϕ. *A non deterministic computable aggregation P is said to be robust in ϕ, if and only if, for any $l \in L<[0,1]>$, and for any $\epsilon > 0$, there exists n_0 such that the absolute difference between $\phi(\pi(DP_{l,n}))$ and $\phi(l)$ is lower than ϵ for any $n \geq n_0$, where $(\pi(DP_{l,n}))$ is the vector representation of the set $DP_{l,n}$.*

Proposition 6 *If $p \in (0,1]$, the following holds:*

- *$P_{M,p}$ is Consistent in mean.*
- *$P_{Max,p}$ is Consistent in Max.*
- *$P_{Min,p}$ is Consistent in Min.*

Proof. For random sample theory, it can be seen that the convergence of $M(\pi(DP_{l,n})$ is $M(l)$, $Max(\pi(DP_{l,n})$ is $Max(l)$ and $Min(\pi(DP_{l,n})$ is $Min(l)$.

It would be also important to consider how the dispersion of $DP_{l,n}$ evolves with n. Ideally we would like the dispersion to reduce when n increases.

Definition 18 Concentrate. *A non deterministic computable aggregation P is said to concentrate when the dispersion of $DP_{l,n}$ decreases (is non increasing) with n.*

Now let us define what we understand as monotonicity in non deterministic computable aggregation, since for the deterministic the definition can be reproduced in the same way. A definition of monotony for any function $f : X \longrightarrow Y$ required first the definition of an order in the spaces X and Y, in such a way if we have two inputs $l, l' \in X$ with $l \leq_X l'$ then monotonicity implies that $f(l) \leq_Y f(l')$.

Taking into account this, if we want to define some class of monotonicity in the case of non-deterministic computable aggregations we have to define first an order in the input space (the set of possible lists $L<[0,1]>$) and also an order relation in the space of finite subsets of $[0,1]$. Let us denote by $\widetilde{\leq}_{L<[0,1]>}$ an order between lists and let us denote by $\widetilde{\leq}_{P_F[0,1]}$ an order between finite sets contained in $[0,1]$. Once these two order relations be explicitly defined, the monotony definition is fixed as follow:

Definition 19 *Let $\widetilde{\leq}_{L<[0,1]>}$ be a partial order on the list set and let $\widetilde{\leq}_{P_F[0,1]}$ be a partial order on the sets of finite sets contained in $[0,1]$, then a computable aggregation P is monotone if and only if given any pair of lists l and l' with $l \widetilde{\leq}_{L<[0,1]>} l'$ this implies that $DP_{l,n} \widetilde{\leq}_{P_F[0,1]} DP_{l',n}$ with n large enough.*

Although this definition is perfectly valid, it would be interesting to study in deep different possibilities that will produce different ideas of monotonicity. In particular, in this article we provide a possible order but others could be defined.

The problem of establishing a possible order among the set of ordered lists does not seem very complex if we focus on the case in which the lists presents equal size, since the order relationship coincides with the natural order relationship in $[0,1]^k$, so we will consider that natural order. The case of order over finite sets (even if they have the same size) is much more complex.

In this paper, we propose a possible idea of order that is related with a majority rule. A finite set $S \leq_T S'$ if and only if the following holds:

$$|\{(x,y) \in S \times S' \; / \; x \leq y\}| \geq |\{(x,y) \in S \times S' \; / \; y \leq x\}|.$$

Just to put an example we have that the set $S = \{0.1, 0.3, 0.5\}$ is lower than $S' = \{0.2, 0.6, 0.7\}$ since from the 9 pairwise comparison the elements of S' win in 7 of the cases. Taking into account this consideration the following monotonicity definition is given:

Definition 20 Tournament monotonicity. *A non deterministic computable aggregation P over the domain T is Tournament monotonic if and only if for any pair of lists l,l' of T with the same cardinal such that $l \leq l'$ there exist n_0 in which the following holds*

$$DP_{n,l} \leq_T DP_{n,l'} \quad for \; n \geq n_0.$$

6 Conclusions

The definition of Computable aggregation implies a rupture between functions and aggregation processes allowing the incorporation of a new class of aggregations: the non deterministic computable aggregation. This new class of computable aggregations is characterized by aggregations where the result of the process could change even if we fix the information that has to be aggregated. Obviously, no function can model this class of aggregation process since by definition the output of a function is always the same for a fixed input value.

It is important to emphasize that there are many real situation in which the information is aggregated in a non deterministic way. For example, any inference based on random sampling is a well-known example of this. If the population is huge, it is very frequent to obtain a sample and operate over it. But this is not the only case, any aggregation process in which randomness appear could be understood as a non deterministic process. Many artificial intelligence or machine learning aggregation process could be classified also as non deterministic.

In this paper, we have tried to define some desirable properties for these new class of aggregation process as robustness, stability, boundary conditions, and monotonicity. This questions open the gate for further research.

Acknowledgment. This research has been partially supported by the Government of Spain (grant PGC2018-096509-B-I00), the Government of Madrid (grant S2013/ICCE-2845), Complutense University (UCM Research Group 910149) and Universidad Politeécnica de Madrid.

References

1. Bouchon-Meunier, B.: Aggregation and Fusion of Imperfect Information, vol. 12. Physica, Heidelberg (2013)
2. Bustince, H., Herrera, F., Montero, J. (eds.): Fuzzy Sets and Their Extensions: Representation, Aggregation and Models. Studies in Fuzziness and Soft Computing, vol. 220. Springer, Heidelberg (2008). https://doi.org/10.1007/978-3-540-73723-0
3. Gómez, D., Montero, J.: A discussion on aggregations operators. Kybernetika **40**, 107–120 (2004)
4. Gómez, D., Rojas, K., Montero, J., Rodríguez, J., Beliakov, G.: Consistency and stability in aggregation operators, an application to missing data problems. Int. J. Comput. Intell. Syst. **7**, 595–604 (2014)
5. Montero, J., del Campo, R.G., Garmendia, L., Gómez, D., Rodríguez, J.: Computable aggregations. Inf. Sci. **460**, 439–449 (2018)
6. González-del-Campo, R., Garmendia, L., Montero, J.: Complexity of increasing phi-recursive computable aggregations. In: XIX Congreso Español sobre Tecnologías y Lógica Fuzzy (CAEPIA-ESTYLF), pp. 187–191 (2018)
7. Magdalena, L., Garmendia, L., Gómez, D., del Campo, R.G., Rodríguez, J.T., Montero, J.: Types of recursive computable aggregations. In: 2019 IEEE International Conference on Fuzzy Systems. FUZZ-IEEE 2019, pp. 1–6. IEEE (2019)

8. Calvo, T., Kolesárová, A., Komorníková, M., Mesiar, R.: Aggregation operators: properties, classes and constructions methods. In: Calvo, T., Mayor, G., Mesiar, R. (eds.) Aggregation Operators. Studies in Fuzziness and Soft Computing, vol. 97, pp. 3–104. Physica, Heidelberg (2002). https://doi.org/10.1007/978-3-7908-1787-4_1

9. Rojas, K., Gómez, D., Montero, J., Rodríguez, J.: Strictly stable families of aggregation operators. Fuzzy Sets Syst. **228**, 44–63 (2013)

10. Beliakov, G., Gómez, D., James, S., Montero, J., Rodríguez, J.: Approaches to learning strictly-stable weights for data missing values. Fuzzy Sets Syst. **325**, 97–113 (2017). https://doi.org/10.1016/j.fss.2017.02.003

11. Olaso, P., Rojas, K., Gómez, D., Montero, J.: A generalization of stability for families of aggregation operators. Fuzzy Sets Syst. **378**, 68–78 (2020)

12. Beliakov, G., Pradera, A., Calvo, T.: Aggregations Functions: A Guide for Practitioners. Springer, Heidelberg (2007). https://doi.org/10.1007/978-3-540-73721-6

13. Cutello, V., Montero, J.: Recursive families of OWA operators. In: Proceedings of the IEEE International Conference on Fuzzy Systems (FUZZIEEE 1994), Orlando, USA, pp. 1137–1141 (1994)

14. Cutello, V., Montero, J.: Hierarchical aggregation of owa operators: basic measures and related computational problems. Uncertainty Fuzziness Knowl.-Based Syst. **3**, 17–26 (1995)

15. Cutello, V., Montero, J.: Recursive connective rules. Int. J. Intell. Syst. **14**, 3–20 (1999)

16. Lucca, J.G., Dimuro, G.P., Bedregal, B.R.C., Mesiar, R., Kolesárová, A., Bustince, H.: Preaggregation functions: construction and an application. IEEE Trans. Fuzzy Syst. **24**(2), 260–272 (2016)

17. Magdalena, L., Gómez, D., Montero, J., Cubillo, S., Torres, C.: Generalized pre-aggregations. In: Kearfott, R.B., Batyrshin, I., Reformat, M., Ceberio, M., Kreinovich, V. (eds.) IFSA/NAFIPS 2019 2019. AISC, vol. 1000, pp. 362–370. Springer, Cham (2019). https://doi.org/10.1007/978-3-030-21920-8_33

18. Minsky, M.: Computation: Finite and Infinite Machines. Prentice-Hall, Englewood Cliffs (1967)

19. Wirth, N.: Algorithms + Data Structures = Programs. Prentice-Hall, Englewood Cliffs (1976)

20. Apt, K.R., Olderog, E.-R.: Deterministic programs. In: Apt, K.R., Olderog, E.-R. (eds.) Verification of Sequential and Concurrent Programs. Graduate Texts in Computer Science, pp. 47–99. Springer, New York (1997). https://doi.org/10.1007/978-1-4757-2714-2_3

Dissimilarity Based Choquet Integrals

Humberto Bustince[1], Radko Mesiar[2], Javier Fernandez[1], Mikel Galar[1],
Daniel Paternain[1], Abdulrahman Altalhi[3], Graçaliz P. Dimuro[4],
Benjamín Bedregal[5], and Zdenko Takáč[6]($\boxtimes$)

[1] Departamento de Estadistica Informatica y Matematicas, Universidad Pública
de Navarra, Campus Arrosadía s/n, P.O. Box 31006, Pamplona, Spain
{bustince,fcojavier.fernandez,mikel.galar,daniel.paternain}@unavarra.es
[2] Department of Mathematics and Descriptive Geometry,
Faculty of Civil Engineering, Slovak University of Technology, Bratislava, Slovakia
radko.mesiar@stuba.sk
[3] Faculty of Computing and Information Technology, King Abdulaziz University,
Jeddah, Saudi Arabia
ahaltalhi@kau.edu.sa
[4] Centro de Ciências Computacionais, Universidade Federal do Rio Grande,
Rio Grande 96077540, Brazil
gracalizdimuro@furg.br
[5] Departamento de Informática e Matemática Aplicada,
Universidade Federal do Rio Grande do Norte, Natal 59078-970, Brazil
bedregal@dimap.ufrn.br
[6] Institute of Information Engineering Automation and Mathematics,
Faculty of Chemical and Food Technology, Slovak University of Technology
in Bratislava, Radlinského 9, Bratislava, Slovakia
zdenko.takac@stuba.sk

Abstract. In this paper, in order to generalize the Choquet integral,
we replace the difference between inputs in its definition by a restricted
dissimilarity function and refer to the obtained function as d-Choquet
integral. For some particular restricted dissimilarity function the corresponding d-Choquet integral with respect to a fuzzy measure is just
the 'standard' Choquet integral with respect to the same fuzzy measure. Hence, the class of all d-Choquet integrals encompasses the class
of all 'standard' Choquet integrals. This approach allows us to construct
a wide class of new functions, d-Choquet integrals, that are possibly,
unlike the 'standard' Choquet integral, outside of the scope of aggregation functions since the monotonicity is, for some restricted dissimilarity
function, violated and also the range of such functions can be wider than
$[0, 1]$, in particular it can be $[0, n]$.

Keywords: Choquet integral · d-Choquet integral · Dissimilarity ·
Pre-aggregation function · Aggregation function · Monotonicity ·
Directional monotonicity

Supported by the project VEGA 1/0545/20.

M.-J. Lesot et al. (Eds.): IPMU 2020, CCIS 1238, pp. 565–573, 2020.
https://doi.org/10.1007/978-3-030-50143-3_44

1   Introduction

The Choquet integral [4] can be regarded as a generalization of additive aggregation functions replacing the requirement of additivity by that of comonotone additivity. In recent years it was shown that, in some cases, additive aggregation functions are not appropriate to model even quite simple situations, which, on the other hand, can be treated with Choquet integrals [1,5,8,9,14,15]. In the literature, some generalizations of the Choquet integral appeared: in [11,12,16] the product operator was replaced by a more general function; in the same pattern, using the distributivity of the product operator and then replacing its two instances by two different functions under some constraints, in [6,17], generalizations of the Choquet integral were obtained; some Choquet-like integrals defined in terms of pseudo-addition and pseudo-multiplication are studied in [18]; a fuzzy t-conorm integral that is a generalization of Choquet integral is introduced in [19]; a non-linear integral that need not be increasing is introduced in [20]; a concave integral generalizing the Choquet integral is introduced in [13]; and a level dependent Choquet integral was also introduced in [10]. An overview of some recent extensions of the Choquet integral can be found in [7].

Our aim is to replace the difference between the inputs in the definition of the Choquet integral by a restricted dissimilarity function [2,3] in order to generalize the Choquet integral. We refer to the obtained function as d-Choquet integral. This approach allows us to construct a wide class of new functions, d-Choquet integrals, which, unlike the "standard" Choquet integral, may be possibly outside of the scope of aggregation functions, since the monotonicity may be violated for some restricted dissimilarity function, and also the range of such functions can be wider than $[0, 1]$. Our work can be seen as the first step to the generalization of the Choquet integral to various settings where the difference causes problems (for example, intervals).

The structure of the paper is as follows. First, we present some preliminary concepts. In Sect. 3, we introduce the notion of d-Choquet integral, describe its construction in terms of automorphisms and study its monotonicity and directional monotonicity. Conclusions and future research are described in Sect. 4.

2   Preliminaries

The necessary basic notions and terminology are recalled in this section.

A function $\delta : [0, 1]^2 \to [0, 1]$ is called a restricted dissimilarity function on $[0, 1]$ if it satisfies, for all $x, y, z \in [0, 1]$, the following conditions:

1. $\delta(x, y) = \delta(y, x)$;
2. $\delta(x, y) = 1$ if and only if $\{x, y\} = \{0, 1\}$;
3. $\delta(x, y) = 0$ if and only if $x = y$;
4. if $x \leq y \leq z$, then $\delta(x, y) \leq \delta(x, z)$ and $\delta(y, z) \leq \delta(x, z)$.

An n-ary aggregation function is a mapping $A : [0, 1]^n \to [0, 1]$ satisfying the following properties:

(A1) A is increasing in each argument;

(A2) $A(0,\ldots,0) = 0$ and $A(1,\ldots,1) = 1$.

An automorphism of $[0,1]$ is a continuous, strictly increasing function φ : $[0,1] \to [0,1]$ such that $\varphi(0) = 0$ and $\varphi(1) = 1$. Moreover, the identity on $[0,1]$ is denoted by Id.

It is well-known that a function $f : [0,1]^n \to [0,1]$ is additive if

$$f(x_1 + y_1, \ldots, x_n + y_n) = f(x_1, \ldots, x_n) + f(y_1, \ldots, y_n) \tag{1}$$

for all $(x_1, \ldots, x_n), (y_1, \ldots, y_n) \in [0,1]^n$ such that $(x_1 + y_1, \ldots, x_n + y_n) \in [0,1]^n$. From now on, $[n]$ denotes the set $\{1, \ldots, n\}$. Vectors $(x_1, \ldots, x_n), (y_1, \ldots, y_n) \in [0,1]^n$ are comonotone if there exists a permutation $\sigma : [n] \to [n]$ such that $x_{\sigma(1)} \leq \ldots \leq x_{\sigma(n)}$ and $y_{\sigma(1)} \leq \ldots \leq y_{\sigma(n)}$. A function $f : [0,1]^n \to [0,1]$ is called comonotone additive if Equality (1) holds for all comonotone vectors $(x_1, \ldots, x_n), (y_1, \ldots, y_n) \in [0,1]^n$ such that $(x_1 + y_1, \ldots, x_n + y_n) \in [0,1]^n$.

A function $\mu : 2^{[n]} \to [0,1]$ is called a fuzzy measure on $[n]$ if $\mu(\emptyset) = 0$, $\mu([n]) = 1$ and $\mu(A) \leq \mu(B)$ for all $A \subseteq B \subseteq [n]$.

Let $\boldsymbol{r} = (r_1, \ldots, r_n)$ be a real n-dimensional vector such that $\boldsymbol{r} \neq \boldsymbol{0}$. A function $f : [0,1]^n \to [0,1]$ is $\boldsymbol{r}$-increasing if, for all $(x_1, \ldots, x_n) \in [0,1]^n$ and for all $c \in {]0,1]}$ such that $(x_1 + cr_1, \ldots, x_n + cr_n) \in [0,1]^n$, it holds

$$f(x_1 + cr_1, \ldots, x_n + cr_n) \geq f(x_1, \ldots, x_n).$$

A function $f : [0,1]^n \to [0,1]$ is said to be an n-ary pre-aggregation function if $f(0, \ldots, 0) = 0$, $f(1, \ldots, 1) = 1$ and f is $\boldsymbol{r}$-increasing for some real n-dimensional vector $\boldsymbol{r} = (r_1, \ldots, r_n)$ such that $\boldsymbol{r} \neq \boldsymbol{0}$ and $r_i \geq 0$ for every $i = 1, \ldots, n$. In this case, we say that f is an $\boldsymbol{r}$-pre-aggregation function.

3 d-Choquet Integral

A new approach to generalization of Choquet integral based on dissimilarity functions is introduced in this section and its monotonicity is studied.

The discrete Choquet integral on the unit interval with respect to a fuzzy measure $\mu : 2^{[n]} \to [0,1]$ is defined as a mapping $C_\mu : [0,1]^n \to [0,1]$ such that

$$C_\mu(x_1, \ldots, x_n) = \sum_{i=1}^{n} (x_{\sigma(i)} - x_{\sigma(i-1)}) \mu\left(A_{\sigma(i)}\right) \tag{2}$$

where σ is a permutation on $[n]$ satisfying $x_{\sigma(1)} \leq \ldots \leq x_{\sigma(n)}$, with the convention $x_{\sigma(0)} = 0$ and $A_{\sigma(i)} = \{\sigma(i), \ldots, \sigma(n)\}$.

In order to generalize the Choquet integral, we replace the difference $x_{\sigma(i)} - x_{\sigma(i-1)}$ by a restricted dissimilarity function $\delta : [0,1]^2 \to [0,1]$ and refer to the obtained function as d-Choquet integral.

Definition 1. *Let n be a positive integer and $\mu : 2^{[n]} \rightarrow [0,1]$ be a fuzzy measure on $[n]$. Let $\delta : [0,1]^2 \rightarrow [0,1]$ be a restricted dissimilarity function. An n-ary discrete d-Choquet integral on $[0,1]$ with respect to μ and δ is defined as a mapping $C_{\mu,\delta} : [0,1]^n \rightarrow [0,n]$ such that*

$$C_{\mu,\delta}(x_1,\ldots,x_n) = \sum_{i=1}^{n} \delta(x_{\sigma(i)}, x_{\sigma(i-1)})\mu\left(A_{\sigma(i)}\right) \tag{3}$$

where σ is a permutation on $[n]$ satisfying $x_{\sigma(1)} \leq \ldots \leq x_{\sigma(n)}$, with the convention $x_{\sigma(0)} = 0$ and $A_{\sigma(i)} = \{\sigma(i),\ldots,\sigma(n)\}$.

It is easy to check that, in general, the range of $C_{\mu,\delta}$ need not be a subset of $[0,1]$, but it is a subset of $[0,n]$. The following condition assures that the outputs of d-Choquet integral $C_{\mu,\delta}$ would be from $[0,1]$, which is often a desired property for some applications:

(P1) $\delta(0,x_1) + \delta(x_1,x_2) + \ldots + \delta(x_{n-1},x_n) \leq 1$ for all $x_1,\ldots,x_n \in [0,1]$ where $x_1 \leq \ldots \leq x_n$.

Under the condition we obtain $C_{\mu,\delta} : [0,1]^n \rightarrow [0,1]$ so we have the following straightforward result.

Proposition 1. *Let $C_{\mu,\delta} : [0,1]^n \rightarrow [0,n]$ be an n-ary discrete d-Choquet integral on $[0,1]$ with respect to μ and δ given by Definition 1. If δ satisfies the condition (P1), then*

$$C_{\mu,\delta}(x_1,\ldots,x_n) \in [0,1]$$

for all $x_1,\ldots,x_n \in [0,1]$ and for any measure μ.

Example 1. Let μ be a fuzzy measure on $\{1,2,3\}$ defined by $\mu(\{1\}) = \mu(\{2\}) = \mu(\{3\}) = 0.3$, $\mu(\{1,2\}) = 0.75$, $\mu(\{2,3\}) = 0.55$ and $\mu(\{1,3\}) = 0.6$.
 (i) Then

$$C_\mu(0.2, 0.9, 0.6) = 0.2 \cdot 1 + 0.4 \cdot 0.55 + 0.3 \cdot 0.3 = 0.51.$$

It is easy to see that for $\delta(x,y) = |x - y|$ it holds $C_{\mu,\delta} = C_\mu$ for any possible inputs and any measure μ.
 (ii) However, if $\delta(x,y) = (x-y)^2$ we have

$$C_{\mu,\delta}(0.2, 0.9, 0.6) = 0.04 \cdot 1 + 0.16 \cdot 0.55 + 0.09 \cdot 0.3 = 0.155.$$

(iii) Finally, taking

$$\delta(x,y) = \begin{cases} 0, & \text{if } x = y; \\ \frac{|x-y|+1}{2}, & \text{otherwise,} \end{cases}$$

we obtain

$$C_{\mu,\delta}(0.2, 0.9, 0.6) = 0.6 \cdot 1 + 0.7 \cdot 0.55 + 0.65 \cdot 0.3 = 1.18,$$

where we can see that $C_{\mu,\delta}(0.2, 0.9, 0.6) > 1$. This may happen since the condition (P1) is not satisfied as can be seen if we take, for instance, $(0, 0.5, 1)$.

The construction of d-Choquet integrals is directly connected with the construction of restricted dissimilarity functions with desired properties. In [2], a construction method for restricted dissimilarity functions in terms of automorphisms was introduced.

Proposition 2 *[2]. If* φ_1, φ_2 *are two automorphisms of* $[0, 1]$*, then the function* $\delta : [0, 1]^2 \to [0, 1]$ *defined by*

$$\delta(x, y) = \varphi_1^{-1}\Big(|\varphi_2(x) - \varphi_2(y)| \Big)$$

is a restricted dissimilarity function.

We write $C_{\mu, \varphi_1, \varphi_2}$ instead of $C_{\mu, \delta}$ if the restricted dissimilarity function δ is given in terms of automorphisms φ_1, φ_2 as in the previous proposition.

Proposition 3. *Let* n *be a positive integer,* $\delta : [0, 1]^2 \to [0, 1]$ *be a restricted dissimilarity function given in terms of automorphisms* φ_1, φ_2 *as in Proposition 2. If* $\varphi_1 \geq Id$ *for all* $x \in [0, 1]$*, then* δ *satisfies (P1).*

Example 2. The restricted dissimilarity functions $\delta(x, y) = (x - y)^2$, $\delta(x, y) = |\sqrt{x} - \sqrt{y}|$, $\delta(x, y) = |x^2 - y^2|$ and $\delta(x, y) = (\sqrt{x} - \sqrt{y})^2$ satisfy the condition $(P1)$ which means that the corresponding d-Choquet integrals have the ranges in $[0, 1]$. However, for the restricted dissimilarity function $\delta(x, y) = \sqrt{|x - y|}$, the condition $(P1)$ is violated. For instance, $\delta(0, 0.1) + \delta(0.1, 1) = 1.2649 > 1$.

Clearly, d-Choquet integral $C_{\mu, \delta}$ for the restricted dissimilarity function $\delta(x, y) = |x - y|$ recovers the "standard" Choquet integral.

Theorem 1. *Let* n *be a positive integer,* $\mu : 2^{[n]} \to [0, 1]$ *be a fuzzy measure on* $[n]$*,* $\delta : [0, 1]^2 \to [0, 1]$ *be the function* $\delta(x, y) = |x - y|$*,* $C_{\mu, \delta} : [0, 1]^n \to [0, 1]$ *be an* n-ary *discrete* d-Choquet integral on $[0, 1]$ *with respect to* μ *and* δ *given by Definition 1 and* $C_\mu : [0, 1]^n \to [0, 1]$ *be an* n-ary *discrete Choquet integral on* $[0, 1]$ *with respect to* μ *given by Eq. (2). Then*

$$C_{\mu, \delta}(x_1, \ldots, x_n) = C_\mu(x_1, \ldots, x_n)$$

for all $x_1, \ldots, x_n \in [0, 1]$*.*

Corollary 1. *Let* n *be a positive integer,* $\mu : 2^{[n]} \to [0, 1]$ *be a fuzzy measure on* $[n]$*. Then*

$$C_{\mu, Id, Id}(x_1, \ldots, x_n) = C_\mu(x_1, \ldots, x_n)$$

for all $x_1, \ldots, x_n \in [0, 1]$*.*

3.1   Monotonicity of d-Choquet Integrals

In general, d-Choquet integrals are not monotone, hence we study conditions under which a d-Choquet integral is increasing in each component. Note that, since the boundary conditions are satisfied, any increasing d-Choquet integral is an aggregation function.

Theorem 2. *Let n be a positive integer, δ be a restricted dissimilarity function and $C_{\mu,\delta}$ be an n-ary d-Choquet integral with respect to μ and δ. Then the following assertions are equivalent:*

(i) For any fuzzy measure μ on $[n]$, $C_{\mu,\delta}(x_1,\ldots,x_n) \leq C_{\mu,\delta}(y_1,\ldots,y_n)$ whenever $x_1 \leq \ldots \leq x_n$, $y_1 \leq \ldots \leq y_n$, $x_1 \leq y_1,\ldots,x_n \leq y_n$.

(ii) $\delta(0,x_1)+\delta(x_1,x_2)+\ldots+\delta(x_{m-1},x_m) \leq \delta(0,y_1)+\delta(y_1,y_2)+\ldots+\delta(y_{m-1},y_m)$ for all $m \in [n]$ and $x_1,\ldots,x_m,y_1,\ldots,y_m \in [0,1]$ where $x_1 \leq \ldots \leq x_m$, $y_1 \leq \ldots \leq y_m$, $x_1 \leq y_1,\ldots,\leq x_m \leq y_m$.

Corollary 2. *Let n be a positive integer, δ be a restricted dissimilarity function and $C_{\mu,\delta}$ be an n-ary d-Choquet integral with respect to μ and δ. If for all $m \in [n]$ there exists an increasing function $f_m : [0,1] \to [0,1]$ such that $\delta(0,x_1) + \delta(x_1,x_2) + \ldots + \delta(x_{m-1},x_m) = f_m(x_m)$ for all $x_1,\ldots,x_m \in [0,1]$ where $x_1 \leq \ldots \leq x_m$, then for any fuzzy measure μ on $[n]$, $C_{\mu,\delta}(x_1,\ldots,x_n) \leq C_{\mu,\delta}(y_1,\ldots,y_n)$ whenever $x_1 \leq y_1,\ldots,x_n \leq y_n$.*

Corollary 3. *Let n be a positive integer, $\delta : [0,1]^2 \to [0,1]$ be a restricted dissimilarity function given in terms of automorphisms φ_1,φ_2 as in Proposition 2. Let $C_{\mu,\varphi_1,\varphi_2}$ be an n-ary d-Choquet integral with respect to μ and δ. If $\varphi_1 = Id$, then for any fuzzy measure μ on $[n]$, $C_{\mu,\delta}(x_1,\ldots,x_n) \leq C_{\mu,\delta}(y_1,\ldots,y_n)$ whenever $x_1 \leq y_1,\ldots,x_n \leq y_n$.*

It is easy to see that for $\varphi_1 = Id$ we have:

$$C_{\mu,Id,\varphi_2}(x_1,\ldots,x_n) = C_\mu(\varphi_2(x_1),\ldots,\varphi_2(x_n)),$$

i.e. C_{μ,Id,φ_2} is fully determined by a "standard" Choquet integral C_μ. It also means that C_{μ,Id,φ_2} is an aggregation function.

Since the monotonicity is not always satisfied, we also study directional monotonicity. From the previous results it is clear that, in general, an n-ary d-Choquet integral is not r-increasing for a vector $r = (r_1,\ldots,r_n)$ such that there exists $k \in \{1,\ldots,n\}$ with $r_i \neq 0$ if and only if $i = k$. In what follows we focus on the directional monotonicity with respect to the vector $r = (1,\ldots,1)$.

Theorem 3. *Let n be a positive integer and $C_{\mu,\delta} : [0,1]^n \to [0,n]$ be an n-ary d-Choquet integral with respect to a fuzzy measure μ and a restricted dissimilarity function δ. Then*

(i) $C_{\mu,\delta}$ is 1-increasing for any fuzzy measure μ whenever

$$\delta(x+c,y+c) \geq \delta(x,y)$$

for all $x,y,c \in [0,1]$ such that $x+c,y+c \in [0,1]$;

(ii) $C_{\mu,\delta}$ is 1-increasing for any fuzzy measure μ whenever for all $m \in [n]$ there exists an increasing function $f_m : [0,1] \to [0,1]$ such that

$$\delta(0,x_1) + \delta(x_1,x_2) + \ldots + \delta(x_{m-1},x_m) = f_m(x_m)$$

for all $x_1,\ldots,x_m \in [0,1]$ where $x_1 \leq \ldots \leq x_m$.

Corollary 4. *Let n be a positive integer, $\delta : [0,1]^2 \to [0,1]$ be a restricted dissimilarity function given in terms of automorphisms φ_1, φ_2 as in Proposition 2. Let $C_{\mu,\varphi_1,\varphi_2} : [0,1]^n \to [0,n]$ be an n-ary d-Choquet integral with respect to μ and δ. Then $C_{\mu,\delta}$ is 1-increasing for any fuzzy measure μ whenever at least one of the following conditions is satisfied:*

(i) φ_2 is convex;
(ii) $\varphi_1 = Id$.

The conditions under which an n-ary d-Choquet integral is a 1-pre-aggregation function directly follow from the previous results.

Corollary 5. *Let n be a positive integer, $\delta : [0,1]^2 \to [0,1]$ be a restricted dissimilarity function given in terms of automorphisms φ_1, φ_2 as in Proposition 2. Then the n-ary d-Choquet integral with respect to μ and δ is a 1-pre-aggregation function for any fuzzy measure μ on $[n]$ whenever at least one of the following conditions is satisfied:*

(i) $\varphi_1 = Id$;
(ii) $\varphi_1 > Id$ and φ_2 is convex.

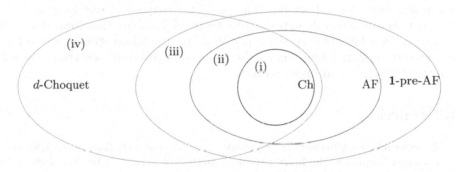

Fig. 1. The relations among the classes of all standard Choquet integrals (Ch), aggregation functions (AF), 1-pre-aggregation functions (1-pre-AF) and d-Choquet integrals (d-Choquet).

Taking the following restricted dissimilarity functions, we obtain an example of d-Choquet integral which is (see Fig. 1):

(i) a standard Choquet integral, if $\delta(x,y) = |x - y|$;
(ii) an aggregation function which is not a standard Choquet integral, if $\delta(x,y) = |\sqrt{x} - \sqrt{y}|$ or $\delta(x,y) = |x^2 - y^2|$;
(iii) an 1-pre-aggregation function which is not an aggregation function, if $\delta(x,y) = (x - y)^2$;
(iv) a d-Choquet integral which is not an 1-pre-aggregation function, if $\delta(x,y) = \sqrt{|\sqrt{x} - \sqrt{y}|}$.

4 Conclusions

In this paper we have introduced a generalization of the Choquet integral replacing the difference by a restricted dissimilarity function. Our approach results in a wide class of d-Choquet integrals that encompasses the class of all "standard" Choquet integrals. We have shown that, based on the choice of a restricted dissimilarity function, the d-Choquet integral is or is not an aggregation function or pre-aggregation function.

The class of d-Choquet integrals can be useful in applications where Choquet integrals have shown themselves valuable, for instance in image processing, decision making or classification. In particular, these new functions can be used in the settings where the difference causes problems, for example for intervals. In this sense, we intend to make a research of possibilities to apply our results in image processing and classification problems in the interval-valued setting in such a way that classical fuzzy algorithms which make use of the Choquet integral can be appropriately formulated.

The Choquet integral can be equivalently expressed in terms of differences of capacity weights on a nested family of sets or in terms of the Moebius transform. We intend to develop an analysis similar to this one regarding the former; nevertheless the equivalence between the two representations will be most probably lost in our more general setting. The possible extension of our generalization idea to Choquet integrals expressed in terms of Moebius transform does not seem so easily achievable, since it is not clear how the information provided by a restricted dissimilarity function can be included in such representation. We will analyze this question in future works.

References

1. Barrenechea, E., Bustince, H., Fernandez, J., Paternain, D., Sanz, J.A.: Using the Choquet integral in the fuzzy reasoning method of fuzzy rule-based classification systems. Axioms **2**(2), 208–223 (2013)
2. Bustince, H., Barrenechea, E., Pagola, M.: Relationship between restricted dissimilarity functions, restricted equivalence functions and normal en-functions: image thresholding invariant. Pattern Recogn. Lett. **29**(4), 525–536 (2008)
3. Bustince, H., Jurio, A., Pradera, A., Mesiar, R., Beliakov, G.: Generalization of the weighted voting method using penalty functions constructed via faithful restricted dissimilarity functions. Eur. J. Oper. Res. **225**(3), 472–478 (2013)
4. Choquet, G.: Theory of capacities. In: Annales de l'Institut Fourier, vol. 5, pp. 131–295 (1953–1954)
5. Dimuro, G.P., Lucca, G., Sanz, J.A., Bustince, H., Bedregal, B.: CMin-Integral: a Choquet-like aggregation function based on the minimum t-norm for applications to fuzzy rule-based classification systems. In: Torra, V., Mesiar, R., De Baets, B. (eds.) AGOP 2017. AISC, vol. 581, pp. 83–95. Springer, Cham (2018). https://doi.org/10.1007/978-3-319-59306-7_9
6. Dimuro, G.P., et al.: Generalized $c_{F_1 F_2}$-integrals: from Choquet-like aggregation to ordered directionally monotone functions. Fuzzy Sets Syst. **378**, 44–67 (2020)

7. Dimuro, G.P., et al.: The state-of-art of the generalizations of the Choquet integral: from aggregation and pre-aggregation to ordered directionally monotone functions. Inf. Fusion **57**, 27–43 (2020)
8. Grabisch, M., Labreuche, C.: A decade of application of the Choquet and Sugeno integrals in multi-criteria decision aid. Ann. Oper. Res. **175**(1), 247–286 (2010)
9. Grabisch, M., Marichal, J., Mesiar, R., Pap, E.: Aggregation Functions. Cambridge University Press, Cambridge (2009)
10. Greco, S., Matarazzo, B., Giove, S.: The Choquet integral with respect to a level dependent capacity. Fuzzy Sets Syst. **175**(1), 1–35 (2011)
11. Horanská, Ľ., Šipošová, A.: A generalization of the discrete Choquet and Sugeno integrals based on a fusion function. Inf. Sci. **451–452**, 83–99 (2018)
12. Horanská, Ľ., Šipošová, A.: Generalization of the discrete Choquet integral. In: Uncertainty Modelling 2015, pp. 49–54. Jednota slovenských matematikov a fyzikov (2016)
13. Lehrer, E.: A new integral for capacities. Econ. Theory **39**(1), 157–176 (2009)
14. Lucca, G., Dimuro, G. P., Mattos, V., Bedregal, B., Bustince, H., Sanz, J.A.: A family of Choquet-based non-associative aggregation functions for application in fuzzy rule-based classification systems. In: 2015 IEEE International Conference on Fuzzy Systems (FUZZ-IEEE), pp. 1–8. Los Alamitos (2015)
15. Lucca, G., et al.: CC-integrals: Choquet-like copula-based aggregation functions and its application in fuzzy rule-based classification systems. Knowledge-Based Syst. **119**, 32–43 (2017)
16. Lucca, G., Sanz, J.A., Dimuro, G.P., Bedregal, B., Bustince, H., Mesiar, R.: CF-integrals: a new family of pre-aggregation functions with application to fuzzy rule-based classification systems. Inf. Sci. **435**, 94–110 (2018)
17. Lucca, G., Dimuro, G.P., Fernández, J., Bustince, H., Bedregal, B., Sanz, J.A.: Improving the performance of fuzzy rule-based classification systems based on a nonaveraging generalization of CC-integrals named $C_{F_1 F_2}$-integrals. IEEE Trans. Fuzzy Syst. **27**(1), 124–134 (2019)
18. Mesiar, R.: Choquet-like integrals. J. Math. Anal. Appl. **194**(2), 477–488 (1995)
19. Murofushi, T., Sugeno, M.: Fuzzy t-conorm integral with respect to fuzzy measures: generalization of Sugeno integral and Choquet integral. Fuzzy Sets Syst. **42**(1), 57–71 (1991)
20. Wang, Z., Leung, K.-S., Wong, M.-L., Fang, J.: A new type of nonlinear integrals and the computational algorithm. Fuzzy Sets Syst. **112**(2), 223–231 (2000)

Aggregation: Aggregation of Different Data Structures

A S-QFD Approach with Bipolar Fuzzy Hamacher Aggregation Operators and Its Application on E-Commerce

Esra Çakır$^{(\boxtimes)}$ ⓘ and Ziya Ulukan$^{(\boxtimes)}$ ⓘ

Department of Industrial Engineering, Galatasaray University, İstanbul, Turkey
{ecakir,zulukan}@gsu.edu.tr

Abstract. In a global competitive environment, companies' ability to develop products and respond to customer demands is crucial to their success. Quality Function Deployment is an approach that companies use to meet customer needs and expectations in the product design process. It provides competitive advantage to the firm by shortening the development period of products that meet customer expectations. The purpose of this research is to explore the feasibility of the QFD approach in the design of an e-commerce web site. In this context, the QFD application has been implemented to meet customer expectations and increase competitive power of the site to be designed. Bipolar fuzzy numbers are used to express customers' decisions. This research contributes to the literature with a new QFD approach with bipolar fuzzy numbers and practice by expanding QFD's application field in software sector.

Keywords: Bipolar Fuzzy Set · Hamacher aggregation operator · House of Quality · Software Quality Function Deployment

1 Introduction

Due to globalization, companies must use innovation in order to be able to withstand competition and customer-oriented production has become mandatory. Thus, customer-oriented approaches such as Quality Function Deployment (QFD) used in product design have become popular in determining product features [1]. For this reason, it is important to understand which product features are meaningful to the customer and reflect the feedback from the customer to the product design and development process [2].

In terms of competition in the market, the effectiveness of the new product design process emerges as one of the most important success conditions for companies. QFD has been working on designing new products or improving existing products to ensure that customer requirements are accurately reflected in the product. The correct reflection of the customers' desires and requirements on the product means the satisfaction of the customers, which is one of the objectives of the firms.

In this study, QFD has been implemented in order to increase customer satisfaction in designing an e-commerce site and to create a user-oriented site. The contribution of this paper is to examine front-end and back-end features of a music streaming web site

© Springer Nature Switzerland AG 2020
M.-J. Lesot et al. (Eds.): IPMU 2020, CCIS 1238, pp. 577–587, 2020.
https://doi.org/10.1007/978-3-030-50143-3_45

and relate this features with customer expectations by using for the first time Software-QFD with bipolar fuzzy Hamacher aggregation operators. The paper's overview is as follows. Section 2 illustrates the literature highlights. Section 3 presents proposed methodology, which is formalized as bipolar fuzzy House of Quality. A music streaming website design is selected as the application area in Sect. 4. Finally, the comments and discussion are given in the last section.

2 Literature Review

The House of Quality is a set of matrices that relate customer wants and quality characteristics determined to meet them, compares quality characteristics based on objective measures, and determines positive or negative correlations between them [3]. Graphical presentation, known as "House of Quality" used in QFD, is a rich and easily accessible information bank. This clear communication mechanism enables the basic facts to occur more timely and more accurately than traditional development documents.

QFD-based product development provides significant competitive advantages by prioritizing more customer satisfaction, product reaching the market in a shorter time, and improved product performance compared to other conventional methods. With this method, the priorities of the customer requests and requirements are determined and according to the voice of customer, the features related to the products are ordered according to their importance. Therefore, the product is designed and manufactured to include the most desirable features.

Software Quality Function Deployment (S-QFD) is a type of QFD (used in the manufacturing industry) used for software products. S-QFD has a significant positive impact on: user engagement, management support and participation, and techniques to shorten the software development life cycle.

In recent years, fuzzy set theory has evolved to solve real world problems. In many studies, fuzzy logic was included in mathematical modeling. Pandey [4] performed evaluating the strategic design parameters of airports in Thailand to meet service expectations of Low-Cost Airlines using the Fuzzy-based QFD method. He demonstrates and signifies that the Fuzzy based QFD method is a promising and pragmatic decision-making tool for customer-oriented airport strategic planning. Aijun et al. [5] studied a fuzzy three-stage multi-attribute decision-making approach based on customer needs for sustainable supplier selection. They use interval-valued intuitionistic trapezoidal fuzzy numbers with QFD and combined with VIKOR. Pooya [6] stressed about an extended approach for manufacturing strategy process based on organization performance through Fuzzy QFD. His proposed model is implemented in an automotive rubber profile manufacturer in Iran. Efe [7] applied fuzzy cognitive map based quality function deployment approach to dishwasher machine selection. His paper contributes to the literature by integrating QFD approach and fuzzy cognitive map approach. Wei et al. [8] presented pythagorean hesitant fuzzy Hamacher aggregation operators in multiple-attribute decision making process. Garg [9] performed intuitionistic fuzzy Hamacher aggregation operators. He used this operators on entropy weight and their applications to multi-criteria decision-making problems. Abtahi et al.

[10] studied to design cloud computing service using QFD. They used flexibility, scalability, high accessibility and cost-effectiveness as features of their software. Prasad et al. [11] proposed a QFD-based decision making model for computer-aided design software selection. They stressed on design and development of a QFD-based decision making model in Visual BASIC 6.0 for selection of computer- aided design software for manufacturing organizations.

3 Methodology

In this section, the preliminaries and definitions of the proposed method with bipolar fuzzy information [12, 13] are given. Then it is shown step by step how to apply.

3.1 Preliminaries

Definition: [12, 13] Let X be a fix set. A bipolar fuzzy set (BFS) is an object having the form

$$B = \{\langle x, (\mu_B^+(x), v_B^-(x))\rangle | x \in X\} \tag{1}$$

where the positive membership degree function $\mu_B^+(x) : X \to [0, 1]$ denotes the satisfaction degree of an element x to the property corresponding to a BFS B and the negative membership degree function $v_B^-(x) : X \to [-1, 0]$ denotes satisfaction degree of an element x to some implicit counter-property corresponding to a BFS B, respectively, and for every $x \in X$.

Let $\tilde{b} = (\mu^+, v^-)$ be a bipolar fuzzy number (BFN). Now we can define a score function and an accuracy function for $\tilde{b}$.

Definition: The score function S of $\tilde{b} = (\mu^+, v^-)$ is evaluated as

$$S(\tilde{b}) = \frac{1}{2}(1 + \mu^+ + v^-), S(\tilde{b}) \in [0, 1].$$

Definition: The accuracy function H of $\tilde{b} = (\mu^+, v^-)$ is evaluated as

$$H(\tilde{b}) = \frac{1}{2}(\mu^+ - v^-), H(\tilde{b}) \in [0, 1].$$

It is evident that $S(\tilde{b}) \in [0, 1]$ and $H(\tilde{b}) \in [0, 1]$. Note that $H(\tilde{b})$ assesses the degree of accuracy of $\tilde{b}$. A larger value of $H(\tilde{b})$ implies a higher degree of accuracy of the BFN $\tilde{b}$. Appling the score function S and the accuracy function H, the next definition is an ordered relation between two BFNs $\tilde{b}_1 = (\mu_1^+, v_1^-)$ and $\tilde{b}_2 = (\mu_2^+, v_2^-)$.

Definition: If $S(\tilde{b}_1) < S(\tilde{b}_2)$, or $S(\tilde{b}_1) = S(\tilde{b}_2)$ but $H(\tilde{b}_1) < H(\tilde{b}_2)$, then $\tilde{b}_1$ is smaller than $\tilde{b}_2$, denoted by $\tilde{b}_1 < \tilde{b}_2$; If $S(\tilde{b}_1) = S(\tilde{b}_2)$ and $H(\tilde{b}_1) = H(\tilde{b}_2)$, then $\tilde{b}_1 = \tilde{b}_2$.

Hamacher [14] proposed a more generalized t-norm and t-conorm. Hamacher operations, i.e., Hamacher product and Hamacher sum, are respective instances of t-norms and t-conorms.

Hamacher product $\otimes$ is a t-norm and Hamacher sum $\oplus$ is a t-conorm, where

$$T(a,b) = a \otimes b = \frac{ab}{\gamma + (1-\gamma)(a+b-ab)}, \gamma > 0.$$

$$T^*(a,b) = a \oplus b = \frac{a+b-ab-(1-\gamma)ab}{1-(1-\gamma)ab}, \gamma > 0.$$

Let $\tilde{b}_1 = (\mu_1^+, v_1^-)$, $\tilde{b}_2 = (\mu_2^+, v_2^-)$ and $\tilde{b} = (\mu^+, v^-)$ denote BFNs. Basic Hamacher operators of BFNs with $\gamma > 0$ can be define as follows:

$$\tilde{b}_1 \oplus \tilde{b}_2 = \left(\frac{\mu_1^+ + \mu_2^+ - \mu_1^+ \mu_2^+ - (1-\gamma)\mu_1^+ \mu_2^+}{1 - (1-\gamma)\mu_1^+ \mu_2^+}, \frac{-v_1^- v_2^-}{\gamma + (1-\gamma)(v_1^- + v_2^- - v_1^- v_2^-)} \right)$$

$$\tilde{b}_1 \otimes \tilde{b}_2 = \left(\frac{\mu_1^+ \mu_2^+}{\gamma + (1-\gamma)(\mu_1^+ + \mu_2^+ - \mu_1^+ \mu_2^+)}, \frac{v_1^- + v_2^- - v_1^- v_2^- - (1-\gamma)v_1^- v_2^-}{1 - (1-\gamma)v_1^- v_2^-} \right)$$

$$\lambda\tilde{b} = \left(\frac{(1+(\gamma-1)\mu^+)^\lambda - (1-\mu^+)^\lambda}{(1+(\gamma-1)\mu^+)^\lambda + (\gamma-1)(1-\mu^+)^\lambda}, \frac{-\gamma|v^-|^\lambda}{(1+(\gamma-1)(1+v^-))^\lambda + (\gamma-1)|v^-|^\lambda} \right), \lambda > 0$$

$$(\tilde{b})^\lambda = \left(\frac{\gamma(\mu^+)^\lambda}{(1+(\gamma-1)(1-\mu^+))^\lambda + (\gamma-1)(\mu^+)^\lambda}, \frac{(1+(\gamma-1)|v^-|)^\lambda - (1+v^-)^\lambda}{(1+(\gamma-1)|v^-|)^\lambda + (\gamma-1)(1+v^-)^\lambda} \right), \lambda > 0$$

Let $\tilde{b}_j = \left(\mu_j^+, v_j^-\right)$ $(j = 1, 2, ..., n)$ be a collection of BFNs. Bipolar fuzzy Hamacher arithmetic aggregation operators can be established as follows:

Definition: The bipolar fuzzy Hamacher weighted average (BFHWA) operator is

$$\text{BFHWA}_\omega(\tilde{b}_1, \tilde{b}_2, ..., \tilde{b}_n) = \bigoplus_{j=1}^{n} (\omega_j \tilde{b}_j)$$

where $\omega = (\omega_1, \omega_2, ..., \omega_n)^T$ denotes the weight vector associated with $\tilde{b}_i (j = 1, 2, ..., n)$, and $\omega_j > 0$, $\sum_{j=1}^{n} \omega_j = 1, \gamma > 0$.

Theorem: [16] The BFHWA operator returns a BFN with

$$\mathrm{BFHWA}_\omega(\tilde{b}_1, \tilde{b}_2, \ldots, \tilde{b}_n) = \overset{n}{\underset{j=1}{\oplus}}(\omega_j \tilde{b}_j)$$

$$= \left(\frac{\prod_{j=1}^n \left(1+(\gamma-1)\mu_j^+\right)^{\omega_j} - \prod_{j=1}^n \left(1-\mu_j^+\right)^{\omega_j}}{\prod_{j=1}^n \left(1+(\gamma-1)\mu_j^+\right)^{\omega_j} - (\gamma-1)\prod_{j=1}^n \left(1-\mu_j^+\right)^{\omega_j}}, \frac{-\gamma \prod_{j=1}^n \left|v_j^-\right|^{\omega_j}}{\prod_{j=1}^n \left(1+(\gamma-1)\left(1+v_j^-\right)\right)^{\omega_j} - (\gamma-1)\prod_{j=1}^n \left|v_j^-\right|^{\omega_j}} \right)$$

3.2 Proposed Model

In this section, we introduce the steps of new S-QFD approach with bipolar fuzzy Hamacher aggregation operators. HoQ is essentially a four-stage process [15]. Detailed information of stages is given in the application section.

Stage 0: Planning
Stage 1: Collection of "Voice of Customer"
Stage 2: Creating House of Quality.

- Determination of customer requirements and importance levels with

- Step 1: Applying the BFHWA operator to process the information in CR relationship matrix $\widetilde{CR}$, derive the overall values $\tilde{cr}_i (i = 1, 2, \ldots, m)$ of the CR$_i$.

$$\tilde{cr}_i = (\mu_i^+, v_i^-) = \mathrm{BFHWA}_\omega(\tilde{cr}_{i1}, \tilde{cr}_{i2}, \ldots, \tilde{cr}_{im}) = \overset{m}{\underset{j=1}{\oplus}}(\omega_j \tilde{cr}_{ij})$$

$$= \left(\frac{\prod_{j=1}^m \left(1+(\gamma-1)\mu_j^+\right)^{\omega_j} - \prod_{j=1}^m \left(1-\mu_j^+\right)^{\omega_j}}{\prod_{j=1}^m \left(1+(\gamma-1)\mu_j^+\right)^{\omega_j} - (\gamma-1)\prod_{j=1}^m \left(1-\mu_j^+\right)^{\omega_j}}, \frac{-\gamma \prod_{j=1}^m \left|v_j^-\right|^{\omega_j}}{\prod_{j=1}^m \left(1+(\gamma-1)\left(1+v_j^-\right)\right)^{\omega_j} - (\gamma-1)\prod_{j=1}^m \left|v_j^-\right|^{\omega_j}} \right)$$

- Step 2: Calculate the scores $S(\tilde{cr}_i)(i = 1, 2, \ldots, m)$ and importance levels.

- Determination of technical requirements
- Determination of the relationships between customer requirements and technical requirements with bipolar fuzzy sets.
  - Step 1: Applying the BFHWA operator to process the information in interrelationship matrix $\widetilde{IR}$, derive the overall values $\tilde{ir}_i (i = 1, 2, \ldots, n)$ of the IR$_i$.

$$\tilde{ir}_i = (\mu_i^+, v_i^-) = \mathrm{BFHWA}_\omega(\tilde{ir}_{i1}, \tilde{ir}_{i2}, \ldots, \tilde{ir}_{in}) = \overset{n}{\underset{j=1}{\oplus}}(\omega_j \tilde{ir}_{ij})$$

$$= \left(\frac{\prod_{j=1}^n \left(1+(\gamma-1)\mu_j^+\right)^{\omega_j} - \prod_{j=1}^n \left(1-\mu_j^+\right)^{\omega_j}}{\prod_{j=1}^n \left(1+(\gamma-1)\mu_j^+\right)^{\omega_j} - (\gamma-1)\prod_{j=1}^n \left(1-\mu_j^+\right)^{\omega_j}}, \frac{-\gamma \prod_{j=1}^n \left|v_j^-\right|^{\omega_j}}{\prod_{j=1}^n \left(1+(\gamma-1)\left(1+v_j^-\right)\right)^{\omega_j} - (\gamma-1)\prod_{j=1}^n \left|v_j^-\right|^{\omega_j}} \right)$$

- Step 2: Calculate the scores $S(\tilde{ir}_i)(i = 1, 2, \ldots, n)$.
- Step 3: Rank all the alternatives $A_i(i = 1, 2, \ldots, n)$ in terms of $S(\tilde{ir}_i)(i = 1, 2, \ldots, n)$. If there is no difference between two scores $S(\tilde{ir}_i)$ and $S(\tilde{ir}_j)$,

then calculate the accuracy degrees H($\widetilde{ir}_i$) and H($\widetilde{ir}_j$) to rank the alternatives A$_i$ and A$_j$.

Stage 3: Analysis and interpretation of results.

4   Case Study

In this study QFD technique was applied to design a music stream website. Thus, customer needs, technical requirements and their importance are determined for music streaming website design.

4.1   Stage 0: Planning

Music streaming services do not have a specific target group. People of all ages and professions can use these services. Although the general tendency is thought to be the young people generation, we did not limit the age in survey. Many professions and age groups are included in the research. Based on the literature review and the opinions of experts, 10 types of customer requirements were identified.

4.2   Stage 1: Collection of "Voice of Customer"

For survey, 50 people were randomly selected and no personal information was received. The goal is to make the assessment of the group more meaningful because this service is used by people of all ages and professions. The profile of the customer participating in the study is shown in Fig. 1.

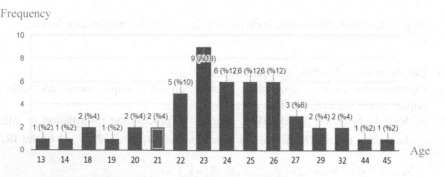

Fig. 1. The profile of the customer participating

82% of the participants are between 20 and 30 years old. The gender distribution of participants is shown in Fig. 2.

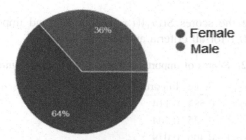

Fig. 2. The gender distribution of participants

4.3   Stage 2: Creating House of Quality

Determination of Customer Requirements and Importance Levels
The QFD application to music delivery services is to show the points that need to be improved in order to improve customer satisfaction. 10 customer requirements are shown in Table 1.

Table 1. Customer requirements

Criteria code	Criteria
C1	Song capacity
C2	User-friendly interface
C3	Free version
C4	Fast music download
C5	Music list properties
C6	Sound quality
C7	Mobile application
C8	Access without internet on mobile application
C9	Fast search engine
C10	No adds between songs

- Step 1: Applying the BFHWA operator to process the information in CR relationship matrix $\widetilde{CR}$. Criteria are evaluated using bipolar fuzzy numbers by 50 customers. The aggregated ratings are presented in the following matrix.

$$\widetilde{CR} = \begin{bmatrix} (0.5,-0.5) & (0.5,-0.5) & (0.7,-0.3) & (0.6,-0.4) & (0.8,-0.2) & (0.4,-0.6) & (0.8,-0.2) & (0.4,-0.6) & (0.7,-0.3) & (0.2,-0.8) \\ (0.5,-0.5) & (0.5,-0.5) & (0.6,-0.4) & (0.5,-0.5) & (0.7,-0.3) & (0.4,-0.6) & (0.6,-0.4) & (0.5,-0.5) & (0.6,-0.4) & (0.4,-0.6) \\ (0.3,-0.7) & (0.4,-0.6) & (0.5,-0.5) & (0.4,-0.6) & (0.6,-0.4) & (0.2,-0.8) & (0.6,-0.4) & (0.2,-0.8) & (0.5,-0.5) & (0.3,-0.7) \\ (0.4,-0.6) & (0.5,-0.5) & (0.6,-0.4) & (0.5,-0.5) & (0.7,-0.3) & (0.3,-0.7) & (0.7,-0.3) & (0.3,-0.7) & (0.3,-0.7) & (0.3,-0.7) \\ (0.2,-0.8) & (0.3,-0.7) & (0.4,-0.6) & (0.3,-0.7) & (0.5,-0.5) & (0.3,-0.7) & (0.4,-0.6) & (0.2,-0.8) & (0.6,-0.4) & (0.1,-0.9) \\ (0.6,-0.4) & (0.6,-0.4) & (0.8,-0.2) & (0.7,-0.3) & (0.7,-0.3) & (0.5,-0.5) & (0.7,-0.3) & (0.4,-0.6) & (0.5,-0.5) & (0.4,-0.6) \\ (0.2,-0.8) & (0.4,-0.6) & (0.4,-0.6) & (0.3,-0.7) & (0.6,-0.4) & (0.3,-0.7) & (0.5,-0.5) & (0.2,-0.8) & (0.6,-0.4) & (0.2,-0.8) \\ (0.6,-0.4) & (0.5,-0.5) & (0.8,-0.2) & (0.7,-0.3) & (0.8,-0.2) & (0.6,-0.4) & (0.8,-0.2) & (0.5,-0.5) & (0.9,-0.1) & (0.5,-0.5) \\ (0.3,-0.7) & (0.4,-0.6) & (0.5,-0.5) & (0.7,-0.3) & (0.4,-0.6) & (0.5,-0.5) & (0.4,-0.6) & (0.1,-0.9) & (0.5,-0.5) & (0.2,-0.8) \\ (0.8,-0.2) & (0.6,-0.4) & (0.7,-0.3) & (0.7,-0.3) & (0.9,-0.1) & (0.6,-0.4) & (0.8,-0.2) & (0.5,-0.5) & (0.8,-0.2) & (0.5,-0.5) \end{bmatrix}$$

- Step 2: Calculate the scores $S(\tilde{cr}_i)(i = 1, 2, \ldots, m)$ and importance levels. (The weight vector is 0.1 for all criteria and $\gamma = 3$) (Table 2).

Table 2. Scores of importance levels of customer requirements

$S(\tilde{cr}_{Ci})$	Score	Importance levels (percentage of score)
$S(\tilde{cr}_{C1})$	0,584	0,114
$S(\tilde{cr}_{C2})$	0,535	0,104
$S(\tilde{cr}_{C3})$	0,406	0,079
$S(\tilde{cr}_{C4})$	0,472	0,092
$S(\tilde{cr}_{C5})$	0,334	0,065
$S(\tilde{cr}_{C6})$	0,604	0,118
$S(\tilde{cr}_{C7})$	0,376	0,073
$S(\tilde{cr}_{C8})$	0,697	0,136
$S(\tilde{cr}_{C9})$	0,408	0,080
$S(\tilde{cr}_{C10})$	0,714	0,139

Determination of Technical Requirements

Three experts defined the specifications of this software. The specifications are divided into two categories: front-end and back-end in Table 3.

Table 3. Technical measures of music streaming services.

Code	Front-end	Code	Back-end
T1	No advertising while listening to music	T7	High speed search
T2	Capability of colors and font	T8	Detailed product information
T3	Personalization	T9	Link with other social media
T4	Video animation	T10	Update list
T5	See comments of other users	T11	Storage of user information
T6	High speed access and high quality music	T12	Good conception and organization
		T13	Security keys for purchase

Determination of the Relationship Between Customer Requirements and Technical Requirements

Three experts participated in the creation of the interrelational matrix in Table 4. Their shared opinions are expressed in order to establish the relationship between customer and technical requirements. ($\gamma = 3$, no relationship = $(0, -1)$).

Table 4. The final version of the House of Quality for Music Streaming Website Design using bipolar fuzzy Hamacher aggregation operators

CR	Weights	Technical measures of music streaming services												
		T_1	T_2	T_3	T_4	T_5	T_6	T_7	T_8	T_9	T_{10}	T_{11}	T_{12}	T_{13}
C_1	0,114				(0.2, −0,6)		(0.9, −0,2)	(0.8, −0,1)						
C_2	0,104	(0.9, −0,3)	(0.4, −0,3)	(0.3, −0,1)		(0.2, −0,5)		(0.7, −0,2)	(0.5, −0,4)			(0.4, −0,6)	(0.8, −0,3)	
C_3	0,079				(0.4, −0,4)								(0.9, −0,3)	(0.3, −0,8)
C_4	0,092				(0.6, −0,8)		(0.7, −0,2)	(0.8, −0,5)			(0.5, −0,1)		(0.4, −0,7)	
C_5	0,065				(0.9, −0,4)	(0.4, −0,2)		(0.3, −0,8)	(0.6, −0,8)	(0.3, −0,5)	(0.5, −0,6)			
C_6	0,118						(0.9, −0,4)							
C_7	0,073						(0.4, −0,7)			(0.4, −0,3)			(0.5, −0,8)	(0.2, −0,1)
C_8	0,136	(0.3, −0,8)		(0.5, −0,6)									(0.8, −0,5)	
C_9	0,080				(0.4, −0,5)		(0.9, −0,1)	(0.7, −0,4)			(0.4, −0,5)		(0.3, −, −0,1)	
C_{10}	0,139	(0.9, −0,5)												(0.4, −0,1)
$\tilde{i}r_i$		(0,336, −0,101)	(0,039, −0,564)	(0,207, −0,027)	(0,107, −0,110)	(0,044, −0,247)	(0,446, −0,021)	(0,351, −0,008)	(0,088, −0,252)	(0,018, −0,595)	(0,106, −0,084)	(0,039, −0,603)	(0,381, −0,009)	(0,075, −0,222)
$S(\tilde{i}r_i)$		0,617	0,237	0,590	0,499	0,399	0,712	0,672	0,418	0,212	0,511	0,218	0,686	0,426
Rank		4	11	5	7	10	1	3	9	13	6	12	2	8

4.4 Stage 3: Analysis and Interpretation of Results

The aim of this study is to specify priority orders for designing customer requirements and technical requirements. Ratings are calculated using the relationship between customer requirements and technical requirements. The final version of the House of Quality is shown in Table 4. According to the final table, *"high-speed access and high quality music-good conception and organization-high speed search"* are the three most important technical features that will increase customer satisfaction in music streaming site design.

5 Conclusions and Future Work

Quality function deployment-based product development offers significant competitive advantages by emphasizing more importance to customer satisfaction and by improving product performance compared to other conventional methods. The priorities of the customer requirements are determined and according to the customer's voice, the features related to the products are ordered according to their importance. Thus, the designer has to choose one of customer demands due to technical or aesthetic reasons in the product design. In addition, determining the importance levels of technical requirements reveals which technical specifications should be studied more carefully during design.

In this study, QFD with bipolar fuzzy set is used in designing a music streaming website considering customer satisfaction. Ten customer expectations were taken into account and three experts established the relation of these specifications. At the end of the application, *"high-speed access and high quality music-good conception and organization-high speed search"* are selected as most desirable features in e-commerce site design.

In future studies, this methodology can be applied to other web service design areas with criteria weighting techniques and QFD matrix can be constructed by other fuzzy set types.

Acknowledgements. Galatasaray University Research Fund financially supports this research.

References

1. Bergquist, K., Abeysekera, J.: Quality function deployment (QFD)-a means for developing usable products. Int. J. Ind. Ergon. **18**, 269275 (1996)
2. Cristiano, J.J., Liker, J.K., White, C.C.: Customer-driven product development through quality function deployment in the US and Japan. J. Prod. Innov. Manag. **17**(4), 228–308 (2000)
3. Morris, L.J., Morris, J.S.: Introducing quality function deployment in the marketing classroom. J. Mark. Educ. **21**(2), 131–137 (1999)
4. Pandey, M.M.: Evaluating the strategic design parameters of airports in Thailand to meet service expectations of low-cost airlines using the fuzzy-based QFD method. J. Air Transp. Manag. **82**, 101738 (2020)
5. Aijun, L., Xiao, Y., Lu, H., Tsai, S., Song, W.: A fuzzy three-stage multi-attribute decision-making approach based on customer needs for sustainable supplier selection. J. Clean. Prod. **239**, 118043 (2019)
6. Pooya, A., Moghadam, E.V.T.: An extended approach for manufacturing strategy process base on organization performance through fuzzy QFD. Oper. Res. Appl. Ind. **53**, 1529–1550 (2019)
7. Efe, B.: Fuzzy cognitive map based quality function deployment approach for dishwasher machine selection. Appl. Soft Comput. **83**, 105660 (2019)
8. Wei, G., Lu, M., Tang, X., Wei, Y.: Pythagorean hesitant fuzzy Hamacher aggregation operators and their application to multiple attribute decision making. Int. J. Intell. Syst. **33**, 1197–1233 (2018)
9. Garg, H.: Intuitionistic fuzzy Hamacher aggregation operators with entropy weight and their applications to multi-criteria decision-making problems. Iran. J. Sci. Technol. Trans. Electr. Eng. **43**, 597–613 (2019). https://doi.org/10.1007/s40998-018-0167-0
10. Abtahi, A.R., Abdi, F.: Designing software as a service in cloud computing using quality function deployment. Int. J. Enterp. Inf. Syst. (IJEIS) **14**(4), 16–27 (2020)
11. Prasad, K., Chakraborty, S.: A QFD-based decision making model for computer-aided design software selection. Manag. Sci. Lett. **6**(3), 213–224 (2018)
12. Zhang, W.R.: Bipolar fuzzy sets and relations: a computational frame work for cognitive modelling and multiagent decision analysis. In: Proceedings of IEEE Conference, pp. 305–309 (1994)
13. Zhang, W.R.: Bipolar fuzzy sets. In: Proceedings of Fuzzy-IEEE Conference, pp. 835–840 (1998)

14. Hamacher, H.: Uber logische Vernüpfungen unscharfer Aussagen und deren Zugehörige Bewertungsfunctionen. In: Trappl, R., Klir, G.J., Riccardi, L. (eds.) Progress in Cybernatics and Systems Research, vol. 3, pp. 276–288 (1978)
15. Cohen, L.: Quality Function Deployment: How to Make QFD Work for You, pp. 11, 32–33, 210. Addison Wesley, Reading (1995)
16. Wei, G., Alsaadi, F.E., Hayat, T., Alsaedi, A.: Bipolar fuzzy Hamacher aggregation operators in multiple attribute decision making. Int. J. Fuzzy Syst. **20**(1), 1–12 (2018). https://doi.org/10.1007/s40815-017-0338-6

An Undesirable Behaviour of a Recent Extension of OWA Operators to the Setting of Multidimensional Data

Raúl Pérez-Fernández[1,2](✉)

[1] Department of Statistics and O.R. and Mathematics Didactics,
University of Oviedo, Oviedo, Spain
perezfernandez@uniovi.es
[2] KERMIT, Department of Data Analysis and Mathematical Modelling,
Ghent University, Ghent, Belgium

Abstract. OWA operators have been ubiquitous in many disciplines since they were introduced by Yager in 1988. Aside of some other intuitive properties (e.g. monotonicity and idempotence), OWA operators are known to be continuous and, for some carefully constructed weighing vectors, very robust in the presence of outliers. In a recent paper, a natural extension of OWA operators to the setting of multidimensional data has been proposed based on the use of a linear extension of the product order by means of several weighted arithmetic means. Unfortunately, OWA operators constructed in such a way focus too strongly on the level sets of one of the weighted arithmetic means. It is here shown that this focus ultimately results in a forfeit of the properties of continuity and robustness in the presence of outliers.

Keywords: OWA operator · Multidimensional data · Linear extension · Weighted arithmetic mean

1 Introduction

Back in 1988, Yager introduced OWA operators in the context of decision making [12] as a family of functions that lie in between the 'and' and the 'or' operators. Formally, an OWA operator simply is the result of applying a symmetrization process to a weighted arithmetic mean [7] in which the weighted arithmetic mean is applied to the order statistics of the values to be aggregated rather than to the values themselves. Aside of symmetry, OWA operators satisfy very natural properties such as (increasing) monotonicity, idempotence and continuity.

Several families of OWA operators have been studied in the literature [13], probably centered OWA operators being the most prominent family [14]. Interestingly, some centered OWA operators have been studied in the field of statistics due to their robustness in the presence of outliers, e.g., the median, trimmed means and winsorized means.

This research has been partially supported by the Spanish MINECO project (TIN2017-87600-P) and the Research Foundation of Flanders (FWO17/PDO/160).

The field of multivariate statistics has studied for a long time how to extend the notion of median to the multidimensional setting [10]. The field of aggregation theory is also experimenting an increasing interest in this multidimensional setting. For instance, one can find recent works on penalty-based aggregation of multidimensional data [5] and on the property of monotonicity for multidimensional functions [6,9]. It is no surprise then that an extension of OWA operators to the multidimensional setting has been recently proposed by De Miguel et al. [2] by making use of a linear extension of the product order. Unfortunately, as we shall see in the upcoming sections, the consideration of a linear extension of the product order extends OWA operators to the multidimensional setting at the cost of losing continuity and, if applicable, the robustness in the presence of outliers.

The remainder of the paper is structured as follows. In Sect. 2, it is discussed how a linear extension of the product order can be defined by means of several weighted arithmetic means. Section 3 presents the extension of OWA operators to the multidimensional setting by De Miguel et al. The strange behaviour of such extension is discussed in Sect. 4. We end with some concluding remarks in Sect. 5.

2   Linear Extensions of the Product Order by Means of Weighted Arithmetic Means

Consider n points $\mathbf{x}_1, \ldots, \mathbf{x}_n \in \mathbb{R}^m$. The j-th component of the point $\mathbf{x}_i$ is denoted by $\mathbf{x}_i(j)$. The product order $\leq_m$ on $\mathbb{R}^m$ is defined as $\mathbf{x}_{i_1} \leq_m \mathbf{x}_{i_2}$ if $\mathbf{x}_{i_1}(j) \leq \mathbf{x}_{i_2}(j)$ for any $j \in \{1, \ldots, m\}$. Obviously, $\leq_m$ is not a linear order on $\mathbb{R}^m$.

As discussed in [2] (see Proposition 2), a linear extension of $\leq_m$ can be defined by means of m linearly independent weighted arithmetic means $M_1, \ldots, M_m : \mathbb{R}^m \to \mathbb{R}$. More precisely, the linear extension $\preceq_\mathbf{M}$ of $\leq_m$ based on $\mathbf{M} = (M_1, \ldots, M_m)$ is defined by $\mathbf{x}_{i_1} \preceq_\mathbf{M} \mathbf{x}_{i_2}$ if $\mathbf{x}_{i_1} = \mathbf{x}_{i_2}$ or there exists $k \in \{1, \ldots, m\}$ such that

$$M_j(\mathbf{x}_{i_1}) \leq M_j(\mathbf{x}_{i_2}), \text{ for any } j \in \{1, \ldots, k-1\},$$
$$M_k(\mathbf{x}_{i_1}) < M_k(\mathbf{x}_{i_2}).$$

The most prominent such linear extensions of $\mathbb{R}^m$ are the lexicographic orders $\preceq_\sigma$ [4], where a permutation σ of $\{1, \ldots, n\}$ serves for establishing a sequential order in which the different components are considered. Formally, for any $j \in \{1, \ldots, k\}$, M_j is defined as $M_j(\mathbf{x}_i) = \mathbf{x}_i(\sigma(j))$.

In the two-dimensional case, Xu and Yager's linear order $\preceq_{XY}$ on $\mathbb{R}^2$ [11] (induced by $M_1(\mathbf{x}_i) = \frac{1}{2}\mathbf{x}_i(1) + \frac{1}{2}\mathbf{x}_i(2)$ and $M_2(\mathbf{x}_i) = \mathbf{x}_i(2)$) is also very well-known[1] in the context of intervals and intuitionistic fuzzy sets.

---

[1] It is admittedly more common to find an equivalent definition of the order in which M_2 is defined as $M_2(\mathbf{x}_i) = \mathbf{x}_i(2) - \mathbf{x}_i(1)$. This equivalent definition is here abandoned in order to guarantee M_2 to be monotone increasing.

Example 1. Consider $\mathbf{x}_1 = (3, 1)$, $\mathbf{x}_2 = (1, 3)$ and $\mathbf{x}_3 = (3, 3)$. It obviously holds that $\mathbf{x}_1 \leq_2 \mathbf{x}_3$ and $\mathbf{x}_2 \leq_2 \mathbf{x}_3$, however, $\mathbf{x}_1$ and $\mathbf{x}_2$ are not comparable with respect to $\leq_2$.

If one considers Xu and Yager's linear order $\preceq_{XY}$ on $\mathbb{R}^2$, it holds that $\mathbf{x}_1 \preceq_{XY} \mathbf{x}_2$ due to the fact that

$$M_1(\mathbf{x}_1) = \frac{1}{2}\mathbf{x}_1(1) + \frac{1}{2}\mathbf{x}_1(2) = 2 \leq 2 = \frac{1}{2}\mathbf{x}_2(1) + \frac{1}{2}\mathbf{x}_2(2) = M_1(\mathbf{x}_2),$$
$$M_2(\mathbf{x}_1) = \mathbf{x}_1(2) = 1 < 3 = \mathbf{x}_2(2) = M_2(\mathbf{x}_2).$$

It is concluded that $\mathbf{x}_1 \preceq_{XY} \mathbf{x}_2 \preceq_{XY} \mathbf{x}_3$.

An illustration of this procedure is given in Fig. 1.

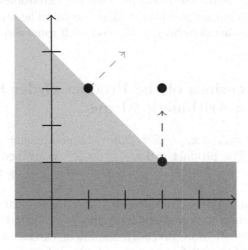

Fig. 1. Graphical representation of the linear extension of the product order based on $M_1(\mathbf{x}_i) = \frac{1}{2}\mathbf{x}_i(1) + \frac{1}{2}\mathbf{x}_i(2)$ and $M_2(\mathbf{x}_i) = \mathbf{x}_i(2)$ for the points $\mathbf{x}_1 = (3, 1)$, $\mathbf{x}_2 = (1, 3)$ and $\mathbf{x}_3 = (3, 3)$. The green area represents the points that lead to values of M_1 smaller than those given by $\mathbf{x}_1$ and $\mathbf{x}_2$. The blue area represents the points that lead to values of M_2 smaller than that given by $\mathbf{x}_2$. The green and blue dashed arrows respectively represent the direction in which M_1 and M_2 increase.

3   Extending OWA Operators to the Setting of Multidimensional Data by Means of a Linear Extension of the Product Order

OWA operators as defined by Yager [12] for the aggregation of unidimensional data are characterized by a weighing vector $\mathbf{w} = (w_1, \ldots, w_n)$ with $w_i \geq 0$ for any

$i \in \{1, \ldots, n\}$ and $\sum_{i=1}^{n} w_i = 1$. In particular, the OWA operator $f_{\mathbf{w}} : \mathbb{R}^n \to \mathbb{R}$ associated with $\mathbf{w}$ is defined as

$$f_{\mathbf{w}}(x_1, \ldots, x_n) = \sum_{i=1}^{n} w_i \, x_{(i)},$$

where $x_{(i)}$ denotes the i-th largest value among $x_1, \ldots, x_n$.

Typical examples of OWA operators are

- the minimum, where $\mathbf{w} = (0, \ldots, 0, 1)$;
- the mid-range, where $\mathbf{w} = (\frac{1}{2}, 0, \ldots, 0, \frac{1}{2})$;
- the arithmetic mean, where $\mathbf{w} = (\frac{1}{n}, \ldots, \frac{1}{n})$;
- the median, where,
    if n is odd, $\mathbf{w} = (0, \ldots, 0, 1, 0, \ldots, 0)$, where the 1 appears at the middle position, or,
    if n is even, $\mathbf{w} = (0, \ldots, 0, \frac{1}{2}, \frac{1}{2}, 0, \ldots, 0)$, where the two $\frac{1}{2}$ appear at the middle positions;
- and the maximum, where $\mathbf{w} = (1, 0, \ldots, 0)$.

When moving to the setting of multidimensional data, since $\leq_m$ is not a linear order on $\mathbb{R}^m$, it is possible that one cannot simply identify the i-th largest point among $\mathbf{x}_1, \ldots, \mathbf{x}_n$. De Miguel et al. proposed to consider a linear extension of $\mathbb{R}^m$ in order to straightforwardly extend OWA operators to the setting of multidimensional data. More precisely, given a linear extension $\preceq$ of the product order $\leq_m$ on $\mathbb{R}^m$, the OWA operator $f_{\mathbf{w}, \preceq} : (\mathbb{R}^m)^n \to \mathbb{R}^m$ associated with a weighing vector $\mathbf{w} = (w_1, \ldots, w_n)$ and $\preceq$ is defined as follows:

$$f_{\mathbf{w}, \preceq}(\mathbf{x}_1, \ldots, \mathbf{x}_n) = \sum_{i=1}^{n} w_i \, \mathbf{x}_{(i)},$$

where $\mathbf{x}_{(i)}$ denotes the i-th largest point among $\mathbf{x}_1, \ldots, \mathbf{x}_n$ according to $\preceq$.

Example 2. Continue with Example 1. Consider $\mathbf{w} = (\frac{1}{4}, \frac{1}{2}, \frac{1}{4})$. Due to the fact that $\mathbf{x}_1 \preceq_{\mathrm{XY}} \mathbf{x}_2 \preceq_{\mathrm{XY}} \mathbf{x}_3$, it holds that

$$f_{\mathbf{w}, \preceq_{\mathrm{XY}}}(\mathbf{x}_1, \mathbf{x}_2, \mathbf{x}_3) = \frac{1}{4}\mathbf{x}_3 + \frac{1}{2}\mathbf{x}_2 + \frac{1}{4}\mathbf{x}_1 = \frac{1}{4}(3, 3) + \frac{1}{2}(1, 3) + \frac{1}{4}(3, 1) = \left(2, \frac{5}{2}\right).$$

4   A Strange Behaviour

A further look reflects that the computation of the OWA operator $f_{\mathbf{w}, \preceq_{\mathbf{M}}} : (\mathbb{R}^m)^n \to \mathbb{R}^m$ associated with a weighing vector $\mathbf{w} = (w_1, \ldots, w_n)$ and a linear extension $\preceq_{\mathbf{M}}$ of the product order by means of m weighted arithmetic means $\mathbf{M}$ works as follows. At first $\mathbb{R}^m$ is reduced into the unidimensional quotient space spanned by the level sets of M_1. The unidimensional OWA operator $f_{\mathbf{w}} : \mathbb{R}^n \to \mathbb{R}$ associated with $\mathbf{w}$ is computed within this unidimensional space and it is only in case the order of the points $\mathbf{x}_1, \ldots, \mathbf{x}_n \in \mathbb{R}^m$ is not uniquely determined that $M_2, \ldots, M_m$ are further considered.

This oversimplification of an m-dimensional space into a unidimensional space leads to two main issues. Firstly, unlike in the unidimensional setting, OWA operators as defined in the previous section are no longer continuous functions.

Example 3. Continue with Example 2. Consider now $\mathbf{x}_2' = (1 - \varepsilon, 3)$ for some $\varepsilon > 0$. It then holds that $\mathbf{x}_2' \preceq_{XY} \mathbf{x}_1 \preceq_{XY} \mathbf{x}_3$, and, thus,

$$f_{\mathbf{w}, \preceq_{XY}} (\mathbf{x}_1, \mathbf{x}_2', \mathbf{x}_3) = \frac{1}{4}\mathbf{x}_3 + \frac{1}{2}\mathbf{x}_1 + \frac{1}{4}\mathbf{x}_2' = \frac{1}{4}(3,3) + \frac{1}{2}(3,1) + \frac{1}{4}(1 - \varepsilon, 3)$$
$$= \left(\frac{5}{2} - \frac{\varepsilon}{4}, 2 \right).$$

It is concluded that $f_{\mathbf{w}, \preceq_{XY}}$ is not continuous.

Even worse, points have such an undesirable freedom of movement within its own level set of the first weighted arithmetic mean that even very robust OWA operators in the presence of outliers in the unidimensional setting become non-robust in higher dimensions.

Example 4. Continue with Example 2. Consider now the weighing vector $\mathbf{w}' = (0, 1, 0)$ associated with the median. Note that, for any $a > 0$, it holds that $\mathbf{x}_1 \preceq_{XY} (\mathbf{x}_2 + (-a, a)) \preceq_{XY} \mathbf{x}_3$. For instance, let $a = 100$, it holds that

$$f_{\mathbf{w}', \preceq_{XY}} (\mathbf{x}_1, \mathbf{x}_2, \mathbf{x}_3) = \mathbf{x}_2 = (1, 3),$$
$$f_{\mathbf{w}', \preceq_{XY}} (\mathbf{x}_1, (\mathbf{x}_2 + (-a, a)), \mathbf{x}_3) = \mathbf{x}_2 + (-a, a) = (-99, 103).$$

It is concluded that $f_{\mathbf{w}', \preceq_{XY}}$ is not robust. Actually, this example implies that the finite sample breakdown point [3,8] of the median is $\frac{1}{n}$ in the multidimensional setting, rather than $\frac{1}{2}$ as in the unidimensional setting. This lack of robustness might not be a big deal in the context of De Miguel et al. [2] since the use of these OWA operators is restricted to a unit hypercube, however, it definitely becomes a major problem if dealing with an unbounded domain (as typically is the case in multivariate statistics [5]).

5   Concluding Remarks

In this paper, a recent extension of OWA operators to the setting of multidimensional data is discussed. As natural as said extension sounds, it is proven to lead to functions that are neither continuous, nor robust. This is due to the fact that the use of a linear extension of the product order is inherently linked to a unidimensional behaviour, and should definitely be abandoned in the multidimensional setting. The use of geometric quantiles [1] instead of linear extensions of the product order in the construction of OWA operators for multidimensional data is encouraged by the author. This direction will be further explored in future work.

References

1. Chaudhuri, P.: On a geometric notion of quantiles for multivariate data. J. Am. Stat. Assoc. **91**(434), 862–872 (1996)
2. De Miguel, L., Sesma-Sara, M., Elkano, M., Asiain, M., Bustince, H.: An algorithm for group decision making using n-dimensional fuzzy sets, admissible orders and OWA operators. Inf. Fus. **37**, 126–131 (2017)
3. Donoho, D.L., Huber, P.J.: The notion of breakdown point. In: A Festschrift for Erich L. Lehmann, pp. 157–184. Wadsworth, Belmont, USA (1983)
4. Fishburn, P.C.: Lexicographic orders, utilities and decision rules: a survey. Manag. Sci. **20**(11), 1442–1471 (1974)
5. Gagolewski, M.: Penalty-based aggregation of multidimensional data. Fuzzy Sets Syst. **325**, 4–20 (2017)
6. Gagolewski, M., Pérez-Fernández, R., De Baets, B.: An inherent difficulty in the aggregation of multidimensional data. IEEE Trans. Fuzzy Syst. **28**, 602–606 (2020)
7. Grabisch, M., Marichal, J.L., Mesiar, R., Pap, E.: Aggregation functions: means. Inf. Sci. **181**, 1–22 (2011)
8. Hampel, F.R.: A general qualitative definition of robustness. Ann. Math. Stat. **42**, 1887–1896 (1971)
9. Pérez-Fernández, R., De Baets, B., Gagolewski, M.: A taxonomy of monotonicity properties for the aggregation of multidimensional data. Inf. Fus. **52**, 322–334 (2019)
10. Small, C.G.: A survey of multidimensional medians. Int. Stat. Rev. **58**(3), 263–277 (1990)
11. Xu, Z., Yager, R.R.: Some geometric aggregation operators based on intuitionistic fuzzy sets. Int. J. Gen. Syst. **35**(4), 417–433 (2006)
12. Yager, R.R.: On ordered weighted averaging aggregation operators in multicriteria decisionmaking. IEEE Trans. Syst. Man Cybern. **18**(1), 183–190 (1988)
13. Yager, R.R.: Families of OWA operators. Fuzzy Sets Syst. **59**, 125–148 (1993)
14. Yager, R.R.: Centered OWA operators. Soft Comput. **11**, 631–639 (2007)

Combining Absolute and Relative Information with Frequency Distributions for Ordinal Classification

Mengzi Tang[1(✉)], Raúl Pérez-Fernández[1,2], and Bernard De Baets[1]

[1] KERMIT, Department of Data Analysis and Mathematical Modelling,
Ghent University, Coupure links 653, 9000 Ghent, Belgium
{mengzi.tang,raul.perezfernandez,bernard.debaets}@ugent.be
[2] Department of Statistics and O.R. and Mathematics Didactics, University
of Oviedo, C/ Federico García Lorca 18, 3307 Oviedo, Spain

Abstract. A large amount of labelled data (absolute information) is usually needed for an ordinal classifier to attain a good performance. As shown in a recent paper by the present authors, the lack of a large amount of absolute information can be overcome by additionally considering some side information in the form of relative information, thus augmenting the method of nearest neighbors. In this paper, we adapt the method of nearest neighbors for dealing with a specific type of relative information: frequency distributions of pairwise comparisons (rather than a single pairwise comparison). We test the proposed method on some classical machine learning datasets and demonstrate its effectiveness.

Keywords: Ordinal classification · Nearest neighbors · Absolute information · Relative information · Frequency distributions

1 Introduction

Typically for ordinal classification there is only absolute information available, i.e., examples with an associated class label of a fixed ordinal scale. Unfortunately, in real-life applications, it is often the case that the amount of absolute information available is limited, thus largely impacting the performance of an ordinal classifier. Fortunately, different types of side information can be additionally collected and make up for the limitation regarding the little amount of absolute information available [1,2]. A popular type of such side information is relative information, i.e., couples of examples without an explicitly given class label but with an order relation between them.

Interestingly, in real-life applications, relative information with frequency distributions arises quite commonly. For instance, the emergence of online transaction platforms such as Amazon Mechanical Turk offers some possibilities to distribute evaluation tasks to consumers and collect a large amount of relative information. However, the order relations from relative information are usually

© Springer Nature Switzerland AG 2020
M.-J. Lesot et al. (Eds.): IPMU 2020, CCIS 1238, pp. 594–602, 2020.
https://doi.org/10.1007/978-3-030-50143-3_47

less informative than the class labels from absolute information. It is also quite common for customers to have contradictory order relations for the same couple of examples. Due to these facts, it is better to consult several customers for collecting the preference among two examples, thus obtaining for each couple of examples a frequency distribution of order relations (hereinafter referred to as relative information with frequency distributions). Hence, how to combine a small amount of absolute information and a large amount of relative information with frequency distributions becomes our main goal.

Some related works [3,4] have shown the effectiveness of fusing absolute and relative information. In the field of ordinal classification, for example, Sader et al. [5] proposed an ordinal classification method for combining both absolute and relative information to perform prediction tasks. This method needs to learn many parameters for solving a constrained convex optimization problem. In a similar direction, our previous work [6] incorporated both types of information into the method of nearest neighbors, and proposed an augmented method for ordinal classification that is non-parametric and easy to explain. However, this method was designed to deal with just one order relation for each couple of examples and not with a frequency distribution of order relations. An immediate extension of this method to the latter setting reduces the study of the nearest couples of examples to just the nearest couple of examples, thus impacting its overall performance. To properly address our problem setting, where there is a small amount of absolute information and a large amount of relative information with frequency distributions available, we propose a method to incorporate both types of information into the method of nearest neighbors for ordinal classification on the basis of our previous work [6].

The remainder of this paper is organized as follows. In Sect. 2, we formulate our problem. We propose our method in Sect. 3. In Sect. 4, we perform experiments and analyze the performance. Some conclusions are presented in Sect. 5.

2   Problem Setting

The available data is composed of absolute and relative information. We denote the input examples by $\mathcal{D} = \{\mathbf{x}_1, \mathbf{x}_2, \ldots, \mathbf{x}_n\}$. The input examples $\mathbf{x}_i = (x_{i1}, \ldots, x_{id})$ belong to the input space $\mathcal{X} \subseteq \mathbb{R}^d$ and their corresponding class labels y_i belong to the output space $\mathcal{Y} = \{C_1, C_2, \ldots, C_r\}$, where the class labels are supposed to be ordered as follows: $C_1 \prec C_2 \prec \ldots \prec C_r$. Absolute information is collected in a set $\mathcal{A} = \{(\mathbf{x}_1, y_1), (\mathbf{x}_2, y_2), \ldots, (\mathbf{x}_n, y_n)\}$.

Although for some examples there is no explicitly given class label, it is still possible to have some side information in the form of relative information. Relative information is typically expressed for a set of couples of examples $\mathcal{C} = \{(\mathbf{a}^i, \mathbf{b}^i), \ldots, (\mathbf{a}^m, \mathbf{b}^m)\} \in \mathcal{X}^2$. With each couple $(\mathbf{a}^i, \mathbf{b}^i)$, a frequency distribution (α^i, β^i) is associated, α^i representing the proportion of times that $\mathbf{a}^i$ is preferred to $\mathbf{b}^i$ and β^i representing the proportion of times that $\mathbf{b}^i$ is preferred to $\mathbf{a}^i$. Obviously, $\alpha^i + \beta^i = 1$. Relative information is collected in a set $\mathcal{R} = \{((\mathbf{a}^1, \mathbf{b}^1), (\alpha^1, \beta^1)), ((\mathbf{a}^2, \mathbf{b}^2), (\alpha^2, \beta^2)), \ldots, ((\mathbf{a}^m, \mathbf{b}^m), (\alpha^m, \beta^m))\}$.

If $((\mathbf{a}^i, \mathbf{b}^i), (\alpha^i, \beta^i))$ belongs to $\mathcal{R}$, then $((\mathbf{b}^i, \mathbf{a}^i), (\beta^i, \alpha^i))$ is supposed also to belong to $\mathcal{R}$. Note that here we do not consider the case in which $\mathbf{a}^p$ and $\mathbf{b}^p$ are equally preferred. The main characteristic of our problem is that the amount of absolute information is typically smaller than the amount of relative information, i.e., $n \ll m$.

3 Proposed Method

3.1 Existing Method: Fusing Absolute and Relative Information for Augmenting the Method of Nearest Neighbors

In this subsection, we recall the method proposed in our previous work [6]. Firstly, according to a fixed distance metric d, we find the k nearest neighbor examples $\mathcal{D}_k = \{\mathbf{x}_{i_j}\}_{j=1}^k$ of the test example $\mathbf{x}^*$. We see each couple $(\mathbf{x}^*, \mathbf{x}_{i_j})$ as a new object and look for the ℓ nearest neighbor couples $\mathcal{C}_\ell^j = \{(\mathbf{a}_q^j, \mathbf{b}_q^j)\}_{q=1}^\ell$ of this new object $(\mathbf{x}^*, \mathbf{x}_{i_j})$. For this process, we compute the distance between couples according to the product distance metric (see [7], page 83, with $p{=}1$), which is defined as

$$d_*((\mathbf{u}, \mathbf{v}), (\mathbf{w}, \mathbf{t})) = d(\mathbf{u}, \mathbf{w}) + d(\mathbf{v}, \mathbf{t}). \tag{1}$$

Secondly, we rely on the assumption that a couple and its nearest neighbor couples have similar order relations. More in detail, for the new object $(\mathbf{x}^*, \mathbf{x}_{i_j})$, we focus on the nearest neighbor couple and get its corresponding order relation. For instance, if the nearest neighbor couple of $(\mathbf{x}^*, \mathbf{x}_{i_j})$ is $(\mathbf{a}_1^j, \mathbf{b}_1^j)$ and its given order relation is $\mathbf{a}_1^j \succ \mathbf{b}_1^j$, then we assume the same order relation $\mathbf{x}^* \succ \mathbf{x}_{i_j}$ for $(\mathbf{x}^*, \mathbf{x}_{i_j})$. Since the class label of $\mathbf{x}_{i_j}$ is known to be, for instance, C_{c_j}, the class label of $\mathbf{x}^*$ is expected to be at least C_{c_j}. The same applies to the other $\ell - 1$ neighbor couples. For each among these ℓ relations, we obtain an interval of potential class labels for $\mathbf{x}^*$. We denote this interval as $I_{jq} = [l_{I_{jq}}, r_{I_{jq}}]$, where $j \in \{1, \ldots, k\}$ and $q \in \{1, \ldots, \ell\}$. For instance, if the given class label of $\mathbf{x}_{i_j}$ is C_{c_j} and we obtain that the relation for the couple $(\mathbf{x}^*, \mathbf{x}_{i_j})$ is $\mathbf{x}^* \succ \mathbf{x}_{i_j}$ according to its q-th nearest neighbor couple, then the interval of possible values of y^* is $I_{jq} = [l_{I_{jq}}, r_{I_{jq}}] = [C_{c_j}, C_r]$. Similarly, if the relation is $\mathbf{x}^* \prec \mathbf{x}_{i_j}$, then the interval of possible values of y^* is $I_{jq} = [l_{I_{jq}}, r_{I_{jq}}] = [C_1, C_{c_j}]$.

Finally, we denote by $\mathbf{I} = (I_{jq})_{j \in \{1,\ldots,k\}, q \in \{1,\ldots,\ell\}}$ the list of all the obtained intervals. We consider the penalty function associated with the median for intervals (see, for instance, Beliakov et al. [8]):

$$P(\mathbf{I}, y) = \sum_{j=1}^k \sum_{q=1}^\ell (|l_{I_{jq}} - y| + |r_{I_{jq}} - y|), \tag{2}$$

where $|C_i - C_j|$ denotes the L_1-distance between two class labels C_i and C_j. Note that the L_1-distance metric treats all class labels of the ordinal scale as if they were equidistant, something that is not always advisable depending on the nature of $\mathcal{Y}$. The class label y^* of $\mathbf{x}^*$ is then determined using the corresponding penalty-based (aggregation) function:

$$y^* = f(\mathbf{y}^*) = \arg\min_{y \in \mathcal{Y}} P(\mathbf{I}, y). \qquad (3)$$

3.2 New Method: Combining Absolute and Relative Information with Frequency Distributions

The method above only focuses on how to deal with couples provided with only one order relation. However, in our problem setting, we have relative information with frequency distributions. More specifically, we have a frequency distribution of order relations for each $(\mathbf{a}^j, \mathbf{b}^j)$. We use (α^j, β^j) to characterize this frequency distribution. In the following, we explain how to deal with such information.

Firstly, we repeat the process above to look for the k nearest neighbor examples of the test example $\mathbf{x}^*$. We see each couple $(\mathbf{x}^*, \mathbf{x}_{i_j})$ as a new object, search for the ℓ nearest neighbor couples, and get their frequency distributions of order relations. We rely on the same assumption above that a couple and its nearest neighbor couples have similar order relations. More in detail, if the nearest neighbor couple of $(\mathbf{x}^*, \mathbf{x}_{i_j})$ is $(\mathbf{a}_q^j, \mathbf{b}_q^j)$ and its frequency distribution is (α_q^j, β_q^j), which implies that a proportion α_q^j of times the order relation of $(\mathbf{a}_q^j, \mathbf{b}_q^j)$ is $\mathbf{a}_q^j \succ \mathbf{b}_q^j$ and a proportion β_q^j of times the order relation of $(\mathbf{a}_q^j, \mathbf{b}_q^j)$ is $\mathbf{a}_q^j \prec \mathbf{b}_q^j$, then in case the couple $(\mathbf{x}^*, \mathbf{x}_{i_j})$ needs to be labelled, we would expect that a proportion α_q^j of times the order relation of $(\mathbf{x}^*, \mathbf{x}_{i_j})$ is $\mathbf{x}^* \succ \mathbf{x}_{i_j}$ and a proportion β_q^j of times the order relation of $(\mathbf{x}^*, \mathbf{x}_{i_j})$ is $\mathbf{x}^* \prec \mathbf{x}_{i_j}$.

For the new couple $(\mathbf{x}^*, \mathbf{x}_{i_j})$ in which the given class label of $\mathbf{x}_{i_j}$ is C_{c_j}, we get α_q^j times the interval $[C_{c_j}, C_r]$ and β_q^j times the interval $[C_1, C_{c_j}]$ for the potential class label y^* of $\mathbf{x}^*$. We repeat this process for the other $\ell - 1$ nearest neighbor couples. Exploiting all k nearest neighbors and ℓ nearest neighbor couples, we get a list of intervals of potential class labels for $\mathbf{x}^*$. We denote by $\mathbf{I}$ the list of all gathered intervals.

Finally, differently to the previous section, here we do not use the notation $I_{jq} = [l_{I_{jq}}, r_{I_{jq}}]$ to represent the interval. More specifically, we now have a proportion α_q^j of times the interval $[C_{c_j}, C_r]$ and a proportion β_q^j of times the interval $[C_1, C_{c_j}]$ for each nearest neighbor couple of $(\mathbf{x}^*, \mathbf{x}_{i_j})$. Thus, the penalty function associated with the median reads as follows:

$$P(\mathbf{I}, y) = \sum_{j=1}^{k} \sum_{q=1}^{\ell} \beta_q^j (|C_1 - y| + |C_{c_j} - y|) + \alpha_q^j (|C_{c_j} - y| + |C_r - y|)$$

$$= \sum_{j=1}^{k} \sum_{q=1}^{\ell} (\beta_q^j |C_1 - y| + |C_{c_j} - y| + \alpha_q^j |C_r - y|), \qquad (4)$$

where $(\mathbf{a}_q^j, \mathbf{b}_q^j)$ is the q-th nearest neighbor couple of the couple $(\mathbf{x}^*, \mathbf{x}_{i_j})$, (α_q^j, β_q^j) is the corresponding frequency distribution and the given class label of the j-th

Table 1. Description of the benchmark datasets.

Dataset	#Examples	#Features	#Classes
Real ordinal classification datasets			
Tae (TA)	151	54	3
Automobile (AU)	205	26	6
Balance-scale (BS)	625	4	3
Eucalyptus (EU)	736	91	5
Red-wine (RW)	1599	12	6
Car (CA)	1728	21	4
Discretized regression datasets			
Housing5 (HO5)	506	14	5
Abalone5 (AB5)	4177	11	5
Bank1-5 (BA1-5)	8192	8	5
Bank2-5 (BA2-5)	8192	32	5
Computer1-5 (CO1-5)	8192	12	5
Computer2-5 (CO2-5)	8192	21	5
Housing10 (HO10)	506	14	10
Abalone10 (AB10)	4177	11	10
Bank1-10 (BA1-10)	8192	8	10
Bank2-10 (BA2-10)	8192	32	10
Computer1-10 (CO1-10)	8192	12	10
Computer2-10 (CO2-10)	8192	21	10

nearest neighbor $\mathbf{x}_{i_j}$ of $\mathbf{x}^*$ is C_{c_j}. The class label y^* of $\mathbf{x}^*$ is then determined using the corresponding penalty-based (aggregation) function:

$$y^* = f(\mathbf{y}^*) = \arg\min_{y \in \mathcal{Y}} P(\mathbf{I}, y). \tag{5}$$

4   Experiments

4.1   Datasets

We perform the experiments on some classical machine learning datasets from some typical repositories [9–11]. The detailed characteristics of these datasets can be found in Table 1, including the number of examples, features and classes. All the features have been properly normalized (by making all the features to have zero mean and unit standard deviation) to avoid the impact of the scale of features. Note that the datasets do not contain relative information with frequency distributions. Based on a similar generation process of relative information as the one described in [6], we generate couples with frequency distributions of order relations.

More in detail, when comparing two examples for generating a couple $(\mathbf{a}^i, \mathbf{b}^i)$, we randomly sample α^i or β^i from a uniform distribution. For example, if the real order relation of these two examples is $\mathbf{a}^i \succ \mathbf{b}^i$, then we sample α^i from a uniform distribution on $[0.5, 1]$ and set $\beta^i = 1 - \alpha^i$. Similarly, if the real order relation of these two examples is $\mathbf{a}^i \prec \mathbf{b}^i$, then we sample β^i from a uniform distribution on $[0.5, 1]$ and set $\alpha^i = 1 - \beta^i$. Thus, we generate a couple $((\mathbf{a}^i, \mathbf{b}^i), (\alpha^i, \beta^i))$.

To test our method, we construct two different datasets for each original dataset. Based on a similar generation process as in our previous work [6], we fix 10% of the data that will be shared by both datasets for testing. The remaining 90% is used for generating the data for training. We keep 5% of the remaining 90% as absolute information. We use the remaining 95% for generating relative information following the aforementioned description. *Dataset 1* includes just absolute information. *Dataset 2* not only includes the same absolute information as *Dataset 1*, but also includes relative information with frequency distributions. By comparing the performance on these two datasets, we test the impact of incorporating relative information with frequency distributions.

4.2 Performance Measures

We use the three most common performance measures to evaluate ordinal classification models [13,14]: the Mean Zero-one Error (MZE), the Mean Absolute Error (MAE) and the C-index.

The MZE describes the error rate of the classifier and is computed as

$$ \text{MZE} = \frac{1}{T} \sum_{i=1}^{T} \delta(y_i^* \neq y_i) = 1 - \text{Acc}, \tag{6} $$

where T is the number of test examples, y_i is the real class label and y_i^* is the predicted class label. Acc is the accuracy of the classifier. The value of MZE ranges from 0 to 1. It describes the global performance, but it neglects the relations among the class labels.

The MAE is the average absolute error between y_i and y_i^*. If the class labels are represented by numbers, the MAE is computed as:

$$ \text{MAE} = \frac{1}{T} \sum_{i=1}^{T} |y_i - y_i^*|. \tag{7} $$

The value of MAE ranges from 0 to $r - 1$ (maximum absolute error between classes). Because the real distances among the class labels are unknown, the numerical representation of the class labels has a big impact on the MAE performance.

In order to avoid this impact, one could consider the relations between the real class label and the predicted class label. Here we use the concordance index

Table 2. Performances on the two newly constructed datasets for each original dataset. The best results are highlighted in boldface.

Dataset	MZE		MAE		1 − C-index	
	Dataset 1	Dataset 2	Dataset 1	Dataset 2	Dataset 1	Dataset 2
TA	0.5959	**0.5839**	0.7664	**0.6889**	0.4549	**0.3963**
AU	0.6047	**0.5395**	0.9410	**0.7522**	0.3804	**0.2749**
BS	0.2366	**0.2080**	0.3549	**0.3039**	0.1817	**0.1547**
EU	0.6383	**0.5693**	1.0267	**0.8329**	0.3275	**0.2539**
WR	0.4997	**0.4914**	0.5747	**0.5608**	0.3338	**0.3180**
CA	0.2269	**0.2205**	0.2778	**0.2744**	0.2440	**0.1976**
HO5	0.5496	**0.5089**	0.7456	**0.6837**	0.2085	**0.1871**
AB5	**0.5937**	0.6007	**0.8547**	0.8750	**0.2515**	0.2548
BA1-5	0.7187	**0.6979**	1.1602	**1.0993**	0.3622	**0.3372**
BA2-5	0.7761	**0.7682**	1.4203	**1.4043**	0.4615	**0.4520**
CO1-5	0.4159	**0.4039**	0.4856	**0.4707**	0.1243	**0.1207**
CO2-5	0.3727	**0.3552**	0.4217	**0.3948**	0.1064	**0.0990**
HO10	0.7921	**0.7468**	1.6814	**1.4935**	0.2367	**0.2091**
AB10	0.7750	**0.7732**	**1.7603**	1.8014	**0.2531**	0.2571
BA1-10	0.8699	**0.8576**	2.4688	**2.3383**	0.3796	**0.3522**
BA2-10	0.8850	**0.8827**	2.8525	**2.8630**	0.4535	**0.4497**
CO1-10	0.6141	**0.6072**	1.0158	**0.9876**	0.1294	**0.1256**
CO2-10	0.5853	**0.5787**	0.8784	**0.8555**	0.1080	**0.1051**
Median difference	−0.0120		−0.02755		−0.01860	
p-value	0.00053		0.00329		0.00074	

or C-index to represent these relations. The C-index is computed as the proportion of the number of concordant pairs to the number of comparable pairs (see [15], page 50):

$$\text{C-index} = \frac{1}{\sum_{C_p \prec C_q} T_{C_p} T_{C_q}} \sum_{y_i \prec y_j} (\delta(y_i^* \prec y_j^*) + \frac{1}{2}\delta(y_i^* = y_j^*)), \qquad (8)$$

where T_{C_p} and T_{C_q} are respectively the numbers of test examples with the class label C_p and C_q, $\{y_i, y_j\}$ is the real pair from the test examples, while $\{y_i^*, y_j^*\}$ is the corresponding predicted pair. When there are only two different class labels, the C-index amounts to the area under the Receiver Operating Characteristic (ROC) curve [16] and ranges from 0.5 to 1. A lower MZE or MAE means a better performance, while a higher C-index means a better performance. Here, we replace C-index by (1 − C-index) to keep an analogy with the other performance measures and facilitate the discussion of the results.

4.3   Performance Analysis

In this subsection, we analyze the performance of the proposed method on the different datasets listed in Subsect. 4.1. All the experimental results are obtained by applying ten-fold cross validation. We perform experiments on all datasets, setting the number k of nearest neighbor examples to 5. Table 2 shows the performance on *Dataset 1* and *Dataset 2*. It is clear that the performance on *Dataset 2* is better than the performance on *Dataset 1* for almost all original datasets except for *AB5* and *AB10*.

In order to test whether there is a significant difference in performance on these two datasets, we perform the Wilcoxon signed-rank test [12] at a significance level of $\alpha = 0.05$. If the p-value is smaller than the fixed significance level of α, then it means that there exists a statistically significant difference between these two datasets. In Table 2, it can be seen that the p-values for MZE, MAE and $1 - $C-index are smaller than α, which means that there exists a statistically significant difference between the performance on these two datasets obtained from all original datasets. The experimental results, together with the obtained p-values and associated point estimates (median differences), show that using relative information with frequency distributions is meaningful.

5   Conclusions and Future Work

Based on our previous work [6], we have proposed an augmented method for ordinal classification for the setting in which there exists a small amount of absolute information and a large amount of relative information with frequency distributions. Specifically, we adapt the method of nearest neighbors for dealing with relative information with frequency distributions. We have carried out experiments on some classical ordinal classification or regression datasets. The experimental results show that the performance improves when relative information with frequency distributions is considered, which validates the usefulness of taking into account relative information with frequency distributions.

We see several interesting future directions for extending this work. On the one hand, absolute information with frequency distributions is also common. How to combine both absolute and relative information with frequency distributions for ordinal classification is still an open problem. On the other hand, in case the amount of relative information is large, it might be necessary to explore how to select the most informative pairwise comparisons for relative information in order to reduce the computational complexity of the proposed method.

Acknowledgements. Mengzi Tang is supported by the China Scholarship Council (CSC). Raúl Pérez-Fernández acknowledges the support of the Research Foundation of Flanders (FWO17/PDO/160) and the Spanish MINECO (TIN2017-87600-P). This research received funding from the Flemish Government under the "Onderzoeksprogramma Artificiële Intelligentie (AI) Vlaanderen" programme.

References

1. Chen, J., Liu, X., Lyu, S.: Boosting with side information. In: Lee, K.M., Matsushita, Y., Rehg, J.M., Hu, Z. (eds.) ACCV 2012. LNCS, vol. 7724, pp. 563–577. Springer, Heidelberg (2013). https://doi.org/10.1007/978-3-642-37331-2_43
2. De Bie, T., Momma, M., Cristianini, N.: Efficiently learning the metric with side-information. In: Gavaldá, R., Jantke, K.P., Takimoto, E. (eds.) ALT 2003. LNCS (LNAI), vol. 2842, pp. 175–189. Springer, Heidelberg (2003). https://doi.org/10.1007/978-3-540-39624-6_15
3. Ovadia, S.: Ratings and rankings: reconsidering the structure of values and their measurement. Int. J. Soc. Res. Methodol. 7(5), 403–414 (2004)
4. van Herk, H., van de Velden, M.: Insight into the relative merits of rating and ranking in a cross-national context using three-way correspondence analysis. Food Qual. Prefer. 18(8), 1096–1105 (2007)
5. Sader, M., Verwaeren, J., Pérez-Fernández, R., De Baets, B.: Integrating expert and novice evaluations for augmenting ordinal regression models. Inf. Fusion 51, 1–9 (2019)
6. Tang, M., Pérez-Fernández, R., De Baets, B.: Fusing absolute and relative information for augmenting the method of nearest neighbors for ordinal classification. Inf. Fusion 56, 128–140 (2020)
7. Deza, M.M., Deza, E.: Encyclopedia of Distances, pp. 1–583. Springer, Heidelberg (2009). https://doi.org/10.1007/978-3-642-30958-8
8. Beliakov, G., Bustince, H., James, S., Calvo, T., Fernández, J.: Aggregation for Atanassov's intuitionistic and interval valued fuzzy sets: the median operator. IEEE Trans. Fuzzy Syst. 20(3), 487–498 (2011)
9. Asuncion, A., Newman, D.: UCI Machine Learning Repository (2007). http://www.ics.uci.edu/~mlearn/MLRepository.html
10. PASCAL: Pattern Analysis, Statistical Modelling and Computational Learning. Machine Learning Benchmarks Repository (2011). http://mldata.org/
11. Chu, W., Ghahramani, Z.: Gaussian processes for ordinal regression. J. Mach. Learn. Res. 6(7), 1019–1041 (2005)
12. Demšar, J.: Statistical comparisons of classifiers over multiple data sets. J. Mach. Learn. Res. 7, 1–30 (2006)
13. Cruz-Ramírez, M., Hervás-Martínez, C., Sánchez-Monedero, J., Gutiérrez, P.A.: Metrics to guide a multi-objective evolutionary algorithm for ordinal classification. Neurocomputing 135, 21–31 (2014)
14. Baccianella, S., Esuli, A., Sebastiani, F.: Evaluation measures for ordinal regression. In: Proceedings of the 9th IEEE International Conference on Intelligent Systems Design and Applications, pp. 283–287. IEEE, Pisa (2009)
15. Waegeman, W., De Baets, B., Boullart, L.: Learning to rank: a ROC-based graph-theoretic approach. Pattern Recogn. Lett. 29(1), 1–9 (2008)
16. Cortes, C., Mohri, M.: AUC optimization vs. error rate minimization. In: Proceedings of the 16th International Conference on Neural Information Processing Systems, pp. 313–320. MIT Press, Cambridge (2003)

A Bidirectional Subsethood Based Fuzzy Measure for Aggregation of Interval-Valued Data

Shaily Kabir$^{(\boxtimes)}$ and Christian Wagner

School of Computer Science, University of Nottingham, Nottingham NG8 1BB, UK
{shaily.kabir,christian.wagner}@nottingham.ac.uk

Abstract. Recent advances in the literature have leveraged the fuzzy integral (FI), a powerful multi-source aggregation operator, where a fuzzy measure (FM) is used to capture the worth of all combinations of subsets of sources. While in most applications, the FM is defined either by experts or numerically derived through optimization, these approaches are only viable if additional information on the sources is available. When such information is unavailable, as is commonly the case when sources are unknown a priori (e.g., in crowdsourcing), prior work has proposed the extraction of valuable insight (captured within FMs) directly from the evidence or input data by analyzing properties such as specificity or agreement amongst sources. Here, existing agreement-based FMs use established measures of similarity such as Jaccard and Dice to estimate the source agreement. Recently, a new similarity measure based on bidirectional subsethood was put forward to compare evidence, minimizing limitations such as *aliasing* (where different inputs result in the same similarity output) present in traditional similarity measures. In this paper, we build on this new similarity measure to develop a new instance of the agreement-based FM for interval-valued data. The proposed FM is purposely designed to support aggregation, and unlike previous agreement FMs, it degrades gracefully to an average operator for cases where no overlap between sources exists. We validate that it respects all requirements of a FM and explore its impact when used in conjunction with the Choquet FI for data fusion as part of both synthetic and real-world datasets, showing empirically that it generates robust and qualitatively superior outputs for the cases considered.

Keywords: Data aggregation · Fuzzy measures · Fuzzy integrals · Subsethood · Similarity measure · Interval-valued data

1 Introduction

Data aggregation from multiple sources has become more prevalent in many applications including sensor fusion [8], and crowdsourcing [20]. In such aggregation contexts, the fuzzy integral (FI) which is specified in respect to a fuzzy measure (FM) [9] is often used to capture the importance of information arising

© Springer Nature Switzerland AG 2020
M.-J. Lesot et al. (Eds.): IPMU 2020, CCIS 1238, pp. 603–617, 2020.
https://doi.org/10.1007/978-3-030-50143-3_48

from different combinations of sources. Generally, FMs are defined by experts or generated through algorithms, such as the Sugeno λ-measure [19] and the Decomposable measure [7] which leverage the 'worth' of the singletons (individual sources), a.k.a. the densities. Another approach to generating FMs is optimization based on tuning an FM in respect to the behaviour of an aggregation function such as the FI and training data [2,4]. If training data or information on the densities is limited or missing, specifying a FM is a challenging task, even though such a situation arises often, for example in aggregating crowdsourced data. To deal with this, Wagner and Anderson [21] first extracted FMs directly from the input data (the evidence) by analyzing and extracting key properties such as agreement, and specificity. Later, Havens et al. [11,12] introduced more data-driven FMs which refined the established agreement FM in particular to leveraging a generic similarity measure (SM) to extract the property of 'agreement' amongst evidence from combinations of sources. This paper focuses on a recently introduced SM – bidirectional subsethood based SM [14,15] which has been shown to address a number of limitations in common existing SMs such as Jaccard [13] and Dice [6], and explores the impact of its use in conjunction with agreement-based FMs.

So far, three agreement-based FMs have been proposed—the *FM of Agreement (AG)* [21], the *FM of Generalized Accord (GenA)* [12], and the *Additive Measure of Agreement (AA)* [11]. The *AG* FM captures the source agreement by using the intersection operation which considers only the overlap amongst multi-source data without tracking changes in their cardinality/size. This limitation of the intersection operation causes the *AG* FM to generate the same agreement and thus worth for very different subsets of sources. Figure 1 shows such a situation for interval-valued data with the *AG* FM. On the other hand, the use of the Jaccard or Dice SM with the *GenA* and *AA* FMs to estimate the source agreement makes the resulting FM susceptible to limitations of these measures, in particular *aliasing*–returning the same similarity for very different sets of intervals [14,15]. Figure 2 presents such a case where the *GenA* and *AA* FMs produce identical agreement values and thus worth for different sets.

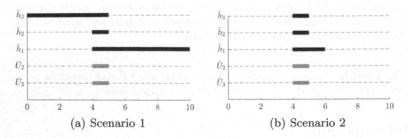

(a) Scenario 1   (b) Scenario 2

Fig. 1. Example highlighting the behaviour of AG FM [21], where $\overline{U}_2$ and $\overline{U}_3$ capture the union of the intersections of all of two and three source combinations as per (8).

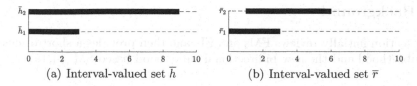

(a) Interval-valued set $\overline{h}$ (b) Interval-valued set $\overline{r}$

Fig. 2. Two different interval-valued sets $\overline{h} = [\overline{h}_1, \overline{h}_2]$ and $\overline{r} = [\overline{r}_1, \overline{r}_2]$ with equal Jaccard similarity of 0.33 and Dice similarity of 0.50 respectively. Clearly, the intervals within $\overline{h}$ and $\overline{r}$ do not appear to be in equal agreement to each other.

Given this context, this paper focuses on developing a new instance of an agreement FM to avoid the limitations of the existing ones. The proposed FM leverages the bidirectional subsethood based SM [14,15] to minimize *aliasing* in the inter-source agreement and worth calculation. The proposed FM is designed following the concept of the *GenA* FM [12], and considers both cases where sources are overlapping (some agreement) or non-overlapping (no agreement). When sources are non-overlapping, the proposed FM in combination with the FI gracefully degrades to an average operator, whereas existing agreement FMs are not designed to deal with such cases. Beyond developing this FM, this paper also demonstrates its behaviour against the existing agreement FMs in aggregating interval datasets when used in combination with the Choquet FI (CFI) [5].

The paper is structured as follows: Sect. 2 reviews FMs and FIs along with a brief discussion of subsethood and the bidirectional subsethood based SM [14,15]. Section 3 discusses existing agreement FMs. Section 4 develops a new instance of the agreement-based FM exploiting the bidirectional subsethood based SM. Section 5 demonstrates the behaviour of the proposed FM against the existing agreement FMs in aggregating interval-valued datasets when used with an FI for both synthetic and real-world datasets. Finally, Sect. 6 concludes the paper with suggestions and future work (Table 1).

Table 1. Acronyms and notation

FM	Fuzzy Measure	FI	Fuzzy Integral
CFI	Choquet Fuzzy Integral	SFI	Sugeno Fuzzy Integral
SM	Similarity Measure	AG	Fuzzy Measure of Agreement
$GenA$	Fuzzy Measure of Generalized Accord	AA	Additive Measure of Agreement
AS_h	Proposed Agreement Fuzzy Measure		
g^{AG}	AG	g^{GenA}	$GenA$
g^{AA}	AA	g^{AS_h}	AS_h
S_h	Subsethood	S_{S_h}	Bidirectional Subsethood based SM
a	Crisp set	$\overline{a}$	Interval $\{\overline{a} \subseteq \mathbb{R} : \overline{a} = [a^-, a^+],\ a^- \leq a^+\}$

2   Background

This section initially reviews FMs and FIs and then provides a short discussion on subsethood and the new bidirectional subsethood based SM [14,15].

2.1   Fuzzy Measures

FMs are defined as a hierarchical weighting structures (lattices) that capture the worth of all subsets in a set of sources, including that of the singletons, also referred to as the densities. Mathematically, an FM, g defined on a finite set of sources, $X = \{x_1, ..., x_n\}$ is a function $g : 2^X \rightarrow [0,1]$ satisfying the properties [9]:

(P1) $g(\emptyset) = 0$ and $g(X) = 1$ (Boundedness)
(P2) If $a \subseteq b \subseteq X$ then $g(a) \subseteq g(b) \subseteq g(X)$ (Monotonicity)

Here, $g(a)$ is the worth of a subset a of X. Property (P1) states that the worth of empty set $(\emptyset)$ is 0 and the worth of universal set (X) is 1. We note that the worth of the universal set is not always required to be 1, but this convention is adopted here. Property (P2) shows the monotonicity of g, stating that if a is a subset of b $(a \subset b)$, the worth of a is smaller or equal to the worth of b. There is a third property of continuous FMs, which is not applicable to discrete FMs, as used in this paper and most practical applications.

In practice, the FMs are defined in various ways, such as expert-defined, or derived by algorithms or optimization based on existing data and in conjunction with an aggregation functions such as the FI; for more details, please see [11,22]. This paper focuses only on algorithmically derived FMs leveraging the evidence data arising from multiple sources. Section 3 reviews such FMs that are derived on the concept of source agreement.

2.2   Fuzzy Integrals

FIs have been efficiently used as powerful non-linear aggregation operators in evidence fusion [3,9]. They aggregate multi-source data (evidence) by combining it with the worth information of all subsets of sources (captured by an FM). Two well-known FIs are the Sugeno FI (SFI) [19] and the Choquet FI (CFI) [5]. In practice, discrete SFI and CFI are commonly used [17] and in this paper, we focus on the discrete CFI as it is most popular for evidence aggregation.

Let $h : X \rightarrow [0, \infty)$ be a real-valued function that presents the evidence from a source. The discrete CFI is defined as

$$\int_{CFI} h \circ g = CFI_g(h) = \sum_{i=1}^{n} h(x_{\pi(i)})[g(A_i) - g(A_{i-1})], \qquad (1)$$

where π is a permutation of X arranged like $h(x_{\pi(1)}) \geq h(x_{\pi(2)}) \geq ... \geq h(x_{\pi(n)})$. $A_i = \{x_{\pi(1)}, x_{\pi(2)}, ..., x_{\pi(i)}\}$ is a subset of sources. g is the FM where $g(A_i)$ is the worth of the subset A_i with $g(A_0) = 0$.

In most cases, the multi-source data h is provided in a numeric form. However, in some applications h is better represented by interval-valued or fuzzy set-valued data. Considering this, FIs have been generalized for non-numeric evidence [1, 10, 16]. Let $\bar{h} : X \to I(\mathbb{R})$ be a set of interval-valued data where $I(\mathbb{R})$ is the set of all closed intervals over the real numbers and $\bar{h}_i = \bar{h}(x_i) = [h_i^-, h_i^+]$ be the ith interval (where h_i^- and h_i^+ are the left and right endpoints respectively). Following the notation in [12], the CFI on $\bar{h}$ is defined as

$$\int_{CFI} \bar{h} \circ g = CFI_g(\bar{h}) = [CFI_g(h^-), CFI_g(h^+)], \tag{2}$$

where the output $CFI_g(\bar{h})$ is itself interval-valued [7]. In other words, the CFI for interval-valued data is computed by applying the CFI for the numeric case of the left and right interval endpoints separately. Please see [11, 12, 21] for more detail about the interval aggregation using the FM and the CFI.

2.3   Subsethood

The subsethood between two sets a and b is a relation, indicating the degree to which a is a subset of b [18]. It is defined as

$$S_h(a, b) = \frac{|a \cap b|}{|a|}, \tag{3}$$

where $|a \cap b|$ is the cardinality of the intersection of a and b, and $|a|$ is the cardinality of a. It is always bounded on the interval $[0, 1]$, where 1 means that a is a subset of b ($a \subseteq b$) and 0 means that a and b are disjoint ($a \not\subseteq b$).

Similarly, the degree of subsethood of two intervals $\bar{a}$ and $\bar{b}$ can be defined as

$$S_h(\bar{a}, \bar{b}) = \frac{|\bar{a} \cap \bar{b}|}{|\bar{a}|}, \tag{4}$$

where $|\bar{a} \cap \bar{b}|$ is the size of the intersection between $\bar{a}$ and $\bar{b}$ and $|\bar{a}| \neq 0$.

2.4   Bidirectional Subsethood Based Similarity Measure

A new SM was introduced in [14, 15] which uses the reciprocal subsethoods of intervals to capture their similarity. This measure for two intervals $\bar{a}$ and $\bar{b}$ is,

$$S_{S_h}(\bar{a}, \bar{b}) = \bigstar \left(S_h(\bar{a}, \bar{b}), S_h(\bar{b}, \bar{a}) \right), \tag{5}$$

where $\bigstar$ is a *t-norm*. We can rewrite (5) using the definition of S_h at (4) as

$$S_{S_h}(\bar{a}, \bar{b}) = \bigstar \left(\frac{|\bar{a} \cap \bar{b}|}{|\bar{a}|}, \frac{|\bar{a} \cap \bar{b}|}{|\bar{b}|} \right). \tag{6}$$

3   Existing Agreement Fuzzy Measures

Here, we briefly recapture the AG [21], $GenA$ [12], and AA [11] FMs with respect to a set of intervals $\overline{h} = \{\overline{h}_1, \overline{h}_2, ..., \overline{h}_n\}$ arising from n individual sources.

3.1   Fuzzy Measure of Agreement

Wagner and Anderson [21] proposed the AG FM by extracting it from the interval-valued data with no prior knowledge about sources. The AG FM is defined as

$$\tilde{g}^{AG}\left(\overline{A}_i\right) = \begin{cases} 0 & \text{for } i = 0,1 \quad (7\text{a}) \\ \sum\limits_{K=2}^{i} \left|\overline{U}_K\left(\overline{A}\right)_i\right| z_K & \text{for } i = [2:n] \quad (7\text{b}) \end{cases}$$

where $\overline{A}_i = \{\overline{h}_{\pi(1)}, \overline{h}_{\pi(2)} ..., \overline{h}_{\pi(i)}\}$ is the permuted set of intervals with $\overline{A}_0 = \emptyset$, $z_i = \frac{i}{n}$ and $|.|$ refers to the cardinality/size of the interval. Here, $\overline{U}_K(\overline{A}_i)$ unites the intersections of the K-tuples in $\overline{A}_i \subseteq \overline{h}$ as defined in (8) [11,12].

$$\overline{U}_K(\overline{A}_i) = \bigcup_{k_1=1}^{i-K+1} \bigcup_{k_2=k_1+1}^{i-K+2} ... \bigcup_{k_K=k_{K-1}+1}^{i} \left(\overline{h}_{\pi(k_1)} \cap \overline{h}_{\pi(k_2)} \cap ... \cap \overline{h}_{\pi(k_K)}\right) \quad (8)$$

Further, the $\tilde{g}^{AG}(\overline{A}_i)$ is normalized by $\tilde{g}^{AG}(\overline{h})$ to satisfy the property of the FM, i.e., $g^{AG}(\overline{A}_i) = \frac{\tilde{g}^{AG}(\overline{A}_i)}{\tilde{g}^{AG}(\overline{h})}$.

3.2   Additive Measure of Agreement

Havens et al. [11] proposed the AA FM in order to alleviate the asymmetry issue of agreement FMs. This FM utilizes the SMs for determining the source agreement. The AA FM is expressed in (9).

$$\tilde{g}^{AA}(\overline{A}_i) = \tilde{g}^{AA}(\overline{A}_{i-1}) + \sum_{\substack{j=1 \\ j \neq i}}^{n} S^p(\overline{h}_j, \overline{h}_{\pi(i)}), \quad i = [n], p \geq 0 \quad (9)$$

where p is a tuning parameter and S is the SM. Further, $\tilde{g}^{AA}(\overline{A}_i)$ is normalized by $\tilde{g}^{AA}(\overline{A}_n)$ like $g^{AA}(A_i) = \frac{\tilde{g}^{AA}(\overline{A}_i)}{\tilde{g}^{AA}(\overline{A}_n)}$.

3.3   Fuzzy Measure of Generalized Accord

Havens et al. [12] proposed the $GenA$ FM leveraging a generic SM to estimate the agreement (accord) of subsets of sources. The $GenA$ FM is defined as

$$g^{Gen}\left(\overline{A}_i\right) = \begin{cases} 0 & \text{for } i = 0,1 \quad (10\text{a}) \\ \alpha_{\overline{h}} \sum\limits_{K=2}^{i} S_K\left(\overline{A}_i\right) & \text{for } i = [2:n] \quad (10\text{b}) \end{cases}$$

where $\overline{A}_i = \{\overline{h}_{\pi(1)}, \overline{h}_{\pi(2)}, ..., \overline{h}_{\pi(i)}\}$ is the permuted set of intervals with $\overline{A}_0 = \emptyset$, and $S_K(\overline{A}_i)$ is defined in (11).

$$S_K(\overline{A}_i) = \binom{n}{K}^{-1} \sum_{k_1=1}^{i-K} \sum_{k_2=k_1+1}^{i-K+1} \cdots \sum_{k_K=k_{K-1}+1}^{i} S(\{\overline{h}_{\pi(k_1)}, \overline{h}_{\pi(k_2)}, ..., \overline{h}_{\pi(k_K)}\})$$

$$(11)$$

Here, $\binom{n}{K}$ is the number of possible K-tuples in $\overline{h}$ and S is the SM. The quantity $S_K(\overline{A}_i)$ is the sum of similarities of the K-tuples in $\overline{A}_i \subseteq \overline{h}$, weighted by $\binom{n}{K}^{-1}$. Further, the constant $\alpha_{\overline{h}}$ is defined in (12) so that $g^{GenA}(\overline{h}) = 1$.

$$\alpha_{\overline{h}} = \left(\sum_{K-2}^{n} S_K(\overline{A}_n) \right)^{-1}$$

$$(12)$$

In [11,12], the *GenA* and *AA* FMs are explored in respect to the popular SMs (within (11) and (9)). As detailed in [14,15], we note however that Jaccard or Dice SMs are liable to *aliasing*, thus making the *GenA* and *AA* FMs to generate the same worth for very different subsets of sources which in turn affects the quality of the overall aggregation. To avoid this, in the next section, we leverage the recently introduced bidirectional subsethood based SM (minimizing *aliasing*), designing a new instance of the *GenA* FM.

4   A New Instance of the Agreement Fuzzy Measure Based on Bidirectional Subsethood

Here, we develop a new instance of agreement FM following the concept of the *GenA* FM and exploit the new bidirectional subsethood based SM for computing the source agreement. As the new SM minimizes *aliasing*, it helps the proposed FM avoid generating the same agreement and worth for different subsets of sources. This section first defines the subsethood for a set of intervals. Then, the new SM at (5) is revisited to enable it to compute similarity for a set of intervals. Finally, the new instance of agreement FM involving the new SM is introduced.

4.1   Defining Subsethood for a Set of Intervals

The subsethood of an interval, $\overline{h}_r$ as regards to a set of intervals $\overline{A}_i \subseteq \overline{h}$ is defined as a mean of its subsethood to each interval $\overline{h}_t$ in $\overline{A}_i$. It is expressed as

$$S_h(\overline{h}_r, \overline{A}_i) = \frac{1}{|\overline{A}_i|} \sum_{\overline{h}_t \in \overline{A}_i} S_h(\overline{h}_r, \overline{h}_t) = \frac{1}{|\overline{A}_i|} \sum_{\overline{h}_t \in \overline{A}_i} \frac{|\overline{h}_r \cap \overline{h}_t|}{|\overline{h}_r|},$$

$$(13)$$

where $S_h(\overline{h}_r, \overline{A}_i) \rightarrow [0,1]$ such that $S_h(\overline{h}_r, \overline{A}_i) = 1$ when $\overline{h}_r \subset \overline{h}_t$, for all $\overline{h}_t \in \overline{A}_i$ and $S_h(\overline{h}_r, \overline{A}_i) = 0$ when $\overline{h}_r \not\subset \overline{h}_t$ for any of $\overline{h}_t \in \overline{A}_i$.

4.2 Defining Bidirectional Subsethood Based Similarity Measure for a Set of Intervals

The bidirectional subsethood based SM, S_{S_h} for $\overline{h}$ is the *t-norm* ($\bigstar$) of their reciprocal subsethoods, i.e.,

$$
\begin{aligned}
S_{S_h}\left(\overline{h}\right) &= \bigstar\left(S_h(\overline{h}_1, \{\overline{h}_2, ..., \overline{h}_n\}), ..., S_h(\overline{h}_n, \{\overline{h}_1, ..., \overline{h}_{n-1}\})\right) \\
&= \bigstar\left(S_h(\overline{h}_1, \overline{h}\backslash\overline{h}_1), ..., S_h(\overline{h}_n, \overline{h}\backslash\overline{h}_n)\right)
\end{aligned}
\tag{14}
$$

where $\overline{h}\backslash\overline{h}_i$ is the nonempty subset of intervals excluding $\overline{h}_i$, $i \in \{1, ..., n\}$. In this paper, we use the minimum *t-norm* ($\bigstar$) as it is the most common in practice.

4.3 Bidirectional Subsethood Based Agreement Fuzzy Measure

Consider again the set of n intervals, $\overline{h}$. For any nonempty subset $\overline{A}_i \in \overline{h}$, $1 \le i \le n$, the new FM, $\tilde{g}^{AS_h}$ using the new SM (14) is defined as follows (which is later normalized to a proper FM, g^{AS_h}):

$$
\tilde{g}^{AS_h}(\overline{A}_0) = 0,
\tag{15a}
$$

$$
\tilde{g}^{AS_h}(\overline{A}_1) = \binom{n}{1}^{-1} \times \sum_{k_1=1}^{1} S_{S_h}\left(\overline{h}_{k_1}, \overline{h}_{k_1}\right) = \frac{1}{n},
\tag{15b}
$$

$$
\tilde{g}^{AS_h}(\overline{A}_i) = i \times \tilde{g}^{AS_h}(\overline{A}_1) + \binom{n}{2}^{-1} \sum_{k_1=1}^{i-1}\sum_{k_2=k_1+1}^{i} S_{S_h}\left(\overline{h}_{k_1}, \overline{h}_{k_2}\right) + ...
\tag{15c}
$$

$$
+ \binom{n}{i}^{-1} S_{S_h}\left(\overline{h}_1, ..., \overline{h}_i\right),
$$

where $\overline{A}_0 = \emptyset$, $\overline{A}_1$ is a singleton subset, and $\overline{A}_i$ is a non-singleton subset with i sources, $1 < i \le n$. $\binom{n}{K}$ is total number of K-tuples in the set, $\overline{h}$, where $1 \le K \le n$. (15a) is the worth of $\overline{A}_0$, which is always 0. (15b) is the worth of $\overline{A}_1$, which is the similarity of 1, weighted by $\binom{n}{1}^{-1}$. (15c) is the worth of $\overline{A}_i$, which is the sum of the similarities of all K-tuples in $\overline{A}_i$, $1 \le K \le i$, weighted by $\binom{n}{K}^{-1}$.

Remark 1. (15b) captures the worth of singleton subsets $(\overline{A}_1)$ which is, $\tilde{g}^{AS_h}(\overline{A}_1)$ $= \frac{1}{n}$, where $n = |\overline{h}|$. For a non-singleton subset consisting of all disagreeing sources, the inclusion of the worth of the singleton subsets in (15c) enables it to generate the worth information for this set.

Following [11,12], (15c) is rewritten as follows,

$$
\tilde{g}^{AS_h}(\overline{A}_i) = \frac{i}{n} + \sum_{K=2}^{i}\left[\binom{n}{K}^{-1} Z_K(\overline{A}_i)\right], \quad i \ge 1,
\tag{16}
$$

where the first part of (16) is the sum of the worth of all singletons in $\overline{A}_i$. The other part gives summation of the similarities of all K-tuples in $\overline{A}_i$ ($K \ge 2$), weighted by $\binom{n}{K}^{-1}$. $Z_K(\overline{A}_i)$ captures the cumulative similarity for all K-tuples in A_i $(K \geq 2)$ using (14) and is defined in (17).

$$Z_K\left(\overline{A}_i\right) = \sum_{k_1=1}^{i-K+1} \sum_{k_2=\atop k_1+1}^{i-K+2} \cdots \sum_{k_K=\atop k_{K-1}+1}^{i} \bigstar \left(S_h(\overline{h}_{k_1}, \overline{A}_i \backslash \overline{h}_{k_1}), ..., S_h(\overline{h}_{k_K}, \overline{A}_i \backslash \overline{h}_{k_K})\right)$$

(17)

Finally, $\tilde{g}^{AS_h}(\overline{A}_i)$ is normalized by $\tilde{g}^{AS_h}(\overline{h})$ in (18) so that $g^{AS_h}(\overline{A}_i) \leq 1$ and $g^{AS_h}(\overline{h}) = 1$, which maintains the bounded property of the FM.

$$g^{AS_h}\left(\overline{A}_i\right) = \frac{\tilde{g}^{AS_h}(\overline{A}_i)}{\tilde{g}^{AS_h}(\overline{h})}, \quad 1 \leq i \leq n.$$

(18)

In the following **Example 1** demonstrates that unlike the g^{GenA} and g^{AA} FMs, the new instance agreement FM, g^{AS_h} avoids generating the same agreement and worth for different sets of sources. In addition, **Example 2** presents the interval aggregation using the g^{AS_h} FM and the CFI.

Example 1: Consider two interval-valued datasets, $\overline{h}$ and $\overline{r}$, as shown in Fig. 3. Their corresponding FM lattices using the g^{AS_h}, g^{AG}, g^{GenA}, and g^{AA} FMs are also shown in Fig. 3 (we skip showing the FM values for $\emptyset$ and $\overline{h}$). Due to *aliasing* of the Jaccard SM, both g^{GenA} and g^{AA} FMs generate the same FM lattices for these sets whereas the g^{AS_h} and g^{AG} FMs generate distinct FM lattice.

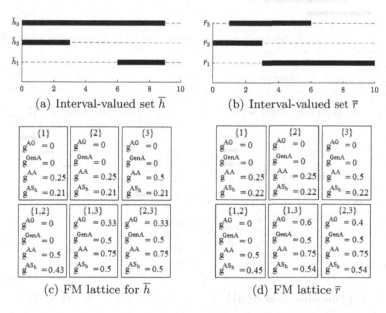

(a) Interval-valued set $\overline{h}$ (b) Interval-valued set $\overline{r}$

(c) FM lattice for $\overline{h}$ (d) FM lattice $\overline{r}$

Fig. 3. Example showing avoidance of generating same FM lattice for different subsets of sources by the g^{AS_h} FM. Any subset $\{\overline{h}_1, \overline{h}_2\}$ or $\{\overline{r}_1, \overline{r}_2\}$ is presented as $\{1, 2\}$.

Example 2: Consider the interval-valued dataset, $\bar{r}$ in Fig. 3(b) and its corresponding g^{AS_h} FM lattice in Fig. 3(d). Using (1), the aggregation of left interval endpoints is, $CFI_g(h^-) = 3 \times [g^{AS_h}(\{1\}) - g^{AS_h}(\{\emptyset\})] + 1 \times [g^{AS_h}(\{1,3\}) - g^{AS_h}(\{1\})] + 0 \times [g^{AS_h}(\{1,2,3\}) - g^{AS_h}(\{1,3\})] = 3 \times [0.22 - 0] + 1 \times [0.54 - 0.22] + 0 \times [1 - 0.54] = 0.98$. Similarly, the aggregation of right interval endpoints is, $CFI_g(h^+) = 10 \times [0.22 - 0] + 6 \times [0.54 - 0.22] + 3 \times [1 - 0.54] = 5.5$. Finally, using (2) the interval aggregation is, $CFI_g(\bar{h}) = [CFI_g(h^-), CFI_g(h^+)] = [0.98, 5.5]$.

5   Demonstration

This section demonstrates the behaviour of the new FM against the AG, $GenA$, and AA FMs for two synthetic datasets and a real-world example. For convenience, the new instance of agreement FM is denoted as AS_h and the CFI is used throughout. Further, the Jaccard SM is used for the $GenA$ and AA FMs, and AVG represents the arithmetic mean of the left and right endpoints of the intervals respectively. In all experiments, we follow the assumption that no worth information of sources is available (e.g. as in crowdsourcing). If there was such information, it could be captured and a meta-measure could be created (see [21]).

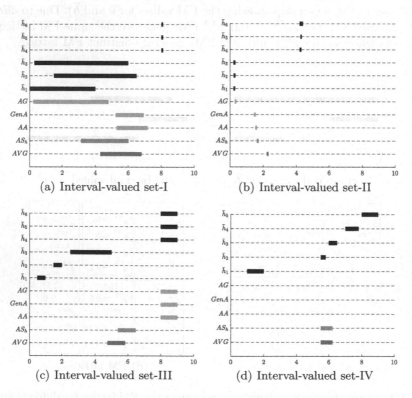

Fig. 4. Comparison of aggregation results from the CFI with the AS_h, AG, $GenA$ and AA FMs for four different interval-sets.

5.1   Demonstration with Synthetic Dataset-1

Figure 4 shows four examples of synthetic datasets together with aggregated results based on the CFI using the AS_h, AG, $GenA$, and AA FMs.

(1) The interval-valued set-I shown in Fig. 4(a) consists of three smaller intervals $\bar{h}_4$, $\bar{h}_5$, and $\bar{h}_6$ that agree completely and three larger intervals $\bar{h}_1$, $\bar{h}_2$ and $\bar{h}_3$ agreeing to a certain degree. The aggregation results (Fig. 4(a)) show that the AG FM gives importance only to the subset of larger intervals, whereas the $GenA$ and AA FMs are influenced by the subset of smaller intervals as they agree totally. However, the AS_h FM not only gives more importance to the subset of smaller intervals having a complete agreement, but also considers other subsets, $\{\bar{h}_1, \bar{h}_3\}$ and $\{\bar{h}_2, \bar{h}_3\}$ with agreement to a certain degree.

(2) For the interval-valued set-II shown in Fig. 4(b), there are three intervals $\bar{h}_1$, $\bar{h}_2$ and $\bar{h}_3$ having higher agreement than three other intervals $\bar{h}_4$, $\bar{h}_5$ and $\bar{h}_6$. Here, the AG FM is greatly influenced by the subset $\{\bar{h}_1, \bar{h}_2, \bar{h}_3\}$, whereas the $GenA$, AA, and AS_h FMs show more balanced aggregation by considering the two subsets ($\{\bar{h}_1, \bar{h}_2, \bar{h}_3\}$ and $\{\bar{h}_4, \bar{h}_5, \bar{h}_6\}$) when used with the CFI.

(3) The interval-valued set-III shown in Fig. 4(c) includes three intervals agree to each other completely and the other three wholly disagrees. Here, the AG, $GenA$ and AA FMs are completely influenced by the subset of agreed intervals, i.e., $\{\bar{h}_4, \bar{h}_5, \bar{h}_6\}$. Like other FMs, the AS_h FM shows the influence of the subset $\{\bar{h}_4, \bar{h}_5, \bar{h}_6\}$, concurrently, it also considers disagreed singletons, $\{\bar{h}_1, \bar{h}_2, \bar{h}_3\}$.

(4) The interval-valued set-IV shown in Fig. 4(d) consists of five intervals where all intervals are completely non-overlapped. At this situation, the AG, $GenA$, and AA FMs are not designed to generate the worth information for the subsets of sources and hence do not provide aggregation when combined with the CFI. Contrarily, the AS_h FM, by its construction, assigns worth to all singletons, which is later normalized by $\tilde{g}^{AS_h}(\bar{h})$. Even though there is no agreement amongst the sources regarding their intervals, the AS_h FM still can estimate the worth of other subsets by utilizing the worth of singletons. Table 2 shows the normalized worth of all subsets of intervals for the dataset-IV (in Fig. 4(d)) using the AS_h FM, where all intervals are in complete disagreement. Intuitively, when there is no overlap between intervals and all intervals are unique, then all sources should be treated with an equal worth and the aggregation should be equal to the average. In Fig. 4(d), only the AS_h FM with the CFI generates the aggregation results accordingly (i.e., performs like an average operator).

5.2   Demonstration with Synthetic Dataset-2

Here, we investigate how the FMs in combination with the CFI behave in producing the aggregation result when the overlap between intervals are gradually decreased. Five different sets of two intervals $\bar{h}_1$ and $\bar{h}_2$ are considered in Fig. 5(a)

Table 2. The normalized worth of subsets of intervals using the AS_h FM (g^{AS_h})

$|\{\overline{A}_i\}| = 1$	$|\{\overline{A}_i\}| = 2$	$|\{\overline{A}_i\}| = 3$
$g^{AS_h}(\{\overline{h}_1\})=0.2$	$g^{AS_h}(\{\overline{h}_1,\overline{h}_2\})=0.4$	$g^{AS_h}(\{\overline{h}_1,\overline{h}_2,\overline{h}_3\})=0.6$
$g^{AS_h}(\{\overline{h}_2\})=0.2$	$g^{AS_h}(\{\overline{h}_1,\overline{h}_3\})=0.4$	$g^{AS_h}(\{\overline{h}_1,\overline{h}_2,\overline{h}_4\})=0.6$
$g^{AS_h}(\{\overline{h}_3\})=0.2$	$g^{AS_h}(\{\overline{h}_1,\overline{h}_4\})=0.4$	$g^{AS_h}(\{\overline{h}_1,\overline{h}_2,\overline{h}_5\})=0.6$
$g^{AS_h}(\{\overline{h}_4\})=0.2$	$g^{AS_h}(\{\overline{h}_1,\overline{h}_5\})=0.4$	$g^{AS_h}(\{\overline{h}_1,\overline{h}_3,\overline{h}_4\})=0.6$
$g^{AS_h}(\{\overline{h}_5\})=0.2$	$g^{AS_h}(\{\overline{h}_2,\overline{h}_3\})=0.4$	$g^{AS_h}(\{\overline{h}_1,\overline{h}_3,\overline{h}_5\})=0.6$
	$g^{AS_h}(\{\overline{h}_2,\overline{h}_4\})=0.4$	$g^{AS_h}(\{\overline{h}_1,\overline{h}_4,\overline{h}_5\})=0.6$
	$g^{AS_h}(\{\overline{h}_2,\overline{h}_5\})=0.4$	$g^{AS_h}(\{\overline{h}_2,\overline{h}_3,\overline{h}_4\})=0.6$
	$g^{AS_h}(\{\overline{h}_3,\overline{h}_4\})=0.4$	$g^{AS_h}(\{\overline{h}_2,\overline{h}_3,\overline{h}_5\})=0.6$
	$g^{AS_h}(\{\overline{h}_3,\overline{h}_5\})=0.4$	$g^{AS_h}(\{\overline{h}_2,\overline{h}_4,\overline{h}_5\})=0.6$
	$g^{AS_h}(\{\overline{h}_4,\overline{h}_5\})=0.4$	$g^{AS_h}(\{\overline{h}_3,\overline{h}_4,\overline{h}_5\})=0.6$

$|\{\overline{A}_i\}| = 4$	$|\{\overline{A}_i\}| = 5$	
$g^{AS_h}(\{\overline{h}_1,\overline{h}_2,\overline{h}_3,\overline{h}_4\})=0.8$	$g^{AS_h}(\{\overline{h}_2,\overline{h}_3,\overline{h}_4,\overline{h}_5\})=1.0$	
$g^{AS_h}(\{\overline{h}_1,\overline{h}_2,\overline{h}_3,\overline{h}_5\})=0.8$		
$g^{AS_h}(\{\overline{h}_1,\overline{h}_2,\overline{h}_4,\overline{h}_5\})=0.8$		
$g^{AS_h}(\{\overline{h}_1,\overline{h}_3,\overline{h}_4,\overline{h}_5\})=0.8$		
$g^{AS_h}(\{\overline{h}_2,\overline{h}_3,\overline{h}_4,\overline{h}_5\})=0.8$		

with 100%, 75%, 50%, 25%, and 0% overlap respectively. Note that $\overline{h}_1$ is set to $[0, 1]$ in all five sets, while $\overline{h}_2$ is altered depending on the % of overlap. Figure 5(b) shows that all FMs (used with the CFI) aggregates the intervals equally (i.e., $[0, 1]$) when 100% overlap exists. However, despite degrading overlap, the AG and $GenA$ FMs continue to show the same aggregation (i.e., $[0, 1]$), whereas the AA and AS_h FMs follow the overlap degradation and aggregate the intervals accordingly. Finally, when the intervals are in complete disagreement (i.e., 0% overlap), the AS_h FM with the CFI performs like an average operator, whereas the other FMs do not support aggregation.

5.3   A Real-World Example

This experiment uses the outcome of different ageing methods (Pubic Symphysis (PS), Auricular Surface (AS), Ectocranial Suture-Vault (ESV), and Ectocranial Suture-Lateral Anterior($ESLA$)) to estimate the age-at-death of an individual skeleton [3] which is useful for forensic and biological anthropologists. Each of them provides an estimated age range for the individual skeleton. Considering the worth information of the aging methods are unknown, here our aim is to fuse their estimated age range directly to get a combined view of the skeletal age-at-death. In this aggregation experiment, the more intuitive aggregation outcome is likely to be a narrow age range capturing the actual age-at-death. Figure 6 presents the estimated age range of each aging methods for three

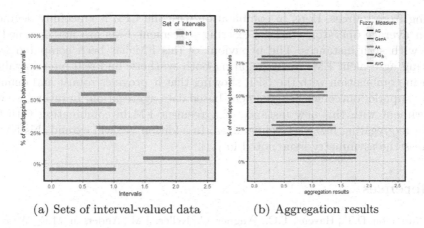

(a) Sets of interval-valued data   (b) Aggregation results

Fig. 5. (a) Five sets of interval-valued data with degrading interval-overlap (b) Aggregation results of the AG, $GenA$, AA and AS_h FMs with the CFI for Fig. 5(a).

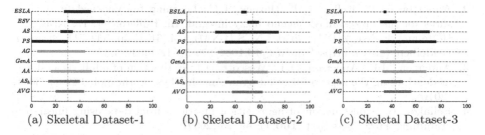

(a) Skeletal Dataset-1   (b) Skeletal Dataset-2   (c) Skeletal Dataset-3

Fig. 6. Aggregation of estimated age range of four different ageing methods using the agreement FMs with the CFI. The vertical line shows chronological age-at-death.

individual skeletons together with their true chronological age-at-death. Figure 6 also shows the aggregation results for all agreement FMs when used with the CFI. The results reveal that the g^{AS_h} FM specifies the age range more narrowly (while also capturing the true chronological age-at-death) compared to other agreement-based FMs. While this is only one example and not an extensive study, it demonstrates the interesting potential robustness in aggregation outcome of the proposed agreement FM.

6   Conclusions

As the agreement calculation of agreement FMs are affected by the limitations of popular SMs, this paper has developed a new instance of an evidence-driven agreement FM for interval-valued datasets building on the structure of $GenA$ FM, and leveraging a recently introduced SM [14, 15] to provide better capture of the inter-source agreement and worth estimation. Further, the proposed FM is designed to deal with cases where no agreement exists amongst the evidence arising from sources. Here, in combination with the CFI, it gracefully degrades to an average operator, whereas existing agreement FMs are not designed to deal with such instances. The behaviour of this FM has been compared with existing agreement FMs by aggregating both synthetic and real interval-valued data in combination with the CFI, showing that it provides robust and qualitatively superior outcomes in agreement-based data aggregation. In future, we will experiment with this new instance of agreement FM in combination with the FI for aggregating fuzzy set-valued data. In addition, we will extend this FM to address the asymmetry issue noted in [11].

References

1. Anderson, D.T., Havens, T.C., Wagner, C., Keller, J.M., Anderson, M.F., Wescott, D.J.: Sugeno fuzzy integral generalizations for sub-normal fuzzy set-valued inputs. In: Proceedings of the IEEE International Conference on Fuzzy Systems, pp. 1–8 (2012)
2. Anderson, D.T., Keller, J.M., Havens, T.C.: Learning fuzzy-valued fuzzy measures for the fuzzy-valued sugeno fuzzy integral. In: Hüllermeier, E., Kruse, R., Hoffmann, F. (eds.) IPMU 2010. LNCS (LNAI), vol. 6178, pp. 502–511. Springer, Heidelberg (2010). https://doi.org/10.1007/978-3-642-14049-5_52
3. Anderson, M.F., Anderson, D.T., Wescott, D.J.: Estimation of adult skeletal age-at-death using the sugeno fuzzy integral. Am. J. Phys. Anthropol. **142**(1), 30–41 (2010)
4. Beliakov, G.: Construction of aggregation functions from data using linear programming. Fuzzy Sets Syst. **160**(1), 65–75 (2009)
5. Choquet, G.: Theory of capacities. Ann. de l'institut Fourier **5**, 131–295 (1954)
6. Dice, L.R.: Measures of the amount of ecologic association between species. Ecology **26**(3), 297–302 (1945)
7. Dubois, D., Prade, H.: Fuzzy Sets and Systems: Theory and Applications. Academic Press, New York (1980)
8. Feng, Y., Chen, Y., Wang, M.: Multi-sensor data fusion based on fuzzy integral in AR system. In: Pan, Z., Cheok, A., Haller, M., Lau, R.W.H., Saito, H., Liang, R. (eds.) ICAT 2006. LNCS, vol. 4282, pp. 155–162. Springer, Heidelberg (2006). https://doi.org/10.1007/11941354_17
9. Grabisch, M., Murofushi, T., Sugeno, M.: Fuzzy Measures and Integrals-Theory and Applications, vol. 40. Physica Verlag, Heidelberg (2000)
10. Havens, T.C., Anderson, D.T., Keller, J.M.: A fuzzy choquet integral with an interval type-2 fuzzy number-valued integrand. In: Proceedings of the IEEE International Conference on Fuzzy Systems, Barcelona, Spain, pp. 1–8 (2010)
11. Havens, T.C., Anderson, D.T., Wagner, C.: Data-informed fuzzy measures for fuzzy integration of intervals and fuzzy numbers. IEEE Trans. Fuzzy Syst. **23**(5), 1861–1875 (2015)
12. Havens, T.C., Anderson, D.T., Wagner, C., Deilamsalehy, H., Wonnacott, D.: Fuzzy integrals of crowd-sourced intervals using a measure of generalized accord. In: Proceedings of the IEEE International Conference on Fuzzy Systems, pp. 1–8 (2013)
13. Jaccard, P.: Nouvelles recherches sur la distribution florale. Bull. de la socit vaudoise des Sci. Nat. **44**, 223–270 (1908)

14. Kabir, S., Wagner, C., Havens, T.C., Anderson, D.T.: A similarity measure based on bidirectional subsethood for intervals. IEEE Trans. Fuzzy Syst. https://doi.org/10.1109/TFUZZ.2019.2945249
15. Kabir, S., Wagner, C., Havens, T.C., Anderson, D.T., Aickelin, U.: Novel similarity measure for interval-valued data based on overlapping ratio. In: Proceedings of the IEEE International Conference on Fuzzy Systems, pp. 1–6 (2017)
16. Meyer, P., Roubens, M.: On the use of the choquet integral with fuzzy numbers in multiple criteria decision support. Fuzzy Sets Syst. **157**(7), 927–938 (2006)
17. Murofushi, T., Sugeno, M.: An interpretation of fuzzy measures and the choquet integral as an integral with respect to a fuzzy measure. Fuzzy Sets Syst. **29**(2), 201–227 (1989)
18. Nguyen, H.T., Kreinovich, V.: Computing degrees of subsethood and similarity for interval-valued fuzzy sets: fast algorithms. In: Proceedings of the 9th International Conference on Intelligent Technologies, pp. 47–55 (2008)
19. Sugeno, M.: Theory of fuzzy integrals and its applications. Tokyo Institute of Technology (1974)
20. Quoc Viet Hung, N., Tam, N.T., Tran, L.N., Aberer, K.: An evaluation of aggregation techniques in crowdsourcing. In: Lin, X., Manolopoulos, Y., Srivastava, D., Huang, G. (eds.) WISE 2013. LNCS, vol. 8181, pp. 1–15. Springer, Heidelberg (2013). https://doi.org/10.1007/978-3-642-41154-0_1
21. Wagner, C., Anderson, D.T.: Extracting meta-measures from data for fuzzy aggregation of crowd sourced information. In: Proceedings of the IEEE International Conference on Fuzzy Systems, pp. 1–8 (2012)
22. Wagner, C., Havens, T.C., Anderson, D.T.: The arithmetic recursive average as an instance of the recursive weighted power mean. In: Proceedings of the IEEE International Conference on Fuzzy Systems, pp. 1–6 (2017)

Fuzzy Methods in Data Mining and Knowledge Discovery

Hybrid Model for Parkinson's Disease Prediction

Augusto Junio Guimarães[1][iD], Paulo Vitor de Campos Souza[1,2(✉)][iD], and Edwin Lughofer[2][iD]

[1] UNA University Center, Av. Gov. Valadares, 640 - Centro, Betim, MG 32510-010, Brazil
augustojunioguimaraes@gmail.com
[2] Department Knowledge-Based Mathematical Systems, Johannes Kepler University Linz, Linz, Austria
paulo.de_campos_souza@jku.at, Edwin.Lughofer@jku.at

Abstract. Parkinson's is a chronic, progressive neurological disease with no known cause that affects the central nervous system of older people and compromises their movement. This disorder can impair daily aspects of people and therefore identify their existence early, helps in choosing treatments that can reduce the impact of the disease on the patient's routine. This work aims to identify Parkinson's traces through a voice recording replications database applied to a fuzzy neural network to identify their patterns and enable the extraction of knowledge about situations present in the data collected in patients. The results obtained by the hybrid model were superior to state of the art for the theme, proving that it is possible to perform hybrid models in the extraction of knowledge and the classification of behavioral patterns of high accuracy Parkinson's.

Keywords: Parkinson's disease · Fuzzy neural network · Hybrid models

1 Introduction

Parkinson's disease - PD [12], progressively affects specific areas of the central nervous system composed of the brain and spinal cord, caused by the intense loss of nerve cells in parts of the basal ganglia known as the black substance, located mainly in a small region of the brain mass. These neurotransmitters are responsible for carrying out the voluntary movements of the body automatically, that is, all those in which there is no need to think to perform them, the muscles perform them to the presence of this substance in the brain. Dopamine is one of the main neurotransmitters in the basal ganglia, and its primary function is to intensify nerve impulses to the muscles [12]. In the absence of it, the individual's control is lost, causing characteristic signs and symptoms. The initial phase of the disease presents subtle symptoms such as variation in speech and other

© Springer Nature Switzerland AG 2020
M.-J. Lesot et al. (Eds.): IPMU 2020, CCIS 1238, pp. 621–634, 2020.
https://doi.org/10.1007/978-3-030-50143-3_49

notable ones, such as extensive and rhythmic tremors in the hands, stiffness in the muscles and joints making movement difficult, balance, and posture are also gradually compromised. PD is the second most common degenerative disease of the central nervous system, after Alzheimer's disease [6], so early diagnosis can provide the patient with a significant improvement in the quality of life and control of PD progression.

In order to carry out a diagnosis in an efficient and less invasive way to patients, methods that evaluate voice recordings are being used to perform an acoustic analysis on speech signals and tones to identify various diseases ([18, 21, 29, 33, 38]). Many works already consider this sound recording technique useful for the creation of intelligent systems or using smart approaches to diagnose the disease, and these approaches allow discriminating healthy people from those with PD ([2, 3, 7, 29, 30, 36, 41]). These structures facilitate the decision making of specialists in the correct diagnosis of patients, with a reduction in the risk of failures and a decrease in false-positive diagnoses, and taking into account the time provided to the patient in a medical follow-up. The classification is only possible by Naranjo et al. [28] that made available a database with 195 recordings consisting of people with PD and healthy people. This work proposes the use of a hybrid model, a Fuzzy Neural Network (FNN), in order to extract diffuse rules and maintain a high level of precision for the results obtained, this interpretability allows the extraction of knowledge from the database. The WEKA tool [14] presents the leading machine learning tools, and it will be possible to compare the results of the FNN network with the main machine learning techniques.

In addition to the introduction, this article presents the following sections: right after that, the related works (Sect. 2) and the main concepts that guide this research are present. In Sect. 3, concepts of fuzzy neural networks and the architecture used in this paper are presented to the reader. Also, in Sect. 4, Parkinson's detection tests are presented, and finally, in Sect. 5 it presents the conclusions obtained in the paper and future works that may expand what was accomplished in this paper.

2   Related Works

2.1   Parkinson's Disease

It can be said that the cause of PD is considered uncertain (first), with more than one factor involved in triggering the disease. These factors can be genetic or environmental. Studies indicate that due to the abnormal accumulation of synuclein (a protein in the brain that helps the communication of nerve cells). These neurotransmitters, called Lewy bodies, can accumulate in various regions of the brain, mainly in the black substance, and consequently interfere with brain function. When performing an early diagnosis, the impacts on quality of life are reduced by monitoring and treatment by specialists [39]. Medication and physical activities are the main assets for reducing the progression of PD.

2.2   Intelligent Approaches in Parkinson Disease Detection

The treatment and early identification of Parkinson's disease have stimulated several scientists to produce renowned academic works. In this paper, the database developed by Naranjo et al. [28] obtained excellent results using Bayesian models. Their studies generated other proposals and approaches with the same database. A summary of these works is listed in Table 1.

Table 1. Pattern Classification Models.

Models			
Applications	Accuracy	Sens	Spec.
Variable selection and classification approach for Parkinson's disease [29]	0.779 (0.080)	0.765 (0.135)	0.792 (0.150)
Bayesian approach [28]	0.752 (0.086)	0.786 (0.135)	0.718 (0.132)

3   Fuzzy Neural Networks

Fuzzy neural networks are an example of hybrid models that act in the synergy between interpretability (provided by fuzzy systems) and the ability to generalize training (provided by artificial neural networks) [24]. These models are seen as the union of a fuzzy inference system and a neural aggregation network responsible for carrying out actions of different natures, such as solving problems in the software area [8], astronomy [10], and the time series prediction [9,34]. It should be noted that these models have been working efficiently to become a reference in solving problems in the area of health and human behavior, such as in solving problems related to immunotherapy [20], breast cancer [32], ECG [23], autism detection [13], in addition to helping in the detection of cognitive and motor problems in children and adolescents [35]. The FNN model presented in this paper acts with three main layers, a fuzzification technique based on data density [17], training following the concepts of Extreme Learning Machine [16], and the classification of patterns performed by a singleton neuron that uses the ReLU [27] approach as a function of activation. Its architecture can be seen in Fig. 1, and its layers and training methodology are explained below.

3.1   First Layer- Data Density Fuzzification

The first layer of the model is responsible for the fuzzification process and the formation of Gaussian-type fuzzy neurons that will compose the model's input structure. All information will be fuzzified using a technique based on data clusters due to its density. Thus, the neurons formed in this layer, represent the data cluster and, in turn, assist in the construction of a more compact FNN architecture and with neurons more significant to the problem.

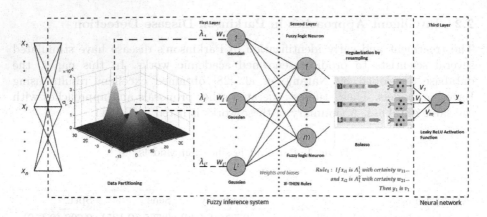

Fig. 1. Fuzzy neural network architecture

The objective of the clustering technique proposed by Hyde and Angelov [17] is to autonomously determine the clusters to be formed by the data, thus allowing the use of a simple fuzzification process and inspired by the behavior of the problem data. For this, the following concepts are fundamental, such as data distribution data density, radii and distances between reference points, among others.

The DDC works initially with the average of all the data to be evaluated. They are recursively calculated in order to decrease the complexity of the fuzzification approach and can be expressed by [17]:

$$\mu_0 = \frac{1}{N} \sum_{i=1}^{N} x_i \tag{1}$$

where x_i represents a sample from the problem, and N is the number of samples [17]. The next step is to calculate the sample density recursively and is expressed by [17]:

$$X_o = \frac{1}{N} \sum_{i=1}^{N} x_i^2 \tag{2}$$

The concepts reported in the Eqs. 1 and 2 provide the fundamental calculation in the construction of clusters. The density, as a result of the previous calculations, is also recursively calculated. Thus, it is expressed by [17]:

$$D_i = \frac{1}{1 + \|x_i - \mu_0\|^2 + X_0 - \|\mu_0\|^2} \tag{3}$$

Finally, in order to adjust the centers found in the grouping procedure, the following training algorithm is used [17]:

$$r_j^2 = \alpha r_0^2 + (1 - \alpha)\frac{1}{N_j} \sum_{i=1}^{N} \|x_i = \mu_j\|^2 \tag{4}$$

where α it is seen as a learning parameter that can be set by the user or using data propagation. In this paper, the second option was defined so that the clustering approach was autonomous, based on the data essence.

Therefore, the clustering algorithm in this paper works by defining the sample with the highest density, transforming it into a center, and starting from there, defining the forms of clustering with the adjacent samples. Therefore, cluster C will have centers that will allow the construction of Gaussian neurons in the first layer of the model. They will be constructed based on the final centers found by the DDC approach and a sigma value defined in the unit interval, that is, between 0 and 1. For each input variable x_{ij}, L neurons are defined A_{lj}, $l = 1$, ... L, whose activation functions are composed of membership functions in the corresponding neurons. Therefore, it is represented by:

$$a_{jl} = \vartheta_{A_l}, \ j = 1... \ n, \ l = 1 \ ... \ L \tag{5}$$

where ϑ is the membership degree. The number of Gaussian neurons in the first layer will be equal to the number of centers found by the fuzzification technique and these neurons can be expressed by:

$$gauss = (x_{ij}, c_l, \sigma_l) = \sum_{j=1}^{n} e^{-\frac{1}{2}\left(\frac{x_{ij} - c_l}{\sigma_l}\right)^2}, \ for \ l = 1 \ ... \ L \tag{6}$$

3.2 Second Layer- Fuzzy Rules

The second layer is composed of fuzzy neurons capable of expressing knowledge relationships on the database through IF-THEN rules. For this purpose, fuzzy logical neurons are constructed through the aggregation of the Gaussian neurons of the first layer of the model. To perform the proper calculations, operators such as t-norm (product) and s-norm (probabilistic sum) [5] are applied to the aggregation operators, which in the specific case of this paper, is called uninorm [40].

The uninorm [40] is an aggregator that allows functions of t-norm and s-norm to be used within the same context, because depending on a term, sometimes the operator can perform calculations with the product, sometimes with the probabilistic sum. These factors facilitate the construction of more interpretable, contextual rules that represent the problem domain in a clear and precise way. The uninorm format used in this paper is defined by:

$$U(x,y) = \begin{cases} o \, T(\frac{x}{o}, \frac{y}{o}), if \ y \in [0, o] \\ o + (1-o)S(\frac{x-o}{1-o}, \frac{y-o}{1-o}), if \, y \in [o, 1] \\ \delta(x,y), otherside \\ \qquad\qquad and \\ \delta(x,y) = \begin{cases} max(x,y) - if \ g \in [0, \ 0.5] \\ min(x,y) \ - if \ g \in (0.5, \ 1] \end{cases} \end{cases} \tag{7}$$

where T is a *t-norm*, S is a *s-norm* and o is the activation of the fuzzy neuron randomly set between [0, 1]. In order for the uninorm to act on fuzzy neurons, two preliminary steps are required. The first procedure is the transformation of the neuron input (a_i) (which in the case of this paper is represented by the Gaussian neuron of the first layer) together with a weight value (w_i), defined in a random between 0 and 1 with Eq. 7, for $i = 1, \dots L$. The second step aggregates all the values resulting from the first step through the following Equation:

$$p(w, a, o) = wa + \bar{w}o \tag{8}$$

where $\bar{w}$ represents the complement of w. The fuzzy neuron $(\mathbf{z})$ that uses uninorm to aggregate neurons in the first layer with weights is called a unineuron [22] and can be described by:

$$z = UNI(w; a; o) = U_{i=1}^n p(w_i, a_i, o) \tag{9}$$

The $\mathbf{z}$ neurons are seen as an aggregation of the first layer to obtain knowledge about the database. In this case, these neurons can be interpreted as a set of fuzzy rules of the IF-THEN type that can be expressed as:

$$\begin{aligned} Rule_z : \ IF \ x_{i1} \ is \ A_1^1 \ with \ certainty \ w_{11}... \\ THEN \ z_z \ = \ v_z \end{aligned} \tag{10}$$

The concept of certainty can also be extended to the rule consequent in Eq. 10.

$$THEN \ z = Class \ \Upsilon \ with \ certainty \ w_\Upsilon \ (in[0,1]) \tag{11}$$

In binary problems, Eq. 11 could be seen as an auxiliary process to interpret fuzzy rules. Therefore, in a hypothetical example where $z = Class \ 0$ with certainty 0.8 would mean that this rule has certainty 0.8 in $Class \ 0$ and 0.2 in $Class \ 1$.

As the relationship of $\mathbf{z}$ neurons occurs through the fuzzification process and the determination of centers, it is concluded that the model that makes up the first two layers of the model can be seen as a fuzzy inference system based on the density of the data evaluated in the model. Where the values of $\mathbf{v}$ are defined by training the model and represent the weights that connect the fuzzy inference system with the third layer. Another view for the value of $\mathbf{v}$ is that it represents the weight of that rule in the context of the expected output.

3.3   Third Layer- Artificial Neural Network

The third layer of the model merges the fuzzy rules and provides the expected answers to the problem. It is composed of an artificial neuron (can also be seen as a Singleton), which receives the $\mathbf{z}$ neurons of the second layer as input and performs the due calculations with a set of weights $\mathbf{v}$ obtained analytically. The neuron activation function, responsible for the necessary calculations for the final responses of the model, uses the Leaky ReLU approach [25]. Therefore, the neurons of the third layer can be expressed by:

$$y = sign \sum_{j=0}^{l} f_\omega(z_l, v_l) \qquad (12)$$

where ω is defined by the leaky ReLU activation function and the sign function represents the signal function. This activation function introduces a β factor to prevent neurons from being discarded when analyzing the problem. Its function is expressed by [25]:

$$f_{LeakyReLU}(\mathbf{z}, \beta) = max(\beta \mathbf{z}, \mathbf{z}) \qquad (13)$$

where, in this paper, $\beta = 0.001$.

The process carried out in the third layer can also be seen as the stage of defuzzification of the model, allowing fuzzy relations to return to crisp values.

3.4 Training

The training of the model is based on the concepts of Extreme Learning Machine, where the parameters of the hidden layer are defined randomly, and the weights used by the neural aggregation network (third layer) are calculated analytically, through the Moore Penrose pseudoinverse. Thus, we can express the obtainment of the weights that connect the fuzzy inference system to the neural aggregation network by the Equation:

$$v = Z^+ y \qquad (14)$$

where Z^+ is the Moore Penrose pseudoinverse of $\mathbf{Z}$, which is defined by:

$$Z = \begin{bmatrix} UNI(w_1, a_1 + o_1) & ... & UNI(w_l, a_1 + o_l) \\ UNI(w_1, a_2 + o_1) & ... & UNI(w_l, a_2 + o_l) \\ UNI(w_1, a_n + o_1) & ... & UNI(w_l, a_N + o_l) \end{bmatrix}_{N \times l} \qquad (15)$$

The model proposed in this paper needs the initial radii value in the fuzzification process and the sigma value of neurons in the first layer. The other values are defined according to the training algorithm. That makes the model simple and easy to adapt to solve the Parkinson's problem.

The fuzzification technique can generate some redundant neurons. To avoid this problem, a resampling technique linked to the LARS [11] model is used to select the best neurons, which consequently will be the best rules. This technique, called bolasso [4], was proposed by Bach and has been widely used in hybrid models for this purpose.

It combines several random replications with fuzzy rules and assesses their relevance to the model's expected outputs. Thus it is possible to define a subgroup of fuzzy candidate rules. At each replication, a different number of candidate rules are selected, and a consensus threshold defines the final selection of fuzzy rules. For example, if in 16 replications with the base, four rules were the most significant in 50% of the replications (that is, 8), they are selected to compose the final model. This selection is made before the weights are generated, so it is guaranteed that the weights that connect the cloudy inference system and the model's output layer are generated with the significant rules of the problem.

4   Parkinson's Detection Test

In order to verify the capacity of the model proposed in this paper, standard classification tests will be performed with a database that was worked on by Naranjo et al. [28]. Initially, the database has 48 features. However, for these experiments, the ID dimension, which is responsible for identifying the people in the experiment, was discarded. The collected records were 240, and the database is balanced, as there are 120 people identified with Parkinson's and 120 others without the disease. Figure 2 presents some of the dimensions present in the database, where the blue colors represent healthy people, and the red color identifies a person with Parkinson's.

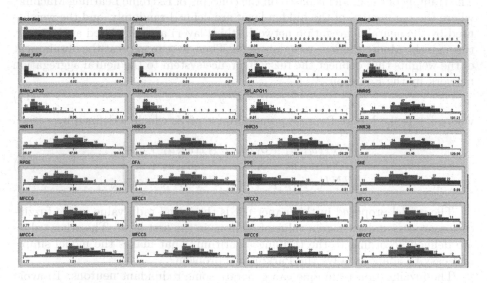

Fig. 2. Parkinson dataset- example (Color figure online)

To verify the model's ability to identify people with Parkinson's disease, the dataset will be divided into 70% for training the model, and the remaining 30% will be used to evaluate the results. All samples were normalized and were selected at random. To avoid trends in the test results, 30 repetitions were performed for the model that is part of the test.

As it is a database with binary outputs, the evaluation criteria of an intelligent model are well known by the academic community. The following criteria assess the model's ability to correct the diagnosis, the number of false positives and false negatives.

$$accuracy = \frac{TP + TN}{TP + FN + TN + FP} \tag{16}$$

$$sensitivity = \frac{TP}{TP + FN} \tag{17}$$

$$specificity = \frac{TN}{TN + TP} \tag{18}$$

$$AUC = \frac{1}{2} \left(sensitivity + specificity\right) \tag{19}$$

where, $TP =$ true positive, $TN =$ true negative, $FN =$ false negative and $FP =$ false positive.

The model proposed in this paper used the following values: $\sigma = 1.9$ and $radii = 0.08$ defined through a preliminary 10 k-fold procedure for the following value range of $\sigma = [1.5, 1.6, 1.7, 1.8, 1.9, 2.0]$ and $radii = [0.05, 0.06, 0.07, 0.08, 0.09]$. For the experiments in this paper, the replication values were defined as 16, consensus threshold $= 60\%$ after cross-validation and 10 -fold tests with the database for training the model[1]. The other models used in the test are listed below:

SVM - Support vector machine algorithm [37] is to find a hyperplane in an N-dimensional space that distinctly classifies the data points.[2]

MLP - Multilayer Perceptron [26]. It uses training based on the backpropagation technique and has a hidden layer.[3]

NB - The Naive Bayes [19] algorithm is a probabilistic classifier based on the Bayes Theorem.[4]

C4.5 - Generating a pruned or unpruned C4.5 decision tree [31].[5]

RNT - Random Tree [1] is use for constructing a tree that considers K randomly chosen attributes at each node.[6]

---

[1] For replication values between 8 and 32 and for the consensus threshold between 40 to 80%.

[2] In all tests that the model was used, one should consider the linear kernel. For more information on the code used, visit https://la.mathworks.com/help/stats/support-vector-machines-for-binary-classification.html.

[3] In all tests the following settings were used: batch size $= 100$, hidden layers $= 1$, learning rate $= 0.3$, momentum $= 0.2$, validation Threshold $= 20$.

[4] In all tests the following settings were used: batch size $= 100$, use kernet Estimator $=$ false, Supervised Discretization $=$ false.

[5] In all tests the following settings were used: batch size $= 100$, confidence factor $= 0.25$, min. num. obj $= 2$, num. Folds $= 3$, seed $= 1$, reduce Error pruning $=$ false, unpruned $=$ false, laplace $=$ false.

[6] In all tests the following settings were used: batch size $= 100$, breakTiesRandomly $=$ false, KValue $= 0$, num. Folds $= 3$, seed $= 1$, minNum $= 1.0$, allowUnclassifiedInstances $=$ false, maxDepth $= 0$.

The values in parentheses are standard deviations. Simulations were performed on a Core (TM) 2 Duo CPU, 2.27 GHz, with 3-GB RAM. Time is represented by the sum of training time and test (seconds) in each of the models. Neurons represent the most representative neurons after the pruning or regularization of the models.

Table 2. Results Optical Interconnection database.

Models	Accuracy	AUC	Sens.	Spec.	Time
This Paper	**80.88 (4.43)***	0.8204 (0.05)*	**0.8534 (0.65)***	0.7875 (0.09)	10.11 (0.42)
SVM	79.93 (4.20)*	0.7969 (0.10)	0.8244 (0.04)*	0.7695 (0.06)	129.94 (16.22)
MLP	77.69 (4.95)*	0.8599 (0.04)*	0.7935 (0.07)	0.9263 (0.10)*	5.63 (0.06)
NB	78.51 (4.56)*	0.8693 (0.04)*	0.8213 (0.06)*	0.9173 (0.05)*	**0.01 (0.01)**
C-4.5	73.47 (4.97)	0.7386 (0.06)	0.7426 (0. 08)	0.7346 (0.65)	**0.01 (0.01)**
RNT	79.81 (3.32)*	**0.8807 (0.04)***	0.8343 (0.06)*	**0.9271 (0.21)***	0.12 (0.01)

The results presented in the Table 2 present the total accuracy more significant than the other models used in the test and also surpass the original dataset. All test results were evaluated using a statistical test (ANOVA) [15]. With a 95% probability, we can say that all results marked with an '*' are statistically equal concerning the equitable performance of the factors collected between the models analyzed in the test. All premises (normality of residues, homoscedasticity, and independence) were not violated.

Therefore, it proves the efficiency of fuzzy neural networks in solving problems. It should be noted that the total fuzzy rules generated initially revolved around 46 fuzzy relations. After the regularization technique, around 5 to 6 rules are used in the 30 executions performed during the procedures reported in the tests. Another relevant factor reported in the results is that the Random Tree algorithm has results very close to the hybrid model, also standing out in AUC and specificity. The model proposed in this paper had better results in sensitivity (the ability of the diagnostic test to detect truly positive individuals, that is, to correctly diagnose patients with Parkinson's disease). So we can say that the model has the best ability to identify people who have the disease. The following fuzzy rule was extracted from the dataset in an experiment with a final accuracy of 84% probability of defining a patient with a Parkinson's diagnosis.

IF *Gender* is female with certainty 0.33 AND/OR *Jitterrel* is very high with certainty 0.90 AND/OR *Jitterabs* is very small with certainty 0.12 AND/OR *JitterRAP* is very small with certainty 0.81 AND/OR *JitterPPQ* is very small with certainty 0.12 AND/OR *Shimloc* is small with certainty 0.15 AND/OR *ShimdB* is very high with certainty 0.94 AND/OR *ShimAPQ3* is very small with certainty 0.04 AND/OR *ShimAPQ5* is very small with certainty 0.74 AND/OR *ShiAPQ11* is very small with certainty 0.73 AND/OR *HNR05* is very small with certainty 0.72 AND/OR *HNR15* is very small with certainty 0.90 AND/OR *HNR25* is very small with certainty 0.15 AND/OR *HNR35* is very small with certainty 0.83 AND/OR *HNR38* is very small with certainty 0.04 AND/OR *RPDE* is medium with certainty 0.72 AND/OR *DFA* is medium with certainty 0.07 AND/OR *PPE* is medium with certainty 0.05 AND/OR *GNE* is very small with certainty 0.76 AND/OR *MFCC0* is medium with certainty 0.87 AND/OR *MFCC1* is medium with certainty 0.98 AND/OR *MFCC2* is small with certainty 0.65 AND/OR *MFCC3* is very small with certainty 0.32 AND/OR *MFCC4* is very small with certainty 0.03 AND/OR *MFCC5* is high with certainty 0.04 AND/OR *MFCC6* is small with certainty 0.65 AND/OR *MFCC7* is very high with certainty 0.45 AND/OR*MFCC8* is very high with certainty 0.62 AND/OR *MFCC9* is very small with certainty 0.05 AND/OR *MFCC10* is small with certainty 0.08 AND/OR *MFCC11* is very small with certainty 0.14 AND/OR *MFCC12* is medium with certainty 0.11 AND/OR *Delta0* is medium with certainty 0.10 AND/OR *Delta1* is extremely high with certainty 0.64 AND/OR *Delta2* is very high with certainty 0.77 AND/OR *Delta3* is very high with certainty 0.65 AND/OR *Delta4* is very small with certainty 0.75 AND/OR *Delta5* is small with certainty 0.66 AND/OR *Delta6* is very small with certainty 0.15 AND/OR *Delta7* is medium with certainty 0.54 AND/OR *Delta8* is medium with certainty 0.02 AND/OR *Delta9* is extremely high with certainty 0.32 AND/OR *Delta10* is very high with certainty 0.16 AND/OR *Delta11* is very high with certainty 0.34 AND/OR *Delta12* is extremely high with certainty 0.16 THEN *Status* is *Parkinson* with certainty 0.18.

5   Conclusion

The results of the tests carried out corroborate that the fuzzy neural network proposed in this paper can act efficiently in the identification of patients with Parkinson's. That corroborates the high accuracy of the model in identifying intricate patterns within a database, and the fuzzification technique based on the data essence demonstrates that the constructed fuzzy neurons can efficiently represent characteristics of the problem. This paper encourages new works to be elaborated for the construction of rules more representative of the Parkinson's problem, at the same time that it provides the evolution of comparative techniques using all dimensions of the problem. In future work, it is expected to expand the techniques to be used in hybrid models, such as new fuzzification, training, and defuzzification techniques.

Acknowledgment. The authors acknowledge the support by the Austrian Science Fund (FWF): contract number P32272-N38, acronym IL-EFS.

References

1. Aldous, D.: The continuum random tree. I. Ann. Probab. 1–28 (1991)
2. Ali, L., Zhu, C., Zhou, M., Liu, Y.: Early diagnosis of Parkinson's disease from multiple voice recordings by simultaneous sample and feature selection. Expert Syst. Appl. **137**, 22–28 (2019)

3. Arora, S., Baghai-Ravary, L., Tsanas, A.: Developing a large scale population screening tool for the assessment of Parkinson's disease using telephone-quality voice. J. Acoust. Soc. Am. **145**(5), 2871–2884 (2019)
4. Bach, F.R.: Bolasso: model consistent lasso estimation through the bootstrap. In: Proceedings of the 25th International Conference on Machine Learning, pp. 33–40 (2008)
5. Balasubramaniam, J., Rao, C.J.M.: On the distributivity of implication operators over T and S norms. IEEE Trans. Fuzzy Syst. **12**(2), 194–198 (2004)
6. Boller, F., Mizutani, T., Roessmann, U., Gambetti, P.: Parkinson disease, dementia, and Alzheimer disease: clinicopathological correlations. Ann. Neurol.: Off. J. Am. Neurol. Assoc. Child Neurol. Soc. **7**(4), 329–335 (1980)
7. Brabenec, L., Mekyska, J., Galaz, Z., Rektorova, I.: Speech disorders in Parkinson's disease: early diagnostics and effects of medication and brain stimulation. J. Neural Transm. **124**(3), 303–334 (2017)
8. de Campos Souza, P.V., Guimaraes, A.J., Araujo, V.S., Rezende, T.S., Araujo, V.J.S.: Incremental regularized data density-based clustering neural networks to aid in the construction of effort forecasting systems in software development. Appl. Intell. **49**(9), 3221–3234 (2019)
9. de Campos Souza, P.V., Torres, L.C.B.: Regularized fuzzy neural network based on or neuron for time series forecasting. In: Barreto, G.A., Coelho, R. (eds.) NAFIPS 2018. CCIS, vol. 831, pp. 13–23. Springer, Cham (2018). https://doi.org/10.1007/978-3-319-95312-0_2
10. de Campos Souza, P.V., Torres, L.C.B., Guimarães, A.J., Araujo, V.S.: Pulsar detection for wavelets soda and regularized fuzzy neural networks based on and neuron and robust activation function. Int. J. Artif. Intell. Tools **28**(01), 1950003 (2019)
11. Efron, B., Hastie, T., Johnstone, I., Tibshirani, R., et al.: Least angle regression. Ann. Stat. **32**(2), 407–499 (2004)
12. Gelb, D.J., Oliver, E., Gilman, S.: Diagnostic criteria for Parkinson disease. Arch. Neurol. **56**(1), 33–39 (1999)
13. Guimarães, A.J., Araujo, V.J.S., Araujo, V.S., Batista, L.O., de Campos Souza, P.V.: A hybrid model based on fuzzy rules to act on the diagnosed of Autism in adults. In: MacIntyre, J., Maglogiannis, I., Iliadis, L., Pimenidis, E. (eds.) AIAI 2019. IAICT, vol. 559, pp. 401–412. Springer, Cham (2019). https://doi.org/10.1007/978-3-030-19823-7_34
14. Hall, M., Frank, E., Holmes, G., Pfahringer, B., Reutemann, P., Witten, I.H.: The WEKA data mining software: an update. ACM SIGKDD Explor. Newslett. **11**(1), 10–18 (2009)
15. Hastie, T., Tibshirani, R., Friedman, J.: The Elements of Statistical Learning: Data Mining, Inference, And Prediction. SSS. Springer, New York (2009). https://doi.org/10.1007/978-0-387-84858-7
16. Huang, G.B., Zhu, Q.Y., Siew, C.K.: Extreme learning machine: theory and applications. Neurocomputing **70**(1–3), 489–501 (2006)
17. Hyde, R., Angelov, P.: Data density based clustering. In: 2014 14th UK Workshop on Computational Intelligence (UKCI), pp. 1–7. IEEE (2014)
18. Ishihara, L., Brayne, C.: A systematic review of depression and mental illness preceding Parkinson's disease. Acta Neurol. Scand. **113**(4), 211–220 (2006)
19. John, G.H., Langley, P.: Estimating continuous distributions in Bayesian classifiers. In: Proceedings of the Eleventh Conference on Uncertainty in Artificial Intelligence, pp. 338–345. Morgan Kaufmann Publishers Inc. (1995)

20. Junio Guimarães, A., Vitor de Campos Souza, P., Jonathan Silva Araújo, V., Silva Rezende, T., Souza Araújo, V.: Pruning fuzzy neural network applied to the construction of expert systems to aid in the diagnosis of the treatment of cryotherapy and immunotherapy. Big Data Cogn. Comput. **3**(2) (2019). https://doi.org/10.3390/bdcc3020022, https://www.mdpi.com/2504-2289/3/2/22
21. Karan, B., Sahu, S.S., Mahto, K.: Parkinson disease prediction using intrinsic mode function based features from speech signal. Biocybern. Biomed. Eng. **40**(1), 249–264 (2020)
22. Lemos, A., Caminhas, W., Gomide, F.: New uninorm-based neuron model and fuzzy neural networks. In: 2010 Annual Meeting of the North American Fuzzy Information Processing Society, pp. 1–6. IEEE (2010)
23. Lin, C.T., et al.: Adaptive EEG-based alertness estimation system by using ICA-based fuzzy neural networks. IEEE Trans. Circuits Syst. I Regul. Pap. **53**(11), 2469–2476 (2006)
24. Lin, C.T., Lee, C.: Neural Fuzzy Systems: A Neuro-fuzzy Synergism to Intelligent Systems. Prentice-Hall, Inc. (1996)
25. Maas, A.L., Hannun, A.Y., Ng, A.Y.: Rectifier nonlinearities improve neural network acoustic models. In: Proceedings of ICML, vol. 30, p. 3 (2013)
26. McClelland, J.L., Rumelhart, D.E., Group, P.R., et al.: Parallel Distributed Processing, vol. 2. MIT Press, Cambridge (1987)
27. Nair, V., Hinton, G.E.: Rectified linear units improve restricted Boltzmann machines. In: Proceedings of the 27th International Conference on Machine Learning (ICML 2010), pp. 807–814 (2010)
28. Naranjo, L., Pérez, C.J., Campos-Roca, Y., Martín, J.: Addressing voice recording replications for Parkinson's disease detection. Expert Syst. Appl. **46**, 286–292 (2016)
29. Naranjo, L., Pérez, C.J., Martín, J., Campos-Roca, Y.: A two-stage variable selection and classification approach for Parkinson's disease detection by using voice recording replications. Comput. Methods Programs Biomed. **142**, 147–156 (2017)
30. Nilashi, M., bin Ibrahim, O., Ahmadi, H., Shahmoradi, L.: An analytical method for diseases prediction using machine learning techniques. Comput. Chem. Eng. **106**, 212–223 (2017)
31. Quinlan, J.R.: C4. 5: programs for machine learning. Elsevier (2014)
32. Silva Araújo, V.J., Guimarães, A.J., de Campos Souza, P.V., Rezende, T.S., Araújo, V.S.: Using resistin, glucose, age and BMI and pruning fuzzy neural network for the construction of expert systems in the prediction of breast cancer. Mach. Learn. Knowl. Extr. **1**(1), 466–482 (2019). https://doi.org/10.3390/make1010028, https://www.mdpi.com/2504-4990/1/1/28
33. Smith, M.E., Ramig, L.O., Dromey, C., Perez, K.S., Samandari, R.: Intensive voice treatment in Parkinson disease: laryngostroboscopic findings. J. Voice **9**(4), 453–459 (1995)
34. Soares, E., Costa, P., Costa, B., Leite, D.: Ensemble of evolving data clouds and fuzzy models for weather time series prediction. Appl. Soft Comput. **64**, 445–453 (2018). https://doi.org/10.1016/j.asoc.2017.12.032, http://www.sciencedirect.com/science/article/pii/S1568494617307573
35. Souza, P.V.C., et al.: Using hybrid systems in the construction of expert systems in the identification of cognitive and motor problems in children and young people. In: 2019 IEEE International Conference on Fuzzy Systems (FUZZ-IEEE), pp. 1–6, June 2019. https://doi.org/10.1109/FUZZ-IEEE.2019.8858906
36. Vaiciukynas, E., Verikas, A., Gelzinis, A., Bacauskiene, M.: Detecting Parkinson's disease from sustained phonation and speech signals. PloS One **12**(10) (2017)

37. Vapnik, V.: The Nature of Statistical Learning Theory. Springer, Heidelberg (2013). https://doi.org/10.1007/978-1-4757-3264-1
38. Viswanathan, R., et al.: Complexity measures of voice recordings as a discriminative tool for Parkinson's disease. Biosensors **10**(1), 1 (2020)
39. Wood-Kaczmar, A., Gandhi, S., Wood, N.: Understanding the molecular causes of Parkinson's disease. Trends Mole. Med. **12**(11), 521–528 (2006)
40. Yager, R.R., Rybalov, A.: Uninorm aggregation operators. Fuzzy Sets Syst. **80**(1), 111–120 (1996)
41. Zhang, H.H., et al.: Classification of Parkinson's disease utilizing multi-edit nearest-neighbor and ensemble learning algorithms with speech samples. Biomed. Eng. Online **15**(1), 122 (2016)

A Word Embedding Model for Mapping Food Composition Databases Using Fuzzy Logic

Andrea Morales-Garzón[✉], Juan Gómez-Romero, and M. J. Martin-Bautista

Department of Computer Science and Artificial Intelligence, Universidad de Granada, Granada, Spain
andreamgmg@correo.ugr.es, {jgomez,mbautis}@decsai.ugr.es

Abstract. This paper addresses the problem of mapping equivalent items between two databases based on their textual descriptions. Specifically, we will apply this technique to link the elements of two food composition databases by calculating the most likely match of each item in another given database. A number of experiments have been carried by employing different distance metrics, some of them involving Fuzzy Logic. The experiments show that the mappings are highly accurate and Fuzzy Logic improves the precision of the model.

Keywords: Word embedding · Fuzzy distance · Database alignment

1 Introduction

Nutrition and health organizations offer specialized and curated resources describing food and food composition, often under open access licenses. The most widely used resource is the database of the United States Department of Agriculture (USDA), which collects and harmonizes food facts from academic and industrial sources [10]. In Europe, the primary reference is the European Food Information Resource Network (EuroFIR), which compiles data from different European countries' databases [15]. There are also private initiatives such as i-Diet [22], an information system addressed to nutritionists to create personalized diets and focused on Spanish cuisine. Along with their nutritional information, i-Diet includes food item labels in Spanish and English.

These resources differ in scope and focus and usually struggle to capture the peculiarities of regional cuisines and the specificity of local products. At the same time, diet recommendation systems must be localized to the patients' context and need to be effective. Based on this principle, the Stance4Health project[1] aims

---

[1] Stance4Health (Smart Technologies for Personalised Nutrition and Consumer Engagement) is a project funded by the European Union under the Horizon 2020 research and innovation programme. More information: https://www.stance4health.com.

Supported by the European Union under grant agreement No. 816303.

© Springer Nature Switzerland AG 2020
M.-J. Lesot et al. (Eds.): IPMU 2020, CCIS 1238, pp. 635–647, 2020.
https://doi.org/10.1007/978-3-030-50143-3_50

at developing a personalized and localized nutrition service that will optimize the gut microbiota activity and long-term consumer commitment. The absence of wide-scope large databases, including regional and local products at the European and Spanish levels [16], makes it necessary in Stance4Health to combine resources mentioned above, e.g., USDA and i-Diet. However, this task is not trivial since these databases have significant differences in structure, semantics, and coverage. The latter, along with the vagueness associated to the language (e.g., a mapping of two equivalent items with different level of specialization) calls for flexible approaches to calculate the mappings.

In this paper, we propose a methodology based on a word embedding model to map food items' databases from their respective short descriptions in English. Similarity between items is calculated by using a (fuzzy) distance metric. In particular, we use this methodology to map the i-Diet and USDA databases: given an i-Diet food item, we calculate the most similar USDA item by measuring the distance between their embedding representations, obtained after encoding the short text associated with each of them with the learnt model.

In contrast to similar works, we use a larger corpus to train the language model and consider the complete recipes instead of just the ingredient list. This approach allows us to find matches between items that are different but have a similar role in several preparations, e.g., hazelnut and almond butter. This contribution could be used to cross-link food items used in different regional cuisines and to propose ingredient substitutions (or even new fusion dishes). More importantly, we expect that the mapped databases will support personalized nutrition in Stance4Health, as well as other Food Computing applications such as recipe nutrients calculation before and after cooking.

The remainder of this paper is structured as follows. In the following section, we contextualize our work within the recent literature on food item mapping and Food Computing. In Sect. 3, we further describe the data sources used in the study: USDA, i-Diet, and the corpus of recipes. Afterward, we describe the methodological approach (Sect. 4) and the experiments carried out (Sect. 5). In the last section, we analyze the results and interpret them. The paper finishes unfolding the conclusions of the work and hinting some promising directions for future work.

2   Related Work

Food Computing researchers have long acknowledged the need for a standard and open food and food components resource considering regional cuisines and cultural differences [20]. Given the effort required for such development and the absence of a central organization, the usual procedure is to extend the USDA database according to application needs [11]. In this regard, database and ontology merging and alignment techniques can be applied to find similarities and links between item registries automatically [21].

Food databases' principal elements are meals and ingredients. Therefore, it is possible to leverage ingredient detection and cuisine prediction methods to match food items based on their constituents. For instance, in [28] and [25], we can find algorithms to classify recipes by country from their ingredients. Similarly, in [19], the authors identified cuisine by using topics extracted from the recipe's text. Predictive models have also been used to translate typical dishes from one region to another by applying an encoder-decoder Deep Learning architecture [12].

From a broader perspective, other research works studied the relation among ingredients and cooking methods from food data descriptions, as in [1]. These relationships can be reused to match food items in different databases. Our work follows the same strategy, but we learn a language model based on embeddings instead of a network of ingredients that appear together in recipes. Our approach has some advantages over the latter, such as avoiding the need for precise ingredient identification in the texts. The latter problem has been extensively addressed in the literature, mostly by applying customized parsers statistical natural language processing, with limited results, e.g., [6,7,29].

Regarding the use of Deep Learning for recipe text processing, Food2Vec used a word embedding model trained only with the list of ingredients included in recipes [2]. In contrast, we also use the text describing the cooking instructions. Therefore, we obtain close encodings for ingredients that appear together in recipes (as in Food2Vec), but also for those that are involved in similar preparations (which is useful for cross-cultural item matching). The Recipe2Vec [5] tool does encode the whole text, although it focuses on recipe comparison and retrieval and not publicly available. Food images were used in [26] to enhance the embedding model. Since we do not have image information in the recipes of our corpus, analyzing the possible improvement after incorporating images remains as future work.

Furthermore, we must take into account that the food text includes the use of food brands, often replacing ingredients themselves. Moreover, brand information also appears in the USDA database. Consequently, our language model must be able to deal with such terms. We follow the guidelines of [8], which identified semantically-related terms with an embedding model, including brands.

Finally, we can use several metrics to measure the distance between two words encoded according to the model [9] and, more interestingly, between two short texts [13]. In this context, similarity techniques combining token-based similarity and Fuzzy Logic [30] can be applied to obtain the mappings. We leverage and validate these approaches to formulate a fuzzy distance metric to tackle both vagueness of the language and syntactic/semantic content within the tokens.

3   Data

We used the English recipe corpus published by archive.org[2] to build the word embedding model. This corpus collates recipes extracted from several websites, e.g., BBC Food Recipe, Epicurious, Cookstr, and AllRecipes. The final corpus includes 267,071 texts. The records corresponding to each recipe source can be seen in Table 1.

---

[2] https://archive.org/download/recipes-en-201706.

Table 1. Recipe corpus: sources and number of records

Data source	Records
BBC Food Recipe	10,679
Epicurious	20,111
Cookstr	225,602
AllRecipes	10,679
Total of records	**267,071**

As mentioned in the introduction, the databases used in this work are i-Diet and the USDA Food Composition Databases. i-Diet is a proprietary database that provides nutritional content of food items usually found in Spanish diets. The USDA database, in turn, contains more extensive and more detailed data, since its scope goes beyond the use in diet recommendations. Examples of their structure and fields are respectively shown in Tables 2 and 3. Due to the nature of the databases, item descriptions have a substantial variability.

Each register in the i-Diet Food Composition Database corresponds to a food item, which can be a complete meal or an ingredient. A food item register consists of an identification number, a description of the item in Spanish, the corresponding translation of the description into English, and the food group to which the item belongs in Spanish. Translations in i-Diet have been performed manually by nutritionists. Additionally, each register includes numerical fields corresponding to the nutritional values of the item. The mapping procedure only uses the English description field; others are discarded.

Table 2. Example of food items in the i-Diet Food Composition Database

ID	Description (ENG)	Group	...
96	Onion	HORTALIZAS BULBOSAS[a]	...
290	Apple	FRUTAS[b]	...

[a] Bulbous vegetables
[b] Fruits

The structure of the USDA Food Composition Database is similar. Each food item register in USDA encompasses an identification number, a short description of the item, a food group category, and the category description. The rest of the fields are related to the item nutritional facts (mostly major and minor nutrient values). The mapping only uses the description field; the others are discarded.

4   Methods

Our methodology is organized into four main steps: (1) data preprocessing, (2) word embedding model training and parameter tuning, (3) distance metrics,

Table 3. Example of food items in the USDA Food Composition Database

Food code	Main food description	WWEIA Category code	WWEIA Category description	...
75117020	Onions, mature, raw	6414	Onions	...
63101210	Apple, cooked or canned, with syrup	6002	Apples	...

(4) calculation of mappings by computing the Word Mover's Distance between pairs of short texts from the encodings obtained with the trained model, and (5) validation of the mappings. These steps are further described in the following sections.

4.1 Data Preprocessing

Although the recipe corpus was already collated and published on the web in a readable format, an extra preprocessing stage was required to prepare the data to train the model:

1. We extracted the data from the text files, i.e., the ingredient list and the cooking instructions. (Note that we did not consider ingredients and instruction separately.) These two pieces of data were filtered and saved in text files, one per recipe.
2. We performed a typical text cleaning process: conversion to lowercase; removal of punctuation marks, digits and special characters; removal of stop words; and lemmatization.
3. The clean data was used to train a bigram model to detect compound words. For this step, we used the Software Framework for Topic Modelling with Large Corpora [24]. English stop words were also imported from this module.

The steps above were applied to the cooking instructions presented in the recipes, e.g., the recipe text *"Combine nutritional yeast, salt, cumin, garlic powder, onion powder, paprika, chili powder, and cayenne pepper in a small bowl."* is turned into *"combin nutrit yeast salt cumin garlic_powder onion powder paprika_chili powder cayenn pepper small_bow"* after the preprocessing phase.

4.2 Model Training and Parameter Tuning

We built the language model from a corpus of text recipes by using Word2Vec [17, 18], an unsupervised Deep Learning algorithm for the creation of word embeddings. An embedding is a set of numeric vectors, each one coding a feature, which represents a language unit preserving its semantics [3]. That is, two related language units (e.g., words) will have encodings located closely in the embeddings space. Therefore, they allow us to operate with the embeddings in a meaningful way; e.g., $\langle$ king $\rangle$ - $\langle$ man $\rangle$ + $\langle$ woman $\rangle$ = $\langle$ queen $\rangle$. There are other algorithms for learning word embeddings that can be used with the same purpose, such as GloVe [23] and *fasttext* [4].

Since a generic word embedding model does not encompass such a specific domain as food from a nutritional context, a Word2Vec model was trained on the preprocessed corpus by using the Continuous Bag Of Words (CBOW) implementation, also provided by the Software Framework for Topic Modelling with Large Corpora [24]. We trained the model using the cooking instructions as a whole entry to the training model, instead of processing every sentence from each recipe separately. The nature of the text of the corpus, with short sentences and frequent anaphora, suggests that this is the most suitable approach. Experimental work and comparison to other works confirmed this assumption [27].

4.3   Distance Metrics

Let S_i be the textual representation of an item, and let $T_i = \{t_1, ..., t_n\}$ be the token set obtained as a result of the preprocessing task of such item; e.g., consider the item k whose textual representation is $S_k =$ "*Canned fish, average*", the corresponding T_k would be {"*can*", "*fish*", "*averag*"}.

We formulate the mapping problem between two items as finding the minimal distance of an item token set against every item token set from the other database. For that purpose, the different distance metrics listed below were compared.

Crisp Distance Metrics

- *Jaccard Distance: JACCARD* is a token-based distance metric which quantifies the distance based on the lexical difference between the token sets [30]:

$$JACCARD(S_1, S_2) = 1 - \frac{|T_1 \cap T_2|}{|T_1| + |T_2| - |T_1 \cap T_2|} \tag{1}$$

- *Word Mover's Distance: WMD* treats a text document as a cloud of words; each word represented as a point in the vector embeddings space [14]. The distance between two clouds is quantified by the minimum cumulative distance that words from one text document need to travel to match exactly the point cloud of the other text document. To calculate the distance between two single words, an Euclidean Distance between the corresponding vector representation is used. Therefore, WMD takes advantage from the semantic information provided by the word embedding model.
- *Hybrid Distance:* Preliminary studies within this work showed that using a unique distance measure, either lexical or semantic, strongly reduces the precision of the model. Therefore, we propose a hybrid distance measure formulated as a weighted combination of Jaccard and Word Mover's Distances.

$$HDISTANCE(t_1, t_2) = wJACCARD(t_1, t_2) + (1 - w)WMD(t_1, t_2) \tag{2}$$

$$\text{where } w \in \mathbb{R} \text{ and } 0 \leq w \leq 1$$

Fuzzy Distance Metrics

- *Fuzzy Jaccard Distance* [30]: This metric consists of a combination of token-based similarity and character-based similarity to determine the fuzzy overlap set. The Jaccard Distance described above is used to measure the distance between tokens, and a threshold determines which ones belong to the fuzzy overlap set. This latter parameter has been empirically tuned to 0.2.

$$FJACCARD_\delta(S_1, S_2) = \frac{\left| T_1 \widetilde{\cap}_\delta T_2 \right|}{|T_1| + |T_2| - \left| T_1 \widetilde{\cap}_\delta T_2 \right|} \tag{3}$$

$$\delta = 0.2$$

- *Fuzzy Document Distance:* We propose a fuzzy approach of the distance between short documents, considering each document as a token set. The distance between two sets is calculated as the Euclidean Distance between the vectors' tokens in both sets. These vectors correspond to the numerical representation obtained from the Word Embedding model previously trained. The fuzzy function is described as follows:

$$FDIST(S_1, S_2) = \frac{\sum_{x \in T_1 \cup T_2} min(\mu S_1) x \times min(\mu T_2) x}{\sum_{x \in T_1}(\mu T_1)(x) + \sum_{x \in T_2}(\mu T_2)(x) - \sum_{x \in T_1 \cup T_2} min(\mu S_1) x \times min(\mu T_2) x}$$

$$\mu_{T_i}(x) = \begin{cases} sigmoid(\frac{1}{distance(t_i, x)}) & 0 < distance(t_i, x) < \infty \\ 1 & distance(t_i, x) = 0 \\ 0 & distance(t_i, x) = \infty \end{cases} \tag{4}$$

where $distance(t_i, x)$ is the Euclidean distance between t_i and x

Noted that the membership of a token x to a set S_i is defined as the minimum distance of x to every token in S_i.

4.4 Mapping Food Items

Once the embedding model is available, it can be used to compare the similarity of two words. To this aim, as already introduced, we tested different metrics to get the most accurate results. Our mapping procedure calculated item mappings for each i-Diet register. That is, for each i-Diet item, we obtained the distance between its English description and the description of every USDA item. The algorithm finally returns the USDA item that minimizes the distance, i.e., the most likely match. Let us mention that we tackle the mapping as a multilabel classification problem, where there are many labels as USDA items apart from the *"No matches"* label (which represents the case where there is not a possible matching for an item between the databases).

4.5 Validation

A nutrition expert validated the quality of the mappings by verifying their exactness. Note that, in some cases, there may be more than one best candidate mapping (i.e., with the same quality). This situation typically happens when items in one database are more general (hypernym) than the corresponding items in the other one (hyponyms). In these cases, the validation labels the mapping as *correct* as long as one of the possible best mappings is retrieved.

Different flexibility levels have been considered to detect the robustness of the model. We obtained the number of i-Diet items where the best possible matching is achieved. We also calculate a less restrictive accuracy value, that allows us to determine the number of items whose best matching is reached between the first and the tenth candidate from the whole USDA database.

5 Experiments

The embedding model was trained during 30 epochs with vector dimensionality set to 300 and a window of size 5. Words that appear less than three times in the whole corpus are ignored. The final model yielded a vocabulary of 11,288 words. Mappings were calculated for every i-Diet food item (735 items). One human expert manually assigned the validation label of each mapping.

The results of the validation of the mappings with the different metrics are showed in Table 4. The first column "Top 1" shows, for each metric, the percentage of items whose best possible matching is achieved by the model. The rest of columns show, respectively, the percentage of items in which the best matching is found in the 2,3,5 or 10 best candidates. The weight parameter of (3) was empirically tuned to achieve the optimal performance ($w = 0.2$).

Table 4. Accuracy of the model (%) obtained with the different metrics

Distance metric	Top 1	Top 2	Top 3	Top 5	Top 10
(1) Jaccard Distance	16.75	20.16	22.20	25.20	27.52
(2) Word Mover's Distance	30.65	35.55	36.92	40.87	44.82
(3) Hybrid Distance	32.15	37.12	40.19	43.05	47.41
(4) Fuzzy Jaccard Distance	23.84	29.70	33.37	39.23	45.64
(5) Fuzzy Document Distance	35.55	40.46	43.46	47.00	53.26

A sample of the final results is provided in Tables 5 and 6. In both cases, matching are carried out using the distance metric with the best performance (see Table 4). Both tables have the same structure. In the first column, we show the original i-Diet item name. Columns from 2 to 4 show the results of the mapping: from left to right, the English description of the source i-Diet item, the description of the mapped USDA item, and the distance between both of them. The last column corresponds to the most accurate mapping identified manually.

6   Discussion

Table 5 shows a selection of successful mappings between i-Diet and USDA, i.e., mappings labeled as correct. Rows (1) to (3) show that when equivalent items had a similar text description in both databases, the model was able to match them properly. Note that a lower distance value of a mapping with respect to another one does not necessarily entail that it is better. The relative values of the distance metric are useful to select the best match for a given item, but not to compare different mappings.

Table 5. Selected examples of correct mappings

	Mapping			
	◇	△	▽	○
1	Pickle, cucumber, sour	Cucumber pickles, sour	0.0	Cucumber pickles, sour
2	All Bran Kellogg's	(Kellogg's All-Bran)	0.25	(Kellogg's All-Bran)
3	Sunflower seeds	Sunflower seeds, NFS	0.333	Sunflower seeds, NFS
4	Meat extract 'Bovril'	Meat, NFS	0.578	Meat, NFS
5	Blue Cheese	Cheese, Blue or Roquefort	0.333	Cheese, Blue or Roquefort
6	Pears canned	Pear, cooked or canned, in light syrup	0.571	Pear, cooked or canned, in light syrup
7	Canned fish, average	Fish, NS as to type, canned	0.6	Fish, NS as to type, canned
8	Chicken giblets	Chicken liver, fried	0.5	Chicken liver, fried
9	Pate liver not specified	Liver paste or pate, chicken	0.6	Liver paste or pate, chicken

◇ i-Diet text (ENG) △ USDA mapped item ▽ WMD ○ Best USDA mapping

Rows (4) to (6) illustrate more difficult mappings that were correctly solved by the procedure. In these cases, the model was capable of matching item descriptions even though one of them was slightly less specific than the other. In particular, row (4) includes a commercial brand. Rows (5) and (6) correspond to cases in which the model can map a broad (i-Diet) description with a more precise one (in USDA). Last but not least, the rows (7) to (9) show correct mappings that were not as obvious as the previous ones.

We also found some limitations to our approach, as depicted in Table 6, largely due to the coverage of the corpus and errors in translations in i-Diet from the original Spanish item description into the English one. First, in rows (1) and (2), we can see that items with no real translation and that are never used in the English recipe corpus were not mapped. We expected this behavior since there is no proper embedding for the terms used in the description. Accordingly, a more diverse corpus should be used, including recipes for local cuisines.

Table 6. Selected examples of not found, acceptable, approximate, and wrong mappings

		Mapping		
	◇	△	▽	○
1	Salchichon	No matches	1.0	Sausage, NFS
2	Morcilla asturiana (38,5%H)	No matches	0.793	Blood Sausage
3	Fondu cheese	Cheese fondue	0.0	Cheese fondue
4	Cottage cheese	Cheese, cottage, NFS	0.0	No matches
5	cocoa and hazelnut butter, Nocilla, Nutela	Almond butter	0.8	No matches
6	Wine Special	Wine, nonalcoholic	0.666	No Wine, table, white
7	Strawberry mermelade	Strawberries, raw	0.666	Jam, preserve, all flavors
8	Low fat sausage	Buttermilk, low fat (1%)	0.5	Sausage, NFS / Pork sausage
9	Scallop	Potato, scalloped, NFS	0.66	Scallops, cooked, NS as to cooking method

◇ i-Diet text (ENG) △ USDA mapped item ▽ WMD ○ Best USDA mapping

Besides, rows (3) and (4) illustrate mappings where the Spanish text is poorly translated, and therefore the mapped item has a slightly different meaning. In these cases, mappings are marked as acceptable because, despite their similar semantics, there is a better match in USDA. These problems could be addressed by manually editing the translations or by using a (more accurate) machine translation system.

Rows (5) to (7) show approximate mappings in which the link USDA is semantically related, but the association is not correct or can be improved. It is interesting to highlight that row (5) include a food brand that is correctly identified. Also, row (5) shows a case of mapping a local item and a replacement with similar usage.

Finally, rows (8) and (9) depict incorrect mappings due to the limitations of the corpus and the (unfrequent) case that the i-Diet item is more specific than the possible USDA candidates. The last column of (9) shows that dealing with hypernym and hyponyms is difficult, and can lead to several possible candidate mappings in USDA for one item in i-Diet.

As shown in Table 4, the fuzzy metrics improved the outcomes obtained with crisp approaches. From the obtained results we can draw the conclusion that vagueness of the language can make Fuzzy Logic a suitable option to tackle the matching task. Given the dimensionality and complexity of the problem, the results are reasonably accurate.

7   Conclusions and Future Work

This research work was motivated by the need for mapping two food composition databases with different scopes. This problem poses additional obstacles when the food items correspond to different regions and local cuisines. We created a word embedding model to address these issues and showed that this technique has the potential to facilitate working with non-overlapping data resources in the Food Computing domain. Our model worked well with regional brands and was able to some extent to identify substitute items used in similar preparations. Fuzzy distance metrics showed better performance than crisp alternatives.

For the future, we plan to improve the mappings by training the embedding model with a larger-scale recipe corpus and by improving the translations of Spanish item descriptions into English in i-Diet. A relevant aspect of our approach that can be further explored is the capability for finding ingredient replacements in recipes, which also entails using more imprecise knowledge. These replacements can either refer to the same item expressed differently, or to similar ingredients more often used in a particular region or cuisine. This kind of situation cannot be addressed by more traditional techniques –e.g., regex and concordances– without resorting to a specialized and comprehensive knowledge base. The absence of such resources is indeed the original motivation for our work. This same idea can be applied to recipe retrieval and automatic generation of recipes.

This work only considered English text recipes from the web. Consequently, some bias is introduced, since the popular dishes from other countries could not have sufficient representation in the collected corpus. Nevertheless, since international dishes have been introduced in cuisines from all over the world we consider that this corpus is suitable to generate useful word embeddings. We acknowledge that including typical recipes from other cuisines would help to improve the model performance. As well, more sophisticated measures can be added as well as combined with the implemented ones. Additionally, Machine Translation techniques can be applied to the Spanish text descriptions in order to reduce the errors generated by the manual translations. Also, we plan to research a multi-modal extension of this work, combining short text embeddings with the numerical fields from Food Composition Databases and other media resources, e.g., images.

Acknowledgements. This work was supported by the European Union under grant agreement No. 816303 (Stance4Health).

References

1. Ahn, Y.Y., Ahnert, S.E., Bagrow, J.P., Barabási, A.L.: Flavor network and the principles of food pairing. Sci. Rep. **1**, 96 (2011)
2. Altosaar, J.: Augmented cooking with machine intelligence. https://jaan.io/food2vec-augmented-cooking-machine-intelligence/
3. Bengio, Y., Ducharme, R., Pascal Vincent, C.J.: A neural probabilistic language model. J. Mach. Learn. Res. **3**, 1137–1155 (2003)
4. Bojanowski, P., Grave, E., Joulin, A., Mikolov, T.: Enriching word vectors with subword information. Trans. Assoc. Comput. Linguist. **5**, 135–146 (2017)
5. BuzzFeed: Recipe2vec: How word2vec helped us discover related tasty recipes. https://pydata.org/nyc2017/schedule/presentation/65/
6. Chang, M., Guillain, L.V., Jung, H., Hare, V.M., Kim, J., Agrawala, M.: RecipeScape: an interactive tool for analyzing cooking instructions at scale. In: Proceedings of the 2018 CHI Conference on Human Factors in Computing Systems, p. 451. ACM (2018)
7. Chen, Y.: A Statistical Machine Learning Approach to Generating Graph Structures from Food Recipes, August 2017
8. Fan, Y., Pakhomov, S., McEwan, R., Zhao, W., Lindemann, E., Zhang, R.: Using word embeddings to expand terminology of dietary supplements on clinical notes. JAMIA Open (2), 246–253 (2019). https://doi.org/10.1093/jamiaopen/ooz007
9. Farouk, M.: Measuring sentences similarity: a survey. ArXiv (2019)
10. Gebhardt, S., et al.: USDA national nutrient database for standard reference, release 21. United States Department of Agriculture Agricultural Research Service (2008)
11. Ispirova, G., Eftimov, T., Korošec, P., Koroušić Seljak, B.: MIGHT: statistical methodology for missing-data imputation in food composition databases. Appl. Sci. (19), 4111 (2019). https://doi.org/10.3390/app9194111
12. Kazama, M., Sugimoto, M., Hosokawa, C., Matsushima, K., Varshney, L.R., Ishikawa, Y.: A neural network system for transformation of regional cuisine style. Front. ICT **5**, 14 (2018). https://doi.org/10.3389/fict.2018.00014, https://www.frontiersin.org/article/10.3389/fict.2018.00014
13. Kenter, T., Rijke, M.D.: Short text similarity with word embeddings. In: International Conference on Information and Knowledge Management, Proceedings, pp. 1411–1420 (2015). https://doi.org/10.1145/2806416.2806475
14. Kusner, M.J., Sun, Y., Kolkin, N.I., Weinberger, K.Q.: From word embeddings to document distances, pp. 957–966 (2015)
15. Laboratory, U.N.D., Consumer, U., Institute, F.E.: USDA nutrient database for standard reference (1999)
16. Lupiañez-Barbero, A., Blanco, C., De Leiva, A.: Spanish food composition tables and databases: need for a gold standard for healthcare professionals (review). Endocrinología, Diabetes y Nutrición (English ed.), July 2018. https://doi.org/10.1016/j.endien.2018.05.011
17. Mikolov, T., Chen, K., Corrado, G., Dean, J.: Efficient estimation of word representations in vector space. In: Proceedings of Workshop at ICLR, January 2013

18. Mikolov, T., Sutskever, I., Chen, K., Corrado, G., Dean, J.: Distributed representations of words and phrases and their compositionality. In: Advances in Neural Information Processing Systems, October 2013
19. Min, W., Bao, B.K., Mei, S., Zhu, Y., Rui, Y., Jiang, S.: You are what you eat: exploring rich recipe information for cross-region food analysis. IEEE Trans. Multimed. **20**(4), 950–964 (2018). https://doi.org/10.1109/TMM.2017.2759499
20. Min, W., Jiang, S., Liu, L., Rui, Y., Jain, R.: A survey on food computing, August 2018. http://arxiv.org/abs/1808.07202
21. Noy, N.F., Musen, M.A.: Prompt: algorithm and tool for automated ontology merging and alignment. In: Proceedings of the Seventeenth National Conference on Artificial Intelligence and Twelfth Conference on Innovative Applications of Artificial Intelligence, pp. 450–455. AAAI Press (2000). http://dl.acm.org/citation.cfm?id=647288.721118
22. Gestión de Salud y Nutrición, S.: I-diet food composition database, updated from original version of g. martín peña fcd (2019)
23. Pennington, J., Socher, R., Manning, C.D.: Glove: global vectors for word representation. In: Empirical Methods in Natural Language Processing (EMNLP), pp. 1532–1543 (2014). http://www.aclweb.org/anthology/D14-1162
24. Řehůřek, R., Sojka, P.: Software framework for topic modelling with large corpora. In: Proceedings of the LREC 2010 Workshop on New Challenges for NLP Frameworks, ELRA, Valletta, Malta, pp. 45–50, May 2010
25. Sajadmanesh, S., et al.: Kissing cuisines: exploring worldwide culinary habits on the web. In: 26th International World Wide Web Conference 2017, WWW 2017 Companion, pp. 1013–1021 (2019). https://doi.org/10.1145/3041021.3055137
26. Salvador, A., et al.: Learning cross-modal embeddings for cooking recipes and food images. In: 2017 IEEE Conference on Computer Vision and Pattern Recognition (CVPR), pp. 3068–3076, July 2017. https://doi.org/10.1109/CVPR.2017.327
27. Sauer, C.R., Haigh, A.: Cooking up food embeddings understanding flavors in the recipe-ingredient graph (2017)
28. Singh, R., Arora, H.: CSE 255 assignment 2 cuisine prediction/classification based on ingredients (2015)
29. Takahashi, J., Ueda, T., Nishikawa, C., Ito, T., Nagai, A.: Implementation of automatic nutrient calculation system for cooking recipes based on text analysis. In: Anthony, P., Ishizuka, M., Lukose, D. (eds.) PRICAI 2012. LNCS (LNAI), vol. 7458, pp. 789–794. Springer, Heidelberg (2012). https://doi.org/10.1007/978-3-642-32695-0_74
30. Wang, J., Li, G., Fe, J.: Fast-join: an efficient method for fuzzy token matching based string similarity join. In: 2011 IEEE 27th International Conference on Data Engineering, pp. 458–469. IEEE (2011)

Mining Text Patterns over Fake and Real Tweets

Jose A. Diaz-Garcia[1]($\boxtimes$) ⓘ, Carlos Fernandez-Basso[1] ⓘ, M. Dolores Ruiz[2] ⓘ,
and Maria J. Martin-Bautista[1] ⓘ

[1] Department of Computer Science and A.I., University of Granada, Granada, Spain
`joseangeldiazg@ugr.es`, {`cjferba,mbautis`}`@decsai.ugr.es`
[2] Department of Statistics and O.R., University of Granada, Granada, Spain
`mariloruiz@ugr.es`

Abstract. With the exponential growth of users and user-generated content present on online social networks, fake news and its detection have become a major problem. Through these, smear campaigns can be generated, aimed for example at trying to change the political orientation of some people. Twitter has become one of the main spreaders of fake news in the network. Therefore, in this paper, we present a solution based on Text Mining that tries to find which text patterns are related to tweets that refer to fake news and which patterns in the tweets are related to true news. To test and validate the results, the system faces a pre-labelled dataset of fake and real tweets during the U.S. presidential election in 2016. In terms of results interesting patterns are obtained that relate the size and subtle changes of the real news to create fake news. Finally, different ways to visualize the results are provided.

Keywords: Association rules · Social media mining · Fake news · Text Mining · Twitter

1 Introduction

With the rise of social networks and the ease with which users can generate content, publish it and share it around the world, it was only a matter of time before accounts and people would appear to generate and share fake news. These fake news can be a real problem as it usually includes content that can go viral and be taken as true by a large number of people. In this way, political orientations, confidence in products and services, etc. can be conditioned. The textual nature of these news, has made it perfectly approachable by Data Mining techniques such as Text Mining, a sub area of Data Mining that tries to obtain relevant information from unstructured texts.

Because of the potential of these techniques in similar problems, in this paper we address the analysis of tweets that deal with fake content and real content by

© Springer Nature Switzerland AG 2020
M.-J. Lesot et al. (Eds.): IPMU 2020, CCIS 1238, pp. 648–660, 2020.
https://doi.org/10.1007/978-3-030-50143-3_51

using text mining by means of association rules. With this, we intend to prove that through these techniques relevant information can be obtained that can be used for the detection of patterns related to fake news. The contribution to the state of the art of paper is twofold:

- A reusable workflow that can get patterns on fake and real news, that can be the input of a posterior classification algorithm in order to discern between both types of news.
- A comprehensive analysis of patterns related to fake and real news during the 2016 US presidential election campaign.

In order to test and validate the system a tweet dataset has been used in which the tweets have been previously labelled as fake and real. The dataset [4] corresponds to tweets from the 2016 presidential elections in the United States. On this dataset, very interesting conclusions and patterns have been drawn, such as the tendency of fake news to slightly change real news to make it appear real. Different visualization methods are also offered to allow a better analysis of the patterns obtained.

The paper is structured as follows: Sect. 2 reviews some of the related theoretical concepts that allow to understand the following sections. Section 3 describes the related work. Section 4 explains the methodology followed. Finally Sect. 5 includes the experimentation carried out. The paper concludes with an analysis of the proposed approach and the future lines that this work opens.

2   Preliminar Concepts

In this section we will see the theoretical background of the Data Mining techniques that will be mentioned throughout the paper and that were used for the experimental development.

2.1   Association Rules

Association rules belong to the Data Mining field and have been used and studied for a long time. One of the first references to them dates back to 1993 [1]. They are used to obtain relevant knowledge from large transactional databases. A transactional database could be for example, a shopping basket database, where the items would be the products, or a text database, as in our case, where the items are the words. In a more formal way, let $t = \{A, B, C\}$ be a transaction of three items *(A, B and C)*, and any combination of them forms an itemset. Examples of differents itemsets are $\{A,B,C\}$, $\{A,B\}$, $\{B,C\}$, $\{A,C\}$, $\{A\}$, $\{B\}$ *and* $\{C\}$. According to this, an association rule would be represented in the form $X \rightarrow Y$ where X is an itemset that represents the antecedent and Y an itemset called consequent. As a result, we can conclude that consequent items have a co-occurrence relationship with antecedent items. Therefore, association rules can be used as a method of extracting hidden relationships between items or elements within transactional databases, data warehouses or other types of data storage from which it is interesting to extract information to help in decision-making processes. The classical way of measuring the goodness of association rules regarding a given problem is with two measures: support and confidence. To these metrics, new metrics have been added over time, among which the certainty factor [5] stands out, which we have used in our experimental process and we will define together with the support and confidence in the following lines.

- Support of an itemset. It is represented as $supp(X)$, and is the proportion of transactions containing item X out of the total amount of transactions of the dataset (D). The equation to define the support of an itemset is:

$$supp(X) = \frac{|t \in D : X \subseteq t|}{|D|} \tag{1}$$

- Support of an association rule. It is represented as $supp(X \rightarrow Y)$, is the total amount of transactions containing both items X and Y, as defined in the following equation:

$$supp(X \rightarrow Y) = supp(X \cup Y) \tag{2}$$

- Confidence of an association rule. It is represented as $conf(X \rightarrow Y)$ and represents the proportion of transactions containing item X which also contains Y. The equation is:

$$conf(X \rightarrow Y) = \frac{supp(X \cup Y)}{supp(X)} \tag{3}$$

- Certainty factor. It is used to represent uncertainty in rule-based expert systems. It has been shown to be one of the best models for measuring the fit of rules. Represented as $CF(X \rightarrow Y)$, a positive CF measures the decrease of probability that Y is not in a transaction when X appears. If we have a negative CF, the interpretation will be analogous. It can be represented mathematically as follows:

$$CF(X \rightarrow Y) = \begin{cases} \dfrac{conf(X \rightarrow Y) - supp(Y)}{1 - supp(Y)} & \text{if } conf(X \rightarrow Y) > supp(Y) \\ \dfrac{conf(X \rightarrow Y) - supp(Y)}{supp(Y)} & \text{if } conf(X \rightarrow Y) < supp(Y) \\ 0 & \text{otherwise} \end{cases} \tag{4}$$

The most widespread approach to obtain association rules is based on two stages using the downward-closure property. The first of these stages is the generation of frequent itemsets. To be considered frequent the itemset have to exceed the minimum support threshold. In the second stage the association rules are obtained using the minimum confidence threshold. In our approach, we will employ the certainty factor to extract more accurate association rules due to the goo properties of this assessment measure (see for instance [9]). Within this category we find the majority of the algorithms for obtaining association rules, such as Apriori, proposed by Agrawal and Srikant [2] and FP-Growth proposed by Han et al. [10]. Although these are the most widespread approaches, there are other frequent itemset extraction techniques such as vertical mining or pattern growth.

2.2   Association Rules and Text Mining

Since association rules demonstrated their great potential to obtain hidden co-occurrence relationships within transactional databases, they have been increasingly applied in different fields. One of the fields is Text Mining [14]. In this field, text entities (paragraphs, tweets, ...) are handled as a transaction in which each of the words is an item. In this way, we can obtain relationships and metrics about co-occurrences in large text databases. Technically, we could define a text transaction as:

Definition 1. *Text transaction: Let W be a set of words (items in our context). A text transaction is defined as a subset of words, i.e. a word will be present or not in a transaction.*

In a text database, in which each tweet is a transaction, it will be composed of each of the terms that appear in that tweet once the cleaning processes have been carried out. So the items will be the words. The structure will be stored in a matrix of terms in which the terms that appear will be labelled with 1 and those that are not present as 0. For example for the transactional database $D = \{t1, t2\}$ being $t1 = (just, like, emails, requested, congress)$ and $t2 = (just, anyone, knows, use, delete, keys)$ the representation of text transactions would be as we can see in Table 1.

Table 1. Example of a database with two textual transactions.

Transactionn\Item	Anyone	Congress	Delete	Emails	Just	Keys	Knows	Like	Requested	Use
t1	0	1	0	1	1	0	0	1	1	0
t2	1	0	1	0	1	1	1	0	0	1

3   Related Work

In this section, we will see in perspective the use of Data Mining techniques applied in the field of fake news. This is a thriving area within Data Mining and more specifically Text Mining, in which there are more and more related articles published.

Within the field of text analysis or Natural Language Processing for the detection of fake news, solutions based on Machine Learning and concretely classification problems stand out. This is corroborated in the paper [7], where the authors make a complete review of the approaches to address the problem of analysing fakes news and clearly highlight the problems of classification either by traditional techniques or by deep learning. According to the traditional techniques we find works like [17], in which Ozbay and Alatas, apply 23 different classification algorithms over a set previously labelled fake news coming from the political scene. With this same approach we find the paper [8] in which, the authors apply again a battery of different classification methods that go from the traditional decision trees to the neural networks, all of them with great results. If we look at the branch of deep learning, we also find some works [13,15,16] in which the authors try to train neural network models to classify texts in fake news or real. If we look at other Machine Learning methods, another interesting work that focuses on selecting which features are interesting to classify fake news is the paper [18]. On the other hand, we also find solutions based on linear regression as presented by Luca Alfaro et al. in the paper [3]. These works, despite being at the dawn of their development, work quite well but are difficult to generalize to other domains in which they have not been trained.

Because of this, within the aspect of textual entities based on fake news, another series of studies appear that try to address the problem from the descriptive and unsupervised perspective of Text Mining. A very interesting work in this sense, because it combines NLP metrics with a rule-based system is [11], in which in a very descriptive way a solution is provided that is based on the combination of a rule-based system with metrics such as the length of the title, the % of stop-words or the proper names. In the same line there is the proposal in [6] in which authors try to improve the behaviour of a random forest classifier using Text Mining metrics like bigrams, or word frequencies. Finally, in this more descriptive aspect that combines classification and NLP or Text Mining techniques, we also find the social network analysis aspect [12], where the authors classify fake or real news in twitter according to network topologies, information dissemination and especially patterns in retweets.

As far as we know, this is the first work that applies association rules in the field of fakes news. By using this technique we will try to find out which patterns are related to fake news within our domain and try to generalize to possible general patterns related to fake news in other domains of the political field. Due to the impossibility of confronting the system against a similar one, we will carry out in the next sections a descriptive study of the obtained rules.

4   Our Proposal

In this section we will depict the procedure followed in our proposal. For that we will detail the pre-processing carried out on the data. We will also look at the pattern mining process on the textual transactions. For a better understanding we can look at Fig. 1. In it we can see how the first part of the process passes

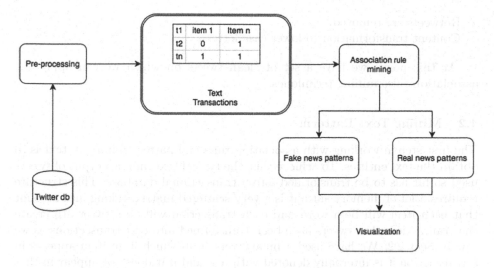

Fig. 1. Process flow for association rule extraction in Twitter transactions

through pre-processing the data, then the textual transactions are obtained, the association rules are applied and results are obtained for fake and real news.

Through this processing flow, we offer a system that discovers patterns on fake and real news that can set the basis of new interesting input values for a latter system to, for instance, obtain and classify new coming patterns into real or fake news. In this first approach the system is able to obtain, in a very friendly and interpretable way for the user, which patterns or rules can be related to fake and/or real news.

4.1   Pre-processing

The data obtained from Twitter are often very noisy so it is necessary a pre-processing step before working with them. The techniques used have been:

- Language detection. We are only interested in English tweets.
- Removal of links, removal of punctuation marks, non-alphanumeric characters, and missing values (empty tweets).
- Removal of numbers.
- Removal of additional white spaces.
- Elimination of empty words in English. We have eliminated empty English words, such as articles, pronouns and prepositions. Empty words from the problem domain have been also added, such as, the word via or rt, which can be considered empty since in Twitter it is common to use this word to reference some account from which information is extracted.
- Hashtags representing readable and interpretable terms are taken as normal words, and longer words which do not represent an analysable entity are eliminated.

- Retweets are removed.
- Content transformation to lower case letters.

At this point, we have a set of clean tweets on which we can apply the association rules mining techniques.

4.2 Mining Text Patterns

The first step in working with association rules and pattern mining in text is to obtain the text entities. To achieve this, the typical text mining corpus of tweets used so far has to be transformed into a transactional database. This structure requires a lot of memory since it is a very scattered matrix, taking into account that each item will be a word and each transaction will be a tweet. To create the transactions, the tweets have been transformed into text transactions as we saw in Sect. 2.2. We have used a binary version in which if an item appears in a transaction it is internally denoted with a 1, and if it does not appear in that transaction the matrix will have a 0.

The association rule extraction algorithm described in [1] has been used for the results. For this purpose, the parameters of minimum support threshold of 0.005 and minimum certainty factor of 0.7 have been chosen. For experimentation, we have varied the support value from 0.05 to 0.001, with fixed values of confidence and certainty factor.

5 Experimentation

In this section we will go into detail on the experimental process. We will study the dataset, the results obtained according to the input thresholds for the Apriori algorithm and finally the visualization methods used to interpret the operation of the system.

5.1 Dataset

In order to compare patterns from fake news and on the other from real news we have divided the dataset [4] into two datasets depending on whether they are labelled as fake news or not.

After this, we have two datasets, which will be analysed together but being able to know which patterns correspond to each one. The fake news dataset is composed of 1370 transactions (tweets), on the other hand, the real news dataset is composed of 5195 transactions.

5.2 Results

The experimentation has been carried out with different values of supports aiming to obtain interesting patterns within the two sets of data. It is possible to observe in Fig. 2 how the execution time is greater as the support decreases, due to the large set of items that we find with these support values.

In the Fig. 3 we can see the number of rules generated for the different support values. According to the comparison of both graphs we could draw a correlation between this graph and the previous runtime graph. As for the volume of rules generated and also the time in generating them (that as we have seen offers a graph of equal tendency), it is necessary to emphasize as the dataset fake offers more time and rules, in spite of having less transactions something that comes offered by the variability of the items inside this dataset.

Moreover, in the Figures we see how the AprioriTID algorithm has an exponential increase in the number of rules and execution time when it is executed with low support values or with more transactions. This would rule out in versions based on Big Data, where the volume of input data increases and support must be lowered.

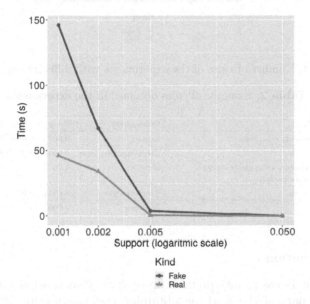

Fig. 2. Execution time of the experiments with different supports

This variability and an interpretation of the obtained patterns can be seen attending to the Table 2 where we have the strongest rules of both datasets. If we pay attention to its interpretation, it is curious how for both datasets we can find very similar rules but with some differences. This may be due to the fact that fake news are usually generated with real news to which some small element is changed. This is something that the rules of association discover for example in the rules {*sexism, won*} → {*electionnight, hate*} for fake news and the rule {*sexism, won*} → {*electionnight*} for real news. We can also observe how the tendency is to discover more items in the rules corresponding to fake news, probably caused by these sensationalist adornments that are usually charged to fake news.

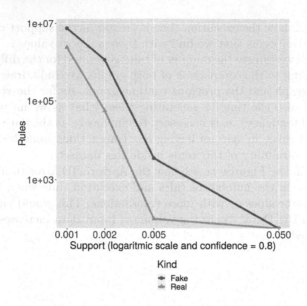

Fig. 3. Number of rules of the experiments with different supports

Table 2. Example of rules obtained in the experiments

Antecedent	Consequent	Supp	Conf	Dataset
Electionnight, won, hate	Sexism	0.0075	0.909	Fake
Sexism, won	Electionnight; hate	0.0075	0.9	Fake
Didn't, trump, won, electionnight, sexism, win, racism	Hate	0.06	1	Fake
Sexism, won	Electionnight	0.0051	0.97	Real
Projects	Foxnews	0.005	1	Real

5.3 Visualization

A system that is easily interpretable must have visualization methods so we have focused part of the work on obtaining and interpreting interesting and friendly graphics on the fake and real news. We can observe the results obtained through the graphics of the Figures. In Fig. 4 we can see the rules obtained for the fake news, where we can appreciate that the resulting rules associate in great quantity of occasions to trump with sexist, winning or racist. But some of them are interesting because the indicate the opposite, like the rule that relates*racist, trump, didnt and sexist.*

On the other hand, in Fig. 5 we can see the rules obtained for the real news. Here we can see how fewer rules are obtained for experimentation and that the terms that appear in them encompass media such as *fox, news, usa* or *winning.* Studying the terms that appear in both examples we can see racist that in this case is associated with *fox* and *donald.*

Finally, a graph has been generated, which can be seen in Fig. 6, with the results of the fake news filtering the 80 rules with a higher certainty factor. It

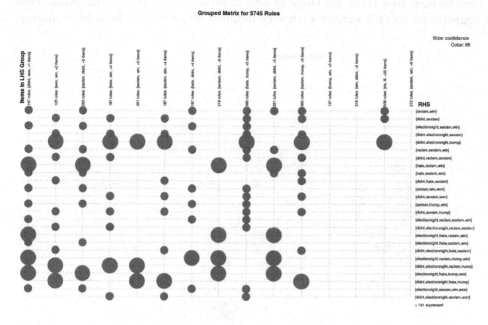

Fig. 4. Example of rules in fake news

Fig. 5. Example of rules in real news

can be seen that there are three groups of terms, one with very interconnected negative terms and another with very frequent terms due to the subject matter.

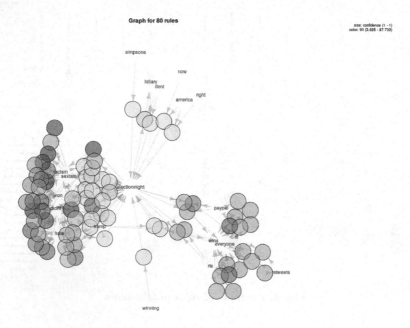

Fig. 6. Example of rules in fake news

6   Conclusions and Future Work

In conclusion, we can see how the application of Data Mining on this kind of data allows us to extract hidden patterns. These patterns allow us to know better the terms more used in each type of news according to if it is false or real in addition to the interrelations between them.

Data mining techniques and, in particular, association rules have also been corroborated as techniques that can provide relevant and user-friendly information in Text Mining domains such as this.

In future works we will extend this technique in order to classify new tweets using the information provided after the application of association rule mining. Another application would be the use of the extracted patterns in order to create a knowledge base that can be applied in real time data.

Acknowledgement. This research paper is part of the COPKIT project, which has received funding from the European Union's Horizon 2020 research and innovation programme under grant agreement No 786687. Finally the project is also partially supported by the Spanish Ministry of Education, Culture and Sport (FPU18/00150) and the program of research initiation for master students of the University of Granada.

References

1. Agrawal, R., Imieliński, T., Swami, A.: Mining association rules between sets of items in large databases. In: ACM SIGMOD Record, vol. 22, pp. 207–216. ACM (1993)
2. Agrawal, R., Srikant, R., et al.: Fast algorithms for mining association rules. In: Proceedings of 20th International Conference on Very Large Data Bases, VLDB, vol. 1215, pp. 487–499 (1994)
3. de Alfaro, L., et al.: Identifying fake news from Twitter sharing data: a large-scale study. coRR abs 1902.07207 (2019)
4. Amador Diaz Lopez, J., Oehmichen, A., Molina-Solana, M.: Fakenews on 2016 US elections viral tweets (November 2016–March 2017), November 2017. https://doi.org/10.5281/zenodo.1048826
5. Berzal, F., Blanco, I., Sánchez, D., Vila, M.A.: Measuring the accuracy and interest of association rules: a new framework. Intell. Data Anal. **6**(3), 221–235 (2002)
6. Bharadwaj, P., Shao, Z.: Fake news detection with semantic features and text mining. Int. J. Nat. Lang. Comput. (IJNLC) **8** (2019)
7. Bondielli, A., Marcelloni, F.: A survey on fake news and rumour detection techniques. Inf. Sci. **497**, 38–55 (2019)
8. Cordeiro, P.R.D., Pinheiro, V., Moreira, R., Carvalho, C., Freire, L.: What is real or fake?-Machine learning approaches for rumor verification using stance classification. In: IEEE/WIC/ACM International Conference on Web Intelligence, pp. 429–432 (2019)
9. Delgado, M., Ruiz, M.D., Sanchez, D.: New approaches for discovering exception and anomalous rules. Int. J. Uncertainty Fuzziness Knowl.-Based Syst. **19**(02), 361–399 (2011)
10. Han, J., Pei, J., Yin, Y.: Mining frequent patterns without candidate generation. In: ACM SIGMOD Record, vol. 29, pp. 1–12. ACM (2000)
11. Ibrishimova, M.D., Li, K.F.: A machine learning approach to fake news detection using knowledge verification and natural language processing. In: Barolli, L., Nishino, H., Miwa, H. (eds.) Advances in Intelligent Networking and Collaborative Systems, vol. 1035, pp. 223–234. Springer, Cham (2020). https://doi.org/10.1007/978-3-030-29035-1_22
12. Jang, Y., Park, C.H., Seo, Y.S.: Fake news analysis modeling using quote retweet. Electronics **8**(12), 1377 (2019)
13. Kaliyar, R.K.: Fake news detection using a deep neural network. In: 2018 4th International Conference on Computing Communication and Automation (ICCCA), pp. 1–7. IEEE (2018)
14. Martin-Bautista, M., Sánchez, D., Serrano, J., Vila, M.: Text mining using fuzzy association rules. In: Loia, V., Nikravesh, M., Zadeh, L.A. (eds.) Fuzzy Logic and the Internet, vol. 137, pp. 173–189. Springer, Heidelberg (2004). https://doi.org/10.1007/978-3-540-39988-9_9
15. Molina-Solana, M., Amador Diaz Lopez, J., Gomez, J.: Deep learning for fake news classification. In: I Workshop in Deep Learning, 2018 Conference Spanish Association of Artificial Intelligence, pp. 1197–1201 (2018)

16. Monti, F., Frasca, F., Eynard, D., Mannion, D., Bronstein, M.M.: Fake news detection on social media using geometric deep learning. arXiv preprint arXiv:1902.06673 (2019)
17. Ozbay, F.A., Alatas, B.: Fake news detection within online social media using supervised artificial intelligence algorithms. Phys. A **540**, 123174 (2020)
18. Reis, J.C., Correia, A., Murai, F., Veloso, A., Benevenuto, F., Cambria, E.: Supervised learning for fake news detection. IEEE Intell. Syst. **34**(2), 76–81 (2019)

Computational Intelligence for Logistics and Transportation Problems

A Genetic Approach to the Job Shop Scheduling Problem with Interval Uncertainty

Hernán Díaz[1] , Inés González-Rodríguez[2] , Juan José Palacios[1] ,
Irene Díaz[1]([✉]) , and Camino R. Vela[1]

[1] Department of Computing, University of Oviedo, Oviedo, Spain
{diazhernan,palaciosjuan,sirene,crvela}@uniovi.es
[2] Department of Maths, Stats and Computing, University of Cantabria,
Santander, Spain
gonzalezri@unican.es

Abstract. In this paper we tackle a variant of the job shop scheduling problem where task durations are uncertain and only an interval of possible values for each task duration is known. We propose a genetic algorithm to minimise the schedule's makespan that takes into account the problem's uncertainty during the search process. The behaviour of the algorithm is experimentally evaluated and compared with other state-of-the-art algorithms. Further analysis in terms of solution robustness proves the advantage of taking into account interval uncertainty during the search process with respect to considering only the expected processing times and solving the problem's crisp counterpart. This robustness analysis also illustrates the relevance of the interval ranking method used to compare schedules during the search.

Keywords: Job shop scheduling · Interval processing time · Genetic algorithms · Robustness

1 Introduction

Scheduling plays an important role in most manufacturing and production systems as well as in most information processing environments, transportation and distribution settings and, more generally, in service industries [24]. One of the most relevant problems in scheduling is the job shop, both because it is considered to be a good model for many practical applications and because it poses

Supported by the Spanish Government under research grants TIN2016-79190-R and TIN2017-87600-P and by the Principality of Asturias Government under grant IDI/2018/000176.

a challenge to the research community due to its complexity. This complexity is the reason why approximate methods and, in particular, metaheuristic search techniques, are especially suited for solving the job shop [28].

Traditionally, it has been assumed that scheduling problems are deterministic. However, for many real-world problems design variables such as processing times may be subject to perturbations or changes, dependent on human factors, etc. The most common approach to handling uncertainty is that of stochastic scheduling, modelling the duration of tasks by probability distributions [24]. However, not only does this present some tractability issues, but also, probability distributions are better suited to model variability of repetitive tasks instead of uncertainty due to lack of information [8]. Alternatively, fuzzy scheduling models uncertain durations as fuzzy numbers or fuzzy intervals, that is, possibility distributions representing more or less plausible values, in an approach that is computationally more appealing and presupposes less knowledge [9]. A third and simpler way of representing uncertainty for activity durations are intervals. Interval uncertainty is present as soon as information is incomplete and it does not assume any further knowledge. Also, it represents a first step towards solving problems in the other uncertain frameworks. Indeed, an interval can be seen as a uniform probability distribution or the support of an unknown probability distribution [1]. Also, an interval not only is a particular case of a fuzzy interval, but also the α-cuts of fuzzy intervals are intervals, so a fuzzy scheduling problem can be decomposed in multiple interval scheduling problems.

Despite its interest, research on the job shop scheduling problem with interval activity durations is still scarce. In [15] a job shop scheduling problem with interval processing times is considered and a population-based neighborhood search (PNS) is presented to optimize the makespan. A genetic algorithm is proposed in [16], but with the objective of minimising the total tardiness with respect to job due dates. Different variants of multiobjective interval job shop problems are considered in [17] and [19]. The former incorporates non-resumable jobs and flexible maintenance to the problem and proposes a multi-objective artificial bee colony algorithm to minimise both the makespan and the total tardiness. The latter considers a dual-resource constrained job shop with heterogeneous resources, and a dynamical neighbourhood search is proposed for lexicographic minimisation of carbon footprint and makespan. Finally, we find a flexible job shop problem with interval processing times in [18], where a shuffled frog-leaping algorithm is adopted to minimise the makespan. Uncertain activity durations are also modelled as intervals in scheduling problems other than the job shop and its variants in [1, 11, 20, 26]. At a more theoretical level, several attempts have been made to study how to compute earliest and latest starting times of all activities and, therefore, critical paths, over all duration scenarios in an activity-on-node network where the duration of every activity is an interval. This is essential to devise successful local search methods, as shown in deterministic job shop. A summary of the main results together with a thorough literature review can be found in [3].

This paper constitutes a starting point for a systematic study of solving methods for the interval job shop scheduling problem with makespan minimisation. We propose a solution for Interval Job Shop Scheduling Problem using genetic algorithms. The paper is organised as follows. Section 2 briefly describes the interval job shop scheduling problem. In Sect. 3 the concept of ϵ-robustness is introduced to measure the error of the prediction made by the a-priori makespan compared to the executed makespan with respect to its expected value used in this work is introduced. Section 4 details the basic schema of a genetic algorithm and the coding approach. Finally, in Sect. 5 an experimental study is developed to check the performance of the genetic algorithm in solving this problem. In addition, we also test if modelling uncertainty as intervals for processing times during the search process is worth the while and carry out some preliminary analysis of the influence of the different interval rankings. Some conclusions are drawn in Sect. 6.

2   The Job Shop Problem with Interval Durations

The classical *job shop scheduling problem*, or *JSP* in short, consists in scheduling a set of jobs $J = \{J_1, \ldots, J_n\}$ on a set of physical resources or machines $M = \{M_1, \ldots, M_m\}$, subject to a set of constraints. There are *precedence constraints*, so each job J_j, $j = 1, \ldots, n$, consists of $m_j \leq m$ tasks $(o(j,1), \ldots, o(j,m_j))$ to be sequentially scheduled. There are also *resource constraints*, whereby each task $o(j,l)$ requires the uninterrupted and exclusive use of a machine $\nu_{o(j,l)} \in M$ for its whole processing time $p_{o(j,l)}$. We assume w.l.o.g. that tasks are indexed from 1 to $N = \sum_{j=1}^{n} m_j$, so we can refer to a task $o(j,l)$ by its index $o = \sum_{i=1}^{j-1} m_i + l$ and simply write ν_o, p_o to refer respectively to its machine and processing time. The set of all tasks is denoted $O = \{1, \ldots, N\}$.

A solution to this problem is a *schedule s*, i.e. an allocation of starting times for each task, which, besides being *feasible* (in the sense that all precedence and resource constraints hold), is *optimal* according to some criterion, most commonly minimising the makespan C_{max}, that is, the completion time of the last operation (and therefore, of the whole project).

2.1   Interval Durations

In real-life applications, it is often the case that the time it takes to process a task is not exactly known in advance; instead, only some uncertain knowledge about the duration is available. If only an upper and a lower bound of each duration are known, an uncertain processing time can be represented as a closed interval of possible values denoted $\mathbf{a} = [\underline{a}, \overline{a}] = \{x \in \mathbb{R} : \underline{a} \leq x \leq \overline{a}\}$.

Let $\mathbb{IR}$ denote the set of closed intervals. The job shop problem with makespan mimisation essentially requires two arithmetic operations on $\mathbb{IR}$: addition and maximum. These are defined by extending the corresponding operations on real numbers [21], so given two intervals $\mathbf{a} = [\underline{a}, \overline{a}], \mathbf{b} = [\underline{b}, \overline{b}] \in \mathbb{IR}$,

$$\mathbf{a} + \mathbf{b} = [\underline{a} + \underline{b}, \overline{a} + \overline{b}], \tag{1}$$

$$\max(\mathbf{a}, \mathbf{b}) = [\max(\underline{a}, \underline{b}), \max(\overline{a}, \overline{b})]. \tag{2}$$

Another issue to be taken into account when processing times take the form of intervals is that of comparisons. Indeed, if several schedules are available, the "best" one would be the one with "minimal" value of the makespan (an interval). However, there is no natural total order in the set of intervals, so an interval ranking method needs to be considered among those proposed in the literature [7,14].

In [7], the authors highlight the following three total orders in $\mathbb{R}$ with certain nice behaviour (called admissibility in that work):

$$\mathbf{a} \leq_{Lex1} \mathbf{b} \Leftrightarrow \underline{a} < \underline{b} \vee (\underline{a} = \underline{b} \wedge \overline{a} < \overline{b}) \tag{3}$$

$$\mathbf{a} \leq_{Lex2} \mathbf{b} \Leftrightarrow \overline{a} < \overline{b} \vee (\overline{a} = \overline{b} \wedge \underline{a} < \underline{b}) \tag{4}$$

$$\mathbf{a} \leq_{YX} \mathbf{b} \Leftrightarrow \underline{a} + \overline{a} < \underline{b} + \overline{b} \vee (\underline{a} + \overline{a} = \underline{b} + \overline{b} \wedge \overline{a} - \underline{a} \leq \overline{b} - \underline{b}) \tag{5}$$

Both (3) and (4) are derived from a lexicographical order of interval extreme points while the last one is proposed in [27]. Obviously, all three linear orders can be used to rank intervals. In [15], a different ranking method is used for the interval job shop:

$$\mathbf{a} \leq_{pd} \mathbf{b} \Leftrightarrow P(\mathbf{b} \geq \mathbf{a}) \geq 0.5 \vee P(\mathbf{a} \geq \mathbf{b}) \leq 0.5 \tag{6}$$

where $P(\mathbf{a} \geq \mathbf{b})$ is the possibility degree that $\mathbf{a}$ is greater or equal than $\mathbf{b}$ as introduced in [13]:

$$P(\mathbf{a} \geq \mathbf{b}) = \begin{cases} 0 & \overline{a} \geq \overline{b} \\ 0.5 \cdot \frac{\overline{b}-\underline{a}}{\overline{a}-\underline{a}} \cdot \frac{\overline{b}-\underline{a}}{\overline{b}-\underline{b}} & \underline{b} \leq \underline{a} < \overline{b} \leq \overline{a} \\ \frac{\overline{b}-\underline{a}}{\overline{a}-\underline{a}} + 0.5 \cdot \frac{\overline{b}-\underline{b}}{\overline{a}-\underline{a}} & \underline{a} < \underline{b} < \overline{b} \leq \overline{a} \\ \frac{\overline{b}-\underline{a}}{\overline{a}-\underline{a}} + \frac{\overline{a}-\underline{b}}{\overline{a}-\underline{a}} \cdot \frac{\overline{b}-\underline{a}}{\overline{b}-\underline{b}} + 0.5\frac{\overline{a}-\underline{b}}{\overline{a}-\underline{a}} \cdot \frac{\overline{a}-\underline{b}}{\overline{b}-\underline{b}} & \underline{a} < \underline{b} \leq \overline{a} < \overline{b} \\ \frac{\overline{b}-\overline{a}}{\overline{b}-\underline{b}} + 0.5 \cdot \frac{\overline{a}-\underline{a}}{\overline{b}-\underline{b}} & \underline{b} \leq \underline{a} < \overline{a} < \overline{b} \\ 1 & \overline{a} \leq \underline{b} \end{cases} \tag{7}$$

It can be easily shown that this ranking is equivalent to the one induced by the interval midpoint:

$$\mathbf{a} \leq_{MP} \mathbf{b} \Leftrightarrow m(\mathbf{a}) \leq m(\mathbf{b}) \tag{8}$$

where $\forall \mathbf{a} \in \mathbb{R}$, $m(\mathbf{a}) = \frac{(\underline{a}+\overline{a})}{2}$. It coincides with the classical Hurwicz criterion for interval comparison with $\alpha = 1/2$ [12], used for interval scheduling in [1]. Also, since the interval's midpoint is the expected value of the uniform probability distribution in that interval, using the midpoint for comparing interval-valued objective functions is also closely related to the stochastic dominance based on expectation used in stochastic scheduling [24].

2.2   Interval Schedules

A schedule s establishes an order π among tasks requiring the same machine. Conversely, given a task processing order π, the schedule s (starting times of all tasks) may be computed as follows. For every task $o \in O$ with processing time $\mathbf{p_o}$, let $\mathbf{s_o}(\pi)$ and $\mathbf{c_o}(\pi)$ denote respectively the starting and completion times of o, let $PM_o(\pi)$ and $SM_o(\pi)$ denote the predecessor and successor tasks of o in the machine ν_o according to π, and let PJ_o and SJ_o denote respectively the predecessor and successor tasks of o in its job $(PM_o(\pi) = 0$ or $PJ_o = 0$ if o is the first task to be processed in its machine or its job). Then the starting time $\mathbf{s_o}(\pi)$ of o is an interval given by $\mathbf{s_o}(\pi) = \max(\mathbf{s_{PJ_o}} + \mathbf{p_{PJ_o}}, \mathbf{s_{PM_o}(\pi)} + \mathbf{p_{PM_o}(\pi)})$. Clearly, $\mathbf{c_o}(\pi) = \mathbf{s_o}(\pi) + \mathbf{p_o}(\pi)$. If there is no possible confusion regarding the processing order, we may simplify notation by writing $\mathbf{s_o}$ and $\mathbf{c_o}$. The completion time of the last task to be processed according to π thus calculated will be the makespan, denoted $\mathbf{C_{max}}(\pi)$ or simply $\mathbf{C_{max}}$. We obtain an *interval-valued schedule* in the sense that the starting and completion times of all tasks and the makespan are intervals, interpreted as the possible values that the times may take. However, notice that the task processing ordering π that determines the schedule is crisp; there is no uncertainty regarding the order in which tasks are to be processed.

2.3   Problem Formulation

We are now in a position to formulate the *Interval Job Shop Scheduling Problem* or *IJSP* in short, as follows:

$$\min_R \mathbf{C_{max}} \tag{9}$$

$$\text{subject to:} \quad \mathbf{C_{max}} = \max_{1 \leq j \leq n} \{\mathbf{c_{o(j,m_j)}}\} \tag{10}$$

$$\underline{c}_o = \underline{s}_o + \underline{p}_o, \; \forall o \in O \tag{11}$$

$$\overline{c}_o = \overline{s}_o + \overline{p}_o, \; \forall o \in O \tag{12}$$

$$\underline{s}_{o(j,l)} \geq \underline{c}_{o(j,l-1)}, \; 1 \leq l \leq m_j, 1 \leq j \leq n \tag{13}$$

$$\overline{s}_{o(j,l)} \geq \overline{c}_{o(j,l-1)}, \; 1 \leq l \leq m_j, 1 \leq j \leq n \tag{14}$$

$$\underline{s}_o \geq \underline{c}_{o'} \; \vee \; \underline{s}_{o'} \geq \underline{c}_o, \forall o \neq o' \in O : \nu_o = \nu_{o'} \tag{15}$$

$$\overline{s}_o \geq \overline{c}_{o'} \; \vee \; \overline{s}_{o'} \geq \overline{c}_o, \forall o \neq o' \in O : \nu_o = \nu_{o'} \tag{16}$$

where the minimum $\min_R \mathbf{C_{max}}$ in (9) is the smallest interval according to a given ranking R in the set of intervals $\mathbb{IR}$. Constraint (10) defines the makespan as the maximum completion time of the last task of each job. Constraints (11) and (12) establish the relationship between the starting and completion time of each task. Constraints (13) and (14) correspond to precedence relations between tasks within each job and constraints (15) and (16) establish that the execution of two tasks requiring the same machine cannot overlap. Notice that the completion time of each job J_j in the resulting schedule s is the completion time of the last task in that job, given by $\mathbf{C_j} = \mathbf{c_{o(j,m_j)}}$.

The resulting problem will be denoted $J|\underline{p}_o \leq p_o \leq \overline{p}_o|C_{max}$, following the three-field notation schema for scheduling problems. Clearly, the IJSP is NP-hard, since setting all processing times to crisp numbers yields the classical JSP, which is itself NP-hard [24].

3 Robust Schedules

A solution to the IJSP provides an interval of possible values for the starting time of each task and, hence, an interval of possible values for the makespan. In fact, it is impossible to predict what the exact starting and completion times will be until the project is actually executed. This idea is the basis for a semantics for fuzzy schedules from [10] by which solutions to a job shop problem with uncertainty should be understood as a-priori solutions. Only when tasks are executed according to the ordering π provided by the schedule we shall know their real duration and, hence, obtain an a-posteriori solution with deterministic times $p_o \in [\underline{p}_o, \overline{p}_o]$ for all tasks $o \in O$.

It would be expected that the predictive schedule does not differ much from the actual executed one. This is strongly related to the idea of robust schedule as one that minimises the effect of executional uncertainties on its performance [4]. This high-level definition is subject to many different interpretations when it comes to specifying robustness measures [25]. Here, we adapt the concept of ϵ-robustness first proposed for fuzzy scheduling problems in [22] inspired by the work on stochastic scheduling from [5].

The rationale behind this concept is to measure the predictive error of the a-priori makespan, the interval $\mathbf{C_{max}}$, compared to the actual makespan C^{ex}_{max} obtained after execution. Notice that C^{ex}_{max} is a real number that corresponds to a specific realisation of task processing times $P^{ex} = \{p^{ex}_o \in [\underline{p}_o, \overline{p}_o], o \in O\}$, usually called a *configuration* in the literature. Assuming that tasks are executed without unnecessary delays at their earliest possible starting times (as explained in Sect. 2.2), it is clear that $C^{ex}_{max} \in \mathbf{C_{max}}$. Thus, the prediction is always accurate in terms of bounds for the possible makespan values after execution. Now, if we are to give a single value as predicted makespan based on the interval $\mathbf{C_{max}}$, in the absence of further information it seems natural to consider the expected or mean value of the uniform distribution on that interval, $E[\mathbf{C_{max}}] = (\overline{C}_{max} - \underline{C}_{max})/2$. We can then measure the error of the prediction made by the a-priori makespan as the (relative) deviation of the executed makespan with respect to this expected value. In consequence, for a given $\epsilon \geq 0$, a predictive schedule with makespan interval value $\mathbf{C_{max}}$ will be considered to be ϵ-*robust* if the relative error made by $E[\mathbf{C_{max}}]$ with respect to the makespan C^{ex}_{max} of the executed schedule is bounded by ϵ, that is:

$$\frac{|C^{ex}_{max} - E[\mathbf{C_{max}}]|}{E[\mathbf{C_{max}}]} \leq \epsilon. \tag{17}$$

Clearly, the smaller the bound ϵ, the more accurate the a-priori prediction is or, in other words, the more robust the interval schedule is.

Although the expression for the expected value $E[\mathbf{C_{max}}]$ is the same as the interval's midpoint used in the ranking criterion $\leq_{MP}$, this is just a mere coincidence. In general, a robustness measure must be independent of the ranking method used to compare schedules. In particular, $E[\mathbf{C_{max}}]$ represents a prediction based on $\mathbf{C_{max}}$ in the absence of further knowledge on how values are distributed in that interval, whereas the midpoint $m(\mathbf{C_{max}})$ is a weighted average of the optimistic $\overline{C}_{max}$ and pessimistic $\underline{C}_{max}$ makespan values, representing a decision maker's equilibrium between those two extreme attitudes.

Finally, this measure of robustness is dependent on a specific configuration P^{ex} of task processing times obtained upon execution of the predictive schedule s. In the absence of real data regarding executions of the project, as is the case with the usual synthetic benchmark instances for job shop, we may resort to Monte-Carlo simulations. The idea is to simulate K possible configurations $P^k = \{p_o^k \in [\underline{p}_o, \overline{p}_o], o \in O\}$ for task processing times, using uniform probability distributions to sample possible durations for every task. For each configuration $k = 1, \ldots, K$, let C_{max}^k denote the exact makespan obtained after executing tasks according to the ordering provided by s. Then, the average ϵ-robustness of the predictive schedule across the K possible configurations, denoted $\overline{\epsilon}$, can be calculated as:

$$\overline{\epsilon} = \frac{1}{K} \sum_{k=1}^{K} \frac{|C_{max}^k - E[\mathbf{C_{max}}]|}{E[\mathbf{C_{max}}]}. \tag{18}$$

This value provides an estimate of how robust is the predictive schedule s across different processing times configurations. Again, the lower $\overline{\epsilon}$, the better.

4   A Genetic Algorithm for the IJSP

Genetic algorithms have proved to be a very useful tool for solving job shop problems, either on their own or combined with other metaheuristics [28]. Roughly speaking, a genetic algorithm starts by building a set of initial solutions or initial population P_0. This population is then evaluated and the algorithm begins an iterative process until a stopping criterion is met, typically a fixed number of iterations or consecutive iterations without improvement. At each step i, individuals from the population P_i are selected and paired for mating and, recombination operators of crossover and mutation are applied to each pair with probability p_{cross} and p_{mut} respectively, creating a new population of offspring solutions Off_i. The new population is evaluated and a replacement operator is applied to merge P_i and Off_i into the new population P_{i+1} for the next iteration. Once the stopping criterion is met, the best solution according to the interval ranking is selected and returned from the last population. Algorithm 1 summarises these steps.

In this work, several well-known selection, recombination and replacement operators for Job Shop Scheduling problems are tried in order to find the best setup for the genetic algorithm. The set of operators and their impact on solving this problem are detailed in Sect. 5. A crucial part in designing algorithms is how to encode and decode solutions. Following [6], we encode a solution as a permutation with repetition. This is a permutation of the set of tasks, where each task $o(j, l)$ is represented by its job number j. For example, a topological order $(o(2, 1), o(1, 1), o(2, 2), o(3, 1), o(3, 2), o(1, 2))$ is encoded as $(2\,1\,2\,3\,3\,1)$.

Require: An IJSP instance
Ensure: A schedule
  Generate a pool P_0 of random solutions.
  Evaluate P_0
  $i \leftarrow 0$;
  while stop condition not satisfied **do**
    $Off_i \leftarrow$ pairs of individuals selected from P_i;
    for each pair of individuals in Off_i **do**
      Apply crossover operator with probability p_{cross};
      Apply mutation operator with probability p_{mut};
    Evaluate Off_i
    $P_{i+1} \leftarrow$ Apply replacement operator in (P_i, Off_i);
    $i \leftarrow i + 1$;
  $Best \leftarrow$ Best solution in P_i based on the order on intervals;
  return $Best$

Algorithm 1: Main steps of the genetic algorithm

The decoding is done using an insertion strategy: we iterate along the chromosome and for each task $o(j,l)$ we schedule it at its earliest feasible insertion position as follows. Let η_k be the number of tasks scheduled on machine $k = \nu_{o(j,l)}$ and let $\sigma_k = (0, \sigma(1,k), ..., \sigma(\eta_k, k))$ denote the partial processing order of tasks already scheduled in machine k. Then a feasible insertion position $q, 0 \leq q < \eta_k$ for $o(j,l)$ is a position such that $\max\{\underline{c}_{\sigma(q,k)}, \underline{c}_{o(j,l-1)}\} + \underline{p}_{o(j,l)} \leq \underline{s}_{\sigma(q+1,k)}$ and $\max\{\overline{c}_{\sigma(q,k)}, \overline{c}_{o(j,l-1)}\} + \overline{p}_{o(j,l)} \leq \overline{s}_{\sigma(q+1,k)}$, so the earliest feasible insertion position is the smallest value q^* verifying these inequalities. We set $\mathbf{s}_{o(j,l)} = \max\{\mathbf{c}_{\sigma(q^*,k)}, \mathbf{c}_{o(j,l-1)}\}$ if q^* exists, and $\mathbf{s}_{o(j,l)} = \max\{\mathbf{c}_{\sigma(\eta_k,k)}, \mathbf{c}_{o(j,l-1)}\}$ otherwise.

5   Experimental Study

The purpose of the experimental study is threefold: assess the proposed genetic algorithm, see if considering the uncertainty in processing times during the search process is worth the while and carry out a preliminary analysis of the influence of the different interval rankings.

    To test the algorithm, we consider 12 very well-known instances for the job shop problem: classical instances FT10 (size 10×10) and FT20 (20×5), and instances La21, La24, La25 (15×10), La27, La29 (20×10), La38, La40 (15×15), and ABZ7, ABZ8, ABZ9 (20×15) that form the set of 10 problems identified in [2] as hard to solve for classical JSP. The processing times are modified to be intervals as follows: given the original crisp processing time of an operation p_o, the interval time is generated as $\mathbf{p_o} = [p_o - \delta, p_o + \delta]$, where δ is a random value in $[0, 0.15p_o]$. The resulting IJSP instances are available online[1]. All the experiments reported in this section have been run on a PC with Intel Xeon

---

[1] Repository section at http://di002.edv.uniovi.es/iscop.

Gold 6132 processor at 2.6 Ghz and 128 Gb RAM with Linux (CentOS v6.9), using a C++ implementation.

Regarding the algorithm's parameter configuration, we have run several tests to find the best setup. A first batch of experiments are conducted to test different recombination operations and probabilities as well as selection strategies. The considered operators are given in Table 1, with the best setup values in bold. In all cases the stopping criterion is set to 500 iterations.

Table 1. Parameter tuning with the best configuration in bold

Parameter	Tested values
Crossover operator	Generalised Order Crossover (GOX)
	Job-Order Crossover (JOX)
	Precedence Preservative Crossover (PPX)
Crossover probability	0.5, 0.75, **1**
Mutation operator	**Insertion**, Inversion, Swap
Mutation probability	0, **0.15**, 0.25
Selection operator	Roulette
	Tournament (t = 3)
	Shuffle
	Stochastic Universal Sampling (SUS)
Replacement	Generational replacement with elitism (k = 1, 12, 25)
	Tournament 2/4 parents-offspring (allowing repetition)
	Tournament 2/4 parents-offspring (no repetition)

A second test based on convergence demonstrates that the best population size is 250. Figure 1 shows the average evolution of the expected value of makespan across 30 runs of the algorithm on instance FT10. The dotted line corresponds to the expected makespan of the best solution in the population and the continuous line to the average of the whole population. It is clear that within 500 iterations, the algorithm reaches a convergence point. The behaviour on the remaining instances is similar, so we adopt this number of iterations as stopping criterion for the algorithm.

To asses the performance of the genetic algorithm (GA in the following), we compare it with the PNS algorithm proposed in [15], which to our knowledge constitutes the state-of-the art in the IJSP with makespan minimisation. The authors use $\leq_{pd}$ to rank different intervals in PNS. Since $\leq_{pd}$ is equivalent to $\leq_{MP}$ and for the sake of a fair comparison, we also adopt the same ranking. GA is run on the same set of 17 instances as PNS, which are adapted versions of the well-known crisp instances ORB1–5, LA16–25 and ABZ5–6, and the stopping criterion is set to 25 consecutive iterations without improvement. Table 2 shows for each algorithm, the average expected makespan across all the runs (20 runs

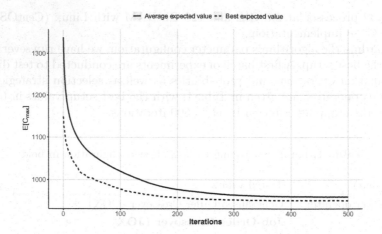

Fig. 1. Evolution of the best and average individual for FT10 instance.

for PNS and 30 for GA) together with runtimes, as well as a column with the relative difference between the performances of GA and PNS. We can see that, despite not having a neighbourhood search component, GA outperforms PNS in 14 out of the 17 instances, and is marginally worse in the remaining 3 instances (0.4% worse in ORB2, 0.5% in La16 and 0.01% in La17). Overall, GA obtains an average improvement of 1.2% compared to PNS. Additionally, a t-test for paired samples is run to compare the results of GA and PNS (after both samples pass a Kolmogorov-Smirnov normality test), confirming that there are indeed significant differences between both algorithms for a significance level of 0.05. Regarding runtime, GA is 93.8% faster than PNS. Notice however, that runtimes of PNS are those provided by the authors using their own machine and therefore comparisons in this sense must be done with caution.

We have used the set of 17 instances considered in [15] in Table 2 to compare GA with the state-of-the-art. However, in the deterministic case, the original crisp instances have already been solved to optimality and their fuzzy counterparts offer little room for improvement, as shown in [23]. For this reason we will now switch to the set of more challenging instances introduced at the beginning of this section for the remaining experimental results.

One may wonder if solving the crisp problem that results from considering only the midpoint of the interval processing times yields similar results to using intervals with the added advantage of having all the available tools for deterministic JSP. Including uncertainty in the search process adds some difficulty to the problem: different concepts need to be adapted or redefined and solving methods tailored to handle the uncertainty need to be proposed, usually with an increased complexity. It is also natural to see if the choice of a ranking method in the interval setting has any influence on the outcome. To try to answer these questions, we carry out a new set of experiments. For every IJSP instance we run GA 30 times considering each of the four different ranking methods and 30

Table 2. Computational results and times of PNS and GA

Instance	PNS		GA		Relative diff.
	Avg. $E[C_{max}]$	Runtime (s)	Avg. $E[C_{max}]$	Runtime (s)	
ORB1	1187.00	6.7	1171.12	0.5	−1.3%
ORB2	968.25	6.8	971.93	0.4	0.4%
ORB3	1145.23	3.5	1117.23	0.7	−2.4%
ORB4	1110.85	6.8	1087.13	0.6	−2.1%
ORB5	974.60	6.9	955.02	0.4	−2.0%
ABZ5	1308.45	5.3	1296.58	0.3	−0.9%
ABZ6	1012.40	6.3	998.95	0.3	−1.3%
La16	1019.40	6.2	1024.52	0.4	0.5%
La17	834.45	6.5	834.50	0.3	0.0%
La18	912.95	6.7	900.75	0.3	−1.3%
La19	919.65	6.0	904.95	0.5	−1.6%
La20	966.50	6.4	952.87	0.3	−1.4%
La21	1173.45	16.9	1150.73	1.0	−1.9%
La22	1036.05	16.8	1019.98	1.1	−1.6%
La23	1105.45	16.8	1083.27	0.9	−2.0%
La24	1047.55	16.9	1038.40	0.9	−0.9%
La25	1089.15	17.1	1077.05	1.0	−1.1%

times on the instance's crisp counterpart. Notice that the objective function is an interval in the first four cases and a crisp value in the last one, so they are not directly comparable. Instead, we measure the $\bar{\epsilon}$-robustness of the 30 solutions obtained by GA in each case using $K = 1000$ possible realisations, to compare the resulting solutions in terms of their quality as predictive schedules.

Figure 2 depicts for each instance the boxplots with the $\bar{\epsilon}$ values with the schedules that result from the 30 runs of GA in each case. We can see that, regardless of the ranking considered, solutions are more robust when intervals are taken into account during the search process. This is confirmed by several t-tests, showing that the $\bar{\epsilon}$-robustness of the interval schedules, regardless of the ranking, is significantly better than the one of the crisp schedule for a significance level of 0.05 on all instances expect La25. Regarding the choice of ranking method, according to the t-tests there is no significant difference between $\leq_{MP}$ and $\leq_{YX}$ on any instance. This is actually not surprising, since $\leq_{YX}$ can be understood as a refinement of $\leq_{MP}$, but it shows that this refinement does not necessarily translate into more robust schedules. Also, there are no significant differences between $\leq_{MP}$ and $\leq_{Lex1}$ on any instance except ABZ9, La21 and La24. More interestingly, the ranking $\leq_{Lex2}$ yields solutions significantly more robust than those obtained using any other ranking on all instances except FT20, La24 and La40, where it is not significantly better than $\leq_{MP}$ and $\leq_{YX}$. We may conclude

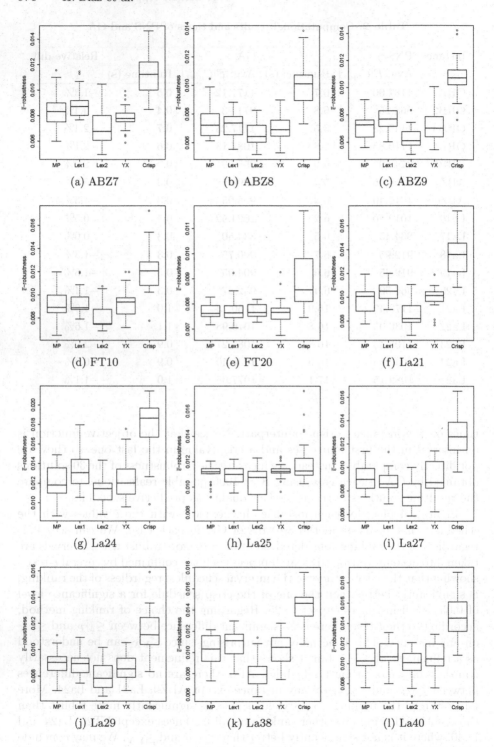

Fig. 2. $\bar{\epsilon}$-robustness of solutions obtained with four different rankings and solving the crisp counterpart

that solving the interval JSP results in more robust schedules than solving a simpler deterministic counterpart and that the choice of interval ranking method does have an influence on the outcome.

6 Conclusions

In this work we have developed an approach to solving the IJSP using a GA. Results show that GA is competitive with the existing methods from the literature. In addition, incorporating the interval uncertainty in the search process yields more robust solutions than solving an alternative crisp problem. On the other hand, the choice of interval ranking method plays an important role in the final solution's performance. Further work needs to be done to obtain more powerful search methods specifically designed for handling interval uncertainty and to thoroughly analyse the influence of different ranking methods in order to make a proper choice.

References

1. Allahverdi, A., Aydilek, H., Aydilek, A.: Single machine scheduling problem with interval processing times to minimize mean weighted completion time. Comput. Oper. Res. **51**, 200–207 (2014). https://doi.org/10.1016/j.cor.2014.06.003
2. Applegate, D., Cook, W.: A computational study of the job-shop scheduling problem. ORSA J. Comput. **3**, 149–156 (1991)
3. Artigues, C., Briand, C., Garaix, T.: Temporal analysis of projects under interval uncertainty. In: Schwindt, C., Zimmermann, J. (eds.) Handbook on Project Management and Scheduling. IHIS, vol. 2, pp. 911–927. Springer, Cham (2015). https://doi.org/10.1007/978-3-319-05915-0_11
4. Aytung, H., Lawley, M.A., McKay, K., Shantha, M., Uzsoy, R.: Executing production schedules in the face of uncertainties: a review and some future directions. Eur. J. Oper. Res. **161**, 86–110 (2005)
5. Bidot, J., Vidal, T., Laboire, P.: A theoretic and practical framework for scheduling in stochastic environment. J. Sched. **12**, 315–344 (2009)
6. Bierwirth, C.: A generalized permutation approach to jobshop scheduling with genetic algorithms. OR Spectr. **17**, 87–92 (1995)
7. Bustince, H., Fernandez, J., Kolesárová, A., Mesiar, R.: Generation of linear orders for intervals by means of aggregation functions. Fuzzy Sets Syst. **220**, 69–77 (2013)
8. Dubois, D., Prade, H., Smets, P.: Representing partial ignorance. IEEE Trans. Syst. Man Cybern. Part A **26**(3), 361–377 (1996)
9. Dubois, D., Prade, H., Sandri, S.: On possibility/probability transformations. In: Fuzzy Logic, Theory and Decision Library, vol. 12, pp. 103–112. Kluwer Academic (1993)
10. González Rodríguez, I., Puente, J., Vela, C.R., Varela, R.: Semantics of schedules for the fuzzy job shop problem. IEEE Trans. Syst. Man Cybern. Part A **38**(3), 655–666 (2008)
11. Han, Y., Gong, D., Yaochu, J., Pan, Q.K.: Evolutionary multi-objective blocking lot-streaming flow shop scheduling with interval processing time. Appl. Soft Comput. **42**, 229–245 (2016)

12. Hurwicz, L.: A class of criteria for decision-making under ignorance. Cowles Commission Discussion Paper. Statistics **370** (1951)
13. Jiang, C., Han, X., Liu, G., Liu, G.: A nonlinear interval number programming method for uncertain optimization problems. Eur. J. Oper. Res. **188**(1), 1–13 (2008)
14. Karmakar, S., Bhunia, A.K.: A comparative study of different order relations of intervals. Reliable Comput. **16**, 38–72 (2012)
15. Lei, D.: Population-based neighborhood search for job shop scheduling with interval processing time. Comput. Ind. Eng. **61**, 1200–1208 (2011). https://doi.org/10.1016/j.cie.2011.07.010
16. Lei, D.: Interval job shop scheduling problems. Int. J. Adv. Manuf. Technol. **60**, 291–301 (2012). https://doi.org/10.1007/s00170-011-3600-3
17. Lei, D.: Multi-objective artificial bee colony for interval job shop scheduling with flexible maintenance. Int. J. Adv. Manuf. Technol. **66**, 1835–1843 (2013). https://doi.org/10.1007/s00170-012-4463-y
18. Lei, D., Cao, S.: A novel shuffled frog-leaping algorithm for flexible job shop scheduling with interval processing time. In: Proceedings of the 36th Chinese Control Conference, pp. 2708–2713 (2017)
19. Lei, D., Guo, X.: An effective neighborhood search for scheduling in dual-resource constrained interval job shop with environmental objective. Int. J. Prod. Econ. **159**, 296–303 (2015). https://doi.org/10.1016/j.ijpe.2014.07.026
20. Matsveichuk, N., Sotskov, Y., Egorova, N., Lai, T.C.: Schedule execution for two-machine flow-shop with interval processing times. Math. Comput. Modell. **49**(5), 991–1011 (2009). https://doi.org/10.1016/j.mcm.2008.02.004
21. Moore, R.E., Kearfott, R.B., Cloud, M.J.: Introduction to Interval Analysis. Society for Industrial and Applied Mathematics (2009)
22. Palacios, J.J., González-Rodríguez, I., Vela, C.R., Puente, J.: Robust swarm optimisation for fuzzy open shop scheduling. Nat. Comput. **13**(2), 145–156 (2014). https://doi.org/10.1007/s11047-014-9413-1
23. Palacios, J.J., Puente, J., Vela, C.R., González-Rodríguez, I.: Benchmarks for fuzzy job shop problems. Inf. Sci. **329**, 736–752 (2016). https://doi.org/10.1016/j.ins.2015.09.042
24. Pinedo, M.L.: Scheduling. Theory, Algorithms, and Systems, 5th edn. Springer, Heidelberg (2016). https://doi.org/10.1007/978-3-319-26580-3
25. Roy, B.: Robustness in operational research and decision aiding: a multi-faceted issue. Eur. J. Oper. Res. **200**, 629–638 (2010)
26. Sotskov, Y.N., Egorova, N.G.: Single machine scheduling problem with interval processing times and total completion time objective. Algorithms **11**(5), 66 (2018)
27. Xu, Z., Yager, R.R.: Some geometric aggregation operators based on intuitionistic fuzzy sets. Int. J. Gen. Syst. **35**(4), 417–433 (2006)
28. Zhang, J., Ding, G., Zou, Y., Qin, S., Fu, J.: Review of job shop scheduling research and its new perspectives under Industry 4.0. J. Intell. Manuf. **30**(4), 1809–1830 (2017). https://doi.org/10.1007/s10845-017-1350-2

A Fuzzy Goal Programming Approach to Fully Fuzzy Linear Regression

Boris Pérez-Cañedo[1]($\boxtimes$) (iD), Alejandro Rosete[2] (iD), José Luis Verdegay[3] (iD), and Eduardo René Concepción-Morales[1] (iD)

[1] University of Cienfuegos, Cienfuegos 55100, Cuba
bpcanedo@gmail.com, econcepm@gmail.com
[2] Technological University of Havana, Havana, Cuba
rosete@ceis.cujae.edu.cu
[3] University of Granada, Granada, Spain
verdegay@ugr.es

Abstract. Traditional linear regression analysis aims at finding a linear functional relationship between predictor and response variables based on available data of a given system, and, when this relationship is found, it is used to predict the future behaviour of the system. The difference between the observed and predicted data is supposed to be due to measurement errors. In fuzzy linear regression, on the other hand, this difference is supposed to be mainly due to the indefiniteness of the system. In this paper, we assume that predictor and response variables are LR-type fuzzy numbers, and so are all regression coefficients; this is known as fully fuzzy linear regression (FFLR) problem. We transform the FFLR problem into a fully fuzzy multiobjective linear programming (FFMOLP) problem. Two fuzzy goal programming methods based on linear and Chebyshev scalarisations are proposed to solve the FFMOLP problem. The proposed methods are compared with a recently published method and show promising results.

Keywords: Fully fuzzy linear regression · Fully fuzzy multiobjective linear programming · Fuzzy goal programming · Linear scalarisation · Chebyshev scalarisation

1 Introduction and Preliminaries

Traditional linear regression is one of the most frequently applied technique for finding functional relationships between predictor and response variables, and for making predictions. However, decision problems arising in ever-changing environments are difficult to describe or formulate with precise terms. Expert knowledge then gains a special value, and the need for its introduction into classical decision-making techniques has motivated the appearance and development of several mathematical theories dealing with uncertainty and vagueness. Among those theories, Fuzzy Sets Theory [31] has succeeded in numerous practical situations and is now an established research field.

© Springer Nature Switzerland AG 2020
M.-J. Lesot et al. (Eds.): IPMU 2020, CCIS 1238, pp. 677–688, 2020.
https://doi.org/10.1007/978-3-030-50143-3_53

Fuzzy linear regression is a natural extension of the classical regression analysis and allows to predict the future behaviour of systems whose structure is not well defined and/or is influenced by subjectivity. It is particularly useful to forecast, e.g., future demands, resource availability and prices that could then be used to set up fuzzy optimisation problems in areas such as production, transportation, project management and so forth. Fuzzy linear regression has been used to forecast airport demand [21], oil consumption [1], house prices [32], sales [5] and short-term load in power distribution systems [24]. Several other applications are reported in [6].

Numerous fuzzy linear regression models and methods have been developed since the 1980s. Tanaka et al. [25] introduced fuzzy linear regression analysis and formulated a regression problem with crisp predictor variables, fuzzy response variable and fuzzy coefficients as a conventional linear programming problem. A modified version of Tanaka et al.'s [25] fuzzy linear regression method allowing negative spreads in the parameters was proposed in [3]. Chang and Lee [2] proposed fuzzy least square deviation and least absolute deviation models based on ranking functions. A multiobjective approach was proposed by Sakawa and Yano [23] by simultaneous consideration of the model fit and fuzziness. Recent methods for fuzzy linear regression have been presented in [14,18,22]. A comprehensive review until year 2019 is provided by Chukhrova and Johannssen [6].

So far, fuzzy linear regression methods mainly resort to the minimisation of crisp-valued distance functions between fuzzy numbers, either by direct generalisations of known crisp distance functions or by the use of linear ranking functions to defuzzify response observations and model predicted values, and then taking the absolute value of the difference as the distance between the two fuzzy numbers. A simulation study, considering distance functions from both approaches, was conducted in [14] to determine the best distance function in fuzzy linear regression using Monte Carlo methods. Notably, Voxman [26] has argued that the distance between two fuzzy numbers should also be a fuzzy number, and proposed a fuzzy-valued distance function. However, to the best of our knowledge, fuzzy-valued distance functions have not been used in fuzzy regression analysis.

In this paper, we seek to evaluate other models and methods for fuzzy linear regression analysis, which do not rely on crisp-valued distance functions. We propose two methods based on FFMOLP for fuzzy linear regression analysis, in which the predictor variables, response variable and regression coefficients are LR-type fuzzy numbers. The proposed methods rely on the lexicographic approach to fully fuzzy linear programming (FFLP) with inequality constraints recently proposed in [19]. The rest of the paper is organised as follows. Section 1.1 presents some fundamental definitions on LR-type fuzzy numbers. Section 1.2 outlines the lexicographic method [19] for solving FFLP problems. In Sect. 2, we formulate the FFLR problem as a FFMOLP problem, and propose two fuzzy scalarisation methods based on classical goal programming to solve it. Section 3 discusses a numerical example. Lastly, Sect. 4 presents the conclusions and remarks for future work.

1.1 LR-type Fuzzy Numbers

Dubois and Prade [8] defined the concept of LR-type fuzzy number and proposed simple formulae for arithmetic operations. In this section, we present some definitions concerning LR-type fuzzy numbers taken from reference [8].

Definition 1. *A fuzzy number $\tilde{a} = (m, \alpha, \beta)_{LR}$ is said to be an LR-type fuzzy number if its membership function is given by:*

$$\mu_{\tilde{a}}(x) = \begin{cases} L\left(\frac{m-x}{\alpha}\right) & m - \alpha \leq x \leq m, \ \alpha > 0 \\ R\left(\frac{x-n}{\beta}\right) & m \leq x \leq m + \beta, \ \beta > 0 \end{cases}$$

where m is the modal value; L and R (called left and right reference functions, respectively) are non-increasing functions $[0, +\infty) \rightarrow [0,1]$, fulfilling $L(0) = R(0) = 1$; α and β are the left and right spreads of $\tilde{a}$, respectively. The set of all LR-type fuzzy numbers, defined on $\Re$, is denoted by $\mathcal{F}(\Re)$.

Definition 2. *Let $\tilde{a}_1 = (m_1, \alpha_1, \beta_1)_{LR}$ and $\tilde{a}_2 = (m_2, \alpha_2, \beta_2)_{LR}$ be any LR-type fuzzy numbers, then $\tilde{a}_1 = \tilde{a}_2$ if and only if $m_1 = m_2$, $\alpha_1 = \alpha_2$ and $\beta_1 = \beta_2$.*

Definition 3. *An LR-type fuzzy number $\tilde{a} = (m, \alpha, \beta)_{LR}$ is said to be non-negative (resp. non-positive) if $m - \alpha \geq 0$ (resp. $m + \beta \leq 0$). This is denoted by $\tilde{a} \geq 0$ (resp. $\tilde{a} \leq 0$).*

Definition 4. *An LR-type fuzzy number $\tilde{a} = (m, \alpha, \beta)_{LR}$ is said to be unrestricted if m is an arbitrary real number.*

Definition 5. *Let $\tilde{a}_1 = (m_1, \alpha_1, \beta_1)_{LR}$ and $\tilde{a}_2 = (m_2, \alpha_2, \beta_2)_{LR}$ be two LR-type fuzzy numbers, then fuzzy addition is given by $\tilde{a}_1 \oplus \tilde{a}_2 = (m_1 + m_2, , \alpha_1 + \alpha_2, \beta_1 + \beta_2)_{LR}$.*

Definition 6. *The product of two non-negative LR-type fuzzy numbers $\tilde{a}_1 = (m_1, \alpha_1, \beta_1)_{LR}$ and $\tilde{a}_2 = (m_2, \alpha_2, \beta_2)_{LR}$ is given by $\tilde{a}_1 \odot \tilde{a}_2 = (m_1 m_2, m_1 \alpha_2 + \alpha_1 m_2 - \alpha_1 \alpha_2, n_1 \beta_2 + \beta_1 n_2 + \beta_1 \beta_2)_{LR}$.*

The reader is referred to [15] for the definition of the product of unrestricted LR-type fuzzy numbers.

1.2 FFLP Problem and Lexicographic Solution Method

Due to the vast number of practical situations where fuzzy quantities must be compared, ranking fuzzy numbers is still recognised as a fundamental research problem in Fuzzy Sets Theory. Many ranking methodologies have been proposed in the literature [29,30]. However, several researchers have noticed that most existing ranking methodologies cannot yield a total order of fuzzy numbers in a strict sense. To resolve this issue, lexicographic ranking criteria have been proposed as an alternative [11,27,28].

The integration of lexicographic ranking criteria into FFLP methods started with [12,13] and has been recently investigated in [7,9,10,16,19,20]. In particular, the use of lexicographic ranking criteria for handling fuzzy inequality constraints has been proposed in [12,19,20]. In this section, we present the lexicographic method [19] for solving FFLP problems with inequality constraints. This method constitutes the basis of the results presented in the following sections.

Firstly, we need to introduce an order relation on $\mathcal{F}(\Re)$. Let $\tilde{a} = (m, \alpha, \beta)_{LR}$ be an arbitrary LR-type fuzzy number, and suppose we have three linear functions of the parameters of $\tilde{a}$, $f_k(\tilde{a}) := w_{k1}m + w_{k2}\alpha + w_{k3}\beta$ for $k = 1, 2, 3$. If each w_{kr} is chosen such that matrix $[w_{kr}]$ is non-singular, then $\tilde{a}_1 = \tilde{a}_2$ if and only if $f_k(\tilde{a}_1) = f_k(\tilde{a}_2)$ for $k = 1, 2, 3$.

Based on the above idea, we may consider the following criterion for ranking LR-type fuzzy numbers.

Definition 7. *Let $\leq_{lex}$ denote the lexicographic order relation on $\Re^3$. For any $\tilde{a}_1, \tilde{a}_2 \in \mathcal{F}(\Re)$, the strict inequality $\tilde{a}_1 \prec \tilde{a}_2$ holds, if and only if $(f_k(\tilde{a}_1))_{k=1,2,3} <_{lex} (f_k(\tilde{a}_2))_{k=1,2,3}$. The weak inequality $\tilde{a}_1 \preceq \tilde{a}_2$ holds, if and only if $(f_k(\tilde{a}_1))_{k=1,2,3} <_{lex} (f_k(\tilde{a}_2))_{k=1,2,3}$ or $(f_k(\tilde{a}_1))_{k=1,2,3} = (f_k(\tilde{a}_2))_{k=1,2,3}$.*

It can be shown that $\preceq$ satisfies the total order properties. That is, for all $\tilde{a}$, $\tilde{b}$ and $\tilde{c}$ in $\mathcal{F}(\Re)$:

- $\tilde{a} \preceq \tilde{a}$ (reflexivity);
- $\tilde{a} \preceq \tilde{b}$ or $\tilde{b} \preceq \tilde{a}$ (comparability);
- if $\tilde{a} \preceq \tilde{b}$ and $\tilde{b} \preceq \tilde{c}$, then $\tilde{a} \preceq \tilde{c}$ (transitivity);
- if $\tilde{a} \preceq \tilde{b}$ and $\tilde{b} \preceq \tilde{a}$, then $\tilde{a} = \tilde{b}$ (anti-symmetry).

Next, we present the lexicographic method proposed in [19] for solving FFLP problems with equality and inequality constraints.

The FFLP problem can be formulated as follows, where $\tilde{c}_j$, $\tilde{a}_{ij}$ and $\tilde{b}_i$ are LR-type fuzzy parameters, $\tilde{x}_j$ denote the LR-type fuzzy decision variables, and $\preceq$ is an order relation on $\mathcal{F}(\Re)$; here, we assume that $\preceq$ is given by Definition 7.

$$\min \sum_{j=1}^{n} \tilde{c}_j \odot \tilde{x}_j$$

$$\text{s.t.} \sum_{j=1}^{n} \tilde{a}_{ij} \odot \tilde{x}_j \{\preceq, =, \succeq\} \tilde{b}_i; \ i = 1, 2, \ldots, m$$

$$\tilde{x}_j \in \mathcal{F}(\Re); \ j = 1, 2, \ldots, n \tag{1}$$

By using Definitions 2 and 7, FFLP problem (1) is transformed into problem (2), which is then transformed into problem (3). To carry out these transformations, we have assumed that $\tilde{z} = \sum_{j=1}^{n} \tilde{c}_j \odot \tilde{x}_j$, $\tilde{a}_i = (m_i^a, \alpha_i^a, \beta_i^a)_{LR} = \sum_{j=1}^{n} \tilde{a}_{ij} \odot \tilde{x}_j$, $\tilde{b}_i = (m_i^b, \alpha_i^b, \beta_i^b)_{LR}$ and $\tilde{x}_j = (m_j^x, \alpha_j^x, \beta_j^x)_{LR}$. In addition, I_e, I_{le} and I_{ge} denote the index sets of the fuzzy equality, less-than-or-equal-to and greater-than-or-equal-to constraints of FFLP problem (1), respectively; ϵ and M are positive real numbers sufficiently small and large, respectively.

$$\text{lexmin } (f_k(\tilde{z}))_{k=1,2,3}$$

$$\text{s.t. } (f_k(\tilde{a}_i))_{k=1,2,3} \{\leq_{lex}, \geq_{lex}\} \left(f_k\left(\tilde{b}_i\right)\right)_{k=1,2,3} ; i \in I_{le} \cup I_{ge} \qquad (2)$$

$$m_i^a = m_i^b, \alpha_i^a = \alpha_i^b, \beta_i^a = \beta_i^b; i \in I_e$$

$$\alpha_j^x \geq 0, \beta_j^x \geq 0; j = 1, 2, \ldots, n$$

$$\text{lexmin } (f_k(\tilde{z}))_{k=1,2,3}$$

$$\text{s.t. } -M \sum_{p=1}^{k-1} y_{ip} + \epsilon y_{ik} \leq f_k\left(\tilde{b}_i\right) - f_k(\tilde{a}_i) \leq M y_{ik}; i \in I_{le}, k = 1,2,3$$

$$-M \sum_{p=1}^{k-1} y_{ip} + \epsilon y_{ik} \leq f_k(\tilde{a}_i) - f_k\left(\tilde{b}_i\right) \leq M y_{ik}; i \in I_{ge}, k = 1,2,3$$

$$m_i^a = m_i^b, \alpha_i^a = \alpha_i^b, \beta_i^a = \beta_i^b; i \in I_e \qquad (3)$$

$$y_{ik} \in \{0,1\}; i \in I_{le} \cup I_{ge}, k = 1,2,3$$

$$\alpha_j^x \geq 0, \beta_j^x \geq 0; j = 1, 2, \ldots, n$$

Theorem 1. *FFLP problem* (1) *is equivalent to problem* (3).

Proof. See [19].   □

Remark 1. In order to solve FFLP problem (1), we must choose a lexicographic criterion for ranking LR-type fuzzy numbers. There are several such criteria in the literature (see, e.g., [11,27]). Notably, the solution method outlined here is general enough so as to allow a decision-maker to use the criterion that best fits the decision-making problem at hand.

2   FFLR: Proposed Methods

Let $\tilde{x}_j$, $\tilde{A}_j$ $(j = 0, 2, \ldots, n)$ and $\tilde{y}$ be LR-type fuzzy numbers. Then the FFLR model is formulated as in Eq. (4).

$$\tilde{y} = \tilde{A}_0 \oplus \tilde{A}_1 \odot \tilde{x}_1 \oplus \tilde{A}_2 \odot \tilde{x}_2 \oplus \cdots \oplus \tilde{A}_n \odot \tilde{x}_n = \tilde{A}_0 \oplus \sum_{j=1}^{n} \tilde{A}_j \odot \tilde{x}_j \qquad (4)$$

In Eq. (4), each $\tilde{x}_j$ is termed fuzzy predictor variable, $\tilde{y}$ fuzzy response variable and $\tilde{A}_j$ fuzzy regression coefficient. Now, let us consider a sample of LR-type fuzzy numbers $\left(\tilde{X}|\tilde{Y}\right)$, where $\tilde{X} = (\tilde{x}_{ij})_{\substack{i=1,2,\ldots,m \\ j=1,2,\ldots,n}}$ contains the observations corresponding to each fuzzy predictor variable $\tilde{x}_j$, and the column vector

$\tilde{Y} = (\tilde{y}_i)_{i=1,2,\ldots,m}$ contains the observations of the fuzzy response variable $\tilde{y}$. We wish to determine the estimates of $\tilde{A}_j$ so as to obtain the best fitting model given the available data.

In what follows, we formulate the FFLR problem as a FFMOLP problem. To this aim, we introduce two non-negative fuzzy deviation variables $\tilde{S}p_i$ and $\tilde{S}n_i$ for each sample. Thus, the following set of fuzzy equalities is obtained.

$$\tilde{A}_0 \oplus \sum_{j=1}^{n} \tilde{A}_j \odot \tilde{x}_{ij} \oplus \tilde{S}p_i = \tilde{y}_i \oplus \tilde{S}n_i; \ i = 1, 2, \ldots, m$$

Therefore, we may consider the following FFMOLP problem:

$$\min \left(\tilde{S}p_1 \oplus \tilde{S}n_1, \tilde{S}p_2 \oplus \tilde{S}n_2, \ldots, \tilde{S}p_m \oplus \tilde{S}n_m \right)$$

$$\text{s.t: } \tilde{A}_0 \oplus \sum_{j=1}^{n} \tilde{A}_j \odot \tilde{x}_{ij} \oplus \tilde{S}p_i = \tilde{y}_i \oplus \tilde{S}n_i; \ i = 1, 2, \ldots, m \qquad \text{(P1)}$$

$$\tilde{A}_j \text{ unrestricted}; \ j = 0, 1, \ldots, n$$

$$\tilde{S}p_i \geq 0, \ \tilde{S}n_i \geq 0; \ i = 1, 2, \ldots, m$$

In order to solve (P1), we resort to two known classical scalarisation methods based on goal programming, which are extended to the fuzzy case: *linear scalarisation method* and *Chebyshev (minimax) scalarisation method*.

2.1   Linear Scalarisation Method

In this method, each objective function is multiplied by a positive weighting factor and the resulting expressions are added together. Thus, we have,

$$\min \sum_{i=1}^{m} w_i \left(\tilde{S}p_i \oplus \tilde{S}n_i \right)$$

$$\text{s.t: } \tilde{A}_0 \oplus \sum_{j=1}^{n} \tilde{A}_j \odot \tilde{x}_{ij} \oplus \tilde{S}p_i = \tilde{y}_i \oplus \tilde{S}n_i; \ i = 1, 2, \ldots, m \qquad \text{(l-P1)}$$

$$\tilde{A}_j \text{ unrestricted}; \ j = 0, 1, \ldots, n$$

$$\tilde{S}p_i \geq 0, \ \tilde{S}n_i \geq 0; \ i = 1, 2, \ldots, m$$

Hereafter, $w_i = 1$ for $i = 1, 2, \ldots, m$ since no particular preference for the objective functions shall be considered.

2.2   Chebyshev Scalarisation Method

In this case the scalarising function is $\tilde{S} = \max \left(\tilde{S}p_1 \oplus \tilde{S}n_1, \tilde{S}p_2 \oplus \tilde{S}n_2, \ldots, \tilde{S}p_m \oplus \tilde{S}n_m \right)$; hence, we have that $\tilde{S}p_i \oplus \tilde{S}n_i \preceq \tilde{S}$ for $i = 1, 2, \ldots, m$. By substituting into (P1), we obtain,

$$\min \ \tilde{S}$$

$$\text{s.t: } \tilde{A}_0 \oplus \sum_{j=1}^{n} \tilde{A}_j \odot \tilde{x}_{ij} \oplus \tilde{S}p_i = \tilde{y}_i \oplus \tilde{S}n_i; \ i = 1, 2, \ldots, m$$

$$\tilde{S}p_i \oplus \tilde{S}n_i \preceq \tilde{S}; \ i = 1, 2, \ldots, m \qquad\qquad \text{(ch-P1)}$$

$$\tilde{A}_j \text{ unrestricted}; \ j = 0, 1, \ldots, n$$

$$\tilde{S} \geq 0, \ \tilde{S}p_i \geq 0, \ \tilde{S}n_i \geq 0; \ i = 1, 2, \ldots, m$$

thus, the objective is to minimise the maximal deviation.

2.3   Steps of the Proposed FFLR Methods

The whole procedure can be summarised in the following six steps.

1. Input: Sample data $\tilde{X}$ and $\tilde{Y}$;
2. choose a lexicographic criterion for ranking LR-type fuzzy numbers;
3. set up FFMOLP problem (P1);
4. choose either of the proposed scalarisation methods, and set up FFLP problem (l-P1) or FFLP problem (ch-P1);
5. solve the FFLP problem chosen in Step 4 by using the lexicographic method outlined in Sect. 1.2;
6. output: $\tilde{A}_j$ for $j = 0, 1, \ldots, n$ as the estimated regression parameters.

3   Numerical Example

The example in this section is taken from references [4, 17, 18]. The dataset contains 30 samples, each having four predictor variables and one response variable (see Table 1). It is a real-life dataset comprising triangular fuzzy numbers used to subjectively evaluate employee's performance according to work quality, inability to endure job stress, frequency of delays, and communication and coordination ability. As part of the solution procedure from Sect. 2.3, the functions $f_1(\tilde{a}) := 3m + \beta - \alpha$, $f_2(\tilde{a}) := m + \beta$ and $f_3(\tilde{a}) := \alpha + \beta$ were used to define a lexicographic order relation on $\mathcal{F}(\Re)$.

We applied the proposed methods and obtained the following two models. The estimated responses of both models are shown in Table 2.

$$\tilde{y}_{\text{linear}} = (0.85684, 0, 0)_{LR} \odot \tilde{x}_1 \oplus (-0.12989, 0, 0)_{LR} \odot \tilde{x}_2$$
$$\oplus (-0.17893, 0.01697, 0)_{LR} \odot \tilde{x}_3 \oplus (0.04995, 0, 0)_{LR} \odot \tilde{x}_4$$
$$\oplus (12.60258, 0.14101, 0)_{LR}$$

$$\tilde{y}_{\text{chebyshev}} = (0.85662, 0, 0)_{LR} \odot \tilde{x}_1 \oplus (-0.15074, 0.03749, 0)_{LR} \odot \tilde{x}_2$$
$$\oplus (-0.14948, 0, 0)_{LR} \odot \tilde{x}_3 \oplus (0.09081, 0, 0)_{LR} \odot \tilde{x}_4$$
$$\oplus (10.03612, 0, 0)_{LR}$$

Table 1. Sample dataset.

$\tilde{x}_1$	$\tilde{x}_2$	$\tilde{x}_3$	$\tilde{x}_4$	$\tilde{y}$
$(50,8,8)_{LR}$	$(98,6,2)_{LR}$	$(71,9,11)_{LR}$	$(70,11,13)_{LR}$	$(30,11,9)_{LR}$
$(29,8,8)_{LR}$	$(76,6,2)_{LR}$	$(61,9,11)_{LR}$	$(46,11,13)_{LR}$	$(20,13,10)_{LR}$
$(41,8,8)_{LR}$	$(88,6,2)_{LR}$	$(73,9,11)_{LR}$	$(58,11,13)_{LR}$	$(25,11,12)_{LR}$
$(60,9,7)_{LR}$	$(62,9,10)_{LR}$	$(79,9,6)_{LR}$	$(66,8,9)_{LR}$	$(45,12,10)_{LR}$
$(49,9,7)_{LR}$	$(50,9,10)_{LR}$	$(75,9,6)_{LR}$	$(54,8,9)_{LR}$	$(38,12,8)_{LR}$
$(59,9,7)_{LR}$	$(60,9,10)_{LR}$	$(85,9,6)_{LR}$	$(64,8,9)_{LR}$	$(43,11,9)_{LR}$
$(61,9,11)_{LR}$	$(77,8,6)_{LR}$	$(85,5,8)_{LR}$	$(18,7,13)_{LR}$	$(40,17,11)_{LR}$
$(58,9,11)_{LR}$	$(75,8,6)_{LR}$	$(82,5,8)_{LR}$	$(16,7,13)_{LR}$	$(38,11,12)_{LR}$
$(55,9,11)_{LR}$	$(72,8,6)_{LR}$	$(79,5,8)_{LR}$	$(13,7,13)_{LR}$	$(37,12,12)_{LR}$
$(66,8,7)_{LR}$	$(59,17,11)_{LR}$	$(39,8,9)_{LR}$	$(83,14,11)_{LR}$	$(60,11,12)_{LR}$
$(69,8,7)_{LR}$	$(63,17,11)_{LR}$	$(49,8,9)_{LR}$	$(87,14,11)_{LR}$	$(59,10,9)_{LR}$
$(59,8,7)_{LR}$	$(53,17,11)_{LR}$	$(39,8,9)_{LR}$	$(77,14,11)_{LR}$	$(54,11,8)_{LR}$
$(74,4,6)_{LR}$	$(89,11,5)_{LR}$	$(70,12,13)_{LR}$	$(82,14,10)_{LR}$	$(61,14,3)_{LR}$
$(41,4,6)_{LR}$	$(57,11,5)_{LR}$	$(58,12,13)_{LR}$	$(50,14,10)_{LR}$	$(34,10,8)_{LR}$
$(49,4,6)_{LR}$	$(65,11,5)_{LR}$	$(66,12,13)_{LR}$	$(58,14,10)_{LR}$	$(38,9,9)_{LR}$
$(76,8,7)_{LR}$	$(75,10,8)_{LR}$	$(37,8,11)_{LR}$	$(75,5,10)_{LR}$	$(64,16,9)_{LR}$
$(57,8,7)_{LR}$	$(56,10,8)_{LR}$	$(18,8,11)_{LR}$	$(56,5,10)_{LR}$	$(56,13,7)_{LR}$
$(72,8,7)_{LR}$	$(71,10,8)_{LR}$	$(33,8,11)_{LR}$	$(71,5,10)_{LR}$	$(63,11,9)_{LR}$
$(78,7,8)_{LR}$	$(65,6,6)_{LR}$	$(82,11,11)_{LR}$	$(64,8,12)_{LR}$	$(66,16,5)_{LR}$
$(58,7,8)_{LR}$	$(45,6,6)_{LR}$	$(62,11,11)_{LR}$	$(44,8,12)_{LR}$	$(49,12,9)_{LR}$
$(72,7,8)_{LR}$	$(59,6,6)_{LR}$	$(76,11,11)_{LR}$	$(58,8,12)_{LR}$	$(55,10,12)_{LR}$
$(90,8,5)_{LR}$	$(95,13,3)_{LR}$	$(80,11,8)_{LR}$	$(72,7,13)_{LR}$	$(67,11,14)_{LR}$
$(68,8,5)_{LR}$	$(73,13,3)_{LR}$	$(58,11,8)_{LR}$	$(50,7,13)_{LR}$	$(53,10,9)_{LR}$
$(71,8,5)_{LR}$	$(76,13,3)_{LR}$	$(61,11,8)_{LR}$	$(53,7,13)_{LR}$	$(54,9,10)_{LR}$
$(92,8,6)_{LR}$	$(76,6,9)_{LR}$	$(78,10,6)_{LR}$	$(27,9,15)_{LR}$	$(70,13,7)_{LR}$
$(94,8,6)_{LR}$	$(78,6,9)_{LR}$	$(80,10,6)_{LR}$	$(29,9,15)_{LR}$	$(68,9,10)_{LR}$
$(87,8,6)_{LR}$	$(71,6,9)_{LR}$	$(73,10,6)_{LR}$	$(22,9,15)_{LR}$	$(65,10,9)_{LR}$
$(94,6,5)_{LR}$	$(51,9,8)_{LR}$	$(30,9,11)_{LR}$	$(29,9,16)_{LR}$	$(75,5,14)_{LR}$
$(95,6,5)_{LR}$	$(52,9,8)_{LR}$	$(31,9,11)_{LR}$	$(30,9,16)_{LR}$	$(84,10,7)_{LR}$
$(86,6,5)_{LR}$	$(43,9,8)_{LR}$	$(22,9,11)_{LR}$	$(21,9,16)_{LR}$	$(80,12,6)_{LR}$

The obtained models' predicted values were compared with the ones reported by Li et al. [18], according to the overall absolute distance from the observed responses, using Eqs. (5) and (6).

$$d\left(\tilde{y}_i^{obs}, \tilde{y}_i^{pred}\right) = \left|m_i^{obs} - m_i^{pred}\right| + \left|m_i^{obs} - \alpha_i^{obs} - \left(m_i^{pred} - \alpha_i^{pred}\right)\right|$$
$$+ \left|m_i^{obs} + \beta_i^{obs} - \left(m_i^{pred} + \beta_i^{pred}\right)\right| \tag{5}$$

Table 2. Estimated values reported by Li et al. [18] and the ones obtained by using the proposed methods. The observed values have been kept to aid visual comparison.

Observed response	Li et al. [18]	Chebyshev Scalarisation	Linear Scalarisation
$(30, 11, 9)_{LR}$	$(32.053, 10.999, 9.352)_{LR}$	$(33.837, 13.547, 10.283)_{LR}$	$(33.508, 11.166, 9.894)_{LR}$
$(20, 13, 10)_{LR}$	$(17.915, 10.916, 9.352)_{LR}$	$(18.480, 12.722, 10.283)_{LR}$	$(18.962, 10.996, 9.894)_{LR}$
$(25, 11, 12)_{LR}$	$(25.053, 10.962, 9.352)_{LR}$	$(26.247, 13.172, 10.283)_{LR}$	$(26.138, 11.199, 9.894)_{LR}$
$(45, 12, 10)_{LR}$	$(45.501, 12.046, 8.795)_{LR}$	$(46.271, 13.540, 9.516)_{LR}$	$(45.121, 12.068, 9.227)_{LR}$
$(38, 12, 8)_{LR}$	$(37.998, 11.999, 8.795)_{LR}$	$(38.166, 13.090, 9.516)_{LR}$	$(37.371, 12.000, 9.227)_{LR}$
$(43, 11, 9)_{LR}$	$(43.947, 12.038, 8.795)_{LR}$	$(44.638, 13.465, 9.516)_{LR}$	$(43.350, 12.170, 9.227)_{LR}$
$(40, 17, 11)_{LR}$	$(39.767, 11.304, 12.074)_{LR}$	$(39.611, 13.558, 12.557)_{LR}$	$(40.558, 11.992, 12.009)_{LR}$
$(38, 11, 12)_{LR}$	$(37.874, 11.296, 12.074)_{LR}$	$(37.609, 13.483, 12.557)_{LR}$	$(38.684, 11.941, 12.009)_{LR}$
$(37, 12, 12)_{LR}$	$(36.090, 11.285, 12.074)_{LR}$	$(35.668, 13.370, 12.557)_{LR}$	$(36.890, 11.890, 12.009)_{LR}$
$(60, 11, 12)_{LR}$	$(59.973, 12.457, 9.485)_{LR}$	$(59.387, 13.753, 10.754)_{LR}$	$(58.658, 11.549, 10.187)_{LR}$
$(59, 10, 9)_{LR}$	$(60.580, 12.472, 9.485)_{LR}$	$(60.222, 13.903, 10.754)_{LR}$	$(59.120, 11.719, 10.187)_{LR}$
$(54, 11, 8)_{LR}$	$(54.633, 12.434, 9.485)_{LR}$	$(53.750, 13.528, 10.754)_{LR}$	$(53.140, 11.549, 10.187)_{LR}$
$(61, 14, 3)_{LR}$	$(56.998, 8.863, 8.283)_{LR}$	$(56.992, 10.919, 9.500)_{LR}$	$(56.019, 8.653, 9.217)_{LR}$
$(34, 10, 8)_{LR}$	$(34.060, 8.741, 8.283)_{LR}$	$(32.435, 9.720, 9.500)_{LR}$	$(32.449, 8.449, 9.217)_{LR}$
$(38, 9, 9)_{LR}$	$(38.818, 8.772, 8.283)_{LR}$	$(37.613, 10.020, 9.500)_{LR}$	$(37.233, 8.585, 9.217)_{LR}$
$(64, 16, 9)_{LR}$	$(64.026, 11.115, 9.292)_{LR}$	$(65.113, 13.269, 9.608)_{LR}$	$(65.107, 11.068, 9.228)_{LR}$
$(56, 13, 7)_{LR}$	$(52.726, 11.043, 9.292)_{LR}$	$(52.817, 12.557, 9.608)_{LR}$	$(53.745, 10.745, 9.228)_{LR}$
$(63, 11, 9)_{LR}$	$(61.647, 11.100, 9.292)_{LR}$	$(62.525, 13.119, 9.608)_{LR}$	$(62.715, 11.000, 9.228)_{LR}$
$(66, 16, 5)_{LR}$	$(59.633, 10.203, 9.923)_{LR}$	$(60.608, 11.934, 10.492)_{LR}$	$(59.518, 10.865, 10.202)_{LR}$
$(49, 12, 9)_{LR}$	$(47.738, 10.127, 9.923)_{LR}$	$(47.664, 11.184, 10.492)_{LR}$	$(47.558, 10.526, 10.202)_{LR}$
$(55, 10, 12)_{LR}$	$(56.064, 10.180, 9.923)_{LR}$	$(56.725, 11.709, 10.492)_{LR}$	$(55.930, 10.763, 10.202)_{LR}$
$(67, 11, 14)_{LR}$	$(68.175, 12.176, 6.479)_{LR}$	$(67.391, 12.811, 9.068)_{LR}$	$(66.661, 10.661, 8.591)_{LR}$
$(53, 10, 9)_{LR}$	$(55.091, 12.092, 6.479)_{LR}$	$(53.152, 11.986, 9.068)_{LR}$	$(53.505, 10.287, 8.591)_{LR}$
$(54, 9, 10)_{LR}$	$(56.877, 12.104, 6.479)_{LR}$	$(55.094, 12.099, 9.068)_{LR}$	$(55.299, 10.338, 8.591)_{LR}$
$(70, 13, 7)_{LR}$	$(68.807, 10.989, 8.026)_{LR}$	$(68.181, 13.111, 8.901)_{LR}$	$(68.952, 11.114, 8.459)_{LR}$
$(68, 9, 10)_{LR}$	$(69.995, 10.996, 8.026)_{LR}$	$(69.475, 13.186, 8.901)_{LR}$	$(70.148, 11.148, 8.459)_{LR}$
$(65, 10, 9)_{LR}$	$(65.834, 10.970, 8.026)_{LR}$	$(64.945, 12.923, 8.901)_{LR}$	$(65.962, 11.029, 8.459)_{LR}$
$(75, 5, 14)_{LR}$	$(82.242, 9.475, 7.819)_{LR}$	$(81.020, 11.020, 8.438)_{LR}$	$(82.602, 9.435, 7.863)_{LR}$
$(84, 10, 7)_{LR}$	$(82.837, 9.479, 7.819)_{LR}$	$(81.667, 11.057, 8.438)_{LR}$	$(83.200, 9.452, 7.863)_{LR}$
$(80, 12, 6)_{LR}$	$(77.484, 9.445, 7.819)_{LR}$	$(75.842, 10.720, 8.438)_{LR}$	$(77.818, 9.299, 7.863)_{LR}$
Overall distance	113.448	142.3720	**109.9628**

$$d_T = \sum_{i=1}^{m} d\left(\tilde{y}_i^{obs}, \tilde{y}_i^{pred}\right) \tag{6}$$

In order to compare the predicted values, first Li et al.'s [18] solution is converted to LR representation of fuzzy numbers, since the authors used a different representation. From the last row of Table 2, it can be seen that the model obtained by the Linear Scalarisation Method (FFLP problem (l-P1)) has the smallest overall distance value, followed by Li et al.'s [18] model and the model obtained by using the proposed Chebyshev Scalarisation Method.

4   Concluding Remarks

In this paper, we proposed two methods for FFLR analysis. Contrary to existing methodologies that use crisp-valued distance functions, we formulated the FFLR problem as a FFMOLP problem. Fuzzy linear and Chebyshev scalarisations were proposed to solve the FFMOLP problem using a lexicographic method for FFLP.

The proposed methods were compared with a recently published method and showed promising results. In a future work, we plan to conduct an extensive simulation study and consider real-world applications to gain more insights into the performance of the proposed methods. In addition, the use of fuzzy-valued distance functions for FFLR analysis will be investigated.

Acknowledgements. The research of José Luis Verdegay is supported in part by project TIN2017-86647-P (Spanish Ministry of Economy and Competitiveness and FEDER funds from the European Union).

References

1. Azadeh, A., Khakestani, M., Saberi, M.: A flexible fuzzy regression algorithm for forecasting oil consumption estimation. Energy Policy **37**(12) (2009). https://doi.org/10.1016/j.enpol.2009.08.017
2. Chang, P.T., Lee, E.S.: Fuzzy least absolute deviations regression based on the ranking of fuzzy numbers. In: Proceedings of 1994 IEEE 3rd International Fuzzy Systems Conference, pp. 1365–1369. IEEE (1994). https://doi.org/10.1109/FUZZY.1994.343613
3. Chang, P.T., Lee, E.S.: Fuzzy linear regression with spreads unrestricted in sign. Comput. Math. Appl. **28**(4), 61–70 (1994). https://doi.org/10.1016/0898-1221(94)00127-8
4. Chen, L.h., Hsueh, C.C.: Fuzzy regression models using the least-squares method based on the concept of distance. IEEE Trans. Fuzzy Syst. **17**(6), 1259–1272 (2009). https://doi.org/10.1109/TFUZZ.2009.2026891
5. Chen, T., Wang, M.J.J.: Forecasting methods using fuzzy concepts. Fuzzy Sets Syst. **105**(3), 339–352 (1999). https://doi.org/10.1016/S0165-0114(97)00265-0
6. Chukhrova, N., Johannssen, A.: Fuzzy regression analysis: systematic review and bibliography. Appl. Soft Comput. J. **84**, 105708 (2019). https://doi.org/10.1016/j.asoc.2019.105708
7. Das, S.K., Mandal, T., Edalatpanah, S.A.: A mathematical model for solving fully fuzzy linear programming problem with trapezoidal fuzzy numbers. Appl. Intell. 1–11 (2016). https://doi.org/10.1007/s10489-016-0779-x
8. Dubois, D., Prade, H.: Operations on fuzzy numbers. Int. J. Syst. Sci. **9**(6), 613–626 (1978). https://doi.org/10.1080/00207727808941724
9. Ebrahimnejad, A.: An effective computational attempt for solving fully fuzzy linear programming using MOLP problem. J. Ind. Prod. Eng. **36**(2), 59–69 (2019). https://doi.org/10.1080/21681015.2019.1585391
10. Ezzati, R., Khorram, E., Enayati, R.: A new algorithm to solve fully fuzzy linear programming problems using the MOLP problem. Appl. Math. Model. **39**(12), 3183–3193 (2015). https://doi.org/10.1016/j.apm.2013.03.014

11. Farhadinia, B.: Ranking fuzzy numbers based on lexicographical ordering. Int. J. Appl. Math. Comput. Sci. **5**(4), 248–251 (2009)
12. Hashemi, S.M., Modarres, M., Nasrabadi, E., Nasrabadi, M.M.: Fully fuzzified linear programming, solution and duality. J. Intell. Fuzzy Syst. **17**(1), 253–261 (2006)
13. Hosseinzadeh Lotfi, F., Allahviranloo, T., Alimardani Jondabeh, M., Alizadeh, L.: Solving a full fuzzy linear programming using lexicography method and fuzzy approximate solution. Appl. Math. Model. **33**, 3151–3156 (2009). https://doi.org/10.1016/j.apm.2008.10.020
14. İçen, D., Cattaneo, M.E.G.V.: Different distance measures for fuzzy linear regression with Monte Carlo methods. Soft Comput. **21**(22), 6687–6697 (2016). https://doi.org/10.1007/s00500-016-2218-7
15. Kaur, J., Kumar, A.: Mehar's method for solving fully fuzzy linear programming problems with L-R fuzzy parameters. Appl. Math. Model. **37**(12–13), 7142–7153 (2013). https://doi.org/10.1016/j.apm.2013.01.040
16. Kaur, J., Kumar, A.: A new method to find the unique fuzzy optimal value of fuzzy linear programming problems. J. Optim. Theory Appl. **156**, 529–534 (2013). https://doi.org/10.1007/s10957-012-0132-4
17. Kelkinnama, M., Taheri, S.M.: Fuzzy least-absolutes regression using shape preserving operations. Inf. Sci. **214**, 105–120 (2012). https://doi.org/10.1016/j.ins.2012.04.017
18. Li, J., Zeng, W., Xie, J., Yin, Q.: A new fuzzy regression model based on least absolute deviation. Eng. Appl. Artif. Intell. **52**, 54–64 (2016). https://doi.org/10.1016/j.engappai.2016.02.009
19. Pérez-Cañedo, B., Concepción-Morales, E.R.: A method to find the unique optimal fuzzy value of fully fuzzy linear programming problems with inequality constraints having unrestricted L-R fuzzy parameters and decision variables. Expert Syst. Appl. **123**, 256–269 (2019). https://doi.org/10.1016/j.eswa.2019.01.041
20. Pérez-Cañedo, B., Verdegay, J.L., Miranda Pérez, R.: An epsilon-constraint method for fully fuzzy multiobjective linear programming. Int. J. Intell. Syst. **35**(4), 600–624 (2020). https://doi.org/10.1002/int.22219
21. Profillidis, V.: Econometric and fuzzy models for the forecast of demand in the airport of Rhodes. J. Air Transp. Manag. **6**(2), 95–100 (2000). https://doi.org/10.1016/S0969-6997(99)00026-5
22. Rabiei, M.R., Arghami, N.R., Taheri, S.M., Gildeh, B.S.: Least-squares approach to regression modeling in full interval-valued fuzzy environment. Soft Comput. **18**(10), 2043–2059 (2013). https://doi.org/10.1007/s00500-013-1185-5
23. Sakawa, M., Yano, H.: Multiobjective fuzzy linear regression analysis and its application. Electron. Commun. Jpn. (Part III: Fundam. Electron. Sci.) **73**(12), 1–10 (1990). https://doi.org/10.1002/ecjc.4430731201
24. Song, K.B., Baek, Y.S., Hong, D., Jang, G.: Short-term load forecasting for the holidays using fuzzy linear regression method. IEEE Trans. Power Syst. **20**(1), 96–101 (2005). https://doi.org/10.1109/TPWRS.2004.835632
25. Tanaka, H., Uejima, S., Asai, K.: Linear regression analysis with fuzzy model. IEEE Trans. Syst. Man Cybern. **12**(6), 903–907 (1982). https://doi.org/10.1109/TSMC.1982.4308925
26. Voxman, W.: Some remarks on distances between fuzzy numbers. Fuzzy Sets Syst. **100**(1–3), 353–365 (1998). https://doi.org/10.1016/S0165-0114(97)00090-0
27. Wang, M.L., Wang, H.F., Chih-Lung, L.: Ranking fuzzy number based on lexicographic screening procedure. Int. J. Inf. Technol. Decis. Making **4**(4), 663–678 (2005). https://doi.org/10.1142/S0219622005001696

28. Wang, W., Wang, Z.: Total orderings defined on the set of all fuzzy numbers. Fuzzy Sets Syst. **243**, 131–141 (2014). https://doi.org/10.1016/j.fss.2013.09.005
29. Wang, X., Kerre, E.E.: Reasonable properties for the ordering of fuzzy quantities (I). Fuzzy Sets Syst. **118**(3), 375–385 (2001). https://doi.org/10.1016/S0165-0114(99)00062-7
30. Wang, X., Kerre, E.E.: Reasonable properties for the ordering of fuzzy quantities (II). Fuzzy Sets Syst. **118**(3), 387–405 (2001). https://doi.org/10.1016/S0165-0114(99)00063-9
31. Zadeh, L.: Fuzzy sets. Inf. Control **8**(3), 338–353 (1965). https://doi.org/10.1016/S0019-9958(65)90241-X
32. Zhou, J., Zhang, H., Gu, Y., Pantelous, A.A.: Affordable levels of house prices using fuzzy linear regression analysis: the case of Shanghai. Soft Comput. **22**(16), 5407–5418 (2018). https://doi.org/10.1007/s00500-018-3090-4

Planning Wi-Fi Access Points Activation in Havana City: A Proposal and Preliminary Results

Cynthia Porras[1], Jenny Fajardo[2], Alejandro Rosete[1], and David A. Pelta[3(✉)]

[1] Universidad Tecnológica de la Habana "José Antonio Echeverría" (CUJAE),
La Habana, Cuba
{cporrasn,rosete}@ceis.cujae.edu.cu
[2] University of Deusto, Bilbao, Spain
fajardo.jenny@deusto.es
[3] Universidad de Granada, Granada, Spain
dpelta@decsai.ugr.es

Abstract. The availability of Wi-Fi connection points or hotspots in places such as parks, transport stations, libraries, and so on is one of the key aspects to allow people the usage of Internet resources (to study, work or meet). This is even more important in Central America and Caribbean countries where the deployment of huge cost infrastructure (like optical fiber) to provide Internet access at home is not envisaged neither in the short or mid term. And this is clearly the case in Havana, Cuba.

This contribution presents the problem of planning the Wi-Fi access points activation, where each point can have different signal power levels and availability along the time. Due to power consumption constraints, it is impossible to have all the points activated simultaneously with maximum signal strength.

The problem is modelled as a dynamic maximal covering location one with facility types and time dependant availability. A metaheuristic approach is used to solve the problem by using an Algorithm portfolio and examples on how solutions can be analyzed (beyond the coverage provided) are shown.

Keywords: Signal levels · Wi-Fi access points

1 Introduction

In a recent report from the International Telecommunication Union (ITU) (United Nations specialized agency for information and communication technologies) [8], one can read:

... Internet use continues to grow globally, with 4.1 billion people now using the Internet, or 53.6% of the global population. However, an estimated 3.6

© Springer Nature Switzerland AG 2020
M.-J. Lesot et al. (Eds.): IPMU 2020, CCIS 1238, pp. 689–698, 2020.
https://doi.org/10.1007/978-3-030-50143-3_54

billion people remain offline, with the majority of the unconnected living in the Least Developed Countries where an average of just two out of every ten people are online.

In all regions of the world, households are more likely to have Internet access at home than to have a computer because Internet access is also possible through other devices. While 93% of the world's population lives within reach of a mobile broadband (or Internet) service, just over 53% actually uses the Internet.

According to the World Bank' DataBank (an analysis and visualisation tool that contains collections of time series data on a variety of topics)[1] Cuba has the following figures:

	2010	2011	2012	2013	2014	2015	2016	2017	2018
Individuals using the Internet (% of population)	15,90	16,02	21,20	27,93	29,07	37,31	42,98	57,15	...
Fixed broadband subscriptions (per 100 people)	0,03	0,04	0,04	0,05	0,07	0,07	0,13	0,30	0,87
Mobile cellular subscriptions (per 100 people)	8,93	11,70	14,94	17,69	22,38	29,45	35,18	40,69	47,39

Although Internet usage and mobile cellular subscriptions are steadily increasing, they are still quite low. Also in the rest of Central America and Caribbean countries the deployment of huge cost infrastructure (like optical fiber) to provide Internet access at home is not envisaged neither in the short or mid term. So deploying wireless connection is one of the steps needed to enlarge Internet usage among the people.

These Wi-Fi access points (WAPs), provide signal to the users if they are within its radius of signal. That means, the location of the WAP is very important for the satisfactory access to Internet. With this approach, the questions of where to locate the WAPs and how to plan their activation/deactivation can be treated as special cases of location problems.

A classic location problem is the maximal covering location problem (MCLP) proposed by Church and ReVelle in [3]. The MCLP aims to locate a limited number of facilities in order to maximize the coverage over a set of demand nodes. The term "facility" should be understood in a wide sense, ranging from warehouses, bus stops or, as in our case, WAPs.

Several examples of the application of MCLP for placing wireless devices may be found in the literature. In [12], one the first contributions in telecommunications scenarios was proposed. A greedy-add method was used over the MCLP and the set covering location problem (SCLP) to design a network of cellular mobile communication system.

---

[1] https://databank.worldbank.org/.

In [9] the MCLP was used to locate wireless routers, taking into account a central tower with a main signal. The routers share its signal, therefore they must be located at a certain distance from each other to maintain the signal strength. In [11] a variant of MCLP was proposed to locate routers and gateways. The model has capacity constraint to guarantee the correct flow on the connections. In [1] a multi-objective MCLP was proposed to locate the nodes in a wireless network with bandwidth capacity constraints. The objectives were: to maximize the coverage, to minimize the bandwidth usage, and to minimize the cost of the network and the interference signal. The NSGA-II algorithm was used to obtain the Pareto-front of the problem. In [2] an extension of MCLP was used to locate a set of access points in a wireless network in order to maximize the total offloaded traffic in an area. A greedy adding algorithm with substitution was used to solve the problem. In [7] a fuzzy MCLP was proposed to locate Wi-Fi antennas in Havana City. The fuzzy constraint describe uncertain information about the availability of antennas due to meteorological events near the coast. A parametric approach was used to solve the fuzzy problem. In [4] a variant of DMCLP to locate flying base stations was proposed. These bases move according to the change in population density over time. The bases have user connectivity capacity constraints.

Let's now consider the following situation. Havana City is exploring ways to provide Internet access to their citizens. There is information about the population density and how they may move. During a day the number of users requiring an Internet connection can change. For example, in the early morning there are less users than the afternoon. Besides, there is a set of deployed WAPs that provide a wireless signal, which provide coverage up to a maximal distance, named the coverage radius. A WAP has different signal levels or strength (e.g. strong, medium and low), that affects the coverage radius and just one level can be activated.

Due to energy saving policies, it is impossible to activate all of the available WAPs with a strong signal level (which would be the obvious solution to the problem). For example, during the early morning all signal levels may be available, since it is when the overall power consumption in the city is lower. However, at night, only the lowest signal level can be used, since it is where the highest energy consumption is perceived. The previous situation can be modelled using the location problem presented in [10]: the dynamic maximal covering location problem with facilities types and time dependant availability (DMCLP-FT).

In this contribution we apply the DMCLP-FT to manage the WAPs activation in Havana, Cuba. The paper is organized as follows: Sect. 2 describes the DMCLP-FT. Section 3 presents the DMCLP-FT to activate a set of WAPs in Havana City, Cuba and comments the principals results and analysis. Finally, we present the conclusions of our contribution.

2   Dynamic Maximal Covering Location Problem with Facility Types and Time Dependant Availability

Several features should be considered when solving the problem but we highlight two of them: 1) WAPs may not have available all the different signal levels and 2) the demand at a geographic area may change over the time. For example, the number of users can decreased or increased depending on the schedule, in the morning there more users on-line than the early morning.

These features can be modelled using a generalization of DMCLP presented in [10]: the dynamic maximal covering location problem with facility types and time dependent availability (DMCLP-FT). The mathematical formulation of the DMCLP-FT, applied to the WAPs activation, is the follows:

- i, I: the index and set of user nodes.
- j, J: the index and set of WAPs.
- t, T: the index and number of time periods.
- k, K: index and number of signal levels.
- a_{it}: population or number of users at node i in period t.
- d_{ij}: the shortest distance (or time) from demand node i to Wi-Fi access j.
- p_k: number of WAPs to be activated with level k.
- S_k: coverage radius provided by a signal level k. It is the minimum distance (or time) between a node and WAP to be considered as covered.
- $W_{jtk} \in \{0;1\}$: a binary variable, 1 if the WAP j has available the level k in period t, 0 otherwise.
- $N_{itk} = \{j | d_{ij} \leq S_k \text{ and } W_{jtk} = 1\}$: set of potential WAPs that can cover the users at node i if a WAP with level k is available in the location j in period t.
- $X_{jtk} \in \{0;1\}$: a binary variable, 1 if a WAP at location j is activated with level k in period t, 0 otherwise.
- $Y_{it} \in \{0;1\}$: a binary variable, 1 if the node i is covered by one or more WAPs in period t, 0 otherwise.

The objective function is:

$$Maximize : Z = \sum_{t \in T} \sum_{i \in I} a_{it} Y_{it} \tag{1}$$

Subject to:

$$Y_{it} \leq \sum_{k \in K} \sum_{j \in N_{itk}} X_{jtk}, \forall i \in I, \forall t \in T \tag{2}$$

$$\sum_{t \in T} \sum_{j \in J} X_{jtk} = p_k, \forall k \in K \tag{3}$$

$$\sum_{k \in K} p_k < J * T \tag{4}$$

$$\sum_{t \in T} \sum_{j \in J} W_{jtk} > p_k, \forall k \in K \tag{5}$$

$$\sum_{t \in T} \sum_{k \in K} W_{jtk} \geq 1, \forall j \in J \tag{6}$$

$$\sum_{k \in K} X_{jtk} \leq 1, \forall j \in J, \forall t \in T \tag{7}$$

$$\sum_{k \in K} X_{jtk} \leq W_{jtk}, \forall j \in J, \forall t \in T \tag{8}$$

The objective function (1) is aimed at maximizing the coverage of the nodes (sets of users). Constraint (2) shows that a node can be covered if an activated WAP with level k in period t belong to set N_{itk}. Constraint (3) shows that the number of WAPs with level k to activate must be equal to p_k. Constraint (4) shows that the total of WAPs with level k that will be activated should be less than the total of available WAPs.

Constraint (5) shows that the availability of WAP with level k must be greater than the number of WAPs with level k that will be activated in all time periods. Constraint (6) shows that a WAP must have at least one available signal level in some period t. Constraint (7) shows that a WAP j can be activated with only one level k. Finally, constraint (8) shows that a WAP can be activated with an available level k.

3   The Case in Havana City, Cuba

In this section we describe the problem of planning the WAPs activation for a whole day in Havana City, Cuba. The WAPs can adjust their signal levels (signal power), thus affecting the covered area. The stronger the signal level is, the higher the covered area (the higher the number of users served). Due to energy saving policies in Cuba, it is not possible to have the already deployed WAPs working simultaneously with a strong signal. Thus, in short, the problem is to determine which WAP (and when) should be activated and with which signal level.

There are several factors that may influence the availability of a given signal level in a schedule. Firstly, the consumption of electrical energy. At night, the energy consumption is higher and therefore, it is not possible to use strong signal levels, since it consumes more energy.

Secondly, geographical features during working hours (morning and afternoon). During these schedules of the day, activities are carried out that need WAPs with a certain signal level. For example, cultural or economic activities, where the signal levels suitable for a WAP are known.

Finally, at early morning, the energy consumption is low, so there are no restrictions on the signal levels.

As the user may notice, within a day the availability of the signal levels per access point would change, depending on the access point location.

In this case study we aim at planning the activation just for one day divided in four stages.

3.1   Problem Information

Part of the data available for the problem is taken from [7]. We include $T = 4$ time periods that represent times of a day: $t = 1$ is early morning, $t = 2$ is morning, $t = 3$ is afternoon and $t = 4$ is night. There are $I = 956$ areas or user nodes and for each node i and time period t there is a demand estimation or number of users to be covered a_{it}. At period $t = 1$, the values a_{i1} are taken from the data set used in [7]. Then, the demand a_{it} for $t = 2, 3, 4$ was randomly generated using values between $[MIN(a_{i1}) - 1; MAX(a_{i1}) + 1]$. The total demand/users to be covered for each time period are $\{35350, 146820, 145646, 144771\}$ for to $t = 1$ (early morning), $t = 2$ (morning), $t = 3$ (afternoon) and $t = 4$ (night) respectively.

There are $J = 200$ WAPs deployed, and $K = 3$ signal levels for each WAP j are considered. The availability of the signal levels is as follows: at $t = 1$ every WAPs can be activated with any signal level. At $t = 2$ and $t = 3$, 70% of J have available the low power level $k = 1$, 50% with medium power level $k = 2$ and 30% with strong power level $k = 3$. Please note that a WAP may have available more than one type of signal during a time period.

Finally, at $t = 4$ the WAPs can only be activated with a low power signal level $k = 1$.

For each signal level k, the corresponding coverage radius S_k is calculated as follows. Firstly, it is computed the $maxD = max(d_{ij})$. Then, for each signal level, the coverage radii are defined as: $S_1 = maxD \times 0.02, S_2 = maxD \times 0.03$ and $S_3 = maxD \times 0.04$.

In a day, we can activate $p_1 = 30$ WAPs with low signal, $p_2 = 18$ with medium signal and $p_3 = 12$ with a strong signal.

3.2   Solving Strategy

As we assume that imprecision can be present in the problem' information, we propose to solve it by means of an approximate method. Instead of choosing just a single strategy (like an isolated metaheuristic), we focus in an Algorithm portfolio (AP) presented in [6]. The AP is composed by different metaheuristics, that are used in a cooperative way to obtain good solutions for optimization problems. Here, the portfolio is composed by: a Hill climbing, an Evolutionary strategy and a basic Genetic algorithm, both with a population size $= 10$ and a Simulated annealing with and initial temperature $t_{initial} = 10000$, a cooling rate $\alpha = 0.9$, a number of iterations at each temperature $iter = 50$ and a final temperature $t_{final} = 0.005$ [5].

A solution of the DMCLP-FT can be represented as an integers' list where the value 0 means that the WAP is off, k means that the WAP is activated with level k and -1 means that the WAP has not any signal level in that period.

As we are not aiming at performing comparisons among several metaheuristics method, we omit more details about the implementation used.

3.3 Results

The AP method was run 30 times. Figure 1 displays the best solution obtained. The overall coverage was 370314 (78.36%). The activated WAPs have a circle that represents the radius corresponding to the signal level used. It is interesting to observe the number of users covered at each time period.

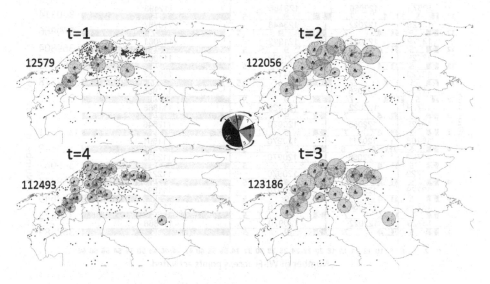

Fig. 1. Description of the best solution after 30 runs.

At the center of Fig. 1 a pie chart appears showing the number of WAPs activated per time period. The black/grey/white series indicates the level $k = 1, 2, 3$ respectively.

At period $t = 1$ there are 6 WAPs activated: 1 WAP with level $k = 1$ and 5 with level $k = 2$. At the period $t = 2$, 12 WAPs were activated: 1 WAP with level $k = 1$, 4 with $k = 2$ and 7 with $k = 3$. At period $t = 3$, 17 WAPs are activated: 3 WAPs with level $k = 1$, 9 with level $k = 2$ and 5 with level $k = 3$ were activated. Finally, at period $t = 4$, a higher number of WAPs is used because they can only use a low level signal strength.

Figure 1 also shows the geographical distribution of the WAPs, together with their coverage radii (associated with the circle radii). In the solution, we can see that the WAPs were activated in areas where there is high concentration of users to be connected. It can be observed that at period $t = 1$, just a few WAPs were activated. This period corresponds with the early morning, where the demand is lower. At $t = 2$, WAPs with medium and strong signals are used. At $t = 3$, the covered area is quite similar but with a different configuration for the WAPs. By the end of the day $(t = 4)$, the covered demand is reduced due to the use of WAPs with low level signals.

As we made 30 runs of the portfolio, we analyse a set of different solutions with a similar value of demand covered. Figure 2 shows a graphic where we can see the behaviour in the number of activated WAPs during a day. We denoted each level using the previous scheme. For visualization purposes, just 14 solutions (out of 30) achieving different coverage values are selected. The solutions are sorted by the coverage value (solution 14 is the one showed in Fig. 1).

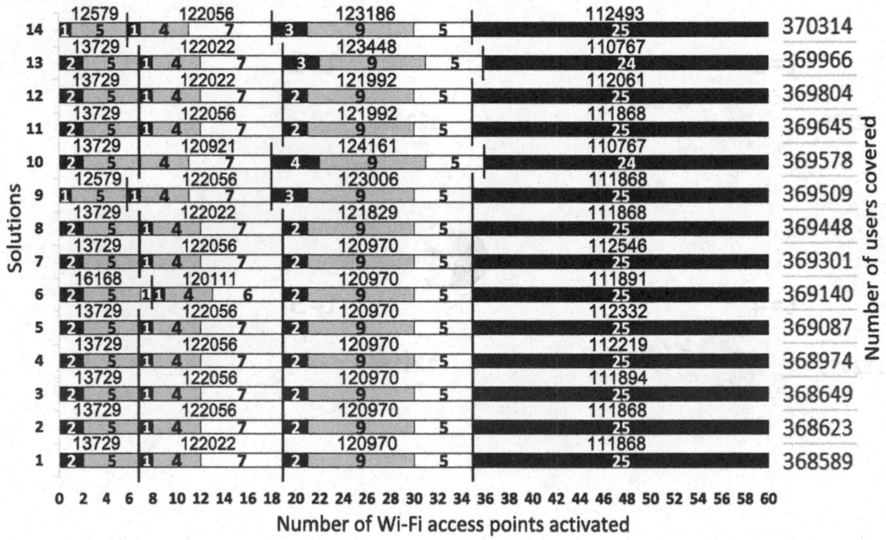

Fig. 2. Distribution of activated WAPs and their types for a day. Solutions have a different value of coverage.

For each solution we indicate the number of users covered. The end of the periods are marked by lines and for each one we show the number of users covered. For each series we can see the number of WAPs that were activated.

In general, we can see that the variability among the solutions is low. The graph provides insight on how the coverage is attained. The use of WAPs in the period $t = 1$ was not prioritized. The algorithm makes use of WAPs with level $k = 1$ in the last period $t = 4$, and not in the previous periods. An example is the best solution (14), which provides new and varied combinations that resulted in a better planning. The activated WAPs with level $k = 3$ were the most used in the period $t = 2$.

It is interesting to note that solution 9 uses the same distribution of levels as the best solution. When we explore the location of the WAPs we observed that both solutions differ in 16 activated WAPs. The selection used by the best solution allowed to obtain a greater number of users covered.

We can also note that solution 13 has the same number of WAPs activated in the period $t = 3$ as the solution 14. However, solution 13 has more users covered in that period than the solution 14. We could obtain an even better solution than solution 14, if we keep the distribution of WAPs activated in the periods $t = 1$, $t = 2$ and $t = 4$ of solution 14, and replace the distribution of $t = 3$ obtained in solution 13.

In general, in the period $t = 1$, we can see few difference between the solutions, in terms of the number of users covered. You can also note that the solution 6 was the only one that used a WAP with level $k = 3$ at $t = 1$, but it does not yield to better results. In the best solutions, the use of low power level $k = 1$ is almost neglected in the first periods. From solutions 2 to 5, we can see that the main difference was in the period $t = 4$, where the quality of the distribution of the WAPs with signal level $k = 1$ made the difference.

4   Conclusions and Future Research

In this contribution we propose a model and an algorithm to deal with the planning of WAPs activation in Havana. A dynamic maximal covering location problem with facility types and time dependant availability were used to model the situation.

As we consider the data available as imprecise, we discard solving the problem with an exact solver. Instead, we applied an Algorithm portfolio composed by classic metaheuristics that allows us to obtain *a set of good solutions*. We explore several ways to analyse them, ranging from simple coverage values, to the type of signal levels used in the WAPs, and to the WAPs geographical distribution.

Several venues are opened now. One is regarding how the quality of a solution should be assessed. In our opinion, a sort of max-min approach can be useful: to maximize the minimum level of coverage provided at any time of the day. Second is to consider some activation/deactivation costs for the WAPs. Both aspects are under study.

Acknowledgments. D. Pelta acknowledges support of Project TIN2017-86647-P (Spanish Ministry of Economy and Competitiveness, includes FEDER funds from the European Union). J. Fajardo, acknowledges support received funding from the European Union Horizon 2020 research and innovation programme under grant agreement No. 769142.

References

1. Abdelkhalek, O., Krichen, S., Guitouni, A.: A genetic algorithm based decision support system for the multi-objective node placement problem in next wireless generation network. Appl. Soft Comput. **33**, 278–291 (2015). https://doi.org/10.1016/j.asoc.2015.03.034

2. Bulut, E., Szymanski, B.K.: Rethinking offloading WiFi access point deployment from user perspective. In: 2016 IEEE 12th International Conference on Wireless and Mobile Computing, Networking and Communications (WiMob), pp. 1–6. IEEE (2016). https://doi.org/10.1109/WiMOB.2016.7763179

3. Church, R., ReVelle, C.: The maximal covering location problem. Pap. Reg. Sci. Assoc. **32**(1), 101–118 (1974). https://doi.org/10.1111/j.1435-5597.1974.tb00902.x

4. Cicek, C.T., Gultekin, H., Tavli, B., Yanikomeroglu, H.: UAV base station location optimization for next generation wireless networks: overview and future research directions. In: 2019 1st International Conference on Unmanned Vehicle Systems-Oman (UVS), pp. 1–6. IEEE (2019). https://doi.org/10.1109/UVS.2019.8658363

5. Eydi, A., Mohebi, J.: Modeling and solution of maximal covering problem considering gradual coverage with variable radius over multi-periods. RAIRO-Oper. Res. **52**(4), 1245–1260 (2018). https://doi.org/10.1051/ro/2018026

6. Calderín, J.F., Masegosa, A.D., Pelta, D.A.: An algorithm portfolio for the dynamic maximal covering location problem. Memetic Comput. **9**(2), 141–151 (2016). https://doi.org/10.1007/s12293-016-0210-5

7. Fajardo, J., Lamata, M.T., Pelta, D.A., Porras, C., Rosete, A., Verdegay, J.L.: Placing Wi-Fi hotspots in Havana with locations availability based on fuzzy constraints. In: 2018 IEEE International Conference on Fuzzy Systems (FUZZ-IEEE), pp. 1–6. IEEE (2018). https://doi.org/10.1109/FUZZ-IEEE.2018.8491568

8. ITU: Measuring digital development: Facts and figures. Technical report, International Telecommunication Union, Place des Nations CH-1211 Geneva Switzerland (2019). https://www.itu.int/en/mediacentre/Pages/2019-PR19.aspx

9. Lee, G., Murray, A.T.: Maximal covering with network survivability requirements in wireless mesh networks. Comput. Environ. Urban Syst. **34**(1), 49–57 (2010). https://doi.org/10.1016/j.compenvurbsys.2009.05.004

10. Porras, C., Fajardo, J., Rosete, A., Álvarez, R., Pelta, D.: Dynamic maximal covering location problem with facility types and time dependent availability. Computational Intelligence and Mathematics for Tackling Complex Problems (2020, to appear)

11. Shillington, L., Tong, D.: Maximizing wireless mesh network coverage. Int. Reg. Sci. Rev. **34**(4), 419–437 (2011). https://doi.org/10.1177/0160017610396011

12. Tutschku, K.: Demand-based radio network planning of cellular mobile communication systems. In: Proceedings of IEEE INFOCOM 1998, The Conference on Computer Communications. Seventeenth Annual Joint Conference of the IEEE Computer and Communications Societies. Gateway to the 21st Century (Cat. No. 98), vol. 3, pp. 1054–1061. IEEE (1998). https://doi.org/10.1109/INFCOM.1998.662915

Fuzzy Set Based Models Comparative Study for the TD TSP with Rush Hours and Traffic Regions

Ruba Almahasneh[1,2(✉)], Tuu-Szabo[1], Peter Foldesi[3], and Laszlo T. Koczy[1,2]

[1] Department of Information Technology, Széchenyi István University, Gyor, Hungary
{mahasnehr, koczy}@tmit.bme.hu,
tuu.szabo.boldizsar@sze.hu
[2] Telecommunications and Media Informatics, Budapest University of Technology and Economics, Budapest, Hungary
[3] Department of Logistics Technology, Széchenyi István University, Gyor, Hungary
foldesi@sze.hu

Abstract. This study compares three fuzzy based model approaches for solving a realistic extension of the Time Dependent Traveling Salesman Problem. First, the triple Fuzzy (3FTD TSP) model, where the uncertain costs between the nodes depend on time are expressed by fuzzy sets. Second, the intuitionistic fuzzy (IFTD TSP) approach, where including hesitation was suitable for quantifying the jam regions and the bimodal rush hour periods during the day. Third, the interval-valued intuitionistic fuzzy sets model, that calculates the interval-valued intuitionistic fuzzy weighted arithmetic average (IIFWAA) of the edges' confirmability degrees and non-confirmability degrees, was contributing in minimizing the information loss in cost (delay) calculation between nodes.

Keywords: Rush hours · Jam regions · Interval-valued fuzzy sets · Intuitionistic fuzzy set · Fuzzy set

1 Introduction

The Traveling Salesman Problem (TSP) is one of the extensively studied NP-hard graph search problems [1]. Various approaches are known for finding the optimum or semi optimum solution. The Time Dependent Traveling Salesman Problem (TD TSP) is a more realistic extension of the TSP, where the costs of edges vary in time, depending on the jam regions and rush hours. In the TD TSP, the edges are assigned higher weights if they are traveled within the traffic jam regions during rush hour periods, and lower weights otherwise [2]. The information on the rush hour periods and jam regions is uncertain and vague (fuzzy), hence, representing them by crisp numbers in the classic TD TSP does not quantify the effects of traffic jams accurately [2]. This limitation of simulating real life cases was to the point in constructing three novel fuzzy models capable of addressing the TD TSP with jam regions and rush hours more efficiently. In

© Springer Nature Switzerland AG 2020
M.-J. Lesot et al. (Eds.): IPMU 2020, CCIS 1238, pp. 699–714, 2020.
https://doi.org/10.1007/978-3-030-50143-3_55

the Triple Fuzzy (3FTD TSP) model, the costs between the nodes may depend on time and location; and are expressed by fuzzy sets [3]. Here, fuzzy values represent the uncertainties of the costs caused by the fuzzy extensions of the traffic jam areas and rush hour times, which depend on several vague or non-deterministic factors. Rush hour time was represented as a bimodal piecewise linear normal fuzzy set, the jam areas as fuzzy oblongs, and the costs as trapezoidal sets. This model expressed the uncertain costs affected by the jam situations, and calculated the overall tour length quantitatively [3]. Second, the Intuitionistic Fuzzy IFTD TSP approach, involves a hesitation part expressing the effects of membership and non-membership values allowing a higher level of uncertainty [4]. In the IFTD TSP, the use of intuitionistic fuzzy sets ensured an even more realistic cost estimate of the TD TSP problem. By successfully representing simultaneously the higher degrees of association and the lower degrees of non-association of the jam factor and rush hours and lower degrees of hesitation to edges cost, resulted in a more accurate cost of the tours [4]. Third, we proposed the interval-valued intuitionistic fuzzy set model (IVIFTD TSP) [5]. In the IVIFTD TSP, additional uncertainty was modeled, and by using an aggregation of the costs rather than using the max-min composition of the fuzzy factors resulted in an even more adequate model. In this paper, the three models are briefly presented and examined, from the point of view of realistical representation and results.

2 Solution of the Classic TSP

The original TSP was first formulated in 1930 [6]. A salesman starts the journey from the headquarters and visits each city or shop exactly once then returns to the starting point. The task is to find the route with minimum overall travelled distance visiting all destination points. TSP is a graph search problem with edge weights in Eq. 1. In the symmetric case with n nodes $c_{ij} = c_{ji}$, so, in the graph, there is only one edge between every two nodes. Let $x_{ij} = \{0, 1\}$ be the decision variable $(i, j = 1, 2, ..., n)$, and $x_{ij} = 1$, if edge e_{ij} between $node_i$ and $node_j$ is part of the tour. Let $x_{ii} = 0$ $(i = 1, 2, ..., n)$, $G_{TSP} = (N_{cities}, E_{conn})$, $C: N_{cities} \times N_{cities} \rightarrow R, C = (C_{ij})_{n \times n}$ C is called cost matrix, where C_{ij} represent the cost of going from city i to city j. then:

$$N_{cities} = \{n_1, n_2, ..., n_n\}, E_{conn} \subseteq \{(n_i, n_j) | i \neq j\} \tag{1}$$

$$\left(\sum_{i=1}^{n-1} C_{P_i, P_{i+1}}\right) + C_{P_n, P_1} \tag{2}$$

The goal is to find the directed Hamiltonian cycle with minimal total length.

3 The Time Dependent TSP (TD TSP)

Despite TD TSP's good results in determining the overall cost for a trip under realistic traffic conditions, yet one major drawback is the crisp values used for the proportional jam factors [2]. The total cost of any trip consists of two main elements: costs proportional to physical distances and costs increased by traffic jams occurring in rush hour periods or in certain areas between the pairs of nodes (such as in city center areas). The first can be looked at as constant; although transit times are subject to external and unexpected environmental factors. Thus, even they should be treated as uncertain variables, in particular, as fuzzy cost coefficients. In the TD TSP, the edges have fixed costs, which may be multiplied by a rush hour factor). This representation of the traffic jam effects is too rigid for real life circumstances.

4 The Triple Fuzzy TD TSP (3FTD TSP)

In the 3FTD TSP approach [Put here a citation of our paper] two parameters modify the fuzzy edge costs, the actual jam factor calculated from the membership degree of being in the jam region, and the degree of membership being in the rush hour period in the given moment. We proposed to use a simple Mamdani rule base [7] in the form: If N_i is in the traffic jam region J and t_j is in the rush hour time R then the cost is C_k. Here, membership functions from the unit interval [0, 1] help describe the uncertainty of the jam region and the rush hour period, more efficiently. In this model, the distance between cities is also expressed in terms of the elapsed time. Here, we introduced a velocity (v) as a new parameter of the TSP route. The costs were represented by asymmetrical triangular fuzzy numbers. The total cost of the tour was calculated as follows:

$$C'(t) = \left(C'_L, C'_C, C'_R\right) = \left(C_L, C_C, C_R\right) \times \left(1 + \left(jam_{factor} - 1\right) \times \mu_1 \times \mu_2\right) \quad (3)$$

where μ_1 and μ_2 are obtained from the Fuzzy Jam Region (J) and Fuzzy Traffic Rush Hours (R) membership functions. A valid solution for the problem is a permutation of the nodes:

$$P_1, P_2, \ldots, P_n, P_1 \quad (4)$$

where p_1 is the starting and end node of the tour. The time needed for visiting the first city from the start node is:

$$t_{P_1, P_2} = \frac{COG\left(C'_{P_1, P_2}(t = 0)\right)}{v} \quad (5)$$

where P_1 is the starting node, P_2 is the first visited one and v is the velocity. The calculation of travel time is necessary as the costs are time dependent, and the actual cost between two cities can be determined by Eq. 3, thus, the time dependency in the cost matrix is represented by virtual distance values. The cost of the trip is calculated from

$$t_{P_k, P_{k+1}} = \frac{COG\left(C'_{P_k, P_{k+1}}\left(t_{elapsed_k}\right)\right)}{v} \quad (6)$$

The total cost is:

$$S = \sum_{i=1}^{n-1} COG\left(C'_{P_i,P_{i+1}}\left(t_{elapsed_i}\right)\right) + COG\left(C'_{P_n,P_1}\left(t_{elapsed_n}\right)\right) \qquad (7)$$

where p_n is the location last visited, and $t_{elapsed_k}$ is the total time elapsed from the beginning of the tour till the salesman arrives in city p_k. In the implementation, the three fuzzy elements used were triangular fuzzy costs between the edges, fuzzy oblong type membership function(s) of the fuzzy jam region(s) J; and bimodal normal piecewise linear membership function(s) for the traffic rush hour time period(s) R – see next sections.

4.1   Triangular Fuzzy Costs for the "Distances"

The uncertain costs between the nodes, is expressed by triangular fuzzy numbers. Triangular fuzzy numbers may be expressed by the support $C = [C_L, C_R]$, and the peak value C_C is, so it is denoted by $\tilde{C} = (C_L, C_C, C_R)$. To calculate the overall distance of the tour, these fuzzy values are summed up. The calculation of the total length of the tour was done by the defuzzified values of the fuzzy numbers using Center of Gravity (COG) [8].

4.2   The Membership Function of the Jam Regions

The fuzzy extensions of the city center areas (the degree of belonging to the jam region J) are expressed by fuzzy borders as in Fig. 1. Thus, μ_1 is simply calculated as $\mu_1 = \frac{f(d1) + f(d2)}{2}$, where $d1$, $d2$ are the distances from the peak of J. This approach sophisticates Schneider's original model (cf. [2]), so that the breakpoints are: [0, 1000, 5000, 6000], (see Fig. 1).

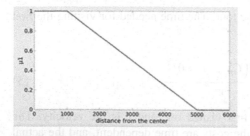   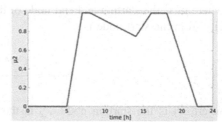

Fig. 1. Jam regions membership function (J)   **Fig. 2.** Rush hours membership function (R)

4.3   Membership Function of the Traffic Rush Hour Periods

The model uses the bimodal membership function in Fig. 2 for representing the Traffic Rush Hour Time (μ_2). In this example, the two peak rush hour periods are from 7 to 8 a.m. and from 4 to 6 p.m. Between the two periods the traffic is lower. (We used the traffic data base …). The breakpoints of J are {0, 5, 7, 8, 14, 16, 18, 22, 24}, and its membership value at 14 h is 0.75. For illustration, assuming a jam factor of 5, a sample calculation is run to clarify the approach. The peak point of the fuzzy triangular cost for each edge is the Euclidean distance between the end-points, namely, the left-side and right-side points were determined randomly (0–50% lower and higher than the middle point) in this test.

Table 1. Computational results for jam factor equal 5.0.

	Run 1	Run 2	Run 3	Run 4	Run 5
Elapsed time	22.2021	22.1314	22.3246	22.3067	22.0133
Low	16.5611	16.2926	16.5862	16.642	16.8112
Middle	22.4831	22.6905	22.4858	23.0948	22.3641
High	27.562	27.4111	27.9018	27.2832	27.3146

Table 1 was calculated by averaging five times runs for jam factor 5.0 for the three areas of the triangular membership functions (low, medium and high). Table 2 shows the middle values of the supports of the triangular fuzzy numbers for that specific case (for jam factor 5). Depending on Table 1 the average elapsed time is 22.19562.

Table 2. Support value for jam factor 5

Run 1	Run 2	Run 3	Run 4	Run 5	Average elapsed time
11.0009	11.1185	11.3156	10.6412	10.5034	10.91592

Applying the same approach results in Table 3, which contains the total time in hours required to visit each location with different jam factors by applying the same concept explained in the previous section

Table 3. Computational results for the 3FTD TSP.

Jam factor	Average elapsed time
1.00	19.5
1.05	19.6
1.20	19.7
1.50	20.3
2.00	20.9
3.00	21.6
5.00	22.2
10.00	23.1
20.00	23.5
50.00	24.2
100.00	24.7

With such high jam factors, the tour lengths were longer for the 3FTD TSP problem, because the traffic jam period is longer compared to the classic TD TSP.

5 Intuitionistic Fuzzy TD TSP (IFTD TSP)

In this model, we moved one step further and extended the model by using intuitionistic fuzzy set theory. First some related definitions as introduced [4].

5.1 Basic Definitions of Intuitionistic Fuzzy Sets (IFS)

Let a universal set E be fixed and $0 \le \mu_A(x) + v_A(x) \le 1$. An *intuitionistic fuzzy set* or IFS A in E is an object having the form

$$A = \{\langle x, \mu_A(x), v_A(x)\rangle | x\epsilon\, E\} \tag{8}$$

The amount $\pi_A(x) = 1 - (\mu_A(x) + v_A)$ is called the *hesitation part*, which may cater to either the membership value or to the non-membership value, or to both [9, 10]. If A is an IFS of X, the max-min-max composition of the If Relation (IFR) $R\ (X \to Y)$ with A is an IFS B of Y denoted by $(B = R \circ A)$ and is defined by the *membership function*

$$\mu_{R\circ A}(y) = \vee_x[\mu_A(x) \wedge \mu_R(x, y)] \tag{9}$$

and the *non-membership function*

$$v_{R\circ A}(y) = \wedge_x[v_A(x) \vee v_R(x, y)] \tag{10}$$

The previous formulas hold for all Y. Let $Q(X \to Y)$ and $R(Y \to Z)$ be two IFRs. *The max-min-max composition* $(R \circ Q)$ is the intuitionistic fuzzy relation from X to Z, defined by the membership function

$$\mu_{R\circ Q}(x, z) = \vee_y[\mu_Q(x, y) \wedge \mu_R(y, z)] \tag{11}$$

and the non-membership function $\forall(x, z) \in X \times Z\, and\, \forall y \in Y$ is given by

$$V_{R\circ Q}(x, z) = \wedge_y[v_Q(x, y) \vee v_R(y, z)] \tag{12}$$

Let A be an IFS of the set J, and R be an IFR from J to C. Then the max-min-max composition B of IFS A with the IFR $R\ (J \to C)$ denoted by $B = A \circ R$ gives the cost of the edges as an IFS B of C with the membership function given by

$$\mu_B(c) = \vee_{j\epsilon J}[\mu_A(j) \wedge \mu_R(j, c)] \tag{13}$$

and the non-membership function given as:

$$v_B(c) = \wedge_{j \in J}[v_A(j) \vee v_R(j,c)] \tag{14}$$

$$\forall c \in C. \; (\textit{Here } \wedge = \textit{min and } \vee \textit{ max})$$

If the state of the edge E is described in terms of an IFS A of J; then E is assumed to be the assigned cost in terms of IFSs B of C, through an IFR R from J to C, which is assumed to be given by a knowledge base directory (given by experts) on the destination cities and the extent (membership) to which each one is included in the jam region. This will be translated to the degrees of association and non-association, respectively, between jam and cost.

5.2   IFTD TSP Applied on the TD TSP Case

Let there be n edges E_i; $i = 1; 2;..., 16$ as in Fig. 3; in a trip. Thus $e_i \in E$. Let R be an IFR $(J \rightarrow C)$ and construct an IFR Q from the set of edges E to the set of jam factors J. Clearly, the composition T of IFRs R and $(T = R \circ Q)$ give the cost for each edge from E to C by the membership function given as:

$$\mu_T(e_i, c) = \vee_{j \in J}[\mu_Q(e_i, j) \wedge \mu_R(j, c)] \tag{15}$$

and the non-membership function $\forall e_i \in E$ and $c \in C$ given as:

$$v_T(e_i, c) = \wedge_{j \in J}[v_Q(e_i, j) \vee v_R(j, c)] \tag{16}$$

For given R and Q, the relation $T = R \circ Q$ can be computed. From the knowledge of Q and T, an improved version of the IFR R can be computed, for which the following holds valid:

(i) $J_R = \mu_R - v_R \cdot \pi_R$ is greatest
(ii) The equality $T = R \circ Q$ is retained

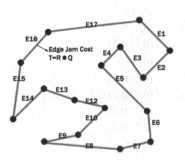

Fig. 3. Tour for a simple example

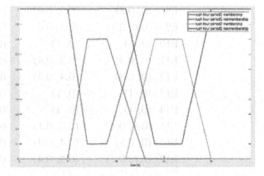

Fig. 4. Fuzzy membership and non- membership functions of the rush hour periods

Table 4 shows each edge of the tour and the jam factors associated. The ultimate goal is to be able to calculate the total tour jam factor which will be multiplied by the physical distances between two nodes. The intuitionistic fuzzy relations $Q(E \rightarrow J)$ are given as shown in Table 4, and $R(J \rightarrow C)$ as in Table 5, and the composition ($T = R \circ Q$) as in Table 6. Then we calculated the jam region cost factors (c_{jam}) (see Table 7), where the four cost factors are $c_1 = 1.2, c_2 = 1.5, c_3 = 2, c_4 = 5$ with weighted average calculations:

$$c_{jam_e} = \frac{\sum_i j_i \times c_i}{\sum_i j_i} \quad (17)$$

The rush hour cost factors of each tour edge (c_{rush}) are determined in a similar intuitionistic model. The relations between the tour time and the rush hour periods ($\dot{Q}$) are described with intuitionistic fuzzy functions in Fig. 4. An IFR ($\dot{R}$) is given between the rush hour periods and the cost factors similarly, as it was done for the jam regions in Table 5. Then the composition ($\dot{T} = \dot{Q} \circ \dot{R}$ is calculated. Finally, rush hour cost factors were calculated with weighted averaging. The cost of the edges is calculated taking into account the two cost factors ($dist_e$ is the Euclidean distance):

IF $c_{jam_e} > 0$ **AND** $c_{rush_e} > 0$ [*the edge belongs to at least one of the jam regions and is passed during rush hour periods*] **THEN** $c_e = c_{jam_e} \times c_{rush_e} \times dist_e$ **ELSE** $C_e = dist_e$.

Table 4. Route1 = (Edge 1 ... Edge 17)

(Q)	Jam Region1	Jam Region2	Jam Region3	Jam Region4
E1	(0.8, 0.1)	(0.6, 0.1)	(0, 1)	(0, 1)
E2	(0, 1)	(0, 1)	(0.2, 0.8)	(0.6, 0.1)
E3	(0.8, 0.1)	(0.8, 0.1)	(0, 1)	(0, 1)
E4	(0, 1)	(0, 1)	(0, 0.6)	(0.2, 0.7)
E5	(0.8, 0.1)	(0.8, 0.1)	(0, 0.6)	(0.2, 0.7)
E6	(0, 0.8)	(0.4, 0.4)	(0, 1)	(0, 1)
E7	(0, 1)	(0, 1)	(0.6, 0.1)	(0.1, 0.7)
E8	(0, 0.8)	(0.4, 0.4)	(0.6, 0.1)	(0.1, 0.7)
E9	(0.6, 0.1)	(0.5, 0.4)	(0, 1)	(0, 1)
E10	(0, 1)	(0, 1)	(0.3, 0.4)	(0.7, 0.2)
E11	(0, 0.8)	(0.4, 0.4)	(0.6, 0.1)	(0.1, 0.7)
E12	(0, 0.8)	(0.4, 0.4)	(0, 1)	(0, 1)
E13	(0, 1)	(0, 1)	(0.2, 0.8)	(0.6, 0.1)
E14	(0, 1)	(0, 1)	(0.6, 0.1)	(0.1, 0.7)
E15	(0.8, 0.1)	(0.8, 0.1)	(0, 0.6)	(0.2, 0.7)
E16	(0, 0.8)	(0.4, 0.4)	(0.6, 0.1)	(0.1, 0.7)
E17	(0.6, 0.1)	(0.5, 0.4)	(0, 1)	(0, 1)

Table 5. Jam factors

Jam area (R)	Cost factor 1 (c1)	Cost factor 2 (c2)	Cost factor 3 (c3)	Cost factor 4 (c4)
Jam Region1	(0.4, 0)	(0.7, 0)	(0.3, 0.3)	(0.1, 0.7)
Jam Region2	(0.3, 0.5)	(0.2, 0.6)	(0.6, 0.1)	(0.2, 0.4)
Jam Region3	(0.1, 0.7)	(0, 0.9)	(0.2, 0.7)	(0.8, 0)
Jam Region4	(0.4, 0.3)	(0.4, 0.3)	(0.2, 0.6)	(0.2, 0.7)

Table 6. $T = R \circ Q$

Jam cost (T)	Cost factor 1	Cost factor 2	Cost factor 3	Cost factor 4
E1	(0.4, 0.1)	(0.7, 0.1)	(0.6, 0.1)	(0.2, 0.4)
E2	(0.4, 0.3)	(0.4, 0.3)	(0.2, 0.6)	(0.2, 0.2)
E3	(0.4, 0.1)	(0.7, 0.1)	(0.6, 0.1)	(0.2, 0.4)
E4	(0.2, 0.7)	(0.2, 0.7)	(0.2, 0.7)	(0.2, 0.6)
E5	(0.3, 0.1)	(0.7, 0.1)	(0.6, 0.1)	(0.2, 0)
E6	(0.3, 0.5)	(0.2, 0.6)	(0.4, 0.4)	(0.2, 0.4)
E7	(0.1, 0.7)	(0.1, 0.7)	(0.2, 0.7)	(0.6, 0.1)
E8	(0.3, 0.5)	(0.2, 0.6)	(0.4, 0.4)	(0.6, 0.1)
E9	(0.4, 0.1)	(0.6, 0.1)	(0.5, 0.3)	(0.2, 0.4)
E10	(0.4, 0.3)	(0.2, 0.6)	(0.2, 0.6)	(0.3, 0.4)
E11	(0.3, 0.5)	(0.2, 0.6)	(0.4, 0.4)	(0.6, 0.1)
E12	(0.3, 0.5)	(0.2, 0.6)	(0.4, 0.4)	(0.2, 0.4)
E13	(0.4, 0.3)	(0.4, 0.3)	(0.2, 0.6)	(0.2, 0.2)
E14	(0.1, 0.7)	(0.1, 0.7)	(0.2, 0.7)	(0.6, 0.1)
E15	(0.3, 0.1)	(0.7, 0.1)	(0.6, 0.1)	(0.2, 0)
E16	(0.3, 0.5)	(0.2, 0.6)	(0.4, 0.4)	(0.6, 0.1)
E17	(0.4, 0.1)	(0.6, 0.1)	(0.5, 0.3)	(0.2, 0.4)

Table 7. Intuitionistic jam region costs for edges

J_R	J_R1	C1	J_R2	C2	J_R3	C3	J_R4	C4	Total jam regions cost
E1	0.35	1.2	0.68	1.5	0.57	2	0.04	5	1.695
E2	0.31	1.2	0.31	1.5	0.08	2	0.08	5	1.791
E3	0.35	1.2	0.68	1.5	0.57	2	0.04	5	1.695
E4	0.13	1.2	0.13	1.5	0.13	2	0.08	5	2.151
E5	0.24	1.2	0.68	1.5	0.57	2	0.2	5	2.04
E6	0.2	1.2	0.08	1.5	0.32	2	0.36	5	2.97
E7	0	1.2	0	1.5	0.13	2	0	5	2
E8	0.2	1.2	0.08	1.5	0.32	2	0.57	5	3.291
E9	0.35	1.2	0.57	1.5	0.44	2	0.04	5	1.682
E10	0.31	1.2	0.08	1.5	0.08	2	0.18	5	2.388

(*continued*)

Table 7. (*continued*)

J_R	J_R1	C1	J_R2	C2	J_R3	C3	J_R4	C4	Total jam regions cost
E11	0.2	1.2	0.08	1.5	0.32	2	0.57	5	3.291
E12	0.2	1.2	0.08	1.5	0.32	2	0.36	5	2.917
E13	0.31	1.2	0.31	1.5	0.08	2	0.08	5	1.791
E14	0	1.2	0	1.5	0.13	2	0	5	2
E15	0.24	1.2	0.68	1.5	0.57	2	0.2	5	2.04
E16	0.2	1.2	0.08	1.5	0.32	2	0.57	5	3.291
E17	0.35	1.2	0.57	1.5	0.44	2	0.04	5	1.682

Clearly, the improved version of R in the IFTD TSP model is more adequate in translating the higher degrees of association and lower degrees of non-association of the jam factors and rush hours as well as lower degrees of hesitation to any cost C; If almost equal values in T are obtained, then we consider the case for which hesitation is least. From a refined version of R one may infer cost from jam factors in the sense of a paired value, one being the degree of association and other the degree of non-association. Ultimately, this model offers more realistic costs calculation for the traveled routes under real traffic conditions.

6   The Interval Valued IFTD TSP (IVIFTD TSP)

First, some basic definitions are overviewed [11, 12]. In a *type-2 fuzzy set*, the uncertain values of the membership function $\tilde{A}$ in Eq. 18 consists of a rounded region called "*footprint of uncertainty*" (FOU). It is the union of all primary memberships

$$FOU(\tilde{A}) = \bigcup_{x \in X} J_x \tag{18}$$

FOUs emphasize the distribution that sits on top of the primary membership function of the type-2 fuzzy set. The shape of this distribution depends on the choice made for the secondary grades. When they are equal between two bounds, it gives an *interval type-2 fuzzy set* as given in Eq. 19.

$$\tilde{A} = \{\langle (x,u), \mu_A(x,u) \rangle | \forall x \in X, \forall u \in J_x \subseteq [0,1]\} \tag{19}$$

For discrete universe of discourse X and U, an embedded type-2 set $\tilde{A}_e$ has N elements, where $\tilde{A}_e$ contains *exactly one* element from set $J_{x1,x2........xN}$, namely $U_{1,2...N}$, with its associated secondary grade $f_{x1}(u_1)$, $f_{x2}(u_2)$.....$f_{xN}(u_N)$, which equals to Eq. 20.

$$\tilde{A}_e(x) = \sum_{i=1}^{N} [f_{xi}(u_i)/u_i]/x_i | u_i \in J_{xi} \subseteq U = [0,1] \tag{20}$$

As we discussed previously, the jam factor costs on the edges in a tour were represented as fuzzy relations between the jam factors and the predicted costs (delays). Let $J = \{J_1, J_2, J_3, \ldots, J_m\}$, $C = \{C_1, C_2, C_3, \ldots, C_n\}$ and $E = \{E_1, E_2, E_3, \ldots, E_q\}$ denote the sets of jam factors, costs and edges, respectively. Two fuzzy relations (FR) Q and R are defined in Eqs. 21 and 22.

$$Q = \{\langle (e,j), \mu_Q(e,j), v_Q(e,j) \rangle \mid (e,j) \in E \times J\} \tag{21}$$

$$R = \{\langle (j,c), \mu_R(j,c), v_R(j,c) \rangle \mid (j,c) \in J \times C\} \tag{22}$$

where $\mu_Q(e,j)$ and $v_Q(e,j)$ indicates jam factors degrees for edges. The degree is the relationship between the edges and the jam factors (rush hours or jam regions). Hence, $\mu_Q(e,j)$ indicates the degree to which jam factor j affects edge E and $v_Q(e,j)$ indicates the degree to which jam factor j does not affect the same edge. Similarly, $\mu_R(j,c)$ and $v_R(j,c)$ are the relationships between the jam factors and the respective costs. (This is called *confirmability degree* in the coming sections). (j,c) represents the degree to which jam factor j confirms, and $v_R(j,c)$ the degree to which jam factor j does not confirm the presence of cost c, respectively [5]. Since Q is defined on set $E \times J$ and R on set $J \times C$, the composition T of R and Q ($T = R \circ Q$) for the prediction of the cost for a specific edge in terms of the cost can be represented by FR from E to C, given the membership function in Eq. 23 and non membership function in Eq. 24 for all $e \in E$ and $c \in C$

$$\mu_T(e,c) = max_j\{min\left[\mu_Q(e,j), \mu_R(j,c)\right]\} \tag{23}$$

$$v_T(e,c) = min_j\{max[v_Q(e,j), v_R(j,c)]\} \tag{24}$$

Let any two IVIFS $A = \{\langle x_i, M_A(x_i), N_A(x_i) \rangle\} i = (1, 2, \ldots, n)$ be a collection of interval-valued intuitionistic fuzzy degrees. Then, an IIFWAA operator is defined in Eq. (25)

$$IFFWAA(A) = \begin{array}{l} [1 - \prod_{n=1}^{n} (1 - M_{AL}(x_i))^{w_i}, \\ 1 - \prod_{n=1}^{n} (1 - M_{AU}(x_i))^{w_i})], \\ \prod_{n=1}^{n} (1 - N_{AL}(x_i))^{w_i}, \prod_{n=1}^{n} (1 - N_{AU}(x_i))^{w_i})] \end{array} \tag{25}$$

In the next section, we explain the IVIFTD TSP by simulating a simple real life TD TSP cost problem. The approach consists of four main steps:

Step 1. Prediction for the rush hours and jam regions of the edges, in the sense that if the trip between the two cities happens during the rush hours and within the jam regions, both will be taken into consideration and none of the factors will be neglected. Table 8 identifies the cost of each jam factor, which is supposed to be predefined by experts in this domain, according to the rush hours and the jam regions.

Step 2. Calculation of the interval-valued intuitionistic fuzzy weighted arithmetic average (IIFWAA) of the edges' confirmability and non-confirmability degrees, respectively with the chosen aggregation [13]. where $(w_1, w_2, \ldots, w_i)^T$ are the weight vectors of A. Further, $w_i > 0$ and $\sum_{i=1}^{n} w_i = 1$. In our model we propose giving equal weights to the factors by $w_i = 1/n$. For finding the final jam factor cost, we first calculate the IIFWAA from the degrees given in Table 8 and then use a measure based on distance between IVIFS.

Step 3. Calculating the distance between the IFSs using the IIFWAA obtained in Step 2.

To calculate the jam factor cost based on the Park distance model between IVIFS [14]. Particularly, we consider the hesitation part to modify the Park distances. The normalized Hamming distance considering the hesitation part, is defined as below for any $A = \{\langle x_i, M_A(x_i), N_A(x_i)\rangle\}i = (1, 2\ldots\ldots n)$ and $= \{\langle x_i, M_B(x_i), N_B(x_i)\rangle\}$ $i = (1, 2\ldots\ldots n)$. The normalized Hamming distance considering the hesitation part (where H is the hesitation part) is defined as:

$$l_h(A, B) = (1/4n) \sum [|M_{AL}(x_i) - M_{BL}(x_i)| + |M_{AU}(x_i) - M_{BU}(x_i)| + N_{AL}(x_i)$$
$$- N_{BL}(x_i)| + |N_{AU}(x_i) - N_{BU}(x_i)| + |H_{AL}(x_i) - H_{BL}(x_i)| + |H_{AU}(x_i) - H_{BU}(x_i)|]$$

$$(26)$$

Step 4. Determination of the final jam factor affected costs based on the distance (assuming equal weights for all factors).

Table 8. Knowledge base for rush hours and jam regions costs

Traffic factor	IF degree					
	Cost1		Cost2		Cost3	
	μ_R	v_R	μ_R	v_R	μ_R	v_R
Factor 1.1	*[0.6, 0.7]*	*[0.1, 0.2]*	[0.2, 0.3]	[0.5, 0.6]	[0.1, 0.3]	[0.4, 0.6]
Factor 1.2	*[0.6, 0.7]*	*[0.2, 0.3]*	[0.2, 0.4]	[0.4, 0.6]	[0.4, 0.6]	[0.1, 0.2]
Factor 1.3	*[0.5, 0.6]*	*[0.1, 0.2]*	[0.1, 0.2]	[0.6, 0.7]	[0.3, 0.4]	[0.3, 0.5]
Factor 1.4	*[0.7, 0.8]*	*[0.1, 0.2]*	[0.1, 0.2]	[0.6, 0.8]	[0.1, 0.2]	[0.7, 0.8]
...	...	...	...	...	...	...
Factor 2.1	*[0.5, 0.6]*	*[0.2, 0.3]*	[0.2, 0.3]	[0.4, 0.6]	[0.2, 0.3]	[0.5, 0.6]
Factor 2.2	*[0.7, 0.8]*	*[0.1, 0.2]*	[0.1, 0.2]	[0.6, 0.7]	[0.1, 0.2]	[0.6, 0.7]
...	...	...	...	...	...	...
Factor n.1	[0.6, 0.8]	[0.1, 0.2]	[0.1, 0.2]	[0.6, 0.7]	[0.2, 0.3]	[0.6, 0.7]
Factor n.n	$[\mu_R U, \mu_R L]$	$[v_R U, v_R L]$	$[\mu_R U, \mu_R L]$	$[v_R U, v_R L]$	$[\mu_R U, \mu_R L]$	$[v_R U, v_R L]$

6.1 The IVIFTD TSP Model Applied on the TD TSP Case

To illustrate how to apply the proposed new model on a TD TSP case study with double uncertain jam regions and rush hours, let us consider that $E1$ is the first edge of a tour as in Fig. 4. There are different traffic factors assumed, they may represent jam regions or rush hours Factors (1.1, 1.2, 1.3, 1.4, 2.1 and 2.2 in bold Table 8) affecting $E1$ simultaneously. Here, we use $\langle M_Q(e, j), N_Q(e, j)\rangle$ assigned by domain experts, to indicate the degrees how a jam factor j affects edge e as in Eq. 23, and the confirmability degree as in Eq. 24 is given by $\langle M_R(j,c), N_R(j,c) \rangle$.

Step 1. Table 9 shows the confirmability and non-confirmability degrees of the jam factors assigned to $E1$, according to their degree of belonging to the rush hour periods and jam regions

Table 9. $E1$ degrees of jam factors. $M_Q(E_1, J)$, $N_Q(E_1, J)$

E1 traffic factors	1.1	1.2	1.3	1.4	2.1	2.2
$M_Q(E_1)$	[0.5, 0.6]	[0.5, 0.6]	[0.4, 0.6]	[0.7, 0.8]	[0.5, 0.6]	[0.5, 0.7]
$N_Q(E_1)$	[0.2, 0.3]	[0.1, 0.3]	[0.1, 0.2]	[0.1, 0.2]	[0.1, 0.2]	[0.2, 0.3]

Step 2. Based on Tables 9 and 8, calculate the results in Tables 10 and 11 by applying the IIFWAA operator (see Eq. 26). For example, [0.61, 0.71], an IIFWAA M_R of Table 11, is calculated as follows: The confirmability membership degrees of the edge jam factors (1.1, 1.2, 1.3 and 1.4) are ([0.6, 0.7], [0.6, 0.7], [0.5, 0.6], [0.7, 0.8]) respectively, the first edge, for example, belongs to four jam factors, then $w_i = 1/n$, and the distributed weight for n = 4 is $w = \frac{1}{4}, \frac{1}{4}, \frac{1}{4}, \frac{1}{4}$ then; $0.61 = 1 - \left\{ \left(1 - 0.6 \right)^{\frac{1}{4}} \right\} *$
$\left\{ \left(1 - 0.6 \right)^{\frac{1}{4}} \right\} * \left\{ \left(1 - 0.5 \right)^{\frac{1}{4}} \right\} * \left\{ \left(1 - 0.7 \right)^{\frac{1}{4}} \right\}$ and $0.71 = 1 - \left\{ (1 - 0.7)^{\frac{1}{4}} \right\} *$
$\left\{ (1 - 0.7)^{\frac{1}{4}} \right\} * \left\{ (1 - 0.6)^{\frac{1}{4}} \right\} \left\{ (1 - 0.8)^{\frac{1}{4}} \right\}$. N_R in Table 11 is calculated by taking the confirmability values for the non-membership degrees of the jam factors [0.1, 0.2] [0.2, 0.3] [0.1, 0.2] [0.1, 0.2] and applying IIFWAA. $0.12 = \left\{ (0.1^{1/4}) * (0.2^{1/4}) * (0.1^{1/4}) * (0.1^{1/4}) \right\}$ and $0.22 = \left\{ (0.2^{1/4}) * (0.3^{1/4}) * (0.2^{1/4}) * (0.2^{1/4}) \right\}$

Table 10. $E1$ IVIF degrees (IIFWAA M_Q, IIFWAAN$_Q$)

Q	Factor1	Factor2
Edge 1	([0.54, 0.66], [0.12, 0.24])	[0.5, 0.65], [0.14, 0.24]

Table 11. $E1$ confirmability degrees (IIFWAA M_R, IIFWAAN$_R$)

R	Cost1	Cost2	Cost3
Factor1 IIFWAA	[0.61, 0.71], [0.12, 0.22]	[0.15, 0.28], [0.52, 0.67]	[0.24, 40], [0.30, 0.47]
Factor2 IIFWAA	[0.37, 0.51], [0.24, 0.39]	[0.21, 0.31], [0.35, 0.46]	[0.55, 70], [0.10, 0.24]

Step 3. Calculate the distance by applying Eq. 25, taking values from Tables 11 and 12.

Table 12. Distance for *E1* with traffic factors l_h

T	Cost1	Cost2	Cost3
Edge 1	0.16	0.26	0.24

Step 4. The lowest distance points of the traffic costs that affect the edge the most, will cause the most extreme delays In our case, this is 0.16 as shown in Table 12. Carrying out the same calculations for all the edges, we end up with Table 13. It contains the jam factor costs for all edges, depending on their confirmabilities ("–" indicates the absence of confirmability).

Table 13. Distances for *E1, 2, …, E16* with traffic factors l_h

Edge	Cost1	Cost2	Cost3
Edge 1	0.16	0.26	0.24
Edge 2	0.13	0.20	–
Edge 3	0.15	–	0.24
Edge 4	0.16	0.14	0.44
Edge 5	0.3	0.6	0.2
Edge 6	0.16	–	0.24
Edge 7	0.16	0.30	0.20
Edge 8	0.16	0.06	0.44
Edge 9	0.04	0.36	–
Edge 10	0.1	0.23	0.34
Edge 11	0.6	–	0.3
Edge 12	–	0.27	0.23
Edge 13	0.13	0.20	–
Edge 14	0.16	0.30	0.20
Edge 15	0.3	0.6	0.2
Edge 16	0.16	0.06	0.44

The results indicate that this model effectively simulates the real-life conditions and successfully quantifies the traffic delays without information loss [5]. It gives more tangible conditions for such intangible factors as vagueness and non-determinnistic effects with better accuracy than all previous models.

7   Conclusions

In this paper, we constructed a comparison of three different fuzzy extensions of the Time Dependent Traveling Salesman Problem, namely, the 3FTD TSP, the IFTD TSP, and the IVIFTD TSP. These models offer alternative extensions of the abstract TD TSP with crisp traffic regions and time dependent rush hour periods. The 3FTD TSP represents the jam regions and rush hour costs by fuzzy sets. The IFTD TSP offers higher degrees of association and lower degrees of non-association of the jam factors and rush hours as well as lower degrees of hesitation to any edge cost. Lastly, the IVIFTD TSP decreases the information loss by employing the IIFWAA operator to aggregate interval-valued fuzzy information from the jam factors in order to measure the final cost based on the distance between IVIFS(s) for the TD TSP.

The results of the examples indicate that our models effectively simulate real-life conditions and successfully quantify the traffic jam regions and rush hours with minimum information loss. After fuzzification of the jam regions and rush hours each model slightly differs in the optimal solution it suggests, including the best tour and the total cost. Although each one of those proposed approaches uniquely contributes to a more adequate calculation of the jam regions and rush hours under vague and uncertain circumstances, yet it is hard to choose one as the unambiguously best solution. In our future work, we are eager to simulate those approaches on more complicated examples, with larger instances and to compare the results with other models, to test their capability and efficiency.

Acknowledgement. This research was supported by the "Higher Education Institutional Excellence Program – Digital Industrial Technologies Research at University of Győr UDFO/4 7138-1/2019-ITM)", and M. F. acknowledges the financial support of the DE Excellence Program. L. T. K. is supported by NKFIH K124055 grants.

References

1. Zadeh, L.A.: Fuzzy sets. Inf. Control **8**(3), 338–353 (1965)
2. Schneider, J.: The time-dependent traveling salesman problem. Physica A **314**, 151–155 (2002)
3. Kóczy, L.T., Földesi, P., Tüü-Szabó, B., Almahasneh, R.: Modeling of fuzzy rule-base algorithm for the time dependent traveling salesman problem. In: Proceedings of the IEEE International Conference on Fuzzy Systems (FUZZ-IEEE) (2019)
4. Almahasneh, R., Tüü-Szabó, B., Földesi, P., Kóczy, L.T.: Intuitionistic fuzzy model of traffic jam regions and rush hours for the time dependent traveling salesman problem. In: Kearfott, R.B., Batyrshin, I., Reformat, M., Ceberio, M., Kreinovich, V. (eds.) IFSA/NAFIPS 2019 2019. AISC, vol. 1000, pp. 123–134. Springer, Cham (2019). https://doi.org/10.1007/978-3-030-21920-8_12
5. Almahasneh, R., Tüü-Szabó, B., Földesi, P., Kóczy, L.T.: Optimization of the time dependent traveling salesman problem using interval-valued intuitionistic fuzzy sets. Axioms Special Issue (2020, under review)
6. Applegate, D.L., Bixby, R.E., Chvátal, V., Cook, W.J.: The Traveling Salesman Problem: A Computational Study, pp. 1–81. Princeton University Press, Princeton (2006)
7. Mamdani, E.H.: Application of fuzzy algorithms for control of simple dynamic plant. IEEE Proc. **121**(12), 1585–1588 (1974)
8. Voskoglou, M.Gr.: Comparison of the COG defuzzification technique and its variations to the GPA index. Am. J. Comput. Appl. Math. **6**(5), 187–193 (2016)
9. Atanassov, R.K.: Intuitionistic fuzzy sets. Fuzzy Sets Syst. **20**, 87–96 (1986)

10. Biswas, R.: On fuzzy sets and intuitionistic fuzzy sets. Notes Intuitionistic Fuzzy Sets **3**, 3–11 (1997)
11. Mendel, J.M., Bob John, R.I., Liu, F.: Interval type-2 fuzzy logic systems made simple. IEEE Trans. Fuzzy Syst. **14**(6) (2006)
12. Mendel, J.M., Bob John, R.I.: Type-2 fuzzy sets made simple. IEEE Trans. Fuzzy Syst. **10** (2) (2002)
13. Xu, Z.S.: Methods for aggregating interval-valued intuitionistic fuzzy information and their application to decision making. Control Decis. **22**, 215–219 (2007)
14. Park, J.H., Lim, K.M., Park, J.S., Kwun, Y.C.: Distances between interval-valued intuitionistic, fuzzy sets. J. Phys.: Conf. Ser. **96** (2008)

Fuzzy Greedy Randomized Adaptive Search Procedure and Simulation Model to Solve the Team Orienteering Problem with Time Windows

Airam Expósito-Márquez[✉][iD], Christopher Expósito-Izquierdo[iD],
Belén Melián-Batista[iD], and José Marcos Moreno-Vega[iD]

Universidad de La Laguna, Avenida Astrofísico s/n, La Laguna, Spain
{aexposim,cexposit,mbmelian,jmmoreno}@ull.edu.es

Abstract. Tourism is a relevant economic activity that provides important resources (income, employment,...) to countries. When a tourist visits a country or city, he/she wants to know their points of interest. To do this, he/she must select some of the places according to his/her preferences and design routes to visit them. This problem can be adequately modeled as a Team Orienteering Problem with Time Windows (TOPTW). In this paper we propose a fuzzy GRASP and a multi-agent simulation model to solve the TOPTW. Our proposal incorporates two criteria to build the restricted candidate list. The computational results obtained show the validity of the proposal.

Keywords: Tourism · GRASP · Fuzzy · Simulation model

1 Introduction

Tourism plays an important role in the economy of the countries and represents an important source of income. According to the World Tourism Organization[1], 1400 million people moved around the world in 2019. A significant percentage of these tourists want to visit the points of interest of the country or city they visit. Travel agencies and specialized companies offer routes to visit these points that generally do not satisfy the preferences of all tourists. An interesting route for one tourist may not be suitable for another. It is therefore necessary to design tools that provide routes adapted to the preferences of each tourist. In addition to tourist preferences, other factors must be considered to build customized routes. Some of the most relevant are the importance of the points of interest, the duration of the routes and the opening hours.

The previous problem can be modeled as a Team Orienteering Problem with Time Windows (TOPTW). The TOPTW is an extension of the Team Orienteering Problem (TOP) [4], in which, given a set of visiting points with a score

---

[1] https://www.unwto.org.

© Springer Nature Switzerland AG 2020
M.-J. Lesot et al. (Eds.): IPMU 2020, CCIS 1238, pp. 715–727, 2020.
https://doi.org/10.1007/978-3-030-50143-3_56

and service time, the objective is to maximize the sum of the collected scores in a given number of routes, while guarantying a maximum travel time in each route. For solving the TOP, different exact an heuristic algorithms have been proposed [6]. In the TOPTW, each visiting point has to be visited within a given time window.

In this paper we present a Fuzzy Greedy Randomized Adaptive Search Procedure (GRASP) to solve the TOPTW. GRASP is a metaheuristic which consists of two phases. In the constructive phase, a solution is built by iteratively selecting an element of a restricted candidate list. In the improvement phase, the above solution is improved by applying a local search. These steps are repeated until a stopping rule is met. In our case, the restricted list of candidates is formed by the appropriate points of interest to be included in some route. Unlike the standard GRASP, Fuzzy GRASP uses two criteria to build this restricted candidate list. The first criterion evaluates the point of interest by attending to time increment of the route. The second criterion uses the score to evaluate the point of interest. On the other hand, we also propose a multi-agent simulation model aimed at handling the randomness of the optimization problem in practical scenarios. We consider that the scores associated to the points of interest are not deterministic, but can have low, moderate and high levels of variability instead. This means that some points have been clearly bad or good for most visitors, but others have not.

We use a classic set of instances to test our proposal. The obtained computational results show its good behavior. Fuzzy GRASP efficiently identifies high quality solutions in insignificant computational time.

2   Problem Description

The problem addressed in this paper is modelled as a multiple route-planning problem, specifically as a TOPTW. It consists of a set I of n points of interests (POIs), each with a given score, and a set K of m routes with the aim of visiting some of them in a given limited time. Each POI has associated a score or profit, a visit time, and a time windows. The starting and ending points are fixed, and represent tourists place of stay. However, not all POIs have to be visited. The number of routes corresponds to the number of stay days at destination. The objective function is to maximize the total collected score.

The POIs are identified by an index i, $i = 1, 2, ..., n$. Indexes 1 and n represent the starting and ending POIs. Each route is represented by an index k, $k = 1, 2, ..., m$. The score obtained in each POI is s_i, $i = 1, 2, ..., n$. The time spent by the tourist when visiting POI i is r_i, $i = 1, 2, ..., n$. The travel time from POI i to POI j is denoted by t_{ij}, $i, j = 0, 1, ..., n$. T_{max}^k is the limit time for route k, $k = 1, ..., m$, considering travel, visit and waiting times. m is the number of tourist stay days. The time windows are represented by e_i, l_i, the opening and closing times of a POI i, respectively; $i = 1, 2, ..., n$.

The variable decisions of the mathematical model are: x_{ij}^k, y_i^k, and a_i^k, $i, j = 1, ..., n$, $k = 1, 2, ..., m$. x_{ij}^k is a binary variable that takes value 1 if route k goes from POI i to j, and 0 otherwise. The binary variable y_i^k is equal to 1 if POI i is visited in route k and $y_j^k = 0$ otherwise. The arrival time at POI i in route k is contained in variable a_i^k.

On the other hand, the TOPTW can be formulated as a linear programming problem (LP) as follows:

Maximize:

$$\sum_{k=1}^{m}\sum_{i=1}^{n} s_i y_i^k \tag{1}$$

$$\sum_{k=1}^{m}\sum_{j=1}^{n} x_{0j}^k = \sum_{k=1}^{m}\sum_{i=1}^{n} x_{i0}^k = m \tag{2}$$

$$\sum_{k=1}^{m} y_i^k \leq 1 \ i \in I \tag{3}$$

$$\sum_{j=0}^{n} x_{ji}^k = \sum_{j=0}^{n} x_{ij}^k = y_i^k \ k \in K, i \in I \tag{4}$$

$$a_i^k + r_i + t_{ij} - a_j^k \leq M(1 - x_{ji}^k) \ k \in K, i,j \in I^0 \tag{5}$$

$$a_i^k + t_{i0} \leq T_{max}^k \ k \in K, i \in I \tag{6}$$

$$e_i \leq a_i^k \leq l_i \ k \in K, i \in I \tag{7}$$

$$x_{ij}^k \in \{0,1\} \ k \in K, i,j \in I^0 \tag{8}$$

$$y_i^k \in \{0,1\} \ k \in K, i \in I \tag{9}$$

$$a_i^k \geq 0 \ k \in K, i \in I \tag{10}$$

The objective function is represented in Eq. (1). The (2) constraints guarantee that routes start and end at point 0. The constraints (3) ensure that every location is visited at most once. The subtour elimination and flow conservation rule in constraints (4) and (5) establish the connectivity and time of each tour, where M is a large constant. The constraints (6) guarantee the maximum time for each tour. The time windows constraints are identified in (7). The condition variables are defined in (8), (9) and (10).

The values of x_{ij}^k, $k = 1, 2, ..., m$, $i, j = 0, 1, ..., n$ variables define the routes. With these variables, the selected POIs can be obtained by Eqs. (4). The combination of (5) and (7) leads to the arrival time at each POI j is iteratively given by

$$a_j^k = \max\{a_i^k, e_i\} + r_i + t_{ij} \tag{11}$$

for the previously point i in this day k; i.e., such that $x_{ij}^k = 1$.

3 Literature Review

The Team Orienteering Problem with Time Windows (TOPTW) [18] has been extensively addressed in the literature over the last decade. For a comprehensive survey on variants of the orienteering problem, we refer the reader to the paper by Gunawan et al. [6].

Due to the complexity of the TOPTW, most papers in the scientific literature tackle it by means of heuristic algorithms. Among the metaheuristic algorithms proposed for solving the TOPTW, there are implementations of Iterated Local Search, Local Search algorithm, Simulated Annealing, Variable Neighborhood Search, Large Neighborhood Search and Greedy Randomized Adaptive Search (GRASP).

A simple, fast and effective iterated local search that combines insert and shake steps to escape from local optima is presented in [18]. A variable neighborhood search algorithm for a multiple time windows extension has been addressed in [17]. A hybrid algorithm that combines GRASP with the evolutionary local search is designed in [10]. A Local Search algorithm is proposed in [9]. Two versions of a simulated annealing algorithm are developed in [11]. A variable neighborhood search procedure based on exploring granular instead of complete neighborhoods is presented in [9]. Granular exploration allows to obtain an efficient and effective algorithm. A Simulated Annealing with restart strategy is proposed in [12] for a multiple time windows extension. A Capacitated TOPTW, in which customers have associated also demands is presented in [2]. An iterated local search and a hybrid algorithm based on simulated annealing and iterated local search are proposed in [7]. A new variant of the TOPTW that includes mandatory visits is considered in [13]. In order to solve this problem, a multistart simulated annealing is developed. A large neighborhood search algorithm is proposed in [15]. A multi-objective version of the TOPTW, in which the score has to be maximized and the time needed for the itinerary of the tourist is minimized, is presented in [8]. It is developed an iterated local search into adjustment iterated local search, in which the construction phase is performed heuristically. The uncertainty theory is used in [19] to consider uncertain travel times when solving the TOPTW. In this regard, simulation is a powerful tool to handle these scenarios. They have been successfully applied to a wide variety of environments. This is illustrated in [5] and [3], among others. Due to its versatility and capability to model specific features of the environment under analysis, simulation based on multiple agents [1] has become in one of the most relevant paradigms. For this reason, this is the approach considered in this paper.

In this work, we propose a Fuzzy GRASP for solving the TOPTW, in which the restricted candidate list used in the construction phase of GRASP is the fuzzy set of the high quality points of interest according to their travel time and score. Furthermore, we propose a multi-agent simulation model to deal with the randomness associated with the optimization problem in practical scenarios.

4   Solution Approach

4.1   Solver

A Fuzzy GRASP (Greedy Randomized Adaptive Search Procedure) is proposed in order to solve the proposed problem and obtain high-quality solutions in reasonable time. The GRASP metaheuristic [14], in its standard version, consists of two phases; a first construction phase, and a second local search improvement phase.

The construction phase obtains a feasible solution by iteratively selecting at random a new location from a Restricted Candidate List (RCL) with size given by the parameter $RCLsize$. Finally, the neighborhood of the solution is explored until a local minimum is found in the second phase. The Algorithm 1 shows the GRASP phases cited above with $maxIterations$ as the maximum number of iterations of the GRASP procedure.

Algorithm 1. GRASP

1: **function GRASP**($maxIterations$, $RCLsize$)
2:   readInput()
3:   **for** i:=1 **to** $maxIterations$ **do**
4:       $solution$ = **GRASPConstructionPhase**($RCLsize$)
5:       $solution$ = **localSearch**($solution$)
6:       **updateSolution**($solution$, $bestSolution$)
7:   **end for**
8:   **return** $bestSolution$
9: **end GRASP**

In Algorithm 2 the construction phase of GRASP is shown. The initial solution has $m = |K|$ empty routes. The procedure starts with the initial solution, and subsequently the construction mechanism adds step-by-step a new POI to the current partial solution guaranteeing the solution feasibility.

The mechanism inserts each new no visited POI i in the best possible position guaranteeing the feasibility of the partial route. The several feasible positions in which insert a new POI i is defined as the (i, j, k) triplet with $y_j^k = 1$. For the partial route k, the mentioned insertion consists in inserting the POI i after j. The variables y_i^k, x_{ji}^k and x_{ih}^k are set to 1 in order to get the new solution. Then for the only index h such that x_{jh}^k is 1; this variable x_{jh}^k is also set to 0.

The variables a_i^k for this route k have to be computed using (11) and feasibility tested by the corresponding time limit constraints (6) and time window constraints (7). A greedy travel time function f computes the travel time increase in the route produced by an insertion represented by the triplet defined previously. The f function is defined as $f(i, j, k) = t_{ji}^k + t_{ih}^k - t_{jh}^k$ for the above index h.

A Restricted Candidate List (RCL) with candidate POIs to be visited in the solution is constructed using the greedy function f. The standard version of GRASP designs the RCL introducing $RCLsize$ feasible insertion triplets (i, j, k) with the best quality values for function f. In this paper a fuzzy version of the GRASP is proposed.

The proposed fuzzy mechanism considers the fuzzy set of the high quality POIs of the RCL based on their score. $\mu(\cdot)$ is the membership function of the

---

Algorithm 2. Construction Phase of the GRASP

1: **function** **GRASPConstructionPhase**($RCLsize$)
2: **Initialize** the *partialSolution* with m empty routes
3: **while** (it is possible to visit new locations) **do**
4:     $CL = \emptyset$
5:     **for all** location $i \in I$ **do**
6:         **Find** the best feasible triplet (i, j, k) to insert this new location i in
            partialSolution according to greedy time function $f(i, j, k)$
7:         **Add** the feasible triplets (i, j, k) to the Candidate List CL
8:     **end for**
9:     **Create** the Restricted Candidate List, RCL, with the best $RCLsize$ triplets
        (i, j, k) from CL according to f
10:    **Select** a random triplet (i, j, k) from RCL
11:    **Update** the variables of route k by inserting the location i at position j
12: **end while**
13: **return** *partialSolution*
14: **end GRASPConstructionPhase**

---

set stated by $\mu(i) = s_i / s_{max}$ with s_i the score related with POI i and s_{max} the highest score in RCL. RCL^* is the set with the best locations in RCL created by the α-cut of RCL related with $\mu(\cdot)$, with $\alpha \in [0, 1]$.

$$RCL^* = \{(i, j, k) \in RCL : \mu(s_i) \geq \alpha\}\}$$

The new POI inserted in the solution is randomly selected from RCL^*. The decision maker is responsible of fixing the α parameter. The choice of this parameter influences the quality of the POI that will be inserted into the solution. The Algorithm 3 shows the fuzzy GRASP construction phase.

Basically, the selection of the best insertion position of POIs in a tour composes the candidate list. The greedy function f evaluates the increase in time due to the POI insertion in partial route. The value of f determine the position of candidate in the list. The best position to insert a candidate POI for all routes is located by the procedure. The goal of the best position is minimize the total time after the insertion of POI. Finally the list is ordered ascending by the increase in time such that lowest time increase candidates are placed at candidate list top. The candidate list size is settled in $RCLsize$ due to build the RCL. Subsequently, the α-cut, that recognizes the best score POIs, conforms RCL^*. The decision maker fix the α value.

The proposed GRASP mix the minimization of travel times in the RCL composition and maximization score in the RCL^* composition. Therefore the RCL^* contains the best POIs that combines the two criteria mentioned. The POIs that will be insert in current partial solution are those with low travel time increase and high score. Later, in RCL^* a POI is randomly selected and inserted into partial solution. The inserted POI is banned for future candidate lists. Finally, the feasible solution is obtained and construction phase ends.

The next phase is an improvement phase that uses a local search for this purpose. Commonly, a local search procedure performances iteratively by replacing

---

Algorithm 3. Construction Phase of the Fuzzy GRASP

1: **function GRASPFuzzyConstructionPhase**($RCLsize$, α)
2: **Initialize** the $partialSolution$ with m empty routes
3: **while** (it is possible to visit new locations) **do**
4:    $CL = \emptyset$
5:    **for all** location $i \in I$ **do**
6:       **Find** the best position j to insert i in a partial route k of the $partialSolution$
         according to greedy time function $f(i, j, k)$
7:       **Add** the feasible triplets (i, j, k) to the Candidate List CL
8:    **end for**
9:    **Create** the Restricted Candidate List, RCL, with the top $RCLsize$ triplets
      (i, j, k) from CL
10:   **Get** $RCL^* = \{(i, j, k) \in RCL : \mu(s_i) \geq \alpha\}$
11:   **Select** a random triplet (i, j, k) from RCL^*
12:   **Update** the variables of route k by inserting the location i at position j
13: **end while**
14: **return** $partialSolution$
15: **end GRASPFuzzyConstructionPhase**

---

the current solution with a improved solution achieved in the neighborhood. The stop criteria of local search is not to find a better solution in the neighborhood. The Algorithm 4 presents a standard local search procedure. The proposed local search applies exchange movements between POIs in order to minimize the total travel time. The POIs can be selected in the same or different routes. A best-improving strategy is used, so that the search explores all neighborhoods in order to replace the current solution by the best solution found. Once the total travel time of the solution has been improved, the procedure tries to insert a new POI in the solution with the aim of maximize the solution score.

Algorithm 4. FUZZY GRASP improvement phase

1: **function localSearch**($solution$)
2: $s = solution$
3: **repeat**
4:    **Find** the best neighbor n of current solution s according the total time
5:    **if** ($TotalTime(n) \leq TotalTime(s)$) **then**
6:       $s = n$
7:    **end if**
8: **until** $TotalTime(n) \geq TotalTime(s)$ for all neighbor n
9: **return** s
10: **end localSearch**

Overall, the aim of the proposed solver is to maximize the total solution score. The procedure is an two-phases iterated algorithm that progress until the stop criteria is met. The introduced fuzzy mechanism in the standard GRASP allows to consider two criteria in construction of the RCL list; the minimization of travel time and the maximization of score.

4.2 Simulation Model

In this paper, we also propose a simulation model aimed at handling randomness associated with the optimization problem in practical scenarios. This randomness is derived from the potential changes in the preferences of the visitors, inadequate treatment of the staff in the points or unadvertised changes in the time windows, among others. Thus, the scores of the points of interest are considered as random variables instead of constant values, as done in previous works found in the scientific literature so far.

We design a multi-agent simulation approach that uses asynchronous time and receives a feasible solution of the problem. This solution is provided by the solver previously described. It also receives the random variables that defines the behaviour of the scores. The model is composed of a hierarchy of agents, The base agent is governed by the state machine shown in Fig. 1 and is aimed at simulating a route of the solution. The nodes in the state machine represent states of the visitor during its route, whereas the arcs represent transitions between pairs of consecutive states. This way, a transition is applied when the triggering event occurs. As can be checked, the visitor moves to the next point of interest and waits until it is open. This is visited once the point is open. Then, next point is checked or the visitor goes to its end point. It is worth mentioning that the score obtained when visiting a point of interest is derived from sampling the random variable that defines it. At the same time, there is a manager agent which is responsible for assign routes to the base agents and managing the performance metrics (*i.e.*, number of points visited, time in each point, etc.). Thus, the model is composed of one manager agent and m base agents.

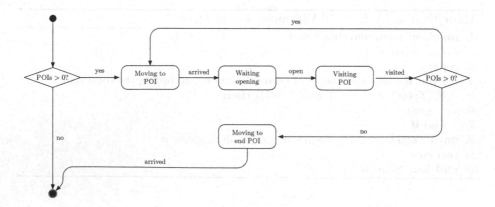

Fig. 1. State machine of an agent aimed at simulating a route

5   Computational Experiments

5.1   Solver Computational Experiments

This section describes the results from the first computational experiment carried out in our study. The aim of this experiment is to evaluate the accuracy of the Fuzzy GRASP metaheuristic proposed to solve the TOPTW and get high-quality solutions.

The benchmark suite tackled along this section is based on the classical set of instances introduced by [16] for Vehicle Routing Problem with Time Windows. Vansteenwegen et al. [18] provide a set of locations with a given score which can be visited. A time windows and service time have been defined for each location. The limit number of routes match with the number of vehicles of the original Solomon data set. The number of locations used is equal to 100 and the limit time per route varies according to the specific instance. This benchmark considers travel times equal to euclidean distances. Regarding the geographical data, there are three kinds of instances: instances with clustered customers (C), instances with random customers (R) and a mixture of random and clustered customers (RC).

The instances data set used are described in Table 1 in more detail.

The parameter of the proposed Fuzzy GRASP are:

1. The α value of α-cut of $\mu(s_j)$
2. The size $sRCL$ of the GRASP restricted candidate list

The parameter values of the proposed solver assess during the computational experiments are shown in Table 2.

The GRASP procedure was run 1000 times for each instance and parameter combination used in computational experiments. The results are presented in Table 3 and show the comparison between the best solutions obtained by standard GRASP and Fuzzy GRASP, and the optimal solution. The score of the optimal solutions is the sum of scores for all locations. Column one and two give the name of the instance and the score of the optimal solution, respectively. The third and fourth columns show the best solutions scores of standard and Fuzzy GRASP, respectively. The gap between the best solutions of the standard and Fuzzy GRASP and the optimal solution are given by the fifth and sixth columns, respectively. Finally, the last column shows the execution time of Fuzzy GRASP. The computational experiments confirm that Fuzzy GRASP is highly efficient when solving the instances under analysis.

5.2   Simulation Analysis

We have carried out several experiments dedicated to check the suitability of the simulation model proposed in Sect. 4.2. In this case, the score associated with the points of interest is modelled as a random variable, S, that does not follow any arbitrary non-negative probability distribution. In particular, a Log-Normal

Table 1. Instances used in experimentation

Instance	Size	Max. routes	Max. time
c101	100	10	1236
c102	100	10	1236
c103	100	10	1236
c104	100	10	1236
c105	100	10	1236
c106	100	10	1236
c107	100	10	1236
c108	100	10	1236
c109	100	10	1236
r101	100	19	230
r102	100	17	230
r103	100	13	230
r104	100	9	230
r105	100	14	230
r106	100	12	230
r107	100	10	230
r108	100	9	230
r109	100	11	230
r110	100	10	230
r111	100	10	230
r112	100	9	230
rc101	100	14	240
rc102	100	12	240
rc103	100	11	240
rc104	100	10	240
rc105	100	13	240
rc106	100	11	240
rc107	100	11	240
rc108	100	10	240

Table 2. Parameters used by Fuzzy GRASP

Parameter	Values
α value of α-cut	$\alpha \in \{0.1,\ 0.2,\ 0.3,\ 0.4,\ 0.5,\ 0.6,\ 0.7,\ 0.8,\ 0.9\}$
Size of RCL	$sRCL \in \{3, 4, 5, 6, 7, 8, 9, 10, 11, 12, 13, 14, 15, 16, 17, 18, 19, 20\}$

Table 3. Comparative results

Instance	Opt.	GRASP	Fuzzy GRASP	Gap (%) Fuzzy GRASP	Gap (%) GRASP	Fuzzy GRASP Execution time (ms)
c101	1810	1600	1700	6.1	11.6	7.07
c102	1810	1730	1780	1.7	4.4	4.17
c103	1810	1800	1810	0.0	0.6	5.32
c104	1810	1810	1810	0.0	0.0	71.45
c105	1810	1710	1770	2.8	5.5	4.66
c106	1810	1710	1770	2.8	5.5	3.93
c107	1810	1800	1810	0.0	0.6	3.84
c108	1810	1810	1810	0.0	0.0	4.18
c109	1810	1810	1810	0.0	0.0	4.94
r101	1458	1429	1447	0.7	2.0	4.01
r102	1458	1441	1444	0.8	1.2	4.95
r103	1458	1421	1443	1.0	2.5	5.45
r104	1458	1352	1403	3.8	7.3	5.32
r105	1458	1415	1435	1.6	2.9	3.93
r106	1458	1379	1418	2.6	5.4	4.77
r107	1458	1366	1405	3.8	6.3	5.51
r108	1458	1391	1418	2.0	4.6	5.55
r109	1458	1386	1403	3.3	4.9	4.38
r110	1458	1356	1401	3.5	7.0	5.29
r111	1458	1377	1421	3.6	5.6	5.87
r112	1458	1372	1414	3.6	5.9	5.58
rc101	1724	1658	1640	4.5	3.8	3.88
rc102	1724	1566	1677	4.4	9.2	4.48
rc103	1724	1628	1688	2.6	5.6	4.86
rc104	1724	1693	1696	1.8	1.8	5.70
rc105	1724	1636	1665	2.9	5.1	4.50
rc106	1724	1595	1631	5.1	7.5	4.02
rc107	1724	1652	1688	2.1	4.2	4.86
rc108	1724	1633	1674	3.0	5.3	5.77

distribution with $Var[S] = k \cdot E[S]$. $k = \{0.25, 1, 2\}$ to represent scenarios with low, moderate, and high levels of variability.

Table 4 shows the average computational results obtained by the proposed model when simulating the behaviour of 10 solutions selected at random. In this case, the score, travelled distance, total time, minimum time in the routes, maximum time in the routes, and slack time used to satisfy the time windows are shown when changing the level of variability in the scenarios. Due to the fact that the variability impacts only on the score of the points of interest, the remaining metrics are not altered in the different scenarios and their values are only provided for further analysis. In this case, the results indicate that the score obtained when visiting the involved points of interest in the routes increases when the variability of the scenarios is higher.

Table 4. Computational results obtained by the simulation model

	$k = 0.25$	$k = 1$	$k = 2$
Score	1302.606	1304.991	1308.624
Distance	1100.106	1100.106	1100.106
Time	1970.106	1970.106	1970.106
Min. time	201.583	201.583	201.583
Max. time	226.534	226.534	226.534
Slack	6388.068	6388.068	6388.068

6   Conclusions and Future Research

This work proposes a Fuzzy GRASP metaheuristic and a multi-agent simulation model to solve the Team Orienteering Problem with Time Windows, in which the scores associated to the points of interest are not deterministic. In the first place, we develop a Fuzzy GRASP to solve the optimization problem with deterministic scores to obtain a set of high-quality solutions. Then, these solutions are used by the designed multi-agent simulation model to consider low, moderate and high levels of variability in the scores.

The computational experiments carried out in this work corroborate that the GRASP metaheuristic can reach a set of high-quality solutions in reasonable computational time. Moreover, the simulation model is a tool able to perform experiments about the impact of changes in the score of the points of interest on the quality of the solutions in realistic scenarios. However, there are still several promising lines for further research. In particular, uncertainty can be associated with travelling times between pairs of points of interest, and therefore the optimization and simulation approaches must be modified appropriately to handle this. At the same time, it is highly interesting to develop recommendation systems to propose visiting routes to the tourists according to their individual features.

Acknowledgment. Airam Expósito-Márquez would like to thank the Canary Government for the financial support he receives through his post-graduate grant.

References

1. Abar, S., Theodoropoulos, G.K., Lemarinier, P., O'Hare, G.M.P.: Agent based modelling and simulation tools: a review of the state-of-art software. Comput. Sci. Rev. **24**, 13–33 (2017)

2. Aghezzaf, B., El Fahim, H.: Solving the capacitated team orienteering problem with time windows through variable neighborhood search. Int. Rev. Comput. Softw. **10**(11), 1134–1142 (2015)
3. Amaran, S., Sahinidis, N.V., Sharda, B., Bury, S.J.: Simulation optimization: a review of algorithms and applications. Ann. Oper. Res. **240**(1), 351–380 (2015). https://doi.org/10.1007/s10479-015-2019-x
4. Chao, I.M., Golden, B.L., Wasil, E.A.: The team orienteering problem. Eur. J. Oper. Res. **88**(3), 464–474 (1996)
5. Günal, M.M., Pidd, M.: Discrete event simulation for performance modelling in health care: a review of the literature. J. Simul. **4**(1), 42–51 (2010)
6. Gunawan, A., Lau, H.C., Vansteenwegen, P.: Orienteering problem: a survey of recent variants, solution approaches and applications. Eur. J. Oper. Res. **255**(2), 315–332 (2016)
7. Gunawan, A., Lau, H.C., Vansteenwegen, P., Lu, K.: Well-tuned algorithms for the team orienteering problem with time windows. J. Oper. Res. Soc. **68**(8), 861–876 (2017)
8. Hapsari, I., Surjandari, I., Komarudin, K.: Solving multi-objective team orienteering problem with time windows using adjustment iterated local search. J. Ind. Eng. Int. **15**(4), 679–693 (2019). https://doi.org/10.1007/s40092-019-0315-9
9. Labadie, N., Mansini, R., Melechovsky, J., Wolfler Calvo, R.: The team orienteering problem with time windows: an LP-based granular variable neighborhood search. Eur. J. Oper. Res. **220**(1), 15–27 (2012)
10. Labadie, N., Mansini, J., Melechovsky, J., Wolfler Calvo, R.: Hybridized evolutionary local search algorithm for the team orienteering problem with time windows. J. Heuristics **17**(6), 729–753 (2011)
11. Lin, S.-W., Yu, V.F.: A simulated annealing heuristic for the team orienteering problem with time windows. Eur. J. Oper. Res. **217**(1), 94–107 (2012)
12. Lin, S.-W., Yu, V.F.: A simulated annealing heuristic for the multiconstraint team orienteering problem with multiple time windows. Appl. Soft Comput. J. **37**, 632–642 (2015)
13. Lin, S.-W., Yu, V.F.: Solving the team orienteering problem with time windows and mandatory visits by multi-start simulated annealing. Comput. Ind. Eng. **114**, 195–205 (2017)
14. Resende, M.G.C., Ribeiro, C.C.: Optimization by GRASP: Greedy Randomized Adaptive Search Procedures. Springer, New York (2016)
15. Schmid, V., Ehmke, J.F.: An effective large neighborhood search for the team orienteering problem with time windows. In: Bektaş, T., Coniglio, S., Martinez-Sykora, A., Voß, S., et al. (eds.) ICCL 2017. LNCS, vol. 10572, pp. 3–18. Springer, Cham (2017). https://doi.org/10.1007/978-3-319-68496-3_1
16. Solomon, M.M.: Algorithms for the vehicle routing and scheduling problems with time window constraints. Oper. Res. **35**(2), 254–265 (1987)
17. Tricoire, F., Romauch, M., Doerner, K.F., Hartl, R.F.: Heuristics for the multi-period orienteering problem with multiple time windows. Comput. Oper. Res. **37**(2), 351–367 (2010)
18. Vansteenwegen, P., Souffriau, W., Vanden Berghe, G., Van Oudheusden, D.: Iterated local search for the team orienteering problem with time windows. Comput. Oper. Res. **36**, 3281–290 (2009)
19. Wang, J., Guo, J., Chen, J., Tian, S., Gu, T.: Uncertain team orienteering problem with time windows based on uncertainty theory. IEEE Access **7**, 63403–63414 (2019)

General-Purpose Automated Machine Learning for Transportation: A Case Study of Auto-sklearn for Traffic Forecasting

Juan S. Angarita-Zapata[1]([⊠]) [iD], Antonio D. Masegosa[1,2] [iD],
and Isaac Triguero[3] [iD]

[1] DeustoTech, Faculty of Engineering, University of Deusto,
Av. Universidades, 24, 48007 Bilbao, Spain
{js.angarita,ad.masegosa}@deusto.es
[2] IKERBASQUE, Basque Foundation for Science, 48011 Bilbao, Spain
[3] Computational Optimisation and Learning (COL) Lab,
School of Computer Science, University of Nottingham, Nottingham, UK
Isaac.Triguero@nottingham.ac.uk

Abstract. Currently, there are no guidelines to determine what are the most suitable machine learning pipelines (i.e. the workflow from data pre-processing to model selection and validation) to approach Traffic Forecasting (TF) problems. Although automated machine learning (AutoML) has proved to be successful dealing with the model selection problem in other applications areas, only a few papers have explored the performance of general-purpose AutoML methods, purely based on optimisation, when tackling TF. In this paper, we provide a thorough exploration of the benefits of Auto-sklearn for TF, as a general-purpose AutoML method that follows a hybrid search strategy combining optimisation with meta-learning and ensemble learning. Particularly, we focus on how well Auto-sklearn is able to recommend competitive machine learning pipelines to forecast traffic, modelled as a TF multi-class imbalanced classification problem, along different time horizons at two spatial scales (point and road segment) and two environments (freeway and urban). Concretely, we test the following scenarios: I) a hybrid search strategy with the three components (optimisation, meta-learning, ensemble learning), II) a strategy based on meta-learning and ensemble learning, and III) a strategy based on the estimation of the best performing pipeline from those suggested by the meta-learning. Experimental results show that the meta-learning component of Auto-sklearn does not work properly on TF problems, and on the other hand, that the optimisation does not contribute too much to the final performance of predictions.

Keywords: Traffic Forecasting · Transportation · Supervised learning · Machine learning · Automated machine learning · Computational intelligence

© Springer Nature Switzerland AG 2020
M.-J. Lesot et al. (Eds.): IPMU 2020, CCIS 1238, pp. 728–744, 2020.
https://doi.org/10.1007/978-3-030-50143-3_57

1 Introduction

A well-established strategy to tackle congestion is the design, development and implementation of TF systems. TF can be defined as the prediction of near-future traffic conditions (e.g. travel time) [16]. The recent emergence of telecommunications technologies integrated into transportation infrastructure generates vast volumes of traffic data. This unprecedented data availability and growing computational capacities have incremented the use of Machine Learning (ML) to address TF. From a ML perspective, TF is focused on building a predictive model using historical data to make predictions of traffic measures based on new and unseen data.

In spite of the aforementioned progress, different ML algorithms and preprocessing approaches may be more appropriate for different kinds of traffic data. Determining the best pipeline (sequence of data preprocessing techniques and a learning algorithm) for making traffic predictions is not a trivial task. In the ML area, this challenge is known as the Model Selection Problem (MSP) and Automated Machine Learning (AutoML) has been one of the most successful approaches addressing it so far. AutoML aims at automatically finding the best combination of preprocessing techniques, ML algorithm and hyperparameters that maximise a performance measure on given data without being specialized in the problem domain where this data comes from. The search strategy to find the mentioned combination can be based either on a "pure" optimisation process that tests different promising combinations from a predefined base of preprocessing and learning algorithms [10]; or it can be based on a hybrid search where the optimisation is complemented with learning strategies such as meta-learning [3]. In the latter case, the learning approach is in charge of systematically observing how different ML pipelines perform on a wide range of tasks to take advantage of this experience to learn new tasks faster [14]. Roughly speaking, it can be seen as using ML for designing ML algorithms.

AutoML methods have successfully approached the MSP in other areas [8,18], however, it has hardly been explored in TF [1]. In the latter area, the current progress is focused only on AutoML methods designed purely on optimisation approaches; thus, leaving aside the study of AutoML methods that have hybrid search strategies. Having this idea in mind, the contribution of this paper is to study the benefits in terms of performance and computational cost of hybrid AutoML for TF. We use Auto-sklearn [3], a state-of-the-art hybrid AutoML method whose search strategy of pipelines uses bayesian optimisation, meta-learning and ensemble learning. To accomplish this objective, we use as a benchmark a multi-class imbalanced classification problem for different time horizons and for freeway and urban environments. Under these traffic forecasting settings, we explore the performance of the Auto-sklearn's components through three scenarios: I) a hybrid search strategy that uses its three components (optimisation, meta-learning, ensemble learning), II) a meta-learning strategy combined with ensemble learning, and III) a strategy based on the estimation of the best performing pipeline from those suggested by the meta-learning.

The rest of this paper is structured as follows. Sections 2 and 3 present background and related work about AutoML methods in TF. Section 4 exposes the methodology followed in this paper. Then, Sect. 5 analyzes the main results obtained. Finally, conclusions are discussed in Sect. 6.

2   Background

This section reviews literature related to AutoML in the context of TF. We start presenting the foundations of general-purpose AutoML methods and finally, Sect. 2.2 reviews Auto-sklearn, the state-of-the-art hybrid AutoML method used in this research.

2.1   Automated Machine Learning

According to [18], a ML pipeline P can be defined as a combination of algorithms A that transforms input data X into target values Y. Let A be defined as

$$A = \{A_{preprocessing} \cup A_{feature} \cup A_{algorithm}\} \tag{1}$$

wherein $A_{preprocessing}$ is a subset of preprocessing techniques, $A_{feature}$ a subset of feature engineering methods, and $A_{algorithm}$ a ML algorithm with configuration of hyperparameters $\lambda^i \in \Lambda$. In order to build a ML pipeline with this structure, human effort and high computational capacities are needed because there is no pipeline that can achieve good performance on every learning problem [6, 17]. This usually is done by means of a trial and error approach in an iterative manner, which causes that the success of ML comes at a great price [17].

AutoML is an emerging sub-area in ML that seeks to automatise the ML workflow from data preprocessing to model validation [5]. It allows reducing human bias and improving computational costs by making the construction of ML applications more efficient. The process consists of identifying the most promising combination P_{A^i,λ^i} that satisfies a given performance metric or condition when P_{A^i,λ^i} is trained on training data $D_{train}^{(i)}$ and evaluated on test data $D_{test}^{(i)}$.

Current literature [5, 18] reports a variety of general-purpose AutoML methods. According to Chen et al. [17], there are two types of taxonomies that can categorise these methods. First, a "what" taxonomy that determines which stages of a ML pipeline are going to be automated (e.g., data preprocessing and algorithm selection, algorithm selection and hyperparameters, or even the entire pipeline). Within this taxonomy, the most common case is the CASH [13] (Combined Algorithm Selection and Hyperparameter) problem wherein AutoML is focused on finding the best combination of ML algorithm and its hyperparameters setting, leaving the data preprocessing up to the human user. In this paper, we are focused on the automation of fixed-size ML pipelines composed of data preprocessing techniques and a classifier algorithm with their respective hyperparameters configurations.

In contrast, the second taxonomy proposed by [17] classifies how the automation process to find the most promising pipeline is done. On the one hand, some AutoML methods use only an optimisation strategy wherein the ML pipeline is built testing multiple possible combinations from a predefined search space of preprocessing methods, ML algorithms and hyperparameters configurations. From this perspective, the ML pipeline building problem consists of finding a pipeline structure $P_{A,\lambda}$ that minimises a cross-validation loss function

$$P_{A^*,\lambda^*} = \underset{A^{(i)} \in A, \lambda^{(i)} \in \Lambda}{argmin} \frac{1}{K} \sum_{i=1}^{k} \gamma \left(P_{A^{(i)},\lambda^{(i)}}, D_{train}, D_{test} \right) \qquad (2)$$

As shown in Eq. 2, this search process can be considered as a black-box optimisation problem that is not easily solvable as the search space can be large and complex. This equation is usually non-smooth and derivative-free, and convergence speed is a critical problem for building ML pipelines. Some methods to solve this equation are grid search, random search, bayesian optimisation and sequential Model-Based Optimisation.

On the other hand, regarding the second taxonomy proposed by [17], there are other AutoML methods that use the aforementioned optimisation in combination with learning strategies to constitute a hybrid search strategy with the purpose of reducing computational costs. In this case, the focus is on applying a ML algorithm at the meta-level to learn meta-knowledge that guides the AutoML process; this approach is known as meta-learning [7,14]. Meta-learning is the data-driven task of systematically observing how different learning algorithms or pipelines perform on different learning tasks and then learning from this experience to warm-start the optimisation process in a new and unknown ML task. This warm-start consists of promising pipelines that are used by the optimisation as starting points to be evaluated in the first place before trying pipelines extracted from a predefined search space.

Meta-learning can extract meta-knowledge using three different strategies [14]. First, there is the "Learning from prior evaluations" strategy wherein a set of known-previous learning tasks $t_j \in T$, a set of configurations $\lambda_i \in \Lambda$ (e.g., hyperparameter settings), and a set of all prior evaluations $P_{i,j}$ coming from applying the configurations Λ over the tasks T, have to be given. Having this knowledge, the objective is to train a meta-learner L able to recommend promising hyperparameters configurations λ_i for an unseen and new task t_{new}. In contrast, the second approach is known as "Learning from task properties". It is based on characterising the known-previous learning tasks $t_j \in T$ using meta-features $m_j \in M$ (e.g, number of instances and features, class imbalance), then extracting the configurations $\lambda_i \in \Lambda$ of the learners associated with these prior tasks, and finally collecting the performance $P_{i,j}$ of the trained models given the meta-features m_j and the configurations λ_i. Having this meta-knowledge, the objective is to train a meta-leaner L that predicts the performance of pipelines or recommend them for an unseen and new task $task_{new}$.

Lastly, the third approach is "Learning from prior models". In this case, the focus is on training a meta-learner given the parameters $m_j \in M$ (e.g. model parameters, features) of prior learnt models, their configurations λ_i, and the performance $P_{i,j}$ of these learners over the previous and known tasks. Then, the objective is to train a meta-learner L that transfers trained models to save computational costs at the moment of approaching a new task t_{new}.

Within this paper, we focus on Auto-sklearn, the AutoML method with a hybrid search strategy that includes optimisation and meta-learning based on the approach "Learning from task properties". This method is presented with more details in the following section.

2.2 Auto-sklearn

Auto-sklearn is an AutoML method that uses meta-learning, bayesian optimisation and ensemble selection to find promising ML pipelines composed of preprocessing methods and ML classifiers. Here we provide a brief description of the method. The interested reader is referred to [3] for further details.

In an off-line phase, for a repository of 121 data-sets, bayesian optimisation is used to determine an optimised ML pipeline with high performance on every data-set. These pipelines are generated from a search space of 15 classifiers, 14 feature preprocessing methods, and 4 data preprocessing methods. Then, for each data-set, a set of 38 meta-features is extracted to characterise every set of data; these meta-features include simple, information-theoretic and statistical information such as statistics about the number of data points, features, the number of classes, data skewness, the entropy of the targets, among others. Later on, instead of storing the 121 data-sets, their meta-features and the ML pipelines are saved in a meta-knowledge base wherein each instance contains the set of meta-features describing every data-set and the optimised pipeline that works well on it.

In the online phase, that is, when a new data-set D_{new} is given, Auto-sklearn computes its meta-features, ranks all the data-sets stored in the meta-knowledge base (stored in the form of meta-features and not the data itself) by their $L1$ distance w.r.t. D_{new}, and selects the stored ML pipelines for the k nearest data-sets (by default $k = 25$). The assumption is that these selected pipelines are likely to perform quite well in D_{new} as they performed well on data-sets with similar meta-features (pipelines closer to the first position of the ranking would expect higher performance on D_{new}). This selection of K most promising pipelines is used then to seed the bayesian optimisation component as a warm-start approach, which boosts the performance of the optimisation. In addition to the recommendations done by the meta-learning component, the bayesian optimisation process (under a time budget constraint) generates and tests new pipeline structures from the same aforementioned search space. In the final step of Auto-sklearn's workflow, the best pipelines identified during the bayesian search process are used to construct an ensemble. This automated ensemble construction avoids to commit itself to a single hyper-parameter setting, and it is more robust than only using the best pipeline found with the optimisation component.

3   Related Work

Within the most representative AutoML methods are Auto-WEKA [13], Auto-sklearn [3], TPOT [10], ATM [12], and ML-Plan [9]. In the case of the first two methods, they are focused on the construction of fixed ML pipelines in which the pipeline structure is a linear sequence of data preprocessing and algorithm learning. The other methods work by building pipeline structures that can be more complex and diverse. In the case of Auto-WEKA, TPOT, ATM and ML-Plan, they use an optimisation approach to find pipeline structures; meanwhile, Auto-sklearn is the state-of-the-art method to generate ML pipelines using a hybrid search strategy. As a common denominator, all these AutoML methods are agnostic w.r.t. the problem domains in which they have been applied; in this sense, they are general-purpose methods that have shown competitive performance in different applications areas [5].

In the transportation area, to the best authors' knowledge, only three papers have used AutoML methods for TF [1,2,15]. The first research carried out by Vlahogianni et al. [15] proposed a meta-modelling technique that, based on surrogate modelling and a genetic algorithm with an island model, optimises both the algorithm selection and the hyper-parameter setting. The AutoML task is performed from an algorithms base of three ML methods (Neural Network, Support Vector Machine and Radial Base Function) that forecast average speed in a time horizon of 5 min, using a regression approach. After that, Angarita et al. in [1] and [2] used Auto-WEKA, an AutoML method that applies sequential model-based bayesian optimisation [4] to find optimal ML pipelines. Both papers compared the performance of Auto-WEKA w.r.t. the general approach, which consists of selecting by trial and error the best of a set of algorithms to predict traffic. In the case of [2], the paper was centred in forecasting traffic LoS at a fixed freeway location through multiple time horizons. On the other hand, in [1], the authors were focused on predicting traffic speed on a subset of families of TF regression problems focused on making predictions at the point and the road segment levels within the freeway and urban environments.

The main differences between this research and the three aforementioned papers lay on the typology of AutoML method used and the addressed TF problems. Whereas the previous three focused on "pure" AutoML optimization approaches, in this research, we centre on a hybrid strategy based on meta-Learning, optimization and ensemble learning with the purpose of evaluating the benefits that the former has on TF within three scenarios: by its own (without optimization and ensemble learning), in combination solely with ensemble learning, and integrated to optimization plus ensemble learning.

4   Methodology

This research seeks to keep exploring the benefits that meta-learning within AutoML of hybrid search strategies can bring to TF. To accomplish such purpose, we analyse to what extent Auto-sklearn, the state-of-the-art AutoML

Table 1. Data-sets

Type	Data-sets	# Instances	# Attributes	# Instances per class	Imbalance ratios
Type I	Fw_T+CD in time horizons of 5, 15, 30, 45, 60 min	[10927-9906]	13	A = 4533, B = 3640 C = 893, D = 850	IR (A/D) = 5,07 IR (A/B) = 1,24 IR (A/C) = 5,33
	Fw_TS+CD in time horizons of 5, 15, 30, 45, 60 min	[10927-9906]	28	A = 5983, B = 4580, C = 363	IR (A/B) = 1,30 IR (A/C) = 16,48
Type II	Fw_T+CD in time horizons of 5, 15, 30, 45, 60 min	[10927-9906]	13	A = 4533, B = 4023, C = 136	IR (A/B) = 1,12 IR (A/C) = 3,33
	Fw_TS+CD in time horizons of 5, 15, 30, 45, 60 min	[10927-9906]	28	B = 7782, C = 2125, A = 101	IR (B/C) = 3,66 IR (B/A) = 7,63
Type I	Ub_T+CD in time horizons of 15, 30, 45, 60 min	[2684-2634]	13	A = 1337, B = 1188, C = 111	IR (A/B) = 1,12 IR (A/C) = 12,04
	Ub_TS+CD in time horizons of 15, 30, 45, 60 min	[2684-2634]	28	B = 1659, A = 691, C = 33	IR (B/A) = 2,40 IR (B/C) = 4
Type II	Ub_T+CD in time horizons of 15, 30, 45, 60 min	[2684-2634]	13	A = 1337 B = 1299	IR (A/B) = 1,02
	Ub_TS+CD in time horizons of 15, 30, 45, 60 min	[2684-2634]	28	B = 1561 A = 1122	IR (B/A) = 1,39

method for this category of search strategies, is able to recommend competitive ML pipelines for TF. In this context, the following parts of these sections are devoted to giving more details about the data-sets used for the experimentation (Sect. 4.1); and the experimental set-up of this study (Sect. 4.2).

4.1 Data-Sets

For experimentation, we considered two TF environments: freeway and urban. For the freeway environment, the data was collected from the Caltrans Performance Measurement System[1] whereas for the urban one, the data was collected from the Madrid Open Data Portal[2]. In both cases, the traffic measure used was thee months of speed in aggregation times of 5 and 15 min, respectively. For more details about the raw data used to generate the data-sets employed in this research, the interested reader is referred to [1].

Concretely, we approach two types of TF classification problems with two problem instances for each of them. In both problems, the objective is to predict

---

[1] http://pems.dot.ca.gov.
[2] https://datos.madrid.es/portal/site/egob/.

a categorical measure named LoS as a multi-class classification problem based on continuous traffic speed. LoS is used to categorise the quality levels of traffic through letters from A to E in a gradual way[3] [11].

The first TF problem corresponds to the prediction of LoS at a target location in a freeway environment. The first instance of this problem is based only on past traffic speed data of the target location (temporal data, T); meanwhile, the second instance considers historical traffic data coming from the target location and from four downstream positions (temporal and spatial, TS). It is important to clarify that these two instances of the first TF problem are correlated because they share the same target location.

The second kind of TF problem is focused on forecasting LoS within an urban context independent of the freeway data described above. Repeatedly, the two correlated instances of this problem are: predict LoS for a single target location considering exclusively historical data of this spot, and forecasting LoS taking into account past traffic speed of the target location together with other four downstream positions.

For the two TF problems described above, we generated 36 data-sets (20 for freeway data and 16 for urban data). In the freeway case, the time horizons wherein LoS is predicted are 5, 15, 30, 45, and 60 min using data granularity of 5 min (granularity means how often and how long the traffic measure is aggregated). Unlike the previous one, for the urban TF problem, the forecasting time steps are 15, 30, 45, and 60 min with data granularity of 15 min. To better identify the data-sets, they are named following the next structure: *Context_InputData_TimeHorizon*.

Attributes of the freeway and urban data-sets where the input is composed of only temporal traffic data from the target location and calendar data are: 1) Day of the week; Minute of the day, 2) Traffic speed of the objective spot at past 5, 10, 15, 20, 25, 30, 35, 40, 45 min for freeway and 15, 30, 45, 60, 75, 90, 105, 120, 135 min for urban, and 3) LoS in the target location. In the case of the freeway and urban data-sets where the input consists of historical speed taken from the target location and from four downstream detectors, the attributes are the same mentioned above for the target location and include also attributes of traffic speed of the four downstream locations at the same past times.

Table 1 presents a summary of the 36 data-sets that includes the number of instances, the number of attributes, the number of instances per class and the Imbalance Ratios (IRs) of each data-set. The IR is calculated by dividing the number of instances of the majority class over the instances of each of all the other classes. IR values show that the generated data-sets have a different imbalance degree. Some data-sets do not contain all the possible classes because on some occasions some of the classes had an extremely low presence (e.g. 20 samples) which introduced noise in the results. Samples of these classes where tagged as classes of the closest label with the lowest number of samples. Moreover, the differences between freeway and urban data-sets of Type I and Type II

---

[3] Category *A* indicates light to moderate traffic, whereas a category *E* means extended delays.

are their class distributions. Within each type, the class distribution is the same for all time horizons. In this sense, we can explore the capacity of Auto-sklearn when approaching different degrees of imbalanced data-sets.

4.2 Experimental Set-Up

Considering the traffic forecasting setting presented, we explore the performance of the Auto-sklearn's components through three scenarios using the data-sets presented above. First, a default scenario in which the hybrid search strategy of the AutoML method uses its three components to find pipelines. In this case, we considered three execution times for Auto-sklearn (ET): 15, 60 and 120 min. They correspond to the time that the bayesian optimisation can take to find the best pipelines and their hyper-parameter configuration for a given data-set. For assessing the performance of this scenario, the data-sets are partitioned in training (80%) and test (20%), keeping the chronological order of the data.

In the second scenario, we probe an alternative approach in which the recommendations done via meta-learning are combined in two ensembles based on weighted-voting without using the optimisation process. First, we extract the 25 best ML pipelines (default value used by Auto-sklearn) suggested by the meta-learning component, which then are combined in the weighted-voting ensemble named MetaEns25. To test this ensemble, the data-sets are partitioned in the same way as Auto-sklearn, such as was described above. For the second ensemble, we extract the complete list of 121 ML pipelines that can be suggested by the meta-learning component and choose again 25 best pipelines, based on their validation error, to generate the ensemble MetaEn25-121. In this case, we do the following procedure: the data-sets are partitioned in training (60%), validation (20%) and test (20%). To select the 25 best pipelines, these are trained on the training set and their performance is assessed on the validation set. Then, the ensemble is built with the 25 pipelines with the best validation error. Finally, the ensemble is trained on training+validation partitions (same number of instances as previous strategies, that is, 80%) and validated on the test set.

Lastly, for the third test scenario, we consider the meta-learning component in isolation. We follow a similar approach to that of MetaEn25-121, that is, we split the data-sets in training (60%), validation (20%) and test (20%). This means that for every data-set, we train the 121 pipelines suggested by meta-learning on the training set and based on their error on the validation set, we choose the best pipeline. The latter is then trained on training+validation (that is, over an 80% of the instances) and assessed over the test set.

To evaluate the experimental set-up presented, we use the metric *G-measure* *(mGM)* that is applied for multi-class imbalanced data in classification problems. Its calculation is expressed as $mGM = \sqrt[M]{\prod_{i=1}^{M} \text{specificity}_i \cdot \text{recall}_i}$ where M is the total number of classes.

Table 2. Mean mGM values and their standard deviations (in brackets) obtained by the three Auto-sklearn's ET, two weighted-voting ensembles and the BestPipe_Val approach.

Type	Data-set	BestPipe_Val	MetaEn25	MetaEn25-121	AutoS_15ET	AutoS_60ET	AutoS_120ET	Winner
Freeway Type I	T+CD.5	**0.70 (0.00)**	0.67 (0.00)	0.67 (0.00)	0.67 (0.01)	0.68 (0.01)	0.67 (0.01)	(Pipe 5, 2.4) - (0.9, 2.0, 11.6)
	T+CD.15	**0.48 (0.01)**	0.41 (0.01)	0.43 (0.01)	0.32 (0.01)	0.35 (0.02)	0.34 (0.01)	(Pipe 114, 5.2) - (0.9, 2.0, 11.6)
	T+CD.30	**0.29 (0.00)**	0.24 (0.01)	0.25 (0.01)	0.22 (0.01)	0.22 (0.01)	0.22 (0.02)	(Pipe 114, 5.2) - (0.9, 2.0, 11.6)
	T+CD.45	**0.85 (0.01)**	0.17 (0.01)	0.15 (0.00)	0.17 (0.01)	0.16 (0.01)	0.18 (0.00)	(Pipe 67, 2.8) - (0.8, 2.0, 11.6)
	T+CD.60	**0.84 (0.00)**	0.18 (0.01)	0.18 (0.01)	0.19 (0.01)	0.19 (0.00)	0.19 (0.01)	(Pipe 67, 2.8) - (0.8, 2.0, 11.6)
	TS+CD.5	**0.71 (0.00)**	0.70 (0.00)	0.70 (0.00)	0.54 (0.04)	0.60 (0.05)	0.61 (0.08)	(Pipe 72, 2.8) - (1.1, 1.9, 11.1)
	TS+CD.15	**0.50 (0.01)**	0.47 (0.01)	0.48 (0.01)	0.36 (0.03)	0.26 (0.14)	0.31 (0.08)	(Pipe 67, 2.7) - (1.1, 1.9, 11.1)
	TS+CD.30	**0.45 (0.00)**	0.00 (0.00)	0.00 (0.00)	0.00 (0.00)	0.00 (0.00)	0.00 (0.00)	(Pipe 72, 3.2) - (1.6, 1.9, 11.1)
	TS+CD.45	**0.35 (0.00)**	0.00 (0.00)	0.17 (0.30)	0.00 (0.00)	0.00 (0.00)	0.00 (0.00)	(Pipe 72, 3.2) - (1.6, 1.9, 11.1)
	TS+CD.60	0.00 (0.00)	0.00 (0.00)	0.00 (0.00)	0.00 (0.00)	0.00 (0.00)	0.00 (0.00)	—
Freeway Type II	T+CD.5	**0.97 (0.00)**	0.87 (0.00)	0.88 (0.00)	0.87 (0.00)	0.88 (0.00)	0.88 (0.00)	(Pipe 77, 3.0) - (1.2, 1.8, 11.5)
	T+CD.15	**0.96 (0.00)**	0.76 (0.00)	0.76 (0.00)	0.75 (0.01)	0.75 (0.01)	0.76 (0.01)	(Pipe 23, 1-8) - (1.2, 1.8, 11.5)
	T+CD.30	**0.94 (0.00)**	0.66 (0.00)	0.66 (0.00)	0.67 (0.00)	0.67 (0.01)	0.67 (0.01)	(Pipe 58, 2.4) - (1.2, 1.8, 11.5)
	T+CD.45	0.60 (0.00)	0.61 (0.00)	**0.61 (0.00)**	0.60 (0.00)	0.61 (0.00)	0.61 (0.00)	MetaEns25-121
	T+CD.60	0.56 (0.00)	0.57 (0.00)	0.57 (0.00)	**0.58 (0.00)**	0.58 (0.00)	0.57 (0.00)	AutoS_15ET
	TS+CD.5	**0.91 (0.00)**	0.65 (0.00)	0.66 (0.01)	0.66 (0.01)	0.65 (0.02)	0.65 (0.01)	(Pipe 2, 1.0) - (0.9, 1.9, 10.9)
	TS+CD.15	**0.47 (0.00)**	0.41 (0.00)	0.42 (0.00)	0.41 (0.03)	0.42 (0.01)	0.42 (0.02)	(Pipe 41, 2.2) - (0.9, 1.9, 11.0)
	TS+CD.30	**0.46 (0.00)**	0.31 (0.00)	0.34 (0.00)	0.32 (0.02)	0.32 (0.01)	0.31 (0.02)	(Pipe 84, 3.5) - (0.9, 1.9, 11.0)
	TS+CD.45	**0.44 (0.00)**	0.25 (0.00)	0.39 (0.25)	0.27 (0.01)	0.28 (0.03)	0.29 (0.01)	(Pipe 84, 3.4) - (0.9, 1.9, 11.0)
	TS+CD.60	**0.41 (0.00)**	0.22 (0.00)	0.23 (0.01)	0.27 (0.02)	0.25 (0.02)	0.27 (0.02)	(Pipe 83, 3.4) - (0.9, 1.9, 11.0)

(continued)

Table 2. (*continued*)

Type	Data-set	BestPipe_Val	MetaEn25	MetaEn25-121	AutoS_15ET	AutoS_60ET	AutoS_120ET	Winner
Urban Type I	T+CD_15	**0.78 (0.03)**	0.24 (0.00)	0.24 (0.00)	0.32 (0.17)	0.33 (0.11)	0.35 (0.07)	(Pipe 4, 1.3) - (0.8, 2.0, 11.7)
	T+CD_30	**0.70 (0.00)**	0.15 (0.02)	0.20 (0.02)	0.25 (0.09)	0.15 (0.14)	0.12 (0.10)	(Pipe 85, 3.3) - (0.8, 2.0, 11.7)
	T+CD_45	0.71 (0.01)	0.12 (0.07)	0.18 (0.02)	0.30 (0.05)	0.29 (0.06)	0.21 (0.05)	(Pipe 101, 4-1) - (0.9, 2.0, 11.7)
	T+CD_60	**0.40 (0.03)**	0.19 (0.00)	0.21 (0.02)	0.23 (0.06)	0.27 (0.06)	0.26 (0.02)	(Pipe 108, 4.3) - (0.9, 2.0, 11.7)
	TS+CD_15	**0.88 (0.01)**	0.54 (0.00)	0.55 (0.01)	0.56 (0.03)	0.54 (0.01)	0.53 (0.02)	(Pipe 107, 4.3) - (1.1, 1.9, 11.9)
	TS+CD_30	**0.67 (0.01)**	0.50 (0.01)	0.53 (0.02)	0.56 (0.03)	0.58 (0.04)	0.54 (0.05)	(Pipe 94, 3.7) - (1.1, 1.9, 11.8)
	TS+CD_45	**0.69 (0.01)**	0.47 (0.00)	0.51 (0.01)	0.56 (0.02)	0.57 (0.02)	0.56 (0.01)	(Pipe 98, 4.0) - (1.1, 1.9, 11.8)
	TS+CD_60	**0.70 (0.00)**	0.50 (0.00)	0.54 (0.01)	0.57 (0.02)	0.61 (0.04)	0.58 (0.03)	(Pipe 73, 2.8) - (1.1, 1.9, 11.9)
Urban Type II	T+CD_15	**0.91 (0.01)**	0.69 (0.01)	0.69 (0.01)	0.69 (0.00)	0.69 (0.01)	0.68 (0.01)	(Pipe 1, 0.3) - (0.3, 1.3, 12.6)
	T+CD_30	**0.92 (0.01)**	0.66 (0.00)	0.68 (0.01)	0.68 (0.01)	0.68 (0.01)	0.69 (0.01)	(Pipe 48, 2.4) - (0.3, 1.3, 12.6)
	T+CD_45	**0.79 (0.01)**	0.69 (0.00)	0.70 (0.01)	0.69 (0.0)	0.69 (0.01)	0.68 (0.01)	(Pipe 39, 1.9) - (0.3, 1.3, 12.6)
	T+CD_60	**0.89 (0.00)**	0.68 (0.00)	0.68 (0.01)	0.69 (0.013)	0.69 (0.00)	0.70 (0.01)	(Pipe 18, 1.0) - (0.3, 1.3, 12.6)
	TS+CD_15	**0.92 (0.00)**	0.66 (0.00)	0.66 (0.00)	0.66 (0.00)	0.67 (0.01)	0.67 (0.01)	(Pipe 109, 5.0) - (0.5, 1.2, 12.3)
	TS+CD_30	**0.91 (0.00)**	0.64 (0.01)	0.64 (0.01)	0.62 (0.00)	0.64 (0.01)	0.64 (0.01)	(Pipe 109, 5.1) - (0.6, 1.2, 12.4)
	TS+CD_45	**0.90 (0.01)**	0.64 (0.01)	0.63 (0.01)	0.62 (0.01)	0.62 (0.15)	0.64 (0.01)	(Pipe 109, 5.1) - (0.6, 1.2, 12.4)
	TS+CD_60	**0.92 (0.00)**	0.67 (0.01)	0.70 (0.01)	0.63 (0.01)	0.65 (0.03)	0.68 (0.02)	(Pipe 2, 0.6) - (0.6, 1.2, 12.4)

5   Results

This section presents the results obtained with the experimental set-up proposed in the previous section. Table 2 shows the mean mGM values obtained by the three execution times (ET) of Auto-sklearn (AutoS_ET), the two voting ensembles (MetaEns25 and MetaEn25-121) and the best pipeline in validation from the meta-learning component (BestPipe_Val). These mGM values were calculated by carrying out five repetitions for each approach on every data-set. mGM values in bold indicate the best result achieved in every data-set. Besides, the last column of Table 2 shows which is the winner approach in terms of performance on each data-set.

In the cases wherein the best performing is obtained by the $BestPipe\_Val$ approach, we indicate the following information: the first pair between brackets indicates the ranking position of the winner pipeline according to the similarity metric used by Auto-sklearn, and the difference between the value of this metric and the one from the pipeline in the first position of the ranking, whereas in the second tuple between brackets, the value of the metric for the pipelines in positions 1, 25 and 121 (this information appears in the same order in the column named "Winner" of Table 2). In this way, we can observe whether there is a positive correlation between the ranking positions and the actual performance of the pipelines. The assumption of Auto-sklearn is that pipelines closer to position 1 (distances near 0) are likely to perform better on the input data.

From Table 2, the following highlights can be extracted regarding the behaviour of the methods AutoML methods compared. The BestPipe_Val component is by far the best performing approach when making traffic predictions. Concretely, it is able to suggest the best pipeline in 33 out of 36 data-sets, performing even better than the longer ET (120 min) of Auto-sklearn. However, these results also show that the distance measure in which Auto-sklearn is based it is not well correlated with performance as we explain below.

If we check carefully the winner pipelines in the last column of Table 2, only in 5 cases (data-sets: Fw_TS+CD_5 - Type II, Ub_T+CD_15 - Type I, Ub_T+CD_15 - Type II, Ub_T+CD_60 - Type II, Ub_TS+CD_60 - Type II) the pipelines are located in a position higher than 25. As it was stated in Sect. 2.1, this is the default value that Auto-sklearn uses to recommend the 25 pipelines that are more likely to perform well on the input data. Such recommendation is made by the similarity metric that compares the meta-features of the input data-sets against the meta-features stored in the meta-knowledge base. Considering such comparison, the similarity metric chooses the best pipelines found in the off-line Auto-sklearn's phase for each of the 25 most similar data-sets w.r.t. the one at hand. Based on results of Table 2, the meta-features used for the comparison are not working properly and they are providing information to the similarity metric that makes it leaving out competitive pipelines located beyond position 25. In conclusion, the majority of pipelines in the column winner of Table 2 are associated to data-sets, that with the current Auto-sklearn's meta-features comparison, are no being categorised as similar w.r.t the TF data-sets.

Table 3. Execution times in minutes of the BestPipe_Val approach and the two weighted voting ensembles. Values in bold indicates an execution time that is between 60 and 120 min which are the two longer execution times of Auto-sklearn

Type	Data-sets	BestPipe_Val	MetaEns25	MetaEns25-121	Type	BestPipe_Val	MetaEns25	MetaEns25-121
Freeway Type I	T+CD_5	9	1	6	Freeway type II	9	1	5
	T+CD_15	**90**	25	76		**80**	17	52
	T+CD_30	**86**	26	77		**83**	18	45
	T+CD_45	8	2	10		8	1	9
	T+CD_60	8	1	10		8	1	6
	TS+CD_5	33	4	22		**129**	25	22
	TS+CD_15	38	4	23		31	8	23
	TS+CD_30	36	7	30		35	5	38
	TS+CD_45	39	7	23		37	6	28
	TS+CD_60	40	7	22		37	5	5
Urban Type I	T+CD_15	11	7	13	Urban type II	10	2	7
	T+CD_30	15	7	12		7	3	7
	T+CD_45	16	2	9		9	2	6
	T+CD_60	16	2	12		7	2	6
	TS+CD_15	46	7	43		38	15	20
	TS+CD_30	46	7	43		40	16	24
	TS+CD_45	47	7	30		38	15	27
	TS+CD_60	46	7	28		35	17	22

For the default scenario in which Auto-sklearn uses its three components to find competitive pipelines, longer ET are supposed to improve the final results of predictions. However, the improvements only rank from 0.01 to 0.07, approximately, in the best of the cases (e.g., Fw_TS+CD_5 - Type I). This could be due to the fact that the meta-learning component is suggesting low-performance pipelines for the warm-start process of the optimisation component. Opposite to this tendency are data-sets $Fw\_TS + CD\_15$ - Type I, $Fw\_T + CD\_15$ - Type II and $Ub\_TS + CD\_15$ - Type I wherein the best mGM value is found by an ET shorter than 120m ET. We observed that this worsening is due to the over-fitting produced by the hyperparameters tuning of Auto-sklearn on the recommended pipelines. This result indicates that it is necessary to introduce mechanisms in the hybrid search strategy of AutoML to deal with over-fitting, especially when execution times of the optimisation are higher.

Regarding the performance of the two ensembles approaches based on weighted-voting (MetaEns25 and MetaEns25-121), the results of MetaEns25-121 are quite similar w.r.t. the results obtained when the optimisation component is taken into account. Concretely, in data-sets of freeway Types I-II and urban Type II, the MetaEns25-121 is able to outperform Auto-sklearn in multiple cases. In particular, in the data-sets $Fw\_TS + CD\_45$ - Type II, $Fw\_TS + CD\_45$ - Type I and $Fw\_TS + CD\_5$ - Type I the performance of MetaEns25-121 is better than any of the Auto-sklearn's ET. This can be explained because this latter ensemble is built using already optimised pipelines located beyond position 25 of the ranking, and as it was stated before, in those positions are competitive pipelines whose performance is boosted by the ensemble without the need of doing optimisation. For the case of MetaEns25, its performance is lower than MetaEns25-121 and the three ET of Auto-sklearn. However, it is interesting to note that these ensembles are not better than the best pipeline suggested via meta-learning; in this sense, it would be interesting to explore why the ensembles obtains a performance worse than the best pipeline in isolation.

As the computational cost is a key factor in AutoML, Table 3 shows the execution times in minutes that the BestPipe_Val and the two meta-ensembles took to make predictions on every data-set. As can be seen, in the majority of the cases, the three approaches spent less than 60 min, which is the second longer ET of Auto-sklearn.

Finally, additional results that are observed regardless of what approach is the one with the highest mGM values are discussed below. In the cases of data-sets freeway Types I-II and urban Type I, as the time horizon of predictions increases, the performance of all approaches decreases. For these data-sets, the ones that have a time horizon of five minutes are the TF problems in which the six approaches perform better. Besides, in data-sets $Fw\_TS + CD\_30$ and $Fw\_TS + CD\_60$, most of them have problems predicting the minorities classes and therefore their mGM values in these cases are equal to zero.

Regarding urban data-set of Type I with only temporal traffic data (T), the three ET of Auto-sklearn and the two ensembles have the lowest performance. This is due to the fact that these data-sets have the highest IRs $(\text{IR}(A/B) = 1.12$

$IR(A/C) = 12.04$). This demonstrates that Auto-sklearn does not incorporate in its inner structure mechanisms to deal with high imbalanced classification data-sets. Meanwhile, in the case of urban data-sets of Type I with spatial and temporal data (TS) and all urban data-sets of Type II, the performance of the six approaches is quite acceptable and homogeneous across them. This behaviour can be argued as these 12 data-sets are the most balanced of the 36 data-sets ($IR(B/A) = 2.40$, $IR(B/C) = 4.98$; $IR(A/B) = 1.02$, $IR(B/A) = 1.39$).

6 Conclusions

In this paper, we studied the benefits in terms of performance and computational cost of hybrid AutoML for TF. We use Auto-sklearn, a state-of-the-art hybrid AutoML method whose search strategy of pipelines uses bayesian optimisation, meta-learning and ensemble learning. We focused on how well Auto-sklearn is able to recommend competitive ML pipelines to forecast traffic, modelled as a multi-class imbalanced classification problem, along different time horizons in urban and freeway environments.

From the results, we drew interesting conclusions. A simple approach based on estimating the best pipeline from Auto-sklearn's meta-learning component is able to suggest competitive pipelines that perform better than the results obtained by the three ET of Auto-sklearn considered and the two weighted-voting ensembles. However, these winner pipelines usually were not included in the 25 suggestions done by default by the Auto-sklearn's meta-learning component. Instead, they were located in lower positions, which could lead to thinking that the meta-features and the similarity metric in charge of recommending pipelines are not performing as expected for these data-sets. As a result, the ranking positions are not directly related to the performance that the pipelines could have on the TF data-sets.

Another interesting conclusion is that the optimisation component is not adding too much to the final mGM values. Higher execution times for Auto-sklearn not always lead to better results as we can expect; this was also corroborated by previous research that approached the use of Auto-WEKA (another AutoML method) for TF [1,2]. In spite of this, the performance of the optimisation process could be improved if the ranking recommended by the meta-learning component was re-organized using the validation error of these pipelines on the input data. Thus, the optimisation would only be fed by pipelines that already are corroborated for having high performance on the data-set at hand. However, caution needs to be taken to check the computational cost consumed when calculating the validation error of the 121 pipelines in the meta-knowledge base.

Further research lines that we aim to explore in the future are: I) improving the synergy between meta-learning and ensemble learning; II) determining the TF problems in which the optimisation is strictly necessary to improve the results obtained via meta-learning.

Acknowledgments. This project has received funding from the European Union's Horizon 2020 research and innovation programme under the grant agreement No. 815069 and the Marie Sklodoska-Curie grant agreement No. 665959.

References

1. Angarita-Zapata, J.S., Masegosa, A.D., Triguero, I.: Evaluating automated machine learning on supervised regression traffic forecasting problems. In: Llanes Santiago, O., Cruz Corona, C., Silva Neto, A.J., Verdegay, J.L. (eds.) Computational Intelligence in Emerging Technologies for Engineering Applications. SCI, vol. 872, pp. 187–204. Springer, Cham (2020). https://doi.org/10.1007/978-3-030-34409-2_11
2. Angarita-Zapata, J.S., Triguero, I., Masegosa, A.D.: A preliminary study on automatic algorithm selection for short-term traffic forecasting. In: Del Ser, J., Osaba, E., Bilbao, M.N., Sanchez-Medina, J.J., Vecchio, M., Yang, X.-S. (eds.) IDC 2018. SCI, vol. 798, pp. 204–214. Springer, Cham (2018). https://doi.org/10.1007/978-3-319-99626-4_18
3. Feurer, M., Klein, A., Eggensperger, K., Springenberg, J., Blum, M., Hutter, F.: Efficient and robust automated machine learning. In: Cortes, C., Lawrence, N.D., Lee, D.D., Sugiyama, M., Garnett, R. (eds.) Advances in Neural Information Processing Systems, pp. 2962–2970. Curran Associates, Inc. (2015)
4. Hutter, F., Hoos, H.H., Leyton-Brown, K.: Sequential model-based optimization for general algorithm configuration. In: Coello, C.A.C. (ed.) LION 2011. LNCS, vol. 6683, pp. 507–523. Springer, Heidelberg (2011). https://doi.org/10.1007/978-3-642-25566-3_40
5. Hutter, F., Kotthoff, L., Vanschoren, J. (eds.): Automated Machine Learning: Methods, Systems, Challenges. Springer, Cham (2019). https://doi.org/10.1007/978-3-030-05318-5
6. Kerschke, P., Hoos, H., Neumann, F., Trautmann, H.: Automated algorithm selection: survey and perspectives. CoRR (2018)
7. Lemke, C., Budka, M., Gabrys, B.: Metalearning: a survey of trends and technologies. Artif. Intell. Rev. **44**(1), 117–130 (2013). https://doi.org/10.1007/s10462-013-9406-y
8. Luo, G.: A review of automatic selection methods for machine learning algorithms and hyper-parameter values. Netw. Model. Anal. Health Inform. Bioinform. **5**(1), 18 (2016). https://doi.org/10.1007/s13721-016-0125-6
9. Mohr, F., Wever, M., Hüllermeier, E.: ML-Plan: automated machine learning via hierarchical planning. Mach. Learn. **107**(8), 1495–1515 (2018). https://doi.org/10.1007/s10994-018-5735-z
10. Olson, R.S., Bartley, N., Urbanowicz, R.J., Moore, J.H.: Evaluation of a tree-based pipeline optimization tool for automating data science. In: Proceedings of the Genetic and Evolutionary Computation Conference 2016, pp. 485–492 (2016)
11. Skycomp, I.B.M.: Major High- way Performance Ratings and Bottleneck Inventory. Maryland State Highway Administration, the Baltimore Metropolitan Council and Maryland Transportation Authority, State of Maryland (2009)
12. Swearingen, T., Drevo, W., Cyphers, B., Cuesta-Infante, A., Ross, A., Veeramachaneni, K.: ATM: a distributed, collaborative, scalable system for automated machine learning. In: 2017 IEEE International Conference on Big Data, pp. 151–162 (2017)

13. Thornton, C., Hutter, F., Hoos, H.H., Leyton-Brown, K.: Auto-WEKA. In: Proceedings of the 19th International Conference on Knowledge Discovery and Data Mining, pp. 847–855 (2013)
14. Vanschoren, J.: Meta-learning: a survey. arXiv preprint arXiv:1810.03548 (2018)
15. Vlahogianni, E.I.: Optimization of traffic forecasting: intelligent surrogate modeling. Transp. Res. Part C Emerg. Technol. **55**, 14–23 (2015). http://www.sciencedirect.com/science/article/pii/S0968090X15000959. Engineering and Applied Sciences Optimization (OPT-i) - Professor Matthew G. Karlaftis Memorial Issue
16. Vlahogianni, E.I., Karlaftis, M.G., Golias, J.C.: Short-term traffic forecasting: where we are and where we're going. Transp. Res. Part C: Emerg. Technol. **43**, 3–19 (2014)
17. Yao, Q., et al.: Taking human out of learning applications: a survey on automated machine learning. CoRR (2018)
18. Zöller, M.A., Huber, M.F.: Survey on automated machine learning. CoRR (2019)

Fuzzy Implication Functions

An Initial Study on Typical Hesitant (T,N)-Implication Functions

Monica Matzenauer[1] (ID), Renata Reiser[1] (ID), Helida Santos[2(✉)] (ID),
Jocivania Pinheiro[3] (ID), and Benjamin Bedregal[4] (ID)

[1] Centro de Desenvolvimento Tecnológico, Universidade Federal de Pelotas,
Pelotas, Brazil
{monica.matzenauer,reiser}@inf.ufpel.edu.br
[2] Centro de Ciências Computacionais, Universidade Federal do Rio Grande,
Rio Grande, Brazil
helida@furg.br
[3] Centro de Ciências Exatas e Naturais, Universidade Federal Rural do Semi-Árido,
Mossoró, Brazil
vaniamat@ufersa.edu.br
[4] Departamento de Informática e Matemática Aplicada,
Universidade Federal do Rio Grande do Norte, Natal, Brazil
bedregal@dimap.ufrn.br

Abstract. In the theory of Hesitant Fuzzy Sets (HFS), the membership
degree of an element is characterized by a membership function which
always returns a fuzzy set. This approach enables one to express, for
example, the hesitance of several experts in the process of decision making based on multiple attributes and multiple criteria. In this work, we
focus on the study of a class of implication functions for typical hesitant
fuzzy sets (THFS). The novelty of our proposal lies on the fact that it
is the first time that an admissible order is used to define operators on
hesitant fuzzy setting. Thus, we introduce typical hesitant fuzzy negations, typical hesitant t-norms and typical hesitant implication functions
considering an admissible order, which allows the comparison of typical
hesitant fuzzy elements with different cardinalities.

Keywords: Hesitant Fuzzy Sets · Admissible orders on THFS ·
Typical Hesitant Implication Functions · (T,N)-implication functions

1 Introduction

In situations where there are conflicts among the several experts in the process of
decision making based on multiple attributes and multiple criteria, it is common
to use Hesitant Fuzzy Sets (HFS). In the HFS theory [25], one considers, as the
membership degree of an element, a membership function which always returns
a fuzzy set expressing this hesitance. Since its introduction in 2010, relevant
research in decision making has used HFS theory, for example, the studies found

© Springer Nature Switzerland AG 2020
M.-J. Lesot et al. (Eds.): IPMU 2020, CCIS 1238, pp. 747–760, 2020.
https://doi.org/10.1007/978-3-030-50143-3_58

in [12,28–30]. In particular, several weighted average and ordered weighted average (OWA)-like operators have been proposed to be used in decision making, as we can see in [5,29,34].

A frequent issue in the context of decision making is that it is not always possible to find a consensus between a group of experts. So, it seems more appropriate to consider a set of possible values taking into account everyone's opinion. For instance, in order to provide a membership degree for an element of the universe, HFS can be useful to express this membership degree through a set of typical hesitant fuzzy elements (THFE), which will consider each opinion given by everyone in the group of experts.

On the other hand, it is common sense the importance of fuzzy implication functions which have been widely investigated and applied in many fields, as for example in decision making [9,17,23] and clustering [27]. In order to have a better understanding of logical connectives, one must know their properties and main characteristics. There are many different ways to model implication-like operators. In [6,18,19], a class of implication functions named (T,N)-implications was investigated and in their definition it is used a t-norm and a fuzzy negation. In this context, this work presents the definition of Typical Hesitant Implication Functions, including the class of (T,N)-implications, and the correspondent analysis of their main properties. Besides, and important contribution of the present work is that an admissible order on a HFS is provided allowing the comparison of hesitant fuzzy elements with different cardinalities. This novelty corroborates with the meaning of hesitant implication functions, providing semantic interpretation for implications setting found in multi-valued fuzzy logics. Therefore, the main properties proposed in the literature for fuzzy implications were studied and extended to HFS, which we discuss in this paper and present the properties of what we call Typical Hesitant (T,N)-Implication Functions (THIF).

This work is organized as follows: some preliminary and necessary concepts are given in Sect. 2, which allow us to provide in Sect. 3 an admissible order for the HFS elements and also allow us to introduce some operators, such as the typical hesitant fuzzy negations and typical hesitant t-norms. Then, Sect. 4 presents typical hesitant implication functions and discusses their main properties, including an incipient study on typical hesitant (T,N)-implication functions. Finally, Sect. 5 concludes the study.

2   Preliminaries

We start with some basic concepts of aggregation functions on the unit interval $[0,1]$, and then we recall triangular norms, fuzzy negations and fuzzy implication functions, for more details refer to [1,3,7,8,14,15].

Definition 1. *A function $A : [0,1]^n \rightarrow [0,1]$ is an n-ary aggregation function (AF) if it verifies, respectively, the isotonicity and boundary conditions, as follows:*

(A1) If $x_i \leq y_i$ for each $i = 1, \ldots, n$, then $A(x_1, \ldots, x_n) \leq A(y_1, \ldots, y_n)$;
(A2) $A(0, \ldots, 0) = 0$ and $A(1, \ldots, 1) = 1$.

Definition 2. *A function* $A : \bigcup_{n=1}^{\infty} [0,1]^n \rightarrow [0,1]$ *is an extended aggregation function (EAF) if the following condition holds:*

(A3) For each natural number $n \geq 2$, $A \upharpoonright [0,1]^n : [0,1]^n \rightarrow [0,1]$ is an AF and $A(x, \ldots, x) = x$, for each $x \in [0,1]$.

Definition 3. *A function $T : [0,1]^2 \rightarrow [0,1]$ is a t-norm if, for each $x, y, z \in [0,1]$, it satisfies:*

(T1) $T(x,y) = T(y,x)$ (commutativity);
(T2) $T(x, T(y,z)) = T(T(x,y), z)$ (associativity);
(T3) If $x \leq y$ then $T(x,z) \leq T(y,z)$ (isotonicity);
(T4) $T(x,1) = x$ (neutrality of 1-element).

Observe that each t-norm is a bivariate aggregation function.

Definition 4. *A function $N : [0,1] \rightarrow [0,1]$ is a fuzzy negation if*

(N1) $N(0) = 1$ and $N(1) = 0$;
(N2) If $x \leq y$ then $N(y) \leq N(x)$, for all $x, y \in [0,1]$.

A fuzzy negation N is strict if it is continuous and $N(x) < N(y)$ when $y < x$ and additionally, it is strong if it is involutive, i.e.

(N3) $N(N(x)) = x, \forall x \in [0,1]$.

The most common strong fuzzy negation is $N_S(x) = 1 - x$, also known as the standard or Zadeh negation. Each strong fuzzy negation is strict but the converse does not hold. For example, the negation $N(x) = 1 - \sqrt{x}$ is strict but it is not strong.

An important notion in our work is the concept of implication functions, in the sense of Fodor and Roubens, see [2,3,10,20]) for additional information.

Definition 5. *A fuzzy implication is a function $I : [0,1]^2 \rightarrow [0,1]$ such that, for every $x, y, z \in [0,1]$:*

(I1) If $x \leq y$ then $I(y,z) \leq I(x,z)$ (first place antitonicity);
(I2) If $y \leq z$ then $I(x,y) \leq I(x,z)$ (second place isotonicity);
(I3) $I(0,y) = 1$ (left boundary);
(I4) $I(x,1) = 1$ (right boundary);
(I5) $I(1,0) = 0$ (corner condition).

Finally, let us recall the notions of partial ordering. Let P be a non-empty set, we say that a partial order $\preceq$ on the set P is a binary relation on P which satisfies, respectively, the reflexivity, antisymmetry and transitivity properties:

(P1) $p \preceq p$, for each $p \in P$,
(P2) If $p \preceq q$ and $q \preceq p$, then $p = q$ for all $p, q \in P$,
(P3) If $p \preceq q$ and $q \preceq r$, then $p \preceq r$ for all $p, q, r \in P$.

Note that we say $a \prec b$ when (a,b) is in a relation $\preceq$ but $a \neq b$. A set P with a partial order $\preceq$ is referred to as a partially ordered set (poset) and denoted by $(P, \preceq)$. If any two elements a, b are comparable in a poset $(P, \preceq)$, i.e. either $a \preceq b$ or $b \preceq a$, then the partial order $\preceq$ is said to be a linear (or total) order (and then P is a chain).

2.1 Typical Hesitant Fuzzy Sets

Hesitant Fuzzy Sets (HFS) were introduced by Torra in [24] and Torra and Narukawa in [25]. In their work, the membership degree of an element that belongs to a set was represented by means of a subset of $[0, 1]$. In the process of decision-making, HFS can be useful to handle situations where there is indecision among many possible values for the preferences over objects. Formally, let $\wp([0, 1])$ be the power set of $[0, 1]$. A HFS A defined over U, where U is a non-empty set, is given by:

$$A = \{(x, \mu_A(x)) | x \in U\}. \tag{1}$$

and $\mu_A \colon U \to \wp([0, 1])$, where μ_A is the membership function. There is a particular case when $\mu_A(x)$ is finite and non-empty for each $x \in U$, and in this case we have Typical Hesitant Fuzzy Sets (THFS).

Definition 6. [5] *Let $\mathbb{H} = \{X \subseteq [0, 1] | X$ is finite and $X \neq \emptyset\}$. A THFS A defined over U is given by Eq. (1), where $\mu_A \colon U \to \mathbb{H}$.*

Each $X \in \mathbb{H}$ is named Typical Hesitant Fuzzy Element (THFE) of $\mathbb{H}$ and the cardinality of X, i.e. the number of elements of X, is referred to as $\#X$. The ith smallest element of a THFE X will be denoted by $X^{(i)}$.

Some examples of THFS are $X = \{0.1, 0.4, 0.7\}$ and $Y = \{0.1, 0.6, 0.9\}$ where $\#X = \#Y = 3$. In those examples, $X^{(1)} = 0.1$ and $Y^{(2)} = 0.6$.

Definition 7. *From every EAF A, and knowing that the least and the greatest elements are $\mathbf{0}_{\mathbb{H}} = \{0\}$ and $\mathbf{1}_{\mathbb{H}} = \{1\}$, respectively, we define the function $f_A \colon \mathbb{H} \to [0, 1]$ as:*

$$f_A(X) = \begin{cases} 0, & \text{if } X = \mathbf{0}_{\mathbb{H}} \\ 1, & \text{if } X = \mathbf{1}_{\mathbb{H}} \\ k \cdot A(X^{(1)}, \dots, X^{(\#X)}) + \frac{1-k}{2}, & \text{otherwise.} \end{cases}$$

where $0 < k < 1$.

For example, if A is the arithmetic average, $k = 0.8$ and $X = \{0.1, 0.2, 0.4, 0.9\}$ then $f_A(X) = 0.8 \cdot 0.4 + \frac{0.2}{2} = 0.42$.

In the literature, one can find many proposals of orders for THFE, such as the ones found in [5, 13, 26, 31–33]. The unique consensus among all these orders is that all of them refine[1] the following order on $\mathbb{H}$:

$$X \preceq_{\mathbb{H}} Y \text{ iff } X = \mathbf{0}_{\mathbb{H}} \text{ or } Y = \mathbf{1}_{\mathbb{H}} \text{ or } (\#X = \#Y \text{ and } X^{(i)} \leq Y^{(i)}, \forall i = 1, \dots, \#X) \tag{2}$$

considered in [5]. However, this is very restrictive, since for two THFE to become comparable, it is required that both have the same cardinality.

---

[1] A partial order $\leq_1$ on a set S refines another partial order $\leq_2$ on S if $(S, \leq_2) \subseteq (S, \leq_1)$, i.e. for each $x, y \in S$ such that $x \leq_2 y$ we have that $x \leq_1 y$.

Our aim in the present work is to establish an admissible order to allow comparisons between THFE without this restriction. The idea of admissible order was presented in [11] for interval-valued fuzzy sets and after in [16] for interval-valued Atanassov's intuitionistic fuzzy sets. And in [21], the study of admissible total orders on hesitant fuzzy sets was included as a challenge. We acknowledge that some efforts have already been made in order to establish an admissible ordering for hesitant fuzzy sets, as seen in [26]. However, their proposal requires that both THFE must have the same cardinality.

In the next section, we present an admissible order in the typical hesitant fuzzy setting, which will allow us to introduce the notion of some typical hesitant connectives.

3   Admissible Orders for Typical Hesitant Fuzzy Elements

Take $\mathbb{H}^{(m)} = \{X \subseteq [0,1] | \#X = m\}$, we start by defining an admissible ordering for the typical hesitant fuzzy elements with cardinality m.

Definition 8. [26] *A total order $\leq_{\mathbb{H}^{(m)}}$ on $\mathbb{H}^{(m)}$ is said to be admissible if for all $X, Y \in \mathbb{H}^{(m)}$, we have that $X \leq_{\mathbb{H}^{(m)}} Y$ if and only if $X^{(i)} \leq Y^{(i)}$ for each $1 \leq i \leq m$.*

Example 1. (i) At first, take $\mathbb{H}^{(m)}$ for $m \geq 1$, and then consider the lexicographical order (with respect to the first variable) [11]. So, we have that $X \leq_{\mathbb{H}^{(m)}} Y$, if $X = Y$ or exists an i such that $X^{(i)}$ is strictly less than $Y^{(i)}$ and for all $j < i, X^{(j)} = Y^{(j)}$. For instance, $X = \{0.1, 0.4, 0.7\} \leq_{\mathbb{H}^{(3)}} \{0.1, 0.6, 0.9\} = Y$.

Definition 9. *A total order $\leq_{\mathbb{H}}$ on $\mathbb{H}$ is said to be admissible if, for all $X, Y \in \mathbb{H}$, we have that $X \leq_{\mathbb{H}} Y$ whenever $X \preceq_{\mathbb{H}} Y$.*

Observe that $(\mathbb{H}, \leq_{\mathbb{H}})$ is a bounded chain with the least and the greatest elements $\mathbf{0}_{\mathbb{H}} = \{0\}$ and $\mathbf{1}_{\mathbb{H}} = \{1\}$, respectively.

Remark 1. Note that for all admissible orders $\leq_{\mathbb{H}}$ on $\mathbb{H}$, their restriction to $\mathbb{H}^{(m)}$ is an admissible order on $\mathbb{H}^{(m)}$.

Next, we provide a method to generate admissible order for THFE based on an indexed family of admissible order $\leq_{\mathbb{H}^{(m)}}$, where $m \in \mathbb{N}^+$.

Theorem 1. *Let $(\leq_{\mathbb{H}^{(m)}})_{m \in \mathbb{N}^+}$ be family of indexed admissible orders and an EAF operator $A \colon \bigcup_{n=1}^{\infty} [0,1]^n \to [0,1]$. Then, the binary relation*

$$X \leq_{\mathbb{H}}^{A} Y \Leftrightarrow \begin{cases} f_A(X) < f_A(Y), \text{ or} \\ f_A(X) = f_A(Y) \text{ and } \#Y < \#X, \text{ or} \\ f_A(X) = f_A(Y) \text{ and } \#Y = \#X = m \text{ and } X \leq_{\mathbb{H}^{(m)}} Y \end{cases} \quad (3)$$

is an admissible order on $\mathbb{H}$.

Proof. It is straightforward the prove that the binary relation $\leq_{\mathbb{H}}^A$ is reflexive and antisymmetric. In addition, the relation $\leq_{\mathbb{H}}^A$ is also a transitive relation on $\mathbb{H}$, which is shown as follows:

(*i*) If $X \leq_{\mathbb{H}}^A Y$ and $Y \leq_{\mathbb{H}}^A Z$ for a given $X, Y, Z \in \mathbb{H}$, then $f_A(X) \leq f_A(Y) \leq f_A(Z)$. In case $f_A(X) < f_A(Y)$ and $f_A(Y) \leq f_A(Z)$ or $f_A(X) \leq f_A(Y)$ and $f_A(Y) < f_A(Z)$, it follows that $f_A(X) < f_A(Z)$. In case $f_A(X) = f_A(Y) = f_A(Z)$, we need to consider the four situations as described below:

Case 1:

$$\left. \begin{array}{l} f_A(X) = f_A(Y) \text{ and } \#Y < \#X \\ f_A(Y) = f_A(Z) \text{ and } \#Z < \#Y \end{array} \right\} \Rightarrow f_A(X) = f_A(Z) \text{ and } \#Z < \#X;$$

Case 2:

$$\left. \begin{array}{l} f_A(X) = f_A(Y) \text{ and } \#Y < \#X \\ f_A(Y) = f_A(Z) \text{ and } \#Z = \#Y = m \text{ and } Y \leq_{\mathbb{H}(m)} Z \end{array} \right\} \Rightarrow f_A(X) = f_A(Z) \text{ and } \#Z < \#X;$$

Case 3:

$$\left. \begin{array}{l} f_A(X) = f_A(Y) \text{ and } \#X = \#Y = m \text{ and } X \leq_{\mathbb{H}(m)} Y \\ f_A(Y) = f_A(Z) \text{ and } \#Z < \#Y \end{array} \right\} \Rightarrow f_A(X) = f_A(Z) \text{ and } \#Z < \#X;$$

Case 4:

$$\left. \begin{array}{l} f_A(X) = f_A(Y) \text{ and } \#X = \#Y = m \text{ and } X \leq_{\mathbb{H}(m)} Y \\ f_A(Y) = f_A(Z) \text{ and } \#Y = \#Z = m \text{ and } Y \leq_{\mathbb{H}(m)} Z \end{array} \right\} \Rightarrow \begin{array}{l} f_A(X) = f_A(Z) \text{ and } \#X = \#Z = m \\ \text{and } X \leq_{\mathbb{H}(m)} Z. \end{array}$$

For any of the above cases, $X \leq_{\mathbb{H}}^A Z$ and, therefore, the $\leq_{\mathbb{H}(m)}$-transitivity holds.

(*ii*) Besides, we also have to prove that either $X \leq_{\mathbb{H}}^A Y$ or $Y \leq_{\mathbb{H}}^A X$. There are three possible situations: (1) $f_A(X) < f_A(Y)$ and therefore $X \leq_{\mathbb{H}}^A Y$. (2) $f_A(Y) < f_A(X)$ and therefore $Y \leq_{\mathbb{H}}^A X$. (3) $f_A(X) = f_A(Y)$ and so, we also have three cases:

(3a) $\#X < \#Y$, so $Y \leq_{\mathbb{H}}^A X$.
(3b) $\#Y < \#X$, so $X \leq_{\mathbb{H}}^A Y$.
(3c) $\#X = \#Y = m$, so since $\leq_{\mathbb{H}(m)}$ is admissible, then $X \leq_{\mathbb{H}(m)} Y$ or $Y \leq_{\mathbb{H}(m)} X$.

Hence, $X \leq_{\mathbb{H}}^A Y$ or $Y \leq_{\mathbb{H}}^A X$.

(*iii*) Finally, let $X, Y \in \mathbb{H}$ and suppose $X \prec_{\mathbb{H}} Y$, then by Eq. (2), we have three possibilities: (1) $X = \mathbf{0}_{\mathbb{H}}$, and in this case, $f_A(X) = 0$ and $f_A(Y) > \frac{k}{2}$ and so, $f_A(X) < f_A(Y)$, i.e. $X <_{\mathbb{H}}^A Y$. (2) $Y = \mathbf{1}_{\mathbb{H}}$, which is analogous to (1). At last, (3) ($\#X = \#Y = m$ and $X^{(i)} \leq Y^{(i)}, \forall i = 1, \ldots, m$) then, because A is an EAF, we have $A(X^{(1)}, \ldots, X^{(m)}) \leq A(Y^{(1)}, \ldots, Y^{(m)})$ and therefore, $f_A(X) \leq f_A(Y)$. Hence, $X \leq_{\mathbb{H}(m)} Y$. Therefore, Theorem 1 holds.

Example 2. Considering the THFS $X = \{0.1, 0.4, 0.7\}, Y = \{0.1, 0.6, 0.9\}$ and $Z = \{0, 1\}$ the following EAF operators:

(1) $A_1(x_1, \ldots, x_n) = \sum_{i=1}^{n} \frac{x_i}{n}$;

(2) $A_2(x_1, \ldots, x_n) = \max\{x_i\}_{1 \leq i \leq n}$;

(3) $A_3(x_1, \ldots, x_n) = \sqrt[n]{\prod_{i=1}^{n} x_i}$.

Thus, one can easily observe the following relations:

i. $X \leq_{\mathbb{H}}^{A_1} Z \leq_{\mathbb{H}}^{A_1} Y$.
ii. $X \leq_{\mathbb{H}}^{A_2} Y \leq_{\mathbb{H}}^{A_2} Z$.
iii. $Z \leq_{\mathbb{H}}^{A_3} X \leq_{\mathbb{H}}^{A_3} Y$.

In the sequence, some operators are given regarding admissible ordering on $\mathbb{H}$.

3.1 Typical Hesitant Fuzzy Negations

In [4,22], different definitions of Typical Hesitant Fuzzy Negations (THFN) were provided, both using partial orders. Now we introduce the concept of $\langle \mathbb{H}, \leq_{\mathbb{H}} \rangle$-negations, which consider an admissible order $\leq_{\mathbb{H}}$.

Definition 10. *Let $\mathcal{N} : \mathbb{H} \to \mathbb{H}$ be a function. $\mathcal{N}$ is said to be a THFN with respect to an admissible order $\leq_{\mathbb{H}}$, $\langle \mathbb{H}, \leq_{\mathbb{H}} \rangle$-negation in short, if the following conditions hold:*

($\mathcal{N}$1) $\mathcal{N}(\mathbf{0}_{\mathbb{H}}) = \mathbf{1}_{\mathbb{H}}$ *and* $\mathcal{N}(\mathbf{1}_{\mathbb{H}}) = \mathbf{0}_{\mathbb{H}}$.
($\mathcal{N}$2) *If* $X \leq_{\mathbb{H}} Y$ *then* $\mathcal{N}(Y) \leq_{\mathbb{H}} \mathcal{N}(X)$.

Additionally, we state that the $\langle \mathbb{H}, \leq_{\mathbb{H}} \rangle$-negation $\mathcal{N}$ is strong if it is involutive, i.e. if for each $X \in \mathbb{H}$, it satisfies a third property, namely:

($\mathcal{N}$3) $\mathcal{N}(\mathcal{N}(X)) = X$.

Example 3. Consider an admissible order $\leq_{\mathbb{H}}$ on the EAF A_1, A_2 and A_3, given in Example 2. Now take the function $\mathcal{N}_S : \mathbb{H} \to \mathbb{H}$, defined as follows:

$$\mathcal{N}_S(X) = \{1 - x | x \in X\}$$

It is easy to see that $\mathcal{N}_S$ is a $\langle \mathbb{H}, \leq_{\mathbb{H}} \rangle$-negation for $\leq_{\mathbb{H}}^{A_1}$ and $\leq_{\mathbb{H}}^{A_3}$, but not for $\leq_{\mathbb{H}}^{A_2}$.

Remark 2. $\mathcal{N}_S$ is a trivial example of a strong THFN with respect to the admissible orders $\leq_{\mathbb{H}}^{A_1}$ and $\leq_{\mathbb{H}}^{A_3}$.

3.2 Typical Hesitant Triangular Norms

The extension of the notion of t-norms for typical hesitant fuzzy elements was presented in [5], taking into account the partial order proposed in that paper. The following definition generalizes this notion by considering admissible orders on $\mathbb{H}$.

Definition 11. *Let* $\mathcal{T} \colon \mathbb{H}^2 \to \mathbb{H}$ *and let* $\leq_{\mathbb{H}}$ *be an admissible order on* $\mathbb{H}$. $\mathcal{T}$ *is a typical hesitant triangular norm with respect to* $\leq_{\mathbb{H}}$, *or* $\langle \mathbb{H}, \leq_{\mathbb{H}} \rangle$-*t-norm in short, if*

(*T*1) *It is commutative:* $\mathcal{T}(X, Y) = \mathcal{T}(Y, X)$;
(*T*2) *It is associative:* $\mathcal{T}(X, \mathcal{T}(Y, Z)) = \mathcal{T}(\mathcal{T}(X, Y), Z)$;
(*T*3) *It is monotonic, i.e., if* $X \leq_{\mathbb{H}} Y$ *then* $\mathcal{T}(X, Z) \leq_{\mathbb{H}} \mathcal{T}(Y, Z)$; *and*
(*T*4) $\mathbf{1}_{\mathbb{H}}$ *is the neutral element:* $\mathcal{T}(X, \mathbf{1}_{\mathbb{H}}) = X$.

Remark 3. Observe that each $\langle \mathbb{H}, \leq_{\mathbb{H}} \rangle$-t-norm also verifies the following property:

(*T*5) $\mathcal{T}(X, \mathbf{0}_{\mathbb{H}}) = \mathbf{0}_{\mathbb{H}}$, $\forall X \in \mathbb{H}$.

In fact, $\mathcal{T}(X, \mathbf{0}_{\mathbb{H}}) \leq_{\mathbb{H}} \mathcal{T}(\mathbf{1}_{\mathbb{H}}, \mathbf{0}_{\mathbb{H}}) = \mathbf{0}_{\mathbb{H}}$, for all $X \in \mathbb{H}$.

Other additional property is reported below:

(*T*6) $\mathcal{T}(X, \mathcal{N}(X)) = \mathbf{0}_{\mathbb{H}}$, $\forall X \in \mathbb{H}$.

Example 4. Consider an admissible order $\leq_{\mathbb{H}}$ on the EAF A_1, A_2 and A_3, given in Example 2. Now take the functions $\mathcal{T}_P, \mathcal{T}_M, \mathcal{T}_L \colon \mathbb{H}^2 \to \mathbb{H}$, defined as follows:

  i. $\mathcal{T}_P(X, Y) = \{x \cdot y \mid x \in X, y \in Y\}$
  ii. $\mathcal{T}_M(X, Y) = \{\min\{x, y\} \mid x \in X, y \in Y\}$
  iii. $\mathcal{T}_L(X, Y) = \{\max\{x + y - 1, 0\} \mid x \in X, y \in Y\}$

It is possible to prove that $\mathcal{T}_P$, $\mathcal{T}_M$ and $\mathcal{T}_L$ are $\langle \mathbb{H}, \leq_{\mathbb{H}} \rangle$-t-norms for $\leq_{\mathbb{H}}^{A_1}$, $\leq_{\mathbb{H}}^{A_2}$ and $\leq_{\mathbb{H}}^{A_3}$.

4   Typical Hesitant Implication Functions

Here we introduce the notion of $\langle \mathbb{H}, \leq_{\mathbb{H}} \rangle$-typical hesitant implication functions, $\langle \mathbb{H}, \leq_{\mathbb{H}} \rangle$-THIF in short, considering an admissible order $\leq_{\mathbb{H}}$, discussing their main properties.

The typical hesitant fuzzy approach for a fuzzy implication is conceived as an extension of axioms in Definition 5.

Definition 12. *Let* $\mathcal{I} \colon \mathbb{H}^2 \to \mathbb{H}$ *and let* $\leq_{\mathbb{H}}$ *be an admissible order.* $\mathcal{I}$ *is a typical hesitant fuzzy implication function with respect to* $\leq_{\mathbb{H}}$, $\langle \mathbb{H}, \leq_{\mathbb{H}} \rangle$-*THIF in short, if for each* $X, Y, Z \in \mathbb{H}$, *the following properties are verified:*

(*I*1) *If* $X \leq_{\mathbb{H}} Y$ *then* $\mathcal{I}(Y, Z) \leq_{\mathbb{H}} \mathcal{I}(X, Z)$ *(first place antitonicity);*
(*I*2) *If* $Y \leq_{\mathbb{H}} Z$ *then* $\mathcal{I}(X, Y) \leq_{\mathbb{H}} \mathcal{I}(X, Z)$ *(second place isotonicity);*
(*I*3) $\mathcal{I}(\mathbf{0}_{\mathbb{H}}, \mathbf{0}_{\mathbb{H}}) = \mathbf{1}_{\mathbb{H}}$ *(corner condition 1);*
(*I*4) $\mathcal{I}(\mathbf{1}_{\mathbb{H}}, \mathbf{1}_{\mathbb{H}}) = \mathbf{1}_{\mathbb{H}}$ *(corner condition 2); and*
(*I*5) $\mathcal{I}(\mathbf{1}_{\mathbb{H}}, \mathbf{0}_{\mathbb{H}}) = \mathbf{0}_{\mathbb{H}}$ *(corner condition 3).*

Table 1. Typical Hesitant Implication Functions

$\langle \mathbb{H}, \leq_{\mathbb{H}} \rangle$-THIF	Restrictions
$\mathcal{I}_{FD}(X,Y) = \begin{cases} \mathbf{1}_{\mathbb{H}}, & \text{if } X \leq_{\mathbb{H}} Y, \\ \max(\mathcal{N}_S(X), Y), & \text{otherwise} \end{cases}$	$\mathcal{N}_S$ is a $\langle \mathbb{H}, \leq_{\mathbb{H}} \rangle$-negation
$\mathcal{I}_{GD}(X,Y) = \begin{cases} \mathbf{1}_{\mathbb{H}}, & \text{if } X \leq_{\mathbb{H}} Y, \\ Y, & \text{otherwise} \end{cases}$	–
$\mathcal{I}_{WB}(X,Y) = \begin{cases} \mathbf{1}_{\mathbb{H}}, & \text{if } X \leq_{\mathbb{H}} \mathbf{1}_{\mathbb{H}}, \\ Y, & \text{if } X = \mathbf{1}_{\mathbb{H}}. \end{cases}$	–
$\mathcal{I}_{GR}(X,Y) = \begin{cases} \mathbf{1}_{\mathbb{H}}, & \text{if } X \leq_{\mathbb{H}} Y, \\ \mathbf{0}_{\mathbb{H}}, & \text{otherwise} \end{cases}$	–

Proposition 1. *If $\mathcal{I}$ is an $\langle \mathbb{H}, \leq_{\mathbb{H}} \rangle$-THIF then it also satisfies the following properties:*

(I6a) $\mathcal{I}(\mathbf{0}_{\mathbb{H}}, Y) = \mathbf{1}_{\mathbb{H}}$ (left boundary);
(I6b) $\mathcal{I}(X, \mathbf{1}_{\mathbb{H}}) = \mathbf{1}_{\mathbb{H}}$ (left and right boundary).

Proof. Straightforward.

There are other properties that some $\langle \mathbb{H}, \leq_{\mathbb{H}} \rangle$-THIF can verify as the listed ones presented in the following.

(I7) $\mathcal{I}(\mathbf{1}_{\mathbb{H}}, X) = X$ (left neutrality property);
(I8) $\mathcal{I}(X, X) = \mathbf{1}_{\mathbb{H}}$ (identity principle);
(I9) $\mathcal{I}(X, \mathcal{I}(Y, Z)) = \mathcal{I}(Y, \mathcal{I}(X, Z))$ (exchange principle);
(I10) $\mathcal{I}(X, \mathcal{N}(Y)) = \mathcal{I}(Y, \mathcal{N}(X))$, if $\mathcal{N}$ is a strong $\langle \mathbb{H}, \leq_{\mathbb{H}} \rangle$-negation (right contraposition or contrapositive symmetry w.r.t. $\mathcal{N}$);
(I11) $\mathcal{I}(X, Y) = \mathcal{I}(\mathcal{N}(Y), \mathcal{N}(X))$, if $\mathcal{N}$ is a strong $\langle \mathbb{H}, \leq_{\mathbb{H}} \rangle$-negation (law of contraposition w.r.t. $\mathcal{N}$).

See in Table 1 examples illustrating the extension of important $\langle \mathbb{H}, \leq_{\mathbb{H}} \rangle$-THIF, namely: Fodor ($\mathcal{I}_{FD}$), Gödel ($\mathcal{I}_{GD}$), Weber ($\mathcal{I}_{WB}$) and Gaines-Rescher ($\mathcal{I}_{GR}$), with respect to the admissible $\leq_{\mathbb{H}}$-order.

4.1 Obtaining $\langle \mathbb{H}, \leq_{\mathbb{H}} \rangle$-THIF from $\langle \mathbb{H}, \leq_{\mathbb{H}} \rangle$-t-norms and $\langle \mathbb{H}, \leq_{\mathbb{H}} \rangle$-negations

Inspired in [6,18,19], which introduced a family of implication functions constructed from fuzzy negations and a triangular norm, in the following proposition we present a method to construct a $\langle \mathbb{H}, \leq_{\mathbb{H}} \rangle$-THIF from a $\mathbb{H}$-t-norm and a $\langle \mathbb{H}, \leq_{\mathbb{H}} \rangle$-negation.

Theorem 2. *Let $\mathcal{T}$ be a $\langle \mathbb{H}, \leq_{\mathbb{H}} \rangle$-t-norm and let $\mathcal{N}$ be a $\langle \mathbb{H}, \leq_{\mathbb{H}} \rangle$-negation. The function $\mathcal{I}_{\mathcal{T}}^{\mathcal{N}} : \mathbb{H}^2 \to \mathbb{H}$ defined by*

$$\mathcal{I}_{\mathcal{T}}^{\mathcal{N}}(X, Y) = \mathcal{N}(\mathcal{T}(X, \mathcal{N}(Y))) \tag{4}$$

is a typical hesitant implication function, denoted as $\langle \mathbb{H}, \leq_{\mathbb{H}} \rangle$-implication.

Proof. We have to prove that $\mathcal{I}_T^{\mathcal{N}}$ satisfies the five properties of Definition 12.

($\mathcal{I}1$) If $X \leq_{\mathbb{H}} Y$ then by the monotonicity of T, we have $T(X, \mathcal{N}(Z)) \leq_{\mathbb{H}} T(Y, \mathcal{N}(Z))$, and by ($\mathcal{N}1$), $\mathcal{N}(T(Y, \mathcal{N}(Z))) \leq_{\mathbb{H}} \mathcal{N}(T(X, \mathcal{N}(Z)))$. So, $\mathcal{I}_T^{\mathcal{N}}(Y, Z) \leq_{\mathbb{H}} \mathcal{I}_T^{\mathcal{N}}(X, Z)$.

($\mathcal{I}2$) If $Y \leq_{\mathbb{H}} Z$ then, by ($\mathcal{N}1$), $\mathcal{N}(Z) \leq_{\mathbb{H}} \mathcal{N}(Y)$. Therefore, by the monotonicity of T, we have $T(X, \mathcal{N}(Z)) \leq_{\mathbb{H}} T(X, \mathcal{N}(Y))$ and $\mathcal{N}(T(X, \mathcal{N}(Y))) \leq_{\mathbb{H}} \mathcal{N}(T(X, \mathcal{N}(Z)))$. Thus, $\mathcal{I}_T^{\mathcal{N}}(X, Y) \leq_{\mathbb{H}} \mathcal{I}_T^{\mathcal{N}}(X, Z)$.

($\mathcal{I}3$) $\mathcal{I}_T^{\mathcal{N}}(0_{\mathbb{H}}, 0_{\mathbb{H}}) \overset{Eq.(4)}{=} \mathcal{N}(T(0_{\mathbb{H}}, \mathcal{N}(0_{\mathbb{H}}))) \overset{(\mathcal{N}1)}{=} \mathcal{N}(T(0_{\mathbb{H}}, 1_{\mathbb{H}})) \overset{(T4)}{=} \mathcal{N}(0_{\mathbb{H}}) \overset{(\mathcal{N}1)}{=} 1_{\mathbb{H}}$.

($\mathcal{I}4$) $\mathcal{I}_T^{\mathcal{N}}(1_{\mathbb{H}}, 1_{\mathbb{H}}) \overset{Eq.(4)}{=} \mathcal{N}(T(1_{\mathbb{H}}, \mathcal{N}(1_{\mathbb{H}}))) \overset{(\mathcal{N}1)}{=} \mathcal{N}(T(1_{\mathbb{H}}, 0_{\mathbb{H}})) \overset{(T4)}{=} \mathcal{N}(0_{\mathbb{H}}) \overset{(\mathcal{N}1)}{=} 1_{\mathbb{H}}$.

($\mathcal{I}5$) $\mathcal{I}_T^{\mathcal{N}}(1_{\mathbb{H}}, 0_{\mathbb{H}}) \overset{Eq.(4)}{=} \mathcal{N}(T(1_{\mathbb{H}}, \mathcal{N}(0_{\mathbb{H}}))) \overset{(\mathcal{N}1)}{=} \mathcal{N}(T(1_{\mathbb{H}}, 1_{\mathbb{H}})) \overset{(T4)}{=} \mathcal{N}(1_{\mathbb{H}}) \overset{(\mathcal{N}1)}{=} 0_{\mathbb{H}}$.

Therefore, Theorem 2 is verified.

Definition 13. *Let T be a $\langle \mathbb{H}, \leq_{\mathbb{H}} \rangle$-t-norm and let $\mathcal{N}$ be a $\langle \mathbb{H}, \leq_{\mathbb{H}} \rangle$-negation. The function $\mathcal{I}_T^{\mathcal{N}}$ defined by Eq. (4) is called a typical hesitant (T,N)-implication function.*

Now, it is shown that a $\langle \mathbb{H}, \leq_{\mathbb{H}} \rangle$-t-norm can be constructed from a $\langle \mathbb{H}, \leq_{\mathbb{H}} \rangle$-THIF.

Proposition 2. [6] *Let $\mathcal{N}$ be a strong $\langle \mathbb{H}, \leq_{\mathbb{H}} \rangle$-negation and let T be a $\langle \mathbb{H}, \leq_{\mathbb{H}} \rangle$-t-norm. Then, for each $X, Y \in \mathbb{H}$,*

$$T(X, Y) = \mathcal{N}(\mathcal{I}_T^{\mathcal{N}}(X, \mathcal{N}(Y))).$$

Proof. Straightforward.

Proposition 3. *Let $\mathcal{I}_T^{\mathcal{N}}$ be a typical hesitant (T,N)-implication function, let T be a $\langle \mathbb{H}, \leq_{\mathbb{H}} \rangle$-t-norm and let $\mathcal{N}$ be a strong $\langle \mathbb{H}, \leq_{\mathbb{H}} \rangle$-negation, then:*

(i) *$\mathcal{I}_T^{\mathcal{N}}$ satisfies the left neutrality property ($\mathcal{I}7$);*
(ii) *$\mathcal{I}_T^{\mathcal{N}}$ satisfies the exchange principle ($\mathcal{I}9$);*
(iii) *$\mathcal{I}_T^{\mathcal{N}}$ satisfies the law of right contraposition w.r.t. $\mathcal{N}$ ($\mathcal{I}10$);*
(iv) *$\mathcal{I}_T^{\mathcal{N}}$ satisfies the law of contraposition w.r.t. $\mathcal{N}$ ($\mathcal{I}11$).*

Proof. ($\mathcal{I}7$) Bearing in mind that T is a $\langle \mathbb{H}, \leq_{\mathbb{H}} \rangle$-t-norm, then for any $X \in \mathbb{H}$, we have: $\mathcal{I}_T^{\mathcal{N}}(1_{\mathbb{H}}, X) \overset{Eq.(4)}{=} \mathcal{N}(T(1_{\mathbb{H}}, \mathcal{N}(X))) \overset{(T1)/(T4)}{=} \mathcal{N}(\mathcal{N}(X)) \overset{(\mathcal{N}3)}{=} X$.

($\mathcal{I}9$) Once $\mathcal{N}$ is a strong $\langle \mathbb{H}, \leq_{\mathbb{H}} \rangle$-negation and T is a $\langle \mathbb{H}, \leq_{\mathbb{H}} \rangle$-t-norm, we have

$$\mathcal{I}_T^{\mathcal{N}}(X, \mathcal{I}_T^{\mathcal{N}}(Y, Z)) \overset{Eq.(4)}{=} \mathcal{N}(T(X, \mathcal{N}(\mathcal{N}(T(Y, \mathcal{N}(Z))))))$$
$$\overset{(\mathcal{N}3)}{=} \mathcal{N}(T(X, T(Y, \mathcal{N}(Z))))$$
$$\overset{(T1)}{=} \mathcal{N}(T(X, T(\mathcal{N}(Z), Y)))$$
$$\overset{(T2)}{=} \mathcal{N}(T(T(X, \mathcal{N}(Z)), Y))$$
$$\overset{(T1)}{=} \mathcal{N}(T(Y, T(X, \mathcal{N}(Z))))$$
$$\overset{(\mathcal{N}3)}{=} \mathcal{N}(T(Y, \mathcal{N}(\mathcal{N}(T(X, \mathcal{N}(Z)))))) \overset{Eq.(4)}{=} \mathcal{I}_T^{\mathcal{N}}(Y, \mathcal{I}_T^{\mathcal{N}}(X, Z)).$$

($\mathcal{I}$10) Due to the commutativity property of $\mathcal{T}$ and since $\mathcal{N}$ is a strong $\langle \mathbb{H}, \leq_{\mathbb{H}} \rangle$-negation, it follows that

$$\mathcal{I}_{\mathcal{T}}^{\mathcal{N}}(X, \mathcal{N}(Y)) \overset{Eq.\,(4)}{=} \mathcal{N}(\mathcal{T}(X, \mathcal{N}(\mathcal{N}(Y))))$$

$$\overset{(\mathcal{N}3)}{=} \mathcal{N}(\mathcal{T}(X, Y))$$

$$\overset{(\mathcal{T}1)}{=} \mathcal{N}(\mathcal{T}(Y, X))$$

$$\overset{(\mathcal{N}3)}{=} \mathcal{N}(\mathcal{T}(Y, \mathcal{N}(\mathcal{N}(X)))) \overset{Eq.\,(4)}{=} \mathcal{I}_{\mathcal{T}}^{\mathcal{N}}(Y, \mathcal{N}(X)).$$

($\mathcal{I}$11) Analogously, from the commutativity of $\mathcal{T}$ and as $\mathcal{N}$ is a strong $\langle \mathbb{H}, \leq_{\mathbb{H}} \rangle$-negation, we have the next results

$$\mathcal{I}_{\mathcal{T}}^{\mathcal{N}}(\mathcal{N}(Y), \mathcal{N}(X)) \overset{Eq.\,(4)}{=} \mathcal{N}(\mathcal{T}(\mathcal{N}(Y), \mathcal{N}(\mathcal{N}(X))))$$

$$\overset{(\mathcal{N}3)}{=} \mathcal{N}(\mathcal{T}(\mathcal{N}(Y), X))$$

$$\overset{(\mathcal{T}1)}{=} \mathcal{N}(\mathcal{T}(X, \mathcal{N}(Y))) \overset{Eq.\,(4)}{=} \mathcal{I}_{\mathcal{T}}^{\mathcal{N}}(X, Y).$$

Proposition 4. *Let $\mathcal{I}_{\mathcal{T}}^{\mathcal{N}}$ be a typical hesitant (T,N)-implication and $\mathcal{T}$ be a $\langle \mathbb{H}, \leq_{\mathbb{H}} \rangle$-t-norm satisfying $\mathcal{T}6$. Then $\mathcal{I}_{\mathcal{T}}^{\mathcal{N}}$ satisfies the identity principle ($\mathcal{I}8$).*

Proof. Suppose that $\mathcal{I}_{\mathcal{T}}^{\mathcal{N}}$ is a typical hesitant (T,N)-implication, $\mathcal{T}$ is a $\langle \mathbb{H}, \leq_{\mathbb{H}} \rangle$-t-norm such that $\mathcal{T}(X, \mathcal{N}(Y)) = \mathbf{0}_{\mathbb{H}}$. Then, the following results are verified:

$$\mathcal{I}_{\mathcal{T}}^{\mathcal{N}}(X, X) \overset{Eq.\,(4)}{=} \mathcal{N}(\mathcal{T}(X, \mathcal{N}(X)) \overset{(\mathcal{T}6)}{=} \mathcal{N}(\mathbf{0}_{\mathbb{H}}) = \mathbf{1}_{\mathbb{H}}.$$

Therefore, Proposition 4 is verified.

Concluding, three examples illustrating such methodology are presented in the following:

Example 5. Based on the methodology established in Theorem 2 and main operators presented in Examples 2 and 4, we construct new $\langle \mathbb{H}, \leq_{\mathbb{H}} \rangle$-implication functions in the class of typical hesitant (T,N)-implication functions. Meaning that, the next three functions $\mathcal{I}_{\mathcal{T}_P}^{\mathcal{N}_S}, \mathcal{I}_{\mathcal{T}_M}^{\mathcal{N}_S}, \mathcal{I}_{\mathcal{T}_L}^{\mathcal{N}_S} : \mathbb{H}^2 \to \mathbb{H}$, respectively expressed as follows

$$\mathcal{I}_{\mathcal{T}_P}^{\mathcal{N}_S}(X, Y) = \mathcal{N}_S(\mathcal{T}_P(X, \mathcal{N}_S(Y))),$$

$$\mathcal{I}_{\mathcal{T}_M}^{\mathcal{N}_S}(X, Y) = \mathcal{N}_S(\mathcal{T}_M(X, \mathcal{N}_S(Y))),$$

$$\mathcal{I}_{\mathcal{T}_L}^{\mathcal{N}_S}(X, Y) = \mathcal{N}_S(\mathcal{T}_L(X, \mathcal{N}_S(Y)));$$

are $\langle \mathbb{H}, \leq_{\mathbb{H}} \rangle$-THIF with respect to the admissible linear $\leq_{\mathbb{H}}^{A_1}$-order and $\leq_{\mathbb{H}}^{A_3}$-order:

5   Final Remarks

Regarding many extensions of multi-valued fuzzy logics, this paper introduces the definition of the class of (T,N)-implications in the context of Typical Hesitant Implication Functions, extending such analysis in order to consider their main properties: left neutrality property, right boundary, law of contraposition and its corresponding right contraposition based on $\langle \mathbb{H}, \leq_{\mathbb{H}} \rangle$-negations, also including the identity and exchange principles. As another important contribution, we investigate the conditions under which the use of admissible orders based on aggregation operators, performed on $\langle \mathbb{H}, \leq_{\mathbb{H}} \rangle$-lattice, allows a comparison of THFS with different cardinalities. Additionally, among several partial orders defined over $\langle \mathbb{H}, \leq_{\mathbb{H}} \rangle$, the discussed admissible orders on $\langle \mathbb{H}, \leq_{\mathbb{H}} \rangle$ promote comparisons even between THFS with different cardinalities.

Our results in the class of $\langle \mathbb{H}, \leq_{\mathbb{H}} \rangle$-implication functions, named (T,N)-implications extend the previous study presented in [19]. Thus, this novelty methodology corroborates with the meaning of $\langle \mathbb{H}, \leq_{\mathbb{H}} \rangle$-implication functions, providing semantic interpretation for implications setting found in multi-valued fuzzy logics.

As ongoing work, we are considering to prove some other properties of $\langle \mathbb{H}, \leq_{\mathbb{H}} \rangle$-THIF operators as a support to generate hesitant fuzzy subsethood measures, based on the studied class of (T,N)-implications. This study needs to consider the discussion about how can we have any (generalized) property of typical hesitant (T,N)-implication, for which the standard property is not satisfied by the standard (T,N)-implication.

One can easily observe that fuzzy implications have been used in preference computations also including ordering relations in related works, see e.g. [9]. Following such research approach, further work also intends to apply the present results on typical hesitant (T,N)-implication in order to achieve new results on hesitant-based fuzzy preferences relations.

Acknowledgments. This work was partially supported by CNPq, PQ(309160/2019-7) and PQ(307781/2016-0), and PqG/FAPERGS 02/2017(17/2551-0001207-0).

References

1. Alsina, C., Maurice, F., Schweizer, B.: Associative Functions: Triangular Norms And Copulas. World Scientific, Singapore (2006)
2. Baczyński, M., Beliakov, G., Bustince, H., Pradera, A.: Advances in Fuzzy Implications Functions. STUDFUZZ, vol. 300. Springer, Heidelberg (2013). https://doi.org/10.1007/978-3-642-35677-3
3. Baczyński, M., Jayaram, B.: Fuzzy Implications. STUDFUZZ, vol. 231. Springer, Heidelberg (2008). https://doi.org/10.1007/978-3-540-69082-5
4. Bedregal, B., Santiago, R., Bustince, H., Paternain, D., Reiser, R.: Typical hesitant fuzzy negations. Int. J. Intell. Syst. **29**(6), 525–543 (2014). https://doi.org/10.1002/int.21655

5. Bedregal, B., Reiser, R., Bustince, H., Lopez-Molina, C., Torra, V.: Aggregation functions for typical hesitant fuzzy elements and the action of automorphisms. Inf. Sci. **255**, 82–99 (2014). https://doi.org/10.1016/j.ins.2013.08.024

6. Bedregal, B.C.: A normal form which preserves tautologies and contradictions in a class of fuzzy logics. J. Algorithms **62**(3), 135–147 (2007). https://doi.org/10.1016/j.jalgor.2007.04.003

7. Beliakov, G., Bustince Sola, H., Calvo Sánchez, T.: A Practical Guide to Averaging Functions. SFSC, vol. 329. Springer, Cham (2016). https://doi.org/10.1007/978-3-319-24753-3

8. Beliakov, G., Pradera, A., Calvo, T.: Aggregation Functions: A Guide for Practitioners. STUDFUZZ, vol. 221. Springer, Heidelberg (2007). https://doi.org/10.1007/978-3-540-73721-6

9. Biba, V., Hlinená, D.: Generated fuzzy implicators and fuzzy preference structures. Kybernetika **48**(3), 453–464 (2012)

10. Bustince, H., Burillo, P., Soria, F.: Automorphisms, negations and implication operators. Fuzzy Sets Syst. **134**(2), 209–229 (2003). https://doi.org/10.1016/S0165-0114(02)00214-2

11. Bustince, H., Fernandez, J., Kolesárová, A., Mesiar, R.: Generation of linear orders for intervals by means of aggregation functions. Fuzzy Sets and Syst. **220**, 69–77 (2013). https://doi.org/10.1016/j.fss.2012.07.015

12. Farhadinia, B.: A novel method of ranking hesitant fuzzy values for multiple attribute decision-making problems. Int. J. Intell. Syst. **28**(8), 752–767 (2013). https://doi.org/10.1002/int.21600

13. Garmendia, L., del Campo, R.G., Recasens, J.: Partial orderings for hesitant fuzzy sets. Int. J. Approx. Reason. **84**, 159–167 (2017). https://doi.org/10.1016/j.ijar.2017.02.008

14. Grabisch, M., Marichal, J., Mesiar, R., Pap, E.: Aggregation Functions (Encyclopedia of Mathematics and Its Applications), 1st edn. Cambridge University Press, New York (2009)

15. Klement, E., Mesiar, R., Pap, E.: Triangular Norms, Trends in Logic - Studia Logica Library, vol. 8. Kluwer Academic Publishers, Dordrecht (2000)

16. Miguel, L.D., Bustince, H., Fernandez, J., Induráin, E., Kolesárová, A., Mesiar, R.: Construction of admissible linear orders for interval-valued Atanassov intuitionistic fuzzy sets with an application to decision making. Inf. Fusion **27**, 189–197 (2016). https://doi.org/10.1016/j.inffus.2015.03.004

17. Paternain, D., Jurio, A., Barrenechea, E., Bustince, H., Bedregal, B., Szmidt, E.: An alternative to fuzzy methods in decision-making problems. Expert Syst. Appl. **39**(9), 7729–7735 (2012). https://doi.org/10.1016/j.eswa.2012.01.081

18. Pinheiro, J., Bedregal, B., Santiago, R.H.N., Santos, H.: (T, N)-implications. In: 2017 IEEE International Conference on Fuzzy Systems (FUZZ-IEEE), Naples, pp. 1–6, July 2017. https://doi.org/10.1109/FUZZ-IEEE.2017.8015568

19. Pinheiro, J., Bedregal, B., Santiago, R.H., Santos, H.: A study of (T, N)-implications and its use to construct a new class of fuzzy subsethood measure. Int. J. Approx. Reason. **97**, 1–16 (2018). https://doi.org/10.1016/j.ijar.2018.03.008

20. Pradera, A., Beliakov, G., Bustince, H., Baets, B.D.: A review of the relationships between implication, negation and aggregation functions from the point of view of material implication. Inf. Sci. **329**, 357–380 (2016). https://doi.org/10.1016/j.ins.2015.09.033

21. Rodríguez, R.M., et al.: A position and perspective analysis of hesitant fuzzy sets on information fusion in decision making. Towards high quality progress. Inf. Fusion **29**, 89–97 (2016). https://doi.org/10.1016/j.inffus.2015.11.004

22. Santos, H.S., Bedregal, B.R.C., Santiago, R.H.N., Bustince, H.: Typical hesitant fuzzy negations based on Xu-Xia-partial order. In: IEEE Conference on Norbert Wiener in the 21st Century (21CW), Boston, pp. 1–6 (2014). https://doi.org/10.1109/NORBERT.2014.6893951

23. Shi, Y., Gasse, B.V., Kerre, E.: The role a fuzzy implication plays in a multi-criteria decision algorithm. Int. J. Gen. Syst. **42**(1), 111–120 (2013). https://doi.org/10.1080/03081079.2012.710441

24. Torra, V., Narukawa, Y.: On hesitant fuzzy sets and decision. In: Proceedings of the FUZZ-IEEE 2009, IEEE International Conference on Fuzzy Systems, Jeju Island, Korea, 20–24 August 2009, pp. 1378–1382 (2009). https://doi.org/10.1109/FUZZY.2009.5276884

25. Torra, V.: Hesitant fuzzy sets. Int. J. Intell. Syst. **25**, 529–539 (2010). https://doi.org/10.1002/int.20418

26. Wang, H., Xu, Z.: Admissible orders of typical hesitant fuzzy elements and their application in ordered information fusion in multi-criteria decision making. Inf. Fusion **29**(C), 98–104 (2016). https://doi.org/10.1016/j.inffus.2015.08.009

27. Wang, Z., Xu, Z., Liu, S., Yao, Z.: Direct clustering analysis based on intuitionistic fuzzy implication. Appl. Soft Comput. **23**, 1–8 (2014). https://doi.org/10.1016/j.asoc.2014.03.037

28. Wei, G.: Hesitant fuzzy prioritized operators and their application to multiple attribute decision making. Knowl.-Based Syst. **31**, 176–182 (2012). https://doi.org/10.1016/j.knosys.2012.03.011

29. Xia, M., Xu, Z.: Hesitant fuzzy information aggregation in decision making. Int. J. Approx. Reason. **52**(3), 395–407 (2011). https://doi.org/10.1016/j.ijar.2010.09.002

30. Xia, M., Xu, Z., Chen, N.: Some hesitant fuzzy aggregation operators with their application in group decision making. Group Decis. Negot. **22**(2), 259–279 (2013). https://doi.org/10.1007/s10726-011-9261-7

31. Xu, Z., Xia, M.: Distance and similarity measures for hesitant fuzzy sets. Inf. Sci. **181**(11), 2128–2138 (2011). https://doi.org/10.1016/j.ins.2011.01.028

32. Zhang, H.Y., Yang, S.Y.: Typical hesitant fuzzy rough sets. In: 2015 International Conference on Machine Learning and Cybernetics (ICMLC), vol. 1, pp. 328–333, July 2015. https://doi.org/10.1109/ICMLC.2015.7340943

33. Zhang, H., Yang, S.: Inclusion measure for typical hesitant fuzzy sets, the relative similarity measure and fuzzy entropy. Soft Comput. **20**(4), 1277–1287 (2015). https://doi.org/10.1007/s00500-015-1851-x

34. Zhu, B., Xu, Z., Xia, M.: Hesitant fuzzy geometric Bonferroni means. Inf. Sci. **205**, 72–85 (2012). https://doi.org/10.1016/j.ins.2012.01.048

Is the Invariance with Respect to Powers of a t-norm a Restrictive Property on Fuzzy Implication Functions? The Case of Strict t-norms

Raquel Fernandez-Peralta[1,2], Sebastia Massanet[1,2(✉)], and Arnau Mir[1,2]

[1] Soft Computing, Image Processing and Aggregation (SCOPIA) Research Group, Department Mathematics and Computer Science, University of the Balearic Islands, 07122 Palma, Spain
{r.fernandez,s.massanet,arnau.mir}@uib.es
[2] Balearic Islands Health Research Institute (IdISBa), 07010 Palma, Spain

Abstract. The invariance with respect to powers of a t-norm has emerged as an important property for fuzzy implication functions in approximate reasoning. Recently, those fuzzy implication functions satisfying this property where fully characterized leading to seemingly new families of these operators. In this paper, the additional properties of the family of fuzzy implication functions which are invariant with respect to powers of a strict t-norm are analyzed. In particular, properties such as the exchange principle, the law of importation with respect to a t-norm or the left neutrality principle, among others, can be fulfilled by some members of this family. This study allows to characterize the intersection of these operators with the most important families of fuzzy implication functions.

Keywords: Fuzzy implication function · Invariance · Powers of t-norms · Exchange principle

1 Introduction

In the last decades, dozens of families of fuzzy implication functions have been proposed in the literature (see [1,2,5] and references therein). Although some of these families have boosted some applications in which fuzzy implication functions play a key role, other families struggle to stand out since they do not satisfy any differentiating additional property with respect to the rest. Moreover, this vast number of families is starting to cause some major problems in the research of the field [6]. Therefore, it is necessary to analyze in depth those families of fuzzy implication functions which provide uncommon but useful properties and to analyze their relationship and intersections with other well-known families. This will allow the community to disclose more about the structure of these operators and to open new potentially useful lines of research.

© Springer Nature Switzerland AG 2020
M.-J. Lesot et al. (Eds.): IPMU 2020, CCIS 1238, pp. 761–774, 2020.
https://doi.org/10.1007/978-3-030-50143-3_59

In this direction, the invariance property with respect to powers of a continuous t-norm was proposed in [8] as an additional property of fuzzy implication functions with applications in approximate reasoning. As it is stated in [8], the fulfillment of this property ensures that the following fuzzy propositions from the classical example given in [10]:

If the tomato is red, then it is ripe.
If the tomato is very red, then it is very ripe.
If the tomato is little red, then it is little ripe.

have the same truth value whenever the linguistic modifiers "very" and "little" are modeled using powers of continuous t-norms. This additional property is not satisfied in general by the most usual families of fuzzy implication functions. Therefore, in [8], the so-called T-power based implications are introduced as a family of fuzzy implication functions satisfying the invariance for many t-norms (see the corrigendum [7] also). Later, in [9], the complete characterization of all fuzzy implication functions satisfying the invariance property with respect to powers of a continuous t-norm is achieved. Indeed, the characterization depends on the type of continuous t-norm and provides the expression of the family of fuzzy implication functions fulfilling the property. However, in [9], only this property is studied and up to now, it is unknown which other additional properties can be satisfied by the members of these families of invariant implications. Thus, the goal of this paper is to study which well-known additional properties these fuzzy implication functions satisfy and under which conditions. As a first approach to this problem, this paper deals with the family of fuzzy implication functions which are invariant with respect to powers of a strict t-norm. This study will encourage the use of fuzzy implication functions in approximate reasoning where other additional properties may be required in addition to the invariance property. Moreover, as a straightforward consequence, the study of the additional properties of this family allows to determine the intersection of this family with some of the most important families of fuzzy implication functions, namely (S, N), R, QL and Yager's f and g generated implications (see [1]).

The paper is organized as follows. In the next section we recall some basic definitions and properties on fuzzy implication functions. In Sect. 3, the family of fuzzy implication functions which are invariant with respect to powers of strict t-norms is recalled and its definition is revisited. Then, in Sect. 4, the additional properties of the family are deeply analyzed and the conditions under which this family fulfills them are determined. After that, in Sect. 5, the intersections of this family with some well-known families is derived from the study carried out in the previous section. The paper ends with some conclusions and future work.

2   Preliminaries

To make this work self-contained, we recall here some of the concepts and results which will be used throughout the paper. Although we will suppose the reader is familiar with basic results on t-norms (see [4, 12] for more details), we recall the definition of a strict t-norm and the expression of its powers.

Definition 1 ([4]). *A function* $T : [0, 1]^2 \rightarrow [0, 1]$ *is called a* strict t-norm *if there exists a continuous, strictly decreasing function* $t : [0, 1] \rightarrow [0, +\infty]$ *with* $t(0) = +\infty$ *and* $t(1) = 0$, *which is uniquely determined up to a positive multiplicative constant, such that* T *is given by*

$$T(x, y) = t^{-1}(t(x) + t(y))$$

for all $x, y \in [0, 1]$.

Powers of a t-norm T, which are defined in detail in [12] and will be denoted by $x_T^{(r)}$ with $x \in [0, 1]$ and $r \in [0, +\infty]$, can be expressed for strict t-norms T in terms of an additive generator of the t-norm.

Proposition 1 ([12]). *Let* T *be a strict t-norm with additive generator* t. *Then*

$$x_T^{(r)} = t^{-1}(rt(x)) \text{ for all } x \in [0, 1] \text{ and } r \in [0, +\infty]$$

with the convention that $+\infty \cdot 0 = 0$.

We start now with the definition of a fuzzy implication function.

Definition 2 ([1, 3]). *A binary operator* $I : [0, 1]^2 \rightarrow [0, 1]$ *is said to be a* fuzzy implication function *if it satisfies:*

(I1) $I(x, z) \geq I(y, z)$ *when* $x \leq y$, *for all* $z \in [0, 1]$.
(I2) $I(x, y) \leq I(x, z)$ *when* $y \leq z$, *for all* $x \in [0, 1]$.
(I3) $I(0, 0) = I(1, 1) = 1$ *and* $I(1, 0) = 0$.

From the definition, it can be easily derived that $I(0, x) = 1$ and $I(x, 1) = 1$ for all $x \in [0, 1]$. On the other hand, the symmetrical values $I(x, 0)$ and $I(1, x)$ are not predetermined from the definition.

Along the history of fuzzy implication functions, additional properties of these functions have been postulated (see [1, 3, 11] for more details). Among the most important and those that are relevant for this work we stand out the following ones:

- The *identity principle*

$$I(x, x) = 1, \quad x \in [0, 1]. \tag{IP}$$

- The *ordering property*

$$I(x, y) = 1 \Leftrightarrow x \leq y, \quad x, y \in [0, 1]. \tag{OP}$$

- The *exchange principle*

$$I(x, I(y, z)) = I(y, I(x, z)), \quad x, y, z \in [0, 1]. \tag{EP}$$

- The *law of importation* with respect to a t-norm T

$$I(T(x,y),z) = I(x,I(y,z)), \quad x,y,z \in [0,1]. \tag{$\mathbf{LI}$}_\mathbf{T}$$

- The *left neutrality principle*

$$I(1,y) = y, \quad y \in [0,1]. \tag{NP}$$

- The *iterative boolean law*

$$I(x,y) = I(x,I(x,y)), \quad x,y \in [0,1]. \tag{IB}$$

In addition to the previous additional properties, the invariance property with respect to t-norms was recently proposed in [8] in order to deal with the classical problem of the tomato recalled in the introduction.

Definition 3 ([8]). *Let I be a fuzzy implication function and T a continuous t-norm. It is said that I is* invariant with respect to T-powers, *or simply that it is T-power invariant when*

$$I(x,y) = I\left(x_T^{(r)}, y_T^{(r)}\right), \tag{$\mathbf{PI}_\mathbf{T}$}$$

holds for all real number $r > 0$ and for all $x,y \in [0,1]$ such that $x_T^{(r)}, y_T^{(r)} \neq 0,1$.

3 Strict T-power Invariant Implications

In [9] all fuzzy implication functions which are invariant with respect to T-powers when T is a strict t-norm were characterized in the following theorem.

Theorem 1 ([9, **Theorem 8**]). *Let T be a strict t-norm and t an additive generator of T. A mapping $I : [0,1]^2 \to [0,1]$ is a fuzzy implication function invariant with respect to T-powers if and only if there exists an increasing mapping $\varphi : [0,+\infty] \to [0,1]$ with $\varphi(0) = 0$, $\varphi(+\infty) = 1$ and such that I is given by*

$$I(x,y) = \varphi\left(\frac{t(x)}{t(y)}\right), \quad \text{for all } (x,y) \in [0,1]^2 \setminus \{(x,0),(1,y)|0 < x,y < 1\}, \tag{1}$$

with the convention that $\frac{0}{0} = \frac{+\infty}{+\infty} = +\infty$, and such that the remaining values $I(x,0)$ and $I(1,y)$ preserve the monotonicity conditions.

The aim of this paper is to study the family of fuzzy implication functions described in the theorem above. In order to do so, we first provide a concrete definition of such family and we establish the conditions which ensure that these functions are indeed fuzzy implication functions.

Let us consider T a strict t-norm, t an additive generator of T and $I : [0,1]^2 \to [0,1]$ a fuzzy implication function invariant with respect to T-powers.

From Theorem 1 we know that there exists an increasing mapping $\varphi : [0, +\infty] \to [0, 1]$ with $\varphi(0) = 0$, $\varphi(+\infty) = 1$ and such that I is given by (1).

First of all, notice that since the additive generator of a t-norm is unique up to a positive multiplicative constant, the definition of I is independent from the considered additive generator of T. Let us define $f(x) = I(x, 0)$ for all $x \in (0, 1)$ and $g(y) = I(1, y)$ for all $y \in (0, 1)$. Now, in order for I to be a fuzzy implication function, the functions f and g need to respect the monotonicity conditions (**I1**) and (**I2**). From (**I2**) we get that g has to be an increasing function and, for a fixed $x \in (0, 1)$ the following condition must hold

$$f(x) = I(x, 0) \leq I(x, y) = \varphi\left(\frac{t(x)}{t(y)}\right), \quad \text{for all } y \in (0, 1). \qquad (2)$$

On the other hand, from (**I1**) we get that f has to be a decreasing function and, for a fixed $y \in (0, 1)$ the following condition must hold

$$g(y) = I(1, y) \leq I(x, y) = \varphi\left(\frac{t(x)}{t(y)}\right), \quad \text{for all } x \in (0, 1). \qquad (3)$$

Now, it is easy to prove that inequalities (2) and (3) hold if and only if

$$\inf_{w \in (0, +\infty)} \varphi(w) \geq \max\left\{ \sup_{y \in (0, 1)} g(y), \sup_{x \in (0, 1)} f(x) \right\}.$$

Having said this, we provide the following definition of the family of fuzzy implication functions that are T-power invariant with respect to a strict t-norm.

Definition 4. *Let T be a strict t-norm and t an additive generator of T. Let $f : (0, 1) \to [0, 1]$ be a decreasing function and $\varphi : [0, +\infty] \to [0, 1]$, $g : (0, 1) \to [0, 1]$ increasing functions such that $\varphi(0) = 0$, $\varphi(+\infty) = 1$ and*

$$\inf_{w \in (0, +\infty)} \varphi(w) \geq \max\left\{ \sup_{y \in (0, 1)} g(y), \sup_{x \in (0, 1)} f(x) \right\}. \qquad (4)$$

The function $I_{\varphi, f, g}^T ; [0, 1]^2 \to [0, 1]$ defined by

$$I_{\varphi, f, g}^T(x, y) = \begin{cases} f(x) & \text{if } x \in (0, 1) \text{ and } y = 0, \\ g(y) & \text{if } x = 1 \text{ and } y \in (0, 1), \\ \varphi\left(\frac{t(x)}{t(y)}\right) & \text{otherwise,} \end{cases} \qquad (5)$$

with the understanding $\frac{0}{0} = \frac{+\infty}{+\infty} = +\infty$, is called a strict T-power invariant implication.

In Fig. 1 we can see the structure of fuzzy implication functions given by Expression (5).

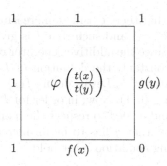

Fig. 1. Structure of the family of strict T-power invariant implications.

Notice that Condition (4) imposes that the function φ is bounded below by any possible value of f and g (see Example 1). Although the structure of strict T-power invariant implications may seem flexible since it depends of three unknown functions, as a matter of fact, Condition (4) severely restricts the choices of functions φ, f and g for which $I^T_{\varphi,f,g}$ is a fuzzy implication function. Indeed, the following proposition studies the continuity of strict T-power implications and shows that certain desired properties of $I^T_{\varphi,f,g}$ lead to impose that φ, f or g are constant functions.

Proposition 2. Let $I^T_{\varphi,f,g}$ be a strict T-power invariant implication. The following statements hold:

(i) If φ is continuous at $w = 0$, then $f(x) = g(y) = 0$ for all $x, y \in (0,1)$.

(ii) If $\lim_{x \to 0^+} f(x) = 1$ or $\lim_{y \to 1^-} g(y) = 1$, then $\varphi(w) = 1$ for all $w \in (0,1)$.

(iii) $I^T_{\varphi,f,g}$ is continuous at $(x,0)$ for $x \in (0,1)$ if and only if $f(x) = \lim_{w \to 0^+} \varphi(w)$.

(iv) $I^T_{\varphi,f,g}$ is continuous at $(1,y)$ for $y \in (0,1)$ if and only if $g(y) = \lim_{w \to 0^+} \varphi(w)$.

(v) $I^T_{\varphi,f,g}$ is continuous at $(x,1)$ and $(0,x)$ for all $x \in [0,1]$ if and only if $\lim_{w \to +\infty} \varphi(w) = 1$.

(vi) $I^T_{\varphi,f,g}$ is continuous at (x_0, y_0) with $x_0, y_0 \in (0,1)$ if and only if φ is continuous at $\frac{t(x_0)}{t(y_0)}$. In this case, $I^T_{\varphi,f,g}$ is also continuous at the following points

$$\left(x, t^{-1}\left(\frac{t(x)t(y_0)}{t(x_0)} \right) \right), \quad \text{for all } x \in (0,1).$$

Moreover, from (ii) in the previous result it is easy to see that $I^T_{\varphi,f,g}$ is never a continuous function.

Corollary 1. Let $I^T_{\varphi,f,g}$ be a strict T-power invariant implication. Then at least one of the following conditions holds:

(i) $I^T_{\varphi,f,g}$ is discontinuous at (1,0).

(ii) $I^T_{\varphi,f,g}$ is discontinuous at (0,0) or (1,1).

Example 1. Let us show two examples of strict T-power invariant implications.

(i) Let us consider $t(s) = \frac{1-s}{s}$ for all $s \in [0,1]$, $f(x) = \frac{1-x}{3}$ for all $x \in (0,1)$, $g(y) = \frac{y}{3}$ for all $y \in (0,1)$ and

$$\varphi(w) = \begin{cases} 0 & \text{if } w = 0, \\ \frac{w+1}{w+3} & \text{otherwise.} \end{cases}$$

The corresponding strict T-power invariant implication is given by

$$I_1(x,y) = \begin{cases} 1 & \text{if } (x,y) \in \{(1,1),(0,0)\}, \\ \frac{1-x}{3} & \text{if } x \in (0,1] \text{ and } y = 0, \\ \frac{y}{3} & \text{if } x = 1 \text{ and } y \in (0,1), \\ \frac{y-2xy+x}{y-4xy+3x} & \text{otherwise.} \end{cases}$$

(ii) Let us consider $t(s) = \frac{1-s}{s}$ for all $s \in [0,1]$, $f(x) = g(y) = 0$ for all $x, y \in (0,1)$ and

$$\varphi(w) = \begin{cases} w \text{ if } w < 1, \\ 1 \text{ otherwise.} \end{cases}$$

The corresponding strict T-power invariant implication is given by

$$I_2(x,y) = \begin{cases} 0 & \text{if } (x \in (0,1] \text{ and } y = 0) \text{ or } (x = 1 \text{ and } y \in (0,1)), \\ \frac{(1-x)y}{(1-y)x} & \text{if } 0 < x < y < 1, \\ 1 & \text{otherwise.} \end{cases}$$

These two fuzzy implication functions are displayed in Fig. 2.

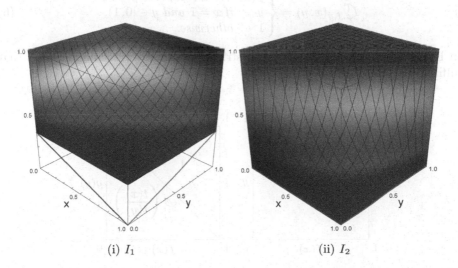

(i) I_1 (ii) I_2

Fig. 2. Plots of fuzzy implication functions given in Example 1.

4   Additional Properties

In [8] it was shown that T-power based implications do not satisfy most of the main additional properties of fuzzy implication functions such as the left neutrality principle, the exchange property or the law of importation with respect to any t-norm. This situation represents a problem if we want to consider fuzzy implication functions that are T-power invariant but also satisfy other common additional properties. In this section, we generalize the study made in [8] by considering strict T-power invariant implications and we show that, in this case, there are choices for φ, f and g which ensure that $I^T_{\varphi,f,g}$ can also satisfy the properties considered.

First of all, we consider the identity principle and the ordering property. These two properties were already studied in [9] to show that although $I^T_{\varphi,f,g}(x,x)$ is constant for all $x \in (0,1)$, (**IP**) is not guaranteed.

Proposition 3 ([9, **Theorem 9**]). *Let $I^T_{\varphi,f,g}$ be a strict T-power invariant implication. Then $I^T_{\varphi,f,g}$ satisfies* (**IP**) *if and only if $\varphi(1) = 1$. In this case, $I^T_{\varphi,f,g}$ satisfies* (**OP**) *if and only if $\varphi(w) < 1$ for all $w < 1$.*

Next, we show that if a strict T-power invariant implication satisfies the left neutrality principle then it is constant to 1 for all $x,y \in (0,1)$.

Proposition 4. *Let $I^T_{\varphi,f,g}$ be a strict T-power invariant implication. Then $I^T_{\varphi,f,g}$ satisfies* (**NP**) *if and only if $g(y) = y$ for all $y \in (0,1)$. Moreover, in this case $I^T_{\varphi,f,g}$ is given by*

$$I^T_{\varphi,f,g}(x,y) = \begin{cases} f(x) & \text{if } x \in (0,1) \text{ and } y = 0, \\ y & \text{if } x = 1 \text{ and } y \in [0,1), \\ 1 & \text{otherwise.} \end{cases} \tag{6}$$

In Fig. 3 we can see the structure of strict T-power invariant implications that fulfill (**NP**) and (**OP**).

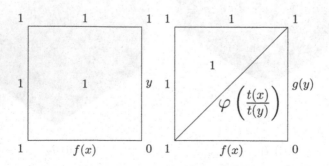

Fig. 3. Structure of strict T-power invariant implications that satisfy (**NP**) and (**OP**), respectively.

Now, let us consider the exchange principle. The next result shows that there are five possible configurations of strict T-power invariant implications that result in functions that satisfy (**EP**).

Proposition 5. *Let $I_{\varphi,f,g}^T$ be a strict T-power invariant implication. Then $I_{\varphi,f,g}^T$ satisfies (**EP**) if and only if one of the following conditions hold:*

(i) *Let $C \in (0,+\infty)$, then $\varphi(w) = t^{-1}\left(\frac{C}{w}\right)$ for all $w \in (0,+\infty)$ and $f(x) = g(y) = 0$ for all $x,y \in (0,1)$.*

(ii) *Let $k \in [0,1]$, then $\varphi(w) = k$ for all $w \in (0,+\infty)$ and $f(x) = g(y) = 0$ for all $x,y \in (0,1)$.*

(iii) *Let $k \in (0,1]$, then $\varphi(w) = k$ for all $w \in (0,+\infty)$ and one of the following conditions holds:*

(a) $f(x) = \begin{cases} k & \text{if } x \in A, \\ 0 & \text{if } x \in (0,1) \setminus A, \end{cases}$ *where A is $(0,a]$ or $(0,a)$ with $a \in (0,1)$ or $A = \emptyset$ and $\operatorname{Im} g \subseteq (0,k]$.*

(b) *$f(x) = k$ for all $x \in (0,1)$ and $\operatorname{Im} g \subseteq [0,k]$.*

(c) *$\operatorname{Im} f \subseteq (0,k]$, $\operatorname{Im} g \subseteq (0,k]$ and $g(y) = y$ for all $y \in \operatorname{Im} f \setminus \{1\}$. Moreover, if $k < 1$, g must additionally satisfy $g(k) = k$.*

In Fig. 4 we summarize the possible configurations of strict T-power invariant implications that fulfill (**EP**). Notice that only configuration (i) corresponds to a fuzzy implication function that is not constant in $(0,1)^2$.

Example 2. Let us consider $t(s) = \frac{1-s}{s}$ for all $s \in [0,1]$. The corresponding strict T-power invariant implication that satisfies (**EP**) and is non-constant in $(0,1)^2$ is given by

$$I_3(x,y) = \begin{cases} 1 & \text{if } x = 0 \text{ and } y = 1, \\ 0 & \text{if } (y = 0 \text{ and } x \in (0,1)) \text{ or } (x = 1 \text{ and } y \in [0,1)), \\ \frac{(1-x)y}{Cx - Cxy + y - xy} & \text{otherwise,} \end{cases}$$

where $C \in (0,+\infty)$. Note that I_3 corresponds to the solution given in Proposition 5-(i). In Fig. 5 we can see the plots of some members of this family of fuzzy implication functions for $C = 1$, $C = 10$ and $C = 100$.

Next, we study the law of importation with respect to a t-norm T^*. The following result establishes the three possible configurations of strict T-power invariant implications that satisfy the law of importation with respect to some t-norm.

Proposition 6. *Let $I_{\varphi,f,g}^T$ be a strict T-power invariant implication and T^* a t-norm. Then $I_{\varphi,f,g}^T$ satisfies (**LI**) with respect to T^* if and only if one of the following conditions hold:*

(i) *$\varphi(w) = 0$ for all $w \in (0,+\infty)$, $f(x) = g(y) = 0$ for all $x,y \in (0,1)$ and T^* is a positive t-norm.*

(ii) *Let $k \in (0,1]$, then $\varphi(w) = k$ for all $w \in (0,+\infty)$, $g(y) = y$ for all $y \in \operatorname{Im} g \setminus \{0,1\}$, and one of the following conditions hold:*

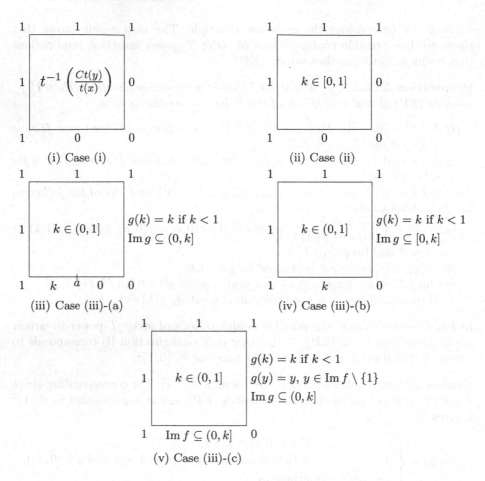

Fig. 4. Structure of strict T-power invariant implications that satisfy (**EP**) defined in Proposition 5.

(a) $f(x) = \begin{cases} k \text{ if } x \in A, \\ 0 \text{ if } x \in (0,1) \setminus A, \end{cases}$ *where A is $(0,a]$ or $(0,a)$ with $a \in (0,1)$ or $A = \emptyset$, T^* satisfies the following property:*

$$T^*(x,y) \in (0,1] \setminus A \text{ if and only if } x,y \in (0,1] \setminus A,$$

and $\operatorname{Im} g \subseteq (0,k]$.

(b) $0 \notin \operatorname{Im} f$, $f(x) = k$ *for all* $x \in \operatorname{Im} T^*|_{(0,1)^2} \setminus \{0\}$, $g(y) = y$ *for all* $y \in \operatorname{Im} f \setminus \{1\}$ *and* $\operatorname{Im} g \subseteq [0,k]$ *but* $g(y) > 0$ *for all* $y \in (0,1)$ *when f is not a function constant to k.*

Moreover, if $k < 1$, g must additionally satisfy $g(k) = k$ and T^* must be a positive t-norm.

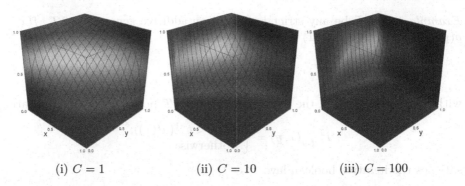

(i) $C = 1$ (ii) $C = 10$ (iii) $C = 100$

Fig. 5. Plots of fuzzy implication function given in Example 2 for $C = 1$, $C = 10$ and $C = 100$.

Example 3. Let us consider $\varphi(w) = \frac{1}{2}$ for all $w \in (0,1)$, $f(x) = \frac{1}{2}$ for all $x \in (0,1)$ and

$$g(y) = \begin{cases} y & \text{if } y \in (0, \frac{1}{4}), \\ \frac{1}{2} & \text{if } y \in [\frac{1}{4}, \frac{1}{2}], \\ 1 & \text{if } y \in (\frac{1}{2}, 1). \end{cases}$$

The corresponding strict T-power invariant implication

$$I_{\varphi,f,g}^{T}(x,y) = \begin{cases} y & \text{if } x = 1 \text{ and } y \in [0, \frac{1}{4}), \\ 1 & \text{if } x = 0 \text{ or } y = 1 \text{ or } (x = 1 \text{ and } y \in (\frac{1}{2}, 1)), \\ \frac{1}{2} & \text{otherwise,} \end{cases}$$

satisfies the law of importation with respect to any positive t-norm, for instance the minimum t-norm $T_M(x,y) = \min(x,y)$ or the product t-norm $T_P(x,y) = xy$.

Remark 1. Let us consider a strict T-power invariant implication under the conditions of (iii)-(a) in Proposition 6 where $A = (0, a)$ with $a \in (0,1)$ and $k \in (0,1)$. Then, this fuzzy implication function satisfies the law of importation with respect to any positive t-norm T^* such that $T^*(x,y) \in [a, 1]$ if and only if $x, y \in [a, 1]$. For instance, T^* can be a continuous positive t-norm with a as an idempotent element. However, a further study must be made in order to characterize all t-norms that fulfill such condition.

Finally, we analyze under which conditions strict T-power invariant implications satisfy the iterative boolean law. In this case, we see that φ needs to be a constant function or its only possible values are 0 and 1.

Proposition 7. *Let $I_{\varphi,f,g}^{T}$ be a strict T-power invariant implication. Then $I_{\varphi,f,g}^{T}$ satisfies (**IB**) if and only if one of the following conditions hold:*

(i) $\operatorname{Im}\varphi \subseteq \{0,1\}$, φ *is not constant to 1 and* $f(x) = g(y) = 0$ *for all* $x, y \in (0,1)$.
(ii) Let $k \in (0,1]$, *then* $\varphi(w) = k$ *for all* $w \in (0, +\infty)$, $\operatorname{Im} f \subseteq \{0, k\}$, $\operatorname{Im} g \subseteq [0, k]$ *and* $g(y) = y$ *for all* $y \in \operatorname{Im} g \setminus \{0, 1\}$.

Example 4. Let T be any strict t-norm, t an additive generator of T, $f(x) = g(y) = 0$ for all $x, y \in (0, 1)$ and

$$\varphi(w) = \begin{cases} 0 \text{ if } w < \frac{1}{a}, \\ 1 \text{ otherwise,} \end{cases}$$

with $a \in (0, +\infty)$. Then, the corresponding strict T-power invariant implication

$$I_{\varphi, f, g}^T(x, y) = \begin{cases} 0 \text{ if } y < t^{-1}(at(x)), \\ 1 \text{ otherwise,} \end{cases}$$

satisfies the iterative boolean law.

Remark 2. According to Propositions 4, 5, 6 and 7 all strict T-power invariant implications that satisfy (**NP**), (**EP**), (**IB**) or (**LI**)$_T$ except case (i) in Proposition 5 and case (i) in Proposition 7, are given by a φ which is constant in $(0, +\infty)$. Therefore, these fuzzy implication functions are given by an expression which is independent from the generator of the corresponding t-norm. Then, they are also T-power invariant with respect to any strict t-norm. This fact reflects that imposing the T-power invariance with some additional property is very restrictive and usually results in degenerated solutions.

5   Intersection with the Main Families of Fuzzy Implication Functions

It is well-known that an important step when studying a new family of fuzzy implication functions is to study their intersection among other families of fuzzy implication functions. In this section we investigate whether the family of strict T-power invariant implications intersects with five of the most well-known families of fuzzy implication functions. Let us denote the following families of fuzzy implication functions:

$\mathbb{I}_{\varphi, f, g}^T$ — the family of all strict T-power invariant implications;

$\mathbb{I}_{S, N}$ — the family of all (S, N)-implications;

$\mathbb{I}_T$ — the family of all R-implications;

$\mathbb{I}_{QL}$ — the family of all QL-implications;

$\mathbb{I}_F$ — the family of all f-generated implications;

$\mathbb{I}_G$ — the family of all g-generated implications.

In [8] it was pointed out that T-power based implications have no intersection with all the above families because they do not satisfy the left neutrality principle. However, we have seen that there are choices for f, g and φ such that the corresponding strict T-power invariant implication satisfies (**NP**). For instance, the following fuzzy implication function satisfies both (**NP**) and (**EP**):

$$I^*(x, y) = \begin{cases} f(x) \text{ if } x \in (0, 1) \text{ and } y = 0, \\ y \quad\;\; \text{if } x = 1 \text{ and } y \in [0, 1), \\ 1 \quad\;\; \text{otherwise,} \end{cases}$$

where $f : (0,1) \to [0,1]$ is any decreasing function with $\operatorname{Im} f \subseteq (0,1]$. Notice that if we choose $f(x) = 1$ for all $x \in (0,1)$ we obtain the well-known Weber implication

$$I_{\mathbf{WB}}(x,y) = \begin{cases} y \text{ if } x = 1 \text{ and } y \in [0,1], \\ 1 \text{ otherwise.} \end{cases}$$

Thanks to the study of the previous section, we are able to prove that strict T-power invariant implications have non-empty intersection with R, QL and (S,N)-implications. Indeed, the following proposition provides the complete characterization of the intersections of interest.

Proposition 8. *The following equalities are true:*

- $\mathbb{I}_{\varphi,f,g}^{T} \cap \mathbb{I}_{\mathrm{T}} = I_{\boldsymbol{WB}}.$
- $\mathbb{I}_{\varphi,f,g}^{T} \cap \mathbb{I}_{\mathrm{S,N}} = \mathbb{I}_{\varphi,f,g}^{T} \cap \mathbb{I}_{\mathrm{QL}} = I^*.$
- $\mathbb{I}_{\varphi,f,g}^{T} \cap \mathbb{I}_{\mathrm{F}} = \mathbb{I}_{\varphi,f,g}^{T} \cap \mathbb{I}_{\mathrm{G}} = \emptyset.$

Notice that although the intersection of strict T-power invariant implications and R, QL and (S,N)-implications is not empty, the fuzzy implication functions that belong to this intersection are constant to 1 in $(0,1)^2$. Therefore, we can conclude that the T-power invariance property with respect to a strict t-norm is not satisfied for almost all fuzzy implication functions that belong to the most well-known families of fuzzy implication functions. In this sense, it is relevant to study strict T-power invariant implications as a new family of fuzzy implication functions.

6   Conclusions and Future Work

In this paper, the most usual additional properties of fuzzy implication functions have been studied for the family of strict T-power invariant implications, those fuzzy implication functions which are invariant with respect to powers of a strict t-norm. The results show that members of this family can satisfy other additional properties in addition to the invariance property. While for some properties such as (**NP**) or (**LI**)$_{\mathrm{T}}$, the only solutions are fuzzy implication functions which are constant in $(0,1)^2$, other properties such as (**IB**), (**OP**) or (**EP**) provide richer non-constant solutions. Note that these results are in fact characterizations of all fuzzy implication functions which satisfy both the invariance with respect to powers of a strict t-norm and the corresponding other additional property. Finally, from these results, the intersections with the most well-known families is fully determined leading to some R, QL and (S,N)-implications which are also invariant with respect to powers of a strict t-norm.

As future work, we want to perform a similar study for the family of nilpotent T-power invariant implications and also for those fuzzy implication functions which are invariant with respect to powers of an ordinal sum t-norm.

Acknowledgment. This paper has been partially supported by the Spanish Grant TIN2016-75404-P, AEI/FEDER, UE. Raquel Fernandez-Peralta benefits from the fellowship FPU18/05664 granted by the Spanish Ministry of Science, Innovation and Universities within the Training university lecturers (FPU) program.

References

1. Baczyński, M., Jayaram, B.: Fuzzy Implications. Studies in Fuzziness and Soft Computing, vol. 231. Springer, Heidelberg (2008). https://doi.org/10.1007/978-3-540-69082-5
2. Baczynski, M., Jayaram, B., Massanet, S., Torrens, J.: Fuzzy implications: past, present, and future. In: Kacprzyk, J., Pedrycz, W. (eds.) Springer Handbook of Computational Intelligence, pp. 183–202. Springer, Heidelberg (2015). https://doi.org/10.1007/978-3-662-43505-2_12
3. Fodor, J.C., Roubens, M.: Fuzzy Preference Modelling and Multicriteria Decision Support. Kluwer Academic Publishers, Dordrecht (1994)
4. Klement, E., Mesiar, R., Pap, E.: Triangular Norms. Kluwer Academic Publishers, Dordrecht (2000)
5. Mas, M., Monserrat, M., Torrens, J., Trillas, E.: A survey on fuzzy implication functions. IEEE Trans. Fuzzy Syst. **15**(6), 1107–1121 (2007)
6. Massanet, S.: Tidying up the mess of classes of fuzzy implication functions. In: Halas, R., Gagolewski, M., Mesiar, R. (eds.) Book of Abstracts of the 10th International Summer School on Aggregation Operators, pp. 1–2. Olomouc University (2019)
7. Massanet, S., Recasens, J., Torrens, J.: Corrigendum to "Fuzzy implication functions based on powers of continuous t-norms" [Int. J. Approx. Reason. **83**, 265–279 (2017)]. Int. J. Approx. Reason. **104**, 144–147 (2019)
8. Massanet, S., Recasens, J., Torrens, J.: Fuzzy implication functions based on powers of continuous t-norms. Int. J. Approx. Reason. **83**, 265–279 (2017)
9. Massanet, S., Recasens, J., Torrens, J.: Some characterizations of T-power based implications. Fuzzy Sets Syst. **359**, 42–62 (2019)
10. Mizumoto, M., Zimmermann, H.J.: Comparison of fuzzy reasoning methods. Fuzzy Sets Syst. **8**, 253–283 (1982)
11. Pradera, A., Beliakov, G., Bustince, H., De Baets, B.: A review of the relationships between implication, negation and aggregation functions from the point of view of material implication. Inf. Sci. **329**, 357–380 (2016)
12. Walker, C.L., Walker, E.A.: Powers of t-norms. Fuzzy Sets Syst. **129**(1), 1–18 (2002)

Some Remarks on Approximate Reasoning and Bandler-Kohout Subproduct

Katarzyna Miś[ID] and Michał Baczyński[✉][ID]

Faculty of Science and Technology, University of Silesia in Katowice,
Bankowa 14, 40-007 Katowice, Poland
{katarzyna.mis,michal.baczynski}@us.edu.pl

Abstract. In our contribution we give some remarks and conclusions regarding reasoning schemas used in approximate reasoning. Based on created computer tool for image customization we give some advices regarding FITA. Also, we show some facts regarding Bandler-Kohout subproduct and we present results for several inference schemas.

Keywords: Bandler-Kohout subproduct · Compositional rule of inference · Fuzzy implications · Fuzzy connectives

1 Introduction

Approximate reasoning is an often used concept when it is required to obtain meaningful results from imprecise data. One idea comes from Zadeh [15], where we use some inference schemas to receive such results. These schemas are based on their classic versions like:

- hypothetical syllogism

$$\frac{A \to B \land B \to C}{\therefore \quad A \to C}$$

- modus ponens

$$\frac{A \to B \land A}{\therefore \quad B}$$

- modus tollens

$$\frac{A \to B \land \neg B}{\therefore \quad \neg A}$$

- reduction to absurdity

$$\frac{\neg A \to B \land \neg B}{\therefore \quad A}$$

where A, B, C are any statements. The one which is the most commonly applied is generalized modus ponens (GMP):

© Springer Nature Switzerland AG 2020
M.-J. Lesot et al. (Eds.): IPMU 2020, CCIS 1238, pp. 775–787, 2020.
https://doi.org/10.1007/978-3-030-50143-3_60

RULE: IF x is A, THEN y is B.
FACT: x is A'.

CONCLUSION: y is B',

where x, y are some objects and A, B, A', B' are some properties. In our case these properties are represented by fuzzy sets so we are able to compute the values of B' using the following method:

$$B' := A' \star R,$$

where R is a fuzzy relation and $\star$ is a composition of such R and a fuzzy set A'. More precisely, if we use the composition proposed by Zadeh - the compositional rule of inference (CRI), which is the most popular, we have the following formula for our result:

$$B'(y) := \sup_{x \in X} T(A'(x), I(A(x), B(y))), \quad y \in Y, \tag{1}$$

where T is a t-norm (or any generalization of classical conjunction) and I is a fuzzy implication (or any generalization of classical implication). Another kind of a composition $\star$ is the Bandler-Kohout subproduct (BKS) [3,13]. In this case we usually use different model of a relation R and we consider the following formula:

$$B'(y) := \inf_{x \in X} I(A'(x), T(A(x), B(y))), \quad y \in Y. \tag{2}$$

For both of them we almost always require a minimal property of interpolativity. It is nothing else but satisfying a classic version of modus ponens, i.e., if we take $A' = A$, then we should obtain $B' = B$. Therefore, if we take into account all possible values of fuzzy sets A, B (i.e., the unit interval $[0, 1]$), then from (1) and (2) we receive the following two functional equations:

$$y = \sup_{x \in [0,1]} T(x, I(x, y)), \quad y \in [0, 1], \tag{CRI-GMP}$$

$$y = \inf_{x \in [0,1]} I(x, T(x, y)), \quad y \in [0, 1], \tag{BK-GMP}$$

where T is a t-norm and I is a fuzzy implication. Both functional equations (or their counterparts written as inequalities) were studied extensively in the literature for different classes of operations (see [2,14]).

In many applications of generalized modus ponens, the rule base is not consisted of only one rule IF - THEN, but is contains more of them. Hence we have the following scheme:

$$\text{IF } x \text{ is } A_1, \text{ THEN } y \text{ is } B_1, \tag{3}$$

$$\text{IF } x \text{ is } A_2, \text{ THEN } y \text{ is } B_2,$$

$$\cdots$$

$$\text{IF } x \text{ is } A_n, \text{ THEN } y \text{ is } B_n,$$

FACT: x is A'.

To obtain the conclusion: y is B' (which is based on all given rules) there are two main strategies (see [2,10]):

- FATI - First Aggregate, Then Infer:

$$\text{Aggregate: } R(x,y) = G(R_1(x,y), \ldots, R_n(x,y))$$
$$= G(I(A_1(x), B_1(y)), \ldots, I(A_n(x), B_n(y))),$$
$$\text{Infer: } B'(y) := A'(x) \circ R(x,y);$$

- FITA - First Infer, Then Aggregate:

$$\text{Infer: } B_i'(y) = A'(x) \overset{T}{\circ} R_i(x,y) = A'(x) \overset{T}{\circ} I(A_i(x), B_i(y)),$$
$$\text{for all } i \in \{1, \ldots, n\},$$
$$\text{Aggregate: } B'(y) = G(B_1'(y), \ldots, B_n'(y)),$$

where G is some aggregation function (it can be a t-norm).

Later we will present some conclusions regarding usage of FITA. Moreover, in our paper we would like to present some other results and corollaries that we obtained during work on a computer tool for image customization. We prepared it in order to test if our theoretical results concerning some reasoning schemas have an impact on applications of approximate reasoning (and what type). We used GMP scheme and this is a starting point for some other investigations.

The paper is organized as follows. Section 2 contains briefly description of the mentioned tool and some preliminaries which will be used in a sequel. In Sect. 3 we present some observations based on exemplary images. In Sect. 4 we conclude our paper and give some plans for future work.

2   Basic Notions and Facts

2.1   Preliminaries

To make this paper self-contained we placed here some basic definitions regarding fuzzy connectives and some aggregation operators.

Definition 2.1 (see [2,9]). *A non-increasing function* $N \colon [0,1] \to [0,1]$ *is called a fuzzy negation, if* $N(0) = 1$, $N(1) = 0$. *Moreover, a fuzzy negation* N *is called*

(i) strict if it is strictly decreasing and continuous,
(ii) strong if it is an involution, i.e., $N(N(x)) = x$ *for all* $x \in [0,1]$.

Definition 2.2 ([8]). *Let* $n \in \mathbb{N}$. *An aggregation function in* $[0,1]^n$ *is a function* $A^{(n)} \colon [0,1]^n \to [0,1]$ *which satisfies the following conditions:*

(i) it is nondecreasing (in each variable),
(ii) $A^{(n)}(0, \ldots, 0) = 0$ *and* $A^{(n)}(1, \ldots, 1) = 1$.

Definition 2.3 ([7]). *A function* $S: [0,1]^2 \to [0,1]$ *is called a semicopula if it satisfies the following conditions:*

(i) $S(x,1) = S(1,x) = x, \quad x \in [0,1],$
(ii) S *is non-decreasing with respect to each variable.*

Definition 2.4 ([8]). *An n-ary mean in* $[0,1]^n$ *is an aggregation function* $M^{(n)}$ *which is internal, i.e.,*

$$\min\{x_1, \ldots, x_n\} \leq M^{(n)}(x_1, \ldots, x_n) \leq \max\{x_1, \ldots, x_n\},$$

for all $(x_1, \ldots, x_n) \in [0,1]^n.$

Definition 2.5 ([9, **Definition 1.1**]). *A function* $T: [0,1]^2 \to [0,1]$ *is called a triangular norm (shortly t-norm) if it satisfies the following conditions, for all* $x, y, z \in [0,1]:$

(T1) $T(x,y) = T(y,x)$, *i.e.,* T *is commutative,*
(T2) $T(x,T(y,z)) = T(T(x,y),z)$, *i.e.,* T *is associative,*
(T3)]T *is non-decreasing with respect to each variable,*
(T4) $T(x,1) = x.$

Definition 2.6 ([1]). *The diagonal of a t-norm* T *is the function* $\delta_T: [0,1] \to [0,1]$ *defined by*

$$\delta_T(x) = T(x,x), \quad x \in [0,1]. \tag{4}$$

For more detailed facts about diagonals of t-norms see [11].

Definition 2.7 ([8, **Definition 2.45**]). *A function* $F: [0,1]^n \to \overline{\mathbb{R}}$ *is idempotizable if its diagonal* δ_F *is strictly increasing and satisfies* $ran(\delta_F) = ran(F).$

Definition 2.8 ([8, **Definition 4.2**]). *A function* $M: [0,1]^n \to [0,1]$ *is an average in* $[0,1]^n$ *if there exists a nondecreasing and idempotizable function* $F: [0,1]^n \to \mathbb{R}$ *such that*

$$F = \delta_F \circ M.$$

In this case, we say that M *is as average associated with* F *in* $[0,1]^n.$

Definition 2.9 ([2, **Definition 1.1.1**]). *A function* $I: [0,1]^2 \to [0,1]$ *is called a fuzzy implication if it satisfies the following conditions:*

(I1) I *is non-increasing with respect to the first variable,*
(I2) I *is non-decreasing with respect to the second variable,*
(I3) $I(0,0) = I(1,1) = 1$ *and* $I(1,0) = 0.$

The set of all fuzzy implications will be denoted by $\mathcal{FI}.$

Definition 2.10 (see [2]). *We say that a fuzzy implication* I *satisfies*

(i) the left neutrality property, if

$$I(1,y) = y, \quad y \in [0,1], \tag{NP}$$

(ii) the ordering property, if

$$x \leq y \Longleftrightarrow I(x,y) = 1, \quad x, y \in [0,1]. \tag{OP}$$

Definition 2.11 ([2, **Definition 2.5.1**]). *A function $I: [0,1]^2 \to [0,1]$ is called an R-implication if there exists a t-norm T such that*

$$I(x,y) = \sup\{t \in [0,1] : T(x,t) \leq y\}, \quad x, y \in [0,1].$$

If I is generated from a t-norm T, then it will be denoted by I_T.

Definition 2.12 ([2, **Definition 1.4.15**]). *Let I be a fuzzy implication. A function $N_I: [0,1] \to [0,1]$ given by*

$$N_I(x) = I(x,0), \quad x \in [0,1], \tag{5}$$

is called the natural negation of I.

Example 2.13. Some basic t-norms and generated R-implications are presented in Fig. 1.

t-norm T		R-implication I_T	
$T_{\mathbf{D}}(x,y) = \begin{cases} 0, & x, y \in [0,1) \\ \min\{x,y\}, & \text{otherwise} \end{cases}$		$I_{\mathbf{WB}}(x,y) = \begin{cases} 1, & x < 1 \\ y, & x = 1 \end{cases}$	
$T_{\mathbf{LK}}(x,y) = \max\{0, x+y-1\}$		$I_{\mathbf{LK}}(x,y) = \min\{1, 1-x+y\}$	
$T_{\mathbf{P}}(x,y) = x \cdot y$		$I_{\mathbf{GG}}(x,y) = \begin{cases} 1, & x \leq y \\ \frac{y}{x}, & x > y \end{cases}$	
$T_{\mathbf{M}}(x,y) = \min\{x,y\}$		$I_{\mathbf{GD}}(x,y) = \begin{cases} 1, & x \leq y \\ y, & x > y \end{cases}$	

Fig. 1. Examples of basic R-implications.

2.2 Description of a Computer Tool

Let us take any picture where colors are given in RGB romat. Any pixel of such image can be seen as an object where we could adjust properties according to user choice/preferences. For instance, the user wants to change the color palette like below:

$$[76, 153, 0] \to [128, 255, 0]$$

from green to lighter between yellow and green

$$[255, 255, 0] \to [255, 178, 102]$$

from yellow to one with orange tones.

We can use GMP to compute values of such "vectors" of all pixels.

Let us consider some examples. Firstly, let us take the pair $(T_{\mathbf{LK}}, I_{\mathbf{LK}})$. In the phase of aggregating (here for FITA) we need to apply an aggregation function. It turns out the same t-norm will not give satisfying results. Let us compare images presented in Fig. 2. Better results (according to our subjective opinion) are obtained in Fig. 2(b) and worse are presented in Fig. 2(c). For this pair we used different aggregation methods – the arithmetic mean and the Łukasiewicz t-norm. In the next part of our paper we give some possible justification which shows why it is better to apply some means instead of t-norms or other class of aggregation functions.

(a) The original image

(b) Image obtained with $(T_{\mathbf{LK}}, I_{\mathbf{LK}})$ and arithmetic mean for aggregating

(c) Image obtained with $(T_{\mathbf{LK}}, I_{\mathbf{LK}})$ and $T_{\mathbf{LK}}$ for aggregating

Fig. 2. Examples of images used in our programme – CRI.

3 Main Results

3.1 Case of FITA

As we mentioned in Introduction, the basic property for compositions like CRI or BKS is interpolativity. Sometimes it is also called consistency [5] (in fact in [4,5] the authors showed that FATI is not consistent). However, when we are thinking about this property we focus mainly on IF - THEN rules. Nevertheless, for the cases of more than one rule it is important to choose a proper aggregation operator. Note that for the rule base (3) and the fact x *is* A' and taking $A' = A_i$, $i \in \{1, \ldots, n\}$, we can check interpolativity separately for each rule. So for the aggregation operator G we would obtain $G(B_1, \ldots, B_n)$. Based on intepolativity, an intuitive approach is when $B_i = B, i \in \{1, \ldots, n\}$ and then $G(B, \ldots, B)$ should be equal to B, which means G is idempotent. Such functions are for instance OWA functions, medians, averaging aggregation functions. Among them there are also means.

Remark 3.1 (cf [8]). For many t-norms and means there is the following relationship between M (a mean) and T (a t-norm):

$$\delta_T \circ M = T.$$

Example 3.2. Some classical means associated with t-norms are presented in Fig. 3.

M	T	δ_T
$T_\mathbf{M}$	$T_\mathbf{M}$	x
$\frac{x+y}{2}$	$T_\mathbf{LK}$	$\max\{0, 2x - 1\}$
$\sqrt{xy}$	$T_\mathbf{P}$	x^2
$\frac{2}{\frac{1}{x}+\frac{1}{y}}$	$T_\mathbf{H}(x,y) = \begin{cases} 0, & x = 0 \land y = 0, \\ \frac{xy}{x+y-xy}, & \text{otherwise.} \end{cases}$	$\frac{x^2}{2x - x^2}$

Fig. 3. Classical means, corresponding t-norms and their diagonals.

Motivated by the above usage of means and t-norms we can define the following class of aggregation functions.

Proposition 3.3. *Let* $G \colon [0,1]^2 \to [0,1]$ *be an idempotent function and let* T_1, T_2 *be t-norms. Then a function* $G_{T_1, T_2} \colon [0,1]^2 \to [0,1]$ *defined by a formula*

$$G_{T_1, T_2}(x, y) = G(T_1(x, y), T_2(x, y)), \quad x, y \in [0, 1], \tag{6}$$

is a commutative semicopula.

Proof. It is easy to see that all properties are satisfied. For instance an existence of the neutral element 1 is as follows:

$$G_{T_1,T_2}(1,y) = G(T_1(1,y), T_2(1,y)) = G(y,y) = y, \quad y \in [0,1].$$

□

Proposition 3.4. *Let* $G, T_1, T_2 \colon [0,1]^2 \to [0,1]$, *let* G *be an idempotent and let* T_1, T_2 *be t-norms. If* $T_1 = T_2$ *then a function* G_{T_1,T_2} *defined by a formula (6) is a t-norm.*

Proof. It is clear since for $T_1 = T_2$ and $x, y \in [0,1]$ we have

$$G_{T_1,T_2}(x,y) = G(T_2(x,y), T_2(x,y)) = T_2(x,y).$$

However usually G_{T_1,T_2} is not a t-norm.

Example 3.5. Let G be the arithmetic mean and $T_1 = T_M$, $T_2 = T_{LK}$. Then for $x = 0.8$, $y = 0.8$ and $z = 0.5$ we have

$$G_{T_M,T_{LK}}(G_{T_M,T_{LK}}(x,y), z) = G_{T_M,T_{LK}}(G(0.8, 0.6), 0.5)$$
$$= G_{T_M,T_{LK}}(0.7, 0.5) = G(0.5, 0.2) = 0.35,$$

while

$$G_{T_M,T_{LK}}(x, G_{T_M,T_{LK}}(y, z)) = G_{T_M,T_{LK}}(0.8, G(0.5, 0.3))$$
$$= G_{T_M,T_{LK}}(0.8, 0.4) = G(0.4, 0.2) = 0.3.$$

3.2   Case of Usage BK-subproduct

Firstly we cite the following result, which justify the reason of choice of pairs of functions for the generalized hypothetical syllogism and CRI.

Theorem 3.6 ([12, **Theorem 3.2**]).   *Let* T *be a t-norm. Then the following statements are equivalent:*

(i) T *is left-continuous.*
(ii) The pair (T, I_T) *satisfies the following functional equation*

$$\sup_{z \in [0,1]} (T(I(x,z), I(z,y))) = I(x,y), \qquad x, y \in [0,1]. \qquad \text{(CRI-GHS)}$$

In fact, we have the similar result for (CRI-GMP).

Proposition 3.7 *For a t-norm* T *the following statements are equivalent:*

(i) T *is left-continuous.*
(ii) The pair (T, I_T) *satisfies (CRI-GMP).*

Proof $(i) \implies (ii)$ If T is left continuous, then from Theorem 3.6 it satisfies (CRI-GHS). Putting $x = 1$ in (CRI-GHS) we obtain

$$\sup_{z \in [0,1]} (T(I(1,z), I(z,y))) = I(1,y), \quad y \in [0,1].$$

But I_T satisfies the left neutrality property (NP) (see [2, Theorem 2.5.4]), so

$$\sup_{z \in [0,1]} (T(z, I(z,y))) = y, \quad y \in [0,1],$$

and thus (ii) is satisfied.

$(ii) \implies (i)$ Let T be a t-norm and let us assume that the pair (T, I_T) satisfies (CRI-GMP), but T is not left-continuous. Based on [2, Proposition 2.5.5] we know that residual principle is not valid, so there exist $x, y, z \in [0,1]$ such that $I_T(x,y) \geq z$ and $T(x,z) > y$. Then, using the fact that (CRI-GMP) is valid for these x, y, z, we have

$$y < T(x,z) \leq T(x, I_T(x,y)) \leq \sup_{t \in [0,1]} T(t, I_T(t,y)) = y;$$

a contradiction. □

Therefore, we know which pairs can be used in (CRI-GMP), like we considered earlier the pair $(T_{\mathbf{LK}}, I_{\mathbf{LK}})$. Moreover, note that if we use it we don't have to take care about the commutativity, i.e., the same result we would obtain for the functional equation

$$y = \sup_{x \in [0,1]} T(I_T(x,y), x), \quad y \in [0,1].$$

However, if we consider BK-subproduct where there is a different composition and a different relation R, then the situation is unlike. Usually, we consider the Eq. (BK-GMP). Nevertheless, when in case of CRI we can change the order of arguments, here we would obtain:

$$y = \inf_{x \in [0,1]} I(T(x,y), x), \quad y \in [0,1]. \qquad \text{(BK-GMP-Rev)}$$

Of course, it makes a big difference since we have a fuzzy implication I. This is in fact so called supercomposition $\triangleright$ or actually a superdirect image of a fuzzy set A under a relation R [6]:

$$R_\triangleright(A)(y) := \inf_{x \in X} I(R(x,y), A(x)), \quad y \in Y.$$

Now, let us mention about another functional equation obtained from the generalized hypothetical syllogism and based on the Bandler-Kohout subproduct:

$$I_2(x,y) = \inf_{z \in [0,1]} I_1(C(x,z), C(z,y)), \quad x, y \in [0,1], \qquad \text{(BK-GHS)}$$

which is a generalization of the following equation, where A, B, C are fuzzy sets,

$$I(A(x), B(y)) = \inf_{z \in Z} I(T(A(x), C(z)), T(C(z), B(y))), \quad x \in X, y \in Y. \quad (7)$$

The next two results can be proven as corollaries from our previous article [12].

Proposition 3.8. *Let* $I \in \mathcal{FI}$ *satisfies* (NP) *and let* C *be a semicopula. If the triplet* (C, I, I) *satisfies* (BK-GHS), *then the pair* (C, I) *satisfies* (BK-GMP).

Proposition 3.9. *For any t-norm* T *the pair* (T, I_T) *satisfies* (BK-GMP).

However, in applications except interpolativity property we know that we have discrete cases, so the infimum is in fact the minimum. Moreover, these results are also valid for such cases.

Proposition 3.10. *Let* A, B, C *be fuzzy sets and let* T *be a t-norm. If* C *is normal, then the triplet* (T, I_T, I_T) *satisfies* (7).

Proposition 3.11. *Let* A, B *be fuzzy sets and let* T *be a t-norm. If* A *is normal, then the pair* (T, I_T) *satisfies*

$$B(y) = \inf_{x \in X} I(A(x), T(A(x), B(y))), \quad y \in Y.$$

Note that for any fuzzy implication I with (OP) we have the following equality $T(x, y) \leq x \iff I(T(x, y), x) = 1$, $x, y \in [0, 1]$, and then $I(x, T(x, y)) < 1$. Hence, for such fuzzy implications I Eq. (BK-GMP-Rev) does not have any solutions. But as we can see on Fig. 4, usage of such order gave in our tool results which cannot be called useless. This could be a reason why we should consider these composition (BKS and supercomposition $\triangleright$) in different frameworks then CRI and check some other properties besides interpolativity.

(a) Image obtained from BKS with $(T_{\mathbf{P}}, I_{\mathbf{GG}})$

(b) Image obtained with supercomposition $\triangleright$ $(T_{\mathbf{P}}, I_{\mathbf{GG}})$

Fig. 4. Examples of images used in our programme – BK-subproduct.

Now let us give some other examples of using BK-subproduct for other reasoning schemas which does and does not have solutions. Firstly, let us consider generalized modus tollens:

$$N(x) = \inf_{y \in [0,1]} I(N(y), T(x,y)), \quad x \in [0,1]. \qquad \text{(BK-GMT)}$$

Remark 3.12. Note that (BK-GMT) have no solutions. Indeed,

$$\inf_{y \in [0,1]} I(N(y), T(x,y)) \leq I(N(0), T(x,0)) = I(1,0) = 0 \neq N(x),$$

for all $x \in [0,1]$.

Hence, now let us focus on the following functional equation

$$N_1(x) = \inf_{y \in [0,1]} I(T(x,y), N_2(y)), \quad x \in [0,1], \qquad \text{(BK-GMT-Rev)}$$

where $I \in \mathcal{FI}$, T is a t-norm and N_1, N_2 are some fuzzy negations.

Proposition 3.13. *Let $I \in \mathcal{FI}$, T be a t-norm and N_1, N_2 be any fuzzy negations. Then the following statements are equivalent:*

(i) The quadruple (T, I, N_1, N_2) satisfies (BK-GMT-Rev).
(ii) $N_1 = N_I$.

Proof. It is enough to see that

$$\inf_{y \in [0,1]} I(T(x,y), N_2(y)) = I(T(x,1), N(1)) = I(x,0) = N_I(x).$$

$\square$

Next, let us consider the functional equation obtained from the reduction to absurdity:

$$x = \inf_{y \in [0,1]} I(N(y), T(N(x), y)), \quad x \in [0,1]. \qquad \text{(BK-GRA)}$$

Here note that there is an analogous case as in (BK-GMT) equation.

Remark 3.14. Let $I \in \mathcal{FI}$, T be a t-norm and N be a fuzzy negation. Then there does not exist any triplet (T, I, N) that satisfies (BK-GRA). Indeed, let us take $x \in [0,1]$. Then

$$\inf_{y \in [0,1]} I(N(y), T(N(x), y)) \leq I(N(0), T(N(x), 0)) = I(1,0) = 0 \neq x,$$

for any $x > 0$.

Hence, similarly as for (BK-GMT), when we consider the reverse order of arguments we can write the following functional equation:

$$x = \inf_{y \in [0,1]} I(T(N(x), y), N(y)), \quad x \in [0, 1]. \qquad \text{(BK-GRA-Rev)}$$

Proposition 3.15. *Let* $I \in \mathcal{FI}$, T *be a t-norm,* N *be a negation and let* N_I *be a strict negation. Then the following statements are equivalent:*

(i) The triplet (T, I, N) *satisfies* (BK-GRA-Rev).
(ii) $N_I^{-1} = N$.

Proof. Take $x \in [0, 1]$. Then

$$\inf_{y \in [0,1]} I(T(N(x), y), N(y)) = I(T(N(x), 1), N(1)) = I(N(x), 0),$$

so from the fact x is arbitrarily fixed we have

$$x = N_I(N(x)) \iff N_I^{-1}(x) = N(x),$$

because N_I is strict. $\qquad\qquad\qquad\qquad\qquad\qquad\qquad\qquad\qquad\qquad\Box$

4   Conclusions

In this paper we briefly described the tool which we created. We can conclude that in case of FITA it may be worth to use means associated with t-norms for the aggregation part. Also, we noticed that a different look on BK-subproduct may give various results. Moreover, we presented some results regarding different functional equations. For the future work we would like to examine BK-subproduct deeply and investigate differences, which we discovered from our tool, between it and CRI.

Acknowledgment. The work on this paper was supported by the National Science Centre, Poland, under Grant No. 2015/19/B/ST6/03259.

References

1. Alsina, C., Frank, M., Schweizer, B.: Associative Functions: Triangular Norms and Copulas. World Scientific Publishing, Singapore (2006). https://doi.org/10.1142/6036
2. Baczyński, M., Jayaram, B.: Fuzzy Implications. In: Studies in Fuzziness and Soft Computing, vol. 231, Springer, Heidelberg (2008). https://doi.org/10.1007/978-3-540-69082-5
3. Bandler, W., Kohout, L.: Semantics of implication operators and fuzzy relational products. Int. J. Man-Mach. Stud. **12**(1), 89–116 (1980). https://doi.org/10.1016/S0020-7373(80)80055-1

4. Buckley, J.J.: Erratum to "Can approximate reasoning be consistent?": [Fuzzy sets and systems 65 (1994) 13–18]. Fuzzy Sets Syst. **103**(1), 189–190 (1999). https://doi.org/10.1016/S0165-0114(98)00443-6

5. Buckley, J.J., Hayashi, Y.: Can approximate reasoning be consistent? Fuzzy Sets Syst. **65**(1), 13–18 (1994). https://doi.org/10.1016/0165-0114(94)90243-7

6. De Baets, B.: Analytical solution methods for fuzzy relational equations. In: Dubois, D., Prade, H. (eds.) Fundamentals of Fuzzy Sets. FSHS, vol. 7, pp. 291–340. Springer, Boston (2000). https://doi.org/10.1007/978-1-4615-4429-6_7

7. Durante, F., Sempi, C.: Semicopulae. Kybernetika **41**(3), 315–328 (2005)

8. Grabisch, M., Marichal, J., Mesiar, R., Pap, E.: Aggregation Functions (Encyclopedia of Mathematics and its Applications). Cambridge University Press, Cambridge (2009). https://doi.org/10.1017/CBO9781139644150

9. Klement, E.P., Mesiar, R., Pap, E.: Triangular Norms. Kluwer Academic Publishers, Dordrecht (2000). https://doi.org/10.1007/978-94-015-9540-7

10. Klir, G.J., Bo, Y.: Fuzzy Sets and Fuzzy Logic Theory and Applications. Prentice Hall, New Jersey (1995)

11. Mesiar, R., Navara, M.: Diagonals of continuous triangular norms. Fuzzy Sets Syst. **104**(1), 35–41 (1999). https://doi.org/10.1016/S0165-0114(98)00256-5

12. Miś, K., Baczyński, M.: Different forms of generalized hypothetical syllogism with regard to R-implications. In: Rutkowski, L., Scherer, R., Korytkowski, M., Pedrycz, W., Tadeusiewicz, R., Zurada, J.M. (eds.) ICAISC 2019. LNCS (LNAI), vol. 11508, pp. 304–313. Springer, Cham (2019). https://doi.org/10.1007/978-3-030-20912-4_29

13. Pedrycz, W.: Applications of fuzzy relational equations for methods of reasoning in presence of fuzzy data. Fuzzy Sets Syst. **16**, 163–175 (1985). https://doi.org/10.1016/S0165-0114(85)80016-6

14. Trillas, E., Alsina, C., Pradera, A.: On MPT implication functions for fuzzy logic. Rev. R. Acad. Cien. Ser. A Mat. **98**, 259–271 (2004)

15. Zadeh, L.A.: Outline of a new approach to the analysis of complex systems and decision processes. IEEE Trans. Syst. Man Cyber. **3**, 28–44 (1973). https://doi.org/10.1109/TSMC.1973.5408575

Modus Ponens Tollens
for RU-Implications

Isabel Aguiló[1,2], Sebastia Massanet[1,2], Juan Vicente Riera[1,2],
and Daniel Ruiz-Aguilera[1,2(✉)]

[1] Department of Mathematics and Computer Science,
University of the Balearic Islands, 07122 Palma, Spain
{isabel.aguilo,s.massanet,jvicente.riera,daniel.ruiz}@uib.es
[2] Health Research Institute of the Balearic Islands (IdISBa), 07010 Palma, Spain

Abstract. In fuzzy rules based systems, fuzzy implication functions
are usually considered to model fuzzy conditionals and to perform forward and backward inferences. These processes are guaranteed by the
fulfilment of the Modus Ponens and Modus Tollens properties by the
fuzzy implication function with respect to the considered conjunction
and fuzzy negation. In this paper, we investigate which residual implications derived from uninorms satisfy both Modus Ponens and Modus
Tollens properties with respect to the same t-norm and a fuzzy negation
simultaneously. The most usual classes of uninorms are considered and
many solutions are obtained which allow to model the fuzzy conditionals in a fuzzy rules based systems (and perform backward and forward
inferences) with a unique residual implication derived from a uninorm.

Keywords: Fuzzy implication function · Modus Ponens · Modus
Tollens · Uninorm

1   Introduction

Fuzzy implication functions have been extensively studied in the last decades
(see [3,4,18] and references therein). There exist two main reasons to support the
great effort made by the scientific community in this field. First, fuzzy implication
functions have proved useful in many applications ranging from approximate reasoning to image processing, including fuzzy control, fuzzy relational equations,
fuzzy DI-subsethood measures or computing with words, among other fields. The
second reason is a direct consequence of their definition, which imposes only some
monotonicities and corner conditions to ensure that they generalize the binary
implication when restricted to $\{0,1\}$. This fact opens a plethora of additional
properties which, although they are studied from a theoretical point of view, are
useful to obtain feasible and more adequate fuzzy implication functions in the
applications.

Two of such additional properties are the (generalized) Modus Ponens and
Modus Tollens. These properties are of paramount importance in approximate

© Springer Nature Switzerland AG 2020
M.-J. Lesot et al. (Eds.): IPMU 2020, CCIS 1238, pp. 788–801, 2020.
https://doi.org/10.1007/978-3-030-50143-3_61

reasoning. Indeed, in any fuzzy rules based system, the fuzzy conditionals are usually modelled by fuzzy implication functions. However, in order to perform backward and forward inferences, the considered fuzzy implication functions must satisfy the aforementioned (generalized) Modus Ponens and Modus Tollens properties with respect to the conjunction and fuzzy negation considered in the system. These properties are usually carried out through the Compositional Rule of Inference (CRI) of Zadeh, based on the sup $-T$ composition, where T is a t-norm (see for instance [5] or Chapter 7 in [3]). Applying this approach, the (generalized) Modus Ponens and Modus Tollens are usually expressed by the following two functional inequalities:

$$T(x, I(x,y)) \leq y, \text{ for all } x, y \in [0,1],$$

$$T(N(y), I(x,y)) \leq N(x), \text{ for all } x, y \in [0,1],$$

where T is a t-norm, I a fuzzy implication function and N a fuzzy negation. These properties have been studied in the literature for the most usual families of fuzzy implication functions such as (S, N), R, QL and D-implications derived from t-norms and t-conorms [2,3,16,18,23–25] or from uninorms [14,15]. Even recently, a whole new line of research has been proposed in which the t-norm T is generalized to a more general conjunction such as a conjunctive uninorm [19] or an overlap function [8], leading to the so-called U-Modus Ponens or O-Modus Ponens.

Although the functional inequalities of Modus Ponens and Modus Tollens have quite similar expressions, it is well-known that both properties are not equivalent. Thus, in [23,24], the simultaneous fulfillment of both properties was studied for the first time for some restricted classes of (S, N), R, QL and D-implications. It was proved that when the fuzzy negation N is a strong negation, both properties are equivalent if the fuzzy implication function satisfies the contrapositive symmetry with respect to N. The importance of the disposal of fuzzy implication functions satisfying both properties lies on the possibility of considering a unique implication to model fuzzy conditionals regardless of whether backward or forward inference processes have to be performed.

Following this line of research, in this paper, we analyze which residual implications derived from uninorms, or RU-implications for short, satisfy both the Modus Ponens and the Modus Tollens properties with respect to the same t-norm T and a fuzzy negation N (continuous, but not necessarily strong). It has to be said that the Modus Ponens property had been already studied for this family of uninorms in [15] and the Modus Tollens property was analyzed in [14]. However, while in [15], the results were given in terms of the t-norm T_U associated to the uninorm U, in [14] the results were presented for each class of uninorms separately. This fact makes it difficult to coordinate the results of both studies in order to find RU-implications satisfying both properties. This is the main goal of this paper in addition to find some cases for which the fulfillment of one property implies the fulfillment of the other one.

The paper is organized as follows. In the next section we recall some basic definitions and properties on fuzzy implication functions and uninorms. In Sect. 3, we introduce the so-called Modus Ponens Tollens property and we discuss its fulfillment when the fuzzy implication function satisfies the contrapositive symmetry with respect to a strong negation. After that, in Sect. 4, the Modus Ponens Tollens property is studied in depth for RU-implications depending on the class of the uninorm U considered. The paper ends with some conclusions and future work.

2   Preliminaries

We will suppose the reader to be familiar with the theory of t-norms, t-conorms and fuzzy negations (all necessary results and notations can be found in [11]). We also suppose that some basic facts on uninorms are known (see for instance [9]) as well as their most usual classes (see [13] for a complete survey), that is, uninorms in $\mathcal{U}_{\min}$ ([9]), representable uninorms ([9]), idempotent uninorms ([6,12,22]) and uninorms continuous in the open unit square ([10]).

We recall here only some facts on implications and uninorms in order to establish the necessary notation that we will use along the paper.

Definition 1 ([3]). *A binary operator $I : [0,1] \times [0,1] \to [0,1]$ is said to be a* fuzzy implication function, *or a* fuzzy implication, *if it satisfies:*

(I1) $I(x,z) \geq I(y,z)$ *when $x \leq y$, for all $z \in [0,1]$.*
(I2) $I(x,y) \leq I(x,z)$ *when $y \leq z$, for all $x \in [0,1]$.*
(I3) $I(0,0) = I(1,1) = 1$ *and $I(1,0) = 0$.*

Note that, from the definition, it follows that $I(0,x) = 1$ and $I(x,1) = 1$ for all $x \in [0,1]$ whereas the symmetrical values $I(x,0)$ and $I(1,x)$ are not derived from it.

Definition 2 ([3]). *A function $N : [0,1] \to [0,1]$ is called a* fuzzy negation *if it is decreasing, $N(0) = 1$ and $N(1) = 0$. If N is a fuzzy negation that is strictly decreasing and continuous, it will be called* strict, *and if it is involutive, $N(N(x)) = x$ for all $x \in [0,1]$, then it will be called* strong.

Definition 3 ([3]). *Let T be a t-norm. A function $N_T : [0,1] \to [0,1]$ defined as*

$$N_T(x) = \sup\{y \in [0,1] \mid T(x,y) = 0\}, \quad x \in [0,1]$$

is called the natural negation *of T or the* negation induced *by T.*

Definition 4 ([9]). *A uninorm is a two-place function $U : [0,1]^2 \to [0,1]$ which is associative, commutative, increasing in each place and such that there exists some element $e \in [0,1]$, called* neutral element, *such that $U(e,x) = x$ for all $x \in [0,1]$.*

Evidently, a uninorm with neutral element $e = 1$ is a t-norm and a uninorm with neutral element $e = 0$ is a t-conorm. For any other value $e \in]0,1[$ the operation works as a t-norm in the $[0,e]^2$ square, as a t-conorm in $[e,1]^2$ and its values are between the minimum and the maximum in the set of points $A(e)$ given by

$$A(e) = [0, e[\times]e, 1] \cup]e, 1] \times [0, e[.$$

We will usually denote a uninorm with neutral element e and underlying t-norm T_U and t-conorm S_U by $U \equiv \langle T_U, e, S_U \rangle$. For any uninorm it is satisfied that $U(0, 1) \in \{0, 1\}$ and a uninorm U is called *conjunctive* if $U(1, 0) = 0$ and *disjunctive* when $U(1, 0) = 1$. On the other hand, let us recall the most usual classes of uninorms in the literature that will be used along the paper. We start with the class of uninorms in $\mathcal{U}_{\min}$.

Definition 5. *Let U be a conjunctive uninorm with neutral element $e \in]0, 1[$. If the mapping $x \mapsto U(x, 1)$ is continuous except in $x = e$, then it is said that U is a uninorm in $\mathcal{U}_{\min}$.*

Theorem 1 ([9]). *Let $U : [0, 1]^2 \to [0, 1]$ be a function. Then U is a uninorm in $\mathcal{U}_{\min}$ if and only if U is given by*

$$U(x, y) = \begin{cases} eT_U\left(\frac{x}{e}, \frac{y}{e}\right) & \text{if } (x, y) \in [0, e]^2, \\ e + (1 - e)S_U\left(\frac{x-e}{1-e}, \frac{y-e}{1-e}\right) & \text{if } (x, y) \in [e, 1]^2, \\ \min(x, y) & \text{if } (x, y) \in A(e), \end{cases}$$

where T_U is a t-norm, and S_U is a t-conorm. We will denote a uninorm in $\mathcal{U}_{\min}$ with underlying t-norm T_U, underlying t-conorm S_U and neutral element e as $U \equiv \langle T_U, e, S_U \rangle_{\min}$.

The class of idempotent uninorms, that satisfy $U(x, x) = x$ for all $x \in [0, 1]$, was characterized first in [6] for those uninorms with a lateral continuity and in [12] for the general case. An improvement of this last result was done in [22] as follows.

Theorem 2 ([22]). *U is an idempotent uninorm with neutral element $e \in [0, 1]$ if and only if there exists a non increasing function $g : [0, 1] \to [0, 1]$, symmetric with respect to the identity function, with $g(e) = e$, such that*

$$U(x, y) = \begin{cases} \min(x, y) & \text{if } y < g(x) \text{ or } (y = g(x) \text{ and } x < g^2(x)), \\ \max(x, y) & \text{if } y > g(x) \text{ or } (y = g(x) \text{ and } x > g^2(x)), \\ x \text{ or } y & \text{if } y = g(x) \text{ and } x = g^2(x), \end{cases}$$

being commutative in the points (x, y) such that $y = g(x)$ with $x = g^2(x)$.

Any idempotent uninorm U with neutral element e and associated function g will be denoted by $U \equiv \langle g, e \rangle_{\text{ide}}$ and the class of idempotent uninorms will be denoted by $\mathcal{U}_{\text{ide}}$. Obviously, for any of these uninorms, the underlying t-norm is the minimum and the underlying t-conorm is the maximum.

Definition 6 ([9])**.** *Let e be in $]0, 1[$. A binary operation $U : [0, 1]^2 \rightarrow [0, 1]$ is a representable uninorm if and only if there exists a strictly increasing function $h : [0, 1] \rightarrow [-\infty, +\infty]$ with $h(0) = -\infty$, $h(e) = 0$ and $h(1) = +\infty$ such that*

$$U(x, y) = h^{-1}(h(x) + h(y))$$

for all $(x, y) \in [0, 1]^2 \setminus \{(0, 1), (1, 0)\}$ and $U(0, 1) = U(1, 0) \in \{0, 1\}$. The function h is usually called an additive generator of U.

Any representable uninorm U with neutral element e and additive generator h will be denoted by $U \equiv \langle h, e \rangle_{\text{rep}}$ and the class of representable uninorms will be denoted by $\mathcal{U}_{\text{rep}}$. For any of these uninorms the underlying t-norm and t-conorm are always strict. For all representable uninorm U, a strong negation can be defined from U as $N_U(x) = h^{-1}(-h(x))$ for all $x \in [0, 1]$.

Another studied class of uninorms is $\mathcal{U}_{\cos}$, composed by all uninorms continuous in $]0, 1[^2$. They were introduced and characterized in [10] as follows.

Theorem 3 ([10])**.** *Suppose U is a uninorm continuous in $]0, 1[^2$ with neutral element $e \in]0, 1[$. Then one of the following cases is satisfied:*

(a) There exist $u \in [0, e[$, $\lambda \in [0, u]$, two continuous t-norms T_1 and T_2 and a representable uninorm R such that U can be represented as

$$U(x, y) = \begin{cases} \lambda T_1(\frac{x}{\lambda}, \frac{y}{\lambda}) & \text{if } (x, y) \in [0, \lambda]^2, \\ \lambda + (u - \lambda)T_2(\frac{x-\lambda}{u-\lambda}, \frac{y-\lambda}{u-\lambda}) & \text{if } (x, y) \in [\lambda, u]^2, \\ u + (1 - u)R(\frac{x-\lambda}{1-\lambda}, \frac{y-\lambda}{1-\lambda}) & \text{if } (x, y) \in [\lambda, u]^2, \\ 1 & \text{if } \min(x, y) \in]\lambda, 1] \\ & \text{and } \max(x, y) = 1, \\ \lambda \text{ or } 1 & \text{if } (x, y) \in \{(\lambda, 1), (1, \lambda)\}, \\ \min(x, y) & \text{elsewhere.} \end{cases}$$

(b) There exist $v \in]e, 1]$ $\omega \in [v, 1]$, two continuous t-conorms S_1 and S_2 and a representable uninorm R such that U can be represented as

$$U(x, y) = \begin{cases} vR(\frac{x}{v}, \frac{y}{v}) & \text{if } (x, y) \in]0, v[^2, \\ v + (\omega - v)S_1(\frac{x-v}{\omega-v}, \frac{y-v}{\omega-v}) & \text{if } (x, y) \in [v, \omega]^2, \\ \omega + (1 - \omega)S_2(\frac{x-\omega}{1-\omega}, \frac{y-\omega}{1-\omega}) & \text{if } (x, y) \in [\omega, 1]^2, \\ 0 & \text{if } \max(x, y) \in [0, \omega[\\ & \text{and } \min(x, y) = 0, \\ \omega \text{ or } 0 & \text{if } (x, y) \in \{(0, \omega), (\omega, 0)\}, \\ \max(x, y) & \text{elsewhere.} \end{cases}$$

Now we will recall residual implications from uninorms: RU-implications.

Definition 7 ([7])**.** *Let U be a uninorm. The residual operation derived from U is the binary operation given by*

$$I_U(x, y) = \sup\{z \in [0, 1] \mid U(x, z) \le y\} \text{ for all } x, y \in [0, 1].$$

Proposition 1 ([7]). *Let U be a uninorm and I_U its residual operation. Then I_U is a fuzzy implication if and only if the following condition holds*

$$U(x,0) = 0 \text{ for all } x < 1.$$

In this case I_U is called an RU-implication.

This includes all conjunctive uninorms but also many disjunctive ones, for instance in the classes of representable uninorms (see [7]) and idempotent uninorms (see [20]).

Some properties of RU-implications have been studied involving the main classes of uninorms, those previously stated: uninorms in $\mathcal{U}_{\min}$, idempotent uninorms and representable uninorms (for more details see [1,3,7,17,20,21]).

3   Modus Ponens Tollens

First of all, let us recall the definition of the Modus Ponens and the Modus Tollens in the framework of fuzzy logic.

Definition 8. *Let I be a fuzzy implication function and T a t-norm. It is said that I satisfies the* Modus Ponens *property with respect to T if*

$$T(x, I(x,y)) \leq y \text{ for all } x,y \in [0,1]. \tag{MP}$$

Definition 9. *Let I be a fuzzy implication function, T a t-norm and N a fuzzy negation. It is said that I satisfies the* Modus Tollens *property with respect to T and N if*

$$T(N(y), I(x,y)) \leq N(x) \text{ for all } x,y \in [0,1]. \tag{MT}$$

The Modus Ponens and Modus Tollens properties have been studied for different types of implications, usuallly taking into account continuous t-norms T and continuous fuzzy negations N. If we consider RU-implications, (**MP**) and (**MT**) have been studied in depth in [15] and [14], respectively. These properties are not equivalent in general, as it is stated in the following examples. First, we have a fuzzy implication function that satisfies (**MP**) but not (**MT**).

Example 1. Consider $U \equiv \langle h, \frac{3}{4} \rangle_{\text{rep}}$ a representable uninorm with $T_U = T_{\mathbf{P}}$, the product t-norm (with additive generator $t_U(x) = -\ln(x)$ up to a multiplicative constant) and S_U any strict t-conorm. Let us consider its residual implication I_U which will be given later in Proposition 10. Let us also consider $T = T_{\mathbf{P}}$ and the negation $N(x) = \frac{1-x}{1+10x}$ which belongs to the family of Sugeno negations with $\lambda = 10$. In this case, I_U safisfies (**MP**) with respect to T by using Proposition 9 in [15]. However, I_U does not satisfy (**MT**) with respect to T and N (just taking $x = 0.7$ and $y = 0.5$ in Eq. (**MP**)).

Next example provides an RU-implication that satisfies (**MT**) but not (**MP**).

Example 2. Let $U \equiv \langle h, \frac{1}{2} \rangle_{\mathrm{rep}}$ be a representative uninorm with additive generator $h(x) = \ln\left(\frac{x}{1-x}\right)$ for all $x \in [0,1]$. Let T be a t-norm whose expression is given by the ordinal sum $T \equiv (\langle 0, \frac{1}{2}, T_{\mathbf{P}} \rangle, \langle \frac{1}{2}, 1, T_1 \rangle)$ with T_1 any continuous t-norm and let us consider the continuous fuzzy negation N given by

$$N(x) = \begin{cases} 1 - x & \text{if } x \leq \frac{1}{2}, \\ \sqrt{x - x^2} & \text{otherwise.} \end{cases}$$

In this case, I_U is given by

$$I_U(x,y) = \begin{cases} 1 & \text{if } (x,y) \in \{(0,0),(1,1)\}, \\ \frac{(1-x)y}{x+y-2xy} & \text{otherwise.} \end{cases}$$

According to Proposition 5.3.20-(ii) in [14], I_U satisfies (**MT**) with respect to T and N. However, a simple computation shows that $g : [0,1] \to [0,1]$ given by $g(x) = -\ln(\frac{2}{1+e^x})$ is not subadditive (for instance, take $x = 0.3$ and $y = 0.2$) and therefore, by using Proposition 10 in [19], I_U does not satisfy (**MP**) with respect to T.

Then, as we have seen, (**MT**) and (**MP**) are not equivalent in general, and the question about which fuzzy implication functions satisfy both properties with respect to the same t-norm T and fuzzy negation N is worthy to study.

Definition 10. *Let I be a fuzzy implication function, T a t-norm and N a fuzzy negation. It is said that I satisfies the* Modus Ponens Tollens (**MPT**) *property with respect to T and N whenever Eqs. (**MP**) and (**MT**) are satisfied simultaneously.*

Remark 1. Note that, when $x \leq y$ we have $N(y) \leq N(x)$ and then (**MT**) trivially holds in these cases. Similarly, (**MP**) is satisfied in these cases. Thus, both properties need to be checked only in points $(x,y) \in [0,1]^2$ where $y < x$.

Anyway, a special case that can be considered is when I satisfies the contrapositive symmetry with respect to N. Contrapositive symmetry is a well known property, which is related to the Modus Ponens Tollens as it is stated in the following results.

Definition 11. *Consider I a fuzzy implication function and N a fuzzy negation. Then I satisfies the* contrapositive symmetry *with respect to N if*

$$I(x,y) = I(N(y), N(x)) \text{ for all } x, y \in [0,1]. \tag{CP}$$

Theorem 4 ([24])**.** *Consider I a fuzzy implication function, T a t-norm and N a strong negation. If I satisfies the contrapositive symmetry with respect to N, then I satisfies (**MP**) with respect to T if and only if I satisfies (**MT**) with respect to N and T.*

From the result above, in the case that a fuzzy implication function I satisfies (**CP**) with respect to N, only one of (**MP**) or (**MT**) needs to be checked in order to satisfy (**MPT**). Now, let us recall the result on contrapositive symmetry for residual implications derived from idempotent uninorms.

Proposition 2 ([20]). *Consider $U \equiv \langle g, e \rangle_{\text{ide}}$ an idempotent uninorm with $g(0) = 1$, I_U its residual implication and N a strong negation. Then I_U satisfies (**CP**) with respect to N if and only if $g = N$.*

As a consequence of the previous result, we have infinite RU-implications that satisfy (**MPT**) for any t-norm T, by using Proposition 5.3.14 in [14].

Corollary 1. *Let N be a strong negation, $U \equiv \langle N, e \rangle_{\text{ide}}$ an idempotent uninorm, I_U its residual implication, and T a t-norm. Then I_U satisfies (**MPT**) with respect to T and N.*

Coming up next, we recall the case of (**CP**) for uninorms continuous in $]0, 1[^2$.

Proposition 3 ([21]). *Let U be a uninorm in $\mathcal{U}_{\cos}$ such that $U(0, x) = 0$ for all $x < 1$, I_U its residual implication and N a strong negation. Then I_U satisfies (**CP**) with respect to N if and only if U is representable and $N = N_U$.*

4   Modus Ponens Tollens for Implications Derived from Different Classes of Uninorms

In this section we investigate the Modus Ponens Tollens property (**MPT**) for fuzzy implication functions derived from three well known classes of uninorms.

4.1   Case When U is a Uninorm in $\mathcal{U}_{\min}$

In this section we will deal with RU-implications derived from uninorms in $\mathcal{U}_{\min}$, that is, uninorms $U \equiv \langle T_U, e, S_U \rangle_{\min}$ with neutral element $e \in]0, 1[$. Recall that for this kind of uninorms, RU-implications have the following structure.

Proposition 4 (Theorem 5.4.7 in [3]). *Let $U \equiv \langle T_U, e, S_U \rangle_{\min}$ a uninorm in $\mathcal{U}_{\min}$ and I_U its residual implication. Then*

$$
I_U(x, y) = \begin{cases}
1 & \text{if } x \leq y < e, \\
e I_{T_U}\left(\frac{x}{e}, \frac{y}{e}\right) & \text{if } y < x \leq e, \\
e + (1 - e) I_{S_U}\left(\frac{x-e}{1-e}, \frac{y-e}{1-e}\right) & \text{if } e \leq x \leq y, \\
e & \text{if } e \leq y < x, \\
y & \text{elsewhere.}
\end{cases}
$$

For this family of RU-implications we have the following result.

Proposition 5. *Let $U \equiv \langle T_U, e, S_U \rangle_{\min}$ a uninorm in $\mathcal{U}_{\min}$ and I_U its residual implication. Let T be a continuous t-norm, and N be a continuous fuzzy negation with fixed point $s \in]0, 1[$. Then, it holds that:*

– If I_U satisfies (**MPT**) with T and N, then T is nilpotent with normalized additive generator $t : [0,1] \to [0,1]$ and associated negation $N_T(x) = t^{-1}(1 - t(x))$ for all $x \in [0,1]$ such that $N(y) \le N_T(y)$ for all $y \le e$.

Thus, from now on, let us consider T satisfying the previous conditions. In this case,

(i) If $T_U = \min$ then I_U always satisfies (**MPT**) with respect to T and N.

(ii) If T_U is a strict t-norm with additive generator t_U and either $s \ge e$ or $N(y) = N_T(y)$ for all $y \le e$, then I_U satisfies (**MPT**) with respect to T and N if and only if the following condition holds:

($\star_1$) Function $g : [0, t(0)] \to [t(e), 1]$ given by the formula $g(x) = t(et_U^{-1}(x))$ is subadditive.

(iii) If T_U is a strict t-norm with additive generator t_U, $s < e$ and $N(y) < N_T(y)$ for some $y \le e$, then I_U satisfies (**MPT**) with respect to T and N if and only if Property ($\star_1$) is fulfilled and the following condition holds:

($\star_2$) For all $y < x < e$,

$$et_U^{-1}\left(t_U\left(\frac{y}{e}\right) - t_U\left(\frac{x}{e}\right)\right) \le t^{-1}(t(N(x)) - t(N(y))).$$

(iv) If T_U is a nilpotent t-norm with additive generator t_U and either $s \ge e$ or $N = N_T$, then I_U satisfies (**MPT**) with respect to T and N if and only if Property ($\star_1$) and the following property holds:

($\star_3$) For all $x \le e$

$$e \cdot N_{T_U}\left(\frac{x}{e}\right) \le N_T(x).$$

(v) If T_U is a nilpotent t-norm with additive generator t_U, $s < e$ and $N(y) < N_T(y)$ for some $y \le e$, then I_U satisfies (**MPT**) with respect to T and N if and only if Properties ($\star_1$), ($\star_2$) and ($\star_3$) hold.

Example 3. Let us consider the uninorm $U \equiv \langle T_\mathbf{M}, e, S_U \rangle_{\min}$ with $T_\mathbf{M}$ is the minimum t-norm and S_U any t-conorm. Let $T_\mathbf{L}$ be the Łukasiewicz t-norm and $N = N_c$ the classical negation given by $N_c(x) = 1 - x$ for all $x \in [0,1]$. Thus, from the previous proposition, taking into account that $N_c(x) = 1 - x \le N_{T_\mathbf{L}}(x) = 1 - x$, I_U satisfies (**MPT**) with respect to $T_\mathbf{L}$ and N_c.

Example 4. Let us take now $U \equiv \langle T_\mathbf{P}, \frac{1}{2}, S_U \rangle_{\min}$ with S_U any t-conorm. Let $T_\mathbf{L}$ be the Łukasiewicz t-norm (with additive generator $t(x) = 1 - x$ up to a multiplicative constant), and $N = N_c$. In this example, we are under the conditions of Case (ii) from the previous result, and it remains only to prove that $g : [0,1] \to [\frac{1}{2}, 1]$ given by $g(x) = 1 - \frac{1}{2}e^{-x}$ is subadditive. A straightforward computation ensures this fact. Therefore, I_U satisfies (**MPT**) with respect to $T_\mathbf{L}$ and N_c.

Example 5. Let us consider $U \equiv \langle T_\mathbf{P}, \frac{1}{2}, S_U \rangle_{\min}$ with S_U any t-conorm. Let $T = T_\mathbf{L}$ and let us take

$$N(x) = \begin{cases} 1 - 2x & \text{if } x \le \frac{1}{2}, \\ 0 & \text{elsewhere,} \end{cases}$$

a continuous fuzzy negation with fixed point $s = \frac{1}{3}$. Note that in this case $N_T = N_c(x) = 1 - x$ and $N(\frac{1}{4}) < N_T(\frac{1}{4})$. Thereby, we are under the conditions of Case (iii) of the previous proposition, and then Properties $(\star_1)$ and $(\star_2)$ must be checked. A simple computation shows that $g : [0, 1] \to [\frac{1}{2}, 1]$ given by $g(x) = 1 - \frac{1}{2}e^{-x}$ is subadditive and so Property $(\star_1)$ is fulfilled. With respect to Property $(\star_2)$, we obtain the inequality $\frac{y}{2x} \le 1 - 2(x - y)$ which is valid for $y < x < \frac{1}{2}$. Consequently, I_U satisfies (**MPT**) with respect to $T_{\mathbf{L}}$ and N.

Example 6. Let us consider $U \equiv \langle T_{\mathbf{L}}, \frac{1}{2}, S_U \rangle_{\min}$ with S_U any t-conorm. Let us consider $T = T_{\mathbf{L}}$ and the same negation N used in the previous example. In this case, again it holds that $N_T = N_c$ and it follows that $N(x) < N_T(x)$ (take for instance $x = \frac{1}{3}$, $N(\frac{1}{3}) = \frac{1}{3} < N_{T_{\mathbf{L}}}(\frac{1}{3}) = \frac{2}{3}$). Now, we are under the conditions of Case (v) from the previous result, and then we must check Properties $(\star_1)$, $(\star_2)$ and $(\star_3)$. Property $(\star_1)$ follows directly. With respect to Property $(\star_2)$, we have $g : [0, 1] \to [\frac{1}{2}, 1]$ given by $g(x) = \frac{1+x}{2}$ is obviously subadditive. Finally a simple computation shows Property $(\star_3)$. Consequently, I_U satisfies (**MPT**) with respect to $T_{\mathbf{L}}$ and N.

4.2   Case When U is an Idempotent Uninorm

In this section we will deal with RU-implications derived from idempotent uninorms, that is, uninorms $U \equiv \langle g, e \rangle_{\text{ide}}$ with neutral element $e \in [0, 1]$ and such that $g(0) = 1$. Recall that for this kind of uninorms, the corresponding RU-implications have the following structure.

Proposition 6 ([20]). *Let $U = \langle g, e \rangle_{ide}$ be an idempotent uninorm with neutral element $e \in]0, 1[$ and such that $g(0) = 1$. Then I_U is given by*

$$I_U(x, y) = \begin{cases} \max(g(x), y) & \text{if } x \le y, \\ \min(g(x), y) & \text{if } x > y. \end{cases} \tag{1}$$

From results in [15], for an idempotent uninorm $U \equiv \langle g, e \rangle_{\text{ide}}$ with $g(0) = 1$, as $T_U = \min$, I_U satisfies (**MP**) with respect to any t-norm. Therefore, we can write the following result.

Proposition 7. *Let $U \equiv \langle g, e \rangle_{ide}$ be an idempotent uninorm with neutral element $e \in]0, 1[$ and such that $g(0) = 1$, T a t-norm and N a fuzzy negation. Then I_U satisfies (**MPT**) with respect to T and N if and only if I_U satisfies (**MT**) with respect to T and N.*

Thus, in the rest of the section, all the conditions in the results will be related to the fulfillment of (**MT**) (that was studied in [14]), which will imply the fulfillment of (**MPT**). Now we will distinguish two cases depending on the value of $g(1)$. We will start with the case $g(1) > 0$.

Proposition 8. *Let $U \equiv \langle g, e \rangle_{ide}$ with $g(0) = 1$ and $g(1) > 0$ and I_U its residual implication. Let T be a t-norm and N a continuous fuzzy negation. If I_U satisfies the (**MPT**) property with respect to T and N, then the following statements are true:*

(i) $T(N(y), y) = 0$ for all $y \leq g(1)$.

(ii) If T is a continuous t-norm then T must be nilpotent with normalized additive generator $t : [0,1] \to [0,1]$ and associated negation N_T, which is given by $N_T(x) = t^{-1}(1 - t(x))$, such that $N(y) \leq N_T(y)$ for all $y \leq g(1)$.

Although this result provides only necessary conditions on T, the following example gives infinite cases of residual implications I_U from U an idempotent uninorm such that satisfy (**MPT**) with respect a t-norm T and a strong negation N.

Example 7. Consider $U \equiv \langle g, e \rangle_{\text{ide}}$ any idempotent uninorm, $T = T_{\mathbf{L}}$ the Lukasiewicz t-norm and $N = N_c$ the classical negation. Take $x, y \in [0,1]$ such that $y < x$, then we have:

$$T_{\mathbf{L}}(N_c(y), \min(g(x), y)) = \max(0, \min(g(x), y) - y) = 0,$$

and, by Remark 1, (**MT**) is satisfied and by Proposition 7, I_U satisfies (**MPT**) with respect to $T_{\mathbf{L}}$ and N_c. If $g \neq N$, I_U satisfies (**MPT**) with respect to $T_{\mathbf{L}}$ and N_c but I_U does not have (**CP**) with respect to N_c.

When $g(1) = 0$ we have a first result which can be applied for any fuzzy negation N.

Theorem 5. *Let* $U \equiv \langle g, e \rangle_{\text{ide}}$ *be an idempotent uninorm with neutral element* $e \in {]}0, 1{[}$ *and* $g(0) = 1$, $g(1) = 0$ *and* I_U *its residual implication. Let* T *be a t-norm and* N *a continuous fuzzy negation. Then* I_U *satisfies* (**MPT**) *with respect to* T *and* N *if and only if*

$$\min(T(N(y), y), T(N(y), g(x))) \leq N(x) \text{ for all } y < x.$$

Example 8. Let us consider $U \equiv \langle N_c, \frac{1}{2} \rangle_{\text{ide}}$ an idempotent uninorm, $T = T_{\mathbf{L}}$ and $N = N_c$. Similarly to the previous Example 7 we have

$$\min(T_{\mathbf{L}}(N_c(y), y), T_{\mathbf{L}}(N_c(y), g(x))) = 0$$

and then I_U satisfies (**MPT**) with respect to $T_{\mathbf{L}}$ and N_c.

When N is strict, the following result provides an easier condition in order to verify the fulfillment of (**MPT**).

Proposition 9. *Let* T *be a t-norm,* N *a strict fuzzy negation, and* $U \equiv \langle g, e \rangle_{\text{ide}}$ *be an idempotent uninorm with neutral element* $e \in {]}0, 1{[}$ *with* $g(0) = 1$, $g(1) = 0$ *and* I_U *its residual implication. Then* I_U *satisfies* (**MPT**) *with respect to* T *and* N *if and only if* $g(x) \leq N(x)$ *for all* $x \geq e$.

Example 9. Let us consider $U \equiv \langle g, \frac{1}{4} \rangle_{\text{ide}}$ an idempotent uninorm where

$$g(x) = \begin{cases} 1 - 3x & \text{if } x \leq \frac{1}{3}, \\ 0 & \text{otherwise}, \end{cases}$$

$T = T_{\mathbf{P}}$ and $N = N_c$. It is straightforward to prove that $g(x) \leq N(x)$ for all $x \geq \frac{1}{4}$. Then, we are under the conditions of the previous result. Thus, I_U satisfies (**MPT**) with respect to T and N.

4.3 Case When U is a Representable Uninorm

In this section we will deal with RU-implications derived from representable uninorms, that is, from uninorms $U \equiv \langle h, e \rangle_{\mathrm{rep}}$ with neutral element $e \in]0, 1[$. Let us recall in this case the expression of the residual implication derived from U.

Proposition 10 (Theorem 5.4.10 in [3]). *Let* $U \equiv \langle h, e \rangle_{\mathrm{rep}}$ *be a representable uninorm with neutral element* $e \in]0, 1[$. *Then* I_U *is given by*

$$I_U(x, y) = \begin{cases} 1 & \text{if } (x, y) \in \{(0, 0), (1, 1)\}, \\ h^{-1}(h(y) - h(x)) & \text{otherwise.} \end{cases}$$

For this kind of uninorms we will consider only continuous t-norms which are not an ordinal sum, namely, the minimum t-norm and continuous Archimedean t-norms.

Proposition 11. *Let* $U \equiv \langle h, e \rangle_{\mathrm{rep}}$ *be a representable uninorm with neutral element* $e \in]0, 1[$ *and* I_U *its residual implication. Let* T *be a continuous non-ordinal sum t-norm and* N *a continuous fuzzy negation. Then, it holds that:*

– *If* I_U *satisfies* (**MPT**) *with* T *and* N, *then* T *is continuous Archimedean with additive generator* $t : [0, 1] \to [0, +\infty]$, *up to a multiplicative constant.*

Thus, in this case, the following statements are true:

(i) I_U *satisfies* (**MPT**) *with respect to* T *and* N *if and only if Property* $(\star_1)$ *is fulfilled and the following property holds:*
($\bullet_1$) For all $y \leq x$,

$$h^{-1}(h(y) - h(x)) \leq t^{-1}(t(N(x)) - t(N(y))).$$

(ii) *If* T *is nilpotent and* $N = N_T$, I_U *satisfies* (**MPT**) *with respect to* T *and* N *if and only if the following property holds:*
($\bullet_2$) Function $\phi : [0, 1] \to [-\infty, +\infty]$ *given by* $\phi(x) = h(t^{-1}(x))$ *for all* $x \in [0, 1]$ *is subadditive.*

Example 10. Let us consider $T = T_{\mathbf{L}}$ and $N = N_c$. Let U be the representable uninorm given by

$$U(x, y) = \begin{cases} 0 & \text{if } (x, y) \in \{(0, 0), (1, 1)\}, \\ \frac{xy}{xy + (1-x)(1-y)} & \text{otherwise.} \end{cases}$$

which has $e = \frac{1}{2}$ as neutral element and additive generator $h(x) = \ln(\frac{x}{1-x})$. In this case, $\phi(x) = h(t^{-1}(x)) = \ln\left(\frac{1-x}{x}\right)$ which is clearly subadditive. By applying Case (ii) of the previous proposition, we conclude that I_U satisfies (**MPT**) with respect to T and N.

5   Conclusions and Future Work

In this paper, we have studied the fulfillment of the so-called Modus Ponens Tollens property (**MPT**) by the family of RU-implications, i.e., we have analyzed which RU-implications satisfy at the same time the Modus Ponens and the Modus Tollens properties with respect to a t-norm T and a negation N. From this study, many solutions are available. On the one side, all RU-implications which satisfy the Modus Ponens property with respect to a t-norm T and the contrapositive symmetry with respect to a strong negation N are solutions of (**MPT**). On the other side, when N is not strong or the contrapositive symmetry is not satisfied, other solutions exist within RU-implications derived from uninorms in $\mathcal{U}_{\min}$, representable uninorms and idempotent uninorms. For most of these families, necessary and sufficient conditions are presented and in some cases, it is shown that the fulfillment of the Modus Tollens property implies the fulfillment of the Modus Ponens property.

As future work, we want to complete the results presented in this paper by considering also continuous ordinal sum t-norms as T in some of the results presented in Sect. 4 and to deepen the study in the particular case of idempotent uninorms with $g(0) = 1$ and $g(1) > 0$.

Acknowledgments. This paper has been supported by the Spanish Grant TIN2016-75404-P AEI/FEDER, UE.

References

1. Aguiló, I., Suñer, J., Torrens, J.: A characterization of residual implications derived from left-continuous uninorms. Inf. Sci. **180**(20), 3992–4005 (2010)
2. Alsina, C., Trillas, E.: When (S, N)-implications are (T, T_1)-conditional functions? Fuzzy Sets Syst. **134**, 305–310 (2003)
3. Baczyński, M., Jayaram, B.: Fuzzy Implications. Studies in Fuzziness and Soft Computing, vol. 231. Springer, Heidelberg (2008). https://doi.org/10.1007/978-3-540-69082-5
4. Baczynski, M., Jayaram, B., Massanet, S., Torrens, J.: Fuzzy implications: past, present, and future. In: Kacprzyk, J., Pedrycz, W. (eds.) Springer Handbook of Computational Intelligence, pp. 183–202. Springer, Heidelberg (2015). https://doi.org/10.1007/978-3-662-43505-2_12
5. Bustince, H., Burillo, P., Soria, F.: Automorphisms, negations and implication operators. Fuzzy Sets Syst. **134**, 209–229 (2003)
6. De Baets, B.: Idempotent uninorms. Eur. J. Oper. Res. **118**, 631–642 (1999)
7. De Baets, B., Fodor, J.C.: Residual operators of uninorms. Soft. Comput. **3**, 89–100 (1999)
8. Dimuro, G.P., Bedregal, B.R.C., Fernández, J., Sesma-Sara, M., Pintor, J.M., Bustince, H.: The law of O-conditionality for fuzzy implications constructed from overlap and grouping functions. Int. J. Approx. Reason. **105**, 27–48 (2019)
9. Fodor, J.C., Yager, R.R., Rybalov, A.: Structure of uninorms. Int. J. Uncertainty Fuzziness Knowl.-Based Syst. **5**, 411–427 (1997)
10. Hu, S., Li, Z.: The structure of continuous uninorms. Fuzzy Sets Syst. **124**, 43–52 (2001)

11. Klement, E.P., Mesiar, R., Pap, E.: Triangular Norms. Kluwer Academic Publishers, Dordrecht (2000)
12. Martín, J., Mayor, G., Torrens, J.: On locally internal monotonic operators. Fuzzy Sets Syst. **137**, 27–42 (2003)
13. Mas, M., Massanet, S., Ruiz-Aguilera, D., Torrens, J.: A survey on the existing classes of uninorms. J. Intell. Fuzzy Syst. **29**(3), 1021–1037 (2015)
14. Mas, M., Monreal, J., Monserrat, M., Riera, J.V., Torrens Sastre, J.: Modus Tollens on fuzzy implication functions derived from uninorms. In: Calvo Sánchez, T., Torrens Sastre, J. (eds.) Fuzzy Logic and Information Fusion. SFSC, vol. 339, pp. 49–64. Springer, Cham (2016). https://doi.org/10.1007/978-3-319-30421-2_5
15. Mas, M., Monserrat, M., Ruiz-Aguilera, D., Torrens, J.: RU and (U, N)-implications satisfying Modus Ponens. Int. J. Approx. Reason. **73**, 123–137 (2016)
16. Mas, M., Monserrat, M., Torrens, J.: Modus Ponens and Modus Tollens in discrete implications. Int. J. Approx. Reason. **49**, 422–435 (2008)
17. Mas, M., Monserrat, M., Torrens, J.: A characterization of (U, N), RU, QL and D-implications derived from uninorms satisfying the law of importation. Fuzzy Sets Syst. **161**, 1369–1387 (2010)
18. Mas, M., Monserrat, M., Torrens, J., Trillas, E.: A survey on fuzzy implication functions. IEEE Trans. Fuzzy Syst. **15**(6), 1107–1121 (2007)
19. Mas, M., Ruiz-Aguilera, D., Torrens, J.: Uninorm based residual implications satisfying the Modus Ponens property with respect to a uninorm. Fuzzy Sets Syst. **359**, 22–41 (2019)
20. Ruiz, D., Torrens, J.: Residual implications and co-implications from idempotent uninorms. Kybernetika **40**, 21–38 (2004)
21. Ruiz-Aguilera, D., Torrens, J.: S- and R-implications from uninorms continuous in $]0, 1[^2$ and their distributivity over uninorms. Fuzzy Sets Syst. **160**, 832–852 (2009)
22. Ruiz-Aguilera, D., Torrens, J., De Baets, B., Fodor, J.: Some remarks on the characterization of idempotent uninorms. In: Hüllermeier, E., Kruse, R., Hoffmann, F. (eds.) IPMU 2010. LNCS (LNAI), vol. 6178, pp. 425–434. Springer, Heidelberg (2010). https://doi.org/10.1007/978-3-642-14049-5_44
23. Trillas, E., Alsina, C., Pradera, A.: On MPT-implication functions for fuzzy logic. Revista de la Real Academia de Ciencias. Serie A. Matemáticas (RACSAM) **98**(1), 259–271 (2004)
24. Trillas, E., Alsina, C., Renedo, E., Pradera, A.: On contra-symmetry and MPT-conditionality in fuzzy logic. Int. J. Intell. Syst. **20**, 313–326 (2005)
25. Trillas, E., Valverde, L.: On Modus Ponens in fuzzy logic. In: Proceedings of the 15th International Symposium on Multiple-Valued Logic, Kingston, Canada, pp. 294–301 (1985)

Author Index